PLANT-ASSOCIATED BACTERIA

Plant-Associated Bacteria

Edited by

SAMUEL S. GNANAMANICKAM

Novozymes Biologicals, Salem, VA/USA &
formerly at University of Madras, Chennai, India

Springer

A C.I.P. Catalogue record for this book is available from the Library of Congress.

ISBN 978-1-4020-4537-0 (PB)
ISBN 978-1-4020-4536-3 (HB)
ISBN 978-1-4020-4538-7 (e-book)

Published by Springer,
P.O. Box 17, 3300 AA Dordrecht, The Netherlands.

www.springer.com

Printed on acid-free paper

TABLE OF CONTENTS

PREFACE

The idea of developing a comprehensive volume on PLANT-ASSOCIATED BACTERIA was born in my mind about 5 years back. I decided therefore that the proposed volume on Plant-Associated Bacteria will be a complete volume on plant bacteriology. There are several books on symbiotic and beneficial bacteria. There are also books and volumes on plant pathogenic bacteria. In recent years, the plant-growth promoting rhizobacteria (PGPR) has emerged as an important group which has significant applications to crop production and biological disease control. There is a vital need for a comprehensive volume on all plant-associated bacteria that also includes epiphytic and endophytic bacteria. In my mind, the proposal appeared worthwhile and timely.

In the present era of genomics, there has been an enhanced interest in the genomics of the plant-associated bacteria (Example: the Plant-Associated Microbe Genome Initiative, American Phytopathological Society–www.apsnetorg/ media/ ps/top.asp). Traditional methods of identification of bacteria have been replaced by molecular methods for diagnostics and their phylogeny. This volume tries to carefully document both the traditional and recent methods. To achieve this, the contributors have been chosen from among the best experts to provide updated information for a reader who has specific expectations in a volume of this nature. Today, readers want specific and detailed information in a narrow area and there are those who want information on broad themes with adequate information on all different groups of bacteria with links and citations. There are several volumes (perhaps there are several others in the making) which could fulfill the first category. This volume belongs to the second category, a resource manual and justifiably, a reference volume.

It is hard to find volumes on broad themes. The volume of Plant-Associated Bacteria has been designed to cover the whole theme of plant bacteriology. Recent advances on all the different groups of bacteria that are associated with the phylloplane and rhizosphere have been dealt with and described whether they are beneficial (symbiotic/diazotrophic), epiphytes, endophytes or pathogens of the plants. It is my hope that a reader who is looking for information either on beneficial or pathogenic bacteria of plants, he/she finds the information in this volume with adequate details. If the reader is a researcher who needs detailed coverage of individual groups of bacteria, their strains, molecular biology and improvement, the volume devotes substantial portions of information on each aspect. This volume can be a reference material which chronicles both the traditional information on aspects such as methods of identification and the new/recent molecular methods. Therefore, it should fill the need for a good resource volume both for research students and scientists.

SAM S. GNANAMANICKAM
Editor

ACKNOWLEDGEMENT

I am deeply grateful to all the contributors. Each one brings a great deal of expertise and yet it was a very cordial experience in relating to work that went into developing this volume by assembling eighteen contributions made by these able men and women.

I am very thankful to my research associates who took a heavy share of the work. In particular, Dr. Boney Kuriakose, Dr. V. Arun and Dr. J. Ebenezar Immanuel are acknowledged for the help they rendered.

A great deal of encouragement was afforded to me by Dr. Gwyn Beattie, Dr. Tim Denny and Dr. Cindy Morris. I appreciate these sources of strength and encouragement.

Dr. Hari B. Krishnan USDA-ARS at U. of Missouri-Columbia readily lent his picture that is featured on the book's cover to represent a nodulating *Sinorhizobium* on *Erythrina* sp. I am thankful for this courtesy.

GWYN A. BEATTIE

PLANT-ASSOCIATED BACTERIA: SURVEY, MOLECULAR PHYLOGENY, GENOMICS AND RECENT ADVANCES

Abstract. Bacteria that associate with plants are diverse in the habitats they occupy, their phylogeny, and their effects on plant and environmental health. The spermosphere, rhizosphere, phyllosphere, vascular tissue and endophytic regions offer distinctive habitats for bacteria. The phylogenetic diversity of bacteria in these habitats extends across the Bacterial and Archaeal Domains; however, bacteria that are known phytosymbionts or phytopathogens are classified into only four Bacterial phyla, the Cyanobacteria, Proteobacteria, Firmicutes, and Actinobacteria. Most plant-associated prokaryotes are commensals that have no detectable effect on plant growth or physiology; these are found primarily on plant surfaces. Mutualistic bacteria include the legume symbionts, *Frankia*, and cyanobacterial symbionts, which form nitrogen-fixing symbioses, as well as associative nitrogen-fixing bacteria and plant growth-promoting rhizobacteria, which can enhance plant growth directly by increasing nutrient availability or producing plant growth-enhancing products, and indirectly by biologically controlling plant diseases. Phytopathogenic bacteria are diverse in the symptoms they induce as well as in their invasion strategies, mechanisms of pathogenesis, culturability, and even genome structure and fluidity. Other organisms that can detrimentally influence plant health are the ice nucleating bacteria, bacteria that overproduce plant growth regulators, and the deleterious rhizobacteria, which can help control weeds. Effects of plant-associated bacteria on environmental health include contributions to the remediation of soilborne pollutants, the decomposition of organic matter, and soil aggregation. Recent advances in genomic sequencing, functional genomics, and microbial ecology are dramatically changing our research approaches to these organisms and the questions we can address about their biology and their interactions with plants.

1. INTRODUCTION

Bacteria that associate with plants are diverse in their ability to affect plant health, their genotypic and phenotypic characteristics, and their phylogeny. These bacteria are typically members of complex microbial communities, with only a few establishing pure clonal populations within a plant. Although the majority of research on plant-associated bacteria has focused on phytopathogens and diazotrophic (nitrogen-fixing) phytosymbionts, interest in the diversity of organisms associated with plants has increased as the tools to assess diversity have advanced. It is clear that many plant-associated microbes, even those that comprise only a small proportion of a community, can have functions that are of agricultural or environmental importance. Technical advances in microbial ecology and genomics have been paralleled by advances in our understanding of the structure and dynamics of these microbial communities and in the molecular basis of plant-microbe and microbe-microbe interactions. The aim of this book is to provide comprehensive coverage of all types of bacteria that are found in association with plants.

S.S. Gnanamanickam (ed.), Plant-Associated Bacteria, 1-56.

2. OVERVIEW OF PLANT-ASSOCIATED BACTERIA AND THEIR HABITATS

Prokaryotes, and primarily members of the Bacterial Domain, are the numerically dominant component of most microbial communities on plants. These prokaryotes, collectively referred to as bacteria, can attain densities as high as 10^9 cells per gram of plant tissue of roots based on culturing, and 10^{10} cells per g based on cultivation-independent methods. The eukaryotic microflora can include filamentous fungi, yeasts, algae, protozoa, and nematodes, but usually at densities many orders of magnitude lower than the prokaryotes. For example, the rhizosphere may contain, in culturable numbers per gram of root, 10^6-10^9 non-actinomycetous bacteria, 10^7 actinomycetes, 10^5-10^6 fungi, 10^3 algae and 10^2-10^3 protozoa. Bacteriophage and viruses can also be members of these communities and, although the populations in the rhizosphere have not been extensively examined, they have been found at densities as high as 10^8 to 10^9 per gram of soil (Ashelford et al., 2003; Williamson et al., 2005). Although many interesting interactions occur within these communities, and some are being exploited for the control of plant pathogens, this chapter is focused exclusively on the associations between bacteria and their plant hosts.

2.1. Plants as habitats for bacteria

Plants offer a wide range of habitats that support microbial growth. These include sites that are moist and rich in nutrients, and thus ideal for fostering bacterial growth, as well as sites that are nutrient-poor or exposed to stressful environmental conditions. The surfaces of seeds, roots, leaves, and fruits often harbor large, diverse bacterial communities, whereas blossoms, stems, vascular tissue, and the intercellular spaces within plant tissues often are free of microorganisms or support only limited bacterial communities.

2.1.1. The spermosphere

The *spermosphere* is the zone that is influenced by a seed; it often extends 1 to 10 mm from the seed surface. Nutrients that support microorganisms are released when the seed imbibes, with nutrient release being greatest from the embryo end, i.e., the end closest to the emerging radicle, and from seeds that are cracked or damaged. Bacteria that establish populations on seeds can colonize roots as they emerge.

2.1.2. The rhizosphere

The *rhizosphere* is the zone that is influenced by the root, although experimentally it is often defined as the soil adhering to the root. Root growth changes the physical and chemical properties of the soil, including the mineral and organic content, the water potential, the pH, and the salinity. Whereas most soils are low in nutrients, the rhizosphere is nutrient rich. This region contains root exudates (low molecular weight (MW) compounds released during normal root cell metabolism), root secretions (low and high MW compounds that are synthesized for secretion), and compounds released by plant cell lysis. A major component of root secretions is mucilage, which contains hydrated polysaccharides, organic acids, vitamins, and amino acids and thus is an excellent substrate for microbial growth. Mucilage binds water and thus helps to form a well-hydrated environment for the roots and rhizosphere microorganisms. During root growth, the root cap continuously sloughs off cells, called peripheral or border cells, that secrete large amounts of mucilage to lubricate the roots, and after living for about 3 weeks, die and lyse. These cells are thus a major source of the mucilage and lysates available to microbial communities in the rhizosphere. The root tip itself is usually devoid of microorganisms due to the rapid rate at which border cells are shed and the rapid growth of roots, which together prevent the continuous presence of bacteria at the root tip. In contrast, the region between the root cap and the mature root, where the border cells accumulate, supports large communities. The epidermal cells in this region release additional exudates and secretions during cell division, elongation, and differentiation into root hairs. Emerging lateral roots promote the release of even more nutrients via plant cell lysis. Because of the outward diffusion of nutrients and the inward movement of salts and minerals during transpiration, chemical gradients form around the root and create a range of distinct microbial habitats. Furthermore, mature roots produce less mucilage and fewer cell lysates, due to the absence of border cells and emerging lateral roots, and leak less water due to the deposition of a water-impermeable suberin layer around epidermal cells. Consequently, the developing roots generally support fast-growing microorganisms like bacteria, whereras mature roots support slower-growing microorganisms like fungi and actinomycetous bacteria.

2.1.3. The phyllosphere

The external regions of the above-ground parts of plants, including leaves, stems, blossoms, and fruits, are collectively referred to as the *phyllosphere.* Leaves are the dominant tissue in the phyllosphere based on the surface area available for colonization. The waxy plant cuticle ensures that water loss occurs primarily through the stomata, an adaptation that probably evolved to allow plants to live on land and that has major implications for the leaf surface microflora. Unlike the rhizosphere, the phyllospehre is subject to large and rapid fluctuations in temperature, solar radiation, and water availability, the latter of which increases with rain, dew, or fog and usually decreases with wind. These changes in environmental conditions are somewhat buffered by the *boundary layer,* an air layer that surrounds the leaf. In contrast to the rhizosphere, which often supports at least 10^9 bacteria per gram of root, leaf surfaces typically support fewer than 10^7 bacteria per gram of leaf, although these community sizes vary greatly with plant species, physiology, age, and environmental conditions. Plant-derived nutrients on leaves probably originate from mesophyll and epidermal cell exudates leaking onto the surface as well as lysates from wounds and broken trichomes. Several lines of evidence indicate that the distribution of these nutrients is highly heterogeneous (Leveau & Lindow, 2001), as would be predicted for those originating from leaf damage or exogenous sources such as the honeydew of visiting insects. Blossoms and fruits offer unique habitats for microbial growth. Blossoms are short-lived, contain sugar-rich nectar (e.g., 10-50% sucrose), and are attractive to insects, which are excellent vectors for bacteria. Fruits generally have a thick cuticle on their epidermis, which probably minimizes the leakage of water and nutrients to the fruit surface.

2.1.4. Endophytic sites

Endophytic sites include any region internal to the plant epidermis, although the vascular system is usually considered separately. Endophytic microorganisms are usually found within the intercellular, or apoplastic, spaces. Intercellular air spaces comprise a significant fraction of the tissue inside roots and leaves. For example, the spaces between the cortical root cells can comprise as much as 30% of the root volume, and those between the mesophyll leaf cells can comprise as much as 70% of the leaf volume. Microorganisms that reach these intercellular regions must contend with plant defense responses, which are triggered when bacteria are in close proximity to the plant cells. Bacteria in endophytic sites may access nutrients and water more easily than those on the surface, particularly if the plant cells lyse or leak nutrients, as occurs during pathogenesis. Bacteria in endophytic sites may be buffered from the environmental fluctuations characteristic of the phyllosphere and from the intense competition for nutrients characteristic of the rhizosphere. Bacterial entry into plants occurs at sites of epidermal damage, of lateral root or radicle emergence, through natural openings such as stomata and lenticels (pores for gas exchange), hydathodes (water pores), nectarthodes (openings in the nectary of blossoms), and into progency plants through infected seeds. Some symbiotic bacteria have evolved sophisticated entry mechanisms that include directing the plant to form a channel, called an infection thread, that promotes bacterial penetration into the plant tissue.

2.1.5. Vascular tissue

The two elements of the vascular system, the xylem and the phloem, offer distinct habitats for microbial colonization. Xylem vessels function in the transport of water and minerals and thus contain a highly dilute solution of minerals and simple organic compounds. The vessels consist of dead cells that do not contain cytoplasm, lignified secondary cell walls, and lateral wall openings, called pits, with membranes that must be crossed to leave the xylem. Some bacterial xylem colonists can live on only the nutrients in the xylem; these fastidious organisms exhibit complex requirements for growth indicative of a high level of adaptation to this habitat. Others live on nutrients that are released following destruction of the xylem vessel walls. In contrast, the phloem functions in the transport of sugars from the leaves to the rest of the plant and contains a concentrated sucrose solution (15-30%). The phloem consists of living cells, including sieve tube elements and their associated companion cells, which load sucrose into the sieve tubes. The plates between adjacent sieve elements have pores with a diameter of 1 to 15 μm, which is large enough to allow bacterial passage. Bacteria that can colonize the phloem are highly adapted to this habitat, as evidenced by the low cultivation rate of these organisms and the unique presence of phytoplasmas, or cell wall-less bacteria, in the phloem. Phloem colonists, all of which are believed to be pathogens, are also unique among plant-associated bacteria in being located intracellularly. Bacteria gain access to the xylem tissue through a variety of routes, including through stomata or hydathodes that lead to the open ends of xylem vessels in leaves, through the sites of lateral root emergence in roots, through xylem-feeding insects and wounds, and via active dissolution of the xylem vessel walls. In contrast, bacteria gain access to the phloem primarily by transmission from phloem-feeding insects or by cultivation practices such as grafting, which involves cutting stems.

2.2. Current bacterial taxonomy

Prokaryotes that live in the multitude of habitats in and on plants are classified into a limited number of taxonomic groups. The current taxonomic outline of the prokaryotes, as presented in *Bergey's Manual of Systematic Bacteriology*, is shown in Fig 1. At present, all named prokaryotes fall within 2 phyla that comprise the Archaeal Domain and 24 phyla that comprise the Bacterial Domain. This outline requires continuous updating because of the current pace of change in bacterial taxonomy. These changes are fueled by improved methods for characterizing cultivated organisms, particularly at the genotypic and molecular level, as well as exponential increases in the discovery of new organisms that have not yet been cultivated. Uncultivated organisms are included in the taxonomic overview compiled by *Bergey's Manual of Systematic Bacteriology* only when they meet at least a minimal level of characterization criteria and are published in, or approved by, the *International Journal of Systematic and Evolutionary Microbiology*.

The Archaea and the Bacteria are the currently accepted names for the two distinct lineages of prokaryotes. These lineages were deduced primarily based on the sequence of the 16S ribosomal RNA (rRNA) subunit found in all prokaryotic cells. The current, standard polyphasic approach to bacterial taxonomy involves constructing a phylogenetic, or evolutionary, tree based on 16S rRNA sequence information, using it as the initial basis for classification, and then validating the tree multidimensionally by examining many phenotypic characteristics (physiological and biochemical traits), genotypic characteristics (G+C content, DNA-DNA similarity, and DNA-rRNA similarity), and other chemotaxonomic characteristics (e.g., cellular fatty acids). In general, hierarchical classifications at the levels of Domain, Phylum, Class, Order, Family, and Genus are mainly based on phylogenetic analyses using 16S rRNA sequences, whereas classification at the Species level is mainly based on DNA-DNA hybridization analyses. Although the resulting bacterial taxa are not "official" in that an official bacterial classification system does not exist, the taxa that are presented by *Bergey's Manual* represent the best consensus available.

2.3. Plant-bacterial associations

Plant-associated prokaryotes can be grouped based on the nature of their interaction with a host plant. Bacteria that are not known to affect the plant, at least directly, can be classified as *commensals*, those that have a beneficial effect on the plant can be called *mutualists*, and those that have an adverse effect on the plant are usually called *pathogens* or *parasites*. These terms are premised on the assumption that, in all cases, the plant benefits the bacteria by providing nutrients. Although many bacteria are clearly commensals, mutualists or pathogens, others fall along a continuum among these groups (Hentschel et al., 2000; Hirsch, 2004). For example, one species may utilize nutrients only passively released by a plant, another may induce a low level of active nutrient release with no detectable impact on the host, and yet another may induce the release of sufficient nutrients to cause a slight, but detectable, decrease in host fitness. A single organism can be a commensal on one host and a pathogen on another, or a commensal for months preceding entry into a pathogenic phase. All plant-associated bacteria are likely to exhibit a commensalistic relationship with a host at some stage during their lifecycle. Thus, in the following description of plant-associated bacteria, the designation of an organism as a mutualist or a pathogen indicates only the *potential* for an organism to have a mutualistic or pathogenic association with one or more plant species.

⟶

Figure 1. Taxonomic overview of the prokaryotes. Taxonomic classifications within phyla are shown only for those phyla that contain phytopathogens or phytosymbionts. Phytopathogens are found only in those taxa that are in bold; plant symbionts are found only in those taxa indicated with an asterisk. This classification is from Bergey's Manual of Systematic Bacteriology, 2nd Edition, Release 5.0 (March 2004) http://dx.doi.org/10.1007/ bergeysoutline200310. The names in quotes have no standing in nomenclature, but are useful as indicators of the taxon.

Domain: Archaea
Phylum: Crenarchaeota
Phylum: Euryarchaeota

Domain: Bacteria
Phylum: Aquificae
Phylum: Thermotogae
Phylum: Thermodesulfobacteria
Phylum: Deinococcus-Thermus
Phylum: Chrysiogenetes
Phylum: "Chloroflexi"
Phylum: Thermomicrobia
Phylum: "Nitrospiria"
Phylum: Deferribacteres
Phylum: Cyanobacteria*
Phylum: Chlorobi
Phylum: Proteobacteria*
Phylum: "Firmicutes"
Phylum: Actinobacteria*
Phylum: "Planctomycetes"
Phylum: "Chlamydiae"
Phylum: Spirochaetes
Phylum: "Fibrobacteres"
Phylum: "Acidobacteria"
Phylum: "Bacteroidetes"
Phylum: "Fusobacteria"
Phylum: "Verrucomicrobia"
Phylum: "Dictyoglomi"
Phylum: Gemmatimonadetes

Class: α -Proteobacteria
Order: Rhodospirillales
Order: Rickettsiales
Order: Rhodobacterales
Order: Sphingomonadales
Order: Caulobacterales
Order: Rhizobiales*
Order: Parvularculales
Class: β -Proteobacteria
Order: Burkholderiales*
Order: Hydrogenophilales
Order: Methylophilales
Order: Neisseriales
Order: Nitrosomonadales
Order: Rhodocyclales
Order: Procabacteriales
Class: γ -Proteobacteria
Order: Chromatiales
Order: Acidithiobacillales
Order: Xanthomonadales
Order: Cardiobacteriales
Order: Thiotrichales
Order: Legionellales
Order: Methylococcales
Order: Oceanospirillales
Order: Pseudomonadales
Order: "Alteromonadales"
Order: Vibrionales
Order: Aeromonadales
Order: Enterobacteriales
Order: Pasteurellales
Class: δ -Proteobacteria
Class: ε -Proteobacteria

Class: Actinobacteria
Subclass: Acidimicrobidae
Subclass: Rubrobacteridae
Subclass: Coriobacteridae
Subclass: Sphaerobacteridae
Subclass: Actinobacteridae
Order: Actinomycetales
Suborder: Actinomycineae
Suborder: Micrococcineae
Suborder: Corynebacterineae
Suborder: Micromonosporineae
Suborder: Propionibacterineae
Suborder: Pseudonocardineae
Suborder: Streptomycineae
Suborder: Streptosporangineae
Suborder: Frankineae*
Suborder: Glycomycineae
Order: Bifidobacteriales

Class: "Clostridia"
Order: Clostridiales
Order: Thermoanaerobacteriales
Order: Halanaerobiales
Class: Mollicutes
Order: Mycoplasmatales
Order: Entomoplasmatales
Order: Acholeplasmatales
Order: Anaeroplasmatales
Order: Incertae sedia
Class: "Bacilli"
Order: Bacillales
Order: "Lactobacillales"

3. PHYTOBACTERIA WITH A COMMENSALISTIC ASSOCIATION WITH PLANTS

3.1. Rhizosphere and phyllosphere bacteria

The vast majority of plant-associated bacteria live on the nutrients that are leaked onto plant surfaces and do not detectably alter the growth or physiology of the plant; thus, most plant-associated bacteria have a commensalistic association with their host. The distinct habitats on a plant support distinct microbial communities. For example, cultured isolates from leaves are commonly pigmented whereas those from the rhizosphere are not; pigmentation is a trait often associated with tolerance to solar radiation. In general phyllosphere communities are composed of a few taxa that have a relatively large number of individuals and many taxa with a small number of individuals (Hirano & Upper, 2000). For example, among 1,701 cultivated isolates from olive leaves, *Pseudomonas syringae* comprised the majority of the isolates whereas each of the other taxa comprised less than 10% (Table 1), and among isolates from sugar beet leaves that were collected over a complete growing season, no taxon formed a majority (Table 1).

Among the cultivated isolates from leaves, Gram$^-$ bacteria are more common than Gram$^+$ bacteria. The size and composition of phyllosphere communities, however, varies greatly with plant species, leaf physiology, leaf age, and environmental conditions, making it difficult to make generalizations about these communities. As an illustration of their taxonomic diversity, isolates from spring wheat leaves collected over a grown season represented 37 genera and 88 species (Legard et al., 1994). Rhizosphere communities are similarly, or perhaps even more, complex. Microbial communities in the rhizosphere clearly differ in composition from those in non-rhizosphere (bulk) soils, and rhizospere diversity has been found to be influenced by a wide range of factors, including plant species, plant disease, plant nutrition, fertilizer application, soil type, and environmental conditions (Kent & Triplett, 2002).

Cultivation-independent methods for examining microbial communities in natural habitats have suggested that cultivated isolates represent less than 1% of the bacterial taxa present (Torsvik & Øvreås, 2002). The finding that many organisms identified by cultivation-independent methods do not belong to known taxa further suggests that there is a rich, uncultured microbial world to be explored. The richness of this uncultured world is reflected in the fact that including candidate phyla that are monophyletic lineages based on 16S rRNA sequence and that are distinct from the existing phyla would likely double the number of recognized phyla in the domain Bacteria (Rappe & Giovannoni, 2003). It is clear that the breadth of bacterial diversity in the environment is significantly wider than is reflected in our current taxonomic outline (Fig 1). Cultivation-independent methods have indicated that plant-associated communities are more complex than was suggested by the cultivated isolates. For example, of 17 organisms identified as the dominant organisms in a phyllosphere community using cultivation-independent methods, 13 had not been described in previous cultivation-based phyllosphere studies and 5 were novel species (C. H. Yang et al., 2001). A large number of novel species were also found in a microbial community from an aquatic leaf (Weidner et al., 2000). Cultivation-independent methods led to the exciting discovery that members of the Archaeal Domain, and specifically of the phylum *Crenarcheae*, can be found in great abundance on plant roots (Simon et al., 2000). Subsequent studies have shown that roots select for a crenarchael consortium that is distinct from that the bulk soil (Sliwinski & Goodman, 2004).

3.2. Endophytic bacteria

Endophytic bacteria, or microorganisms that live within the tissues of living plants without causing substantive damage to the host. They can be isolated from surface-disinfected plant tissue or extracted from internal plant tissue. Although endophytic fungi have been extensively studied, endophytic bacteria have not. These bacteria, however, are of increasing interest due in part to their potential uses in agriculture

Table 1. Bacterial community composition on leaves.

Olive leaves[1]	%	Sugar beet leaves[2,3]	%
Pseudomonas syringae	51.0	*Pantoea agglomerans*	11.2
Xanthomonas campestris	6.7	*Pseudomonas aureofaciens*	9.6
Pantoea agglomerans	6.0	*Pseudomonas syringae*	6.2
Acetobacter aceti	4.7	*Arthrobacter oxydans*	4.8
Gluconobacter oxydans	4.3	*Raoultella terrigena*	3.2
Pseudomonas fluorescens	3.9	*Kocuria lylae*	2.9
Bacillus megaterium	3.8	*Kurthia zopfi*	2.8
Leuconostoc mesenteroides	3.1	*Aeromonas caviae*	2.7
Lactobacillus plantarum	2.8	*Pseudomonas marginalis*	2.2
Curtobacterium plantarum	2.2	*Serratia liquefaciens*	2.1
Micrococcus luteus	2.2	*Micrococcus kristinae*	2.1
Arthrobacter globiformis	1.4	*Total Bacillus spp.*	2.0
Raoultella planticola	1.2	*Stenotrophomonas maltophilia*	1.9
Enterococcus faecium	1.2	*Pseudomonas putida*	1.8

Clavibacter spp.	0.98	*Microbacterium saperdae*	1.8
Micrococcus spp.	0.82	*Flavobacterium johnsonae*	1.3
Serratia marcescens	0.81	*Arthrobacter globiformis*	1.3
Bacillus subtilis	0.57	*Pseudomonas fluorescens*	1.1
Cellulomonas flavigena	0.40	*Brevundimonas vesicularis*	1.0
Erwinia spp.	0.37	*Pseudomonas pseudoalcaligenes*	0.9
Zymomonas mobilis	0.30	*Curtobacterum flaccumfaciens*	0.9
Bacillus spp.	0.29	*Microbacterium lacticum*	0.8
Alcaligenes faecalis	0.27	*Arthrobacter mysorens*	0.7
Erwinia carotovora	0.08	*Arthrobacter viscosus*	0.7
Pseudomonas aeruginosa	0.04	*Bacillus pumilus*	0.7
		Pseudomonas cichorii	0.6
		Pseuodmonas viridiflava	0.6
		Raoultella planticola	0.6
		Yersina enterocolitica	0.6
		Pseudomonas chlororaphis	0.5
		Pseduomonas mendocina	0.5
		Chryseobacterium indologenes	0.5
		Erwinia rhapontici	0.4
		Hafnia alvei	0.4
		Serratia plymuthica	0.4
		Agrobacterium rubi	0.4

[1] 1,701 isolates from olive leaves were identified based on phenetic traits (Ercolani, 1991)
[2] 1,236 isolates from sugar beet leaves were identified based on fatty acid methyl ester analysis (Thompson, 1993)
[3] Additional species were identified that collectively represented less than 0.4% of the community: Acinetobacter calcoaceticus, Ac. lwoffii, Aeromonas salmonicida, Ae. sobria, Agrobacterium radiobacter, Ag. tumefaciens, Arthrobacter atrocyaneus, Ar. crystallopoietes, Ar. protophormiae, Ar. ureafaciens, Aureobacterium esteraromaticum, Cellulomonas turbata, Chryseobacterium balustinum, Clavibacter michiganensis, Comamonas acidovorans, Com. testosteroni, Corynebacterium bovis, Enterobacter amnigenus, Ent.intermedius, Erwinia chrysanthemi, Hydrogenophaga pseudoflava, Janthinobacterium lividum, Kocuria roseus, Microbacterium barkeri, Mi. liquefaciens, Micrococcus luteus, Morganella morganii, Ochrobacterium antropi, Pseudomonas corrugata, Ps. oryzihabitans, Ps. stutzeri, Rhodococcus rhodochrous, Sphingobacterum

(Kobayashi & Palumbo, 2000). Endophytic bacteria have been isolated from a wide variety of healthy plant species and tissue types. Kobayashi and Palumbo (2000) compiled a list of 55 bacterial genera and 144 species isolated as endophytes from roots, stems, flowers, tubers, seeds, or fruit of the many plant species examined; this included almost 50 bacterial endophytes that were isolated from only two plant species (McInroy & Kloepper, 1995). Approximately 15% of the endophytes identified could also be classified as phytopathogenic species, but the isolates may represent nonpathogenic strains or may have been present in a host or environmental context that was not appropriate for disease. Some of the most commonly isolated endophytes from the rhizosphere include species of *Bacillus, Enterobacter, Serratia, Stenotrophomonas,* and fluorescent pseudomonads (Kobayashi & Palumbo, 2000), whereas isolates showing promising levels of colonization and persistence in various agronomic crops and prairie plants included species of *Cellulomonas, Clavibacter, Curtobacterium* and *Microbacterium* (Zinniel et al., 2002). All of these isolates fall in the Proteobacteria, Firmicutes, Actinobacteria and Bacteroidetes phyla of the domain Bacteria (Fig 1).

4. PHYTOBACTERIA WITH A MUTUALISTIC ASSOCIATION WITH PLANTS

Bacteria that can benefit plants have been of great biological and agricultural interest since 1888, when Martinus Beijerink discovered that root nodule bacteria are critical to the ability of legumes to obtain nitrogen from the air. In the last two decades, much work has focused on identifying the breadth and extent of the beneficial effects of bacteria on plants with the goal of exploiting these benefits to improve plant productivity. Mutualisms between bacteria and plants can be divided into those that involve close associations between the bacteria and plant, often called symbioses, and those that involve only loose associations, often called associative interactions. In the former, the bacterial mutualist is often called a symbiont. These mutualists include some who provide a clear and direct benefit to the plant, such as the nitrogen-fixing mutualists, some who are more complex in their benefit, such as those that produce phytohormones that beneficially alter plant physiology, and some whose benefit is indirect, such as those that reduce the negative effects of pathogens.

4.1. Symbiotic N_2-fixing organisms

Biological nitrogen fixation, or the ability to convert dinitrogen to ammonia, is limited to prokaryotes but is widespread among prokaryotic genera. Among the approximately 100 nitrogen-fixing, or *diazotrophic,* genera, only a few exhibit the highly specific, intimate interactions needed to induce a new organ or organ-like structure in the host. The intimacy of such associations maximizes the transfer of fixed nitrogen to the plant, thus providing a nutrient that is often in short supply. It also provides the prokaryote with a protected environment and a supply of energy-rich compounds, which are necessary to support the high energy demands of nitrogen fixation. Symbiotic diazotrophs include legume symbionts, which are Gram$^-$ bacteria that form nodules on leguminous plants, and members of the actinomycetous genus *Frankia,* which are Gram$^+$ bacteria that form nodules on woody, dicotyledonous trees and shrubs. In contrast to these symbioses with higher plants, cyanobacteria are symbiotic diazotrophs that form mutualisms with primitive plants, described below, as well as with fungi in mutualisms called lichens; these cyanobacterial symbionts are often called *cyanobionts*. Collectively, these mutualisms result in the efficient fixation of large amounts of nitrogen and are responsible for a significant proportion of the global conversion of atmospheric dinitrogen to ammonia, nitrate and organic compounds.

4.1.1. Legume symbionts

The most well-studied plant-bacterial mutualisms are those between members of the Order Rhizobiales and members of the Leguminoseae (Fabaceae) family of plants. This plant family includes crops such as peas, beans, lentils and alfalfa, as well as trees such as locust, acacia and mimosa. All legume symbionts were originally classified into the genus *Rhizobium*, hence these symbionts are often referred to generically as rhizobia and the symbioses as *Rhizobium*-legume symbioses. Today, these symbionts are classified into several genera, with most species belonging to the genera *Rhizobium, Sinorhizobium* (*Ensifer*), *Mesorhizobium,* and *Bradyrhizobium* (Table 2). These organisms are phylogenetically clustered with non-phytosymbionts (Fig 1). Biological nitrogen fixation by legume symbionts occurs only within nodules, which are specialized outgrowths of plant tissue formed specifically in response to the bacterial symbionts. The agricultural importance of this mutualism was realized during the many centuries of crop rotation practice in which a rotation with legumes was used to rejuvenate the soil; the biological basis of this agricultural practice, however, was understood only with the recognition of the symbiosis.

Table 2. Plant-pathogenic bacteria and plant symbionts[1].

Order/Family (Fam)	**Genus**	**Species[2]**
Class: Cyanobacteria		
Subsection 4 Fam: 4.1	***Anabaena***	**spp.**
	Nostoc	**spp.**
Class: α-proteobacteria		
Order: Rhodospirillales Fam: Acetobacteraceae	*Acetobacter*	*aceti, pasteurianus*
	Gluconobacter	*oxydans*
Order: Sphingomonadales Fam: Sphingomonadaceae	*Sphingomonas*	*suberifaciens* (prev. *Rhizomonas suberifaciens*)
Order: Rhizobiales Fam: Rhizobiaceae	*Rhizobium*	***etli*[3] (prev. *Rh. legminosarum* bv. *phaseoli* type I), *galagae* (prev. *Rh. leguminosarum*), *gallicum*[3], *giardinii*[3], *hainanense, huautlense, indigoferae, leguminosarum*[3], *lupini, mongolense, sullae, tropici* (prev. *Rh. leguminosarum* bv. *phaseoli* type II), *yanglingense***

	Agrobacterium/ Rhizobium[4]	*larrymoorei, radiobacter, rhizogenes, rubi, tumefaciens (= radiobacter), vitis*
	Sinorhizobium/ Ensifer[6]	***adhaerens, arboris, fredii***[5] **(prev. *Rh. fredii*), *kostiensis, kummerowiae, medicae, meliloti* (= *Ens. meliloti*, prev. *Rh. meliloti*), *morelense, saheli, terangae***[3], ***xinjiangense***
Fam: Phyllobacteriaceae	*Mesorhizobium*	***amorphae, chacoense, ciceri* (prev. *Rh. ciceri*), *huakuii* (prev. *Rh. huakuii*), *loti* (prev. *Rh. loti*), *mediterraneum* (prev. *Rh. mediterraneum*), *plurifarium, tianshanense* (prev. *Rh. tianshanense*)**
	"Candidatus Liberibacter"	*asiaticus, africanus* (subsp. *africanus, capensis*)
Fam: Bradyrhizobiaceae	*Bradyrhizobium*	*betae*[7], ***canariense***[7], ***elkanii, japonicum* (prev. *Rh. japonicum*), *liaoningense, yuanmingense***
	Blastobacter	***denitrificans***
Fam: Methylobacteriaceae	*Methylobacterium*	***nodulans***[7]
Fam: Hyphomicrobiaceae	*Azorhizobium*	***caulinodans***
	Devosia	***neptuniae* (prev. *Rh. neptunii*)**
Class: β-proteobacteria		
Order: Burkholderiales Fam: Burkholderiaceae	*Burkholderia* (prev *Pseudomonas*)	*andropogonis, caryophylli, cepacia, gladioli*[8], *glumae, plantarii*, ***phymatum, tuberum***
	Ralstonia	*solanacearum* (prev. *Bu. solanacearum*), *syzgii* (prev. *Ps. syzgii*)
	Wautersia	*taiwanensis* (prev. *Ra. taiwanensis*)
Fam: Oxalobacteraceae	*Herbaspirillum*	*rubrisubalbicans* (prev. *Ps. rubrisubalbicans*)
Fam: Comamonadaceae	*Acidovorax*	*anthurii, avenae* (subsp. *avenae, citrulli, cattleyae*) (prev. *Ps. avenae* subsp. *avenae, Ps. avenae* subsp. *citrulli, Ps. cattleyae*), *konjaci* (prev. *Ps. avenae* subsp. *konjaci*), *valerianellae*
	Xylophilus	*ampelinus* (prev. *Xa. ampelinus*)
Class: γ-proteobacteria		
Order: Xanthomonadales Fam: Xanthomonadaceae	*Xanthomonas*	*albilineans, arboricola*[8,9] (= *Xa. juglandis*), *axonopodis*[8], *bromi*[9] (= *Xa. campestris* pv. *bromi*), *campestris*[8], *cassavae*[9] (prev. *Xa. campestris* pv. *cassavae*), *citri*[9]

		(= *Xa. campestris* pv. *citri* = *Xa. axonopodis* pv. *citri*), *codiaei*[9] (= *Xa. campestris* pv. *poinsettiicola*), *cucurbitae*[9] (prev. *Xa. campestris* pv. *cucurbitae*), *cynarae, fragariae, hortorum*[8,9] (= *Xa. hederae*), *hyacinthi*[9] (prev. *Xa. campestris* pv. *hyacinthi*), *melonis*[9] (= *Xa. campestris* pv. *melonis*), *oryzae*[8], *phaseoli*[9] (=*Xa. campestris* pv. *phaseoli* = *Xa. axonopodis* pv. *phaseoli*), *pisi*[9] (prev. *Xa. campestris* pv. *pisi*), *populi, sacchari*[9] (=*Xa. albilineans*), *theicola*[9] (= *Xa. campestris* pv. *theicola*), *translucens*[8] (prev. *Xa. campestris* pv. *hordei*), *vasicola*[8,9] (=*Xa. campestris* pvs. *holcicola, vasculorum*), *vesicatoria*[9] (= *Xa. campestris* pv. *vesicatoria* or *Xa. exitiosa*)
	Xylella	*fastidiosa*
Order: Pseudomonadales Fam: Pseudomonadaceae	*Pseudomonas*	*agarici, amygdali, asplenii, avellanae, beteli, cannabina, caricapapayae, cichorii, cissicola, corrugate, costantinii, flectens, fuscovaginae, marginalis*[8], *mediterranea, orrugata, palleroniana, salomonii, savastanoi*[8,9] (= *Ps. syringae* pv. *savastanoi* = *Ps. syrinage* subsp. *savastanoi*), *syringae*[8], *tolaasii, tremae, viridiflava*
Order: Enterobacteriales Fam: Enterobacteriaceae	*Brenneria* (prev. *Erwinia*)	*alni, nigrifluens, paradisiaca, quercina, rubrifaciens, salicis*
	Enterobacter	*cancerogenus* (prev. *Er. cancerogena*), *dissolvens, nimipressuralis* (prev. *Er. nimipressuralis*), *pyrinus*
	Erwinia	*amylovora, carnegieana, cypripedii* (= *Pe. cypripedii*), *mallotivora, papayae*[7], *persicina, psidii, pyrifoliae, rhapontici, tracheiphila*
	Pantoea (prev. *Erwinia*)	*agglomerans* (= *Er. herbicola* = *En. agglomerans*), *ananatis*[8], *citrea, stewartii* (subsp. *stewartii, indologenes*)
	Pectobacterium (prev. *Erwinia*)	*atrosepticum* (prev. *Er. carotovora* subsp. *atroseptica*),

		betavasculorum (prev. *Er. carotovora* subsp. *betavasculorum*), *cacticida*, *carotovorum* (subsp. *carotovorum*, *odoriferum*), *chrysanthemi* (= *Er. chrysanthemi*), *cypripedii* (= *Er. cypripedii*), *wasabiae* (prev. *Er. carotovora* subsp. *wasabiae*)
	"Candidatus Phlomobacter"	*fragariae*
	Samsonia	*erythrinae*
	Serratia	*proteamaculans* (subsp. *proteamaculans*) (prev. *Er. proteamaculans*)
Class: "Clostridia"		
Order: Clostridiales Fam: Clostridiaceae	*Clostridium*	*puniceum*
Class: Mollicutes		
Order: Entomoplasmatales Fam: Spiroplasmataceae	*Spiroplasma*	*citri, kunkelii, phoeniceum*
Order: Acholeplasmatales Fam: Acholeplasmataceae	*"Candidatus Phytoplasma"*[7]	*allocasuarinae, asteris, aurantifolia, australiense, australasia, brasiliense, castaneae, cyondontis, fraxini, japonicum, mali, oryzae, phoenicium, pini, prunorum, pyri, rhamni, spartii, trifolii, ulmi, ziziphi*
Class: "Bacilli"		
Order: Bacillales Fam: Bacillaceae	*Bacillus*	*megaterium, pumilus*
Class: Actinobacteria Subclass:Actinobacteridae		
Order: Actinomycetales Suborder: Micrococcinaea Fam: Micrococcaceae	*Arthrobacter*	*ilicis* (prev. *Corynebacterium ilicis*)[10]
Fam: Microbacteriaceae	*Clavibacter*	*michiganensis* (subsp. *insidiosus, michiganensis, nebraskensis, sepedonicus, tessellarius)* (prev. *Corynebacterium michiganense* subsps. *insidiosum, michiganense, nebraskense, sepedonicum tessellarius*)
	Curtobacterium	*flaccumfaciens*[8] (prev. *Corynebacterium flaccumfaciens* subsp. *flaccumfaciens*)
	Leifsonia (prev. *Clavibacter*)	*xyli* (subsp. *xyli, cynodontis*)
	Rathyibacter	*iranicus, rathayi,tritici*

	(prev. *Clavibacter*)	
Suborder:Corynebacterineae Fam: Nocardiaceae	*Nocardia*	*vaccinii*
	Rhodococcus	*fascians* (prev. *Co. fascians*)
Suborder: Streptomycineae Fam: Streptomycineae	*Streptomyces*	*acidiscabies, albidoflavus, aureofaciens, cacaoi* (subsp. *cacaoi, asoensis*), *candidus, caviscabies, clavifer, collinus, europaeiscabiei, fimbriatus, globisporus* (subsp. *globisporus, caucasicus*), *griseus* (subsp. *alpha, cretosus, griseus, solvifaciens*), *intermedius, ipomoeae, longisporus, parvulus, praecox, reticuliscabiei, rimosus* (subsp. *rimosus, paromomycinus*), *sampsonii, scabiei* (= *St. scabies*), *setonii, sparsogenes, stelliscabiei, tricolor, turgidiscabies, violaceus, wedmorensis*
Suborder: Frankineae Fam: Frankiaceae	*Frankia*	*alni*

[1] *Organisms were identified as plant pathogens (normal text) as indicated by the International Society for Plant Pathology (Young, Bull et al., 2004; Young et al., 1995) as plant symbionts (bold text) as indicated by Sawada et al. (2003).*

[2]*Prev., previous name; "=" indicates synonyms (included only homotypic synonyms and senior heterotypic synonyms); pv., pathovar; bv., biovar; subsp, subspecies.*

[3]*This species has been further divided into biovars.*

[4]*Young et al. (2001; 2003) have proposed to amalgamate the genera Rhizobium, Allorhizobium, and Agrobacterium into one genus named Rhizobium, although this emendation is not fully supported within the scientific community (Farrand et al., 2003).*

[5]*This species has been further divided into chemovars.*

[6]*Recent evidence supports the amalgamation of the genera Sinorhizobium and Ensifer into one genus, although the name of the resulting genus has been under debate (Willems et al., 2003; Young, 2003).*

[7]*These species have been proposed since the most recent release of the Bergey's Taxonomic Outline (May 2004): Bradyrhizobium betae (Rivas et al., 2004), Bradyrhizobium canariense, Methylobacterium nodulans (Jourand et al., 2004), Er. papayae (Gardan et al., 2004), and all of the "Candidatus Phytoplasma" species except ulmi (IRPCM, 2004; Lee et al., 2004; Schneider et al., 2005)*

[8]*This species has been further divided into pathovars.*

[9]*These taxonomic classifications have been subject to evaluation by Schaad et al. (2000).*

[10]*Young et al. (2004) have proposed that pathogenic strains of this species be reclassified as Curtobacterium flaccumfaciens pv. ilicis.*

Most *Rhizobium*-legume symbioses involve the development of nitrogen-fixing nodules on roots. Root nodulation involves signal exchange between the bacteria and the plant: the bacteria sense specific plant-derived signals, often flavanoids, and in response produce lipochitooligosaccharide signal molecules called Nod factors. These Nod factors induce root hair curling, localized hydrolysis of the plant cell wall within the curl, invagination of the plasma membrane, and deposition of new plant cell wall material to create an infection thread, which serves as a passageway into the plant tissue. The bacteria divide within the infection thread as it develops through the many layers of cortical cells in the root. Alternatively, some legume symbionts enter the plant through gaps in the epidermis and move through the root intercellularly, with or without the induction of an infection thread. In some host plants, such as alfalfa and pea, the bacteria induce mitotic division of the inner cortical cells and induce indeterminate nodules, which continuously elongate due to a persistent meristem at the apex. In other host plants, such as soybean, the bacteria induce mitotic division of the outer cortical cells and induce determinate nodules, which are spherical and reach a terminal size due to only transient meristemmatic activity. In two of the three legume subfamilies (Mimosoideae and Papilionoideae), bacteria move from the infection thread into the cytoplasm of the host cells by endocytosis, allowing the bacteria to become enveloped by the plant membrane. These intracellular, enveloped bacteria, called bacteroids, divide until they fill the host cell and differentiate into a form that is rich in nitrogenase, a key enzyme for nitrogen fixation. In the Caesalpinoideae legume subfamily, the bacteria do not leave the infection threads; instead, the infection threads branch until they fill the host cell. Nitrogen fixation by legume symbionts is induced only when the concentration of free oxygen in the nodule is low enough to permit activity of the oxygen-sensitive enzyme nitrogenase. The oxygen-binding plant-derived protein leghemoglobin, which is functionally related to hemoglobin, plays a key role in reducing the free oxygen levels while making oxygen available for bacterial respiration. The resemblance of nitrogen-fixing, intracellular, bacterial-derived structures to organelles such as chloroplasts have led to their designation as *symbiosomes.*

At least one legume symbiont, *Azorhizobium caulinodans,* can induce the formation of stem nodules. These have been found on the tropical host *Sesbania rostrata. A. caulinodans* initiates infection by entering through the gaps in the epidermis on the stem and moving through the intercellular spaces, with subsequent induction of an infection thread. Stem nodules have also been found on *Aeschynomene* species, and interestingly, some of these symbionts are not only diazotrophic, but also photosynthetic (Giraud & Fleischman, 2004), a relationship that could alleviate some of the energy demand by the bacteria in the nodules. Legume symbionts have been found to form nodules on a non-legume, the tropical tree *Parasponia* in the family Ulmaceae (Vessey et al., 2004). The nodules on this plant species are indeterminate and branched, similar to those of actinorhizal nodules, described below, and are induced by bacteria in the genera *Bradyrhizobium* and *Rhizobium.*

Nodulation is a strongly species-specific process, with individual bacterial strains often forming nodules only on specific plant species. In fact, species-specificity was central to the development of the taxonomy of these organisms. Whereas all legume symbionts were originally classified into the genus *Rhizobium*, the classification at the species level was based primarily on host specificity. As more isolates were examined on a wider range of legumes, it became clear that many strains can nodulate more than one host species, and distinct legume symbionts can be isolated from a single legume species, although individual nodules typically contain a single strain. Thus, host specificity gradually lost its reliability as a taxonomic criterion (Sawada et al., 2003). The application of new taxonomic criteria, including phenotypic and genotypic characteristics, chemotaxonomic markers (e.g., fatty acids) and molecular phylogenetic data (16S rRNA gene sequences and protein-coding gene sequences) along with advances in systematics, resulted in a revised classification. In the early 1990's, legume symbionts were classified into four genera: the fast-growing strains into *Rhizobium, Sinorhizobium*, and *Azorhizobium*, and the slow-growing strains into *Bradyrhizobium.* A group of fast-growing strains was subsequently recognized as a distinct taxon and the genus *Mesorhizobium* was created, with "meso" referring to its intermediate position between the other groups based on its phylogenetic position and the intermediate growth rate of its members. Strains of *Mesorhizobium* are also intermediate in having 2 rRNA operons versus 3 in the fast growers and 1-2 in *Bradyrhizobium*, and in having their nodulation and fixation genes on a megaplasmid or on the chromosome versus only on a megaplasmid like the fast growers and only on the chromosome like *Bradyrhizobium*. Today, the majority of legume symbionts are classified in the genera *Rhizobium*, *Sinorhizobium* (*Ensifer*), *Mesorhizobium*, and *Bradyrhizobium*, with host species specificity often reflected at the species level as well as at the subspecies level by divisions known as *biovars,* a term used to indicate that an organism is capable of nodulating one or more host species (Table 2). Speciation of these organisms, however, is still experiencing rapid changes as illustrated by the classification of symbiotic diazotrophs in these genera into 15 species in 1996 (Young & Haukka, 1996) but into 37 species in 2004 (Young, Park et al., 2004). Phylogenetic analyses of *B. japonicum* strains indicate that they are too diverse to be grouped into one species, so modifications to the species classification in this genus can be expected in the future. Root-nodulating diazotrophs have recently been isolated and classified as members of the genera *Blastobacter, Devosia, Methylobacterium*, and *Ochrobactrum*, all of which are also in the Rhizobiales order, as well as *Ralstonia,* which is in the Burkholderiales order (Table 2) (Rivas et al., 2004; Rivas et al., 2003; Sy et al., 2001; Trujillo et al., 2005; van Berkum & Eardly, 2002). Interestingly, tripartite mutualisms have been reported that involve mycorrhizal fungi incorporated into root nodules, or bacteria closely related to *Burkholderia* located intracellularly in mycorrhizal fungi (Bonfante, 2003).

4.1.2. Frankia

The actinomycete *Frankia*, which is a Gram$^+$ prokaryote that grows as branched hyphae, can form root nodules on over 200 species of woody, dicotyledonous trees and shrubs, such as *Alnus* (alders), *Casuarina*, *Ceanothus* and *Myrica*, as well as on two non-woody species of *Datisca*. These plants are collectively referred to as actinorhizal plants. They are distributed throughout tropical and temperate climates, and are often the pioneer species on nitrogen-poor soils, making them useful for the reclamation of poor soils. Similar to legumes, actinorhizal plants are likely of global importance to nitrogen cycling due to their widespread occurrence.

Frankia induce root nodules on actinorhizal plants via an infection process similar to that of the legume symbionts (Vessey et al., 2004). *Frankia* can infect intracellularly by inducing root hair deformation, infection thread development, and encapsulation of hyphae by a host-derived membrane. *Frankia* can also infect intercellularly via hyphal colonization of the cortical intercellular spaces. Both infection routes ultimately lead to mitotic activation of the root pericycle cells and nodules that are similar to indeterminate legume nodules in their elongated and persistent growth; actinorhizal nodules, however, are branched. *Frankia* cells do not form bacteroids in the host cell cytoplasm, but rather differentiate the terminal regions of their hyphae into thick-walled, lipid-encapsulated vesicles. The multiple lipid layers on the vesicles collectively function as a gas-diffusion barrier, and when the oxygen concentration is sufficient low, nitrogen fixation occurs. This ability to form vesicles permits *Frankia* to fix nitrogen in a free-living state. Like most legume symbionts, *Frankia* strains induce nodules in a host-specific manner.

Compared to the *Rhizobium*-legume symbioses, relatively little is known of actinorhizal associations; this is due, at least in part, to the fact that both *Frankia* and their hosts grow relatively slowly. Furthermore, obtaining and maintaining pure culture isolates of *Frankia* is extremely difficult. *Frankia* was isolated in pure culture in 1959, with the culture subsequently lost (Pawlowski, 2002), and again in 1978 (Callaham et al., 1978), 90 years after the first isolation of *Rhizobium* from root nodules. Genetic tools that work in *Frankia* have also been difficult to develop. To date, *Frankia* strains have been isolated from 20 of the 24 genera of actinorhizal plants, but these isolates represent only a small number of the over 200 actinorhizal plant species. Although attempts to differentiate the genus into species have been hindered by insufficient numbers of isolates, a recent study has identified at least four clades within the *Frankia* phylogenetic tree (Normand et al., 1996).

4.1.3. Cyanobionts

Mutualisms between plants and cyanobacteria primarily involve diazotrophic cyanobacteria in the genus *Nostoc*. These mutualisms involve plant structures that develop in the absence of the cyanobiont, but that are changed when colonized by the cyanobiont. Features that are similar to all of these mutualisms are the presence of mucilage in the cavities that support the cyanobiont, the sole presence of the cyanobiont in these cavities, and the formation in response to the cyanobiont of long, finger-like plant cells that extend into these cavities. These extensions likely increase the surface area of the cyanobiont-host interface and thus the opportunities for nutrient exchange. The cyanobacterial association with the angio- sperm *Gunnera* is the only cyanobacterial-plant mutualism that induces the development of a symbiotic "organ", as in the nitrogen-fixing nodules.

Cyanobionts on liverworts and hornworts. Cyanobacteria form mutualisms with only two of the 330 known liverwort genera, and with four of the six genera of hornworts (Rai et al., 2000). These cyanobionts, classified as *Nostoc*, respond to plant signals by differentiating into structures called hormogonia that are specialized for taxis, and moving toward and into mucilage-filled cavities located on the ventral surface of the gametophyte. A plant signal that induces these behaviors has been isolated and named HIF, for hormogonia inducing factor. Only compatible strains are able to successfully enter these cavities, indicating host-specificity among *Nostoc* strains, and each cavity contains only one strain of *Nostoc,* like in nitrogen-fixing nodules. The infecting cyanobiont strain multiplies within the cavities and induces plant production of multicellular filaments (hairs) that extend throughout the cavity. The host eventually releases HRF, a hormogonia repressing factor, which promotes the formation of heterocysts, which are structures that are specific to nitrogen fixation. Subsequent growth of the host and the cyanobiont is synchronized, thus maintaining a similar cyanobiont:plant ratio throughout the life of the plant.

Cyanobionts on Azolla. Cyanobacterial mutualisms with the small water fern *Azolla* have attracted much agricultural interest because of their potential to enhance rice production (Vaishampayan et al., 2001). Nitrogen limitation is a major factor limiting rice productivity, and this limitation is poorly addressed by nitrogen application due, in part, to losses via NH_3 volatilization. Studies in the last few decades have shown that biofertilizers based on the *Azolla*-cyanobacterial symbiosis can significantly enhance rice cultivation. The cyanobiont, which has historically been called *Anabaena azollae*, was recently reported to be *Nostoc* (Rai et al., 2000) or a novel organism that is neither *Anabaena* or *Nostoc* (Baker et al., 2003). The infection process is similar to that described for liverworts and hornworts except that the cyanobiont is transmitted vertically (through the germ line) via the megaspore rather than being acquired by horizontal transmission from the environment (Rai et al., 2000). It enters cavities in the cotyledonary leaves via movement from an inoculum chamber during *Azolla* development, thus avoiding the need for host production of HIF and HRF signals, and like the cyanobionts on liverworts and hornworts, becomes embedded in mucilage, fixes nitrogen, and grows in relative synchrony with the host.

Cyanobionts on cycads. Cyanobacterial mutualisms with gymnosperms are slightly more sophisticated than those with the bryophytes and pteridophyte, described above, in that they involve greater differentiation of the plant tissue (Vessey et al., 2004). *Nostoc* colonizes a mucilage-filled zone, called the *cyanobacterial zone*, between the inner and outer cortical cells of specialized cycad roots. *Nostoc* colonization induces these precoralloid roots to become geotropic, lose a papillose sheath, and develop prominent lenticels, thus transforming them into coralloid, or coral-like, roots. The plant develops elongated host cells that specifically traverse the cyanobacterial zone; similar to in other cyanobacterial mutualisms, this increases the cyanobiont-host surface contact. Subsequent growth of the host and the cyanobiont is synchronized, with only one strain of *Nostoc* present in any single coralloid root.

Cyanobionts on Gunnera. Cyanobacteria can form mutualisms with only one higher plant family, the subtropical angiosperm *Gunnera* in the Gunneraceae family. In this mutualism the cyanobiont, *Nostoc,* is located intracellularly rather than extracellularly. *Nostoc* is attracted to and enters a specific gland that is present at base of each petiole along the stem; this entry may be facilitated by entrapment in mucilage that is secreted through the glands. *Nostoc* cells move through this mucilage to a cavity at the base of the gland. If the strain is compatible with the host, it enters into the thin-walled meristemmatic cells lining the cavity. This entry is mediated by localized hydrolysis of the cell wall, entry of the cyanobiont while being enveloped by the host membrane, and formation of new cell wall around the cyanobacterial cell. Subsequent differentiation and growth of the cyanobiont is accompanied by division of the infected host cells, demonstrating that like the rhizobia, cyanobacteria can stimulate mitosis in host cells.

4.2. Associative nitrogen-fixing organisms

Many diazotrophic plant-associated bacteria benefit the plant via nitrogen fixation but do not enter into host species-specific mutualisms. These organisms, as with non-diazotrophic bacteria, vary in how closely they interact with their plant host (Table 3). For example, individual diazotrophic strains may be classified anywhere along a continuum that extends from establishing intracellular, and thus highly evolved, interactions with a host to living exclusively on a plant's surface. It is tempting to speculate that increasing closeness in this association is paralleled by increasing availability of fixed nitrogen to the plant. Although such a correlation may be generally true, it has not been demonstrated and is in fact contradicted by the inability of many strains of legume symbionts and *Frankia* to effectively fix nitrogen following nodule formation. However, this spectrum of closeness is useful for discussing the range of organisms capable of *associative nitrogen fixation*, i.e., asymbiotic plant-associated nitrogen fixation. Furthermore, it is reasonable to believe that nitrogen fixed by free-living diazotrophs on plant surfaces is transferred less efficiently to the plant host and is subject to greater losses than nitrogen fixed by endophytic bacteria.

Table 3. The spectrum of diazotroph-plant interactions.

Increasing closeness in the diazotroph-plant association →	**Endosymbionts:** bacteria are located intracellularly Example: Nodule-forming bacteria on legumes and actinorhizal plants; *Nostoc* on *Gunnera*
	Ectosymbionts: bacteria are located extracellularly Example: cyanobacterial associations with primitive plants
	Endophytes: bacteria are within the host plant tissue Example: Associations between *Gluconacetobacter diazotrophicus* and *Herbaspirillum* spp. in sugarcane
	Plant surface colonists: bacteria that grow only on plant surfaces Example: diazotrophic rhizosphere colonists such as *Azospirillum* and *Klebsiella*

Many important crops, such as maize, rice, wheat, and root and tuber crops, have not evolved mutualisms that involve the synthesis of specific nitrogen-fixing organs, but some, particularly grasses, often detectably benefit from the presence of diazotrophs. The extent to which associative diazotrophs transfer fixed nitrogen directly to their host versus indirectly via their death and decomposition is not clear and has been experimentally challenging to determine. The growth-enhancing effects of some of these diazotrophs is due, at least in part, to the production of phytohormones, which can not only stimulate growth directly, but also increase the efficiency with which fixed nitrogen is taken into the plant. A primary research goal of the last century has been to extend the benefit of biological nitrogen fixation to nonleguminous plants. Although research in the 1960s, 1970s and 1980s focused on rhizosphere bacteria as sources of fixed nitrogen, more recent efforts have focused on using endophytes to provide fixed nitrogen to plants, and particularly to agronomic grasses such as wheat, rice and maize. This switch in research focus was motivated, in part, by the hypothesis described above, namely that the closer the association between a diazotroph and its host, the better the transfer of fixed nitrogen and thus the greater the benefit to the host.

4.2.1. Diazotrophic endophytes

Diazotrophic endophytes that do not form nodules have been isolated primarily from grasses such as maize, rice, sorghum, sugar cane and wheat. The most well-studied diazotrophic endophytes are *Gluconacetobacter diazotrophicus* (previously *Acetobacter diazotrophicus*) and *Herbaspirillum* spp. in sugarcane, *Azoarcus* spp.

in Kallar grass, and *Alcaligenes, Azospirillum, Bacillus, Enterobacter, Herbaspirillum, Klebsiella, Pseudomonas* and *Rhizobium* in rice and maize (James, 2000). Inoculation of these bacterial species onto roots can result in colonization of the root cortex and sometimes even entry into the vascular system with subsequent translocation to the aerial regions of the plant. Populations in the intercellular spaces and xylem vessels have been found to range from 10^1 to 10^7 cells per gram of fresh weight. N-balance, ^{15}N isotope dilution and ^{15}N natural abundance studies have provided evidence that some plants can obtain at least part of their nitrogen from associative nitrogen fixation, although whether the fixed nitrogen is transferred directly to the plant or is transferred by the death and subsequent mineralization of the cells is not clear. In many cases involving inoculation of diazotrophic endophytes, nitrogen fixation activity was detected only after amendment of roots with a carbon source, suggesting that the high energy requirement of nitrogen fixation may not be met during endophytic root colonization. Recently, however, a *Klebsiella pneumoniae* strain was shown to transfer fixed nitrogen to wheat during endophytic growth in the roots without carbon amendment, and to be capable of endophytic colonization in a broad range of hosts (Iniguez et al., 2004). Other studies have shown that up to 70% of the nitrogen in sugarcane grown with *G. diazotrophicus* can originate from biological nitrogen fixation (James, 2000). Significant efforts are being made to identify endophytic diazotrophs that promote rice growth, and these have resulted in the identification of many that enhance rice growth (Reddy et al., 2002), although whether the bacteria directly transfer the fixed nitrogen to plants has not been clearly established in most cases.

4.2.2. Diazotrophic plant surface colonists

Most nitrogen-fixing prokaryotes fix nitrogen in a free-living state, presumably for their own benefit. Nitrogen fixation in the rhizosphere is likely more common than in bulk soil because the high energy demands of nitrogen fixation can be met by the carbohydrate-rich plant exudates in the rhizosphere. *Azospirillum* species were the first diazotrophs to be isolated from the rhizosphere of plants and remain the best characterized plant-surface associated diazotrophic species. Unlike some of the endophytic diazotrophs, such as *Gluconacetobacter diazotrophicus* and *Herbaspirillum* spp., which appear to be obligate endophytes, *Azospirillum* spp. are facultative endophytes, i.e., they can live either endophytically or in the rhizosphere. *Azospirillum* spp. have been the subject of extensive inoculation efforts due to their plant growth promotion activity, although it is now generally accepted that their primary benefit to plants is due to activities such as phytohormone production or altering root morphology, as discussed below, rather than nitrogen fixation. *Azotobacter* spp. have also been the subject of extensive investigations into the role of nitrogen fixation in promoting plant growth, with similar conclusions that plant growth benefits may be due, at least in part, to activities other than nitrogen fixation. In addition to *Azospirillum* and *Azotobacter,* which are obligate aerobes, nitrogen-fixing enteric bacteria such as *Klebsiella* and *Enterobacter*, which are facultative anaerobes, are common in the rhizosphere. In the phyllosphere, the extent to which bacteria fix nitrogen is not well known, although nitrogen-fixing genera such as *Beijerinkia* and *Azotobacter* are common and nitrogen fixation activity on leaves has been reported (Jones, 1970).

4.3. Plant growth-promoting bacteria

Bacteria can stimulate plant growth by mechanisms other than by providing fixed nitrogen. Identifying and characterizing these mechanisms is challenging because of the diversity of mechanisms by which bacteria can promote plant growth, the difficulty in estimating enhanced growth or improved health, particularly with small increases, and the narrow range of abiotic and biotic conditions under which growth promotion may occur. For example, even the diazotrophs may only promote detectable increases in plant growth when soils are nitrogen poor. Bacterial phytostimulation may result from bacteria 1) increasing the availability of nutrients to the plant, 2) producing growth-stimulating phytohormones, 3) enhancing the positive effects of symbionts, or 4) reducing the negative effects of pathogens (Bloemberg & Lugtenberg, 2001; Vessey, 2003). Organisms that increase the growth of plants by increasing nutrient availability or increasing plant access to nutrients, such as by increasing root surface area, are sometimes referred to as *biofertilizers*. In contrast, organisms that promote plant growth by controlling deleterious organisms are called *biocontrol agents* or *biopesticides*. The majority of bacteria that promote plant growth are rhizosphere colonists; such organisms have been designated *plant growth-promoting rhizobacteria,* or PGPR.

4.3.1. Bacteria that increase the availability of nutrients to plants

In addition to bacteria that generally contribute to plant nutrition by decomposing soil organic matter, some bacteria provide specific nutrients to plants. As discussed above, nitrogen limitation can be lessened or ameliorated by some diazotrophic bacteria. The importance of these bacteria was recognized as early as the 1890s when symbiotic nitrogen-fixing bacteria were first marketed as agricultural inoculants; they continue to be sold as inoculants today. Phosphorous is second to nitrogen among the mineral nutrients that limit terrestrial plant growth. Phosphorous is often abundant in soil, but most is in an insoluble form. Many rhizosphere bacteria can solubilize inorganic phosphates by releasing organic acids, such as gluconic acid and 2-ketogluconic acid, or can mineralize organic phosphates by secreting extracellular phosphatases. Examples of phosphate-solubilizing and phosphate-mineralizing bacteria that can enhance plant growth include *Azotobacter chroococcum, Bacillus* spp., *Enterobacter agglomerans, Pseudomonas chlororaphis, Pseudomonas putida,* and *Rhizobium* and *Bradyrhizobium* spp. (Vessey, 2003). Like phosphorous, iron is often in an insoluble form in soil. Many bacteria produce organic compounds, called siderophores, that bind Fe^{3+} and make it available for conversion to the preferred form, Fe^{2+}. Bacterial Fe^{3+}-siderophore complexes may facilitate uptake of iron not only into bacteria, but also into plants that are able to recognize the bacterial Fe^{3+}-siderophore complex (Sharma et al., 2003). Whether these siderophore complexes can significantly contribute to the iron requirements of the plant, however, remains equivocal.

4.3.2. Bacterial products that stimulate plant growth

Bacteria may induce changes in root morphology or physiology that increase the root surface area or respiration rate, thus influencing nutrient uptake and plant growth. Following the inoculation of plants with PGPR, plants often exhibit increases in root weight, root length, and number of root hairs, all of which are associated with increased root surface area.

Phytohormones. Phytostimulation often results from bacterial production of phytohormones. The most common phytohormone produced by PGPR, the auxin indole-3-acetic acid (IAA), has been suggested or demonstrated to contribute to plant growth promotion by *Azospirillum brasilense, Aeromonas veronii, Agrobacterium* spp., *Alcaligenes piechaudii, Bradyrhizobium* spp., *Comamonas acidovorans, Enterobacter* spp., and *Rhizobium leguminosarum* (Vessey, 2003). IAA production can increase root growth and root length, and has been associated with root hair proliferation and root hair elongation. Some plant growth-promoting bacteria produce gibberellin or cytokinins (de Salamone et al., 2001), although extent to which these compounds have a role in plant growth promotion is not known.

Enzymes. Bacterial production of 1-aminocyclopropane-1-carboxylate (ACC) deaminase contributes to plant growth promotion by some PGPR (Li et al., 2000). Specifically, ACC deaminase is made by PGPR strains of *Alcaligenes* spp. *Bacillus pumilus, Enterbacer cloacae, Burkholderia cepacia, Pseudomonas putida, Pseudomonas* spp. and *Variovorax paradoxus* (Vessey, 2003). The mode of action is believed to be via cleavage of ACC, which is the immediate precursor to ethylene during ethylene biosynthesis in roots. Among its many effects on plants, ethylene inhibits root growth; thus, reductions in ethylene via ACC deaminase-mediated reductions should increase root growth.

Lumichrome. Some PGPR have been found to increase root respiration. Lumichrome, a degradation product of riboflavin, is a compound that contributes to this activity in *Sinorhizobium meliloti.* At concentrations of only 3 nM, lumichrome was found to increase root respiration by as much as 21% (Phillips et al., 1999). Increased root respiration is associated with increased CO_2 in the rhizosphere, which favors bacterial growth, as well as increased net carbon assimilation in plants. The finding that at least 50% of the rhizosphere isolates examined release riboflavin indicates that lumichrome may be a common bacterial-derived, plant growth-promoting signal in the rhizosphere (Phillips et al., 1999).

Volatile compounds. Recently, PGPR strains of *Bacillus subtilis* and *Bacillus amyloliquefaciens* were shown to produce volatile compounds that promote plant growth (Ryu et al., 2003). In particular, the compounds 2,3-butanediol and acetoin promoted the growth of *Arabidopsis thaliana* seedlings. The involvement of such compounds in plant growth promotion by other bacteria is not known and until now has not been rigorously examined, in part due to the experimental challenges of studying volatile compounds.

4.3.3. Bacteria that enhance the positive effects of symbionts

Some bacteria enhance the beneficial effects of bacterial-legume and fungal-plant symbioses. Enhancement of legume symbioses is often manifest as an increase in the number or total mass of the nodules on a plant or in the nitrogen content of the plant. Strains of at least 12 bacterial species in the genera *Aeromonas, Azospirillum, Bacillus*, *Pseudomonas,* and *Serratia* have been identified as active in such nodulation or nitrogen fixation enhancement (Vessey, 2003). In many of these strains, this activity is attributable to the production of IAA, which can lead to more root growth and more sites for infection. The enhancement activity of *Azospirillum brasilense,* however, results from its ability to stimulate flavonoid production by the host (Burdman et al., 1996), which consequently stimulates bacterial nodule initiation activities. The enhancement activity of other bacteria involves bacterial secretion of B vitamins (Marek-Kozaczuk & Skorupska, 2001) or even a phytotoxin that inactivates a specific enzyme in the nodule (Knight & Langston-Unkefer, 1988). Co-inoculation of plants with PGPR and beneficial fungi, such as arbuscular mycorrhizae (AM), can measurably enhance plant growth. This enhancement activity appears to occur, at least in part, from increased phosphorous content in the plant. Examples of such tripartite relationships include *Bacillus polymyxa* or *Azospirillum brasilense* and the AM fungus *Glomus aggregatum* on a grass (Ratti et al., 2001), and *Enterobacter agglomerans* and the AM fungus *Glomus etunicatum* on tomato plants (Kim et al., 1998). Interestingly, an endocellular bacterium, a β-proteobacterium called "Candidatus *Glomeribacter gigasporarum*" has recently been recognized within the AM fungus *Gigaspora margarita* (Jargeat et al., 2004), suggesting potentially complex and highly evolved tripartite mutualisms on plants.

4.3.4. Bacteria that reduce the negative effects of pathogens

Bacteria that reduce the incidence or severity of plant diseases are often referred to as biological control agents, biocontrol agents, or, if they exhibit antagonistic activity toward a pathogen, antagonists. At least 20 bacterial biocontrol products have been marketed for use in the rhizosphere; these include *Pseudomonas* species (*chlororaphis, fluorescens, putida*), *Bacillus* species (*amyloliquefaciens, cereus, licheniformis, pumilus, subtilis*), and *Streptomyces griseoviridis* (Haas & Defago, 2005). Products for use aboveground include *Agrobacterium* strains and *Pseudomonas fluorescens*. Additional bacteria reported to exhibit biocontrol activity but that are not yet commercialized include *Burkholderia cepacia* complex strains and strains of *Paenibacillus polymyxa, Enterobacter cloacae, Lysobacter enzymogenes, Pantoea agglomerans, Pseudomonas* spp. (*aeruginosa, aureofaciens, corrugate, syringae*), *Serratia marcescens*, and *Streptomyces* spp. (Thomashow & Weller, 1996; van Loon et al., 1998), as well as nonpathogenic variants of pathogens such as *Xanthomonas campestris* (Rukayadi et al., 2000). Among these, biocontrol-active *Pseudomonas* species have been best characterized at the molecular level. Some biocontrol bacteria have been identified based on their presence in disease suppressive soils, which are soils that foster only low levels of one or more soilborne diseases. Investigations into the mechanisms of this suppressiveness have sometimes indicated a microbial basis. For example, disease suppressive soils that reduce take-all of wheat, a fungal disease caused by *Gaeumannomyces graminis* var. *tritici* (*Ggt*), exhibit a decine in take-all during consecutive seasons of wheat cultivation, and this decline is paralleled by a shift in the microbial community composition in which the populations of *Pseudomonas* species that produce an antibiotic, 2,4-diacetylphloroglucinol, effective against *Ggt*, increase. These findings are suggestive of bacteria that reduce the negative effects of a pathogen by directly affecting the pathogen. Biocontrol agents may also control pathogens by influencing the plant or the other members of the microbial community. In fact, bacteria that provide effective disease protection probably typically employ multiple modes of action, several of which are described below.

Production of antimicrobial compounds. The first well-developed biocontrol agent, *Agrobacterium radiobacter* strain K84, produces an antibiotic called agrocin 84 that contributes to its ability to control the pathogen *A. tumefaciens*. Most known biocontrol agents in the genus *Pseudomonas* produce antifungal metabolites that contribute to biocontrol activity. These fall into six classes: phenazines, pyrrolnitrin, pyoluteorin, phloroglucinols, cyclic lipopeptides, and hydrogen cyanide (Haas & Defago, 2005). The biosynthesis, regulation, and mode of action of these compounds, as well as the extent to which they contribute to biocontrol activity, has been extensively investigated (Chin-A-Woeng et al., 2003; Haas & Defago, 2005). Antifungal compounds have also been isolated from biocontrol-active *Streptomyces* and *Bacillus* species. Recent studies have shown that signals from other members of the microbial community, and even from transgenic and natural plants, can influence the production of antimicrobial compounds (Chin-A-Woeng et al., 2003; Fray et al., 1999; Gao et al., 2003). This illustrates the complexity of predicting a strain's behavior in a community versus in pure culture.

Production of enzymes that interfere with fungal pathogenesis. Production of chitinases, glucanases, cellulases, lipases and other lytic enzymes may also contribute to effective biocontrol activity. For example, β-1,3-glucanases contribute to the ability of *Lysobacter enzymogenes* to control *Biopolaris* leaf spot (Palumbo et al., 2005). Such enzymes may be involved in bacterial mycoparasitism, a form of antagonism in which bacteria directly colonize the hyphae. Other examples are enzymes that can degrade fusaric acid, which is required for the phytopathogenicity of *Fusarium oxysporum* (Toyoda et al., 1988; Utsumi et al., 1991), enzymes that degrade volatile seed exudates, which are critical to the initiation of pathogenesis by *Pythium ultimum* (Paulitz, 1991), and enzymes that degrade fatty acids from seeds and roots (McKellar & Nelson, 2003).

Competition for resources. Under conditions in which particular resources are limited, bacteria that are superior competitors can restrict the growth of poorer competitors. For example, iron is an important limiting resource in many terrestrial habitats. Successful scavenging of iron can result from producing siderophores and siderophore receptors, or alternatively, from producing only siderophore receptors if these receptors can mediate iron uptake from siderophores from other microorganisms in the community. Although iron competition can contribute to the biocontrol activity of many organisms, particularly for the fluorescent pseudomonads, iron competition is not the sole or even primary factor contributing to biocontrol in the bacteria that have been characterized. In contrast, nutrient limitation appears to be involved in the control of *Erwinia amylovora* by *Pantoea agglomerans* (*Erwinia herbicola*) on flowers (Giddens et al., 2003). Competition for sites, also called *niche exclusion,* has been implicated as a mode of control almost exclusively in the phyllosphere, where sites harboring sufficient nutrients for growth may be low. An excellent example is the ability of *Pseudomonas syringae* mutants that were deficient in the production of a single protein, the icenucleation protein, to competitively exclude the wild-type *P. syringae* strain when subsequently introduced onto leaves; this phenomenon was termed *pre-emptive exclusion* (Lindow, 1987).

Induction of plant resistance. Some bacteria induce a state of systemic resistance in plants that protects the plants against a broad spectrum of pathogens, including fungal, bacterial and viral pathogens. At least in the *Arabidopsis thaliana* model system, non-pathogenic bacteria elicit an induced systemic resistance (ISR) response that involves a distinct signal transduction pathway from that involved in systemic acquired resistance (SAR), which is elicited by pathogens (Bloemberg & Lugtenberg, 2001; Ping & Boland, 2004). Some bacteria, however, activate both ISR and SAR pathways simultaneously. Reported elicitors of ISR include whole cells and cellular components, such as lipopolysaccharides and siderophores, as well as salicylic acid and the volatile compound butanediol. Successful elicitation of ISR has been most extensively characterized for strains of *Pseudomonas* species, but also occurs with other organisms such as *Bacillus* species, and appears to occur in a bacterial strain- and plant species-specific manner.

Interactions among members of the microbial community. Interactions among members of a microbial community may also contribute to pathogen control. For example, some bacterial traits involved in pathogenesis or in pathogen control are induced when the population density exceeds a certain threshold, i.e., when a quorum is reached. This occurs in response to extracellular bacterial signal molecules, such as acyl-homoserine lactones (AHLs) in $Gram^-$ bacteria or small peptides in $Gram^+$ bacteria (Winans & Bassler, 2002). Interception, degradation, or production of similar signal molecules by one population can alter signal-based cell-cell communication among cells of another population. For example, the rhizosphere colonist *Pseudomonas aureofaciens* produces phenazines in response to AHLs produced by other rhizosphere bacteria, including phenazine-sensitive bacteria that are killed by the resulting phenazine (Pierson & Pierson, 1996). This phenomenon of sharing identical or similar autoinducers is called *cross talk.* AHL degradation can minimize the expression of AHL-induced pathogenic traits in plant pathogens, as has been shown for *Erwinia carotovora* and *Agrobacterium tumefaciens* (Zhang & Dong, 2004), but may also reduce the effectiveness of AHL-dependent biocontrol activities. This phenomenon, called *quorum quenching,* may be common in the rhizosphere based on the large number of AHL-degraders, but its implications for biocontrol have not yet been fully explored. *Cross-feeding*, which is the sharing of nutrients such as by assimilating iron that is complexed to siderophores produced by other bacteria, likely occurs in plant-associated communities but has not been extensively documented. Lastly, the introduction of a biocontrol-active *Bacillus cereus* strain onto roots was discovered change the rhizosphere community to one resembling that of a soil community; this community shift was proposed to help *camouflage* the plant from the pathogen (Gilbert et al., 1994). This finding illustrates the complex interactions within microbial communities and the difficulty in identifying the mode(s) of action of a given biocontrol agent.

5. PHYTOBACTERIA WITH A PARASITIC ASSOCIATION WITH PLANTS

5.1. Phytopathogens

The major group of bacteria that adversely affect plants are the bacterial plant pathogens. Although the changes in plant physiology induced by these organisms are generally considered to be detrimental to plant health, some have been exploited as favorable from a horticultural perspective. Symptoms induced by phytopathogenic bacteria range from local areas of cell death such as leaf spots, cankers, and scabby lesions, to wilts, yellowing, tissue liquefication, and tumor formation. A given symptom may be caused by pathogens in multiple genera, and organisms within a give genus may incite diseases with diverse symptoms.

5.1.1. Taxonomy of phytopathogenic bacteria

Only a limited diversity of prokaryotes can cause diseases in plants. Among the 26 phyla that currently comprise the prokaryotes (Fig 1), phytopathogenic bacteria are found in only 3 Bacterial phyla and none of the Archaeal phyla. As with the phytosymbionts, the phytopathogens are phylogenetically clustered with non-pathogens (Fig 1). The majority of bacterial plant pathogens are Gram$^-$ bacteria that fall within the α-, ß-, and γ- classes of the Proteobacteria (Table 2). This phylum contains the largest number of Bacterial genera and the greatest diversity in metabolic and morphological diversity among the Bacteria. The remaining phytopathogens fall into either the Firmicutes or Actinobacteria phyla. The Firmicutes consists of both low G+C content Gram$^+$ bacteria, i.e., the clostridia and bacilli, as well as the cell wall-less genera known collectively as the Mollicutes. The Actinobacteria, or actinomycetes, are Gram$^+$ bacteria with a high G+C content that form branched filaments or hyphae. Collectively, these plant pathogens are diverse in their approach to invading plants, their mechanisms of pathogenicity, their effect on plant growth and vigor, their epidemiology, their ability to be cultured, their requirements for growth and survival, and even the very structure and fluidity of their genomes.

Historically, the names of phytopathogenic species were based on the host from which they were isolated. As common in all fields of bacteriology, this taxonomy has been revised using polyphasic approaches that consider phylogenetic and phenetic information. At present, the taxonomy of bacterial phytopathogens is reviewed and regularly updated by the International Society for Plant Pathology. In addition to the standard taxonomic ranks of family, genus and species, phytopathogens are commonly divided into subspecies designated with descriptors such as subspecies, pathovars, and races. *Pathovar* indicates that an organism is a pathogen on one or more host species, and is usually applied to pathogens with a fairly narrow host range. In 1991, the term pathovar was extended to organisms that induce distinct symptoms on a single plant species, and most notably to distinguish *Xanthomonas oryzae* pv. *oryzae* from *X. oryzae* pv. *oryzicola*, which cause leaf blight and leaf streak of rice, respectively. *Race* is used to distinguish among strains that infect distinct sets of cultivars, or other identifiable germplasm, within a host species. As an exception, the term *race* is also used to divide *Ralstonia solanacearum* into subspecies based on the ability to infect distinct plant species, such as most solanaceous plants (race 1) versus only a single species, bananas (race 2). The term *biovar* has also been used to divide *R. solanacearum* into subspecies based on differences in carbohydrate utilization profiles.

5.1.2. Economic importance of phytopathogenic bacteria

Bacterial diseases are generally of lower economic importance than fungal and viral diseases, but their economic impact can be large. Many bacterial diseases cause direct crop losses. For example, leaf blight of rice, caused by *Xanthomonas oryzae*, reduces the growth, maturation, and quality of grain, with yield losses up to 30% in southeast Asia. Similarly, fireblight, caused by *Erwinia amylovora,* can kill entire pear and apple trees if not controlled, and soft rots can cause devastatingly large losses of fleshy vegetables in the field, in transit and during storage. Some bacterial diseases decrease crop quality but not necessarily crop yield; examples of this include potato scab (*Streptomyces scabies*) and bacterial brown spot of beans (*Pseudomonas syringae* pv. *syringae*), which cause visible lesions that decrease produce marketability. Simply the threat of a bacterial disease can reduce the potential for crop production, as illustrated by the lack of production of pears in the eastern U. S. and parts of California because of a fireblight threat, and the absence of grape production in the southeastern U. S. due to the possibility of Pierce's disease of grape. Control measures, which also place an economic burden on productivity, include the generation of pathogen-free seeds, sterilization of seedbeds, use of sanitation and controlled conditions for storage of produce, and application of bactericides such as antibiotics and copper sprays. The control of citrus canker, alone, has incurred losses of over 6 million citrus trees in Florida in the most recent epidemic (1995-2005); this is due to disease control via eradication of citrus trees to prevent the spread of the causal agent *Xanthomonas citri* (syn *X. campestris* pv. *citri, X. axonopodis* pv. *citri*).

Interestingly, the economic impact of phytopathogenic bacteria also includes increases in the quality of at least one ornamental, poinsettia. In the early 1900's, most poinsettias were sold as cut flowers; however, the discovery of a variant that was highly branched and thus bushy led to its marketing as a potted ornamental plant. These plants are the foundation of an industry that was worth $325 million by 1997. This bushy phenotype is now known to result from infection with a phytoplasma, a pathogen that often causes fasciation of plants.

5.1.3. Pathogens that cause necrotic diseases of plants

Bacteria that cause plant cell death, or necrosis, are often referred to as *necrogens.* In these diseases, most of the bacterial growth precedes plant cell death. Bacterial induced necroses may remain localized or may spread through the plant, thus becoming systemic. The most common symptoms are necrotic spots, blights and cankers on leaves, stems, blossoms and fruits. Spots usually begin as translucent or watersoaked areas, suggestive of water leaking into the intercellular spaces; these areas eventually turn brown, indicative of necrosis, and may dry and even fall out, leaving a hole. In some plants, the spots are restricted by the large veins and thus appear angular, as in dicots, or as stripes or streaks, as in monocots. The presence of a halo of chlorosis, or yellowing, around a spot is often caused by the diffusion of a toxin; such phytotoxins are produced primarily by pathovars of *Pseudomonas syringae*. In contrast to distinct spots, necroses may appear as a darkening of the vascular tissue, as a coalescence of spots over a large area of a leaf, stem or blossom, called a bligtht, or as a sunken area surrounded by living tissue, called a canker.

Bacteria in the genera *Xanthomonas* and *Pseudomonas* cause most bacterial spots and blights, as well as many streaks and stripes diseases. Strains in these genera typically exhibit a narrow host range and thus are classified into many pathovars. For example, *P. syringae* has been subdivided into at least 50 pathovars and *X. axonopodis* into at least 34. In fact, prior to a major reclassification of *Xanthomonas* in 1995 (Vauterin et al., 1995), the species *X. campestris*, alone, was subdivided into over 140 pathovars. The induction of necrotic spots or blights in leaves and fruits generally involves bacterial invasion through the stomata, the hydathodes or wounds, bacterial multiplication in the intercellular regions of the tissue, and eventual plant cell collapse and plant cell death in the infected site. Humid or wet weather tends to favor invasion into plant tissue as well as the release of masses of bacteria; these bacteria can then spread to new tissues or plants. Examples of such necrotic diseases include wildfire of tobacco (*P. syringae* pv. *tabaci*), angular leaf spot of cucumber (*P. syringae* pv. *lachrymans*), bacterial streak of barley and black chaff of wheat (*X. translucens*), bacterial leaf blight (*X. oryzae* pv. *oryzae*) and leaf streak (*X. oryzae* pv. *oryzicola*) of rice, bacterial spot (*X. vesicatoria*) and bacterial speck (*P. syringae* pv. *tomato*) of tomato, and bacterial brown spot (*P. syringae* pv. *syringae*), common blight (*X. phaseoli*), and halo blight (*P. syringae* pv. *phaseolicola*) of bean.

Phytopathogenic bacteria also induce other types of necrotic diseases. *Erwinia amylovora*, causal agent of fireblight, was the first bacterial species to be identified as a plant pathogen. Fireblight symptoms often progress from watersoaking to wilting and blackening of the tissue, with characteristic symptoms of shriveled black flowers and darkened twigs with hanging leaves. *E. amylovora* infection occurs primarily through blossoms and secondarily through wounds, and is followed by multiplication in the intercellular spaces, host cell collapse and necrosis, bacterial spread into older tissue, and induction of plant-directed deposition of cork layers, forming cankers that support overwintering of *E. amylovora*. In the spring, *E. amylovora* cells multiply and ooze out of the canker margins, spreading via wind and attachment to insects. Canker of stone and pome fruits (*P. syringae* pvs. *morsprunorum* and *syringae*) and citrus canker (*X. citri*) are among the few canker diseases caused by bacteria. Like fireblight, citrus canker involves multiplication in the intercellular spaces, as well as induction of plant cell growth, pustule formation, and eventual epidermal rupturing and release of bacterial "ooze" that promotes spreading. Yet another necrotic disease, common scab of potato (*Streptomyces scabies*) involves entry through natural openings or wounds in the belowground parts of plants, growth in the intercellular spaces, and induction of layers of corky cells that yield the appearance of scab-like lesions. Necrotic diseases comprise the majority of distinct diseases induced by phytopathogenic bacteria.

5.1.4. Pathogens that cause vascular diseases

Bacteria that decrease vascular function are also generally necrogens. As might be predicted by the distinct functions of the xylem versus the phloem, bacteria that reduce xylem function are different from those that reduce phloem function (Bové & Garnier, 2002). Typical symptoms of xylem diseases are wilts; these result from blocked transport of water and nutrients and can lead to accelerated leaf senescence, a decline in general health, stunting, and death. Although not well understood, this dysfunction may be caused by occlusion of the pit membranes or from damage to the xylem walls and pit membranes, particularly damage that introduces air into the xylem and thus disrupts the hydrostatic pressure needed for water transport. Some bacteria, including *Ralstonia solanacearum, Pantoea stewartii,* and *Xanthomonas campestris*, can destroy the xylem vessel walls, spread and multiply in the intercellular spaces, and possibly eventually egress onto the leaf surface through stomata or wounds. Examples of this type of xylem disease include black rot of crucifers (*X. campestris* pv. *campestris*), which appears as V-shaped chlorotic blotches on leaf margins that progress to blackening of the veins and stems, Southern bacterial wilt (*Ralstonia solanaearum* race 1), which manifests as sudden wilt and death of solanaceous plants, Moko disease of bananas (*R. solanaearum* race 2), potato ring rot (*Clavibacter michiganensis* subsp. *sepedonicus*), and bacterial canker and wilt of tomato (*C. michiganensis* subsp. *michiganensis*). In contrast to these non-fastidious bacteria, fastidious bacteria, or bacteria that require complex media for growth, invade only the xylem and cause wilt entirely by vascular occlusion. Only three known species are pathogens of this type: *Xylella fastidiosa*, *Leifsonia xyli*, and *Ralstonia syzygii.* An example of this type of disease is Pierce's disease of grape (*X. fastidiosa*), which appears as sudden scalding and drying of leaves, wilting of vines, and halting of grape production.

A typical symptom of phloem diseases is yellowing of the plant tissue. Like xylem diseases, the mechanism of phloem dysfunction is not well understood, but has been proposed to result from blocked transport of photosynthetic products, resulting in nutrient starvation, gradual yellowing or reddening of leaves, stunting, fasciation (shoot proliferation or witches' broom), and sometimes death. *Spiroplasma citri,* the only phloem-restricted bacteria that have been cultivated, appear to impair sucrose loading into sieve tubes. The cell wall-less phytoplasmas, previously called mycoplasma-like organisms (MLOs), and the Gram^{-} liberibacters and phlomobacters are all phloem-restricted and have resisted cultivation. Xylem invaders may be spread by mechanical transfer, including by contamination of the mouthparts of feeding insects, whereas phloem invaders can be ingested by insects, subsequently infect the salivary glands, and be spread during feeding events for the remainder of the insect's life. Citrus stubborn is an example of a disease caused by *S. citri*; it involves the formation of an excessive numbers of shoots, stunting, mottled leaves, and small, lopsided fruits that have a bad odor and flavor. In contrast, aster yellows, a disease caused by a phytoplasma, involves an excessive numbers of shoots and dwarfing as well as flower sterility and organ deformation. Collectively, the phytoplasmas cause over 200 plant diseases.

5.1.5. Pathogens that cause soft rots

Bacteria that cause plant cell separation leading to tissue collapse, i.e., maceration of plant tissue, are called *macergens*. Tissue maceration, the most characteristic symptom of soft rot diseases, begins as a small watersoaked lesion that expands and intensifies until the tissue turns soft and watery. Externally, the surface may remain intact, although brown and depressed, or become covered in an oozing bacterial slime layer. Foul odors are common due to the release of volatile compounds during tissue degradation. Most bacterial growth occurs after plant cell lysis in these diseases. Soft-rotting bacteria are notable for the speed at which they promote soft rot: stored produce may liquefy in only a few hours. These pathogens typically invade through wound sites or natural openings such as lenticels and remain in the intercellular spaces and vascular tissue until the environmental conditions become suitable for disease development. At this time, they coordinately produce large amounts of exoenzymes, including cellulolytic enzymes, pectate lyases and pectin methylesterases; these dissolve the plant cell walls and pectin holding the plant cells together and cause tissue collapse and plant cell lysis. The primary species that cause soft rots are *Erwinia chrysanthemi* and *E. carotovora* subsp. *carotovora,* which exhibit a broad host range, and *E. carotovora* subsp. *atroseptica*, which infects primarily potatoes. These organisms have been renamed *Pectobacterium chrysanthemi, P. carotovorum* subsp. *carotovorum,* and *P. atro-septicum,* respectively; however, these names have not yet been widely adopted. A range of opportunistic pathogens can also cause soft rot under some conditions, including *Bacillus* spp., *Clostridium* spp., *Pseudomonas marginalis* and *Pantoea agglomerans*.

5.1.6. Pathogens that cause tumor diseases

Bacteria that cause uncontrolled host cell proliferation in roots, stems or leaves are called *tumorigens* or *oncogens.* These bacteria induce both hyperplasia (enhanced cell division) and hypertrophy (cell enlargement). Although phytopathogens cause only a limited number of tumor diseases, one such disease, crown gall, has been extensively studied and will be discussed more below. The major symptom of crown gall, caused by *Agrobacterium* species, is a mass of plant cells that often occurs at the crown of the plant. Another tumor disease is leaf gall caused by *Rhodococcus fascians.* This pathogen secretes molecules that interfere with the hormone balance in the plant, thus inducing fasciation and sometimes fusing of stems. Galls on olive trees (called olive knot), oleander, ash, and forsythia can be formed by *Pseudomonas savastanoi,* a pathogen that enters wounds, proliferates in the intercellular spaces, and induces plant cell proliferation and enlargement in the cambium at least partly through the production of the plant hormone IAA. Other tumor diseases include crown and root gall on *Gypsophila* (known in the floral industry as Baby's Breath), which is caused by *Pantoea agglomerans* pv. *gypsophilae*, and carrot gall, which is caused by *Rhizobacter dauci*.

Agrobacterium tumefaciens causes crown gall on a wide range of plant species, whereas the closely related species *A. vitis* causes crown gall on grape, *A. rubi* causes cane gall on raspberries and blackberries, and *A. rhizogenes* causes hairy root disease on apple. The galls starts as small, white, spongy masses but sometimes grow to have convoluted dark brown surfaces that may even lignify, and thus become hard and woody. The extent to which they affect the health of the host varies. During the development of crown gall, agrobacterial cells enter wounds, such as those made by insects or grafting. They bind to plant cells, transfer tumorigenic DNA (T-DNA) into the plant cells via a pilus much like that used during bacterial conjugation, and direct the plant cell to release nutrients called opines. These opines are utilized specifically by the infecting strain, which remains on the surface of the developing tumor. The ability of agrobacteria to mediate this interkingdom transfer of DNA has been widely exploited as a tool for genetically modifying plants. These *Agrobacterium* species contain a large plasmid, designated Ti (tumor-inducing) in *A. tumefaciens* and Ri (root-inducing) in *A. rhizogenes*, which harbors genes involved in phytopathogenicity, just as symbiotic strains of *Rhizobium, Sinorhizobium,* and *Azorhizobium* contain a Sym (symbiosis) plasmid that harbors genes involved in nodulation and nitrogen fixation. The Sym, Ti and Ri plasmids are closely related and likely originated from a common ancestor. In fact, *Agrobacterium* strains that acquire a Sym plasmid also acquire the ability to form root nodules and fix nitrogen symbiotically (Brom et al., 1988; Martinez et al., 1987; Novikova & Safronova, 1992), whereas *Rhizobium* strains that acquire a Ti plasmid acquire the ability to incite galls on plants (Hooykaas et al., 1977). Genome fluidity due to the horizontal transfer of plasmids was not fully appreciated when these genera were first named, and the ability to symbiotically fix nitrogen or cause plant disease was used as the primary criterion for taxonomic classification. The intermingling of the members of these genera on a phylogenetic tree based on 16S rRNA, combined with the historical classification of these organisms into distinct genera, has led to current debates over the taxonomy of these organisms, including a recent proposal to transfer the six known pathogenic *Agrobacterium* species to the genus *Rhizobium* (Table 2).

5.2. Ice nucleating bacteria

Ice nucleating bacteria can also adversely affect plants (Lindow, 1983). Because of their ability to produce proteins that nucleate ice formation in supercooled water, the presence of these bacteria on plant surfaces increases the probability of frost injury when the plants are under low temperature conditions. This is particularly important for front-sensitive plants, which lack significant mechanisms for frost tolerance and thus must avoid ice formation to avoid frost injury. An ice nucleation event on a frost-sensitive plant can result in the rapid intracellular and intercellular spread of ice. All of the known ice nucleation active (ina) bacteria are Gram$^-$, plant-associated bacterial species. The first ina species identified was *Pseudomonas syringae*, which is commonly found on leaves throughout the world. Some *P. syringae* strains produce ice nuclei that are active at temperatures as high as -1.5°C, as compared to those of other species that are active at -3 to -8°C, and silver iodide ions, which are active at -8 to -10°C, making them among the most active inorganic ice nuclei known. Other bacterial species that have strains that are capable of ice nucleation are *Pseudomonas viridiflava*, *Ps. fluorescens, Xanthomonas campestris, Pantoea ananatis,* and *Pa. agglomerans*, although the proportion of strains within each species that are ina varies from 100% for *Ps. viridiflava* to 50% for *Ps. syringae*, 10% for *Pa. agglomerans*, and <5% for *Ps. fluorescens*.

5.3. Bacteria that produce high amounts of IAA

Although bacterial production of the plant growth hormone IAA can promote plant growth, as discussed above, and is a factor contributing to gall formation by some tumor-inducing phytopathogens, some bacteria produce sufficiently high levels of IAA that the IAA negatively impacts plant development. A few studies have shown that some nonphytopathogenic bacteria that produce high amounts of IAA in culture can inhibit or reduce root growth (Barbieri & Galli, 1993; Patten & Glick, 2002), although the significance of this in a field setting has not been explored. IAA production by epiphytic bacteria such as *Pseudomonas fluorescens* has also been shown to contribute to increased fruit russet, a condition in which brown, corky regions are present on fruits such as apples and pears (Lindow et al., 1998). This condition reduces the quality and storage life of the fruit.

5.4. Deleterious rhizobacteria

Bacteria that inhibit plant growth without causing obvious disease symptoms have been referred to as deleterious rhizobacteria (DRB) (Suslow & Schroth, 1982). Over the last decade, the potential for these bacteria to help control weeds has been actively investigated (Boyetchko, 1997; Kremer & Kennedy, 1996). Studies have reported DRB-mediated suppression of weed growth and reduction in seed germination, seedling vigor, root elongation, and weed density. Although the mechanisms of plant growth suppression have not been rigorously evaluated, DRB are thought to inhibit plant growth in part through the production of harmful metabolites, such as cyanide, phytotoxins, and high concentrations of phytohormones. DRB may also inhibit plant growth by enhancing the effects of pathogens or attenuating the effects of beneficial microorganisms, such as mycorrhizae. Like with the PGPR, bacteria in the genus *Pseudomonas* are common among isolates recognized as DRB and are particularly common among the DRB that produce cyanide. Interestingly, individual isolates have been shown to promote plant growth under one set of conditions or on one host genotype, but to suppress it under other conditions or on other genotypes. This illustrates at least one of the challenges to the effective exploitation of these bacteria for weed control.

6. FUNCTIONS OF PLANT-ASSOCIATED BACTERIA THAT INFLUENCE PLANT OR ENVIRONMENTAL HEALTH

6.1. Decomposition of organic matter

Bacteria on roots function as a reservoir of nutrients for plants as well as degrade and transform nutrients deposited by the roots (Beare et al., 1997; K. E. Lee & Pankhurst, 1992). Bacteria are critical to the decomposition of most organic matter in soils, including plant exudates, secretions, and residues, and to nutrient cycling, immobilization, and uptake by plants. Bacteria are the primary agents for cycling nitrogen because of their ability to perform all of the major nitrogen transformations (nitrification, dissimilatory nitrate and nitrite reduction, denitrification, and symbiotic and asymbiotic N_2 fixation). They also promote the release of available phosphorus, as described above, and contribute to sulfur transformations, including sulfur oxidation, which can release sulfuric acid that promotes phosphorus release via acidification, and sulfate reduction, which actually decreases sulfur availability to plants. Bacteria are major players in iron cycling due to the ability of various species to oxidize, reduce and/or precipitate iron. Many of these transformations depend on available carbon and nitrogen and thus likely occur preferentially in the rhizosphere. Bacteria can also enhance decomposition processes performed by other microorganisms, as illustrated by the ability of free-living diazotrophs to provide fixed nitrogen to wood-rotting fungi and thus promote fungal composting of the wood (Postgate, 1998). Interestingly, the ability of rhizosphere bacteria to sequester carbon has been proposed to contribute to carbon immobilization, and thus may be one factor helping to mitigate the rising CO_2 levels associated with global climate change.

6.2. Degradation of xenobiotic compounds (e.g., synthetic pesticides and pollutants)

Plants can enhance the rates of microbial degradation of herbicides, pesticides and other pollutants in soils. *Phytoremediation*, which is the use of plants and their associated microbes for environmental cleanup, offers a relatively low cost, environmentally friendly form of bioremediation that can be performed *in situ*, thus not requiring the physical movement of contaminated soils. Plants, themselves, can mediate the cleanup of some pollutants, such as by taking them up from the soil and releasing them into the air (phytovolatilization), accumulating them in tissues (phytoextraction), or degrading them (phytodegradation). Plants can also promote microbial degradation of pollutants, a process known as rhizosphere remediation, phytostimulation, or rhizodegradation (Pilon-Smits, 2005). In rhizosphere remediation, plant roots promote bacterial access to soil contaminants by moving them toward the roots in the transpiration stream and providing bacterial access to soil layers that may otherwise be impermeable. Bacterial degradation activities are likely stronger in the rhizosphere than in bulk soil because of 1) high bacterial densities and metabolic activity, due in part to the exudation of up to 35% of the plant photosynthate into the rhizosphere, 2) the induction of bacterial degradation pathways by secondary plant compounds released from roots, and 3) potential co-metabolism of the secondary plant compounds and the pollutants. Pollutant-degrading bacteria often exhibit a diversity of catabolic pathways, allowing degradation by a single strain as well as by consortia of bacteria that are each capable of distinct steps in a catabolic pathway (Kuiper et al., 2004). Rhizosphere microorganisms have been found to mediate the cleanup of many hydrophobic organic pollutants, including polychlorinated biphenyls and polycyclic aromatic hydrocarbons (Pilon-Smits, 2005). Surprisingly few studies have focused on identifying bacterial degraders in rhizosphere communities or introducing pure cultures of pollutant-degrading strains to promote rhizosphere remediation. Interestingly, bacteria that can degrade specific pollutants were enriched among endophytic bacteria in some plants as compared to in the rhizosphere or soil communities (Siciliano et al., 2001), suggesting that endophytic populations, in particular, may contribute to the degradation of soil contaminants. The degradation of compounds inhibitory to plants, such as herbicides, can contribute not only to environmental cleanup, but also to the protection of plants from the phytotoxic effects of those compounds.

6.3. Production of extracellular polymers that contribute to soil aggregation

Microbes help maintain soil structure by producing organic polymers that contribute to soil aggregation; this production is greatest in the rhizosphere where nutrients are abundant. Bacterial polysaccharides, in particular, adsorb onto clay particles and help bind the particles together (Chotte, 2005). Soil aggregation is important to the water holding capacity of a soil, the movement of water, the growth of roots, and the resistance of soil to erosion. The hygroscopic nature of bacterial polysaccharides, and perhaps other biopolymers, contributes to soil water retention. The central role of microorganisms in soil aggregation is illustrated by the steps leading to stable aggregate formation: first, the plant residues are broken into large particles by the macrofauna; second, the microorganisms decompose the organic matter into smaller particles; third, the microorganisms, minerals, and organic polymers associate with one another to form aggregates; and fourth, the microorganisms eventually die, leaving mineral-organic polymer complexes, or microaggregates, that protect the organic carbon against further decomposition.

6.4. Contribution of atmospheric freezing nuclei to climatic processes

The production of ice nucleating proteins appears to be an exclusive ability of bacteria that are common to plant surfaces. These bacteria are readily disseminated into the atmosphere, due in part to an upward flux of bacteria from plant canopies under dry conditions (Lindemann & Upper, 1985) and to deposition during rain (Constantinidou et al., 1990). Although these bacteria have been found in clouds (Szyrmer & Zawadski, 1997), the extent to which they function in bioprecipitation is not known. The fact that the ice nuclei from bacteria are the only organic or inorganic ice nuclei that function at near-zero temperatures, as opposed to at the low temperatures of -12° to -15°C, suggests that they may be of particular importance to cloud properties when the temperatures are near zero (Szyrmer & Zawadski, 1997). The role of ice-nucleating bacteria in seeding clouds is an exciting area of investigation that should benefit from developments in agricultural sciences, microbial ecology, and meteorology (Morris et al., 2004).

7. RECENT ADVANCES IN PLANT-BACTERIAL INTERACTIONS

7.1. Genomic sequencing

The generation of complete genomic sequences for many plant-associated bacteria is exponentially increasing our information base on these organisms and is changing how we study them and even how we think about their biology. To date, the complete genome sequence has been published for 13 phytopathogens and 3 phytosymbionts (Table 4). The complete genome has been sequenced, but the data not yet published, for 4 more organisms (*Pantoea citri, Pseudomonas syringae* pv. *phaseolicola,* Candidatus *Phytoplasma asteris,* and *Spiroplasma citri*), and the genome has been sequenced to a draft stage or is being completed for at least 24 more. Complete genomic sequence data is also available for species that associate with plants, including *Bacillus subtilus, Pseudomonas aeruginosa, Ps. putida,* and *Ps. fluorescens.* Among the sequenced strains, the obligate vascular pathogens *Xylella fastidiosa, Leifsonia xyli,* and Candidatus *Phytoplasma asteris* have the smallest genomes (Table 4), possibly due to the lower metabolic flexibility required for growth in the relatively invariant vascular environment. In addition to genome size, the genome structure of these organisms varies. Most have a single, circular chromosome; however, *Agrobacterium tumefaciens* contains two chromosomes, one linear and one circular. Several have plasmids that comprise a significant portion of their genome, as illustrated by *Ralstonia solanacearum,* which has a 3.72-Mb chromosome and a 2.09-Mb megaplasmid, and by *Sinorhizobium meliloti,* which has a 3.65-Mb chromosome and 1.35-Mb and 1.68-Mb megaplasmids. The strains shown in Table 4 also vary in their %G+C content and in the number of plasmids and rRNA operons in their genomes. Many of them have been reported to contain genomic islands, including pathogenicity islands, symbiosis islands, fitness islands, and other regions that were putatively acquired by horizontal transfer (Finan, 2002).

Such horizontal movement may have been promoted by the many insertion elements (25-200 per genome) and transposon and bacteriophage genes (appx 2-200 per genome) that are present in the genomes of these organisms. Surprisingly, the functions of 40% of the protein-coding sequences, on average, have yet to be predicted, and the majority of genes have not yet been experimentally defined, thus illustrating the challenges that lie ahead.

Despite these challenges, the sequence data have provided some insights into the interactions of these organisms with their plant hosts. For example, many of the organisms contain genes that encode a type III secretion system (TTSS), a system that has a central role in the interactions of many pathogens with plants due to its ability to direct the movement of bacterial proteins into plant cells. This TTSS is present in some phytopathogens and phytosymbionts, but not all (Table 5). The demonstration that the TTSS system is critical to the ability of many phytopathogens to elicit a defense response in resistant and nonhost plants, and to cause disease on susceptible plants, caused a major paradigm shift in our understanding of plant-pathogen interactions over the last decade. At present, elucidating the functions of the translocated proteins is a major research thrust, the fruits of which are hoped to greatly improve our understanding of molecular plant-bacterial interactions. The complete genomic sequences have also provided insights into the production of 1) a type IV secretion system, which secretes macromolecules such as single-stranded DNA from *A. tumefaciens* to host cells and is present in most of the organisms examined, 2) cell-wall degrading enzymes, the complement of which varies greatly among the organisms, 3) type IV pili, which aid in movement across surfaces and may be critical to the development of communities in biofilms on plants, 4) phytotoxin and phytohormones, and 5) enzymes involved in nitrogen fixation (Table 4).

Table 4. Phytopathogens and phytosymbionts for which complete genomic sequence information is available[1].

Year	Organism	Strain	Habitat	Mb	GC	Prot	Pl	rRNA	% Fn
Phytopathogenic bacteria									
2000	*Xylella fastidiosa*	9a5c	xylem	2.732	51	2,832	2	2	47
2001	*Agrobacterium tumefaciens*	C58	soil, rhizophere	5.674	59	5,402	2	4	64
2002	*Ralstonia solanacearum*	GMI1000	soil, rhizosphere	5.811	67	5,116	1	4	44
2002	*Xanthomonas axonopodis* pv. *citri*	306	phyllosphere, mesophyll, xylem	5.274	64	4,427	2	2	63
2002	*Xanthomonas campestris* pv. *campestris*	ATCC-33913	phyllosphere, mesophyll, xylem	5.076	64	4,818	0	2	65
2003	*Pseudomonas syringae* pv. *tomato*	DC3000	phyllosphere, mesophyll	6.538	58	5,608	2	5	61
2003	*Xylella fastidiosa*	Temecula 1	xylem, insects	2.521	51	2,036	1	2	57
2004	*Candidatus Phytoplasma asteris (Onion yellows phytoplasma)*	OY-M	phloem, insects	0.861	28	754	0	2	59
2004	*Erwinia carotovora subsp. atroseptica*	SCRI1043	soil, rhizosphere, endophyte	5.064	51	4,472	0	7	NR[2]
2004	*Leifsonia xyli* subsp. *xyli*	CTCB07	xylem, insects	2.584	68	2,030	0	1	58
2005	*Pseudomonas syringae* pv. *syringae*	B728a	phyllosphere, mesophyll	6.094	59	5,090	0	5	74

Year	Organism	Strain	Habitat	Mb	GC	Prot	P1	rRNA	%Fn
2005	*Xanthomonas campestris* pv. *campestris*	8004	phyllosphere, mesophyll, xylem	5.149	65	4,273	0	2	63
2005	*Xanthomonas oryzae* pv. *oryzae*	KACC10331	xylem, insect vectors	4.941	64	4,637	0	2	72
Phytosymbionts									
2000	*Mesorhizobium loti*	MAFF303099	soil, rhizosphere, nodule	7.596	62	7,272	2	2	54
2001	*Sinorhizobium meliloti*	1021	soil, rhizosphere, nodule	6.692	63	6,205	2	3	60
2002	*Bradyrhizobium japonicum*	USDA 110	soil, rhizosphere, nodule	9.106	64	8,317	0	1	52

[1] *Year, year in which sequenced was published; Mb, total size of the genome; GC, %G+C content; Prot, number of predicted proteins; Pl, number of plasmids; rRNA, number of rRNA operons; % Fn, % ORFs with a putative function at the initial time of the report; data is from The Institute for Genomic Research (TIGR) or from the references shown in Table 5.*
[2] *NR, not reported.*

7.2. Advances in functional genomics

The availability of increasing amounts of genomic data enables comparative genomic studies, which are useful for addressing evolutionary questions, as well as global studies of gene and protein function. High throughput approaches, in which all or most of the genes, proteins, or even metabolites in an organism are subjected to functional analyses, are required to maximally utilize genomic data. For example, identifying the full collection of genes that is expressed during symbiosis, pathogenesis, or simply growth *in planta* should provide insights into the developmental and physiological adaptations that allow bacteria to invade and grow in and on plants. At one time, the identification of such genes required screening individual strains containing reporter gene fusions for differential expression in culture versus in plants (Beaulieu & Van Gijsegem, 1990; Lindow & Cirvilleri, 1994; Osbourn et al., 1987). More recent studies have employed promoter-trapping techniques such as *in vivo* expression technology (IVET), in which allows for the selection of promoters expressed in a specified niche, and differential fluorescence induction (DFI), which allows for screening for such promoters based on fluorescence (Rediers et al., 2005). These techniques have enabled the identification of genes that are expressed specifically during plant infection (Boch et al., 2002; Brown & Allen, 2004; Oke & Long, 1999; Osbourn et al., 1987; S. Yang et al., 2004) and during colonization of the phyllosphere (Marco et al., 2005), rhizosphere (Rainey, 1999; Ramos-Gonzalez et al., 2005), and endophytic sites (Rediers et al., 2003). Global gene expression studies that employ microarrays, or microchips, that contain high density arrays of oligonucleotides are becoming more common. Microarray studies have thus far identified bacterial and plant genes expressed during infection by virulent or avirulent pathogens (de Torres et al., 2003; Gibly et al., 2004; Navarro et al., 2004; Okinaka et al., 2002), during symbiosis (Colebatch et al., 2004; Colebatch et al., 2002), during bacterial colonization of plants (Cartieaux et al., 2003; Verhagen et al., 2004; Wang et al., 2005), and in response to specific plant signals (Barnett et al., 2004).

Functional genomics approaches are also being employed to characterize the proteomes expressed during bacterial infection (A. M. Jones et al., 2004; Natera et al., 2000; Rosen et al., 2003). Global analyses of metabolites, such as the metabolites produced in response to symbioses, have also been performed (Desbrosses et al., 2005). Among the most informative studies performed to date are those that combine computational methods with functional genomics approaches (Fouts et al., 2002; Zwiesler-Vollick et al., 2002). Experimental strategies for evaluating gene and protein function have traditionally relied on correlating phenotypic alterations with gene inactivation. Whereas insertional inactivation or deletion of individual genes has been a foundational molecular tool for decades, methods for large-scale inactivation of known genes are now being developed. For example, Holeva et al. (2004) generated a mutation grid of *Erwinia carotovora* transposon-insertion mutants that enables the rapid identification of mutants altered in any given gene. Similarly, a rapid PCR-mediated inactivation strategy was recently applied to a phytosymbiont (Sukdeo & Charles, 2003). Other inactivation strategies, such as signature-tagged mutagenesis (STM) (Hensel et al., 1995), have yet to be applied to plant-associated bacteria.

7.3. Advances in microbial ecology

Advances in the techniques for characterizing microbial communities are exponentially increasing our awareness of which organisms are present in plant-associated microbial communities and how the structure of these communities is influenced by various biotic and abiotic forces. Techniques that provide detailed information on community composition are currently enjoying widespread use (Kent & Triplett, 2002). Most of these techniques detect sequence differences among specific genes, usually those in the ribosomal RNA operon, using fingerprinting techniques; these techniques are laden with acronyms, such as DGGE (denaturing gradient gel electrophoresis), RAPD (random amplified polymorphic DNA), ARDRA (amplified rDNA restriction analysis), and RISA (ribosomal intergenic spacer analysis), with T-RFLP (terminal restriction fragment length polymorphism) analysis directly providing phylogenetic information. Although still early in their development, microarrays are also being adapted for phylogenetic analyses of microbial communities in which whole genomic community DNA is applied to phylogeny-based oligonucleotide arrays. Metagenomic analyses, in which whole community genomic DNA is collected, cloned and evaluated for particular functions, may also provide insights into specific functions in plant-associated microbial communities, although these analyses have thus far not been performed with microbial communities from plants.

Table 5. Putative presence or absence of specific traits in the phytopathogens and phytosymbionts from Table 4[1].

Organism	TTSS	TFSS	CWDE	Pili	Toxin	Horm	N_2 fix	Ref[c]
Phytopathogenic bacteria								
Xylella fastidiosa	no	no	low	yes	no	no	no	1
Agrobacterium tumefaciens	no	yes	low	yes	no	yes	no	2,3
Ralstonia solanacearum	yes	no	high	yes	yes	yes	no	4
Xanthomonas axonopodis pv. *citri*	yes	yes	high	yes	no	no	no	5
Xanthomonas campestris pv. *campestris*	yes	yes	high	yes	no	no	no	5
Pseudomonas syringae pv. *tomato*	yes	no	low	yes	yes	yes	no	6
Xylella fastidiosa	no	no	low	yes	no	no	no	7
Candidatus Phytoplasma asteris (Onion yellows phytoplasma)	no	no	low	no	no	no	no	8
Erwinia carotovora subsp. atroseptica	yes	yes	high	yes	yes	yes	yes	9
Leifsonia xyli subsp. *xyli*	no	no	low	no	no	no	no	10
Pseudomonas syringae pv. *syringae*	yes	no	low	yes	yes	yes	no	11
Xanthomonas campestris pv. *campestris*	yes	yes	high	yes	no	no	no	12
Xanthomonas oryzae pv. *oryzae*	yes	yes	high	yes	no	no	no	13

Phytosymbionts								
Mesorhizobium loti	yes[2]	yes	low	no	no	yes	yes	14
Sinorhizobium meliloti	no	yes	low	no	no	no	yes	15
Bradyrhizobium japonicum	yes[2]	yes	low	yes	mo	yes	yes	16

[1] *TTSS, type III secretion system; TFSS, type IV secretion system; CWDE, cell wall-degrading enzymes (low or high, low or high total number of CWDE); Pili, type IV pili; Toxin, phytotoxin production; Horm, phytohormone production; N_2 fix, nitrogen fixation; Ref, references.*

[2] *The structure is distinct from that in the phytopathogens.*

[3] *1 (Simpson et al., 2000), 2,3 (Goodner et al., 2001; Wood et al., 2001), 4 (Salanoubat et al., 2002), 5 (da Silva et al., 2002), 6 (Buell et al., 2003), 7 (Van Sluys et al., 2003), 8 (Oshima et al., 2004), 9 (Bell et al., 2004), 10 (Monteiro-Vitorello et al., 2004), 11 (Feil et al., 2005), 12 (Qian et al., 2005), 13 (B. M. Lee et al., 2005), 14 (Kaneko et al., 2000), 15 (Capela et al., 2001), 16 (Kaneko et al., 2002).*

Advances in microbial ecology techniques, particularly when coupled with increasing genomic sequence information, are providing novel approaches for exploring the behavior and functional roles of bacteria in natural communities on plants. Improvements in fluorescent protein cellular markers and fluorescence microscopy techniques are enabling unprecedented viewing of bacteria as they colonize and invade plants. Furthermore, uncultured cells in natural communities can now be identified using phylogenetic probes and fluorescence *in situ* hybridization (FISH). Recently, microautoradiography (MAR), which can detect cellular accumulation of a radiolabeled compound, has been coupled with FISH to directly link a specific metabolic activity to a phylogenetically-identifiable organism; however, this technique, called FISH-MAR, has not yet been applied to bacteria on plants. Stable isotope probing (SIP) also links functional activity to specific community members. For example, by providing $^{13}CO_2$ to plants, SIP has been used to follow carbon movement into rhizosphere communities (Manefield et al., 2002; Radajewski et al., 2003) and has the potential to identify which members of the population are active in the subsequent transformation of root exudates. Information on community functions may also be obtained using microarrays of functional genes or even isotope arrays, in which nucleic acids are labeled when cells utilize a radiolabeled substrate prior to microarray hybridization. For example, Singh (2004) has proposed that, following incubation of plants with $^{14}CO_2$, an isotope microarray analysis of nucleic acids from the rhizosphere could be used with a phylogenetic probe-based microarray to identify those taxa that utilize root-derived compounds. Although still in the early days of their application, these tools have the potential to dramatically increase our understanding of the ecology and functional roles of bacteria in plant-associated bacterial communities.

7.4. *Future directions*

Up until recently, we have practiced primarily reductionist science, looking at individual genes in an organism and individual organisms in a community. The plethora of tools for global analyses, including functional genomics for the bacteria and their hosts, and for analyzing organisms within natural communities, including community structure analyses and *in situ* coupling of phylogenetic and functional information, strongly complements the powerful reductionist tools of molecular biology and traditional microbiology. The application of these tools to plant-associated bacteria is really still in its infancy, but is widely predicted to catapult our current knowledge of the biology of these organisms and their interactions with plants to new heights.

8. ACKNOWLEDGEMENTS

This work was supported by the Cooperative State Research, and Extension Services, U.S. Department of Agriculture, under Award no. 2005-35319-15300.

9. REFERENCES

Ashelford, K. E., Day, M. J., & Fry, J. C. (2003). Elevated abundance of bacteriophage infecting bacteria in soil. *Appl. Environ. Microbiol., 69*, 285-289.

Baker, J. A., Entsch, B., & McKay, D. B. (2003). The cyanobiont in an *Azolla* fern is neither *Anabaena* nor *Nostoc*. *FEMS Microbiol. Lett., 229*, 43-47.

Barbieri, P., & Galli, E. (1993). Effect on wheat root development of inoculation with an *Azospirillum brasilense* mutant with altered indole-3-acetic acid production. *Res. Microbiol., 144*, 69-75.

Barnett, M. J., Toman, C. J., Fisher, R. F., & Long, S. R. (2004). A dual-genome Symbiosis Chip for coordinate study of signal exchange and development in a prokaryote-host interaction. *Proc. Natl. Acad. Sci. U. S. A., 101*, 16636-16641.

Beare, M. H., Reddy, M. V., Tian, G., & Srivastava, S. C. (1997). Agricultural intensification, soil biodiversity and agroecosystem function in the tropics: the role of decomposer biota. *Appl. Soil Ecol., 6*, 87-108.

Beaulieu, C., & Van Gijsegem, F. (1990). Identification of plant-inducible genes in *Erwinia chrysanthemi* 3937. *J. Bacteriol., 172*, 1569-1575.

Bell, K. S., Sebaihia, M., Pritchard, L., Holden, M. T., Hyman, L. J., Holeva, M. C., et al. (2004). Genome sequence of the enterobacterial phytopathogen *Erwinia carotovora* subsp. *atroseptica* and characterization of virulence factors. *Proc. Natl. Acad. Sci. U. S. A., 101*, 11105-11110.

Bloemberg, G. V., & Lugtenberg, B. J. (2001). Molecular basis of plant growth promotion and biocontrol by rhizobacteria. *Curr. Opin. Plant. Biol., 4*, 343-350.

Boch, J., Joardar, V., Gao, L., Robertson, T. L., Lim, M., & Kunkel, B. N. (2002). Identification of *Pseudomonas syringae* pv. *tomato* genes induced during infection of *Arabidopsis thaliana*. *Mol. Microbiol., 44*, 73-88.

Bonfante, P. (2003). Plants, mycorrhizal fungi and endobacteria: a dialog among cells and genomes. *Biolog. Bull., 204*, 215-220.

Bové, J. M., & Garnier, M. (2002). Phloem- and xylem-restricted plant pathogenic bacteria. *Plant Sci., 163*, 1083-1098.

Boyetchko, S. M. (1997). Principles of biological weed control with microorganisms. *Hortscience, 32*, 201-205.

Brom, S., Martinez, E., Davila, G., & Palacios, R. (1988). Narrow- and broad-host-range symbiotic plasmids of *Rhizobium* spp. strains that nodule *Phaseolus vulgaris*. *Appl. Environ. Microbiol., 54*, 1280-1283.

Brown, D. G., & Allen, C. (2004). *Ralstonia solanacearum* genes induced during growth in tomato: an inside view of bacterial wilt. *Mol. Microbiol., 53*, 1641-1660.

Buell, C. R., Joardar, V., Lindeberg, M., Selengut, J., Paulsen, I. T., Gwinn, M. L., et al. (2003). The complete genome sequence of the *Arabidopsis* and tomato pathogen *Pseudomonas syringae* pv. *tomato* DC3000. *Proc. Natl. Acad. Sci. U. S. A., 100*, 10181-10186.

Burdman, S., Volpin, H., Kapulnik, Y., & Okon, Y. (1996). Promotion of *nod* gene inducers and nodulation in common bean (*Phaseolus vulgaris*) root inoculated with *Azospirillum brasilense* Cd. *Appl. Environ. Microbiol., 62*, 3030-3033.

Callaham, D., Del Tredici, P., & Torrey, J. G. (1978). Isolation and cultivation *in vitro* of the actinomycete causing root nodulation in *Comptonia*. *Science, 199*, 899-902.

Capela, D., Barloy-Hubler, F., Gouzy, J., Bothe, G., Ampe, F., Batut, J., et al. (2001). Analysis of the chromosome sequence of the legume symbiont *Sinorhizobium meliloti* strain 1021. *Proc. Natl. Acad. Sci. U. S. A., 98*, 9877-9882.

Cartieaux, F., Thibaud, M. C., Zimmerli, L., Lessard, P., Sarrobert, C., David, P., et al. (2003). Transcriptome analysis of *Arabidopsis* colonized by a plant-growth promoting rhizobacterium reveals a general effect on disease resistance. *Plant J., 36*, 177-188.

Chin-A-Woeng, T. F. C., Bloemberg, G. V., & Lugtenberg, B. J. J. (2003). Phenazines and their role in biocontorl by *Pseudomonas* bacteria. *New Phytol., 157*, 503-523.

Chotte, J.-L. (2005). Importance of microorganisms for soil aggregation. In F. Buscot & A. Varma (Eds.), *Microorganisms in soil: roles in genesis and functions* (Vol. 3, pp. 107-119). Berlin: Springer-Verlag.

Colebatch, G., Desbrosses, G., Ott, T., Krusell, L., Montanari, O., Kloska, S., et al. (2004). Global changes in transcription orchestrate metabolic differentiation during symbiotic nitrogen fixation in *Lotus japonicus*. *Plant J., 39*, 487-512.

Colebatch, G., Kloska, S., Trevaskis, B., Freund, S., Altmann, T., & Udvardi, M. K. (2002). Novel aspects of symbiotic nitrogen fixation uncovered by transcript profiling with cDNA arrays. *Mol. Plant-Microb. Interact., 15*, 411-420.

Constantinidou, H. A., Hirano, S. S., Baker, L. S., & Upper, C. D. (1990). Atmospheric dispersal of ice nucleation-active bacteria: the role of rain. *Phytopathology, 80*, 934-937.

da Silva, A. C., Ferro, J. A., Reinach, F. C., Farah, C. S., Furlan, L. R., Quaggio, R. B., et al. (2002). Comparison of the genomes of two *Xanthomonas* pathogens with differing host specificities. *Nature, 417*, 459-463.

de Salamone, I. E. G., Hynes, R. K., & Nelson, L. M. (2001). Cytokinin production by plant growth promoting rhizobacteria and selected mutants. *Can. J. Microbiol., 47*, 404-411.

de Torres, M., Sanchez, P., Fernandez-Delmond, I., & Grant, M. (2003). Expression profiling of the host response to bacterial infection: the transition from basal to induced defence responses in *RPM1*-mediated resistance. *Plant J., 33*, 665-676.

Desbrosses, G. G., Kopka, J., & Udvardi, M. K. (2005). *Lotus japonicus* metabolic profiling. Development of gas chromatography-mass spectrometry resources for the study of plant-microbe interactions. *Plant Physiol., 137*, 1302-1318.

Ercolani, G. L. (1991). Distribution of epiphytic bacteria on olive leaves and the influence of leaf age and sampling time. *Microb. Ecol., 21*, 35-48.

Farrand, S. K., Van Berkum, P. B., & Oger, P. (2003). *Agrobacterium* is a definable genus of the family *Rhizobiaceae*. *Int. J. Syst. Evol. Microbiol., 53*, 1681-1687.

Feil, H., Feil, W. S., Chain, P., Larimer, F., Dibartolo, G., Copeland, A., et al. (2005). Comparison of the complete genome sequences of *Pseudomonas syringae* pv. *syringae* B728a and pv. *tomato* DC3000. *Proc. Natl. Acad. Sci. U. S. A., 102*, 11064-11069.

Finan, T. M. (2002). Evolving insights: symbiosis islands and horizontal gene transfer. *J. Bacteriol., 184*, 2855-2856.

Fouts, D. E., Abramovitch, R. B., Alfano, J. R., Baldo, A. M., Buell, C. R., Cartinhour, S., et al. (2002). Genomewide identification of *Pseudomonas syringae* pv. *tomato* DC3000 promoters controlled by the HrpL alternative sigma factor. *Proc. Natl. Acad. Sci. U. S. A., 99*, 2275-2280.

Fray, R. G., Throup, J. P., Daykin, M., Wallace, A., Williams, P., Stewart, G. S., et al. (1999). Plants genetically modified to produce N-acylhomoserine lactones communicate with bacteria. *Nat. Biotechnol., 17*, 1017-1020.

Gao, M., Teplitski, M., Robinson, J. B., & Bauer, W. D. (2003). Production of substances by *Medicago truncatula* that affect bacterial quorum sensing. *Mol. Plant-Microb. Interact., 16*, 827-834.

Gardan, L., Christen, R., Achouak, W., & Prior, P. (2004). *Erwinia papayae* sp. nov., a pathogen of papaya (*Carica papaya*). *Int. J. Syst. Evol. Microbiol., 54*, 107-113.

Gibly, A., Bonshtien, A., Balaji, V., Debbie, P., Martin, G. B., & Sessa, G. (2004). Identification and expression profiling of tomato genes differentially regulated during a resistance response to *Xanthomonas campestris* pv. *vesicatoria*. *Mol. Plant-Microb. Interact., 17*, 1212-1222.

Giddens, S. R., Houliston, G. J., & Mahanty, H. K. (2003). The influence of antibiotic production and pre-emptive colonization on the population dynamics of *Pantoea agglomerans* (*Erwinia herbicola*) Eh1087 and *Erwinia amylovora* in planta. *Environ. Microbiol., 5*, 1016-1021.

Gilbert, G. S., Handelsman, J., & Parke, J. L. (1994). Root camouflage and disease control. *Phytopathology, 84*, 222-225.

Giraud, E., & Fleischman, D. (2004). Nitrogen-fixing symbiosis between photosynthetic bacteria and legumes. *Photosynthesis Res., 82*, 115-130.

Goodner, B., Hinkle, G., Gattung, S., Miller, N., Blanchard, M., Qurollo, B., et al. (2001). Genome sequence of the plant pathogen and biotechnology agent *Agrobacterium tumefaciens* C58. *Science, 294*, 2323-2328.

Haas, D., & Defago, G. (2005). Biological control of soil-borne pathogens by fluorescent pseudomonads. *Nat. Rev. Microbiol., 3*, 307-319.

Hensel, M., Shea, J. E., Gleeson, C., Jones, M. D., Dalton, E., & Holden, D. W. (1995). Simultaneous identification of bacterial virulence genes by negative selection. *Science, 269*, 400-403.

Hentschel, U., Steinert, M., & Hacker, J. (2000). Common molecular mechanisms of symbiosis and pathogenesis. *Trends Microbiol., 8*, 226-231.

Hirano, S. S., & Upper, C. D. (2000). Bacteria in the leaf ecosystem with emphasis on *Pseudomonas syringae* - a pathogen, ice nucleus, and epiphyte. *Microbiol. Mol. Biol. Rev., 64*, 624-653.

Hirsch, A. M. (2004). Plant-microbe symbioses: a continuum from commensalism to parasitism. *Symbiosis, 37*, 345-363.

Holeva, M. C., Bell, K. S., Hyman, L. J., Avrova, A. O., Whisson, S. C., Birch, P. R., et al. (2004). Use of a pooled transposon mutation grid to demonstrate roles in disease development for *Erwinia carotovora* subsp. *atroseptica* putative type III secreted effector (DspE/A) and helper (HrpN) proteins. *Mol. Plant-Microb. Interact., 17*, 943-950.

Hooykaas, P. J. J., Klapwijk, P. M., Nuti, M. P., Shilperoort, R. A., & Rorsch, A. (1977). Transfer of *Agrobacterium tumefaciens* Ti plasmid to avirulent agrobacteria and to *Rhizobium ex-planta*. *J. Gen. Microbiol., 98*, 477-484.

Iniguez, A. L., Dong, Y., & Triplett, E. W. (2004). Nitrogen fixation in wheat provided by *Klebsiella pneumoniae* 342. *Mol. Plant-Microb. Interact., 17*, 1078-1085.

IRPCM Phytoplasma/Spiroplasma Working Team – Phytoplasma taxonomy group (2004). 'Candidatus *Phytoplasma*', a taxon for the wall-less, non-helical prokaryotes that colonize plant phloem and insects. *Int. J. Syst. Evol. Microbiol., 54*, 1243-1255.

James, E. K. (2000). Nitrogen fixation in endophytic and associative symbiosis. *Field Crops Res., 65*, 197-209.

Jargeat, P., Cosseau, C., Ola'h, B., Jauneau, A., Bonfante, P., Batut, J., et al. (2004). Isolation, free-living capacities, and genome structure of "Candidatus *Glomeribacter gigasporarum*," the endocellular bacterium of the mycorrhizal fungus *Gigaspora margarita*. *J. Bacteriol., 186*, 6876-6884.

Jones, A. M., Thomas, V., Truman, B., Lilley, K., Mansfield, J., & Grant, M. (2004). Specific changes in the Arabidopsis proteome in response to bacterial challenge: differentiating basal and R-gene mediated resistance. *Phytochemistry, 65*, 1805-1816.

Jones, K. (1970). Nitrogen fixation in the phyllosphere of the Douglas fir, *Pseudotsuga douglasii*. *Ann. Bot., 34*, 239-244.

Jourand, P., Giraud, E., Bena, G., Sy, A., Willems, A., Gillis, M., et al. (2004). *Methylobacterium nodulans* sp. nov., for a group of aerobic, facultatively methylotrophic, legume root-nodule-forming and nitrogen-fixing bacteria. *Int. J. Syst. Evol. Microbiol., 54*, 2269-2273.

Kaneko, T., Nakamura, Y., Sato, S., Asamizu, E., Kato, T., Sasamoto, S., et al. (2000). Complete genome structure of the nitrogen-fixing symbiotic bacterium *Mesorhizobium loti*. *DNA Res., 7*, 331-338.

Kaneko, T., Nakamura, Y., Sato, S., Minamisawa, K., Uchiumi, T., Sasamoto, S., et al. (2002). Complete genomic sequence of nitrogen-fixing symbiotic bacterium *Bradyrhizobium japonicum* USDA110. *DNA Res., 9*, 189-197.

Kent, A. D., & Triplett, E. W. (2002). Microbial communities and their interactions in soil and rhizosphere ecosystems. *Annu. Rev. Microbiol., 56*, 211-236.

Kim, K. Y., Jordan, D., & McDonald, G. A. (1998). Effect of phosphate-solubilizing bacteria and vesicular-arbuscular mycorrhizae on tomato growth and soil microbial activity. *Biol. Fertil. Soils, 26*, 79-87.

Knight, T. J., & Langston-Unkefer, P. J. (1988). Enhancement of symbiotic dinitrogen fixation by a toxin-releasing plant pathogen. *Science, 241*, 951-954.

Kobayashi, D. Y., & Palumbo, J. D. (2000). Bacterial endophytes and their effects on plants and uses in agriculture. In C. W. Bacon & J. F. White, Jr. (Eds.), *Microbial Endophytes* (pp. 199-233). New York: Marcel Dekker, Inc.

Kremer, R. J., & Kennedy, A. C. (1996). Rhizobacteria as biocontrol agents of weeds. *Weed Technol., 10*, 601-609.

Kuiper, I., Lagendijk, E. L., Bloemberg, G. V., & Lugtenberg, B. J. J. (2004). Rhizoremediation: a beneficial plant-microbe interaction. *Mol. Plant-Microb. Interact., 17*, 6-15.

Lee, B. M., Park, Y. J., Park, D. S., Kang, H. W., Kim, J. G., Song, E. S., et al. (2005). The genome sequence of *Xanthomonas oryzae* pathovar *oryzae* KACC10331, the bacterial blight pathogen of rice. *Nucl. Acids Res., 33*, 577-586.

Lee, I. M., Gundersen-Rindal, D. E., Davis, R. E., Bottner, K. D., Marcone, C., & Seemuller, E. (2004). 'Candidatus *Phytoplasma asteris*', a novel phytoplasma taxon associated with aster yellows and related diseases. *Int. J. Syst. Evol. Microbiol., 54*, 1037-1048.

Lee, K. E., & Pankhurst, C. E. (1992). Soil organisms and sustainable productivity. *Aust. J. Soil Res., 30*, 855-892.

Legard, D. E., McQuilken, M. P., Whipps, J. M., Fenlon, J. S., Fermor, T. R., Thompson, I. P., et al. (1994). Studies of seasonal changes in the microbial populations on the phyllosphere of spring wheat as a prelude to the release of a genetically modified microorganism. *Agric. Ecosys. Environ., 50*, 87-101.

Leveau, J. H., & Lindow, S. E. (2001). Appetite of an epiphyte: quantitative monitoring of bacterial sugar consumption in the phyllosphere. *Proc. Natl. Acad. Sci. U. S. A., 98*, 3446-3453.

Li, J., Ovakim, D. H., Charles, T. C., & Glick, B. R. (2000). An ACC deaminase minus mutant of *Enterobacter cloacae* UW4 no longer promotes root elongation. *Curr. Microbiol., 41*, 101-105.

Lindemann, J., & Upper, C. D. (1985). Aerial dispersal of epiphytic bacteria over bean plants. *Appl. Environ. Microbiol., 50*, 1229-1232.

Lindow, S. E. (1983). The role of bacterial ice nucleation in frost injury to plants. *Annu. Rev. Phytopathol., 21*, 363-384.

Lindow, S. E. (1987). Competitive exclusion of epiphytic bacteria by ice$^-$ *Pseudomonas syringae* mutants. *Appl. Environ. Microbiol., 53*, 2520-2527.

Lindow, S. E., & Cirvilleri, G. (1994). Differential expression of genes of *Pseudomonas syringae* on leaves and in culture evaluated with random genomic lux fusions. *Mol. Ecol., 3*, 249-257.

Lindow, S. E., Desurmont, C., Elkins, R., McGourty, G., Clark, E., & Brandl, M. T. (1998). Occurrence of indole-3-acetic acid-producing bacteria on pear trees and their association with fruit russet. *Phytopathology, 88*, 1149-1157.

Manefield, M., Whiteley, A. S., Griffiths, R. I., & Bailey, M. J. (2002). RNA stable isotope probing, a novel means of linking microbial community function to phylogeny. *Appl. Environ. Microbiol., 68*, 5367-5373.

Marco, M. L., Legac, J., & Lindow, S. E. (2005). *Pseudomonas syringae* genes induced during colonization of leaf surfaces. *Environ. Microbiol., 7*, 1379-1391.

Marek-Kozaczuk, M., & Skorupska, A. (2001). Production of B-group vitamins by plant-growth-promoting *Pseudomonas fluorescens* strain 267 and the importance of vitamins in the colonization and nodulation of red clover. *Biol. Fertil. Soils, 33*, 146-151.

Martinez, E., Palacios, R., & Sanchez, F. (1987). Nitrogen-fixing nodules induced by *Agrobacterium tumefaciens* harboring *Rhizobium phaseoli* plasmids. *J. Bacteriol., 169*, 2828-2834.

McInroy, J. A., & Kloepper, J. W. (1995). Survey of indigenous bacterial endophytes from cotton and sweet corn. *Plant Soil, 173*, 337-342.

McKellar, M. E., & Nelson, E. B. (2003). Compost-induced suppression of *Pythium* damping-off is mediated by fatty-acid-metabolizing seed-colonizing microbial communities. *Appl. Environ. Microbiol., 69*, 452-460.

Monteiro-Vitorello, C. B., Camargo, L. E., Van Sluys, M. A., Kitajima, J. P., Truffi, D., do Amaral, A. M., et al. (2004). The genome sequence of the gram-positive sugarcane pathogen *Leifsonia xyli subsp. xyli*. *Mol. Plant-Microb. Interact., 17*, 827-836.

Morris, C. E., Georgakopoulos, D. G., & Sands, D. C. (2004). Ice nucleation active bacteria and their potential role in precipitation. *J. Phys. IV France, 121*, 87-103.

Natera, S. H., Guerreiro, N., & Djordjevic, M. A. (2000). Proteome analysis of differentially displayed proteins as a tool for the investigation of symbiosis. *Mol. Plant-Microb. Interact., 13*, 995-1009.

Navarro, L., Zipfel, C., Rowland, O., Keller, I., Robatzek, S., Boller, T., et al. (2004). The transcriptional innate immune response to flg22. Interplay and overlap with *avr* gene-dependent defense responses and bacterial pathogenesis. *Plant Physiol., 135*, 1113-1128.

Normand, P., Orso, S., Cournoyer, B., Jeannin, P., Chapelon, C., Dawson, J., et al. (1996). Molecular phylogeny of the genus *Frankia* and related genera and emendation of the family *Frankiaceae*. *Int. J. Syst. Bacteriol., 46*, 1-9.

Novikova, N., & Safronova, V. (1992). Transconjugants of *Agrobacterium radiobacter* harboring *sym* genes of *Rhizobium galegae* can form an effective symbiosis with *Medicago sativa*. *FEMS Microbiol. Lett., 93*, 261-268.

Oke, V., & Long, S. R. (1999). Bacterial genes induced within the nodule during the *Rhizobium*-legume symbiosis. *Mol. Microbiol., 32*, 837-849.

Okinaka, Y., Yang, C. H., Perna, N. T., & Keen, N. T. (2002). Microarray profiling of *Erwinia chrysanthemi* 3937 genes that are regulated during plant infection. *Mol. Plant-Microb. Interact., 15*, 619-629.

Osbourn, A. E., Barber, C. E., & Daniels, M. J. (1987). Identification of plant-induced genes of the bacterial pathogen *Xanthomonas campestris* pathovar *campestris* using a promoter-probe plasmid. *Embo. J., 6*, 23-28.

Oshima, K., Kakizawa, S., Nishigawa, H., Jung, H. Y., Wei, W., Suzuki, S., et al. (2004). Reductive evolution suggested from the complete genome sequence of a plant-pathogenic phytoplasma. *Nat. Genet., 36*, 27-29.

Palumbo, J. D., Yuen, G. Y., Jochum, C. C., Tatum, K., & Kobayashi, D. Y. (2005). Mutagenesis of beta-1,3-glucanase genes in *Lysobacter enzymogenes* strain C3 results in reduced biological control activity toward *Bipolaris* leaf spot of tall fescue and *Pythium* damping-off of sugar beet. *Phytopathology, 95*, 701-707.

Patten, C. L., & Glick, B. R. (2002). Role of *Pseudomonas putida* indoleacetic acid in development of the host plant root system. *Appl. Environ. Microbiol., 68*, 3795-3801.

Paulitz, T. C. (1991). Effect of *Pseudomonas putida* on the stimulation of *Pythium ultimum* by seed volatiles of pea and soybean. *Phytopathology, 81*, 1282-1287.

Pawlowski, K. (2002). Actinorhizal symbioses. In G. J. Leigh (Ed.), *Nitrogen Fixation at the Millennium* (pp. 167-189). Boston: Elsevier Science.

Phillips, D. A., Joseph, C. M., Yang, G. P., Martinez-Romero, E., Sanborn, J. R., & Volpin, H. (1999). Identification of lumichrome as a sinorhizobium enhancer of alfalfa root respiration and shoot growth. *Proc. Natl. Acad. Sci. U. S. A., 96*, 12275-12280.

Pierson, L. S. I., & Pierson, E. A. (1996). Phenazine antibiotic production in *Pseudomonas aureofaciens*: role in rhizosphere ecology and pathogen suppression. *FEMS Microbiol. Lett., 136*, 101-108.

Pilon-Smits, E. (2005). Phytoremediation. *Annu. Rev. Plant Biol., 56*, 15-39.

Ping, L., & Boland, W. (2004). Signals from the underground: bacterial volatiles promote growth in *Arabidopsis*. *Trends Plant Sci., 9*, 263-266.

Postgate, J. (1998). *Nitrogen Fixation* (3rd edition ed.). Cambridge: Cambridge University Press.

Qian, W., Jia, Y., Ren, S. X., He, Y. Q., Feng, J. X., Lu, L. F., et al. (2005). Comparative and functional genomic analyses of the pathogenicity of phytopathogen *Xanthomonas campestris* pv. *campestris*. *Genome Res., 15*, 757-767.

Radajewski, S., McDonald, I. R., & Murrell, J. C. (2003). Stable-isotope probing of nucleic acids: a window to the function of uncultured microorganisms. *Curr. Opin. Biotechnol., 14*, 296-302.

Rai, A. N., Söderbäck, E., & Bergman, B. (2000). Cyanobacterium-plant symbioses. *New Phytol., 147*, 449-481.

Rainey, P. B. (1999). Adaptation of *Pseudomonas fluorescens* to the plant rhizosphere. *Environ. Microbiol., 1*, 243-257.

Ramos-Gonzalez, M. I., Campos, M. J., & Ramos, J. L. (2005). Analysis of *Pseudomonas putida* KT2440 gene expression in the maize rhizosphere: *in vivo* expression technology capture and identification of root-activated promoters. *J. Bacteriol., 187*, 4033-4041.

Rappe, M. S., & Giovannoni, S. J. (2003). The uncultured microbial majority. *Annu. Rev. Microbiol., 57*, 369-394.

Ratti, N., Kumar, S., Verma, H. N., & Gautam, S. P. (2001). Improvement in bioavailability of tricalcium phosphate to *Cymbopogon martinii* var. *motia* by rhizobacteria, AMF and *Azospirillum* inoculation. *Microbiol. Res., 156*, 145-149.

Reddy, P. M., James, E. K., & Ladha, J. K. (2002). Nitrogen fixation in rice. In G. J. Leigh (Ed.), *Nitrogen Fixation at the Millennium* (pp. 421-445). Boston: Elsevier Science.

Rediers, H., Bonnecarrere, V., Rainey, P. B., Hamonts, K., Vanderleyden, J., & De Mot, R. (2003). Development and application of a dapB-based *in vivo* expression technology system to study colonization of rice by the endophytic nitrogen-fixing bacterium *Pseudomonas stutzeri* A15. *Appl. Environ. Microbiol., 69*, 6864-6874.

Rediers, H., Rainey, P. B., Vanderleyden, J., & De Mot, R. (2005). Unraveling the secret lives of bacteria: use of *in vivo* expression technology and differential fluorescence induction promoter traps as tools for exploring niche-specific gene expression. *Microbiol. Mol. Biol. Rev., 69*, 217-261.

Rivas, R., Willems, A., Palomo, J. L., Garcia-Benavides, P., Mateos, P. F., Martinez-Molina, E., et al. (2004). *Bradyrhizobium betae* sp. nov., isolated from roots of *Beta vulgaris* affected by tumour-like deformations. *Int. J. Syst. Evol. Microbiol., 54*, 1271-1275.

Rivas, R., Willems, A., Subba-Rao, N. S., Mateos, P. F., Dazzo, F. B., Kroppenstedt, R. M., et al. (2003). Description of *Devosia neptuniae* sp. nov. that nodulates and fixes nitrogen in symbiosis with *Neptunia natans*, an aquatic legume from India. *Syst. Appl. Microbiol., 26*, 47-53.

Rosen, R., Matthysse, A. G., Bedher, D., Blran, D., Yura, T., Hecker, M., et al. (2003). Proteome analysis of plant-induced proteins of *Agrobacterium tumefaciens*. *FEMS Microbiol. Ecol., 44*, 355-360.

Rukayadi, Y., Suwanto, A., Tjahjono, B., & Harling, R. (2000). Survival and epiphytic fitness of a nonpathogenic mutant of *Xanthomonas campestris* pv. glycines. *Appl. Environ. Microbiol., 66*, 1183-1189.

Ryu, C. M., Farag, M. A., Hu, C. H., Reddy, M. S., Wei, H. X., Pare, P. W., et al. (2003). Bacterial volatiles promote growth in *Arabidopsis*. *Proc. Natl. Acad. Sci. U. S. A., 100*, 4927-4932.

Salanoubat, M., Genin, S., Artiguenave, F., Gouzy, J., Mangenot, S., Arlat, M., et al. (2002). Genome sequence of the plant pathogen *Ralstonia solanacearum*. *Nature, 415*, 497-502.

Sawada, H., Kuykendall, L. D., & Young, J. M. (2003). Changing concepts in the systematics of bacterial nitrogen-fixing legume symbionts. *J. Gen. Appl. Microbiol., 49*, 155-179.

Schaad, N. W., Vidaver, A. K., Lacy, G. H., Rudolph, K., & Jones, J. B. (2000). Evaluation of proposed amended names of several pseudomonads and xanthomonads and recommendations. *Phytopathology, 90*, 208-213.

Schneider, B., Torres, E., Martin, M. P., Schroder, M., Behnke, H. D., & Seemuller, E. (2005). 'Candidatus *Phytoplasma pini*', a novel taxon from *Pinus silvestris* and *Pinus halepensis*. *Int. J. Syst. Evol. Microbiol., 55*, 303-307.

Sharma, A., Johri, B. N., Sharma, A. K., & Glick, B. R. (2003). Plant growth-promoting bacterium *Pseudomonas* sp, strain GFP(3) influences iron acquisition in mung bean (*Vigna radiata* L. Wilzeck). *Soil Biol. Blochem., 35*, 887-894.

Siciliano, S. D., Fortin, N., Mihoc, A., Wisse, G., Labelle, S., Beaumier, D., et al. (2001). Selection of specific endophytic bacterial genotypes by plants in response to soil contamination. *Appl. Environ. Microbiol., 67*, 2469-2475.

Simon, H. M., Dodsworth, J. A., & Goodman, R. M. (2000). *Crenarchaeota* colonize terrestrial plant roots. *Environ. Microbiol., 2*, 495-505.

Simpson, A. J., Reinach, F. C., Arruda, P., Abreu, F. A., Acencio, M., Alvarenga, R., et al. (2000). The genome sequence of the plant pathogen *Xylella fastidiosa*. *Nature, 406*, 151-157.

Singh, B. K., Millard, P., Whiteley, A. S., & Murrell, J. C. (2004). Unravelling rhizosphere - microbial interactions: opportunities and limitations. *Trends Microbiol., 12*, 386-393.

Sliwinski, M. K., & Goodman, R. M. (2004). Comparison of crenarchaeal consortia inhabiting the rhizosphere of diverse terrestrial plants with those in bulk soil in native environments. *Appl. Environ. Microbiol., 70*, 1821-1826.

Sukdeo, N., & Charles, T. C. (2003). Application of crossover-PCR-mediated deletion-insertion mutagenesis to analysis of the *bdhA-xdhA2-xdhB2* mixed-function operon of *Sinorhizobium meliloti*. *Arch. Microbiol., 179*, 301-304.

Suslow, T. V., & Schroth, M. N. (1982). Role of deleterious rhzobacteria as minor pathogens in reducing crop growth. *Phytopathology, 72*, 111-115.

Sy, A., Giraud, E., Jourand, P., Garcia, N., Willems, A., de Lajudie, P., et al. (2001). *Methylotrophic methylobacterium* bacteria nodulate and fix nitrogen in symbiosis with legumes. *J. Bacteriol., 183*, 214-220.

Szyrmer, W., & Zawadski, I. (1997). Biogenic and anthropogenic sources of ice-forming nuclei: a review. *Bull. Amer. Meterol. Soc., 78*, 209-228.

Thomashow, L. S., & Weller, D. M. (1996). Current concepts in the use of introduced bacteria for biological disease control: mechanisms and antifungal metabolites. In G. Stacey & N. T. Keen (Eds.), *Plant-microbe Interactions* (Vol. 1, pp. 187-227). New York: Chapman & Hall.

Thompson, I. P., Bailey, M. J., Fenlon, J. S., Fermor, T. R., Lilley, A. K., Lynch, J. M., et al. (1993). Quantitative and qualitative seasonal changes in the microbial community from the phyllosphere of sugar beet (*Beta vulgaris*). *Plant Soil, 150*, 177-191.

Torsvik, V., & Øvreås, L. (2002). Microbial diversity and function in soil: from genes to ecosystems. *Curr. Opin. Microbiol., 5*, 240-245.

Toyoda, H., Hashimoto, H., Utsumi, R., Kobayashi, H., & Ouchi, S. (1988). Detoxification of fusaric acid by a fusaric acid-resistant mutant of *Pseudomonas solanacearum* and its application to biological control of *Fusarium* wilt of tomato. *Phytopathology, 78*, 1307-1311.

Trujillo, M. E., Willems, A., Abril, A., Planchuelo, A. M., Rivas, R., Ludena, D., et al. (2005). Nodulation of *Lupinus albus* by strains of *Ochrobactrum lupini* sp. nov. *Appl. Environ. Microbiol., 71*, 1318-1327.

Utsumi, R., Yagi, T., Katayama, S., Katsuragi, K., Tachibana, K., Toyoda, H., et al. (1991). Molecular cloning and characterization of the fusaric acid-resistance gene from *Pseudomonas cepacia*. *Agric. Biol. Chem., 55*, 1913-1918.

Vaishampayan, A., Sinha, R. P., Hader, D. P., Dey, T., Gupta, A. K., Bhan, U., et al. (2001). Cyanobacterial biofertilizers in rice agriculture. *Bot. Rev., 67*, 453-516.

van Berkum, P., & Eardly, B. D. (2002). The aquatic budding bacterium *Blastobacter denitrificans* is a nitrogen-fixing symbiont of *Aeschynomene indica*. *Appl. Environ. Microbiol., 68*, 1132-1136.

van Loon, L. C., Bakker, P. A. H. M., & Pieterse, C. M. J. (1998). Systemic resistance induced by rhizosphere bacteria. *Annu. Rev. Phytopathol., 36*, 453-483.

Van Sluys, M. A., de Oliveira, M. C., Monteiro-Vitorello, C. B., Miyaki, C. Y., Furlan, L. R., Camargo, L. E., et al. (2003). Comparative analyses of the complete genome sequences of Pierce's disease and citrus variegated chlorosis strains of *Xylella fastidiosa*. *J. Bacteriol., 185*, 1018-1026.

Vauterin, L., Hoste, B., Kersters, K., & Swings, J. (1995). Reclassification of *Xanthomonas*. *Int. J. Syst. Bacteriol., 45*, 472-489.

Verhagen, B. W., Glazebrook, J., Zhu, T., Chang, H. S., van Loon, L. C., & Pieterse, C. M. (2004). The transcriptome of rhizobacteria-induced systemic resistance in *Arabidopsis*. *Mol. Plant-Microb. Interact., 17*, 895-908.

Vessey, J. K. (2003). Plant growth promoting rhizobacteria as biofertilizers. *Plant Soil, 255*, 571-586.

Vessey, J. K., Pawlowski, K., & Bergman, B. (2004). Root-based N_2-fixing symbioses: legumes, actinorhizal plants, *Parasponia* sp. and cycads. *Plant Soil, 266*, 205-230.

Wang, Y., Ohara, Y., Nakayashiki, H., Tosa, Y., & Mayama, S. (2005). Microarray analysis of the gene expression profile induced by the endophytic plant growth-promoting rhizobacteria, *Pseudomonas fluorescens* FPT9601-T5 in *Arabidopsis*. *Mol. Plant-Microb. Interact., 18*, 385-396.

Weidner, S., Arnold, W., Stackebrandt, E., & Puhler, A. (2000). Phylogenetic analysis of bacterial communities associated with leaves of the seagrass *Halophila stipulacea* by a culture-independent small-subunit rRNA gene approach. *Microb. Ecol., 39*, 22-31.

Willems, A., Fernandez-Lopez, M., Munoz-Adelantado, E., Goris, J., De Vos, P., Martinez-Romero, E., et al. (2003). Description of new *Ensifer* strains from nodules and proposal to transfer *Ensifer adhaerens* Casida 1982 to *Sinorhizobium* as *Sinorhizobium adhaerens* comb. nov. Request for an opinion. *Int. J. Syst. Evol. Microbiol., 53*, 1207-1217.

Williamson, K. E., Radosevich, M., & Wommack, K. E. (2005). Abundance and diversity of viruses in six Delaware soils. *Appl. Environ. Microbiol., 71*, 3119-3125.

Winans, S. C., & Bassler, B. L. (2002). Mob psychology. *J. Bacteriol., 184*, 873-883.

Wood, D. W., Setubal, J. C., Kaul, R., Monks, D. E., Kitajima, J. P., Okura, V. K., et al. (2001). The genome of the natural genetic engineer *Agrobacterium tumefaciens* C58. *Science, 294*, 2317-2323.

Yang, C. H., Crowley, D. E., Borneman, J., & Keen, N. T. (2001). Microbial phyllosphere populations are more complex than previously realized. *Proc. Natl. Acad. Sci. U. S. A., 98*, 3889-3894.

Yang, S., Perna, N. T., Cooksey, D. A., Okinaka, Y., Lindow, S. E., Ibekwe, A. M., et al. (2004). Genome-wide identification of plant-upregulated genes of *Erwinia chrysanthemi* 3937 using a GFP-based IVET leaf array. *Mol. Plant-Microb. Interact., 17*, 999-1008.

Young, J. M. (2003). The genus name *Ensifer* Casida 1982 takes priority over *Sinorhizobium* Chen et al., 1988, and *Sinorhizobium morelense* Wang et al. 2002 is a later synonym of *Ensifer adhaerens* Casida 1982. Is the combination "*Sinorhizobium adhaerens*" (Casida 1982) Willems et al., 2003 legitimate? Request for an Opinion. *Int. J. Syst. Evol. Microbiol., 53*, 2107-2110.

Young, J. M., Bull, C. T., De Boer, S. H., Firrao, G., Gardan, L., Saddler, G. E., et al. (2004). Names of plant pathogenic bacteria published since 1995. *Intl. Soc. Plant Pathol., www.isppweb.org/names_bacterial.asp.*

Young, J. M., Kuykendall, L. D., Martinez-Romero, E., Kerr, A., & Sawada, H. (2001). A revision of *Rhizobium* Frank 1889, with an emended description of the genus, and the inclusion of all species of *Agrobacterium* Conn 1942 and *Allorhizobium undicola* de Lajudie et al. 1998 as new combinations: *Rhizobium radiobacter*, *R. rhizogenes*, *R. rubi*, *R. undicola* and *R. vitis*. *Int. J. Syst. Evol. Microbiol., 51*, 89-103.

Young, J. M., Kuykendall, L. D., Martinez-Romero, E., Kerr, A., & Sawada, H. (2003). Classification and nomenclature of *Agrobacterium* and *Rhizobium*. *Int. J. Syst. Evol. Microbiol., 53*, 1689-1695.

Young, J. M., Park, D. C., & Weir, B. S. (2004). Diversity of 16S rDNA sequences of *Rhizobium* spp. implications for species determinations. *FEMS Microbiol. Lett., 238*, 125-131.

Young, J. M., Saddler, G. S., Takikawa, Y., De Boer, S. H., Vauterin, L., Gardan, L., et al. (1995). Names of plant pathogenic bacteria, 1864-1995. *Intl. Soc. Plant Pathol., www.isppweb.org/names_bacterial.asp.*

Young, J. M., Watson, D. R., & Dye, D. W. (2004). Reconsideration of *Arthrobacter ilicis* (Mandel et al., 1961) Collins et al., 1982 as a plant-pathogenic species. Proposal to emend the authority and description of the species. Request for an opinion. *Int. J. Syst. Evol. Microbiol., 54*, 303-305.

Young, J. P. W., & Haukka, K. E. (1996). Diversity and phylogeny of rhizobia. *New Phytol., 133*, 87-94.

Zhang, L.-H., & Dong, Y.-H. (2004). Quorum sensing and signal interference: diverse implications. *Mol. Microbiol., 53*, 1563-1571.

Zinniel, D. K., Lambrecht, P., Harris, N. B., Feng, Z., Kuczmarski, D., Higley, P., et al. (2002). Isolation and characterization of endophytic colonizing bacteria from agronomic crops and prairie plants. *Appl. Environ. Microbiol., 68*, 2198-2208.

Zwiesler-Vollick, J., Plovanich-Jones, A. E., Nomura, K., Bandyopadhyay, S., Joardar, V., Kunkel, B. N., et al. (2002). Identification of novel hrp-regulated genes through functional genomic analysis of the *Pseudomonas syringae* pv. *tomato* DC3000 genome. *Mol. Microbiol., 45*, 1207-1218.

10. AFFILIATION

Gwyn A. Beattie, *Department of Plant Pathology, Iowa State University in Ames, Iowa, U.S.A.*

Beneficial Bacteria

HARI B. KRISHNAN AND JOHN O. BENNETT

RHIZOBIUM-LEGUME SYMBIOSES: MOLECULAR SIGNALS ELABORATED BY RHIZOBIA THAT ARE IMPORTANT FOR NODULATION

Abstract. Symbiotic nitrogen fixation has traditionally been considered the domain of rhizobia belonging to α-proteobacteria and leguminous plants. Recent discovery of members of β-proteobacteria capable of nitrogen reduction indicates a greater diversity among nitrogen fixing symbionts than previously realized. Intense research on nitrogen-fixing bacteria conducted during the past few decades has shed light on the molecular signals and thye intricate biochemical events that occur when the rhizobia interact with legumes. In this paper, we have carefully documented these advances on the present understanding of nodule formation which is an orchestrated sequence of events evidenced by changes in root morphology at the site of infection. Attachment of the bacteria to the legume roots results in root hair curling and deformation forming a pocket, which serves to entrap the rhizohia thus facilitating subsequent events of nodulation. Initiation and continuation of the nodulation process is dependent upon a constant exchange of chemical signals between the host legume and the bacteria. Several bacterial and plant-derived components play a crucial role in establishing symbiotic association. Flavonoids released by the legumes and lipochiooligosaccharides (LCO) or nodulation factors (Nod factors) secreted by the rhizobia are the two principal signal molecules involved in establishing symbioses. Nodulation genes set apart the rhizobia from other nitrogen fixing and endophytic bacteria. They encode proteins essential for the production of a unique set of molecules, known as Nod factors and other complex organic molecules integral to the symbiotic relationship between these bacteria and the legumes. Divided into three groups according to their respective functions, *nod* genes are designated as regulatory, common, or host specific. We observe that our increased knowledge and understanding of the symbiotic process, however, has resulted in minimal transfer of technology for use in commercial agricultural production and suggest that further research on the manufacture and use of flavonoids and Nod-factors to enhance nodulation in cultivated legumes such as soybean can be a promising area.

1. INTRODUCTION

Rhizobia, derived from the Greek nouns '*riza*' (root) and '*bios*' (life), are soil bacteria that express an enzymatic system capable of reducing diatomic nitrogen after having formed a symbiotic relationship with specific legume hosts. In the third century B.C, Theophrastus noted in a treatise *Enquiry into Plants* that leguminous plants reinvigorated the soil (Fred et al., 1932). This was reiterated by the Roman agriculturalist Cato in the second century B.C. who wrote of using lupines to manure the land (Fred et al., 1932). It was not until the 18th century that scientific strides began to unravel the mystery between the structures on the plant roots and the organism residing within these structures. Experiments conducted in the 1880's showed conclusively that the ability of the leguminous plant to utilize molecular nitrogen depended upon the formation of swellings on the roots, which harbored a particular organism (Hellriegel, 1886; Hellriegel & Wilfarth, 1888). Beijerinck (1888) isolated bacteria from the swellings on the roots of *Vica faba* and designated them *Bacillus radiocola*, a name subsequently changed to *Rhizobium*

S.S. Gnanamanickam (ed.), Plant-Associated Bacteria, 57–104.

leguminosarum (Frank, 1889). He also showed that these organisms were capable of eliciting the formation of root nodule, nascent plant organs in which differentiated bacteria supply reduced nitrogen to the plant (Beijerinck, 1890). Interestingly after this work was published, little time elapsed before inoculation of agricultural legumes was the practice. Prazmowski (1889; 1890) noted that the bacterium entered the plant through root hair. Further observation revealed cell free filtrates induced morphological alterations in root hair structure (Hiltner, 1900). Ljunggren and Fahraeus (1959) found that rhizobia produce a water-soluble non-dialyzable substance that is involved in nodule formation. The elusive compound responsible for nodule formation was determined 30 years after Ljunggren and Fabraeus observation to be Nod factors, which are modified chitin molecules (Lerouge et al., 1990).

Fred et al. (1932) noted that since the seminal work conducted in the latter 19th century, thousands of papers had been published on the matter of the symbiotic nitrogen fixation. Some eight decades later, we are still unraveling the mysteries of the intricate orchestration of communication between the plant and bacteria. The advent of techniques capable of facilitating investigation on the molecular level and more sophisticated analytical machinery have accelerated discoveries relating to this intriguing relationship as evidenced by the myriad of scientific publications over the last three decades. Research centered upon elucidation of this symbiotic relationship has provided not only information specific to the rhizobia-legume relationship but also a wealth of data concerning plant microbe interactions and mitogenic signal transduction.

2. CLASSIFICATION

Rhizobial classification has evolved over the decades and the bacteria are presently divided among six principal genera (Table 1), which include *Mesorhizobium, Sinorhizobium, Rhizobium, Allorhizobium, Bradyrhizobium* and *Azorhizobium* (Chen et al., 1988; Young & Haukka, 1996; Jarvis et al., 1997; De Lajudie et al., 1998). Rhizobia from these genera are placed among three groups in the α-2 subclass of proteobacteria (Sy et al., 2001). The first group consists of *Mesorhizobium, Sinorhizobium, Rhizobium,* and *Allorhizobium,* the second *Bradyrhizobium* and the third *Azorhizobium.*

A myriad of proposals dealing with classification had been made by the first third of the 20th century (Hiltner & Stormer, 1903). Previously, Nobbe et al. (1895) had noted that bacteria isolated from *Pisum sativum* did not elicit nodule formation on other legumes, thus a possible method of classification based on host-specificity was realized. This host plant specificity has given rise to the establishment of cross-inoculation groups in which the bacterial species are classified based on their host-range. In another general classification method initiated during this period, rhizobia were also considered either fast or slow growing (Lohnis & Hansen, 1921). Fast growing organisms include the genus *Rhizobium* while *Bradyrhizobium* were considered slow growing bacteria. Since the inception of these early classification methods, some bacteria especially those from the tropical regions such as the *Rhizobium* sp. NGR234 have been found to be quite promiscuous having legume hosts representing 232 species from over hundred genera (Pueppke & Broughton, 1999; Perret et al., 2000). Other rhizobia are more discriminating, forming a symbiotic relationship with an individual legume or a small group of legumes. *Sinorhizobium meliloti* nodulates species of *Medicago, Melilotus and Trigonella. R. leguminosarum bv viciae* nodulates plants of the genera *Pisum, Vicia, Lathyrus* spp *and Lens* while another strain of *R. leguminosarum* bv. *trifolii,* only nodulates plants of the genera *Trifolium*. Although host specificity has continued to play a role in classification through the years, it became evident that host-based classification alone was not sufficient (Wilson, 1939; Lim & Burton, 1982; Laguerre et al., 1993; Michiels et al., 1998; Perret et al., 2000).

With the development of DNA sequencing methods, another mode of classifying bacteria was envisioned using nucleotide base sequences of specific genes to distinguish between species (Graham et al., 1991; Vandamme et al., 1996; Young, 1996; Young & Haukka, 1996; Terefework et al., 1998). Ribosomal RNA (rRNA) became a benchmark for classification of bacteria (Young et al., 1991; Young et al., 2004). Databases of 16S rRNA base sequences facilitated identification and classification of bacteria (Maidak et al., 1996; Ludwig et al., 1998; De Rijk et al., 2000; Van de Peer et al., 2000; Kwon et al., 2005). The use of

Table 1. Rhizobium-legume associations.

Rhizobia	Common Host	Reference
Allorhizobium undicola	*Neptunia natans*	De Lajudie et al., 1998
Azorhizobium caulinodans	*Sesbania rastrata*	Dreyfus et al., 1988
Bradyrhizobium japonicum	*Glycine max*	Jordan, 1980
B. elkanii	*Glycine max*	Kuykendall et al., 1992
Mesorhizobium loti	*Lotus species*	Jarvis et al., 1982
		Jarvis et al., 1997
M. ciceri	*Cicer arietinum* (chickpea)	Nour et al., 1994
		Jarvis et al., 1997
M. mediterraneum	*Cicer arietinum* (chickpea)	Nour et al., 1995
		Jarvis et al., 1997
M. huakuii	*Astragalus sinicus*	Chen et al., 1991
		Jarvis et al., 1997
Rhizobium leguminosarum		Frank, 1879

biovar *viciae*	*Pisum sativum*	Laguerre et al., 1996
biovar *trifolii*	*Trifolium repens,*	Laguerre et al., 1996
	T. subteraneum, T. pratense	Laguerre et al., 1996
biovar *phaseoli*	*Phaseolus vulgaris*	Laguerre et al., 1996
R. tropici	*Phaseolus vulgaris*	Martínez-Romero et al., 1991
R. galegae	*Galega orientalis*	Lindström, 1989
R. giardinii	*Phaseolus vulgaris*	Amarger et al., 1997
Sinorhizobium meliloti	*Medicago sativa,*	de Lajudie et al., 1994
	Melilotus alba	
Sinorhizobium fredii	*Glycine max*	Scholla &Elkan, 1984; Chen, et al., 1988
Sinorhizobium sp. NGR234	*Vigna unguiculata*	Trinick, 1973

the 16S rRNA was expanded to include the 23S rRNA and the internally transcribed space (ITS) region in an attempt to determine the viability of rRNA sequences as a tool for classification. Other gene sequences, such as of nodulation (*nod*), nitrogen fixation (*nif*), ATP synthase (*atpD)* and recombinase (*recA)* genes, have been used in conjunction with the 16S rRNA in an attempt to classify the rhizobia (Dobert et al., 1994; Gaunt et al., 2001; Laguerre et al., 2001). Research has shown use of the 16S rRNA gene and possibly other sequences alone for distinguishing and classifying bacteria may not be entirely adequate. Sequence heterogeneity of genes encoding 16S rRNA has been noted (Nübel et al., 1996; Ludwig et al., 1998). Multiple copies of rRNA genes in the same organism have been identified (Boros et al., 1979; Carbon et al., 1979; Amann et al., 2000; Conville & Witebsky, 2005). Lateral gene transfer and recombination among alleles of 16S rRNA across genera was detected suggesting the possibility of similar occurrences in other genes (Cilia et al., 1996; Eardly et al., 1996; Yap et al., 1999; van Berkum & Fuhrmann, 2000; van Berkum et al., 2003). Sequencing has shown that multiple copies of rRNA, which differ by more than 5%, conventionally considered necessary to distinguish between genera, exist within the same organism (Broughton, 2003). In total, these findings indicate that rRNA genes in an organism are not entirely the result of a vertical transfer and thus should not be used as the sole criteria for classification (Broughton, 2003). It has been suggested that classification should be based in part on several loci and other parameters, not only on rRNA sequence (Broughton, 2003). These include growth rate on laboratory media, carbohydrate utilization, metabolic capacities, antibiotic resistance, total protein fingerprinting, fatty acid analysis, DNA GC content, DNA:DNA and DNA:RNA hybridization, ribosome make-up, composition of cell walls, and 16S rRNA sequences (Graham et al., 1991; Vandamme et al., 1996).

Although nodulation of legumes was thought the exclusive domain of rhizobia belonging to the α-proteobacteria, recent work has shown that members of β-proteobacteria can also initiate nodules on host legumes (Moulin et al., 2001). Bacteria identified as *Burkholderia tuberum* and *B. phymatum,* respectively (Vandamme et al., 2002) were isolated from *Aspalathus carnosa* and *Machaerium lunatum* (Moulin et al., 2001). A conserved *nod box* was identified in front of *nodAB* and *nodC* in *Burkholderia* and mutational analysis showed that these genes were necessary for nodulation (Moulin et al., 2001). Additionally, *nifH* which encodes dinitrogenase was isolated from *B. tuberum* lending support to the inference that the organism embodied characteristics of rhizobia (Moulin et al., 2001). *Ralstonia taiwanensis* sp. nov. was recently isolated from nodules on *Mimosa pudica* and *M. diplotricha* thereby adding another member of β-proteo bacteria to the expanding list of bacterial symbionts. Sequencing revealed the presence of *nodA* and *nodBC* (Chen et al., 2001; Chen et al., 2003a). Since the initial discoveries, several strains of *Burkholderia* and *R. taiwanensis* have been shown to be nitrogen fixing symbionts (Chen et al., 2003b). A number of other α-proteo bacteria capable of nitrogen fixation, in addition to those generally classified as rhizobia, have been identified. *Methylobacterium* spp. which nodulate *Crotalaria* and *Lotononis* (Sy et al., 2001; Jaftha et al., 2002; Jourand et al., 2004), *Blastobacter denitrificans* IFAM 1005, a freshwater bacterium which forms nodules on *Aeschynomene indica* (van Berkum & Eardly, 2002), and *Devosia* strains that nodulate the aquatic legume *Neptunia natans* (Rivas et al., 2002; 2003) all belong to α-proteobacteria.

3. NODULATION PROCESS

Symbiotic nitrogen fixation provides legumes with nutrients for growth and development, obviating application of manufactured forms of reduced nitrogen, whose synthesis consumes vast amounts of fossil fuel. In itself, biological nitrogen fixation is energy intensive requiring 16 molecules of plant derived ATP for every diatomic nitrogen reduced to ammonia. Two morphologically distinct nodule types, determinate and indeterminate, have been identified (Fig. 1). Determinate nodules, derived from cells of the outer root cortex are viable for a short duration and are found predominantly associated with legumes of tropical origin such as soybean and cowpea. Reduced nitrogen produced in these globular shaped determinate nodules is delivered to the plant as glutamine and asparagine. In contrast, the indeterminate nodules are associated with legumes such as clover and alfalfa, plants of temperate origin. These nodules derived from root inner cortical cells display persistent meristematic tissue. The presence of an apical meristem generates nascent tissue that becomes infected with rhizobia, allowing the nodules to remain functional for longer duration. The constant growth of tissue imparts an elongated shape to the nodule, which exhibits distinct functional zones. The distal meristematic zone is followed by a region consisting of mature infected cells which are actively involved in fixing nitrogen that is to be delivered to the plant in the form of the ureides, allantoin and allantoic acid. The zone proximal to the plant root is comprised of senescing cells, which are at various stages of degeneration.

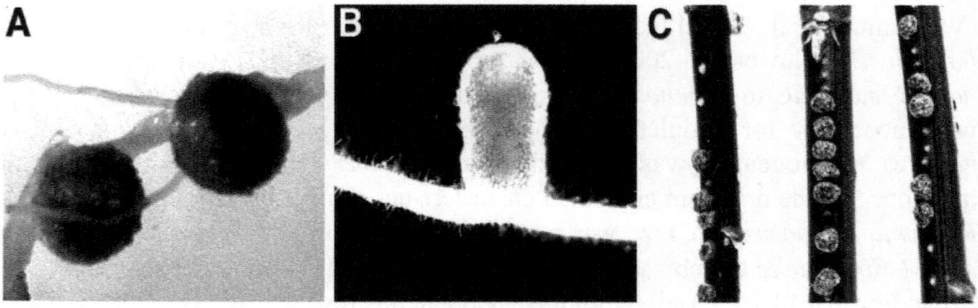

Fig. 1. Photomicrographs of determinate soybean nodules (A), indeterminate alfalfa nodule (B) and *Sesbania rostrata* stem nodules (C).

Nodule formation is an orchestrated sequence of events evidenced by changes in root morphology at the site of infection. Attachment of the bacteria to the legume roots results in root hair curling and deformation forming a pocket, which serves to entrap the rhizobia thus facilitating subsequent events of nodulation (Fig. 2). Concurrently, changes occur within the cortical region of the root tissue, which undergoes pronounced cell division leading to the formation of nodule primordia. Subsequently, the infection thread, a conduit between the root surface and cortical cells, is initiated and begins to grow towards the cortical region (Fig. 3). Within the expanding infection thread bacteria multiply and move to the awaiting cortical cell derived nodule primordia. When the bacterial laden thread reaches the nodule primodia, plant membrane bound structures containing rhizobia are released into the plant cytoplasm where the bacteria differentiate into morphologically distinct bacteroids. The bacteroids are surrounded by plant derived peribacteriod membrane and the entire unit is refered as the symbiosome. Within the symbiosome, the bacteroids enzymatically reduce dinitrogen to ammonia.

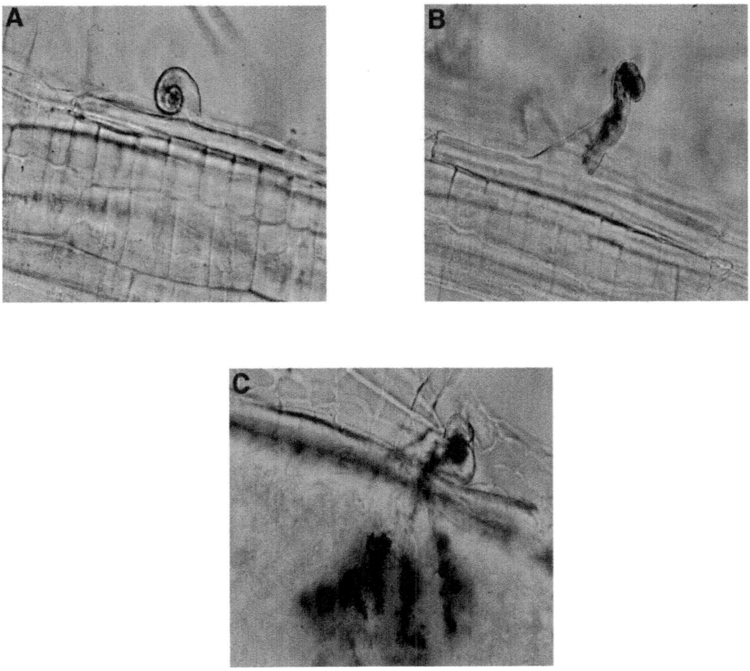

Fig. 2. Early events in soybean nodulation. *Sinorhizobium fredii* USDA257 entrapped in the curled root hairs (Panel A) enter the root hairs and form infection threads (Panel B) which ramify into the root cortex resulting in the formation of nodule primordia (Panel C).

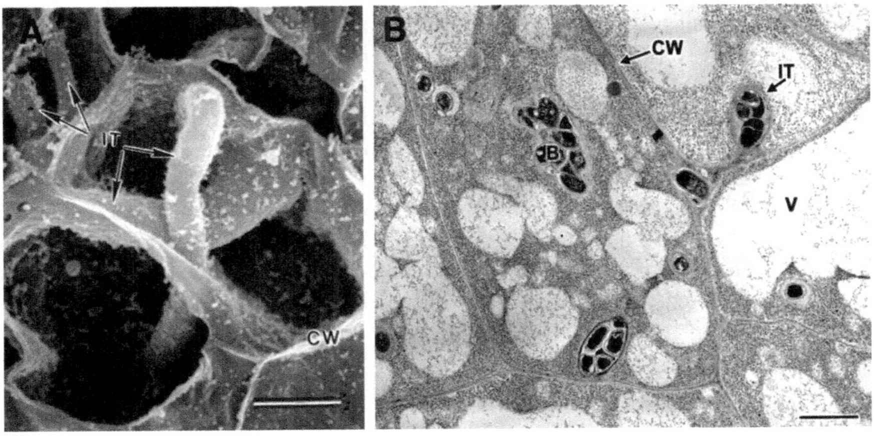

Fig. 3. Electron micrographs of infection threads. Scanning electron micrograph of infection thread in a *Sesbania rostrata* stem nodule. Bar = 4 μm (Panel A). Transmission electron micrograph showing a cross-section of infection thread in a soybean nodule. Bar = 2 μm.

4. SIGNAL EXCHANGE BETWEEN LEGUMES AND RHIZOBIUM

Initiation and continuation of the nodulation process is dependent upon a constant exchange of chemical signals between the host legume and the bacteria (Dénarié & Cullimore, 1993; Dénarié et al., 1996; Parniske & Downie, 2003). Several bacterial and plant-derived components play a crucial role in establishing symbiotic association. Flavonoids released by the legumes and lipochitooligosaccharides (LCO) or nodulation factors (Nod factors) secreted by the rhizobia are the two principal signal molecules involved in establishing symbisoses. In addition, other bacterial components such as secreted exopolysaccharides (EPS), membrane bound lipopolysaccharides (LPS), cyclic β-glucans, and nodulation outer proteins (Nops) play an integral role in nodule formation. The plethora of Nod factors, LPS, EPS, and Nops associated with effective nodulation on different leguminous plants indicate, that it is an expression of a specific combination of these chemical signals, that allows the final union of bacteria and plant to occur.

Legumes are dispersed from tropical to polar latitudes and have phenotypes as wide-ranging as the environmental conditions in which they thrive. The success of these plants has ostensibly been enhanced by their association with symbiotic nitrogen fixing bacteria (Long, 1989). Formation of this symbiotic relationship is dependent upon a molecular dialogue that is exchanged between the bacteria and plant (Dénarié & Cullimore, 1993). The plant's contribution to the dialogue begins with secretion of flavonoids into the rhizosphere (Redmond et al., 1986). These secondary metabolites act in conjunction with the product of constitutively expressed rhizobial *nodD*, a transcriptional activator of sundry nodulation genes which are abbreviated *nod*, *nol*, and *noe*. Since NodD combines with flavonoid signals from specific potential host legumes only, it also serves in host range determination (Györgypal et al., 1991a; 1991b). Luteolin is among the flavonoids from *Medicago sativa* (alfalfa) which induce the *S. meliloti nod* genes while apigenin and eriodictyol from *Pisum sativum* (pea) are inducers of *R. leguminosarum* bv. *viciae nod* genes (Firmin et al., 1986; Peters et al., 1986; Peters & Long, 1988). *Glycine max* (soybean) produces isoflavonoids, among others, genistein and daidzein which are potent inducers of *nod* genes in *Bradyrhizobium japonicum* and *S. fredii* (Kosslak et al., 1987; Krishnan & Pueppke, 1991). The bacterial contribution to the molecular dialogue with the legumes is through Nod factors, chitin-like chemical messengers released by the rhizobia in response to plant flavonoids (Redmond et al., 1986; Firmin et al., 1986; Peters et al., 1986; Kosslak et al., 1987; Peters & Long, 1988; Dénarié et al., 1996; Spaink, 1996). Other genes are instrumental in the production extracellular polysaccharides and extracellular proteins, which function in symbiosis.

In most rhizobia, the *nod* genes are carried on symbiotic (*sym*) plasmids or megaplasmids. Exceptions are the *Bradyrhizobium spp. Azorhizobium caulinodans,* and *M. loti* in which the genes are located within symbiotic islands of the bacterial chromosome. Regardless of the location of these loci, horizontal transfer of genes among species is thought to have promulgated the symbiotic phenomena. Today, sequencing of rhizobial genomes is providing not only information concerning the symbiotic relationship but also plant-microbe interactions as well. Complete genome sequences of *S. meliloti, M. loti* and *B. japonicum* along with the symbiotic regions of *Rhizobium* sp. NGR234 are currently known (Freiberg et al., 1997; Kaneko et al., 2000; Galibert et al., 2001; Kaneko et al., 2002). Accrual of information by the evolving tools of genetics and proteomics will continue to augment our knowledge and ultimately lead to a thorough understanding of the unique relationship between the rhizobia and legumes.

5. NODULATION GENES

Nodulation genes set apart the rhizobia from other nitrogen fixing and endophytic bacteria. They encode proteins essential for the production of a unique set of molecules, known as Nod factors (Fig. 4), and other complex organic molecules integral to the symbiotic relationship between these bacteria and the legumes. Divided into three groups according to their respective functions, *nod* genes are designated as regulatory, common, or host specific. Conserved sequences, *nod* boxes, preceding common and host specific *nod* operons, are a defining characteristic of most *nod* genes. The regulatory *nodD* and common *nodABCIJ* are ubiquitous to rhizobia while the host specific genes, in accordance with their function, vary among the species. Recognition and discrimination of the host signal begins with the interaction of the *nodD* product and a plant-borne inducer (Burn et al., 1987). Constitutively expressed in Rhizobia, NodD varies in response to the type and quantity of host flavonoids (Spaink et al., 1987; Györgypal et al., 1991a; Spaink, 2000). This omnipresent rhizobial protein is related to LysR family of DNA binding proteins, which serve as transcriptional activators in prokaryotes (Henikoff et al., 1988). Three functional domains in LysR type transcription activators have been determined, an N-terminal DNA binding domain, a coinducer binding domain, and a conserved C- terminal domain (Schell, 1993; Lochowska et al., 2001).

In the presence of plant inducing compounds, the cytoplasmic membrane localized NodD protein binds to conserved *nod* boxes in the promoter regions and activates the transcription of down stream genes (Fisher et al., 1988; Györgypal et al., 1991; Schlaman et al., 1992; Fisher & Long, 1993). NodD has also been shown to bind the promoter region of *nod* genes in the absence of inducer molecules (Rostas et al., 1986; Goethals et al., 1992; Fisher & Long, 1993). Within each *nod* box is a highly conserved 32 base pair consensus sequence, which contains the imperfect palindromic structure ATC-N9-GAT that is the NodD binding site (Spaink et al., 1987; Goethals et al., 1992).

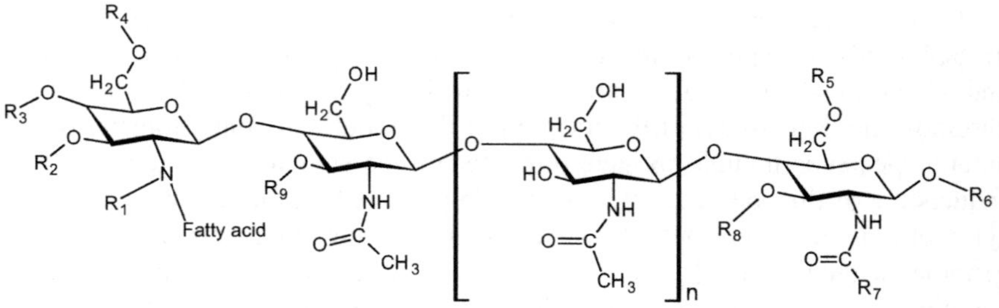

Fig. 4. General structure of Nod factors of rhizobia. For information on substitutions at R1 to R9 and the nature of the fatty acids refer to Table 3.

The mechanism by which the NodD in conjunction with flavonoid inducers promotes transcription of *nod* genes has not been fully determined. However, modulation of DNA tertiary structure due to binding to conserved 47 bp *nod* box sequences is thought to play a role in transcriptional activation (Feng et al., 2003). DNA symmetry and position of *cis* elements indicate that NodD attaches as a dimer or tetramer at two sites on the same face of the DNA inducing a bend in the helical structure (Goethals et al., 1992; Fisher & Long, 1993). Recently, DNaseI footprint and *nod* box deletion analysis have shown that that homotetremeric NodD is the biologically active form, which binds tandem subsites or half sites of the *nod* boxes. The distal and proximal binding sites are 75 to 25 base pairs and 50 to 25 base pairs upstream from the transcriptional start site, respectively (Feng et al., 2003; Chen et al., 2005). This molecule is a V-shape with four DNA binding domains at the end of the bifurcation which interact with the proximal and distal *nod* box sequences (Feng et al., 2003; Chen et al., 2005).

Small organic molecules are known to participate in protein-protein interactions and protein-nucleic acid interactions both of which are involved in formation of the transcriptional machinery (Berg, 2003; Mapp, 2003). Increasingly, evidence indicates that individual proteins often function as a part of a complex, acting in conjuction with other proteins to exert a biological function (Gavin et al., 2002; Ho et al., 2002). Small organic molecules such as the flavonoids are capable of both negatively and positively influencing protein-protein interactions. There are then several interfaces at which the flavonoids could influence the transcription of the *nod* genes such as protein-protein interactions and interactions between the machinery and the nucleic acid. In the LysR-type transcriptional activators conformational change has been noted in response to these signal or inducer molecules (Wang & Winans, 1995; Akakura & Winans, 2002; Feng et al., 2003; Chen et al., 2005). In the presence of inducer molecule naringenin, an increased bending of DNA, which was bound at *nod* boxes by NodD was noted (Chen et al., 2005). Mutation of the *nod* box resulted in sharper bends in the DNA structure and transcription was initiated in the absence of both the naringenin and the NodD indicating that this contortion of the DNA structure facilitated transcription (Chen et al., 2005). Another aspect of protein function is the specific conformation required for biological activity. The chaperone protein complex GroESL is necessary for proper folding and assembly of the NodD and thus its function as a transcriptional activator (Yeh et al., 2002). Biochemical and genetic evidence has shown that the *groESL* is required for expression of the *nod* genes and GroESL copurifies with the NodD and DNA binding activity decreases in the absence of the chaperone (Ogawa & Long, 1995).

Several variations in the design of the transcriptional activators occur among the rhizobia including copy number, and control mechanisms. *nodD* is present in all the rhizobia known and some species contain more than one copy of the gene (Honma & Ausubel, 1987; Davis et al., 1990; Pueppke, 1996). At least one functional NodD is required in the initiation of nodulation (Schlaman et al., 1992). Rhizobia species such as *Bradyrhizobium japonicum, Rhizobium* sp. NGR234, *R. meliloti* and *R. tropici* contain two to five copies of the *nodD* genes while *R. leguminosarum* bv. *viciae* and *R. leguminosarum* bv. *trifolii* have but one. *R. meliloti* has three copies of the transcriptional activator, *nodD1, nodD2* and *nodD3* and elimination of nodulation by this bacterium requires mutation of all three (Honma & Ausubel, 1987). *S. meliloti* expresses *syrM*, another member of the LysR family, which is independent of plant flavonoid influence (Mulligan & Long, 1989; Barnett & Long, 1990; Barnett et al., 1998). The genes *syrM* and *nodD3* with their encoded proteins mutually enhance transcription of each other thus generating amplification of the signal (Swanson et al., 1993).

6. *NOD* GENES INVOLVED IN SYNTHESIS AND DECORATION OF NOD FACTORS

Three of the genes common to all rhizobia, *nodABC*, encode proteins responsible for the synthesis of the basic structure of the Nod factors, an oligomer of β-1, 4 linked *N*-acetyl glucosamine (Fig.4) which varies in length from three to five glucosyl residues (Table 2). NodC, a UDP-*N*-acetyl glucosaminyltransferase, catalyzes the first step in the synthesis of the Nod factor from precursor molecules UDP-*N*-acetyl glucosamine, generating the oligomer (Geremia et al., 1994). Anchored to the inner bacterial membrane by three C-terminal transmembrane domains NodC provides the epicenter for Nod factor synthesis (Barny et al., 1996). The active site of NodC is exposed to the cytoplasm, where ostensibly the substrates UDP-*N*-acetylglucosamine, acetyl CoA, S-adenosyl L-methionine and the acyl-acyl carrier protein (ACP) complex are located. This possibly results in the formation of a multienzyme complex based around the membrane bound NodC facilitating channeling of substrates from one enzyme to the next in the synthetic pathway (Norris et al., 1996; Mergaert et al., 1997b). Such multienzyme systems increase the efficiency of production as substrates for the sequential series of reactions are held in proximity. The second step in the process involves deacetylation of the acetamido group on the nonreducing sugar moiety of the oligomer by NodB, a chito-oligosaccharide deacetylase (John et al., 1993). NodA, an acyltransferase catalyzes acylation of the acetamido group on the nonreducing sugar moiety, the final step in the formation of the basic oligomer (Atkinson et al.,

1994; Röhrig et al., 1994). Interestingly, allelic forms of both *nodA* and *nodC* exist, the former showing a preference for acyl groups of varying characteristics (Debellé et al., 1996) and the latter giving rise to different lengths of the oligomer backbone (Kamst et al., 1999). Allelic variations in *nodC* results in synthesis of a tetrameric lipooligosaccharide in *R. meliloti* and a pentameric backbone in *R. tropici*. As shown by Roche et al. (1996) a genetically modified bacteria carrying the tetrameric backbone was more efficacious in the nodulation of the host alfalfa than one carrying the pentameric backbone, thus allelic variation establishes a modicum of host specificity at this level. Additionally, a similar situation arises with respect to *nodA* alleles as the encoded protein is specific for the acyl group that it carries (Debellé et al., 1996). Exchanging *nodA* genes with a homologue rendered the recipient bacteria incapable of nodulation of its host plant (Ritsema et al., 1996; Debellé et al., 1996). The *nodA* mutant of *R. leguminosaurm* biovar *viciae*, when complemented with its homologue from *Bradyrhizobium* was incapable of nodulating its host plant, indicating a level of host specificity is generated by the allelic variation (Ritsema et al., 1996). Another factor to consider is the order in which the modifications occur as this determines the substrate of the enzyme catalyzing the subsequent reactions. Some modifications at the reducing terminus of the oligomer occur before acylation. If these structures are part of the recognition sites of the NodA enzyme, allelic variation could play a role in acylation as well as in other reactions, which place substituent molecules on the Nod factor backbone (Bloemberg et al., 1995; Spaink et al., 1991; Jabbouri et al., 1995; Quinto et al., 1997). The remaining common genes, *nodIJ*, are involved in

Table 2. Nodulation genes involved in Nod factor synthesis and transport.

Nod gene	Function	Reference
Regulation genes		
nodD	LysR-transcriptional activator	Egelhoff & Long, 1985; Downie et al., 1985
nodV	Two component regulation	Loh et al., 1997
nodW	Two component regulation	Loh et al., 1997
nolA	Transcriptional regulation	Garcia et al., 1996; Loh et al., 1999
nolR	LysR type regulator	Kondorosi et al., 1991
syrM	LysR type regulator	Maillet et al., 1990
Synthesis of glucosamine backbone		
nodB	Deactylase	John et al., 1993
nodC	UDP-GlcNAc transferase	John et al., 1988
nodM	D-glucosamine synthetase	Baev et al., 1991 Marie et al., 1992

Fatty acid synthesis and transfer to nonreducing terminus		
nodA	*N*-acyltransferase	Röhrig et al., 1994
nodE	β-ketoacyl synthase	Spaink et al., 1989
nodF	Acyl carrier protein	Demont et al., 1993
Modifications to nonreducing terminus		
nodS	S-adenosyl methionine methyl transferase	Göttfert et al., 1990; Krishnan et al., 1992
nodL	6-*O*-acetyltransferase	Downie, 1989; Berck et al., 1999
nodU	6-*O*-carbamoyltranserase	Jabbouri et al., 1995
nolO	3-*O*-carbamoyltransferase	Jabbouri et al., 1998
Modifications to reducing terminus		
nodP	ATP sulfurylase	Schwedock & Long, 1989 Schwedock et al., 1994
nodQ	ATP sulfurylase, APS kinase	Cervantes et al., 1989
nodH	Sulfotransferase	Roche et al., 1991a
nodZ	Fucosyl transferase	Stacey et al., 1994
nolK	Epimerase	Mergaert et al., 1997b
nodX	Acetyl transferase	Davis et al., 1988
noeE	Sulfotransferase	Hanin et al., 1997
nolL	*O*-acetyltransferase	Berek et al., 1999
noeI	2-*O*-methyltransferase	Jabbouri et al., 1998
Secretion of nod factors		
nodI	ATP-binding protein	Spaink et al., 1995a; Cardenas et al., 1996
nodJ	membrane protein	Spaink et al., 1995a; Cárdenas et al., 1996
nodT	membrane protein	Lewis-Henderson & 1991, Rivilla et al., 1995
Djordjevic		
nolF,G,H,I	membrane protein	Saler et al., 1994

secretion of the completed Nod factors. Rhizobia bearing mutatations of *nodIJ* are still capable of secreting the Nod factors indicating duplicity of genes or the presence of counterparts in the housekeeping genes (Spaink et al., 1995a; Cárdenas et al., 1996; Fernández-López et al., 1996).

Non-allelic variation in *nod* genes among the rhizobia is the crux of diversity in the enzymatic production of the Nod factors. Modifications to the core structure depend upon genes encoding transferases and enzymes in the synthetic pathways of the substituent entities. After the carbohydrate based oligomer has been completed, Nod factors encoded by host specific nodulation (*hsn*) genes in different rhizobia decorate the structure by addition of modified and unmodified monosaccharides, sulfate esters, methyl, carbamoyl groups, and other tags to generate unique structures (Dénarié et al., 1996; Perret et al., 2000; Spaink, 2000). Modifications made to this basic structure by *hsn* genes result in the variety of Nod factors, which define host specificity.

7. STRUCTURE OF NOD FACTORS

The chemical structure of Nod factors (Table 3) was first determined in *S. meliloti* and found be a sulfated β-1,4- tetrasaccharide of D-glucosamine in which three of the amino groups were acetylated and the fourth acylated with a polyunsaturated 16 carbon fatty acid (Lerouge et al., 1990; Roche et al., 1991b). Genetic and biochemical evidence show that structural variations of the Nod factors due to substituent groups contribute to host recognition and subsequent events in the nodulation process (Dénarié et al., 1996; Broughton et al., 2000). Different rhizobial strains each produce a spectrum of Nod factors unique to the particular strain (D'Haeze & Holsters, 2002). To facilitate investigation and communication of findings, a number of formats of Nod factor nomenclature have been developed (Roche et al., 1991b; Spaink, 1992). Each proposed method of designation includes an abbreviaton of the *Rhizobium* from which the Nod factor arises, the length of the lipochitooligosaccharide backbone, the length and degree of unsaturation of the ubiquitous *N*-linked acyl group on the nonreducing terminal and the various substituents on the terminal residues.

Although Nod factors are necessary to determine host plant specificity, they are not alone sufficient to produce and maintain the physiological and morphological changes resulting in formation of a functional nodule (Poupot et al., 1993; Cárdenas et al., 1995; Orgambide et al., 1995; Pueppke & Broughton, 1999; Perret et al., 2000). There are three basic variations in the overall structure of the Nod factors, length of the degree of polymerization, the identity of the *N*-acyl group on the nonreducing terminus, and the various and sundry reactions that occur at the hydroxyl groups on the terminal glucosaminyl residues. The number of individual lipochitooligosaccharides synthesized by a species varies from two in *R. etli* FFN42 (Poupot et al., 1995) to 60 in *R. galegae* HAMBI1207 (Yang et al., 1999).

Nod factors are divided into two general groups based on their structure (Spaink, 1996). One group produces a mixture of tetrameric or pentameric backbones with N-acylated polyunsaturated fatty acids on the nonreducing terminal residue while the other group has primarily a saturated or monounsaturated fatty acid and the acyl group shares the *N*-linking nitrogen with a methyl group (Meregaert et al., 1997a; Spaink, 1996). The reducing terminal residue is most often modified with a sulfate ester, L-fucosyl, 2-*O* Me-fucosyl, 4-*O*-Ac-fucosyl, or acetyl group while the nonreducing terminal carry *N*-methyl, *O*-acetyl or *O*-carbamoyl groups (D'Haeze & Holsters, 2002). Other monosaccharides such as mannose and arabinose or their acetylated or methylated derivatives are also utilized in the modification of the backbone.

Table 3. Structural details of Nod factors from various rhizobia.

Rhizobial Strain	Fatty acids	Substituents	References
A. caulinodans ORS571	C16:0, C18:0, C18:1	R1-Me, R4-Cb, R5 Fuc or Ara, R7-Me R8-Ara	Mergaert et al., 1997a
B. japonicum USDA110	C16:0, C16:1, C18:1	R5-MeFuc	Sanjuan et al., 1992
B. elkanii USDA61	C16:0, C18:1	R1-Me, R3-Cb or Ac,	Stokkermans et al., 1996 R4-Ac, R5- Fuc or MeFuc R6-Glo, R7-Me
M. loti NZP2037	C18:0, C18:1	R1-Me, R2-Cb, R3-Cb	López-Lara et al., 1995 R4- Cb, R5-AcFuc, R7-Me
M. huakuii	C18:4	R5-S, R7-G, -Me, -CH_2OH	Yang et al., 1999
R. etli	C18:0, C18:1	R1-Me, R3-Cb, R5-AcFuc, R7-Me	Cárdenas et al., 1995; Segovia et al., 1993
R. tropici CIAT899	C18:1	R1-Me. R5-S, R6-Man, R7-Me	Folch-Mallol et al., 1996 Poupot et al., 1993
R. leguminosaurm bv. *trifolii* ANU843	3OH-C14:0, 16:0, 3OH-C16:0 C16:1, C18:0	R3-Ac, R4-Ac, R5-Ac, R6-Et, R7-Me	Philip-Hollingsworth et al., 1991 Philip-Hollingsworth et al., 1997 Orgambide et al., 1995, van der Drift et al., 1996
	3OH-C18:0, C18:1 C18:2, C18:3, C18:4 C20:3, C20:4		
bv. *viciae* RBL5560	C16:0, C16:1 C18:0, C18:1 C18:4	R4-Ac, R7-Me	Spaink et al., 1995b

R. galegae	C14:0, C16:0, C18:0 C18:1, C18:2, C18:3, C18:4, 3OH- C18:3, 3OH-C20:1, C20:2, C20:3	R4-Cb, R7-Me, R9-Ac	Yang et al., 1999
S. meliloti	C16:1, C16:2, C16:3, C18, 20, 22, 24, & 26:(ω-1)OH	R4-Ac, R5-S	Ardourel et al., 1994, Demont et al., 1993 Lerouge et al., 1990; Schultze et al., 1992; Truchet et al., 1991
S. fredii USDA191	C16:1, C18:0, C18:1		Bec-Ferté et al., 1996
S. fredii USDA 257	C18:1	R5-Fuc, R5-MeFuc, R7-Me	Bec-Ferte et al., 1994
S. fredii NGR234	C16:0, C16:1, C18:0 C18:0, C18:1	R1-Me; R2-Cb, R3-Cb R4-Cb, R5-MeFuc; R5-AcMeFuc, R5-SMeFuc, R7-Me	Price et al., 1992
S. teranga			
bv. *sesbaniae*	C18:1	R1-Me, R3-Cb, R4-Cb,	Lorquin et al., 1997
bv. *acaciae* ORS1602	C16:0, C18:0, C18:1	R1-Me, R2 or R3-Cb, R5-S R5-Fuc, R8-Ara	Lorquin et al., 1997

The substituents given are not present in all Nod factors. Abbreviations: Ara, arabinosyl; Ac, *O*-acetyl; AcMeFuc, 4-*O*-acetyl-2-*O*-methylfucosyl; Cb, *O*-carbamoyl; Et, ethyl; Fuc, α-linked fucosyl; G, *N*-glycolyl; Glo, glycerol; Man, mannosyl; Me, *N*-methyl; MeFuc, 2-*O*-methylfucosyl; S, *O*-sulfuryl; SMeFuc, 3-*O*-sulfate-2-*O*-methylfucosyl.

The broad host range of *Rhizobium* sp. NGR234 has been attributed to the diversity of Nod factors synthesized (Price, 1992). However, it has been noted that rhizobia with a more limited host range also produce not one but several Nod factors varying in the length and substituents (Spaink et al., 1987). This combination of Nod factors produced by each organism has been proposed to function in a cooperative manner to assure proper matching of bacteria to host (Minami et al., 1996). A host plant can harbor rhizobia from different genera but the Nod factors will have similar structures thus reflecting adaptation of the bacteria to the plant (Lorquin et al., 1997). This is in essence a redundancy of selection. All the keys must fit all the locks before the symbiotic process can proceed (Parniske & Downie, 2003).

Some unusual modifications to the basic structure result in atypical Nod factors which ostensibly places restrictions on the host range of the bacterium. Generally, the oligosaccharide is a tetramer or pentamer but *Rhizobium* sp. GRH2 has a hexameric backbone (López-Lara et al., 1995). *S. fredii* USDA191 synthesizes a Nod factor in which one of the nonterminal glucosaminyl residues is replaced by a glucosyl residue (Bec-Ferté et al., 1996). In *M. loti* a dimeric lipochitooligosaccharide and a Nod factor with a fucose residue α-1,3 linked to the *N*-acetylglucosamine residue next to the nonreducing terminus, have been found (Olsthoorn et al., 1998). In two strains of *R. galegae* the glucosamine residue next to the nonreducing terminus was acetylated on C-3 (Yang et al., 1999). *S. meliloti* RCR2001, *R. galegae* HAMBI1207, *R. leguminoarum* bv. *trifolii* ANU843 and some strains of *Rhizobium* synthesize fatty acids whose carbon chain carries hydroxyl groups (Fraysse et al., 2003). An open ring structure on the reducing terminal residue exists in *Rhizobium* sp BR816 (Snoeck et al., 2001) An arctic bacterium, *Mesorhizobium* sp strain N33, exhibits a methyl group at the alpha and beta carbon of the *N*-linked acyl chain (Poinsot et al., 2001). Unusual modifications are possibly an adaptation to a particular set of environmental conditions. The modified structure in some way may have given the symbiont an advantage over competing organisms.

Nano-and pico-molar concentrations of Nod factors are sufficient to elicit changes in root hair cytoskeleton and subsequent root hair deformation (Lerouge et al., 1990; Cárdenas et al., 1998; Timmers et al., 1999; Weerasinghe et al., 2005). Physiological changes include root hair membrane depolarization and oscillation of intracellular calcium levels referred to as calcium spiking (Ehrhardt et al., 1992; Gehring et al., 1997; Felle et al., 1998; Wais et al., 2002a; 2002b). Other manifestations of the plant response to Nod factors include preinfection thread formation and localized mitogenesis in the root cortex resulting in the formation nodule primordia (Spaink, 1992; van Brussel et al., 1992).

Elevated levels of Nod factors, however, can elicit defense response that interfere with rhizobial infection and cause defective nodulation (Knight et al., 1986; Savore et al., 1997; Hogg et al., 2002; Ramu et al., 2002). Interestingly, the Nod factors at nanomolar concentrations in an autoregulation mechanism appear to function as inducers of this hydrolytic cleavage by enhancing activity of chitinolytic enzymes (Staehelin et al., 1994a; 1994b; Staehelin et al., 1995; Ovtsyna et al., 2000) The self induced cleavage after host recognition is possibly required for effective signal transduction and induction of the plant genes involved or to avoid initiation of the pathogen response by the plant (Staehelin et al., 1995; Savoure et al., 1997; D'Haeze & Holsters, 2002). A sulfate ester on the reducing terminal residue of Nod factors produced by *R. meliloti* prevents hydrolytic cleavage by purified chitinases from the host plant *Medicago*. The resulting cleavage products were less active as measured by root hair deformation (Staehelin et al., 1994b; Schultze et al., 1992). This is indicative that the biological activity of Nod factors may be in part controlled by chitinase.

8. THE ROLE OF POLYSACCHARIDES IN SYMBIOSIS

Currently, at least seven types of carbohydrate polymers, lipo-, exo-, capsular, gelling polysaccharides, glucomannan, β-1,2 glucan, and cellulose, associated with nodulation have been identified. A plethora of mutational analyses has linked rhizobial carbohydrates to aberrant nodulation phenotypes. Current investigations into the roles of these compounds suggest they function in invasion, infection thread initiation and expansion, subversion of plant defense mechanisms, bacterial release from infection thread, bacteroid development, and initiation of plant gene expression (González et al., 1996; Dazzo et al., 1991; Perroto et al., 1994; Parniske et al., 1994; Becker & Pühler, 1998; Broughton et al., 2000; Perret et al., 2000; Fraysse et al., 2003; D'Haeze & Holsters, 2004, Mathis et al., 2005). Determination of structure and function of these complex molecules continues to be a painstaking adventure. Data garnered from work in the various disciplines suggests these complex molecules can function as both structural entities and signal molecules. The demarcation between these roles is rather subtle. The surface polysaccharides facilitate adherence to the root surface, nutrient acquisition, buffering environmental stresses, and circumventing or suppressing host plant defense mechanisms. In the capacity of signal molecules the surface polysaccharides emulate the Nod factors as they generate a plant response with respect to initiation of infection thread development and release of rhizobia from infection threads, and plant defense systems and development of the bacteriods. Elucidating the function of a particular polysaccharide is made difficult as the monomers from which they are synthesized often originate from the same biochemical pathways (Cedergren et al., 1995; Geiger et al., 1998; de Rudder et al., 1999, Price, 1999; López-Lara et al., 2003; Sohlenkamp et al., 2003). As an example, galactose is ubiquitous among the various polysaccharides, thus an interruption of its synthesis through a mutation in *exoB* which encodes UDP-glucose 4'-epimerase will affect several polysaccharides making determination of function difficult (Canter Cremers et al., 1990). Although functional complementation among glucosyl transferases appears to occur (Gluckmann et al., 1993) two genes *pssE* and *pssD* putatively encoding a glycosyl transferase of *R. leguminosarum* bv *trifolii* RBL 5599 are somewhat unique in that they appear essential for the formation of the exopolysaccharides (van Workum et al., 1997).

9. EXOPOLYSACCHARIDES

The chemical structure of exopolysaccharides secreted by *Rhizobium* is strain and biovar dependent, varying in monosaccharide composition, glycosidic linkage, polymer length, and substituent molecules (van Workum & Kijne, 1998). Glucuronic and galacturonic acids in the chain along with acetyl, pyruvyl, and succinyl substituents contribute to the acidic nature of these polymers. Essentially a preponderance of the knowledge concerning EPS synthesis, structure and function has been gleaned from work using the *R. leguminosarum* and *S. meliloti* as models. EPS deficient mutants reveal various phenotypes such as failure to form nodules or infection threads, formation of empty nodules, or defects in the bacterial entry into nodule (Finan et al., 1985; Leigh et al., 1985, Borthakur et al., 1986; Leigh et al., 1987; Rolfe et al., 1996). Complementation by the addition of an exopolysaccharide fraction to EPS deficient mutants *R. leguminosarum* bv. *trifolii* and *S. meliloti* restored nodulation effeciency (Djordjevic et al., 1987; Battisti et al., 1992). However, host specificity was noted in the rescue of nodulation indicating that the host plant has requirements for specific ESP structures (Djordjevic et al. 1987; Gray et al., 1991; Battisti et al., 1992; van Workum et al., 1998). In a general sense, EPS are required for infection thread formation since mutants exhibiting particular deficiencies in EPS are incapable of infecting the plant root hair. Additionally, the EPS are involved in modulation of plant defense mechanisms. Whether the EPS function as antagonists in the signal cascade eliciting plant defense mechanisms or disguise epitotic structures capable of eliciting the reponse is not at present known (Niehaus et al., 1993; Rolfe et al., 1996). *S. meliloti* produces two types of exopolysaccharides, EPS I and EPS II. EPSI or succinoglycan encoded by the plasmid borne *exo* gene cluster (Long et al., 1988; Reinhold et al., 1994), is composed of repeating units which consist of one galactosyl and seven glucosyl residues connected by β-1,4, β-1,3 and β-1,6 linkages (Reinhold et al., 1994). Acetyl, succinyl, and pyruvyl groups decorate this backbone structure. Mutations in the various *exo* genes affect nodulation at different stages such as infection thread elongation and cortical cell invasion (Cheng & Walker, 1998). Succinoglycan does not appear to be essential for colonization of the root hair or infection thread induction but is required for infection thread elongation. Modifications to the succinoglycan structure by the various mutations had noted effects on infection thread initiation and elongation. (Cheng & Walker, 1998). EPSII, encoded by the *exp* gene cluster, is formed by alternating glucosyl and galactosyl residues in α-1,3 and β-1,3 linkages (Glazebrook & Walker 1989; Becker et al., 1997). Adducts to this basic polymeric structure include acetyl and pyruvyl groups (Her et al., 1990). Both the ESPI and II are synthesized as either high or low molecular weight entities (Gonzalez et al., 1996). Synthesis of these acidic polysaccharides begins by transfer to a

C55-isoprenylphosphate carrier by glycosyl-isoprenyl transferases, a large family of related membrane proteins (van Workum et al., 1997; Pollock et al., 1998)

Although the composition of the EPS has been shown to vary with carbon source used to support growth in culture medium, the varying EPS structure however does not affect the soybean nodulation efficacy. This is in accordance with the observation that legumes exhibiting indeterminate nodules require EPS while those expressing determinate nodules do not require this polysaccharide (Borthakur et al., 1986; Djordjevic et al., 1987; Diebold & Noel, 1989; Hotter & Scott, 1991; Karr et al., 2000). However, analysis of *S. meliloti* expolysaccharides and capsular polysaccharides mutants has suggested a redundancy exists in the functioning of these compounds (Pellock, 2000; Fraysse et al., 2003). It is possible then that the exopolysaccharides are also involved in determinate nodulation but other polysaccharides can serve in their absence (Fraysse et al., 2003). Another aspect attributed to the ESP is masking the rhizobia toward plant defense. Mutants deficient in exopolysaccharide synthesis are subject to growth inhibition by plant phytoalexin and additionally elicit plant defense mechanisms (Vasse et al., 1993; Parniske et al., 1994; Perotto et al., 1994; Rolfe et al., 1996; Niehaus et al., 1998). The redundancy in the EPS indicates that these compounds have a role in both determinate and indeterminate nodule formation (Fraysse et al., 2003).

10. LIPOPOLYSACCHARIDES

Lipopolysaccharides (LPS), acidic polysaccharides displayed on the rhizobial cell surface, are comprised of three distinct entities, the core polysaccharide, O-antigen and lipid A. The core polysaccharide structure has been determined for several rhizobia and although generally conserved, differences in composition exist (Carlson et al., 1989; Bhat et al., 1994; Carlson & Krishnaiah, 1992; Reuhs et al., 1998). The core is bridged to the O-antigen and the lipid A by 3-deoxy-D-manno-2-octulosonic acid (Kdo). Determined for several LPS, the O-antigen portion comprised of modified monosaccharides, heptose, uronic acid residues is the most variable portion of the LPS and can differ between strains of the same organism (Reuhs, et al., 1994; Wang & Hollingsworth, 1994; Gil-Serrano, 1995; Forsberg et al., 2000). Lipid A, comprised of a glucosamime disaccharide linked to four to six long chain 27OH-C:28 hydroxy fatty acids unique to α-proteobacteria, anchors the entire LPS to the bacterial outer membrane (Hollingsworth & Carlson, 1989; Bhat et al., 1991a; Bhat et al., 1991b).

Overall, LPS appear to be involved in the later stages of symbiosis such as infection thread development, bacterial release from the infection thread and suppression of plant defense response (Perotto et al., 1994; Albus et al., 2001; Gao et al., 2001). Complementation studies involving LPS defective mutants supplemented with purified LPS have suggested that the LPS also function as signal molecules in nodulation (Mathis et al., 2005). Mutants with alterations in the LPS tend to compromise the symbiotic process at different stages of infection and nodule formation (Kannenberg et al., 1998). Evidence that these carbohydrates serve as signal molecules is supplemented by determination of variations in LPS structure, which appear during the progressive stages of symbiosis suggesting that these structures are required at defined stages for the process to continue (Sindhu et al., 1990; Goosen-de Roo et al., 1991; Duelli et al., 1997; Reuhs, et al., 1999). Lipopolysaccharides may function to amend the characteristics of the cell surface facilitating the interaction between plant cells and bacteria which is necessary for symbiosome formation. Additionally, the use monoclonal antibodies raised against the O-antigen to monitor the presence of the different LPS compositions indicated that the structure is differentially regulated during symbiosis (Kannenberg et al., 1994; Brewin et al., 1998). The progressive change in the structure of LPS leads to an increased hydrophobicity of the rhizobial membrane ostensibly to facilitate interaction with plant cell membranes in the formation of the peribacteroid membrane that finally encompasses the nitrogen reducing bacteroids (Kannenberg & Carlson, 2001).

11. CYCLIC GLUCANS

Cyclic glucans are β-linked glucose polymers found predominantly in the periplasmic space and are variously decorated with charged substituents such as phosphoglycerol, succinic acid, and phosphocholine (Batley et al., 1987). Synthesis of cyclic β-glucans from UDP-glucose is controlled by *ndv* loci, a chromosomal loci similar to *chv* loci of *Agrobacterium* (Dylan et al., 1986; Ielpi et al., 1990; Bhagwat, et al., 1999). The degree of polymerization, type of glycosidic linkage, presence of branching and substituent groups varies with the rhizobial species (Gore & Miller, 1992; Inon de Iannino & Ugdale, 1993). Examination of the mutations in the *ndvA* and *ndvB* loci encoding cytoplasmic membrane proteins involved in β-1,2 glucan synthesis from UDP glucose showed that these mutants developed ineffective nodules (Dylan et al., 1990; Ielpi et al., 1990). A multitude of roles for the cyclic polysaccharides have been suggested including promoting growth under hypoosmotic conditions, suppressing plant defense response and facilitating transport of molecules (Chen et al., 1985; Tully et al., 1990; Gore & Miller, 1993; Dunlap et al., 1996; Ingram-Smith & Miller, 1998; Bhagwat et al., 1999). Accumulation of significant quantities of the cyclic glucans in contrasting environments occupied by free bacteria and bacteroids within the root nodules supports the notion that the polymers function as an osmoprotectant facilitating adjustment to the changing conditions (Rolin et al., 1992; Gore & Miller, 1993; Breedveld & Miller, 1994; Fraysse et al., 2003). Cyclic glucans apparently function to suppress host plant defense response. Structural similarities exist between cyclic glucans from soybean fungal pathogens and those of the symbiont *B. japonicum*. However, the plant response to exposure to these glucans is antithetical to that of the pathogen derived carbohydrate. Non cyclic β-1,6 and β-1,3 glucan fragments from the fungus elicit production of the isoflavonoid derived plant defense phytoalexins, while those from the symbiont elicits production of daidzein, an inducer of nodulation genes (Miller et al., 1994). Subsequent work has shown that the symbiont-produced glucan suppresses induction of phytoalexin induced by the fungal pathogen (Mithöfer et al., 1996). When site-specific mutations in *ndvB* and *ndvC* were generated in *B. japonicum,* loss of β-glucan synthesis and production of an aberrant structure (cyclodecakis-(1→3)- β-D-glucosyl) resulted, respectively. The phytoalexin, glyceollin, was three to five fold higher in the nodules or pseudonodules induced by the mutants (Bhagwat et al., 1999). In competitive binding tests, when opposed to a model of the fungal β-glucan, wild type β-glucan was 40 fold more effective than the aberrant structure in binding to the putative soybean β-glucan receptor (Ebel & Cosio, 1994; Bhagwat et al., 1999). Mutants of *R. leguminosarum* defective in cyclic β-1,2-glucan secretion showed a deficiency in nodulation and induced synthesis of chalcone synthase, an enzyme expressed early in the isoflavonoid pathway (Yang et al., 1992). Up regulation of the gene could have resulted in increased production of plant defense compounds thus thwarting nodulation.

12. CAPSULAR POLYSACCHARIDES

Capsular polysaccharides bear structural similarity to a group of K-antigens found in *Escherichia coli* thus the designation (Reuhs et al., 1993). These polysaccharides possibly serve an active role as signal molecules and a passive role in protection against plant defense. Serving as a signal molecule in *S. meliloti*, capsular polysaccharides were shown to induce mRNA of genes involved in isoflavanoid pathway (Becquart-de-Kozak et al., 1997). The hydrated capsular polysaccharides form a matrix around the rhizobia providing protection from dessication and bacteriophage invasion (Campbell et al., 1998). Similarly, the capsule could play a role in subverting host defense mechanisms thus allowing the process of nodulation to proceed. These compounds can vary from strain to strain and consist of a repeating unit comprised of hexose or modified hexose linked to Kdo or similar carbohydrate such as neuraminic acid.

Gelling surface polysaccharide is a neutral polymer of unknown function. Consisting of repeating units of glucose, mannose and galactose, this insoluble polymer thus far isolated only from *R. leguminosarum* may function as a protectant against adverse environmental conditions (Laus & Kijne, 2004). Another neutral polysaccharide recently isolated from *R. leguminosaum* RBL5523 is glucomannan. This polymer was shown to be present at the bacterial pole, which attaches to the root hair cell membrane (Laus & Kijne, 2004).

13. EXTRACELLUAR PROTEINS

Rhizobial extracellular proteins appear to function in part as signal molecules in nodulation. Since the flavonoid inducible protein, NodO was first isolated from the extracellular medium of *R. leguminosarum* (de Maagd et al., 1989; Economou et al., 1990), a number of other proteins have been found to be secreted by rhizobia. Among them, a group of extracellular proteins, have recently been shown secreted by the type III secretion system (TTSS) (Krishnan et al., 1995; Viprey et al., 1998; Marie et al., 2001; Krause et al., 2002). Since first identified in *Yersinia enterocolitica* as the mechanism by which *Yersinia* outer proteins (Yops) were secreted, TTS systems have now been found in many animal and plant pathogenic bacteria, insect endosymbionts and nitrogen reducing rhizobia (Forsberg et al., 1988; Michiels et al., 1990; Freiberg, 1997; Lee, 1997; He, 1998; Hueck, 1998; Heddi et al., 1999; Cornelis & Van Gijsegem, 2000; Dale et al., 2002). Comprised of membrane spanning protein complexes, type III secretion systems serve in the translocation of effector proteins into host cells, secretion of proteins required for development of various symbiotic relationships, and construction of bacterial flagella (Freiberg et al., 1997; He, 1998; Hueck, 1998; Galán & Collmer, 1999; Macnab, 1999; Dale et al., 2002). A family of conserved genes encodes the transmembrane protein complex through which the proteins are secreted and a cytosolic ATPase, which putatively energizes the process. Proteins which move through the complex include effectors and accompanying translocators, some of which comprise extracellular portions of the complex (Büttner & Bonas, 2002; Jin & He, 2001).

Although distinct TTS systems have been identified, there is commonality among them (Young et al., 1999). While a variety of proteins are secreted by TTS systems, sequence similarity exists among the genes encoding the secretion apparatus indicating the mechanics of operation are similar among the various bacterial species, which embody this system (Galán & Collmer, 1999). Genes encoding TTS system are located either on a plasmid or within a pathogenicity island of the bacterial chromosome. Secretion of a protein does not involve cleavage of an N-terminal signal peptide but does require the presence of chaperones (Michiels, 1988; Page & Parsot, 2002; Ghosh, 2004). TTS systems consist of approximately 20-25 proteins, many of which are similar in sequence and have morphological and mechanistic characteristics akin to the flagellar proteins (He, 1998; Macnab, 1999; Aizawa, 2001; Blocker et al., 2003). Since the flagella are both ancient and widely distributed among bacteria whereas TTS systems are more restricted in dispersion, phylogenetic analyses suggests that the latter evolved from the flagella (Macnab, 1999; Nguyen et al., 2000.) Other phylogenetic analyses, however, suggests that flagella and TTS systems are both ancient structures and have evolved from a common ancestor (Gophna et al., 2003).

14. IDENTIFICATION OF TTS SYSTEMS IN RHIZOBIA

Investigation of the molecular basis of host specificity of *Sinorhizobium fredii* USDA257 provided the first evidence for the existence of a TTS system in rhizobia. *Sinorhizobium fredii* USDA257 enters into symbiosis with primitive soybean cultivars but is unable to form such a relationship with modern cultivars (Keyser et al., 1982; Annapurna & Krishnan, 2003). The key to cultivar-specific nodulation of *S. fredii* USDA257 lays in the gene cluster, *nolXWBTUV*, which is located on a *sym* plasmid (Meinhardt et al., 1993; Kovacs et al., 1995). A comparison of sequences between open reading frames at this locus and that of the genes encoding structural components of plant and animal pathogen TTS systems revealed a striking homology. Furthermore, amino acid sequences of NolW and NolT showed significant homology to the HrcC and HrcJ of plant pathogenic bacteria (Meinhardt et al., 1993; Kovacs et al., 1995) corroborating the possibility that a TTS system was present in USDA257. Sequence analysis of the symbiotic plasmid pNGR234a from *Rhizobium* sp. NGR234 provided the complete identification of a rhizobial TTS system (Freiberg et al., 1997). The TTS system also has been found in *M. loti* MAFF303099 (Kaneko et al., 2000), *B. japonicum* USDA 110 (Göttfert et al., 2001), *S. fredii* USDA191 (Bellato et al., 1997) and HH103 (Lamrabet, et al., 1999; Marie et al., 2001), *B. elkanii* (Viprey et al., 1998), and *R. etli* CFN42 (National Center for Biotechnology Information [NCBI] database, accession number U80928). However, the TTS system has not been identified in the genomes of *R. meliloti*, *M. loti* strain R7a or *S. meliloti* 1021 (Viprey et al., 1998; Galibert et al., 2001; Sullivan et al., 2002).

15. GENETIC ORGANIZATION OF TTS SYSTEM IN SYMBIOTIC BACTERIA

Secretion of proteins through the TTS system is flavonoid dependent, however the typical regulatory *nod* box sequences do not appear in the promoter regions of either the conserved *rhc* or the *nops* which comprise the *tts* cluster (Krishnan et al., 1995; Viprey et al., 1998; Marie et al., 2001; Krause et al., 2002). However, a *nod* box was found in front of *ttsI,* which encodes a transcriptional activator that binds to the promoter regions of certain genes containing *cis* regulatory elements, the *tts* boxes. Identification of the *tts* boxes, upstream from *rhc* and *nops* has brought to light another element in the hierarchy of regulation, which appears to control assembly of the machinery necessary to secrete Nops (Krause et al., 2002; Marie et al., 2004).

Genes encoding TTS systems of rhizobia, as in other bacteria, are contained within 25 to 50 kb regions either on a symbiotic plasmid or within symbiotic islands on the bacterial chromosome. In *S. fredii* USDA257 and *Rhizobium* sp. NGR234, for example, these genes are located on the *sym* plasmid, while those from *B. japonicum* and *M. loti* MAFF303099 are located on the bacterial chromosome. These loci contain not only the genes encoding the secretion machinery but those of the secreted proteins as well (Marie et al., 2001). A subset of 10 genes of the 27 genes identified, which comprise plasmid borne TTS system clusters of NGR234 and *S. fredii* USDA257 encode the structural components of the TTSS (Freiberg et al., 1997).

Nomenclature of the rhizobial TTS system parallels that of the animal pathogen *Y. entercolita* in which TTS system genes are designated as *ysc* (*Yersinia* conserved). Accordingly, the rhizobial TTSS genes are designated as *rhc* (*Rhizobium* conserved). Eleven protein components of *Yersinia* TTS system, YscC, YscD, YscJ, -YscJ, -YscN, YscQ, YscR, YscS, -YscT, YscU, YscV, are commonly found in TTS systems of plant pathogenic and symbiotic bacteria (Hueck, 1998). Based on the structure of the TTS system apparatus of *P. syringae*

(Baker et al., 1997) a model has been proposed for the *rhc* encoded TTS system (Viprey et al., 1998; Marie et al., 2003). RhcC1 and RhcC2, which are analogous to the N-terminal and C-terminal domains of HrcC of plant pathogenic bacteria, are located in the outer membrane. RhcJ codes for an outer membrane lipoprotein and RhcR, RhcS, RhcT, RhcU, and RhcQ encode inner membrane proteins. Two cytoplasmic proteins, RhcQ and RhcN, are also part of the TTS system machinery of NGR234 and USDA257. The functions of these proteins are assumed similar to those ascribed to the Hrc proteins. The TTS system of *B. japonicum* USDA110 contains several open-reading frames that are not present in other symbiotic bacteria (Göttfert et al., 2001). Interestingly, the TTSS of NGR234 and that of USDA257 are 98% identical but only the later rhizobia can form nodules on soybeans (Krishnan et al., 2003). Antithetically, the TTS systems of *S. fredii* USDA257 and *B. japonicum* USDA110, both capable of soybean nodulation, have only limited sequence homology. In spite of these differences, the *rhc* genes appear to be highly conserved among all the symbiotic bacteria expressing a TTS system.

16. SURFACE APPENDAGES OF SYMBIOTIC BACTERIA

Electron microscopy has revealed the fine structure of surface appendage associated with TTS system present in animal and plant pathogens (Roine et al., 1997; Kubori et al., 1998; Galán & Collmer, 1999; Blocker et al., 2001; Daniell et al., 2001). *Salmonella typhimurium* elaborates a structure that consists of a cylindrical base embedded in the membrane with a needle-like projection appearing to emanate from the bacterial surface. Visualization under the electron microscope shows these needle-like structures to be hollow and approximately 120 nm length. Presumably, these appendages serve to deliver effector proteins into the host cells (Kubori et al., 1998). Other types of surface appendages have been reported. *Escherichia coli* produce filamentous appendages about 50 nm in diameter (Daniell et al., 2001), while the plant pathogen *P. syringae*, elaborates a 6-8 nm diameter pilus whose presence is dependent upon *hrp* locus (Roine et al., 1997). The *hrp* dependent pilus is a conduit for the delivery of effector proteins into the host cell (Jin & He, 2001). Surface appendages similar in structure and dimension to those of *P. syringae* have also been observed in *S. fredii* and *Rhizobium* sp. NGR234 (Pueppke et al., 1999; Krishnan et al., 2003; Saad et al., 2005). Immunocytochemical localization revealed the presence of type III secreted proteins throughout the entire length of the pili, indicating that these structures may function in the delivery of proteins to host cells (Pueppke et al., 1999). Additionally, NopX, which has been identified as a type III secreted protein was found to be present in the infection threads of soybean nodules (Krishnan, 2002). Although the presence of surface appendages has not been confirmed in other rhizobia, conservation of *rhc* genes would suggest that these structures are elaborated by the other species of symbiotic nitrogen fixing bacteria.

17. TYPE III-SECRETED PROTEINS OF SYMBIOTIC BACTERIA

In contrast to animal and plant pathogenic bacteria, only a few rhizobial TTS system proteins have been identified. Prior to the elucidation of the presence of TTS system in rhizobia, *Sinorhizobium fredii* USDA257 was found to secrete several extracellular proteins when induced by flavonoids (Krishnan & Pueppke, 1993; Krishnan et al., 1995). Mutations in *hrcC* and *hrcJ*, which encode the outer membrane proteins of the type III transport complex, negate the ability of the mutants to secrete several proteins. In accordance with the accepted nomenclature used to define TTS system secreted *Yersinia* outer proteins, it has been suggested that Rhizobia TTS system secreted proteins be designated as nodulation outer proteins (Nops) (Marie et al., 2001, Krishnan et al., 2003). At present, six nodulation outer proteins have been identified and their function characterized to a varying extent (Fig. 5). These proteins include NopA, NopB, NopC, NopL, NopP, and NopX (Marie et al., 2003, Krishnan et al., 2003; Lorio et al., 2004; Deakin et al., 2005; Saad et al., 2005). NopA a secreted protein associated with the pili-like surface appendages, which form under flavonoid induction in *S. fredii* and NGR234 is believed to be a structural component of the system and is necessary for the secretion of other proteins (Krishnan et al., 2003; Marie et al., 2003;

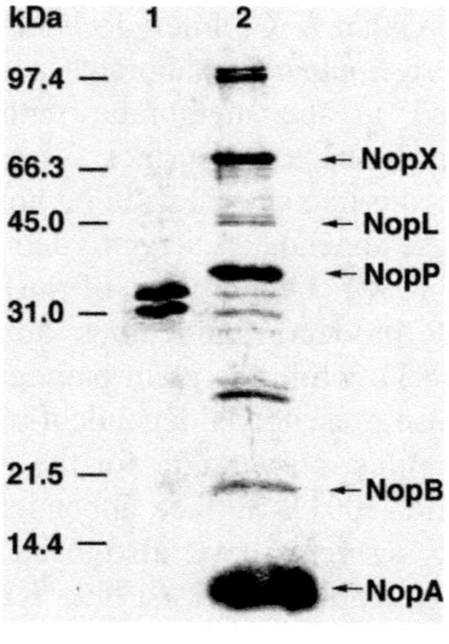

Fig. 5. Proteins associated with surface appendages of *S. fredii* USDA257. Rhizobia grown in the presence of flavonoids secrete several Nops (Lane 2). In the absence of flavonoids, USDA 257 elaborates two abundant flagellar proteins (Lane 1).

Deakin et al., 2005). Similarly, NopB an 18-kDa protein has also been associated with pili like structures on USDA257 (Fig. 6) and NGR234. Mutats of *nopB* were unable to form pili or secrete other Nops including NopX, NopB and NopA (Lorio et al., 2004; Saad et al., 2005). Another protein, NopL, was shown to be a substrate for plant protein kinases. Phosphorylation of this protein was inhibited by mitogen activated protein kinase inhibitors suggesting that NopL could influence the host

MAP kinase signal pathway (Bartsev et al., 2004). Effector proteins of plant pathogens often subvert signal transduction pathways thus NopL is possibly involved in abrogating cellular defense mechanisms (Büttner & Bonas, 2003; Bartsev et al., 2004). Another TTS secreted protein, NopX is involved in regulation of cultivar specific nodulation and thought to function as an effector or translocator protein (Krishnan, 2002; Marie et al., 2003). Interestingly, NopX, which was initially identified as a soybean cultivar-specificity protein in *S. fredii* USDA257 is not found in *B. japonicum* USDA110, a classical symbiont of soybean. Antithetically, this protein is present in *Rhizobium* sp. NGR234, a strain that does not form nitrogen-fixing nodules on soybean. NopX is homologous to HrpF of *Xanthomonas campestris* pv. *vesicatoria* which functions in the translocation of effector proteins into the host cells (Huguet & Bonas, 1997). NopC has been identified as a 10-kDa secreted protein of *Rhizobium* NGR234 with homologues identified in USDA257, *B. japonicum* USDA110 and *M. loti* MAFF303099 (Deakin et al., 2005). NopP is a 31.2-kDa secreted protein isolated from cultures of NGR234 and USDA257. Neither NopP nor NopL appear to influence secretion of other Nops. Various affects with respect to host specificity were noted in the presence or absence of NopP and NopL (Ausmees et al., 2004; Bartsev et al., 2004).

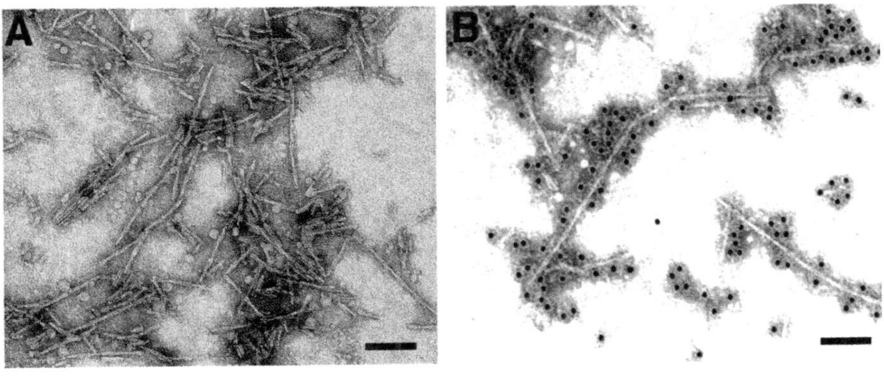

Fig. 6. Transmission electron micrographs of *S. fredii* USDA257 pili. USDA257 when grown in presence of flavonoid produces narrow tubular appendages (Panel A). Immunocytochemical localization studies reveal that NopB, a type three secreted protein, is associated with these surface appendages (Panel B). Bars = 100 nm.

A few other proteins encoded by genes outside the TTS system locus of *Rhizobium* sp. NGR234 and *B. japonicum* also show sequence homology with effector proteins of pathogenic bacteria. These include y4I0, y4IR, and y4zC from *Rhizobium* sp. NGR234 and Id431 and Id797 from *B. japonicum* USDA110. y4fR and Id431 are homologous to the IpaH family of *Shigella flexneri*, the YopM of

Yersinia spp. and the SspH of *S. typhimurium*. Additionally y4I0, y4zC, and Id797 appear homologous to plant avirulence proteins (White et al., 2000). y4I0 shows sequence homology to YopJ which functions as an ubiquitin-like protease (Orth et al., 2000). Whether these rhizobial proteins exhibit functions analogous to their pathogenic counterparts or are secreted by the TTS system is not known.

18. FUNCTION OF TYPE III-SECRETED PROTEINS IN SYMBIOSIS

The role of type III secreted proteins in nodulation was first demonstrated in mutants of *S. fredii* USDA257. Although USDA257 forms nodules on the primitive soybean cv. Peking it does not nodulate the modern cv. McCall. Unlike the wild type strain an HrcC and HrcJ mutant of *S. fredii* USDA257 was able to form nitrogen-fixing nodules on both McCall soybean and *Erythrina spp.* (Meinhardt et al., 1993; Krishnan & Pueppke, 1994). This indicates that the type III secreted proteins are involved in both soybean cultivar specificity and host range extension. Negation of protein secretion can reduce, enhance, or have no effect on the number of nodules formed depending upon the host involved. An *rhcN* mutant of NGR234 had no effect of *Vigna unguiculta* nodulation but increased the nodule number on *Pachyrhizus tuberosus* while reducing nodulation on *Tephrosia vogelii* (Viprey et al., 1998). Apparently, some hosts require type III secreted proteins for optimal nodulation, while in other hosts these proteins have a deleterious effect on nodulation. Plant proteins homologous to those of the TTS system are thought to be involved in these contrasting responses. It appears that a functional complementation occurs when a plant lacking the homologue of the secreted protein is inoculated with rhizobia mutated in the same gene (Broughton et al., 2000). Certainly, further work in this area is required to verify this proposition.

19. CONCLUDING REMARKS

Symbiotic nitrogen fixation has traditionally been considered the domain of rhizobia belonging to α-proteobacteria and leguminous plants. Recent discovery of members of β-proteobacteria capable of nitrogen reduction indicates a greater diversity among nitrogen fixing symbionts than previously realized. Additionally, the presence of *nod* genes in *Methylobacterium* and *Devosia* gives the thought that horizontal transfer of genes has introduced the capacity for substantiation to symbiotic nitrogen fixation to very diverse bacteria. Various genera of bacteria adapted to diverse environments harbor species capable of symbiotic nitrogen fixation. Although commonality among these organisms such as *nod* genes appear to exist, the discovery of new symbiotic pairs enhances the possibility of uncovering subtle and novel aspects in the communication process which could be exploited in our attempt to increase biological nitrogen fixation among legumes and ultimately extend the characteristic to graminaceous plants.

Intense research on nitrogen-fixing bacteria conducted during the past few decades has shed light on the molecular signals and the intricate biochemical events that occur when the rhizobia interact with legumes. Our increased knowledge and understanding of the symbiotic process however has resulted in minimal transfer of technology for use in commercial agricultural production. One promising area is the manufacture and use of flavonoids and Nod-factors to enhance nodulation in cultivated legumes such as soybean. Since nodulation involves other host-specific signal molecules such as polysaccharides and bacterial type III secreted proteins, inclusion of these factors may be necessary to realize the full potential of nodulation enhancement. However, the cost benefit ratio associated with the use of these synthetic biological signal molecules is a matter that will require careful consideration. Competition among rhizobia for nodulation, especially between the endogenous poor nitrogen fixers and the introduced efficient nitrogen fixers for host plants is another factor, which limits biological nitrogen fixation. It is exasperating that the poor nitrogen fixers are often the most effective nodulators. Ultimately shifting the nodule inhabitant population toward the efficient nitrogen fixers could result in an increase of the total nitrogen fixation in the leguminous crop. This shifting of populations is likely to involve an alteration of the biochemical signals which are responsible for the communication pathway between the rhizobia and host legume. Undoubtedly, basic research will continue to unravel the plethora the mysteries surrounding rhizobia-legume symbioses but we also need to focus on applied research to improve agricultural production.

20. REFERENCES

Aizawa, S.-I. (2001). Bacterial flagella and type III secretion systems. *Federation of European Microbiological Societies Microbiology, 36,* 1436-1446.

Akakura, R., & Winans, S. C. (2002). Constitutive mutations of the OccR regulatory protein affect DNA bending in response to metabolites released from plant tumors. *Journal of Biological Chemistry, 277,* 5866-5874.

Albus, U., Baier, R., Holst, O., Pühler, A., & Niehaus, K. (2001). Suppression of an elicitor-induced oxidative burst in *Medicago sativa* cell cultures by *Sinorhizobium meliloti* lipopolysaccharides. *New Phytologist, 151,* 597-606.

Amann, G., Stetter, K. O., Llobete-Brossa, E., Amann, R., & Anton, J. (2000). Direct proof for the presence and expression of two 5% different 16S rRNA genes in individual cells of *Haloarcula marismortui*. *Extremeophiles, 4,* 373-376.

Amarger, N., Macheret, V., & Laguerre, G. (1997). *Rhizobium gallicum* sp. nov. and *Rhizobium giardinii* sp. nov., from *Phaseolus vulgaris* nodules. *International Journal of Systematic Bacteriology, 47,* 996-1006.

Annapurna, K., & Krishnan, H. B. (2003). Molecular aspects of soybean cultivar-specific nodulation by *Sinorhizobium fredii* USDA257. *Indian Journal of Experimental Biology, 41,* 1114-1123.

Ardourel, M., Demont, N., Debellé, F., Maillet, F., deBilly, F., Promé, J.-C., et al. (1994). *Rhizobium meliloti* lipooligosaccharide nodulation factors: different structural requirements for bacterial entry into target root hair cells and induction of plant symbiotic developmental responses. *Plant Cell, 6,* 1357-1374.

Atkinson, E. M., Palcic, M. M., Hindsgaul, O., & Long, S. R. (1994). Biosynthesis of *Rhizobium meliloti* lipooligosaccharide nod factors: *nodA* is required for an *N*-acyltransferase activity. *Proceedings of the National Academy of Sciences of the United States of America, 91,* 8418-8422.

Ausmees, N., Kobayashi, H., Deakin, W. J., Marie, C., Krishnan, H. B., Broughton, W. J., et al. (2004). Characterization of NopP, a type III secreted effector of *Rhizobium* sp. strain NGR234. *Journal of Bacteriology, 186,* 4774-4780.

Baev, N., Endre, G., Petrovics, G., Banfalvi, Z., & Kondorosi, Á. (1991). Six nodulation genes of *nod* box locus 4 in *Rhizobium meliloti* are involved in nodulation signal production: *nodM* codes for D-glucosamine synthetase. *Molecular & General Genetics, 228,* 113-124.

Baker, B., Zambryski, P., Staskawicz, B., & Dinesh-Kumar, S. P. (1997). Signaling in plant-microbe interactions. *Science, 276,* 726-733.

Barnett, M. J., Swanson, J. A., & Long, S. R. (1998). Multiple genetic controls on *Rhizobium meliloti syrA,* a regulator of exopolysaccharide abundance. *Genetics, 148,* 19-32.

Barnett, M. J., & Long, S. R. (1990). DNA sequence and translational product of a new nodulation-regulatory locus: SyrM has sequence similarity to NodD proteins. *Journal of Bacteriology, 172,* 3695-3700.

Barny, M.-A., Schoonejans, E., Economou, A., Johnston, A. W.B., & Downie, J. A. (1996). The C-terminal domain of the *Rhizobium leguminosarum* chitin synthase NodC is important for function and determines the orientation of the N-terminal region in the inner membrane. *Molecular Microbiology, 19,* 443-453.

Bartsev, A. V., Deakin, W. J., Boukli, N. M., McAlvin, C. B., Stacey, G., Malnoe, P., et al. (2004). NopL, an effector protein of *Rhizobium sp.* NGR234, thwarts activation of plant defense reactions. *Plant Physiology, 134,* 871-879.

Batley, M., Redmond, J. W., Djordjevic, S. P., & Rolfe, B. G. (1987). Characterization of glycerophosphorylated cyclic β-1,2-glucans from a fast growing *Rhizobium* species. *Biochimica et Biophysica Acta, 901,* 119-126.

Battisti, L., Lara, J. C., & Leigh, J. A. (1992). Specific oligosaccharide form of the *Rhizobium meliloti* exopolysaccharide promotes nodule invasion in alfalfa. *Proceedings of the National Academy of Sciences of the United States of America, 89,* 5625-5629.

Bec-Ferté, M. P., Krishnan, H. B., Promé, D., Savagnac, A., Pueppke, S. G., & Promé, J.-C. (1994). Structures of nodulation factors from the nitrogen fixing-fixing symbiont *Rhizobium fredii* USDA257. *Biochemistry, 33,* 11782-11788.

Bec-Ferté, M. P., Krishnan, H. B., Savagnac, A., Pueppke, S. G., & Promé, J.-C. (1996). *Rhizobium fredii* synthesizes an array of lipooligosaccharides, including a novel compound with glucose inserted into the backbone of the molecule. *Federation of European Biological Societies Letters, 393,* 273-279.

Becker, A., Kleickmann, A., Arnold, W., & Pühler, A. (1993). Analysis of the *Rhizobium meliloti exoH/exoK/exoL* fragment: ExoK shows homology to excreted endo-β-1,3,-1,4-glucanases and ExoH resembles membrane proteins. *Molecular and General Genetics, 238,* 145-154.

Becker, A., & Pühler, A. (1998). Production of exopolysaccharides, In H. P. Spaink, Á. Kondorosi, & P. J. J. Hooykaas (Eds.), *The Rhizobiaceae: molecular biology of model plant-associated bacteria* (97-118). Dordrecht, The Netherlands: Kluwer Academic Publishers.

Becker A., Rüberg S., Küster, H., Roxlau, A. A., Keller, M., Ivashina, T., et al. (1997). The 32-kilobase *exp* gene cluster of *Rhizobium meliloti* directing the biosynthesis of galactoglucan: genetic organization and properties of the encoded gene products. *Journal of Bacteriology, 179,* 1375-1384.

Becquart-de-Kozak, I., Reuhs, B. L., Buffard, D., Breda, C., Kim, J. S., Esnault, R., & Kondorosi, Á. (1997). Role of the K-antigen sub-group of capsular polysaccharide in the early recognition process between *Rhizobium meliloti* and alfalfa leaves. *Molecular Plant-Microbe Interactions, 10,* 114-123.

Beijerinck, M. W. (1888). Die Bacterien der Papilionaceen-Knöllchen. *Botanische Zeitung, 46,* 725-804.

Beijerinck, M. W. (1890). Künstliche Infection von *Vicia faba* mit *Bacillus radicicola*: Ernährungsbedingungen dieser Bacterie. *Botanische Zeitung 48,* 837-843.

Bellato, C., Krishnan, H. B., Cubo, T., Temprano, F., & Pueppke, S. G. (1997). The soybean cultivar specificity gene *nolX* is present, expressed in a nodD-dependent manner, and of symbiotic significance in cultivar- nonspecific strains of *Rhizobium (Sinorhizobium) fredii. Microbiology, 143,* 1381-1388.

Berck, S., Perret, X., Quesada-Vincens, D., Promé, J.-C., Broughton, W. J., & Jabbouri, S. (1999). NolL of *Rhizobium* sp. strain NGR234 is required for *O*-acetyltransferase activity. *Journal of Bacteriology, 181*, 957-964.

Berg, T. (2003). Modulation of protein-protein interactions with small organic molecules. *Angewandte Chemie International Edition, 42,* 2462-2481.

Bhat, U. R., Carlson, R. W., Busch, M., & Mayer, H. (1991b). Distribution and phylogenetic significance of 27-hydroxy-octacosanoicacid in lipopolysaccharides from bacteria belonging to the alpha-2 subgroup of *Proteobacteria*. *International Journal of Systematic Bacteriology, 41,* 213-217.

Bhat, U. R., Forsberg, S. L., & Carlson, R. W. (1994). Structure of lipid A component of *Rhizobium leguminosarum* bv. *phaseoli* lipopolysaccharide. *Journal of Biological Chemistry, 269,* 14402-14410.

Bhat U. R., Mayer, H., Yokota, A., Hollingsworth, R. I., & Carlson, R W. (1991a). Occurrence of lipid A variants with 27-hydroxyoctacosanoic acid in lipopolysaccharides from members of the family of *Rhizobiaceae*. *Journal of Bacteriology, 173,* 2155-2159.

Bhagwat, A. A., Mithöfer, A., Pfeffer, P. E., Kraus, C., Spickers, N., Hotchkiss, A., et al. (1999). Further studies of the role of cyclic β-glucans in symbiosis. An *ndvC* mutant of *Bradyrhizobium japonicum* synthesizes cyclodecakis-(1→3)-β-glucosyl. *Plant Physiology, 119,* 1057-1064.

Blocker, A., Komoriya, K., & Aizawa, S.-I. (2003). Type III secretion systems and bacterial flagella: Insights into their function from structural similarities. *Proceedings of the National Academy of Sciences of the United States of America, 100,* 3027-3030.

Blocker, A., Noureddine, J., Eric, L., Pierre, G., Claude, E. F. P., Philippe, S., & Abdelmounaaim, A. (2001). Structure and composition of the *Shigella flexneri* needle complex, a part of its type III secretion. *Molecular Microbiology, 39,* 652-663.

Bloemberg, G. V., Kamst, E., Harteveld, M., van der Drift, K. M. G. M., Haverkamp, J., Thomas-Oates, J. E., et al. (1995). A central domain of *Rhizobium* NodE protein mediates host specificity by determining the hydrophobicity of fatty acyl moieties of nodulation factors. *Molecular Microbiology, 16,* 1123-1136.

Boros, I., Kiss, A., & Venetianer, P. (1979). Physical map of the seven ribosomal RNA genes of *Escherichia coli*. *Nucleic Acids Research, 6,* 1817-1830.

Borthakur, D., Barber, C. E., Lamb, J. W., Daniels, M. J., Downie, J. A., & Johnston, A. W. B. (1986). A mutation that blocks exopolysaccharide synthesis prevents nodulation of peas by *Rhizobium leguminosarum* but not of beans by *R. phaseoli* and is corrected by cloned genes from *Rhizobium* or the phytopathogen *Xanthomonas*. *Molecular and General Genetics, 203,* 320-323.

Breedveld, M. W., & Miller, K. J. (1994). Cyclic β-glucans of members of the family, *Rhizobiaceae*. *Microbiology Review, 58,* 145-151.

Brewin, N. J. (1998). Tissue and cell invasion by *Rhizobium*: the structure and development of infection threads and symbiosomes. In H. P. Spaink, Á. Kondorosi, & P. J. J. Hooykaas (Eds.), *The Rhizobiaceae: molecular biology of model plant-associated bacteria* (417-429). Dordrecht, The Netherlands: Kluwer Academic Publishers.

Broughton, W. J. (2003). Roses by other names: Taxonomy of the *Rhizobiaceae*. *Journal of Bacteriology, 185,* 2975-2979.

Broughton, W. J., Jabbouri, S., & Perret, X. (2000). Keys to symbiotic harmony. *Journal of Bacteriology, 182,* 5641-5652.

Broughton, W. J., Zhang, F., Perret, X., & Staehelin, C. (2003). Signals exchanged between legumes and *Rhizobium*: agricultural uses and perspectives. *Plant & Soil, 252,* 129-137.

Burn, J., Rossen, L., & Johnston, A. W. B. (1987). Four classes of mutations in the *nodD* gene of *Rhizobium leguminosaurm* biovar *viciae* that affect its ability to autoregulate and/or activate other *nod* genes in the presence of flavonoid inducers. *Genes and Development, 1*, 456-464.

Büttner, D., & Bonas, U. (2002). Port of entry - the type III secretion translocon. *Trends in Microbiology, 10,* 186-192.

Büttner, D., & Bonas, U. (2003) Common infection strategies of plant and animal pathogenic bacteria. *Current Opinion in Plant Biology, 6*, 312-319.

Campbell, G. R. O., Reuhs, B. L., & Walker, G. C. (1998). Different phenotypic classes of *Sinorhizobium meliloti* mutants defective in synthesis of K antigen. *Journal of Bacteriology, 180,* 5432-5436.

Canter Cremers, H. C. J., Batley, M., Redmond, J., Eydems, L., Breedveld, M. W., Zevenhuizen, L. P. T., M., et al. (1990). *Rhizobium leguminosarum exoB* mutants are deficient in the synthesis of UDP-glucose 4-epimerase. *Journal of Biological Chemistry, 265,* 21122-21127.

Carbon, C., Philips, J., Fu, Z. Y., Squires, C., & Squires, C. L. (1979). The complete nucleotide sequence of ribosomal 16S rRNA from *Escherichia coli. European Molecular Biology Organization Journal, 11,* 4175-4185.

Cárdenas, L., Dominguez, J., Quinto, C., López-Lara, I. M., Lugtenberg, B. J. J., Spaink, H. P., et al. (1995). Isolation, chemical structure, and biological activity of the lipo-chitin oligosaccharide nodulation signals from *Rhizobium etli. Plant Molecular Biology, 29,* 453-464.

Cárdenas, L., Dominguez, J., Santana, O., & Quinto, C. (1996). The role of the *nodI* and *nodJ* genes in the transport of nod metabolites in *Rhizobium etli. Gene, 173,* 183-187.

Cárdenas, L., Vidali, L., Domínguez, J., Pérez, H., Sánchez, F., Hepler, P. K., & Quinto, C. (1998). Rearrangement of actin microfilaments in plant root hairs responding to *Rhizobium etli* nodulation signals. *Plant Physiology, 116,* 871-877.

Carlson, R. W., Garci, F., Noel, D., & Hollingsworth, R. (1989). The structures of the LPS core components from *Rhizobium leguminosarum* biovar *phaseoli* and two of its symbiotic mutants, CE 109, and CE 309. *Carbohydrate Research, 195,* 101-110.

Carlson, R. W., & Krishnaiah, B. S. (1992). Structures of the oligosaccharides obtained from the core regions of the lipopolysaccharides of *Bradyrhizobium japonicum* 61A101c and its symbiotically defective lipopolysachharide mutant, JS314. *Carbohydrate Research, 231,* 205-219.

Cedergren, R. A., Lee, J. G., Ross, K. L., & Hollingsworth, R. I. (1995). Common links in the structure and cellular-localization of *Rhizobium* chitolipooligosaccharides and general *Rhizobium* membrane phospholipids and glycolipid components. *Biochemistry, 34,* 4467-4477.

Cervantes E., Sharma, S. B. Maillet, F., Vasse, J., Truchet, G., Rosenberg, C. (1989). The *Rhizobium-meliloti* host range *nodQ* gene encodes a protein which shares homology with translation elongation and initiation factors. *Molecular Microbiology, 3,* 745-756.

Chen, H., Batley, M., Redmond, J., & Rolfe, B. G. (1985). Alteration of effective nodulation properties of a fast-growing broad host range *Rhizobium* due to changes in exopolysaccharide synthesis. *Journal of Plant Physiology, 120,* 331-349.

Chen, W.-M., James, E. K., Prescott, A. R., Kierans, M., & Sprent, J. I. (2003). Nodulation of *Mimosa* spp. by the β-proteobacterium *Ralstonia taiwanensis. Molecular Plant-Microbe Interactions, 16,* 1051-1061.

Chen, W.-M., Laevens, S., Lee, T. M., Coenye, T., de Vos, P., Mergeay, M., et al. (2001). *Ralstonia taiwanensis* sp. nov., isolated from root nodules of *Mimosa* species and sputum of a cystic fibrosis patient. *International Journal of Systematic and Evolutionary Microbiology, 51,* 1729-1735.

Chen, W.-M., Moulin, L., Bontemps, C., Vandamme, P., Béna, G., Boivin-Masson, C. (2003). Legume symbiotic nitrogen fixation by β-proteobacteria is widespread in nature. *Journal of Bacteriology, 185,* 7266-7272.

Chen, W.-X., Li, G.-S., Qi, Y.-L., Wang, E.-T., Yuan, H.-L., & Li, J.-L. (1991). *Rhizobium huakuii* sp. nov., isolated from the root nodules of *Astragalus sinicus. International Journal of Systematic Bacteriology, 41,* 275-280.

Chen, W.-X., Yan, G.-H., & Li, J.-L. (1988). Numerical taxonomic study of fast-growing soybean rhizobia and a proposal that *Rhizobium fredii* be assigned to *Sinorhizobium* gen. nov. *International Journal of Systematic Bacteriology, 38,* 392-397.

Chen, X.-C., Feng, J., Hou, B.-H., Li, F.-Q., Li, Q., & Hong, G.-F. (2005). Modulating DNA bending affects NodD-mediated transcriptional control in *Rhizobium leguminosarum. Nucleic Acids Research, 33,* 2540-2548.

Cheng, H.-P. & Walker, G. C. (1998). Succinoglycan is required for initiation and elongation of infection threads during nodulation of alfalfa by *Rhizobium meliloti. Journal of Bacteriology, 180,* 5183-5191.

Cilia, V., Lafay, B., & Christen, R. (1996). Sequence heterogeneities among 16S ribosomal RNA sequences, and their effect on phylogenetic analyses at the species level. *Molecular Biology and Evolution, 13,* 451-461.

Conville, P. S., & Witebsky, F. G. (2005). Multiple copies of the 16S rRNA gene in *Nocardia nova* isolates and implications for sequence-based identification procedures. *Journal of Clinical Microbiology, 43,* 2881-2885.

Cornelis, G. R., & Van Gijsegem, F. (2000). Assembly and function of type III secretory systems. *Annual Review of Microbiology, 54,* 735-774.

Dale, C., Plague, G. R., Wang, B., Ochman, H., & Moran, N. (2002). Type III secretion systems and the evolution of mutualisic endosymbiosis. *Proceedings of the National Academy of Sciences of the United States of America, 99,* 12397-12402.

Daniell, S., Takahashi, N., Wilson, R., Friedberg, D., Rosenshine, I., Booy, F. P., et al. (2001). The filamentous type III secretion translocon of enteropathogenic *Escherichia coli. Cellular Microbiology, 3,* 865-871.

Davis, E. O., Evans, I. J., & Johnston, A. W. B. (1988). Identification of *nodX,* a gene that allows *Rhizobium leguminosarum* biovar *viciae* strain TOM to nodulate Afghanistan peas. *Molecular and General Genetics, 212,* 531-535.

Davis, E. O., & Johnston, A. W. B. (1990). Analysis of three *nodD* genes in *Rhizobium leguminosarum* biovar *phaseoli*; *nodD1* is preceded by *nolE*, a gene whose product is secreted from the cytoplasm. *Molecular Microbiology, 4,* 921-932.

Dazzo, F. B., Truchet, G. L., Hollingsworth, R. I., Hrabak, E. M., Pankratz, H. S., Philip-Hollingsworth, S., et al. (1991). *Rhizobium* lipopolysaccharide modulates infection thread development in white clover root hairs. *Journal of Bacteriology, 173,* 5371-5384.

Deakin W. J., Marie, C., Saad, M. M., Krishnan, H. B., & Broughton, W. J. (2005). NopA Is associated with cell surface appendages produced by the type III secretion system of *Rhizobium* sp. strain NGR234. *Molecular Plant-Microbe Interactions, 18,* 499-507.

Debellé, F., Plazanet, C., Roche, P., Pujol, C., Savagnac, A., Rosenberg, C., et al. (1996). The NodA proteins of *Rhizobium meliloti* and *Rhizobium tropici* specify the *N*-acylation of Nod factors by different fatty acids. *Molecular Microbiology, 22,* 303-314.

Diebold, R., & Noel, K. D. (1989). *Rhizobium leguminosarum* exopolysaccharide mutants: biochemical and genetic analyses and symbiotic behavior on three hosts. *Journal of Bacteriology, 171,* 4821-4830.

De Lajudie, P., Laurent-Fulele, E., Willems, A., Torck, U., Coopman, R., Collins, M. D., et al. (1998). *Allorhizobium undicola* gen. nov. sp. nov., nitrogen-fixing bacteria that efficiently nodulate *Neptunia natans* in Senegal. *International Journal of Systematic Bacteriology, 48,* 1277-1290.

De Lajudie, P., Willems, A., Pot, B., Dewettinck, D., Maestrojuan, G., Neyra, M., et al. (1994). Polyphasic taxonomy of *Rhizobia*: emendation of the genus *Sinorhizobium* and description of *Sinorhizobium meliloti* comb. nov., *Sinorhizobium saheli* sp. nov., and *Sinorhizobium teranga* sp. nov. *International Journal of Systematic Bacteriology, 44,* 715-733.

de Maagd, R. A., Wijfjes, A. H. M., Spaink, H. P., Ruiz-Sainz, J. E., Wijffelman, C. A., Okker, R. J. H., et al. (1989). *NodO* a new nod gene of the *Rhizobium leguminosarum* biovar *viciae* sym plasmid PRL1JI encodes a secreted protein. *Journal of Bacteriology, 171,* 6764-6770.

Demont, N., Debellé, F., Aurelle, H., Dénarié, J., & Promé, J.-C. (1993). Role of the *Rhizobium meliloti nodF* and *nodE* genes in the biosynthesis of lipo-oligosaccharidic nodulation factors. *Journal of Biological Chemistry*, *268,* 20134-20142.

Dénarié, J., & Cullimore, J. (1993). Lipo-oligosaccharide nodulation factors: a new class of signalling molecules mediating recognition and morphogenesis. *Cell, 74,* 951-954.

Dénarié, J., Debellé, F., & Promé, J.-C. (1996). *Rhizobium* lipo-chitooligosaccharide nodulation factors: Signaling molecules mediating recognition and morphogenesis. *Annual Review of Biochemistry, 65,* 503-535.

De Rijk, P., Wuyts, J., Van de Peer, Y., Winkelmans, T., & De Wachter, R. (2000). The European large subunit ribosomal RNA database. *Nucleic Acids Research, 28,* 177-178.

de Rudder, K. E. E., Sohlenkamp, C., & Geiger, O. (1999). Plant-exuded choline is used for rhizobial membrane lipid biosynthesis by phosphatidylcholine synthase. *Journal of Biological Chemistry, 274,* 20011-20016.

D'Haeze, W., & Holsters, M. (2002). Nod factor structures, responses, and perception during initiation of nodule development. *Glycobiology, 12,* 79-105.

D'Haeze, W., & Holsters, M. (2004). Surface polysaccharides enable bacteria to evade plant immunity. *Trends in Microbiology, 12*, 555-561.

Djordjevic, S. P., Chen, H., Batley, M., Redmond, J. W., & Rolfe, B. G. (1987). Nitrogen fixation ability of exopolysaccharide mutants of *Rhizobium sp.* NGR234 and *Rhizobium trifolii* is restored by the addition of homologous exopolysaccharides. *Journal of Bacteriology, 169*, 53-60.

Dobert, R. C., Breil, B. T., & Triplett, E. W. (1994). DNA sequence of the common nodulation genes of *Bradyrhizobium elkanii* and their phylogenetic relationship to those of other nodulating bacteria. *Molecular Plant-Microbe Interactions, 7,* 564-572.

Downie, J A. (1989). The *nodL* gene from *Rhizobium leguminosarum* is homolgous to the acetyl transferases encoded by *lacA* and *cysE. Molecular Microbiology, 3,* 1649-1651.

Downie, J. A., Knight, C. D., Johnston, A. W. B., & Rossen, L. (1985). Identification of genes and gene products involved in the nodulation of peas by *Rhizobium leguminosarum. Molecular and General Genetics, 198,* 255-262.

Dreyfus, B., Garcia, J. L., & Gillis, M. (1988). Characterization of *Azorhizobium caulinodans* gen. nov. a stem nodulating nitrogen-fixing bacterium isolated from *Sesbania rostrata. International Journal of Systematic Bacteriology, 38,* 89-98.

Duelli, D. M., & Noel, K. D. (1997). Compounds exuded by *Phaseolus vulgaris* that induce a modification of *Rhizobium etli* lipopolysaccharide. *Molecular Plant-Microbe Interactions, 10,* 903-910.

Dunlap, J., Minami, E., Bhagwat, A. A., Keister, D. L., Stacey, G. (1996). Nodulin gene expression by nodule development (*ndv*) mutants of *Bradyrhizobium japonicum. Molecular Plant-Microbe Interactions, 9,* 546-555.

Dylan, T., Ielpi, L., Stanfield, S., Kashyap, L., Douglas, C., Yanofsky, M., et al. (1986). *Rhizobium meliloti* genes required for nodule development are related to chromosomal virulence genes in *Agrobacterium tumefaciens. Proceedings of the National Academy of Sciences United States of America, 83,* 4403-4407.

Dylan, T, Nagpal, P., Helinski, D. R., & Ditta, G. S. (1990) Symbiotic pseudorevertants of *Rhizobium meliloti ndv* mutants. *Journal of Bacteriology, 172,* 1409-1417.

Eardly, B. D., Wang, F.-S., van Berkum, P. (1996). Corresponding 16S rRNA gene segments in *Rhizobiaceae* and *Aeromonas* yield discordant phylogenies. *Plant and Soil, 186,* 69-74.

Ebel, J., & Cosio, E. G. (1994). Elicitors of plant defense responses. *International Review of Cytology, 148,* 1-36.

Economou, A., Hamilton, W. D. O., Johnston, A. W. B., & Downie, J. A. (1990). The *Rhizobium* modulation gene *nodO* encodes a Ca^{+2} binding protein that is exported without N-terminal cleavage and is homologous to hemolysin and related proteins. *European Molecular Biology Organization Journal, 9,* 349-354.

Egelhoff, T. T., & Long, S. R. (1985). *Rhizobium meliloti* nodulation genes: Identification of *nodDABC* gene products, purification of *nodA* protein, and expression of *nodA* in *Rhizobium meliloti. Journal of Bacteriology, 164,* 591-599.

Ehrhardt, D. W., Atkinson, E. M., & Long, S. R. (1992). Depolarization of alfalfa root hair membrane potential by *Rhizobium meliloti* Nod factors. *Science, 256,* 998-1000.

Felle, H. H., Kondorosi É., Kondorosi Á., & Schultze, M. (1998). The role of ion fluxes in Nod factor signalling in *Medicago sativa. The Plant Journal, 13,* 455-463.

Feng, J., Li, Q., Hu, H.-L., Chen, X.-C., & Hong, G.-F. (2003). Inactivation of the *nod* box distal half-site allows tetrameric NodD to activate *nodA* transcription in an inducer-independent manner. *Nucleic Acids Research, 31,* 3143-3156.

Fernández-López, M., D'Haeze, W., Mergaert, P., Verplancke, C., Promé, J.-C., van Montagu, M., et al. (1996). Role of *nodI* and *nodJ* in lipo-oligosaccharide secretion in *Azorhizobium caulinodans* and *Escherichia coli*. *Molecular Microbiology, 20,* 993-1000.

Finan, T. M., Hirsch, A. M., Leigh, J. A., Johansen, E., Kuldau, G. A., Deegan, S., et al. (1985). Symbiotic mutants of *Rhizobium meliloti* that uncouple plant from bacterial differentiation. *Cell, 40,* 869-877.

Firmin, J. L., Wilson, K. E., Rossen, L., & Johnston, A. W. B. (1986). Flavonoid activation of nodulation genes in *Rhizobium* reversed by other compounds present in plants. *Nature 324,* 90-92.

Fisher, R. F., Egelhoff, T. T., Mulligan, J. T., & Long, S. R. (1988). Specific binding of proteins from *Rhizobium meliloti* cell-free extracts containing NodD to DNA sequences upstream of inducible nodulation genes. *Genes and Development, 2,* 282-293.

Fisher, R. F., & Long, S. R. (1993). Interactions of NodD at the *nod* box. NodD binds to two distinct sites on the same face of the helix and induces a bend in the DNA. *Journal of Molecular Biology, 233,* 336-348.

Folch-Mallol, J. L., Marroqui S., Sousa, S., Manyani, H., López-Lara, I. M., van der Drift, K. M. G. M., et al. (1996). Characterization of *Rhizobium tropicii* CIAT899 nodulation factors: The role of *nodH* and *nodPQ* genes in their sulphation. *Molecular Plant-Microbe Interactions, 9,* 151-63.

Forsberg, A., & Wolf-Watz, H. (1988). The virulence protein Yop5 of *Yersinia pseudotuberculosis* is regulated at transcriptional level by plasmid-pIB1-encoded trans-acting elements controlled by temperature and calcium. *Molecular Microbiology, 2,* 121-133.

Forsberg, L. S., Bhat, U. R., & Carlson, R. W. (2000). Structural characterization of the O-antigenic polysaccharide of the lipopolysaccharide from *Rhizobium etli* strain CE3 a unique *O*-acetylated glycan of discrete size, containing 3-*O*-methyl-6-deoxy-L-talose and 2,3,4-tri-*O*-methyl-L-fucose. *Journal of Biological Chemistry, 275,* 18851-18863.

Fox, G. E., Stackebrandt, E., Hespell, R. B. Gibson, J., Maniloff, J., Dyer, T. A., et al. (1980). The phylogeny of prokaryotes. *Science, 209,* 457-463.

Frank, B. (1879). Über die Parasiten in den Wurzelanschwillungen der Papilionaceen. *Botanische Zeitung, 37,* 376-387.

Frank, B. (1889). Über die Pilzsymbiose der Leguminosen. *Berichte der Deutschen Botanischen Gesellschaft, 7,* 332-346.

Fraysse, N., Couderc, F., & Poinsot, V. (2003). Surface polysaccharide involvement in establishing the rhizobium–legume symbiosis *European Journal of Biochemistry, 270,* 1365-1380.

Fraysse, N., Jabbouri, S., Treilhou, M., Couderc, F., Poinsot, V. (2002). Symbiotic conditions induce structural modifications of *Sinorhizobium* sp. NGR234 surface polysaccharides. *Glycobiology 12,* 741-748.

Fred, E. B., Baldwin, I. L., & McCoy, E. (1932). Root nodule bacteria and leguminous plants. University of Wisconsin Studies in Science 5, Madison, Wisconsin.

Freiberg, C., Fellay, R., Bairoch, A., Broughton, W. J., Rosenthal, A., & Perret, X. (1997). Molecular basis of symbiosis between *Rhizobium* and legumes. *Nature, 387,* 394-401.

Galán, J. E., & Collmer. A. (1999). Type III secretion machines: bacterial devices for protein delivery into host cells. *Science 284,* 1322-1328.

Galibert, F., Finan, T. M., Long, S. R., Pühler, A., Abola, P., Ampe, F., et al. (2001). The composite genome of the legume symbiont *Sinorhizobium meliloti*. *Science, 293,* 668-672.

Gao, M., D'Haeze, W., De Rycke, R., Wolucka, B., & Holsters, M. (2001). Knockout of an Azorhizobial dTDP-L-rhamnose synthase affects lipopolysaccharide and extracellular polysaccharide production and disables symbiosis with *Sesbania rostrata*. *Molecular Plant-Microbe Interactions, 14,* 857-866.

Garcia, M., Dunlap, J., Loh, J., & Stacey, G. (1996). Phenotypic characterization and regulation of the *nolA* gene of *Bradyrhizobium japonicum*. *Molecular Plant-Microbe Interactions, 9,* 625-636.

Gaunt, M. W., Turner, S. L., Rigottier-Gois, L., Lloyd-Macgilp, S. A., & Young, J. P. W. (2001). Phylogenies of *atpD* and *recA* support the small subunit rRNA-based classification of rhizobia. *International Journal of Systematic and Evolutionary Bacteriology, 51,* 2037-2048.

Gavin, A. C., Bösche, M., Krause, R., Grandi, P., Marzioch, M., Bauer, A. et al. (2002) Functional organization of the yeast proteome by systematic analysis of protein complexes. *Nature, 415,* 141-147.

Gehring, C. A., Irving, H. R., Kabbara, A. A., Parish, R. W., Boukli, N. M., & Broughton, W. J. (1997). Rapid, plateau-like increases in intracellular free calcium are associated with nod-factor-induced root hair deformation. *Molecular Plant-Microbe Interactions, 10,* 791-802.

Geiger, O., Glushka, J., Lugtenberg, B. J. J., Spaink, H. P., & Thomas-Oates J. E. (1998). Rhizobial NodFE-dependent fatty acids which lack an α-β unsaturation are subject to differential transfer, leading to novel phospholipids. *Molecular Plant-Microbe Interactions, 11,* 33-44.

Geremia, R. A., Mergaert, P., Geelen, D., Van Montagu, M., & Holsters, M. (1994). The NodC protein of *Azorhizobium caulinodans* is an *N*-acetylglucosaminyltransferase. *Proceedings of the National Academy of Sciences of the United States of America, 91,* 2669-2673.

Ghosh, P. (2004). Process of protein transport by the type III secretion system. *Microbiology and Molecular Biology Review, 68,* 771-795.

Gil-Serrano, A. M., González-Jiménez, I., Mateo, P. T., Bernabé, M., Jiménez-Barbero, J., Megias, M., et al. (1995). Structural analysis of the O-antigen of the lipopolysaccharide of *Rhizobium tropici* CIAT899. *Carbohydrate Research, 275,* 285-294.

Glazebrook, J., & Walker, G. C. (1989). A novel exopolysaccharide can function in place of calcofluor-binding exopolysaccharides in nodulation of alfalfa by *Rhizobium meliloti. Cell, 56,* 661-672.

Gluckman, M. A., Reuber, T. L., & Walker, G. C. (1993). Family of glycosyltransferases needed for the synthesis of succinoglycan by *Rhizobium meliloti. Journal of Bacteriology, 175,* 7033-7044.

Goethals, K., van Montagu, M., & Holsters, M. (1992). Conserved motifs in a divergent *nod* box of *Azorhizobium caulinodans* ORS571 reveal a common structure in promoters regulated by the LysR-type protein. *Proceedings of the National Academy of Sciences of the United States of America, 89,* 1646-1650.

González, J. E., York, G. M, & Walker, G. C. (1996). *Rhizobium meliloti* exopolysaccharides: synthesis and symbiotic function. *Gene, 179,* 141-146.

Goosen-de Roo, L., de Maagd, R. A., & Lugtenberg, B. J. J. (1991). Antigenic changes in lipopolysaccharide I of *Rhizobium leguminosarum* bv. *viciae* in root nodules of *Vicia sativa* subsp. *nigra* occur during release from infection threads. *Journal of Bacteriology, 173,* 3177-3183.

Gophna, U., Ron, E. Z., & Graur, D. (2003). Bacterial type III secretion systems are ancient and evolved by multiple horizontal-transfer events. *Gene, 312,* 151-163.

Gore, R. S., & Miller K. J. (1992). Cell surface carbohydrates of microaerobic, nitrogenase-active, continous cultures of *Bradyrhizobium* sp. strain 32H1. *Journal of Bacteriology, 174,* 7838-7840.

Gore, R. S., & Miller, K. J. (1993). Cyclic β-1,6 -1,3 glucans are synthesized by *Bradyrhizobium japonicum* bacteroids within soybean (*Glycine max*) root nodules. *Plant Physiology, 102,* 191-194.

Göttfert, M., Hitz, S., & Hennecke, H. (1990). Identification of *nodS* and *nodU* two inducible genes inserted between the *Bradyrhizobium japonicum nodYABC* and *nodIJ* genes. *Molecular Plant-Microbe Interactions, 3,* 308-316.

Göttfert, M., Röthlisberger, S., Kündig, C., Beck, C., Marty, R., & Henneke, H. (2001). Potential symbiosis-specific genes uncovered by sequencing a 410-kilobase DNA region of the *Bradyrhizobium japonicum* chromosome. *Journal of Bacteriology, 183,* 1405-1412.

Graham, P. H., Sadowsky, M. J., Keyser, H. H., Barnet, Y. M., Bradley, R. S., Cooper, J. E., et al. (1991). Proposed minimal standards for the description of new genera and species of root- and stem-nodulating bacteria. *International Journal of Systematic Bacteriology, 41,* 582-587.

Gray, J. X., Zhan, H., Levery, S. B., Battisti, L., Rolfe, B. G., & Leigh, J. A. (1991). Heterologous exopolysaccharide production in *Rhizobium* sp. strain NGR234 and consequences for nodule development. *Journal of Bacteriology, 173,* 3066-3077.

Györgypál, Z., Kiss, G. B., & Kondorosi, Á. (1991a). Transduction of plant signal molecules by the *Rhizobium* NodD proteins. *BioEssays, 13,* 575-581.

Györgypal, Z., Kondorosi, É., & Kondorosi, Á. (1991b). Diverse signal sensitivity of NodD protein homologs from narrow and broad host range rhizobia. *Molecular Plant-Microbe Interactions, 4,* 356-364.

Hanin, M., Jabbouri, S., Quesada-Vincens, D., Freiburg, C., Perret, X., Promé, J.-C., et al. (1997). Sulphation of *Rhizobium* sp. NGR234 Nod factors is dependent on *noeE*, a new host-specificity gene. *Molecular Microbiology, 24,* 1119-1129.

He, S. Y. (1998). Type III protein secretion systems in plant and animal pathogenic bacteria. *Annual Review of Phytopathology, 36,* 363-392.

Heddi, A., Grenier, A. M., Khatchadourian, C., Charles, H., & Nardon, P. (1999). Four intracellular genomes direct weevil biology: nuclear, mitochondrial, principal endosymbion and *Wolbachia. Proceedings of the National Academy of Sciences of the United States of America, 96,* 6814-6819.

Hellriegel, H (1886). Welche Stickstoffquellen stehen der Pflanze zu Gebote? Zeitschrift des Vereins für Rübenzucker-Industrie des Deutschen Reichs, 38, 863-877.

Hellriegel, H, & Wilfarth, H. (1888). Untersuchungen über die Stickstoffnahrung der Gramineen und Leguminosen. Beilageheft zu der Zeitschrift des Vereins fur Rübenzucker-Industrie Deutschen Reichs, pp. 234, Buchdruckerei der Post, Kayssler & Co.

Henikoff, S., Haughn, G. W., Calvo, J. M., & Wallace, J. C. (1988). A large family of bacterial activator proteins. *Proceedings of the National Academy of Sciences of the United States of America, 85,* 6602-6066.

Her, G.-R., Glazebrook, J., Walker, G. C., & Reinhold, V. N. (1990). Structural studies of a novel exopolysaccharide produced by a mutant of *Rhizobium meliloti* strain Rm1021. *Carbohydrate Research, 198,* 305-312.

Hiltner, L. (1900). Über die ursachen, welche die Grösse, Zahl, Stellung und Wirkung der Wurzelknöllchen der Leguminosen bedingen. Arbeiten aus der Biologischen Abteilung für Land und Forstwirtschaft am Kaiserlichen Gesndheitsamte, 1, 177-222. Berlin, Germany.

Hiltner, L. & Stormer, K. (1903). Neue Untersuchungen über die wurzelknöllchen der Leguminosen und deren Erreger. Arbeiten aus der Biologischen Abteilung für Land und Forstwirtschaft am Kaiserlichen Gesndheitsamte, 3, 151-307. Berlin, Germany.

Ho, Y, Gruhler, A, Heilbut, A, Bader, G. D., Moore, L., Adams, S. L., et al. (2002). Systematic identification of protein complexes in *Saccharomyces cerevisiae* by mass spectrometry. *Nature, 415,* 180-183.

Hogg, B., Davies, A. E., Wilson, K. E., Bisseling, T., & Downie, J.A. (2002). Competitive nodulation blocking of cv. Afghanistan pea is related to high levels of nodulation factors made by some strains of *Rhizobium leguminosarum* bv. *viciae. Molecular Plant-Microbe Interactions, 15,* 60-68.

Hollingsworth R. I., & Carlson, R. W. (1989). 27-Hydroxyoctacosanoic acid is a major structural fatty acyl component of the lipopolysaccharide of *Rhizobium trifolii* ANU 843. *Journal of Biological Chemistry, 264,* 9300-9303.

Honma, M. A., & Ausubel, F. M. (1987). *Rhizobium meliloti* Has three functional copies of the *nodD* symbiotic regulatory gene. *Proceedings of the National Academy of Science of the United States of America, 84,* 8558-8562.

Hotter, G. S., & Scott, B. (1991). Exopolysaccharide mutants of *Rhizobium loti* are fully effective on a determinate nodulating host but are ineffective on an indeterminate nodulating host. *Journal of Bacteriology, 173,* 851-859.

Hueck, C. J. (1998). Type III protein secretion system in bacterial pathogens of animal and plants. *Microbiology and Molecular Biology Reviews, 62,* 379-433.

Huguet, E., & Bonas, U. (1997). *hrpF* of *Xanthomonas campestris* pv. *vesticatoria* encodes an 87-kDa protein with homology to NolX of *Rhizobium fredii. Molecular Plant-Microbe Interactions 10,* 488-498.

Ielpi, L., Dylan, T., Ditta, G. S., Helinski, D. R., & Stanfield, S. W. (1990). The *ndvB* locus of *Rhizobium meliloti* encodes a 319-kDa protein involved in the production of β-(1→2)-glucan. *Journal of Biological Chemistry, 265,* 2843-2851.

Ingram-Smith, C., & Miller, K. J. (1998). Effects of ionic and osmotic strength on the glucosyltransferase of *Rhizobium meliloti* responsible for cyclic β-(1,2)-glucan biosynthesis. *Applied and Environmental Microbiology, 64,* 1290-1297.

Inon de Iannino, N., & Ugdale, R.A. (1993). Biosynthesis of cyclic beta-(1-3), beta-(1-6) glucan in *Bradyrhizobium* spp. *Archives of Microbiology, 159,* 30-38.

Jabbouri, S., Fellay, R., Talmont, F., Kamalaprija, P., Burger, U., Relic, B., et al. (1995). Involvement of *nodS* in *N*-methylation and *nodU* in 6-*O*-carbamoylation of *Rhizobium* sp. NGR234 Nod factors. *Journal of Biological Chemistry, 270,* 22968-22973.

Jabbouri, S., Relic, B., Hanin, M., Kamalaprija, P., Burger, U., Promé, D., et al. (1998). *nolO* and *noeI* (HsnIII) of *Rhizobium sp.* NGR234 are involved in 3-*O*-carbamoylation and 2-*O*-methylation of nod factors. *Journal of Biological Chemistry, 273,* 12047-12055.

Jaftha, J. B., Strijdom, B. W., & Steyn, P. L. (2002). Characterization of pigmented methylotrophic bacteria which nodulate *Lotononis bainesii. Systematic and Applied Microbiology, 25,* 440-449.

Jarvis, B. D. W., Pankhurst, C. E., & Patel, J. J. (1982). *Rhizobium loti*, a new species of legume root bacteria. *International Journal of Systematic Bacteriology, 32,* 378-380.

Jarvis, B. D. W., van Berkum, P., Chen, W.-X., Nour, S. M., Fernández, M. P., Cleyet-Marel, J. C., et al. (1997). Transfer of *Rhizobium loti, Rhizobium huakuii, Rhizobium ciceri, Rhizobium mediterraneum,* and *Rhizobium tianshanense* to *Mesorhizobium* gen. nov. *International Journal of Systematic Bacteriology*, *47,* 895-898.

Jin, Q., & He, S. Y. (2001). Role of the Hrp pilus in type III protein secretion in *Pseudomonas syringae. Science, 294,* 2556-2558.

John, M., Röhrig, H., Schmidt, J., Wieneke, U., & Schell, J. (1993). *Rhizobium* NodB protein involved in modulation signal synthesis is a chitooligosacccharide deacetylase. *Proceedings of the National Academy of Sciences of the United States of America, 90,* 625-629.

John, M, Schmidt, J., Wieneke, U., Kruessmann, H.-D., & Schell, J. (1988). Transmembrane orientation and receptor-like structure of the *Rhizobium meliloti* common nodulation protein NodC. *European Molecular Biology Organization Journal, 7,* 583-588.

Jordan, D. C. (1980). Transfer of *Rhizobium japonicum* Buchanan to *Bradyrhizobium* gen. nov., a genus of slow-growing, root nodule bacteria from leguminous plants. *International Journal of Systematic Bacteriology, 32,* 136-139.

Jourand, P., Giraud, E., Béna, G., Sy, A., Willems, A., Gillis, M., et al. (2004). *Methylobacterium nodulans* sp. nov., for a group of aerobic, facultatively methylotrophic, legume root-nodule-forming and nitrogen-fixing bacteria. *International Journal of Systematic and Evolutionary Microbiology, 54,* 2269-2273.

Kamst, E., Bakkers, J., Quaedvlieg, N. E., Pilling, J., Kijne, J. W., Lugtenberg, B. J., Spaink, H. P. (1999). Chitin oligosaccharide synthesis by Rhizobia and Zebrafish embryos starts by glycosyl transfer to O4 of the reducing-terminal residue. *Biochemistry, 38,* 4045-52

Kaneko, T., Nakamura, Y., Sato, S., Asamizu, E., Kato, T., Sasamoto, S., et al. (2000). Complete genome structure of the nitrogen-fixing symbiotic bacterium *Mesorhizobium loti. DNA Research, 7,* 331-338.

Kaneko, T., Nakamura, Y., Sato, S., Minamisawa, K., Uchiumi, T., Sasmoto, S., et al. (2002). Complete genomic sequence of nitrogen-fixing symbiotic bacterium *Bradyrhizobium japonicum* USDA110. *DNA Research, 9,* 189-197.

Kannenberg, E. L., Perotto, S., Bianciotto, V., Rathbun, E. A., & Brewin, N. J. (1994). Lipopolysaccharide epitope expression of *Rhizobium* bacteroids as revealed by *in situ* immunolabelling of pea root nodule sections. *Journal of Bacteriology, 176,* 2021-2032.

Kannenberg, E. L., Reuhs, B.L., Forsberg, L. S., & Carlson, R.W. (1998). Lipopolysaccharides and K-antigens: their structures, biosynthesis, and functions. In H. P. Spaink, Á. Kondorosi, P. J. J. Hooykaas (Eds.) *The Rhizobiaceae: Molecular biology of model plant-associated bacteria* (119-54). Dordrecht, The Netherlands: Kluwer Academic Press.

Kannenberg, E. L., & Carlson, R. W. (2001). Lipid A and O-chain modifications cause *Rhizobium* lipopolysaccharides to become hydrophobic during bacteroid development. *Molecular Microbiology, 39,* 379-391.

Karr, D. B., Liang, R. T., Reuhs, B L., & Emerich, D. W. (2000). Altered exopolysaccharides of *Bradyrhizobium japonicum* mutants correlate with impaired soybean lectin binding, but not with effective nodule formation. *Planta, 211,* 218-226.

Keyser, H. H., Bohlool, B. B., Hu, T. S., & Weber, D. F. (1982). Fast-growing rhizobia isolated from root nodules of soybean. *Science, 215,* 1631-1632.

Knight, C. D., Rossen, L., Robertson, J. G., Wells, B., & Downie, J. A. (1986). Nodulation inhibition by *Rhizobium leguminosarum* multicopy *nodABC* genes and analysis of early stages of plant infection. *Journal of Bacteriology, 166,* 552-558.

Kondorosi, É., Pierre, M., Cren, M., Haumann, U., Buire, M., Hoffmann, B., et al. (1991). Identification of NolR, a negative transacting factor controlling the *nod* regulon in *Rhizobium meliloti. Journal of Molecular Biology, 222,* 885-896.

Kosslak, R. M., Bookland, R., Barkei, J., Paaren, H. E., & Appelbaum, E. R. (1987). Induction of *Bradyrhizobium japonicum* common *nod* genes by isoflavones isolated from *Glycine max. Proceedings of the National Academy of Sciences of the United States of America, 84,* 7428-7432.

Kovacs, L. G., Balatti, P. A., Krishnan, H. B., & Pueppke, S. G. (1995). Transcriptional organization and expression of *nolXWBTUV*, a locus that regulates cultivar-specific nodulation of soybean by *Rhizobium fredii* USDA257, *Molecular Microbiology, 17,* 923-933.

Krause, A., Doerfel, A., Göttfert, M., (2002). Mutational and transcriptional analysis of the type III secretion system of *Bradyrhizobium japonicum. Molecular Plant-Microbe Interactions, 15,* 1228-1235.

Krishnan, H. B. (2002). NolX of *Sinorhizobium fredii* USDA257, a type III-secreted protein involved in host range determination, is localized in the infection threads of cowpea (*Vigna unguiculata* [L.] Walp) and soybean (*Glycine max* [L.] Merr.) nodules. *Journal of Bacteriology, 184,* 831-839.

Krishnan, H. B., Kuo, C.-L., & Pueppke, S. G. (1995). Elaboration of flavonoid-induced proteins by the nitrogen-fixing soybean symbiont *Rhizobium fredii* is regulated by both *nodD1* and *nodD2*, and is dependent on the cultivar-specificity locus, *nolXWBTUV. Microbiology, 141,* 2245-2251.

Krishnan, H. B., Lewin, A., Fellay, R., Broughton, W. J., & Pueppke, S. G. (1992). Differential expression of *nodS* accounts for the varied abilities of *Rhizobium fredii* USDA257 and *Rhizobium sp.* strain NGR234 to nodulate *Leucaena spp. Molecular Microbiology, 6,* 3321-3330.

Krishnan, H. B., Lorio, J., Kim, W. S., Guoqiao, J., Kim, K.Y., DeBoer, M., et al. (2003). Extracellular proteins involved in soybean cultivar-specific nodulation are associated with pilus-like surface appendages and exported by a type III secretion system in *Sinorhizobium fredii* USDA257. *Molecular Plant-Microbe Interactions, 16,* 617-625.

Krishnan, H. B., & Pueppke, S. G. (1991). Sequence and analysis of the *nodABC* region of *Rhizobium fredii* USDA257, a nitrogen-fixing symbiont of the soybean and other legumes. *Molecular Plant-Microbe Interactions, 5,* 512-520.

Krishnan, H. B., & Pueppke, S. G. (1993). Flavonoid inducers of nodulation genes stimulate *Rhizobium fredii* USDA257 to export proteins into the environment. *Molecular Plant-Microbe Interactions, 6,* 107-113.

Krishnan, H. B., & Pueppke. S. G. (1994). Cultivar-specificity genes of the nitrogen-fixing soybean symbiont, *Rhizobium fredii* USDA257, also regulate nodulation of *Erythrina* spp. *American Journal of Botany, 81,* 38-45.

Kubori, T., Matsushima, Y., Nakamura, D., Uralili, J., Lara-Tejero, M., Sukhan A., et al. (1998). Supramolecular structure of *Salmonella typhimurium* type III protein secretion system. *Science, 280,* 602-605.

Kuykendall, L. D., Saxena, B., Devine, T. E., & Udell, S. E. (1992). Genetic diversity in *Bradyrhizobium japonicum* Jordan 1982 and a proposal for *Bradyrhizobium elkanii* sp. nov. *Canadian Journal of Microbiology, 38,* 501-505.

Kwon, S.-W., Park, J.-Y., Kim, J.-S., Kang, J.-W., Cho, Y.-H., Lim, C.-K., et al. (2005). Phylogenetic analysis of the genera *Bradyrhizobium, Mesorhizobium, Rhizobium* and *Sinorhizobium* on the basis of 16S rRNA gene and internally transcribed spacer region sequences. *International Journal of Systematic and Evolutionary Microbiology, 55,* 263-270.

Laguerre, G., Geniaux, E., Mazurier, S. I., Rodríguez-Casartelli, R., & Amarger, N. (1993). Conformity and diversity among field isolates of *Rhizobium leguminosarum* bv. *viciae*, bv. *trifolii*, and bv. *phaseoli* revealed by DNA hybridization using chromosome and plasmid probes. *Canadian Journal of Microbiology, 39,* 412-419.

Laguerre, G., Mavingui, P., Allard, R. R., Charnay, M. P., Louvrier, R., Mazurier, S. I., et al. (1996). Typing of rhizobia by PCR DNA fingerprinting and PCR-restriction fragment length polymorphism analysis of chromosomal and symbiotic gene regions: application to *Rhizobium leguminosarum* and its different biovars. *Applied and Environmental Microbiology, 62,* 2029-2036.

Laguerre, G., Nour, S. M., Macheret, V., Sanjuan, J., Drouin, P., & Amarger, N. (2001). Classification of rhizobia based on *nodC* and *nifH* gene analysis reveals a close phylogenetic relationship among *Phaseolus vulgaris* symbionts. *Microbiology, 147,* 981-993.

Lamrabet, Y., Bellogin, R. A., Cubo, M. T., Espuny, R., Gil, A., Krishnan, H. B., et al. (1999). Mutation in GDP-fucose synthesis genes of *Sinorhizobium fredii* alters nod factors and significantly decreases competitiveness to nodulate soybeans. *Molecular Plant-Microbe Interactions, 12,* 207-217.

Laus, M. C., & Kijne, J. W. (2004). A fixer's dress code: surface polysaccharides and host-plant-specificity in the root nodule symbiosis. *Trends in Glycoscience and Glycotechnology*, 16, 281-290.

Lee, C. A. (1997). Type III secretion systems: machines to deliver bacterial proteins into eukaryotic cells? *Trends in Microbiology, 5,* 148-156.

Leigh, J. A., Signer, E. R., & Walker, G. C. (1985). Exopolysaccharide-deficient mutants of *Rhizobium meliloti* that form ineffective nodules. *Proceedings of the National Academy of Sciences of the United States of America, 82,* 6231-6235.

Leigh, J. A., Reed, J. W., Hanks, J. F., Hirsch, A. M., & Walker, G. C. (1987). *Rhizobium meliloti* mutants that fail to succinylate their calcofluor-binding exopolysaccharide are defective in nodule invasion. *Cell, 51,* 579-588.

Lerouge, P., Roche, P., Faucher, C., Maillet, F., Truchet, G., Promé, J.-C., et al. (1990). Symbiotic host specificity of *Rhizobium meliloti* is determined by a sulphated and acylated glucosamine oligosaccharide signal. *Nature, 344,* 781-784.

Lewis-Henderson, W. R., & Djordjevic, M. A. (1991). NodT, a positively-acting cultivar specificity determinant controlling nodulation of *Trifolium subterraneum* by *Rhizobium leguminosarum* biovar *trifolii. Plant Molecular Biology, 16,* 515-526.

Lim, G., & Burton, J. C. (1982). Nodulation status of the Leguminosae. In: W. J. Broughton, (Ed.), *Nitrogen Fixation.* II. *Rhizobium* (1-34). Oxford, United Kingdom: Oxford University Press.

Lindström, K. (1989). *Rhizobium galegae*, a new species of legume root nodule bacteria. *International Journal of Systematic Bacteriology,* 39, 365-367.

Ljunggren, H., & Fahraeus, G. (1959). Effect of *Rhizobium* polysaccharide on the formation of polygalacturonase in lucerne and clover. *Nature, 184 (supplement 20),* 1578-1580.

Lochowska, A., Iwanicka-Nowicka, R., Plochocka, D., & Hryniewicz, M. M. (2001). Functional dissection of the LysR-type CysB transcriptional regulator. *Journal of Biological Chemistry, 276,* 2098-2107.

Loh, J., Garcia, M., & Stacey, G. (1997). NodV and NodW, a second flavonoids recognition system regulating *nod* gene expression in *Bradyrhizobium japonicum. Journal of Bacteriology, 179,* 3013-3020.

Loh, J., Stacey, M. G., Sadowsky, M. J., & Stacey, G. (1999). The *Bradyrhizobium japonicum nolA* gene encodes three functionally distinct proteins. *Journal of Bacteriology, 181,* 1544-1554.

Lohnis, F., & Hansen, R. (1921). Nodule bacteria of leguminous plants. *Journal of Agricultural Research, 20,* 543-556.

Long, S. R. (1989). *Rhizobium*-legume nodulation: Life together in the underground. *Cell, 56,* 203-214.

Long, S., Reed, J. W., Himawan, J., & Walker, G. C. (1988). Genetic analysis of a cluster of genes required for synthesis of the calcofluor-binding expolysaccharide of *Rhizobium meliloti. Journal of Bacteriology, 170*, 4239-4248.

López-Lara, I. M., Sohlenkamp, C., & Geiger, O. (2003). Membrane lipids in plant-associated bacteria: their biosyntheses and possible functions. *Molecular Plant-Microbe Interactions, 16*, 567-579.

López-Lara, I. M., van der Drift, K. M. G. M., van Brussel, A. A. N., Haverkamp, J., Lugtenberg, B. J. J., Thomas-Oates, J. E., et al. (1995). Induction of nodule primordia on *Phaseolus* and *Acacia* by lipo-chitin oligosaccharide nodulation signals from broad-host-range *Rhizobium* strain GRH2. *Plant Molecular Biology*, *29,* 465-477.

Lorio, J. C., Kim, W. S., & Krishnan, H. B. (2004). NopB, a soybean cultivar-specificity protein from *Sinorhizobium fredii* USDA257, is a type III secreted protein. *Molecular Plant-Microbe Interactions, 17,* 1259-1268.

Lorquin, J., Lortet, G., Ferro, M., Méar, N., Dreyfus, B., Promé, J.-C., et al. (1997). Nod factors from *Sinorhizobium saheli* and *S. teranga* bv. *sesbaniae* are both arabinosylated and fucosylated, a structural feature specific to *Sesbania rostrata* symbionts. *Molecular Plant-Microbe Interactions, 10,* 879-90.

Ludwig W., Amann, R., Martínez -Romero, E., Schönhuber, W., Bauer, S., Neef, A., et al. (1998). rRNA based identification and detection systems for rhizobia and other bacteria. *Plant and Soil, 204,* 1-19.

Macnab, R. M. (1999). The bacterial flagellum: reversible rotary propeller and type III export apparatus. *Journal of Bacteriology, 181,* 7149-7153.

Maidak, B. L. Olsen, G. J., Larsen, N., Overbeek, R., McCaughey, M. J., & Woese, C. R. (1996). The ribosomal database project (RDP). *Nucleic Acids Research, 24,* 82-85.

Maillet, F., Debellé, F., & Dénarié, J. (1990). Role of the *nodD* and *syrM* genes in the activation of the regulatory NodD3 and of the common and host-specific *nod* genes of *Rhizobium meliloti*. *Molecular Microbiology, 4,* 1975-1984.

Mapp, A. K. (2003). Regulating transcription: a chemical perspective. *Organic & Biomolecular Chemistry, 1,* 2217-2220.

Marie, C., Barny, M.-A., & Downie, J. A. (1992). *Rhizobium-leguminosarum* has two glucosamine synthases Glms and NodM required for nodulation and development of nitrogen fixing nodules. *Molecular Microbiology, 6,* 843-851.

Marie, C., Broughton, W. J., & Deakin, W. J. (2001). Rhizobium type III secretion systems: legume charmers or alarmers? *Current Opinion in Plant Biology, 4,* 336-342.

Marie, C., Deakin, W. J., Ojanen-Reuhs, T., Diallo, E., Reuhs, B., Broughton, W. J., et al. (2004) TtsI, a key regulator of *Rhizobium* species NGR234 is required for type III-dependent protein secretion and synthesis of rhamnose-rich polysaccharides. *Molecular Plant-Microbe Interactions, 17,* 958-966.

Marie, C., Deakin, W. J., Viprey, V., Kopciñska, J., Golinowski, W., Krishnan, H. B. et al. (2003). Characterization of Nops, nodulation outer proteins, secreted via the type III secretion system of NGR234. *Molecular Plant-Microbe Interactions, 16,* 743-751.

Mathis, R., Van Gijsegem, F., De Rycke, R., D'Haeze, W., Van Maelsaeke, E., Anthonio E., et al. (2005). Lipopolysaccharides as a communication signal for progression of legume endosymbiosis. *Proceedings of the National Academy of Sciences of the United States of America, 102,* 2655-2660.

Martínez-Romero, E., Segovia, L., Marcante, F. M., Franco, A. A., Graham, P., Pardo, M. A. (1991). *Rhizobium tropici*, a novel species nodulating *Phaseolus vulgaris* L. beans and *Leucaena* sp. trees. *International Journal of Systematic Bacteriology, 41,* 417-426.

Meinhardt, L. W., Krishnan, H. B., Balatti, P. A., & Pueppke, S. G (1993). Molecular cloning and characterization of a *sym* plasmid locus that regulates cultivar-specific nodulation of soybean by *Rhizobium fredii* USDA257. *Molecular Microbiology, 9,* 17-29.

Mergaert, P., Ferro, M., D'Haeze, W., Van Montagu, M., Holsters, M., & Promé, J.-C. (1997a). Nod factors of *Azorhizobium caulinodans* strain ORS571 can be glycosylated with an arabinosyl group, a fucosyl group, or both. *Molecular Plant-Microbe Interactions, 10,* 683-687.

Mergaert, P., Van Montagu, M., & Holsters, M. (1997b). Molecular mechanisms of Nod factor diversity. *Molecular Microbiology, 25,* 811-817.

Mergaert, P., Van Montagu, M., & Holsters, M. (1997c). The nodulation gene *nolK* of *Azorhizobium caulinodans* is involved in the formation of GDP-fucose from GDP-mannose. *Federation of European Biochemical Societies Letters, 409,* 312-316.

Michiels, J., Dombrecht, B., Vermeiren, N., Xi., C.-W., Luyten, E., & Vanderleyden, J. (1998). *Phaseolus vulgaris* is a non-selective host for nodulation. *Federation of European Microbiological Societies Microbiology Ecology, 26,* 193-205.

Michiels, T., & Cornelis, G. (1988). Nucleotide sequence and transcription analysis of *yop51* from *Yersinia enterocolitica* W22703. *Microbial Pathogenesis, 5,* 449-459.

Michiels, T., Wattiau, P., Brasseur, R., Ruysschaert, J.-M., Cornelis, G. (1990). Secretion of Yop proteins by *Yersiniae*. *Infection and Immunity, 58,* 2840-2849.

Miller, K. J., Hadley, J. A., & Gustine, D. L. (1994). Cyclic β-1,6-1,3-glucans of *Bradyrhizobium japonicum* USDA 110 elicit isoflavonoid production in the soybean (*Glycine max*) host. *Plant Physiology, 104,* 917-923.

Minami, E., Kouchi, H., Carlson, R. W., Cohn, J. R., Kolli, V. K., Day, R. B., et al. (1996). Cooperative action of lipo-chitin nodulation signals on the induction of the early nodulin, ENOD2, in soybean roots. *Molecular Plant-Microbe Interactions, 9,* 574-583.

Mithöfer, A. (2002). Suppression of plant defence in rhizobia-legume symbiosis. *Trends in Plant Science, 7,* 440-444.

Mithöfer, A., Bhagwat, A. A., Feger, M., & Ebel, J. (1996). Suppression of fungal β-glucan-induced plant defense in soybean (*Glycine max* L.) by cyclic 1,3-1,6-β-glucans from the symbiont *Bradyrhizobium japonicum*. *Planta, 199,* 270-275.

Moulin, L., Munive, A, Dreyfus, B., & Boivin-Masson, C. (2001). Nodulation of legumes by members of the β-subclass of Proteobacteria. *Nature, 411,* 948-950.

Mulligan, J. T., & Long, S. R. (1989). A family of activator genes regulates expression of *Rhizobium meliloti* nodulation genes. *Genetics, 122,* 7-18.

Niehaus, K., Kapp, D., & Pühler, A. (1993). Plant defence and delayed infection of alfalfa pseudonodules induced by an exopolysaccharide (EPS I)-deficient *Rhizobium meliloti* mutant. *Planta, 190,* 415-425.

Niehaus, K., Lagares, A., & Pühler, A. (1998). A *Sinorhizobium meliloti* lipopolysaccharide mutant induces effective nodules on the host plant *Medicago sativa* (alfalfa) but fails to establish a symbiosis with *Medicago truncatula*. *Molecular Plant-Microbe Interactions, 11,* 906-914.

Nguyen, L., Paulsen, I. T., Tchieu, J., Hueck, C. J., Saier, M. H. Jr. (2000). Phylogenetic analyses of the constituents of type III protein secretion systems. *Journal of Molecular Microbiology and Biotechnology 2,* 125-144.

Nobbe, F., Hiltner, L., & Schmid, E. (1895). Versuche über die Biolgie der Knöllchenbakterien der Leguminosen, insbesondere über die Frage der Arteinheit derselben. Bericht über die Tätigkeit der Landwirtschaftlichen Abteilung der Königlichen Versuchsstation für Pflanzenkultur zu Dresden, 45, 1-27.

Nour, S. M., Fernandez, M. P., Normand, P., & Cleyet-Marel, J. C. (1994). *Rhizobium ciceri* sp. nov., consisting of strains that nodulate chickpeas (*Cicer arietinum* L.). *International Journal of Systematic Bacteriology*, *44,* 511-522.

Nour, S. M., Cleyet-Marel, J. C., Normand, P., & Fernandez, M. P. (1995). Genomic heterogeneity of strains nodulating chickpeas (*Cicer arietinum* L.) and description of *Rhizobium mediterraneum* sp. nov. *International Journal of Systematic Bacteriology, 45,* 640-648.

Norris, V., Turnock, G. & Sigee, D. (1996). The *Escherichia coli* enzoskeleton. *Molecular Microbiology, 19,* 197-204.

Nübel, U., Engelen, B., Felske, A., Snaidr, J., Wieshuber, A., Amann, R. I., et al. (1996). Sequence heterogeneities of genes encoding 16S rRNAs in *Paenibacillus polymyxa* detected by temperature gradient gel electrophoresis. *Journal of Bacteriology, 178,* 5636-5643.

Ogawa, J., & Long, S. R. (1995) The *Rhizobium meliloti groELc* locus is required for regulation of early nod genes by the transcription activator NodD. *Genes and Development, 9,* 714-729.

Olsthoorn, M. M. A., López-Lara, I. M., Petersen, B. O., Bock, K., Haverkamp, J., Spaink, H. P., et al. (1998). Novel branched Nod factor structure results from α-(1→3) fucosyl transferase activity: The major lipo-chitin oligosaccharides from *Mesorhizobium loti* strain NZP2213 bear an α-(1→3) fucosyl substituent on a nonterminal backbone residue. *Biochemistry, 37,* 9024-9032.

Orgambide, G. C., Li, J., Hollingsworth, R. I., & Dazzo, F. B. (1995). Structurally diverse chitolipooligosaccharide Nod factors accumulate primarily in membranes of wild-type *Rhizobium leguminosarum* biovar *trifolii*. *Biochemistry, 34,* 3832-3840.

Orth, K., Xu, Z., Mudgett, M. B., Bao, Z. Q., Palmer, L. E., Bliska, J. B., et al. (2000). Disruption of signaling by *Yersinia* effector YopJ, a ubiquitin-like protein protease. *Science, 290,* 1594-1597.

Ovtsyna, A. O., Schultze, M., Tikhonovich, I. A., Spaink, H. P., Kondorosi, É., Kondorosi, Á., et al. (2000). Nod factors of *Rhizobium leguminosarum* bv. *viciae* and their fucosylated derivatives stimulate a Nod factor cleaving activity in pea roots and are hydrolyzed in vitro by plant chitinases at different rates. *Molecular Plant-Microbe Interactions, 13,* 799-807.

Page, A. L., & Parsot, C. (2002). Chaperones of the type III secretion pathway: jacks of all trades. *Molecular Microbiology, 46,* 1-11.

Parniske, M., & Downie, J. A. (2003). Locks, keys and symbioses. *Nature, 425,* 569-570.

Parniske, M., Schmidt, P. E., Kosch, K., & Müller, P. (1994). Plant defense response of host plants with determinate nodules induced by EPS-defective *exoB* mutants of *Bradyrhizobium japonicum. Molecular Plant-Microbe Interactions, 7,* 631-638.

Pellock, B. J., Cheng, H. P., & Walker, G. C. (2000). Alfalfa root nodule invasion efficiency is dependent on *Sinorhizobium meliloti* polysaccharides. *Journal of Bacteriology, 182,* 4310-4318

Perotto, S., Brewin, N. J., & Kannenberg, E. L. (1994). Cytological evidence for a host defence response that reduces cell and tissue invasion in pea nodules by lipopolysaccharides-defective mutants of *Rhizobium leguminosarum* strain 3841. *Molecular Plant-Microbe Interactions, 7,* 99-112.

Perret, X., Staehelin, C., & Broughton, W. J. (2000). Molecular basis of symbiotic promiscuity. *Microbiology and Molecular Biology Reviews, 64,* 180-201.

Peters, N. K., Frost, J. W., & Long, S. R. (1986). A plant flavone, luteolin, induces expression of *Rhizobium meliloti* nodulation genes. *Science, 233,* 977-980.

Peters, N. K., & Long, S. R. (1988). Alfalfa root exudates and compounds which promote or inhibit induction of *Rhizobium meliloti* nodulation genes. *Plant Physiology, 88,* 396-400.

Philip-Hollingsworth, S., Dazzo, F. B., & Hollingsworth, R. I. (1997). Structural requirements of *Rhizobium* chitolipooligosaccharides for uptake and bioactivity in legume roots as revealed by synthetic analogs and fluorescent probes. *Journal of Lipid Research, 38,* 1229-1241.

Philip-Hollingsworth, S., Hollingsworth, R. I., & Dazzo, F. B. (1991). *N*-acetylglutaminic acid: an extracellular *nod* signal of *Rhizobium trifolii* ANU843 that induces root hair branching and nodule-like primordia in white clover roots. *Journal of Biological Chemistry, 266,* 16854-16858.

Poinsot, V., Bélanger, E., Laberge, S., Yang, G.-P., Antoun, H., Cloutier, J., et al. (2001). Unusual methyl-branched α,ß-unsaturated acyl chain substitutions in the Nod factors of an Arctic *Rhizobium, Mesorhizobium* sp. strain N33 (*Oxytropis arctobia*). *Journal of Bacteriology, 183,* 3721-3728.

Pollock T. J., van Workum, W. A. T., Thorne, L., Mikolajczak, M. J., Yamazaki, M., Kijne, J. W., et al. (1998). Assignment of biochemical functions to glycosyl transferase genes which are essential for biosynthesis of exopolysaccharides in *Sphingomonas* strain S88 and *Rhizobium leguminosarum. Journal of Bacteriology, 180,* 586-593.

Poupot, R., Martínez-Romero, E., Gautier, N., & Promé, J.-C. (1995). Wild type *Rhizobium etli,* a bean smbiont, produces acetyl-fucosylated, *N*-methylated and carbamoylated nodulation factors. *Journal of Biological Chemistry, 270,* 6050-6055.

Poupot, R., Martínez-Romero, E., & Promé, J.-C. (1993). Nodulation factors from *Rhizobium tropici* are sulfated or nonsulfated chitopentasaccharides containing an *N*-methyl-*N*-acylglucosamine terminus. *Biochemistry, 32,* 10430-10435.

Prazmowski, A. (1889). Das Wesen und die biologise Bedeutung der Wurzelknöllchen der Erbse. *Berichte Botanisches Centralblatt, 39,* 356-362.

Prazmowski, A. (1890). Die Wurzelknöllchen der Erbse. Bericht über die Tätigkeit der Landwirtschaftlichen Abteilung der Königlichen Versuchsstation für Pflanzenkultur zu Berlin, *37,* 161-238.

Price, N. P. (1999). Carbohydrate determinants of *Rhizobium*-legume symbioses. *Carbohydrate Research, 317,* 1-9.

Price, N. P. J., Relic, B., Talmont, F., Lewin, A., Promé , D., Pueppke, S. G., et al. (1992). Broad-host-range *Rhizobium* species strain NGR234 secretes a family of carbamoylated, and fucosylated, nodulation signals that are *O*-acetylated or sulphated. *Molecular Microbiology, 6,* 3575-84.

Pueppke, S. G. (1996). The genetic and biochemical basis for nodulation of legumes by rhizobia. *Critical Reviews in Biotechnology*, 16, 1-51.

Pueppke, S. G., & Broughton, W. J. (1999). *Rhizobium* sp. NGR234 and *R. fredii* USDA 257 share exceptionally broad, nested host-ranges. *Molecular Plant-Microbe Interactions, 12,* 293-318.

Pueppke, S. G., DeBoer M., He, X.-T., &. Krishnan, H. B. (1999). An isoflavone from soybean roots induces *Sinorhizobium fredii* strain USDA257 to produce pilus-like surface appendages. In E. Martinez & G. Hernandez (Eds.), *Highlights of Nitrogen Fixation Research* (131-134). New York: Plenum Press.

Quinto, C., Wijfjes, A. H. M., Bloemberg, G. V., Blok-Tip, L., López-

Lara, I. M., Lugtenberg, B. J. J., et al. (1997). Bacterial nodulation protein NodZ is a chitin oligosaccharidefucosyltransferase which can also recognize related substrates of animal origin. *Proceedings of the National Academy of Sciences of the United States of America, 94,* 4336-4341.

Ramu, S. K., Peng, H. M., & Cook, D. R. (2002). Nod factor induction of reactive oxygen species production is correlated with expression of the early nodulin gene *rip1* in *Medicago truncatula. Molecular Plant-Microbe Interactions, 15,* 522-528.

Redmond, J. W., Batley, M., Djordjevic, M. A., Innes, R. W., Kuempel, P. L., & Rolfe, B. G. (1986). Flavones induce expression of nodulation genes in *Rhizobium. Nature 323,* 632-636.

Reinhold, B. B., Chan, S. Y., Reuber, T. L., Marra, A., Walker, G. C., & Reinhold, V. N. (1994). Detailed structural characterization of succinoglycan, the major exopolysaccharide of *Rhizobium meliloti* Rm1021. *Journal of Bacteriology, 176,* 1997-2002.

Reuhs, B. L., Carlson, R. W., & Kim, J. S. (1993). *Rhizobium fredii* and *Rhizobium meliloti* produce 3-deoxy-D-manno-2-octulosonic acid-containing polysaccharides that are structurally analogous to group II K antigens (capsular polysaccharides) found in *Escherichia coli. Journal of Bacteriology, 175,* 3570-3580.

Reuhs, B. L., Geller, D. P., Kim, J. S., Fox, J. E., Kolli, V. S., & Pueppke, S. G. (1998). *Sinorhizobium fredii* and *Sinorhizobium meliloti* produce structurally conserved lipopolysaccharides and strain-specific K antigens. *Applied and Environmental Microbiology, 64,* 4930-4938.

Reuhs, B. L., Kim, J. S., Badgett, A., & Carlson, R. W. (1994). Production of cell-associated polysaccharides of *Rhizobium fredii* USDA205 is modulated by apigenin and host root extract. *Molecular Plant-Microbe Interactions, 7,* 240-247.

Reuhs, B. L., Stephens, S. B., Geller, D. P., Kim, J. S., Glenn, J., Przytycki, J., et al. (1999). Epitope identification for a panel of anti-*Sinorhizobium meliloti* monoclonal antibodies and application to the analysis of K-antigen and lipopolysaccharides from bacteroids. *Applied and Environmental Microbiology, 65,* 5186-5191.

Ritsema, T., Wijfjes, A. M. H., Lugtenberg, B. J. J., & Spaink, H. P. (1996). *Rhizobium* nodulation protein NodA is a host specific determinant of the transfer of fatty acids in Nod factor biosynthesis. *Molecular and General Genetics, 251,* 44-51.

Rivas, R., Velazquez, E., Willems, A., Vizcaíno, N., Subba-Rao, N. S., Mateos, P., et al. (2002). A new species of *Devosia* that forms a unique nitrogen-fixing root-nodule symbiosis with the aquatic legume *Neptunia natans* (L.f.) Druce. *Applied and Environmental Microbiology, 68,* 5217-5222.

Rivas, R., Willems, A., Subba-Rao, N. S., Mateos, P. F., Dazzo, F. B., Kroppenstedt, R. M., et al. (2003). Description of *Devosia neptuniae* sp. nov. that nodulates and fixes nitrogen in symbiosis with *Neptunia natans*, an aquatic legume from India. *Systematic and Applied Microbiology, 26,* 47-53.

Rivilla, R., Sutton, J. M., & Downie, J. A. (1995). *Rhizobium leguminosarum* NodT is related to a family of outer-membrane transport proteins that includes TolC, PrtF, CyaE and AprF. *Gene , 161,* 27-31.

Roche, P., Debellé, F., Maillet, F., Lerouge, P., Faucher, C., Truchet, G., et al. (1991a). Molecular basis of symbiotic host specificity in *Rhizobium meliloti*: *nodH* and *nodPQ* genes encode the sulfation of lipo-oligosaccharide signals. *Cell, 67,* 1131-1143.

Roche, P., Lerouge P., Ponthus, C., & Promé, J.-C. (1991b). Structural determination of bacterial nodulation factors involved in the *Rhizobium meliloti* alfalfa symbiosis. *Journal of Biological Chemistry, 266,* 10933-10940.

Roche, P., Maillet, F., Plazanet, C., Debellé, F., Ferro, M., Truchet, G., et al. (1996). The common *nodABC* genes of *Rhizobium meliloti* are host-range determinants. *Proceedings of the National Academy of Sciences of the United States of America, 93,* 15305-15310.

Röhrig, H., Schmidt, J., Wieneke, U., Kondorosi, É., Barlier, I., Schell, J., et al. (1994). Biosynthesis of lipooligosaccharide nodulation factors: *Rhizobium* NodA protein is involved in *N*-acylation of the chitooligosaccharide backbone. *Proceedings of the National Academy of Sciences of the United States of America. 91,* 3122-3126.

Roine, E., Wei, W., Yuan, J., Nurmiaho-Lassila, E.-L., Kalkkinen, N., Romantschuk, M., et al. (1997). Hrp pilus: an *hrp*-dependent bacterial surface appendage produced by *Pseudomonas syringae* pv. tomato DC3000. *Proceedings of the National Academy of Sciences of the United States of America, 94,* 3459-3464.

Rolfe, B. G., Carlson, R. W., Ridge, R. W., Dazzo, F. B., Mateos, P. F., & Pankhurst, C. E. (1996). Defective infection and nodulation of clovers by exopolysaccharide mutants of *Rhizobium leguminosarum* bv. *trifolii. Australian Journal of Plant Physiology, 23*, 285-303.

Rolin D. B., Pfeffer, P. E., Osman, S. F., Szwergold, B. S., Kappler, F., & Benesi, A. J. (1992) Structural studies of a phosphocholine substituted β-(1,3);(1,6) macrocyclic glucan from *Bradyrhizobium japonicum* USDA 110. *Biochimica et Biophysica Acta, 1116,* 215-225.

Rostas, K., Kondorosi, É., Horvath, B., Simoncsits, A., & Kondorosi, Á. (1986). Conservation of extended promoter regions of nodulation genes in *Rhizobium. Proceedings of the National Academy of Science of the United States of America, 83,* 1757-1761.

Saad, M. M., Kobayashi, H., Marie, C., Brown, I. R., Mansfield, J. W., Broughton, W. J., et al. (2005). NopB, a Type III secreted protein of *Rhizobium* sp. strain NGR234, is associated with pilus-like surface appendages. *Journal of Bacteriology, 187,* 1173-1181.

Saler, M. H., Tam, R., Reizer, A., & Reizer, J. (1994). Two novel families of bacterial membrane proteins concerned with nodulation, cell division, and transport. *Molecular Microbiology, 11,* 841-847.

Sanjuan, J., Carlson, R. W., Spaink, H. P., Bhat, U. R., Barbour, W. M., Glushka, J., et al. (1992). A 2-*O*-methylfucose moiety is present in the lipo-oligosaccharide nodulation signal of *Bradyrhizobium japonicum. Proceedings of the National Academy of Sciences of the United States of America, 89,* 8789-8793.

Savouré, A., Sallaud, C., El-Turk, J., Zuanazzi, J., Ratet, P., Schultze, M., et al. (1997). Distinct response of *Medicago* suspension cultures and roots to Nod factors and chitin oligomers in the elicitation of defense-related responses. *Plant Journal, 11,* 277-287.

Schell, M. A. (1993). Molecular biology of the LysR family of transcriptional regulators. *Annual Review of Microbiology, 47,* 597-626.

Schlaman, H. R. M., Okker, R. J. H., & Lugtenberg, B. J. J. (1992). Regulation of nodulation gene expression by NodD in Rhizobia. *Journal of Bacteriology, 174,* 5177-5182.

Scholla, M. H., & Elkan, G. H. (1984). *Rhizobium fredii* sp. nov. a fast-growing species that effectively nodulates soybeans. *International Journal of Systematic Bacteriology, 34,* 484-486.

Schultze, M., Quiclet-Sire, B., Kondorosi, É., Virelizier, H., Glushka, J. N., Endre, G., et al. (1992). *Rhizobium meliloti* produces a family of sulfated lipo-oligosaccharides exhibiting different degrees of plant host specificity. *Proceedings of the National Academy of Sciences of the United States of America, 89,* 192-196.

Schwedock, J. S., Liu, C., Leyh, T. S., & Long, S. R. (1994). *Rhizobium meliloti* NodP and NodQ form a multifunctional sulfate-activating complex requiring GTP for activity. *Journal of Bacteriology, 176,* 7055-7064.

Schwedock, J., & Long, S. R. (1989). Nucleotide sequence and protein products of two new nodulation genes of *Rhizobium meliloti, nodP* and *nodQ. Molecular Plant-Microbe Interactions. 2,* 181-194.

Segovia, L., Young, J. P. W., & Martínez-Romero, E. (1993). Reclassification of American *Rhizobium leguminosarum* biovar *phaseoli* type I strains as *Rhizobium etli* sp. nov. *International Journal Systematic Bacteriology, 43,* 374-377.

Sindhu, S. S., Brewin, N. J., & Kannenberg, E. L. (1990). Immunochemical analysis of lipopolysaccharides from free-living and endosymbiotic forms of *Rhizobium leguminosarum. Journal of Bacteriology, 172,* 1804-1813.

Snoeck, C., Luyten, E., Poinsot, V., Savagnac, A., Vanderleyden, J., & Promé, J.-C. (2001) *Rhizobium* sp. BR816 produces a complex mixture of known and novel lipochitooligosaccharide molecules. *Molecular Plant-Microbe Interactions, 14,* 678-684.

Sohlenkamp, C., López-Lara, I. M., & Geiger, O. (2003). Biosynthesis of phosphatidylcholine in bacteria. *Progress in Lipid Research, 42,* 115-62.

Spaink, H. P. (1992). Rhizobial lipo-oligosaccharide: answers and questions. *Plant Molecular Biology, 20,* 977-986.

Spaink, H. P. (1996). Regulation of plant morphogenesis by lipo-chitin oligosaccharides. *Critical Reviews in Plant Sciences, 15,* 559-82.

Spaink, H. P. (2000). Root nodulation and infection factors produced by rhizobial bacteria. *Annual Review of Microbiology, 54,* 257-288.

Spaink, H. P., Bloemberg, G. V., van Brussel, A. A. N., Lugtenberg, B. J. J., van der Drift, K. M. G. M., Haverkamp, J., et al. (1995a). Host specificity of *Rhizobium leguminosarum* is determined by the hydrophobicity of highly unsaturated fatty acyl moieties of the nodulation factors. *Molecular Plant-Microbe Interactions, 8,* 155-164.

Spaink, H. P., Sheeley, D. M., van Brussel, A. A. N., Glushka, J., York, W. S., Tak, T., et al. (1991). A novel highly unsaturated fatty acid moiety of lipo-oligosaccharide signals determines host specificity of *Rhizobium*. *Nature, 354,* 125-130.

Spaink, H. P., Weinman, J., Djordjevic, M. A., Wijffelman, C. A., Okker, R. J. H., & Lugtenberg, B. J. J. (1989). Genetic analysis and cellular localization of the *Rhizobium* host-specificity-determining NodE protein *European Molecular Biology Organization Journal, 8,* 2811-2818.

Spaink, H. P., Wijffelman, C. A., Pees, E., Okker, R. J. H., & Lugtenberg, B. J. J. (1987). *Rhizobium* nodulation gene *nodD* as a determinant of host specificity. *Nature, 328,* 337-340.

Spaink, H. P., Wijfjes, A. H. M., & Lugtenberg, B. J. J. (1995b). *Rhizobium* NodI and NodJ proteins play a role in the efficiency of secretion of lipochitin oligosaccharides. *Journal of Bacteriology, 177,* 6276-6281.

Stacey, G., Luka, S., Sanjuan, J., Banfalvi, Z., Nieuwkoop, A. J., Chun, J. Y., et al. (1994). *nodZ*, a unique host-specific nodulation gene, is involved in the fucosylation of the lipooligosaccharide nodulation signal of *Bradyrhizobium japonicum*. *Journal of Bacteriology, 176,* 620-633.

Staehelin, C., Granado, J., Müller, J., Wiemken, A., Mellor, R. B., Felix, G., et al. (1994). Perception of *Rhizobium* nodulation factors by tomato cells and inactivation by root chitinases. *Proceedings of the National Academy of Sciences of the United States of America, 91,* 2196-2200.

Staehelin, C., Schultze, M., Kondorosi, É., & Kondorosi, Á. (1995). Lipo-chitooligosaccharide nodulation signals from *Rhizobium meliloti* induce their rapid degradation by the host plant alfalfa. *Plant Physiology, 108,* 1607-1614.

Staehelin, C., Schultze, M., Kondorosi, É., Mellor, R. B., Boller, T., & Kondorosi, Á. (1994). Structural modifications in *Rhizobium meliloti* Nod factors influence their stability against hydrolysis by root chitinases. *Plant Journal, 5,* 319-330.

Stokkermans, T. J. W., Orlando, R., Kolli, V. S. K., Carlson, R. W., & Peters, N. K. (1996). Biological activities and structures of *Bradyrhizobium elkanii* low abundance lipo chitin-oligosaccharides. *Molecular Plant-Microbe Interactions, 9,* 298-304.

Sullivan, J. T., Trzebiatowski, J. R., Cruickshank, R. W., Gouzy, J., Brown, S. D., Elliot, R. M., et al. (2002). Comparative sequence analysis of the symbiosis island of *Mesorhizobium loti* strain R7A. *Journal of Bacteriology, 184,* 3086-3095.

Swanson, J. A., Mulligan, J. T., & Long, S. R. (1993). Regulation of syrM and nodD3 in *Rhizobium meliloti. Genetics, 134,* 435-444.

Sy, A., Giraud, E., Jourand, P., Garcia, N. Willems, A., De Lajudie, P., et al. (2001). Methylotrophic *Methylobacterium* bacteria nodulate and fix nitrogen in symbiosis with legumes. *Journal of Bacteriology, 183,* 214-220.

Terefework, Z., Paulin, L., Suomalainen, S., Nick, G., & Lindström, K. (1998). Phylogeny of *Rhizobium galegae* compared with other rhizobia and agrobacteria. *International Journal of Systematic Bacteriology, 48,* 349-356.

Timmers, A. C. J., Auriac, M. C., & Truchet, G. (1999). Refined analysis of early symbiotic steps of the *Rhizobium medicago* interaction in relationship with microtubular cytoskeleton rearrangements. *Development, 126,* 3617-3628.

Trinick, M. J. (1973). Symbiosis between *Rhizobium* and the non-legume, *Trema aspera. Nature 244,* 459-460.

Truchet, G., Roche, P., Lerouge, P., Vasse, J., Camut, S., de Billy, F., et al. (1991). Sulphated lipo-oligosaccharide signals of *Rhizobium meliloti* elicit root nodule organogenesis in alfalfa. *Nature, 351,* 670-673.

Tully, R. E., Keister, D. L., & Gross, K. C. (1990). Fractionation of the β-linked glucans of *Bradyrhizobium japonicum* and their response to osmotic potential. *Applied and Environmental Microbiology, 56,* 1518-1522.

Viprey, V., Del Greco, A., Golinowski, W., Broughton, W. J., & Perret, X. (1998). Symbiotic implications of type III protein secretion machinery in *Rhizobium. Molecular Microbiology, 28,* 1381-1389.

van Berkum, P., & Eardly, B. D. (2002). The aquatic budding bacterium *Blastobacter denitrificans* is a nitrogen fixing symbiont of *Aeschynomene indica. Applied and Enviornmental Microbiology, 68,* 1132-1136.

van Berkum, P., & Fuhrmann, J. J. (2000). Evolutionary relationships among the soybean *Bradyrhizobia* reconstructed from 16S rRNA gene and internally transcribed spacer region sequence divergence. *International Journal of Systematic and Evolutionary Microbiology, 50,* 2165-2172.

van Berkum, P., Terefework, Z., Paulin, L., Suomalainen, S., Lindström, K., & Eardly, B. D. (2003). Discordant phylogenies within the *rrn* loci of rhizobia. *Journal of Bacteriology, 185,* 2988-2998.

van Brussel, A. A. N., Bakhuizen, R., Van Spronsen, P. C., Spaink, H. P., Tak, T., Lugtenberg, B. J. J., et al. (1992). Induction of preinfection thread structures in the leguminous host plant by mitogenic lipo-oligosaccharides of *Rhizobium. Science, 257,* 70-72.

Vandamme, P., Goris, J., Chen, W. M., & Willems, A. (2002). *Burkholderia tuberum* sp. nov. and *Burkholderia phymatum* sp. nov., nodulate the roots of tropical legumes. *Systematic and Applied Microbiology, 25,* 507-12.

Vandamme, P., Pot, B., Gillis, M., de Vos, P., Kersters, K., & Swings, J. (1996). Polyphasic taxonomy, a consensus approach to bacterial systematics. *Microbiological Reviews, 60,* 407-438.

Van de Peer, Y., De Rijk, P., Wuyts, J., Winkelmans, T., De Wachter, R. (2000). The European small subunit ribosomal RNA database. *Nucleic Acids Research, 28,* 175-176.

van der Drift, K. M. G. M., Spaink, H. P., Bloemberg, G. V., van Brussel, A. A. N., Lugtenberg B. J. J., Haverkamp, J., et al. (1996). *Rhizobium leguminosarum* bv. *trifolii* produces lipo-chitin oligosaccharides with *nodE*-dependent highly unsaturated fatty acyl moieties: an electrospray ionisation and collision induced dissociation tandem mass spectrometric study. *Journal of Biological Chemistry, 271,* 22563-22569.

van Workum, W. A. T., Canter Cremers, H. C. J., Wijfjes, A. H. M., van der Kolk, C., & Kijne, J. W. (1997). Cloning and characterization of four genes of *Rhizobium leguminosarum* bv. *trifolii* involved in exopolysaccharide production and nodulation. *Molecular Plant-Microbe Interactions, 10,* 290-301.

van Workum, W. A. T., & Kijne, J. W. (1998). Biosynthesis of rhizobial exopolysaccharides and their role in the root nodule symbiosis of leguminous plants. In J. T. Romeo, K. R. Downum, R. Verpoorte, *Recent Advances in Phytochemistry (vol 32), Phytochemical Signals and Plant-Microbe Interactions* (139-66). New York: Plenum Press.

van Workum, W. A. T., van Slageren, S., van Brussel, A. A. N. & Kijne, J.W. (1998) Role of exopolysaccharides of *Rhizobium leguminosarum* bv. *viciae* as host plant-specific molecules required for infection thread formation during nodulation of *Vicia sativa. Molecular Plant-Microbe Interactions, 11,* 1233-1241.

Vasse, J., de Billy, F., & Truchet, G. (1993). Abortion of infection during the *Rhizobium meliloti*-alfalfa symbiotic interactions is accompanied by a hypersensitive reaction. *Plant Journal, 4,* 555-566.

Viprey, V., Del Greco, A., Golinowski, W., Broughton, W. J., & Perret, X. (1998). Symbiotic implications of type III protein secretion machinery in *Rhizobium*. *Molecular Microbiology, 28,* 1381-1389.

Wais, R. J., Galera, C., Oldroyd, G., Catoira, R., Penmetsa, R. V., Cook, D., et al. (2002a). Genetic analysis of calcium spiking responses in nodulation mutants of *Medicago truncatula. Proceedings of the National Academy of Science of the United States of America, 97, 13407-13412.*

Wais, R. J., Keating, D. H., & Long, S. R. (2002b). Structure-function analysis of Nod factor-induced root hair calcium spiking in Rhizobium-legume symbiosis. *Plant Physiology, 128,* 1-14.

Wang, Y., & Hollingsworth, R. I. (1994) The structure of the O-antigenic chain of the lipopolysaccharide of *Rhizobium trifolii* 4s. *Carbohydrate Research, 260,* 305-317.

Wang L., & Winans S. C. (1995). The sixty nucleotide of OccR operator contains a subsite essential and sufficient for OccR binding and a second subsite required for ligand-responsive DNA bending. *Journal of Molecular Biology, 253,* 691-702.

Weerasinghe, R., Bird, D. M., & Allen, N. S. (2005). Root-knot nematodes and bacterial Nod factors elicit common signal transduction events in *Lotus japonicus. Proceedings of the National Academy of Science of the United States of America, 102,* 3147-3152.

White, F. F., Yang, B., & Johnson, L. B. (2000). Prospects for understanding avirulence gene function. *Current Opinion in Plant Biology, 3,* 291-298.

Wilson, J. K., 1939. Leguminous plants and their associated organisms. *Cornell University Agricultural Experiment Station Memorandum 221.* Ithaca, New York: Cornell University Press.

Yang, W.-C., Canter-Cremers, H. C. J., Hogendijk, P., Katinakis, P., Wijffelman, C. A., Franssen, H., et al. (1992). In-situ localization of chalcone synthase mRNA in pea root nodule development. *Plant Journal, 2,* 143-151.

Yang, G.-P., Debellé, F., Savagnac, A., Ferro, M., Schiltz, O., Maillet, F. et al. (1999). Structure of the *Mesorhizobium huakuii* and *Rhizobium galegae* Nod factors: a cluster of phylogenetically related legumes are nodulated by rhizobia producing Nod factors with α, ß-unsaturated *N*-acyl substitutions. *Molecular Microbiology, 34,* 227–237.

Yap, W. H., Zhang, Z., & Wang, Y. (1999). Distinct types of rRNA operons exist in the genome of the actinomycete *Thermomonospora chromogena. Journal of Bacteriology, 181,* 5201-5209.

Yeh, K.-C., Peck, M. C., & Long, S. R. (2002). Luteolin and GroESL modulate in vitro activity of NodD. *Journal of Bacteriology, 184,* 525-530.

Young, J. P. W. (1996). Phylogeny and taxonomy of rhizobia. *Plant and Soil, 186,* 45-52.

Young, J. M., Park, D.-C., & Weir, B. S. (2004). Diversity of 16S rDNA sequences of *Rhizobium* spp. implications for species determinations. *Federation of European Microbiological Societies Microbiology Letters, 238,* 125-131.

Young, J. P. W., Downer, H. L., & Eardly, B. D. (1991). Phylogeny of the phototrophic *Rhizobium* strain BTAi1 by polymerase chain reaction-based sequencing of a 16S rRNA gene segment. *Journal of Bacteriology, 173,* 2271-2277

Young, J. P. W., & Haukka, K. E. (1996). Diversity and phylogeny of rhizobia. *New Phytologist, 133,* 87-94.

Young, G. M., Schmiel, D. H., & Miller, V. L. (1999). A new pathway for the secretion of virulence factors by bacteria: the flagellar export apparatus functions as a protein secretion system. *Proceedings of the National Academy of Science of the United States of America, 96,* 6456-6461.

21. AFFILIATIONS

Hari B. Krishnan, *Plant Genetics Research Unit, Agricultural Research Service, U.S. Department of Agriculture, and [2]Department of Agronomy, University of Missouri, Columbia, MO 65211, USA, Phone: 573-882-8151, Fax: 573-884-7850, E-mail: KrishnanH@missouri.edu*

John O. Bennett, *Department of Agronomy, University of Missouri, Columbia, MO 65211, USA.*

DAVID A. DALTON AND SASHA KRAMER

NITROGEN-FIXING BACTERIA IN NON-LEGUMES

Abstract. The process of biological nitrogen fixation, in which bacteria pass fixed nitrogen on to a plant host, is well known in the legume-rhizobia system. Less well known, but equally intriguing, is the ability of some grasses to harbor nitrogen-fixing bacteria. The plant-bacteria association is less distinct in the case of grasses, since the bacteria may be dispersed in low numbers in or near roots (associative fixation) or ensconced within cell walls inside the plant (endophytic fixation). It has been difficult to define the significance of nitrogen fixation in these systems. The most studied system in this regard is the association between sugarcane and endophytic bacteria such as *Gluconoacetobacter diazotrophicus*. This association appears to be responsible for the ability of sugarcane to produce large crops for many years without the addition of nitrogen fertilizers. Experiments comparing inoculated and uninoculated sugarcane plants show that nitrogen-fixing bacteria provide a strong benefit for the host plant. However, some of those benefits may arise from the ability of the bacteria to produce phytohormones. Other similar systems include: 1) various tropical grasses such as *Digitaria* and *Paspalum* with bacteria such as *Azospirillum*; 2) Kallar grass (*Leptochloa fusca*) with *Azoarcus*; and 3) a wide range of host plants (coffee, maize, dune grasses) with the widespread bacteria *Burkholderia*. PCR-based screens of DNA isolated directly from environmental samples suggest that associative/endophytic nitrogen fixation is more widespread than generally acknowledged.

1. INTRODUCTION

1.1. Biological Nitrogen Fixation

Despite the fact that 78% of the earth's atmosphere is composed of nitrogen, it is often a limiting factor in plant growth. This is because atmospheric nitrogen exists as dinitrogen (N_2), a form of the element that is inaccessible to all but a handful of specially adapted prokaryotic organisms including some eubacteria, cyanobacteria, and actinomycetes. For nitrogen to be incorporated into living organisms it must first be converted to ammonia or nitrate, a process that cannot be carried out by higher plants and animals in the absence of endosymbiotic bacteria.

Biological nitrogen fixation is the process by which dinitrogen is reduced to ammonia by a specialized group of prokaryotic organisms called diazotrophs. The nitrogen fixation reaction is catalyzed by the nitrogenase enzyme.

$$N_2 + 8e^- + 8H^+ + 16ATP \rightarrow 2NH_3 + H_2 + 16ADP + 16\ P_i$$

This reaction is energetically demanding, requiring 16 molecules of ATP for each molecule of N_2 reduced. Additional energy is required to elevate the electrons for nitrogenase to a sufficiently large redox potential, so the actual energy requirement for this reaction is in the range of 25 molecules of ATP per molecule of N_2 fixed. In terms of overall cellular energy the energetic requirements are exacerbated by the fact that the ATP used for nitrogen fixation would otherwise have been available for other energy-requiring processes. Microbial nitrogen

S.S. Gnanamanickam (ed.), Plant-Associated Bacteria, 105–130.

fixation rates are highest in carbon rich environments and many diazotrophs are frequently concentrated in the rooting zone of plants.

Environmental restrictions on nitrogen fixation are largely governed by the physiological properties of the enzyme nitrogenase. Nitrogenase is a large enzyme that requires a considerable investment on the part of the organism, up to 30% of total cell protein (Haaker and Klugkist, 1987). This also contributes to the need for available carbon for efficient diazotrophy.

Nitrogen fixation is complicated by the complex nature of nitrogenase that simultaneously requires and is inactivated by oxygen. Oxygen is required for the process of oxidative phosphorylation that is responsible for the production of ATP needed to fuel nitrogen fixation. Both protein components of the nitrogenase enzyme are inactivated in the presence of atmospheric oxygen levels, with half lives of 4.5 seconds to 60 minutes (Robeson and Postgate, 1980). In order for nitrogen fixation to proceed at optimal efficiency it is necessary that very low levels of oxygen be maintained in the vicinity of nitrogenase at all times.

Diazotrophic organisms have developed a wide variety of strategies for overcoming the sensitivity of nitrogenase as well as the energy intensive nature of nitrogen fixation. This chapter will focus on nitrogen-fixing organisms that associate with plants as a strategy for addressing their physiological demands.

1.2. Nitrogen-Fixing Organisms

Diazotrophic organisms are loosely classified into three subgroups: symbiotic, free-living, and associative. The distinctions between these groups, particularly the free-living and associative diazotrophs, are not clearly delineated and several organisms could fall under the heading of more than one group. A symbiosis is defined as the living together in intimate association of two organisms, the collaboration being mutually beneficial (Paul and Clark, 1989). The most well-studied diazotrophs are those that have formed a tight symbiotic relationship with host plants from the legume family. These organisms are valued for their importance for agricultural fertility. This chapter will focus explicitly on another group of diazotrophs that live in close association with plants but do not inhabit specialized growth structures on their host plants.

The distinction between associative and symbiotic diazotrophs is a question of the degree of association. Associative nitrogen fixation is commonly defined as nitrogen fixation by a free-living diazotroph under the direct influence of a host. Associative nitrogen-fixers engage in a protocooperative relationship with a plant host in which some exchange of nutrients takes place but no differentiated root structures are formed (Elmerich *et al.*, 1992; You and Zhou, 1989). Plants are provided with some excess nitrogen and the microbes can utilize plant derived carbon compounds to fuel the nitrogen fixation reaction. In an associative relationship both partners benefit but the relationship is more casual than cooperative (Klucas, 1991).

Associative nitrogen fixation constitutes an ecological niche midway between symbiotic and asymbiotic diazotrophy. The organisms involved represent a continuum from free-living within the vicinity of the plant to endogenous colonization of plant tissue. In order for fixation to be termed associative the following requirements must be met: 1) a plant host; 2) a diazotroph; 3) an adequate supply of substrate provided by the plant; 4) an appropriate microenvironment for nitrogenase activity; and 5) a net transfer of fixed nitrogen from the diazotroph to the host (Klucas, 1991). Important bacterial genera that enter into associative relationships with plant hosts include *Azospirillum*, *Burkholderia*, *Enterobacter*, *Gluconoacetobacter*, *Herbaspirillum*, and *Klebsiella*.

1.3. The Rhizosphere: A Solution to the Carbon and Oxygen Problems

The soil under the immediate influence of a plant's root system is called the rhizosphere (Döbereiner and Pedrosa, 1987). The rhizosphere is composed of soil particles, microorganisms, polymeric secretions, water-soluble exudates, lysates, and gases. Root effects are mainly due to the exudation of carbon substrate, creating a gradient that decreases with distance from the root (Rovira, 1969). Other factors that vary within the rhizosphere include pH and concentrations of CO_2 and O_2. The outer edge is determined by the diffusion limits of water-soluble, plant-derived compounds.

The availability of carbon exudates in the rhizosphere leads to the selective enrichment of diazotrophs, a phenomenon called the rhizosphere effect (Rovira, 1965, 1969; Kuzyakov 2002). Enrichment of diazotrophs in rhizosphere soil as compared to non-rhizosphere soil is termed the R/S ratio and is generally calculated to be between 1 and 20 (Balandreau *et al.*, 1978). This preferential enrichment of diazotrophs is influenced by the following selective factors: competitive advantage in carbon-rich, nitrogen-poor soil, chemotactic attraction to root exudates, and aerotactic attraction to decreased pO_2 in the root zone. Reduced oxygen supply in the root zone has been shown to enhance nitrogenase activity in rhizosphere organisms (Döbereiner *et al.*, 1972), and the highest levels of nitrogenase activity are achieved in soils with high water content, as this restricts oxygen diffusion.

The presence of microorganisms in the rhizosphere can enhance carbon exudation (Barber and Martin, 1977; Barber and Lynch, 1976; Kraffcyzk *et al.*, 1984; Heulin et al., 1987) and, in the case of *Azospirillum*, root permeability (Venkateswarlu and Rao, 1985). Other factors that influence carbon exudation include plant species, age, temperature, light intensity, nutrition, soil moisture, and root damage (Rovira, 1969). Studies indicate that from 14-25% of carbon assimilated by wheat and barley is released into the soils (Martin, 1977), and values as high as 80-90% have been obtained for some species (Lynch, 1990). The amount of exudate is regulated by the rate of photosynthesis and, in several studies, nitrogenase activity of rhizosphere diazotrophs has been shown to be tightly correlated to the photosynthetic activity of the host plant (Dommergues *et al.*,

1973; Day *et al.*, 1975). The correlation of nitrogenase activity and photosynthate flux indicates that carbon exudates are a major regulatory factor in diazotrophic activity in the rhizosphere.

Certain diazotrophs are capable of endogenous colonization of plant tissue, called the histosphere or endorhizosphere (Watanabe *et al.*, 1979; Patriquin and Döbereiner, 1978). Endophytic diazotrophs tend to colonize the intercellular spaces of the inner cortex, stele, and lumen of xylem vessels without disrupting the outermost cortex of the endodermis. Initial entry into the roots probably occurs through disrupted cortical tissues where lateral branches emerge from the main roots (Döbereiner and Pedrosa, 1987). The infection then proceeds via active penetration of the middle lamella and through lytic holes in the cell wall (Patriquin *et al.*, 1983).

Endophytic bacteria such as *Gluconoacetobacter*, *Azoarcus*, and *Alcaligenes* exhibit high densities in or surrounding the xylem in grasses, rice and sugarcane (Hurek *et al.*, 1994; You and Zhou, 1989; Oliveira et al., 2002). Vascular tissue provides an ideal environment for colonization because it has a relatively low pO_2 (Crawford, 1976) and high concentrations of photosynthate (Butz and Long, 1979). Association within the vascular elements (both xylem and phloem) facilitates nutrient exchange between the diazotroph and the host plant (Patriquin *et al.*, 1983). Additionally, the xylem can provide a means of bacterial transport via the transpiration stream, which allows for the longitudinal and lateral spread of non-pathogenic diazotrophs without repeated ingress via the rhizoplane (root surface) (James *et al.*, 1994; Hurek *et al.*, 1994). Once the diazotrophic infection is established the bacteria can be transmitted via an insect vector, vesicular-arbuscular mycorrhizae (Varma *et al.*, 1981), or normal vegetative reproduction of the host plant (Vose, 1983).

Originally it was believed that associative nitrogen fixers were confined to intercellular spaces, with no observed colonization of healthy plant cells (Döbereiner and Pedrosa, 1987). In 1989 You and Zhou reported the first evidence of intracellular colonization by *Alcaligenes faecalis* in rice seedlings. Further support for the role of intracellular colonization in associative nitrogen fixation was provided by Hurek *et al.* (1994) who reported that *Azoarcus* is able to colonize roots of gnotobiotically grown Kallar grass both intra- and intercellularly.

It is estimated that endorhizosphere contributions to plant nitrogen are more substantial than rhizosphere contributions due to the lack of competition from other rhizosphere organisms and the increased availability of carbon substrate. Endogenous bacteria also have the benefit of relatively small fluctuations in pO_2 when compared to rhizosphere fluctuations caused by changes in soil moisture (Boddey and Döbereiner, 1995).

There are several other ways in which diazotrophic associations can enhance plant productivity such as the role of bacterial nitrate reductase in enhanced nitrate assimilation by the host plant (Boddey *et al.*, 1986) and the softening of the cell walls which can facilitate germination of spores of vesicular-arbuscular mycorrhizae (VAM) fungi, stimulating hyphal extension and enhancing subsequent colonization (Will and Sylvia, 1990). VAM fungi have numerous functions in plant growth the most important of which is enhancing nutrient uptake. Many rhizosphere bacteria are also able to solubilize phosphorous making it more readily available to plants.

1.4. Seasonal and Diurnal Variations in Nitrogenase Activity

Nitrogen fixation by associative diazotrophs appears to be highly regulated by factors such as host plant life cycle, temperature, moisture, and availability of photosynthate. In some plants, microbial nitrogenase activity is either absent or very low during early vegetative growth and reaches a maximum at some stage during reproductive development, a period when physiological N demands are highest but when root growth has slowed or stopped (Evans and Wardlaw, 1976). This tends to vary from system to system and is not a general rule.

Seasonal variation is well documented and appears to be similar for all associative systems studied. In general, the highest rates of fixation are obtained in moist soil during the warm summer months (Döbereiner and Day, 1975). Nitrogenase activity becomes insignificant when night temperatures drop below 15°C for a prolonged period. This effect could be due to insufficient photosynthesis and lack of growth during the winter or the result of decreased photosynthesis due to chloroplast damage (Döbereiner and Day, 1975). As a result of these general requirements for optimal nitrogenase activity, the tropics are much more likely to benefit from associative nitrogen fixation than temperate zones, although surprisingly little research has been conducted in temperate zones (Nelson *et al.*, 1975; Dalton *et al.*, 2004). The tropics are more conducive to associative nitrogen fixation because of optimal temperatures, high light intensity resulting in an increased supply of photosynthate, and efficient use of the C_4-dicarboxylic acid photosynthetic pathway by most grasses (Döbereiner *et al.*, 1972).

Diurnal variations are also frequently observed in associative fixation systems and can in fact be used to determine the extent of the association. Generally, it is observed that nitrogenase activity is substantially reduced several hours after cessation of photosynthesis (Nelson *et al.*, 1975). *Azotobacter paspali* actually shows two peaks in nitrogenase activity, one at midday and one during the night (Döbereiner and Day, 1975). These diurnal variations could be due to either the availability of carbohydrate or changes in stomatal conductance, probably a combination of both factors is involved (Vose, 1983).

1.5. Agricultural Significance

For many years the main goal of nitrogen fixation research has been the manipulation of biological diazotrophy to benefit agricultural systems. Numerous strategies have been suggested but in terms of associative nitrogen fixation the two main approaches are enhancement of natural diazotrophic associations with crop species and inoculation with non-native diazotrophs in an attempt to induce novel associations. The importance of associative nitrogen fixation to agriculture remains controversial, due in part to methodological limitations. The benefits of associative nitrogen fixation to agriculture are largely confined to the tropics where the environment is more conducive to optimal levels of nitrogenase activity. Additionally associative nitrogen fixation is only beneficial to plants in nitrogen-stressed environments, otherwise there can be decreases in crop yield due to substrate diversion (Klucas, 1991).

Diazotrophs, both endogenous and rhizospheric, have been found in association with most tropical crop species. Certainly the most well-studied system is that of sugar cane, a crop species that has an integral system of associated bacteria including: *Azotobacter*, *Beijerinckia*, *Clostridium*, *Azospirillum*, *Klebsiella*, *Bacillus*, *Derxia*, *Pseudomonas*, *Vibrio*, and. most importantly, *Gluconoacetobacter* (Cavalcante and Döbereiner, 1988; Döbereiner, 1961; Ruschel, 1981). Bacteria associated with the root systems of sorghum and millet include: *Klebsiella*, *Azospirillum*, *Bacillus*, *Derxia*, and *Erwinia* (Dart and Wani, 1982). Additionally, 29 bacterial genera have been isolated from stems and roots of field-grown maize of which the following were found to exhibit diazotrophy: *Bacillus*, *Enterobacter*, *Erwinia*, *Klebsiella*, *Herbaspirillum*, and *Xanthobacter* (Young, 1992; McInroy and Kloepper, 1995; Baldini *et al.*, 1986, 1992). Corn tissue is estimated to contain 10^4 bacterial cells per gram fresh weight (Triplett, 1996). Collectively these data indicate that many natural diazotrophic associations are present in crop species and further research in this area could provide the knowledge base necessary for enhancement of diazotrophy.

Many attempts have been made to inoculate crop species with non-native diazotrophs in an effort to establish new associations. The most frequently used organism for inoculation studies has been *Azospirillum*. Results of inoculation trials have been variable and will be discussed in more detail in a later section. Successful inoculation results are difficult to interpret as they are complicated by the contribution of factors other than nitrogen fixation, which promote increased growth and yield. The major growth-promoting factor in plant-microbe interactions is bacterial production of substances, which resemble plant hormones (Döbereiner and Pedrosa, 1987). This phenomenon will also be covered in detail in later sections.

2. METHODOLOGY

2.1. Overview

Several different methods have been used verify the presence of nitrogen-fixing bacteria associated with plants. It is important to understand the significance as well as the limitations of each of these various methods to avoid overly optimistic or misleading interpretations. The most compelling studies often involve a synthesis of methodological approaches that provide independent evidence of diazotrophy. While it may be relatively straightforward to find evidence of nitrogen accretion in a system, it is considerably more challenging to demonstrate that 1) the "new" nitrogen has come from fixation; 2) the fixing organism is properly identified; and 3) the amounts of nitrogen fixed are ecologically relevant.

2.2. Nitrogen budget studies

Carefully crafted experiments in which the total nitrogen content of all plant components is monitored over time can produce strong evidence for the presence of biological nitrogen fixation (BNF). The technique is most often applied in a large scale, ecological setting, but it can also be useful for smaller, greenhouse experiments. Quantification of total nitrogen regardless of its chemical form is routinely achieved with the Kjeldahl procedure that requires only modest equipment and expertise. While straightforward in concept, it is difficult to control and measure all the diverse forms of transfer and flow of components within the nitrogen budget. Inputs that may be particularly difficult to measure include uptake from surface flow and wet and dry N deposition. Losses may occur from removal of plant or animal material, denitrification (loss of NO_x and N_2), volatilization of NH_3, soil erosion, and leaching.

App *et al.* (1984) provided a particularly useful demonstration of the N balance method for evaluation of BNF by rice under field conditions. After accounting for inputs and outputs, the accretion of N was determined to be as much as 103 kg $N \cdot ha^{-1} \cdot yr^{-1}$, an amount comparable to that observed in a typical legume crop. A later study (App *et al.*, 1986) extended these observations to rice grown in pots in a greenhouse. Nitrogen accretion varied substantially depending on the variety of rice and was also observed for two aquatic weeds. The rates observed were estimated to be equivalent to up to 70 kg $N \cdot ha^{-1}$, thus roughly supporting the conclusions from the earlier field study. Similar results were obtained with long-term experiments with sugar cane (Oliveira *et al.*, 1994) where the total N gain in the soil and plant was estimated to be 38 kg $N \cdot ha^{-1} \cdot yr^{-1}$.

The N balance method can also be used in greenhouse experiments with plants in pots. This has the advantage of increased control of inputs and outputs so that subsequent results are more reliable. Iniguez *et al.* (2004) used such an approach with wheat plants to show that inoculation with nitrogen-fixing *Klebsiella* led to a greater than 300% increase in total N concentration in roots and shoots compared to uninoculated controls.

2.3. Acetylene reduction assay

The study of nitrogen fixation was revolutionized by the discovery of Schollhorn and Burris (1967) that nitrogenase reduces acetylene to ethylene as shown in the following reaction:

$$C_2H_2 + 2e^- + 2H^+ \rightarrow C_2H_4$$

Acetylene can be conveniently generated by reacting calcium carbide with water. Ethylene can be easily detected with a gas chromatograph equipped with a flame ionization detector. The technique is extremely sensitive and adaptable to a range of applications both in the field and the laboratory. This technique has been used successfully to detect nitrogen fixation in hundreds of biological systems. It remains as one of the most valuable techniques in this regard, but it does have some substantial limitations.

One drawback arises from the difficulty of converting the rates of acetylene reduction to rates of dinitrogen reduction. The equation for nitrogen fixation shown below suggests a theoretical ratio of 4:1 based on the number of electrons involved

$$N_2 + 8e^- + 8H^+ \rightarrow 2NH_3 + H_2$$

(i.e. it takes 4 times as many electrons to reduce one mole of N_2 as it does to reduce one mole of acetylene.) However, many factors, both physiological and ecological, can severely influence the actual ratio. Values have been reported to range from as high as 20:1 to as low as 0.3:1 (Liengen, 1999, Nohrstedt, 1983). Consequently, it is generally unreliable to try to make such a conversion unless the ratio is confirmed experimentally with ^{15}N. The technique is even less reliable in predicting nitrogen accretion over time (e.g. kg N fixed . $ha^{-1}.yr^{-1}$) since the additional factors of spatial, diurnal, and seasonal variation in rates of acetylene reduction introduce so much uncertainty that any calculations are essentially worthless.

Another issue of concern is the long lag period before acetylene reduction fully develops. A common practice in dealing with grass-bacteria symbioses is to seal the plant and/or soil material in an airtight chamber for 8-24 hours before the addition of acetylene (Bergersen, 1980; Giller, 1987). This has been generously been interpreted as allowing the bacteria to re-adjust to disturbances (especially in oxygen tension) resulting from digging or agitation as the samples are inserted into a chamber. However, it is equally likely that such long "pre-incubation" allows for bacteria to benefit from fermentation of roots or soil carbon and thus to proliferate to artificially high numbers. A related complication is how to best regulate the oxygen level during incubation since rates of nitrogen fixation are highly sensitive to oxygen, with the highest activity usually observed at O_2 levels < 0.05 atm. Fig. 1 shows typical data in which a pre-incubation without acetylene and low oxygen tensions were used to generate substantial activity.

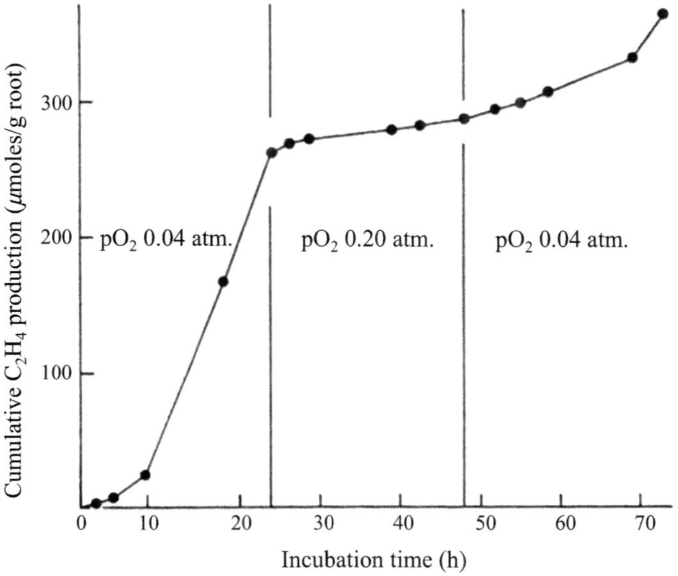

Fig. 1. Effect of varying pO_2 on acetylene reduction activity by roots of *Paspalum notatum*. The roots were pre-incubated under N_2 for 6 h without acetylene at pO_2 0.04 atm. From Döbereiner *et al.*, 1972.

Traces of "endogenous" ethylene may be produced in the soil or by plant material. Acetylene has variously been reported to block the production of this ethylene or to inhibit the oxidation of it. (Giller 1987, Giller and Merckx, 2003). These concerns are less troublesome than those described earlier because they can be minimized by use of proper controls, but such precautions are frequently ignored.

In conclusion, the acetylene reduction technique is highly useful due to its simplicity and flexibility. It is generally reliable as a qualitative or relative indicator of nitrogen fixation, but is of dubious merit for quantitative determinations.

2.4. ^{15}N studies

^{15}N is a stable isotope of nitrogen that provides an extremely reliable marker for nitrogen fixation. There are at least four different approaches to using this isotope in such studies: direct use of $^{15}N_2$, isotope dilution, natural abundance and ^{15}N-aided N balances. The simplest method is to expose the plant material to an atmosphere enriched with $^{15}N_2$ gas and to use a mass spectrometer to check for incorporation of labeled nitrogen into plant-derived compounds. This has worked well to demonstrate N_2 fixation in sugar cane (Ruschel *et al.*, 1975), rice (Eskew *et al.*, 1981), and sorghum and millet (Giller *et al.*, 1988.) Care must be taken to insure that all traces of contaminating $^{15}NH_3$ are removed from the $^{15}N_2$ gas source. Other drawbacks include the high cost of $^{15}N_2$ gas and the technical challenges of growing plants in a sealed atmosphere, but there are many successful applications of this technique.

The next two $^{15}N_2$-based techniques are based on the principle that atmospheric N_2 has an isotopic signature that distinguishes it from other forms of N that are available to the plant. These techniques have been reviewed in detail by Boddey (1987). Although the principle of these techniques is sound there are many potential complications, mostly having to do with incomplete control and understanding of ^{15}N distribution, the difficulty in selecting an appropriate control (non-fixing) plant for reference., and the fact that ^{15}N cycling in any system is controlled by many other factors besides just N_2 fixation. The natural abundance technique has been used to suggest that sugar cane in commercial plantations may obtain between 0 and 60% of its total N from plant-associated nitrogen fixation (Boddey *et al.*, 2003).

The fourth application of ^{15}N involves using ^{15}N to label the soil over long periods in order to reach a stable background level of enrichment. Any deviation of the ^{15}N content of plant N below that of the labeled soil indicates a contribution from BNF. This technique also suffers from complications relating to possibly unidentified sources of N from irrigation water or from deep soil sources, but in general the data are reliable. It has been used most successfully to demonstrate BNF in sugarcane (Boddey *et al.*, 2003).

One particular concern with using ^{15}N analysis of material from grasses with putative endophytic nitrogen-fixing bacteria is the difficulty in establishing that the isotope signature is actually a feature of a plant compound and not simply present in the contaminating bacteria that is intermixed with plant material. In other words, the bacteria may be endophytic and they may be fixing N_2, but how is one to know that this N is actually making it to the plant? This problem was solved elegantly by Iniguez *et al.* (2004) by measuring incorporation of fixed ^{15}N into chlorophyll of wheat plants that had been inoculated with N_2-fixing *Klebsiella*.

Unfortunately, there are no convenient radioisotopes of nitrogen. In theory $^{13}N_2$ could be used in radiotracer studies but this isotope has a half-life of only 10 min. and so is of no practical use for studying nitrogen fixation in grasses where the rates are likely to be low and detectable only by long-running experiments.

2.5. PCR-based techniques

The hunt for N_2 fixation in grasses has a long history based on culturing and identification of putative N_2-fixing bacteria isolated from plant material. Indeed the modern burst of research activity on this subject was instigated by the discovery of N_2-fixing bacteria such as *Azospirillum* from tropical grasses in Brazil by Döbereiner and associates in the early 1970's. This approach continues to be valuable but it is limited by the prerequisite for proper knowledge of culture media and conditions. It is seldom appreciated that the vast majority of bacterial species in many natural ecosystems can not be cultured by known techniques and thus these bacteria remain essentially unknown to microbiologists. For the skeptic trained in Koch postulates, this always leaves open the possibility that any genuine agent of BNF might avoid detection and identification. This problem has been somewhat ameliorated by the application of PCR-based techniques in which genes for nitrogenase (usually *nifH* or *nifK*) can be screened for in samples without the need for culturing. DNA is first isolated directly from soil or from plant material and then screened with the proper *nif* primers. PCR products are cloned and sequenced to confirm that these correspond to authentic *nif* genes. The sequences can then be compared with known sequences to assign a tentative taxonomic affiliation for the bacteria. Despite the great number of known *nif* sequences, those sequences obtained by this technique often do not match any known sequence, thus indicating the presence of unknown (and perhaps unculturable) bacterial taxa. The most relevant application of this technique to grass-bacteria associations was its use with rice roots to detect 23 *nif* clones, none of which was identical to any known published sequence (Ueda *et al.*, 1995). This implies that natural populations of N_2-fixating bacteria associated with rice may be much larger and diverse than previously thought and that the bacteria are largely unidentified.

It is also possible to use a wider PCR screen based on 16S rDNA primers in order to identify bacteria (nitrogen-fixing or not) associated with plant tissue. Chelis and Triplett (2001) used this approach to identify 74 phylotypes of bacteria associated with *Zea mays* roots. Although some of these bacteria resembled the proteobacterial population of a typical soil community, there was also a high number of potential diazotrophs such as *Rhizobium*-, *Burkholderia*- and *Herbaspriliium*-related phylotypes. In addition, a range of other phylotypes with surprisingly diversity was also detected, including representatives from uncultured hot spring communities and marine archaea. Thus the situation with maize is similar to that of rice (and presumably other grasses) in that the plant harbors a diverse and highly complex microbial community that is poorly understood.

3. PHYTOHORMONES AND ASSOCIATIVE BACTERIA

Once a putative N_2-fixing grass symbiosis has been identified and the bacterial symbiont cultured, a common next step has been to link the two partners together under carefully-controlled conditions in order to demonstrate a beneficial effect. Does the plant+bacteria combination do better than the plant alone? However, an answer in the affirmative does not always mean that BNF is involved because bacteria are capable of interacting with plants in many ways besides simply BNF. Of special concern is the ability of many bacteria to produce phytohormones such as indole-3-acetic acid (IAA), gibberellins, or cytokinins all of which could have positive effects on associated plants. This complication has been demonstrated particularly well with two examples: *Azoarcus*/Kallar grass and *Glucono*-

acetobacter/sugar cane. Mutants of *Azoarcus* unable to fix N_2 (*nif*$^-$) can still stimulate the growth of rice plants to an extent equal to that of the wild type *nif*$^+$ (Hurek *et al.*, 1994). Similarly, *nif*$^-$ mutants of *Gluconoacetobacter* can stimulate growth of sugarcane as long as N is not limiting (Sevilla *et al.,* 2001). Under N-limiting conditions, only the wild type can stimulate growth, an observation consistent with the central role of nitrogen fixation. The strain of *Gluconoacetabacter* used (PA15) was found by Fuentes-Ramirez *et al.*, (1993) to produce large amounts of IAA, thus providing the likely explanation for the stimulating effect of the *nif*$^-$ mutant. The production of IAA appears to be widespread in associative nitrogen-fixing bacteria and has since been confirmed in a number of other genera including *Azospirillum*, *Herbaspirillum*, and *Pseudomonas* (Radwan *et al.,* 2002; Pderaza *et al.*, 2004). In addition to hormone production, associative bacteria may also benefit hosts plants in a variety of ways including improved nutrient cycling or uptake (especially through production of siderophores for iron uptake) or by inhibition of potential pathogenic or competing microorganisms (Dobbelaere *et al.*, 2003).

4. *AZOSPIRILLUM*

Azospirillum is one of the most extensively studied genera of diazotrophs. It was first isolated in the Netherlands in 1925 (Beijerinck, 1925) but its potential significance to plant growth was not appreciated until half a century later when Döbereiner and Day (1975) noted that grasses associated with *Azospirillum* were not as nitrogen deficient as neighboring grasses that lacked *Azospirillum.* Many bacterial members of this genus are capable of fixing nitrogen and promoting plant growth through a variety of other mechanisms, stimulating extensive physiological, genetic, and ecological research (Michiels *et al.*, 1989; Holguin *et al.*, 1999; Vande Broek and Vanderleyden, 1995; Somers *et al.*, 2004). Seven species have been characterized within the genus: *A. brasilense, A. lipoferum, A. amazonense, A. halopreferens, A. irakense, and A. doebereinerae* (Tarrand *et al.*, 1978; Magalhaes *et al.*, 1984; Reinhold *et al.*, 1987; Khammas *et al.*, 1989; Sly and Stackebrandt, 1999; Eckert *et al.*, 2001).

Although the growth promoting effects of *Azospirillum* have been well documented, the exact mechanism of growth promotion goes beyond nitrogen fixation to include nitrate reduction, phytohormone production, production of undefined signal molecules that can interfere with plant metabolism and enhancement of mineral uptake by plants in response to root elongation (Okon and Itzigsohn, 1995). All of these mechanisms interact to promote plant growth and the relative contribution of each process is influenced by soil environmental parameters, plant and bacterial growth phases, and bacterial interactions (Bashan and Holguin, 1997; Okon and Labandera-Gonzalez, 1994).

Azospirillum survives as a typical aerobe in the presence of combined nitrogen. Its capacity as a diazotroph only emerges in microaerobic, N-deficient conditions.

It appears that microaerobic conditions are maintained in part by encapsulation, which provides a barrier to O_2 diffusion (Berg *et al.*, 1980). *Azospirillum* is preferentially enriched in the rhizosphere because of its ability to fix nitrogen, reduce nitrate, produce plant hormones and siderophores and its ability to anchor itself to roots with fibrillar material (Patten and Glick, 1996; Bashan and Holguin, 1997). Following a field inoculation trial, 10 to 100 times more cells of *Azospirillum* were found in rhizosphere soil as compared to surrounding bulk soil (De Connick *et al.*, 1988).

Azospirillum has been isolated from the roots of numerous wild and cultivated plants around the world, from the moist tropics to temperate regions (Döbereiner *et al.*, 1976; Bally *et al.*, 1983; Ladha *et al.*, 1987; Kirchhof *et al.*, 1997; Gunarto *et al.*, 1999). This genus is essentially ubiquitous in tropical ecosystems and is associated with the roots of 50% of crop plants and grasses studied, including sugar cane, corn, Kallar grass, and many others. It is found at densities as high as 10^5-10^7 bacteria per gram soil or root in the tropics (Döbereiner and Pedrosa, 1987), though it is also found in temperate zones at lower densities (De Connick *et al.*, 1988). *Azospirillum* has also been described as a facultative endophyte, capable of inhabiting living plant tissue (Döbereiner *et al.*, 1995). It is presumed to enter the plant through cortical wounds and cracks (Ramos *et al.*, 2002) although entrance into plant tissue may be facilitated by pectinolytic activity as well (Faure *et al.*, 1999; Faure *et al.*, 2001; Bekri *et al.*, 1999).

Its widespread distribution and ability to persist in appreciable numbers in association with a host plant make *Azospirillum* an attractive candidate for inoculation studies, and indeed many attempts have been made. Results of inoculation trials have been variable although there have been multiple reports of significant increases in crop yields of 20-60% following inoculation with *Azospirillum* (Bouton, 1979; Baldini *et al.*, 1983; Okon and Vanderleyden, 1997; Okon and Itzigsohn, 1995; Dobbelaere *et al.*, 2001). There have also been reports of no significant increase in yield following inoculation (Barber *et al.*, 1976) or even reduced yield when *Azospirillum* is applied with organic matter (Freitas and Stamford, 2002). Generally inoculations have been reported to be most successful when the plants were supplemented with intermediate levels of fertilizer (Smith *et al.*, 1976, Nath *et al.*, 2002). Most inoculation studies are carried out in the greenhouse and extrapolations to field conditions are difficult because of abnormal restrictions imposed on the plants during the trials, such as severe nitrogen deficiency (Klucas, 1991).

Root segments inoculated with *Azospirillum* show enhanced uptake of NO_3^-, P_2O_5, and K^+ (Jain and Patriquin, 1984). This is likely a combined effect of *Azospirillum's* ability to soften the middle lamella using pectinolytic enzymes thereby promoting mineral uptake (Tien *et al.*, 1981; Umali-Garcia *et al.*, 1980) and the production of root-elongating hormones like cytokinins, gibberellins, and auxins (Tien *et al.*, 1979; Bottini *et al.*, 1989; Somers *et al.*, 2004). *Azospirillum* has also been shown to enhance colonization of vesicular arbuscular mycorrhizae in chickpea and sorghum (Saini *et al.*, 2004).

5. *GLUCONOACETOBACTER* AND SUGARCANE

Of the various grass/bacteria systems suspected of N_2 fixation, the sugarcane /*Gluconoacetobacter* team has probably attracted the most attention. There are several reasons for this, the primary one being that the evidence for a positive effect due to N_2 fixation is compelling, much more so than for other grass crops such as rice, maize, or wheat. Second, sugarcane is a very valuable and profitable crop.

The original discovery of N_2 fixation in sugarcane was precipitated by the observation that high yields of this crop have been obtained for many years in Brazil without the use of any N fertilizer. Brazil is the world's leading produced of sugarcane. Not only is sugarcane an export commodity in high demand, but it is also used to produce ethanol to fuel about one half of the automobiles in Brazil (Triplett, 1996). Early work by Cavalcante and Döbereiner (1988) and Gillis *et al.*, (1989) led to the isolation of a novel diazotrophic bacteria from surface-sterilized stem and root tissue of sugarcane. This bacterium, originally called *Acetobacter* (now *Gluconoacetobacter*) *diazotrophicus*, has a number of features that make it well-suited for its role as a plant-associated diazotrophic. It can use sucrose, which is abundant in sugarcane stems, as its sole carbon source. Second, it can grow at the low pH values likely to be present in plant tissues. This bacterium is an aerobe, but is capable of N_2 fixation under microaerobic conditions. Lastly, NO_3^- and NH_4^+ show little or no inhibition of nitrogenase activity thus allowing the bacteria to remain a potentially unending source that leaks out fixed N that is not subject to any inherent down regulation.

Prior to this time, plant associated diazotrophs were assumed to be located primarily in the rhizosphere i.e., on the surface of roots or in nearby soil. In contrast, *Gluconoacetobacter* is present in large numbers (up to 10^7 cfu per g fresh weight) within the root and stem tissues. It first colonizes the epidermis of stems and roots and then becomes established in xylem vessels (James *et al.*, 1994, Fig. 2). In addition to *Gluconoacetobacter*, several other endophytic diazotrophic bacteria – including *Herbaspirillum seropedicae*, *H. rubrisubalbicans*, and *Burkholderia* sp. – have also been discovered in sugarcane and it is not clear what the relative contributions of each of there bacteria might be to BNF within the plant (Boddey *et al.*, 2003). The complexity of BNF in sugarcane is further evidenced by the range of diazotrophic bacteria associated with the surface of roots. These include species of *Beijerinckia*, *Azospirillum*, *Azotobacter*, *Bacillus*, *Derxia*, *Enterobacter*, and *Erwinia* and *Pantoea* (Boddey *et al.*, 2003; Loiret *et al.*, 2004).

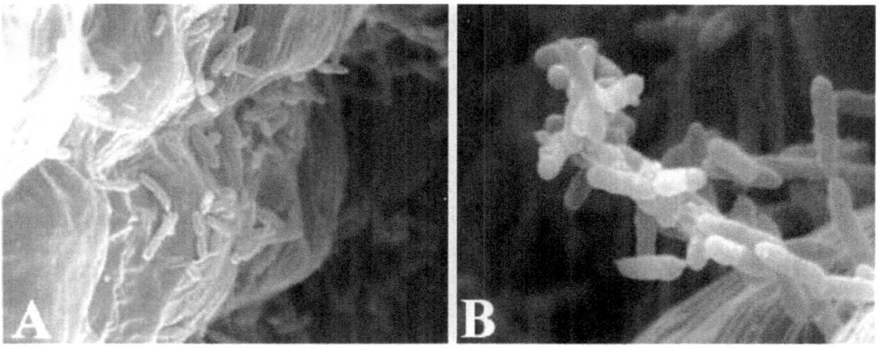

Fig. 2. A) Scanning electron micrograph of a sugarcane stem and *Gluconoacetobacter diazotrophicus* colonization. B) SEM of *A. diazotrophicus* cells held together by a mucilage-type material inside the sugarcane tissue. Photos provided by Christina Kennedy, University of Arizona.

Numerous studies have shown that sugarcane obtains substantial N from fixation, although it is not always clear which bacteria are involved. For instance, a survey of 10 different varieties of sugarcane showed that between 35 to 79% of the total N needs of the plant came from fixation in field situations relying entirely upon indigenous populations of bacteria (Boddey and Döbereiner, 1995). In the best performing varieties, N accumulated in the soil at over 230 kg N ha^{-1} $year^{-1}$ even though all plant material was removed at harvest and no N fertilizer was used at all during the 3-year duration of the experiment. These are impressive rates that would seldom be matched even by a prolific legume crop. In carefully controlled greenhouse with axenic sugarcane plants, inoculation with *Gluconoacetobacter* led to significant increases in weight and in N content and in weight of both shoots and roots (Sevilla *et al.*, 2001). The inoculated plants showed uptake of $^{15}N_2$ after 24 hours incubation, whereas uninoculated or Nif^--inoculated plants did not.

Nitrogenase contains molybdenum and the supply of this element in soils is sometimes insufficient to fully support BNF. In such cases, fertilization with Mo might be expected to enhance crop performance by facilitating BNF. This was demonstrated convincingly by field studies in Brazil in which soils were treated with 4 levels of Mo (0, 100, 200, and 400 g ha^{-1}, Boddey *et al.*, 2003). The yield of sugarcane responded proportionally to the amount of Mo added. The highest treatment (400 g Mo ha^{-1}) actually yielded more fresh cane than did the treatment receiving 120 kg N ha^{-1}, but no Mo. This ability of Mo to stimulate yields is an independent confirmation that BNF is important in this crop.

6. *AZOARCUS* AND KALLAR GRASS

Another endophytic association that has received considerable attention is the association between *Azoarcus*, a member of the β-subclass of proteobacteria, and

Kallar grass (*Leptochloa fusca*), a C_4 grass that is pervasive in nutrient-deficient, waterlogged, or alkaline soils in Pakistan (Reinhold *et al.*, 1986; Reinhold-Hurek and Hurek, 1993; Reinhold-Hurek *et al.*, 1993b). Kallar grass also contains a variety of well-known rhizoplane-based diazotrophic bacteria, including *Klebsiella*, *Beijerinckia*, and *Azospirillum* but the discovery of *Azoarcus* in 1993 was distinctive because the bacteria were endophytic and the first representatives of a new genus (Reinhold-Hurek and Hurek, 1997). Other species of *Azoarcus* have since been isolated from various (often polluted) soils (Hurek and Reinhold-Hurek, 1995). The ability of these species to degrade aromatic hydrocarbons anaerobically with nitrate as an electron acceptor has generated interest in their potential for bioremediation.

Azoarcus is able to colonize both inter- and intracellularly and populations are preferentially localized in stellar tissue (Hurek *et al.*, 1994). *Azoarcus* is also able to colonize rice where colonization has been observed to be concentrated in the lower parts of the roots, especially the zone of elongation and differentiation just above the root tips (Reinhold-Hurek and Hurek, 1998). The plant-associated species of *Azoarcus* are unusual in that they are unable to utilize carbohydrates; however, they do possess enzymes that degrade cellulose, an ability, which may facilitate the infection process by localized digestion of plant cell walls (Reinhold-Hurek *et al.*, 1993).

The oxygen metabolism of endophytic diazotrophic bacteria is critical because of the rapid inactivation of nitrogenase by oxygen. *Azoarcus* is an aerobe that fixes nitrogen only under low (μM) concentrations of oxygen. If the oxygen concentration should decline to the nM range, the cells of *Azoarcus* enter a "hyperinduced" state in which a complex of intracytoplasmic membranes is formed (Hurek *et al.*, 1995). Immunohistochemical studies show that the iron protein of nitrogenase is highly enriched on these membranes. It is likely that these "diazosomes" concentrate components of the respiratory electron transport to facilitate nitrogen fixation.

The ability of *Azoarcus* to contribute fixed nitrogen to the host plant has been examined in careful greenhouse experiments (Hurek *et al.*, 2002). In nitrogen-deficient conditions, inoculated plants grew better and accumulated more nitrogen than did plants inoculated with a *nifK*$^-$ mutant. The inability of researchers to re-isolate *Azoarcus* from infected plants has long been an issue that raises uncertainty about the significance of the relationship. However, in this study abundant PCR-amplified *nif* transcripts were retrieved from test (but not control plants) even though it was not possible to reisolate the bacteria from infected plants. Abundant *Azoarcus nifH* transcripts were detected also in uninoculated plants taken from the natural environment, from which *Azoarcus* also could not be isolated. The implications of this are significant since the results indicate that bacteria in an unculturable state can contribute combined nitrogen to the plant. The extent to which this applies in the other systems is unknown, but it is likely a major, complicating factor that has contributed to the uncertainties surrounding our understanding of endophytic nitrogen fixation in general.

7. *BURKHOLDERIA*

Burkholderia is one of the more intriguing genera of diazotrophic plant-associated bacteria. There are at least 19 recognized species of *Burkholderia* most of which are normally present in soil and rhizosphere (Estrada-de Los Santos *et al.*, 2001). Some species are pathogens of plants and are also opportunistic pathogens of humans. Agronomic interest in this genus began to emerge in 1995 with its detection as a diazotroph associated with rice roots and has since expanded considerably (Gillis *et al.*, 1995). It appears to be widely distributed in a number of plants including maize (*Zea*), coffee (*Coffea*), pineapple (*Ananas*), sorghum, dune grasses (e.g. *Ammophila arenaria*), and almost certainly many others that have not yet been described. The type genus, *B. cepacia*, is especially common and is recognized for its ability to promote growth and suppress pathogens of maize and other crops (Estrada-De Los Santos *et al.*, 2001).

The extent of distribution of *Burkholderia* was recently verified by an extensive survey in which 51 nitrogen-fixing strains representing 15 different 16s rDNA genotypes were isolated from coffee and maize plants from throughout Mexico (Estrada-De Los Santos *et al.*, 2001). There is nothing particularly unique about Mexico in this regard as *Burkholderia* has been found from many parts of the world, often but not necessarily from tropical latitudes. In the study by Estrada-De Los Santos, *Burkholderia* was readily recovered from the rhizospheres and root surfaces of maize and coffee plants, as well as from within the tissue (endophytically) of maize. The occurrence of *Burkholderia* in maize raises some interesting questions with respect to the potential contribution of BNF towards maize yield, particularly in remote parts of Mexico such as in the state of Oaxaca where native peoples have grown maize for centuries with high yields and no N input (Estrada *et al.*, 2002; Fig. 3A and B). Although it is clear that *Burkholderia* is a common endophyte in maize (at least in Mexico), the evidence for any substantial contribution to the N needs of the plant is largely anecdotal. This situation will likely remain ambiguous until comprehensive N budget and ^{15}N studies are applied to this system. *Burkholderia* has also been found in teosinte, the wild ancestor of maize that still grows wild in Oaxaca. This has led to the suggestion that *Burkholderia* may have established an early symbiosis with teosinte that has persisted in modern varieties of maize in the region (Estrada *et al.*, 2002). Even if true, this proposed symbiosis is probably not present in areas of intensive maize cultivation in developed nations because of the heavy use of chemical fertilizers and the inadvertent selection of varieties that lack the ability. Maize responds very well to massive fertilization with N and this remains the standard practice in most areas.

Burkholderia also holds the distinction of being able to form nitrogen-fixing nodules on at least some legumes (Moulin *et al.*, 2001). This is surprising considering that *Burkholderia* is a member of the β-subclass of Proteobacteria, whereas traditional rhizobia (*Rhizobium, Bradyrhizobium* and close relatives) are members of the α-subclass of Proteobacteria. This ability to nodulate legumes is made possible by the presence in some isolates of *Burkholderia* of the *nod* genes that code for Nod factors that act as molecular signals involved in host-bacteria recognition. To date only a few tropical legumes (such as *Macroptilium atropurpureum*) have been shown to nodulate with *Burkholderia*.

Fig. 3A. Typical landscape in the Mexican state of Oaxaca where primitive land races of maize are grown on subsistence farms throughout remote mountainous regions. Most of the fields visible in this view are maize. B. Oaxacan maize plants in excess of 4 meters high, grown without addition of N fertilizer.

Burkholderia also appears to be a factor in BNF in natural ecosystems but here the situation is even less well understood than is the case in crops. One of the more studied examples involves sand dune grasses such as European beach grass (*Ammophila arenaria*) and American dune grass (*Elymus mollis*) that harbor endophytic, diazotrophic *Burkholderia* (Dalton *et al.*, 2004). These grasses (especially *A. arenaria*) are very prolific on bare sand on the Pacific Northwest coast of the United States. The rhizomes and other plant tissues harbor a large and complex bacterial community of up to 10^8 cfu g^{-1} FW. These bacteria have been cultured and identified by 16S rRNA gene sequence. In addition to *Burkholderia*, other genera identified included *Allorhizobium*, *Enterobacter, Hydrocarbophaga, Ochrobactrum, Pantoea, Pseudomonas, Sphingomonas,* and *Stenotrophomonas*. Shoots and rhizomes reduced acetylene at rates comparable to those reported for tropical grasses such as *Digitaria* and *Paspalum*. The *Burkholderia* isolate (but none of the other isolates) reduced acetylene in culture under microaerobic conditions and contained *nifH* and *nifD* genes. Immunolocalization with fluorescent antibodies to nitrogenase indicated that the bacteria were present within the cell walls of rhizomes and shoots (Fig. 4). It is likely that similar symbioses remain to be discovered in other grasses, particularly those growing on nutrient-poor sand.

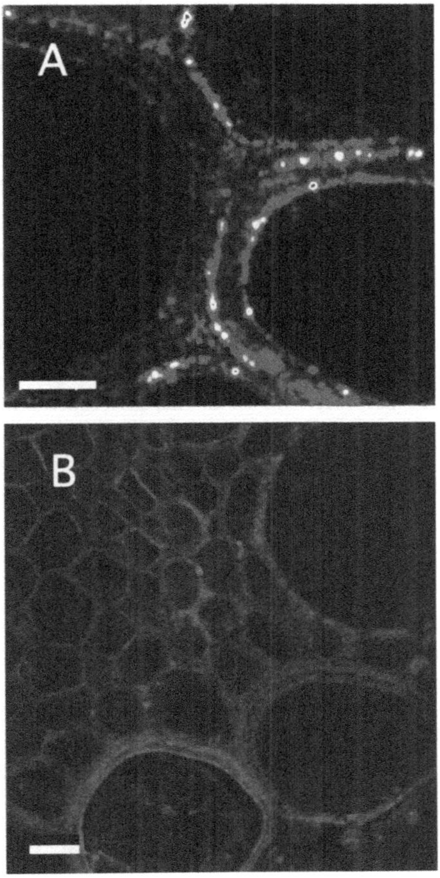

Fig. 4. A. Fluorescence immunolocalization of nitrogenase in the cell walls of rhizomes of *Elymus mollis*. Bar = 5 µm. B. Negative control with rabbit normal serum. Bar = 10 µm. From Dalton *et al.*, 2004.

8. CONCLUSIONS

Plants do not exist in a sterile environment. Plants share their environment as well as their own tissues with a host of microorganisms, often in prolific numbers of staggering diversity. Some of these microorganisms appear to benefit the plant in a number of ways, including especially nitrogen fixation. The important genera of bacteria in this regard include *Azoarcus, Azospirillum, Burkholderia, Gluconoacetobacter, Klebsiella*, and *Herbaspirillum*, though there are probably many others that have not been cultured and are perhaps unculturable by known methods.

The field of associative and endophytic nitrogen fixation is both contentious and elusive because the evidence for significant contribution of fixed nitrogen to the host plant is difficult to obtain. With respect to plant-bacteria interactions, there is a gradient from symbiotic to associative to commensal to coincidental with boundaries that are difficult to establish. At one extreme is the legume-rhizobia symbiosis in which the benefits of the symbiosis have been obvious to researchers for over 100 years. However, this is a poor model for the less rigorous associations between grasses and various diazotrophic bacteria. In the latter case, the processes are diffused throughout the plant body and not localized into concentrated packages of metabolic industry (nodules). Furthermore, the exchange of materials (especially carbon and nitrogen) may be of a smaller scale that is difficult to quantify. Although the amounts of nitrogen accrued in a system may be small (< 5 kg N ha^{-1} yr^{-1}) compared to legumes, they could be significant in the long term, especially in natural ecosystems with nitrogen-deficient conditions. For crop plants, in which high yields and short growth cycles are the norm, associative and endophytic nitrogen fixation has not yet been shown to have a significant impact. The major exception to this is the *Gluconoacetobacter*/sugar cane system for which a preponderance of evidence indicates that substantial nitrogen fixation occurs. The potential for other crops remains as an area of active research worldwide.

9. REFERENCES

App, A., Watanabe, I., Ventura, T., Bravo, M., & Jurey, C.D. (1986). The effect of cultivated and wild rice varieties on the nitrogen balance of flooded soil. *Soil. Sci.* , 141, 448-452.

App, A., Santiago, T., Daez, C., Menguito, C., Ventura, W., Tirol, A., Po, J., Watanabe, I., De Datta, S.K. & Roger, P. (1984). Estimation of the nitrogen balance for irrigated rice and the contribution of phototrophic nitrogen fixation. *Field Crops Res.,* 9, 17-27.

Bally, R.D. Thomas-Bauzon, Heulin, T. & Balandreau, J. (1983). Determination of the frequent N2-fixing bacteria in a rice rhizosphere. *Can. J. Microbiol.*, 29, 881-887. Barber, D.A. & Martin, J.K. (1976). The release of organic substances by cereal roots into soil. *New Phytol.*, 16, 69-80.

Barber, D.A., & Lynch, J.M. (1977). Microbial growth in the rhizosphere. *Soil Biol. Biochem.*, 9, 305-308.

Barber, L.E., Tjepkema. J.D., Russell, S.A., & Evans, H.J. (1976). Acetylene reduction associated with corn inoculated with *Spirillum. Appl. Environ. Microbiol.,* 32, 108-113. Balandreau, J. (1986). Ecological factors and adaptive processes in N_2-fixing bacterial populations of the plant environment. *Plant & Soil,* 90, 73-92.

Balandreau, J., Ducerf, P., Hamad-Fares, P., Weinhard, G., Rinaudo, G., Miller, C., & Dommergues, Y. (1978). Limiting factors in grass nitrogen fixation. pp. 275-302. In: *Limitations and Potentials for Biological Nitrogen Fixation in the Tropics*, Döbereiner, J., Burris, R.H., & Hollaender, A., (Eds.), New York and London: Plenum Press.

Baldini, J.I., Baldini, V.L.D., Seldin, L., & Döbereiner, J. (1986). Characterization of *Herbaspirillum seropedicae* gen. nov. sp. nov. a root-associated, nitrogen-fixing bacterium. *Internat. J. Systematic Bacteriol.,* 36, 86-93. Baldini, V.L.D., Baldini, J.I., & Döbereiner, J. (1983). Effects of *Azospirillum* inoculation on root infection and nitrogen incorporation in wheat. *Can. J. Microbiol.,* 29, 924-929.

Baldini, V.L.D., Baldini, J.I., Olivares, F., & Döbereiner, J. (1992). Identification and ecology of *Herbaspirillum seropedicae* and the closely related *Pseudomonas rubrisubalbicans*. *Symbiosis,* 13, 65-73.

Bashan, Y. & Holguin, G. (1997). *Azospirillum*/plant relationships: environmental and physiological advances (1990-1996). *Can. J. Microbiol.*, 43, 103-121. Beijerinck, M.W. (1925). Uber ein *Spirillum*, welches freien Stickstoff bindin kann? *Zentralb l Bakteriol Parasitenkd Infektionskr Abt,* 63, 353-359.

Bekri, M.A., Desair, J., Keijers, V., Proost, P., Searle-van Leeuwen, M., Vanderleyden, J. & Vande Broek, A. (1999). *Azospirillum irakense* produces a novel type of pectate lyase. *J. Bacteriol.*, 181, 2440-2447.

Berg, R.H., Tyler, M.E., Novick, N.J., Vasil, V., & Vasil, I.K. (1980). Biology of *Azospirillum*-sugarcane association: enhancement of nitrogenase activity. *Appl. Environ, Microbiol.*, 39, 642-649.

Boddey, R.M. (1987). Methods for quantification of nitrogen fixation associated with gramineae. *Crit. Rev. Plant Sci.*, 6, 209-266.

Boddey, R.M. & J. Döbereiner. (1995). Nitrogen fixation associated with grasses and cereals: recent progress and perspectives for the future. *Fert. Res.*, 42, 241-250.

Boddey, R.M., Urquiaga, S.B., Alves, J.R., & Reis, V. (2003). Endophytic nitrogen fixation in sugarcane: present knowledge and future applications. *Plant & Soil,* 252, 139-149.

Boddey, R.M., Chalk, P.M., Victoria, R.L., Matsui, E., & Döbereiner, J. (1983). The use of the ^{15}N isotope dilution technique to estimate the contribution of associated nitrogen fixation nutrition of *Paspalum notatum* cv. *batatais*. *Can. J. Microbiol.*, 29, 1036-1045.
Boddey, R.M., Baldini, V.L.D., Baldini, J.I., & Döbereiner, J. (1986). Effect of inoculation of *Azospirillum* spp. on the nitrogen assimilation of field grown wheat. *Plant & Soil,* 95, 109-121.

Bottini, R., Fulchieri, M., Pearce, D., & Pharis, R.P. (1989). Identification of gibberellins A1, A3, and iso-A3 in cultures of *Azospirillum lipoferum*. *Plant Physiol.*, 90, 45-47.

Bouton, J.H., Smith, R.L., Schank, S.C., Burton, G.W., Tyler, M.E., Littell, R.C., Gallaher, R.N., & Queensberry, K.H. (1979). Responses of pearl millet inbreds and hybrids to inoculation with *Azospirillum brasilense*. *Crop Science*, 19, 12-16.

Butz, R.G., & Long, R.C. (1979). L-malate as an essential component of the xylem fluid of corn seedling roots. *Plant Physiol.*, 64, 684-689.

Cavalcante, V.A., & Döbereiner, J. (1988). A new acid-tolerant nitrogen-fixing bacterium associated with sugarcane. *Plant Soil,* 108, 23-31.

Chelius, M. K., & Triplett, E.W. (2001). The diversity of archaea and bacteria in association with the roots of *Zea mays L. Microb. Ecol.*, 41, 252-263.

Crawford, R.M.M. (1976). Mineral nutrition. pp. 197-253. In: *Plant, Function, Structure, and Adaptation*, Hall, M.A. (Ed.). London: MacMillan.

Dalton, D.A., Kramer, S.B., Azios, N., Fusaro, S., Cahill, E. & Kennedy, C. (2004). Endophytic nitrogen fixation in dune grasses (*Ammophila arenaria* and *Elymus mollis*) from Oregon. *FEMS Microbiol. Ecol.*, 49, 469-479.

Dart, P.J., & Wani, S.P. (1982). Non-symbiotic nitrogen fixation and soil fertility. pp.3-27. In: *Proceedings of the 12th International Congress on Soil Science*, Symposium papers. Vol 1. New Delhi: International Society of Soil Science.

Day, J.M., Neves, M.C.P., & Döbereiner, J. (1975). Nitrogenase activity on the roots of tropical forage grasses. *Soil Biol. Biochem.*, 7, 107-112.

DeConnick, K., Horemans, S., Randombage, S., & Vlassak, K. (1988). Occurrence and survival of *Azospirillum* spp. in temperate regions. *Plant & Soil,* 110, 213-218.

Döbereiner, J. (1961). Nitrogen-fixing bacteria of the genus *Beijerinckia* in the rhizosphere of sugarcane. *Plant & Soil,* 15, 211-216.

Döbereiner, J., & Day, J.M. (1975). Nitrogen fixation in the rhizosphere of tropical grasses. pp. 39-55. In: *Nitrogen Fixation by Free-living Microorganisms*, Stewart, W.D.P., (Ed.), Cambridge: Cambridge University Press.

Döbereiner, J., & Pedrosa, F.O. (1987). *Nitrogen-Fixing Bacteria in Nonleguminous Crop Plants*. Madison, Wisconsin: Science-Tech Publishers.

Döbereiner, J., Day, J.M., & Dart, P.J. (1972). Nitrogenase activity and oxygen sensitivity of the *Paspalum notatum-Azotobacter paspali* association. *J. Gen. Microbiol.,* 71, 103-116.

Döbereiner, J., Marriel, I.E., & Nery, M. (1976). Ecological distribution of *Spirillum lipoferum* Beijerinck. *Can. J. Microbiol.,* 22, 1464-1474.

Döbereiner, J., Baldini, V.L.D., & Reis, V.M. (1995). Endophytic occurrence of diazotrophic bacteria in non-leguminous crops in *Azospirillum VI and Related Microorganisms: Genetics, Physiology, Ecology*, Fendrick, I., Del Gallo, M., Vanderleyden, J. & de Zamoroczy, M. (Eds.). Berlin Heidelberg.

Dobbelaere, S., Croonenborghs, A., Thys, A., Ptacek, D., Vanderleyden, J., Dutto, P., Labandera-Gonzalez, C., Caballero-Mellado, J., Aguirre, J.F., Kapulnik, Y., Brener, S., Burdman, S., Kadouri, D., Sarig, S. & Okon, Y. (2001). Responses of agronomically important crops to inoculation with *Azospirillum. Austr. J. Plant Physiol.,* 28, 871-879.

Dobbelaere, S., Vanderleyden, J. & Okon, Y. (2003). Plant growth-promoting effects of diazotrophs in the rhizosphere. *Crit. Rev. Plant Sci.* 22, 107-149. Dommergues, J., Balandreau, J., Rinaudo, G., & Weindard, P. (1973). Non-symbiotic nitrogen fixation in the rhizospheres of rice, maize, and different tropical grasses. *Soil Biol. Biochem.,* 5, 83-89.

Eckert, B., Weber, O.B., Kirchhof, G., Halbritter, A., Hartmann, A., & Hartmann, A. (2001). *Azospirillum doebereinerae* sp. nov., a nitrogen-fixing bacterium associated with the C4-grass *Miscanthus. Int. J. Syst. Evol. Microbiol.,* 51, 17-26.

Elmerich, C., Zimmer, W., & Vieille, C. (1992). Associative nitrogen-fixing bacteria. pp. 212-245. In: *Biological Nitrogen Fixation*, Stacey, G., Burris, R.H., & Evans, H. J. (Eds.). New York: Chapman and Hall.

Eskew, D.L., Eaglesham, A.R.J. & App, A.A. (1981). Heterotrophic $^{15}N_2$ fixation and distribution of newly fixed nitrogen in a rice-flooded soil system. *Plant Physiol.,* 68, 48-52.

Estrada-De Los Santos, P., Bustillos-Cristales, R., & Caballero-Mellado, J. (2001). *Burkholderia*, a genus rich in plant-associated nitrogen fixers with wide environmental and geographic distribution. *Appl. Environ. Microbiol.,* 67, 2790-2798.

Estrada, P., Mavingui, P., Cournoyer, B., Fontaine, F., Balandreau, J. & Caballero-Mellado, J. (2002). A N_2-fixing endophytic *Burkholderia* sp. associated with maize plants cultivated in Mexico. *Can. J. Microbiol.,* 48, 285-294.

Evans, L.T. & Wardlaw, I.F. (1976). Aspects of the comparative physiology of grain yields in cereals. *Adv. Agron.,* 28, 301-359.

Faure, D., Desair, J., Keijers, V., Bekri, M.A., Proost, P., Henrissat, B., & Vanderleyden, J. (1999). Growth of *Azospirillum irakense* KBC1 on the *aryl* beta glucoside salicin requires either *salA* or *salB. J. Bacteriol.,* 181, 3003-3009.

Faure, D., Henrissat, B., Ptacek, D., Bekri, M.A., Vanderleyden, J. (2001). The celA gene, encoding a glycosyl hydrolase family 3 beta-glucosidase in *Azospirillum irakense*, is required for optimal growth on cellobiosides. *Appl. Environ. Microbiol.,* 67, 2380-2383.

Freitas, A.D.S. & Stamford, N.P. (2002). Associative nitrogen fixation and growth of maize in a Brazilian rainforest soil as affected by *Azospirillum* and organic materials. *Tropical Grasslands,* 36, 77-82.

Fuentes-Ramirez, L.E., Jiminez-Salgado, T., Abarca-Ocampo, I.R., & Caballero-Mellado, J. (1993). *Acetobacter diazotrophicus*, an indoleacetic acid producing bacterium isolated from sugarcane cultivars of Mexico. *Plant & Soil,* 154, 145-150.

Giller, K.E. (1987). Use and abuse of the acetylene reduction assay for measurement of "associative" nitrogen fixation. *Soil Biol. Biochem.,* 19, 783-784.

Giller, K.E., Wani, S.P., Day, J.M., & Dart, P.J. (1988). Short term measurements of uptake of nitrogen fixed in the rhizospheres of sorghum and millet using $^{15}N_2$. *Biol. Fert. Soil.,* 7, 11-15.

Gillis, M., Kersters, K., Hoste, B., Janssens, D., Kroppenstedt, R.M., Stephan, M.P., Teizeira, K.R.S., Döbereiner, J. & De Lay, J. (1989) *Acetobacter diazotrophicus* sp. nov., a nitrogen-fixing acetic acid bacteria associated with sugarcane. *Int. J. Syst. Bacteriol.*, 39, 361-364.

Gillis, M., Tran Van, V., Bardin, R., Goor, M., Hebbar, P., Willems, A., Segers, P., Kersters, K., Heulin, T., & Fernandez, M.P. (1995). Polyphasic taxonomy in the genus *Burkholderia* leading to an emended description of the genus and proposition of *Burkholderia vietnamiensis* sp. nov. for N_2-fixing isolates from rice in Vietnam. *Internat. J. Syst. Bacteriol.*, 45, 274-289.

Gunarto, L., Adachi, K., & Senboku, T. (1999). Isolation and selection of indigenous *Azospirillum* spp. From a subtropical island, and effect of inoculation on growth of lowland rice under several levels of N application. *Biol Fertil. Soils,* 28, 129-136.

Haaker, H., & Klugkist, J. (1987). The bioenergetics of electron transfer to nitrogenase. *FEMS Microbiol Rev.* 46, 57-71.

Heulin, T., Guckert, A. & Balandreau, J. (1987). Stimulation of root exudation of rice seedlings by Azospirillum strains: Carbon budget under gnotobiotic conditions. *Biol. Fert. Soils,* 4, 9-14.

Hill, S. (1978). Factors influencing the efficiency of nitrogen fixation in free-living bacteria. pp. 130-136. In: *Environmental Role of Nitrogen-Fixing Blue-Green Algae and Asymbiotic Bacteria,* Granhall, U. (Ed.). Stockholm: Swedish Natural Science Research Council.

Holguin, G., Patten, C.L. & Glick, B.R. (1999). Genetics and molecular biology of *Azospirillum. Biol Fertil Soils,* 29, 10-23.

Hurek, T., Reinhold-Hurek, B., Van Montagu, M., & Kellenberger, E. (1994). Root colonization and systemic spreading of *Azoarcus* sp. strain BH72 in grasses. J. Bact. 176, 1913-1923.

Hurek, T., & Reinhold-Hurek, B. (1995). Identification of grass-associated and toluene-degrading diazotrophs, *Azoarcus* spp., by analyses of partial 16S ribosomal DNA sequences. *Appl. Environ. Microbiol.* , 61, 2257-2261.

Hurke, T., Van Montagu, M., Kellenberger, E., & Reinhold-Hurek, B. (1995). Induction of complex intracytoplasmic membranes related to nitrogen fixation in *Azoarcus* sp. BH72. *Mol. Microbiol.,* 18, 225-236.

Hurek, T., Handley, L.L., Reinhold-Hurek, B., & Piche, Y. (2002). *Azoarcus* grass endophytes contribute fixed nitrogen to the plant in an unculturable state. *Mol. Plant-Microbe Interactions,* 15, 233-242.

Iniguez, A. L., Doong, Y., & Triplett, E.W. (2004). Nitrogen fixation in wheat provided by *Klebsiella pneumoniae* 342. *Mol. Plant-Microb Interact.,* 17, 1078-1085.Jain, D.K. & Patriquin, D.G. (1984). Root hair deformation, bacterial attachment, and plant growth in wheat-*Azospirillum* associations. *Appl. Environ. Microbiol.,* 48, 1208-1213.

James, E. K., Reis, V.M., Olivares, F.L., Baldani, J.I., & Döbereiner, J. (1994). Infection of sugar cane by the nitrogen-fixing bacterium *Acetobacter diazotrophicus. J. Exp. Bot.* , 45, 757-766.

Khammas, K.M., Ageron, E., Grimont, P.A., & Kaiser, P. (1989). *Azospirillum irakense* sp. nov., a nitrogen-fixing bacterium associated with rice roots and rhizospheric soil. *Res. Microbiol.,* 140, 679-694.

Kirchhof, G., Reis, V.M., Baldini, J.I., Eckert, B., Döbereiner, J., & Hartmann, A. (1997). Occurrence, physiological and molecular analysis of endophytic diazotrophic bacteria in graminaceous energy plants. *Plant & Soil,* 194, 45-55.

Klucas, R.V. (1991) Associative nitrogen fixation in plants. pp. 187-198. In: *Biology and Biochemistry of Nitrogen Fixation,* Dilworth, M.J. & Glenn, A.R. (Eds.). New York: Elsevier.

Kraffczyk, I., Trolldenier, G., & Beringer, H. (1984). Soluble root exudates of maize: influence of potassium supply and rhizospheric organisms. *Soil Biol. Biochem.,* 16, 315-322.

Kuzyakov, Y. (2002). Factors affecting rhizosphere priming effects. *J. Plant Nutr. Soil Sci.,* 165, 382-396.

Ladha, J.K., So, R.B., Wattanabe, I. (1987). Composition of *Azospirillum* species associated with wetland rice plants grown in different soils. *Plant & Soil,* 102, 127-129.

Liengen. T. (1999). Conversion factor between acetylene reduction and nitrogen fixation in free-living cyanobacteria from high arctic habitats. *Can. J. Microbiol.,* 45, 223-229.

Loiret, F. G., Ortega, E., Kleiner, D., Ortega-Rodes, P., Rodes, R., & Dong, Z. (2004). A putative new endophytic nitrogen-fixing bacterium *Pantoea* sp. from sugarcane. *J. Appl. Microbiol.,* 97, 504-511.

Lynch, J.M. (1990). *The Rhizosphere*. Chichester, England: John Wiley and Sons.

Magalhaes, F.M., Baldini, J.I., Souto, S.M., & Döbereiner, J. (1984). A new-acid tolerant *Azospirillum* species. *An. Acad. Brasil. Cienc.,* 55, 417.

Martin, J.K. (1977). Factors affecting the loss of organic carbon from wheat roots. *Soil Biol. Biochem.,* 9, 1-8.

McInroy, J.A., & Kloepper, J.W. (1995). Survey of indigenous bacterial endophytes from cotton and sweet corn. *Plant & Soil,* 173, 337-342.

Michiels, K., Vanderleyden, J., & Van Gool, A. (1989). *Azospirillum*-plant root associations: A review. *Biol Fertil Soils,* 8, 356-368.

Moulin, L., Munive, A., Dreyfus, B., & Boivin-Masson, C. (2001). Nodulation of legumes by members of the β-subclass of proteobacteria. *Nature*, 411, 948-950.

Nath, D.J., Ahmed, A.A., Patgiri, S.R., & Bhattacharjee, R.N. (2002). Associative influence of *Azospirillum lipferum* and *Azospirillum brasilense* inoculation on rice (*Oryza sativa*) crop. *Adv. Plant Sci.,* 15, 171-175.

Nelson, A.D., Barber, L.E., Tjepkema, J., Russell, S.A., Powelson, R., Evans, H.J., & Seidler, R.J. (1975). Nitrogen fixation associated with grasses in Oregon. *Can. J. Microbiol.,* 22, 523-530.

Nohrstedt, H. (1983). Conversion factor between acetylene reduction and nitrogen fixation in soil: effect of water content and nitrogenase activity. *Soil Biol. Biochem.,* 15, 275-279.

Okon, Y., & Labandera-Gonzalez, C.A. (1994). Agronomic applications of *Azospirillum*: an evaluation of 20 years worldwide field inoculation. *Soil Biol. Biochemg.,* 26, 1591-1601.

Okon, Y., & Itzigsohn, R. (1995). The development of *Azospirillum* as a commercial inoculant for improving crop yields. *Biotechnol. Adv.,* 13, 365-374.

Okon, Y., & Vanderleyden, J. (1997). Root-associated *Azospirillum* species can stimulate plants. *ASM News,* 63, 366.

Oliviera, A.L.M., Urquiaga, S., & Boddey, R.M. (1994). Burning cane: the long term effects. *Int. Sug. J.,* 96, 272-275.

Oliviera, A.L.M., Urquiaga, S., Döbereiner, J., & Baldini, J.I. (2002). The effect of inoculating endophytic N2-fixing bacteria on micropropagated sugarcane plants. *Plant & Soil,* 242, 205-215.

Patten, C., & Glick, B.R. (1996). Bacterial biosynthesis of indole-3-acetic acid. *Can. J. Microbiol.,* 42, 207-220.

Patriquin, D.G., & Döbereiner, J. (1978). Light microscopy observations of tetrazolium-reducing bacteria in the endorhizosphere of maize and other grasses in Brazil. *Can. J. Microbiol.,* 24, 734-742.

Patriquin, D.G., Döbereiner, J., and Jain, D.K. 1983. Sites and processes of association between diazotrophs and grasses. *Can. J. Microbiol.,* 29, 900-915.

Paul, E.A., & Clark, F.E. (1989). *Soil Microbiology and Biochemistry*. San Diego: Academic Press.

Pedraza, R.O., Ramirez, A., Xiqui, M.L., & Baca, B.E. (2004). Aromatic amino acid amino transferase activity and indole-3-acetic acid production by associative nitrogen-fixing bacteria. *FEMS Microbiol. Lett.,* 233, 15-21.

Radwan. T., Mohamed, Z.K., & Reis, V.M. (2002). Production of indole-3-acetic acid by different strains of *Azospirillum* and *Herbaspirillum* spp. *Symbiosis,* 32, 39-53.

Ramos, H.J., Roncato-Maccari, L.D., Souza, E.M., Soares-Ramos, J.R., Hungria, M. & Pedrosa, F.O. (2002). Monitoring *Azospirillum*-wheat interactions using *gfp* and *gusA* genes constitutively expressed from a new broad range host vector. *J. Biotechnol.* , 97, 243-252.

Reinhold, B., Hurek, T., Niemann, E.G., & Fendrik, I. (1986). Close association of *Azospirillum* and diazotrophic rods with different root zones of Kallar grass. *Appl. Environ. Microbiol.,* 52, 520-526.

Reinhold, B., Hurek, T., Fendrik, I., Pot, B., Gillis, M., Kesters, K., Thielemans, S., & De Ley, J. (1987). *Azospirillum halopraeferens* sp. nov., a nitrogen-fixing organism associated with roots of Kaller grass (*Leptochloa fusca* (L.) Kunth). *Int. J. Syst. Bacteriol.,* 37, 43-51.

Reinhold-Hurek, B., Hurek, T., Claeyssens, M., &Van Montagu, M. (1993a.). Cloning, expression in *Escherichia coli*, and characterization of cellulolytic enzymes of *Azoarcus* sp., a root-invading diazotroph. *J. Bacteriol.*, 175, 7056-7065.

Reinhold-Hurek, B., Hurek, T., Gillis, M., Hoste, B., Vancanneyt, M., Kersters, K., & de Ley, J. (1993b.). *Azoarcus* gen. nov. Nitrogen-fixing proteobacteria associated with roots of Kaller grass (*Leptochloa fusca* (L.) Kunth), and description of two species, *Azoarcus indignes* sp. nov. and *Azoarcus communis* sp. nov. *Internat. J. Systematic Bacteriol.*, 43, 574-584.

Reinhold-Hurek, B., & Hurek, T. (1997). *Azoarcus* spp. and their interactions with grass roots. *Plant & Soil,* 194, 57-64.

Reinhold-Hurek, B. & Hurek, T. (1998). Interactions of graminaceous plants with *Azoarcus* spp. and other diazotrophs: identification, localization, and perspectives to study their function. *Crit. Rev. Plant Sci.,* 17, 29-54.

Robson, R.L., & Postgate, J.R. (1980). Oxygen and hydrogen in biological nitrogen fixation. *Ann. Rev. Microbiol.*, 34, 183-207.

Rovira, A.D. (1965). Interaction between plant roots and soil microorganisms. *Ann. Rev. Microbiol.,* 19, 241-266.

Rovira, A.D. (1969). Diffusion of carbon compounds away from wheat roots. *Austr. J. Biol. Sci.,* 22, 1287-1290.

Ruschel, A.P., Henis, Y., & Salai, E. (1975). Nitrogen-15 tracing of N-fixation with soil-grown sugar cane seedlings. *Soil. Biol. Biochem.,* 7, 181-182.

Ruschel, A.P. (1981). Associative nitrogen-fixation by sugarcane. pp. 81-90. In: *Associative N_2-Fixation*, Vol. 2, Vose, P.B., & Ruschel, A.P. (Eds.). Boca Raton, Florida: CRC Press.

Saini, V.K., Bhandari, S.C., & Tarafdar, J.C. (2004). Comparison of crop yield, soil microbial C, N, and P, N-fixation, nodulation and mycorrhizal infection in inoculated and non-inoculated sorghum and chickpea crops. *Field Crops Res.,* 89, 39-47.

Schollhorn, R., & Burris, R.H. (1967). Acetylene as a competitive inhibitor of N2 fixation. Proc. Natl. Acad. Sci. USA, 58, 213.

Sevilla, M., De Oliveira, A., Baldani. I., & Kennedy, C. (1998). Contributions of the bacterial endophyte *Acetobacter diazotrophicus* to sugarcane nutrition: a preliminary study. *Symbiosis,* 25, 181-191.

Sevilla, M., Burris, R.H., Gunapala, N. & Kennedy, C. (2001). Comparison of benefit to sugarcane plant growth and $^{15}N_2$ incorporation following inoculation of sterile plants with *Acetobacter diazotrophicus* wild-type and *nif* mutant strains. *Mol. Plant-Microbe. Interact.,* 14, 358-366.

Sly, L.I., & Stackebrandt, E. (1999) Description of *Skermanella parooensis* gen. nov., sp. nov. to accommodate *Conglomeromonas largomobilis* subsp. *parooensis* following the transfer of *Conglomeromonas largomobilis* subsp. *largomobilis* to the genus *Azospirillum. Int. J. Syst. Bacteriol.,* 49, 541-544.

Smith, R.L., Bouton, J.H., Schank, S.C., Queensberry, K.H., Tyler, M.E., Milam, J.R., Gaskins, M.H., & Littell, R.C. (1976). Nitrogen fixation in grasses inoculated with *Spirillum lipoferum. Science,* 193, 1003-1005.

Somers, E., Vanderleyden, J. & M. Srinivasan. (2004). Rhizosphere bacterial signaling: A love parade beneath our feet. *Crit. Rev. Microbio.*, 30, 205-240.

Tarrand, J.J., Döbereiner, J. & Döbereiner, J. (1978). A taxonomic study of the *Spirillum lipoferum* group with descriptions of a new genus, *Azospirillum* gen. nov. and two species *Azospirillum lipoferum* (Beijerinck) comb. Nov. and *Azospirillum brasilense* sp. nov. *Can. J. Microbiol.,* 24, 967-974.

Tien, T.M., Gaskins, M.H., & Hubbell, D.H. (1979.) Plant growth substances produced by *Azospirillum brasilense* and their effect on the growth of pearl millet (*Pennisetum americanum* L.) *Appl. Environ. Microbiol.,* 37, 1016-1024.

Tien, T.M., Diem, H.G., Gaskins, M.H., & Hubbell, D.H. (1981). Polygalacturonic acid transeliminase production by *Azospirillum* species. *Can. J. Microbiol.,* 27, 426-431.

Triplett, E.W. (1996) Diazotrophic endophytes: progress and prospects for nitrogen fixation in monocots. *Plant & Soil,* 186, 29-38.

Ueda, T., Suga, Y., Yahiro, N., & Matsuguchi, T. (1995). Remarkable N_2-fixing bacterial diversity detected in rice roots by molecular evolutionary analysis of *nifH* gene sequences. *J. Bacteriol.,* 177, 1414-1417.

Umali-Garcia, M., Hubbell, D.H., Gaskins, M.H., & Dazze, F.H. (1980). Association of *Azospirillum* with grass roots. *Appl. Environ. Microbiol.,* 39, 219-226.

Urquiaga, S., Cruz, K.H.S., & Boddey, R.M. (1992). Contribution of nitrogen fixation to sugar cane nitrogen-15 and nitrogen balance estimates. *Soil Sci. Soc. Amer. J.,* 56, 105-114.

Vande Broek, A., & Vanderleyden, J. (1995). Review: Genetics of the *Azospirillum*-plant root association. *Crit. Rev. Plant Sciences,* 14, 445- 466.

Varma, A.K., Singh, K., & Lall, V.K. (1981). Lumen bacteria from endomycorrhizal spores. *Curr. Microbiol.,* 6, 207-211.Volpon, A.G.T., De-Polli, H. & Döbereiner, J. (1981). Physiology of nitrogen fixation in *Azospirillum lipoferum* Br17 (ATCC 29709). *Arch. Microbiol.,* 128, 371-375.

Vose, P.B. (1983). Developments in non-legume N_2-fixing systems. *Can. J. Microbiol.,* 29, 837-850.

Watanabe, I., Barraquio, W.L., Guzman, M.R., & Cabrera, D.A. (1979). Nitrogen-fixing (acetylene reduction) activity and population of aerobic heterotrophic nitrogen-fixing bacteria associated with wetland rice. *Appl. Environ. Microbiol.,* 37, 813-815.

Will, M.E., & Sylvia, D.M. (1990). Interaction of rhizosphere bacteria, fertilizer, and vesicular-arbuscular mycorrhizal fungi and sea oats. *Appl. Environ. Microbiol.,* 56, 2073-2079.

You, C., & Zhou, F. (1989). Non-nodular endorhizospheric nitrogen fixation in wetland rice. *Can. J. Microbiol.,* 35, 403-408.

Young, J.P.W. (1992). Phylogenetic classification of nitrogen-fixing organisms. pp. 43-86. In: *Biological Nitrogen Fixation*, Stacey, G., Burris, R.H., and Evans, H.J. (Eds.). New York: Chapman and Hall.

10. AFFILIATIONS

David A. Dalton, *Biology Department, Reed College, Portland, OR 97202, USA.*

Sasha Kramer, *Department of Biological Sciences, Stanford University, Stanford, CA 94305-5020, USA.*

SAM S. GNANAMANICKAM AND J. EBENEZAR IMMANUEL

EPIPHYTIC BACTERIA, THEIR ECOLOGY AND FUNCTIONS

Abstract: Bacteria are a major component of the diverse epiphytic microbial flora that are harbored by leaves of plants in temperate, tropical and subtropical climates. In this review, we compile the information on the phyllosphere microbiology of plant pathogenic bacteria, ice nucleation-active bacteria, and bio-film bacteria. In particular, we examine their fitness and evolution of common epiphytic bacteria such as strains of *Pseudomonas fluorescens* and *Bacillus* which function as agents of biological disease suppression and plant growth-promotion in rice (*Oryza sativa*, L).

1. INTRODUCTION

What are epiphytic bacteria? Beattie & Lindow (1995) considered some of the definitions that have been proposed to describe them. These are bacteria that are capable of living (i.e. multiplying) on plant surfaces (Hirano & Upper, 1991; Leben, 1965). In contrast to those that are considered endophytic bacteria, the epiphytes can be removed from leaves by washing (Leben, 1965), or killed by UV irradiation or chemical surface disinfection (Henis & Bashan, 1986). All the bacteria associated with a leaf have been referred to as "phyllobacteria" (Beattie & Lindow, 1999) in parallel to the term, "rhizobacteria" that has been used to refer to all the root-associated bacteria.

2. DIVERSITY OF EPIPHYTIC BACTERIA

Kado has compiled a diverse list (*In* Plant Pathogenic Bacteria: http:// 141. 150. 157.117:8080/prokPUB/chaphtm/027/02_01.htm) of several of the epiphytic types that have associated themselves with the foliar and root surfaces of plants. These are dependent on food material shed by the plant as biproducts of growth and development. They may prefer certain types of plants and certain plant parts. On foliar surfaces many of these epiphytes are rod-shaped, Gram-negative, pigmented, and fermentative (Graham & Hodkiss, 1967; Thomas & McQuillen, 1952; Papavassiliou et al., 1967; Leben et al., 1968). The nonpigmented types often possess high GC contents in their genomic DNA. Because thymine dimerization caused by UV irradiation leads to mutations, the high GC content (low relative thymine content) may serve as a measure of protection from the lethal effects of ultraviolet radiation present in sunlight. Likewise, the pigment in other epiphytes may serve as a protective screen against this radiation. These properties have been suggested to provide selective advantage for these bacteria (Singer & Ames, 1970), but there is also the advantage of having specialized nutritional requirements that favor the growth and survival of these epiphytic bacteria over other saprophytes.

S. S. Gnanamanickam (ed.), Plant-Associated Bacteria, 131–153.

Generally, most epiphytic bacteria do not harm the plant on which they reside, but, in some cases, they can either be beneficial or detrimental. One case in point are those bacteria that promote the formation of ice crystals. Such ice nucleation sites serve as foci for ice formation on plants at temperatures well above normal freezing. Certain strains of common epiphytes such as *Pseudomonas fluorescens* possess ice-nucleating potential (Lindow, 1983). Certain strains of *Pseudomonas syringae* and *Erwinia herbicola* also show ice-nucleating traits, and *P. syringae* is also known to be pathogenic. Indeed *P. syringae* seems to remain as resident among epiphytic populations on grasses and trees (Malvick & Moore, 1988). *Xanthomonas* isolates also have been found to be active in ice nucleation (Goto et al., 1988). Besides ice-nucleation potential, certain pathogenic *Pseudomonas* species seem to associate epiphytically on their respective host plants. Examination of olive and oleander showed that these plants often harbor the olive knot pathogen *Pseudomonas syringae* pathovar *savastanoi* and other pseudomonads, which represent about 33 percent of the population. Other members of the epiphytic community include *Bacillus* (22 percent) and *Xanthomonas* (10 percent), as well as lesser numbers of *Acinetobacter*, *Erwinia*, *Serratia*, *Lactobacillus*, *Corynebacterium*, and *Flavobacterium*, and unidentified nitrogen fixers (Lavermicocca et al., 1987). A similar list of bacteria was compiled for olive leaves, with *Pseudomonas syringae* pathovar *savastanoi* and *Erwinia herbicola* being the major epiphytic organisms present (Ercolani, 1978). Such close association of the pathogen in the epiphytic state may be important for its long term survival.

3. DISTINCTIVES OF EPIPHYTIC BACTERIA

3.1. Survival of epiphytic bacteria

Beatie & Lindow (1999) used the term "phyllobacteria" to refer to all the leaf-associated bacteria regardless of their location. In a previous paper, they listed two major strategies these bacteria use for their growth and survival as foliar bacterial pathogens. These are, 1. A tolerance strategy that requires the ability to tolerate direct exposure to environmental stresses on leaf surfaces and 2. an avoidance strategy by which they seek sites that are protected from these stresses, including endophytic sites. These researchers suggested that their second strategy of avoidance that enables them to grow endophytically is likely to make them potential plant pathogens. According to them, the previously described terms "epiphytic" and "endophytic" bacteria should be viewed to refer to bacteria that are at two ends of a spectrum rather than to two distinct groups of bacteria that colonize external leaf surface and those that colonize internal sites of a leaf. Foliar pathogens can colonize both the surfaces and internal sites of a leaf (Henis & Bashan, 1986; Hirano and Upper, 1990; Leben, 1965; Mew & Vera Cruz, 1986).

3.2. Colonization strategies of epiphytic bacteria

Beattie & Lindow (1999) illustrated the complexity in the ecology of phyllobacteria when they researched beyond survival strategies to a broader perspective of leaf colonization. They suggested that phyllobacteria employed a number of strategies for colonization that included, modification of the leaf habitat, aggregation, ingression, and egression. In describing these strategies, they found strong evidence in recent literature for a density-dependent interaction among bacterial cells (Beck-VonBodman & Farrand, 1995; Greenberg, 1997; Pierson et al., 1998; Swift et al., 1994). Such density-dependent interactions and the ability of bacteria to sense the presence of neighboring cells are often made possible by quorum sensing (QS) which is mediated by the secretion of signal molecules that belong to N-acyl homoserine lactones (HSL) (Greenberg, 1997; Swift et al., 1994). These strategies contribute to enhanced cooperation among the phyllobacteria through enhanced transcription of a number of genes that include, genes for EPS, antibiotic production and virulence traits (Beck-VonBodman & Farrand, 1995; Greenberg, 1997; Pierson et al., 1998, Swift et al., 1994).

Beattie & Lindow (1999) developed a general model of leaf colonization by phyllobacteria and this model had 8 important steps. These are, 1. bacterial immigration, 2. habitat modification, 3. bacterial division, 4. microcolony formation, 5. large aggregate formation, 6. entry into internal spaces, 7. habitat modification and bacterial division, and 8. egression onto the leaf surface. They acknowledged, however, that these steps will vary with distinct bacterial species.

4. IMPORTANT CLASSES OF EPIPHYTIC BACTERIA

It is now known that epiphytic bacteria have developed fitness strategies to survive and multiply on leaf surfaces which help to resist periods of water stress, UV irradiation, fluctuating temperatures and shifts in availability of nutrients. One of the fascinating strategies is the demonstration of biofilm-like formations, harboring multiple bacterial species, as well as yeast and filamentous fungi. Biofilms were demonstrated about 7 years ago by C. Morris and her associates at INRA, France as aggregations of microbial cells attached to a leaf surface and bound in an exopolymetric matrix (Morris et al., 1997, 1998). Equally fascinating are the other major groups of bacterial epiphytes which are either causative agents of major bacterial plant diseases or frost injury to crop plants and are antagonists of such devastating plant pathogens that are beneficial for crop production.

In this paper we make detailed descriptions of four classes of epiphytic bacteria. These are,

I. epiphytic bacteria that are plant pathogens;
II. epiphytic bacteria that are biological disease control agents;
III. epiphytic bacteria that are ice-nucleation active, and
IV. epiphytic bacteria that form biofilms

I. PLANT PATHOGENIC EPIPHYTIC BACTERIA

Bacterial pathogens colonize leaf surfaces of healthy leaves of host plants. This phenomenon was first reported in 1959 by Crosse of Italy by the isolation of large numbers of *Pseudomononas syringae* pv. *morsprunorum* (incitant of bacterial canker of stone fruit trees) from healthy leaves of cherry. In 1960, English and Davis also isolated *P. syringae* in healthy peach and almond leaves and also in symptomless tissues of the plant hosts. From these observations emerged a new concept that was forwarded by Leben (1963, 1965) and it observed that phytopathogenic bacteria have a resident or epiphytic phase of growth on the healthy tissues of the host. Leben proposed that a plant pathogenic bacterium may multiply on the surfaces of a healthy plant and thus provide inoculum in the absence of disease and when suitable conditions prevail, diseases follows. Such pathogens could have a resident and pathogenic phase of growth. This hypothesis provides new insights into the epidemiology of foliar bacterial plant pathogens.

Hirano and Upper (1983) compiled a list of those bacterial plant pathogens that were known or suspected of having the potential for epiphytic growth on aerial plant surfaces. This is perhaps as good a list as it was during the time it was compiled and it included major pathogenic species of *Erwinia*, *Pseudomonas* and *Xanthomonas* that affect host plants of global (both temperate and tropic regions) importance. Out of this we have compiled a list of important epiphytic bacterial pathogens (Table 1).

I. A. Epiphytic bacteria in relation to disease induction.

At times of disease onset and development of severe disease, large population of epiphytic bacterial pathogens have been observed and have been causally linked. Detailed studies on quantitative relationships between foliar disease incidence and population frequencies of bacterial plant pathogens carried out for brown spot disease of bean (Lindeman et al., 1984; Rouse et al., 1985), halo blight of oats

Table 1. A short list of phytopathogenic epiphytic bacteria that are important to agriculture in the tropics, temperate regions or both.

I. Tropical area	
01. *Xanthomonas oryzae* pv. *oryzae*	bacterial blight of rice
02. *Xanthomonas campestriss* pv. *malvacearum*	bacterial blight of cotton
03. *Xanthomonas axonopodis* pv. *manihotis*	bacterial blight of cassava
04. *Xanthomonas campestris* pv. *vesicatoria*	bacterial spot of tomato
05. *Xanthomonas axonopodis* pv. *citri*	canker of citrus
II. Temperate area	
01. *Erwinia amylovora*	fire blight of *Pyrus, Malus*
02. *Erwinia carotovora*	soft rot of vegetables
03. *Erwinia herbicola*	frost injury to crop species
04. *Pseudomonas syringae* pv. syringae	bacterial canker of *Prunus* spp. brown spot of bean and soybean; bacterial speck of tomato; frost injury to many crop species
05. *Pseudomonas syringae* pv. *glycinea*	bacterial blight of soybean
06. *Pseudomonas syringae* pv. *phaseolicola*	halo blight of French bean
07. *Pseudomonas syringae* pv. *tomato*	bacterial speck of tomato
III. Temperate and Tropic areas	
01. *Erwinia carotovora* subsp. *carotovora*	soft rots in vegetable crops
02. *Xanthomonas campestris* pv. *vesicatori*a	leaf spot of tomato

(Hirano et al (1981) and frost-injury by ice-nucleation active bacteria (Lindow et al., 1978) do show that bacterial population size is critical for disease initiation.

Not only the size, it is known that their endophytic populations, and not the epiphytic populations, are responsible for disease induction (Beatie & Lindow, 1995). The former population increases always with increasing latter population suggesting a mechanistic basis for this relationship of the two bacterial populations. While disease induction might be initiated by a large-enough endophytic populations of a pathogen, it should also be understood that it might not be the case always. Thus, a large epiphytic population of a phytopathogenic bacterium does not ensure development of disease, although invariably, it leads to the probability for large endophytic populations. Disease induction is in turn governed by host genotype and other environmental parameters that prevail. When the bacterial population reaches the threshold size, and if there are changes in the virulence of the pathogen or susceptibility of the host genotype, disease incidence occurs. It has been generally accepted that host genotype more than the environment, determines the outcome of disease or no disease (Gnanamanickam & Patil, 1977; Patil & Gnanamanickam, 1976). This is in conformity with speculated view of Beattie & Lindow (1995) that resistant genotypes limit the development of size and spread of endophytic populations of phytopathogenic bacteria.

II. EPIPHYTIC BACTERIA AS AGENTS OF BIOLOGICAL DISEASE SUPPRESSION

II. A. Epiphytic bacteria and suppression of diseases of rice (Oryza sativa,L.): a case study.

Epiphytic bacteria present in the leaf surfaces or those introduced as foliar sprays do suppress plant pathogenic bacteria (and fungal pathogens) of global importance. *Erwinia herbicola,* the epiphytic bacterium present in the leaf surfaces of rice was known to lower the pH of the rice leaf and thus made it difficult for the bacterial pathogen (*Xanthomonas oryzae* pv. *Oryzae*) to grow (Hsieh & Buddenhagen, 1974; Santhi et al., 1987).

The experience of the authors in biocontrol research has been limited to the use of applying bacterial inoculants for the suppression of devastating rice pathogens of the foliage such as rice blast caused by *Magnaporthe grisea*, sheath blight caused by *Rhizoctonia solani* and bacterial blight caused by *Xanthomonas oryzae* pv. *Oryzae* (Gnanamanickam & Mew, 1992; Chatterjee et al., 1996; Kavitha et al., 2005; Vasudevan et al., 2002; Velusamy et al., 2004, 2005). Therefore, in this section of the paper, some of the detailed studies that have been conducted in our laboratory with rice-associated bacteria with emphasis on disease suppression and enhancement of plant growth/yield increase in rice are described as case studies. The diverse kinds of metabolites these epiphytic bacteria produce to suppress rice pathogens or cause enhanced plant growth suggest how well they have evolved in their fitness for performing these vital functions.

In Table 2, we have listed a number of rice-associated strains of *P. fluorescens* and *Bacillus* spp. that have shown good potential for suppression of four major foliar diseases of rice.

Table 2. Bacterial biocontrol agents developed for major diseases of rice. Center for Advanced Studies in Botany, University of Madras, India.

Disease	Pathogen (causal agent)	Biocontrol agent developed	Reference
Blast (Bl)	*Pyricularia grisea* (Teliomorph: *Magnaporthe grisea)*	*Pseudomonas fluorescens, Bacillus spp: B. polymyxa, B. pumulus, B. coagulans, Enterobacter agglomerans*	Gnanamanickam & Mew, 1992; Valasubramanian, 1994; Kavitha, 2002.
Sheath blight (ShB)	*Rhizoctonia solani* (Teliomorph: *Thanetophorus cucumeris)*	*P. fluorescens, P. putida, Bacillus megaterium, B. polymyxa, B. pumulus, B. coagulans, Enterobacter agglomerans*	Vasantha Devi et al., 1989; Thara, 1994; Krishnamurthy & Gnanamanickam, 1996; Kavitha, 2002.
Sheath-rot (Sh-R)	*Sarocladium oryzae*	*P. fluorescens*	Sakthivel, 1987; Sakthivel & Gnanamanickam, 1987, 1989.
Bacterial Blight (BB)	*Xanthomonas oryzae* pv. *Oryzae*	*Bacillus* spp: *B. lentus B. cereus B. circulans,* and *P. fluorescens*	Preeti Vasudevan, 2002; Velusamy & Gnanamanickam, 2003.

The key to consistent good performance by the bacteria (*Pseudomonas fluorescens, Bacillus* spp) in controlling foliar pathogens appears to be the bacterial population size that is introduced by bacterial treatments (bacterization) and maintained throughout the critical phases of crop growth. Production of broad spectrum antibiotics such as 2,4-diacetylphloroglucinol (implicated earlier in wheat "take-all decline") is now known to be involved in the biological suppression of *X. oryzae* pv. *Oryzae* (bacterial blight of rice) (Velusamy et al., 2004, 2006) (Table 3). Since this molecule can also trigger the induction of induced systemic resistance (ISR) in the host at microgram levels, bacterial population size appears to be of lesser importance. This aspect, however, remains to be fully clarified.

Table 3. Evaluation of *Pseudomonas fluorescens* PTB9 and its *Phl⁻* mutants for suppression of bacterial blight (BB) in IR24 rice. Greenhouse experiment, Chennai, southern India, July-November, 2004 (Velusamy et al., 2006).

Bacterial strain	Mean BB lesion length (cm)[1]	Difference in lesion length from control (cm)[2]	Percent BB suppression
Wild type *Pseudomonas fluorescens* PTB9	7.66**	11.27**	59.52
Phl⁻mutants			
PTB9a	15.21^{ns}	3.72^{ns}	19.65
PTB9b	15.69^{ns}	3.24^{ns}	17.11
PTB9c	14.43^{ns}	4.50^{ns}	23.77
PTB9d	14.99^{ns}	3.94^{ns}	20.81
PTB9e	15.50^{ns}	3.43^{ns}	18.11
Check	18.93	0.00	0.00
LSD 5%	*5.8*	*5.8*	*5.8*
1%	*6.6*	*6.6*	*6.6*

**Reduction in lesion length significant at 1% by LSD method of analysis; ^{ns} = not significant.
[1]Mean of three replications.
[2]Mean of normalized lesion lengths (mean lesion length in untreated control/check deducted from mean lesion length in bacteria-treated plants).

To perform the functions of biological disease suppression and growth promotion/yield increase in rice, our analyses show that these bacteria produce a spectrum of biologically active metabolites other than 2,4-DAPG. Additional data presented in Table 4 and 5 show that the biological activity of these metabolites is distinctly target-specific.

Table 4. Suppression of rice bacterial blight in IR24 rice by *Bacillus* treatments in net-house and field experiments (Vasudevan, 2002).

Bacillus strain	Tiller numbers in experiments		Percent bacterial blight (BB) suppression in experiments		Production of secondary metabolites/ growth hormone
	Net-house	Field	Net-house	Field	
B. lentus ALP18 (B1)	5.88	5.78	27.0	55.3	Aminoglycoside antibiotic, kanasomine
B. cereus NGCI15 (B2)	4.33	5.31	21.0	54.20	IAA and GA_3
B. circulans VYI 18 (B3)	5.80	5.81	53.0	52.0	-
Bacillus sp.CAL9 (B4)	4.33	5.18	33.0	52.8	-
Bacillus sp.MON-7	5.43	4.81	38.0	54.9	IAA (5.5 µg/mL)
Check	2.35	2.84	-	-	-

Each value is a mean of 120 observations.
Reduction in lesion lengths were significant at 1% by statistical analysis by LSD.
NT= Not tested because of the limited numbers of treatments

We believe that our data demonstrate how well the rice-associated bacteria when applied as bacterial inoculants to the rice foliage suppress devastating rice diseases by producing an array of biologically active metabolites. The benefits are diseases suppression, enhanced crop growth, tiller numbers and grain yields in rice (Vasudevan et al., 2002). It is tempting to suggest that there must still be another set of uncharacterized compounds and metabolites. Together, the epiphytic bacteria that are associated with rice foliage play an important beneficial role in rice production.

Table 5. Suppression of rice blast in IR50 and CO39 rice by bacterial treatments in the field, RARS, Pattambi, Kerala, southern India (Kavitha, 2002; Kavitha et al., 2005).

Bacterial strain	Diameter of zone of inhibition of *Pyricularia grisea* (cm)	Percent blast incidence	Percent Blast suppression	Production of antifungal metabolite
Bacillus pumilus (IM3)	3.0	51.03	46.22	-
Bacillus polymyxa (KRU22)	4.4	44.31	53.30	-
Bacillus polymyxa (VLB16)	5.0	47.36	50.06	37 kDa antifungal, heat-resistant protein (Kavitha et al., 2005)
Bacillus coagulans (PD7)	3.4	51.22	46.02	-
Enterobacter agglomerans (UPM18)	2.4	41.67	56.09	antifungal glucanase (9.3 units)

III. THE ICE-NUCLEATION ACTIVE EPIPHYTIC BACTERIA

In the year 1976, Arny, Lindow & Upper of University of Wisconsin showed that *Pseudomonas syringae* van Hall incited frost damage to corn (*Zea mays*). In 1978, Lindow et al., demonstrated that strains of *Erwinia herbicola* also had bacterial ice-nucleus activity and its application induced frost damage in corn. With rapid methods developed for quantification of ice-nucleation active bacteria, a large number of plant species were found to harbor INA bacteria. And these resembled either *P. syringae* or *E. herbicola* (Lindow et al., 1978). Leaf surface populations of these bacterial species limit supercooling in the plant parts in which they reside by initiating damaging ice formation at temperatures of –2 to –4°C. Plants do not have intrinsic ice nuclei active at these relatively warm temperatures, INA bacteria have a primary role in limiting supercooling and thus, inciting frost damage to plants in nature, if they were present in frost-sensitive plants at the time of frost hazard.

Several elegant reports and detailed analyses emerged from the studies of Lindow and his associates from the "Ice Lab" at the University of California, Berkeley.

Our present understanding of INA bacteria with emphasis on their potential role in precipitation was recently summarized by Morris et al. (2004) who said that the ice-nucleation active (INA) strains of bacteria as certain bacteria that are commonly found on plants and have the capacity to catalyze the freezing of supercooled water at temperatures as warm as -1°C. This capacity is conferred by a protein present in the outer membrane of the bacterial cell. Because of the abundance of these bacteria and the warm temperature at which they function as ice nuclei, they are considered to be among the most active of the naturally-occurring ice nuclei. These researchers proposed that these bacteria also play a role in atmospheric processes leading to rain, given that they are readily disseminated into the atmosphere and have been found in clouds at altitudes of several kilometers. That they participate in a sort of biological cycle of precipitation – whereby they are transported into clouds from plant canopies and incite rain thereby causing favorable conditions for their growth on plant surfaces – was proposed about 20 years ago. Today, sufficient evidence and meteorological tools have emerged to re-ignite interest in bioprecipitation and in the ways in which plants play a role as cloud seeders.

In this section of the review, we describe the work that led to uncovering the important role of bacterial ice nucleation to frost injury to plants.

Frost injury, a serious abiotic stress of crop plants, had been described as a major production constraint in many areas of the temperate zone. In the US alone annual crop losses due to frost injury were estimated at over one billion dollars (White & Haas, 1975). In the absence of information about what caused the problem, frost injury to growing crop plants was considered an avoidable result of physical stress. However, since 1976, it has become known that this problem results from an interaction between certain leaf-surface bacteria and low temperature stress. Such bacteria residing on frost-sensitive plants make them more susceptible to freezing damage by initiating the formation of ice that is required for frost injury (Arny et al., 1976; Lindow et al., 1975, 1976, 1978, 1982, 2002; Yankofsky et al., 1981). With regard to ice formation within plant tissues, the frost-sensitive plants are distinguished from frost-tolerant ones by their relative inability to tolerate such ice formation.

III. A. Bacterial ice nucleation and frost injury.

A single ice nucleus is understood to be sufficient to initiate ice formation and subsequent frost injury to an entire leaf, fruit, flower or even groups of leaves or flowers depending on the degree of restriction of ice propagation within a plant

(Single & Olien, 1967). Therefore, to avoid frost damage, frost-sensitive plants must avoid ice formation. Lindow and his associates (1978, 1982) explained the basis for frost-sensitivity of crop plants by the fact that they harbored very large epiphytic populations of ice nucleation active bacteria, which limit their supercooling ability. Under greenhouse conditions, plants were cooled to as low as –7°C for several hours and there was no apparent damage or internal ice formation due to the absence of nucleation active bacteria (Anderson et al., 1982; Arny et al., 1976; Kozloff et al., 1983; Lindow et al., 1978, 1982). In the field, however, the presence of these bacteria on plant surfaces always led to ice formation to occur on and in the plants with subsequent injury at temperatures above –5°C (Henderschott, 1962; Lindow, 1983; Yelenosky, 1991). Studies have shown that at least 95% (and perhaps all) of ice nuclei on leaf surfaces active at –5°C or above are of bacterial origin (Lindow, 1983; Lindow et al., 1978, 1982). The extent of frost damage at any given temperature increases with increasing populations of ice nucleation active bacteria (Lindow, 1983). This population is supposed to represent only about 0.1 to 10% of the total bacterial population found on leaf surfaces.

III. B. Antagonistic bacteria and biological control of frost injury.

From what was mentioned above, there is still a large bacterial population resident on plant surfaces which are non nucleation active and this can result in competition or other forms of antagonism between the nucleation active and non active epiphytic populations. Application of selected strains of bacterial antagonists are known to have reduced the populations of nucleation active bacteria and frost injury (Lindow, 1979, 1981, 1982). Mechanisms determining effective biological control of frost injury with the use of non nucleation active bacteria were not well known and have been searched subsequently through careful studies. One of them is the use of genetically engineered strains of bacteria for biological suppression of frost injury (Lindow & Panoupoulos, 1988).

III. C. Bacterial ice nucleation active plant pathogenic bacteria.

In this category, *P. syringae* has been extensively studied. Many pathogenic strains of *P. syringae* have been reported to survive in large numbers as epiphytes on a variety of symptomless host plants, including stone fruits, olive, bean and soybean (Crosse, 1959, 1963; Ercolani, 1969; Ercolani et al., 1974; Hirano et al., 1978, 1982; Leben, 1965; Leben et al., 1973). Frost injury was considered a disposing factor for infection of some of these host plants by *P. syringae* (Klement, 1974; Panagopoulos & Crosse, 1964). Panagopulos & Crosse (1964) reported that pear blossoms supercooled to –2°C and if the flowers were sprayed with a bacterial suspension after freezing infection *by P. syringae* was severe than in uninoculated, unfrozen pear flowers. Lindow (1983) suggested that with the development of new knowledge that was emerging, new and better methods can be developed to manage the harmful associations of ice nucleation active bacteria with plants. This has now become possible because of a great volume of impressive research in INA bacteria.

III. D. Molecular basis of ice nucleation.

Several research papers that appeared in late 1970s and early 1980s suggested that the ice nucleating material in bacteria such as, *Pseudomonas syringae, Erwinia herbicola and P. fluorescens* is associated with the outer cell membrane of these gram-negative bacteria (Sprang & Lindow, 1981). Soon it became clear that the determinant is an outer membrane protein (Kozloff et al., 1983). The genes for ice nucleation in strains of *P. syringae* and *E. herbicola* were cloned subsequently by Lindow and his collaborators (Orser et al., 1982, 1985) and expressed in *Escherichia coli.* These researchers maintained that ice nucleation in transformants of *E.coli* was qualitatively and quantitatively similar to what was observed in nucleation-active *P. syringae/E. herbicola.* The cloned DNA sequences conferring ice nucleation complemented ice^- mutants of *P. syringae* derived by chemical mutagenesis thus confirming that the cloned genes were determinants of ice nucleation activity (Lindow & Staskawicz, 1981).

The molecular basis of ice nucleation has been reviewed by Wolber (1992), Warren (1987), Wolber & Green (1990), Gurian-Sherman & Lindow (1992, 1993). These have contributed to our understanding of the process and it is known now that a single protein is responsible for the ice-nucleation phenotype. Single genes are known to confer bacterial ice nucleation in all ice-nucleation-active bacteria. Structurally, these genes vary slightly in length but in overall structure, they are similar and they encode a protein with a small unique amino terminus made up of a tandemly repeated 16 amino acid motif with probable functions in export of the protein to the outer membrane and in aggregation of monomers, and a carboxyl terminus which was proposed to be involved in protein folding (Wolber, 1992; Warren, 1987). Gurian-Sherman & Lindow (1993) discussed at length all the possible applications of bacterial ice nuclei including biological frost control. *Agrobacterium tumefaciens*-mediated transformation of a frost-hardy *Solanum commersonii* and expression of bacterial ice nucleation genes made the transgenic plants tolerate ice at $-3^{\circ}C$ (Baertlein et al., 1992). Ice nucleation activity could also serve as a new reporter gene system when it is fused with a gene of interest (eg. *Lux* operon) to monitor its transcriptional activity with 10,000 to 10,00,000-fold enhanced sensitivity to detect target gene activity (Lindgren et al., 1989). In spite of the known information about the molecular basis of the ice nucleation process, deducing the structure of the ice nucleation protein had theoretical constraints imposed by the function of the protein (Gurian-Sherman & Lindow, 1993).

IV. THE BIOFILM-FORMING EPIPHYTIC BACTERIA

Biofilms have been defined as frequently observed assemblages made by plant-associated bacteria that have been referred to as aggregates, microcolonies, and symplasmata (Morris & Monier, 2003). These are formed on leaves and on root surfaces and also within intercellular spaces of plant tissues. Morris & Monier (2003) reviewed how these biofilms of bacteria associated with leaves, roots and intercellular spaces influence the ecology of the bacteria they harbor and the relationship of bacteria with plants.

We now know that the first observations of microcolonies or biofilm-like structures were made in the early 1960s and early 1970s for leaves (Ruinen, 1961) and roots (Rovira & Campbell, 1974), respectively. Their occurrence and locations have been reported over the last 40 years by several researchers. However, only recently there has been an effort to relate the phenomena of bacterial aggregation on leaves to the biofilm state and its multitude of functions giving rise to novel questions about the biology and ecology of plant-associated bacteria (Morris et al., 2002) also in medical, aquatic and industrial environments.

IV. A. Biofilms: structure and properties.

Basically biofilms are assemblages of microorganisms adherent to each other and are embedded in a matrix of exopolymers. According to Wimpenny & Colasanti (1997), there are three well known models of hydrated biofilm structures. These are, (i) *the water-channel model* where biofilms have mushroom towers attached by stalks of extracellular polysaccharides (EPS) and interspersed with a net-work of water channels (like that of a primitive circulatory system); (ii) *the mosaic biofilm model*, occurs usually on water-distribution pipes, are similar to the first model but the stacks or towers of cell masses are not interconnected, and (iii) the dental plaque biofilm model. This third model of biofilms have high cell densities and arise in a high nutrient environment where they are bathed continuously in a fluid which has a limited fluid flow. Besides these, there exist biofilms that occur at non-water-saturated conditions at ranges of matric water potentials as low as –1.5 Mpa (Auerbach et al., 2000; Holden et al., 1997).

While the first three saturated models have been observed in aquatic plants and plants that are raised in hydroponic systems, the unsaturated biofilms have been observed on roots of terrestrial plants by Auerbach et al. (2000). It is known that the densities of microorganisms occurring in leaf surface biofilms is much lower than those observed in water-unsaturated systems. This is an indication, perhaps for low nutrient availability.

IV. B. Biofilms in the phyllosphere.

Methods which indirectly study the spatial distribution of bacteria on leaf surfaces which include leaf imprints or estimation of population sizes in individual leaf segments, provided the first evidence that bacteria on leaf surfaces are not distributed uniformly across the leaf. These were from observations made by Lindow and his associates (Leveau & Lindow, 2001; Kinkel et al., 2002; Lindow et al., 2002). Direct microscopic observations have revealed the presence of large complexes of aggregated microbial cells of several microns in length and in a distinct exopolymeric matrix (Carmichael et al., 1999; Fett, 2000; Gras et al., 1994; Morris et al.,1997). These biofilm-like structures have been observed on a wide range of plant species including vegetable crops, herbs and trees (Morris et al., 1997). Individual biofilms harbored numerous species of Gram-positive and Gram-negative bacteria, yeasts and filamentous fungi. Plant pathogens, however, did not constitute a measurable component of the epiphytic microorganisms in these biofilms (Riffaud et al., 2003).

Costerton et al., (1995) suggested that biofilms will have an effect on bacterial ecology/biology in ways such as (a) protection from desiccation, uv, predation and bactericides, (b) increased genetic exchange, (c) enhanced synergistic interactions, and also perhaps, (4) fostering the expression of density-dependent phenotypes through quorum sensing. These are regulated by the properties of the biofilms which in turn are influenced by the degree to which the biofilm is favourable or recalcitrant to uptake of gases, nutrients, antimicrobials, etc., and to their movement throughout the biofilm. The boundary layer of the biofilm made up of exopolymeric substances act as molecular sieves or absorbents in determining the properties of the biofilm. Although bacteria were associated in all anatomical features such as veins, hooked trichomes, grooves between epidermal cells) of bean leaves inoculated with *Pseudomonas syringae* B728a, the bacterium formed preferential aggregates at the base of the glandular trichomes (Monier, 2002). This has been observed also with other bacteria and plants such as *Arabidopsis*. Trichomes probably offer the optimal conditions for bacterial growth because of their ability to retain water droplets (Ascensao & Pais, 1998; Simon, 1997).

There are interesting ways in which bacterial biofilms impact the survival and bacterial interactions in phyllosphere. These include, gene exchange, metabolic interactions, changes in phenotypes due to cell signaling and virulence to plants. Monier (2002) examined the effect of aggregation on survival of *P. syringae* on bean leaves exposed to different levels of desiccation stress. It was observed that solitary cells located in small aggregates died due to increased desiccation, whereas cells located in larger aggregates survived. Studies made to compare the population structures of fluorescent pseudomonad and other bacteria in biofilms and other solitary components of epiphytic populations have observed no systematic differences. However, it was agreed that Gram-positive bacteria occurred more frequently in biofilms than outside of biofilms. Do plants control biofilm formation on their surfaces? The answer is yes and the clearest example in support of this came from a sudy of the marine red macroalga, *Delisea pulchra* by Kjellenberg & Steiberg (2002). In the shallow coastal waters of Australasia, this marine plant was surrounded by 10^6-10^7 bacteria ml^{-1} while it was void of any bacterial colonists on its surface. The researchers pointed out that this is due to defenses secreted in the form of furanone molecules such as acylated homoserine lactones (AHLs) involved in quorum sensing by bacteria.

IV. C. Biofilms in the rhizosphere

Bacteria generally form microcolonies or aggregates on root surfaces and they have a patchy, nonuniform distribution. Among the diverse kinds of bacteria that are involved in such colonization, saprophytic strains of fluorescent *Pseudomonas* spp are a major component. This group of bacteria have the potential as biocontrol agents of plant pathogens (Bianciotto et al., 2001; Bloemberg et al., 2000; Dandurang et al., 1997) and or as plant growth-promoting rhizobacteria (PGPR) (Fukui et al., 1994; Wiehe et al., 1996). Several studies have shown that microcolonies of these bacteria are enrobed in a distinct exopolymeric matrix (Achouaket et al., 1994; Briones et al., 2002; Schloter et al., 1993; Toledo et al., 1995). Rovira & Campbell (1974) suggested that these microcolonies are present primarily at sites of root exudation.

In terms of their impact on the biology of plants, the rhizosphere biofilms are known to contribute to the overall soil structure. In sites contaminated by mining, the microbial biofilms on roots may be involved in sequestering metals from the environment and in other aquatic environments, biofilms do play a role in adhesion of plants to rock surfaces. An elegant study carried out by Briones et al. (2002) on the influence of different rice genotypes on populations of ammonia-oxidizing bacteria in the root environment of rice revealed a reciprocal advantages of biofilm formation for both rice and the bacteria. While the ammonia-oxidizing bacteria benefited from their association with rice roots by obtaining a microniche of sufficient oxygen and nutrients in an otherwise anoxic environment, the rice in turn was assured of adequate supply of nitrogen.

Growth of bacteria and their aggregation in plant roots during disease development can be likened to biofilm formations. Vascular pathogens such as *Ralstonia solanacearum, Xyllela fastidiosa and Pantoea stewartii* provoke wilting and conditions of water stress by causing vascular dysfunctions caused, in part, by bacterial aggregates and /or EPS which in turn is regulated by quorum-sensing systems (Denny, 1999). It has been rightly suggested that biofilms are likely to interfere with recognition of pathogenic bacteria by plants and might foster coordinated expression of pathogenicity-related genes by armies of bacterial cells

(Denny, 1999) and therefore it appears that the structure and composition of biofilms are sufficiently important in plant-microbe interaction of different kinds.

5. CONCLUDING REMARKS

In this paper we have reviewed some of the remarkable research advances that have been recorded on four important groups of epiphytic bacteria. Each group has a ecology of its own and functions that are associated with their ecology. It is obvious that they have evolved to survive in their respective habitats by developing unique strategies.

We have treated those strains of *Pseudomonas fluorescens* and *Bacillus* spp. also as epiphytic bacteria which are involved in important interactions with plant pathogens of global importance. Like those ice nucleation active *Pseudomonas fluorescens* and *Erwinia herbicola* that are epiphytic on leaf surfaces, those that afford biological suppression of fungal and bacterial pathogens of crop plants are also epiphytic bacteria. This conviction emerged from our early observation in 1984-87 about the presence of *P. fluorescens* strains (biotype C) on both healthy and cankered citrus leaves. These strains were useful biocontol agents for rice sheath rot (Sakthivel & Gnanamanickam, 1987) and citrus canker (Unnamalai & Gnanamanickam, 1984). Today, there exist exciting sets of possibilities to apply them or their genetically modified/improved strains for crop disease management.

Epiphytic bacteria that incite plant diseases and those cause frost injury continue to remain as extremely important to crop production. Some of the new knowledge that has emerged in recent years should be applied to the management of these harmful interactions. The unfolding results from studies on quorum sensing among these different groups of epiphytic bacteria, including those that suppress plant disease severities and those that form biofilms, offer valuable insights about how they affect bacterial ecology/biology and plant-microbe interactions. Further studies in future should be addressed to developing practical applications to impact their beneficial interactions with plants.

6. ACKNOWLEDGEMENTS

We are greatly indebted to Dr. Cindy Morris, Station de Pathologie Vegetale, Montfavet, France and Dr. Gwyn Beattie, Iowa State University, USA for their generosity in supplying reprints and pdf. files of their published papers on epiphytic bacteria. The Director, Centre for Advanced Studies in Botany, University of Madras is thanked for the encouragement he afforded.

7. REFERENCES

Achouak, W., Heulin, T., Villemin, G. & Balandreau, J. (1994). Root colonization by symplasmata-forming *Enterobacter agglomerans*. *FEMS Microbiol. Ecol.*, *13*, 287-294.

Anderson, J. A., Buchanan, D. W., Stall, R. E. & Hall, C. B. (1982). Frost injury to tender plants increased by *Pseudomonas syringae* van Hall. *J. Am. Soc. Hortic. Sci.*, *107*, 123-125.

Arny, D. C., Lindow, S. E. & Upper, C. D. (1976). Frost sensitivity of *Zea mays* increased by application of *Pseudomonas syringae*. *Nature*, *262*, 282-284.

Ascensao, L. & Pais, M. S. (1998). The leaf capitate trichomes of *Leonotis leonurus*: histochemistry, ultrastructure and secretion. *Ann. Bot.*, *81*, 263-271.

Auerbach, I. D., Sorensen, C., Hansma, H. G. & Holden, P. A. (2000). Physical morphology and surface properties of unsaturated Pseudomoans putida biofilms. *J. Bacteriol.*, *182*, 3809-3815.

Baertlein, D. A., Lindow, S. E., Panoupoulos, N. J., Lee, S. P., Mindrinos, M. N. & Chen, T. H. (1992). Expression of a bacterial ice nucleation gene in plants. *Plant Physiol*, *100*, 1730-1736.

Beattie. G. A. & Lindow, S. E. (1995). The secret life of foliar bacterial pathogens on leaves. *Annu. Rev. Phytopathol.*, *33*, 145-172.

Beattie, G. A. & Lindow, S. E. (1999). Bacterial colonization of leaves: a spectrum of strategies. *Phytopathology*, *89*, 353-359.

Beck-VonBodman, S. & Farrand, S. K. (1995). Capsular polysaccharide biosynthesis and pathogenicity in *Erwinia stewartii* require induction by an N-acylhomoserine lactone autoinducer. *J. Bacteriol.*, *177*, 5000-5008.

Bianciotto, V., Andreotti, S., Balestini, R., Bonfonte, P. & Perotto, S. (2001). Mucoid mutants of the biocontrol strain *Pseudomonas fluorescens* CHA0 show increased ability in biofilm formation on mycorrhizal and nomycorrhizal carrot roots. *Mol. Plant-Microbe Interact.*, *14*, 255-260.

Bloemberg, G. V., Wijfjes, A. H. M., Lamers, G.E.M., Stuurman, N. & Lugtenberg, B. J. J. (2000). Simultaneous imaging of *Pseudomonas fluorescens* WCS365 populations expressing three different autofluorescent proteins in the rhizosphere: new perspectives for studying microbial communities. *Mol. Plant-Microbe Interact.*, *13*, 1170-1176.

Briones, A.M., Okabe, S., Umemiya, Y., Ramsing, N-B., Reichardt, W. & Okuyama, H. (2002). Influence of different cultivars on populations of ammonia-oxidising bacteria in the root environment of rice. *Appl. Environ. Microbiol.*, *68*, 3067-3075.

Carmichel, I., Harper, I. S., Coventry, M. J., Taylor, P.W.J. & Hickey, M. W. (1999). Bacterial colonization and biofilm development on minimally processed vegetables. *J. Appl. Microbiol. Symp. Suppl.*, *85*, 45S-51S.

Chatterjee, A., Valasubramanian, R., Vachani, A., Mau, W-L., Gnanamanickam, S. S. & Chatterjee, A. K. (1996). Biological control of rice diseases with *Pseudomonas fluorescens* 7-14: isolation of ant mutants altered in antibiotic production, cloning of ant^{+} DNA and an evaluation of a role for antibiotic production in the control of blast and sheath blight. *Biological Control*, *7*, 185-195.

Costerton, J. W., Lewandowski, Z., Caldwell, D. E., Korber, D. R. & Lappinscott, H. M. (1995). Microbial biofilms. *Annu. Rev. Microbiol.*, *49*, 711-745.

Crosse, J. E. (1959). Bacterial canker of stone fruits. IV. Investigation of a method for measuring the inoculum potential of cherry trees. *Ann. Appl. Biol.*, *47*, 306-317.

Dandurand, L. M., Schotscko, D. J. & Knudsen, G. R. (1997). Spatial patterns of rhizosphere populations of *Pseudomoans fluorescens*. *Appl. Environ.Microbiol.*, *63*, 3211-3217.

Denny, T. P. (1999). Autoregulator-dependent control of extracellular polysaccharide production in phytopathogenic bacteria. *Eur. J. Plant Pathol.*, *105*, 417-430.

English, H. & Davis, J. R. (1960). The source of inoculum for bacterial canker and blast of stone fruit trees. *Phytopathology*, *50*, 634 (Abstr.)

Ercolani, G. L. (1969). Epiphytic survival of *Pseudomonas morsprunorum* Wormald from cherry and *P. syringae* van Hall from pearon the host and on the non-host plant. *Phytopathol. Mediterr.*, *8*, 197-206.

Ercolani, G. L. (1973). Two hypotheses on the aetiology of response of plants to phytopathogenic bacteria. *J. Gen. Microbiol.*, *74*, 83-95.

Ercolani G. L. (1978). *Pseudomonas savastanoi and other bacteria colonizing the surface of olive leaves in the field. J. Gen. Microbiol.*, *109*, 245-257.

Ercolani, G. L., Hagedorn, D. J., Kelman, A. R. E. 1974. Epiphytic survival of *Pseudomoans syringae* on hairy vetch in relation to epidemiology of bacterial brown spot of bean in Wisconsin. *Phytopathology*, *64*, 1330-1339.

Fett, W. F. (2000). Naturally occurring biofilms on alfalfa and other types of sprouts. *J. Food. Protect.*, *63*, 625-632.

Fukui, R., Poinar, E. I., Bauer, P. H., Schroth, M. N., Hendson, M. (1994). Spatial colonization patterns and interaction of bacteria on inoculated sugar beet seed. *Phytopathology*, *84*, 1338-1345.

Gnanamanickam, S. S., & Patil, S. S. (1977). Phaseotoxin suppresses bacterially-induced hypersensitive reaction and phytoalexin synthesis in bean cultivars. *Physiol. Plant Pathol.*, *10*, 169-179.

Gnanamanickam, S. S. & Mew, T. W. (1992). Biological control of blast disease of rice (*Oryza sativa*, L) with antagonistic bacteria and its mediation by a *Pseudomonas* antibiotic. *Ann. Phytopathol. Soc. Japan*, *58*, 380-385.

Goto, M., Huang, B. L., Makno, T., Goto, T. & Inaba, T. (1988). A taxonomic study on ice nulceation-active bacteria isolated from gemmisphere of tea *(The sinensis L.*), phyllosphere of vegetables and flowers of *Magnolia denudata Desr. Ann. Phytopathol. Soc. Japan*, *54*, 189-197.

Graham, D. C. & Hodgkiss, W. (1967*)*. Identity of gram negative, yellow pigmented, fermentative bacteria isolated from plants and animals. *J. Appl. Bacteriol.*, *30*, 175-189.

Gras, M-H., Druet-Michaud, C. & Cerf, O. (1994). La flore bactrienne des feuilles de salade fraiche. *Sci. Aliments*, *14*, 173-188.

Greenberg, E. P. (1997). Quorum sensing in gram-negative bacteria. *Am. Soc. Microbiol News*, *63*, 371-377.

Gurian-Sherman, D. & Lindow, S. E. (1993). Bacterial ice nucleation: significance and molecular basis. *FASEB J.*, *7*, 1338-1343.

Gurian-Sherman, D. & Lindow, S. E. (1992). Ice nucleation and its application. *Curr. Opin. Biotechnol.*, *3*, 3030-306.

Henderschott, C. H. (1962). The reponse of orange trees and fruit to freezing temperatures. *Proc. Am. Soc. Hortic. Sci.*, *80*, 247-254.

Henis, Y., & Bashan, Y. (1986). Epiphytic survival of bacterial leaf pathogens. In N. J. Fokkema & J. van den Heuvel (eds), *Microbiology of the phyllosphere* (pp. 252-268), New York: Cambridge Univ. Press.

Hirano, S. S. & Upper, C. D. (1983). Ecology and epidemiology of foliar bacterial plant pathogens. *Annu. Rev. Phytopathol.*, *21*, 243-269.

Hirano, S. S. & Upper, C. D. (1990). Population biology and epidemiology of *Pseudomonas syringae*. *Annu. Rev. Phytopathol.*, *28*, 155-177.

Hirano, S. S. & Upper, C. D. (1991). Bacterial community dynamics. In J. H. Andrews, & S. S. Hirano (Eds), *Microbial Ecology of Leaves* (pp. 271-294) New York: Springer-Verlag.

Hirano, S. S., Maher, E. A., Lindow, S. E., Kelma, A. & Upper, C. D. 1978. Types of ice nucleation activity among plant pathogenic fluorescent pseudomonads. In *Phytopathol. News, Proc. Am. Phytopathol. Soc.*, *12*, 176.

Hirano, S. S., Rouse, D. I., Nordheim, E.V. & Upper, C. D. (1981). Epiphytic ice nucleation-active (INA) bacterial population in relation to halo blight incidence in oats. *Phytopathology*, *71*, 881.

Holden, P. A., Hunt, J. R. & Firestone, M. K. (1997). Toluene diffusion and reaction in unsaturated *Pseudomonas putida* biofilms. Biotechnol. Bioeng., *56*, 656-670.

Hsieh, S. P. Y. & Buddenhagen, I. W. (1974). Suppressing effects of *Erwinia herbicola* on infection by *Xanthomonas oryzae* and on symptom development in rice. *Phytopathology*, *64*, 1182-1185.

Kavitha, S. (2002). Strategies for management of rice blast and sheath blight with bacterial biocontrol agents in combination with major genes for disease resistance. Ph. D Thesis, University of Madras, India.

Kavitha, S., Senthilkumar, S., Gnanamanickam, S. S., Inayathullah, M. & Jayakumar, R. (2005). Isolation and partial characterization of antifungal protein from *Bacillus polymyxa* strain VLB16. *Process Biochem.*, *40*, 3236-3243.

Kinkle, L. L., Newton, M. R. & Leonard, K. J. (2002). Resource aggregation in the phyllosphere: implications for microbial dynamics across spatial scales. In S. E. Lindow., Hecht-Poinar, E.I. & Elliot, V.J (eds) *Phyllosphere Microbiology* (pp. 317-340), St.Paul: APS Press.

Kjellenberg, S. & Steiberg, P. D. (2002). Defences against bacterial colonization of marine plants. In S. E. Lindow., Hecht-Poinar, E.I. & Elliot, V.J (eds) *Phyllosphere Microbiology* (pp.157-172), St.Paul: APS Press.

Klement, Z. (1974). Apoplexy of apricots. III. Relationship of winter frost and bacterial canker and die-back of apricots. *Acta Phytopathol. Acad. Sci. Hung.*, *9*, 35-45.

Kozloff, L. M., Schofield, M. A. & Lute, M. (1983). Ice nucleating site of *Pseudomonas syringae and Erwinia herbicola. J. Bacteriol.*, *153*, 222-231.

Krishnamurthy, K. & Gnanamanickam, S. S. (1998). Biocontrol of rice sheath blight with formulated *Pseudomonas putida. Indian Phytopath.*, *51*, 233-236.

Lavermicocca P., Surico, G., Varvaro, L. & Babelegoto, N. M (1987). Plant hormone, cryogenic and antimicrobial activities of epiphytic bacteria of live and oleander. *Phytopathol. Mediterr.*, *26*, 65-72.

Leben, C. (1963). Multiplication of *Xanthomonas vesicatoria* on tomato seedlings. *Phytopathology*, *53*, 778-781.

Leben, C. (1965). Epiphytic microorganisms in relation to plant disease. *Annu. Rev. Phytopathol.*, *3*, 209-230.

Leben, C. & Miller, T. D. (1973). A pathogenic pseudomonad from healthy field grown soybean plants. *Phytopathology*, *63*, 1464-1467.

Leben, C., Daft, G. C., & Schmitthenner, A. F. (1968). Bacterial blight of soybeans: population levels of *Pseudomonas glycinea* in relation to symptom development. *Phytopathology*, *58*, 1143-1146.

Leveau, J. H. J. & Lindow, S. E. (2001). Appetitel of an epiphyte : quantitative monitoring of bacterial sugar consumption in the phyllosphere. *Proc. Natl. Acad. Sci. USA*, *98*, 3446-3453.

Lindeman, J., Arny, D. C. & Upper, C. D. (1984). Epiphytic populations of *Pseudomonas syringae* pv. *syringae* on snap bean and non-host plants and the incidence of bacterial brown spot disease in relation to cropping patterns. *Phytopathology*, *74*, 1329-1233.

Lindgren, P. B., Frederick, R., Govindarajan, A. G., Panoupoulos, N. J., Staskawicz, B. J. & Lindow, S. E. (1989). An ice nucleation reporter gene system: identification of inducible pathogenicity genes in *Pseudomonas syringae* pv. *phaseolicola. EMBO J.*, *8*, 1291-1301.

Lindow, S. E. (1979). Frost damage to potato reduced by bacteria antagonistic to ice nucleation active bacteria. *Phytopathology*, *69*, 1036 (Abstr.).

Lindow, S. E. (1981). Frost damage to pear reduced by antagonistic bacteria, bactericides and ice nucleation ingibitors. *Phytopathology*, *71*, 237 (Abstr.).

Lindow, S. E. (1982). Integrated control of frost injury and fire blight of pear with antagonistic, epiphytic bacteria. *Phytopathology*, *72*, 946 (Abstr.).

Lindow, S. E. (1983). Methods of preventing frost injury through control of epiphytic ice nucleation active bacteria. *Plant Dis.*, *67*, 327-333.

Lindow, S. E. & Panopoulos, N. J. (1988). Field tests of recombinant Ice- *Pseudomonas syringae* for biological frost control in potato. *In* M. Sussman., C. H. Collins., & F. A. Skinner (eds*), Proc. First. Int'l Conf. Release of Genetically Engineered Microorganisms,* (pp. 121-138), London: Academic Press.

Lindow, S. E. & Stakawicz, B. J. (1981). Isolation of ice nucleation deficient mutants of *Pseudomonas syringae* and *Erwinia herbicola* and their transformation with plasmid DNA. *Phytoapthology, 71*, 237.

Lindow, S. E., Arny, D. C. & Upper, C. D. (1975). Increased frost sensitivity of maize in the presence of *Pseudomonas syringae. Proc. Am. Phytopath. Soc,. 2*, 57 (Abstr.).

Lindow, S. E., Arny, D. C., Barchet, W. R. & Upper, C. D. (1976). Bacterial ice nuclei as incitants of warm temperature frost damage. *Am. Phytopathol. Soc. Proc.*, *3*, 224 (Abstr.)

Lindow, S. E., Arny, D. C. & Upper, C. D. (1978). *Erwinia herbicola*: an active ice nucleus incites frost damage to maize. *Phytopathology, 68*, 523-527.

Lindow, S. E., Arny, D. C., Upper, C. D. & Barchet, W. R. (1978). The role of bacterial ice nuclei in frost injury to sensivie plants. In P.H.Li. & A. Sakai (eds), Plant Cold Hardiness and Freezing Stress: mechanisms and crop implications (pp. 249-263), New York: Academic Press.

Lindow, S. E., Arny, D. C. & Upper, C. D. (1982). Bacterial ice nucleation: a factor in frost injury to plants. *Plant Physiol., 70*, 1084-1089.

Lindow, S. E., Hecht-Poinar, E.I. & Elliot, V. J. (2002) (eds). Phyllosphere Microbiology, St. Paul, MN: APS Press.

Malvick, D. K. & Moore, L. W. (1988). Population dynamics and diversity of *Pseudomonas syringae* on apple and pear trees and associated grasses. *Phytopathology, 78,* 1366-1370.

Mew, T. W. & Vera Cruz, C. M. (1986). Epiphytic colonization of host and non-host plants by phytopathogenic bacteria. In N. J. Fokkema & J. van den Heuvel (eds), *Microbiology of the phyllosphere* (pp. 269-282), New York: Cambridge Univ. Press.

Monier, J-M. (2002). Biological significance of bacterial aggregation of leaf surfaces : The social life of epiphytic bacteria. In *Environmetal Science, Policy and Management* (176 pp), Berkeley: Univ. Calif. Press.

Morris, C. E. & Monier, J-M. (2003). The ecological significance of biofilm formation by plant-associated bacteria. *Annu. Rev. Phytopathol.*, 41, 429-453.

Morris, C. E., Monier, J. M. & Jacques, M. A. (1997). Methods for observing microbial biofilms directly on leaf surfaces and recovering them for isolation of culturable microorganisms. *Appl. Environ. Microbiol.*, 63, 1570-1576.

Morris, C. E., Monier, J. M. & Jacques, M. A. (1998). A technique to quantify the population size and composition of the biofilm component in communities of bacteria in the phyllosphere. *Appl. Environ. Microbiol.*, 64, 4789-4795.

Morris, C. E., Barnes, M. B. & McLean, R. J. C. (2002). Biofilms on leaf surfaces: implications for the biology, ecology and management of populations of epiphytic bacteria. *In* S. E. Lindow., E.I. Hecht-Poinar., & V. J. Elliot (eds), *Phyllosphere Microbiology*, (pp. 138-154), St. Paul, MN: APS Press.

Morris C.E., Georgakopoulos D.G. & Sands D.C. (2004). Ice nucleation active bacteria and their potential role in precipitation. *J. Physiques IV France, 121*, 87-103.

Orser, C. S., Staskawicz, J. B., Panopulos, N. J., Dahlbeck, D. & Lindow, S. E. (1985). Cloning and expression of ice nuleation genes in *Escherichia coli. J. Bacteriol., 164*, 359-366.

Orser, C. S., Staskawicz, J. B., Panopulos, N. J., Dahlbeck, D. & Lindow, S. E. (1982). Cloning and expression of ice nucleation genes from *Pseudomonas syringae* and *Erwinia herbicola* in *Escherichia coli.* Phytopatholohy, *71*, 1000 (Abstr.).

Panagopoulos, C. G. & Crosse, J. E. (1964). Frost injury as a predisposing factor in blossom blight of pear caused by *P. syringae* van Hall. *Nature, 202*, 1352.

Papavassiliou, J., Tzannetis, S., Leka, H. & Michopoulos, G. (1967). Coli-aerogenes bacteria on plants. *J. Appl. Bacteriol., 30,* 219-223.

Patil, S. S. & Gnanamanickam, S. S. (1976). Suppression of bacterially-induced hypersensitive reaction and phytoalexin accumulation in bean by phaseotoxin. *Nature*, *259*, 486-487.

Pierson, L. S., Wood, D. W., Pierson, E. A. & Chancey, S. T. (1998). N-acyl-homoserine lactone-mediated gene regulation in biological control by fluorescent pseudomonads: current knowledge and future work. *Eur. J. Plant Pathol.*, *104*, 1-9.

Riffaud, CM-H., Glaux, C., Guilbad, C., Dominguez, H., Prior, P. & Morris, C. E. (2003). Epidemiological clues for developing methods of control of bacterial blight of cantaloupe caused by *Pseudomonas syringae* pv. *aptata.* In N. Iacobellis., A. Collmer., S. W. Hutcheson., J. W. Mansfield., C. E. Morris., J. Murillo., N. W. Schaad., G. Surico., D. E. Stead and M. Ullrich (eds*)* *Pseudomonas syringae and Related Pathogens: Biology, Ecology and Genetics.*, London: Kluwer/Plenum Press.

Rovira, A. D. & Campbell, R. (1974). Scanning electron microscopy of microorganisms on the roots of wheat. *Microbiol. Ecol.*, *1*, 15-23.

Rouse, D. I., Nordheim, E. V., Hirano, S. S. & Upper, C. D. (1985). A model relating the probability of foliar disease incidence to the population frequencies of bacterial plant pathogens. *Phytopathology*, *75*, 505-509.

Ruinen, J. (1961). The phyllosphere. 1. An ecologically neglected milieu. *Plant Soil*, 15, 81-109.

Santhi, D. P., Unnamalai, N. & Gnanamanickam, S. S. (1987). Epiphytic association of *Erwinia herbicola* with rice leaves infected by *Xanthomonas campestris* pv. *oryzae* and its interaction with the pathogen. Indian Phytoapth., *40*, 3270332.

Sakthivel, N. (1997). Biological control of *Sarocladium oryzae* (Sawada) Gams & Hawksworth, sheath-rot pathogen of rice by bacterization with *Pseudomonas fluorescens* Migula. Ph.D Thesis, University of Madras, India.

Sakthivel, N. & Gnanamanickam, S. S. (1987). Evaluation of *Pseudomonas fluorescens* strains for suppression of sheath-rot disease and for enhancement of grain yields in rice, *Oryza sativa*, L. *Appl. Environ. Microbiol.*, *53*, 2056-2059.

Sakthivel, N. & Gnanamanickam, S. S. (1989). Incidence of different biovars of *Pseudomonas fluorescens* in flooded rice rhizospheres of India. *Agriculture, Ecosystem and Environment*, *25*, 287-295.

Schloter, M., Borlinghaus, R., Bode, W. & Hartmann, A. (1993). Direct identification and localization of *Azospirillum* in the rhizosphere of wheat using fluorescence labeled monoclonal antibodies and confocal scanning laser microscopy. *J. Microscopy*, *171*, 173-177.

Singer, C. E., & Ames, B. N. (1970). Sunlight ultraviolet and bacterial DNA base ratios. *Science, 170*, 822-826.

Single, W. V., & Olien, C. R. (1967). Freezing processes in wheat stems. *Aust. J. Biol. Sci., 20*, 1025-1028.

Sprang, M. L., & Lindow, S. E. (1981). Sub-cellular localization and partial characterization of ice nucleation activity of *Pseudomonas syringae* and *Erwinia herbicola*. *Phytopathology, 71*, 256.

Swift, Bainton, N. J., & Winson, M. K. (1994). Gram-negative bacterial communication by N-acyl homoserine lactones: a universal language?. *Trends Microbiol., 2*, 193-198.

Thara, K. V. (1994). Biological control of rice sheath blight by bacterial antagonists: mechanisms of disease suppression. Ph. D. Thesis, University of Madras, India.

Thomas, S., McQuillin, J. (1952). Coli-aerogenes bacteria isolated from grass. *Proc. Soc. Appl. Bacteriol., 15*, 41-52 Single, W. V., & Olien, C. R. (1967). Freezing processes in wheat stems. *Aust. J. Biol. Sci., 20,* 1025-1028.

Toledo, G., Bashan, Y. & Soeldner, A. (1995). In vitro colonization and increase in nitrogen fixation of seedling roots of black mangrove inoculated by a filamentous cyanobacteria. *Can. J. Microbiol.*, *41*, 1012-1020.

Unnamalai, N. & Gnanamanickam, S. S. (1984). *Pseudomonas fluorescens* is an antagonist to *Xanthomonas citri* (Hasse) Dye, the incitant of citrus canker. Curr. Sci., *50*, 703-704.

Valasubramanian, R. (1994). Biological control of rice blast with *Pseudomonas fluorescens* Migula: Role of antifungal antibiotic in disease suppression. Ph. D Thesis, University of Madras, India.

Vasantha devi, T., Malarvizhi, R., Sakthivel, N. & Gnanamanickam, S. S. (1989). Biological control of sheath blight of rice in India with bacterial antagonists. *Plant Soil*, *119*, 121-128.

Vasudevan, P. (2002). Isolation and characterization of *Bacillus* spp. from the rice rhizosphere and their role in biological control of rice caused by *Xanthomonas oryzae* pv. *oryzae*. Ph. D Thesis, University of Madras, India.

Vasudevan, P., Kavitha, S., Brindha Priyadarisini, S., Babujee, L. & Gnanamanickam, S. S. (2002). Biological control of rice diseases. In S. S. Gnanamanickam (ed), *Biological Control of Crop Diseases*, (pp. 11-32) New York: Marcel Dekker.

Velusamy, P. & Gnanamanickam, S. S. (2003). Identification of 2,4-diacetylphloroglucinol production by plant-associated bacteria and its role in suppression of rice bacterial blight in India. *Curr. Sci.*, *85*, 1270-1273.

Velusamy, P., Defago, G., Thomashow, L. S. & Gnanamanickam. S. S. (2004). Role of 2, 4-diacetylphloroglucinol (DAPG) for plant disease control: its importance to rice bacterial blight suppression in India. In C. D. Mayee et al (eds), Biotechnological Approaches to the Integrated Management of Crop Diseases, (pp. 182-191), New Delhi: Daya Publishing House.

Velusamy, P., Ebenezar Immanuel,. Gnanamanickam, S. S. & Thomashow, L. S. (2006). Biological control of rice bacterial blight by 2,4-diacetylphloroglucinol (DAPG)-producing plant-associated bacteria. *Can. J. Microbiol.* 52: 1-10.

Warren, G. J. (1987). Bacterial ice nucleation: molecular biology and applications. *Biotechnol. Gene. Engin. Rev.*, *5*, 107-135.

White, G. F. & Haas, J. E. (1975). *Assessment of Research on Natural Hazards* (pp. 304-312), Cambridge, MA: MIT Press.

Wiehe, W., Schloter, M., Hartmann, A. & Hoflich, G. (1996). Detection of colonization by *Pseudomonas* PsA12 of inoculated roots of *Lupinus albus* and *Pisum sativum* in greenhouse experiments with immunogold techniques. *Symbiosis*, *20*, 129-145.

Wolber, P. K. (1992). Bacterial ice nucleation. *Adv. Microbial. Physiol.*, *34*, 205-237.

Wolber, P. K. & Green, R. L. (1990). Detection of bacteria by transduction of ice nucleation genes. *Trends Biotechnol.*, *8*, 276.

Yankofsky, S. A., Levin, Z. & Moshe, A. (1981). Association with citrus of ice-nucleation bacteria abd their possible role as causative agents of frost damage. *Curr. Microbiol.*, *5*, 213-217.

Yelenosky, G. (1991). Apparent nucleation and freezing in different parts of young citrus trees during controlled freezes. *Hort Science, 26*, 576-579.

Wimpenny, J. W. T. & Colasanti, R. (1997). A unifying hypothesis for the structure of microbial biofilms based on cellular automation models. *FEMS Microbiol. Ecol.*, *22*, 1-16.

8. AFFILIATIONS

Samuel S. Gnanamanickam and J. Ebenezar Emmanuel, *Plant Pathology Group, Centre for Advanced Studies in Botany, University of Madras-Guindy Campus, Chennai-600025, India.*
Tel: 91-44-22350401; Fax: 91-44-22352494; E-mail: gmanick@vsnl.com

C.W. BACON AND D.M. HINTON

BACTERIAL ENDOPHYTES: THE ENDOPHYTIC NICHE, ITS OCCUPANTS, AND ITS UTILITY

Abstract. Endophytic species of bacteria offer an enormous potential for protection and increased agronomic performance of plants. The diversity of bacterial endophytes guarantees that there are endophytes capable of forming compatible association with all agronomically important plants, including monocots and dicots. Fitness of endophytically infected plants is increased both at inter- and intraspecific levels of competition. These bacterial offer the advantages of being sequestered within plants and as long as the plant has a favorable environment to grow, the bacteria contained within it will continue to grow affording the plant continued protection throughout its growth period. The bacteria are endosymbionts, and behave in most instances the relationships are mutualists. The habit is the intercellular spaces of plants, a large area of interconnected spaces that contain high levels of carbohydrates, amino acids, and inorganic nutrients, which serve the purpose of supporting the growth of intercellular bacteria. The intercellular niche is a novel compartment from which the plant can be protected, offering numerous advantages over conventional pesticides. In addition to biocontrol uses, endophytic bacteria offer potential for surrogate transformation of plants resulting in increased nutritional qualities or novel pesticides, as well as the recently emerging use for phytoremediation of soil and water pollutants.

1. INTRODUCTION

Endophytic bacteria are defined as bacteria that colonize healthy plant tissue without causing obvious symptoms in or produce obvious injury to the host. Bacterial endophytes colonize a large number of plants, but because they are symptomless they cannot be detected, and can only be detected by isolating such bacteria from surface sterilized plants placed onto bacterial agar. This procedure foretells the difficulty in providing a valid estimate of the numbers of bacterial endophytes and their roles *in planta*. Bacterial endophytes actively colonize plant tissues, establish long-term associations, actually lifelong natural associations, and are not in generally organ-specific. Thus, they may be isolated from roots, leaves, and stem, and a few from inflorescences and fruits. Excluded from the group of bacterial endophytes are those that are associated with plants as latent infections and or colonize senescent plant tissue as evidenced by the production of macroscopic signs of diseases.

Therefore, a bacterial endophyte is a microbe that lives in association with plants forming a symptomless intercellular biotrophic association (Fig. 1). Endophytic bacteria are distinguished from transient visitors that form plant associations as happenstances, which do not survive long. Bacterial endophytes as used here encompass the broader use of the narrowly defined term of Kado (1992) to include that of Quispel (1992), which defines bacterial endophytes to include

S.S. Gnanamanickam (ed.), Plant-Associated Bacteria, 155–194.

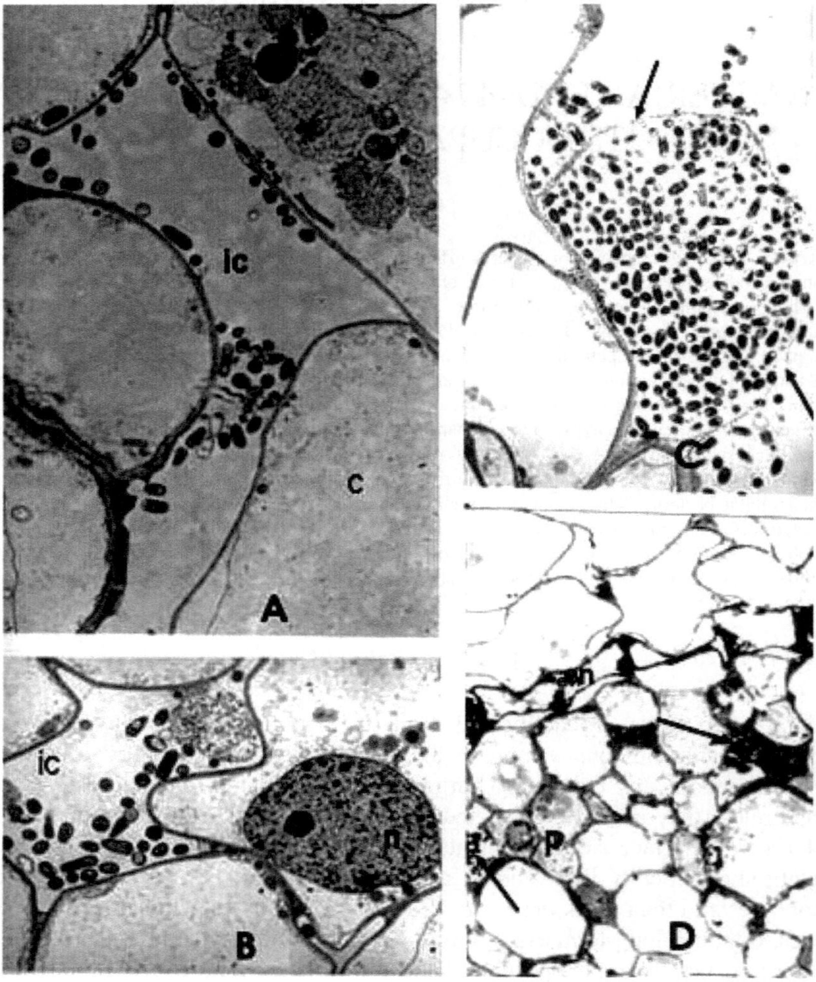

Fig. 1. Transmission electron and light microscopy of the bacterial endophyte *Bacillus mojavensis* in maize seedling cultured in soil. A). A lysigenous lacunae producing a large intercellular space (IC), containing several bacterial cells, x4640. B). Another large intercellular space (IC) containing bacterial cells, and a granulated substance, possibly the remnants of the middle lamellae. C). The association of bacterial cells on the exterior of seedling roots covered by a matrix-like bacterial substance (delimited by arrows) suggestive of a site from which cells may gain entry into young roots, i.e., via an injury. D). Bacterial cells (arrows) next to primary phloem cells, P, located in the stele (x1000 under oil) below the endodermis (en).

those intercellular bacteria that impart an ecological benefit without doing substantive harm to the plant. While this definition is restrictive, it conveys the concepts useful in defining fungal endophytes (Stone, Bacon, & White, Jr., 2000), and will serve as the primary focus for readers of this review as that of using endophytes as biocontrol agents. Inherent in the definition of endophytic bacteria is the concept of a symptomless, non-disease producing infection that results in a series of interactions that range from no effects on the hosts (neutralism), beneficial to the hosts and bacteria (mutualism), or to that of benefiting only one member (commensalism). We, base these interactions on the terminology used for fungal endophytes and their quantitative and qualitative relationships with grasses (Stone et al., 2000). However, the specific interactions used to support the effects of endophytic fungi on their hosts have not yet been completely established for endophytic bacteria.

The definition of endophytic bacteria used here purposely excludes latent and/or pathogenic bacteria and the resulting negative pathological, epidemiological, and physiological effects that these bacteria produce in plants. Also excluded are those bacteria whose interactions with plants are restricted, i.e., epiphytic and rhizoplane or phyllosphere colonizer and dormant or latent stages of endophytic colonizers, as well as diazotrophic bacteria that form root and stem nodules such as the legume species of *Rhizobium, Bradyrhizobium,* etc. However, the diazotrophic endophytic species that are included are those that do not form nodules and these include the species of *Rhizobium, Azorhizobium, Azospirillum,* and others.

This review will center first on the importance of the nature of the endophytic niche since this location is important if we are to ever understand the interactions between endophytes, the number cells that can be supported for optimum expression for the desired outcomes, and for clues useful developing crops varieties that can provide optimal requirements for supporting bacterial endophytes. Other foci of this review include those bacteria found naturally and those that have been experimentally used for various aspects of crop protection, and other technological uses of endophytes, both present and future. Additional review that treat other salient aspects of bacterial endophytes include, but not limited to those of McCully (2001), Chanway (1998), Sturz et al. (2000) and Hallmann et al. (1998), which should be consulted for additional information and perspectives dealing with the exciting world of bacterial endophytes.

2. THE ENDOPHYTIC NICHE

2.1. Evolution of Endophytism in Bacteria

Biological associations of plants by bacteria have occurred in both Gram-positive and Gram-negative species, despite their major differences in cell morphology and biochemical specializations indicating the benefits derived from associations with plants. Further, phylogenetically these associations have occurred in both the diderm or derived group, and the monoderm group, which is viewed currently as being the more primitive group. However, the majority of endophytic species are found within the monoderm bacteria, suggesting that endophytic associations might be an ancestral trait as opposed to the view, a apriori, that it is evolutionarily derived. If indeed evolution among the bacteria occurred in a forward direction, over time the monoderm endophytic bacteria may not be as primitive as the association suggest since over evolutionary periods environmental changes occurred that allowed only those to survive, which lived in protected niches, such as the intercellular spaces of plants. The occurrence of endophytic bacteria as remnants of the primitive forms is strengthened by their low numbers, and the more abundant numbers of β-proteobacteria and γ-proteobacteria, which is indicative of groups at the leading edge of evolution (Gupta, 2002), as for example *Burkolderia* species (McInroy & Kloepper, 1995b). However, this conclusion is based on the total number of culturable bacteria. Another survey of bacteria insolated from maize that was based on 16SrDNA primer analysis isolations directly from roots of maize indicated that the *Proteobacteria* made up the largest group with the α-subdivision predominating, followed by the β-proteobacteria, and the γ-proteobacteria as the third most abundant group (Chelius & Tiplett, 2001). These and other surveys were made on plants grown in soil that is used for commercial production, and how much above observation is true for native forest soils is not known. Finally, the physiological behavior of specific bacteria varies with host. For example, there is specific information indicating that species of bacterial endophytes colonize and are physiologically different on wild native plant compared to their cultivated relatives, especially the endophytic colonizing of grasses and other non leguminous plants by species of rhizobia (Chanintreuil, Giraud, Prin, Lorquin, BA, Gillis, de Lajudie, & Breyfus, 2000; Cavalcante and Dobereiner, 1988; James and Olivares, 1998; Yanni, Rizk, Corich, Squartini, Ninke, Phi, Hollingsworth, Orgambide, De Bruijn, Brckley, Schmidt, Mateos, & Dazzo, 1997).

Since there is such a wide array of bacterial endophytes, evolutionary strategies which led to the observerd diversity are difficult to determine, which led Kobayashi & Palumbo (2000) to conclude that each species of bacterium has evolved toward enhancing its own fitness or survival and each might not express an evolutionary tendency from saprophytism to parasitism, and finally to endophytism. This would result in producing two generally different types, evolved and nonevolved (Kobayashi & Palumbo, 2000). This concept is similar to the evolutionary theory presented by Swensen and Mullin (1997) for the evolution of root nodule symbioses. In their hypothesis data are presented to support a single origin for the predisposition for root nodule symbiosis, but which accounts for the evolution of multiple origins of symbiosis within the nodule symbiotic group (Swensen and Mullin, 1997). In light of this hypothesis, the ability to cohabitate as an endophyte would have a single origin, but several species acquired this trait by one of several sexual recombination means. Present knowledge does not address host requirements for colonization nor is there information concerning the inability of endophytes to colonize hosts. If in fact there are two basic types of bacterial endophytes, i.e. evolved and nonevolved, these should be separated, and the genetics and physiology of the evolved should be studied in terms of basic genetics that govern endophytic colonization of host and nonhosts. Finally, what is needed is a species whose isolates clearly indicate the evolved tendency, reflective of the presence of endophytism in most if not all isolates. Such a species can serve as the model for subsequent studies of most aspects of bacterial endophytes and we present *Bacillus mojavensis*, a species discussed in considerable detail below, as one such model for endophytism (Fig. 2). Additional species might include *Burkholderia, Klebsiella,* and *Herbaspirillum.*

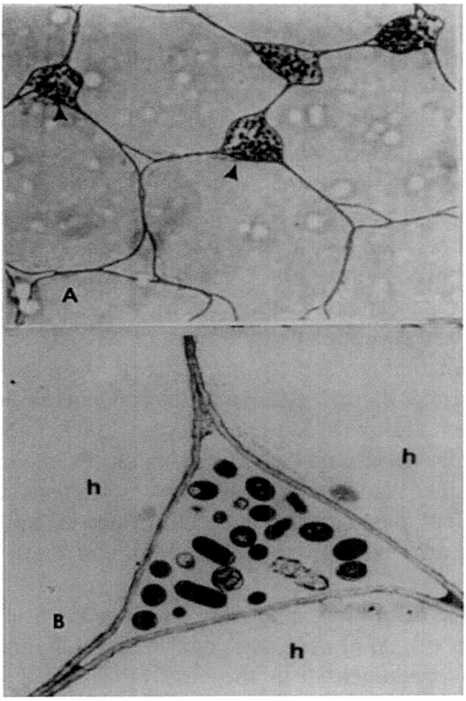

Fig. 2. Light and transmission electron microscopy of *Bacillus mojavensis* (arrowheads) within the fluid filled intercellular spaces of the cortex of corn seedling (A) (x40). In (B) an enlargement (x7250) of a section typified in (A) showing an intercellular space formed by three host cells (h) producing an intercellular space. Note the middle lamella remnants (granular material) observed in all three angles formed by the three plant cells, indicating a biochemical activity of the bacteria to produce the intercellular spaces.

2.2. In Planta Association and Nutrition

Bacterial endophytes are intercellular (Bacon & Hinton, 2002; Chanway, 1998; McCully, 2001), and may colonize plants either exclusively below ground, above ground or both. However, bacterial endophytes are not exclusively intercellular as some form intracellular association as well, e.g., within xylem tissues (Timmer, Brlansky, Lee, & Raju, 1983; Gardner, Feldman, & Zablotowicz, 1982a). Although the xylem is a rich source of nutrients, endophytes that live within it are however, considered not ideal endophytes since such infected vessels are nonfunctional and considered detrimental to the plant (McCully, 2001). Thus, our discussion on xylem endophytes will be rather limited to a specific application of xylem endophytes, phytoremediation, and we will concentrate more on bacterial endophytes that live within the intercellular spaces.

Endophytic bacteria are associated with plants as biotrophic symbionts and these bacterial can be either obligate or facultative, which include thousands of plants hosts, although some endophytes are restricted to plant families. Some are only associated with grasses, while others are symbiotic with a wide range of plant species including both monocots and dicots. The means of infecting host plants intercellular is primarily conjecture as the few studies have demonstrated the nature of the *in planta* association (Bacon & Hinton, 2002; de Souza, Pamphile, Sartori, da Rocha, & Azevedo, 2004; Patriquin & Dobereiner, 1978; James, Olivares, Baldani, & Döbereiner, 1997; James & Olivares, 1998), and there are no direct studies on how the plant is initially infected. Most studies suggest that infection is accomplished through breaks, tares, or wounds of the roots of seedling during either germination or shortly thereafter (Dobereiner & Pedrosa, 1987; Patriquin, Dobereiner, & Jain, 1983; Patriquin & Dobereiner, 1978; Chaintreuil, Giruad, Prin, Lorquin, Gillis, De Lajudie, Dreyfus, 2001; Bacon & Hinton, 2002), which remain open for several days (Moon, Clough, Peterson, & Allaway, 1986). Aerial infection can take place through the stomates of guard cell, and other natural pores but this too is not demonstrated as portals of natural infection. According to McCully (2001), the undamaged root does not offer casual entry by microbes.

As we are dealing with non-pathogenic bacteria, there is very little evidence that they possess the arsenal of hydrolytic enzymes that are used to gain entry into roots and leaves by pathogens. In the case of *Bacillus mojavensis* we have evidence that topical application of the bacterium to the kernel results in seedling infection simultaneously with seedling development (Bacon & Hinton, 2002), although we also know that a low percentage of seed produced from *B. mojavensis*-infected plants are internally infected and such seed give rise to *B. mojavensis*-infected seedlings when planted (Bacon, unpublished). Bacterial endophytes benefit from inhabiting the plant' interior because it is a protected niche in which there is relatively little competition from other microorganisms for a constant and reliable source of nutrition. However, the concentrations of such nutrients have been reported as being sparse, suggesting that intercellular inhabiting bacteria might live under oligotrophic conditions. This however, has not been established as the nutritional requirements of endophytic bacteria reported to date are in fact copiotrophic, which is reinforced in the discussion below where data indicate that the intercellular spaces are in fact rich in organic and inorganic nutrients capable of supporting the cellular concentration of bacteria observed in endophytic infected plants.

2.3. Morphological and Chemical Constituents of the Endophytic Niche

What are the morphological characteristics of the intercellular space and what substances are contained within it? The answers to these questions are important in understanding the nature of interactions that allow endophytic bacteria to exist in this niche. The interactive nature of the intercellular spaces is currently still unresolved but modern views of it suggest that it is an important area of the plants, the nature of which vary for each species although there are generalized information to serve as focus for discussions and there salient points are presented in the discussion that follows.

The intercellular spaces consists of a series of connecting nonliving components of the plant forming what is referred to as the apoplast, which is distinct from the cellular to cellular contact referred to as the symplast. The intercellular spaces are located in the cortical tissue of the root and in the parenchyma tissue of the leaves. The intercellular spaces are formed by the juxtapositions of three to four cells and the dissolution of the middle lamellae and they are significantly different in the leaves and stems as opposed to the roots (Figure 1A-C). The intercellular spaces of all plants have not been fully characterized enough to generalize on their contents, total volume, and relatedness to symplast. However, for the sake of discussion, we do have some information that describes this location as a significant area to the physiology of the plant as it relates to sugarcane and a few other plants. The volume occupied by intercellular spaces consists of a significant portion of the plant axis and in leaves, as much as 6% of the leaf tissue consists of intercellular spaces or the apoplast in sugar canes (Tetlow & Farrar, 1993; Altus & Canny, 1985). The intercellular spaces of roots in most plants are typically schizogenous, although in grasses they may be larger, consisting of lysigenous lacunae formed from the breakdown of surrounding cells (Figure 1A-B). The composition of these spaces varies and some consist of air or dry flaky deposits, but the majority of spaces consists of fluid, which in maize has been shown to contain the inorganic ions potassium, calcium, sulphur, phosphorus, and chlorine (Canny & Huang, 1993). The size of the intercellular lacunae may be greatly reduced in various tissue types, especially in the stele, resulting in lacunae literally filled to capacity with bacterial cells (Figure 1D). The intercellular spaces of the green portion of plants also consist of gases, and fluids. The intercellular spaces in leaves are predominantly located in the extrafascicular plant parenchymatous tissues that consist of oxygen and air-saturated water, and fluid. Evidence that endophytic bacteria may either enlarge or alter the nature of the intercellular space is evidence by apparent remains of middle lamella-type materials of opposing cells (Figure 2A and B).

Nutrient concentration in both the apoplasm and symplasm is interactive with the phloem, dispelling the earlier notion that the apoplasm is relatively free of nutrients. Indeed, current research indicates that nutrient transport within plant tissues are considered to occur through an apoplastic route via the cell wall continuum and via the symplastic route via the plasmodesmata (Canny, 1995; Dong et al., 1994; Kursanov & Brovchenko, 1970; Madore & Webb, 1981). Organic nutrients within the apoplasm consist of the several sugars and related carbohydrates derived from photosynthesis (Tables 1 and 2), although several nitrogenous compounds such as amino acids and amine also occur, as well as key inorganic nutrients (Table 2). The concentrations of sugar available within the apoplasm are dependent on the nature of phloem loading from either the apoplasm or symplasm (Giaquinta 1976), which are both plant species and individually regulated. Additionally, regulation is altered dramatically by the presence of endophytic hyphae (Farrar & Lewis, 1987; Tetlow & Farrar, 1993), resulting in a constant removal, but replacing of nutrients (Madore & Webb, 1981), reflective of uptake by the endophyte. For example, apoplastic sucrose represented 0.4% of the total soluble carbohydrates in healthy leaves, and 0.2% infected tissue, and a similar ratio was reported for total soluble reducing sugars (Tetlow & Farrar, 1993). Further, when expressed as a percentage of tissue water content, the apoplast occupied 8% of infected tissue, compared to 6% in non-infected plants (Tetlow & Farrar, 1993). Therefore, endophyte-infected tissue has a higher apoplastic volume.

Table 1. Sugars in the apoplasm and symplasm of leaves and roots of squash, *Cucurbita pepo* (Madore and Webb, 1981).

Sugar	Concentration, μmol g fresh weight $^{-1}$	
	Apoplasm	Symplasm
Verbascose	Trace	Not measured
Stachyose	0.006	2.7
Raffinose	Trace	0.8
Sucrose	0.023	4.1
Galactose	0.008	0.2
Glucose	0.034	4.5
Fructose	0.005	0.6
Galactinol	0.009	2.5
Myoinositol	0.008	2.0

Table 2. Non-Carbohydrate Metabolites reported to occur in the apoplasm of leaves and roots (Canny & McCully, 1988; Canny & Huang, 1993).

Metabolites
Glutamine
Aspartic acid
Serine
Alanine
Asparagine
Ammonium
γ-aminobutyric acid
Threonine
Valine
Lysine
Tyrosine
Leucine
Glycine
Isoleucine
Arginine
Histidine
Methionine
Malic acid
Tartaric acid
Oxalic acid
Fumaric acid
Citric acid
Cystine
Potassium
Calcium
Sulphur
Phosphorus
Chloride

The nature of nutrients within the apoplast is controversial and complex. The sugars are considered to be compartmentalized in the apoplasm and their concentration within the apoplastic fluid is not controlled by membrane of cells forming the spaces (Kursanov & Brovchenko, 1970; Altus et al., 1985; Canny, 1995; Tetlow &Farrar, 1993; Madore & Webb, 1981). The extent of fluid movement through an apoplastic route occurs (Canny, 1995), but the connection to the symplasm, if any, is complex and uncertain. However, analysis of the contents of the apoplasm indicates that it consists of sucrose, glucose, and fructose in ratios close to those characteristic of surrounding cells, indicating an interaction with compartmentalized sugars within cells, i.e. the symplast. Further, when sugars are removed from the intercellular spaces they are replaced within 60 minutes (Kursanov & Brovcehnko, 1970). Thus, the intercellular space is rich in substances necessary to support the growth of organisms which is reinforces by the large number of bacterial cells observed (Fig. 1). Further, the concentrations of apoplastic sugars reported to occur in non-endophytically-infected apoplasm (Table 1) are sufficient to support bacterial growth in squash and other plants (Canny & McCalley, 1988; Dong et al., 1994; Farrar & Farrar, 1986; Huber & Moreland, 1980; Klement, 1965; Kneale & Farrar, 1985; Kursanov et al., 1970).

Considering the above, the concentration of nutrients within the apoplasm is dynamic as it is cycled between the host and bacterium, although the direction of flow is not considered to be directed by the endophytic microorganisms. However, there is a suggestion that there is a direct correlation between phloem loading of low concentrations of sugars and pH that require a proton gradient across the plasma membrane of the sieve elements and companion elements (Delrot & Bonnemain, 1981; Giaquinta, 1977). Therefore, altering the pH in the apoplast can affect sugar uptake and concentration, which has been documented to occur by the brown rust fungus (Tetlow & Farrar, 1993). It was demonstrated that in host-infected tissue, the pH was increased from 6.6 to 7.3 resulting in a 35 to 40% decrease in total soluble carbohydrate concentration. While this work concentrated on disease tissue, it nevertheless indicates the nature of an interaction between an endophytic system and a host for sugars. It is our opinion that this concentrations gradient would favor flow of continued carbohydrates to an endophyte. Thus, by altering the pH of the apoplasm endophytic organisms can alter directly or indirectly, the activity of specific enzymes, the sugar uptake kinetics of host cells, affecting the activities of host and cell wall invertases, and increasing the concentration of sugars in the apoplasm. It was also noticed that there was an increase in the volume of the intercellular spaces in infected hosts (Tetlow & Farrar, 1993). Therefore, the extent of an endophytic microorganism's alteration on the host occurs both at the physiological and morphological level of expression.

3. BIOLOGY OF THE ASSOCIATION

3.1. Endophyte and Host Diversity

Endophytic bacteria show a tremendous amount of diversity not only in plant hosts but also in bacterial taxa (Table 3). Endophytic bacteria have been isolated from reproductive structures such as fruits, seeds, and nuts, herbaceous plants such as noxious weeds, aquatic weeds, important vegetables, perennial and annual grasses, and woody plants (McInroy & Kloepper, 1995b; Hallmann, 2001; Chanway, 1998; Kuklinsky-Sobral, Araujo, Mendonca, Geran, Piskala, & Azevedo, 2004; Sturz, Christie, & Matheson, 1998; Sturz, Christie, & Nowak, 2000; Hallmann, Quadt-Hallmann, Mahaffee, & Kloepper, 1997; Kobayashi & Palumbo, 2000; Mundt & Hinkle, 1976). The woody plants reported as hosts for bacterial endophytes include such formable tree species as oak, elm, pear, trembling aspen, and pine (Brooks, Gonzales, Appel, & Filer, 1994; Bent & Chanway, 1998; Whitesides & Spotts, 1991; Myers & Strobel, 1983; Knutson, 1973).

Endophytic bacteria consist of both Gram-negative and Gram-positive species, although there is a higher number of Gram negative species that have been reported as agents of biological controls (Kobayashi & Plumbo, 2000), which as discussed above has a larger number of species than the gram positive groups, is considered more evolutionarily diverse and higher than the primitive Firmicutes (Gupta, 2002). However, the number of species reported as endophytic is low compared to other species found on plants as epiphytes, epibiotic, or rhizospheric colonizers, but undoubtedly this number will increase as the search for more efficient endophytic biocontrol agents continues, and as more species are delimited by molecular technology. Table 3 contains additional species added to the Table 4 that was modified by Chanway (1998), who modified the original table of Hallmann et al. (1997). Additional listing for host species of endophytic bacteria can be found in the earlier reviews and should be consulted (Kirchhof et al., 1997; James & Olivares, 1998; Kobayashi & Plumbo, 2000). The point here is not only are there additional hosts and genera listed since these two dates but also the listings are now indicating the specific epithets due to influence of molecular technology to aid in species identification.

Table 3. Common bacterial endophytes isolated from plant parts of different crop species*.

Plant species or organ	Bacterial taxa	References
Alfalfa (*Medicago sativa L.*) roots	*Erwinia-like* sp., *Pseudomonas* sp.	Gagne, Richard, Rousseau, & Antoun, 1987
Coffee (*Coffea Arabica* L.) root, and stem	*Acetobacter* sp.	Jimenez-Salgado et al., 1997
Camerron grass (*Pennisetum purpureum*)	*Acetobacter* sp.	Reis, Olivares, & Bobereiner, 1994
Maize (*Zea mays* L.) root and stem	*Arthrobacter* sp., *Aureobacterium* sp., *Bacillus mojavensis, B. thuringiensis, Bacillus spp. Burkholderia cepacia, Bukholderia* sp., *Corynebacterium* sp., *Enterobacter* sp., *Klebsiella terrigena, K. pneumoniae, Pseudomonas* sp., *Staphylococcus* sp.	Lalande, Bissonnette, Coutlee, & Antoun, 1989; Fisher, Petrini, & Scott, 1992; McInroy & Kloepper, 1995b; McInroy & Kloepper, 1995a; Palus, Borneman, Ludden, & Triplett, 1996; Bacon and Hinton, 2002; Balandreau et al., 2001; Chelius & Triplett, 2000
Cotton (*Gossypium hirsutum* L.) root and stem	*Acinetobacter baumannii, Agrobacterium radiobacter, Bacillus* sp., *Bacillus endophyticus, Burkholderia cepacia, B. gladioli, B. pickettii, Cellulomonas* sp., *Chryseobacterium* sp., *Comamonas testosteroni, Curtobacterium* sp., *Enterobacter cloacae, Enterobacter* sp., *Escherichia coli, Hydrogenophaga* sp., *Klebsiella* sp., *Kluyvera* sp., *Methlobacterium* sp., *Pseudomonas* sp., *Stenotrophomonas* sp., *Ochrobacterum*	Misaghi & Donndelinger, 1990; McInroy & Kloepper, 1995b; McInroy & Kloepper, 1995a; Reva et al., 2002; Gardner, Feldman, & Zablotowicz, 1982b; Mundt & Hinkle, 1976

	anthropi, Pantoea sp., *Phyllobacterium* sp., *Pseudomonas saccharophila, P. stutzeri, P. chloroaphis, Ralstonia japonicum, Rhizobium japonicum, Rhizobium spp.,Serratia* sp., *Sphingomonas paucimobilis, Staphylococcus* sp., *Xanthomonas campestris, Yersinia frederiksenii*	
Cucumber (*Cucumis sativis* L.), root and fruit	*Agrobacterium* sp., *Bacillus* sp., *Burkholderia* sp., *Citrobacter* sp., *Clavibacter* sp., *Erwinia carotovora* sp., *Proteus mirabilis, Serratia* sp., *Xanthomonas* sp.	McInroy & Kloepper, 1995a; Meneley et al., 1975
Grape (*Vitis* spp.)	*Bacillus fastidiosus, B. insolitus, Clavibacter* sp., *Comamonas* sp., *Curtobacterium* sp., *Enterobacter* sp., *Klebsiella ozaenae, K. pneumoniae, K. terrigena,, Moraxella bovis, Pantoea* sp., *Pseudomonas cichorii, Rahnella agquatilis, Rhodococcus luteus, Staphylococcus* sp., *Xanthomonas* sp.	Bell, Dickie, Harvey, & Chan, Meneley 1995
Kaller grass (*Leptochloa fusca*), root	*Azoarcus* sp.	Reinhold-Hurek & Hurek, 1998
Perennial ryegrass (*Lolium perenne*)	*Bacillus* sp.	Chanway et al., 1990
Spring wheat (*Triticum aestivum*), wheat	*Bacillus* sp., *Burkholderia cepacia*	Rennie, De Freitas, Ruschel, & Vose, 1982; Balandreau et al., 2001

Potato (*Solanum tubersum*), tuber	*Acidovorax* sp., *Acinetobacter* sp., *Actinomyces, Agrobacterium* sp., *Alcaligenes* sp., *Arthrobacter ureafaciens, Bacillus alclophialus, B. pasteurii, B. sphaericus, Capnocytophaga* sp., *Comamonas* sp., *Corynebacterium* sp., *Curtobacterium citrenum,C. leteum, Deleya* sp., *Enterobacter* sp., *Erwinia* sp., *Flavobacterium* sp., *Kingella kingae, Klebsiella* sp., *Leuconostoc* sp., *Methylobacterium* sp., *Micrococcus* sp., *Pantoea* sp., *Pasteurella* sp., *Photobacterium* sp., *Pseudomonas tolaasii, Psychrobacter, Serratia liquefaciens* sp., *P. plymuthica, P. proteamaculans, Psychrobacter* sp., *Serratia* sp., *Shewanella* sp., *Sphinogomonas* sp., *Vibrio* sp., *Xanthomonas* sp.	Hollis, 1951; De Freitas & Germida, 1990; Sturz et al., 1998b; Sturz, Christie, & Matheson, 1998a
Red clover (*Trifolium pretense*), leaves, stem and root	*Acidovorax* sp., *Aerobacter cloaceae, Agrobacterium rhizogenes, A. tumefaciens, Arthobacter ilicis, Bacillus brevis,B. megaterium Bordetella* sp., *Cellulomonaas, Comamonas* sp.,	Sturz et al., 1998a

	Curtobacterium citreum sp., *C. luteum, Deleya* sp., *Enterobacter* sp., *Escherichia* sp., *Kingella denitrificans, K. kingae, Klebsiella* sp., *Methylobacterium* sp., *Micrococcus varians, Pantoea agglomerans, Pasteurella* sp., *Phyllobacterium* sp., *Pseudomonas cichorii, P. corrugata, P. fulva, P. syringae, P. tolaasii, Psychrobacter immobilis, Rhizobium* sp., *Serratia* sp., *Sphingomonas* sp., *Variovorax* sp., *Xanthomonas compestris, X. oryzae*	
Wild rice (*Oryza officinalis, O. barthii, O. rufipogon, O. glandiglumis, O. breviligulata*) and cultivated rice (*Oryza sativa*), root and stems	*Agrobacterium* sp., *Azorhizobium* sp., *Azospirillum lipoferum, A. brasilense, Bacillus* sp., *Bradyrhizobium* sp., *Burkholderia graminis, Herbaspirillum rubrisubalbicans, H. seropedicae, Ideonella dechloratans, Enterobacter cancerogenus, Pseudomonas* sp., *Rhizobium leguminosarum*	Reddy et al., 1997; Stolzfus, So, Ladha, & de Bruijn, 1997; Elbeltagy, Suzuki, & Minamisawa, 2001; Yanni et al., 1997; Chaintreuil et al., 2000; Villard et al., 1998
Rough lemon (*Citrus jambhiri*), root, and fruit	*Achromobacter* sp., *Alcaligenes* sp., *Moraxella acinetobacter, Actinomyces* sp., *Arthrobacter* sp., *Bacillus* sp., *Citrobacter* sp., *Coryebacterium* sp., *Enterobacte sakazakiir,*	Gardner et al., 1982a; Araujo, Marcos, Maccheroni, Van Vuurde, & Azevedo, 2002; Gardner et al., 1982a

	E. agglomerans, Flavobacterium sp., *Kelbsiella* sp., *Providencia* sp., *Pseudomonas putida, Serratia liquefaciens, Shigella* sp., *Vibrio* sp., *Yersinia* sp., *Rickettsia-like* sp.	
Sorghum bicolor, shoot	*Herbasspirillum* sp.	James et al., 1997; Olivares, James, Baldani, & Dobereiner, 1992
Sugar beet (*Beta vulgaris*), root	*Bacillus* sp., *Corynebacterium* sp., *Erwinia* sp., *Lactobacillus* sp., *Pseudomonas* sp., *Xanthomonas*sp.	Jacobs et al., 1895
Sugar cane (*Saccharum officinarum*), root and stem	*Acetobacter* sp., *Herbaspirillum* sp., *Gluconacetobacte* sp., *diazotrophicus* sp.	Calvalcante & Bobereiner, 1988; James et al., 1998; Olivares et al., 1992; Rennie et al., 1982
Teosinte (Zea luxurians), stem and root	*Klebsiella* sp.	Palus et al., 1996
Soybean (*Glycine max*), stem, leaves, root	*Pseudomonas citronellolis, P. oryzihabitans, P. staminea, Erwinia* sp., *Klebsiella pneumoniae, K. oxytoca, Agrobacterium, Caulobacter* sp., *Enterobacter agglomerans, Pantoea* sp.	Kuklinsky-Sobral et al., 2004
Tangerine (*Citrus reticulate*) and sweet orange (*C. sinensis*), stem	*Curtobacterium flaccumfaciens, Enterobacter cloacae, Methylobacterium* sp., *M. zatmanii, Nocardia*	Araujo et al., 2002

	sp., *Pantoea agglomerans, Xanthomonas campestris, Bacillus pumilus*	
Tomato (*Lycopersicon esculentum*) stem and fruit	*Pseudomonas syringae, Escherichia coli*	Pillay et al., 1997; Samish, Etinger-Tulezynska, & Bick, 1961
Peanut	*Bacillus* spp.	Kumar, 1996
Wheat (*Triticum aestivum*)	*Bacillus polymyxa, Burkholderia cepacia*	Mavingui et al., 1999; Reddy & Rahe, 1989; Chanway et al., 1988; Balandreau et al., 2001
Brassica spp. (Cauliflower, ect)	*B. polymyxa, Achromobacter* spp.	Pichard & Thouvenst, 1995; Bertrand et al., 2000
Beans, peas (*Phaseolus vulgaris*)	*B. polymyxa, B.* spp.	Petersen et al., 1996; Walker et al., 1998
Crested wheatgrass (*Agropyron cristatum*)	*B. polymyxa*	Holl, 1988
Lupine (*Lupinus* sp.), root	*Burkholderia cepacia*	Balandreau et al., 2001

*Table is a modification of Chanway (1998) and Kobayashi and Palumbo (2000) with current additions by the present authors.

Table 4. Isolations of natural infections or artificial inoculations of bacterial endophytes from trees.

Plant species	Bacterial taxa	References
Lobolly pine (*Pinus taeda*)	*Bacillus subtilis, B. pumilus, B. sphaericus, Paenibacillus macerans, Serratia marcescens*	Enebak, Wei, & Kloepper, 1997
Slash pine (*Pinus elliottii*)	*Bacillus subtilis, B. pumilus, B. sphaericus, Paenibacillus macerans, Serratia marcescens*	Enebak et al., 1997

Black spruce (*Picea mariana*)	*Pseudomonas fluorescens, Arthrobacter, A. citreu*	Beall & Tipping, 1989
White spruce (*Picea glauca*)	*Pseudomonas, Arthrobacter*	Beall & Tipping, 1989
White spruce hybrids (*P. glauca x engelmannii*)	*Bacillus polymyxa Pseudomonas putida, Hydrogenophaga pseudoflava, Staphyloccus hominis*	Bent & Chanway, 1998; Chanway, Radley, & Holl, 1989; Chanway, Shishido, & Holl, 1994; O'Neill, Chanway, Axelrood, Radley, & Holl, 1992
Pinus sylvestris	*Athrobacter, Agrobacterium,*	Pokojska-Burdziej, 1982
Pinus banksiana	*Pseudomonas putida*	Beall & Tipping, 1989; Chanway & Holl, 1991
Lodgepole pine (*Pinus contorta*)	*Bacillus polymyxa*	Chanway & Holl, 1991; Bent & Chanway, 1998; Shishido, Loeb, & Chanway, 1995
Pseudotsuga menziesii	*Arthobacter oxydans, Pseudomonas syringae, Microbacterium omperiale, Pseudomonas aureofaciens, Bacillus polymyxa*	Bent & Chanway, 1998; Chanway & Holl, 1992; Chanway et al., 1994
Tsuga heterophylla	*Bacillus polymyxa*	Chanway, 1995
Quercus serrata	*Azobacter chroococcum, Bacillus megaterium*	Pandey, Bahl, & Rao, 1986
Malus pumila	*Pseudomonas*	Caesar & Burr, 1987

Euclayptus camaldulensis	*Azotobacter chroococcum, bacillus megaterium*	Mohammad & Prasad, 1988
Fagus sylvatica	*Agrobacterium radiobacter*	Leyval & Berthelin, 1989
Casuarina cunninghamiana	*Azospirillum brasilense*	Rodríguez-Barrueco, Cervantes, Subbarao, & Rodríguez-Caceres, 1991
Oak (*Quercus ithaburensis*)	*Azospirillum brasilense*	Zaady, Perevolotsky, & Okon, 1993
Chestnut (*Castanea sativa*)	*Bacillus subtilis*	Wihelm, Arthofer, Schafleitner & Krebs, 1998

Some hosts are reported as having several endophytes and these serve as hosts for a wide range of endophytic taxa. For example, in soybean one analysis of two cultivars of soybeans indicated significant differences in the population of endophytes, which was based on 16S rDNA analysis, and the results indicated that on these cultivars there are endophytic species closely related to epiphytic species and these occur throughout the α-, β-, and γ-proteobacteria subgroups (Kuklinsky-Sobral et al., 2004). This study indicated that while all three subgroups were present, the γ-proteobacteria subgroup was the most dominant with the Enterobacteriaceae showing the highest diversity within this section. Within the primitive bacteria, the Firmicutes, the Gram-positive group, the dominant endophytic species are primarily those belonging to the *Bacillus* species. Additional studies have also demonstrated that a large number of bacterial endophytes can be isolated from the same hosts (Mundt & Hinkle, 1976; McInroy & Kloepper, 1995a). However, it is not know how many of these are actually co-existing as physiologically viable endophytes within the same niche, or are they there as dominant propagules. Competition for intercellular spaces of the same plant must occur and predictions on the dominant species must be determined. For example, competition can be indirectly measured by characterizing the functional potential of each endophyte expressed by either niche overlap indices based on sole carbon source utilization patterns.

The numbers of bacterial endophytes contained within tissues of plants vary but are exceedingly much higher than pathogenic bacteria. Internal concentrations of as large as 10^7 CFU g^{-1} wet weight of plant tissue are reported for several species, while concentrations of 10^2 to 10^6 cfu g^{-1} of wet weight of plant tissue appear the normal. Additionally, in endophytic species that are equally adapt at colonizing the entire plant axis, the distribution of cell of the same species varies within the organs along the axis, with the roots usually containing the highest number of cfu for that species (Bacon & Hinton, 2002; Jacobs, Bugbee, & Gabrielson, 1995). The total number of endophytic bacteria colonizing a plant is related to plant genotype, biotic and abiotic environmental factors and bacterial genotype. Endophyte populations have been observed to fluctuate seasonally (Gardner et al., 1982a; McInroy & Kloepper, 1995a; Whitesides & Spotts, 1991), depending on the host growth temperature (Pillay & Nowak, 1997), genetics of the hosts, either intra- or interspecific (Pillay & Nowak, 1997; Kuklinsky-Sobral et al., 2004), presence of competing endophytic or epiphytic species (Bent & Chanway, 1998), host chemical alterations, and the dryness, heat and salinity of soils (Tejera, Oetega, Gonzalez-Lopex, & Lluch, 2003) and the dryness, heat and salinity of the soil.

3.2. In Planta Secondary Metabolite Production and Affects on Hosts

A primary focus of all studies on endophytic bacteria is based on the use for improvement in plant agriculture. Endophytic bacteria have a wide spectrum of effects on hosts, which is due to the production of secondary metabolites that alter the host's growth and phenotype. The studies are based on the two major uses of bacterial endophytes: Plant growth promotion and plant disease protection and these are discussed below.

Enhanced plant growth is attributed to either or both production of phytoharmones, such as auxins, cytokinins, and the gibberellins (Hall & Chanway, 1992; Lebuhn, Heulin, & Hartmenn, 1997) and antibiotic production (Leifert et al., 1995), by supplying the infected plant with nitrogen, and or solubilizing soil iron and phosphorus for plant growth (Shishido, Breuil, & Chanway, 1999; Sturz et al., 2000), and suppressing endogenous levels of ethylene production (Glick 1995). Of particular important are the diazotrophic endophytes that do not form stem and root nodules, but fix atmospheric dinitrogen into forms that are used by plants resulting in increased growth. These include species of Herbaspirillum, *Azorhizobium, Azoarcus, Azospirillum* and *Acetobacter,* and *Pseudomonas fluorescens*, (James & Olivares, 1998; Kirchhof et al., 1997; Elbeltagy et al., 2001). However, there is concern that the amount of nitrogen produced *in planta* is too small to stimulate growth. They are aptly referred to as the growth promoting bacteria and present information suggests that these have a tendency in being restricted to hosts, which may limit their use as plant growth promotion in general. For example, *Acetobacter diazotrophicus* associates mainly with sugar-rich plants such as sugarcane, sorghum, and sweet potato (James et al., 1997; Kirchhof et al., 1997), but apparently are further restricted to colonizing grasses regardless of their sugar rich status. These bacteria are more in line with the basic concept of endophytes as defined above than are the root and stem nodule diazotrophs and are considered obligate endophytes since they are only found in association with plants.

Biological control of plant disease is perhaps the major effects observed on plants infected by endophytic bacteria. In vitro suppression of plant pathogens is demonstrated in culture, which may not be the mechanisms expressed under natural conditions, but it is these observations that created a potential use of endophytic bacteria. Field data support the suppression of various diseases of plants that correlates with most of the in vitro studies, which drives the interests in bacterial endophytes. Several studies indicate that species of endophytes produce antibiotics, produce siderophores, and lytic enzymes, but very few support the7 production of these substances in *planta*. The current mechanism for disease control by endophytic bacteria in plants is induced systemic resistance (Kloepper & Beauchamp, 1992; Chen, Bauske, Musson, Rodriguez-Kabana, & Kloepper, 1995).

4. TECHNOLOGICAL EXPLOITATION OF THE ENDOPHYTIC NICHE

Historically, endophytic bacteria that did not produce diseases were described well over 50 years ago (Tervet & Hollis, 1948; Hollis, 1951) but bacterial endophytes are rapidly becoming a distinct group of biocontrol organisms as indicated by the recent increase in publications, which reflects an interest in their potential benefits to agriculture (Tables 3 and 4). Usually, endophytic bacteria are represented by nonhost specific strains that include both Gram-positive and Gram-negative species. However, there are some species that apparently have evolved along with plants, exploiting the endophytic habit, and have developed some degree of host specificity (Chanway, Holl, & Turkington, 1990; Rennie & Larson, 1979), suggesting that the endophytic niche is perhaps different for each plant species and would require considerable modifications to accommodate any and all bacteria. This is particularly important since there are some bacteria that do exhibit some degree of specificity as to the species of fungi they control and for that matter any of the future applications of bacterial endophyte for uses in agriculture. Thus, in addition to plant breeding, there must also be a conscious attempt at modifying the bacterium for the endophytic niche.

We have discovered a species of desert dwelling bacteria, *Bacillus mojavensis,* which was subsequently patented as an endophyte to protect plants against diseases (Bacon & Hinton, 1999). All isolates of this species tested to date readily form endophytic associations with plants, and several of these have been reported to be successful in preventing disease development (Bacon, Yates, Hinton, & Meredith, 2001; Bacon & Hinton, 2002; Bacon & Hinton, 2001). Thus, a species of bacteria with an apparent genetic predisposition to colonizing plants endophytically is suggested, indicating a species where endophytism has evolved following the suggestion of Kobayashi & Polumbi (2000). *B. mojavensis* might have all the desired aspects for serving as the model for detailed studies on bacterial endophytes. This species readily colonizes seedlings from the topical application to seed, and there are no ill effects observed over the growing season. The ability of this bacterium to colonize endophytically plants from the external application to seed is one highly desirable characteristic and it is economical since only one application is required. This endophytic bacterium promotes the growth of all plants species colonized, and this is shown in enhanced root growth (Bacon & Hinton, 2002), which also should provide for drought tolerance of *B. mojavensis*-infected plants. One patented isolate of *B. mojavensis* has been used to demonstrate for the first time that a bacterial endophyte can be used to reduce the *in planta* accumulation of a mycotoxin produced by an endophytic fungus (Bacon & Hinton, 1996). However, the association of their endophytic species with the deserts of the world implies an importance to the environment on the uniformity of species to adopt the endophytic habit, i.e., an evolved species for endophytism.

4.1. Biological Control Agents

The endophytic niche offers a unique habitat for the control of pathogens since the endophyte is contained and is not subject to the direct influence of the environment and will multiply within the intercellular spaces as the plant grows, thereby potentially colonizing the entire plant axis. The bacterium is within a stable environment, and in our preliminary trials, there is no genetic modifications of a strain during and after a season's growth, indicating that it is genetically stable. However, the use of multiple endophytic strains of the same species, or genus within the same plant might alter the genetic properties. The association is long-term, as the effects is measured after two or more years from endophyte-infected plants (Chanway, 1998), indicating its application to perennials such as trees (Chanway, 1996) and grasses (Reinhold-Hurek & Hurek, 1998) (Tables 3 and 4). Another benefit is that most endophytic bacteria can be easily applied to the plant or seed before or at planting with natural infection taking place from this one-time application (Bacon & Hinton, 2002). Further, plant hosts appear to be more acceptable of bacterial endophytes so that plant genetic modification is not required for infection.

The list of bacterial endophytes includes a variety of common and uncommon taxa (Tables 3 and 4), which implies that the association with plants has occurred throughout several if not all of the major groups of bacteria, and that the intercellular niche, which may not contain the larger assembly of bacteria as the soil, for example, is nevertheless effective in supporting bacteria.

The diversity of endophytic bacteria might reflect the large number of probable mechanisms of action reported to serve as the basis of disease suppression. Thus, the mechanisms of action for endophytic biocontrol agents include the production of antimicrobial compounds (Schnider-Keel et al., 2000; Maurhofer et al., 1992; Lambert et al., 1987; Lambert et al., 1987; Duffy & Defago, 1997; Bacon & Hinton, 2002; Leifert et al., 1995), nutrient competition and niche overlap (Kloepper, Leong, Tientize, & Schroth, 1980; Wilson & Lindow, 1999), competition for micronutrients such as iron by siderophore production (Bergeron & Weimar, 1990; Buyer & Sikora, 1990; Carrillo & Del Rosario G.Vazquez, 1992; Costa & Loper, 1994; Crowley, Wang, Reid, & Szaniszlo, 1991; Kloepper et al., 1980; Thomashow & Weller, 1990), and more recently systemic acquired host resistance (Chen et al., 1995; Liu, Kloepper, & Tuzun, 1995; Kuc, 1990; Zhang, Reddy, & Kloepper, 2002).

One aspects of the use of bacterial endophytes to control diseases is the possibility of the pathogen producing antibiotics to alter the expression of the biocontrol agent. The use of *Bacillus mojavensis* to control the production of the fumonisin mycotoxin by *Fusarium verticillioides* on maize was demonstrated with experiments conducted under greenhouse and plant growth room conditions (Fig. 3; Bacon et al., 2001). However, the field use of *B. mojavensis* resulted in less than favorable control of *F. verticillioides* and the production of the fumonisin mycotoxins by this fungus. Further, this species and other *Fusarium* species produced fusaric acid, particularly when *Fusarium*-infected plants are grown under abiotic soil stresses (Bacon, Porter, Norred, & Leslie, 1996; Capasso et al., 1996). Fusaric acid is a potent antibiotic for this species as well as several other biocontrol species (Bacon, Hinton, Porter, Glenn, & Kuldau, 2004; Landa, Cachinero-Díaz, Lemanceau, Jiménez-Díaz, & Alabouvette, 2002; Notz, Maurhofer, Dubach, Haas, & Défago, 2002; Schnider-Keel et al., 2000). However, fusaric acid tolerant mutant strains of biocontrol endophytes have been developed that now offer promise for the field use of these species (Arias, 1985; Toyoda, Hashimoto, Utsumi, Kobayashi, & Ouchi, 1988). The occurrence of fusaric acid in most species of *Fusarium* indicates a role for this substance as a wilt toxin, which is produced at concentrations below that which is toxic to plants but sufficient to control the growth of bacterial endophytes. The effects of this antibiotic on bacteria indicate an important concern for the use of endophyte (Bacon, Hinton, & Hinton, Jr., 2005).

4.2. Plant Growth

Plant growth responses to bacteria are well established for several crop plants (Surette, Sturz, Lada, & Nowak, 2003; Lazarovits & Nowak, 1997). What is not known is the direct involvement of the many agents that affect plant growth relative to specific bacteria. However, since there are so many species that can produce this effect, the causes are expected to be varied. The most obvious cause includes the production of phytoharmones such as ethylene, auxins, cytokinins (Arshad & Frankenberger, 1991; Kuklinsky-Sobral et al., 2004; Bashan & Holguin, 1997; Lazarovits & Nowak, 1997), all of which will have a positive effect on plant growth and development.

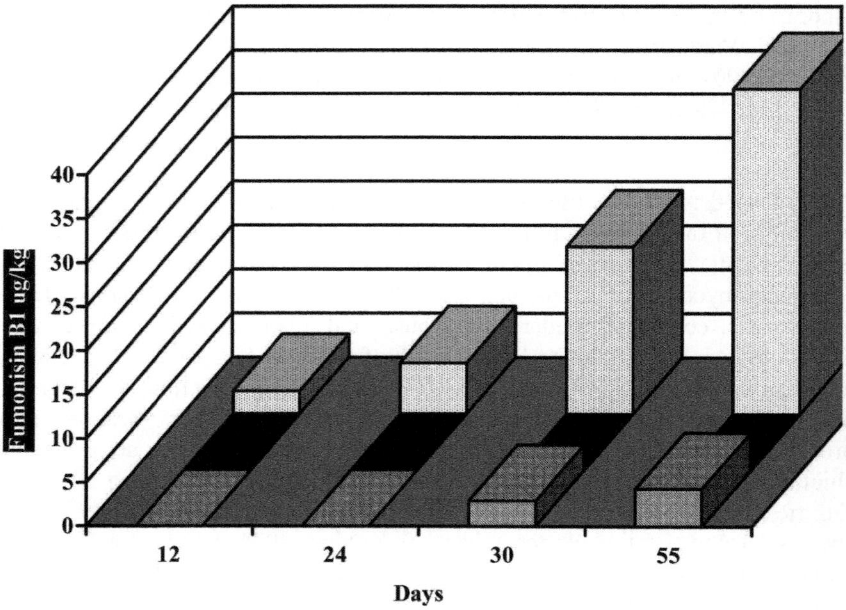

Fig. 3. Effects of *Bacillus mojavensis (B. moj)* on the fumonisin mycotoxin accumulation in maize seedling infected with *Fusarium verticillioides (F. vert)* compared to non-infected bacterium seedlings infected only with the fungus *F. verticillioides* (Control) (Bacon et al., 2001).

Other plant growth effects attributed to endophytic bacteria are enhanced mineral uptake such as the solubilizing of bound soil iron and phosphorus and by providing the plant with nitrogen (Shishido et al., 1999; Sturz et al., 2000; Surette et al., 2003). Additional effects are those that interact with soil bacterial population, prior crop history and the buildup of soil bacteria and the resulting benefits derived from allelopathy (Sturz et al., 1998b; Sturz et al., 2000; Surette et al., 2003). Nitrogen fixation by several endophytic species of bacteria includes *Azospirillum, Enterobacter cloacae, Alcaligenes, Acetobacter diazotrophicus, Herbaspirillum seropedicae, Ideonella dechlorantans,* and *Azoarcus* sp. Recent evidence has indicated the *in planta* fixation of nitrogen in rice (Elbeltagy et al., 2001) and sugarcane (James & Olivares, 1998), which in the past was controversial. Ladha & Reddy (1995) estimated that as much as 200 kg N ha^{-1} $year^{-1}$ can be produced by the use of endophytes in plants. This finding has opened the door to optimize the use of bacterial endophytes for nitrogen fixation in plants either through the use of altered host acceptance of these bacteria, the use of endophytic species yet discovered and genetic alteration of endophytic species to fix nitrogen.

The effects on plant growth can be indirect. The control of pathogenic organisms will maximize plant growth due to lack of parasitisms and diseases, as well as decreased susceptibility to abiotic stresses such as drought tolerance or frost resistance (Xu, Griffith, Patten, & Glick, 1998). Additionally, indirect effects on plant growth through stimulation of the host plant to produce an increased amount of phytoharmones (Lazarovits et al., 1997) and mineral availability.

5. FUTURE TECHNOLOGICAL APPLICATIONS FOR BACTERIAL ENDOPHYTES

An important objective of this chapter is to provide future technological uses of bacterial endophytes, a brief summary of the present and future application of bacteria endophytes follows and include the use of endophytes as on emerging technologies for biocontrol systems, gene delivery system, plant growth- and health promotion, and phytoremediation. This information is intended for the true bacterial endophyte, but since it is presented in general terms, some of the uses discussed may also prove suitable for species of endophytic bacteria with only a limited number of endophytic characteristics listed above.

5.1. Requirements of Endophytic Strains for Exploitations

Bacterial endophytes offer a unique habit for plant protection and other applications: it is a self-contain, self-perpetuating system that is buffered and protected from the many changes in the natural environment, growing as long as the plant grows requiring very little, if any, from the environment and usually having very little competition within the endophytic niche. However, endophytic bacteria intended for use in biotechnology must have specific natural requirements relative to plants and potential uses, which include biocontrol agents, delivery systems for specific pest protecting genes or their products, and to provide enhanced plant nutritional qualities. The first requirement is that the endophyte must not have the potential to induce plant disease. A second requirement is that the endophyte must be capable of being distributed within the plant, especially at locations where pests are expected to reside. Equally important, and in the absence of a uniform plant distribution, and if the intended use is for pesticide delivery, the substance must be translocated from the bacterium, through the plant and to the target. The third requirement is that the organism must be culturalable, easily transformable, and easily manipulated genetically. A fourth requirement, although not absolute is that endophytic colonization is natural and obligate for the species.

It is difficult to find a species of bacteria where endophytism is the basic characteristic of the species. This trait as a characteristic of the species is certainly important as it would not only identify evolved endophytism as discussed above, but it also would suggest a wide application for uses over a wider range of plants with highly successful results for biocontrol. There are some species where a large component of strains are endophytic and we refer you to reviews on the subject where past listings of species can be used to indicate the prevalence of the endophytic trait within a genus (Chanway, 1998; Hallmann et al., 1997; Kobayashi et al., 2000; Reinhold-Hurek & Hurek, 1998; Sturz et al., 2000; Bacon & Hinton, 1997). There are certainly other biotic and abiotic components envolved in endophytism, especially for specific strains of species that are not natural colonizers, and for those stains that co-occur normally as root nodulation bacteria and or as traditional endophytes, which is discussed below. Current reports also suggests that in maize, competitiveness by *Rhizobium etli* and the endophytic colonization are correlated with specific symbiotic plasmids (Brom et al., 1992), and that there is a strict requirement for specific plasmids for the endophytic colonization of both bean and maize (Brom, Girad, Santos, Sanjuan-Pinilla, Olivares, & Sanjuan, 2002; Santos & Brom, 1997). These plasmids are distinct from those required by the nitrogen fixing nodules by *R. leguminosarum,* which also require different additional plasmids for nodulation formation (Hynes & McGregor, 1990). These observations indicates important roles for plasmids in the associations of plants by bacteria, and suggests important potential uses for specific plasmids in the endophytic colonization of plants by bacteria, either in terms of increased efficiency or on plants that are known to be difficult as hosts or non-hosts of endophytes.

Forced experimental use of bacterial endophytes as suggested above may prove difficult. Much can be said for the use of natural endophytes where most strains have a demonstrated ability to colonize most plants. Endophytism by such species is obligate, and free living stains of such species are considered a reduced occurrence, if at all. What is needed is an examination of recent isolates of endophytic genera for such obligate endophytic species. All isolates of *B. mojavensis* examined to date are endophytic (Bacon et al., 2002) and each shows very little host specificity forming compatible relations with peas, beans, maize, wheat, rye, barley, and may other plants, thus its potential utility in agriculture is tremendous. A genetic analysis of several strains of this species indicate a tremendous amount of diversity suggesting additional differences that may define additional benefits for biocontrols (Fig. 4; Bacon et al., 2005). The 14 randomly selected strains, characterized by a ribotyping procedure using 16S rDNA sequence analysis, were isolated from different locations of the great deserts. The analysis showed subspecies or strain variation in ribotype patterns, as well as subspecies variation fragment polymorphisms, independently of a geographic origin (Fig. 4). Our interpretation of these data was that there was no correlation between specific desert origin, and ribotype patterns. Further, the ribotype groups in the area of subspecies or strain variation indicated a considerable overlap of genetically diverse stains and fragment polymorphisms within the world's distribution suggesting a high degree of genetic plasticity, which indicates that these isolates are not clones, a conclusion with high confidence since 16S rDNA analysis allows valid analysis of strain characterization. Thus, there is considerable diversity of the strains presently included within this taxon implying additional potential biocontrol and other uses. Similar molecular diversity for other species suggest that are other specific endophytic colonizers, which include the photosynthetic bradyrhizobia, as well as *Azoarcus* spp., *Rhizobium etli,Acetobacter diazotrophicus, Burholderia cepacia* genomovar III, and other species (Engelhard, Hurek & Reinhold-Hurek, 2000; Fuentess-Ramirez, Caballero-Mellado, Sepulveda, Rartinez-Romero, 1999; Chelium & Triplett, 2001). Selective endophytic colonization may also be a host related event, as here is a specific host interaction and signaling for *Azorhizobium caulinodans*, e.g., host flavonoids and flavonoids that stimulate intercellular colonization (Gough, Vasse, Galera, Batchelor, O'Callaghan, Davey, Kothari, Denarie, & Cocking, 1996; O'Callaghan, Stone, Hu, Griffiths, Davey, & Cocking, 2000).

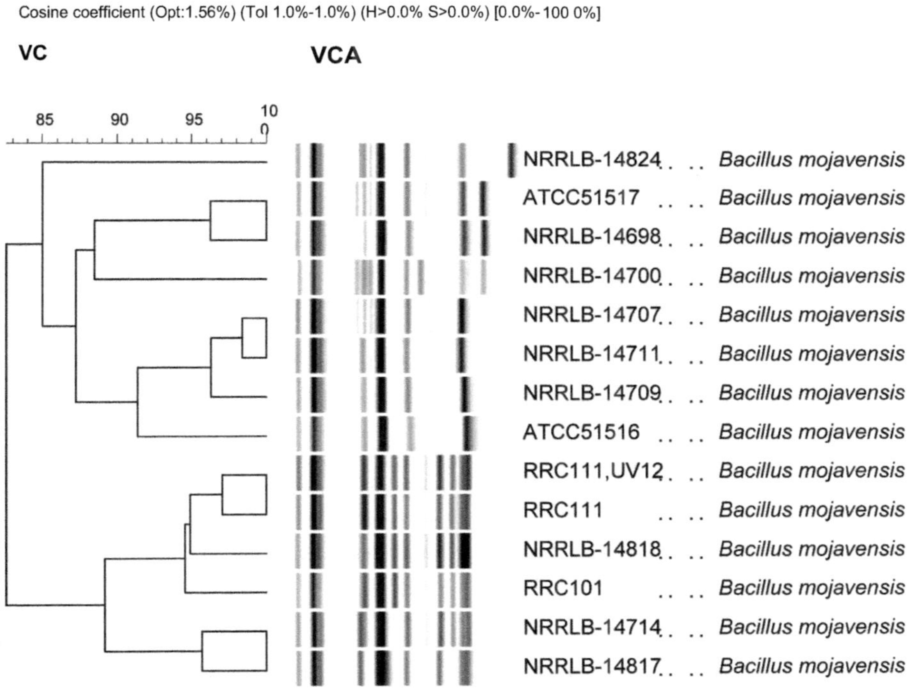

Fig. 4. Phylogenetic dendrogram and ribotype patterns obtained with *Eco*RI restriction enzyme of 16S rDNA sequences of 14 randomly selected *B. mojavensis* strains and other type species of the *Bacillus subtilis*-group. Dendrogram was calculated with Dice coefficient using UPGMA clustering methods. Origin of strains: S, Sahara Desert; M, Mojave Desert; G, Gobi Desert; I, southern Italy; Az, Arizona (Bacon et al., 2005).

Since the discovery of the *B. mojavensis* group of organisms is relatively recent, very little information is available on its natural distribution in plants, desert or otherwise, its role in protecting plants, and it effects on host. However, this group appears as being unique since it is one of few where the seedling is infected after the one-time application to the seed, which can be stored and remain viable in seed coating over dry conditions for months (Bacon & Hinton, 2002). In addition to the unknown areas stated above, we have limited information on the occurrence of this species in other soil-plant locations and knowledge concerning the specifics for genetic manipulations for increased performance for host plant is lacking. The natural endophytic characteristic of so many isolates of this species from all the major desert of the world suggests that endophytism is the rule for this species. It is not established if only the desert isolates of this species have the ability to colonize plants, which implies an environmental induced association. There is a report of an isolate of this species from brine of an oil well (Folmsbee, McInerney, & Nagle, 2004), which has an anticipated use for phytoremediation of oil spills under an earlier name of *B. licheniformis* (Lin, Minton, Sharma, & Georgiou, 1994). The endophytic ability of this strain was not determined but it produces a lipopeptide that acts as a biosurfactant, which is characteristic of most biocontrol *Bacillus* species (Georgiou, Lin, & Sharma, 1992), and may be related to basic biocontrol activity observed on media by strains and species of this genus. Recently another bacterial endophyte, *B. endophyticus*, was identified as being a natural endophyte of cotton plants (Reva, Pettersson, & Priest, 2002). The occurrence of several cryptic species within *Bacillus subtilis* group (Nakamura, Roberts, Cohan, 1999; Nielsen, Fritze, Priest, 1995; Reva et al., 2002), a large heterogeneous and obvious superficial grouping of bacteria, identifies the utility of the bacilli as a rich source for additional biocontrol endophytic species, indicating not only the possibility for other endophytic cryptic species within this.

The use of bacterial endophytes to control other endophytes, fungal or bacterial, requires that the biocontrol species must be highly competitive in order to control the endophytic pathogen. Thus what is needed is a measure of inter- and intraspecific competition to quantify or at least measure the potential for success of niche competition. However most ecological relationships directed for this purpose do not clearly measure nor relate to the intensity of competition. Niche overlap measure can to some extent be used if careful consideration is paid to specifics used in all aspects of the equations developed for niche overlap indices (Pianka 1975; Lawlor and Smith 1976). The most successful of these is the work of Schoener (1970), which was criticized favorably by Abrams (1980).

5.2. Surrogate Transformation or Paratransgenesis of Plants

We adopt surrogate transformation, as we did the definition of endophytism, from studies on endophytic fungi, and consider paratransgenesis a synonym, to describe the use of transgenic technology in crop protection where the bacterial mutualist is transformed with a specific gene or genes and the strain is reinserted in the plant for gene expression. The use of the terms and inherent in them, especially paratransgenesis, is the concept that the bacterial endophyte form a mutualistic association with the plant, or at least the association is neutral, resulting in non host responses to the biological association.

Each specific symbiotic biocontrol agent has several key characteristics that an endophyte must posses in order for it to be an effective agent for surrogate transformation. The major characteristics must include a successful transformation without the loss of endophytic fitness possessed by the non-transformant, i.e. a true biological and any mutualistic relationship must be maintained. Further, the introduced gene must be expressed and not regulated *in planta* and the product or mechanism must be delivered to the appropriate sites, which should be specific, especially if biocontrol is the desired result. The advantages of surrogate transformation over traditional plant transformation include the ease of bacterial transformations compared to the more difficult plant such as grasses, and the horizontal spread of the introduced gene is reduced. Further, the process of bacterial transformation is rapid, reasonably inexpensive, it is easier than eukaryotes, and it is possible for the introduction of large numbers of genes into endophytic bacteria (Rhitu, Prasanta, & Aqbal, 2002; Tomsaino, Leister, Dimock, Beach, & Kelly, 1995; Obukowicz et al., 1987; Sufnit, Barah, Kapuluk, Oppenheim, & Chet, 2000; Downing & Thomson, 2000). Considerable attention must be place on gene regulation *in planta,* which in most instances the transformation should be directed at the mitochondria DNA and not cytoplasmic DNA.

Paratransgenesis or surrogate transformations can be used for specific medicinal aspects for human medicine and other pharmaceutical applications (Strobel & Long, 1998). Most of the uses of surrogately transformed bacterial endophytes have been used for the delivery of pesticides, such as Bt, to various plants as the studied indicate in the references above. However, surrogate transformation has the potential for use as a gene deliver systems in plants for increased plant performance, nitrogen efficiency, enhanced biotic and abiotic stress resistance, accelerated seedling emergence and subsequent plant development, increased or improved nutritional qualities, and increased herbage yield. Endophytic bacteria have the potential as vectors for transformation of useful products that can be expressed *in planta,* augmenting either the natural nutritional qualities of foods or producing value added components of food crops, as well as pesticide living vectors. Surrogately transformed plants might have greater value and public acceptability over transgenic plants, especially if it is demonstrated that the transformed bacterium does not enter the consumed portion of the plant, i.e., seed or fruit.

5.3. Phytoremediation of Pollutants

For almost a decade, plants have been considered to have the potential to remediate polluted water and soils, (Salt, Smith, & Raskin, 1999). Almost concurrently it been shown that bacteria have the potential for remediating polluted soils as well (Walton & Anderson, 1990), but only recently has it been established that bacterial endophytes can also be used to either remediate polluted soils directly or contribute to plants that detoxify polluted soils (Siciliano et al., 2001). The of uses for phytoremediation with endophytic bacteria ranges from reducing petroleum hydrocarbon contaminations in soils to reducing heavy metals levels in soil, including water soluble and volatile organic xenobiotics. Bacteria do a more efficient job of remediating toxins than do plants simply because plants do not degrade many pollutants completely resulting in toxic decomposition products. The pollutants that are reported to be reduced by endophytic bacteria and plants include benzene, toluene, ethylbenzene, high ammonia waste and animal manures, chloroform, dichloromethane, xylene, and other hydrophobic pollutants. Endophytic degradation occurs within the plant or in the rhizosphere if the bacterium is localized in the root.

Phytoremediation may be accomplished by natural isolates within a location (Siciliano et al., 2001) or with transgenic endophytes that may be used along or in combination with plants with natural high tolerance to polluted soils such as yellow lupine (Barac et al., 2004). Transformed endophytic bacteria not only can be used to improve the efficiency of degradation of various compounds, but they can be engineered to degrade pollutants into harmless compounds (Barac et al., 2004), which is a problem of phytoremediation by vascular plants whose degradation products are oftentimes toxic volatiles or the decontaminants are more toxic than the parent pollutants. However, the strains of endophytic bacteria targeted for use in phytoremediations are those that colonize xylem as well as the intercellular spaces. This is because most of the pollutants are translocated in the plants vascular system, therefore to be efficient the bacteria must be able to colonize the plant's vascular system. In the absence of bacterial endophytes, the plant disposes of the decomposed pollutants as volatiles that are released into the environment by evaporation through the leaves. If bacteria are used most of the methylations that are used to make compounds by the plant volatile is prevented, and the amount of toxic volatiles are eliminated, making the process of microbial phytoremediation safer than when done by plants along.

6. REFERENCES

Abrams, A. (1980). Some comments on measuring niche overlap. *Ecology, 61,* 44-49.

Altus, D. P. & Canny, M. J. (1985). Loading of assimilates in wheat laves. *Plant, Cell and Environment, 8,* 275-285.

Araujo, W. L., Marcos, A., Acheron, W., Van Verde, J. W. L. & Azevedo, L. (2002). Diversity of endophytic bacteria populations and their interaction with *Xyletta fastidiosa* in citrus plants. *Applied and Environmental Microbiology, 68,* 4906-4914.

Arias, J. A. (1985). Secretary organelle and mitochondrial alterations induced by fusaric acid in root cells of *Zea mays*. *Physiological Plant Pathology, 27,* 149-158.

Arshad, M. & Frankenberger, W. T. (1991). Microbial production of plant hormones. In B. D. Keister & P. B. Cregan (Eds.), *The rhizosphere and plant growth* (pp. 327-334). Dordrecht, the Netherlands: Kluwer Academic Publishers.

Bacon, C. W. & Hinton, D. M. (2001). Control of seedling blight in wheat by *Bacillus mojavensis*. *Phytopathology, 91(supplement),* S4.

Bacon, C. W. & Hinton, D. M. (1996). Symptomless endophytic colonization of maize by *Fusarium moniliforme*. *Canadian Journal Botany, 74,* 1195-1202.

Bacon, C. W. & Hinton, D. M. (1997). Isolation and culture of endophytic bacteria and fungi. In G.R. Knudsen, M. J. McInerney, L. D. Stetzenbach, & M. V. Walter (Eds.), *Manual of Environmental Microbiology* (pp. 413-421). Washington, D.C.: ASM Press.

Bacon, C. W. & Hinton, D. M. (1999). Use of *Bacillus subtilis* as an endophyte for the control of diseases caused by fungi. *US Patent & Trademark Office, November 30, 1999.*

Bacon, C. W. & Hinton, D. M. (2002). Endophytic and biological control potential of *Bacillus mojavensis* and related species. *Biological Control, 23,* 274-284.

Bacon, C. W., Hinton, D. M. & Hinton, A., Jr. (2005). Growth-inhibiting effects of concentrations of fusaric acid on the growth of *Bacillus mojavensis* and other biocontrol *Bacillus* species. *Journal of Applied Microbiology,* In Press.

Bacon, C. W., Hinton, D. M., Porter, J. K., Glenn, A. E. & Kuldau, G. (2004). Fusaric acid, a *Fusarium verticillioides* metabolite, antagonistic to the endophytic biocontrol bacterium *Bacillus mojavensis*. *Canadian Journal of Botany, 82,* 878-885.

Bacon, C. W., Porter, J. K., Norred, W. P. & Leslie, J. F. (1996). Production of fusaric acid by *Fusarium* species. *Applied and Environmental Microbiology, 62,* 4039-4043.

Bacon, C. W., Yates, I. E., Hinton, D. M. & Meredith, F. (2001). Biological control of *Fusarium moniliforme* in maize. *Environmental Health Perspectives, 109,* 325-332.

Balandreau, J., Viallard, V., Cournoyer, B., Coenye, T., Laevens, S. & Vandamme, P. (2001). *Burkholderia* cepacia Genomovar III is a common plant associated bacterium. *Applied and Environmental Microbiology*, *67,* 982-985.

Barac, T., Taghavi, S., Borremans, B., Provoost, A., Oeyen, L., Colpaert, J. V. et al. (2004). Engineered endophytic bacteria improve phytoremediation of water-soluble, volatile, organic pollutants. *Nature Biotechnolgy, 22,* 583-588.

Bashan, Y. & Holguin, G. (1997). *Azospirillum*-plant relationships: environmental and physiological advances (1990-1996). *Canadian Journal of Microbiology, 43,* 103-121.

Beall, F. & Tipping, B. (1989). Plant growth-promoting rhizobacteria in forestry. Forest Research market Proceedings, Ontario Forest Research Commission, and Toronto Abstract. 117.

Bell, C. R., Dickie, G. A., Harvey, W. L. G. & Chan, J. W. Y. F. (1995). Endophytic bacteria in grapevine. *Canadian Journal of Microbiology, 41,* 46-53.

Bent, E. & Chanway, C. P. (1998). The growth-promoting effects of a bacterial endophyte on lodgepole pine are partially inhibited by the presence of other rhizobacteria. *Canadian Journal of Microbiology, 44,* 980-988.

Bergeron, R. J. & Weimar, W. R. (1990). Kinetics of iron acquisition from ferric siderophores by *Paracoccus denitrificans*. *Journal of Bacteriol., 172,* 2650-2657.

Bertrand, H., Plassard, C., Pinochet, X., Touraine, B., Normand, P. & Clayer-Merel, J.C. (2000). Stimulation of the ionic transport system in *Brassica napus* by a plant growth-promoting rhizobacterium (*Achromobacter* sp.). *Canadian Journal of Microbiology, 46,* 229-236.

Brom, S., Garcia de los Santos, A., Stepkowsky, T., Flores, M., Gavila, G., Romero, D. & Palacios. (1992). Different plasmids of *Rhizobium leguminosarum* bv. Phaseoli are required for optimal symbiotic performance. *Journal of Bacteriology. 174,* 5183-5189.

Brom, S., Girard, Garcia-de los Santos, A., Sanjuan-Pinilla, J. M., Olivares, J., & Sanjuan, J. (2002). Conservation of plasmid-encoded traits among bean-nodulating *Rhizobium* species. Applied Environmental Microbiology. *68,* 2555-2561.

Brooks, D. S., Gonzales, C. F., Appel, D. N. & Filer, T. H. (1994). Evaluation of endophytic bacteria as potential biological control agents for oak wilt. *Biological Control, 4,* 373-381.

Buyer, J. S. & Sikora, L. J. (1990). Rhizosphere interactions and siderophores. *Plant and Soil, 129,* 101-107.

Caesar, A. J. & Burr, T. J. (1987). Growth promotion of apple seedling and rootstocks by specific strains of bacteria. *Phytopathology, 77,* 1583-1588.

Calvalcante, C. P. & Bobereiner, J. (1988). A new acid-tolerant nitrogen-fixing bacterium associated with sugarcane. *Plant and Soil, 108,* 23-31.

Canny, M. J. (1995). Apoplastic water and solute movement: New rules for an old space. *Annual Review of Plant Physiology, 46,* 215-236.

Canny, M. J. & McCully, M. E. (1988). The xylem sap of maize roots: its collection, composition and formation. *Australian Journal of Plant Physiology, 15,* 557-566.

Canny, M. J. & Huang, C. X. (1993). What is in the intercellular spaces of roots? Evidence from the cryo-analytical-scanning electron microscope. *Physiologia Plantarum, 87,* 561-568.

Capasso, R., Evidente, A., Cutignano, A., Vurro, M., Zonno, M. C. & Bottalico, A. (1996). Fusaric and 9, 10-dehydrofusaric acids and their methyl esters from *Fusarium nygamai. Phytochemistry, 41,* 1035-1039.

Carrillo, G. C. & Del Rosario, G. Vazquez, M. (1992). Comparative study of siderophore-like activity of *Rhizobium phaseoli* and *Pseudomonas fluorescens*. *Journal of Plant Nutrition, 15,* 579-590.

Chaintreuil, C., Giraud, E., Prin, Y., Lorquin, J., Ba, A., Gillis, M., Be Lajudie, P. & Dreyfus, B. (2000). Photosynthetic *Bradyrhizobia* are natural endophytes of wild rice *Oryza breviligulata. Applied Environmental Microbiology, 66,* 5437-5447.

Chanway, C. P. (1995). Differential response of western hemlock from low and high elevations to inoculation with plant growth-promoting *Bacillus polymyxa. Soil Biology and Biochemistry, 27,* 767-775.

Chanway, C. P. (1996). Inoculating of tree roots with plant growth promoting soil bacteria: an emerging technology for reforestation. *Forest Science, 43,* 99-112.

Chanway, C. P. (1998). Bacterial endophytes: ecological and practical implications. *Sydowia, 50,* 149-170.

Chanway, C. P. & Holl, F. B. (1991). Biomass increase and associative nitrogen fixation of mycorrhizal *Pinus contorta* Dougl. Seedlings inoculated with a plant growth promoting *Bacillus* strain. *Canadian Journal Botany, 70,* 1025-1031.

Chanway, C. P. & Holl, F. B. (1992). Influence of soil biota on Douglas-fir (*Pseudotsuga menziesii* (Chanway & Holl, 1992) Franco) seedling growth: the role of rhizosphere bacteria. *Canadian Journal Botany, 70,* 1025-1031.

Chanway, C. P., Holl, F. B. & Turkington, R. (1990). Specificity of association between *Bacillus* isolates and genotypes of *Lolium perenne* and *Trifolium repens* from a grass - legume pasture. *Canadian Journal Botany, 68,* 1126-1130.

Chanway, C. P., Holl, F. B. & Turkington, R. (1988). Cultivar specific growth promotion of spring wheat (*Triticum aestivum* L.,) by co-existent *Bacillus* Species. Canadian Journal of Microbiology, *34,* 925-929.

Chanway, C. P., Radley, R. A. & Holl, F. B. (1989). Bacterial inoculation of pine (*Pinus glauca* Voss) and Douglas-fir (*Pseudotsuga menziesii* (Chanway, Radley, & Holl, 1989) Franco) grown in containers. In T.D.Landis (Ed.), *Proceeding of International Foest Nursery Association Annual Meeting* (pp. 93-97). Gen. Tech. Rep. RM–84.

Chanway, C. P., Shishido, M. & Holl, F. B. (1994). Root-endophytic and rhizosphere plant growth promoting *Bacillus* strains causes increased seedling emergence and biomass. In M. H. Ryder (Ed.), *Improving plant productivity with rhizosphere bacteria. Proc. 3rd International Workshop PGPR* (pp. 72-74). Adelaide: CSIRO Div. Soils.

Chelius, M. C. & Triplett, W. (2000). Immunolocalization of dinitrogenase reductase produced by *Klebsiella pneumoniae* in association with *Zea mays. Applied Environmental Microbiology, 66,* 783-787.

Chelius, M. C. & Triplett, W. (2001). The diversity of Archaea and bacteria in association with roots of *Zea mays* L. *Microbial Ecology,* 41, 252-263.

Chen, C., Bauske, E. M., Musson, G., Rodriguez-Kabana, R. & Kloepper, J. W. (1995). Biological control of *Fusarium* wilt of cotton by use of endophytic bacteria. *Biological Control, 6,* 83-91.

Costa, J. M. & Loper, J. E. (1994). Characterization of siderophore production by the biological control agent *Enterobacter cloacae. Molecular Plant-Microbe Interactions, 7,* 440-448.

Crowley, D. E., Wang, Y. C., Reid, C. P. P. & Szaniszlo, P. J. (1991). Mechanisms of iron acquisition from siderophores by microorganisms and plants. *Plant and Soil, 130,* 179-198.

De Freitas, J. R. & Germida, J. J. (1990). Plant growth promoting rhizobacteria for winter wheat. *Canadian Journal of Microbiology, 36,* 265-272.

de Souza, A. O., Pamphile, J. A., Sartori, C. L. D. M., da Rocha, C. & Azevedo, J. L. (2004). Plant-microbe interaction between maize (*Zea mays* L.) and endophytic microorganisms observed by scanning electron microscopy. *Acta Scientiarum. Biol. Sci., 26,* 357-359.

Delrot, S. & Bonnemain, J. L. (1981). Involvement of protons as a substrate for the sucrose carrier during phloem loading in *Vicia fava* leaves. *Plant Physiology, 67,* 560-564.

Dobereiner, J. & Pedrosa, F. O. (1987). *Nitrogen-fixing bacteria in nonleguminous crop plants.* New York: Springer-Verlag.

Dong, Z., Canny, M. J., McCully, M. E., Roboredo, M. R., Cabadilla, C. F., Ortega, E. et al. (1994). A nitrogen-fixing endophyte of sugarcane stems. A new role for the apoplast. *Plant Physiology, 105,* 1139-1147.

Downing, K. & Thomson, J. A. (2000). Introduction of the *Serratia marcescens chiA* gene into an endophytic *Pseudomonas fluorescens* for the biocontrol of phytopathogenic fungi. *Canadian Journal of Microbiology, 46,* 363-369.

Duffy, B. K. & Defago, G. (1997). Zinc improves biocontrol of Fusarium crown and root rot of tomato by *Peudomonas fluorescens* and represses the production of pathogen metabolites inhibitory to bacterial antibiotic biosynthesis. *Phytopathology, 87,* 1250-1257.

Elbeltagy, A., Nishioka, K., Salo, T., Ye, B., Hamada, T. Isawa, T., Mitsam, H. & Minomusawa, K. (2001). Endophytic colonization and in planta nitrogen fixation by a *Herbaspirillum* sp. isolated from wild rice species. *Applied and Environmental microbiology, 67,* 5285-5293.

Enebak, S. A., Wei, G. & Kloepper, J. W. (1997). Effects of plant growth-promoting rhizobacteria on loblolly and slash pine seedlings. *Forest Science, 44,* 139-144.

Engelhard, M., Hurek, T. & Reinhold-Hurek, B. (2000). Preferential occurrence of diazotrophic endophytes, *Azoarcus* spp., in wild rice species and land races of *Oryza stativa* in comparison with modern races. *Environmental Microbiology, 2,* 131-141.

Farrar, S. C. & Farrar, J. F. (1986). Compartmentation and fluxes of sucrose in intact leaf blades of barley. *New Phytologist, 103,* 645-647.

Farrar, S. C. & Lewis, D. H. (1987). Nutrient relations in biotrophic infections. In G.F. Ayres & P.G. Ayres, (Eds.), Fungal Infection of Plants (pp. 92-132). New York: Cambridge University Press.

Fisher, P. J., Petrini, O. & Scott, H. M. L. (1992). The distribution of some fungal and bacterial endophytes in maize (*Zea mays* L.). *New Phytologist, 122,* 299-305.

Folmsbee, M. J., McInerney, M. J. & Nagle, D. P. (2004). Anaerobic growth of *Bacillus mojavensis* and *Bacillus subtilis* requires deoxyribonucleosides or DNA. *Applied and Environmental Microbiology, 70,* 5252-5257.

Fuentes-Ramiez, L., Caballero-Mellado, J., Sepulveda, J. & Romero, D. (1999). Colonization of sugarcane by *Acetobacter diazotrophicus* is inhibited by high N-fertilization. FEEM Microbiology Ecology. *29,* 117-128.

Gagne, S., Richard, H., Rousseau, H. & Antoun, H. (1987). Xylem-residing bacteria in alfalfa roots. *Canadian Journal of Microbiology, 33,* 996-1000.

Gardner, J. M., Feldman, A. W. & Zablotowicz, R. M. (1982a). Identity and behavior of xylem residing bacteria in rough lemon roots of Florida citrus trees. *Applied and Environmental Microbiology, 43,* 1335-1342.

Garcia-de los Santos, A. & Brom, S. (1997). Characterization of two plasmid-borne *lpsβ* loci of *Rhizobium etli* required for lipopolysaccharide synthesis and for optimal interaction with plants. *Molecular Plant-Microbe Interaction, 13,* 1283-1292.

Georgiou, G., Lin, S. G. & Sharma, M. M. (1992). Surface-active compounds from microorganisms. *Bio/technology, 10,* 60-65.

Giaquinta, R.T. (1976). Evidence for phloem loading from the apoplast. Chemical modification of membrane sulhydryl groups. *Plant Physiology*, *57*: 872-875.

Giaquinta, R. T. (1977). Phloem loading of sucrose, pH dependence and selectivity. *Plant Physiology, 59,* 750-753.

Glick, B.R. (1995). The enhancement of plant growth by free-living bacteria. *Canadian Journal of Microbiology, 41,* 109-117.

Gough, G., Webster, G., Vasse, J., Galera, C., Batchelor, C., O'Callaghan, K., Davey, M., Kothari, S., Benarie, J. & Cocking, E. 1996. Specific flavonoids stimulate intercellular colonization of non-legumes by *Azorhizobium caulinodans*. In G. Stacey, B. Mullin, & P.M. Gresshoff (Eds.), *Biology of Plant-Microbe Interactions* (pp. 409-415). St. Paul, MN: International Society for Molecular Plant-Microbe Interactions.

Gupta, R. S. (2002). Phylogeny of Bacteria: Are we now close to understanding it? *ASM News, 68,* 284-291.

Hallmann, J. (2001). Plant interactions with endophytic bacteria. In M.J.Jeger & N. J. Spence (Eds.), *Biotic Interactions in Plant-Pathogen Associations* (pp. 87-119). New York: CABI Publishing.

Hallmann, J., Quadt-Hallmann, A., Mahaffee, W. F. & Kloepper, J. W. (1997). Bacterial endophytes in agricultural crops. *Canadian Journal of Microbiology, 43,* 895-914.

Holl, F. B. & Chanway, C.P. (1992). Rhizosphere colonization and seedling growth promotion of lodgepole pine by *Bacillus polymyxa. Canadian Journal of Microbiology, 38,* 303-308.

Holl, F.B., Chanway, C.P., Turkington, R. & Radley, R. (1988). Growth response of crested wheatgrass (*Agropyron cristatum* L.), white clover (*Trifolium repens* L.), and perennial ryegrass (*Lolium perenne* L.) to inoculation with *Bacillus polymyxa. Soil Biology and Biochemistry, 20,* 19-24.

Hollis, J. P. (1951). Bacteria in healthy tissue. *Phytopathology, 41,* 350-367.

Huber, S. C. & Moreland, D. E. (1980). Translocation efflux of sugars across the plasmalemma of mesophyll protoplasts. *Plant Physiology, 65,* 560-562.

Hynes, M. F. & McGregor, N. F. (1990). Two plasmids other than the nodulation plasmid are necessary for formation of nitrogen-fixing nodules by *Rhizobium leguminosarum*. Molecular Microbioology. *4,* 567-574.

Jacobs, M. J., Bugbee, W. M. & Gabrielson, D. A. (1895). Enumeration, location, and characterization of endophytic bacteria within sugar beet roots. *Canadian Journal Botany, 63,* 1262-1265.

James, E. K. & Olivares, F. L. (1998). Infection and colonization of sugar cane and other graminaceous plants by endophytic diazotrophs. *Critical Reviews in Plant Sciences, 17,* 77-119.

James, E. K., Olivares, F. L., Baldani, J. I. & Döbereiner, J. (1997). Herbaspirillum, an endophytic diazotroph colonizing vascular tissue in leaves of *Sorghum bicolor* L. Moench. *Journal of Experimental Botany, 48,* 785-797.

Jimenez-Salgado, T., Fuentes-Ramirez, L. E., Tapia-Hernandez, A., Mascarua-Esparza, A., Martinez-Romero, E. & Caballero-Mellado, J. (1997). Coffea arabica, a new host plant for *Acetobacter diazotrophicus* and isolation of other nitrogen-fixing acetobacteria. *Applied and Environmental Microbiology, 63,* 3676-3683.

Kado, C. I. (1992). Plant pathogenic bacteria. In A. Balows, H. G. Truper, M. Dworkin, W. Harder, & K.-H. Schleifer (Eds.), *The Prokaryotes* (pp. 660-662). New York: Springer-Verlag.

Kirchhof, G., Reis, V. M., Baldani, J. I., Eckert, B., Döbereiner, J. & Hartmann, A. (1997). Occurrence, physiological and molecular analysis of endophytic diazotrophic bacteria in gramineous energy plants. *Plant and Soil, 194,* 45-55.

Klement, Z. (1965). Method of obtaining fluid from the intercellular spaces of foliage and the fluid's merit as substrate for phytobacterial pathogens. *Phytopathology, 55,* 1033-1034.

Kloepper, J. W. & Beauchamp, C. J. (1992). A review of issues related to measuring colonization of plant roots by bacteria. *Canadian Journal of Microbiology, 38,* 1219-1232.

Kloepper, J. W., Leong, J., Tientize, M. & Schroth, M. N. (1980). Enhanced plant growth by siderophores produced by plant growth promoting rhizobacteria. *Nature, 286,* 885-886.

Kneale, J. & Farrar, J. F. (1985). The localization and frequency of haustoria in colonies of brown rust on barley leaves. *New Phytologist, 101,* 495-505.

Knutson, D. M. (1973). The bacteria in sapwood, wetwood, and heartwood of trembling aspen (*Populus tremuloides*). *Canadian Journal Botany, 51,* 498-500.

Kobayashi, D. Y. & Palumbo, J. D. (2000). Bacterial endophytes and their effects on plants and uses in agriculture. In C.W. Bacon & J. F. White, Jr. (Eds.), *Microbial Endophytes* (pp. 199-233). New York: Marcel Dekker, Inc.

Kuc, J. (1990). Immunization for the control of plant disease. In D. Hornby (Ed.), *Biological control of soil-borne plant pathogens* (pp. 355-373). Wallingford, Oxon, UK: CAB International.

Kuklinsky-Sobral, K., Araujo, W. L., Mendonça, C., Geran, L. C., Pískala, A. & Azevedo, J. L. (2004). Isolation and characterization of soybean-associated bacteria and their potential for plant growth promotion. *Environmental Microbiology, 6,* 1244-1251.

Kumar, B. S. D. (1996). Crop improvement and disease suppression by a *Bacillus* spp. SR2 from peanut rhizosphere. *Indian Joumal of Experimental Biology, 34,* 794-798.

Kursanov, A. L. & Brovchenko, M. I. (1970). Sugars in the free space of leaf plates: their origin and possible involvement in transport. *Canadian Journal Botany, 48,* 1243-1250.

Ladha, J. K. & Reddy, P. M. (1995). Introduction: assessing opportunities for nitrogen fixation in rice - a frontier project. *Plant and Soil, 194,* 1-10.

Lalande, R. N., Bissonnette, N., Coutlee, D. & Antoun, H. (1989). Identification of rhizobacteria from maize and determination of their plant-growth promoting potential. *Plant and Soil, 115,* 7-11.

Lambert, B., Leyns, E., Van Rooyen, L., Gossele, F., Popon, Y. & Swings, J. (1987). Rhizobacteria of maize and their fungal activities. *Applied and Environmental Microbiology, 53,* 1866-1871.

Landa, B. B., Cachinero-Díaz, J. M., Lemanceau, P., Jiménez-Díaz, R. M. & Alabouvette, C. (2002). Effect of fusaric acid and phytoanticipins on growth of rhizobacteria and *Fusarium oxysporum. Canadian Journal of Microbiology, 48,* 971-985.

Lawlor, L.R. & Smith, J. M. 1976. The Coevolution and stability of competing species. *American Naturalist, 110,* 79-99.

Lazarovits, G. & Nowak, J. (1997). Rhizobacteria for improvement of plant growth and establishment. *HortScience, 32,* 188-192.

Lebuhn, M., Heulin, T. & Hartmann, A. (1997). Production of auxin and other indole and phenolic compounds by *Paenibacillus polymyxa* strains isolated from different proximity to plant roots. *FEMS Microbiology Ecology, 22,* 325-334.

Leifert, C., Li, H., Chidburee, S., Hampson, S., Workman, S., Sigee, D. et al. (1995). Antibiotic production and biocontrol activity by *Bacillus subtilis* CL27 and *Bacillus pumilus* CL45. *Journal of Applied Bacteriology, 78,* 97-108.

Leyval, C. & Berthelin, J. (1989). Interactions between *Laccaria laccata, Agrobacterium radiobacter,* and beech roots: Influence of P, K, Mg, and Fe mobilization and ectomycomycorrhizal fungi. *Plant and Soil, 117,* 103-110.

Lin, S. C., Minton, M. A., Sharma, M. M. & Georgiou, G. (1994). Structural and immunological characterization of a biosurfactant produced by *Bacillus licheniformis* JF-2. *Applied and Environmental Microbiology, 60,* 31-38.

Liu, L., Kloepper, J. W. & Tuzun, S. (1995). Induction of systemic resistance in cucumber against bacterial angular leaf spot by plant growth-promoting rhizobacteria. *Phytopathology, 85,* 843-847.

Madore, M. & Webb, J. A. (1981). Lead free space analysis and veins loading in *Curcurbita pepo. Canadian Journal Botany, 59,* 2550-2557.

Maurhofer, M., Keel, C., Schnider, U., Voisard, C., Hass, D. & Defago, G. (1992). Influence of enhanced antibiotic production in *Pseudomonas fluoresces* strain CHAO on its disease suppressive capacity. *Phytopathology, 82,* 190-195.

Mavingui, P., Laguerre, G., Berge, O. & Heulin, T. (1992). Genotypic and phenotypic diversity of *Bacillus polymyxa* in the wheat rhizosphere. *Applied and Environmental Microbiology, 58,* 1894-1903.

McCully, M. E. (2001). Niches for bacterial endophytes in crop plants: a plant biologist's view. *Australian Journal of Plant Physiology, 28,* 983-990.

Mclnroy, J. A. & Kloepper, J. W. (1995a). Population dynamics of endophytic bacteria in field-grown sweet corn and cotton. *Canadian Journal of Microbiology, 41,* 895-901.

Mclnroy, J. A. & Kloepper, J. W. (1995b). Survey of indigenous bacterial endophytes from cotton and sweet corn. *Plant and Soil, 173,* 337-342.

Misaghi, I. J. & Donndelinger, C. R. (1990). Endophytic bacteria in symptom-free cotton plants. *Phytopathology, 80,* 808-811.

Mohammad, G. & Prasad, R. (1988). Influence of microbial fertilizers on biomass accumulation in polypotted *Eucalyptus camaldulensis* Dehn. seedlings. *Journal of Tropical Forestry, 4,* 74-77.

Moon, C. J., Clough, B. F., Peterson, C. A. & Allaway, W. G. (1986). Apoplasmic and symplastic pathways in *Avicennia marina* (Forsk.) Vierh. roots revealed by fluorescent dyes. *Australian Journal of Plant Physiology, 13,* 673-648.

Mundt, J. O. & Hinkle, N. F. (1976). Bacteria within ovules and seeds. *Applied and Environmental Microbiology, 32,* 694-698.

Myers, D. F. & Strobel, G. A. (1983). *Pseudomonas syringae* a microbial antagonist of *Cyratocystis ulmi* in the apoplast of American elm. *Transactions British Mycological Society, 80,* 389-394.

Nakamura, L., Roberts, M.S. & Cohan, F.M. (1990). Relationship of *Bacillus subtilis* clades associated with strains 168 and W23: a proposal for *Bacillus subtilis* subsp. *subtilis* subsp. nov. and *Bacillus subtilis* subsp. *spizizeni* subsp. nov. *International Journal of Systematic Bacteriology, 49,* 1211-1215.

Nielsen, P., Fritze, S. & Priest, F.G. (1995). Phenetic diversity of alkaliphilic *Bacillus* strains: proposal for new species. *Microbiology, 141,* 1745-1761.

Notz, R., Maurhofer, M., Dubach, H., Haas, D. & Défago, G. (2002). Fusaric acid-producing strains of *Fusarium oxysporum* alter 2, 4-diacetylphloroglucinol biosynthetic gene expression in *Pseudomonas fluorescens* CHA0 in vitro and in the rhizosphere of wheat. *Applied and Environmental Microbiology, 68,* 2229-2235.

O'Callaghan, K. J., Stone, P. J., Hu, X., Griffiths, D. W., Davey, M. R. & Cocking, E. C. (2000). Effects of glucosinolates and flavonoids on colonization of the roots of *Brassica napus* by *Azorhizobium caulinnodans* ORS571. *Applied and Environmental Microbiology, 66,* 2185-2191.

O'Neill, G. A., Chanway, C. P., Axelrood, P. E., Radley, R. A. & Holl, F. B. (1992). Growth response specificity of spruce inoculated with coexistent rhizosphere bacteria. *Canadian Journal Botany, 70,* 2347-2353.

Obukowicz, M. G., Perlak, F. J., Bolten, C. L., Kusan-Keetzneis, K., Meyer, E. J. & Watrud, L. S. (1987). IS50L a non-self transposable vector used to integrate the *B. thuringensis* delta endotoxin gene into the chromosome of root colonizing *Pseudomonas*. *Gene, 51,* 95-96.

Olivares, F. L., James, E. K., Baldani, I. & Dobereiner, J. (1992). Infection of mottled stripe disease-susceptible and resistant sugar cane varieties by the endophytic diazotroph *Herbaspirillum*. *New Phytologist, 135,* 723-737.

Palus, J. A., Borneman, J., Ludden, P. W. & Triplett, E. W. (1996). A diazotrophic bacterial endophyte isolated from stems of *Zea mays* L. and *Zea luxurians* Iltis and Doebley. *Plant and Soil, 186,* 135-142.

Pandey, R. K., Bahl, R. K. & Rao, P. R. T. (1986). Growth stimulating effects of nitrogen fixing bacteria (biofertiliser) on oak seedlings. *Industria Forestry, 112,* 75-79.

Patriquin, D. G. & Dobereiner, J. (1978). Light microscopy observations of tetrazolium-reducing bacteria in the endorhizosphere of maize and other grasses in Brazil. *Canadian Journal of Microbiology, 24,* 734-742.

Patriquin, D. G., Dobereiner, J. & Jain, D. K. (1983). Sites and processes of association between diazotrophs and grasses. *Canadian Journal of Microbiology, 29,* 900-915.

Petersen, D. J., Srinivasan, M.S. & Chanway, C.P. (1996). *Bacillus polymyxa* stimulates increased *Rhizobium etli* populations and nodulation when co-resident in the rhizosphere of *Phaseolus vulgaris*. *FEMS Microbiology Letters, 142,* 271-276.

Pianka, E. R. (1975). Niche relations of desert lizards. In J. Disamond & M.L. Cody (Eds.), *Ecology and evolution of communities* (pp. 292-314). Cambridge, Mass: Belknap Press.

Pichart, B. & Thouvenot, D. (1995). Effect of *Bacilus polymyxa* seed treatments on control of black-rot and damping-off of cauliflower. *Seed Science and Technology, 27,* 455-465.

Pillay, V. K. & Nowak, J. (1997). Inoculum density, temperature, and genotype effects on in vitro growth promotion and epiphytic and endophytic colonization of tomato (*Lysopersicon esculentum* L.) seedlings inoculated with a pseudomonad bacterium. *Canadian Journal of Microbiology, 43,* 354-361.

Pokojska-Burdziej, A. (1982). The effect of microorganisms, microbial metabolites and plant growth regulators (IAA and GA) on the growth of pine seedlings (*Pinus sylvestris* L.). *Polish Journal of Soil Science, 15,* 137-143.

Quispel, A. (1992). A search for signals in endophytic microorganisms. In D. P. S. Verma (Ed.), *Molecular signals in plant-microbe communications* (pp. 471-490). Boca Raton, FL: CRC Press.

Reddy, P. M., Ladha, K., So, R., Hernandez, R., Dazzo, F. B., Angeles, O. R. et al. (1997). Rhizobial communication with rice roots: induction of phenotypic changes, mode of invasion and extent of colonization. *Plant and Soil, 194,* 81-98.

Reinhold-Hurek, B. & Hurek, T. (1998). Life in grasses: diazotrophic endophytes. *Trends in Microbiology, 6,* 139-144.

Reis, V. M., Olivares, F. L. & Bobereiner, J. (1994). Improved methodology for isolation of *Acetobacter diazotrophicus* and confirmation of its endophytic habitat. *World journal of Microbiological Biotechnology, 10,* 401-405.

Rennie, R. J., De Freitas, J. R., Ruschel, A. P. & Vose, P. B. (1982). Isolation and identification of N_2-fixing bacteria associated with sugar cane (*Saccharum* sp.). *Canadian Journal of Microbiology, 28,* 462-467.

Rennie, R. J. & Larson, R. L. (1979). Nitrogen fixation associated with disomic chromosome substitution lines of spring wheat. *Canadian Journal Botany, 57,* 2771-2775.

Reva, O. N., Smirnov, V. V., Pettersson, B. & Priest, F. G. (2002). *Bacillus endophyticicus* sp. nov., isolated from the inner tissues of cotton plants (*Gossypium* sp.). *International Journal of Systematic Bacteriology, 52,* 101-107.

Rhitu, R., Prasanta, D. & Aqbal, S. (2002). Cloning of Bt gene in maize endophyte for control of shoot borer, *Chilo partellus* (Swinhoe). In O. Koul, G. S. Dhaliwal, S. S. Marwaha, & J. K. Aroa (Eds.), *Biopesticides and Pest Management* (pp. 116-121). New Delhi: Campus Books International.

Roberts, M.S., Nakamura, L.K. & Cohan, F.M. (1994). *Bacillus mojavensis* sp. Nov., distinguishable from *Bacillus subtilis* by sexual isolation, divergence in DNA sequence, and differences in fatty acid composition. *International Journal of Systematic Bacteriology, 44,* 256-264.

Rodríguez-Barrueco, C., Cervantes, E., Subbarao, N. S. & Rodríguez-Caceres, E. (1991). Growth promoting effect of *Azospirillum brasilense* on *Casuarina cunninghamiana* Miq. seedlings. *Plant and Soil, 135,* 121-124.

Reddy, M. S. & Rahe, J. E. (1989). Growth effects associated with seed bacterization not correlated with populations of *Bacillus subtilis* inoculant in onion seedling rhizospheres. *Soil Biology and Biochemistry, 21,* 373-378.

Ryder, M. H., Yan, Z. Terrace, T. E. Rovira, A. D. Tang, W. & Correll, R. L. (1999). Use of strains of *Bacillus* isolated in China to suppress take-all and rhizoctonia root rot, and promote seedling growth of glasshouse-grown wheat in Australian soils. *Soil Biology and Biochemistry, 31,* 19-29.

Salt, D. E., Smith, R. D. & Raskin, I. (1999). Phytoremediation. *Annual Review of Plant Physiology and Plant Molecular Biology, 49,* 643-668.

Samish, Z., Etinger-Tulezynska, R. & Bick, M. (1961). Microflora within healthy tomatoes. *Applied Microbiology, 9,* 20-25.

Schnider-Keel, U., Seematter, A., Maurhofer, M., Blumer, C., Duffy, B. K., Gigot-Bonnefoy, C. et al. (2000). Autoinduction of 2,4-diacetylphoroglucinol biosynthesis in the biocontrol agent *Pseudomonas fluorescens* CHA0 and repression by the bacterial metabolites salicylate and pyoluteorin. *Journal of Bacteriology, 182,* 1215-1225.

Shishido, M., Breuil, C. & Chanway, C. P. (1999). Endophytic colonization of spruce by plant growth-promoting rhizobacteria. *FEMS Microbiological Ecology, 29,* 191-196.

Shishido, M., Loeb, B. M. & Chanway, C. P. (1995). External and internal root colonization of lodgepole pine seedlings by two growth-promoting *Bacillus* strains originated from different root microsites. *Canadian Journal of Microbiology, 41,* 707-713.

Siciliano, S. D., Fortin, N., Mihoc, A., Wisse, G., Labelle, S., Beaumier, D. et al. (2001). Selection of specific endophytic bacterial genotypes by plants in response to soil contamination. *Applied and Environmental Microbiology, 67,* 2469-2475.

Stolzfus, J. R., So, R. M. P. P., Ladha, J. K. & de Bruijn, F. J. (1997). Isolation of endophytic bacteria from rice and assessment of their potential for supplying rice with biologically fixed nitrogen. *Plant and Soil, 194,* 25-36.

Stone, J. K., Bacon, C. W. & White, J. F., Jr. (2000). An overview of endophytic microbes: endophytism defined. In C.W. Bacon & J. F. White, Jr. (Eds.), *Microbial Endophytes* (pp. 3-30). New York: Marcel Dekker.

Strobel, G. A. & Long, D. M. (1998). Endophytic microbes embody pharmaceutical potential. ASM *News 64,* 263-268.

Sturz, A. V., Christie, B. R. & Matheson, B. G. (1998a). Association of bacterial endophyte populations from red clover and potato crops with potential for beneficial allelopathy. *Canadian Journal of Microbiology, 44,* 162-167.

Sturz, A. V., Christie, B. R. & Matheson, B. G. (1998b). Associations of bacterial endophyte populations from red clover and potato crops with potential for beneficial allelopathy. *Canadian Journal of Microbiology, 44,* 162-167.

Sufnit, Y., Barah, Z., Kapuluk, Y., Oppenheim, A. B. & Chet, I. (2000). Expression of *Serratia marcescens* chitnase genes in *R. meliloti* during symbiosis on alfalfa roots. *Molecular Plant-Microbe Interactions, 6,* 293-298.

Surette, M. A., Sturz, A. V., Lada, R. R. & Nowak, J. (2003). Bacterial endophytes in processing carrots (*Daucus carota* L. var. sativus): their localization, population density, biodiversity and their effects on plant growth. *Plant and Soil, 253,* 381-390.

Swensen, S. M. & Mullin, B. C. 1997. The impact of molecular systematic on hypotheses for the evolution of root nodule symbioses and implications for expanding symbioses to new host plant genera. *Plant and Science, 194,* 185-192.

Tejera, N. A., Oetega, E., Gonzalez-Lopex, J. & Lluch, C. (2003). Effect of some abiotic factors on the biological activity of *Gluconacetobacter diazotrophicus. Journal of Applied Microbiology, 95,* 528-535.

Tervet, I. W. & Hollis, J. P. (1948). Bacteria in the storage organs of healthy plants. *Phytopathology, 38,* 960-967.

Tetlow, I. J. & Farrar, J. F. (1993). Apoplastic sugar concentration and pH in Barley leaves infected with brown rust. *Journal of Experimental Biology, 44,* 929-936.

Thomashow, L. S. & Weller, D. M. (1990). Role of antibiotics and siderophores in biocontrol of take-all disease of wheat. *Plant and Soil, 129,* 93-99.

Timmer, L. W., Brlansky, R. H., Lee, R. F. & Raju, B. C. (1983). A festidious xylem-limited bacterium infecting ragweed. *Phytopathology, 73,* 957-979.

Tomsaino, S. F., Leister, R. T., Dimock, M. B., Beach, R. M. & Kelly, J. W. (1995). Field performance of *Clavibacter xyli* subsp. *cynodontis* expressing the insecticidal protein gene cryIA (c) of *Bacillus thuringiensis* against European corn borer in field corn. *Biological Control, 5,* 442-448.

Toyoda, H., Hashimoto, H., Utsumi, R., Kobayashi, H. & Ouchi, S. (1988). Detoxification of fusaric acid by a fusaric acid-resistant mutant of *Pseudomonas solanacearum* and its application to biological control of *Fusarium* wilt of tomato. *Phytopathology, 78,* 1307-1311.

Viallard, V., Poirier, I., Cournoyer, B., Haurat, J., Wiebkin, S., Ophel-Keller, K. & Balandreau, J. (1998). *Burkholderia graminis* sp. no., a novel species of rhizospheria *Burkholderia* and reassessment of *Pseudomonas phenazinium, P. pyrrocinia,* and *P. glathei* into *Burkholderia. International Journal of Systemic Bacteriology, 48,* 549-563.

Walker, R., Powell, A.A. & Seddon, B. (1998). *Bacillus* isolates from the spermosphere of peas and dwarf French beans with antifungal activity against *Botrytis cinerea* and *Phythium* species. *Journal of Applied Microbiology, 84,* 791-801.

Walton, B. T. & Anderson, T. A. (1990). Microbial degradation of trichloroethylene in the rhizosphere: Potential application to biological remediation of waste sites. *Applied and Environmental Microbiology, 56,* 1012-1016.

Whitesides, S. K. & Spotts, R. A. (1991). Frequency, distribution, and characteristics of endophytic *Pseudomonas syringae* in pear trees. *Phytopathology, 81,* 453-457.

Wilhelm, E., Arthofer, W., Schafleitner, R., & Krebs, B. (1998). *Bacillus subtilis* an endophyte of chestnut (*Castanea sativa)* as antagonist against chestnut blight (*Cryphonectria parasitica*). *Plant Cell, Tissue and Organ Culture, 52,* 105-108.

Wilson, M. & Lindow, S. E. (1999). Location and survival of leaf-associated bacteria in relation to pathogenicity and potential for growth within the leaf. *Applied and Environmental Microbiology, 65,* 1435-1443.

Xu, H., Griffith, M., Patten, C. L. & Glick, B. R. (1998). Isolation of an antifreeze protein with ice nucleation activity from the plant growth promoting rhizobacterium *Pseudomonas putrid* GR12-2. *Canadian Journal Botany, 44,* 64-73.

Yanni, Y. G., Rizk, R. Y., Corich, V., Squartini, A., Ninke, K., Philip-Hollingsworth, S. et al. (1997). Natural endophytic association between *Rhizobium leguminosarum* cv. *trifolii* and rice roots and assessment of its potential to promote rice growth. *Plant and Soil, 194,* 99-114.

Zaady, E. A., Perevolotsky, A. & Okon, Y. (1993). Promotion of plant growth by inoculum with aggregated and single cell suspensions of *Azospirillum brisilense* Cd. *Soil Biology and Biochemistry, 25,* 819-823.

Zhang, S., Reddy, M. S. & Kloepper, J. W. (2002). Development of assays for assessing induced systemic resistance by plant growth-promoting rhizobacteria against blue mold of tobacco. *Biological Control, 23,* 79-86.

7. AFFILIATIONS

Charles W. Bacon and Dorothy M. Hinton, USDA, Agricultural Research Service Toxicology and Mycotoxin Research Unit, Russell Research Center, Athens, Georgia, 30604.

Rhizosphere Bacteria

APPA RAO PODILE AND G. KRISHNA KISHORE

PLANT GROWTH-PROMOTING RHIZOBACTERIA

Abstract. Plant growth-promoting rhizobacteria (PGPR) are a group of free-living bacteria that colonize the rhizosphere and benefit the root growth. Bacteria of diverse genera were identified as PGPR of which *Bacillus* and *Pseudomonas* spp. are predominant. PGPR exert a direct effect on plant growth by production of phytohormones, solubilization of inorganic phosphates, increased iron nutrition through iron-chelating siderophores and volatile compounds that affect the plant signaling pathways. Additionally, by antibiosis, competition for space and nutrients, and induction of systemic resistance in plants against a broad-spectrum of root and foliar pathogens, PGPR reduce the populations of root pathogens and other deleterious microorganisms in the rhizosphere, thus benefiting the plant growth. Root colonization, influenced by a number of biotic and abiotic components, is a limiting factor for the success of PGPR. Diverse reporter genes and nucleic acid-based methods were developed to track the introduced PGPR in the rhizosphere, and also to determine their metabolic status, and their effect on the native rhizosphere microbial communities. Quality of the PGPR formulations, in terms of viability and efficacy, determines their large-scale adoption at the end-user level. We discuss the importance of PGPR in sustainable agriculture with special reference to the mechanisms involved in their action and factors affecting their efficacy, along with the possibilities for their improvement.

1. INTRODUCTION

The rhizosphere, volume of soil surrounding roots and influenced chemically, physically and biologically by the plant root, is a highly favorable habitat for the proliferation of microorganisms and exerts a potential impact on plant health and soil fertility (Sorensen, 1997). Root exudates rich in amino acids, monosaccharides and organic acids, serve as the primary source of nutrients, and support the dynamic growth and activities of various microorganisms within the vicinity of the roots. These root-colonizing microorganisms could be free-living, parasitic or saprophytic and their diversity remains dynamic with a frequent shift in community structure, and species abundance (Kunc & Macura, 1988). An important group of these microbial communities that exerts beneficial effects on plant growth upon root colonization were first defined by Joseph Kloepper and Milton Schroth and termed as plant growth-promoting rhizobacteria (PGPR) (Kloepper & Schroth, 1978). Plant growth enhanced by PGPR is quantified as an increase in seedling emergence, vigor, biomass, proliferation of root system and yield in various plant species. Since their recognition as an important subset of root colonizing microorganisms, in the past three decades, several studies were conducted, at an exponential rate, to identify PGPR in different cropping systems and agro ecological zones (reviewed by Kloepper et al. 1991; Vessey, 2003; Zahir et al. 2004; Ping & Boland, 2004). Based on the results of these studies and release of a few bacterial strains for commercial purposes (Table 1), PGPR gained attention as an important group of agriculturally beneficial bacteria besides symbiotic nitrogen fixing bacteria.

S.S. Gnanamanickam (ed.), Plant-Associated Bacteria, 195–230.

Table 1. Commercially available PGPR strains, that essentially act through direct plant growth promoting mechanisms.

Trade name	PGPR strain	Manufacturer	Recommended application
Bioboost	*Delftia acidovorans*	Brett-young seeds ltd., Manitoba	Seed treatment in canola
Bioplin	*Azotobacter* spp.	Kumar krishi mitra bioproducts pvt. ltd., Pune, India	Soil drenching for sunflower, tomato and other vegetable crops
Bioyield	*Bacillus* spp.	Guftason, LLC, Plano, Tx	Seed treatment in tomato, tobacco, cucumber and pepper
Compete	*Bacillus*, *Pseudomonas* and *Streptomyces* spp.	Plant health care BV, CA Vught	Soil drenching for turfgrass, nursery and greenhouse plantations
Kodiak	*B. subtilis* GB03	Gustafson, LLC, Dallaas, Tx	Seed treatment in fruits and vegetables

The importance of PGPR was realized as an off shoot of biological control of soil-borne pathogens. But, according to the mode of action, PGPRs were divided into two groups viz. biocontrol-PGPBs and PGPBs (Bashan & Holognin, 1997). Protection of bacterial-inoculated seedlings against soilborne pathogens was observed inseparable from the plant growth-promoting activity of several of the reported PGPR (Guo et al. 2004; Manjula & Podile, 2001; Raupach & Kloepper, 2000). As a consequence, PGPR were more emphasized as protectants of soilborne pathogens. However, the need for PGPR as biofertilizers has been rediscovered in recent years with the increasing importance for organic culturing with minimum or no inputs. The need for a threshold level for the initial bacterial inoculum to promote plant growth significantly indicates that quorum sensing by bacteria plays an important role in plant-PGPR interactions (Teplitski et al. 2000). A detailed insight into the mechanisms involved in the action of PGPR and knowledge of the underlying genetics, and biochemical and physiological pathways, has made it possible to use PGPR both for plant growth promotion and disease control. These understandings have also provided an insight for strain improvement. However, there continues a need to overcome the inconsistent performance of PGPR strains especially in more complex natural environments, subjected to differences in physical, chemical and microbial properties in the introduced environments. Innovations in the formulation and delivery systems of PGPR to suit the farmers' needs in various cropping systems enhance their quick adoption. International PGPR workshops, organized every four years, in different parts of the world have been a regular platform to discuss various aspects of PGPR. The most recent 6th International PGPR Workshop was held at Calicut, India during $5\text{-}10^{th}$ October 2003, while the next meeting is scheduled to be held in the Netherlands in 2007. These workshops have given a big boost to the PGPR research in Americas, Africa, Australia, Asia and Europe as evidenced by the regular reports on diversity, mode of action, formulation, and commercialization of PGPRs in these regions.

In this chapter, the importance of PGPR in sustainable agriculture with special reference to their diversity, the basic mechanisms involved in their action, the factors affecting their root colonization and formulation, and the possibilities for their improvement are discussed.

2. DIVERSITY OF PGPR STRAINS

Strains of the genera such as *Aeromonas*, *Azoarcus*, *Azospirillum*, *Azotobacter*, *Arthobacter*, *Bacillus*, *Clostridium*, *Enterobacter*, *Gluconacetobacter*, *Klebsiella*, *Pseudomonas* and *Serratia*, have been identified as PGPR, while the search for additional strains continues (Tripathi et al. 2005; Raj et al. 2004; Dey et al. 2004; Jaizme-vega et al. 2004; Joo et al. 2004; Bpnaterra et al. 2003; Cezon et al. 2003; Esitken et al. 2003; Garica et al. 2003; Munir et al. 2003; Kokalis-Burelle et al. 2002; Khalid et al. 2003; Murphy et al. 2003; Preeti et al. 2002; Gupta et al. 1995; 2002; 2003; Bertand et al. 2001; Hamaouri et al. 2001; Nanda-Kumar et al. 2001, Pan et al. 1999; Arnd et al. 1998; de Freitas et al. 1997; Shishido et al. 1996; Babalola et al. 2003; Mirza et al. 2001). The diversity of PGPR in the rhizosphere largely varies according to the plant type, soil type and nutrients available (recently reviewed by Tilak et al. 2005). Among the diverse range of PGPR identified, *Pseudomonas* and *Bacillus* spp. have a wide distribution and are the most extensively studied. *Azospirillum*, a N_2-fixing genus is an important group of PGPR, since treatment with almost all strains and species of this genus positively affect the root biomass and surface area (Bashan et al. 2004). Recent developments in metagenomics i.e. the study of collective genome of an ecosystem provide insights of bacterial diversity in the rhizosphere including the non-culturable organisms. Though, by definition PGPR are free-living plant-associated bacteria, few *Rhizobium* strains that colonize the roots of non-legume plants such as *Gramineae* and crucifers and promote root growth by mechanisms other than biological N_2 fixation, are also considered as PGPR (Antoun et al. 1998).

Initial explorations for identification of PGPR concentrated on rhizosphere bacteria, and subsequent experiments demonstrated the potential application of root endophytes as PGPR (Vessey, 2003). It is quite possible that majority of the root endophytes have their origin as rhizosphere, and subsequently invaded the root tissues and colonized the inner cortex. Certain epiphytic bacteria were also identified for their characteristics that contribute to plant growth promotion (Kuklinsky-Sorbal et al. 2004; Kishore et al. 2005a). The growth-promoting activity of groundnut or soybean associated bacteria indicated the possible use of phylloplane bacteria as PGPR. Two phylloplane isolates, *Bacillus megaterium* and *Bacillus firmis* colonized the groundnut rhizosphere and increased the plant growth in field (Kishore et al. 2005a). Thus, there exists an opportunity to isolate and use PGPR strains from other plant habitats.

3. MECHANISMS OF GROWTH PROMOTION

According to the mode of action, PGPR have been divided into two groups: biocontrol PGPRs that indirectly benefit the plant growth and PGPRs that directly affect plant growth, seed emergence, or improve crop yields (Glick et al. 1999). Efforts have been made in the past two decades to elucidate both the direct and indirect mechanisms by which PGPR enhance plant growth. Phytohormone production and enhancing plant nutrition were the two prominent mechanisms by which PGPR directly contribute to the plant growth. Enhanced plant nutrition by PGPR is mainly through increased phosphorous uptake by solubilization of inorganic phosphates and iron uptake by production of iron chelating siderophores. Ryu et al. (2003 & 2004) elegantly demonstrated the role of bacterial volatiles in plant growth promotion *in vitro*. PGPR indirectly benefit the plant growth by suppression of deleterious microorganisms that inhibit plant growth or root pathogens through antibiosis, parasitism, competition for nutrients and space within the vicinity of plant roots, and/or activation of host defense responses. Details of these mechanisms and their individual contribution to plant growth promotion are discussed below:

3.1. Direct plant growth promotion

3.1.1. Phytohormones

Plant growth stimulating phytohormones produced by PGPR within the root zone stimulate the density and length of root hairs. The increase in root surface area improves the plant uptake potential of water and mineral nutrients from a large volume of soil (Volkmar & Bremar, 1998). Vessey et al. (2003) provided evidence that different PGPR strains, exhibit their plant growth-promoting activity due to the influence of phytohormones.

Auxin, indole-3-acetic acid (IAA) is a quantitatively important phytohormone produced by PGPR, and treatment with auxin-producing rhizobacteria increased the plant growth (Vessey, 2003). Contrasting observations were made in the studies conducted to determine the specific role of IAA production in PGPR-mediated growth promotion. A mutant strain of *Pseudomonas putida* with four-fold increase in IAA production lost its ability to induce root elongation in canola seedlings, though its growth rate and production of siderophores and 1-aminocyclopropane-1-carboxylate deaminase (ACC) remained unaltered (Xie et al. 1996). In contrast, a positive correlation was observed between L-tryptophan-dependent auxin production by different PGPR strains and their ability to increase the grain yield, and number of branches and pods per plant in *Brassica* spp (Asghar et al. 2002). It was further supported by the positive relation between auxin production by PGPR and the increase in number of branches and oil content in *B. napus* treated with these PGPR (Asghar et al. 2004).

Growth promotion in lodgepole pine seedlings inoculated with auxin-producing *Paenibacillus polymyxa* L6 is supported by the enhanced levels of auxin in roots (Bent et al. 2001). Persello-Cartieaux et al. (2001) used mutants of *Arabidopsis thaliana* affected in root hair development and possible hormone perception, to identify the plant genetic determinants of bacterial colonization. The observation that two auxin-resistant mutants were insensitive to *Pseudomonas thivervalensis* colonization supports the role of bacterial auxin altering the root morphology.

Bacteria like *Azospirillum* and *Pseudomonas* spp. produce cytokinins and gibberellins (gibberellic acid), in addition to IAA (Gaudin et al. 1994). A few PGPR strains were reported to produce cytokinins (Vessey, 2003) and gibberellins (gibberellic acid, GA; Gutierrez-Manero et al. 2001). However, the available information is insufficient to support the role of PGPR-produced cytokinins and gibberellins in plant growth promotion.

Ethylene, a gaseous phytohormone commonly induced by wounding in plants, causes root growth inhibition. Studies focused on indirect promotion of root growth by ethylene inhibition revealed that PGPR produced ACC deaminase cleaves ACC, the immediate precursor molecule of ethylene in plants, with a net result of increase in root growth (Glick et al. 1998). Bacteria containing ACC deaminase activity from different soils stimulated plant growth even in soils with cadmium toxicity (Belimov et al. 2005). *P. fluorescens* strains transformed with ACC deaminase gene, together with its regulatory region increased length of canola plants under gnotobiotic conditions and improved its ability to protect cucumber seedlings and potato tubers against pathogenic fungi (Wang et al. 2000). However, the notable observation is that many but not all PGPR possess ACC deaminase activity (Babalola et al. 2003).

3.1.2. Solubilization of phosphates

Phosphorous, next to nitrogen, is the second important macronutrient required for plant growth. Even in phosphorous rich soils most of the P is in insoluble form - iron and aluminium phosphates in acidic soils, and calcium phosphates in alkaline soils, and only a small proportion (~0.1%) is available to plants (Stevenson & Cole, 1999). Additionally, 3/4th of the phosphate fertilizers applied to soil reprecipitate into insoluble forms thus increasing the P requirement of the crop (Goldstein, 1986). Phosphate solubilizing bacteria (PSB) secrete organic acids and phosphatases to convert the insoluble phosphates into soluble monobasic ($H_2PO_4^-$) and dibasic (HPO_4^{2-}) ions, a process referred to as mineral phosphate solubilization (MPS). MPS in the rhizosphere leads to an increase in the phosphorous available to plants and in turn the plant uptake (Gyaneshwar et al. 2002).

Phosphate solubilizing bacteria are ubiquitous (Gyaneshwar et al. 2002), and *Bacillus*, *Enterobacter, Erwinia* and *Pseudomonas* spp. are among the most potent strains. PSB are common in rhizospheres of crop plants and few examples of beneficial association of phosphate solubilizing PGPR and plants include *Azotobacter chroococcum* and wheat (Kumar & Narula, 1999), *Bacillus circulans* and wheat (Singh & Kapoor, 1998), *Enterobacter agglomerans* and tomato (Kim et al. 1998b), *P. chlororaphis* or *P. putida* and soybean (Cattelan et al. 1999). The phosphate solubilizing ability of PGPR in natural habitats, in turn, depends on the available nutrients - carbon and nitrogen sources, and metal ions (Kim et al. 1998b). In an *in vitro* study Nautiyal et al. (2000) observed that phosphate solubilizing activity of four chickpea rhizobacteria varied with nitrogen source and increased in presence of low levels of Ca^{2+} and EDTA.

Organic acids such as citrate, lactate, and succinate etc., secreted by PSB contribute for phosphate solubilization in the rhizosphere. The most important among the various organic acids that solubilize inorganic phosphates is 2-keto-gluconic acid, a secondary oxidation product of glucose. The oxidation of glucose to gluconic acid and 2-keto-gluconic acid is catalyzed by glucose dehydrogenase (GDH) located on the outer leaf or cytoplasmic membrane. Mutants of phosphate solubilizing *Enterobacter asburiae,* deficient in GDH activity, failed to release phosphate from alkaline soils indicating the essential role of GDH in MPS (Gyaneshwar et al. 1999).

Acid and alkaline phosphatases of PGPR mobilize the organic phosphorous of soil for the benefit of the plants. Phytate (*myo*-inositol hexakisphosphate) accounts for 20-50% of the total soil organic phosphorous. Phytases are involved in the stepwise degradation of phytate to lower phosphate esters of myo-inositol and phosphorous. Phytase has been purified and characterized from selected Gram-positive and Gram-negative soil bacteria including *Bacillus*, *Pseudomonas*, *Klebsiella* and *Enterobacter* spp (Kerovuo et al. 1998; Greiner et al. 1997). Idriss et al. (2002) using a genetic approach, obtained evidence that secreted phytase activity of the PGPR *Bacillus amyloliquefaciens* FZB45 promotes growth of maize seedlings consisting of a phosphate-limited plant nutrient medium in the presence of phytate. A phytase-negative mutant strain, FZB45/M2, whose *phyA* gene is disrupted, and its culture filtrates did not stimulate plant growth.

3.1.3. Increased uptake of iron

Iron is an essential micronutrient of plants as it serves as a cofactor of many enzymes with redox activity. A large portion of iron in soils is in the highly insoluble form of ferric hydroxide, thus iron acts as a limiting factor for plant growth even in iron rich soils. The availability of iron in soil solutions is 10^{-18}M, a concentration which even cannot sustain the microbial growth. Several soil microorganisms produce siderophores, low molecular weight iron chelating compounds that bind Fe^{3+} with very high affinity and helps in iron uptake. It is possible for the rhizosphere microorganisms to use siderophores provided they contain the appropriate uptake protein (Raaijmakers et al. 1995). A great deal of evidence exists that a number of plant species can absorb bacterial Fe^{3+}-siderophore complexes, and this process is vital in absorption of iron by plants, especially in calcareous soils (Wang et al. 1993; Masalha et al. 2000).

Buyer et al. (1993), using monoclonal antibodies, confirmed the production of siderophores by PGPR in rhizosphere under iron-limiting conditions. These siderophores can be of different types: hydroxamates, phenol-catecholates, and carboxylates. Siderophore-mediated iron transport of Gram -ve PGPR is well studied than the Gram +ve PGPR (Guerinot, 1994). *Pseudomonas* spp. are the potent siderophore producers among Gram -ve PGPR and they produce pseudobactin, pyochelin, pyoverdine, quinolobactin and salicylic acid, and the structure of the outer membrane receptor proteins complementary to some of these siderophores has been determined (David et al. 2005). Nutrient availability influences the siderophore production of PGPR. Co^{2+}, fructose, mannitol, and glucose increased *in vitro* production of pyochelin by *P. fluorescens*, while NH_4Mo^{2+}, glycerol and glucose increased the production of its precursor salicylic acid (Duffy & Defago, 1999).

The role of siderophores produced by PGPR in enhancing plant iron nutrition was well established by mutation analysis and siderophore production (Kloepper et al. 1991). Mutant strains of siderophore producing PGPR, deficient in siderophore production lost their plant growth and complementation of the siderophore production restored their activity. A cold-tolerant mutant of *P. fluorescens* with 17-fold increase in siderophore production had an increased rhizosphere colonization and growth-promoting effect on mungbean (Katiyar & Goel, 2004). Seed treatment of maize with siderophore-producing *Pseudomonas* spp. GRP3A, PRS9 and *P. chlororaphis* ATCC 9446 increased seed germination, shoot and root lengths, and dry weight of seedlings (Sharma & Johri, 2003). Siderophore-mediated growth promoting activity of PGPR is associated with the suppression of root pathogens by competitive exclusion and the activity is reversed by the addition of Fe EDTA. A purified pseudobactin from *Pseudomonas* sp. B 10 caused significant increase in the growth of potato seedlings and the activity was reversed by the addition of 50 μm Fe EDTA (Kloepper et al. 1980). Even in acidic soils with pH less than 6.0, iron availability increases and siderophores become less effective (Neilands & Nakamura, 1991).

The genetics and regulation of production and uptake of siderophores are well understood (Leoni et al. 2002; Ravel & Cornelis, 2003). Since the synthesis of each siderophore generally requires the activity of several gene products (Mercade-Blanco et al. 2001), it is difficult to genetically engineer bacteria to produce modified siderophores. Genetic complementation of siderophore-deficient mutants of *P. fluorescens* M114 indicated that at least five separate loci are needed to encode the enzymes involved in the synthesis of the siderophore psuedobactin M114 (O'Sullivan et al. 1990).

3.1.4. Volatiles in growth promotion

Plant growth promotion by volatiles produced by PGPR is the most recently identified mechanism. Ryu et al. (2003) demonstrated that PGPR strains release different volatile blends and the differences in these volatile blends stimulate the plant growth. Volatiles produced by *Bacillus subtilis* and *B. amyloliquefaciens* stimulated the growth of *Arabidopsis thaliana* in *in vitro* experiments as observed by an increase in the total leaf surface area compared to control. Two compounds, 3-hydroxy-2-butanone (acetoin) and 2, 3-butanediol, were produced by both the bacterial strains whereas they are not produced by bacterial strains that did not affect the plant growth. A mutation in *B. subtilis* that affected the production of 2, 3-butanediol, but not acetoin, impaired its growth promotion activity. The observed growth promotion was dose-dependent in the exogenous application of 2, 3-butanediol to *A. thaliana* seedlings. The volatile-mediated growth promotion by PGPR is by activation of cytokinin-signaling pathways.

Ryu et al. (2004) identified that exposure of *A. thaliana* seedlings to the volatile blends from *B. subtilis* and *B. amyloliquifaciens* reduced the disease severity by the bacterial pathogen *Erwinia carotovora* subsp. *carotovora*. The level of disease protection was directly related to the production of volatile organic compounds by *B. subtilis*. It is also observed that the signaling pathway activated by volatiles from *B. subtilis* is dependent on ethylene and independent of the salicylic acid or jasmonic acid signaling pathways. These observations confirm the contribution of bacterial volatile organic compounds to plant growth promotion both by direct and indirect mechanisms. The bacterial acetoin-pathway leading to the production of 2,3-butanediol is triggered by low partial pressures of O_2, a condition that exists in the rhizosphere soil, thus at least a small portion of the PGPRs produce 2,3-butanediol in the rhizosphere above the threshold concentrations required to bring about plant responses (Ping & Boland, 2004).

3.2. Indirect plant growth promotion

3.2.1. Antibiosis

Diverse PGPR antagonize the root pathogens through one or more of the different mechanisms identified, for example by production of volatile or non-volatile antibiotics, siderophores, enzymes and other secondary metabolites like HCN. Production of these compounds is highly influenced by the qualitative and quantitative nutrient availability and is also subjected to quorum sensing (Hass & Keel, 2003).

The antibiotics commonly produced by different antagonistic bacteria include ammonia, butyrolactones, 2,4-diacetyl phloroglucinol (DAPG), kanosamine, oligomycin A, oomycin A, phenazine-1-carboxylic acid, pyoluteorin, pyrrolnitrin, viscosinamide, xanthobaccin, and zwittermycin A (Whipps, 2001). Many of these antibiotics possess a broad-spectrum activity, and DAPG was the most potent and most extensively studied (Raaijmakers et al. 2002). The role of individual antibiotic compounds in suppression of root pathogens has been clearly established using mutation analysis and molecular genetic tools, and purified antibiotic compounds. For example, DAPG overproducing mutants of *P. fluorescens* offered a better protection against take-all in wheat and bacterial wilt in tomato, though the root colonization pattern remained the same (Hongyou et al. 2005). Selection of PGPR based on the presence or absence of the genes encoding for the antibiotic compounds facilitate rapid selection of PGPR from a large number of collections (Raaijmakers et al. 1997). The genes encoding the enzymes responsible for synthesis of few of the above-mentioned antibiotics were isolated and their regulation both at transcriptional and post-transcriptional stages was studied to facilitate the genetic improvement of antibiotic production (Haas & Keel, 2003).

Siderophores produced by PGPR, inhibit the root pathogens by creating iron-limiting conditions in the rhizosphere, as mentioned above. *P. fluorescens* inhibited the *in vitro* growth of *Fusarium culmorum* by producing iron-chelating siderophores, thus competing for Fe (III) (Kurek & Jaroszuk-Scisel, 2003). A siderophore over-producing mutant of *P. putida* was more effective than the wild type in suppression of Fusarium wilt in tomato (Vandenburgh & Gonzalez, 1984), while a siderophore-deficient mutant of *Pseudomonas aeruginosa* lost its biocontrol ability (Buysens et al. 1994). Purified siderophores exhibited disease-suppressive effect similar to the producer strain (Kloepper et al. 1980; Neilands & Leong, 1986). However, the suppression of pathogens by majority of the siderophore-producing PGPR is due to a combination of various other traits (Ongena et al. 1999).

Certain PGPR strains produce volatile antibiotics, of which HCN that inhibit the cytochrome oxidase of many organisms is most important. The producer strains possess an alternate cyanide-resistant cytochrome oxidase and are relatively insensitive to HCN. Mutants of *P. fluorescens* CHAO deficient in HCN production were less effective in control of take-all of wheat and black root rot of tobacco. Complementation of mutant strain by cloned wild type hcn^{+} genes restored the biocontrol ability (Voisard, 1989).

3.2.2. Competition

PGPR compete with deleterious microorganisms and pathogens for limited available nutrients in root exudates and suitable colonization niches, and finally out number them. Populations of PGPR established on the plant roots could act as a sink for the available nutrients and limit the nutrient availability for pathogen stimulation and its subsequent root colonization. This mechanism is most often used by fluorescent pseudomonads due to their nutritional versatility, and because of their high growth rates in the rhizosphere (Walsh et al. 2001). Apart from root colonization, the PGPR should be able to compete for nutrients with native microbial populations in the rhizosphere for successful elimination of the pathogens. Siderophore production by PGPR, sequester most of the available Fe^{3+} in the rhizosphere and force the pathogens for iron starvation, thus is a major contributor for pathogen suppression (O'Sullivan & O'Gara, 1992). Suppression of Fusarium wilt of radish by *Pseudomonas* strain WCS358 through siderophore-mediated competition for iron (Costa & Loper, 1994) is one such example.

3.2.3. Parasitism or lysis

Parasitism of pathogenic fungi by PGPR is facilitated through the production of hydrolytic enzymes that degrade the fungal cell walls. Chitinases, among the hydrolytic enzymes, are of prime importance since chitin, a linear polymer of β- (1, 4)-*N*-acetylglucosamine is a major cell wall constituent in majority of the phytopathogenic fungi. Purified chitinases of *Bacillus subtilis* AF 1 (Manjula et al. 2004), *Serratia marcescens* (Kishore et al. 2005b; Ordentlich et al. 1988) and *S. plymuthica* (Frankowski et al. 2001) were highly antifungal. The role of chitinolysis in root pathogen suppression has been further supported by improved disease control in chitin-supplemented treatments of chitinolytic PGPR (Manjula & Podile, 2001; Hallmann et al. 1999).

Another important group of hydrolytic enzymes, glucanases degrade the β-1, 3-glucans of the fungal cell walls. β-1, 3-glucanase producing strain of *P. cepacia* inhibited the rhizosphere proliferation of various phytopathogenic fungi including *Rhizoctonia solani*, *S. rolfsii* and *Pythium ultimum* (Fridlender et al. 1993). A synergistic action of the two hydrolytic enzymes chitinases and β-1, 3-glucanases was more effective in inhibition of fungal pathogens than either enzyme alone (Tanaka & Watanabe, 1995).

3.2.4. Inhibition of pathogen-produced enzymes or toxins

Pathogenic fungi produce extracellular hydrolytic enzymes, which degrade the polymers present in plant cell walls and facilitate the fungal infection by disintegrating the cell wall. These hydrolytic enzymes include pectolytic enzymes (exo- and endo- polygalacturonases, pectin lyases), cellulases and cutinase. A reduction in the activity of these enzymes correlates with a reduction in virulence (Beraha et al. 1983). *Bacillus megaterium* B 153-2-2 inhibited the activities of extracellular enzymes, like cellulase, pectin lyase and pectinase produced by *R. solani,* by producing an extracellular endoproteinase (Bertagnolli et al. 1996). A groundnut seed endophytic bacterium *Pseudomonas aeruginosa* GSE 18, effective in control of groundnut stem rot disease, when applied as seed treatment or soil amendment, inhibited the production of CWDE by the pathogen *Sclerotium rolfsii* tested *in vitro* (Kishore et al. 2005c).

3.2.5. Induced resistance

A few strains of rhizobacteria activate plant defense responses against a broad spectrum of plant pathogens, termed as induced systemic resistance (ISR). Rhizobacteria-mediated ISR has been demonstrated in many plant-pathogen systems wherein the bacterium and the challenging pathogen remained spatially separated, and these observations indicate that ISR is genetically determined

(Pieterse et al. 2001). Various biochemical pathways of plants that are activated by PGPR were reviewed by Van Loon et al. (1998). PGPR trigger host defense responses through two different signaling pathways; broadly classified as salicylic acid (SA) dependent and independent. The later pathways involved jasmonic acid (JA) and ethylene to trigger defense responses (Pieterse et al. 2001). The application of molecular, genetic and biochemical techniques led to the identification of signaling pathways that activate defense responses in *Arabidopsis*. Signaling molecules like SA, JA, and ethylene accumulate in activated plants and coordinate the defense responses. These signaling molecules activate specific sets of defense-related genes: SA induces genes encoding pathogenesis-related proteins (Uknes et al. 1992) with antimicrobial activity (Kombrink & Somssich, 1995). JA and ethylene induce genes encoding three antimicrobial proteins - *Hel*, a heveine-like protein, *ChiB*, basic chitinase, and *Pdf1.2*, a plant defensin, all three proteins possessing antimicrobial activity (Potter et al. 1993; Samac & Shah, 1994). Interestingly, exogenous application of these compounds also induces resistance (Ryals, 1996).

Induced resistance protected the two legume crops peanut and pigeonpea against soilborne fungal diseases following seed treatment with PGPR. Treatment of peanut seeds with *B. subtilis* AF1 altered the phytoalexin metabolism in seedlings and conferred protection against crown rot caused by *Aspergillus niger* (Sailaja & Podile, 1998). Also, in AF1-treated seedlings, lipoxygenase activity quickly increased compared to *A. niger*-treated control, and the breakdown products of lipoxygenase were inhibitory to *A. niger* (Sailaja et al. 1998). Seed treatment with *P. aeruginosa* resulted in rapid accumulation of defense-related enzymes - chitinase, β-1, 3-glucanase, peroxidase and phenylalanine ammonia lyase in peanut seedlings to protect against *A. niger* infection (Kishore and Podile, Unpublished). A similar rapid accumulation of phenylalanine ammonia lyase was involved in protection of AF1-treated pigeonpea seedlings against Fusarium wilt infection (Podile and Laxmi, 1998).

Bacterial determinants of ISR include lipopolysaccharides (LPS), siderophores and SA, as supported by various findings and also depend on the host plant (Audenaert et al. 2002; Meziane et al. 2005). Cell-free culture filtrate of *P. putida* BTP1 without SA, pyochelin and pyoverdine induced resistance in bean to *Botrytis cinerea*, indicating the possible involvement of other macromolecules (Ongena et al. 2002). The role of LPS in ISR was confirmed by the efficacy of LPS-containing cell wall preparations of WCS417r in activation of *Arabidopsis* defense responses (Van Wees et al. 1997), while the mutant lacking the O-antigenic side chain of LPS, lost its ability to induce ISR (Leeman et al. 1995).

In general, free-living rhizobacteria usually do not rely on a single mechanism of plant growth promotion (Glick et al. 1999), and may involve two or more of the above-listed individual mechanisms. Understanding the environmental factors that regulate the biosynthesis of growth-promoting and antimicrobial compounds by

PGPR is an essential step toward improving the level and reliability of their growth promoting activity.

4. ROOT COLONIZATION BY PGPR

Successful root colonization by the introduced PGPR is an essential criterion to exert beneficial effects on plant growth as identified from different field experiments. Bacterial plant root colonization is a complex multistep process dependent on several bacterial traits, root exudates and signaling between bacteria and plants. Bacterial traits involved in root colonization are motility and surface components such as flagella, pili, and the O antigen of LPS. Other influencing factors are the composition of root exudates, and host-PGPR-environment interactions.

4.1. Bacterial traits

Bacterial flagella, pili, LPS and exopolysaccharides are the major determinants of PGPR root colonization. Bacterial pili, outer membrane proteins, and flagella (through motility), contribute to PGPR adherence to plant root surfaces. A type IV pilus on the surface of *P. fluorescens* is involved in competitive root tip colonization (Lugtenberg et al. 2001). Through active extension and retraction, type IV pili are also involved in other processes such as cell movement (Skerker & Berg, 2005), and biofilm formation (O'toole & Kolter, 1988).

Extracellular polysaccharides were similar to the fibrillar material produced during the anchoring phase of the plant root attachment process (Skvortsov & Ignatov, 1998). Bacterial LPS contribute to growth and survival of the bacteria in planta by aiding in colonization, creating a favorable micro-environment, acting as a barrier to plant defensive compounds, and by modulating host reactions (Dow et al. 2000). Mutants impaired in the synthesis of the O-antigen of LPS of *P. fluorescens* and *P. putida* show impaired colonization ability in the rhizosphere. Pretreatment with *Azospirillum brasilense* LPS significantly promoted plant root attachment of *A. brasilense* (Matora et al. 2001) suggesting its role in the adhesion process.

4.2. Effects of root exudates on PGPR root colonization

Plants secrete species-specific root exudates that contain carbohydrates, proteins, amino acids, organic acids, vitamins and other nutrients that affect the growth and physiology of rhizobacterial populations. The primary colonizers of the microbial community are determined by the composition of root exudates and bacteria that catabolize these diverse nutrients. *Pseudomonas* spp. by virtue of their ability to catabolize diverse nutrients and compete for limited carbon sources are identified as potent root colonizers (Lugtenberg et al. 2001). Simons et al. (1997) demonstrated that amino acid auxotrophic mutants of *P. fluorescens* WCS365 were out competed by wild-type strains. A *P. fluorescens* mutant impaired in the utilization of carbohydrates, the primary carbon source found in tomato exudate retained their root colonization ability through utilization of organic acids

(Lugtenberg et al. 1999; 2001). These results suggest an important role for utilization of root exudates in rhizosphere colonization by PGPR. In contrast, Kuiper et al. (2001) showed that a mutant strain of *Pseudomonas* sp., with increased uptake of putrescine as a source of nitrogen, had less competitive advantage than the wild strain. These observations emphasize the careful upregulation of nutrients by PGPR for effective root colonization (Somers et al. 2004).

4.3. PGPR biofilms

PGPR form biofilms on the root surface, where the cells are densely packed and covered by an exopolysaccharide layer, and facilitate a suitable environment for gene-regulatory systems (Faqua & Greenberg, 2002). The percentage of competent cells has been more in biofilms, compared to individual bacteria. Several of the plant-beneficial phenotypes of PGPR including production of antibiotics and secondary metabolites, are under the control of quorum sensing (Loh et al. 2002; Smoers et al. 2004), a bacterial intercellular communication mechanism that controls gene expression in response to population density.

Quorum sensing has been studied in depth in Gram -ve than in Gram +ve bacteria. In these bacteria, *N*-acyl homoserine lactones (AHLs), usually synthesized by a member of the LuxI protein family, acts as a signal molecule. AHL-mediated cell-cell communication is a widespread phenomenon among plant-associated bacteria (Pierson III, et al. 1998). The AHL molecules act as population density sensors and also facilitate communication between cells of different species colonizing the plant rhizosphere (Pierson et al. 1998). AHL accumulates in the environment surrounding the bacteria, and when reaches a threshold level, trigger specific bacterial responses. AHL binds to its cognate receptor, which is a member of the LuxR protein family, and the signal-receptor complex activates or represses target gene expression. Often quorum-sensing regulatory mechanisms are part of a large complex of regulatory cascades, with the global regulatory GacS/GacA two-component system. This regulatory system influences the target gene expression at post-transcriptional level, involving an mRNA target sequence, two RNA binding proteins (RsmA and RsmE), and a regulatory RNA (RsmZ) capable of binding RsmA (Haas et al. 2002), and is widespread among pseudomonads (Heeb & Haas, 2001).

Quorum sensing through production of AHL is widely detected in *Pseudomonas* spp. that any other root-colonizing bacteria (Juhas et al. 2005). Production of phenazine antibiotics by *Pseudomonas aureofaciens* strain 30-84 in the wheat rhizosphere is regulated by the *phzI* and *phzR* genes, which determine the production of *N*-hexanoyl homoserine lactone (Wood et al. 1997). Subsequently, a second quorum-sensing system, *Csa*R-*Csa*I, determining the rhizosphere competence through biosynthesis of cell surface components was detected in the same strain. Colonization of wheat rhizosphere by strain 30-84 was reduced six-fold by a mutation in the *phzR* gene and forty-fold by a mutation in both *phzR* and *Csa*R genes (Zhang & Pierson, 2001). AHL-mediated quorum sensing has been detected in *P. putida* IsoF, and a quorum sensing locus termed *ppu*, functionally interchangeable with *las*I of *P. aeruginosa* was isolated (Steidle et al. 2002).

A *phzI* mutant strain of *P. aureofaciens* 30-84, with reduced expression of the phenazine biosynthetic operon was restored for phenazine gene expression to wild-type levels in the wheat rhizosphere after coinoculation with an isogenic strain that produced the endogenous AHL signal, supporting the role of AHL in interpopulation signaling (Wood et al. 1997). Through, GFP-based AHL reporter systems *P. putida* and *Serratia liquefaciens* were observed to sense AHLs from each other in tomato rhizosphere (Steidle et al. 2001). The model plant *Medicago truncatula* responded to nanomolar concentrations of AHLs from *Sinorhizobium meliloti* and *P. aeruginosa*, with significant changes in the accumulation of over 150 proteins, including auxin-responsive proteins and flavonoid synthesis proteins (Mathesius et al. 2003). Elasri et al. (2001) observed that certain rhizobacteria degrade AHL, which could be one of the probable reasons for inconsistent field performance of rhizobacterial inoculants that exert beneficial effects through quorum-sensing-regulating processes. Recent technological breakthroughs in molecular biology, such as in the development of auto fluorescent marker proteins, genomics, and proteomics, accelerate the fundamental understanding of biofilm formation, and intra- and inter species bacterial interactions that help in development of bioinoculants with consistent performance.

In addition to the above-mentioned factors affecting root colonization, determination of the bacterial metabolic pathways activated during root colonization using molecular genetic approaches unravels the information needed on the host-PGPR interactions. *In vivo* expression technology (IVET) is a powerful technique to identify promoters that are functional and involved in different steps of root colonization. Using the functional genetic approach rice colonization by *Pseudomonas stutzeri* A15 was studied (Rediers et al. 2003). Resolvase IVET, a derivative of IVET can be used to detect weakly or transiently expressed promoters (Boch et al. 2002). Using microarray analysis, Wang et al. (2005) identified that during the colonization of an endophytic PGPR, genes involved in metabolism, signal transduction and stress response were upregulated in *Arabidopsis*. Upon colonization of the PGPR strain, putative auxin-regulated genes and nodulin-like genes were up-regulated, and some ethylene-responsive genes were down regulated.

5. SYNERGISTIC ACTION OF PGPR, RHIZOBIA AND ARBUSCULAR MYCORRHIZAL FUNGI

Among the different bioinoculants proven as potential alternatives to chemical fertilizers, *Rhizobium* and arbuscular mycorrhizal (AM) fungi are important groups in addition to PGPR. Rhizobia are Gram -ve bacilli that live freely in soil and through strain-specific symbiotic association with the roots of legume plants reduce atmospheric N_2 to ammonia, a suitable form for plant assimilation. AM fungi distributed in almost all the soils form endophytic association with roots of vascular flowering plants. These fungi produce arbuscules, microscopic tree-like structures, within cells of the plant's root and develop hyphae that extend into the soil. The hyphae explore a far greater volume of soil than root hairs can, and assimilate phosphorous, copper and zinc that are difficult to move through the soil solution, thus enhance the plant nutrition. It became necessary to evaluate the specific functional compatibility of PGPR with rhizobia and AM fungi to develop multi-functional based bioinoculants. In the recent past, much research is addressed for better understanding of the diversity, dynamics and significance of rhizosphere microbial populations and their co-operative activities (Barea et al. 2005). In several of these studies a positive co-operation was observed between PGPR, rhizobia and/or AM fungi (Table 2).

Table 2. Examples of the beneficial association of PGPR and *Rhizobium*/arbuscular mycorrhizal (AM) fungi on plant growth.

PGPR strain	*Rhizobium*/AM strain	Benefit as an increase in	Reference
Pseudomonas fluorescens 92rk	*G. mosseae* BEG12	the growth of tomato plants	Gamalero et al. 2004
Pseudomonas striata	*Bradyrhizobium* sp. and *G. fasciculatum*	N_2 and P uptake by greengram	Zaidi et al. 2004
Azospirillum sp.; *P. striata*	*Rhizobium* sp.	nodule number and dry weight, plant height, dry weight and yield in pigeonpea	Devanand et al. 2002
Pseudomonas sp.	*Mesorhizobium* sp. *cicer*	the shoot dry weight by 3-4 fold and also the N_2 uptake	Goel et al. 2002
P. fluorescens; *Pseudomonas aeruginosa*	*Bradyrhizobium japonicum*	the growth of tomato when applied as soil drench	Siddiqui & Shaukat, 2002
Bacillus sp. CECT450	*Rhizobium tropici*	the growth of field-grown bean	Camacho et al. 2001
P. fluorescens	*B. japonicum*	nodule number and acetylene reduction activity in soybean roots	Chebotar et al. 2001

PGPR strain	*Rhizobium*/AM strain	Benefit as an increase in	Reference
Pseudomonas spp.	*Rhizobium leguminosarum* biovar *viceae*	the shoot height, root length and dry weight of pea	Kumar et al. 2001
P. fluorescens DF57	*Glomus intraradices*; *G. caledonium*	P uptake by cucumber	Ravnskov & Jakobsen, 1999
Pseudomonas sp. MRS 13 and MRS 16	*Bradyrhizobium* sp. S24	nodule weight, plant dry weight and total N_2 content in greengram	Sindhu et al. 1999
Bacillus circulans	*Glomus* sp. 88	N_2 and P uptake, and grain and straw yields of wheat	Singh & Kapoor, 1998a
B. circulans	*Glomus fasciculatum*	increased N_2 and P uptake in mungbean	Singh & Kapoor, 1998b
Serratia liquefaciens; *Serratia proteamaculans*	*B. japonicum*	the grain yield and grain protein yield in soybean	Dashti et al. 1997
Aeromonas hydrophila; *S. liquefacines*; *S. proteamaculans*	*B. japonicum*	the number of nodules and amount of nitrogen fixed in soybean rhizosphere	Zhang et al. 1996

Combined inoculation of PGPR and rhizobia was observed to exert positive effects on the growth of legumes including pea, clover, common bean, cowpea and soybean, by increased nodulation. For example, co-inoculation of rhizobacteria with rhizobia increased the number of nodules and nodule occupancy (Cattelan et al. 1999; Podile, 1995; Polonenko et al. 1987). However, the beneficial effects of PGPR and *Rhizobium* co-inoculation are strain dependent. The basic mechanisms involved in this synergistic activity were not completely known and remains a challenge. One possibility is that PGPR, by altering the host secondary metabolism and/or creating antibiosis in the rhizosphere, out compete the pathogens and eliminate competition of *Rhizobium* with deleterious microorganisms for colonization of the plant parts. Alternately, the applied rhizobacteria stimulate formation of additional infection sites that may later be occupied by rhizobia, thus increasing the nodulation. Alteration of the plant flavonoid metabolism was proposed as another mechanism of synergistic activity of PGPR and rhizobia (van Loon & Bakker, 2003). Isolates of *Azospirillum* spp. co-inoculated with *Rhizobium* increased the nodulation in several legume crops. *Azospirillum* sp. produces large quantities of auxins and stimulates the formation of epidermal cells that become root hair cells or additional infection sites for rhizobial colonization. Inoculation with *A. brasilense* also promoted the *nod*-gene inducers in roots of *Phaseolus vulgaris* (Burdman et al. 1996). Certain isolates of PGPR enhance legume nodulation by affecting the signal exchange between plants and rhizobia. These isolates produce signal molecule analogues and/or stimulate the plant to produce more signal molecules (Parmar & Dadarwal, 1999).

The synergistic effect of PGPR and AM fungi on plant growth promotion is well documented (Bhowmik & Singh, 2004), in spite of the scanty information on the impact of mixed inocula on root architecture. Combined application of a rhizobacterium *P. fluorescens* and AM fungal strain *Glomus mosseae* resulted in improved growth of chickpea combined to the application of the two bioinoculants separately, and also reduced the galling and multiplication of the nematode pathogen *Meliodogyne javanica* (Siddiqui & Mahmood, 2001). Co-inoculation of *Pseudomonas* sp. F113 and *G. mosseae* stimulated mycelial development from *G. mosseae* spores germinating in soil and tomato root colonization (Barea et al. 1998). *Vicia faba* plants with the *Rhizobium* and AMF symbiotic associations were observed for higher photosynthetic rates per unit leaf area (Jia et al. 2004). *P. fluorescens* and *G. mosseae* when co-inoculated had a synergistic effect on root fresh weight in tomato and both the organisms, depending on the inoculum combination, strongly affected root architecture. *P. fluorescens* increased mycorrhizal colonization, suggesting its role as a mycorrhization helper bacterium (Gamalero et al. 2004). Two phosphate-solubilizing rhizobacterial isolates *Enterobacter* sp., & *B. subtilis* promoted the establishment of *Glomus intraradices*, which in turn increased biomass and N and P accumulation in plant tissues (Toro et al. 1997). The two bacterial isolates solubilized the rock phosphate in soil and the presence of mycorrhizal strain resulted in increased absorption of phosphorous. The possible mechanisms affected by the rhizobacteria in favor of mycorrhizal establishment are increased receptivity of root, root-fungus recognition, fungal growth, germination of fungal propagules, and modification of the chemistry of the rhizospheric soil (Garbaye, 1994: Johannson et al. 2004). The role of individual mechanisms in the synergistic action of PGPR and AM fungi needs to be determined.

In a global scenario of reaching the nitrogen and phosphorous resource plateau, increasing concerns on environmental effects of chemical fertilizers, as well as their cost, more emphasis is on promotion of biological nitrogen fixation, AM fungi and PGPR technologies. Symbiotic effects of these three major groups of bioinoculants should be exploited for the economical benefit of subsistence farming systems and attempts should be made for complete understanding of the mechanisms of synergistic activity between these three groups of bioinoculants.

6. TRACKING PGPR IN THE RHIZOSPHERE

Since the rhizosphere colonization of PGPR is dependent on a number of biotic and abiotic factors, sensitive and reliable methods are needed for specific detection and quantification of inoculated PGPR in the rhizosphere. Spontaneous or induced antibiotic resistance markers, fluorescently labelled antibodies, immunodiffusion methods and fatty acid profiling have long been used for tracking PGPR. These methods are laborious and unambiguous (Cavigelli et al. 1995), and don't provide information on the metabolic status of introduced PGPR and their effects on the resident microbial communities. Additionally, the possible loss of ecological competence of antibiotic resistance mutants of PGPR, and also transfer of antibiotic resistance to other organisms, limits the extensive use of this method. To overcome these constraints and improve the efficacy of detection, presently several marker gene-based and nucleic acid-based detection methods have been developed for PGPR tracking.

6.1. Marker gene-based detection methods

Introduction of a selectable marker into a particular PGPR strain facilitates its sensitive and selective detection even in the background of a large population of native microorganisms. Currently available marker genes link a specific metabolic activity of the bacterium with the production of a colorimetric or light-induced phenotype, or with specific degradative capability, or resistance to heavy metals. Thus, these methods are convenient to study ecology of the marked strain and provide more insight to understand its interaction with plant roots with minimum alteration of the environment. Additionally, these methods provide information on spatial distribution and metabolic status of the marked strain with respect to environment.

The *lacZY* -based expression of *E. coli lac* operon genes encoding β-galactosidase and lactose permease, is a sensitive, selectable marker system and has been extensively used to monitor the rhizosphere colonization of *Pseudomonas* spp. (Hofte et al. 1990; Krishnamurthy & Gnanamanickam, 1998). Bacterial luciferase, encoded by *luxAB* genes, is a promising marker to determine both the spatial distribution and cellular metabolic activities of bacteria in their native environment (Jannson & Prosser, 1997). The use of *lux* system requires the presence of oxygen, and thus limits its use with anaerobic PGPR strains. This marker gene was used to evaluate the rhizosphere colonization of cucumber, cotton, maize and soybean roots by *Pseudomonas* sp. strains (Beauchamp et al. 1993). Rattray et al. (1995) reported that luminescence activity of *lux*-marked *Enterobacter cloacae* was greater in cells from rhizosphere than in those from non-rhizosphere arable soil, where luminescence was detected only after addition of growth substrate. This bioluminescence-based assay of root colonization and metabolic activity might, therefore, enable rapid identification of suitable host plants for PGPR in place of glasshouse and field trials.

Green fluorescent protein (GFP) gene of the jellyfish *Aequorea victoria* commonly employed address a range of biological questions (Chalfie et al. 1994) is used for non-destructive monitoring of gene expression, protein localization and various other biological phenomena of PGPR. GFP-marked cells can be visualized by using standard microscopes equipped with commonly available fluorescent filter sets. GFP facilitates *in situ* detection of marked strains, but this marker system is less valuable for studies of transient (real-time) gene expression to assess physiological activities in the rhizosphere. Novel *gfp*-based reporter systems, whose expression is cell growth regulated, has recently been developed. A marker system comprising a fusion of the ribosomal *E. coli rrnB* P1 promoter to a gene encoding an unstable variant of the GFP was used to monitor *P. putida* in barley rhizosphere. Recombinant *P. putida* growing exponentially at a rate of $> 0.17\ h^{-1}$ emitted growth rate-dependent green fluorescence detectable at single-cell level. The monitoring system permitted nondestructive *in situ* detection of fast-growing bacterial micro colonies on the sloughing root sheath cells of barley seedlings and assessment of the growth of *P. putida* single cells in the rhizosphere (Ramos et al. 2000).

The *xylE* gene that encodes catechol 2, 3 dioxygenase (E.C. 1.13.11.2), and *gusA* gene that encodes β-glucuronidase are also adopted as marker genes to study PGPR ecology. In all these studies, it is important to ensure that the transformed strain is as active in the environment as the wild-type strain.

6.2. Nucleic acid-based detection methods

Nucleic acid-based techniques i.e. identification of characteristic patterns and sequences of nucleic acids, doesn't interfere with bacterial ecology prior to the moment of assay. Additionally, the genotyping profiles provide discrete information on different microorganisms of the rhizosphere that may also include non-culturable bacteria (Ward et al. 1990). Nucleic acid-based detection methods are insensitive to the bacterial growth conditions and culture age and do not require any prior preparation for individual strains unlike other marker systems, such as preparation of antisera, induction of antibiotic resistance, or introduction of a marker gene etc.

Both PCR-independent nucleic acid (DNA and RNA) hybridization and PCR-dependent amplification techniques are used for PGPR tracking. In DNA hybridization studies, use of specific probes that hybridize unique DNA sequences of the target strain, allows specific detection of introduced PGPR. Strain specific sequences can be identified by subtraction hybridization protocols (Seal et al. 1992), and wherever strain specific DNA sequences are not identified, a specific sequence can be introduced into the bacterial genome and detected by hybridization with corresponding probes. For DNA hybridization, the standard restriction fragment length polymorphism (RFLP) analysis was improved by DNA fragmentation with infrequently cutting restriction enzymes and the separation of large restriction fragments by pulsed field- or field inversion gel electrophoresis (PFGE; FIGE) followed by hybridization with probes.

Fluorescence *in situ* hybridization (FISH), by use of fluorescent labeled rRNA-targeted oligonucleotide probes of 15-20 nucleotide length detects single cells of the target strain, its cellular locations and exact cell numbers (Amann et al. 2001). An actively growing bacterial cell contains 50,000-70,000 ribosomes, thus generate sufficient signal for detection by conventional fluorescence microscopy. The establishment of seed-inoculated *P. fluorescens* DR54 on sugar beet root surfaces was monitored by immunofluorescence and rRNA-targeted FISH techniques. Population density of DR54 reached a constant level after 2 days, and there after high cellular activity was noticed only in few bacteria located as single cells, especially at the root tip (Lubeck et al. 2000). FISH technique determined the endophytic root colonization of wheat roots by *A. brasilense* strains that resulted in plant growth promotion (Rothballer et al. 2003). The limitations of FISH technique are poor cytoplasmic penetration of probes in bacteria with thick cell walls, hindrance of the target site by other ribosomal components and poor detection of metabolically inactive cells.

The PCR-dependent methods of PGPR tracking and identification are based on characteristic genomic fingerprints that require only a minute quantity of DNA. Continued improvement in the methods for direct extraction of microbial DNA from soil (Simmon et al. 2004), boosted the adoption of PCR-based fingerprinting in analysis of soil microbial communities including that of rhizosphere. The primers used in different fingerprint techniques can be either of arbitrary sequence or directed to sequences that are likely to be repeated many times within the targeted genome. The precision and reproducibility of these fingerprints can be increased by use of primers that are complementary to widespread and well-conserved DNA sequences, which often discriminate between different strains of the same species. Various PCR-based methods that can be used for PGPR identification, tracking and rhizosphere community analysis are arbitrarily primed PCR (AP-PCR), sequence characterized amplified region (SCAR) analysis, repetitive extragenic palindromic (REP)-PCR, reverse transcriptase dependent PCR (RT-PCR), amplified fragment length polymorphism (AFLP), amplified ribosomal DNA restriction analysis (ARDRA), amplification of ribosomal intergenic spacer (RIS) regions between the 16S and 23S rRNA genes, terminal restriction fragment length polymorphism (T-RFLP), denatured gradient gel electrophoresis (DGGE), temperature gradient gel electrophoresis (TGGE) and single strand conformation polymorphism (SSCP).

Among the above-mentioned different methods, DGGE and TGGE are also ideal for separation of amplification products of 16S rRNA genes in complex bacterial communities, and have been applied to study the rhizobacterial communities. Yang et al. (2001) used 16S rDNA fingerprints obtained by PCR-DGGE to monitor bacterial community structures in the rhizosphere of avocado trees during infection by *Phytophthora cinnamomi* and repeated bioaugmentation with a disease suppressive *P. fluorescens* st. 513. More diverse bacteria colonized the infected roots with no detectable visible symptoms, and their community structures significantly differ from those of healthy roots. Repeated applications of *P. fluorescens* st. 513 protected the trees from *P. cinnamoni* infection and the bacterial community structures on the roots of these trees were similar to those of non-treated healthy roots. The same technique was also used to study the effects of antagonistic *P. putida* AC14-3-8 and *Serratia grimesii* L16-3-3 on the native bacterial communities in the rhizosphere of native and transgenic potato plants expressing T_4 lysozyme (Zveld et al. 2001). The introduced strains didn't become dominant among the bacterial community and at any sampling time. There were no differences in the DGGE pattern of bacterial communities of the rhizosphere and geocarposphere of both inoculated and non-inoculated plants. RT-PCR amplification of the 16s rRNA genes followed by DGGE was applied to study the presence and activity of bacterial populations in the rhizosphere of *Chrysanthemum* at different growth stages. The majority of DGGE band sequences were related to the genera *Acetobacter*, *Arthobacter*, *Bacillus*, *Comamonas*, *Pseudomonas* and *Variovorax* and the microflora of root tip and base samples from 2-week-old plants were similar.

The establishment of phenazine-1-carboxylic acid producing *P. aureofaciens* TX-1 and its effect on the indigenous microbial community were studied in the turfgrass system. Use of PCR amplification and hybridization of extracted DNA with the *P. aureofaciens* TX-1-specific primer-probe combination indicated that TX-1 is capable of establishing in the rhizosphere and thatch, and also over wintering. Separation of PCR-amplified partial 16S rRNA genes by DGGE indicate the transient displacement of a leaf surface bacterial community member up on repeated application of *P. aureofaciens* TX-1 through irrigation water, but there was no alteration in any of the dominant members of the thatch and rhizosphere communities (Sigler et al. 2001).

In addition to gradient gel eletrophoresis, other PCR-dependent methods were also widely used to study the population dynamics and diversity of rhizobacteria. Using a pair of SCAR primers specific to *P. fluorescens* Pf29A, Chapon et al. (2003) observed that the total populations of Pf29A detected in wheat rhizoplane using antibiotic resistant and GFP marker systems was only 13% compared to that quantified using SCAR markers. The effect of application method of *Burkholderia ambifaria* MCI 7 on microbial community structure in the rhizosphere of maize was determined by ARDRA technique. Seed treatment with MCI 7 decreased the microbial diversity while soil treatment increased the diversity of microorganisms in the maize rhizosphere (Cicillo et al. 2002). The occurrence and diversity of endophytic bacteria in relation to plant growth in potato was studied by T-RFLP analysis of 16S rDNA (Sessitsch et al. 2004)

7. PGPR FORMULATION AND DELIVERY SYSTEMS

Essential pre-requisites for large-scale promotion of the efficient PGPR strains are development of viable, cost-effective and user-friendly formulations that are easy to handle and convenient for application, and simple method of delivery. The ability of a PGPR formulation to consistently deliver a critical number of viable cells is an indicator of its success (Pauu, 1988). Since the active ingredient of the PGPR formulation is live bacterium, maintenance of the bacterium in a metabolically and physiologically competent state, should be the primary concern in order to derive maximum advantage of the formulation. Other components of the formulation, carrier and additives support the bacterial growth and multiplication, and aid in the stabilization and protection of the bacterial cells during storage, transport and at the target zone, respectively. The ideal carrier and conditions for development of formulations are subjected to strain-to-strain variability. Further, for the same organism different formulations are needed to suit different climates, soil-types and end-user preferences. PGPR and their formulations are commonly applied as seed treatment, soil amendment or root dip in bacterial suspension before transplanting. Certain PGPR were also applied on an experimental basis as foliar spray or through drip irrigation.

Different PGPR formulations were developed that either use liquid or solid carrier materials depending on the requirement. Liquid formulations of PGPR are mostly based on vegetable oils such as rapeseed, palm oil and castor oil. Among the different carrier materials used for solid formulations, peat has been identified globally as the most ideal carrier because of its high water holding and buffering capacities, and nutrient content (Burton 1982). However, because of the limited availability and high cost of good quality peat, the search continued for identification of alternate carrier materials. Compost is one such low-cost and easily available alternate carrier that is also rich in organic material, nontoxic and biodegradable like peat, and readily renewable resource unlike peat. Other carrier materials that are frequently used for PGPR formulations are talc, spent compost, vermiculite, and other agricultural wastes and by-products.

Talc-based formulation of a mixture of *Pseudomonas* spp. PF 1 and PF 7 in addition to reduction of sheath blight severity increased the plant height, number of tillers and grain yield in rice (Nandakumar et al. 2001). Following seed treatment with talc-based formulation of a rhizobacterium *P. fluorescens*, the bacterium was recovered from the chickpea seeds even up to 180 days (Vidhyasekaran & Muthamilan, 1995). *B. subtilis* AF 1 formulated in peat and spent compost had a viable populations of ~log 9.0 CFU g-1 up to 180 days, and the increase in bacterial populations was > log 5.0 CFU g^{-1} compared to the initial populations. These two formulations when applied as seed treatment significantly increased the root length, shoot length and dry biomass of peanut and pigeonpea (Manjula & Podile, 2001; 2005). Peat as a carrier material supported the growth of *B. firmis* GRS 123 and *B. megaterium* GPS 55, and the viable bacteria ranged > log 7.0 CFU g^{-1} up to the measured 180 days after inoculation. Peat formulations of these bacteria applied as seed treatment equally supported the growth of peanut in field compared to their mid-log phase cells (Kishore et al. 2005a).

In addition to solid-state fermentation, other formulations useful for PGPR application as seed treatment include the use of cross-linking organic polymers like alginate, carrageenan or polyacrylamide (Bashan, 1986; Fravel et al. 1985). Beads of these polymers with encapsulated bacterial cells can be applied together with the seeds at sowing. Seed encapsulation in a gelatinous pellet containing beneficial microbes provides an interesting variation of the method. *P. fluorescens* cells encapsulated in alginate beads and introduced into non-sterile loamy sand had higher soil populations and colonization of wheat roots better than compared to fresh cells. The addition of skim milk, or skim milk and bentonite clay further enhanced the survival of the encapsulated cells in soil (van Elsas et al. 1992).

Development of PGPR formulations with improved efficacy is a challenging task for popularization of the PGPR-based bioformulations. Addition of nutrients to the carrier material has been one of the major focuses in such studies, to increase the survival of PGPR. Addition of chitin, a liner polymer of *N*-acetyl glucosamine, to the carrier material peat enhanced the survival and multiplication, and also *in vivo* efficacy of chitinolytic *B. subtilis* AF 1. Chitin-supplemented peat formulation of *B. subtilis* AF 1, applied as a seed treatment, was significantly effective than fresh cells of AF 1 or peat formulation of AF 1 in promotion of seedling emergence and dry biomass in groundnut and pigeonpea (Manjula & Podile, 2001; 2005). This formulation also had an improved biocontrol efficacy against collar rot disease in groundnut and Fusarium wilt in pigeonpea, because of the pre-induction of chitinase production (Manjula & Podile, 2001). Formulations of *B. amyloliquefaciens*, *Bacillus pumilus* and *B. subtilis* based on chitosan, a deacetylated form of chitin, applied as soil amendment increased the root and shoot lengths and grain yield of rice by > 50% (Preeti et al. 2002).

8. CONCLUSIONS AND FUTURE OUTLOOK

In spite of significant advancements in understanding the mechanisms of action of PGPR and factors affecting their root colonization and *in situ* efficacy, results in the field environment are inconsistent. Additionally, because of the challenges in laborious screening programs, formulation and application, PGPR have yet to fulfill their promise and potential as commercial inoculants. Unambiguous and economical strain identification facilitates the comparison of the selected PGPR with that of earlier reported strains and helps in decision-making. Use of PGPR for sustainable increase of plant growth in agricultural, horticultural and agro forestry systems is directly related to our understanding of PGPR diversity, host specificity, mechanisms of action, colonization ability, and formulation and application. There should be continued efforts to develop effective, economical and convenient formulations of PGPR, needed for their success. Attempts for strain improvement either through selection or genetic engineering should target enhanced plant growth promoting ability, greater competitiveness, and applicability of strains to a broad range of host cultivars, soil conditions and climates.

There should be special emphasis to study the introduced PGPR in rhizosphere, their dispersal, metabolic status and interaction with the host plant. Identification of genes and traits involved in the process of root colonization will result in detailed knowledge of bacterial rhizosphere ecology, physiology and its interaction with plant roots, which will facilitate more efficient application methods of inoculant strains and strategies to reduce diseases caused by phytopathogens. Elucidation of plant signaling pathways activated in response to PGPR inoculation should be the major objective in investigations of plant-PGPR interactions.

Apart from the basic research for PGPR identification, application and improvement of efficiency, establishment of a strong researcher-entrepreneur partnership and increased awareness for adoption of PGPR among the end-users, are needed for popularization of the technologies generated towards the promotion of bioinoculants and their long-term success. This goal, in turn, is dependent on setting a set of separate regulatory authorities for marketing of PGPR products and regular quality assessment of the final products released for farmer-use. Free exchange of PGPR strains between researchers and countries may promote the untapped potential of PGPR to support the plant growth in diverse agro ecological regions.

9. REFERENCES

Antoun, H., Beauchamp, C. J., Goussard, N., Chabot, R., & Lalande, R. (1998). Potential of *Rhizobium* and *Bradyrhizobium* species as plant growth promoting rhizobacteria on non-legumes: Effect of radishes (*Raphanus sativus* L.). *Plant and Soil, 204,* 57-67.

Arndt, W., Kolle, C., & Buchenauer, H. (1998). Effectiveness of fluorescent pseudomonads on cucumber and tomato plants under practical conditions and preliminary studies on the mode of action of the antagonists. *Zeitschrift fur Pflanzenkrankheiten und Pflanzenschutz, 105,* 198-215.

Asghar, H. N., Zahir, Z. A., & Arshad, M. (2004). Screening rhizobacteria for improving the growth, yield, and oil content of canola (*Brassica napus* L.). *Australian Journal of Agricultural Research, 55,* 187-194.

Asghar, H., Zahir, Z., Arshad, M., & Khaliq, A. (2002). Relationship between *in vitro* production of auxins by rhizobacteria and their growth-promoting activities in *Brassica juncea* L. *Biology and Fertility of Soils, 35,* 231-237.

Audenaert, K., Pattery, T., Cornelis, P., & Hofte, M. (2002). Induction of systemic resistance to *Botrytis cinerea* in tomato by *Pseudomonas aeruginosa* 7NSK2: Role of salicylic acid, pyochelin, and pyocyanin. *Molecular Plant-Microbe Interactions, 15,* 1147-1156.

Babalola, O. O., Osir, E. O., Sanni, A. I., Odhiambo, G. D., & Bulimo, W. D. (2003). Amplification of 1-amino-cyclopropane-1-carboxylic (ACC) deaminase from plant growth promoting rhizobacteria in *Striga*-infested soil. *African Journal of Biotechnology, 2,* 157-160.

Barea, J. M., Andrade, G., Bianciotto, V., Dowling, D., Lohrke, S., Bonfante, P., O'Gara, F., & Azcon-Aguilar, C. (1998). Impact on arbuscular mycorrhiza formation of *Pseudomonas* strains used as inoculants for biocontrol of soil-borne fungal plant pathogens. *Applied and Environmental Microbiology, 64,* 2304-2307.

Barea, J. M., Pozo, M. J., Azcon, R., & Azcon-Aguilar, C. (2005). Microbial co-operation in the rhizosphere. *Journal of Experimental Botany, 56,* 1761-1778.

Bashan, Y. 1986. Alginate beads as synthetic inoculant carriers for slow release of bacteria that affect plant growth. *Applied and Environmental Microbiology, 51,* 1089-1098.

Bashan, Y., Gina, H., & de-Bashan, L. E. (2004). Azospirillum-plant relationships: physiological, molecular, agricultural, and environmental advances (1997-2003). *Canadian Journal of Microbiology, 50*, 521-577.

Bashan, Y., & Holguin, G. (1997). Proposal for division of plant growth promoting rhizobacteria into two classes: biocontrol-PGPBs and PGPB. *Soil Biology and Biochemistry, 30,* 1225-1228.

Beauchamp, C. J., Kloepper, J. W., & Lemke, P. A. (1993). Luminometric analyses of plant root colonization by bioluminescent pseudomonads. *Canadian Journal of Microbiology, 39*, 434-441.

Belimov, A. A., Hfontzeas, N., Safronova, V. I., Demchinskaya, S. V., Piluzza, G., Bullitta, S., & Glick, B. R. (2005). Cadmium-tolerant plant growth-promoting bacteria associated with the roots of Indian mustard (*Brassica juncea* L. Czem.). *Soil Biology and Biochemistry, 37,* 241-250.

Belimov, A. A., Safronova, V. I., Sergeyeva, T. A., Egorova, T. N., Matveyeva, V. A., Tsyganov, V. E., Borisov, A. Y., Tikhonovich, I. A., Kluge, C., Preisfeld, A., Dietz, K. J., & Stepanok, V. V. (2001). Characterization of plant growth promoting rhizobacteria isolated from polluted soils and containing 1-aminocyclopropane-1-carboxylate deaminase. *Canadian Journal of Microbiology, 47,* 642-652.

Bent, E., Tuzun, S., Chanway, C. P., & Enebak, S. (2001). Alterations in plant growth and in root hormone levels of lodgepole pines inoculated with rhizobacteria. *Canadian Journal of Microbiology, 47,* 793-800.

Beraha, L., Wisniewski, V., & Garber, E. D. (1983). Avirulence and reduced extracellular enzyme activity in *Geotrichum candidum*. *Botanical Gazette, 144,* 461-465.

Bertagnolli, B. L., Soglio, F. K. D., & Sinclair, J. B. (1996). Extracellular enzyme profiles of the fungal pathogen *Rhizoctonia solani* isolate 2B-12 and of two antagonists, *Bacillus megaterium* strain B153-2-2 and *Trichoderma harzianum* isolate Th008. I. Possible correlations with inhibition of growth and biocontrol. *Physiological and Molecular Plant Pathology, 48,* 145-160.

Bertand, H., Nalin, R., Bally, R., & Cleyet-Marel, J. C. (2001). Isolation and identification of the most efficient plant growth-promoting bacteria associated with canola (*Brassica napa*). *Biology and Fertility of Soils, 33*, 152-156.

Bhowmik, S. N., & Singh, C. S. (2004). Mass multiplication of AM inoculum: Effect of plant growth-promoting rhizobacteria and yeast in rapid culturing of *Glomus mosseae*. *Current Science, 86,* 705-709.

Boch, J., Joardar, V., Gao, L., Robertson, T. L., Lim, M., & Kunkel, B. N. (2002). Identification of *Pseudomonas syringae* pv. *tomato* genes induced during infection of *Arabidopsis thaliana*. *Molecular Microbiology, 44,* 73-88.

Bonaterra, A., Ruz, L., Badosa, E., Pinochet, J., & Montesinos, E. (2003). Growth promotion of *Prunus* rootstocks by root treatment with specific bacterial strains. *Plant and Soil, 255*, 555-569.

Burdman, S., Volpin, H., Kigel, J., Kapulnik, Y., & Okon, Y. (1996). Promotion of *nod*-gene inducers and nodulation in common bean (*Phaseolus vulgaris*) roots inoculated with *Azospirillum brasilense* Cd. *Applied and Environmental Microbiology, 62,* 3030-3033.

Burton, J. C. (1982). Modern concepts in legume production. In P. H. Graham & S. C. Harris (Eds.). *Biological nitrogen fixation technology for tropical agriculture* (pp. 105-114). Cali, Colombia: CIAT.

Buyer, J. S., Kratzke, M. G., & Sikora, L. J. (1993). A method for detection of pseudobactin, the siderophore produced by a plant-growth-promoting *Pseudomonas* strain, in the barley rhizosphere. *Applied and Environmental Microbiology, 59,* 677-681.

Buysens, S., Poppe, J., & Hofte, M. (1994). Role of siderophores in plant growth stimulation and antagonism by *Pseudomonas aeruginosa* 7NSK2. In M. H. Ryder, P. M. Stephens & G. D. Bowen (Eds.). *Improving plant productivity with rhizosphere bacteria* (pp. 139-141). Adelaide, Australia: Commonwealth Scientific and Industrial Research Organization.

Camacho, M., Santamaria, C., Temprano, F., Rodriguez-Navarro, D. N., & Daza, A. (2001). Co-inoculation with *Bacillus* sp. CECT 450 improves nodulation in *Phaseolus vulgaris* L. *Canadian Journal of Microbiology, 47,* 1058-1062.

Cattelan, A. J., Hartel, P. G., & Fuhrmann, J. J. (1999). Screening for plant growth-promoting rhizobacteria to promote early soybean growth. *Soil Science Society of America Journal, 63,* 1670-1680.

Cavigelli, M. A., Robertson, G. P., & Klug, M. J. (1995). Fatty acid methyl ester (FAME) profiles as a measures of soil microbial community structure. *Plant and Soil, 170,* 99-113.

Cezon, R., Manero, F. J., Gutierrez, P. A., Ramos, B., & Garica, J. A. L. (2003). Effects of two plant growth-promoting rhizobacteria on the germination and growth of pepper seedlings (*Capsicum annum*) cv. Roxy. *Archives of Agronomy and Soil Science, 49,* 593-603.

Chalfie, M., Tu, Y., Euskirchen, G., Ward, W. W., & Prasher, D. C. (1994). Green fluorescent protein as a marker for gene expression. *Science, 263,* 802-805.

Chapon, A., Boutin, M., Rime, D., Delalande, L., Guillerm, A. Y., & Sarniguet, A. (2003). Direct and specific assessment of colonization of wheat rhizoplane by *Pseudomonas fluorescens* Pf29A. *European Journal of Plant Pathology, 109,* 61-70.

Chebotar, V. K., Asis Jr. C. A., & Akao, S. (2001). Production of growth-promoting substances and high root colonization ability of rhizobacteria enhance the nitrogen fixation of soybean when coinoculated with *Bradyrhizobium japonicum. Biology and Fertility of Soils, 34,* 427-432.

Ciccillo, F., Fiore, A., Bevivino, A., Dalmastri, C., Tabacchioni, S., & Chiarini, L. (2002). Effects of two different application methods of *Burkholderia ambifaria* MCI 7 on plant growth and rhizospheric bacterial diversity. *Environmental Microbiology, 4,* 238-245.

Costa, J. M., & Loper, J. E. (1994). Characterization of siderophore production by the biological control agent *Enterobacter cloacae. Molecular Plant-Microbe Interactions, 7,* 440-448.

Dashti, N., Zhang, F., Hynes, R., & Smith, D. L. (1997). Application of plant growth-promoting rhizobacteria to soybean (*Glycine max* (L.) Merr.) increases protein and dry matter yield under short-season conditions. *Plant and Soil, 188,* 33-41.

David, C., Herve, C., Nicolas, F., Isabelle, S. J., Mohamed, A., & Franc. P. (2005). The crystal structure of the pyoverdine outer membrane receptor FpyA from *Pseudomonas aeruginosa* at 3.6°A resolution. *Journal of Molecular Biology, 347,* 121-134.

de Freitas, J. R., Banerjee, M. R., & Germida, J. J. 1997. Phosphate-solubilizing rhizobacteria enhance the growth and yield but not phosphorous uptake of canola (*Brassica napus* L.). *Biology and Fertility of Soils, 24,* 358-364.

Devanand, B. J., Patil, A. B., Kulkarni, J. H., & Algawadi, A. R. 2002. Effect of plant growth promoting rhizobacteria on growth and yield of pigeonpea (*Cajanus cajan* L.) cultivars. *Karnataka Journal of Agricultural Sciences, 15,* 653-656.

Dey, R., Pal, K. K., Bhatt, D. M., & Chauhan, S. M. (2004). Growth promotion and yield enhancement of peanut (*Arachis hypogaea* L.) by application of plant growth-promoting rhizobacteria. *Microbiological Research, 159,* 371-394.

Dow, M., Newman, M. A., & von Roepenack, E. (2000). The induction and modulation of plant defense responses by bacterial lipopolysaccharides. *Annual Review of Phytopathology, 34,* 241-261.

Duffy, B. K., & Defago, G. (1999). Environmental factors modulating antibiotic and siderophore biosynthesis by *Pseudomonas fluorescens* biocontrol strains. *Applied and Environmental Microbiology, 65,* 2429-2438.

Elasri, M., Delorme, S., Lemanceau, P., Stewart, G., Laue, B., Glickmann, E., Oger, P. M., & Dessaux, Y. (2001). Acyl-homoserine lactone production is more common among plant-associated *Pseudomonas* spp. than among soilborne *Pseudomonas* spp. *Applied and Environmental Microbiology, 67,* 1198-1209.

Esitken, A., Karlidag, H., Ercisli, S., Turan, M., & Sahin, F. (2003). The effect of spraying a growth promoting bacterium on the yield, growth and nutrient element composition of leaves of apricot (*Prunus armeniaca* L. cv. Hacihaliloglu). *Australian Journal of Agricultural Research, 54,* 377-380.

Frankowski, J., Lorito, M., Scala, F., Schmid, R., Berg, G., & Bahl, H. (2001). Purification and properties of two chitinolytic enzymes of *Serratia plymuthica* HRO-C48. *Archives of Microbiology, 176,* 421-426.

Fravel, D. R., Marois, M. J., Lumsden, R. D., & Connick, W. J. (1985). Encapsulation of potential biocontrol agents in an alginate-clay matrix. *Phytopathology, 75,* 774-777.

Fridlender, M., Inbar, J., & Chet, I. (1993). Biological control of soilborne plant pathogens by a β-1, 3-glucanase-producing *Pseudomonas cepacia. Soil Biology and Biochemistry, 25,* 1211-1221.

Fuqua, C., & Greenberg, E. P. (2002). Listening in on bacteria: acyl-homoserine lactone signaling. *Nature Reviews: Molecular Cell Biology, 3,* 685-695.

Gamalero, E., Trotta, A., Massa, N., Copetta, A., Martinotti, M. G., & Berta, G. (2004). Impact of two fluorescent pseudomonads and an arbuscular mycorrhizal fungus on tomato plant growth, root architecture and P acquisition. *Mycorrhiza, 14,* 185-192.

Garbaye, J. (1994). Helper bacteria - A new dimension to the mycorrhizal symbiosis. *New Phytologist, 128,* 197-210.

Garica, J. A. L., Schloter, M., Durkaya, T., Hartmann, A., & Manero, F. J. G. (2003). Colonization of pepper roots by a plant growth promoting *Pseudomonas fluorescens* strain. *Biology and Fertility of Soils, 37,* 381-385.

Gaudin, V., Vrain, D., & Jouanin, L. (1994). Bacterial genes modifying hormonal balance in plant. *Plant Physiology and Biology, 32,* 11-29.

Glick, B. R., Patten, C. L., Holguin, G., & Penrose, D. M. (Eds.). (1999). Biochemical and genetic mechanisms used by plant growth-promoting bacteria. London: Imperial College Press.

Glick, B. R., Penrose, D. M., & Li, J. P. (1998). A model for the lowering of plant ethylene concentrations by plant growth-promoting bacteria. *Journal of Theoretical Biology, 190,* 63-68.

Goel, A., Singhu, S., & Dadarwal, K. (2002). Stimulation of nodulation and plant growth of chickpea (*Cicer arietinum* L.) by *Pseudomonas* spp. antagonistic to fungal pathogens. *Biology and Fertility of Soils, 36,* 391-396.

Goldstein, A. H. (1986) Bacterial solubilization of mineral phosphates: Historical perspectives and future prospects. *American Journal of Alternate Agriculture, 1,* 51-57.

Greiner, R., Haller, E., Koniezny, U., & Jany, K. D. (1997). Purification and characterization of a phytase from *Klebsiella terrigena. Archives of Biochemistry and Biophysics, 341,* 201-206.

Guerinot, M. L. (1994). Microbial iron transport. *Annual Review of Microbiology, 48,* 743-772.

Guo, J. H., Qi, H. Y., Guo, Y. H., Ge, H. L., Gong, L. Y., Zhang, L. X., & Sun, P. H. (2004). Biocontrol of tomato wilt by plant growth-promoting rhizobacteria. *Biological Control, 29,* 66-72.

Gupta, S., Arora, D. K., & Srivastava, A. K. (1995). Growth promotion of tomato plants by rhizobacteria and imposition of energy stress on *Rhizoctonia solani. Soil Biology and Biochemistry, 27,* 1051-1053.

Gupta, C., Dubey, R., & Maheshwari, D. (2002). Plant growth enhancement and suppression of *Macrophomina phaseolina* causing charcoal rot of peanut by fluorescent *Pseudomonas. Biology and Fertility of Soils, 35,* 399-405.

Gupta, A., Saxena, A. K., Gopal, M., & Tilak, K. V. B. R. 2003. Effects of co-inoculation of plant growth-promoting rhizobacteria and *Bradyrhizobium* sp. (*Vigna*) on the growth and yield of green gram (*Vigna radiata* (L.) Wilczek). *Tropical Agriculture, 80,* 28-35.

Gutierrez-Manero, F. J., Ramos-Solano, B., Probanza, A., Mehouachi, J., Tadeo, F. R., & Talon, M. (2001). The plant growth-promoting rhizobacteria *Bacillus pumilus* and *Bacillus licheniformis* produce high amounts of physiologically active gibberellins. *Physiologia Plantarum, 111,* 206-211.

Gyaneshwar, P., Kumar, G. N., Parekh, L. J., & Poole, P. S. (2002). Role of soil microorganisms in improving P nutrition of plants. *Plant and Soil, 245,* 83-93.

Gyaneshwar, P., Parekh, L. J., Archana, G., Poole, P. S., Collins, M. D., Hutson, R. A., & Kumar, G. N. (1999). Involvement of a phosphate-starvation inducible glucose dehydrogenase in soil phosphate solubilization by *Enterobacter asburiae*. *FEMS Microbiology Letters, 171,* 223-229.

Haas, D., & Keel, C. (2003). Regulation of antibiotic production in root-colonized *Pseudomonas* spp. and relevance for biological control of plant disease. *Annual Review of Phytopathology, 41,* 117-153.

Haas, D., Keel, C., & Reimmann, C. (2002). Signal transduction in plant-beneficial rhizobacteria with biocontrol properties. *Antonie van Leeuwenhoek, 81,* 385-395.

Hallmann, J., Rodriguez-Kabana, R., & Kloepper, J. W. (1999). Chitin-mediated changes in bacterial communities of the soil, rhizosphere and within roots of cotton in relation to nematode control. *Soil Biology and Biochemistry, 31,* 551-560.

Hamaoui, B., Abbadi, J. M., Burdman, S., Rashid, A., Sarig, S., & Okon, Y. (2001). Effects of inoculation with *Azospirillum brasilense* on chickpeas (*Cicer arietinum*) and faba beans (*Vicia faba*) under different growth conditions. *Agronomie, 21,* 553-560.

Heeb, S., & Haas, D (2001). Regulatory roles of the GasS/GacA two-component system in plant-associated and other gram-negative bacteria. *Molecular Plant-Microbe Interactions, 14,* 1351-1363.

Hofte, M., Mergeay, M., & Verstraete, W. (1990). Marking the rhizopseudomonas strain 7NSK$_2$ with a Mu d (*lac*) element for ecological studies. *Applied and Environmental Microbiology, 56,* 1046-1052.

Hongyou, Z., Hailei, W., Xili, L., Ye, W., Liqun, Z., & Wenhua, T. (2005). Improving biocontrol activity of *Pseudomonas fluorescens* through chromosomal integration of 2,4-diacetyl-phloroglucinol biosynthesis genes. *Chinese Science Bulletin, 50,* 775-781.

Idriss, E. E., Makarewicz, O., Farouk, A., Rosner, K., Greiner, R., Bochow, H., Richter, T., & Borriss, R. (2002). Extracellular phytase activity of *Bacillus amyloliquefaciens* FZB45 contributes to its plant-growth-promoting effect. *Microbiology, 148,* 2097-2109.

Jaizme-Vega, M. D. C., Rodriguez-Romero, A. S., & Guerra, M. S. P. (2004). Potential use of rhizobacteria from the *Bacillus* genus to stimulate the plant growth of micropropagated bananas. *Fruits, 59,* 83-90.

Jannson, J. K., & Prosser, J. (1997). Quantification of the presence and activity of specific microorganisms in nature. *Molecular Biotechnology, 7,* 103-120.

Jia, Y., Gray, V. M., & Straker, C. J. (2004). The influence of *Rhizobium* and arbuscular mycorrhizal fungi on nitrogen and phosphorous accumulation by *Vicia faba*. *Annals of Botany, 94,* 251-258.

Johansson, J. F., Paul, L. R., & Finlay, R. D. (2004). Microbial interactions in the mycorrhizosphere and their significance for sustainable agriculture. *FEMS Microbiology Ecology,* 48, 1-13.

Joo, G. J., Kim, Y. M., Lee, I. J., Song, K. S., & Rhee, I. K. (2004). Growth promotion of red pepper plug seedlings and the production of gibberellins by *Bacillus cereus, Bacillus macroides* and *Bacillus pumilus*. *Biotechnology Letters, 26,* 487-491.

Juhas, M., Eberi, L., & Tummler, B. (2005). Quorum sensing: the power of cooperation in the world of *Pseudomonas*. *Environmental Microbiology,* 7, 459-471.

Katiyar, V., & Goel, R. (2004). Siderophore-mediated plant growth promotion at low temperature by mutant of fluorescent pseudomonad. *Plant Growth Regulation, 42,* 239-244.

Kerovuo, J., Lauraeus, M., Nurminen, P., Kalkkinen, N., & Apajalahti, J. (1998). Isolation, characterization, molecular gene cloning, and sequencing of a novel phytase from *Bacillus subtilis*. *Applied and Environmental Microbiology, 64,* 2079-2085.

Khalid, A., Arshad, M., & Zahir, Z. A. (2003). Growth and yield response of wheat to inoculation with auxin producing PGPR. *Pakistan Journal of Botany, 35,* 483-498.

Kim, K. Y., Jordan, D., & Krishnan, H. B. (1998a). Expression of genes from *Rahnella aquatilis* that are necessary for mineral phosphate solubilization in *Escherichia coli. FEMS Microbiology Letters, 159,* 121-127.

Kim, K. Y., Jordan, D., & McDonald, G. A. (1998b). *Enterobacter agglomerans,* phosphate solubilizing bacteria, and microbial activity in soil: effect of carbon sources. *Soil Biology and Biochemistry, 30,* 995-1003.

Kishore, G. K., Pande, S., & Podile, A. R. (2005a). Phylloplane bacteria increase seedling emergence, growth and yield of field-grown groundnut (*Arachis hypogaea* L.). *Letters in Applied Microbiology, 40,* 260-268.

Kishore, G. K., Pande, S., & Podile, A. R. (2005b). Biological control of late leaf spot of peanut (*Arachis hypogaea* L.) with chitinolytic bacteria. *Phytopathology, 95,* (in press).

Kishore, G. K., Pande, S., & Rao, J. N., & Podile, A. R. (2005c). *Pseudomonas aeruginosa* inhibits the plant cell wall degrading enzymes of *Sclerotium rolfsii* and reduces the severity of groundnut stem rot. *European Journal of Plant Pathology, (in press).*

Kloepper, J. W., & Schroth, M. N. (1978). Plant growth promoting rhizobacteria on radishes. In *Proceedings of the fourth International Conference on Plant Pathogenic Bacteria (Vol. 2).* (pp. 879-892). Argers, France: Station de Pathologie Vegetale et Phytobacteriologyie, INRA.

Kloepper, J. W., Leong, J., Teintze, M., & Schroth, M. N. (1980). Enhanced plant growth by siderophores produced by plant growth-promoting rhizobacteria. *Nature, 286,* 885-886.

Kloepper, J. W., Zablokovicz, R. M., Tipping, E. M., & Lifshitz, R. (1991). Plant growth promotion mediated by bacterial rhizosphere colonizers. In D. L. Keister & P. B. Cregan (Eds.). *The rhizosphere and plant growth* (pp. 315-326). The Netherlands: Kluwer Academic Publishers.

Kokalis-Burelle, N., Vavrina, C. S., Rosskopf, E. N., & Shelby, R. A. (2002). Field evaluation of plant growth-promoting rhizobacteria amended transplant mixes and soil solarization for tomato and pepper production in Florida. *Plant and Soil, 2,* 257-266.

Kombrink, E., & Somssich, I. E. (1995). Defense responses of plants to pathogens. *Advances in Botanical Research, 21,* 1-34.

Krishnamurthy, K., & Gnanamanickam, S. S. (1998). Biological control of rice blast by *Pseudomonas fluorescens* strain Pf7-14: evaluation of a marker gene and formulations. *Biological Control, 13,* 158-165.

Kuiper, I., Bloemberg, G. V., & Lugtenberg, B. J. J. (2001). Selection of a plant-bacterium pair as a novel tool for rhizostimulation of polycyclic aromatic hydrocarbon-degrading bacteria. *Molecular Plant-Microbe Interactions, 14,* 1197-1205.

Kuklinsky-Sobral, J., Araujo, W. L., Mendes, R., Geraldi, I. O., Pizzirani-Kleiner, A. A., & Azevedo, J. L. (2004). Isolation and characterization of soybean-associated bacteria and their potential for plant growth promotion. *Environmental Microbiology, 6,* 1244-1251.

Kumar, V., & Narula, N. (1999). Solubilization of inorganic phosphates and growth emergence of wheat as affected by *Azotobacter chroococcum* mutants. *Biology and Fertility of Soils, 28,* 301-305.

Kumar, B. S. D., Berggren, I., & Martensson, A. M. (2001). Potential for improving pea production by co-inoculation with fluorescent *Pseudomonas* and *Rhizobium*. *Plant and Soil, 229,* 25-34.

Kunc, F., & Macura, J. (1988). Mechanisms of adaptation and selection of microorganisms in the soil. In V. Vancura & F. Kunc (Eds.). *Soil Microbial Associations* (pp. 281-299). Elsevier: Amsterdam.

Kurek, E., & Jaroszuk-Scisel, J. (2003). Rye (*Secale cereale*) growth promotion by *Pseudomonas fluorescens* and their interactions with *Fusarium culmorum* under various soil conditions. *Biological Control, 26,* 48-56.

Leeman, M., van Pelt, J. A., den Ouden, F. M., Heinsbroek, M., Bakker, P. A. H. M., & Schippers, B. (1995). Induction of systemic resistance against fusarium wilt of radish by lipopolysaccharides of *Pseudomonas fluorescens*. *Phytopathology, 85,* 1021-1027.

Leoni, L., Ambrosi, C., Petrucca, A., & Visca, P. (2002). Transcriptional regulation of pseudobactin synthesis in the plant growth-promoting *Pseudomonas* B10. *FEMS Microbiology Letters, 208,* 219-225.

Loh, J., Pierson, E. A., Pierson III, L. S., Stacey, G., & Chatterjee, A. (2002). Quorum sensing in plant-associated bacteria. *Current Opinion in Plant Biology, 5,* 1-6.

Lubeck, P. S., Hansen, M., & Sorensen, J. (2000). Simultaneous detection of the establishment of seed-inoculated *Pseudomonas fluorescens* strain DR54 and native soil bacteria on sugar beet root surfaces using fluorescence antibody and in situ hybridization techniques. *FEMS Microbiology Ecology, 33,* 11-19.

Lugtenberg, B. J. J., Dekkers, L., & Bloemberg, G. V. (2001). Molecular determinants of rhizosphere colonization by *Pseudomonas*. *Annual Review of Phytopathology, 39,* 461-490.

Lugtenberg, B. J. J., Kravchenko, L.V., & Simons, M. (1999). Tomato seed and root exudates sugars: composition, utilization by *Pseudomonas* biocontrol strains and role in rhizosphere colonization. *Environmental Microbiology, 1,* 439-446.

Manjula, K., & Podile, A. R. (2001). Chitin-supplemented formulations improve biocontrol and plant growth promoting efficiency of *Bacillus subtilis* AF 1. *Canadian Journal of Microbiology, 47,* 618-625.

Manjula, K., & Podile, A.R. (2005). Increase in seedling emergence and dry weight of pigeon pea in the field with chitin-supplemented formulations of *Bacillus subtilis* AF 1. *World Journal of Microbiology and Biotechnology,* (in press).

Manjula, K., Kishore, G. K., & Podile, A. R. (2004). Whole cells of *Bacillus subtilis* AF 1 proved effective than cell free and chitinase-based formulations in biological control of citrus fruit rot and groundnut rust. *Canadian Journal of Microbiology, 50,* 737-744.

Masalha, J., Kosegarten, H., Elmaci, O., & Mengel, K. (2000). The central role of microbial activity for iron acquisition in maize and sunflower. *Biology and Fertility of Soils, 30,* 433-439.

Mathesius, U., Mulders, S., Gao, M., Teplitski, M., Caetano-Anolles, G., Rolfe, B. G. & Bauer, W. D. (2003). Extensive and specific responses of a eukaryote to bacterial quorum-sensing signals. *Proceedings of the National Academy of Sciences of the United States of America, 100,* 1444-1449.

Matora, L. Y., Serebrennikova, O. B., & Shchyogolev, S. Y. (2001). Structural effects of the *Azospirillum* lipopolysaccharides in cell suspensions. *Biomacromolecules, 2,* 402-406.

Mercado-Blanco, J., van der Drift, K. M., Olsson, P. E., Thomas-Oates, J. E., van Loon, L. C., & Bakker, P. A. (2001). Analysis of the *pms*CEAB gene cluster involved in biosynthesis of salicylic acid and the siderophore pseudomonine in the biocontrol strain *Pseudomonas fluorescens* WCS374. *Journal of Bacteriology, 183,* 1909-1920.

Meziane, H., van der Sluis, L., van Loon, L. C., Hofte, M., & Bakker, P. A. H. M. (2005). Determinants of *Pseudomonas putida* WCS358 involved in inducing systemic resistance in plants. *Molecular Plant Pathology, 6,* 177-185.

Mirza, M. S., Ahmad, W., Latif, F., Haurat, J., Bally, R., Normajd, P., & Malik, K. A. (2001). Isolation, partial characterization, and the effect of plant growth-promoting bacteria (PGPB) on micro-propagated sugarcane *in vitro*. *Plant and Soil, 237,* 47-54.

Munir, A., Munir, I., Afrasyab, S., & Hasnain, S. (2003). Growth stimulatory effects of *Azospirillum* strains on *Triticum aestivum* and *Vigna radiata*. *Biotechnology, 2,* 198-205.

Murphy, J. F., Reddy, M. S., Ryu, C. M., Kloepper, J. W., & Li, R. (2003). Rhizobacteria-mediated growth promotion of tomato leads to protection against *Cucumber Mosaic Virus*. *Phytopathology, 93,* 1301-1307.

Nanda-Kumar, R., Babu, S., Viswanathan, R., Sheela, J., Raguchander, T., & Samiyappan, R. (2001). A new bio-formulation containing plant growth promoting rhizobacterial mixture for management of sheath blight and enhanced grain yield in rice. *Biocontrol, 46,* 493-510.

Nautiyal, C. S., Bhadauria, S., Kumar, P., Lal, H., Mondal, R., & Verma, D. (2000). Stress-induced phosphate solubilization in bacteria isolated from alkaline soils. *FEMS Microbiology Letters, 182,* 291-296.

Neilands, J. B., & Leong, S. A. (1986). Siderophores in relation to plant growth and disease. *Annual Review of Plant Physiology, 37,* 187-208.

Neilands, J. B., & Nakamura, K. (1991). Detection, determination, isolation, characterization and regulation of microbial iron chelates. In G. Winkelmann (Eds.) *CRC Handbook of Microbial Iron Chelates*. London: CRC Press.

Ongena, M., Daayf, F., Jacques, P., Thonart, P., Benhamou, N., Paulitz, T. C., Cornelis, P., Koedam, N., & Belanger, R. R. (1999). Protection of cucumber against pythium root rot by fluorescent pseudomonads: predominant role of induced resistance over siderophores and antibiosis. *Plant Pathology, 48,* 66-76.

Ongena, M., Giger, A., Jacques, P., Dommes, J., & Thonart, P. (2002). Study of bacterial determinants involved in the induction of systemic resistance in bean by *Pseudomonas putida* BTP1. *European Journal of Plant Pathology, 108,* 187-196.

Ordentlich, A., Elad, Y., & Chet, I. (1988). The role of chitinase of *Serratia marcescens* in biocontrol of *Sclerotium rolfsii*. *Phytopathology, 78,* 84-88.

O'Sullivan, D. J., Morris, J., & O'Gara. (1990). Identification of an additional ferric-siderophore uptake gene clustered with receptor, biosynthesis, and fur-like regulatory genes in fluorescent *Pseudomonas* sp. strain M114. *Applied and Environmental Microbiology, 56,* 2056-2064.

O'Sullivan, D. J., & O'Gara, F. (1992). Traits of fluorescent *Pseudomonas* spp. involved in suppression of plant root pathogens. *Microbiological Reviews, 56,* 662-676.

O'Toole, G. A., & Kotler, R. (1998). Flagellar and twitching motility are necessary for *Pseudomonas aeruginosa* biofilm development. *Molecular Microbiology, 30,* 295-304.

Paau, A. S. (1988). Formulations useful in applying beneficial microorganisms to seeds. *Trends in Biotechnology, 6,* 276-279.

Pan, B., Bai, Y. M., Leibovitch, S., & Smith, D. L. (1999). Plant growth-promoting rhizobacteria and kinetin as ways to promote corn growth and yield in a short-growing-season area. *European Journal of Agronomy, 11,* 179-186.

Parmar, N., & Dadarwal, K. R. (1999). Stimulation of nitrogen fixation and induction of flavonoid-like compounds by rhizobacteria. *Journal of Applied Microbiology, 86,* 36-44.

Persello-Cartieaux, F., David, P., Sarrobert, C., Thibaud, M. C., Achouak, W., Robaglia, C., & Nussaume, L. (2001). Utilization of mutants to analyze the interaction between *Arabidopsis thaliana* and its naturally root-associated *Pseudomonas*. *Planta, 212,* 190-198.

Pierson III, L. S., Wood, D. W., & Pierson, E. A. (1998). Homoserine lactone-mediated gene regulation in plant-associated bacteria. *Annual Review of Phytopathology, 36,* 207-225.

Pierson, E. A., Wood, D. W., Cannon, J. A., Blachere, F. M., & Pierson III, L. S. 1998. Interpopulation signaling via *N*-acyl-homoserine lactones among bacteria in the wheat rhizosphere. *Molecular Plant-Microbe Interactions, 11,* 1078-1084.

Pieterse, C. M. J., Van Pelt, J. A., Van Wees, S. C. M., Ton, J., Leon-Kloosterziel, K. M., Keurentjes, J. J. B., Verhagen, B. W. M., Knoester, M., Sluis, I. V., Bakker, P. A. H. M., & Van Loon, L. C.

(2001). Rhizobacteria-mediated induced systemic resistance: triggering, signaling and expression. *European Journal of Plant Pathology, 107,* 51-61.

Ping, L., & Boland, W. (2004). Signals from the underground, bacterial volatiles promote growth in *Arabidopsis*. *Trends in Plant Sciences, 9,* 263-266.

Podile, A. R. (1995). Seed bacterization with *Bacillus subtilis* AF 1 enhances seedling emergence, growth and nodulation of pigeonpea. *Indian Journal of Microbiology, 35,* 199-204.

Podile, A. R., & Laxmi, V. D. V. (1998). Seed bacterization with *Bacillus subtilis* increases phenylalanine ammonia lyase and reduces the incidence of fusarial wilt of pigeon pea. *Journal of Phytopathology, 146,* 255-259.

Polonenko, D. R., Scher, F. M., Kloepper, J. W., Singleton, C. A., Laliberte, M., & Zaleska, I. (1987). Effects of root colonizing bacteria on nodulation of soybeans roots by *Bradyrhizobium japonicum*. *Canadian Journal of Microbiology, 33,* 498-503.

Potter, S., Uknes, S., Lawton, K., Winter, A. M., Chandler, D., DiMaio, J., Novitzky, R., Ward, E., & Ryals, J. (1993). Regulation of a hevein-like gene in *Arabidopsis*. *Molecular Plant-Microbe Interactions, 6,* 680-685.

Preeti, V., Reddy, M. S., Kavitha, S., Velusamy, P., PaulRaj, R. S. D., Purushothaman, S. M., Priyadarisini, V. B., Bharathkumar, S., Kloepper, J. W., & Gnanamanickam, S. S. (2002). Role of biological preparations in enhancement of rice seedling growth and grain yield. *Current Science, 83,* 1140-1143.

Raj, N. S., Shetty, N. P., & Shetty, H. S. (2004). Seed bio-priming with *Pseudomonas fluorescens* isolates enhances growth of pearl millet plants and induces resistance against downy mildew. *International Journal of Pest Management, 50,* 41-48.

Raaijmakers, J. M., van der Sluis, I., Koster, M., Bakker, P. A. H. M., Welsbeek, P. J., & Schippers, B. (1995). Utilization of heterologous siderophores and rhizosphere competence of fluorescent *Pseudomonas* spp. *Canadian Journal of Microbiology, 41,* 126-135.

Raaijmakers, J. M., Vlami, M., & de Souza, J. T. (2002). Antibiotic production by bacterial biocontrol agents. *Antonie van Leeuwenhoek, 81,* 537-547.

Raaijmakers, J., Weller, D. M., & Thomashow, L. S. (1997). Frequency of antibiotic-producing *Pseudomonas* spp. in natural environments. *Applied and Environmental Microbiology, 63,* 881-887.

Ramos, C., Molbak, L., & Molin, S. (2000). Bacterial activity in the rhizosphere analyzed at the single-cell level by monitoring ribosome contents and synthesis rates. *Applied and Environmental Microbiology, 66,* 801-809.

Rattray, E, A, S., Prosser, J. I., Glovery, L. A., & Killham, K. (1995). Characterization of rhizosphere colonization by luminescent *Enterobacter cloacae* at the population and single-cell levels. *Applied and Environmental Microbiology, 61,* 2950-2957.

Raupach, G. S., & Kloepper, J. W. (2000). Biocontrol of cucumber diseases in the field by plant growth-promoting rhizobacteria with and without methyl bromide fumigation. *Plant Disease, 84,* 1073-1075.

Ravel, J., & Cornelis, P. (2003). Genomics of pyoverdine-mediated iron uptake in pseudomonads. *Trends in Microbiology, 11,* 195-200.

Ravnskov, S., & Jakobsen, I. (1999). Effects of *Pseudomoans fluorescens* DF57 and P uptake of two arbuscular mycorrhizal fungi in symbiosis with cucumber. *Mycorrhiza, 8,* 329-334.

Rediers, H., Bonnecarrere, V., Rainey, P. B., Hamonts, K., Vanderleyden, J., & De Mot, R. (2003). Development and application of a *dapB*-based *in vivo* expression technology system to study colonization of rice by the endophytic nitrogen-fixing bacterium *Pseudomonas stutzeri* A15. *Applied and Environmental Microbiology, 69,* 6864-6874.

Rothballer, M., Schmid, M., & Hartmann, A. (2003). *In situ* localization and PGPR-effect of *Azospirillum brasilense* strains colonizing roots of different wheat varieties. *Symbiosis, 34,* 261-279.

Ryals, J. A., Neuenschwander, U. H., Willits, M. G., Molina, A., Steiner, H. Y., & Hunt, M. D. (1996). Systemic acquired resistance. *Plant Cell, 8,* 1809-1819.

Ryu, C. M., Farag, M. A., Hu, C. H., Reddy, M. S., Kloepper, J. W., & Pare, P. W. (2004). Bacterial volatiles induce systemic resistance in *Arabidopsis*. *Plant Physiology, 134,* 1017-1026.

Ryu, C. M., Farag, M. A., Hu, C. H., Reddy, M. S., Wei, H. X., Pare, P. W., & Kloepper, J. W. (2003). Bacterial volatiles promote growth in *Arabidopsis*. *Proceedings of the National Academy of Sciences of the USA, 100,* 4927-4932.

Sailaja, P. R., & Podile, A. R. (1998). A phytoalexin is modified to less fungistatic substances by crown rot pathogen in groundnut (*Arachis hypogaea* L.). *Indian Journal of Experimental Biology. 36,* 631-634.

Sailaja, P. R., Podile, A. R., & Reddanna, P. (1998). Biocontrol strain of *Bacillus subtilis* AF1 rapidly induces lipoxygenase in groundnut (*Arachis hypogaea* L.) compared to crown rot pathogen *Aspergillus niger*. *European Journal of Plant Pathology, 104,* 125-132.

Samac, D. A., & Shah, D. M. (1994). Effect of chitinase antisense RNA expression on disease susceptibility of *Arabidopsis* plants. *Plant Molecular Biology, 25,* 587-596.

Seal, S. E., Jackson, L. A., & Daniels, M. J. (1992). Isolation of a *Pseudomonas solanacearum*-specific DNA probe by subtraction hybridization and construction of species-specific oligonucleotide primers for sensitive detection by the polymerase chain reaction. *Applied and Environmental Microbiology, 58,* 3751-3758.

Sessitsch, A., Reiter, B., & Berg, G. (2004). Endophytic bacterial communities of field-grown potato plants and their plant-growth-promoting and antagonistic abilities. *Canadian Journal of Microbiology, 50,* 239-249.

Sharma, A., & Johri, B. N. (2003). Growth promoting influence of siderophore-producing *Pseudomonas* strains GRP3A and PRS_9 in maize (*Zea mays* L.) under iron limiting conditions. *Microbiological Research, 158,* 243-248.

Shishido, M., Petersen, D. J., Massicotte, H. B., & Chanway, C. P. (1996). Pine and spruce seedling growth and mycorrhizal infection after inoculating with plant growth promoting *Pseudomonas* strains. *FEMS Microbiology Ecology, 21,* 109-119.

Siddiqui, Z. A., & Mahmood, I. (2001). Effects of rhizobacteria and root symbionts on the reproduction of *Meliodogyne javanica* and growth of chickpea. *Bioresources Technology, 79,* 41-45.

Siddiqui, I. A., & Shaukat, S. S. (2002). Mixtures of plant disease suppressive bacteria enhance biological control of multiple tomato pathogens. *Biology and Fertility of Soils, 36,* 260-268.

Sigler, W. V., Nakatsu, C. H., Reicher, Z. J., & Turco, R. F. (2001). Fate of the biological control agent *Pseudomonas aureofaciens* TX-1 after application to turfgrass. *Applied and Environmental Microbiology, 67,* 3542-3548.

Simmon, K. E., Steadman, D. D., Sarah, D., Baldwin, A., Jeffrey, W. H., Sheridan, P., Horton , R., & Shields, M. S. (2004). Autoclave method for rapid preparation of bacterial PCR-template DNA. *Journal of Microbiological Methods, 56,* 143-149.

Simons, M., Permentier, H. P., de Weger, L. A., Wijffelman, C. A., & Lugtenberg, B. J. J. (1997). Amino acid synthesis is necessary for tomato root colonization by *Pseudomonas fluorescens* strain WCS365. *Molecular Plant-Microbe Interactions, 10,* 102-106.

Sindhu, S. S., Gupta, S. K., & Dadarwal, K. R. (1999). Antagonistic effect of *Pseudomonas* spp. on pathogenic fungi and enhancement of growth of green gram (*Vigna radiata*). *Biology and Fertility of Soils, 29,* 62-68.

Singh, S., & Kapoor, K. K. (1998a). Inoculation with phosphate-solubilizing microorganisms and a vesicular-arbuscular mycorrhizal fungus improves dry matter yield and nutrient uptake by wheat grown in a sandy soil. *Biology and Fertility of Soils, 28,* 139-144.

Singh, S., & Kapoor, K. K. (1998b). Effects of inoculation of phosphate solubilizing microorganisms and arbuscular mycorrhizal fungus on mungbean grown under natural soil conditions. *Mycorrhiza, 7,* 249-253.

Skerker, J. M., & Berg, H. C. (2005). Direct observation of extension and retraction of type IV pili. *Proceedings of the National Academy of Sciences of the USA, 98,* 6901-6904.

Skvortsov, I. M., & Ignatov, V. V. (1998). Extracellular polysaccharides and polysaccharide-containing biopolymers from *Azospirillum* species: properties and the possible role in interaction with plant roots. *FEMS Microbiology Letters, 165,* 223-229.

Somers, E., Vanderleyden, J., & Srinivasan, M. (2004). Rhizosphere bacterial signaling: A love parade beneath our feet. *Critical Reviews in Microbiology, 30,* 205-240.

Sorensen, J. (1997). The rhizosphere as a habitat for soil microorganisms. In J. D. van Elsas, J. T. Trevors & E. M. H. Welington (Eds.). *Modern Soil Ecology* (pp. 21-46). New York: Marcel Dekker, Inc.

Steidle, A., Allesen-Holm, M., Riedel, K., Berg, G., Givskov, M., Molin, S., & Eberl, L. (2002). Identification and characterization of an *N*-acylhomoserine lactone-dependent quorum sensing system in *Pseudomonas putida* strain IsoF. *Applied and Environmental Microbiology, 68,* 6371-6382.

Steidle, A., Sigl, K., Schuhegger, R., Ihring, A., Schmid, M., Ganter, S., Stoffels, M., Riedel, M., Givskov, A., Hartman, C., Langebartels, C., & Eberl, L. (2001). Visualization of *N*-acylhomoserine lactone-mediated cell-cell communication between bacteria colonizing the tomato rhizosphere. *Applied and Environmental Microbiology, 67,* 5761-5770.

Stevenson, F. J., & Cole, M. A. (1999). *Cycles of soil: Carbon, Nitrogen, Phosphorous, Sulfur, Micronutrients, 2nd Edition.* New York: Wiley.

Tanaka, H., & Watanabe, T. (1995). Glucanases and chitinases of *Bacillus circulans* WL-12. *Journal of Industrial Microbiology, 114,* 478-483.

Teplitski, M., Robinson, J. B., Bauer, W. D. (2000). Plants secrete substances that mimic bacterial *N*-acyl homoserine lactone signal activities and affect population density-dependent behaviour in associated bacteria. *Molecular Plant-Microbe Interactions, 13,* 637-648.

Tilak, K. V. B. R., Ranganayaki, N., Pal, K. K., De, R., Saxena, A. K., Nautiyal, C. S., Mittal, S., Tripathi, A. K., & Johri, B. N. (2005). Diversity of plant growth and soil health supporting bacteria. *Current Science, 89,* 136-150.

Toro, M., Azcon, R., & Barea, J. (1997). Improvement of arbuscular mycorrhiza development by inoculation of soil with phosphate-solubilizing rhizobacteria to improve rock phosphate bioavailability (P^{32}) and nutrient cycling. *Applied and Environmental Microbiology, 63,* 4408-4412.

Tripathi, M., Munot, H. P., Schouche, Y., Meyer, J. M., & Goel, R. (2005). Isolation and functional characterization of siderophore-producing lead- and cadmium-resistant *Pseudomonas putida* KNP9. *Current Microbiology, 50,* 233-237.

Uknes, A., Mauch-Mani, B., Moyer, M., Potter, S., Williams, S., Dincher, S., Chandler, D., Slusarenko, A., Ward, E., & Ryals, J. (1992). Acquired resistance in *Arabidopsis*. *Plant Cell, 4,* 645-656.

Van Elsas, J. D., Trevors, J. T., Jain, D., Wolters, A. C., Heijnen, C. E., & van Overbeek, L. S. (1992). Survival of, and root colonization by alginate-encapsulated *Pseudomonas fluorescens* cells following introduction into soil. *Biology and Fertility of Soils, 14,* 14-22.

Van Loon, L. C., & Bakker, P. A. H. M. (2003). Signaling in rhizobacteria-plant interactions. In H. de Kroon & E. J. W. Visser (Eds.). *Ecological Studies. Vol. 168* (pp. 297-330). Berlin Heidelberg: Springer-Verlag.

vanLoon, L. C., Bakker, P. A. H. M., & Pieterse, C. M. J. (1998). Systemic resistance induced by rhizosphere bacteria. *Annual Review of Phytopathology, 36,* 453-483.

Van Schie, B. J., De Mooy, O. H., Linton, J. D., Van Dijken, J. P., & Kuenen, J. G. (1987). PQQ-dependent production of gluconic acid by *Acinetobacter*, *Agrobacterium* and *Rhizobium* species. *Journal of General Microbiology, 133,* 867-875.

Vandenbergh, P. A., & Gonzalez, C. F. (1984). Method for protecting the growth of plants by employing mutant siderophore producing strains of *Pseudomonas putida*. U.S. patent No. 4479 936.

Vessey, J. K. (2003). Plant growth promoting rhizobacteria as biofertilizers. *Plant and Soil, 255,* 571-586.

Vidhyasekaran, P., & Muthamilan, M. (1995). Development of formulations of *Pseudomonas fluorescens* for control of chickpea wilt. *Plant Disease, 79,* 782-786.

Voisard, C., Keel, C., Haas, D., & Defago, G. (1989). Cyanide production by *Pseudomonas fluorescens* helps suppress black root rot of tobacco under gnotobiotic conditions. *EMBO Journal, 8,* 351-358.

Volkmar, K. M., & Bremer, E. (1998). Effects of seed inoculation with a strain of *Pseudomonas fluorescens* on root growth and activity of wheat in well-watered and drought-stressed glass-fronted rhizotrons. *Canadian Journal of Plant Sciences, 78,* 545-551.

Walsh, U. F., Morrissey, J. P., & O'Gara, F. (2001). *Pseudomonas* for biocontrol of phytopathogens: from functional genomics to commercial exploitation. *Current Opinion in Biotechnology, 12,* 289-295.

Wang, Y., Brown, H. N., Crowley, D. E., & Szaniszlo, P. J. (1993). Evidence for direct utilization of a siderophore ferrioxamine B, in axenically grown cucumber. *Plant Cell and Environment, 16,* 579-585.

Wang, C., Knill, E., Glick, B. R., & Defago, G. (2000). Effect of transferring 1-aminocyclopropane-1-carboxylic acid (ACC) deaminase genes into *Pseudomonas fluorescens* strain CHAO and its *gac*A derivative CHA96 on their growth-promoting and disease-suppressive capacities. *Canadian Journal of Microbiology, 46,* 898-907.

Wang, C., Ohara, Y., Nakayashiki, H., Tosa, Y., & Mayama, S. (2005). Microarray analysis of the gene expression profile induced by the endophytic plant growth-promoting rhizobacteria, *Pseudomonas fluorescens* FPT9601-T5 in *Arabidopsis*. *Molecular Plant-Microbe Interactions, 18,* 385-396.

Ward, D. M., Weller, R., & Bateson, M. M. (1990). 16S rRNA sequences reveal numerous uncultured microorganisms in a natural community. *Nature, 345,* 63-65.

Whipps, J. M. (2001). Microbial interactions and biocontrol in the rhizosphere. *Journal of Experimental Biology, 52,* 487-511.

Wood, D. W., Gong, G., Daykin, M. M., Williams, P., & Pierson III, L. S. (1997). *N*-acyl-homoserine lactone-mediated regulation of phenazine gene expression by *Pseudomonas aureofaciens* 30-84 in the wheat rhizosphere. *Journal of Bacteriology, 179,* 7663-7670.

Xie, H., Pasternak, J. J., & Glick, B. R. (1996). Isolation and characterization of mutants of the plant growth-promoting rhizobacterium *Pseudomonas putida* GR12-2 that overproduce indole acetic acid. *Current Microbiology, 32,* 67-71.

Yang, C. H., Crowley, D. E., & Menge, J. A. (2001). 16S rDNA fingerprinting of rhizosphere bacterial communities associated with healthy and *Phytophthora*-infected avocado roots. *FEMS Microbiology Ecology, 35,* 129-136.

Zahir, A., Muhammad, A., & Frankenberger Jr., W. T. (2004). Plant growth promoting rhizobacteria: applications and perspectives in agriculture. *Advances in Agronomy, 81,* 97-168.

Zaidi, A., Khan, S. M., & Aamil, M. (2001). Bioassociative effect of rhizospheric microorganisms on growth, yield, and nutrient uptake of green gram. *Journal of Plant Nutrition, 27,* 601-612.

Zhang, Z., & Pierson III, L. S. (2001). A second quorum-sensing system regulates cell surface properties but not phenazine antibiotic production in *Pseudomonas aureofaciens*. *Applied and Environmental Microbiology, 67*, 4305-4315.

Zhang, F., Dashti, N., Hynes, R. K., & Smith, D. L. (1996). Plant growth promoting rhizobacteria and soybean (*Glycine max* (L.) Merr.) nodulation and nitrogen fixation at suboptimal root zone temperatures. *Annals of Botany*, *77,* 453-459.

10. AFFILIATIONS

Appa Rao Podile, *Department of Plant Sciences, University of Hyderabad, Hyderabad 500 046, India, E-mail: arpsl@uohyd.ernet.in*

G. Krishna Kishore, *Department of Pathology, International Crops Research Institute for the Semi-Arid Tropics, Patancheru 502 324, AP, India.*

PHILIPPE LEMANCEAU, MONIKA MAURHOFER AND GENEVIÈVE DÉFAGO

CONTRIBUTION OF STUDIES ON SUPPRESSIVE SOILS TO THE IDENTIFICATION OF BACTERIAL BIOCONTROL AGENTS AND TO THE KNOWLEDGE OF THEIR MODES OF ACTION

Abstract. In some soils, disease severity of soilborne diseases remains low despite the presence of pathogenic populations and conditions favourable to disease expression. Natural suppressiveness has been reported for a wide range of soilborne diseases and is specific, for a given soil, to a class of disease. Different types of soil suppressiveness have been described according to their origin and stability: (i) acquired with time after a monoculture of a susceptible host-plant in the presence of the pathogen and eliminated by breaking the monoculture by a non-host crop, for take-all, and (ii) native and stable, for fusarium wilts and balck root rot. Natural suppressiveness of soils to diseases has been related to their indigenous microflora. More precisely, suppressiveness has been ascribed to the total soil microflora (general suppressiveness) and to specific antagonistic microbial populations (specific suppressiveness). Among these populations, a major role has been given to fluorescent pseudomonads. Their implication in soil suppressiveness was shown to be related to siderophore-mediated iron competition (suppressive soils to fusarium wilts) and antibiosis (suppressive soils to take-all and tobacco root rot). Progresses in the knowledge of the microbial populations involved in the natural soil suppressiveness have allowed the isolation of biocontrol agents such as *P. fluorescens* CHA0, 2-79, Q8r1-96, C7 from soils naturally suppressive to black root rot, take-all and fusarium wilts, respectively. Modes of action of these biocontrol agents are related to antibiosis, competition and/or induced systemic resistance. Reciprocal feedbacks between studies on mechanisms of natural soil suppressiveness and modes of action of biocontrol agents are described. Prospects are finally given on the application of molecular approaches to identify new bacteria, genes and metabolites involved in soil suppressiveness.

1. INTRODUCTION

A disease results from the intimate interaction between a plant and a pathogen, with its severity influenced by the environmental factors that affect the plant, the pathogen or both. The existence of soils that suppress diseases caused by soilborne pathogens provide an example of the role of biotic and abiotic factors that affect pathogenic populations; in so-called suppressive soils, disease incidence remains low in spite of the presence of pathogenic populations, susceptible host-plants and climate conditions favourable for disease development (Cook and Baker, 1983).

The level of disease incidence varies from one to the other even when the same susceptible host-plant is cultivated under the same climatic conditions after having infested the soils with the same pathogenic inoculum. This observation shows that each soil can be characterized by its level of receptivity (suppressiveness) to the considered soilborne disease (see paragraph 2.2). The level of receptivity of

S.S. Gnanamanickam (ed.), Plant-Associated Bacteria, 231–267.

a soil to a soilborne disease corresponds to its capacity to suppress more or less the saprophytic growth and infectious activity of the corresponding pathogenic populations present in the soil. Soil suppressiveness is a continuum going from the highly conducive soils in which disease incidence is very high to strongly suppressive soils in which the disease is expressed at a very low rate (Alabouvette et al., 1982; Linderman et al., 1983). Even conducive soils have some potential of disease suppression; and disease may occur, although at a very low rate, in suppressive soils indicating that the suppression is not absolute.

Research on soil suppressiveness has contributed to the development of strategies to assess the phytosanitary quality of soils (Alabouvette et al., in press). This assessment is obviously based on the detection of the presence or absence of the main pathogens of the crop to be cultivated. However, this knowledge is clearly not sufficient to predict the risk for the crop to be severely diseased. In the early 70's, Garett (1970) proposed the concept of soil, 'inoculum potential' as being 'the energy of growth of a parasite available for infection of a host at the surface of the host organ to be infected', to specify the capacity of a soil to provoke the disease. Among factors that affect the 'energy of growth from the inoculum', Garett pointed out 'the collective effect of environmental conditions' and noted that 'endogenous nutrients of the inoculum might be augmented by exogenous nutrients from the environment'. The effect of the soil environment on the inoculum potential was further specified by Louvet (1973) who described that potential as being the result of (i) the inoculum density, (ii) the inoculum capacity corresponding to the genetic and physiological ability to infect the host-plant, and (iii) the effect of the soil environment affecting both the inoculum density and its inoculum capacity, characterized by the level of soil receptivity. The assessment of the soil inoculum potential is based on the numeration of diseased plants when cultivating a susceptible cultivar in environmental conditions the most favourable for the disease expression. To quantify this potential, it is necessary to establish a dose-response relationship, and therefore to dilute the naturally infested soil in a disinfested soil in increasing proportions (Bouhot, 1979). A low inoculum potential may result from the combination of a low inoculum density or capacity, and a high level of soil receptivity (corresponding to a low level of soil suppressiveness), in such way that, due to the high level of receptivity of the considered soil, introduction of aggressive pathogenic populations through contaminated seedlings would lead to a high disease incidence despite the low inoculum potential measured from the initial soil samples. Characterization of the level of soil suppressiveness is then required to predict the long term risk of development of soilborne diseases; the best guarantee of a low disease expression, even in the presence of pathogenic populations, relying on a high level of soil suppressiveness. Methodologies to assess the level of soil suppressiveness are described in paragraph 2.2.

Since the soil suppressiveness has been associated to their microflora (see paragraph 2.3), studies have been dedicated to the identifications of populations that might account for this phenomenon. Comparison of the diversity of microbial populations in suppressive soils and in conducive soils is aiming at identifying populations preferentially associated with suppressive soils in order to define bioindicators of the soil suppressiveness. Studies on suppressive soils have allowed identifying microorganisms with a high biocontrol potential (see paragraph 3) and resolving some of their modes of actions in relation to the analysis of the mechanisms accounting for the natural suppressiveness (see paragraph 2). According to the biocontrol agents isolated, there is or not a good congruence between their modes of actions and the mechanisms of soil suppressiveness (see paragraph 4). Despite the attention dedicated to natural soil suppressiveness since the 70's, a lot of microbes and genes involved in this phenomenon remain probably to be discovered and strategies based on new molecular approaches and strategies are finally proposed to unravel them (see paragraph 5).

2. SUPPRESSIVE SOILS: A RESERVOIR OF ANTAGONISTIC MICROORGANISMS

2.1. Different types of suppressive soils

Soils suppressive to diseases induced by the most important soilborne pathogens have been described all over the world; they include bacterial and fungal pathogens but also pests such as nematodes (Cook and Baker, 1983; Kerry and Crump, 1998; Schneider, 1982; Westphal and Becker, 1999). These soils control root rot and wilt diseases induced by: *Aphanomyces euteiches*, *Cylindrocladium* sp., formae speciales of *F. oxysporum*, *Gaeumannomyces graminis*, *Pythium* spp., *Phytophthora* spp., *Ralstonia solanacearum*, *Streptomyces scabies*, *Thielaviopsis basicola* (*Chalara elegans*) and *Vertcillium dahliae*. The large diversity of pathogens and pests controlled by suppressive soils shows that soil suppressiveness is not a rare phenomenon.

Among the different suppressive soils listed here above, soils naturally suppressive to take-all, fusarium wilts and *T. basicola* mediated-black root rot have been intensively studied since the 70's.

Take-all, caused by the fungus *Gaeumannomyces graminis* var. *tritici (Ggt)*, is an important root disease of wheat worldwide (Asher and Shipton 1981; Hornby, 1998). Beside wheat, *Ggt* affects various other Gramineae such as barley, rye and triticale. Suppressiveness to take-all, commonly called Take-All Decline (TAD) has been described in many wheat growing areas worldwide (Hornby, 1998). The similarity of TAD throughout the world is remarkable in view of the broad range of soil types, climates and agronomic conditions under which wheat is cultivated. In contrast with soils suppressive to fusarium wilts and black root rot that show specific physicochemical properties (Höper et al., 1995; Stutz et al., 1989), the soil type seems only to modulate the extent and speed of development of TAD (Shipton, 1975).

Fusarium wilts are provoked by pathogenic *F. oxysporum* which, after a saprophytic stage, allow mycelium to infect roots of the host-plant and to penetrate them. During this parasitic stage of their life cycle, pathogenic *F. oxysporum* induce a tracheomycosis resulting from their invasion of the plant-vascular system. The wilt-inducing populations of *F. oxysporum* are responsible for severe damage on many plants of economic importance. Fusarium wilt pathogens show a high level of host specificity and, based on the plant species and plant cultivars they can infect, they are classified into more than 120 formae speciales and races (Armstrong and Armstrong, 1981). Natural suppressiveness to fusarium wilt was first described in 1892 (Atkinson, 1892) and has been reported since then in different places round the world (Dominguez et al., 2001; Louvet et al., 1976; Peng et al., 1999; Scher and Baker, 1980; Smith and Snyder, 1971). Among them, soils from Châteaurenard (France) and Salinas Valley (California) have been studied extensively (Alabouvette, 1986; Scher and Baker, 1982). These soils share their high clay content and pH (Alabouvette et al., 1996).

Soils, which are naturally suppressive to black root rot, caused by *T. basicola* on tobacco and cherry trees, were found scattered in a region of 32 km^2 at Morens, Switzerland in 1981(Gasser and Défago, 1981; Stutz, 1985; Stutz et al., 1986). The suppressive soils are derived from a weathered ground moraine that contains mostly native vermiculitic clay minerals (rich in iron and other minerals). In contrast, some neighbouring soils, which are derived from a weathered molasse and contain secondary illitic clay minerals (poor in iron and other minerals), are conducive to black root rot (Stutz and Défago, 1985; Stutz et al., 1989).

Usually, soil suppressiveness is specific to a given class of disease. This is clearly illustrated for the suppressive soils to fusarium wilts caused by *Fusarium oxysporum* f. spp. which are not suppressive to any other soilborne disease even not to diseases caused by *F. solani* (Alabouvette et al., 1980; Smith and Snyder, 1971).

Different mechanisms accounting for the soil suppressiveness to diseases have been proposed by Baker and Cook (1974) who have accordingly described three types of suppressive soils: (i) in the first one 'the pathogen does not establish', (ii) in the second one 'it establishes but fails to produce disease', and finally (iii) in the third one 'it establishes and causes disease at first but then disease severity diminishes with continued growing of the same crop'. The second type of suppressive soil is illustrated by soils suppressive to fusarium wilts and those to black root rot: it is well demonstrated that the pathogen is still present and virulent in the soil (Alabouvette et al., 1984; Couteaudier and Alabouvette, 1990b; Stutz et al., 1986) but the environmental conditions prevent disease expression. The third type is illustrated by TAD which is defined as the spontaneous decrease in the incidence and severity of take-all, caused by *Ggt*, which occurs on wheat or other susceptible host-crops after one or more severe outbreaks of the disease (Cook, 1981; Shipton, 1975).

Suppressiveness to take-all is acquired with time after a monoculture of susceptible host-plants in the presence of the pathogen, whereas soil suppressiveness to fusarium wilts and to black root rotof tobacco do not seem to be related to a specific crop management despite the reports made by Larkin et al. (1993) and Sneh et al. (1987) describing the acquisition of suppressiveness to fusarium wilt after a monoculture of water and musk melon, respectively.

TAD is acquired whereas suppressiveness to fusarium wilts and to black root rot appears to be a long-standing suppressiveness. Indeed, soil suppressiveness to these soilborne diseases is very stable with time (Défago et al., unpublished data; Ramette et al., 2003a; Steinberg et al., in press), whereas TAD can be reduced or eliminated by breaking monoculture with a non-host crop (Weller et al., 2002). The stability of suppressiveness to fusarium wilts points to preexisting properties of the soil and associated microbiota (Alabouvette et al., 1996), this explains why the early studies were focused on relationship between disease incidence and soil type (Stotzky and Martin, 1963; Stover, 1962).

The three types of soil suppressiveness described by Baker and Cook (1974) underlay very diverse mechanisms responsible for disease suppression and many studies have been devoted to the study of biotic and abiotic interactions involved in soil suppressiveness (cf. paragraph 2.4).

2.2 Methods to assess the level of soil suppressiveness

In nature, suppressive soils can be detected by the observation that disease incidence in a crop remains low despite the presence of a susceptible host plant, climatic conditions favorable to disease expression and ample opportunity for the pathogen to be present. It is quite easy to experimentally demonstrate that a soil is suppressive to a given disease. The pathogen has to be produced in the laboratory and introduced into the soil at increasing inoculum densities. A susceptible host plant is sown or transplanted and cultivated under standardized conditions favorable to disease expression. Observations of symptom appearance enable to draw disease progress curves with time and inoculum concentrations. Different statistical methods allow the comparison of these curves to those obtained with another soil known to be conducive to the disease. All experimental conditions being similar, differences in disease incidence have to be attributed to differences of the soil environment, ie differences in the level of soil receptivity.

Different standardized methods have been proposed for fusarium wilts (Alabouvette et al., 1982), take-all (Asher and Shipton, 1981; Hornby, 1998; Lucas and Sarniguet, 1990), and to tobacco black root rot (Stutz and Défago, 1985). These methods have been recently extensively described (Alabouvette et al., in press). The corresponding methods are based on bioassays in which the soil samples are infested with increasing concentrations (fusarium wilts) or not (take-all, black root rot) of a pathogenic strain. A cultivar susceptible to the corresponding disease is then cultivated in these infested soil samples under well controlled conditions. Disease incidence is assessed by numeration of the healthy plants (fusarium wilts) or by determination of a disease index based on the combination of plant and disease variables (take-all, tobacco black root rot). Disease incidence and disease index are recorded either regularly in order to establish disease curves in relation to inoculum concentrations (fusarium wilt) or once at the end of the bioassay (take-all, black root rot).

Different statistical procedures may be applied to compare the level of soil suppressiveness. One of them is based on the use of the multiple infection transformation allowing the calculation of regression lines between disease incidence and time (when available) and then the comparison of the slopes of these regression lines and the determination of the time required to reach 50% of diseased plants (Paulitz et al., 1987). Another statistical method, commonly applied to clinical surviving population (Hill et al., 1990), has been proposed to assess the level of soil suppressiveness by comparing the kinetic of healthy plant populations in different treatments (Steinberg et al., in press). The corresponding model allows the calculation of the Mean Survival Time (MST) for each population of plants. It is also possible to calculate the Area Under the Disease Progress Curve (AUDPC) (Steinberg et al., 2004). The MSTs and AUDPCs obtained within treatments are then analyzed by Analysis of Variance (ANOVA) to evaluate the effects of the soil type but also of other treatments on the level of suppressiveness.

2.3 Microbial origin of soil suppressiveness

Early works on natural suppresiveness have pointed out the importance of the indigenous microflora in the natural phenomenon of soil suppressiveness. Indeed, application of biocide treatments, such as heat, chemicals or γ-rays, destroy both the soil microflora and the soil suppressiveness to fusarium wilts (Alabouvette et al., 1977; Rouxel et al., 1977; Scher and Baker, 1980), whereas soil suppressiveness is restored by mixing a small quantity of suppressive soil into the previously heat-treated soil (Komada and Ezuka, 1970; Louvet et al., 1976; Scher and Baker, 1980; Schneider, 1982). The same type of strategy was followed to demonstrate the microbial nature of suppressiveness of soils to take-all (Cook and Baker, 1983) and to black root rot of tobacco (Stutz et al., 1986). These observations have stimulated many studies dedicated to the identification of microbial populations that may specifically be involved in soil suppressiveness in order to possibly use them as biocontrol agents or even make conducive soils, suppressive.

However, in most cases, natural suppressiveness appears to be a complex phenomenon, resulting from the activity of various populations of microorganisms. Indeed, Cook and Baker (1983) distinguished two types of mechanisms the 'general' and 'specific' suppression. General suppression is determined by the global biomass acting as a nutrient sink decreasing resources required for the saprophytic phase of the life cycle of the pathogens. During this phase, they are indeed susceptible to nutrient competition, which is mainly responsible for the soil fungistasis, i.e. inhibition of spore germination in soil (Lockwood, 1977). The highest are the biomass and activity of the soil microflora, the highest is the nutrient sink and then the general suppressiveness. And general suppressiveness was shown to be enhanced after soil organic amendments that increase soil biomass (Steinberg et al., 2004).

In contrast to general suppressiveness, specific supressiveness is ascribed to the activity of specific antagonistic populations (Weller et al., 2002). Specific suppression is expressed on a background of general suppression. Despite the importance of the general soil suppressiveness in some types of suppressive soils, including those to fusarium wilts (Alabouvette et al., 1985b), soil suppressiveness is known to be disease specific. However, general suppressiveness is based on non specific mechanisms and then is expected to act against any pathogen. These observations indicate then that general suppressiveness cannot account for all the suppression.

A lot of attention has been given to identify microbial populations accounting for the specific suppression. A recent review has been published on specific soil suppressiveness to plant pathogens (Weller et al., 2002). The early strategy followed to identify microbial groups involved in the suppressiveness consisted in submitting the suppressive soils to increased temperatures (50°C, 55°C, 60°C, 65°C, and 70°C). For suppressive soils to fusarium wilts, black root rot and to take-all, it appears that the suppressiveness was destroyed at 55-65°C. The next steps consisted in (i) identifying the type of microorganisms the most affected by this lethal temperature, (ii) isolating strains belonging to that group from the suppressive soil prior to the heat treatment, and (iii) finally re-introducing these strains in the heat treated soil to identify those the most able to restore the soil suppressiveness following the Koch's postulate. Following this strategy, fluorescent pseudomonads were showed to be involved in the natural soil suppressiveness to black root rot of tobacco, fusarium wilts and take-all (Cook and Rovira, 1976; Gerlagh, 1968; Scher and Baker, 1980; Stutz et al., 1986). Interestingly, following that same strategy, Rouxel et al. (1977, 1979) came to a different conclusion and identified non-pathogenic *F. oxysporum* as playing a major role in the natural suppressiveness to fusarium wilts. This observation stresses the limits of the above strategy related to (i) the small number of microbial species tested, (ii) the lack of consideration of the microbial diversity within these species, and (iii) the only consideration of culturable isolates. Progresses in molecular biology together with the limits of this strategy have stimulated new approaches described in paragraph 5. Despite these limits, this strategy has allowed the identification and isolation of various bacterial antagonists from suppressive soils including fluorescent pseudomonads such as the strains 2-79 (Weller and Cook, 1983), CHA0 (Stutz et al., 1986), C7 (Lemanceau and Alabouvette, 1991), Q8r1-96 (Raaijmakers and Weller (2001) (see paragraph 3). Fungal antagonists such non-pathogenic *F. oxysporum* Fo47 were also isolated from suppressive soils (Alabouvette et al., 1987) but will not be detailed in the present chapter.

2.4 Mechanisms of soil suppressiveness

Two major types of mechanisms (microbial antagonism and induced systemic resistance), that affect the life cycle of the pathogen, have been described as being involved in the suppression of soilborne diseases. Microbial antagonism results from microbial interactions (competition and antibiosis) leading to a reduced growth of the pathogen during the saprophytic phase of its life cycle and then to a reduced rate of root infections. Induced systemic resistance results from plant microbe interactions leading to a stimulation of the plant defence reaction and then to a reduced growth of the pathogen during the parasitic phase of its life cycle. Studies on suppressive soils have mostly demonstrated the involvement of microbial antagonism in the natural low expression of the diseases. Possible involvement of induced resistance still need to be evaluated and is open to discussion.

2.4.1 Microbial antagonism

2.4.1.1 Competition

Nutrient competition results from the combination of the low offer and high demand of elements which are essential for the microbial physiology. This is the case for organic compounds and for ferric iron. Both are essential for the carbon and energetic metabolism of soil microorganisms which are mostly heterotrophic and aerobic in the plough layer.

Soils are usually oligotrophic, leading to a high competition for carbon in such way that in soil most of the microorganisms, including pathogens, are surviving as resting structures (chlamydospores, microsclerotia, oospores). This competition for organic compounds is mainly responsible for soil fungistasis, i.e. inhibition of spore germination in soil (Lockwood, 1977). Addition of a carbon source, such as glucose, induces an increased germination of chlamydospores of *F. oxysporum* in any soil (Alabouvette et al., 1985a). However, in suppressive soils to fusarium wilts, the microbial biomass is greater than in conducive soils and determine then a stronger competition; therefore a higher concentration of carbohydrate must be added in suppressive soils than in conducive soils to induce the same percentage of chlamydospore germination and *Fusarium* saprophytic growth (Alabouvette et al., 1985; Smith, 1977; Sneh et al., 1984). These observations suggest that fungistasis is expressed with a greater intensity in suppressive than in conducive soils to fusarium wilts. In these suppressive soils, the intensity of carbon competition appears to be directly related to the carbon sink associated to the microbial biomass and activity contributing to the general suppression (see paragraph 2.3).

Although iron is the 4th element of the earth crust, in cultivated aerated soils the concentration of ferric iron available to microorganisms is low (Lindsay, 1979). Iron starvation was showed to lead to a decreased saprophytic growth of *F. oxysporum in vitro* and in soil conditions (Elad and Baker, 1985; Sneh et al., 1984). Beside their high microbial biomass contributing to the general competition for iron, naturally suppressive soils to fusarium wilts are known to have physicochemical properties (high pH and $CaCO_3$ content) contributing to the very low solubility of ferric iron, both properties accounting for the strong iron competition in these soils (Alabouvette et al., 1996).

Increased saprophytic growth of pathogenic *F. oxysporum* resulting from the addition of these limiting compounds, carbohydrates and iron, leads to an increased disease severity, making the suppressive soil conducive, and the conducive soil even more conducive (Alabouvette et al., 1985a; Lemanceau et al., 1988a; Scher and Baker, 1982); whereas the reduction of iron availability leads to a reduced saprophytic growth of pathogenic *F. oxysporum* and to an increased suppressiveness in both conducive and suppressive soils (Kloepper et al., 1980; Lemanceau et al., 1988a; Scher and Baker, 1982). The interaction between carbon and iron competitions was further shown to make the level of suppressiveness more intense (Lemanceau, 1989).

On this background of general suppressiveness, some microbial populations are particularly competitive for a given nutrient and take advantage of this competitive saprophytic ability which is part of their antagonistic modes of action. As an example, non-pathogenic *F. oxysporum* which show very close trophic requirements to pathogenic populations exert against them a very efficient competition for carbohydrates (Couteaudier and Alabouvette, 1990a; Lemanceau et al., 1992). Fluorescent pseudomonads have developed a very efficient iron uptake strategy based on the synthesis of siderophores showing an affinity for iron higher than those of various pathogenic fungi including *F. oxysporum* (Lemanceau et al., 1992; Loper and Buyer, 1991), in such way that these bacterial siderophores contribute to reduce the iron availability for those pathogens (Lemanceau et al., in press).

Trophic competition occurs at primary infection sites and is then expected to be an important mechanism of soil suppressiveness for monocyclic pathogens (pathogens which have only one infection cycle per plant, e.g. *F. oxysporum*). In contrast, competition appears to be poorly effective for polycyclic pathogens (pathogens which have more than one infection cycle per crop, e.g. *Ggt*, *T. basicola*). Indeed, after primary infections, the pathogen establishes inside the host-plant and makes use of nutrients and minerals from the plant for secondary infections. Therefore, reduction of secondary infections is expected to be mostly ascribed to antimicrobial substances suppressing mycelium growth, spores production and/or pathogenicity factors.

2.4.1.2. Antibiosis

Antibiosis is a major mode of action of many antagonistic microorganisms including bacteria and fungi (Fravel, 1988). For example, fluorescent pseudomonads are well known to produce a large diversity of antifungal secondary metabolites responsible for their biocontrol capacity (Cook et al., 1995; Défago and Haas, 1990; Morrissey et al., 2004; Raaijmakers et al., 2002). These compounds include antimicrobial substances, surfactants, and chitinolytic enzymes. Recent studies demonstrated that fluorescent pseudomonads which synthesise phloroglucinol (Phl) play a key role in the suppressiveness of TAD soils in Washington State and in the Dutch polders (de Souza et al., 2003; Raaijmakers et al., 1997, 1999). Thanks to the development of specific primers, the density of the bacterial populations capable of producing 2,4-Phl (Phl^+) was evaluated in comparison in suppressive and conducive soils to take-all (Raaijmakers et al., 1997). Their densities on roots of wheat grown in TAD soils were shown to be higher than the threshold density required for control of take-all but were below the threshold or even not detected on roots from conducive soils (Raaijmakers et al., 1997). Further evidence includes increased densities of Phl^+ pseudomonads with wheat monoculture and increased suppression of take-all, but decrease of these Phl^+ pseudomonads with disruption of the wheat monoculture and destruction of the natural suppressiveness (Raaijmakers and Weller, 1998). Altogether, these observations enabled Weller et al. (2002) to conclude to the major contribution to 2,4-Phl producers in natural soil suppressiveness to take-all. Phl^+ pseudomonads from TAD soils appear to harbour few *phlD* alleles but a high diversity (ERIC-PCR) between suppressive soils and even within a given soil (de Souza et al., 2003; McSpadden Gardener et al., 2000; Weller et al., 2002).

Phl^+ pseudomonads were found at similar high population densities in both Morens suppressive and conducive soils to black root rot (Ramette et al., 2003a). No clear demarcation was found between Phl and hydrogen cyanide (HCN) producing *Pseudomonas* isolates from suppressive and conducive soils in terms of presence of particular biocontrol alleles and biocontrol performances (Ramette et al., 2006). Both in conducive and suppressive soils of Morens, the root-associated Phl^+ fluorescent pseudomonads have several *phlD* alleles, high diversity (ERIC-PCR) among the strains harboring the same *phlD* allele and a substantial grade of endemism (Ramette et al., 2003b, 2006). Phl^+ pseudomonads were shown to be less efficient when inoculated in a reconstituted Morens conducive soil than when inoculated in the reconstituted suppressive counterpart, and in the latter isolates from suppressive and conducive soils appeared to be as efficient (Ramette et al., 2006; Stutz et al., 1985). These data suggest that the expression of biocontrol genes is hampered in conducive soils. This is in agreement with the hypothesis that the suppressiveness of Morens is related both by the presence of Phl^+ root colonizing pseudomonads and conditions promoting expression of these biocontrol genes (see paragraph 2.3).

Some Phl^+ pseudomonads induce the early production of Phl in other pseudomonads (Maurhofer et al., 2004). Others produce bacteriocins which inhibit the growth of several Phl^+ strains (Validov et al., 2005). The antimicrobial activity of Phl is broad (see paragraph 3.1); when Phl+ pseudomonads are added as inoculants, they control diseases caused by several pathogens but the degree of suppression can vary considerably over time and geographic sites (e.g., in Haas and Défago, 2005). However, natural suppressive characteristics of soils seem to be pathogen-specific and consistent over years. This contradiction has not yet been explained. One possible hypothesis is that the specificity of a natural suppressive soil is due to interpopulation cross talk between Phl^+ pseudomonads adapted to different phases of the pathogen-plant interaction. In inoculation experiments, since Phl^+ pseudomonads are added at high concentrations (10^5 to 10^8 cfu per gram of soil) and they would reduce or silence the interpopulation talks.

Although a major has been given to trophic competition in natural soil suppressiveness to fusarium wilts, possible contribution of antibiosis in suppressiveness mechanisms cannot be excluded since isolates of *F. oxysporum* were shown to be susceptible to 2,4-Phl and to phenazine-1-carboxamide (Schouten et al., 2004; Chin-A-Woeng et al., 1998). This hypothesis is currently being evaluated (Mazurier, Lemanceau & Raaijmakers, personal communication).

2.4.2 Induced Resistance

Stimulation of the plant defense reactions by beneficial microbial populations has been shown to be generally plant specific and non-pathogen specific (Van Loon et al., 1998), whereas suppressive soils have usually be shown to be disease specific and not plant specific (Louvet et al., 1976; Weller et al., 2002). Since, there is a high microbial diversity in suppressive soils (Edel et al., 2001; Latour et al., 1996), different populations belonging to a given bacterial species may elicitate defense reactions in different plant species as it has been shown for fluorescent pseudomonads (Van Loon et al., 1998). This suggestion is also supported by the importance of plant genotype in selecting for soil microbial communities which confer disease suppression (Mazzola, 2004; Mazzola et al., 2004). Despite the fact that possible ISR in natural mechanisms of soil suppressiveness cannot be ruled out, so far there is no real evidence to support this hypothesis. To our knowledge, the only report in favor to a possible involvement of ISR to natural suppressiveness was made by Tamietti et al. (1993) who have shown that tomato plant cultivated in a suppressive soil to fusarium wilts expressed higher enzymatic activities such as laminarase, chitinase, N-acetyl-glucosaminidase and β-1,4-glucosidase which are known to be involved in the process of ISR. Further research based on the use of the so-called split-root experimental set up with one side of the root system cultivated in the steamed suppressive soil being infested and the other side in the natural suppressive soil should bring insights on the involvement of ISR in the natural suppressiveness phenomenon.

3. EXAMPLES OF BIOCONTROL AGENTS ISOLATED FROM SUPPRESSIVE SOILS

3.1. Pseudomonas fluorescens *2-79*

P. fluorescens 2-79 was shown to be one of the most efficient biocontrol agent against take-all among the fluorescent pseudomonads strains isolated by Weller and Cook (1983) from wheat roots grown in a TAD soil. *P. fluorescens* 2-79 was also demonstarted to be effective against *Rhizoctonia solani* rot root of wheat seedlings and to have an improved efficacy against *R. solani* root rot and take-all of wheat when combined with the fungizide pencycuron (Duffy, 2000). The ability of 2-79 to reduce take-all severity was confirmed in field conditions (Weller and Cook 1983).

Weller and Cook (1983) further showed that *P. fluorescens* 2-79 is an aggressive root colonizer and can be isolated from wheat roots throughout the growing season. When applied to winter wheat as a seed treatment at 10^8 cfu/seed, 2-79 establishes a large population of 10^7 cfu/roots one month after sowing. Its density declines during the winter down to 3 x 10^4 cfu/g roots, and increases again during the following spring up to 2 x 10^5 cfu/root with renewed growth of the wheat plant (Weller and Cook 1983). *P. fluorescens* 2-79 appears to be very competitive right after its application and represents about 50% of the total fluorescent pseudomonads and 1.4% of the total root bacteria one month after sowing. However half a year later, the proportion of 2-79 declined down to 0.3% of the total pseudomonads and 0.003% of total root bacteria (Weller and Cook 1983). Bankhead et al. (2004) have recently showed that inoculation of *P. fluorescens* 2-79 does not affect significantly the structure of indigenous bacterial populations in wheat rhizosphere. This observation suggests that although *P. fluorescens* 2-79 establishes well in the rhizosphere, possible risk of strong or lasting impact on indigenous bacterial communities appears to be small. Root colonisation by *P. fluorescens* 2-79 is affected by several environmental factors. Root colonization by 2-79 is greater at a slightly acidic pH of 6-6.5 compared to a pH of 7.0 or above. Bacterial growth in the rhizosphere is also affected by a range of biotic factors such as plants and resident microorganisms. Population densities of 2-79 on most supportive wheat cultivars are 100 times higher than on other cultivars, and the presence of *Ggt* lesions can enhance 2-79 populations up to ten-fold (Weller, 1988).

In culture 2-79 produces the antibiotic PCA (Brisbane et al., 1987; Gurusiddaiah et al., 1986). Purified PCA is active against *Ggt* and other wheat root pathogens in vitro (Brisbane et al., 1987; Gurusiddaiah et al., 1986) and is also involved in the suppression of take-all by 2-79. Indeed, Tn-*5* mutants of 2-79 impaired in their PCA production do not inhibit Ggt *in vitro* and have partially lost their ability to protect wheat against take-all. Complementation of the mutants with DNA from a 2-79 genomic library not only restored pathogen inhibition in vitro but also the suppression of take-all on wheat plants (Thomashow and Weller, 1988). The demonstration of PCA in the suppression by 2-79 was further supported by the measurement of this metabolite in the wheat rhizosphere inoculated by the wild-type strain but not by the PCA^- mutants (Thomashow and Weller, 1988).

More generally, PCA contributes to the competitiveness of 2-79 in wheat rhizosphere and seems to be an important feature for competition with resident microflora (Mazzola et al., 1992). Indeed, densities PCA-negative mutants were shown to decline much faster that those of the wild-type strain during the cultivation of several cycles of wheat in natural soil.

The PCA biosynthetic locus of strain 2-79 has been identified and characterized (Mavrodi et al., 1998). It consists of seven genes designated *phzABCDEFG*. PhzC, PhzD and PhzE are similar to enzymes of the shikimic acid pathway and together with PhzF, which apparently is involved in the formation of the phenazine ring system, are indispensable for PCA production (Mavrodi et al. 1998, Blankenfeldt et al., 2004). The PCA biosynthetic operon is preceded by two genes *phzR* and *phzI* that are homologs of the quorum sensing pairs of the *luxR-luxI* family. PhzI catalyzes the production of an acyl-homoserine lactone (HSL) signal, which interacts with PhzR and initiate the expression of the PCA biosynthetic genes, when reaching a critical concentration. Recently Khan et al. (2005) showed that *phzI* of 2-79 encodes three different 3-OH-acyl-HSLs, one of them, the *N*-(3-hydroxy-hexanoyl)–HSL, being the quorum sensing signal interacting with PhzR.

Even if PCA accounts for most the antagonistic activity of *P. fluorescens* 2-79, other metabolites have been proposed to play a role in its biocontrol activity, since PCA mutants did not fully lose their antagonistic activity (Hamdan et al., 1991). Although early works had suggested the possible involvement of pyoverdine-mediated iron competition in the activity of 2-79 (Weller et al., 1988), it was then nicely demonstrated that its relative contribution to the antagonism and disease suppression was minor or even nil compared to those by phenazine antibiotic (Hamdan et al., 1991). Thomashow and Weller (1990) suggested that the effects attributed in the early studies to pyoverdine in the antagonism against *G. graminis* var. *tritici* could be related to an artefact linked to the production of iron-regulated antibiotics, this type of compound being already described by Gill and Warren (1988). An additional antifungal compound was also suggested to contribute to some extent to the antagonistic activity of 2-79 (Thomashow and Weller, 1990). Finally, Slininger et al. (2000) identified a metabolite (2-acetamidophenol, ABP), produced by 2-79 in vitro, sharing a common segment of the PCA biosynthetic pathway. However, the possible contribution of ABP in disease suppression remains to be demonstrated.

3.2. Pseudomonas fluorescens *CHA0*

Pseudomonas fluorescens CHA0 was isolated in the early 1980s from a soil naturally suppressive to black root rot of tobacco and cherry trees caused by *T. basicola* (Stutz et al., 1986; Stutz, 1985). Strain CHA0 can protect tobacco and cherry trees from *T. basicola*-mediated black root rot when inoculated into some but not all conducive soils (Stutz et al., 1986; Stutz, 1985). When artificial soils mimicking suppressive and conducive soils from Morens were used, strain CHA0 protected tobacco from black root rot better in artificial suppressive soil than in conducive soil (Stutz et al., 1989; Keel et al., 1989).

In addition to black root rot, CHA0 controls effectively, in growth chambers and experimental greenhouses, various root and stem diseases caused by major fungal pathogens such as *G. graminis* var. *tritici*, *F. oxysporum*, *Pythium* and *Rhizoctonia* (Défago and Keel, 1995; Duffy and Défago, 1997a,b; Maurhofer et al., 1992, 1994b, 1995). In commercial greenhouses and in several field plots experiments, CHA0 protected effectively wheat against take-all and *Pythium* diseases, poorly tobacco and cotton against *T. basicola* and inconsistently tomato against *F. oxysporum* f. sp. *lycopersici* and f. sp. *radicis* (Défago et al., 1987; Défago et al., unpublished data; Wüthrich 1991; Wüthrich and Défago 1991). CHA0 based formulations are used to protect tomato plants against tomato spotted wilt virus (TSWV) in commercial tomato production (Kandan et al., 2005). Moreover, strain CHA0 protects tomato against nematodes (Siddiqui and Shaukat, 2003, 2005).

The mechanisms of disease suppression and survival of CHA0 in soil have been studied in details (reviewed in Haas and Défago, 2005). Strain CHA0 produces several extracellular metabolites in culture medium (in addition to Phl and HCN). Using gene replacement and other molecular techniques we have shown that HCN, Phl, pyoluteorin (Plt) and salicylic acid (Sal) are of importance for disease suppression (Défago and Haas, 1990; Keel et al., 1992; Laville et al., 1998; Maurhofer et al., 1994b, 1995, 1998; Schmidli-Sachere et al., 1997; Voisard et al., 1989). No role was found for the production of IAA and for extracellular proteases (Beyeler et al., 1999; Sacherer et al., 1994). CHA0 is able to induce systemic resistance in the upper part of tobacco plant and *Arabidopsis* and sugarcane when colonizing only the roots (Maurhofer et al., 1994a, 1998; Iavicoli et al., 2003; Viswanathan and Samiyappan, 2002). And pyoverdine seems to only play a role in ISR (Maurhofer et al., 1994a).

Strain CHA0 is an efficient rhizosphere coloniser, and can also colonise the inner tissues of the root (Troxler et al., 1997a, b). This endophytic behaviour is perhaps of relevance in terms of biocontrol (Troxler et al., 1997a). When colonising plant roots, CHA0 interacted strongly with indigenous fluorescent pseudomonads, as indicated by the modifications of community of indigenous fluorescent pseudomonads and rhizobia induced by CHA0 (Maurhofer et al., 2004; Moenne-Loccoz and Défago, 2004; Natsch et al., 1997; Niemann et al., 1997). CHA0 has no major impact on the abundance and diversity of many other rhizosphere inhabitants including *Bacillus* (Natsch et al., 1998) and *Cytophaga* (Johansen et al., 2002). A positive effect of *Glomus mosseae* on CHA0 densities was recorded in the rhizosphere and reciprocally Phl does not seems to have any deleterious effect on the mycorrhizal symbiosis (Edwards et al., 1998).

The bacterium may also persist over considerable periods at nontarget sites outside the rhizosphere (Natsch et al., 1996; Troxler et al., 1997b, 1998). However, persistence and biocontrol activity of strain CHA0 can be negatively affected under adverse environmental conditions, e.g. when the bacterium is faced to dry stress, hyperosmolarity (Schnider-Keel et al., 2001; Johansen et al., 2002) or lytic bacteriophages (Keel et al., 2002). Under certain stress conditions, CHA0 was found to persist in soil ecosystems as non-culturable cells (Troxler et al.,

1997b, 1998; Mascher et al., 2000, 2002; Johanson et al., 2002), even when associated with plant roots (Hase et al., 2000).

Several biotic and abiotic factors modulate the synthesis of biocontrol metabolites in CHA0. For instance, Plt and Sal are overproduced in pyoverdine-negative mutants (Schmidli-Sacherer et al., 1997). These two compounds act on the *phlA-phlF* gene region and repress the synthesis of Phl in the wild-type (Schnider-Keel et al., 2000). Toxins by *F. oxysporum* (fusaric acid) and by *R. solani* (phenylacetic acid) have also a negative effect on Phl production. Therefore, the production of these toxins allows the pathogen to protect itself against the biocontrol agent (Duffy and Défago, 1997a; Duffy et al., 2003; Notz et al., 2002; Siddiqui and Shaukat, 2005). Abiotic factors (e.g. oxygen tension, metal ions, minerals and carbon sources) influence also the production of biocontrol metabolites in CHA0 (Duffy and Défago, 1997a, 1999; Laville et al., 1998). HCN synthesis requires the *anr* regulatory gene and low oxygen levels typically found in the case of soil compaction and/or water logging (Laville et al., 1998; Hojberg et al., 1999). Addition of Zn^{2+} blocked the production of fusaric acid by *F. oxysporum* f.sp *radicis* and increased the production of Phl by CHA0 and its efficacy against crown root rot of tomato (Duffy and Défago, 1997a, b). Changes in Phl production obtained through modification of the expression of the *phlA*, *phlF*, *phlD*, *rpoD*, *pqq* and *gacS*/*gacA* genes resulted also in changes in the biocontrol efficacy of CHA0 in several pathosystems (e.g. Laville et al., 1992, Maurofer et al., 1992; Schmidli-Sacherer et al., 1997; Schnider et al., 1995a,b).

The regulatory network controlling the expression of biocontrol genes in CHA0 is complex; important decisions are made at a posttranscriptional level and some of the elements involved (e.g. GacS/GacA, three small regulatory RNAs) have been identified in recent years (reviewed in Haas and Keel 2003 and in Haas and Defago, 2005, Reimmann et al., 2005; Valverde et al., 2003, 2004). Interestingly Phl positively autoregulates its own biosynthesis (Schnider-Keel et al., 2000) and inhibits the synthesis of Plt (Baehler et al., 2005). Other Phl producing pseudomonads are able to induce Phl production in CHA0 (Maurhofer et al., 2004). This interpopulation cross talk is probably a key factor of the ecological fitness of Phl^{+} fluorescent pseudomonads at the infection sites.

Phl^{+} fluorescent pseudomonads were found world wide and are effective biocontrol agents in several pathosystems (Cronin et al., 1997; Keel et al., 1996; Sharifi et al., 1998; Ramette et al., 2001; Vincent et al., 1991; Wang et al., 2001). Moreover Phl synthesis is implicated in decline of wheat take-all disease caused by *Ggt* (Raaijmakers and Weller 1998; de Souza et al., 2003). Of special interest is the strain *P. fluorescens* Pf 5, strain very similar to CHA0. The complete genome sequence of this strain is now available (Paulsen et al., 2005) and appear to be very similar to that of CHA0.

3.3 Pseudomonas fluorescens C7

This strain was issued from a collection of strains (71) isolated from the rhizosphere of flax cultivated in the fusarium wilt suppressive soil of Châteaurenard (Lemanceau et al., 1988b). These strains were tested for their ability to suppress fusarium wilt of flax when inoculated alone or in combination with the antagonistic strain of non-pathogenic *F. oxysporum* Fo47 also previously isolated from the Châteaurenard soil (Alabouvette et al., 1987). *P. fluorescens* C7 was identified as being among the strains improving consistently the efficacy of the suppression by Fo47, without being efficient on its own. The ability of this strain to improve the efficacy and consistency of the control by Fo47 of fusarium crown and root of tomato cultivated in rockwool (soilless culture) was also demonstrated in commercial-like conditions (Lemanceau and Alabouvette, 1991). The efficacy and consistency of the biocontrol achieved by the microbial combination were further confirmed by Olivain et al. (2004). This efficacy does not seem be related to antibiosis since it does not produce any known antibiotic, but does synthesize cyanide (Ellis et al., 2000). But, *P. fluorescens* C7R12 appears to be an efficient iron competitor as indicated by its ability to grow in iron stress conditions (high value of Minimal Inhibiting Concentration of the strong iron chelator 8-hydroxyquinoline) (Delorme et al., submitted for publication) and its ability to incorporate seven types of heterologous pyoverdines (Mirleau et al., 2000).

Experiments with a rifampicin mutant (C7R12) of *P. fluorescens* C7 showed that this strain is rhizosphere competent (Eparvier et al., 1991); this competence being related to (i) its efficient pyoverdine mediated iron-uptake (specially its ability to incorporate heterologous pyoverdines) (Mirleau et al., 2000) and to (ii) its ability to switch from aerobic respiration to nitrogen respiration according to the rhizosphere conditions (Mirleau et al., 2001). Furthermore, *P. fluorescens* C7R12 was shown to form colonies at the root surface but also to colonize root tissues intercellularly and intracellularly (Sanchez et al., 2004) like it was described previously with *P. fluorescens* WCS417r (Duijff et al., 1997).

Interestingly, some of the genes shown to be induced in the model plant *Medicago truncatula* during root colonization of the arbuscular mycorrhizal fungus *Glomus mossease* BEG12 appear also to be induced during that of *P. fluorescens* C7R12, suggesting that some plant cell programmes may be shared by these two types of beneficial microorganisms during their root colonization (Sanchez et al., 2004). Unexpectedly, *P. fluorescens* C7 reported to suppress fusarium crown and root rot of tomato (Lemanceau and Alabouvette, 1991) induces a hypersensitive reaction in tomato after leaf infiltration. This reaction was related to the presence of homologues of *hrp* genes (Mazurier et al., 2004),since this reaction was not anymore induced by a mutant impaired in the sequence encoding type three secretion system (Mazurier et al., unpublished data). Possible involvement of this secretion system in the rhizosphere competence and induced systemic resistance is currently evaluated.

3.4. P. fluorescens Q8r1-96

P. fluorescens Q8r1-96 is a representative strain of the dominant RAPD genotype among the 16 identified during a diversity study of indigenous Phl^+ fluorescent pseudomonads (101) isolated from wheat roots in the Quincy TAD soil (Washington). This dominant RAPD genotype represented up to 50% of the total Phl-producing fluorescent pseudomonads of the Quincy soil (Raaijmakers and Weller, 2001). More generally, Mc Spadden et al. (2000) have reported that nearly one-third of the Phl^+ isolates obtained from different wheat-growing areas in the United States were genotypically similar to Q8r1-96.

Rhizosphere competence of *P. fluorescens* Q8r1-96 appears to be very high (Raaijmakers and Weller, 2001). This was specially clearly shown during long term experiments in which Q8r1-96 density maintained, after 7 growth cycles of wheat, at a level above the threshold previously shown by Raaijmakers et al. (1997) to be required for take-all decline. This high rhizosphere competence does not seem to be soil specific since it was expressed in soils showing different physico-chemical properties. In agreement with observations previously made with *P. fluorescens* F113 (Carroll et al., 1995), the rhizosphere competence of Q8r1-96 appears to be not related to Phl production (Raaijmakers and Weller, 2001).

As expected by its density, during long term experiments *P. fluorescens* Q8r1-96 suppressed take-all significantly better than other Phl^+ pseudomonads that showed a lower rhizosphere competence (Raaijmakers and Weller, 2001). Recently, Huang et al. (2004) have introduced introducing in this aggressive root colonizer a seven-gene operon for the synthesis of PCA. The recombinant strains did not gain in the efficacy of suppression of take-all or *Pythium* root rot, but were able to suppress *Rhizoctonia* root rot at a density of one to two order of magnitude lower than that of the wild-type strain.

Taking in account its specially high and long-lasting rhizosphere competence in different soil types and its efficient ability to suppress take-all, *P. fluorescens* Q8r1-96 appears to be a good candidate for biological control and this strain is in the process of being developed for control of take-all and root diseases of wheat (Weller, personal communication).

4. RELATIONS BETWEEN MECHANISMS OF SOIL SUPPRESSIVENESS AND MODES OF ACTION OF THE BIOCONTROL AGENTS

Studies on these relations have mostly relied on two types of strategies. One consisted in raising hypotheses of possible modes of actions of biocontrol agents on the basis of mechanisms identified as being involved in natural soil suppressiveness. This strategy is illustrated for the carbon and iron competition that has been shown to be involved in the natural suppressiveness to fusarium wilts and then in the antagonistic activities of non-pathogenic *F. oxysporum* and fluorescent pseudomonads against pathogenic *F. oxysporum*. On the other way round, the second strategy consisted in (i) isolating strains from microbial groups shown to be involved in the natural soil suppressiveness and (ii) analyzing their antagonistic modes of action, these modes of action being expected to be part of the mechanisms accounting for natural soil suppressiveness. The limit of such approach has stimulated further studies at a population level. Indeed, biocontrol model strains are not necessarily representative of the diversity of the indigenous antagonistic populations and their modes of actions may then not account for the major mechanisms involved in the natural suppressiveness. Several reviews have been published over the last years on the modes of action of biocontrol agents (Bloemberg and Lugtenberg, 2001; Haas and Keel, 2003; Moënne-Loccoz and Défago, 2004; Morrissey et al., 2004; Raaijmakers et al., 2002).

4.1 Competition

From studies on suppressive soils to fusarium wilts, the importance of competition for carbohydrates and iron was clearly demonstrated (see *paragraph 2.4.1.1*). These studies have also ascribed a major role to non-pathogenic *Fusarium* and fluorescent pseudomonads in natural soil suppressiveness (see *paragraph 2.3*).

Then, it was suggested that non-pathogenic and pathogenic *F. oxysporum* showing close trophic requirements would compete together for the scarce carbon sources in soil. This hypothesis was confirmed with the biocontrol agent non-pathogenic *F. oxysporum* Fo47, isolated from the Châteaurenard suppressive soil to fusarium wilts. Indeed, its antagonistic activity was shown to only occur when carbon is a limiting factor, clearly indicating that competition for carbon is a major antagonistic mode of action for this strain (Lemanceau et al., 1993).

Similarly from the early work of Kloepper et al. (1980) reporting that introduction of *Pseudomonas* B10, isolated from a take-all suppressive soil, or its pyoverdine rendered a conducive soil suppressive, it was suggested that antagonistic activity of fluorescent pseudomonads against pathogenic fungi, including *F. oxysporum*, would be related to pyoverdine-mediated iron competition (Baker, 1992). Indeed, this mode of action was shown to be involved in the antagonism achieved by various fluorescent pseudomonads strains (Lemanceau and Alabouvette, 1993) and this contribution of pyoverdine in bacterial antagonism has been reviewed several times (Bakker et al., 1991; Heming, 1986; Lemanceau, 1992; Leong, 1986; Loper and Buyer, 1991, Neilands and Leong, 1986). Demonstrations of the pyoverdine-mediated antagonism determined by fluorescent pseudomonad strains were based on the use of pyoverdine minus mutant (pvd-mutant) and/or purified pyoverdine. As an example *P. putida* WCS358 was shown to suppress fusarium wilts whereas a pvd- mutant did not (De Boer et al., 2003; Leeman et al., 1996; Raaijmakers et al., 1995). Purified pyoverdine of WCS358 reduced saprophytic growth of pathogenic *F. oxysporum* whereas ferri-pyoverdine did not, clearly indicating that the antagonism determined by the bacterial metabolite is related to iron competition (Lemanceau et al., 1993). This observation is in agreement with the lower stability of complex made between iron and fusarinines, the siderophores from *Fusarium* (Emery, 1965; Lemanceau et al.,

1986), than that of ferri-pyoverdine (Meyer and Abdallah, 1978; Scher and Baker, 1982). As indicated above for the naturally suppressive soil (see *paragraph 2.4.1.1.*), competition for iron is only expressed when the demand for this element is high and the offer is low, and increase of iron availability by addition of FeEDDHA was indeed shown to reduce the biocontrol efficacy of the wild-type strain WCS358 (De Boer et al., 2003; Duijff et al., 1994).

Specific combinations of fluorescent pseudomonads and non-pathogenic *F. oxysporum* were tested to make attempt to somehow mimic the complexity of the microbial interactions occurring in suppressive soils in order to improve the disease suppression by the biocontrol agents (Leeman et al., 1996; Lemanceau and Alabouvette, 1991; Lemanceau et al., 1992; Park et al., 1988). Combinations of *P. fluorescens* C7R12 + non-pathogenic *F. oxysporum* Fo47 (see *paragraph 3.3*) or WCS358 + Fo47 (Duijff et al., 1999; Lemanceau et al., 1992) were shown to improve the efficacy and the consistency of the biocontrol compared to the strains inoculated separately. Mechanisms underlying the improved protection of the combination WCS358+Fo47 were further investigated and were related to pyoverdine synthesis. Indeed, the increased efficacy of Fo47 occurred when co-inoculated with the wild-type strain WCS358 but not with its pvd- mutant. It was confirmed with an ice-nucleation reporter gene to an iron-regulated promoter that the iron availability was low enough to allow pyoverdine synthesis by WCS358 (Duijff et al., 1999). Synergistic effect between carbon competition by Fo47 and pyoverdine-mediated iron competition by WCS358 was related to a reduced efficacy of the energetic metabolism of the pathogen in iron stress conditions, making the pathogen more susceptible to the carbon competition determined by the non-pathogenic *F. oxysporum* (Lemanceau et al., 1993).

For other pathosystems, iron competition seems to play little or no role in the modes of action of antagonistic activities of fluorescent pseudomonads. Pvd- mutant of *P. fluorescens* 2-79 appeared to control take-all as effectively as the parental strain (Hamdan et al., 1991). Even more, enough iron availability is prerequisite for HCN production by *P. fluorescens* CHA0 which is a major mode of action of that strain (Keel et al., 1989).

4.2 Antibiosis

Natural suppressiveness to take-all has been related to fluorescent pseudomonads (see *paragraph 2.3*). Efficient biocontrol fluorescent pseudomonads isolated from suppressive soil appeared to be able to synthesize two types of antibiotics 2,4-Phl or PCA. As recently reviewed by Haas and Défago (2005), the strategy followed to show the implication of these antibiotics in the antagonistic modes of actions of the biocontrol pseudomonads was based on (i) the demonstration of the production of the metabolite in vitro and its purification and characterization, (ii) the detection and quantification of the antibiotic in the rhizosphere, (iii) the identification of the genes encoding the antibiotic allowing the construction of non-producing and over-producing mutants, (iv) the observation of the less efficient protection achieved by the non-producing mutant and in some cases the improved protection by the over-producing mutant, (v) the acquisition of the antagonistic activity in poor biocontrol strains by introduction of biosynthetic genes, and finally (vi) the observation of antibiotic biosynthetic genes in the rhizosphere through the use of reporter genes.

This strategy is nicely illustrated the work of Thomashow and co-workers to demonstrate the implication of PCA in the suppression of take-all by *P. fluorescens* 2-79 isolated from a TAD soil (Huang et al., 2004; Mavrodi et al., 1998; Thomashow and Weller, 1988; Thomashow et al., 1990). Phenazines were shown be involved in the antagonistic activity against various soilborne pathogens in several fluorescents pseudomonads (Chin-A-Woeng et al., 2001; 2003; Timms-Wilson et al., 2000). However, search for populations of fluorescent pseudomonads harbouring sequence encoding PCA thanks to the development of specific probes did not allow confirming the involvement of PCA producers in the natural mechanism of TAD, since their density was below the detection limit both in conducive and suppressive soils (Raaijmakers et al., 1997).

Similar strategy was applied to demonstrate the implication of Phl in the antagonistic modes of action of model strains against take-all (Bonsall et al., 1997; Cook et al., 1995; Fenton et al., 1992; Keel et al., 1992). However, in contrast to PCA^+ pseudomonads, use of specific probes allowed Raaijmakers et al (1998) to show that densities 2,4-Phl^+ pseudomonads on wheat roots were above the threshold required to take-all control when cultivated in suppressive soils, but were below the threshold or even not detected on roots from conducive soils (Raaijmakers et al., 1997). Densities of Phl producers were further shown to increase with wheat monoculture and with enhanced suppression of take-all, whereas their densities decreased with disruption of the wheat monoculture and destruction of the natural suppressiveness. Altogether, these observations allowed Weller et al., (2002) to conclude to the major contribution to 2,4-Phl producers in natural soil suppressiveness. Phl synthesis appears then to be then involved both in the natural soil suppressiveness to take-all and in the antagonistic activity of biocontrol agent isolated from these soils. Such congruence could not be made for PCA despite the clear demonstration of the role of this metabolite in the control achieved by model strains of fluorescent pseudomonad including *P. fluorescens* 2-79.

4.3 Induced Systemic Resistance

Induced systemic resistance has been demonstrated for a wide range of microorganisms including fluorescent pseudomonads (Van Loon et al., 1998), non-pathogenic *Fusarium* (Fuchs et al., 1997; Matta et al., 1989), relative importance of microbial antagonism and ISR varying according to the strains (Duijff et al., 1998). These microorganisms are present in suppressive soils and have been shown to be involved in the natural soil suppressiveness. Even if these observations raise the possible involvement of ISR in natural soil suppressiveness, no demonstration of the involvement of this mechanism in natural suppressiveness has been made so far (see *paragraph 2.4.2.*).

5. PROSPECTS FOR DISCOVERING NEW BACTERIA, GENES AND MODES OF ACTION FROM SUPPRESSIVE SOILS

Soils are considered to be a major reservoir of biodiversity. Despite the efforts made, so far this diversity has only been very partially explored. Until recently, we had only access to the microflora that we know how to cultivate. However according to estimations made by Torsvik et al., (1990) this fraction would represent less than 1% of the soil microflora; and as indicated above (see *paragraph 2.3*), only very few microbial groups were considered in the early studies on soil suppressiveness and the diversity of these groups was not taken in account. A better knowledge of the microbial populations associated with soil suppressiveness requires an untargeted approach to characterize the soilborne communities of microorganisms.

The general strategy to discover new bacteria, genes and metabolites involved in the soil suppressiveness consists in comparing structure, diversity and activity of microbial communities in soils showing different level of soil suppressiveness. At that stage, the conclusions made only allow to conclude that these bacteria, genes and metabolites are preferentially associated with the soil suppressiveness. Their real involvement in the soil suppressiveness requires further experiments based on the strategy summarize in paragraph 4.2.

During the last decade, progresses made in the extraction and amplification of DNA from soils (Martin-Laurent et al., 2001) and more generally advances in molecular techniques have allowed the development of cultivation-independent methods to characterize the structure and diversity of microbial communities in soils; even if we have to be cautious about the analysis of the corresponding data (Von Wintzingerode et al., 1997). Application of these methods to characterize the structure and diversity of microbial communities in relation with soil suppressiveness has been reviewed recently (Garbeva et al., 2004; Mazzola, 2004). The molecular methods developed to assess microbial diversity and community structure in soils are mainly based on the extraction of soil DNA and the PCR amplification of part of the genetic information, generally from the ribosomal operon (neutral markers). These direct methods have initially focused on the bacterial communities, by targeting the small subunit ribosomal RNA gene (16S rRNA) which can serve as phylogenetic marker to determine evolutionary relationships between members of the community and to identify them. The growing sequence database of rRNA genes has later allowed for the selection of a range of specific PCR primers for various taxonomic groups, from the species to high taxonomic levels, including eukaryotic organisms. The PCR products, consisting of a mixture of DNA fragments contributed by community members, can be characterized using a cloning and sequencing strategy, or fingerprinting methods. Depending on the primers used in the PCR, these fingerprinting methods can either be applied to the whole bacterial or archaeal community represented in the soil DNA extract, or focus on more specific groups such as α-proteobacteria, β-proteobacteria or actinomycetes (Gomes et al., 2001; Heuer et al., 1997) or even specific genera (Locatelli et al., 2002; Matsuki et al., 2002). Multivariate analysis of DNA fingerprints of the communities associated with the different soils allows the characterization of shifts in the genetic structure of the microbial communities. The markers explaining the shifts can be then excised and sequenced, for their possible identification by comparison of the sequences already available in the data bases and for developing probes for their targeting among cfu if culturable (Ranjard et al., 2000).

Although direct DNA-based methods are powerful tools to investigate microbial diversity and identify predominant microbial populations associated with specific soil processes, they do not inform us about the metabolically active fraction of the microbial community. New developments based on the use of ^{13}C now allow the discrimination of DNA from active and total microbial communities (DNA-Stable Istope Probing) (Radajewski et al., 2000), and then the comparison of the structure of these communities in soils showing different levels of suppressiveness. However, progresses still need to be made to identify more specifically which populations are active and to assess specific activities in soils. These may be addressed using RNA-based approaches to investigate the set of genes that are expressed or transcribed in soil environments, in the so-called transcriptome. Although more difficult than DNA-based approaches, RNA-based ones have already enabled us to gain insight into microbial ecology. In order to identify active metabolic populations, clone libraries strategies or similar profiling methods as those described above were applied to rDNA template obtained by reverse transcriptase (RT) PCR of RNA extracted directly from soil (Duineveld et al., 2001; Nogales et al., 2001). Soil RNA extracts were also investigated with microarrays consisting of either taxa-specific oligonucleotide probes or known metabolic genes in order to detect and quantify microorganisms or functional genes in soils, respectively (Small et al., 2001; Taroncher-Oldenburg et al., 2003; Wu et al., 2001). Although environmental applications of microarrays are for the moment limited, this technology is potentially a powerful tool for the identification of microbial populations and functions involved in key soil processes. Recent advances in DNA-based methods have also offered the opportunity of directly cloning environmental DNA without any previous PCR step. The so-called metagenome approach is based on the construction of environmental genetic libraries in vectors that can carry large DNA fragments of more than 100 kb and express the coding sequences. These clones may be tested for their antagonistic activity against the plant pathogens considered. This type of strategy is currently applied for searching for new functional genes involved in the natural soil suppressiveness (Raaijmakers et al., & Van Elsas et al., personal communications). Metagenomic libraries provide meaningful genetic and functional information to enable for the discovery of novel genes and organisms (Rondon et al., 2000; Sebat et al., 2003; Voget et al., 2003).

After the DNA and RNA levels that investigate gene presence and expression, respectively, the following-uplevel of analysis would be the proteome and metabolome. Indeed, various reports have stressed the importance of environmental conditions on the gene expression (Haas and Keel, 2003). Currents progresses in the optimisation of the extraction of proteins (Maron et al., submitted for publication; Wilmes and Bond, 2004) and of metabolites (Goodacre et al., 2004) are underway. The corresponding information will not be necessarily correlated with the DNA or RNA levels due to post transcriptional regulation and possible proteins rearrangement (Mann, 1999). This information will be relevant since proteins preferentially associated with suppressive soils could possibly allow targeting new functional genes by reverse genetic. Comparison of metabolomes in soils showing different level of suppressiveness should allow the identification of metabolites preferentially associated to suppressive soils which would then possibly be involved in the soil suppressiveness, this latter demonstration requiring the identification and the purification of the putative compounds and the identification of the producing microorganisms (Goodacre et al., 2004).

Even if promising, strategies based on community approach are hampered by a major limitation related to the fact that only dominant populations and proteins can only be tackled by metagenome and metaproteome characterization. A polyphasic strategy based on a combination of approaches including characterization of microbial communities by genotyping and cultivation, DNA arrays, gene expression, metagenome, metaproteome and metabolome should allow us to make further progresses in the knowledge of microorganisms, genes and metabolites involved in the natural soil suppressiveness to soilborne diseases.

6. REFERENCES

Alabouvette, C. (1986). Fusarium wilt suppressive soil from the Châteaurenard region: a review of a 10-year study. *Agronomie, 6*, 273-284.

Alabouvette, C., Rouxel, F., & Louvet, J. (1977). Recherches sur la résistance des sols aux maladies. III. Effets du rayonnement γ sur la microflore d'un sol et sa résistance à la fusariose vasculaire du melon. *Annales de Phytopathologie, 9*, 467-471.

Alabouvette, C., Rouxel, F., & Louvet, J. (1980). Recherches sur la résistance des sols aux maladies. VI. Mise en évidence de la spécificité de la résistance d'un sol vis-à-vis des fusarioses. *Annales de Phytopathologie, 12*, 11-19.

Alabouvette, C., Couteaudier, Y., & Louvet, J. (1982). Comparaison de la réceptivité de différents sols et substrats de culture aux fusarioses vasculaires. *Agronomie, 2*, 1-6.

Alabouvette, C., Couteaudier, Y., & Louvet, J. (1984). Recherches sur la résistance des sols aux maladies. IX. Dynamique des populations de *Fusarium* spp. et de *Fusarium oxysporum* f.sp. *melonis* dans un sol résistant et dans un sol sensible aux fusarioses vasculaires. *Agronomie, 4*, 729-733.

Alabouvette, C., Couteaudier, Y., & Louvet, J. (1985a). Recherches sur la résistance des sols aux maladies. XI. Etude comparative du comportement des *Fusarium* spp. dans un sol résistant et un sol sensible aux fusarioses vasculaires enrichis en glucose. *Agronomie, 5*, 63-68.

Alabouvette, C., Couteaudier, Y., & Louvet, J. (1985b). Recherches sur la résistance des sols aux maladies. XII. Activité respiratoire dans un sol résistant et un sol sensible aux fusarioses vasculaires enrichis en glucose. *Agronomie, 5*, 69-72.

Alabouvette, C., Höper, C., Lemanceau P., & Steinberg, C. (1996). Soil suppressiveness to diseases induced by soilborne plant pathogens. In G. Stotzky & J. M. Bollag (Eds.), *Soil Biochemistry* (pp. 371-413). New York: Marcel Dekker Inc.

Alabouvette, C., De La Broise, D., Lemanceau, P., Couteaudier, Y., & Louvet, J. (1987). Utilisation de souches non pathogènes de *Fusarium* pour lutter contre les fusarioses : situation actuelle dans la pratique. *OEPP/EPPO, 17*, 665-674.

Alabouvette, C., Raaijmakers, J., de Boer, W., Notz, R., Défago, G, Steinberg, C., & Lemanceau P. Concepts and methods to assess phytosanitary quality of soils. In J. Bloem, D. W. Hopkins & A. Benedetti (Eds.), *Plant-microbe interactions and soil quality*. Wallingford: CABI Publishing, in press.

Armstrong, G. M., & Armstrong, J. K. (1981). Formae speciales and races of Fusarium oxysporum causing wilt diseases. In P. E. Nelson, T. A. Toussoun & R. J. Cook (Eds), *Fusarium: Diseases, Biology, and Taxonomy*. Pennsylvania State University Press.

Asher, M. J. C., & Shipton, P. J. (1981). *Biology and Control of Take-all*. New York: Academic Press.

Atkinson, G. F. (1982). Some diseases of cotton. *Alabama Agricultural Experimental Station Bulletin, 41*, 19-29.

Baehler, E., Bottiglieri, M., Pechy-Tarr, M., Maurhofer, M. & Keel, C. (2005). Use of green fluorescent protein-based reporters to monitor balanced production of antifungal compounds in the biocontrol agent *Pseudomonas fluorescens* CHA0. *Journal of Applied Microbiology, 99*, 24-38.

Baker, R. (1992). Biological control of diseases of crops grown in covered and environmentally controlled structures. In E. S. Tjamos et al. (Eds.), *Biological Control of Plant Diseases*. (pp. 231-241). New York: Plenum Press.

Baker, K. F., & Cook, R. J. (1974). *Biological control of plant pathogens*. San Francisco: Freeman.

Bakker, P. A. H. M., Van Peer, R., & Schippers, B. (1991). Suppression of soil-borne plant pathogens by fluorescent pseudomonads: mechanisms and prospects. In A. B. R. Beemster, G. J. Boeller, M. Gerlagh, M. A. Ruissen, B. Schippers & A. Tempei (Eds.), *Biotic interactions and soil-borne diseases* (pp. 217-230). Amsterdam: Elsevier.

Bangera, M. G., & Thomashow, L. S. (1999). Identification and characterization of gene cluster for synthesis of the polyketide antibiotic 2,4-diacetylphloroglucinol from *Pseudomonas fluorescens* Q2-87. *Journal of Bacteriology, 181*, 3155-3163.

Bankhead, S. B., Landa, B. B., Lutton, E., Weller, D. M., & McSpadden Gardener, B. B. (2004) Minimal changes in rhizobacterial population structure following root colonization by wild type and transgenic biocontrol strains. *FEMS Microbiology Ecology, 49*, 307-318.

Berling, C. H., Défago, G, & Kern, H. (1984). Population-dynamics of *Thielaviopsis basicola* in soils conducive and suppressive to black root rot disease. *Phytopathology, 74*, 867-867.

Beyeler, M., Keel, C., Michaux, P., & Haas, D. (1999). Enhanced production of indole-3-acetic acid by a genetically modified strain of *Pseudomonas fluorescens* CHA0 affects root growth of cucumber but does not improve protection of the plant against Pythium root rot. *FEMS Microbiology Ecology, 28*, 225-233.

Blankenfeldt, W., Kuzin, A. P., Skarina, T., Korniyenko, Y., Tong, L., Bayer, P., Janning, P., Thomashow, L. S., & Mavrodi, D. V. (2004). Structure and function of the phenazine biosynthetic protein PhzF from *Pseudomonas fluorescens. Proceedings of the National Academy of Sciences of the United States, 101*, 16431-16436.

Bloemberg, G. V., & Lugtenberg, B. J. J. (2001). Molecular basis of plant growth promotion and biocontrol by rhizobacteria. *Current Opinion in Plant Biology, 4*, 343-350.

Bonsal, R. F., Weller, D. M., & Thomashow, L. S. (1997). Quantification of 2,4-diacetylphloroglucinol produced by fluorescent *Pseudomonas* spp. in vitro and in the rhizosphere of wheat. *Applied and Environmental Microbiology, 65*, 951-955.

Bouhot, D. (1979). Estimation of inoculum density and inoculum potential: techniques and their values for disease prediction. In B. Schippers & W. Gams (Eds.), *Soil-borne plant pathogens* (pp. 21-34). London: Academic Press.

Brisbane, P. G., Janik, L. J., Tate, M. E., & Warren, R. F. O. (1987). Revised structure for the phenazine antibiotic from *Pseudomonas fluorescens* 2-79 (NRRL B-15132). *Antimicrobial Agents and Chemotherapy, 31*, 1967-1971.

Carroll, H., Moënne-Loccoz, Y., Dowling, D. N., & O'Gara, F. (1995). Mutational disruption of the biosynthesis gene coding for the antifungal metabolite 2,4-diacetylphlroglucinol does not influence the ecological fitness of *Pseudomonas fluorescens* F113 in the rhizosphere of sugarbeets. *Applied and Environmental Microbiology, 61*, 3002-3007.

Chin-A-Woeng, T. F. C., Thomas-Oates, J.E., Lugetnberg, B. J. J., & Bloemberg, G. V. (2001). Introduction of *phzH* gene of *Pseudomonas chlororaphis* PCL1391 extends the range of biocontrol ability of phenazine-1-carboxylic acid-producing *Pseudomonas* spp. strains. *Molecular Plant-Microbe Interactions, 14*, 1006-1015.

Chin-A-Woeng, T. F. C., Bloemberg, G. V., & Lugtenberg, B. J. J. (2003). Phenazines and their role in biocontrol by *Pseudomonas* bacteria. *New Phytologist, 157*, 503-523.

Chin-A-Woeng, T. F. C., Bloemberg, G. V., van der Bij, A. J., van der Drift, K. M. G. F., Schripsema, J., Kroon, B., Scheffer, R. J., Keel, C., Bakker, P. A. H. M., Tichy, H. V., de Bruijn, F. J., Thomas-Oates, J. E., & Lugtenberg, B. J. J. (1998). Biocontrol by phenazine-1-carboxamide-producing *Pseudomonas chlororaphis* PCL1391 of tomato root rot caused by *Fusarium oxysporum* f. sp. *radicis-lycopersici*. *Molecular Plant-Microbe Interactions, 11*, 1069-1077.

Cook, R. J. (1981). The influence of rotation crops on take-all decline phenomenon. *Phytopathology, 71*, 189-192.

Cook, R. J., & Baker, K. F. (1983). *The nature and practice of biological control of plant pathogens*. St Paul: The American Phytopathological Society.

Cook, R. J., & Rovira, A. D. (1976). The role of bacteria in the biological control of *Gaeumannomyces graminis* by suppressive soils. *Soil Biology and Biochemistry, 8*, 269-273.

Cook, J. R., Thomashow, L. S., Weller, D. M., Fujimoto, D., Mazzola, M., Bangera, G., & Kim. D. S. (1995). Molecular mechanisms of defense by rhizobacteria against root disease. *Proceedings of the National Academy of Science, U.S.A. 92*, 4197-4201.

Couteaudier, Y., & Alabouvette, C. (1990a). Quantitative comparison of *Fusarium oxysporum* competitiveness in relation to carbon utilization. *FEMS Microbiology Ecology, 74*, 261-267.

Couteaudier, Y., & Alabouvette, C. (1990b). Survival and inoculum potential of conidia and chlamydospores of *Fusarium oxysporum* f.sp. *lini* in soil. *Canadian Journal of Microbiology, 36*, 551-556.

Cronin, D., Moënne-Loccoz, Y., Fenton, A., Dunne, C., Dowling, D. N. & O'Gara, F. (1997). Role of 2,4-diacetylphloroglucinol in the interactions of the biocontrol pseudomonad strain F113 with the potato cyst nematode *Globodera rostochiensis*. *Applied and Environmental Microbiology, 63*, 1357-1361.

De Boer, M., Bom, P., Kindt, F., Keurentjes, J. J. B., van der Sluis, I., Van Loon, L. C., & Bakker, P. A. H. M. (2003). Control of Fusarium wilt of radish by combining *Pseudomonas putida* strains that have different disease-suppressive mechanisms. *Phytopathology, 93*, 626-632.

Défago, G., & Haas, D. (1990). Pseudomonads as antagonists of soilborne plant pathogens: modes of action and genetic analysis. In J. M. Bollag & G. Stotzky (Eds.), *Soil Biochemistry* (pp. 249-291). New York: Marcel Dekker Inc.

Défago, G, & Keel, C. (1995). Pseudomonads as biocontrol agents of diseases caused by soil-borne pathogens. In H. M. Hokkanen & J. M. Lynch (Eds.), *Benefits and Risks of Introducing Biocontrol Agents* (pp. 137-148). Cambridge: University Press.

Défago, G., Berling, C., H., Henggeler, S., Hungerbühler, W., Kern, H., Schleppi, P., Stutz, E. W., & Zürrer, M. (1987). Survie d'un *Pseudomonas fluorescens* dans le sol et protection du blé contre des maladies d'origine fongique. *Sch Landw Fo., 26*, 155-160.

De Souza, J. T., Weller, D. M., & Raaijmakers, J. M. (2003). Frequency, diversity, and activity of 2,4-diacetylphloroglucinol-producing fluorescent *Pseudomonas* spp. in Dutch take-all decline soils. *Phytopathology, 93*, 54-63.

Dominiguez, J., Negrin, M. A., Rodriguez, C. M. (2001). Aggregate water-stability and soil solution properties in conducive and suppressive soils to Fusarium wilt of banana from Canary islands (Spain). *Soil Biology and Biochemistry, 33*, 449-455.

Duffy, B. (2000). Combination fo pencycuron and *Pseudomonas fluorescens* strain 2-79 for integrated control of Rhizoctonia root rot and take-all of spring wheat. *Crop Protection, 19*, 21-25.

Duffy, B. K., & Défago, G. (1997a). Zinc amendment improves the biocontrol of tomato crown and root rot by *Pseudomonas fluorescens* and represses the production of bacterial metabolites inhibitory to bacterial antibiotic biosynthesis. *Phytopathology, 87*, 1250-57.

Duffy, B. K., & Défago, G. (1997b). Environmental signals in biocontrol of tomato root disease by *Pseudomonas fluorescens*. *Med. Fac. Landbouww. Uni. Gent, 62*, 1015-1022.

Duffy, B. K., & Défago, G. (1999). Environmental factors modulating antibiotic and siderophore biosynthesis by *Pseudomonas fluorescens* biocontrol strains. *Applied and Environmental Microbiology, 65*, 2429-2438.

Duffy, B., Schouten, A., & Raaijmakers, J. M. (2003). Pathogen self-defense: mechanisms to counteract microbial antagonism. *Annual Review of Phytopathology, 41*, 501-538.

Duineveld, B. M., Kowalchuk, G. A., Keijzer, A., van Elsas, J. D., & van Veen, J. A. (2001). Analysis of bacterial communities in the rhizosphere of chrysanthemum via denaturing gradient gel electrophoresis of PCR-amplified 16S rRNA as well as DNA fragments coding for 16S rRNA. : *Applied and Environmental Microbiology, 67*, 172-178.

Duijff, B. J., Bakker, P. A. H. M., & Schippers B. (1994). Suppression of Fusarium wilt of carnation by *Pseudomonas putida* WCS358 at different levels of disease incidence and iron availability. *Biocontrol Science and Technology, 4*, 279-288.

Duijff, B J, Gianinazzi-Pearson, V., & Lemanceau, P. (1996). Involvement of the outermembrane lipopolysaccharides in the endophytic colonization of tomato roots by biocontrol *Pseudomonas fluorescens* strain WCS417r. *New Phytologist, 125*, 325-334.

Duijff, B. J., Gianinazzi-Pearson, V., & Lemanceau, P. (1997). Involvement of the outer membrane lipopolysaccharides in the endophytic colonization of tomato roots by biocontrol *Pseudomonas fluorescens* strain WCS417r. *New Phytologist, 135*, 325-334.

Duijff, B. J., Pouhair, D., Olivain, C., Alabouvette, C., & Lemanceau, P. (1998). Implication of systemic induced resistance in the suppression of fusarium wilt of tomato by *Pseudomonas fluorescens* WCS417r and nonpathogenic *Fusarium oxysporum* Fo47. *European Journal of Plant Pathology, 104*, 903-910.

Duijff, B. J., Recorbet, G., Bakker, P. A. H. M., Loper, J., & Lemanceau, P. (1999). Microbial antagonism at the root level is involved in the suppression of fusarium wilt by the combination of nonpathogenic *Fusarium oxysporum* Fo47 and *Pseudomonas putida* WCS358. *Phytopathology, 89*, 1073-1079.

Edel, V., Steinberg, C., Gautheron, N., Recorbet, G., & Alabouvette, C. (2001). Genetic diversity of *Fusarium oxysporum* populations isolated from different soils in France. *FEMS Microbiology Ecology, 36*, 61-71.

Edwards, S. G., Young, J. P. W. & Fitter, A.H. (1998). Interactions between *Pseudomonas fluorescens* biocontrol agents and *Glomus mosseae*, an arbuscular mycorrhizal fungus, within the rhizosphere. *FEMS Microbiology Letters, 166*, 297-303.

Elad, Y., & Baker, R. (1985). The role of competition for iron and carbon in suppression of chlamydospore germination of *Fusarium* spp. by *Pseudomonas* spp. *Phytoptahology, 75*, 1053-1059.

Ellis, R. J., Timms-Wilson T. M, & Bailey M. J. (2000). Identification of conserved traits in fluorescent pseudomonads with antifungal activity. *Environmental Microbioliogy, 2*, 274-284.

Emery, T. (1965). Isolation, characterization, and properties of fusarinine, a δ-hydroxamic acid derivative of ornithine. *Biochemistry, 4*, 1410-1417.

Eparvier, A., Lemanceau, P., & Alabouvette, C. (1991). Population dynamics of non-pathogenic *Fusarium* and fluorescent *Pseudomonas* strains in rockwool, a substratum for soilless culture. *FEMS Microbiology Ecology, 86*, 177-184.

Fenton, A. M., Stephens, P. M., Crowley, J., O'Callaghan, M., & O'Gara, F. (1992). Exploitation of gene(s) involved in 2,4-diacetylphloroglucinol biosynthesis to confer a new biocontrol capability to *Pseudomonas* strain. *Applied and Environmental Microbiology, 58*, 3873-3878.

Fravel, D. R. (1988). Role of antibiosis in the biocontrol of plant diseases. *Annual Review of Phytopathology, 26*, 75-91.

Fuchs, J. G., Moënne-Loccoz Y., & Défago G. (1997). Nonpathogenic *Fusarium oxysporum* strain Fo47 induces resistance to Fusarium wilt in tomato. *Plant Disease, 81*, 492-496.

Garbeva, P., van Veen, J. A., & van Elsas, J. D. (2004). Microbial diversity in soil: selection of microbial populations by plant and soil type and implications in soil suppressiveness. *Annual Review of Phytopathology, 42*, 243-270.

Garrett, S. D. (1970). *Pathogenic Root-Infecting Fungi*. Cambridge: Cambridge University Press.

Gasser, R., & Défago, G. (1981). Mise en évidence de la résistance de certaines terres à la pourriture noire des racines du tabac causée par le *Thielaviopsis basicola*. *Botanica Helvetica, 91*, 75-80.

Gerlagh, M. (1968). Introduction of Ophiobulus graminis into new polders and its decline. *Netherlands Journal of Plant Pathology, 74 (Suppl. 2)*, 1-97.

Gill, P. R., & Warren, G. J. (1988). An iron-antagonized fungistatic agent that is not required for iron assimilation from a fluorescent rhizosphere pseudomonad. *Journal of Bacteriology, 170*, 163-170.

Gomes, N. C. M., Heuer, H., Schonfeld, J., Costa, R., MendoncaHagler, L., & Smalla, K. (2001). Bacterial diversity of the rhizosphere of maize (*Zea mays*) grown in tropical soil studied by temperature gradient gel electrophoresis. *Plant and Soil, 232*, 167-180.

Goodacre, R., Vaidyanathan, S., Dunn, W. B., Harrigan, G. G., & Kell, D. B. (2004). Metabolomics by numbers: acquiring and understanding global metabolite data. *Trends in Biotechnology, 22*, 245-252.

Gurusiddajah, S., Weller, D. M., Sarkar, A., & Cook, R. J. (1986). Characterization of an antibiotic produced by a strain of *Pseudomonas fluorescens* inhibitory to *Gaeumannomyces graminis* var. *tritici* and *Pythium* spp. *Antimicrobial Agents and Chemotherapy, 29*, 488-495.

Hamdan, H., Weller, D. M., & Thomashow, L. S. (1991). Relative importance of fluorescent siderophores and other factors in biological control of *Gaeumannomyces graminis* var. *tritici* by *Pseudomonas fluorescens* 2-79 and M4-80R. *Applied and Environmental Microbiology, 57*, 3270-3277.

Haas, D., & Keel, C. (2003). Regulation of antibiotic production in root-colonizing *Pseudomonas* spp. and relevance for biological control of plant disease. *Annual Review of Phytopathology, 41*, 117-153.

Haas, D., & Défago. G. (2005). Biological control of soil-borne pathogens by fluorescent pseudomonads. *Nature Reviews Microbiology, 3*, 307-319.

Hase, C., Hottinger, G., Moënne-Loccoz, Y., & Défago, G. (2000). Survival and cell culturability of biocontrol *Pseudomonas fluorescens* CHA0 in the rhizosphere of cucumber grown in two soils of contrasting fertility status. *Biology and Fertility of Soils, 32*, 217-221.

Heming, B. C. (1986). Microbial-iron interactions in the rhizosphere. An overview. *Journal of Plant Nutrition, 9*, 502-521.

Heuer, H., Krsek, M., Baker, P., Smalla, K., & Wellington, E. M. H. (1997). Analysis of actinomycete communities by specific amplification of genes encoding 16S rRNA and gel-electrophoretic separation in denaturing gradients. *Applied and Environmental Microbiology, 63*, 3233-3241.

Hill, C., Com-Nougué, C., Kramar, A., Moreau, T., O'Quigley, J., Senoussi, R., & Chastang, C. (1990). *Analyse statistique des données de survie*. Paris: Flammarion.

Hojberg, O., Schnider, U., Winteler, H.V., Sorensen, J., & Haas, D. (1999). Oxygen-sensing reporter strain of *Pseudomonas fluorescens* for monitoring the distribution of low-oxygen habitats in soil. *Applied and Environmental Microbiology, 65*, 4085-4093.

Höper, H., Steinberg, C., & Alabouvette C. (1995). Involvement of clay type and pH in the mechanisms of soil suppressiveness to fusarium wilt of flax. *Soil Biology Biochemistry, 27*, 955-967.

Hornby, D. (1998). *Take-all disease of cereals: a regional perspective.* Wallingford: CAB International.

Huang, Z., Bonsall, R. F., Mavrodi, D. V., Weller, D. M., & Thomashow, L. S. (2004). Transformation of with genes of phenazine-1-carboxylic acid improves biocontrol of rhizoctonia root rot and in situ antibiotic production. *FEMS Microbiology Ecology, 19*, 243-251.

Iavicoli, A., Boutet, E., Buchala, A., & Métraux, J. P. (2003). Induced systemic resistance in *Arabidopsis thaliana* in response to root inoculation with *Pseudomonas fluorescens* CHA0. *Molecular Plant-Microbe Interactions, 16*, 851-858.

Johansen, J. E., Binnerupn S. J., Bang Lejbølle, K., Mascher, F., Sørensen, J., & Keel, C. (2002). Impact of biocontrol strain *Pseudomonas fluorescens* CHA0 on rhizosphere bacteria isolated from barley (*Hordeum vulgare* L.) with special reference to *Cytophaga*-like bacteria. *Journal Applied Microbiology, 93*, 1065-1074.

Kandan, A., Ramiah, M., Vasanthi, V. J., Radjacommare, R., Nandakumar, R., Ramanathan, A., & Samiyappan, R. (2005). Use of *Pseudomonas fluorescens*-based formulations for management of tomato spotted wilt virus (TSWV) and enhanced yield in tomato. *Biocontrol Science and Technology, 15*, 553-569.

Keel, C., Voisard, C., Berling, C. H., Kahr, G., & Défago, G. (1989). Iron sufficiency, a prerequisite for suppression of tobacco black root rot by *Pseudomonas fluorescens* strain CHA0 under gnotobiotic conditions. *Phytopathology, 79*, 584-589.

Keel, C., Ucurum, Z., Michaux, M., Adrian, M., & Haas, D. (2002). Deleterious impact of a virulent bacteriophage on survival and biocontrol activity of *Pseudomonas fluorescens* strain CHA0 in natural soil. *Molecular Plant-Microbe Interactactions, 15*, 567-576.

Keel, C., Weller, D. M., Natsch, A., Défago, G., Cook, R. J., & Thomashow, L. S. (1996). Conservation of the 2,4-diacetylphloroglucinol biosynthesis locus among fluorescent *Pseudomonas* strains from diverse geographic locations. *Applied and Environmental Microbiology, 62*, 552-563.

Keel, C., Schnider, U., Maurhofer, M., Voisard, C., Burger, U., Wirthner, Ph., Haas, D., & Défago, G. (1992). Suppression of root diseases by *Pseudomonas fluorescens* CHA0: importance of the bacterial secondary metabolite 2,4-diacetylphloroglucinol. *Molecular Plant Microbe Interactions, 5*, 4-13.

Kerry, B. R., & Crump, D. H. (1998). The dynamics of the decline of the cereal cyst-nematode, *Heterodera avenae*, in four soils under intensive cereal production. Fund. *Applied Nematology, 21*, 617-625.

Khan, S. R., Mavrodi, D. V., Jog, G. J., Suga, H., Thomashow, L. S., & Farrand, S. K. (2005). Activation of the *phz* operon of *Pseudomonas fluorescens* 2-79 requires the LuxR homolog PhzR,

N-(3-OH-hexanoyl)-L-homoserine lactone produced by the LuxI homolg phzI, and a *cis*-acting *phz* box. *Journal of Bacteriology, 187*, 6517-6527.

Kloepper, J. W., Leong, J., Teintze, M., & Schroth M. N. (1980). *Pseudomonas* siderophores: a mechanism explaining disease suppression soils. *Current Microbiology, 4*, 317-320.

Komada, H., & Ezuka, A. (1970). Ecological study of Fusarium diseases of vegetable crops. I. Survival of pathogenic Fusaria in different soil types. *Research Progress Report.of the Tokai-Kinki National Agricultural Station, 6*, S.1-6.

Larkin, R. P., Hopkins, D. L., & Martin F. N. (1993). Effect of successive watermelon plantings on *Fusarium oxysporum* and other microorganisms suppressive and conducive to fusarium wilt of watermelon. *Phytopathology 83*, 1097-1105.

Latour, X., Corberand, T., Laguerre, G., Allard, F., & Lemanceau, P. (1996). The composition of fluorescent pseudomonad population associated with roots is influenced by plant and soil type. *Applied and Environmental Microbioliogy, 62*, 2449-2556.

Laville, J., Voisard, C., Keel, C., Maurhofer, M., Défago, G., & Haas D., (1992). Global control in Pseudomonas *fluorescens* mediating antibiotic synthesis and suppression of black root rot of tobacco. *Proceedings of the National Academy of Science, U.S.A., 89*, 1562-1566.

Laville, J., Blumer, C., Von Schroetter, C., Gaia, V., Défago, G., Keel, C., & Haas, D. (1998). Characterization of the hcnABC gene cluster encoding hydrogen cyanide synthase and anaerobic regulation by ANR in the strictly aerobic biocontrol agent *Pseudomonas fluorescens* CHA0. *Journal of Bacteriology, 180*, 3187-3196.

Leeman, M., Den Ouden, F. M., Van Pelt, J. A., Cornelissen, C., Bakker P. A. H. M., & Schippers B. (1996). Suppression of Fusarium wilt of radish by co-inoculation of fluorescent *Pseudomonas* spp. and root colonizing fungi. *European Journal of Plant Pathology, 102*, 21-31.

Lemanceau, P. (1992). Effets bénéfiques de rhizobactéries sur les plantes : exemple des *Pseudomonas* spp. fluorescents. *Agronomie, 12*, 413-437.

Lemanceau, P., & Alabouvette, C. (1991). Biological control of fusarium diseases by fluorescent *Pseudomonas* and non-pathogenic *Fusarium*. *Crop Protection, 10*, 279-286.

Lemanceau, P., & Alabouvette, C. (1993) Suppression of fusarium wilts by fluorescent pseudomonads: mechanisms and applications. *Biocontrol Science and Technology, 3*, 219-234.

Lemanceau, P., Alabouvette, C., & Meyer, J. M. (1986). Production of fusarinine and iron assimilation by pathogenic and non-pathogenic *Fusarium*. In T. R. Swinburne (Ed.), *Iron, siderophores, and plant diseases* (pp. 251-253). New York: Plenum.

Lemanceau, P., Alabouvette, C., Couteaudier, Y. (1988a). Recherches sur la résistance des sols aux maladies. XIV. Modification du niveau de réceptivité d'un sol résistant et d'un sol sensible aux fusarioses vasculaires en réponse à des apports de fer ou de glucose. *Agronomie, 8*, 155-162.

Lemanceau, P., Samson, R., Alabouvette, C. (1988b). Recherches sur la résistance des sols aux maladies. XV. Comparaison des populations de *Pseudomonas* fluorescents dans un sol résistant et un sol sensible aux fusarioses vasculaires. *Agronomie, 8*, 243-249.

Lemanceau, P., Bakker, P. A. H. M., De Kogel, W. J., Alabouvette, C., & Schippers, B. (1992). Effect of pseudobactin 358 production by *Pseudomonas putida* WCS358 on suppression of fusarium wilt of carnations by nonpathogenic *Fusarium oxysporum* Fo47. *Applied and Environmental Microbiology, 58*, 2978-2982.

Lemanceau, P., Bakker, P. A. H. M., De Kogel, W. J., Alabouvette, C., & Schippers, B. (1993). Antagonistic effect of nonpathogenic *Fusarium oxysporum* Fo47 and pseudobactin 358 upon pathogenic *Fusarium oxysporum* f. sp. *dianthi*. *Applied and Environmental Microbiology, 59*, 74-82.

Leong, J. (1986) Siderophores: their biochemistry and possible role in the biocontrol of plant pathogens. *Annual Review of Phytopathology, 24*, 187-209.

Linderman, R. G., Moore, L. W., Baker, K. F., & Cooksey, D. A. (1983). Strategies for detecting and characterizing systems. *Plant Disease, 67*, 1058-1064.

Lindsay, W. L. (1979). *Chemical Equilibria in Soils*. New York: John Wiley & Sons.

Locatelli, L., Tarnawski, S., Hamelin, J., Rossi, P., Aragno, M., & Fromin, N. (2002). Specific PCR amplification for the genus *Pseudomonas* targeting the 3' half of 16S rDNA and the whole 16S-23S rDNA spacer. *Systematic and Applied Microbiology, 25*, 220-227.

Lockwood, J. L. (1977). Fungistasis in soils. *Biology Review, 52*, 1-43.

Loper, J. E., & Buyer, J. S. (1991). Siderophores in microbial interactions on plant surfaces. *Molecular Plant-Microbe Interactions, 4*, 5-13.

Louvet, J. (1973). Les perspectives de lutte biologique contre les champignons parasites des organes souterrains des plantes. In *Perspectives de lutte biologique contre les champignons parasites des plantes cultivées et des tissus ligneux* (pp. 48-56). Lausanne: Symposium international.

Louvet, J., Rouxel, F., & Alabouvette, C. (1976). Recherches sur la résistance des sols aux maladies. I - Mise en évidence de la nature microbiologique de la résistance d'un sol au développement de la fusariose vasculaire du melon. *Annales de Phytopathologie, 8*, 425-436.

Lucas, P., & Sarniguet, A. (1990). Soil receptivity to take-all: influence of some cultural practices and soil chemical characteristics. *Symbiosis, 9*, 51-57.

Mann, M. (1999). Quantitative proteomics? *Nature Biotechnology, 17*, 954-955.

Maron, P.-A., Ranjard, L., Mougel, C., & Lemanceau, P. Metaproteomics: a new approach for studying functional microbial ecology. *Microbial Ecology*, submitted for publication.

Martin-Laurent, F., Philippot, L., Hallet, S., Chaussod, R., Germon, J.C., Soulas, G., & Catroux, G. (2001). DNA extraction from soils: old bias for new microbial diversity analysis methods. *Applied and Environmental Microbiology, 67*, 2354-2359.

Mascher, F., Hase, C., Moënne-Loccoz, Y., & Défago, G. (2000). The viable-but-nonculturable state induced by abiotic stress in the biocontrol strain *Pseudomonas fluorescens* CHA0 does not promote strain persistence in soil. *Applied and Environmental Microbiology, 66*, 1662-1667.

Mascher, F., Moënne-Loccoz, Y., Schnider-Keel, U., Keel, C., Haas, D., & Défago, G. (2002). Inactivation of regulatory genes *algU* or *gacA* affects the ability of biocontrol *Pseudomonas fluorescens* CHA0 to persist as culturable cells in non-sterile soil. *Applied and Environmental Microbiology, 68*, 2085-2088.

Matsuki, T., Watanabe, K., Fujimoto, J., Miyamoto, Y., Takad, T., Matsumoo, K., Oyaizu, H., & Tanaka, R. (2002). Development of 16S rRNA-gene-targeted group-specific primers for the detection and identification of predominant bacteria in human feces. *Applied and Environmental Microbiology, 68*, 5445-5451.

Matta, A., Gentile, I. A., & Ferraris, L. (1989). Induced resistance to Fusarium wilt diseases. In E. C. Tjamos & C. H. Beckman (Eds), *Vascular wilt diseases of pants. Basic studies and control* (pp. 175-196). Berlin: Springer-Verlag.

Maurhofer, M., Baehler, E., Notz, R., Martinez, V. & Keel, C. (2004). Cross-talk between 2, 4-diacetylphloroglucinol-producing biocontrol pseudomonads on wheat roots. *Applied and Environmental Microbiology, 70*, 1990-1998.

Maurhofer, M., Keel, C., Schnider, U., Voisard, C., Haas, D., & Défago. D., (1992). Influence of enhanced antibiotic production in *Pseudomonas fluorescens* strain CHA0 on its disease suppressive capacity. *Phytopathology, 82*, 190-195.

Maurhofer, M., Hase, C., Meuwly, Ph., Métraux, J. P., & Défago, G. (1994a). Induction of systemic resistance of tobacco to tobacco necrosis virus by the root colonizing *Pseudomonas fluorescens* strain CHA0: influence of the *gacA* gene and of pyoverdin production. *Phytopathology, 89*,139-146.

Maurhofer, M., Keel, C., Haas, D., & Défago, D. (1994b). Pyoluteorin production by *Pseudomonas fluorescens* strain CHA0 is involved in the suppression of *Pythium* damping-off of cress but not of cucumber. *European Journal Plant Pathology, 100*, 221-232.

Maurhofer, M., Keel, C., Haas, D., & Défago D. (1995). Influence of plant species on disease suppression by Pseudomonas *fluorescens* strain CHA0 with enhanced antibiotic production. *Plant Pathology, 44*, 40-50.

Maurhofer, M., Reimmann, C., Schmidli-Sacherer, P., Heeb, S., Haas, D., & Défago, D. (1998). Salicylic acid biosynthetic genes expressed in *Pseudomonas fluorescens* strain P3 improve the induction of systemic resistance in tobacco against tobacco necrosis virus. *Phytopathology, 88*, 78-684.

Mavrodi, D. V., Ksenzenko, V. N., Bonsall, R. F., Cook, R. J., Boronin, A. M., & Thomashow, L. S. (1998). A seven gene locus for synthesis of phenazine-1-carboxylic acid by *Pseudomonas fluorescens* 2-79. *Journal of Bacteriology, 180*, 1894-1903.

Mazurier, S., Lemunier, M., Siblot, S., Mougel, C., & Lemanceau, P. (2004). Distribution and diversity of type III secretion system like genes in fluorescent pseudomonads. *FEMS Microbiology Ecology, 49*, 455-467.

Mazzola M. (2002). Mechanisms of natural soil suppressiveness to soilborne diseases. *Antonie van Leeuwenhoek, 81*, 557-564

Mazzola, M. (2004). Assessment and management of soil microbial community structure for disease suppression. *Annual Review of Phytopathology, 42*, 35-59.

Mazzola, M., Funnell, D. L., & Raaijmakers, J. M. (2004). Wheat cultivar-specific selection of 2, 4-diacetylphloroglucinol-producing fluorescent *Pseudomonas* species from resident soil populations. *Microbiol. Ecology, 48*, 338-348.

Mazzola, M., Cook, R. J., Thomashow, L. S., Weller, D. M. & Pierson III, L. S. (1992). Contribution of phenazine antibiotic production to the ecological competence of fluorescent pseudomonads in soil habitats. *Applied and Environmental Microbiology, 58*, 2616-2624.

McSpadden Gardener, B. B., Schroeder, K. L., Kalloger, S. E., Raaijmakers, J. M., Thomashow, L. S., & Weller, D. M. (2000). Genotypic and phenotypic diversity of *phlD*-containing *Pseudomonas* strains isolated from the rhizosphere of wheat. *Applied and Environmental Microbiology, 66*, 1939-1946.

Meyer, J. M., & Abdallah, M. A. (1978). The fluorescent pigment of *Pseudomonas fluorescens*: biosynthesis, purification and physico-chemical properties. *Journal of General Microbiology, 107*, 319-328.

Mirleau, P., Philippot, L., Corberand, T., & Lemanceau, P. (2001). Involvement of nitrate reductase and pyoverdine in competitiveness of *Pseudomonas fluorescens* strain C7R12 in soil. *Applied and Environmental Microbiology, 67*, 2627-2635.

Mirleau, P., Delorme, S., Philippot, L., Meyer, J. M., & Mazurier S., Lemanceau, P. (2000). Fitness in soil and rhizosphere of *Pseudomonas fluorescens* C7R12 compared with a C7R12 mutant affected in pyoverdine synthesis and uptake. *FEMS Microbiology Ecology, 34*, 35-44.

Moënne-Loccoz, Y., & Défago, G. (2004). Life as a biocontrol pseudomonad. In J. L. Ramos (Ed.), *Pseudomonas* (Volume 1, pp. 457-476). New York: Kluwer Academic/Plenum Publishers.

Morrissey, J. P., Cullinane, M., Abbas, A., Mark, G. L., & O'Gara, F. (2004). Biosynthesis and regulation of anti-fungal metabolites by pseudomonads. In J. L. Ramos (Ed.), *Pseudomonas* (Volume 3, pp. 637-670). New York: Kluwer Academic, Plenum Publishers.

Natsch, A., Keel, C., Hebecker, N., Laasik, E., & Défago, D. (1997). Influence of biocontrol strain *Pseudomonas fluorescens* CHA0 and its antibiotic overproducing derivative on the diversity of resident root colonizing pseudomonads. *FEMS Microbiology Ecology, 23*, 341-52.

Natsch, A., Keel, C., Hebecker, N., Laasik, E., & Défago, G. (1998). Impact of *Pseudomonas fluorescens* strain CHA0 and a derivative with improved biocontrol activity on the culturable resident bacterial community on cucumber roots. *FEMS Microbiology Ecology, 27*, 365-380.

Natsch, A., Keel, C., Troxler, J., Zala, M., VonAbertini, N., & Défago, G. (1996). Importance of preferential flow and soil management in vertical transport of a biocontrol strain of Pseudomonas fluorescens in structured field soil. *Applied and Environmental Microbiology, 62*, 33-40.

Neilands, J. B., & Leong, S. A. (1986). Siderophores in relation to plant growth and disease. *Annual Review of Plant Physiology, 37*, 187-208.

Niemann, S., Keel, C., Pühler, A., & Selbitschka, W. (1997). Biocontrol strain *Pseudomonas fluorescens* CHA0 and its genetically modified derivative with enhanced biocontrol capability exert comparable effects on the structure of a *Sinorhizobium meliloti* population in gnotobiotic systems. *Biology and Fertility of Soils, 25*, 240-244.

Nogales, B., Moore, E. R., Llobet-Brossa, E., Rossello-Mora, R., Amann, R., & Timmis, K.N. (2001). Combined use of 16S Ribosomal DNA and 16S rRNA to study the bacterial community of polychlorinated biphenyl-polluted Soil. *Applied and Environmental Microbiology, 67*, 1874-1884.

Notz, R., Maurhofer, M., Dubach, H., Haas, D., & Défago, G. (2002). Fusaric acid-producing strains of *Fusarium oxysporum* alter 2,4-diacetylphloroglucinol biosynthetic gene expression in *Pseudomonas fluorescens* CHA0 in vitro and in the rhizosphere of wheat. *Applied and Environmental Microbiology, 68*, 2229-2235.

Notz, R., Maurhofer, M., Schnider-Keel, U., Duffy, B., Haas, D., & Défago, G. (2001). Biotic factors affecting expression of the 2,4-diacetylphloroglucinol biosynthesis gene *phlA* in *Pseudomonas fluorescens* biocontrol strain CHA0 in the rhizosphere. *Phytopathology, 91*, 873-881.

Olivain, C., Alabouvette, C., & Steinberg, C. (2004). Production of a mixed inoculum of *Fusarium oxysporum* Fo47 and *Pseudomonas fluorescens* C7 to control Fusarium diseases. *Biocontrol Sciences and Technology, 14*, 227-238.

Park, C. S., Paulitz, T. C., & Baker, R. (1988). Biocontrol of Fusarium wilt of cucumber resulting from the interactions between *Pseudomonas putida* and non-pathogenic isolates of *Fusarium oxysporum. Phytopathology, 78*, 190-194.

Paulitz, T. C., Park, C. S., & Baker, R. (1987). Biological control of Fusarium wilt of cucumber with nonpathogenic isolates of *Fusarium oxysporum. Canadian Journal of Microbiology, 33*, 349-353.

Paulsen, I. T., Press, C. M., Ravel, J., Kobayashi, D. Y., Myers, G. S. A., Mavrodi, D. V., DeBoy, R. T., Seshadri, R., Ren, Q. H., Madupu, R., Dodson, R. J., Durkin, A. S., Brinkac, L. M., Daugherty, S. C., Sullivan, S. A., Rosovitz, M. J., Gwinn, M. L., Zhou, L. W., Schneider, D. J., Cartinhour, S. W., Nelson, W. C., Weidman, J., Watkins, K., Tran, K., Khouri, H., Pierson, E. A., Pierson, L. S., Thomashow, L. S., Loper, J. E. (2005) Complete genome sequence of the plant commensal *Pseudomonas fluorescens* Pf-5. *Nature Biotechnology, 23*, 873-878.

Peng, H. X., Sivasithamparam, K., & Turner, D. W. (1999). Chlamydospore germination and Fusarium wilt of banana plantlets in suppressive and conducive soils are affected by physical and chemical factors. *Soil Biology and Biochemistry, 31*, 1363-1374.

Raaijmakers, J. M., & Weller, D. M. (1998). Natural plant protection by 2,4-diacetylphloroglucinol-producing *Pseudomonas* spp. in take-all decline soils. *Molecular Plant-Microbe Interactions, 11*, 144-152.

Raaijmakers, J. M., & Weller, D. M. (2001). Exploting diversity of 2,4-diacetylpholoroglucinol-producing *Pseudomonas* spp.: characterization of superior root-colonizing *P. fluorescens* strain Q8r1-96. *Applied and Environmental Microbiology, 67*, 2545-2554.

Raaijmakers, J. M., Weller, D. M., & Thomashow, L. S. (1997). Frequency of antibiotic-producing *Pseudomonas* spp. in natural environments. *Applied and Environmental Microbiology, 63*, 881-887.

Raaijmakers, J. M., Bonsall, R. E., & Weller, D. M. (1999). Effect of population density of *Pseudomonas fluorescens* on production of 2,4-diacetylphloroglucinol in the rhizosphere of wheat. *Phytopathology, 89*, 470-475.

Raaijmakers J. M., Viami, M., & de Souza, J. T. (2002). Antibiotic production by bacterial biocontrol agents. *Antonie van Leeuwenhoek, 81*, 537-547.

Raaijmakers, J. M., Leeman, M., Van Oorschot, M. M. P., Van der Sluis, I., Schippers, B., & Bakker, P. A. H. M. (1995). Dose-response relationships in biological control of fusarium wilt of radish by *Pseudomonas* spp. *Phytopathology, 85,* 1075-1081.

Radajewski, S., Ineson, P., Parekh, N. R., & Murrell, J. C. 2000. Stable-isotope probing as a tool in microbial ecology. *Nature, 403,* 646-649.

Ramette, A., Moënne-Loccoz, Y., & Défago, G. (2001). Polymorphism of the polyketide synthase gene *phl*D in biocontrol fluorescent *Pseudomonas* spp. producing 2,4-diacetylphloroglucinol. *Molecular Plant-Microbe Interactions, 14,* 639-652.

Ramette, A., Moënne-Loccoz, Y. & Défago, G. (2003a). Prevalence of fluorescent pseudomonads producing antifungal phloroglucinols and/or hydrogen cyanide in soils naturally suppressive or conducive to tobacco black root rot. *FEMS Microbiology Ecology, 44,* 35-43.

Ramette, A., Moënne-Loccoz, Y., & Défago, G. (2006) Genetic diversity and biocontrol potential of fluorescent pseudomonads producing phloroglucinols and HCN from Swiss soils naturally suppressive or conducive to *Thielaviopsis basicola*-mediated black root rot of tobacco. *FEMS Microbiology Ecology,* in press.

Ramette, A., Frapolli, M., Moënne-Loccoz, Y., & Défago, G. (2003b) Phylogenetic analysis of HCN synthase-encoding *hcnBC* in biocontrol fluorescent pseudomonads and relationship with bacterial biogeography and HCN synthesis. *Molecular Plant-Microbe Interactions, 16,* 525-535.

Ranjard, L, Brothier, E., & Nazaret, S. (2000). Sequencing bands of ribosomal intergenic spacer analysis fingerprints for characterization and microscale distribution of soil bacterium populations responding to mercury spiking. *Applied and Environmental Microbiology, 66,* 5334-5339.

Reimmann, C., Valverde, C., Kay, E. & Haas, D. 2005 Posttranscriptional repression of GacS/GacA-controlled genes by the RNA-binding protein RsmE acting together with RsmA in the biocontrol strain *Pseudomonas fluorescens* CHA0. *Journal of Bacteriology, 187,* 276-285.

Rondon, M. R., August, P. R., Bettermann, A. D., Brady, S. F., Grossman, T. H., Liles, M. R., Loiacono, K. A., Lynch, B. A., MacNeil, I. A., Minor, C., Tiong, C. L., Gilman, M., Osburne, M. S., Clardy, J., Handelsman, J., & Goodman, R. M. 2000. Cloning the soil metagenome: a strategy for accessing the genetic and functional diversity of uncultured microorganisms. *Applied and Environmental Microbiology 66,* 2541-2547.

Rouxel, F., Alabouvette, C., & Louvet J. (1977). Recherches sur la résistance des sols aux maladies. II - Incidence de traitements thermiques sur la résistance microbiologique d'un sol à la fusariose vasculaire du melon. *Annales de Phytopathologie, 9,* 183-189.

Rouxel, F., Alabouvette, C., & Louvet, J. (1979). Recherches sur la résistance des sols aux maladies. IV - Mise en évidence du rôle des *Fusarium* autochtones dans la résistance d'un sol à la fusariose vasculaire du melon. *Annales de Phytopathologie, 11,* 199-207.

Sacherer, P., Défago, G., & Haas, D., (1994). Extracellular protease and phospholipase C are controlled by the global regulatory gene *gacA* in the biocontrol strain *Pseudomonas fluorescens* CHA0. *FEMS Microbiology Letters, 116,* 155-160.

Sanchez, L., Weidmann, S., Brechenmacher, L., Batoux, M., van Tuinen, D., Lemanceau, P., Gianinazzi, S., & Gianinazzi-Pearson, V. (2004). Common gene expression in *Medicago truncatula* roots in response to *Pseudomonas fluorescens* colonisation, mycorrhiza development and nodulation. *New Phytologist, 161,* 855-863.

Scher, F. M., & Baker R. (1980). Mechanism of biological control in a Fusarium suppressive soil. *Phytopathology, 70,* 412-417.

Scher, F. M., & Baker, R. (1982). Effect of a *Pseudomonas putida* and a synthetic iron chelator on induction of soil suppressiveness to Fusarium wilt pathogens. *Phytopathology, 72,* 1567-1573.

Schmidli-Sacherer, P., Keel, C., & Défago, G. (1997). The global regulator GacA of *Pseudomonas* fluorescens CHA0 is required for suppression of root diseases in dicotyledons but not in Gramineae. *Plant Pathology, 46,* 80-90.

Schneider, R. W. (1982). *Suppressive soils and plant disease*. St Paul: Amer. Phytopathol. Soc.

Schnider-Keel, U., Bang Lejbølle, K., Baehler, E., Haas, D., & Keel, C. (2001). The sigma factor AlgU (AlgT) controls exopolysaccharide production and tolerance towards desiccation and osmotic stress in the biocontrol agent *Pseudomonas fluorescens* CHA0. *Applied and Environmental Microbiology, 67*, 5683-5693.

Schnider, U., Keel, C., Blumer, C., Troxler, J., Défago, G., & Haas, D. (1995a). Amplification of the housekeeping sigma factor in *Pseudomonas fluorescens* CHA0 enhances antibiotic production and improves biocontrol abilities. *Journal of Bacteriology, 177*, 5387-5392.

Schnider, U., Keel, C., Voisard, C., Défago, G., & Haas, D. (1995b). Tn*5*-directed cloning of *pqq* genes from *Pseudomonas fluorescens* CHA0: mutational inactivation of the genes results in overproduction of the antibiotic pyoluteorin. *Applied and Environmental Microbiology, 61*, 3856-3864.

Schnider-Keel, U., Seematter, A., Maurhofer, M., Blumer, C., Duffy, B., Gigot-Bonnefoy, C., Reimmann, C., Notz, R., Défago, G., Haas, D., Keel, C. (2000). Autoinduction of 2,4-diacetylphloroglucinol biosynthesis in the biocontrol agent *Pseudomonas fluorescens* CHA0 and repression by the bacterial metabolites salicylate an pyoluteorin. *Journal of Bacteriology, 182*, 1215-1225.

Schouten, A., van den Berg, G., Edel, V., Steinberg, C., Gautheron, N., Alabouvette, C., de Vos, E., Lemanceau, P., & Raaijsmakers, J. (2004). Defence responses of *Fusarium oxysporum* to 2,4-diacetylphloroglucinol, a broad-spectrum antibiotic produced by *Pseudomonas fluorescens*. *Molecular Plant-Microbes Interactions, 17*, 1201-1211.

Sebat, J. L., Colwell, F. S., & Crawford, R. L. (2003). Metagenomic profiling: microarray analysis of an environmental genomic library. *Applied and Environmental Microbiology, 69*, 4927-4934.

Sharifi-Tehrani, A., Zala, M., Natsch, A., Moënne-Loccoz, Y., & Défago G. (1998). Biocontrol of soil-borne fungal plant diseases by 2,4-diacetylphloroglucinol-producing fluorescent pseudomonads with different restriction profiles of amplified 16S rDNA. *European Journal Plant Pathology, 104*, 631-643.

Shipton, P. J. (1975). Take-all decline during cereal monoculture. In. G. W. Bruehl (Ed), *Biology and Control of Soil-Borne Plant Pathogens* (pp. 137-144). St Paul: Am. Phytopathol. Soc.

Siddiqui, I. A., & Shaukat, S. S. (2003). Suppression of root-knot disease by *Pseudomonas fluorescens* CHA0 in tomato: importance of bacterial secondary metabolite, 2,4-diacetylpholoroglucinol. *Soil Biology and Soil Biochemistry,* 35, 1615-1623.

Siddiqui, I. A., & Shaukat, S. S. (2005). Phenylacetic acid-producing *Rhizoctonia solani* represses the biosynthesis of nematicidal compounds in vitro and influences biocontrol of *Meloidogyne incognita* in tomato by *Pseudomonas fluorescens* strain CHA0 and its GM derivatives. *Journal of Applied Microbiology, 98*, 43-55.

Slininger, F. J., Burkhead, K. D., Schisler, D. A., & Bothast, R. J. (2000). Isolation, identification, and accumulation of 2-acetamidophenol in liquid cultures of the wheat take-all biocontrol agent *Pseudomonas fluorescens* 2-79. *Applied Microbiology and Biotechnology, 54*, 376-381.

Small, J., Call, D. R., Brockman, F. J., Straub, T. M., & Chandler, D. P. (2001). Direct detection of 16S rRNA in soil extracts by using oligonucleotide microarrays. *Applied and Environmental Microbiology, 67*, 4708-4716.

Smith, S. N. (1977). Comparison of germination of pathogenic *Fusarium oxysporum* chlamydospores in host rhizosphere soils conducive and suppressive to wilts. *Phytopathology, 67*, 502-510.

Smith, S. N., & Snyder, W. C. (1971). Germination of *Fusarium oxysporum* in soils favourable and unfavourable to wilt establishment. *Phytopathology, 62*, 273-277.

Sneh, B., Pozniak, D., & Salomon, D. (1987). Soil suppressiveness to fusarium wilt of melon, induced by repeated croppings of resistant varieties of melons. *Journal of Phytopathology, 120*, 347-354.

Sneh, B., Dupler, M., Elad, Y., & Baker, R. (1984). Chlamydospore germination of *Fusarium oxysporum* f.sp. *cucumerinum* as affected by fluorescent and lytic bacteria from Fusarium suppressive soil. *Phytopathology. 74*, 1115-1124.

Steinberg, C., Edel-Hermann, V., Guillemaut, C., Pérez-Piqueres, A., Singh, P., & Alabouvette, C. (2004). Impact of organic amendments on soil suppressiveness to diseases. In R. A. Sikora, S. Gowen, R. Hauschild & S. Kiewnick (Eds), *Multitrophic interactions in soil and integrated control* (pp. 259-266). Bonn: IOBC wprs Bulletin.

Stotzky, G., & Martin, R.T. (1963). Soil mineralogy in relation to the spread of Fusarium wilt of banana in Central America. *Plant and Soil, 18*, 317-337.

Stover, R. H. (1962). *Fusarial wilt (Panama disease) of bananas and other Musa species*. Kew: Commonwealth Mycological Institute.

Stutz, E. W. (1985). *Krankheitsunterdrückende Böden in der Westschweiz*. Zurich ETHZ: Ph.D. Thesis.

Stutz, E. W., & Défago, G. (1985). Effect of parent materials derived from different geological strata on suppressiveness of soils to black root rot of tobacco. In C. A Parker, A. D. Rovira, K. J. More, & P. T. W. Wong (Eds.), *Ecology and management of soilborne plant pathogens* (pp 215-217). St.Paul: American Phytopathological Society.

Stutz, E. W., Défago, G., & Kern, H. (1986). Naturally occurring fluorescent pseudomonads involved in suppression of black root rot of tobacco. *Phytopathology, 76*, 181-185.

Stutz, E. W., Kahr, G., & Défago, G. (1989). Clays involved in suppression of tobacco black root rot by a strain of *Pseudomonas fluorescens*. *Soil Biology and Biochemistry, 21*, 361-362.

Stutz, E. W., Défago, G., Hantke, R., & Kern, H. (1985). Effect of parent materials derived from different geological strata on suppressiveness of soils to black root rot of tobacco. In C. A. Parker, A. D. Rovira, K. J. Moore, P. T. W. Wong & J. F. Kollmorgen (Eds.), *Ecology and Management of Soilborne Plant Pathogens* (pp. 215-217). St. Paul, Minnesota: The American Phytopathological Society.

Tamietti, G., Ferraris, L., Matta, A., & Abbattista Gentile, I. (1993). Physiological responses of tomato plants grown in Fusarium suppressive soil. *Journal of Phytopathology, 138*, 66-76.

Taroncher-Oldenburg, G., Griner, E. M., Francis, C. A., & Ward, B. B. (2003). Oligonucleotide microarray for the study of functional gene diversity in the nitrogen cycle in the environment. *Applied and Environmental Microbiology, 69*, 1159-1171.

Thomashow, L. S., & Weller, D. M. (1988). Role of a phenazine antibiotic from *Pseudomonas fluorescens* in the biological control of *Gaeumanomyces graminis* var. *tritici*. *Journal of Bacteriology, 170*, 3499-3508.

Thomashow, L. S., & Weller, D. M. (1990). Application of fluorescent pseudomonads to control root diseases of wheat and some mechanisms of disease suppression. In D. Hornby (Ed.), *Biological Control of Soil-Borne Plant Pathogens* (pp.109-122). Oxon: CAB International.

Thomashow, l. S., Weller, D. M., Bonsall, R. F. and Pierson III, L. S. (1990). Production of the antibiotic phenzine-1-carboxylic acid by fluorescent *Pseudomonas* species in the rhizosphere of wheat. *Applied and Environmental Microbiology 56*, 908-912.

Timms-Wilson, T. M., Ellis, R. J., Renwick, A., Rhodes, D. J., Mavrodi, D. V., Weller, D. M., Thomashow, L. S., & Bailey, M. J. (2000). Chromosomal insertion of phenazin-1-carboxylic acid biosynthetic pathway enhances efficacy of damping-off disease control by *Pseudomonas fluorescens*. *Molecular Plant-Microbe Interactions, 13*, 1293-1300.

Torsvik, T., Goksoyr, J. & Daae, F. L. (1990) High diversity in DNA of soil bacteria. *Applied and Environmental Microbiology 56*, 782-787.

Troxler, J., Berlin, C. H., Moënne-Loccoz, Y., Keel, C., & Défago, G. (1997a). Interactions between the biocontrol agent *Pseudomonas fluorescens* CHA0 and *Thielaviopsis basicola* in tobacco roots observed by immunofluorescence microscopy. *Plant Patholology, 46*, 62-71.

Troxler, J., Zala, M., Natsch, A., Moënne-Loccoz, Y., & Défago, G. (1997b). Autecology of the biocontrol strain *Pseudomonas fluo-rescens* CHA0 in the rhizosphere and inside roots at later stages of plant development. *FEMS Microbiology Ecology, 23*, 119-130.

Troxler, J., Zala, M., Natsch, A., Nievergelt, J., Keel, C., & Défago, G. (1998). Transport of a biocontrol *Pseudomonas fluorescens* through 2.5-m deep outdoor lysimeters and survival in the effluent water. *Soil Biology and Biochemistry, 30*, 621-631.

Validov, S., Mavrodi, O., De La Fuente, L., Boronin, A., Weller, D. M., Thomashow L. S., & Mavrodi, D. (2005). Antagonistic activity among 2,4-diacetylphloroglucinol-producing fluorescent *Pseudomonas* spp. *FEMS Microbiology Letters, 242*, 249-256.

Valverde, C., Heeb, S., Keel, C. & Haas, D. (2003). *RsmY*, a small regulatory RNA, is required in concert with *RsmZ* for *GacA*-dependent expression of biocontrol traits in *Pseudomonas fluorescens* CHA0. *Molecular Microbiology, 50*, 1361-1379.

Valverde, C., Lindell, M., Wagner, G.H. & Haas, D. (2004). A repeated GGA motif is critical for the activity and stability of the riboregulatore RsmY of *Pseudomonas fluorescens*. *Journal of Biological Chemistry, 279*, 25066-25074.

Van Loon, L. C., Bakker, P. A. H. M., & Pieterse, C. M. J. (1998). Systemic resistance induced by rhizosphere bacteria. *Annual Review of Phytopathology, 36*, 453-483.

Vincent, M. N., Harrison, M. N., Brackin, L. A., Kovacevich, J. M., Mukerji, P. A., Weller, D. M., & Pierson, E.A. (1991). Genetic analysis of the antifungal activity of a soilborne *Pseudomonas aureofaciens* strain. *Applied and Environmental Microbiology, 57*, 2928-2934.

Viswanathan, R., & Samiyappan, R. (2002). Induced systemic resistance by fluorescent pseudomonads against red rot disease of sugarcane caused by *Colletotrichum falcatum*. *Crop Protection, 21*, 1-10.

Voisard, C., Keel, C., Haas, D., & Défago, G. (1989). Cyanide production by *Pseudomonas fluorescens* helps suppress black root rot of tobacco under gnotobiotic conditions. *EMBO Journal, 8*, 351-358.

Voget, S., Leggewie, C., Uesbeck, A., Raasch, C., Jaeger, K.E., & Streit, W. R. (2003). Prospecting for novel biocatalysts in a soil metagenome. *Applied and Environmental Microbiology, 69*, 6235-6242.

Von Wintzingerode, F., Göbel U.B., & Stackebrandt, E. (1997). Determination of microbial diversity in environmental samples: pitfalls of PCR-based rRNA analysis. *FEMS Microbial Reviews, 21*, 213-219.

Wang, C. X., Ramette, A., Punjasamarnwong, P., Zala, M., Natsch, A., Moënne-Loccoz, Y., & Défago, G. (2001). Cosmopolitan distribution of *phlD*-containing dicotyledonous crop-associated biocontrol pseudomonads of worldwide origin. *FEMS Microbiology Ecology, 37*, 105-116.

Weller, D. M. (1988). Biological control of soilborne plant pathogens in the rhizosphere with bacteria. *Annual Review of Phytopathology, 26*, 379-407.

Weller, D. M., & Cook, R. J. (1983). Suppression of take-all of wheat by seed treatments with fluorescent pseudomonads. *Phytopathology, 73*, 463-460.

Weller, D. M., Howie, W. J., Cook, R. J. (1988). Relationship between un vitro inhibition of *Gaeumannomyces graminis* var. *tritici* and suppression of take-all of wheat by fluorescent pseudomonads. *Phytopathology, 78*, 1094-1100.

Weller, D. M., Raaijmakers, J. M., Gardener, B. B. M. & Thomashow, L. S. (2002). Microbial populations responsible for specific soil suppressiveness to plant pathogens. *Annual Review of Phytopathology, 40*, 309-348.

Westphal, A., & Becker, J. O. (1999). Biological suppression and natural population decline of *Heterodera schachtii* in a California field. *Phytopathology, 83*, 581-589.

Wilmes, P., & Bond, P. L. (2004). The application of two-dimensional polyacrylamide gel electrophoresis and downstream analyses to a mixed community of prokaryotic microorganisms. *Environmental Microbiology, 6*, 911-920.

Wu, L., Thompson, D. K., Li, G., Hurt, R. A., Tiedje, J. M. and Zhou, J. (2001). Development and evaluation of functional gene arrays for detection of selected genes in the environment, *Applied and Environmental Microbiology, 67*, 5780-5790.

Wüthrich, B., & Défago, G. (1991). Suppression of wheat take-all and black root rot of tobacco by *Pseudomonas fluorescens* strain CHA0: results of field and pot experiments. *Bulletin–0ILB/SROP, 14*, 57-62.

Wüthrich, B. (1991). *Biologische Bekämpfung der Schwarzbeinigkeit des Weizens durch Einsatz von Pseudomonas fluorescens Stamm CHA0 im Feld und Mechanismen der Krankheitsunterdrückung*. Zurich ETHZ: Ph.D. Thesis.

7. AFFILIATIONS

Philippe Lemanceau, *UMR Microbiologie et Géochimie des Sols, INRA/Université de Bourgogne, CMSE, BP 86510, 21065 Dijon cedex, France.*

Monika Maurhofer and **Geneviève Défago**, *Swiss Federal Institute of Technology, Universitaestr. 2, LFW C22, CH-8092 Zürich, Switzerland.*

L.C. VAN LOON AND P.A.H.M. BAKKER

ROOT-ASSOCIATED BACTERIA INDUCING SYSTEMIC RESISTANCE

Abstract. We describe the nature of rhizobacterially-induced systemic resistance (ISR) and distinquish ISR from the induced resistance triggered by limited pathogen infection, SAR. The progress made in the elucidation of induction, signaling and expression of ISR is documented for *Arabidopsis thaliana*, the model plant species and several other plants. The results confirm that, in general, ISR is not dependent on salycilic acid (SA), but indicate that, instead, it has a variable requirement or dependence on jasmonic acid (JA) and ethylene signaling. Evidence emerging from recent gene expression studies have suggested that plants can be in the state of ISR without expressing any changes in gene transcripts of leaves in such induced state. As usual, only upon a challenge inoculation, defense responses get boosted. This phenomenon points to the importance of priming of resistance mechanisms in ISR. Such an enhanced defensive capacity is suggested as the main characteristic of rhizobaterially-induced systemic resistance.

1. THE NATURE OF RHIZOBACTERIALLY-INDUCED SYSTEMIC RESISTANCE

When growing into the soil, plant roots become quickly colonized by a highly diverse microflora of bacteria and fungi that may have both beneficial and deleterious effects on plant growth (Lavelle and Spain, 2001). More than 80% of all land plants enter into a symbiosis with mycorrhizal fungi, which aid in the uptake of water and minerals, notably phosphate, and offer protection against abiotic and biotic stresses (Barea et al., 2002). Legumes engage in a symbiosis with rhizobium bacteria, which induce root nodules in which they fix atmospheric nitrogen (Albrecht et al., 1999). The interactions between rhizobia and their host plants are extensively studied and have yielded a fascinating picture of mutual exchanges of molecular signals between both partners for root nodule formation and nitrogen fixation to become established. The obligate nature of the symbiosis between mycorrhizal fungi and plants has slowed progress into the elucidation of the mechanisms involved in this type of interaction, but mutant and gene expression analyses are indicating a similar intricate interplay between the plant and its symbiont (Gianinazzi-Pearson, 1996; Mulder et al., 2005). Moreover, some plant genes have been identified that control initial stages of both the rhizobial nodulation response and of vesicular-arbuscular mycorrhizal development (Harrison, 1998; Stracke et al., 2002). In both of these interactions, the microorganisms invade plant roots and appear to suppress host defense reactions (Gianinazzi-Pearson et al., 1996). Thus, initially the plant seems to recognize the foreign invader and activates resistance mechanisms to limit spread and tissue colonization, but subsequently yields at least partially to the fungus or bacterium in order to allow an effective symbiosis.

Plant pathogenic fungi and bacteria exploit the host plant unilaterally by circumventing to trigger defense reactions or by evading the effects of activated defenses sufficiently to be able to grow and cause disease. Yet, plants do recognize the

S.S. Gnanamanickam (ed.). Plant-Associated Bacteria, 269–316.

invader and start mounting a defensive response, but this is either too weak or too slow, or both, to limit the pathogen sufficiently to prevent it from colonizing the plant (Thordal-Christensen, 2003). However, when resistance reactions are triggered sufficiently early, the plant can restrict the pathogen to certain tissues, and when highly effective, to only a few cells around the site of attack (Kuc, 1982). If the plant overcomes the infection and damage is limited, an enhanced defensive capacity is acquired. This so-called induced resistance enables the plant to react more quickly and strongly to a subsequent attacker, and is variously denoted as systemic acquired resistance (SAR) or induced systemic resistance (ISR) (Van Loon, 2000). In the last decade it has become clear that mechanistically different types of induced resistance exist that are effective against different attackers (Ton et al., 2002a,c). To avoid confusion, the term SAR has been adopted predominantly for the type of induced resistance that is triggered by limited pathogen infection (Ryals et al., 1996; Sticher et al., 1997). This type of resistance is induced optimally when the plant withstands the first invader by developing a hypersensitive reaction. Although tissue necrosis contributes to the level of induced resistance attained, activation of defense mechanisms resulting in limitation of the primary infection seems sufficient to elicit induced resistance (Hammerschmidt and Kuc, 1995). ISR encompasses a broader range of induced resistance phenomena elicited mainly by non-pathogenic organisms (Van Loon et al., 1998). Generally, induced resistance is systemic, in that not only in the primary infected plant parts, but also in non-infected distant tissues the defensive capacity is increased. However, induced resistance is not always expressed systemically: localized acquired resistance (LAR) occurs when only tissues exposed to the primary invader become more resistant (Ross, 1961). Phenotypically, localized and systemically induced resistance are similar in that they are effective against various types of pathogens, but in LAR a signal appears to be lacking that is necessary to propagate the enhanced defensive capacity throughout the plant. The nature of the mobile signal in SAR is unknown (Vernooij et al., 1994) but, at least in Arabidopsis and tobacco, the expression of induced resistance is dependent on the accumulation of salicylic acid (SA) (Gaffney et al., 1993), and its level is modulated by ethylene and jasmonic acid (JA) (Pieterse et al., 1998; Verberne et al., 2003). Through the enhancement of the general defensive capacity of the plant, induced resistance is a protective mechanism that increases the basal level of resistance against all types of pathogens. It limits the effects of a subsequent infection by reducing disease incidence and/or severity, but it only rarely prevents disease development.

Like symbiotic and pathogenic microorganisms, other fungi and bacteria are attracted to plant roots by their release of exudates and lysates (Whipps, 2001; Persello-Cartieaux, 2003). Because of this nutrient supply, they grow and multiply in the rhizosphere, and many appear to establish mutualistic interactions with plants. Most rhizobacteria remain confined to the root surface but some enter the root interior and behave as endophytes. Several of these rhizobacteria have been found to increase plant growth and are therefore called "plant growth-promoting rhizobacteria" (PGPR)

(Kloepper et al., 1980). The mechanisms of growth promotion by these rhizobacteria are complex and usually taken to comprise increasing the availability and uptake of mineral nutrients and modulating plant hormone status (Glick et al., 1999; Van Loon and Glick, 2004). In addition, growth promotion can result from suppression of pathogenic fungi and bacteria through microbial antagonism, i.e. competition for nutrients, particularly iron, production of antibiotics, and/or secretion of lytic enzymes such as chitinases, glucanases and proteases (Handelsman and Stabb, 1996; Van Loon and Bakker, 2003). Some of these mechanisms may also affect the plant. At least one strain of non-pathogenic rhizobacteria has been shown to possess a type III secretion system (Preston et al., 2001). This secretory system is used by pathogenic bacteria to inject effector proteins into eukaryotic cells to induce disease (Collmer et al., 2000). In the case of e.g. the pathogenic leaf bacterium *Pseudomonas syringae*, various proteins have been characterized that play a role in virulence and suppression of plant defenses in susceptible hosts and that are recognized as foreign, and initiate defense responses in resistant hosts (He et al., 2002). The presence of a similar type of secretion apparatus in certain non-pathogenic *Pseudomonas* isolates with growth-promoting properties suggests that more intricate interactions between rhizobacteria and host plants may exist.

The expression of virulence factors by pathogenic bacteria is regulated by quorum sensing (Whitehead et al., 2001). Also non-pathogenic rhizobacteria can rely on quorum sensing for the production of e.g. factors for efficient root colonization or antibiotics to antagonize competing microorganisms (Somers et al., 2004). Thus, non-pathogenic rhizobacteria share with pathogens certain molecular mechanisms that can aid in competition in the rhizosphere and in the colonization of plant roots. Hence, it may not be too surprising that, like pathogens, selected strains of non-pathogenic, root-colonizing bacteria can interact with plant roots in ways that result in induced systemic resistance against pathogens. Induction of systemic resistance by rhizobacteria was first demonstrated independently by Van Peer et al. (1991) against fusarium wilt in carnation, and by Wei et al. (1991) against anthracnose in cucumber. Since then, it has been established in several plant species against different types of pathogens by various research groups using different rhizobacterial strains (e.g. Van Loon et al., 1998). It is not entirely clear in how far the plant reacts to inducing rhizobacteria in a way that resembles a defense response against a pathogen, and it cannot be excluded that some rhizobacteria that are considered non-pathogenic on the host species under study, are actually minor pathogens on other plant species (Hynes et al., 1994). Yet, root colonization by resistance-inducing bacteria does not normally cause any visible symptoms and often promotes plant growth. Thus, these resistance-inducing bacteria seem to be of substantial benefit to the plant and help it to defend itself against not only soil-borne, but also foliar pathogens by boosting its capacity to resist a range of pathogenic attackers.

Because rhizobacteria that trigger ISR can at the same time be antagonistic to the pathogen, disease suppression seen upon their application may result from more than one mechanism. Concluding that ISR is involved rather than a direct effect of the bacterium on the pathogen requires rigorous demonstration that disease suppression is plant-mediated. Many studies in which induced resistance is considered as the mechanism responsible for disease reduction, have not specifically addressed this point, leaving doubts that other mechanisms may (also) be involved. In the presence of inducing rhizobacteria that may also have antagonistic properties, local protection as a result of induced resistance is difficult to verify experimentally. Mutants of the pathogen that are insensitive to the antagonism may aid in resolving this issue. Because rhizobacteria are present on the roots, systemic protection against root pathogens must be demonstrated by applying the inducing bacteria to one part of the root system and the challenging pathogen to another part, for instance by making use of split-root systems. Testing for protection against foliar pathogens is easier, because those are naturally separated from the rhizobacteria. However, rhizobacteria applied to seeds, or to soil into which seeds are sown or seedlings are transplanted, can move into the interior of aerial plant tissues and to some extent maintain themselves on the exterior of aerial surfaces (Kluepfel, 1993; Lamb et al., 1996). Thus, in order to prove that resistance is induced and that it is truly systemic, it must be shown that inducing rhizobacteria are absent from the site of challenge with the pathogen and no contact between the two is established.

Direct antagonistic effects of metabolites of the inducing bacterium on the pathogen should also be excluded. This pertains particularly to the production of antibiotics, and makes it very difficult to establish ISR when the bacterium produces an antimicrobial factor that acts directly on the pathogen. Even when the inducing organism itself is shown not to be present at the site of challenge with the pathogen, the toxic factor may have been produced and transported through the plant, contacting the pathogen directly. This does not mean that induced resistance could not be involved, or even that the toxic factor itself did not (also) induce resistance. Some antibiotics produced by rhizobacteria, such as 2,4-diacetylphloroglucinol (DAPG) or phenazine-1-carboxylic acid (PCA), in high concentrations have some toxicity to plants (Maurhofer et al., 1995), and could induce systemic resistance in the same way as a pathogen causing localized necrosis. Under such circumstances, induced resistance remains difficult to prove, and best evidence is obtained when the rhizobacterium does not antagonize the pathogen in in vitro culture. In Arabidopsis, mutants that cannot express ISR can be used to differentiate between antagonism and induced resistance (Van Loon and Bakker, 2005). If such mutants are not available, doubts can remain because conditions in the rhizosphere are different from those on artificial media, and an antibiotic produced exclusively in planta may go undetected. Determining the presence of an antibiotic in situ and comparing levels to effective doses in vitro can indicate whether microbial antagonism could explain disease reduction.

2. INDUCED SYSTEMIC RESISTANCE IN DIFFERENT PLANT SPECIES

2.1 Arabidopsis

Progress in the elucidation of the induction, signalling and expression of rhizobacterially-induced systemic resistance has been greatly aided by the adoption of *Arabidopsis thaliana* as a model plant species. ISR in Arabidopsis was first established against the vascular wilt fungus *Fusarium oxysporum* f.sp. *raphani* (For) upon application of the rhizobacterial strain *Pseudomonas fluorescens* WCS417r (Pieterse et al., 1996), using a bioassay involving rock wool wetted with nutrient solution (Leeman et al., 1995a). Two- to 3-weeks-old seedlings were placed horizontally on rock wool cubes with the distal part of the roots on cubes contained in a plastic bag, adjacent to another bag with cubes supporting the proximal part of the root system. The roots were laid down through an incision in the bags. The distal part of the roots was then treated with a bacterial suspension in talcum emulsion. Three days later the proximal part of the root system was inoculated with the fungus in a peat mixture. Disease started developing after 14 days and leaf yellowing and wilting symptoms were scored routinely for over 3 weeks after inoculation. After challenge with For, symptom development was delayed and disease severity reduced. Bacterial colonization of the root remained confined to the distal part and no fungus was recoverable from this part, demonstrating that the bacteria remained spatially separated from the pathogen for the duration of the experiments (Pieterse et al., 1996).

A simpler and more convenient bioassay was developed by using the foliar pathogen *Pseudomonas syringae* pv. *tomato* strain DC3000 (Pst) for challenge. Two-weeks-old Arabidopsis seedlings were transplanted into autoclaved potting soil into which the rhizobacterial strain was mixed. Three weeks later the plants were challenged by dipping the leaves in a suspension of the pathogen. Typical symptoms of bacterial speck disease, spreading chlorosis progressing into necrosis, developed within a few days. ISR against Pst was expressed as a reduction in the percentage of leaves with symptoms. In addition, growth of the pathogenic bacterium in the infected leaves was strongly inhibited. Colonization by the rhizobacteria remained confined to the roots and no contact between the inducing strain and the pathogen was demonstrable (Pieterse et al., 1996; Van Wees et al., 1997).

Of three rhizobacterial strains tested, *Pseudomonas putida* WCS358r, *Pseudomonas fluorescens* WCS374r and *P. fluorescens* WCS417r, WCS358r and WCS417r triggered ISR to similar extents, whereas WCS374r was ineffective (Van Wees et al., 1997). The level of resistance attained by application of either WCS417r or WCS358r was mostly similar to that of SAR induced by virulent or avirulent Pst, or by the chemical inducers SA or 2,6-dichloroisonicotinic acid (INA) (Pieterse et al., 1996; Van Wees et al., 1997). Development of SAR in response to limited infection by a pathogen is dependent on endogenous signalling through SA and associated with local and systemic expression of defense genes encoding pathogenesis-related proteins (PRs). Several of these PRs have potential antimicrobial activities and are taken to contribute to the enhanced resistance of induced plants (Van Loon, 1997). In Arabidopsis, development of SAR in response to Pst or treatment with SA or INA is associated with accumulation of PR-1, -2, and –5 mRNAs (Uknes et al., 1992).

However, in plants with ISR triggered by WCS417r or WCS358r no *PR*-gene expression or accumulation of PR-proteins was detectable (Pieterse et al., 1996). Therefore, the induction of ISR by WCS417r or WCS358r in Arabidopsis was not coupled to the activation of *PR* genes.

As shown by Zhang et al. (1998), a biocontrol agent-fortified compost mix suppressive to several diseases caused by soil-borne pathogens, induced resistance against the foliar pathogen *Pseudomonas syringae* pv. *maculicola* (Psm), as evidenced by a reduction in both the severity of bacterial speck symptoms en population densities of the pathogen in the leaves. Autoclaving destroyed the effect of the compost mix, and the activity apparently resided in a water extract of the compost. Transgenic Arabidopsis plants containing a fusion protein construct consisting of the *PR-2* (ß-1,3-glucanase) promoter and the ß-glucuronidase (*GUS*) open reading frame as a reporter did express GUS activity after spraying with the compost water extract. Although the disease suppression is reminiscent of ISR, the mechanism involved seems different from that induced by *P. fluorescens* WCS417r or *P. putida* WCS358r, and a minor toxic action of the water extract leading to activation of *PR*-gene expression cannot be excluded.

To investigate whether the induction of ISR by WCS417r or WCS358r was dependent on SA accumulation in the plant, tests were conducted using transgenic NahG plants. These plants harbour the *NahG* gene from *P. putida*, which encodes salicylate hydroxylase (Delaney et al., 1994). As a result, SA applied to, or synthesized by, the plant is converted into the inactive compound catechol. Since SA is required for the expression of both PRs and SAR, NahG plants no longer accumulate PRs or become induced upon a primary infection. In NahG plants, WCS417r and WCS358r were as effective in inducing ISR against Pst as in untransformed plants, demonstrating that these rhizobacteria activate a signalling pathway different from the one that controls SAR (Pieterse et al., 1996), even though WCS417r is capable of producing SA in vitro under iron-limiting conditions (Leeman et al., 1996). Strikingly, *P. fluorescens* WCS374r, a strain capable of producing a large amount of SA in vitro (Leeman et al., 1996; Mercado-Blanco et al., 2001), did not induce resistance in Arabidopsis, whereas WCS358r, incapable of producing SA, did (Van Wees et al., 1997). The latter result provided direct proof that bacterially-produced SA did not play a role in triggering ISR in Arabidopsis. Moreover, the finding that WCS358r was equally capable of inducing systemic resistance in untransformed and in NahG Arabidopsis demonstrated that also plant-derived SA was not involved. Similar to WCS417r and WCS358r, *P. fluorescens* CHA0r, a PGPR strain with multiple mechanisms of disease suppression, and *Pseudomonas aeruginosa* strain 7NSK2, both capable of producing SA in vitro under iron-limited conditions, induced resistance in NahG *Arabidopsis* (Ran et al., 2005b). It must be concluded that SA does not play a role in the ISR triggered by these strains.

Arabidopsis mutant *npr1* was identified as a non-expresser of *PR* genes, and unable to exhibit SAR (Cao et al., 1994). Because rhizobacterially-mediated ISR was found to be independent of SA and not associated with PRs, it was expected that ISR would still be expressed in this mutant. However, *npr1* did not express WCS417r-mediated ISR (Pieterse et al., 1998). This finding implied that NPR1 functions beyond the expression of *PR* genes and is required for both pathogen-dependent and rhizobacterially-mediated induced systemic resistance. The *npr1* gene is allelic to *nim1* (no immunity) (Delaney et al., 1995) and *sai* (SA-insensitive) (Shah et al., 1997) and encodes an ankyrin-repeat family protein structurally resembling the inhibitor IF-κB, which plays a role in animal innate immunity (Cao et al., 1997). Under the influence of SA, a redox change causes oligomers of NPR1 in the cytoplasm to be reduced to monomers. The monomers are transported into the nucleus, where they interact with specific TGA transcription factors to activate the expression of *PR* genes (Mou et al., 2003). The non-responsiveness of the Arabidopsis *npr1* mutant to rhizobacteria demonstrated that SAR and ISR converge at the last part of the signalling pathway. Yet, concomitant with SAR PRs are induced, whereas activation of their genes is not part of the pathway leading to ISR in Arabidopsis. Apparently, NPR1 regulates defense responses in ISR differently from SAR.

Because SA is essential for the expression of SAR but ISR does not involve SA, the signalling pathway of ISR before NPR1 must be different from that of SAR. Ethylene is an important signalling compound in plant defense responses (Van Loon et al., 2005). It enhances the sensitivity of Arabidopsis to SA (Lawton et al., 1994) and activates several defense-related genes, including specific *PRs* (Brederode et al., 1991). Moreover, during a virus-induced hypersensitive reaction, transgenic ethylene-insensitive tobacco plants were fully capable of elevating SA levels and expressing induced resistance locally but, when used as rootstock, failed to transmit the mobile signal to a wild-type scion. As a result, no systemically induced resistance was evident (Verberne et al., 2003). To investigate the role of ethylene in the expression of ISR, the ethylene-insensitive Arabidopsis mutant *etr1,* as well as the ethylene signalling mutants *ein2* – *ein7* and *axr1-12*, and the ethylene-overproducing mutant *eto1-1* were tested (Pieterse et al., 1998; Knoester et al., 1999). WCS417r was not capable of inducing resistance against Pst in these mutants, whereas pathogen-induced SAR was essentially unimpaired. Because WCS417r colonized the roots of the mutants to the same extent as those of wild-type seedlings, these observations implicate ethylene perception as a specific and essential step in the signal-transduction pathway leading to ISR. Indeed, application of the ethylene precursor, 1-aminocyclopropane-1-carboxylic acid (ACC), to wild-type seedlings induced resistance against Pst to the same level as the rhizobacterial strains, not only in untransformed but also in SA non-accumulating NahG Arabidopsis (Pieterse et al., 1998). However, no significant increase in ethylene production was measurable in plants treated with resistance-inducing WCS417r bacteria (Pieterse et al., 2000).

The Arabidopsis mutant *eir1* is insensitive to ethylene in the roots, but not in the shoots (Roman et al., 1995). In accordance with the requirement for ethylene sensitivity, in bioassays the *eir1* mutant did not express ISR upon application of WCS417r to the roots, while it did exhibit ISR when the inducing bacteria were infiltrated into the leaves (Knoester et al., 1999). This result demonstrated that for the induction of ISR in Arabidopsis by WCS417r, ethylene responsiveness is required at the site of application of the inducing rhizobacteria.

In a screening for natural variation within Arabidopsis to express rhizobacteria-mediated ISR, accessions RLD and Ws-0 were found to not develop ISR after root colonization by WCS417r, in contrast to eight other accessions tested. All ten accessions were capable of expressing pathogen-induced SAR, indicating that RLD and Ws-0 were impaired specifically in ISR expression. When RLD and Ws-0 were crossed, no complementation occurred, indicating that in both accessions the same trait was involved (Ton et al., 1999). The non-responsive phenotype was associated with an enhanced susceptibility to infection by Pst, suggesting that induced resistance represents an enhancement of an extant defensive capacity. Both traits – inability to express ISR and enhanced disease susceptibility – were inherited monogenically and genetically linked. The corresponding locus was designated *Isr1* and appears to encode a so far unidentified component of the ethylene response that is required both for basal resistance against Pst and for the expression of ISR upon root bacterization with WCS417r (Ton et al., 2001). All these observations are suggestive of a triggering of ethylene production or modulated perception by the plant in the presence of the inducing bacteria.

The enhanced disease susceptibility phenotype of RLD and Ws-0 resembles that of Arabidopsis *eds* mutants, that have been described to possess an enhanced susceptibility to Psm or the powdery mildew fungus *Erysiphe orontii* (Glazebrook et al., 1996; Volko et al., 1998). Out of 11 of these *eds* mutants tested, three proved non-responsive to expression of ISR upon root colonization by WCS417r, while pathogen-induced SAR was unimpaired. Of the three, *eds4-1* showed reduced sensitivity to ethylene, whereas *eds8-1* was non-responsive to methyl jasmonate (MeJA) (Ton et al., 2002b). The Arabidopsis JA response mutant *jar1* likewise exhibited wild-type levels of pathogen-induced SAR but failed to express rhizobacterially-mediated ISR (Pieterse et al., 1998). JA levels were not increased in plants treated with resistance-inducing WCS417r bacteria (Pieterse et al., 2000). Yet, on wild-type plants, application of MeJA induced resistance, as did ACC. However, MeJA did not induce resistance in the *etr1* mutant, whereas ACC did induce resistance in the *jar1* mutant, indicating that perception of JA is dependent on components of the ethylene response. Neither JA, nor ACC induced resistance in the *npr1* mutant, indicating an action prior to NPR1 (Pieterse et al., 1998). How perception of JA and ethylene are coupled to ISR is still unclear.

The development of ISR against the downy mildew oomycete *Peronospora parasitica* in Arabidopsis in response to root inoculation with *P. aeruginosa* strain CHA0 has likewise been shown to require JAR1, EIR1 and NPR1, and not SA.

However, mutants *etr1-1* and *ein2-1* still expressed ISR against this pathogen (Iavicoli et al., 2003), suggesting that the requirements for ISR against *P. parasitica* overlap only partly with those defined for ISR against Pst, as induced by WCS417r. On the other hand, ISR activated by *Bacillus subtilis* GB03 was found to be independent of SA and dependent on ethylene, yet did apparently not require JA (Ryu et al., 2004b). ISR elicited in Arabidopsis against Pst or Psm by four other rhizobacterial strains was reported to be variably dependent on ethylene and JA (Ryu et al., 2003a). These results confirm that, in general, ISR is not dependent on SA, but indicate that, instead, ISR has a variable requirement for JA and ethylene signalling (Table 1). The latter does not need to be problematic, because hormone sensitivity is still poorly understood and may vary depending on experimental conditions. In a given situation, either sensitivity to JA or to ethylene, or both, might be limiting.

The lack of expression of WCS417r-elicited ISR in accessions RLD and Ws-0, as well as in the *jar1*, ethylene-response and *npr1* mutants, rules out that rhizobacterially-produced antibiotics might have been responsible for, or contributed to, the induced resistance, in agreement with findings that in vitro no antagonistic effects between the rhizobacteria WCS417r and WCS358r and the pathogen Pst were observed (Van Wees et al., 1997). By requiring JA and ethylene signalling, but not SA, the signalling pathway triggered by these resistance-inducing rhizobacteria is different from SAR (Pieterse et al., 1998) (Fig. 1).

Whereas both SAR and ISR in Arabidopsis are effective against different types of pathogens, their effects vary both qualitatively and quantitatively. Activation of ISR provides a significant level of protection against the leaf spotting fungus *Alternaria alternata* and the grey mould fungus *Botrytis cinerea*, whereas SAR is ineffective against these pathogens. Conversely, activation of SAR results in a high level of protection against turnip crinckle virus, whereas ISR reduces neither virus multiplication, nor symptom severity. Compared to SAR, protection by ISR proved less effective against the downy mildew oomycete *Peronospora parasitica*, but almost equally effective against Pst, and at least as effective against For and *Xanthomonas campestris* pv. *armoraciae* (Xca) (Ton et al., 2002a). These relative effectiveness of SAR and ISR correlate with an enhanced susceptibility of Arabidopsis mutants that are impaired in the accumulation of SA, and sensitivity to JA or ethylene, respectively (Thomma et al., 1998, 2001; Oliver and Ipcho, 2004). Thus, SAR is effective against pathogens that in non-induced plants are resisted through SA-dependent basal resistance responses, whereas ISR is effective against pathogens that in non-induced plants are resisted through JA/ethylene-dependent basal resistance mechanisms, supporting the notion that induced resistance represents an enhancement of extant defensive capacity (Van Loon, 1997). In the assays used, treatment of Arabidopsis seedlings with inducing rhizobacteria occurred for a few days to a few weeks before challenge inoculation, and spatial separation between the rhizobacteria and the challenging pathogen was verified.

Table 1. Dependence of rhizobacteria-induced systemic resistance in Arabidopsis on signalling components [a].

Bacterial strain	*Pathogen*	*SA*	*JA*	*Ethy-lene*	*NPR1*
Bacillus amyloliquefaciens IN937a	*Erwinia carotovora* subsp. *carotovora*	-	-	-	-
Bacillus pumilus SE34	*Pseudomonas syringae* pv. *maculicola*	-	+	+	+
	Pseudomonas syringae pv. *tomato*	+	-	-	+
Bacillus pumilus T4	*Pseudomonas syringae* pv. *maculicola*	-	+	-	-
	Pseudomonas syringae pv. *tomato*	-	-	-	-
Bacillus subtilis GB03	*Erwinia carotovora* subsp. *carotovora*	-	-	+	-
Pseudomonas fluorescens CHA0	*Peronospora parasitica*	-	+	-/+	+
Pseudomonas fluorescens WCS417	*Pseudomonas syringae* pv. *tomato*	-	+	+	+
Pseudomonas fluorescens 89B-61	*Pseudomonas syringae* pv. *maculicola*	-	-	+	+
	Pseudomonas syringae pv. *tomato*	-	-	+	+
Serratia marcescens 90-166	Cucumber mosaic virus	-	+	n.d.	-
	Pseudomonas syringae pv. *maculicola*	-	+	+	+
	Pseudomonas syringae pv. *tomato*	-	-	+	+

[a] +: required: when impaired ISR is abolished; -: not required: when impaired, ISR is not abolished; -/+: variably required as deduced from different mutants; n.d.: not determined. Data compiled from Iavicoli et al., 2003; Pieterse et al., 1998; Ryu et al., 2003a, 2004a,b.

Whereas *P. fluorescens* WCS417r-triggered ISR was not effective against a virus, *Bacillus amyloliquefaciens* IN937a, *B. pumilus* SE34, *P. fluorescens* 89B-61 and *S. marcescens* 90-166 substantially reduced symptom severity upon infection with cucumber mosaic virus (CMV). In plants treated with strains SE34, 89B-61 and 90-166 CMV accumulation in systemically infected leaves was significantly less than in non-bacterized control plants. Reductions as a result of treatments with strains SE34 or 90-166 were maintained in NahG plants. A SA^- mutant of the SA-producing strain 90-166 still reduced CMV symptom severity and virus accumulation in NahG plants, ruling out induction of resistance through a SA-dependent mechanism. A siderophore $(sid)^-$ mutant of strain 90-166 was likewise unimpaired in its inducing action. Resistance induction was also maintained in *npr1* mutant plants. In contrast, no reduced CMV accumulation occurred in strain 90-166-treated triple-mutant *fad3,7,8* plants, that are unable to accumulate JA. These data indicated that strain 90-166 elicited a SA-independent, JA-dependent ISR against CMV that was independent of NPR1 (Ryu et al., 2004a).

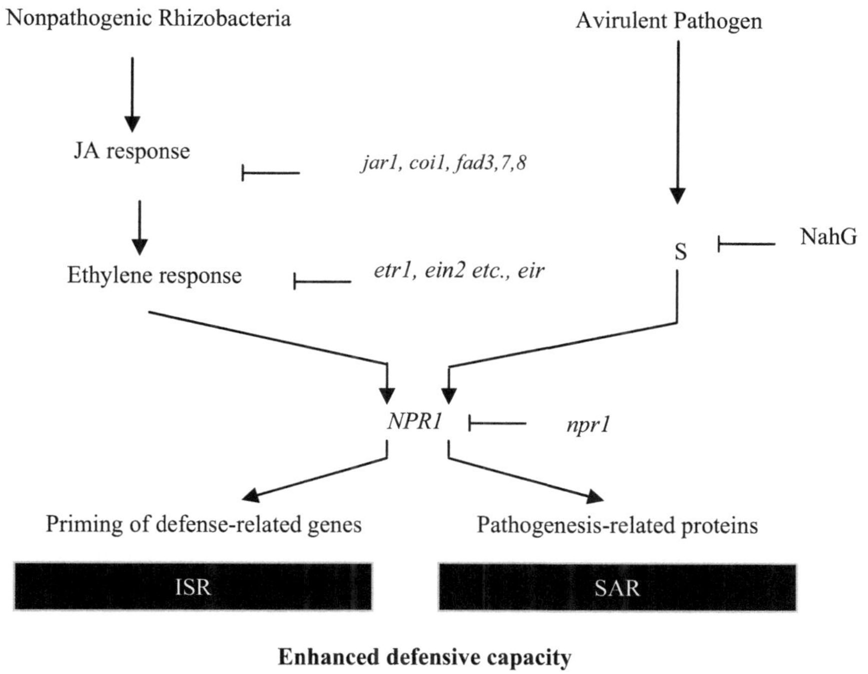

Fig. 1. Signal-transduction pathways leading to pathogen-induced systemic acquired resistance (SAR) and rhizobacterially induced systemic resistance (ISR) (Adapted from Pieterse et al., 1998).

Whereas SAR is associated with accumulation of PRs throughout the plant, no major changes in gene expression occurred either locally in the roots or systemically in the leaves upon induction of ISR by WCS417r (Van Wees et al., 1999). As SA is not involved in WCS417r-mediated ISR, it was not surprising to find that, contrary to SAR, SA-responsive PRs, such as PR-1, PR-2 and PR-5, were not induced. Also the ethylene-inducible hevein gene *Hel*, the ethylene- and JA-responsive genes *ChiB*, encoding a basic chitinase, and *Pdf1.2* (defensin), and the JA-inducible genes *Vsp* (vegetative storage protein), *Lox1*, *Lox2* (lipoxygenase 1 and 2), *Pal1* (phenylalanine ammonia-lyase 1) and *Pin2* (proteinase inhibitor 2) were not expressed, in spite of the involvement of both hormones in ISR. However, pathogen challenge of ISR-expressing plants led to an enhanced level of *Vsp* transcript accumulation.

In contrast, planting Arabidopsis seedlings in soil containing *B. amyloliquefaciens* strain EXTN-1 triggered activation of both *PR-1* and *Pdf1.2* genes. The activation of *PR-1* was abolished in transgenic NahG, and reduced in mutant *npr1* plants, while induction of *Pdf1.2* mRNA was absent in the *jar1*, and reduced in the *etr1* mutants. These observations implied activation of both SA- and JA-dependent signalling pathways, resembling the result of infection by an avirulent pathogen. However, expression of systemically induced resistance was not verified in this study (Ahn et al., 2002).

Recently, transcriptome analyses by cDNA microarrays, RNA differential display, or subtractive hybridization of cDNA libraries have confirmed the notion that rhizobacteria influence plant gene expression to only limited extents. Analysis of the expression of over 8000 genes of Arabidopsis plants with ISR elicited by WCS417r revealed changes in the expression of 102 genes in the roots on which the bacteria were present. In contrast, systemically in the leaves, none of the genes that were expressed to sufficiently high levels to be reliably detected showed a consistent change, in spite of the fact that, when challenge inoculated, these leaves showed a significant ISR response (Verhagen et al., 2004). Clearly, the roots reacted locally to colonization by the bacteria. Within the first week transient changes were observed in the expression of hundreds of genes, but these were not associated with the persistent state of ISR. Of the 102 genes whose expression was changed over a longer period, 39 were up-regulated and 63 were down-regulated. In view of the ethylene dependency of ISR, an increase in expression of a putative ACC oxidase and down-regulation of ethylene response factor 1 (*Erf1*) and ethylene-responsive element binding factors 1 (*Erebp1*) and 2 (*Erebp2*) are particularly interesting.

Upon challenge inoculation of Arabidopsis plants with Pst, SAR-induced plants showed an augmented expression of SA-dependent *PR-1* mRNA, whereas plants with ISR accumulated mRNA of the JA-inducible gene *Vsp* to higher levels than non-induced plants (Van Wees et al., 1999). This “priming” effect indicated that induced plants activate defense-related gene expression earlier and to a greater extent than non-induced plants (Conrath et al., 2002). Indeed, cDNA microarray analysis of WCS417r-induced plants revealed 81 genes showing an augmented expression pattern in ISR-expressing leaves after challenge inoculation with Pst (Verhagen et al., 2004). Of these, 51 genes were expressed at at least 1.5 times higher levels, including *Vsp*, the JA- and ethylene-responsive gene *Pdf1.2*, a thaumatin-like gene, a chitinase gene, and a gene encoding EREBP2. The other 30 genes showed a Pst-induced change in WCS417r-treated plants only, and appear to be ISR-specific. These included genes that are presumably involved in regulating gene transcription and signal transduction. The majority of the genes were predicted to be regulated by JA or ethylene. Thus, the requirement for JA and ethylene sensitivity in ISR seems to be related to the priming action of defense-related gene expression after challenge inoculation of induced plants.

An independent analysis of Arabidopsis plants with induced resistance against Pst as a result of root colonization by *Pseudomonas thivervalensis* led to contrasting results (Cartieaux et al., 2003). cDNA microarray analysis of approximately 14300 Arabidopsis genes revealed that the transcript levels in colonized roots were hardly changed relative to axenic control plants, and none were elevated. In contrast, in shoots the levels of 63 transcripts were modified, including 42 genes that were upregulated. Except for a putative chitinase, no indication of increased *PR* gene expression was evident. An intermediate situation seems to apply to Arabidopsis colonized endophytically by *P. fluorescens* FPT9601-T5, which caused transient suppression of disease symptoms of Pst (Wang et al., 2005). FPT9601-T5 suppressed plant growth in early stages of colonization, but promoted it in later stages. Of 22800 Arabidopsis genes, 95 and 105 genes were up- and down-regulated, respectively, in leaves of bacterized plants as compared to control plants. These up-regulated genes are involved in metabolism, signal transduction and stress responses, whereas genes involved in ethylene signalling were suppressed, including EREBP 1 and 2. Thus, both Verhagen et al. (2004) and Wang et al. (2005) observed that the onset of ISR is associated with a reduction in ethylene signaling, indicating that ISR triggered by FPT9601-T5 may partly shares signalling pathways with ISR elicited by strain WCS417r. Induction of resistance against *Erwinia carotovora* in Arabidopsis by *Paenibacillus polymyxa* was associated with increased tolerance to drought and changes in the abundance of mRNAs encoding drought stress- and biotic stress-responsive proteins, consistent with a mild pathogenic effect of *P. polymyxa* on Arabidopsis (Timmusk and Wagner, 1999). These alterations do not seem typical of ISR.

There is comparatively little information on the bacterial determinants that trigger ISR (Bakker et al., 2003). In searches for the bacterial factor(s) that trigger ISR in Arabidopsis, dead bacterial cells of *P. fluorescens* WCS417r were effective in inducing resistance against Pst, be it to a significantly lower level than live cells (Van Wees et al., 1997). Dead cells of a bacterial mutant that lacked the O-antigenic side-chain (OA^-) of the outer membrane lipopolysaccharide (LPS) were ineffective. These results suggested that the OA of the LPS of WCS417r contributes to elicitation of ISR, but may not be sufficient for full induction. More information is available on the determinants of ISR-eliciting *P. putida* WCS358r (Meziane et al., 2005). As shown by mutant analysis and application of isolated components, at least three factors can be active in eliciting ISR: LPS, the pseudobactin siderophore, and flagella. Due to the redundancy of these inducing factors, their relative importance in the induction of live bacteria was not assessed. *P. fluorescens* WCS374 does not induce systemic resistance in Arabidopsis when grown at 28°C. However, when grown at 33°C or 36°C, it triggered ISR in both wild-type and NahG plants, but not in ethylene-insensitive *ein2* or in non-PR-expressing *npr1* mutant plants, irrespective of the growth temperature of the bacteria (Ran et al., 2005b). These results demonstrate that, when grown at elevated temperatures, WCS374 expresses a, so far unidentified, determinant that triggers the ISR pathway in Arabidopsis.

Other rhizobacteria that trigger ISR in Arabidopsis have been shown to act through other determinants. Both *P. fluorescens* CHA0 and *P. fluorescens* Q2-87 elicit ISR through the production of the antibiotic DAPG (Iavicoli et al., 2003; Weller et al., 2004). Siderophores and antibiotics are important for microbial antagonism in the rhizosphere. These factors can thus function in disease suppression through restriction of growth or activity of soilborne pathogens, as well as in the induction of systemic resistance against both soilborne and foliar pathogens. Recently, it was found that certain bacilli, i.e. *B. amyloliquefaciens* IN937a and *B. subtilis* GB03, trigger ISR through a volatile compound, 2,3-butanediol, which is also active in promoting plant growth (Ryu et al., 2003b, 2004b). Probably, related compounds are also active. These results demonstrate that several different components present on, or released by, rhizobacteria can elicit ISR in Arabidopsis.

2.2 Banana

The causal agent of Panama disease, *Fusarium oxysporum* f.sp. *cubense*, is a major problem in banana cultivation and difficult to control. *Pseudomonas fluorescens* strain IIHRPf12 inhibited growth of the fungus in vitro and, when applied as a soil drench 3 days before inoculation of banana tissue culture explants with the pathogen, reduced colonization of the roots by *F. oxysporum* by 72%. This result could be fully explained by antibiosis, but electron microscopy of treated root parts revealed structural changes in the cortical cells, mainly consisting of densely stained amorphous material and cell wall appositions that were associated with enhanced restriction of fungal growth (Mohandas et al., 2004). In a similar study (Thangavelu et al., 2003), *P. fluorescens* strain Pf10 increased peroxidase, PAL, chitinase and β-1,3-glucanase activities and phenolic content in banana leaves, indicating that at least some defense-related activities were enhanced systemically as a result of bacterial application. Moreover, upon challenge inoculation with the pathogen, these defense-related activities were stimulated to a larger extent in bacterized than in non-bacterized plants. These results suggest that part of the protection might be ascribed to ISR.

2.3 Bean and other legumes

An early report describes how bean seeds were subjected to treatments with suspensions of *P. fluorescens* strain S97 and the emerging seedlings were grown for 3-4 weeks (Alström, 1991). The first trifoliate leaf was then challenged with *P. syringae* pv. *phaseolicola* and lesions due to halo blight were recorded in the third trifoliate leaf. The number of lesions decreased with increasing concentration of the induction inoculum, 4.6 x 10^8 cfu.ml^{-1} reducing the number to 17% of that in non-treated controls. Protection was eliminated when the bacterial suspension was autoclaved, indicating a need for live bacteria for protection to be achieved. S97 exhibited bacteriostatic activity against the bean pathogen and it was suggested that the rhizobacterium might produce substances already during seed germination that were later translocated to the foliage. These inhibitory compounds might accumulate around the site of bacterial multiplication and contribute to their restricted growth.

Thus, antibiosis rather than ISR may have been responsible for the reduction in disease symptoms. Absence of the antagonistic bacteria from the aerial parts of the plants was not checked. However, ISR could have been involved (Alström, 1995). Reduced levels of the challenging pathogen in protected leaves and in the intercellular washing fluid (IWF) from those leaves were correlated with a general rise in protein content in the IWF, and associated with accumulation of certain phenolic compounds. Hence, changes in plant metabolism did occur as a result of prior seed bacterization, but a causal relationship with disease suppression was not established.

Indications for the induction of plant defense responses were also obtained upon root colonization by other fluorescent *Pseudomonas* spp. A *P. aureofaciens* strain induced symptoms of a hypersensitive response on bean cotyledons and in bean suspension cultures, and induced defense proteins resembling the plant's response to pathogens inducing SAR (Zdor and Anderson, 1992). Isolates of *P. tolaasii* and *P. putida* induced only certain acid-soluble proteins. Upon inoculation on cotyledons some accumulation of phytoalexins and phenolics was associated with a slight browning reaction, indicating that the bean plants responded defensively to foliar application of these rhizobacterial species. However, upon root treatment no metabolic changes were apparent, suggesting that the rhizobacteria had only very minor effects on root metabolism. These observations are important in showing that at least some rhizobacterial species have limited pathogenic activity when applied to leaves, and initiate defense reactions also observed in leaves exhibiting SAR. Unfortunately, no attempt was made in this study to relate the metabolic effects seen to protection against pathogens.

The rhizobacterium *P. aeruginosa* 7NSK2 induces resistance to leaf infection by the gray mould fungus *Botrytis cinerea* (De Meyer and Höfte, 1997). Bean seeds were soaked in a bacterial suspension for 5 min prior to planting in soil mixed with the bacterial inoculum also. After plant emergence leaves were wounded by touching with a red hot pinhead to allow *Botrytis* to infect and produce a spreading lesion. Inducing bacteria were never detected in stem and cotyledon extracts of challenged bean leaves, but were present occasionally in extracts of first leaves at levels below 50 cfu.leaf $^{-1}$. 7NSK2 reduced the number of spreading lesions by about one half when suspensions were prepared from iron-limited King's medium B, but not from iron-rich LB medium. Under iron limitation, 7NSK2 produces three siderophores: pyoverdin, pyochelin and SA. By using bacterial mutants deficient in either one or two of these siderophores, it was demonstrated that SA production was essential for the induction of resistance (De Meyer and Höfte, 1997). Because SA is a precursor of pyochelin, a role for pyochelin could not be excluded. Indeed, pyochelin has been shown to be required for 7NSK2-elicited ISR in tomato (Audenaert et al., 2002). 7NSK2-induced resistance was expressed not only against *B. cinerea*, but also against the anthracnose fungus *Colletotrichum lindemuthianum* (Bigirimana and Höfte, 2002). A bacterial mutant, KMPCH, that lacked pyoverdin and pyochelin but still produced SA, induced resistance against anthracnose in a moderately resistant, but not in a susceptible bean cultivar. KMPCH-567, a SA^{-} mutant of KMPCH, failed to induce resistance, demonstrating that SA is important for the induction of systemic resistance by 7NSK2. Production of SA and pyochelin was detected during 7NSK2 colonization of bean, indicating that iron limitation occurs in the rhizosphere. These results demonstrate that induction of systemic resistance in bean by 7NSK2 is dependent on iron nutritional state and confirm that specific siderophores function not only in the acquisition of iron by the bacteria, but also in the induction of systemic resistance in the plant.

In a model system with plants grown in perlite, 7NSK2-induced resistance was equivalent to the inclusion of 1 nM SA in the nutrient solution (De Meyer et al., 1999b). Hydroponic feeding of 1 nM SA solution induced PAL activity in roots and increased free SA levels in leaves. In bacterized roots of soil-grown plants no such increases were associated with the elicitation of ISR by 7NSK2. In contrast, KMPCH was as effective as wild-type 7NSK2 in triggering ISR but did stimulate PAL activity. Thus, although these bacteria can stimulate PAL activity in bean roots, an increase in PAL activity is unlikely to be essential for ISR. Neither increased PAL activity, nor elevated SA levels were found in bean plants in which ISR was elicited by seed treatment with the non-pathogenic *P. putida* strain BTP1. Furthermore, no enhanced fungitoxicity was detected in methanolic leaf extracts, suggesting that accumulation of bean phytoalexins was not part of the stimulated defense mechanisms against *B. cinerea*. However, lipoxygenase and hydroperoxide lyase, two key enzymes in the biosynthetic pathway of JA, were significantly stimulated during the first four days after challenging BTP1-treated plants with the pathogen, suggesting that priming of JA-dependent defenses occurred upon challenge inoculation (Ongena et al., 2004). Pyoverdin, pyochelin or SA were not involved in the systemic resistance triggered by BTP1. Nevertheless, most of the resistance-eliciting activity was present in culture fluid when cells were grown under iron-limited conditions, and not associated with killed bacterial cells. However, the nature of the inducing fraction was not determined (Ongena et al., 2002).

Like 7NSK2, *P. putida* WCS358 and *P. fluorescens* WCS417 have been demonstrated to trigger ISR in bean against gray mould and anthracnose, and anthracnose, respectively (Bigirimana and Höfte, 2002; Meziane et al., 2005). Of strain WCS358, both the LPS and the pseudobactin siderophore were effective in eliciting ISR, whereas flagella were not. In agreement with more than one factor acting as inducing determinant, bacterial mutants lacking either pseudobactin or flagellin were still capable of eliciting ISR. However, the OA^- mutant was not, suggesting an interference with siderophore production.

Few data are available for other legume species. An electron microscopic study showed that infection of in vitro grown, *Agrobacterium rhizogenes* T-DNA-transformed hairy pea roots by *Fusarium oxysporum* f.sp. *pisi* was restricted to the outer root cortex when pretreated with *P. fluorescens* strain 63-28 (Benhamou et al., 1996a) or *B. pumilus* strain SE34 (Benhamou et al., 1996b). Lack of ingress of the fungus toward the vascular stele was associated with deposition onto the inner surface of cell walls of callose-enriched appositions and accumulation of phenolic compounds. Apparently, the rhizobacterial strains enhanced plant defense reactions culminating in the elaboration of physical barriers to fungal spread.

A similar enhancement of the accumulation of defense-related phenolic compounds was observed in pea plants grown from seed bacterized with *P. fluorescens* strain Pf4 or a strain of *P. aeruginosa* and, when 20-25 days old, inoculated with conidia of the powdery mildew fungus *Erysiphe pisi* (Singh et al., 2002). These observations are suggestive of the rhizobacteria acting through a stimulation of a plant defense-related response upon challenge. However, spatial separation of the rhizobacterial inducers and the challenging pathogen and lack of antibiosis between those were not checked. The same two strains, alone or in combination, were also reported to protect chickpea against collar rot caused by the soilborne fungus *Sclerotium rolfsii* (Singh et al., 2003). Protection was associated with increases in specific phenolic acids, including SA, in roots and leaves. Soil drenching with culture filtrate of *S. rolfsii* had similar effects, and a combination of both rhizobacterial strains and culture filtrate induced additive increases. The interpretation of these findings is difficult, but the rhizobacterial strains may have acted as minor pathogens, inducing stress-related phenolic compounds systemically. Seed treatments with seven other *P. fluorescens* strains were likewise found to protect chickpea against collar rot in association with increased accumulation of phenolic acids in the plant (Sarma et al., 2002).

Protection of chickpea against charcoal rot disease caused by the fungus *Macrophomona phaseolina* was obtained upon soil mixing, drenching or root dip with a suspension of *P. fluorescens* isolate 4-92 (Srivastava et al., 2001). Two days were required between induction treatment and challenge inoculation for protection to be manifested. Bacterial treatments increased root glucanase and chitinase activities up to 7-fold after 5 days. Effects were evident already upon bacterization with 10^4 cfu ml $^{-1}$, and reached a plateau at 10^5 cfu ml $^{-1}$, similar to the induction of ISR in e.g. radish (Raaijmakers et al., 1995).

Soil drench of 3-4 weeks old white clover plants with 10^5 -10^6 cfu ml $^{-1}$ of fluorescent *Pseudomonas* spp. strains P29 and P80, or *B. cereus* strain B1, 1-3 days prior to inoculation with the clover cyst nematode *Heterodera trifolii*, caused morhological abnormalities and reduced fecundity of the nematodes (Kempster et al., 2001). Both live and dead cells of P29, but not cell-free culture filtrate, induced these effects, suggesting that some type of ISR was involved. On the basis of the significantly improved performance of the plants, the induced resistance was taken to be expressed also systemically against the blue-green aphid, *Acyrthosiphon kondoi*, in the while clover plants, as well as in annual medic (*Medicago trunculata*) (Kempster et al., 2002), but the mechanism involved remains unclear.

2.4 Carnation

When *P. fluorescens* WCS417r and the soil-borne fungus *Fusarium oxysporum* f.sp. *dianthi* (Fod) were applied to different parts of carnation plants by treating the roots with the bacterium and introducing the fungus one week later in the stem by slashing, a significant reduction in wilting symptoms was obtained (Van Peer et al., 1991). Bacteria and fungus were confirmed to remain spatially separated, indicating that WCS417r protected carnation against Fod by induction of systemic resistance. In contrast, *P. putida* WCS358 did not enhance resistance of carnation against Fod when the fungus was inoculated in the stem.

Resistance of carnation to Fod is polygenic and associated with the accumulation of phytoalexins (Baayen and Niemann, 1989). No increase in phytoalexins was detectable upon induction of systemic resistance by WCS417r prior to challenge, but upon subsequent inoculation with Fod production of phytoalexins was enhanced (Van Peer et al., 1991). Disease incidence was reduced substantially in the moderately resistant carnation cultivar Pallas and less consistently in the susceptible cultivar Lena. These results indicate that genetically determined differences in the level of protection occur in different cultivars, and confirm that priming of the extant defensive capacity is at the basis of the induced resistance expressed.

In the systemic protection of carnation against Fod by WCS417r, heat-killed bacteria or the purified LPS were as effective in inducing resistance as were live bacteria (Van Peer and Schippers, 1992), thus confirming that the protective effect is plant-mediated. Moreover, these results indicated that the bacterial LPS acts as a determinant of resistance induction by WCS417r in carnation.

2.5 Cucumber

Ninety-four rhizobacterial strains previously shown to reduce disease incidence caused by *Pythium ultimum* or *Rhizoctonia solani*, or to promote seedling emergence and crop yield, were screened for induction of systemic resistance in cucumber against the fungus *Colletotrichum orbiculare*, the causal agent of leaf anthracnose (Wei et al., 1991). Bacteria were coated on the seeds and colonized roots to high levels, and none of the strains was recovered from surface-disinfested petioles when the second true leaf of each plant was challenge inoculated 21 days after planting. Out of the 94 strains, six reduced lesion size after challenge inoculation with the fungus by 33-75% relative to the non-bacterized control, compared to a reduction of 84% when SAR was induced by the pathogen itself. Five of the bacterial strains also significantly reduced lesion numbers. Antagonism in vitro to *C. orbiculare* on three media was generally absent with five of the strains, and weak with one strain. Using spontaneous rifampicin-resistant (rif$^+$) mutants for treating seeds, some strains were recovered from inside surface-disinfested roots. However, none of the inducing strains was found in leaves challenged with *C. orbiculare*. Thus, while some bacteria which induced systemic resistance may have limited internal colonization and behave as endophytes, they did not appear to translocate to a significant extent or colonize challenged leaves, suggesting that neither competition nor antibiosis were involved in the observed protection (Wei et al., 1991, 1992).

Subsequently, seed treatment with two different rhizobacterial strains, *P. putida* 89B-27 and *Serratia marcescens* 90-166 was shown to induce systemic resistance against vascular wilt disease, caused *F. oxysporum* f.sp. *cucumerinum* (Foc) (Liu et al., 1995a), angular leaf spot due to the foliar bacterium *P. syringae* pv. *lachrymans* (Psl) (Liu et al., 1995b), bacterial wilt provoked by *Erwinia tracheiphila* (Kloepper et al., 1993), systemic mosaic disease-inducing cucumber mosaic virus (CMV) (Raupach et al., 1996), and *C. orbiculare* (Liu et al., 1995c). In the study on the protection against fusarium wilt, a split-root assay was used in which the inducing bacteria and the pathogen were inoculated on separate halves of seedling roots at the same time, and then planted in separate pots. ISR was expressed by delayed disease symptom development, lesser disease severity and reduced number of dead plants compared to non-bacterized controls, reflecting reduced spread of the pathogen in the plant. Movement of inducing bacteria was monitored with a bioluminescent derivative of 89B-27, which was detected with a charge-coupled device camera. The bacteria showed only limited movement within inoculated pots and did not migrate to the pots in which the pathogen was inoculated, demonstrating that the bacteria and the pathogen remained spatially separated (Liu et al., 1995a). Upon elicitation of ISR by *P. fluorescens* 89B-61 or *S. marcescens* 90-166 and challenge inoculation with *C. orbiculare*, cytological studies using fluorescence microscopy revealed a higher frequency of autofluorescent cells, which are related to accumulation of phenolic compounds at the sites of fungal penetration. Callose deposition was also often observed in epidermal cells. These observations are suggestive of an impediment of fungal penetration in the ISR-expressing plants (Jeun et al., 2004).

Treatment of seeds or cotyledons with either *P. putida* 89B-27 or *Serratia marcescens* 90-166 also resulted in substantial decreases in the number and size of lesions upon challenge inoculation of the second leaves with the bacterium Psl. Upon injection of cotyledons with 0.1 ml of 10^{10} cfu.ml^{-1}, the inducing bacterial strain reached population densities of 10^{11} or 10^{12} cfu.cotyledon^{-1} 3 days later. However, neither bacterial strain was recovered from stems 1 or 2 cm above or below the inoculated cotyledons at any sampling time. Associated with reductions of about 60% in total Psl lesion area pathogen populations in inoculated leaves also were significantly reduced in the rhizobacterial treatments. The level of protection afforded was similar to that upon induction of SAR by the pathogens *C. orbiculare* or Psl itself (Liu et al., 1995b). In field trials involving bacterized seeds, naturally occurring bacterial wilt, caused by *E. tracheiphila*, was also significantly reduced (Kloepper et al., 1993).

Testing for ISR against CMV revealed that both rhizobacterial strains reduced the number of plants with symptoms. Protected plants did not develop mosaic symptoms throughout the experimental period, whereas pathogen-induced SAR was expressed as a delay in symptom development (Bergstrom et al., 1982). No viral antigen was detected by enzyme-linked immunosorbent assay in any asymptomatic bacteria-treated plants (Raupach et al., 1996), indicating that due to the inducing bacteria the plant had become refractory to viral infection.

When strains 89B-27 or 90-166 were inoculated on seeds prior to planting and plants were thereupon challenged with *C. orbiculare*, 89B-27 reduced mean total lesion diameter up to 60%, whereas 90-166 was often significantly less effective. Populations of 89B-27 declined at a consistent rate over time, dropping from over 10^8 cfu.g^{-1} root one week after planting to 10^3 cfu.g^{-1} 3 weeks later. ISR was evident at the first-leaf stage, increased over time, and was maintained at least to the fifth-leaf stage more than 5 weeks after seeding. Strain 90-166 showed population dynamics similar to 89B-27. ISR was somewhat more variable but also generally observed at all leaf stages tested. The decline in population densities of the two strains on the roots did not match the extent of ISR, supporting the notion that once induced, systemic resistance in the plant is maintained (Liu et al., 1995c).

Strain 90-166-mediated ISR was dependent on low-iron conditions, but independent of SA, as four different SA$^-$ mutants were as effective in inducing ISR against *C. orbiculare* as wild-type bacteria, whereas an ISR$^-$ mutant still produced SA (Press et al., 1997). In contrast, a Tn*5-phoA* insertion into the *entA* gene, which encodes an enzyme in the biosynthetic pathway of catechol siderophores, eliminated the ability of strain 90-166 to elicit ISR. Although total population densities of 90-166 and its mutant on cucumber roots were similar, the internal population size of the mutant was significantly lower (Press et al., 2001). However, it is questionable whether internal colonization is a decisive factor in the elicitation of ISR. Rather, the data support a role for a catechol siderophore as the bacterial determinant responsible for triggering ISR in cucumber.

Cultivar specificity of ISR was studied in one resistant and three susceptible cucumber cultivars. Strain 89B-27 induced systemic resistance in all three susceptible cultivars, whereas 90-166 did so in only two. Both strains failed to increase resistance in the resistant cultivar. No cultivar specificity in root colonization patterns by the two strains was observed, ruling out the possibility that failure of the strains to enhance protection was due to impaired root colonization. Apparently, protection in the genetically highly resistant cultivar could not be further enhanced by the rhizobacterium (Liu et al., 1995c). These data confirm that rhizobacterial strains differ in their inducing properties and that plant cultivars differ in inducibility.

The observations that in cucumber two different fungi, the root-infecting Foc and the foliar pathogen *C. orbiculare*, and two leaf-spotting bacteria, Psl and *E. tracheiphila*, were controlled by the same bacterial strains through the induction of systemic resistance indicates that rhizobacterially-mediated ISR in cucumber is effective against different pathogens in a similar way as in Arabidopsis. Moreover, ISR appeared to also reduce insect attack. Cucumber plants growing in the field from seeds treated with *B. pumilus* strain INR7, *Flavomonas oryzihabitans* strain INR5, *P. putida* 89B-61 or *S. marcescens* 90-166 supported lower numbers of the spotted cucumber beetle, *Diabrotica undecimpunctata*, and the striped cucumber beetle, *Acalymna vittatum* (Zehnder et al., 1997a,b). As a result cucumber plant growth and yield were significantly greater. Bacterial treatment also resulted in greater beetle control than weekly applications of insecticide. The inducing bacteria appeared to reduce the level of the secondary compound cucurbitacin in the plants, which acts as a feeding stimulant to the beetles. This observation is particularly interesting because it shows that induction of systemic resistance by these rhizobacterial strains is associated with a defined change in plant metabolism.

Three different rhizobacterial strains, *P. corrugata* 13, *P. fluorescens* C15 and *P. aureofaciens* 63-28 were shown to induce a systemic resistance in cucumber against crown rot-causing *Pythium aphanidermatum* (Zhou and Paulitz, 1994; Chen et al., 1998). The strains were applied to one half of the root system spatially separated from the pathogen-inoculated root part by splitting the root systems in rock wool in two separate pots. Bacterial population densities were maintained at about 10^5 cfu.g^{-1} dry rockwool in the bacterized pots. Systemic resistance was expressed as a significant decrease in the numbers of attaching and germinating zoospores. Occurrence of stem rot was delayed for 4-6 days and disease index was reduced, but application of the bacteria to one pot one week before inoculation of the other pot with the fungus did not reduce disease to the same degree as when both were applied at the same time. This result is puzzling in view of the common observation that time is needed to achieve the induced state. The possibility that the rhizobacteria produced antifungal metabolites that were transported to the opposite root system was discounted on the grounds that these bacterial isolates did not inhibit mycelial growth of *P. aphanidermatum* in vitro. In the split-root system, both *P. corrugata* strain 13 and *P. aureofaciens* strain 63-28 increased PAL, peroxidase and polyphenoloxidase activity both locally and systemically. Moreover, additional peroxidase isoenzymes were induced, indicating that the cucumber roots reacted to colonization by these rhizobacterial strains with increases in defense-related enzymes (Chen et al., 2000).

Both strains were shown to produce SA in liquid medium and also induce cucumber roots to accumulate endogenous SA one day after bacterial treatment. More SA accumulated in roots treated with bacteria than in distant roots on the opposite site of the root system in the first two days, but this difference disappeared after 3-4 days (Chen et al., 1999). It appears unlikely that this SA accumulation was involved in ISR, because exogenously applied SA failed to induce local or systemic resistance against challenging *P. aphanidermatum*, and the mechanisms involved have remained unclear.

P. putida BTP1, as well as a derivative impaired in siderophore production likewise protected cucumber against *P. aphanidermatum* root rot. In analyses of root extracts from split-root experiments, this protection was found to be associated with a systemic accumulation of several phytoalexin-like compounds. The nature of these compounds appeared to be different in leaves as compared to roots, suggesting that the defense response to *Pythium* induced by this strain is not based on a single antifungal factor (Ongena et al., 1999, 2000). These reports are among the few that provide evidence for a defined induced response that is active against the pathogen involved.

In a similar split-root system the suppressive activity of three different types of compost on root rot caused by *P. ultimum* or *P. aphanidermatum* was analysed (Zhang et al., 1996). Mean root rot severity was significantly less for split roots in fungus-infested peat mix paired with compost-amended mixes than in infested peat mix paired with non-infested peat mix without compost. Moreover, 3 weeks after planting, anthracnose due to inoculation of the second leaf with *C. orbiculare* was significantly less severe on plants grown in compost-amended mixes than in peat mix. Plants grown in the compost-amended mixes showed enhanced leaf peroxidase activity, indicative of a stress response. Inoculation with *C. orbiculare* dramatically increased the activity of the major peroxidase isoenzyme in plants grown in peat mix, but this increase was even greater in plants grown in the compost-amended mixes. Initially, β-glucanase activity was low, but when infected with *C. orbiculare*, this activity was induced to significantly higher levels in plants grown in the compost mix than in plants grown in the peat mix (Zhang et al., 1998). This situation is similar to the priming action of defense responses in challenge-inoculated ISR-expressing plants.

ISR triggered by *Pseudomonas chlororaphis* strain O6 upon root colonization of cucumber against target leaf spot, caused by *Corynespora cassiicola*, was associated with a faster and stronger accumulation of transcripts of six distinct genes upon challenge inoculation, as revealed through subtractive hybridization (Kim et al., 2004). Expression of these genes was not induced by O6 colonization alone, and became apparent only after challenge with the pathogen. These results are in line with the priming of defense responses upon challenge inoculation of induced plants, as also observed in e.g. Arabidopsis and carnation. A similar amplification of defense reactions was seen after challenge inoculation with *P. ultimum* of cucumber plants that were endophytically colonized by *Serratia plymuthica*. Light and transmission electron microscopy of root tissues revealed that invading hyphae of the pathogen were trapped in a phenolic-enriched material, which infiltrated and filled moribund hyphae. Cell wall appositions containing callose, pectin, cellulose and phenolics at sites of potential pathogen invasion were suggestive of a coordinated induction of structural and biochemical barriers which adversely affect pathogen growth and development (Benhamou et al., 2000).

Several *Bacillus* spp. have been demonstrated to elicit ISR in cucumber. *B. pumilus* INR7 was reported to elicit ISR against anthracnose and angular leaf spot (Raupach and Kloepper, 1998). When applied as a seed coating singly or in combination with *B. subtilis* GB03 and *Curtobacter flaccumfaciens* ME1 under greenhouse conditions, there was a general trend towards greater suppression and enhanced consistency of protection against anthracnose, angular leaf spot and cucubit wilt when using strain mixtures. In how far ISR contributed to disease suppression was not determined, however. Seed treatments with *B. amyloliquefaciens* IN937a or *B. pumilus* IN937b suppressed symptoms of CMV in cotyledons of challenged plants, suggestive of induced resistance. Mixtures of the two strains or with strains *B. pumilus* INR7, SE34, SE49 or T4, or *B. sphaericus* SE56 were equally or even more effective under both greenhouse (Jetiyanon and Kloepper, 2002) and field conditions (Jetiyanon et al., 2003).

2.6 Pepper

Eighteen isolates of *P. fluorescens* and two of *P. putida* were evaluated for their ability to control damping off caused by *Pythium apanidermatum* in hot pepper. Seeds bacterized with the effective *P. fluorescens* isolate Pf1 and sown in sterilized potting medium mixed with pathogen inoculum developed into substantially more vigorous plants than non-bacterized controls. Disease was reduced by 60%. Earlier and increased activities of PAL, peroxidase and polyphenol oxidase were observed in Pf1-pretreated plants. Moreover, higher accumulation of phenolics was noticed (Ramamoorthy et al., 2002a). These results were interpreted as evidence for ISR, but no spatial separation between the bacterium in the seed coating and the pathogenic fungus was imposed. The increased vigour of the bacterized plants was referred to as increased plant growth promotion, and may have resulted from an action of Pf1 unrelated to disease suppression, or due to suppression of *P. aphanidermatum* through microbial antagonism. The latter was not checked. The increase in the defense-related enzyme activities may have been the consequence of the improved plant development, or of a specific stimulation by the biocontrol agent that may, or may not, be related to ISR. This study exemplifies the difficulties encountered when interpreting mechanisms from studies aimed at the identification of effective disease-suppressing rhizobacterial isolates that were not designed to specifically study involvement of ISR. However, similar results were reported associated with the reduction of chilli fruit rot and die-back incited by *Colletotrichum capsici* as a result of treatments with strain Pf1 or an isolate of *Bacillus subtilis* (Bharathi et al., 2004). Bacteria were applied to the soil, whereas conidia of the pathogen were sprayed on the leaves, suggesting that the effect might indeed be due to ISR.

At least absence of in vitro antibiosis was checked in a study on systemic suppression of pepper anthracnose, caused by *Colletotrichum gloeosporioides*, by seed bacterization with several *Bacillus* spp. Treatments with seven individual strains and eighteen mixtures all caused significant reduction of lesion development on fruits, resulting in smaller lesions than on nontreated control plants (Jetyanon and Kloepper, 2002). Although absence of biocontrol bacteria on the fruits was not checked, it seems likely that in this case ISR was involved.

2.7 Pines and other tree species

Seed treatment of loblolly pine with *Bacillus pumilus* strains INR7, SE34 or SE52, or *Serratia marcescens* 90-166 increased the capacity to resist gall formation by the causal agent of fusiform rust, *Cronartium quercuus* f.sp. *fusiforme*. Averaged over two years, 31% of control seedlings inoculated with the pathogen one month after sowing were infected, whereas treatment with the bacterial strains reduced infection to 15, 13, 16 and 14%, respectively. Although spatial separation of the bacteria and the rust was not checked, it appears that systemic resistance was induced (Enebak and Carey, 2000). This report is of interest because it shows protection in a gymnosperm tree.

Induction of systemic resistance in seedlings of *Eucalyptus urophylla* was found when *P. putida* WCS358 or *P. fluorescens* WCS374 were infiltrated into two lower leaves and the causal agent of bacterial wilt, the soilborne bacterium *Ralstonia solanacearum*, was inoculated 3-7 days later on the decapitated stem. ISR was evident as a reduction in the length of the stem from the top that blackened and decayed (Ran et al., 2005a). However, no resistance was induced when the biocontrol strains were applied to the soil and, thus, ISR was evident only under an unnatural condition.

2.8 Potato

In spite of the importance of potato as a food crop, no data are available on rhizobacteria-mediated ISR against pathogenic fungi, bacteria or viruses. However, for nematode control, in split-root experiments *Bacillus sphaericus* strain B43 and *Rhizobium etli* G12 were found to reduce penetration by the cyst nematode *Globodera pallida* up to 37% (Hasky-Günther et al., 1998; Reitz et al., 2000). Heat-killed bacterial cells were equally active. The ISR-eliciting factor of B43 was present in culture filtrate, wheras G12 was shown to trigger ISR through its LPS. Oligosaccharides of the core region of the LPS were the main factor in triggering ISR (Reitz et al., 2002). ISR was not associated with enhanced accumulation of PRs or increased lignification of root cells (Reitz et al., 2001), leaving the resistance mechanism(s) involved to be determined.

2.9 Radish

Out of 538 rhizobacterial strains isolated from roots of radish seedlings grown in compost-amended potting mixes, eleven suppressed bacterial leaf spot caused by Xca, when applied individually as inoculum in non-inducing potting mixes. Reduction in symptom severity varied between 8 and 22% and was assumed to be the result of ISR (Krause et al., 2003).

Using the two-compartment, separate inoculation rockwool system described for Arabidopsis, *P. fluorescens* strains WCS374 and WCS417 induced ISR against For, reducing disease incidence by up to 50%. ISR was expressed primarily as a reduction in the percentage of diseased plants, apparently resulting from a more frequent failure of the fungus to reach or colonize the vascular tissue (Hoffland et al., 1995). Unlike in

Arabidopsis, *P. putida* WCS358 was not active, but another *P. putida* strain, RE8, was as effective as *P. fluorescens* WCS417 (De Boer et al., 1999). Strain WCS374 induced a similar level of resistance in six cultivars ranging from highly susceptible to relatively resistant, provided fungal inoculum pressure was adjusted to cause between 40 and 80% of the plants in non-bacterized controls to become diseased. Apparently, at higher inoculum doses the resistance induced was insufficient to block disease development, whereas at low disease incidence the effect of the inducing bacteria was too small to be significant (Leeman et al., 1995a). Differences in inducibility would have been recorded, though, if all cultivars had been inoculated with the same dose of the fungus.

At least one day was necessary after application of at least 10^5 cfu.g^{-1} talcum powder for significant resistance to be induced. No increase in resistance was evident when the dosage of bacteria was increased. Root colonization of the bacteria was determined in root macerates using immunofluorescence colony-staining, and found to be maintained at levels above 10^5 cfu.g^{-1} root for at least 3 weeks (Leeman et al., 1995a). A minimum of 10^5 cfu.g^{-1} root were required for effective protection, and no disease suppression was evident when colonization of the roots did not reach this level. However, no relationship was apparent between the remaining rhizosphere population density of WCS374 and disease incidence at later stages, when protection was evident (Raaijmakers et al., 1995). Thus, the initial triggering of the plant leads to its induced state and, once this has been reached, further protection appears independent of the remaining population of the inducing bacterium in the rhizosphere.

Resistance was effective not only against For, but also against avirulent Pst, *Alternaria brassicicola* and a different isolate of *F. oxysporum* (Hoffland et al., 1996). The same results were obtained when the inducing rhizobacterium was applied spatially separated on the roots, or to a single leaf different from the ones used for challenge inoculation. The level of protection achieved was at least as high as that afforded by SAR induced by the pathogen Pst itself. The rhizobacteria remained confined to the induction site and resistance was expressed while bacterial numbers on the induction site decreased with time (Hoffland et al., 1996).

The bacterial LPS appeared to be the main trait responsible for resistance induction (Leeman et al., 1995b). Cell wall extracts of WCS374 or WCS417, or purified LPS were as effective as live bacteria when applied to radish roots. Bacterial OA$^-$ mutants, as well as their cell wall extracts, were ineffective. These results rule out any protective effects resulting from bacterial metabolism and show that the specificity within the structure of the O-antigenic side chain of the LPS determines the induction of systemic resistance by these rhizobacteria in radish. The resistance-inducing OA of WCS374 was effective not only on roots, but also when applied to the cotyledons. Hence, the plant is able to perceive this bacterial determinant not only at the root, but also at the leaf surface.

When similar experiments were conducted under iron-limiting conditions by adjusting the composition of the nutrient solution, the OA$^-$ mutants of WCS374 and

WCS417 were found not to be impaired in their ability to induce resistance (Leeman et al., 1996). Indeed, ISR reached levels equivalent to those induced by the wild type strains. These results indicated that (an)other bacterial determinant(s) expressed only under low-iron conditions induced resistance. Since siderophores are produced by the bacteria under these conditions, the pyoverdin-type pseudobactins of the strains were applied to radish roots. The purified siderophore of WCS374 induced resistance to the same level as the bacterial LPS. In contrast, the siderophore of WCS417 did not. Moreover, the sid$^-$ mutants of both WCS374 and WCS417 induced resistance under conditions of iron deficiency as well as in the presence of excess iron, and again to a level similar to that reached upon application of the wild type bacteria. Although the siderophore of WCS374 could induce resistance, it did not seem to be responsible for the induction under low-iron conditions. Thus, (an)other iron-regulated metabolite(s) appeared to be involved. Apparently, different bacterial determinants are equally able to induce systemic resistance in radish, but these effects seem complementary rather than additive and full induction by one factor masks any contribution by the others.

Unlike *P. aeruginosa* 7NSK2 (De Meyer and Höfte, 1997), WCS374 and WCS417 do not produce an additional siderophore of the pyochelin type. However, at low iron availablity they do produce SA in vitro, amounting to about 50 and 10 μg SA ml^{-1} standard succinate medium (SSM) for WCS374 and WCS417, respectively (Leeman et al., 1996). The production of SA decreased rapidly with increasing iron concentration. WCS374 is able to produce a second, SA-containing siderophore, pseudomonine (Mercado-Blanco et al., 2001), and it is possible that SA produced in the rhizosphere is channelled into pseudomonine, and pseudomonine could act as the additional iron-dependent inducing determinant. However, this hypothesis has not been tested so far.

The non-SA-producing strain *P. putida* WCS358 did not induce resistance under low-iron conditions. Commercial SA induced resistance in radish against Fod when applied to roots at concentrations as low as 100 $fg.g^{-1}$ talcum emulsion. It cannot be excluded, therefore, that low levels of SA released by bacteria in the rhizosphere induce the systemic resistance. Moreover, a fair correlation was found to exist between the capacity of WCS374 and WCS417 to produce SA in vitro and the ability of these strains to trigger ISR in vivo under low-iron conditions (Leeman et al., 1996).

Treatment of radish leaves with avirulent pathogens or millimolar concentrations of SA induced SAR and PR-homologues of the families PR-1, -2 and -5 (Hoffland et al., 1995). However, neither root-applied low doses of SA, nor the rhizobacterial strains triggering ISR induced PRs under either high- or low-iron conditions (Hoffland et al., 1995, 1996). While strong resistance was induced by the rhizobacteria and their OA$^-$-mutants under low-iron conditions, SA was apparently not present in sufficient quantities to induce PRs. Under high-iron conditions, SA is even less likely to be produced and resistance is induced by the wild-type strains only through their LPS.

Therefore, the role of SA in the induction of systemic resistance by these rhizobacterial strains in radish remains to be determined.

2.10 Rice and other monocots

Grain crops are the most widely cultivated plants for world food supply but very little research on ISR in monocots has been documented so far. In rice, two *P. fluorescens* strains, Pf1 and FP7, were reported to inhibit mycelial growth of the sheath blight fungus *Rhizoctonia solani* (Radjacommare et al., 2004) and to increase seedling vigour when applied as a seed coating, by root dipping, by mixing through the soil, or as a foliar spray (Nandakumar et al., 2001). Inoculation with *R. solani* was performed by placing sclerotia in the sheath of the rice plants 40 days after planting. Treatments by the single strains or their combination effectively reduced sheath blight incidence, promoted plant growth, and ultimately increased yield, under glasshouse or field conditions, comparable to the fungicide carbendazim. In those experiments in which the bacteria were applied by root dipping or mixing through the soil, it is possible that the reduction in disease incidence could be ascribed to ISR. In treated plants chitinase and peroxidase activities were enhanced and increased more strongly upon challenge inoculation than in non-treated plants (Nandakumar et al., 2001; Radjacommare et al., 2004). However, the level of increase varied between treatments, *Pseudomonas* strains used and treatment duration, even though disease index was reduced to similar extents. Absence of bacteria on the plants was not checked, the increased vigour of treated plants may have reduced infection by the fungus, and antibiotics produced by the bacteria may have contributed to disease reduction. This makes it difficult to interpret the significance of ISR for the decrease in sheath blight incidence. Similar results were obtained by Madhaiyan et al. (2004) by imbibing rice seeds in a suspension of *Methylobacterium* sp. strain PPFM-Os-7. This bacterium increased plant phenolics content, PAL, chitinase, β-1,3-glucanase and peroxidase activities when sprayed upon the leaves. These conditions are artefactual and should be considered local effects rather than caused by induction of systemically induced resistance.

Seed treatment or root dipping with strains Pf1 and FP7, singly and even more so in combination, also protected rice against damage due to feeding by larvae of the leaffolder *Cnaphalocrocis medinalis*. Larval and pupal weight were reduced and larval mortality was increased. Moreover, increased populations of hymenopterous parasitoids and predatory spider mites acting as natural enemies of the leaffolder, were noticed in *Pseudomonas*-treated plots under field conditions (Radja Commare et al., 2002). Thus, the bacterized plants appeared to express both enhanced direct and indirect defenses against this insect pest.

LacZY-marked *P. fluorescens* strain 7-14 and *P. putida* strain V14i were applied either as a seed treatment or as a root dip and found to suppress rice blast disease, caused by the fungus *Magnaporthe grisea*, by about 25%. Based on the expression of the marker gene, bacteria remained spatially separated from the pathogen, indicating that disease reduction was caused by ISR (Krisnamurthy and Gnanamanickam, 1998).

Similarly, seed treatment, root dipping or soil amendment with either one out of four strains of *P. fluorescens* or various combinations of two out of five *Bacillus* spp. with chitosan as a carrier promoted growth and advanced time of flowering of pearl millet and reduced the incidence of downy mildew caused by *Sclerospora graminicola* under greenhouse and field conditions. Resistance induced by the most effective *P. fluorescens* isolate required a minimum of three days to build up and was sustained throughout the plant's life (Niranjan Raj et al., 2003a,b).

Induced resistance was likewise found to be a mechanism in tall fescue (*Festuca arundinacea*) elicited by the bacterium *Lysobacter enzymogenes* strain C3 and effective against leaf spot development caused by *Bipolaris sorokiniana* and leaf blight caused by *R. solani*. Heat-inactivated cells were likewise active, ruling out an effect of antibiosis. Application of live or heat-killed cells to leaves resulted only in localized resistance, but treatment of roots elicited systemic resistance expressed in the foliage. Both live and heat-killed cells, when applied to the roots, increased peroxidase activity to similar extents in the foliage, demonstrating that the plant reacted systemically to the local application of bacterial components (Kilic-Ekici and Yuen, 2003). Two other strains of *L. enzymogenes* were ineffective in triggering ISR, whereas *Bacillus pumilus* INR7 and *P. fluorescens* 89B-61 and WCS417r elicited ISR against leaf spot rapidly, with reductions being apparent one day after root treatment. *Bacillus* spp. strains SE34, IN937a and IN937b were only slightly active, while strain T4 was ineffective. All bacterial strains colonized tall fescue roots to a similar degree. Increases in peroxidase activity in leaves upon root treatment were too variable to allow any conclusion about a possible correlation between ISR and peroxidase activity (Kilic-Ekici and Yuen, 2004).

In field experiments with sugarcane, seed treatment during planting and two following soil applications with various strains of *P. fluorescens* (Pf1, EP1, VTP4, VTP10, CHA0) or *P. putida* strain KKM1 reduced red rot disease when the pathogen *Colletotrichum falcatum* was inoculated at the third internode from the base of the stalk some time later. Yield of cane and sugar were significantly enhanced. Protection was higher in susceptible sugarcane cultivars than in moderately resistant ones. Four strains were also tested and found suppressive to natural infection under field conditions (Viswanathan and Samiyappan, 2002a). Treated canes had enhanced peroxidase and PAL activities and lowered catalase activity, particularly upon challenge inoculation, and had a higher lignin content (Viswanathan and Samiyappan, 2002b). In addition, increases in chitinases, β-1,3-glucanases and thaumatin-like proteins upon bacterization with strain KKM1 were also recorded (Viswanathan et al., 2003), suggesting that the bacterium induced a systemic resistance associated with the induction of PRs. However, it is not clear whether these differences were related to ISR or to the enhanced growth and development of the plants.

2.11 Tobacco

At least eight of the ten major PRs induced in tobacco in response to pathogens causing hypersensitive necrosis, were found in the IWF of leaves of plants grown in autoclaved soil in the presence of *P. fluorescens* strain CHA0 (Maurhofer et al., 1994). Strain CHA0 was reisolated from the roots of the plants but could not be detected in stems or leaves. Six weeks after the bacteria had been added to the soil, plants were challenged by inoculation with tobacco necrosis virus (TNV). Both numbers of viral lesions and lesion diameter were reduced to the same extent as in plants with SAR due to a previous infection with TNV itself. Strain CHA400, a sid^- mutant of CHA0, was still able to induce PRs but showed only partial resistance to TNV, implicating the involvement of the pyoverdin siderophore of CHA0 in the induction of resistance against TNV. Root colonization of the plants with CHA0 or CHA400, as well as leaf infection with TNV, caused up to five-fold increases in SA in the leaves. These observations suggested that SA was involved in the ISR elicited by CHA0 in tobacco, with induction being at least partly determined by the siderophore of CHA0. However, the transposon insertion generating the sid^- mutation in CHA400 was not localized, and it is not clear whether the loss of pyoverdin is the only mutation in CHA400.

Besides pyoverdin, CHA0 also produces SA under low-iron conditions (Meyer et al., 1992). Therefore, it is not clear whether the increase in SA in the bacterized plants was the result of induction by the bacteria or of synthesis of SA by the plant, or whether the plant takes up bacterial SA and translocates it to the leaves. CHA0 behaves as an endophyte and produces several toxic metabolites, among which are DAPG, pyoluteorin and hydrogen cyanide (Maurhofer et al., 1995). Because tobacco plants are sensitive to these antibiotics, these can cause stress to the plant, resulting in the induction of stress-related defense responses. A bacterial transformant over-producing DAPG and pyoluteorin protected tobacco roots significantly better than the wild type against the black root rot-causing fungus *Thielaviopsis basicola*, but at the same time drastically reduced the growth of the plants. Because CHA0 was seldom found in contact with the mycelium of *T. basicola* despite reducing the extent of black root rot, and the physical integrity of the fungal hyphae in direct contact with cells of CHA0 were not affected, it was inferred that secondary metabolites involved in the antagonism to the fungal pathogen also induced resistance in the roots (Troxler et al., 1997). Although the antibiotics suppressed black root rot through inhibition of fungal growth, there was no correlation between the sensitivity of various pathogens to the synthetic antibiotics and the degree of disease suppression by the transformants, suggesting that the plants exhibited an antibiotic stress-induced resistance resembling SAR (Maurhofer et al., 1995; Troxler et al., 1997). However, protection against TNV and against *T. basicola* may not be caused by the same mechanism. Introduction of the SA biosynthetic gene cluster *pchDCBA* from *P. aeruginosa* PAO1 (Serino et al., 1995) under a constitutive promoter into strain CHA0 did not increase its suppression of TNV or black root rot in tobacco. However, introduction into the non-SA-producing strain P3, leading to the production of 0.8 μg SA.10^9 cfu^{-1}, made this poor biocontrol agent an effective suppressor of TNV lesion formation. In contrast, suppression of black root rot was not improved (Maurhofer et al., 1998).

The SA-producing rhizobacterial strain *S. marcescens* 90-166 induced resistance to the wildfire bacterium *P. syringae* pv. *tabaci*, but mini-*Tn5 phoA* mutants, which did not produce detectable amounts of SA, induced resistance to the same level (Press et al., 1997). Thus, bacterial SA did not appear to be involved in the ISR induced by this strain in tobacco against the wildfire bacterium. Yet, when tested in a microtiter plate assay, strain 90-166 induced the *PR-1a* promoter in transgenic tobacco seedlings containing the *GUS* gene as a reporter, similar to SA (Park and Kloepper, 2000). In this assay, surface-sterilized tobacco seeds were germinated in Murashige & Skoog medium and 7 days later treated with bacteria, effectively bathing the seedlings in the bacterial suspension. *Enterobacter asburiae* strain JM-22, *P. putida* 89B-61 and *Burkholderia gladioli* IN-26 induced *GUS* expression likewise, while *Clavibacter michiganensis* TE-5, *Bacillus pumilus* strains T4, SE-34, SE-49, SE76 and INR7 and *P. fluorescens* 89B-027 did not. Infiltration of greenhouse-grown tobacco leaves with the same strains resulted in significant increases in GUS activity after treatment with all of the tested strains, including T4 and TE5, although the level of GUS activity in plants treated with the latter two strains was lower than in plants treated with strains JM-22, 89B-61, 90-166 and IN-26. When bacteria were applied as a root drench, only T4 and JM-22 induced significant GUS activity in leaves. These data demonstrate that local application of ISR-eliciting bacteria to leaves can induce PRs, whereas ISR elicited by treatment of the roots is not consistently associated with *PR*-gene activation. Indeed, tobacco was significantly protected against *P. syringae* pv. *tabaci* by root treatments with strains T4, JM-22, 89B-61, 90-166 or IN-26. However, these strains did induce stronger GUS expression than non-inducing strains in the microtiter plate assays (Park and Kloepper, 2000).

The relationship between SA and ISR was explored further by assaying strains 90-166, SE34, 89B-61, T4 and *B. pasteurii* C-9 for ISR against blue mould of tobacco, caused by *Peronospora tabacina*, in microtiter plates and detached leaf assays, as well as greenhouse tests. ISR was confirmed by checking for absence of bacteria on challenged leaves. Elicitation of ISR among strains varied in the different types of assays and on different tobacco cultivars, stressing the need for reliable assays for assessing ISR (Zhang et al., 2002a). In the microtiter plate assay, levels of endogenous free SA in tobacco seedlings treated with strains 90-166, 89B-61 and SE 34 increased significantly during the first week after bacterization, but in the second week were substantially lower than in non-treated seedlings. Upon challenge inoculation of plants treated with SE34, levels of SA peaked strongly one day after challenge. Such an increase was absent upon challenge of plants treated with 90-166 or 89B-61. Moreover, the three strains were similarly effective in reducing blue mould severity in both Xanthi-nc and transgenic NahG tobacco. Since SA is effectively broken down in NahG plants, these results indicate that the ISR against *P. tabacina* was SA-independent (Zhang et al., 2002b). In such a case, SA-inducible PRs would not be expected to occur in ISR-expressing plants and the induction of the *PR-1a* promoter by some of the strains remains to be explained.

Abundant induction of PR-like proteins was described in tobacco plants that were treated with LPS from an endophytic strain of *Burkholderia cepacia*, associated with a protective effect against black shank disease, caused by the oomycete *Phytophthora nicotianae*. As a result of the treatment with the LPS, cell permeability was slightly increased and viability decreased, suggesting that at the high concentration used (100 $\mu g\ ml^{-1}$) toxic side effects may have contributed (Coventry and Dubery, 2001). The LPS triggered transient phosphorylation of a 43-kDa (ERK-like) mitogen-activated protein (MAP) kinase within a few minutes, marking a rapid response associated with elicitation (Piater et al., 2004). MAP kinase phosphorylation cascades mediate innate immunity responses in animals, suggesting parallels between induction of defense responses in animals and plants.

No PR-1 expression was associated with ISR against tobacco mosaic virus (TMV) induced by *P. aeruginosa* strain 7NSK2, in spite of the demonstration that resistance was abolished in NahG plants (De Meyer et al., 1999a). A reduction in TMV lesion diameter was also evident after root application of mutant KMPCH, the pyoverdin- and pyochelin-deficient derivative of 7NSK2, ruling out an involvement of the pseudobactin or the SA-containing pyochelin siderophores in triggering ISR. ISR was no longer observed for mutant MPFMI-569, which is additionally deficient in SA production. Thus, SA production is essential for 7NSK2-elicited ISR against TMV and appears to be the only determinant for ISR in KMPCH. However, in grafting experiments in which NahG tobacco was used as rootstock to which bacteria were applied, and non-transformed tobacco plants served as scion for testing ISR, 7NSK2 and KMPCH did elicit ISR against TMV. It must be assumed that any SA produced by the bacteria was converted to inactive catechol by the SA-hydroxylase in NahG plants and, thus, the absence of expression of ISR in intact NahG plants must be due to a dependency of 7NSK2-elicited ISR on plant-derived SA accumulation. The amount of SA produced seems to be sufficient for inducing resistance but too low to induce detectable levels of PR-1 (De Meyer et al., 1999a). Nevertheless, the results remain puzzling and suggest that induced resistance against viruses is expressed differently from induced resistance against fungi and bacteria.

ISR triggered by *B. amyloliquefaciens* EXTN-1 against pepper mild mottle virus was manifested by strong reductions in systemic mosaic symptoms and viral RNA accumulation. Protection was associated with increased transcript levels of *PR-1a*, *PAL* and 3-hydroxy-3-methylglutaryl-CoA reductase (*HMGR*). Upon challenge inoculation with the virus, expression of these genes was enhanced in both inoculated and non-inoculated leaves of induced plants (Ahn et al., 2002), but a dependence on SA was not checked. Together, these data suggest that tobacco is more prone to react to ISR-eliciting rhizobacteria by expressing PRs than Arabidopsis. However, systemic protection induced by *Pseudomonas chlororaphis* strain O6 against *P. syringae* pv. *tabaci* and the soft rot bacterium *Erwinia carotovora* subsp. *carotovora* was maintained in NahG plants (Spencer et al., 2003). In leaves of bacterized plants no induction of PR-1a occurred, but transcripts of JA- and ethylene-inducible *PR-1g*, *Hmgr* and *Lox* accumulated, indicating that the bacterium activated the JA/

ethylene-dependent response pathway. Accordingly, ethylene-insensitive tobacco transformants did not display ISR against *E. carotovora*. However, resistance against *P. syringae* pv. *tabaci* was maintained in this genotype. Hence, the requirements for the expression of ISR against different pathogens are clearly different.

2.12 Tomato

In tomato, many strains of rhizobacteria have been tested for induction of ISR against various diseases. Using a split-root system, *P. fluorescens* WCS417r was found to induce resistance against *F. oxysporum* f.sp. *lycopersici* (Fol) (Duijff et al., 1998), the level of which appeared related to the presence of the OA of the LPS. ISR was associated with a thickening of cortical cell walls in tomato roots if epidermal or hypodermal cells were colonized densely by the bacterium (Duijff et al., 1997).

P. fluorescens 89B-27 and *S. marcescens* 90-166, when inoculated onto seeds, reduced disease severity of CMV challenge inoculated on the first pair of leaves three weeks after planting (Raupach et al., 1996). Combination of *B. subtilis* GB03 and either one of the *B. pumilus* strains SE34, INR7 or T4, or *B. amyloliquefaciens* IN937a or *B. subtilis* IN937b, formulated with the carrier chitosan and mixed through potting medium, all substantially slowed CMV accumulation and reduced the number of plants developing symptoms as well as symptom severity, and promoted plant growth and yield in greenhouse experiments (Murphy et al., 2003). Seed bacterization with three strains (SE34, IN937a and IN937b) was likewise effective under field conditions (Zehnder et al., 2000). Under these conditions, development of tomato mottle virus symptoms was also retarded (Murphy et al., 2000). In view of the intimate relationship between viral multiplication and plant metabolism, it is unlikely that these reductions in viral diseases were caused by a mechanism other than ISR, even though SE34 was reported to move upwards from the soil and colonize the phyllosphere (Yan et al., 2002, 2003).

The three strains, alone or in combination, were also effective in suppressing bacterial wilt, caused by *Ralstonia solanacearum* (Jetiyanon and Kloepper, 2002) and variably reduced southern blight caused by *Sclerotium rolfsii* (Jetiyanon et al., 2003). No in vitro antibiosis was observed, suggesting the mechanism to be ISR. Bacterial wilt was likewise suppressed when transplants at the time of seeding and one week prior to challenge were treated with *P. putida* 89B-61 or a combination of strains IN937a and *B. subtilis* GB03 ("BioYield") (Anith et al., 2004).

When incorporated into soilless growth media, strains 89B-61 as well as SE34 were also active in eliciting ISR against late blight, caused by the oomycete *Phytophthora infestans*. By testing NahG tomato, the ethylene-insensitive mutant *Nr*, and the JA-insensitive mutant *def1*, it was established that the induced protection by both strains was SA-independent, but ethylene- and JA-dependent (Yan et al., 2002). 89B-61 was not detected on leaves of bacterized plants, implying that systemic protection was due to ISR. In contrast, SE34 colonized the leaves to a population density of 10^6 cfu g^{-1}. Presumably, antibiosis was not involved because the bacteria and *P. infestans* did not antagonize each other in vitro. Plant-mediated protection appears to apply because the resistance was abolished in the *Nr* and *def1* genotypes. However, because SE34 was present at high densities on the leaves, protection could have resulted from localized rather than systemically induced resistance (Yan et al., 2003).

Bacterization of tomato seeds with a strain of *Bacillus cereus* retarded development of late blight, early blight (*Alternaria solani*) and septoria leaf spot (*Septoria lycopersici*) under field conditions. Disease severity was reduced by a marginal 13, 19 and 18%, respectively, and fruit yield was increased (Silva et al., 2004b). ISR was inferred, but not proven. Other bacterial strains (*B. cereus* B 101 R and B 212 K and the actinomycete A 068 R) were reported to elicit ISR against Pst, and in greenhouse bioassays reduced lesion formation by *A. solani*, *Corynespora cassiicola* (foliar blight), *Oidium lycopersici* (powdery mildew), *Stemphylium solani* (leaf spot) and *Xanthomonas campestris* pv. *vesicatoria* (Xcv) (bacterial spot) to different extents. Increases in peroxidase and lipoxygenase activities were detected in foliar extracts of bacterized plants, both before and after challenge, whereas no significant increases in PAL activity were evident (Silva et al., 2004a). Another isolate, *B. cereus* UFV101, when grown in modified Simmon's liquid culture produced a factor in the medium that did not inhibit *A. solani*, Pst, *C. cassiicola*, Xcv, *P. syringae* pv. *syringae*, *Pseudomonas corrugata*, *R. solanacearum* or *Clavibacter michiganensis* in vitro, yet when applied to tomato roots, reduced lesion numbers of Pst, Xcv, *A. solani* or *C. cassiicola* inoculated on the leaves (Romeiro et al., 2005). The nature of the elicitor(s) present was not clarified, however.

Earlier and increased activities of PAL, peroxidase and polyphenoloxidase activity, as well as high accumulation of phenolics were recorded in roots from tomato plants that were seed-treated with *P. fluorescens* strain Pf1 and challenged with *Pythium aphanidermatum* or Fol. Similarly, β-1,3-glucanase and chitinase were increased and thaumatin-like proteins accumulated at higher levels in bacterized plants and after challenge with Fol. No spatial separation between the bacteria and the pathogen was present and Pf1 antagonized mycelial growth of *P. aphanidermatum* in vitro. Although bacterized roots reacted more strongly to the pathogen than control roots, the evidence appears insufficient to claim that ISR was involved, as antibiotics can have some toxicity to plant roots. However, a priming effect on defense-related enzyme activities after challenge was apparent and may, or may not, be responsible for the reductions in damping-off and vascular wilt afforded by the bacterial treatment (Ramamoorthy et al., 2002a,b).

Split root trials were performed to demonstrate ISR against damping-off caused by *Rhizoctonia solani* upon root-dip treatment or soil drench with *P. aeruginosa* IE-65 or *P. fluorescens* CHA0. Root infection by *R. solani* as a result of inoculation of the soil one week after bacterial treatment was about halved. Dead bacterial cells were similarly active. Culture filtrate was likewise active, suggesting that the bacterial cells released inducing compounds into the medium. These data provide good proof of ISR being responsible for the disease reduction and imply induction by bacterial components that still need to be defined (Siddiqui and Shaukat, 2002b). Both rhizobacterial strains also induced resistance against the root knot nematode *Meloidogyne javanica* (Siddiqui and Shaukat, 2002a). The culture filtrate of CHA0, as well as of *P. aeruginosa* 7NSK2, caused significant mortality of juveniles of the nematode (Siddiqui and Shaukat, 2004), suggesting that the effect was due to a toxic compound. However, in split-root systems application of the bacterial cell suspension to one half of the root system lowered the population of the root knot nematode in the non-bacterized, nematode-treated other half, indicating that both rhizobacterial strains elicited ISR against attack by the root knot nematode. A similar level of ISR was attained in wild-type and in NahG tomato, confirming observations that mutants of 7NSK2 and CHA0 that lacked or overproduced SA did not influence bacterial efficacy to cause nematode death, and indicating that SA-independent ISR was involved (Siddiqui and Shaukat, 2004). A derivative of CHA0 that overexpressed the antibiotics DAPG and pyoluteorin triggered ISR more strongly than wild-type bacteria, whereas a *GacA* mutant deficient in antibiotic production, did not elicit ISR at all (Siddiqui and Shaukat, 2003). These results implicate antibiotic production by CHA0 as the mechanism of induction of systemic resistance in tomato against the nematode *M. javanica*, similar to the ISR triggered by this strain in Arabidopsis against *Peronospora parasitica* and Pst (Iavicoli et al., 2003; Weller et al., 2004).

In contrast, ISR triggered by 7NSK2 against *Botrytis cinerea* was lost in NahG plants and SA-lacking bacterial mutants were no longer capable of eliciting ISR in tomato (Audenaert et al., 2002). Mutant KMPCH, a producer of SA but not of the SA-containing siderophore pyochelin, induced PAL activity, whereas wild-type 7NSK2 did not, suggesting that in the wild type SA is converted into pyochelin. Pyochelin by itself did not elicit ISR. A mutant lacking the ability to produce the antibiotic pyocyanin was likewise impaired in triggering ISR. However, when the latter mutant and the SA- and pyochelin-lacking 7NSK2-562 were combined, the combination of both mutants did induce ISR. These results indicated that 7NSK2 elicited ISR through the combined action of the SA-containing siderophore pyochelin and the antibiotic pyocyanin, probably through a toxic action of reactive oxygen species (Audenaert et al., 2002). Strain *P. putida* WCS358 also elicited ISR in tomato against *B. cinerea*. Through mutant analysis and application of purified components it was established that this strain triggered ISR through its pseudobactin siderophore and its LPS (Meziane et al., 2005).

Except for the occasional increases in defense-related proteins and enzymes, little information is available on the mechanisms that are responsible for the expression of ISR against different pathogens in tomato. Mpiga et al. (1997) reported that tomato plants treated with *P. fluorescens* strain 63-28 allowed less ingress of *Fusarium oxysporum* f.sp. *radicis-lycopersici*, the causal agent of tomato crown and root rot. Light and electron microscopic analyses showed a restriction of the fungus to the outer root tissues with preferential localization in the intercellular spaces, accumulation of electron-dense material in epidermal and cortical cells and elaboration of callose-enriched wall appositions at sites of attempted fungal penetration. These features indicate a priming of defense responses against the fungus, which appears characteristic of induced resistance. Enhanced defenses against other pathogens in ISR-expressing tomato may similarly depend on priming of basal resistance responses against the various pathogens that have been tested.

3. CONCLUDING REMARKS

In the last few years, many studies have reported on ISR triggered by various root-colonizing bacteria in several dicotyledonous and some monocotyledonous plant species. Due to the diversity of plant species, rhizobacterial strains and pathogens used, it is difficult to draw general conclusions, except that triggering of ISR is not uncommon. This raises the question why, if all plants in nature host high numbers of bacteria on their roots, plants do not seem to be naturally induced already (Tuzun and Kloepper, 1995). However, only selected strains of rhizobacteria appear to be able to trigger ISR. Moreover, in those cases where dose-response relationships for ISR induction have been established, a minimum of 10^5 cfu g^{-1} root appears to be necessary for triggering ISR (Raaijmakers et al., 1995). Because of the enormous diversity of microorganisms on plant roots, this number is unlikely to be reached by any individual strain under natural conditions.

Strains seem to differ considerably in the determinants that are recognized by plant roots and give rise to ISR (Preston, 2004). Depending on bacterial strain and plant species, LPS, different types of siderophores, flagella, antibiotics or volatile alcohols have been shown to be capable of eliciting ISR, sometimes even in a plant species-specific manner (Van Loon and Bakker, 2005). This diversity implies that plants must possess intricate mechanisms for recognizing specific bacterial components. In Arabidopsis, a receptor-like kinase with a structure similar to that of several major resistance (*R*) genes has been shown to bind a common domain in bacterial flagellins (Gómez-Gómez, 2004), but its linkage to systemically induced resistance has not been well established. For none of the other bacterial inducing determinants is it known how perception is achieved.

Some rhizobacterial strains have been shown to elicit ISR in several plant species, others in some but not all species that have been studied. It may be that strains triggering ISR in several plant species possess more than a single determinant that can be recognized, i.e. *P. putida* WCS358r is recognized through at least its LPS, pseudobactin siderophore and flagella in Arabidopsis, bean and tomato (Meziane et al., 2005), but does not elicit ISR in carnation and radish. Otherwise, widely effective strains may possess an evolutionarily conserved determinant that is strongly recognized by many plant species. LPS is often considered to act as such a determinant and, indeed, has been found for many strains to be an important inducing

Table 2. Demonstrated effectiveness of specific rhizobacterial strains in induction of systemic resistance in at least three plant species [a].

Bacterial strain	*Plant species*
Bacillus amyloliquefaciens IN937a	Arabidopsis, cucumber, pearl millet, pepper, tomato
Bacillus pumilus INR7	Cucumber, tall fescue, tomato
Bacillus pumilus SE34	Arabidopsis, cucumber, pearl millet, pepper, pine, tobacco, tomato
Bacillus pumilus T4	Arabidopsis, cucumber, pearl millet, pepper, tobacco, tomato
Bacillus subtilis GB03	Arabidopsis, cucumber, pearl millet, tomato
Bacillus subtilis IN937b	Cucumber, pearl millet, pepper, tomato
Pseudomonas aeruginosa 7NSK2	Arabidopsis, bean, tobacco, tomato
Pseudomonas fluorescens CHA0	Arabidopsis, sugarcane, tobacco, tomato
Pseudomonas fluorescens Pf1	Pepper, rice, sugarcane, tomato
Pseudomonas fluorescens WCS417	Arabidopsis, bean, carnation, radish, tall fescue, tomato
Pseudomonas fluorescens 89B-61	Arabidopsis, cucumber, tall fescue, tobacco, tomato
Pseudomonas putida WCS358	Arabidopsis, bean, tomato
Serratia marcescens 90-166	Arabidopsis, cucumber, pine, tobacco

[a] See main text for references.

factor in more than one plant species. Yet, the structure of the LPS can vary substantially, and strains possessing LPS that is recognized by one plant species but not by another, do occur, e.g. *P. fluorescens* WCS374 elicits ISR in radish through its LPS but under the same conditions does not elicit ISR in Arabidopsis (Van Wees et al., 1997). A listing of strains that have been found to be effective in several plant species is presented in Table 2.

In those systems in which the signal-transduction pathway of ISR has been studied by making use of mutant and transgenic plants, almost all inducing bacteria have been shown to elicit ISR through a SA-independent route (Van Loon and Bakker, 2005). Thus, rhizobacterially-induced ISR is different from SA-dependent SAR. Indeed, whereas SAR is usually most effective against biotrophic pathogens, ISR is often more effective against necrotrophic pathogens, and the combination of SAR and ISR can afford greater protection against a wider range of pathogens (Van Wees et al., 2000; Ton et al., 2002a). Only ISR triggered by *P. aeruginosa* strain 7NSK2 has been demonstrated to be dependent on SA. For ISR in tomato, SA is required for the production of the siderophore pyochelin, which, in conjunction with the antibiotic pyocyanin, triggers ISR (Audenaert et al., 2002). Moreover, results from tobacco indicate that 7NSK2-elicted ISR involves SA accumulation in planta (De Meyer et al., 1999a). So far, these results have not been confirmed for other plant species, and the mechanisms involved remain to be clarified. SA-independent ISR in different rhizobacterium-plant-pathogen combinations has been shown to be variably dependent on ethylene and/or JA perception, and these variations also need to be investigated in more detail.

Recent gene expression studies have revealed a remarkable variation in the number of genes that are activated in plants upon bacterization by different rhizobacterial strains. As shown by Verhagen et al. (2004), Arabidopsis plants can be in the state of ISR without expressing any changes in gene transcripts in induced leaves. Thus, one should be cautious not to link any alterations in gene expression, protein accumulation or enzyme activities to ISR unless a causal connection can be firmly established. Only after challenge inoculation, defense responses are boosted. This has been observed in several studies and points to the importance of priming of resistance mechanisms in ISR. In this way, invading pathogens can be resisted earlier and to a greater extent than in non-induced plants. This enhanced defensive capacity appears the main characteristic of rhizobacterially induced systemic resistance (Conrath et al., 2002). Elucidation of the priming phenomenon will provide further insight into how the recognition of various bacterial determinants by plant roots is channeled through a largely common JA- and ethylene-dependent signalling pathway towards the state of attentive alert in the plant.

4. REFERENCES

Ahn, I.P., Park, K. and Kim, C.H., 2002. Rhizobacteria-induced resistance perturbs viral disease progress and triggers defense-related gene expression. Mol. Cells 13:302-308.

Albrecht, C., Geurts, R. and Bisseling, T., 1999. Legume nodulation and mycorrhizae formation; two extremes in host specificity meet. EMBO J. 18:281-288.

Alström, S., 1991. Induction of disease resistance in common bean susceptible to halo blight bacterial pathogen after seed bacterization with rhizosphere pseudomonads. J. Gen. Appl. Microbiol. 37:495-501.

Alström, S., 1995. Evidence of disease resistance induced by rhizosphere pseudomonad against *Pseudomonas syringae* pv. *phaseolicola*. J. Gen. Appl. Microbiol. 41:315-325.

Anith, K.N., Momol, M.T., Kloepper, J.W., Marois, J.J., Olson, S.M. and Jones, J.B., 2004. Efficacy of plant growth-promoting rhizobacteria, acibenzolar-S-methyl, and soil amendment for integrated management of bacterial wilt on tomato. Plant Dis. 88:669-673.

Audenaert, K., Pattery, T., Cornelis, P. and Höfte, M., 2002. Induction of systemic resistance to *Botrytis cinerea* in tomato by *Pseudomonas aeruginosa* 7NSK2: role of salicylic acid, pyochelin, and pyocyanin. Mol. Plant-Microbe Interact. 15:1147-1156.

Baayen, R.P. and Niemann, G.J., 1989. Correlations between accumulation of dianthramides, dianthalexin and unknown compounds, and partial resistance to *Fusarium oxysporum* f.sp. *dianthi* in eleven carnation cultivars. J. Phytopathol. 126:281-292.

Bakker, P.A.H.M., Ran, L.X., Pieterse, C.M.J. and Van Loon, L.C., 2003. Understanding the involvement of rhizobacteria-mediated induction of systemic resistance in biocontrol of plant diseases. *Can. J. Plant Pathol.* 25:5-9.

Barea, J.M., Azcon, R. and Azcon-Aguilar, C., 2002. Mycorrhizosphere interactions to improve plant fitness and soil quality. Antonie van Leeuwenhoek Int. J. Gen. Mol. Microbiol. 81:343-351.

Benhamou, N., Bélanger, R.R. and Paulitz, T.C., 1996a. Induction of differential host responses by *Pseudomonas fluorescens* in *Ri* T-DNA-transformed pea roots after challenge with *Fusarium oxysporum* f.sp. *pisi* and *Pythium ultimum*. Phytopathology 86:1174-1185.

Benhamou, N., Kloepper, J.W., Quadt-Hallman, A. and Tuzun, S., 1996b. Induction of defense-related ultrastructural modifications in pea root tissues inoculated with endophytic bacteria. Plant Physiol. 112:919-929.

Benhamou, N., Gagné, S., Le Quéré, D. and Dehbi, L., 2000. Bacterial-mediated induced resistance in cucumber: beneficial effect of the endophytic bacterium *Serratia plymuthica* on the protection against infection by *Pythium ultimum*. Phytopathology 90:45-56.

Bergstrom, G.C., Johnson, M.C. and Kuc, J., 1982. Effects of local infection of cucumber by *Colletotrichum lagenarium*, *Pseudomonas lachrymans*, or tobacco necrosis virus on systemic resistance to cucumber mosaic virus. Phytopathology 72:922-926.

Bharathi, R., Vivekananthan, R., Harish, S., Ramanathan, A. and Samiyappan, R., 2004. Rhizobacteria-based bio-formulations for the management of fruit rot infection in chillies. Crop Prot. 23:835-843.

Bigirimana, J. and Höfte, M., 2002. Induction of systemic resistance to *Colletotrichum lindemuthianum* in bean by a benzothiadiazole derivative and rhizobacteria. Phytoparasitica 30:159-168.

Brederode, F.T., Linthorst, H.J.M. and Bol, J.F., 1991. Differential induction of acquired resistance and PR gene expression in tobacco by virus infection, ethephon treatment, UV light and wounding. Plant Mol. Biol. 17:1117-1125.

Cao, H., Bowling, S.A., Gordon, A.S. and Dong, X., 1994. Characterization of an *Arabidopsis* mutant that is nonresponsive to inducers of systemic acquired resistance. Plant Cell 6:1583-1592.

Cao, H., Glazebrook, J., Clarke, J.D., Volko, S. and Dong, X.N., 1997. The Arabidopsis *NPR1* gene that controls systemic acquired resistance encodes a novel protein containing ankyrin repeats. Cell 88:57-63.

Cartieaux, F., Thibaud, M.C., Zimmerli, L., Lessard, P., Sarrobert, C., David, P., Gerbaud, A., Robaglia, C., Somerville, S. and Nussaume, L., 2003. Transcriptome analysis of *Arabidopsis* colonized by a plant-growth promoting rhizobacterium reveals a general effect on disease resistance. Plant J. 36:177-188.

Chen, C., Bélanger, R.R., Benhamou, N. and Paulitz, T.C., 1998. Induced systemic resistance (ISR) by *Pseudomonas* spp. impairs pre- and post-infection development of *Pythium aphanidermatum* on cucumber roots. Eur. J. Plant Pathol. 104:877-886.

Chen, C., Bélanger, R.R., Benhamou, N. and Paulitz, T.C., 1999. Role of salicylic acid in systemic resistance induced by *Pseudomonas* spp. against *Pythium aphanidermatum* in cucumber roots. Eur. J. Plant Pathol. 105:477-486.

Chen, C., Bélanger, R.R., Benhamou, N. and Paulitz, T.C., 2000. Defense enzymes induced in cucumber roots by treatment with plant growth-promoting rhizobacteria (PGPR) and *Pythium aphanidermatum*. Physiol. Mol. Plant Pathol. 56:13-23.

Collmer, A., Badel, J.L., Charkowski, A.O., Deng, W.L., Fouts, D.E., Ramos, A.R., Rehm, A.H., Anderson, D.M., Schneewind, O., Van Dijk, K. and Alfano, J.R., 2000. *Pseudomonas syringae* Hrp type III secretion system and effector proteins. Proc. Natl. Acad. Sci. USA 97:8770-8777.

Conrath, U., Pieterse, C.M.J. and Mauch-Mani, B., 2002. Priming in plant-pathogen interactions. Trends Plant Sci. 7:210-216.

Coventry, H.S. and Dubery, I.A., 2001. Lipopolysaccharides from *Burkholderia cepacia* contribute to an enhanced defensive capacity and the induction of pathogenesis-related proteins in *Nicotianae tabacum*. Physiol. Mol. Plant Pathol. 58:149-158.

Delaney, T.P., Uknes, S., Vernooij, B., Friedrich, L., Weymann, K., Negrotto, D., Gaffney, T., Gut-Rella, M., Kessmann, H., Ward, E. and Ryals, J., 1994. A central role of salicylic acid in plant disease resistance. Science 266:1247-1250.

De Boer, M., Van der Sluis, I., Van Loon, L.C. and Bakker, P.A.H.M., 1999. Combining fluorescent *Pseudomonas* spp. strains to enhance suppression of fusarium wilt of radish. Eur. J. Plant Pathol. 105:201-219.

De Meyer, G. and Höfte, M., 1997. Salicylic acid produced by the rhizobacterium *Pseudomonas aeruginosa* 7NSK2 induces resistance to leaf infection by *Botrytis cinerea* on bean. Phytopathology 87:588-593.

De Meyer, G., Audenaert, K. and Höfte, M., 1999a. *Pseudomonas aeruginosa* 7NSK2-induced systemic resistance in tobacco depends on *in planta* salicylic acid accumulation but is not associated with PR1a expression. Eur. J. Plant Pathol. 105:513-517.

De Meyer, G., Capieau, K., Audenaert, K., Buchala, A., Métraux, J.-P. and Höfte, M. 1999b. Nanogram amounts of salicylic acid produced by the rhizobacterium *Pseudomonas aeruginosa* 7NSK2 activate the systemic acquired resistance pathway in bean. Mol. Plant-Microbe Interact. 12:450-458.

Delaney, T., Friedrich, L. and Ryals, J., 1995. *Arabidopsis* signal transduction mutant defective in chemically and biologically induced disease resistance. Proc. Natl. Acad. Sci. USA 92:6602-6606.

Duijff, B.J., Gianinazzi-Pearson, V. and Lemanceau, P., 1997. Involvement of the outer membrane lipopolysaccharides in the endophytic colonization of tomato roots by biocontrol *Pseudomonas fluorescens* strain WCS417r. New Phytol. 135:325-334.

Duijff, B.J., Pouhair, D., Olivain, C., Alabouvette, C. and Lemanceau, P., 1998. Implication of systemic induced resistance in the suppression of fusarium wilt of tomato by *Pseudomonas fluorescens* WCS417r and by nonpathogenic *Fusarium oxysporum* Fo47. Eur. J. Plant Pathol. 104:903-910.

Enebak, S.A. and Carey, W.A., 2000. Evidence for induced systemic protection to fusiform rust in loblolly pine by plant growth-promoting rhizobacteria. Plant Dis. 84:306-308.

Gaffney, T., Friedrich, L., Vernooij, B., Negrotto, D., Nye, G., Uknes, S., Ward, E., Kessmann, H. and Ryals, J., 1993. Requirement of salicylic acid for the induction of systemic acquired resistance. Science 261:754-756.

Gianinazzi-Pearson, V., 1996. Plant cell responses to arbuscular mycorrhizal fungi: getting to the roots of the symbiosis. Plant Cell 8:1871-1883.

Gianinazzi-Pearson, V., Dumas-Gaudot, E., Gollotte, A, Tahiri-Alaoui, A. and Gianinazzi, S., 1996. Cellular and molecular defence-related root responses to invasion by arbuscular mycorrhizal fungi. New Phytol. 133:45-57.

Glazebrook, J., Rogers, E.E. and Ausubel, F.M., 1996. Isolation of *Arabidopsis* mutants with enhanced disease susceptibility by direct screening. Genetics 143:973-982.

Glick, B.R., Patten, C.L., Holguin, G. and Penrose, D.M., 1999. Biochemical and Genetic Mechanisms Used by Plant Growth Promoting Bacteria. Imperial College Press, London.

Gómez-Gómez, L., 2004. Plant perception systems for pathogen recognition and defence. Mol. Immunol. 41:1055-1062.

Hammerschmidt, R. and Kuc, J., 1995. Induced Resistance to Disease in Plants. Kluwer Acad. Publ., Dordrecht.

Handelsman, J. and Stabb, E.V., 1996. Biocontrol of soilborne plant pathogens. Plant Cell 8:1855-1869.

Harrison, M.J., 1998. Development of the arbuscular mycorrhizal symbiosis. Curr. Opin. Plant Biol. 1:360-365.

Hasky-Günther, K., Hoffmann-Hergarten, S. and Sikora, R.A., 1998. Resistance against the potato cyst nematode *Globodera pallida* systemically induced by the rhizobacteria *Agrobacterium radiobacter* (G12) and *Bacillus sphaericus* (B43). Fundam. Appl. Nematol. 21:511-517.

He, S.Y., Jin, Q.-L., Thilmony, R., Nomura, K., Zwiesler-Vollick, J., Bandyopadhyay, S., Hauck, P., Hu, W., Plovanich-Jones, A., Lu, M. and Bray, E., 2002. The molecular basis of *Pseudomonas syringae* pathogenesis in susceptible *Arabidopsis thaliana* plants. In: Biology of Plant-Microbe Interactions, Vol. 3, Leong, S.A., Allem, C. and Triplett, E.W. (eds), Int. Soc. Mol. Plant-Microbe Interact., St. Paul, MN, pp. 129-134.

Hoffland, E., Pieterse, C.M.J., Bik, L. and Van Pelt, J.A., 1995. Induced systemic resistance in radish is not associated with accumulation of pathogenesis-related proteins. Physiol. Mol. Plant Pathol. 46:309-320.

Hoffland, E., Hakulinen, J. and Van Pelt, J.A., 1996. Comparison of systemic resistance induced by avirulent and non-pathogenic *Pseudomonas* species. Phytopathology 86:757-762.

Hynes, R.K., Hill, J., Reddy, M.S. and Lazarovits, G., 1994. Phytoalexin production by wounded white bean (*Phaseolus vulgaris*) cotyledons and hypocotyls in response to inoculation with rhizobacteria. Can. J. Microbiol. 40:548-554.

Iavicoli, A., Boutet, E., Buchala, A. and Métraux, J.-P., 2003. Induced systemic resistance in *Arabidopsis thaliana* in response to root inoculation with *Pseudomonas fluorescens* CHA0. Mol. Plant-Microbe Interact. 16:851-858.

Jetiyanon, K. and Kloepper, J.W., 2002. Mixtures of plant growth-promoting rhizobacteria for induction of systemic resistance against multiple plant diseases. Biol. Control 24:285-291.

Jetiyanon, K., Fowler, W.D. and Kloepper, J.W., 2003. Broad-spectrum protection against several pathogens by PGPR mixtures under field conditions in Thailand. Plant Dis. 87:1390-1394.

Jeun, Y.C., Park, K.S., Kim, C.H., Fowler, W.D. and Kloepper, J.W., 2004. Cytological observations of cucumber plants during induced resistance elicited by rhizobacteria. Biol. Control 29:34-42.

Kempster, V.N., Davies, K.A. and Scott, E.S., 2001. Chemical and biological induction of resistance to the clover cyst nematode (*Heterodera trifolii*) in white clover (*Trifolium repens*). Nematology 3:35-43.

Kempster, V.N., Scott, E.S. and Davies, K.A., 2002. Evidence for systemic, cross-resistance in white clover (*Trifolium repens*) and annual medic (*Medicago truncatula* var *truncultata*) induced by biological and chemical agents. Biocontrol Sci. Technol. 12:615-623.

Kilic-Ekici, O. and Yuen, G.Y., 2003. Induced resistance as a mechanism of biological control by *Lysobacter enzymogenes* strain C3. Phytopathology 93:1103-1110.

Kilic-Ekici, O. and Yuen, G.Y., 2004. Comparison of strains of *Lysobacter enzymogenes* and PGPR for induction of resistance against *Bipolaris sorokiniana* in tall fescue. Biol. Control 30:446-455.

Kim, M.S., Kim, Y.C. and Cho, B.H., 2004. Gene expression analysis in cucumber leaves primed by root colonization with *Pseudomonas chlororaphis* O6 upon challenge-inoculation with *Corynespora cassiicola*. Plant Biol. 6:105-108.

Kloepper, J.W., Tuzun, S., Liu, L. and Wei, G., 1993. Plant growth-promoting rhizobacteria as inducers of systemic disease resistance. In: Pest Management: Biologically Based Technologies, Lumsden, R.D. and Vaughn, J.L. (eds), Amer. Chem. Soc, Washington DC, pp. 156-165.

Kloepper, J.W., Leong, J., Teintze, M. and Schroth, M.N., 1980. Enhanced plant growth by siderophores produced by plant growth-promoting rhizobacteria. Nature 286:885-886.

Kluepfel, D.A., 1993. The behavior and tracking of bacteria in the rhizosphere. Annu. Rev. Phytopathol. 31:441-472.

Knoester, M., Pieterse, C.M. J., Bol, J.F. and Van Loon, L.C., 1999. Systemic resistance in Arabidopsis induced by rhizobacteria requires ethylene-dependent signaling at the site of application. Mol. Plant-Microbe Interact. 12:720-727.

Krause, M.S., De Ceuster, T.J.J., Tiquia, S.M., Michel Jr., F.C., Madden, L.V. and Hoitink, H.A.J., 2003. Isolation and characterization of rhizobacteria from composts that suppress the severity of bacterial leaf spot of radish. Phytopathology 93:1292-1300.

Krishnamurthy, K. and Gnanamanickam, S.S., 1998. Induction of systemic resistance and salicylic acid accumulation in *Oryza sativa*, L. in the biological suppression of rice blast caused by treatments with *Pseudomonas* spp. World J. Microbiol. Biotechnol. 14:935-937.

Kuc, J., 1982. Induced immunity to plant disease. Bioscience 32:854-860.

Lamb, T.G., Tonkyn, D.W. and Kluepfel, D.A., 1996. Movement of *Pseudomonas aureofaciens* from the rhizosphere to aerial plant tissue. Can. J. Microbiol. 42:1112-1120.

Lavelle, P. and Spain, A.V., 2001. Soil Ecology, Kluwer Acad. Publ., Dordrecht.

Lawton, K.A., Potter, S.L., Uknes, S. and Ryals, J., 1994. Acquired resistance signal transduction in *Arabidopsis* is ethylene independent. Plant Cell 6:581-588.

Leeman, M., Van Pelt, J.A., Den Ouden, F.M., Heinsbroek, M., Bakker, P.A.H.M. and Schippers, B., 1995a. Induction of systemic resistance by *Pseudomonas fluorescens* in radish cultivars differing in susceptibility to fusarium wilt, using a novel bioassay. Eur. J. Plant Pathol. 101:655-664.

Leeman, M., Van Pelt, J.A., Den Ouden, F.M., Heinsbroek, M., Bakker, P.A.H.M. and Schippers, B., 1995b. Induction of systemic resistance against fusarium wilt of radish by lipopolysaccharides of *Pseudomonas fluorescens*. Phytopathology 85:1021-1027.

Leeman, M., Den Ouden, F.M., Van Pelt, J.A., Dirkx, F.P.M., Steijl, H., Bakker, P.A.H.M. and Schippers, B., 1996. Iron availability affects induction of systemic resistance against fusarium wilt of radish by *Pseudomonas fluorescens*. Phytopathology 86:149-155.

Liu, L., Kloepper, J.W. and Tuzun, S., 1995a. Induction of systemic resistance in cucumber against fusarium wilt by plant growth-promoting rhizobacteria. Phytopathology 85:695-698.

Liu, L., Kloepper, J.W. and Tuzun, S., 1995b. Induction of systemic resistance in cucumber against angular leaf spot by plant growth-promoting rhizobacteria. Phytopathology 85:843-847.

Liu, L., Kloepper, J.W. and Tuzun, S., 1995c. Induction of systemic resistance in cucumber by plant growth-promoting rhizobacteria: duration of protection and effect of host resistance on protection and root colonization. Phytopathology 85:1064-1068.

M'piga, P., Bélanger, R.R., Paulitz, T.C. and Benhamou, N., 1997. Increased resistance to *Fusarium oxysporum* f.sp. *radicis-lycopersici* in tomato plants treated with the endophytic bacterium *Pseudomonas fluorescens* strain 63-28. Physiol. Mol. Plant Pathol. 50:301-320.

Madhaiyan, M., Poonguzhali, S., Senthilkumar, M., Seshadri, S., Chung, H., Yang, J., Sundaram, S. and Sa, T., 2004. Growth promotion and induction of systemic resistance in rice cultivar Co-47 (*Oryza sativa* L.) by *Methylobacterium* spp. Bot. Bull. Acad. Sin. 45:315-324.

Maurhofer, M., Hase, C., Meuwly, P., Métraux, J.-P. and Défago, G., 1994. Induction of systemic resistance of tobacco to tobacco necrosis virus by the root-colonizing *Pseudomonas fluorescens* strain CHA0: influence of the *gacA* gene and of pyoverdine production. Phytopathology 84:139-146.

Maurhofer, M., Keel, C., Haas, D. and Défago, G., 1995. Influence of plant species on disease suppression by *Pseudomonas fluorescens* strain CHA0 with enhanced antibiotic production. Plant Pathol. 44:40-50.

Maurhofer, M., Reimmann, C., Schmidli-Sacherer, P., Heeb, S., Haas, D. and Défago, G., 1998. Salicylic acid biosynthetic genes expressed in *Pseudomonas fluorescens* strain P3 improve the induction of systemic resistance in tobacco against tobacco necrosis virus. Phytopathology 88:678-684.

Mercado-Blanco, J., Van der Drift, K.M.G.M., Olsson, P.E., Thomas-Oates, J.E., Van Loon, L.C. and Bakker, P.A.H.M., 2001. Analysis of the *pmsCEAB* gene cluster involved in biosynthesis of salicylic acid and the siderophore pseudomonine in the biocontrol strain *Pseudomonas fluorescens* WCS374. J. Bacteriol. 183:1909-1920.

Meyer, J.M., Azelvandre, P. and Georges, C., 1992. Iron metabolism in *Pseudomonas*: salicylic acid, a siderophore of *Pseudomonas fluorescens* CHA0. BioFactors 4:23-27.

Meziane, H., Van der Sluis, I., Van Loon, L.C., Höfte, M. and Bakker, P.A.H.M., 2005. Determinants of *Pseudomonas putida* WCS358 involved in inducing systemic resistance in plants. Mol. Plant Pathol. 6:177-185.

Mohandas, S., Manamohan, M., Rawal, R.D., Chakraborty, S., Sreekantappa, H., Manjula, R. and Lakshmikantha, H.C., 2004. Interaction of *Fusarium oxysporum* f.sp. *cubense* with *Pseudomonas fluorescens* precolonized to banana roots. World J. Microbiol. Biotechnol. 20:651-655.

Mou, Z., Fan, W.H. and Dong, X.N., 2003. Inducers of plant systemic acquired resistance regulate NPR1 function through redox changes. Cell 113:935-944.

Mulder, L., Hogg, B., Bersoult, A. and Cullimore, J.V., 2005. Integration of signalling pathways in the establishment of the legume-rhizobia symbiosis. Physiol. Plant. 123:207-218.

Murphy, J.F., Zehnder, G.W., Schuster, D.J., Sikora, E.J., Polston, J.E. and Kloepper, J.W., 2000. Plant growth-promoting rhizobacterial mediated protection in tomato against *tomato mottle virus*. Plant Dis. 84:779-784.

Murphy, J.F., Reddy, M.S., Ryu, C.-M., Kloepper, J.W. and Li, R., 2003. Rhizobacteria-mediated growth promotion of tomato leads to protection against *cucumber mosaic virus*. Phytopathology 93:1301-1307.

Nandakumar, R., Babu, S., Viswanathan, R., Raguchander, T. and Samiyappan, R., 2001. Induction of systemic resistance in rice against sheath blight disease by *Pseudomonas fluorescens*. Soil Biol. Biochem. 33:603-612.

Niranjan Raj, S., Deepak, S.A., Basavaraju, P., Shetty, H.S., Reddy, M.S. and Kloepper, J.W., 2003a. Comparative performance of formulations of plant growth promoting rhizobacteria in growth promotion and suppression of downy mildew in pearl millet. Crop Prot. 22:579-588.

Niranjan Raj, S., Shetty, N.P. and Shetty, H.S., 2003b. Seed bio-priming with *Pseudomonas fluorescens* isolates enhances growth of pearl millet plants and induces resistance against downy mildew. Int. J. Pest Manag. 50:41-49.

Oliver, R.P. and Ipcho, S.V.S., 2004. *Arabidopsis* pathology breathes new life into the necrotrophs-vs. biotrophs classification of fungal pathogens. Mol. Plant Pathol. 5:347-352.

Ongena, M., Giger, A., Jacques, P., Dommes, J. and Thonart, P., 2002. Study of bacterial determinants involved in the induction of systemic resistance in bean by *Pseudomonas putida* BTP1. Eur. J. Plant Pathol. 108:187-196.

Ongena, M., Daayf, F., Jacques, P., Thonart, P., Benhamou, N., Paulitz, T.C., Cornélis, P., Koedam, N. and Bélanger, R.R., 1999. Protection of cucumber against Pythium root rot by fluorescent pseudomonads: predominant role of induced resistance over siderophores and antibiosis. Plant Pathol. 48:66-76.

Ongena, M., Daayf, F., Jacques, P., Thonart, P., Benhamou, N., Paulitz, T.C. and Bélanger, R.R., 2000. Systemic induction of phytoalexins in cucumber in response to treatments with fluorescent pseudomonads. Plant Pathol. 49:523-530.

Ongena, M., Duby, F., Rossignol, F., Fauconnier, M.-L., Dommes, J. and Thonart, P., 2004. Stimulation of the lipoxygenase pathway is associated with systemic resistance induced in bean by a nonpathogenic *Pseudomonas* strain. Mol. Plant-Microbe Interact. 17:1009-1018.

Park, K.S. and Kloepper, J.W., 2000. Activation of PR-1a promoter by rhizobacteria that induce systemic resistance in tobacco against *Pseudomonas syringae* pv. *tabaci*. Biol. Control 18:2-9.

Persello-Cartieaux, F., Nussaume, L. and Robaglia, C., 2003. Tales from the underground: molecular plant-rhizobacteria interactions. Plant Cell Environ. 26:189-199.

Piater, L.A., Nürnberger, T. and Dubery, I.A., 2004. Identification of a lipopolysaccharide responsive erk-like MAP kinase in tobacco leaf tissue. Mol. Plant Pathol. 5:331-341.

Pieterse, C.M.J., Van Wees, S.C.M., Hoffland, E., Van Pelt, J.A. and Van Loon, L.C., 1996. Systemic resistance in *Arabidopsis* induced by biocontrol bacteria is independent of salicylic acid accumulation and pathogenesis-related gene expression. Plant Cell 8:1225-1237.

Pieterse, C.M.J., Van Wees, S.C.M., Van Pelt, J.A., Knoester, M., Laan, R., Gerrits, H., Weisbeek, P.J. and Van Loon, L.C., 1998. A novel signaling pathway controlling induced systemic resistance in Arabidopsis. Plant Cell 10:1571-1580.

Pieterse, C.M.J., Van Pelt, J.A., Ton, J., Parchmann, S., Mueller, M.J., Buchala, A.J., Métraux, J.-P. and Van Loon, L.C., 2000. Rhizobacteria-mediated induced systemic resistance (ISR) in *Arabidopsis* requires sensitivity to jasmonate and ethylene but is not accompanied by an increase in their production. Physiol. Mol. Plant Pathol. 57:123-134.

Press, C.M., Wilson, M., Tuzun, S. and Kloepper, J.W., 1997. Salicylic acid produced by *Serratia marcescens* 90-166 is not the primary determinant of induced systemic resistance in cucumber or tobacco. Mol. Plant-Microbe Interact. 10:761-768.

Press, C.M., Loper, J.E. and Kloepper, J.W., 2001. Role of iron in rhizobacteria-mediated induced systemic resistance of cucumber. Phytopathology 91:593-598.

Preston, G.M., 2004. Plant perceptions of plant growth-promoting Pseudomonas. Phil. Trans. Roy. Soc. London B 359:907-918.

Preston, G.M., Bertrand, N. and Rainey, P.B., 2001. Type III secretion in plant growth-promoting *Pseudomonas fluorescens* SBW25. Mol. Microbiol. 41:999-1014.

Raaijmakers, J.M., Leeman, M., Van Oorschot, M.P.M., Van der Sluis, I., Schippers, B. and Bakker, P.A.H.M., 1995. Dose-response relationships in biological control of fusarium wilt of radish by *Pseudomonas* spp. Phytopathology 85:1075-1081.

Radja Commare, R., Nandakumar, R., Kandan, A., Suresh, S., Bharathi, M., Raguchander, T. and Samiyappan, R., 2002. *Pseudomonas fluorescens* based bio-formulation for the management of sheath blight disease and leaffolder insect in rice. Crop Prot. 21:671-677.

Radjacommare, R., Kandan, A., Nandakumar, R. and Samiyappan, R., 2004. Association of the hydrolytic enzyme chitinase against *Rhizoctonia solani* in rhizobacteria-treated rice plants. J. Phytopathol. 152:365-370.

Ramamoorthy, V., Raguchander, T. and Samiyappan, R., 2002a. Enhancing resistance of tomato and hot pepper to Pythium diseases by seed treatment with fluorescent pseudomonads. Eur. J. Plant Pathol. 108:429-441.

Ramamoorthy, V., Raguchander, T. and Samiyappan, R., 2002b. Induction of defense-related proteins in tomato roots treated with *Pseudomonas fluorescens* Pf1 and *Fusarium oxysporum* f.sp. *lycopersici*. Plant Soil 239:55-68.

Ran, L.X., Li, Z.N., Wu, G.J., Van Loon, L.C. and Bakker, P.A.H.M., 2005. Induction of systemic resistance against bacterial wilt in *Eucalyptus urophylla* by fluorescent *Pseudomonas* spp. Eur. J. Plant Pathol., in press.

Ran, L.X., Van Loon, L.C. and Bakker, P.A.H.M., 2005b. No role for bacterially produced salicylic acid in rhizobacterial induction of systemic resistance in *Arabidopsis*. Phytopathology, in press.

Raupach, G.S. and Kloepper, J.W., 1998. Mixtures of plant growth-promoting rhizobacteria enhance biological control of multiple cucumber pathogens. Phytopathology 88:1158-1164.

Raupach, G.S., Liu, L., Murphy, J.F., Tuzun, S. and Kloepper, J.W., 1996. Induced systemic resistance in cucumber and tomato against cucumber mosaic cucumovirus using plant growth-promoting rhizobacteria (PGPR). Plant Dis. 80:891-894.

Reitz, M., Rudolph, K., Schröder, I., Hoffmann-Hergarten, S., Hallmann, J. and Sikora, R.A., 2000. Lipopolysaccharides of *Rhizobium etli* strain G12 act in potato roots as an inducing agent of systemic resistance to infection by the cyst nematode *Globodera pallida*. Appl. Environ. Microbiol. 66:3515-3518.

Reitz, M., Hoffmann-Hergarten, S., Hallmann, J. and Sikora, R.A., 2001. Induction of systemic resistance in potato by rhizobacterium *Rhizobium etli* strain G12 is not associated with accumulation of pathogenesis-related proteins and enhanced lignin biosynthesis. Z. Pflanzenkrankh. Pflanzenschutz 108:11-20.

Reitz, M., Oger, P., Meyer, A., Niehaus, K., Farrand, S.K., Hallmann, J. and Sikora, R.A., 2002. Importance of the O-antigen, core-region and lipid A of rhizobial lipopolysaccharides for the induction of systemic resistance in potato to *Globodera pallida*. Nematology **4**:73-79.

Roman, G., Lubarsky, B., Kieber, J.J., Rothenberg, M., and Ecker, J.R., 1995. Genetic analysis of ethylene signal transduction in *Arabidopsis thaliana*: five novel mutant loci integrated into a stress response pathway. Genetics 139:1393-1409.

Romeiro, R.S., Filho, L., Vieira Junior, J.R., Silva, H.S.A., Baracat-Pereira, M.C. and Carvalho, M.G., 2005. Macromolecules released by a plant growth-promoting rhizobacterium as elicitors of systemic resistance in tomato to bacterial and fungal pathogens. J. Phytopathol. 153:120-123.

Ross, A.F., 1961. Localized acquired resistance to plant virus infection in hypersensitive hosts. Virology 14:329-339.

Ryals, J.A., Neuenschwander, U.H., Willits, M.G., Molina, A., Steiner, H.Y. and Hunt, M.D., 1996. Systemic acquired resistance. Plant Cell 8:1809-1819.

Ryu, C.-M., Hu, C.H., Reddy, M.S. and Kloepper, J.W., 2003a. Different signaling pathways of induced resistance by rhizobacteria in *Arabidopsis thaliana* against two pathovars of *Pseudomonas syringae*. New Phytol. 160:413-420.

Ryu, C.-M., Farag, M.A., Hu, C.H., Reddy, M.S., Wei, H.X., Paré, P.W. and Kloepper, J.W., 2003b. Bacterial volatiles promote growth in *Arabidopsis*. Proc. Natl. Acad. Sci. USA 100:4927-4932.

Ryu, C.-M., Murphy, J.F., Mysore, K.S. and Kloepper, J.W., 2004a. Plant growth-promoting rhizobacteria systemically protect *Arabidopsis thaliana* against *Cucumber mosaic virus* by a salicylic acid and NPR1-independent and jasmonic acid-dependent signaling pathway. Plant J. 39:381-392.

Ryu, C.-M., Farag, M.A., Hu, C.H., Reddy, M.S., Kloepper, J.W. and Paré, P.W., 2004b. Bacterial volatiles induce systemic resistance in Arabidopsis. Plant Physiol. 134:1017-1026.

Sarma, B.K., Singh, D.P., Mehta, S., Singh, H.B. and Singh, U.P., 2002. Plant growth-promoting rhizobacteria-elicited alterations in phenolic profile of chickpea (*Cicer arietinum*) infected by *Sclerotium rolfsii*. J. Phytopathol. 150:277-282.

Serino, L., Reimmann, C., Baur, H., Beyeler, M., Visca, P. and Haas, D., 1995. Structural genes for salicylate biosynthesis from chorismate in *Pseudomonas aeruginosa*. Mol. Gen. Genet. 249:217-228.

Shah, J., Tsui, F. and Klessig, D.F., 1997. Characterization of a salicylic acid-insensitive mutant (*sai1*) of *Arabidopsis thaliana*, identified in a selective screen utilizing the SA-inducible expression of the *tms2* gene. Mol. Plant-Microbe Interact. 10:69-78.

Siddiqui, I.A. and Shaukat, S.S., 2002a. Rhizobacteria-mediated induction of systemic resistance (ISR) in tomato against *Meloidogyne javanica*. J. Phytopathol. 150:469-473.

Siddiqui, I.A. and Shaukat, S.S., 2002b. Resistance against the damping-off fungus *Rhizoctonia solani* systemically induced by the plant-growth-promoting rhizobacteria *Pseudomonas aeruginosa* (IE-6S$^+$) and *P. fluorescens* (CHA0). J. Phytopathol. 150:500-506.

Siddiqui, I.A. and Shaukat, S.S., 2003. Suppression of root-knot disease by *Pseudomonas fluorescens* CHA0 in tomato: importance of bacterial secondary metabolite, 2,4-diacetylpholoroglucinol. Soil Biol. Biochem. 35:1615-1623.

Siddiqui, I.A. and Shaukat, S.S., 2004. Systemic resistance in tomato induced by biocontrol bacteria against the root-knot nematode, *Meloidogyne javanica* is independent of salicylic acid production. J. Phytopathol. 152:48-54.

Silva, H.S.A., Romeiro, R.S., Macagnan, D., Halfield-Vieira, B.A., Pereira, M.C.B. and Mounteer, A., 2004a. Rhizobacterial induction of systemic resistance in tomato plants: non-specific protection and increase in enzyme activities. Biol. Control 29:288-295.

Silva, H.S.A., Romeiro, R.S., Carrer Filho, R., Pereira, J.L.A., Mizubuti, E.S.G. and Mounteer, A., 2004b. Induction of systemic resistance by *Bacillus cereus* against tomato foliar diseases under field conditions. J. Phytopathol. 152:371-375.

Singh, U.P., Sarma, B.K., Singh, D.P. and Bahadur, A., 2002. Plant growth-promoting rhizobacteria-mediated induction of phenolics in pea (*Pisum sativum*) after infection with *Erysiphe pisi*. Curr. Microbiol. 44:396-400.

Singh, U.P., Sarma, B.K. and Singh, D.P., 2003. Effect of plant growth-promoting rhizobacteria and culture filtrate of *Sclerotium rolfsii* on phenolic and salicylic acid contents in chickpea (*Cicer arietinum*). Curr. Microbiol. 46:131-140.

Somers, E., Vanderleijden, J. and Srinivasan, M., 2004. Rhizosphere bacterial signalling: a love parade beneath our feet. Crit. Rev. Microbiol. 30:205-240.

Spencer, M., Ryu, C.-M., Yang, K.Y., Kim, Y.C., Kloepper, J.W. and Anderson, A.J., 2003. Induced defence in tobacco by *Pseudomonas chlororaphis* strain O6 involves at least the ethylene pathway. Physiol. Mol. Plant Pathol. 63:27-34.

Srivastava, A.K., Singh, T., Jana, T.K. and Arora, D.K., 2001. Induced resistance and control of charcoal rot in *Cicer arietinum* (chickpea) by *Pseudomonas fluorescens*. Can. J. Bot. 79:787-795.

Sticher, L., Mauch-Mani, B. and Métraux, J.-P., 1997. Systemic acquired resistance. Annu. Rev. Phytopathol. 35:235-270.

Stracke, S., Kistner, C., Yoshida, S., Mulder, L., Sato, S., Kaneko, T., Tabata, S., Sandal, N., Stougaard, J., Szczyglowski, K. and Parniske, M., 2002. A plant receptor-like kinase required for both bacterial and fungal symbiosis. Nature 417:959-962.

Thangavelu, R., Palaniswami, A., Doraiswamy, S. and Velazhahan, R., 2003. The effect of *Pseudomonas fluorescens* and *Fusarium oxysporum* f.sp. *cubense* on induction of defense enzymes and phenolics in banana. Biol. Plant. 46:107-112.

Thomma, B.P.H.J., Eggermont, K., Penninckx, I.A.M.A., Mauch-Mani, B., Vogelsang, R., Cammue, B.P.A. and Broekaert, W.F., 1998. Separate jasmonate-dependent and salicylate-dependent defense response pathways in Arabidopsis are essential for resistance to distinct microbial pathogens. Proc. Natl. Acad. Sci. USA 95:15107-15111.

Thomma, B.P.H.J., Tierens, K.F.M., Penninckx, I.A.M.A., Mauch-Mani, B., Broekaert, W.F. and Cammue, B.P.A., 2001. Different micro-organisms differentially induce Arabidopsis disease response pathways. Plant Physiol. Biochem. 39:673-680.

Thordal-Christensen, H., 2003. Fresh insights into processes of nonhost resistance. Curr. Opin. Plant Biol. 6:351-357.

Timmusk, S. and Wagner, E.G.H., 1999. The plant-growth-promoting rhizobacterium *Paenibacillus polymyxa* induces changes in *Arabidopsis thaliana* gene expression: a possible connection between biotic and abiotic stress responses. Mol. Plant-Microbe Interact. 12:951-959.

Ton, J., Pieterse, C.M.J. and Van Loon, L.C., 1999. Identification of a locus in Arabidopsis controlling both the expression of rhizobacteria-mediated induced systemic resistance (ISR) and basal resistance against *Pseudomonas syringae* pv. *tomato*. Mol. Plant-Microbe Interact. 12:911-918.

Ton, J., Davison, S., Van Wees, S.C.M., Van Loon, L.C. and Pieterse, C.M.J., 2001. The Arabidopsis *ISR1* locus controlling rhizobacteria-mediated induced systemic resistance is involved in ethylene signaling. Plant Physiol. 125:652-661.

Ton, J., Van Pelt, J.A., Van Loon, L.C. and Pieterse, C.M.J., 2002a. Differential effectiveness of salicylate-dependent and jasmonate/ethylene-dependent induced resistance in Arabidopsis. Mol. Plant-Microbe Interact. 15:27-34.

Ton, J., De Vos, M., Robben, C., Buchala, A.J., Métraux, J.-P., Van Loon, L.C. and Pieterse, C.M.J., 2002b. Characterisation of *Arabidopsis* enhanced disease susceptibility mutants that are affected in systemically induced resistance. Plant J. 29:11-21.

Ton, J., Van Pelt, J.A., Van Loon, L.C. and Pieterse, C.M.J., 2002c. The Arabidopsis *ISR1* locus is required for rhizobacteria-mediated induced systemic resistance against different pathogens. Plant Biol. 4:224-227.

Troxler, J., Berling, C.H., Moënne-Loccoz, Y., Keel, C. and Défago, G. 1997. Interactions between the biocontrol agent *Pseudomonas fluorescens* CHA0 and *Thielaviopsis basicola* in tobacco roots observed by immunofluorescence microscopy. Plant Pathol. 46:62-71.

Tuzun, S. and Kloepper, J., 1995. Practical application and implementation of induced resistance. In: Induced Resistance to Disease in Plants, Hammerschmidt, R. and Kuc, J. (eds), Kluwer Acad. Publ., Dordrecht, pp. 152-168.

Uknes, S., Mauch-Mani, B., Moyer, M., Potter, S., Williams, S., Dincher, S., Chandler, D., Slusarenko, A., Ward, E. and Ryals, J., 1992. Acquired resistance in *Arabidopsis*. Plant Cell 4: 645-656.

Van Loon, L.C., 1997. Induced resistance in plants and the role of pathogenesis-related proteins. Eur. J. Plant Pathol. 103:753-765.

Van Loon, L.C., 2000. Systemic induced resistance. In: Mechanisms of Resistance to Plant Diseases, Slusarenko, A.J., Fraser. R.S.S. and Van Loon, L.C. (eds), Kluwer Acad. Publ., Dordrecht, pp. 521-574.

Van Loon, L.C. and Bakker, P.A.H.M., 2003. Signalling in rhizobacteria-plant interactions. In: Root Ecology, De Kroon, H. and Visser, E. J. W. (eds), Springer-Verlag, Berlin Heidelberg, pp. 297-330.

Van Loon, L.C. and Bakker, P.A.H.M., 2005. Induced systemic resistance as a mechanism of disease suppression by rhizobacteria. In: PGPR: Biocontrol and Biofertilization, Siddiqui, Z.A. (ed.), Springer Science, Dordrecht, pp. 39-66.

Van Loon, L.C. and Glick, B.R., 2004. Increased plant fitness by rhizobacteria. In: Molecular Ecotoxicology of Plants, Sandermann, H. (ed.), Springer-Verlag, Berlin Heidelberg, pp 177-205.

Van Loon, L.C., Bakker, P.A.H.M., and Pieterse, C.M.J., 1998. Systemic resistance induced by rhizosphere bacteria. Annu. Rev. Phytopathol. 36:453-483.

Van Loon, L.C., Geraats, B.P.J. and Linthorst, H.J.M., 2005. Ethylene as a modulator of disease resistance in plants. Trends Plant Sci., in press.

Van Peer, R. and Schippers, B., 1992. Lipopolysaccharides of plant-growth promoting *Pseudomonas* sp. strain WCS417r induce resistance in carnation to fusarium wilt. Neth. J. Plant Pathol. 98:129-139.

Van Peer, R., Niemann, G.J. and Schippers, B., 1991. Induced resistance and phytoalexin accumulation in biological control of fusarium wilt of carnation by *Pseudomonas* sp. strain WCS417r. Phytopathology 81:728-734.

Van Wees, S.C.M., Pieterse, C.M.J., Trijssenaar, A., Van' t Westende, Y., Hartog, F. and Van Loon, L.C., 1997. Differential induction of systemic resistance in *Arabidopsis* by biocontrol bacteria. Mol. Plant-Microbe Interact. 10:716-724.

Van Wees, S.C.M., Luijendijk, M., Smoorenburg, I., Van Loon, L.C. and Pieterse, C.M.J., 1999. Rhizobacteria-mediated induced systemic resistance (ISR) in *Arabidopsis* is not associated with a direct effect on expression of known defense-related genes but stimulates the expression of the jasmonate-inducible gene *Atvsp* upon challenge. Plant Mol. Biol. 41:537-549.

Van Wees, S.C.M., De Swart, E.A.M., Van Pelt, J.A., Van Loon, L.C. and Pieterse, C.M.J., 2000. Enhancement of induced disease resistance by simultaneous activation of salicylate- and jasmonate-dependent defense pathways in *Arabidopsis thaliana*. Proc. Natl. Acad. Sci. USA 97:8711-8716.

Verberne, M.C., Hoekstra, J., Bol, J.F. and Linthorst, H.J.M., 2003. Signaling of systemic acquired resistance in tobacco depends on ethylene perception. Plant J. 35:27-32.

Verhagen, B.W.M., Glazebrook, J., Zhu, T., Chang, H.S., Van Loon, L.C. and Pieterse, C.M.J., 2004. The transcriptome of rhizobacteria-induced systemic resistance in Arabidopsis. Mol. Plant-Microbe Interact. 17:895-908.

Vernooij, B., Friedrich, L., Morse, A., Reist, R., Kolditz-Jawhar, R., Ward, E., Uknes, S., Kessmann, H. and Ryals, J., 1994. Salicylic acid is not the translocated signal responsible for inducing systemic acquired resistance but is required in signal transduction. Plant Cell 6: 959-965.

Viswanathan, R. and Samiyappan, R., 2002a. Induced systemic resistance by fluorescent pseudomonads against red rot disease of sugarcane caused by *Colletotrichum falcatum*. Crop Prot. 21:1-10.

Viswanathan, R. and Samiyappan, R., 2002b. Role of oxidative enzymes in the plant growth promoting rhizobacteria (PGPR) mediated induced systemic resistance in sugarcane against *Colletotrichum falcatum*. Z. Pflanzenkrankh. Pflanzenschutz 109:88-100.

Viswanathan, R., Nandakumar, R. and Samiyappan, R., 2003. Role of pathogenesis-related proteins in rhizobacteria-mediated induced systemic resistance against *Colletotrichum falcatum* in sugarcane. Z. Pflanzenkrankh. Pflanzenschutz 110:524-534.

Volko, S.M., Boller, T. and Ausubel, F.M., 1998. Isolation of new *Arabidopsis* mutants with enhanced disease susceptibility to *Pseudomonas syringae* by direct screening. Genetics 149:537-548.

Wang, Y., Ohara, Y., Nakayashiki, H., Tosa, Y. and Mayama, S., 2005. Microarray analysis of the gene expression profile induced by the endophytic plant growth-promoting rhizobacteria, *Pseudomonas fluorescens* FPT9601-T5 in *Arabidopsis*. Mol. Plant-Microbe Interact. 18:385-396.

Wei, G., Kloepper, J.W. and Tuzun, S., 1991. Induction of systemic resistance of cucumber to *Colletotrichum orbiculare* by select strains of plant growth-promoting rhizobacteria. Phytopathology 81:1508-1512.

Wei, G., Kloepper, J.W. and Tuzun, S., 1992. Internal and external colonization of cucumber by bacteria which induce systemic resistance to *Colletotrichum orbiculare*. Phytopathology 82:1108.

Weller D.M., Van Pelt, J.A., Mavrodi, D.V., Pieterse, C.M.J., Bakker, P.A.H.M. and Van Loon, L.C., 2004. Induced systemic resistance (ISR) in *Arabidopsis* against *Pseudomonas syringae* pv. *tomato* by 2,4-diacetylphloroglucinol (DAPG)-producing *Pseudomonas fluorescens*. Phytopathology 94:S108.

Whitehead, N.A., Barnard, A.M.L., Slater, H., Simpson, N.J.L. and Salmond, G.P.C., 2001. Quorum-sensing in Gram-negative bacteria. FEMS Microbiol. Rev. 25:365-4004.

Whipps, J.M., 2001. Microbial interactions and biocontrol in the rhizosphere. J. Exp. Bot. 52:487-511.

Yan, Z., Reddy, M.S., Ryu, C.-M., McInroy, J.A., Wilson, M. and Kloepper, J.W., 2002. Induced systemic protection against tomato late blight elicited by plant growth-promoting rhizobacteria. Phytopathology 92:1329-1333.

Yan, Z., Reddy, M.S. and Kloepper, J.W., 2003. Survival and colonization of rhizobacteria in a tomato transplant system. Can. J. Microbiol. 49:383-389.

Zdor, R.E. and Anderson, A.J., 1992. Influence of root colonizing bacteria on the defense responses of bean. Plant Soil 140:99-107.

Zehnder, G., Kloepper, J., Tuzun, S., Yao, C., Wei, G., Chambliss, O. and Shelby, R., 1997a. Insect feeding on cucumber mediated by rhizobacteria-induced plant resistance. Entomol. Exp. Appl. 83:81-85.

Zehnder, G., Kloepper, J., Yao, C.B. and Wei, G., 1997b. Induction of systemic resistance in cucumber against cucumber beetles (Coleoptera: *Chrysomelidae*) by plant growth-promoting rhizobacteria. J. Econ. Entomol. 90:391-396.

Zehnder, G.W., Yao, C., Murphy, J.F., Sikora, E.R. and Kloepper, J.W., 2000. Induction of resistance in tomato against cucumber mosaic cucumovirus by plant growth-promoting rhizobacteria. BioControl 45:127-137.

Zhang, S., Reddy, M.S. and Kloepper, J.W., 2002a. Development of assays for assessing induced systemic resistance by plant growth-promoting rhizobacteria against blue mold of tobacco. Biol. Control 23:79-86.

Zhang, S., Moyne, A.-L., Reddy, M.S. and Kloepper, J.W., 2002b. The role of salicylic acid in induced systemic resistance elicited by plant growth-promoting rhizobacteria against blue mold of tobacco. Biol. Control 25:288-296.

Zhang, W., Dick, W.A. and Hoitink, H.A.J., 1996. Compost-induced systemic acquired resistance in cucumber to Pythium root rot and anthracnose. Phytopathology 86:1066-1070.

Zhang, W., Han, D.Y., Dick, W.A., Davis, K.R. and Hoitink, H.A.J., 1998. Compost and compost water extract-induced systemic acquired resistance in cucumber and Arabidopsis. Phytopathology 88:450-455.

Zhou, T. and Paulitz, T.C., 1994. Induced resistance in the biocontrol of *Pythium aphanidermatum* by *Pseudomonas* spp. on cucumber. J. Phytopathol. 142:51-63.

5. AFFILIATION

L.C. van Loon and P.A.H.M. Bakker, *Department of Biology, Section Phytopathology, Utercht University, P.O.Box. 800.84, 3508 TB, UTRECHT, TheNetherlands.*

SANDRA DE WEERT AND GUIDO V. BLOEMBERG

RHIZOSPHERE COMPETENCE AND THE ROLE OF ROOT COLONIZATION IN BIOCONTROL

Abstract. In order to efficiently protect the root system against the attack of phytopathogenic microorganism a biocontrol strain has to establish itself in the rhizosphere. After inoculation the biocontrol agent has to colonize the rhizosphere, in which processes such as growth, chemotaxis, motility, etc play essential roles. This chapter will focus on the mechanisms of *Pseudomonas* biocontrol strains, which are involved in rhizosphere competence and the role of root colonization in biological control. It will address experimental approaches to elucidate colonization genes and give a brief overview of genes and traits that are involved in root colonization.

1. INTRODUCTION

The rhizosphere was first described by Hiltner (1904). It describes the (soil) layer surrounding the root, including the rhizoplane, and the root surface. The rhizosphere is a complex and dynamic environment created by the root. Plant roots are able to exude an extensive range of organic compounds. At least some of these components can function as nutrients or carbon source for micro-organisms. This is one of the reasons why the rhizosphere is inhabited by a wide range of micro-organisms. As a result of release of organic material, such as organic acids and sugars, the rhizosphere becomes an attractive ecological niche, being rich in nutrients as compared to bulk soil. Root exudate is defined as all substances that are released by leakage or secretion from the roots of healthy plants. From several plants root exudate composition has been studied. Detailed analysis of tomato root exudates identified organic acids as the major exudate component (80%) (Lugtenberg et al., 1999). Additionally, sugars (Lugtenberg et al., 1999) and to a minor level amino acids (Simons et al., 1997) have been identified as components of the tomato exudate. In comparison to root-free soil, elevated levels of micro-organisms are found in the plant rhizosphere, since they use these exudate components as a nutrient source.

Various types of interactions can take place in the rhizosphere, including (i) interactions between the root and micro-organisms and (ii) interactions between microorganisms. These interactions can be divided into harmful, neutral and beneficial ones (Lynch and Whipps, 1990). In general beneficial plant-microbe interactions can be grouped in four different classes, e.g. phytostimulation, biofertilisation, bioremediation and biological control. Besides beneficial microbes also harmful micro-organisms are attracted by the root exudate followed by root

S.S. Gnanamanickam (ed.), Plant-Associated Bacteria, 317–333.

colonization, infection and disease development. Plant diseases often result in major crop losses. Plant diseases can be responsible for major crop yield losses and are estimated to be responsible for a world-wide loss of up to 30% (Lugtenberg and de Weger, 1994), illustrating the economical importance of globally controlling these diseases. A wide variety of pathogens are present in the soil, including bacteria and fungi. For example *Fusarium oxysporum,* the causal agent of tomato foot and root rot, and *Pythium ultimum*, which causes damping off.

The most common strategy to suppress plant diseases is the use of chemical pesticides, which besides plant protection in many cases are harmful for the environment by their sustained presence. Therefore policies in the Western world are strongly aiming at reducing the use of chemical pesticides and ban the use of approximately 60% of the chemical pesticide products. Alternative strategies are required to suppress plant diseases to minimize crop losses. A promising alternative is the use of biological (bio) pesticides, using bacteria or fungi occurring naturally in disease suppressive soils. The product list of biocontrol inoculants is growing, but inconsistency of results is hampering their application.

2. GENERAL ASPECTS OF BIOLOGICAL CONTROL

Biocontrol involves the use of micro-organisms able to suppress the actions of phytopathogens and their resulting disease. The mono culture policy of crop growing can lead to an increase in the occurrence of soil-borne plant diseases. This is due to a disturbance of the natural balance in the soil between beneficial and pathogenic micro-organisms (Harman and Lumsden, 1990). However, some soils naturally suppress soil-borne plant pathogens (Schroth and Hancock, 1982). Scher and Baker (1980) reported that conducive soil can be made suppressive by introducing a small amount of disease-suppressive soil. This indicates that micro-organisms are active in this soil (Scher and Baker, 1980; Schroth and Hancock, 1982). These disease-suppressive micro-organisms were designated as plant growth promoting rhizobacteria (PGPR) (Kloepper et al., 1980a). By applying these micro-organisms to seeds growth of various crops could be improved (Burr and Suslow, 1978; Geels and Schippers, 1983a; Geels and Schippers, 1983b; Kloepper et al., 1980b; Suslow and Schroth, 1982). The microbial population present in the rhizosphere consists substantially of gram-negative bacteria such as *Pseudomonas* spp. but can vary and depends on plant species (Shann and Boyle, 1994), plant age, root type and the history of the soil (Anderson et al., 1993). Various mechanisms can contribute to this mechanism of biocontrol such as (i) induction of systemic resistance (ISR), (ii) niche exclusion and competition for nutrients on the root, (iii) parasitism and predation and (iv) production of anti-fungal metabolites.

In order to improve the application of inoculants, an understanding of the functioning of bacterial strains on the root and their biocontrol mechanisms is required. Root colonization is one of the first steps in the interaction between the biocontrol agent and the root necessary for establishing an efficient protection. Inconsistency of biocontrol has been shown to be correlated with variable colonization numbers (Bull et al., 1991). *Pseudomonas* spp. are ubiquitous soil organisms frequently being present on the root. Many biocontrol *Pseudomonas* strains have been isolated showing suppression of a diverse range of plant diseases, which makes *Pseudomonas* a very relevant organism to elucidate genes and traits involved in root colonization. In order to understand the relevance of root colonization for biocontrol, a concise overview of the most common biocontrol mechanism will be given.

3. OVERVIEW OF BIOCONTROL MECHANISMS

In the following sections a concise overview of biocontrol mechanism will be given.

Induction of Systemic Resistance

An immune response of the plant towards a pathogen is known as systemic acquired resistance (SAR) (Metraux JP et al., 2002). This type of defense is characterised by the accumulation of salicylic acid (SA), which in turn induces a set of genes encoding pathogenesis related (PR) proteins in the plant. Some PGPR are also able to induce an immune response in the plant, refered to as ISR, which is different from SAR (van Peer et al., 1991; Schippers et al., 1995; van Loon et al., 1998). Although SAR and ISR both result in resistance against pathogens from the plant there is no indication that these two pathways have an overlap (van Wees et al., 2000). The ISR response is induced by non-pathogenic rhizobacteria and does not result in SA accumulation or in expression of PR proteins. When ISR is induced the plant produces jasmonic acid and ethylene (van Wees et al., 1999). It was shown that several *Pseudomonas* spp. are able to induce ISR in a wide range of plants towards different pathogens. For example, *P. fluorescens* strain WCS417 is able to induce ISR in *A. thaliana* ecotype *Columbia* against the pathogen *P. syringae* pv. *tomato*. The signals of different *Pseudomonas* spp. that induce ISR have not yet been completely elucidated but there are indications that bacterial surface structures such as flagella and lipopolysaccharide (LPS) can be involved (Bakker and Schippers, 1995). Bacterial root colonization seems to be of a limited importance for biocontrol since colonization mutants of the biocontrol strain *P. fluorescens* WCS365 were still able to exhibit biocontrol of tomato foot and root rot (Dekkers et al., 2000)

Niche exclusion and competition for nutrients

In order to protect plants against pathogens micro-organisms must be able to establish themselves in the rhizosphere. After introduction the PGPR must be able to compete with the indigenous microflora, including the pathogen. A rapid establishment of the PGPR on the root and competition for the exuded nutrients has been suggested as an important mechanism for biological control. The cell-cell junctions are the preferable sites on the root where PGPR as well as pathogens colonise the root system (Chin-A-Woeng et al., 1997; Lagopodi et al., 2002; Bolwerk et al., 2003). These sites are assumed to be the locations were nutrients leak from the root. An overlap of preferable niches for colonization is presumably the basis for competition in the rhizosphere. An efficient and fast colonization is necessary to be able to prevent the establishment of the pathogen on the root system and its subsequent infection. Also competition for other components such as iron ions can take place (Buyer and Leong, 1986). Iron is often used as a cofactor in enzymes that are involved in pathways essential for microbial growth. In the rhizosphere there is often a limitation for soluble Fe^{3+}. Fluorescent pseudomonads respond to iron limitation by producing iron chelating compounds, refered to as siderophores. Often siderophores have fluorescent properties resulting in a yellow-green color. Examples of siderophores produced by *Pseudomonas* spp. are pseudobactins or pyoverdins (Leong, 1986; Neilands, 1991). The ability to produce siderophores or to take up related siderophores produced by other microorganims in the rhizosphere results in an competitive advantage and an iron deficiency for pathogens (Raaijmakers et al., 1995). This will inhibit the growth of the pathogen and eventually lead to the suppression of the disease. An efficient suppression of the disease by the production of the siderophores is highly dependent upon the delivery in the rhizosphere, which is determined by the root colonizing ability of the siderophore producing PGPR.

Parasitism and predation

Soil borne bacteria and fungi are able to produce HCN and extracellular enzymes such as chitinases, β-(1,3)-glucanases, lipases, cellulases and proteases. Some of these components, like chitinases, are able to degrade the cell wall of fungi, which results in damage of the fungal cell wall ad leakage of the fungal cells (Lugtenberg et al., 1991; Thomashow and Weller, 1996). Bacteria producing hydrolytic enzymes were shown to posses biocontrol abilities (Markovich and Kononova, 2003). Chitinases produced by biocontrol *Trichoderma* species were shown to be involved in antagonistic activity against and biocontrol of phytopathogen fungi, such as *Botrytis cinerea* (Markovich and Kononova, 2003). For an effective suppression of a plant disease root colonization is important to reach the side tocompete with the pathogen. In addition, colonization of fungal hyphae might form an important trait for an efficient parasitism resulting in an more efficient biocontrol.

Production of anti-fungal metabolites

For various pseudomonads biocontrol is established by the production of secondary metabolites that have anti-fungal properties. These anti-fungal metabolites (AFM) are usually low molecular weight components, which strongly inhibit growth and the metabolism of fungi. The most common AFM found to be produced by Pseudomonas spp. belong to the class of phenazines, di-acetyl phloroglucinol (DAPG) and pyrolnitrin. Most of these AFM affect the growth of a broad range of fungi. Genetic studies have been conducted for several of these strains to analyze the effect of AFM production on their biocontrol ability. For one of these strains we will briefly describe the experimental approach and results.

Pseudomonas chlororaphis PCL1391 was isolated from the rhizosphere of tomato from a commercial field in Andalucia (Spain) and selected for its efficient suppression of tomato root and foot rot caused by *Fusarium oxysporum* f.sp. *radicis lycopersici* (Chin-A-Woeng et al., 1998). *P. chlororaphis* PCL1391 produces and secretes several hydrolytic activities, including protease, chitinase and lipase activities, hydrogen cyanide and the antifungal compound phenazine-1-carboxamide (PCN) (Chin-A-Woeng et al., 1998). A mutant of *P. chlororaphis* PCL1391 deficient in the biosynthesis of phenazine-1-carboxamide lost the biocontrol ability of tomato foot and root rot, showing the essential role of PCN for biocontrol (Chin-A-Woeng et al., 1998). *P. chlororaphis* PCL1391 is an efficient root colonizer of tomato. Using PCL1391 strains impaired in root colonization by mutation of colonization genes or traits, such as amino acid biosynthesis, motility and a functional *sss* gene, it was shown that colonization is essential for the biocontrol ability of *P. chlororaphis* PCL1391(Chin-A-Woeng et al., 2000) These studies indicate that root colonization functions as a delivery system for PCN in the rhizosphere to reach the fungus and protect the root system against infection. Regulation of PCN production by *P. chlororaphis* PCL1391 is subject to quorum sensing, which requires a certain quorum of cells to induce the expression of the *phz* operon, responsible for the biosynthesis of PCN (Chin-A-Woeng et al., 2001). The microcolonies formed by *P. chlororaphis* PCL1391 on the root surface which are encapsidated in an extracellular matrix (Bolwerk et al., 2003) will provide the suitable environment for quorum sensing. In addition, it was shown that PCN production is regulated by a complex cascade of regulators, including *gacA/S*, *psrA* and *rpoS* (Chin-A-Woeng et al., 2005; Girard et al., in press). Environmental factors that affect the level of PCN production include, abiotic factors such as low oxygen and iron concentrations, as well as biotic factors such as the nutrients glucose and phenylalanine (van Rij et al., 2004). Recent studies showed that PCN production can also be repressed by fusaric acid that is secreted by *Fusarium oxysporum* f.sp. *radicis lycopersici* (van Rij et al., 2005). Since pathogenic and non-pathogenic *Fusarium* strains produce fusaric acid this effect might be relevant for the biocontrol efficiency of *P. chlororaphis* PCL1391in the field.

4. MECHANISMS OF *PSEUDOMONAS* ROOT COLONIZATION

When bacterial cells adhere and proliferate on a surface they can form a colony and subsequently a biofilm, in which they are encased in an extracellular polysaccharide matrix. This process of colonization consists of the following steps (i) transport of microbes to the surface, (ii) initial attachment, (iii) formation of microcolonies and (iv) biofilm maturation Bacterial cells are able to colonise different surfaces including plant roots and fungal surfaces (Bolwerk et al., 2003;

de Weert et al., 2002; de Weert et al., 2004a). By formation of micro-colonies structures encased in an extracellular matrix on the root surface (Bloemberg et al., 1997), these colonies are bacterial biofilms.

Root colonization can be the limiting process in biocontrol of *P. chlororaphis* strain PCL1391 (Chin-A-Woeng et al., 2000). Root colonization can be defined as the proliferation of micro-organisms in, on or along the root system (Parke, 1991). Seeds can be inoculated with biocontrol microbes but their success depends on the establishment on and along the growing root system. By migrating down with the root they will have to compete with the indigenous population present in the soil (Parke, 1991). This again emphasizes the extreme importance of root colonization for efficient biocontrol. In case of the production of AFM efficient root colonization is essential for the delivery of the AFM along the root system at the right time and place. Mutants of PCL1391 defective in colonization have completely lost their ability to suppress tomato foot and root rot caused by *Fusarium* (Chin-A-Woeng et al., 2000). In contrast, in the case of ISR, a decrease in root colonization does not clearly result in loss of biological control (Dekkers et al., 2000). Apparently, already a smaller number of micro-organisms inducing ISR on the root can be sufficient to induce the plant response and complete root colonization in this case is not necessary.

5. MODEL SYSTEM TO STUDY ROOT COLONIZATION

To study root colonization in a reproducible manner under controlled laboratory conditions, a gnotobiotic test system was developed by Simons et al. (1996). In this closed system bacterized seedlings are placed in a sterile quartz sand column moistened with plant nutrient solution (PNS) (Hoffland et al., 1989). After 7 days of growth in a climate controlled growth chamber, roots are taken out and bacteria isolated from the root. After plating, numbers of bacteria and ratios between wild type and mutant, which are distinguished from each other by differential marker genes (e.g. ß-galactosidase or antibiotic resistance genes) are determined.

To identify traits involved in root colonization a suitable model organism is required. *Pseudomonas fluorescens* WCS365 (Geels and Schippers, 1983a) was selected out of a pool of eleven *Pseudomonas* strains based on its capacity to efficiently colonise the root system of potato (Brand et al., 1990). Briefly, potato tubers were inoculated with a mixture of eleven *Pseudomonas* strains, being present in equal numbers, and subsequently grown in contained field soil. After two weeks of plant growth, bacteria were isolated from the roots at various depths up to 1.5 meters deep. It was shown that the distribution of bacteria along the root system was not uniform. *P. fluorescens* WCS365 was the only dominant strain that was present on all root parts, including the root tip (Brand et al., 1990; Glandorf, 1992). Subsequently, WCS365 was shown to be an excellent root coloniser of tomato, wheat and radish (Simons et al., 1996).

Different approaches can be used to identify genes involved in competitive root tip colonization after applying a mixture of bacteria on a (germinated) seed. Firstly, genes predicted to have an effect on colonization can be disrupted by homologous recombination in the biocontrol organism. Before testing the mutant strains for their colonizing ability, strains should be tested for an unaltered growth rate as compared to the wt. In this way traits like motility and chemotaxis were identified to be important for colonization. Secondly, a random transposon mutagenesis can be performed usingfor example Tn*5lacZ* (Lam et al., 1990) or Tn*5luxAB* (Wolk et al., 1991) of which single mutants can be tested or pools of mutants can be tested for impaired or enhanced competitive root tip colonization. Applying the transposon approach traits such as uptake of putrescine, proton motive force and the synthesis of the O-antigen of LPS (lipopolysaccharide) as well as hypothetical genes have been identified (for a review see Lugtenberg et al., 2001).

6. TRAITS INVOLVED IN ROOT COLONIZATION

Genes and traits involved in attachment are required for root colonization

Attachment to the plant surface is one of the initial steps in root colonization. Flagella and different types of pili are involved in attachment to diverse surfaces. Flagella of *Pseudomonas* are necessary for biofilm formation on different surfaces (O'Toole and Kolter, 1998a; O'Toole and Kolter, 1998b). Pili (fimbriae) can have different functions among different organisms. For example, type 4 pili are known to be involved in twitching motility and the initial contact between bacteria and epithelial cell surfaces. Camacho (2001) reported that type 4 pili from *P. fluorescens* WCS365 are involved in efficient attachment and biofilm formation on PVC. Type 4 pili are also involved in competitive colonization of the tomato root tip (Camacho, 2001).

The involvement of environmental signals in root colonization is indicated by the identification of the colonizing mutant PCL1210, which was shown to be mutated in the sensor gene of the two-component regulatory system *colR/colS* (Dekkers et al., 1998a). The environmental signal that activates this system is not known. Candidates of genes regulated by *colR* are located downstream of *colR/colS* located in the putative *orf222-inaA/wapQ* operon. *inaA* is involved in maintenance of the intracellular pH (White et al., 1992) and that *wapQ* encodes a putative heptose kinase (Walsh et al., 2000).

Motility

It was described by de Weger et al. (1987) that flagella-less mutants (no flagella present) do not have the ability to colonise the root tip efficiently. On the contrary, Howie et al. (1987) and Scher et al. (1988) stated that flagella-less mutants are not impaired in colonization of wheat and soybean roots, respectively. Motility in relation to root colonization has been reported to depend on the soil type, the plant and bacterial strains used (Weller and Thomashow, 1994). It should be noted that the method of sampling in the above-mentioned experiments is different. Scher et al. (1988) analysed whole roots. Howie et al. (1987) only analysed root parts just below the seed whereas de Weger et al. (1987) analysed only the root tip where the biggest differences in numbers of bacterial cells occur (Chin-A-Woeng et al., 1997; Simons et al., 1996). Apparently, bacteria need functional flagella to migrate down with the growing root and to reach the deeper root parts such as the root tip.

The role of chemotaxis in root colonization

Signal-transduction pathways mediated by histidine kinases play a central role in information processing. Histidine kinases allow bacteria, plants and fungi to sense and respond to their environment. These signal-transduction pathways are sometimes referred to as "two-component systems" and they regulate cellular responses like chemotaxis, sporulation or microbial pathogenesis (Appleby et al., 1996; Parkinson and Kofoid, 1992). Most histidine kinases are homodimers that use ATP to phosphorylate a specific histidine residue on the flanking subunit within the dimer (Swanson et al., 1993; Yang and Inouye, 1991). This phosphoryl group is transferred to a specific aspartyl residue on a response regulator domain. The phosphorylated response regulator acts directly by modifying an effector, which will in turn lead to a change in cellular behaviour. A well-studied signalling pathway that uses a two-component system is bacterial chemotaxis. By rotation of their flagella bacteria are able to swim and tumble. The bacterial chemotaxis pathway is encoded by a number of genes and has been best described for *Escherichia coli*. Genes necessary for chemotaxis and their function are listed in Table 1.

This phosphoryl group is transferred to a specific aspartyl residue on a response regulator domain. The phosphorylated response regulator acts directly by modifying an effector, which will in turn lead to a change in cellular behaviour. A well-studied signalling pathway that uses a two-component system is bacterial chemotaxis. By rotation of their flagella bacteria are able to swim and tumble. The bacterial chemotaxis pathway is encoded by a number of genes and has been best described for *Escherichia coli*. Genes necessary for chemotaxis and their function are listed in Table 1.

Briefly, bacterial chemotaxis acts upon sensing of attractants and repellents. Whenever an attractant is sensed by MCPs these membrane proteins are methylated by CheR. In this case the histidine kinase CheA is not able to phosphorylate its histidine residue. There is no interaction with the flagellar motor by response regulator CheY and flagella will rotate counterclockwise (CCW) causing the bacterial cell to swim smoothly and not change its direction of swimming. However, when the bacterial cell senses a repellent the MCPs are demethylated by CheB. In this case CheA is able to phosphorylate it histidine residue and transfer this phosphoryl group to CheY. Phosphorylated CheY will interact with the flagellar motor causing a conformational change and the flagella will switch from CCW to clockwise (CW) rotation. CW rotation of flagella will result in tumbling of the cell, thereby changing the swimmimg direction of the cell.

Table 1. Genes involved in bacterial chemotaxis and their mode of action

Gene	Protein	Mode of action
cheA	Histidine kinase	Autophosphorylation and transfer of P-group to CheY
cheY	Response regulator	Interaction with flagellar motor
cheR	Methyltransferase	Methylation of methyl accepting chemotaxis proteins (MCPs)
cheW	Coupling protein	Coupling of transmembrane receptors to CheA
cheB	Methylesterase	Demethylation of MCPs
cheZ	Phosphatase	Dephosphorylation of CheY

Tomato root exudate consists mainly of organic acids, sugars and amino acids (Simons et al., 1997; Lugtenberg et al., 1999). All of these components can function as potential nutrient sources for the bacteria present in the rhizosphere (Gamliel and Katan, 1992). It has been shown that *Bradyrhizobium japonicum* shows a chemotactic response to soybean exudates (Barbour et al., 1991). A *P. fluorescens* WCS mutant strain impaired in its organic acid utilisation is also impaired in its competitive colonization abilities (unpublished results). However, a mutant impaired in its sugar utilisation does not show this impaired colonization phenotype, indicating that utilisation of specific nutrients is of importance for rhizosphere establishment of bacteria (unpublished results). It can be hypothesized that a directed movement towards exudate components is necessary for efficient competitive colonising abilities and that specific components of the root exudate function as chemo-attractants. Flagella-driven chemotaxis and its specific chemo-attraction towards individual root exudate components was shown to be essential for competitive root colonization.

Previously, motility was shown to be involved in competitive root colonization of pseudomonads (de Weger et al., 1987). The basis of chemotaxis has been defined as a directed movement towards chemicals. The analysis of *cheA* mutants in four different *P. fluorescens* strains (e.g. F113, WCS365, SBW25 and OE28.3), defective in flagella-driven chemotaxis but still motile, for their competitive root colonising abilities, showed that all *cheA* mutants appeared to be impaired in competitive root tip colonization on tomato (de Weert et al., 2002). It was shown that the differences in competitive colonising abilities between parental strains and *cheA* mutants were significant on all root parts and gradually increased from root base to the root tip. Significant differences at the root tip could already be detected after 2-3 days of inoculation showing that chemotaxis is an important trait for competitive root colonization. Testing compounds previously being identified to be present in tomato root exudates (Lugtenberg et al., 1999; Simons et al., 1997) showed that WCS365 exhibits a chemotactic response towards some organic acids and some amino acids. Interestingly the sugars present in the root exudate did not initiate a chemotactic response from the WCS365 cells. Comparison of the minimal concentrations required for a chemotactic response with concentrations estimated to be present in root exudates suggested that malic acid and citric acid are among major chemo-attractants for *P. fluorescens* WCS365 cells in the tomato rhizosphere (de Weert et al., 2002).

The influence of recombination on root colonization

Via transposon mutagenesis a mutant was selected impaired in competitive root tip colonization and mutated in a gene with homology to the *sss* gene from *P. aeruginosa* (Dekkers et al., 1998b). This gene encodes a protein belonging to the λ-integrase family of site-specific recombinases. The role of the *sss* homologue in colonization is thought to be through genetic rearrangements causing different phenotypes. The *sss* colonization mutant is assumed to be locked in a phase that is not suitable for competitive colonization in the rhizosphere. By introduction of multiple copies of the *sss* gene into a poor colonizer *P. fluorescens* WCS307 and in the good coloniser *P. fluorescens* F113, the competitive root tip colonising ability of these strains was improved (Dekkers et al., 2000). These results show that it is possible to improve colonization through genetic engineering (Dekkers et al., 2000).

7. GENERATION OF ENHANCED ROOT TIP COLONISERS

A criterion for a good root colonizing microorganism is that it can efficiently reach the root tip after seed inoculation. An enrichment procedure for efficient remediating rhizobacteria described by Kuiper et al. (2001) was developed to isolate from roots growing in soil polluted with poly aromatic hydrocarbons efficient rhizoremediating bacteria. These bacteria were selected on the basis of their ability to degrade naphthalene, colonise grass roots and grow rapidly to make production of an inoculant economically feasible (Kuiper et al., 2001; Kuiper et al., 2002).

Using a transposon mutant library of WCS365 in combination with the procedure to enrich for efficient competitive root tip colonisers (Kuiper et al., 2001) it was observed that competitive root tip colonization can be enhanced. It was described that enhanced root tip colonisers, for the dicotyledenous plant tomato as well as for the monocotyledenous plant grass, can be isolated from a Tn*5luxAB* mutant bank of WCS365 (de Weert et al., 2004b). The best colonising mutant was shown to be mutated in a *mutY* homolog. This gene is involved in the repair of A:G mismatches caused by spontaneous oxidation of guanine (Urios et al., 1994; Notley-McRobb et al., 2000). Since a mutant with a disrupted *mutY* gene is defective in repairing its mismatches it is hypothesized that such cells harbour an increased number of mutations and that some of these mutants, with a beneficial combination of mutations, can adapt faster than the wild type to the environment of the root system. By homologous recombination an additional *mutY* mutant was constructed and analysed for its competitive root tip colonization behaviour prior to and after enrichment. The possibility that the enrichment procedure alone altered the colonization ability of the wild type was excluded since wild type cells subjected to this enrichment procedure did not show an altered colonization phenotype. The new *mutY* mutant was even strongly impaired in its colonization ability prior to the enrichment cycles. However, after 3 cycles of enrichment on tomato the PCL1808 derivative strain isolated from the root tip colonised the root tip even significantly better than its wild type (de Weert et al., 2004b). Therefore it can be concluded that both the *mutY* mutation as well as the selection procedure for enrichment of secondary mutants that improve the colonizing ability are required to obtain an enhanced root tip colonising mutant and that an experimental setup is developed to analyze evolutionary events that result in an improved root colonizing ability. The identification of the secondary mutations of the improved root colonizers will be necessary to unravel the processes that are responsible for the enhanced root colonizing ability.

8. COLONIZATION OF FUNGAL HYPHAE

Various microscopic studies showed the close interaction between rhizosphere fungi and bacteria. Bacterial cells were shown to attach to hyphae and colonize the hyphae of beneficial and phytopathogenic fungi. For example, it was shown that arbuscular mycorrhizal (AM) fungi can physically interact with plant growth promoting rhizobacteria (PGPR). Mycorrhizal fungi are able to initiate highly specific interactions with numerous plant species, resulting in an enhanced uptake of phosphate and micro-nutrients by the plant (Reid, 1990). The interaction between mycorhizae and PGPR results in an increased root colonization by PGPRs (Bianciotto et al., 1996). Interestingly, extracellular polysaccharides play an important role in these interactions (Biancotto et al., 1996).

Besides the excellent root colonizing abilities of the biocontrol strains *P. fluorescens* WCS365 and *P. chlororaphis* PCL1391 they were also shown to be able to colonise the hyphae of the phytopathogenic fungus *Fusarium oxysporum* f.sp. *radicis-lycopersici* (FORL) under biocontrol conditions (Lagopodi et al., 2002; Bolwerk et al., 2003). The studies of Biancotto et al also showed that *P. fluorescens* WCS365 colonises the fungal structures of *Gigaspora margarita*, indicating that hyphal colonization is not limited to a single species. Fungi exude compounds which can act as chemo-attractants and growth substrates for other micro-organisms (Arora et al., 1983). Since *P. fluorescens* WCS365 displays a chemotactic response towards root exudate (de Weert et al., 2002) it was hypothesized that they might be attracted to exudate of *F. oxysporum*, which might be a prerequisite for hyphal colonization. By adapting a fungus-bacterium confrontation assay, it was possible to study colonization of hyphae by bacteria *in vitro*, showing that a *cheA* mutant derivative of WCS365 is able to colonise the hyphae of FORL but not as efficient as its wild type WCS365 (de Weert et al., 2004a). This phenomenon was both observed when single cells were applied as well as when a mixture of the parental wt WCS365 and *cheA* mutant was applied. Since the *cheA* mutant was less efficiently colonising fungal hyphae the supernatant of FORL cultures was analyzed for a chemotactic response. It was shown that WCS365 responds to supernatant of FORL grown cultures while its *cheA* mutant did not (de Weert et al., 2004a). A known secreted pathogenicity factor produced by many *Fusarium* spp. is fusaric acid, which is toxic to a wide range of plants, fungi and bacteria (Yabuta et al., 1934). Despite its toxicity, synthetic fusaric acid was able to initiate a chemotactic response in WCS365 cells although WCS365 is not able to use fusaric acid as a carbon source. The importance of fusaric acid as a chemo-attractant was analysed by using supernatants of a number of *Fusarium* strains, secreting different levels of fusaric acid, as chemo-attractants. Amounts of fusaric acid produced by the different fungi positively corresponded to the chemo-attractant activity of the WCS365 cells. No chemotactic response was observed towards the supernatant of the non-fusaric acid producing *Fusarium* strain FO242; however, the hyphae of this fungus were still colonised by WCS365 cells. These results indicate that either low levels of fusaric acid are sufficient for chemotaxis or that other metabolites are produced by the fungus and play a role in chemotaxis. It can be predicted that hyphal colonization is an important trait for biocontrol, especially when compounds or enzymes that inhibit the growth of the fungus are involved in the biocontrol mechanism. The identification of genes involved in the fungal colonization, but which do not affect root colonization, will provide the opportunity to analyze the importance of fungal colonization for biocontrol.

9. FUTURE PERSPECTIVES IN COLONIZATION RESEARCH

Colonization is a complex process in which a range of different factors are involved (Lugtenberg et al., 2001). Colonization finally results in the formation of a biofilm on the root (Bloemberg et al., 1997). To be able to establish themselves in the rhizosphere, bacteria need to sense their surroundings and migrate towards nutrient sources. It is likely that the search for nutrients is the principle behind both colonization of roots and fungal hyphae.

10. REFERENCES

Anderson, T. A., Guthrie, E. A. & Walton, B. T. (1993). Bioremediation in the rhizosphere. *Environ. Sci. Technol. 27*, 2630-2636.

Appleby, J. L., Parkinson, J. S. & Bourret, R. B. (1996). Signal transduction via the multi-step phosphorelay: not necessarily a road less traveled. *Cell 86*, 845-848.

Arora, D. K., Filonow, A. B. & Lockwood, J. L. (1983). Bacterial chemotaxis to fungal propagules in vitro and in soil. *Can. J. Microbiol. 29*, 1104-1109.

Bakker, A. W. & Schippers, B. (1995). Induction of systemic resistance against *Fusarium* wilt of radish by lipopolysaccharides of *Pseudmonas fluorescens*. *Phytopathol. 85*, 1021-1027.

Barbour, W. M., Hattermann, D. R. & Stacey, G. (1991). Chemotaxis of *Bradyrhizobium japonicum* to soybean exudates. *Appl. Environ. Microbiol. 57*, 2635-2639.

Bianciotto, V., Minerdi, D., Perotto, S. & Bonfante, P. (1996). Cellular interactions between arbuscular mycorrhizal fungi and rhizosphere bacteria. *Protoplasma 193*, 123-131.

Bloemberg, G. V., O'Toole, G. A., Lugtenberg, B. J. J. & R. Kolter. (1997). Green fluorescent protein as a marker for *Pseudomonas* spp. *Appl. Environ. Microbiol. 63*, 4543-4551.

Bolwerk, A., Lagopodi, A. L., Wijfjes, A. H. M., Lamers, G.E.M., Chin-A-Woeng, T. F. C., Lugtenberg, B. J. J. & Bloemberg, G. V. (2003). Interactions in the tomato rhizosphere of two *Pseudomonas* biocontrol strains with the phytopathogenic fungus *Fusarium oxysporum* f. sp. *radicis-lycopersici. Mol. Plant-Microbe Interact. 16*, 983-993.

Brand, J., Lugtenberg, B. J. J., Glandorf, D. C. M., Bakker, P. A. H. M., Schippers, B. & Weger, L. A. (1990). Isolation and characterization of a superior potato root-colonizing *Pseudomonas* strain. Pages 350-354 in: Plant growth -promoting rhizobacteria-Progress and Prospects. C. Keel, D. Knoller & G. Defago, eds IOBC/WPRS Bull. XIV-8.Interlaken, Switzerland.

Bull, C. T., Weller, D. M., & Thomashow, L. S. (1991). Relationship between root colonization and suppression of *Gauemannomyces graminis* var. *tritici* by *Pseudomonas* strain 2-79. *Phytopathol. 81*, 954-959.

Burr, T. J. & Suslow, T. V. (1978). Increased potato yields by treatment of seed pieces with specific strains of *Pseudmonas fluorescens* and *P. putida. Phytopathol. 68*, 1377-1383.

Buyer, J. S. & Leong, J. (1986). Iron transport-mediated antagonism between plant growth-promoting and plant-deleterious *Pseudomonas* strains. *J. Biol. Chem. 261*, 791-794.

Camacho, M. M. (2001). Molecular characterisation of the role of Type 4 Pili, NDH-I and PyrR in rhizosphere colonisation of *Pseudomonas fluorescens* WCS365. Thesis, Leiden University.

Chin-A-Woeng, T. F. C., de Priester, W., van der Bij, A. J. & Lugtenberg, B. J. J. (1997). Description of the colonization of a gnotobiotic tomato rhizosphere by *Pseudomonas fluorescens* biocontrol strain WCS365 using scanning electron microscopy. *Mol. Plant-Microbe Interact. 10*, 79-86.

Chin-A-Woeng, T. F. C., Bloemberg, G. V., van der Bij, A. J., van der Drift, K. M. G. M., Schripsema, J., Kroon, B., Scheffer, R. J., Keel, C., Bakker, P. A. H. M., Tichy, H. V., de Bruijn, F. J., Thomas-Oates, J. E., and Lugtenberg, B. J. J. (1998). Biocontrol by phenazine-1-carboxamide-producing *Pseudomonas chlororaphis* PCL1391 of tomato root rot caused by *Fusarium oxysporum* f. sp. *radicis-lycopersici. Mol. Plant-Microbe Interact. 11*, 1069-1077.

Chin-A-Woeng, T. F. C., Bloemberg, G. V., Mulders, I. H. M., Dekkers, L. C. & Lugtenberg, B. J. J. (2000). Root colonization by phenazine-1-carboxamide-producing bacterium *Pseudomonas chlororaphis* PCL1391 is essential for biocontrol of tomato foot and root rot. *Mol. Plant-Microbe Interact. 13*, 1340-1345.

Chin-A-Woeng, T. F. C., van den Broek, D., de Voer, G., van der Drift, K. M. G. M., Tuinman, S., Thomas-Oates, J. E., Lugtenberg, B. J. J. & Bloemberg G.V. (2001). Phenazine-1-carboxamide production in the biocontrol strain *Pseudomonas chlororaphis* PCL1391 is regulated by multiple factors secreted into the growth medium. *Mol. Plant-Microbe Interact. 14*, 969-979.

Chin-A-Woeng, T. F. C., van den Broek, D., Lugtenberg, B. J. J. & Bloemberg, G. V. (2005). The *Pseudomonas chlororaphis* PCL1391 sigma regulator *psrA* represses the production of the antifungal metabolite phenazine-1-carboxamide. *Mol. Plant-Microbe Interact. 18*, 244-253.

de Weert, S., Vermeiren, H., Mulders, I. H. M., Kuiper, I., Hendrickx, N., Bloemberg, G. V., Vanderleyden, J., De Mot, R. & Lugtenberg, B. J. J. (2002). Flagella-driven chemotaxis towards exudate components is an important trait for tomato root colonization by *Pseudomonas fluorescens. Mol. Plant-Microbe Interact. 15*, 1173-1180.

de Weert, S., Kuiper, I., Lagendijk, E. L., Lamers, G. E. M. & Lugtenberg, B. J. J. (2004a). Role of chemotaxis towards fusaric acid in colonization of hyphae of *Fusarium oxysporum* f.sp. *radicis-lycopersici* by *Pseudomonas fluorescens* WCS365. *Mol. Plant-Microbe Interact.17*, 557-566.

de Weert, S., Dekkers, L. C., Kuiper, I., Bloemberg, G. V. & Lugtenberg, B. J. J. (2004b). Generation of enhanced competitive root tip colonizing *Pseudomonas* bacteria through accelerated evolution. *J. Bacteriol. 186*, 3153-3159.

de Weger, L. A., van der Vlugt, C. I., Wijfjes, A. H. M., Bakker, P. A. H. M., Schippers, B. & Lugtenberg B. J. J. (1987). Flagella of a plant-growth-stimulating *Pseudomonas fluorescens* strain are required for colonization of potato roots. *J. Bacteriol. 169*, 2769-2773.

Dekkers, L. C., Bloemendaal, C. J., de Weger, L. A., Wijffelman, C. A., Spaink, H. P. & Lugtenberg, B. J. J. (1998a). A two-component system plays an important role in the root-colonizing ability of *Pseudomonas fluorescens* strain WCS365. *Mol. Plant-Microbe Interact. 11*, 45-56.

Dekkers, L. C., Phoelich, C. C., van der Fits, L. & Lugtenberg, B. J. J. (1998b). A site-specific recombinase is required for competitive root colonization by *Pseudomonas fluorescens* WCS365. *Proc. Natl. Acad. Sci. USA 95*, 7051-7056.

Dekkers, L. C., Mulders, I. H. M., Phoelich, C. C., Chin-A-Woeng, T. F. C., Wijfjes, A. H. M. & Lugtenberg, B. J. J. (2000). The *sss* colonization gene of the tomato-*Fusarium oxysporum* f. sp. *radicis lycopersici* biocontrol strain *Pseudomonas fluorescens* WCS365 can improve root colonization of other wild-type *Pseudomonas* spp. bacteria. *Mol. Plant-Microbe Interact. 13*, 1177-1183.

Gamliel, A. & Katan, J. (1992). Chemotaxis of fluorescent pseudomonads towards seed exudates and germinating seeds in solarized soil. *Phytopathol. 82*, 328-332.

Geels, F. P. & Schippers, B. (1983a). Reduction of yield depressions in high frequency potato cropping soil after seed tuber treatments with antagonistic fluorescent *Pseudomonas* spp.. *Phytopathol. 108*, 207-214.

Geels, F. P. & Schippers, B. (1983b). Selection of antagonistic fluorescent *Pseudomonas* spp. and their root colonization and persistence following treatment of seed potatoes. *Phytopathol.* Z. *108*, 193-206.

Girard, G., van Rij, E., Lugtenberg, B.J.J., & Bloemberg, G.V. Regulatory roles of *psrA* and *rpoS* in phenazine-1-carboxamide production by *Pseudomonas chlororaphis* PCL1391. *Microbiology* In press.

Glandorf, D. C. M. (1992). Root colonization by fluorescent *Pseudomonas*. Ph.D. Thesis. University of Utrecht, Utrecht, The Netherlands.

Harman, G. E. & Lumsden, R. D. (1990). Biological disease control, p. 259-280. In: J. M. Lynch (ed.), *The Rhizosphere.* Wiley & Sons, Chichester.

Hiltner, L. (1904). Über neue Erfahrungen und Probleme auf dem Gebiet der Bodenbacteriologie und unter besonderes Berücksichtigung der Grundugungen und Brauche. *Arb. Dtsch Landwirt Ges Berl 98*, 59-78.

Hoffland, E., Findenegg, G. R. & Nelemans, J. A. (1989). Solubilization of rock phosphate by rape. *Plant Soil 113*, 161-165.

Howie, W. J., Cook, R. J. & Weller, D. M. (1987). Effects of soil matric potential and cell motility on wheat root colonization by fluorescent pseudomonads suppressive to take-all. *Phytopathol. 77*, 286-292.

Kloepper, J. W., Leong, J., Teintze, M. & Shroth, M. N. (1980a). Enhanced plant growth by siderophores produced by plant growth promoting rhizobacteria. *Nature 286*, 885-886.

Kloepper, J. W., Schroth, M. N. & Miller, T. D. (1980b). Effects of rhizosphere colonization by plant growth-promoting rhizobacteria on potato plant development and yield. *Phytopathol. 70*, 1078-1082.

Kuiper, I., Bloemberg, G. V. & Lugtenberg, B. J. J. (2001). Selection of a plant-bacterium pair as a novel tool for rhizostimulation of polycyclic aromatic hydrocarbon-(PAH)-degrading bacteria. *Mol. Plant-Microbe Interact. 14*, 1197-1205.

Kuiper, I., Kravchenko, L., Bloemberg, G. V. & Lugtenberg, B. J. J. (2002). *Pseudomonas putida* strain PCL1444, selected for efficient root colonization and naphthalene degradation, effectively utilizes root exudate components. *Mol. Plant-Microbe Interact. 15*, 734-741.

Lagopodi, A. L., Ram, A. F. J., Lamers, G. E. M., Punt, P. J., Van den Hondel, C. A. M. J. J., Lugtenberg, B. J. J. & Bloemberg, G. V. (2002). Novel apsects of tomato root colonization and infection by *Fusarium oxysporum* f.sp. *radicis-lycopersici* revealed by confocal laser scanning microscopic analysis using the green fluorescent protein as a marker. *Mol. Plant Microbe Interact. 15*, 172-179.

Lam, S., Ellis, T. & Ligon, J. M. (1990). Genetic approaches for studying rhizosphere colonization. *Plant Soil 129*, 11-18.

Leong, J. (1986). Siderophores: their biochemistry and possible role in the biocontrol of plant pathogens. *Ann. Rev. Phytopathol. 24*, 187-209.

Lugtenberg, B. J. J., de Weger, L. A. & Bennett, J. W. (1991). Microbial stimulation of plant growth and protection from disease. *Curr. Opin. Biotechnol. 2*, 457-464.

Lugtenberg, B. J. J. & de Weger, L. A. (1994). Bacterization to protect seed and rhizosphere against disease. *Seed treatment: Progress and prospects 57*, 293-302.

Lugtenberg, B. J. J., Kravchenko, L. V. & Simons, M. (1999). Tomato seed and root exudate sugars: composition, utilization by *Pseudomonas* biocontrol strains and role in rhizosphere colonization. *Environ.Microbiol. 1*, 439-446.

Lugtenberg, B. J. J., Dekkers, L. C. & Bloemberg, G. V. (2001). Molecular determinations of rhizosphere colonization by *Pseudomonas*. *Annu. Rev. Phytopathol. 39*, 461-490.

Lynch, J. M. & Whipps, J. M. (1990). Substrate flow in the rhizosphere. *Plant and Soil 129*, 1-10.

Markovich, N. A. & Kononova, G. L. (2003). Lytic enzymes of *Trichoderma* and their role in plant defense from fungal diseases: A review. *Appl. Biochem. Microbiol. 39*, 341-351.

Metraux, J. P., Nawrath, C. & Genoud, T. (2002). Systemic acquired resistance. *Euphitica* 124, 237-243.

Neilands, J. B. 1991. Brief history of iron metabolism. *Biol. Metals 4*, 1-6.

Notley-McRobb, L. & Ferenci, T. (2000). Experimental analysis of molecular events during mutational periodic selections in bacterial evolution. *Genetics 156*, 1493-1501.

O'Toole, G. A. & Kolter, T. (1998a). Flagellar and twitching motility are necessary for *Pseudomonas aeruginosa* biofilm development. *Mol. Microbiol 30*, 295-304.

O'Toole, G. A. & Kolter, R. (1998b). Initiation of biofilm formation in *Pseudomonas fluorescens* WCS365 proceeds via multiple, convergent signaling pathways: a genetic analysis. *Mol. Microbiol 28*, 449-461.

Parke, J. L. (1991). Root colonization by indigenous and introduced microorganisms., p. 33-42. In: D. L. Keister & P. B. Cregan (eds.), *The Rhizosphere and Plant Growth.* Kluwer Academic Publishers, The Netherlands.

Parkinson, J. S. & Kofoid, E. C. (1992). Communication modules in bacterial signaling proteins. *Annu. Rev. Genet. 26*, 71-112.

Raaijmakers, J. M., van der Sluis, I., Koster, M., Bakker, P. A. H. M., Weisbeek, P. J. & Schippers, B. (1995). Utilization of heterologous siderophores and rhizosphere competence of fluorescent *Pseudmonas* spp. *Can. J. Microbiol. 41*, 126-135.

Reid, C. P. P. (1990). Mycorrhizas, p. 281-315. In: J. M. Lynch (ed.), *The Rhizopshere.* Wiley and Sons, Chichester.

Scher, F. M. & Baker, R. 1980. Mechanisms of biological control in *Fusarium*-suppressive soil. *Phytopathol. 70*, 412-417.

Scher, F. M., Kloepper, J. W., Singleton, C., Zaleski, I. & Laliberte, M. (1988). Colonization, of soybean roots by *Pseudomonas* and *Serratia* species: relationship to bacterial motility, chemotaxis and generation time. *Phytopathol.* 1055-1059.

Schippers, B., Scheffer, R. J., Lugtenberg, B. J. J. & Weisbeek, P. J. (1995). Biocoating of seeds with plant growth-promoting rhizobacteria to improve plant establishment. *Outlook Agricul. 24*,179-185.

Schroth, M. N. & Hancock, J. G. (1982). Disease-suppressive soil and root-colonizing bacteria. *Science 216*, 1376-1381.

Shann, J. R. & Boyle, J. J. (1994). Influence of plant species on in situ rhizosphere degradation, p. 70-81. In: T. A. Anderson and J. R. Coats (eds.), *Bioremediation through Rhizosphere Technology*. American Chemical Society, Washington, DC.

Simons, M., van der Bij, A. J., Brand, I., de Weger, L. A., Wijffelman, C. A. & Lugtenberg, B. J. J. (1996). Gnotobiotic system for studying rhizosphere colonization by plant growth-promoting *Pseudomonas* bacteria. *Mol. Plant-Microbe Interact. 9*, 600-607.

Simons, M., Permentier, H. P., de Weger, L. A., Wijffelman, C. A. & Lugtenberg, B. J. J. (1997). Amino acid synthesis is necessary for tomato root colonization by *Pseudomonas fluorescens* strain WCS365. *Mol. Plant-Microbe Interact. 10*, 102-106.

Suslow, T. V. & Schroth, M. N. (1982). Rhizobacteria of sugarbeets: effects of seed application and root colonization on yield. *Phytopathol. 72*, 199-206.

Swanson, R. V., Bourret, R. B. & Simon, M. I. (1993). Intermolecular complementation of the kinase activity of CheA. *Mol. Microbiol. 8*, 435-441.

Thomashow, L. S. & Weller, D. M. (1996). Current concepts in the use of introduced bacteria for biological disease control: mechanisms and antifungal metabolites., p. 236-271. In: G. Stacey and N. T. Keen (eds.), *Plant-Microbe interactions*. Chapman and Hall, New York.

Urios, A., Herrera, G., Aleixandre, V. & M. Blanco, M. (1994). Processing of MucA protein is required for spontaneous and benzo[a]pyrene-induced reversion of the *Escherichia coli* trpA23 missense mutation by G.C-T.A transversions: effect of a deficiency in the MutY DNA glycosylase. *Mutat. Res. 311*, 257-263.

van Loon, L. C., Bakker, P. A. H. M. & Pieterse, C. M. J. (1998). Systemic resistance induced by rhizosphere bacteria. *Ann. Rev. Phytopathol. 36*, 453-483.

van Peer, R., Nieman, G. J. & Schippers, B. (1991). Induced resistance and phytoalexin accumulation in biological control of *Fusarium* wilt of carnation by *Pseudomonas* sp. strain WCS417r. *Phytopathol. 81*, 728-734.

van Rij, E. T., Wesselink, M., Chin-A-Woeng, T. F. C., Bloemberg, G. V., & Lugtenberg, B. J. J. (2004). Influence of environmental conditions on the production of phenazine-1-carboxamide by *Pseudomonas chlororaphis* PCL1391. *Mol. Plant-Microbe Interact. 17*, 557-566.

van Rij, E.T., Girard, G., Lugtenberg, B. J. J. & Bloemberg G. V. (2005). Influence of fusaric acid on phenazine-1-carboxamide synthesis and gene expression of *Pseudomonas chlororaphis* strain PCL1391. *Microbiol. 151*, 2805-2814.

van Wees, S. C., Luijendijk, M., Smoorenburg, I., van Loon, L. C. & Pieterse, C. M. J. (1999). Rhizobacteria-mediated induced systemic resistance (ISR) in *Arabidopsis* is not associated with a direct effect on expression of known defense-related genes but stimulates the expression of the jasmonate-inducible gene Atvsp upon challenge. *Plant Mol. Biol. 41*, 537-549.

van Wees, S. C., de Swart, E. A., van Pelt, J. A., van Loon, L. C. & Pieterse, C. M. J. (2000). Enhancement of induced disease resistance by simultaneous activation of salicylate- and jasmonate-dependent defense pathways in *Arabidopsis thaliana. Proc. Natl. Acad. Sci. U.S.A 97*, 8711-8716.

Walsh, A. G., Matewish, M. J., Burrows, L. L., Monteiro, M. A., Perry, M. B. & Lam, J. S. (2000). Lipopolysaccharide core phosphates are required for viability and intrinsic drug resistance in *Pseudomonas aeruginosa. Mol.Microbiol. 35*, 718-727.

Weller, D. M. & Thomashow, L. S. (1994). Current challenges in introducing beneficial microorganisms into the rhizosphere., p. 1-17. In: F. O'Gara, D. N. Dowling, and B. Boesten (eds.), *Molecular ecology of rhizosphere organisms*. VCH verlagsgesellschaft mbH., Weinheim, Germany.

White, S., Tuttle, F. E., Blankenhorn, D., Dosch, D. C. & Slonczewski, J. L. (1992). pH dependence and gene structure of *inaA* in *Escherichia coli. J.Bacteriol. 174,* 1537-1543.

Wolk, C. P., Cai, Y. & Panoff, J. M. (1991). Use of a transposon with luciferase as a reporter to identify environmentally responsive genes in cyanobacterium. *Proc. Natl. Acad. Sci. USA 88*, 5355-5359.

Yabuta, T., Kambe, K. & Hayasi, T. (1934). Biochemistry of the bakanae-fungus. I. Fusarinic acid, a new product of the bakanae-fungus. *J. Agric. Chem. Soc. Japan. 10*, 1059-1068.

Yang, Y. & Inouye, M. (1991). Intermolecular complementation between two defective mutant signal-transducing receptors of *Escherichia coli. Proc. Natl. Acad. Sci. U.S.A 88*, 11057-11061.

10. AFFILIATIONS

Sandra de Weert & Guido V. Bloemberg, *Institute of Biology Leiden, Leiden University, Wassenaarseweg 64, 2333AL Leiden, The Netherlands.*
Tel: 31-71-5275066; Fax : 31-71-5275088;
E-mail: Bloemberg@rulbim.leidenuniv.nl

ROBERT J. KREMER

DELETERIOUS RHIZOBACTERIA

Abstract. Deleterious rhizobacteria (DRB) are predominantly saprophytic bacteria that aggressively colonize plant seeds, roots and rhizospheres and readily metabolize organic substances released by plant tissues. Unlike typical phytopathogens, DRB do not invade and parasitize vascular tissues; DRB that inhabit plants endophytically are found intercellularly beneath epidermal cells and in intracellular spaces of root cortical cells without inducing disease symptoms. Because plant growth suppression by DRB is subtle, the origin, distribution and ecophysiology of DRB are reviewed. The development of growth suppression in plants by DRB depends on interaction of several factors including plant host, soil environmental factors, time, and, in agroecosystems, management. The interacting factors may determine the expression and severity of several mechanisms for growth suppression mediated by DRB. Plant growth suppressive mechanisms include production of phytotoxic substances, plant-growth regulators, or interactions with other microorganisms. As new techniques in biotechnology are applied in investigations of DRB and their metabolites, the new information will aid in management of DRB in various ecosystems and in development of DRB as potential beneficial bacteria.

1. INTRODUCTION AND TERMINOLOGY

Regions within the soil ecosystem devoid of plants are nutrient-limited habitats for microorganisms, however, soils that are adjacent to and influenced by seeds (spermosphere) and roots (rhizosphere) provide microenvironments with organic substrates that are readily available for proliferation of a diverse array of soil microorganisms (Kennedy, 2005). Soilborne bacteria adapted to competitive colonization of the spermosphere, rhizosphere, and the root can be grouped under the general term rhizobacteria (Schroth & Hancock, 1982). An expanded definition of rhizobacteria includes bacteria from the rhizosphere, root surface (rhizoplane), and within the root (endorhizal bacteria or endophytes) (Nehl et al., 1997). Although the spermosphere is defined with reference to a seed prior to root emergence, the zone of soil contains substances exuded from the seed that rapidly attract specific microorganisms, which, if the seed survives, often establishes the dominant microbial communities of the longer-lived rhizosphere (Nelson, 2004). Subsequent rhizosphere colonization is often conceptualized as progressing across a continuum comprised of soil adjacent to the root, the rhizoplane, and the interior cells beneath the root epidermis (Sturz et al., 2000). Enhanced growth of bacteria in the rhizosphere is mediated by environmental conditions (soil moisture, temperature, substrate availability, etc.) and can extend 2 mm or more from the rhizoplane. However, microbial growth is most prolific in areas where root exudates and other root-derived organic substances are concentrated, providing a favorable micro-environment (Chao et al., 1986; Newman & Römheld, 2002).

1.1. Rhizosphere and Root Colonization by Rhizobacteria

Rhizobacteria are aggressive rhizosphere colonizers because their growth is stimulated by root-exuded organic compounds, and root cell lysates and secretions (de Weger et al., 1995; Somers et al., 2004). Plants exude high levels of nutrients, many of which act as chemoattractants for bacteria. Indeed, a strong relationship

S.S. Gnanamanickam (ed.), Plant-Associated Bacteria, 335–357.

exists between successful root colonization by rhizobacteria and their high growth rates on root-derived carbon substrates (Somers et al., 2004). Vital characteristics of rhizobacteria necessary for establishment, persistence, and interaction in the rhizosphere include motility, chemotaxis toward root exudates, and colonizing ability (cellular attachment to the rhizoplane) (Kremer et al., 1990; Vande Broeke & Vanderleyden, 1995; Lugtenberg & Dekkers, 1999). After establishment in the rhizosphere or on the rhizoplane, rhizobacteria continue colonization of the developing root system, efficiently compete with and displace weaker root-colonizing organisms, and often persist through the mid-stages of plant ontogeny at population densities $\geq 10^6$ cells/g of root (Kloepper et al., 1991). Endorhizal bacteria, established on the root surface or between epidermal cells, may penetrate the root epidermal layer through natural openings, wounds, micropores, or lateral root-forming sites, and colonize intercellular spaces within the endodermis (Garbeva et al., 2001; Hallman, 2001). Localized release of hydrolytic enzymes (i.e., cellulase or pectinase) by some endorhizal bacteria may aid cellular penetration, however, enzyme production ceases after bacteria successfully establish and begin proliferation in the endodermis (Hallman, 2001).

Root-derived compounds play multiple roles in the rhizosphere-rhizobacteria interaction. The composition and concentration of these compounds and the plant species guide the specific colonization of the root by compatible rhizobacteria. In addition to serving as carbon and energy sources, the compounds may act as signal chemicals in a mechanism that regulates early stages of colonization of the root. Involvement of signaling in colonization is well documented for symbiotic nitrogen-fixing bacteria in *Rhizobiaceae* and for invasive pathogens such as *Agrobacterium* spp. Signaling has also been demonstrated for several non-infective, non-symbiotic rhizobacteria including *Pseudomonas* spp. (Brencic & Winans, 2005; Somers et al., 2004). Synthesis of signal compounds by host plants may increase in the presence of organisms before the onset of root colonization. Establishment of plant-microbe associations may be triggered by bacterial recognition of plant signal molecules, detected by bacterial sensory systems (Brencic & Winans, 2005). Presumably, signaling as an important component of the colonization process may be common among a majority of rhizobacteria.

Colonization of the root surface is essential for rhizobacteria to not only occupy the rhizosphere habitat but also to assure the onset and progression of functional processes carried out by the plant-microbe association. Based on descriptions for the phyllosphere habitat model (Lindow & Brandl, 2003), rhizobacteria may establish a functional association by altering the root surface through release of auxins, which increase root cell permeability and release of soluble carbon, thereby enhancing nutrient availability for sustenance of rhizobacteria on the root surface. Increased available nutrients may lead to exopolysaccharide synthesis by rhizobacteria for formation of cellular aggregates along the root that provide additional protection.

Root colonization patterns often are non-uniformly distributed micro-colonies (Figure 1) or aggregates of rhizobacteria. These may be considered biofilm communities (Somers et al., 2004; Islam et al., 2005) because they provide an ecological advantage for nutrient competition and uptake by the bacterial members and a means of bacterial cell-to-cell communication (quorum sensing), which is required for the community to mutually perform critical functions in the rhizosphere.

1.2. Diversity in Rhizobacteria

In general, the characteristics of bacterial inhabitants of the rhizosphere include the ability to live in a commensal relationship with plants, utilize nutrients exuded by roots, and survive e nvironmental stress by inhabiting protected sites provided by the architecture of the root (Paulsen et al., 2005) and those micro-sites formed by the combined intermingling of root and soil components. The composition of rhizobacterial communities in the rhizosphere is significantly affected by root-borne carbohydrates (Newman & Römheld, 2002) and reflects the metabolic versatility that allows rhizobacteria to out-compete non-rhizosphere bacteria and dominate the rhizosphere environment (Nehl et al., 1997). The rhizosphere inhabitants include a diverse array of rhizobacteria that differ widely in structure or taxonomy and function. Species of *Pseudomonas, Burkholderia, Aeromonas, Agrobacterium, Alcaligenes, Acenitobacter, Azospirillum, Bacillus, Chryseomonas, Klebsiella, Enterobacter, Flavobacterium, Stenotrophomonas, Xanthomonas, Lysobacter, Pantoea*, and *Serratia* are representative of members of the rhizobacterial community that have been identified in rhizospheres of numerous plants. Because similar species of rhizobacteria inhabit the rhizosphere, rhizoplane, and endorhizal cells, and because some rhizobacteria may colonize all three niches, these bacteria will be discussed collectively as suggested in previous reports (Bowen & Rovira, 1999; Nehl et al., 1997; Sturz et al., 2000; Zahir et al., 2004).

Bowen & Rovira (1999) present a thorough discussion on the various broad classes of rhizobacteria including infective and non-infective groups, each of which is further divided into beneficial and detrimental sub-groups. Beneficial, infective bacteria are represented by symbiotic nitrogen-fixing species such as *Rhizobium* and *Bradyrhizobium* with leguminous plants and *Frankia* with a variety of tree and shrub host plants. Detrimental, infective bacteria comprise classical plant pathogenic bacteria (i.e., *Agrobacterium tumefaciens*) that invade root tissues and cause readily distinguishable disease symptoms. Beneficial, non-infective rhizobacteria are broadly termed plant growth promoting rhizobacteria (PGPR) that promote plant growth through numerous mechanisms including synthesis of plant growth regulators, facilitation of nutrient uptake, antagonism of detrimental root pathogens, and inducing disease resistance in the host plant. Other terms often used for this group include yield-increasing bacteria (YIB), biocontrol-PGPR, or biofertilizers. Detrimental, non-infective rhizobacteria that inhibit root growth, suppress overall plant growth, and antagonize beneficial rhizosphere microorganisms are known as deleterious rhizobacteria (DRB) or, in some cases, yield decline (YD) bacteria. Previous reviews consider DRB as "minor pathogens" because interactions with the host plant primarily occur in the rhizosphere and rhizoplane with restricted intercellular penetration of the root cortex; root and shoot growth are inhibited with limited expression of obvious visual symptoms (Nehl et al., 1997). Even though rhizobacteria are typically categorized as PGPR or DRB, no single characteristic or group of characteristics, physiological, morphological or otherwise, distinguishes rhizobacteria as one class or another (Nehl et al., 1997).

A

B

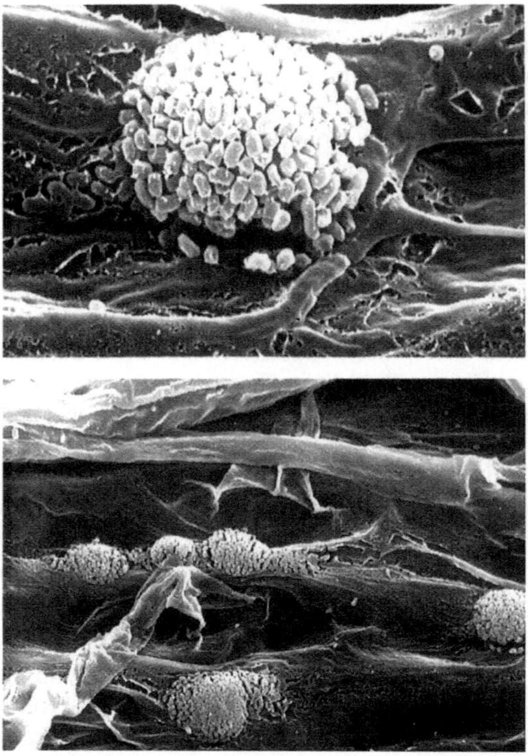

Figure 1. Scanning electron micrographs of colonization pattern of DRB Pseudomonas *on rhizoplane of velvetleaf (Abutilon theophrasti). Individual microcolony characteristics (A) and the random distribution of microcolonies on the rhizoplane (B). From Begonia et al. (1991). Magnification, 5,400X (A); 1,600X (B).*

Deleterious rhizobacteria as major factors in plant health have received less attention than PGPR because of the subtle nature of suppressing growth and difficulty in demonstrating effects of DRB on plants under field conditions.

However, frequent field observations of plant growth suppression and damage symptoms that cannot be linked to typical plant pathogens or soil nutrient imbalances provide substantial evidence to suggest the involvement of other biological entities such as DRB. Therefore, the objectives of this chapter are to provide an introduction to the origin, distribution and ecophysiology of DRB and to survey the natural association between DRB and their plant hosts.

2. DELETERIOUS RHIZOBACTERIA

Deleterious rhizobacteria are predominantly saprophytic bacteria that live on or in plant seeds and roots, surviving on organic compounds released by plant seed or root cells (Schippers et al., 1987). The plant is not parasitized and vascular tissue is not penetrated by DRB like major or true pathogens, but seed tissues, root hairs, and the root tip are externally colonized; DRB that inhabit plants endophytically are found intercellularly beneath epidermal cells and in intracellular spaces of root cortical cells without inducing disease symptoms (Schippers et al., 1987; Hallman, 2001). As pointed out by Paulsen et al. (2005), the plant-commensal microbial species include both beneficial and pathogenic organisms within the rhizosphere; indeed, Desai et al. (2002) suggest that plant pathogens and their antagonists (PGPRs) co-evolved over time to establish specific modes of co-existence. If PGPR are in a commensal association with the pathogens they antagonize, it is reasonable to expect that they are also associated commensally with DRB, which would explain the frequent mutual antagonism reported within these co-existent groups. Indeed, the coexistence of deleterious *Pseudomonas* strains with growth-promoting *Bacillus* strains in the European alder (*Alnus glutinosa*) rhizosphere (Probanza et al., 1996) is an excellent example of a DRB-PGPR commensal association. Furthermore, an early determinant in characterizing rhizobacteria as PGPR was the ability to antagonize or displace DRB and exclude their activity in the rhizosphere (Schippers et al., 1987; Schippers, 1988; Defago & Haas, 1990; Kloepper et al., 1991; Handelsman & Stabb, 1996; Nehl et al., 1997; Bowen & Rovira, 1999; Zahir et al., 2004).

Although the existence of DRB in nature recently has been challenged (Haas & Defago, 2005), the literature continues to acknowledge the presence and influence of DRB in plant-microorganism interactions. In their review of rhizobacteria, Bowen & Rovira (1999) state that strong evidence exists for nonpathogenic rhizosphere bacteria that are detrimental to plant growth. Examples from recent reports describe DRB as significant components of the total rhizosphere community and are often implicated as factors in plant disease complexes (Waschkies et al., 1994; Pankhurst et al., 2003); in plant growth suppression due to continuous cropping or "narrow crop rotation" (Iswandi et al., 1987); or in the interference of plant growth enhancement effects of selected PGPR applied to crops (Berggren et al., 2005). A survey of non-pathogenic bacteria resident in Kennebec potato (*Solanum tuberosum*) tubers revealed that 4.5% of the isolates depressed plant growth and were considered DRB or "plant growth retarding" bacteria (Sturz, 1995). Similarly, a survey of endophytic bacteria colonizing roots of field-grown carrot (*Daucus carota*) showed that 7% of the bacterial isolates inhibited carrot growth and 29% inhibited potato plant growth (Surette et al., 2003), suggesting that DRB-host interactions may occur for different crops under certain conditions. An example of such crop specificity was demonstrated for non-detrimental, endophytic bacteria of red clover (*Trifolium pretense*), which, when released from red clover in crude leachates, performed as DRB of maize (*Zea mays*) by inhibiting emergence rate and growth of seedlings (Sturz & Christie, 1996).

Population densities of DRB are very high in soils continuously cropped to perennial plants (trees, vines) that exhibit "replant disease." Replant disease problems, expressed as poor growth and low fruit production, are associated with monoculture of the same plants for many years. One of the first studies to recognize the important role of DRB in growth suppression was the work with young tree decline of citrus (*Citrus* spp.) in which the diversity and prevalence of DRB in the rhizosphere became the focus for disease management because reduction in citrus growth could not be causally related to typical root phytopathogens (Gardner et al., 1985). Similarly, grapevine (*Vitis* spp.) replant disease was associated with an accumulation of deleterious fluorescent pseudomonads on grapevine roots in soils of old vineyards in which growth of newly-planted grapevine cuttings was severely inhibited (Waschkies et al., 1994). Metabolic profiles (substrate utilization) and molecular analyses (amplification of 16S rDNA gene sequences) of bacterial communities of orchard soils afflicted with peach (*Prunus persica*) replant disease demonstrated the involvement of high densities of phytotoxin-producing DRB in the etiology of the replant disease (Benizri et al., 2005). Apple (*Malus domestica*) replant disease, recognized as a complex disease syndrome, has been attributed primarily to several fungal root pathogens in development of the disease in certain apple-producing areas (i.e., Mazzola, 1998). However, recent research that examined rhizosphere bacterial DNA from apple orchards established in a different geographical region containing different soils showed that high numbers of DRB were associated with specific apple rootstocks and contributed to reduction in tree growth (Yao et al., 2005). Therefore, in addition to the biotic components of the replant disease complex, host plant genotype and soil region must be considered as factors that interact with DRB to influence tree growth and yield.

Other disease complexes in which DRB are likely involved because experimental evidence to date suggests that many detrimental soil organisms remain unidentified include yield decline (YD) of sugarcane and forest decline. Factors that contribute to loss of sugarcane (*Saccharum officinarum*) productivity by YD include extensive monoculture resulting in soils with high compaction and low organic carbon that are favorable for development of detrimental soil organisms, likely including DRB, that affect the health of the cane root system (Pankhurst et al., 2003). Forest decline is characterized by interactions of biotic and abiotic factors that weakens some physiological state of the tree whereby detrimental soil and root organisms invade root tissues and successfully suppress growth (Manion & Lachance, 1992). Several models depict forest decline with a "saprogen" biotic component, opportunistic soil organisms that may include DRB, able to damage tree roots with altered cellular morphology and increase sensitivity to phytopathogens as a result of environmental stress (Houston, 1992; Matschke & Machácková, 2002) or disrupt nutrient uptake by the trees (Devêre et al., 1993).

Seed germination and seedling development are often affected by rhizobacteria that produce phytotoxic metabolites, which may be considered *bacterial allelochemicals* (Compant et al., 2005). Therefore, it has been suggested that the term *allelopathic bacteria* may more accurately describe DRB involved in suppressing plant growth through the production of allelochemicals (Barazani & Friedman, 1999; Kremer, 2005; Sturz & Christie, 2003). Such DRB associate with weed seedlings, are very host specific, and have no detrimental or allelopathic effects on growth of crop or other desirable plant species (Cherrington & Elliott, 1987; Kennedy et al., 2001). The ability to selectively suppress the growth of weeds in the presence of crop plants demonstrates the importance of these DRB as potential weed biological control agents (Kremer & Kennedy, 1996; Kremer, 2002).

Current evidence on the diverse associations of rhizobacteria with plants complicates the distinction between PGPR and DRB (Nehl et al., 1997). Many PGPR selected for beneficial effects of plant growth enhancement or suppression of plant pathogens may also detrimentally affect plants. For example, *Pseudomonas fluorescens* strain CHA0 produces antibiotics to suppress various soilborne fungi that cause plant diseases yet these antibiotic metabolites may also be phytotoxic to crop seedlings (Maurhofer et al., 1992). Other pseudomonads, considered PGPR for producing hydrogen cyanide (HCN) in amounts lethal to a pathogen on one crop, may act as DRB if HCN negatively affects growth of another crop (Defago & Haas, 1990). Another paradoxical relationship of a rhizobacterium with plants was discovered when a gram-negative bacterium, isolated from the rhizosphere of glyphosate-resistant soybean (*Glycine max*), acted as a DRB on the weed ivyleaf morningglory (*Ipomoea hederacea*) because it inhibited seedling growth by producing high concentrations of auxins, yet was identified as *Bradyrhizobium japonicum*, a beneficial, infective PGPR (symbiotic nitrogen fixer) on soybean (Kim & Kremer, 2005). Some rhizobacterial strains may alternately exhibit inhibitory and plant growth promoting properties or may have potential to act as a DRB and a plant growth promoting bacterium (PGPR) at the same time (Nehl et al., 1997). Thus, it is important to define DRB isolate within the context in which the specific host plant, soil, and possible pathogens interact (Defago & Haas, 1990; Preston, 2004).

3. DEVELOPMENT OF GROWTH SUPPRESSION IN PLANTS

Because DRB comprise a diverse group of bacteria and often detrimentally affect plant growth through complex associations with plants in the rhizosphere, the involvement of a number of interacting factors that lead to plant growth suppression, including the ability of the plant to perceive these factors perhaps through signal transduction pathways (Preston, 2004), may be conceptualized in a multi-component model (Figure 2). The model presented is an adaptation of the disease pyramid (Agrios, 1988) used in plant pathology to describe the interaction of components of plant disease to hypothesize the interactions of components specific for development of disease-like symptoms in plants by DRB. Classical plant diseases are explained by the interactions of plant, pathogen, a set of environmental components, and time. Huber (2005) separated the environmental component into biotic and abiotic factors to demonstrate the importance of diverse soil organisms (biotic) interacting with temperature, soil moisture, etc. (abiotic) in critically understanding the determinants of disease severity. In the plant growth suppression model, biotic and abiotic components in the soil environment play critical roles in inducing DRB activity toward host plants, therefore, these are considered as separate environmental factors (Figure 2). Agrios (1988) emphasized the influence of the human component in disease development in cultivated plants. The human component may be represented in symptom development in plants affected by DRB as "management" because many cultural practices followed in agricultural management systems may directly or indirectly be responsible for enhancing or suppressing deleterious effects (Garbeva et al., 2001). The schematic diagram presented in Figure 2 illustrates the host plant, DRB, and the abiotic and biotic soil environments as sides of a pyramid with time as a horizontal line through the center of the pyramid, and management as the center line connecting the components together. This representation thus shows how management interacts with and influences each of the four components of plant growth suppression, the extent of which is affected by time (stage of plant development, time of year, timing of management practice, etc.).

Reference to the model (Figure 2) indicates that growth suppression by DRB is a likely consequence of the association of high populations of bacteria on the root surface of the host plant due to the combined effects of root exudation patterns, soil properties, other soil organisms, and management practices imposed on ecosystem in which the plants occur. It is well documented that interactions of plants and microorganisms in the rhizosphere ultimately depend on the quality and quantity of root exudates (Jones et al., 2003) that define the level of colonization, and the diversity and functions of the rhizosphere inhabitants (Garbeva et al., 2001). Examples of various interacting factors that result in growth suppression by DRB are limited. Soil environmental factors including pH, texture, organic carbon (Li et al., 2002; Howie & Echandi, 1983) and temperature (Fredrickson & Elliott, 1985; Hasegawa et al., 2005) affect root colonization and growth suppressive activity by rhizobacteria. As illustrated later in Section 6, management practices applied in agroecosystems interact with host plants to alter root exudation patterns, which significantly influence the composition, populations, and activities of DRB in the rhizosphere.

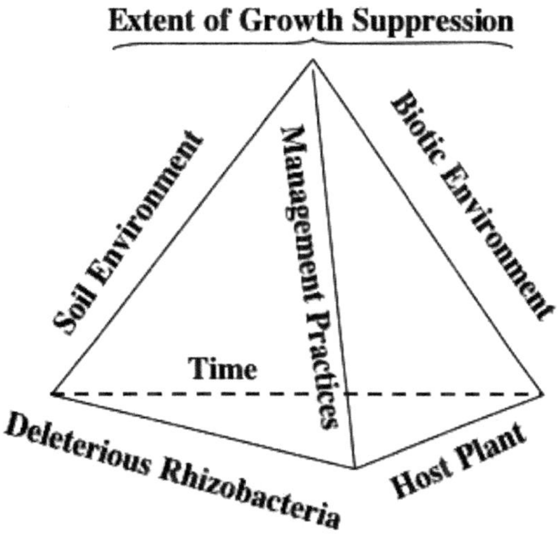

Figure 2. Representation of components and their interactions involved in plant growth suppression associated with deleterious rhizobacteria (modified from Agrios, 1988 and Huber, 2005).

4. SPECIFIC EFFECTS OF DRB

The pathogenicity of DRB is difficult to establish because their effects on plants are usually limited to a general retardation of root and shoot growth. Although distinct symptoms such as root discoloration, necrosis, chlorosis, wilting, and distortions of leaves and roots have been reported, these may have been due to the high inoculum numbers used in laboratory and greenhouse experiments (Schippers et al., 1987).

4.1. Inhibition of Seed Germination, Seedling Emergence, and Seedling Growth

Seedling mortality due to DRB, though rare, has been reported for canola *Brassica napus*) (Campbell et al., 1986). *Pseudomonas* Rp2, the most virulent isolate tested, caused 30% mortality in sprouting seedlings as well as delayed development in 68% of infected plants under laboratory conditions. Field trials supported the laboratory findings and a significant delay in maturation of infected plants compared to controls was observed. Unidentified phytotoxins in culture filtrates of bacteria isolated from rhizospheres of yellow starthistle (*Centaurea solstitialis*) were suppressive to yellow starthistle seed germination (Widmer & Guermache, 2004).

Various rhizobacteria inhibit seedling root and shoot growth of many plants. Growth of field pea (*Pisum sativum*) under gnotobiotic and non-sterile conditions was inhibited by strains of *Pseudomonas putida* (Berggren et al., 2001a). *Pseudomonas* and *Enterobacter* species inhibited white clover (*Trifolium repens*) and perennial ryegrass (*Lolium perenne*) seedling growth (Sarathchandra et al., 1996). Nehl et al. (1997) summarized the growth inhibition by rhizobacteria in many cereal crops, *Brassica* spp., lettuce (*Lactuca sativa*), grape, bean (*Phaseolus vulgaris*), potato, citrus crops, sugarcane, tomato (*Lycopersicon esculentum*), and a number of woody plants. Growth suppression of weed species by DRB is generally observed as stunting of tap and/or lateral roots, reduced hypocotyl or coleoptile length, leaf chlorosis, and reductions in overall plant biomass (Kremer & Kennedy, 1996).

4.2. Interactions with Beneficial Rhizosphere Microorganisms

Deleterious rhizobacteria may inhibit plant growth indirectly by negatively affecting mycorrhizal development through reductions in spore germination and hyphal lengths of arbuscular mycorrhizae (AM), decreased root colonization, or a decline in the metabolic activity of the internal mycelium (Hodge, 2000). Development of structures of vesicular-AM during specific stages of infection and establishment in big bluestem (*Adropogon furcatus*) were inhibited by soil bacteria (Wilson et al., 1988). Low plant available manganese is associated with many AM associations, which may be related to alteration of the rhizosphere and root exudation patterns by AM infection (Posta et al., 1994). The altered rhizosphere environment may favor DRB able to either antagonize Mn-reducing bacteria, which provide plant available Mn in the reduced state, or able to oxidize Mn thereby making it unavailable for plant uptake.

Symbiotic nitrogen fixation in leguminous plants may be inhibited by DRB by reducing nodulation capacity, inhibiting nitrogenase activity, and reducing rhizobial growth in the rhizosphere. Recently, Berggren, et al. (2001b; 2005) demonstrated that growth inhibition of *Rhizobium leguminosarum* bv. *viceae* resulted from its direct contact with, as well as its exposure to extracellular metabolites of the DRB *Pseudomonas putida* Å313. It was suggested that inhibitory substances synthesized by *P. putida* were active toward *R. leguminosarum* in iron-deficient soils. P. putida also colonized the rhizoplane sufficiently to cause deformed root hairs that contributed to reduced rhizobial infection and poor nodulation, suggesting production of phytoinhibitory metabolites possible affecting integrity of root membranes (Berggren et al., 2005). This was supported by observed negative chemotaxis of *R. leguminosarum* in the presence of cell-free filtrate of *P. putida*, indicating that inhibitory substances disrupted early infection of pea by *R. leguminosarum*. This research demonstrated that production of both antimicrobial and phytotoxic metabolites is a mechanism by which DRB are able to reduce or prevent colonization of roots by competing rhizosphere bacteria and disrupt the onset of symbiotic nitrogen fixation.

Deleterious rhizobacteria may be involved in inconsistent performance of selected PGPR used as biocontrol or growth-promoting agents in the field. It is hypothesized that DRB might degrade the signaling compounds, N-acylhomoserine lactones (AHLs), thereby disrupting AHL signaling between PGPR cells that is necessary for optimum root colonization and subsequent production of biocontrol compounds and/or plant-growth regulators (Somers et al., 2004). Also, DRB may be involved in blocking the plant-derived signaling compounds including salicylic acid, jasmonic acid, and ethylene that are important in pathogen defense, which could render plants more susceptible to soilborne pathogens (Somers et al., 2004).

5. MECHANISMS OF PLANT GROWTH SUPPRESSION

The ability of rhizosphere bacteria to produce various metabolites that inhibit plant growth is an important factor in crop productivity and affects the composition of plant communities in natural ecosystems. DRB frequently produce phytotoxic metabolites that inhibit plant growth by disrupting various plant physiological processes (Schippers et al., 1988; Alström and Burns, 1989; Bolton and Elliott, 1989). Frequently, DRB and PGPR produce similar substances that may inhibit or stimulate growth of other microorganisms and/or plants. Whether a metabolite acts as a growth inhibitor or promoter is generally a function of the concentration of the substance produced in the rhizosphere or the relative tolerance of a plant species to the substance. Also, as demonstrated for PGPR (Ryu et al., 2005), various environmental conditions (Figure 2) likely influence the response of DRB to the rhizosphere habitat of a particular host plant and affect amounts and types of metabolites produced and released, and ultimately, the overall extent of plant growth suppression. Phytotoxic compounds that specifically suppress growth of selected plants have been documented for only a few DRB.

5.1. Plant Growth Regulating Substances

Biosynthesis of the plant growth regulating compound indole-3-acetic acid (IAA), considered the most physiologically active auxin-type compound in plants, is widespread among bacterial colonizers of plant surfaces. Also IAA synthesis by bacteria associated with plants enhances leakage of plant cell nutrients to the bacterial colonizers and possibly promotes simultaneous plant growth (Lindow & Brandl, 2003), release of excessive amounts of IAA inhibits plant growth (Persello-Cartoeaix et al., 2003). Auxin-mediated plant growth suppression, as a mechanism used by DRB, is illustrated in several studies in which plant growth suppression of sugarbeet (*Beta vulgaris*) (Loper & Schroth, 1986), maize (Sarwar & Frankenberger, 1994), lettuce (Barazani & Friedman, 1999), sour cherry (*Prunus cerasis*) (Dubeikovsky et al., 1993), and several important weed and crop species (Sarwar & Kremer, 1995) was correlated to elevated IAA levels produced by DRB.

Similarly, plant growth was suppressed when tryptophan, an IAA precursor, was applied with DRB, which enhanced biosynthesis of IAA (Sarwar & Frankenberger, 1994; Sarwar & Kremer, 1995).

Ethylene is a volatile plant growth regulator that facilitates many physiological aspects of plant growth and development. Low levels of ethylene enhance root growth and elongation, however, as concentrations increase, root growth is inhibited. The impacts of microbially derived ethylene on plant growth have been extensively reviewed by Arshad & Frankenberger (1992). The possibility of production of high levels of ethylene by DRB as a mechanism for plant growth suppression has not been fully investigated. Some rhizobacteria capable of producing high levels of ethylene have been investigated as germination biostimulants of seeds of the parasitic weed *Striga* sp. as a potential biological control strategy (Berner et al., 1999). Many PGPR synthesize 1-aminocyclopropane-1-carboxylate (ACC) deaminase, an enzyme which degrades ACC, the precursor of ethylene in plants (Bowen & Rovira, 1999). Subsequently, ethylene concentrations are reduced and inhibition of root growth and elongation is effectively eliminated. The ability of co-existent DRB in the rhizosphere to block ACC deaminase synthesis by PGPR and indirectly suppress growth by maintaining high ethylene levels is yet to be determined.

5.2. Hydrogen Cyanide

The production of hydrogen cyanide (HCN), a volatile metabolite that negatively affects root metabolism and root growth by inhibiting cytochrome oxidase respiration, is common among rhizosphere pseudomonads (Schippers et al., 1990). The rate of HCN synthesis is affected by the availability of precursors such as the amino acids glycine, methionine, and proline, and the cyanogenic glucosides (Knowles & Bunch, 1986; Schippers et al., 1990). The amino acid composition of root exudates as well as environmental factors affecting root exudation (i.e, light intensity, soil water potential, nutrients) may be important as well (Schippers et al., 1990). Bakker & Schippers (1987) were among the first to relate crop yield reductions with damage to potato resulting from high levels of HCN produced by DRB present in crop production fields. HCN-producing DRB have also been implicated in root inhibition of lettuce and bean (Alström & Burns, 1989). The host specificity of cyanogenic DRB was illustrated by *Pseudomonas putida* and *Acidovorax delafieldii* strains, when combined with the glycine precursor of HCN, significantly inhibited seedling growth of velvetleaf but not maize (Owen & Zdor, 2001). Cyanide production by several rhizobacterial strains was a major factor in the inhibition of seedling growth of several weed species and was suggested as a trait for consideration in selecting DRB as potential weed biological control agents (Kremer & Souissi, 2001).

5.3. Phytotoxins

Few reports are available on specific or unique phytotoxins metabolized by DRB that are directly involved in plant growth suppression mechanisms. Fredrickson &

Elliott (1985b) found root-colonizing pseudomonads inhibited growth of winter wheat (*Triticum aestivum*) seedlings by producing a toxin. The phytotoxin was produced under a wide range of pH, growth temperatures, iron and sodium phosphate concentrations and different types of carbon and nitrogen sources (Bolton & Elliott, 1989). A phytotoxin from *Pseudomonas fluorescens* strain D7 was shown to be responsible for root growth inhibition of downy brome (*Bromus tectorum*) (Tranel et al., 1993). Further characterization revealed that the active fraction is a complex of peptides and fatty acid esters in a lipopolysaccharide matrix (Gurusiddaiah et al., 1994). Secondary metabolites isolated from *Pseudomonas syringae* strain 3366 inhibitory to downy brome consisted of phenazine-1-carboxylic acid, 2-amino phenoxazone and 2-amino phenol (Gealy et al., 1996). Electron microscopic observations revealed that disrupted plant cell walls and membranes of leafy spurge (*Euphorbia esula*) roots colonized by DRB were apparently due to production of unidentified phytotoxins and/or enzymes by the bacteria, which consequently inhibited seedling growth (Souissi et al., 1997).

5.4. Exopolysaccharides

A factor that has been overlooked in studies of DRB is the production of exopolysaccharides (EPS), which has traditionally been correlated with virulence of plant-associated, pathogenic bacteria. Further, a diverse range of EPS is often elaborated by *Pseudomonas* spp., which are considered opportunistic plant pathogens (Fett et al., 1989). EPS aids bacteria in systemic colonization of plant tissue and causes wilting in affected plants, interfering with movement of water through xylem vessels. Production of EPS by rhizosphere bacteria may enhance survival under adverse environmental conditions such as high temperatures and moisture stress (Fett et al., 1989). It was recently demonstrated that high EPS production combined with other traits including HCN and siderophore production by DRB were highly correlated with growth suppression of leafy spurge (Kremer et al., 2005a).

5.5. Nutrient Availability

Nutrients influence the expression of plant growth suppression through interaction as part of the soil environment component. Microorganisms play important roles in plant nutrition and may be most critical in mineral-deficient soils with low plant availability of certain nutrients. For example, the amount and composition of root exudates affected by various management practices often select for Mn-oxidizing rhizobacteria, which can immobilize Mn leading to Mn deficiency and render weakened cereal crops susceptible to infection by *Gaeumannomyces graminis*, the take-all disease pathogen (Rengel et al., 1994; Newman & Römheld, 2002; Huber, 2005). In contrast, "Frenching disease" in tobacco (*Nicotiana tobacum*), one of the earliest examples that implicated DRB in disease symptoms rather than classical plant pathogens (Woltz, 1978), was due to the Mn-reducing ability of *Bacillus cereus*, which increased available Mn to toxic levels in plant tissue (Huber, 2005).

Competition for Fe in the rhizosphere between microorganisms is an important factor in determining species distribution in the rhizosphere and can lead to biocontrol of plant pathogens or increased plant growth suppression by DRB or plant disease by pathogens. Siderophores are low molecular weight iron-chelating agents produced by almost all bacteria under iron-limiting conditions and are excreted in order to bind and solubilize extracellular Fe (III), which is then available to microorganisms. Siderophore production in high amounts could be competitive for Fe with plants, affecting Fe availability (Crowley & Gries, 1994). All siderophores have high formation constants for Fe (III), therefore, virtually all siderophore produced would form ferric siderophores; however, most available Fe is in the form of slow-reacting insoluble hydroxides which limit chelating reactions. Thus, bacteria producing high amounts of siderophore or a siderophore that efficiently and rapidly chelates Fe have a *competitive advantage* (Buyer et al., 1994). Also bacteria have a competitive advantage if they require low amounts of Fe for growth or had more efficient transport systems; or if they could use another bacterial siderophore for Fe uptake/transport. Pseudomonads evolved multiple siderophore types in order to produce siderophores that other bacteria cannot use. Thus, competition for iron in the rhizosphere affects microbial community structure and can favor predominance of DRB. This is supported by the siderophore pseudobactin 7SR1 by *Pseudomonas* 7SR1, a DRB on sugarbeet, which is produced in levels surpassing the amounts of siderophores by PGPR thereby allowing the DRB to compete successfully in the rhizosphere (Yang & Leong, 1984).

5.6. Effects of Combined Mechanisms of Action

It is likely that in many cases a single DRB will exhibit several mechanisms for suppressing plant growth. Vessey (2003) reviewed several studies of PGPR that reported many strains stimulated plant growth using two or more different mechanisms (i.e., ACC-deaminase, phosphate solubilization, IAA production). Similar studies are limited for DRB. In a recent survey for DRB as potential biological control agents of leafy spurge, we found that two or more selected traits (HCN, IAA, EPS production) expressed by isolates were highly correlated with plant growth suppression (Kremer et al., 2005a). Thus, DRB may function by multiple mechanisms of action. Also, based on inferences by Vessey (2003) and Cleyet-Marel et al. (2001), DRB strains with single mechanisms of action that co-exist in the rhizosphere may function synergistically to suppress plant growth.

6. MANAGEMENT IMPACTS ON DRB ACTIVITY

6.1. Cropping Systems

Crop and soil management practices imposed in agroecosystems have significant impacts on soil and rhizosphere microbial communities (Sturz & Christie, 2003), which may result in the dominance of either beneficial or detrimental rhizobacterial communities with associated effects on plant health and crop yields. Different plant species select specific rhizosphere communities based on their different root exudate composition (Grayston et al., 1998). However, the effect of several management practices on rhizosphere microorganisms is ultimately linked to altered of root exudation patterns by crop plants subjected to these practices.

Tillage influences the frequency of DRB occurring in soil. Higher proportions of indigenous DRB inhibitory toward the weeds downy brome and jointed goatgrass (*Aegilops cylindrical*) were detected under either conventional or reduced tillage compared to no-tillage (Kremer & Kennedy, 1996). Winter wheat produced in the U.S. Pacific Northwest if frequently heavily colonized by pseudomonad DRB when wheat seed is directly drilled into the residues of the previous crop (Elliott & Lynch, 1984; Fredrickson & Elliott, 1985b). The vegetative residues at or near the soil surface may serve as substrates for DRB allowing production of phytotoxins that inhibit growth of wheat seedlings (Stroo et al., 1988). Alternatively, for wheat grown in Australia, Kirkegaard et al. (1995) suggested that direct drilling in a continuous wheat production system allowed developing roots of wheat seedlings to contact biopores containing high concentrations of DRB from roots of the previous crop.

Repeated planting of the same annual crop for several years can encourage development of inhibitory bacteria on plant roots. Such an effect was described for continuous corn whereby poor seedling vigor was due partly to certain rhizobacteria that developed over time to specifically associate with and suppress growth of corn roots (Turco et al., 1990). Potato yields also decline in continuously cropped soils, however, when wheat or sugarbeet were included in a crop rotation, their yields were not affected, suggesting that inhibitory effects were specific to potato (Schippers et al., 1987). These observations support the general conclusion that in many cases, a community of rhizosphere microorganisms (i.e., DRB) proliferates in continuous monoculture, which enriches for specific rhizobacteria for the particular crop (Vilich & Sikora, 1998).

Cropping systems also influence the occurrence and magnitude of deleterious rhizobacteria from weed seedlings (Li & Kremer, 2000). A study of crop management practices on claypan soils (Epiaqualfs) that involved reduced tillage, maintenance of high soil organic matter, and limited inputs of agrichemicals found increased levels of DRB associated with weed seedlings that may contribute to natural weed suppression. Management systems with soils showing high levels of enzyme activity and a high proportion of water-stable soil aggregates also had greater proportions of weed-suppressive DRB (Kremer & Li, 2003). Agronomic practices that result in relatively high organic matter, such as perennial forage and pasture systems, organic farming, and integrated cropping systems, support higher proportions of weed DRBs.

6.2. Pesticides

Pesticides applied in conventional agricultural and horticultural production systems influence the rhizosphere microbial composition by altering the types and quantities of root exudates compounds and/or through systemic release of the active ingredients via root exudation. An early report demonstrated that the insecticide diazinon applied to bean foliage translocated rapidly to roots and selectively enriched for specific bacteria in the rhizosphere compared to plants receiving no insecticide (Gunner et al., 1966). It was not reported whether the selected bacteria were growth suppressive or not. However, Greaves & Sargent (1986) found that colonization of wheat roots by DRB (inhibitory *Pseudomonas* spp.) was greatly enhanced and resulted in extensive cellular and tissue damage when plants were treated with the herbicide mecoprop. A significant shift in endophytic bacteria detected by molecular analyses in soybean roots was related to cultivation of soybean in soils with a history of pre-plant glyphosate application (Kuklinsky-Sobral et al., 2005). It was suggested that continued or increased application of glyphosate could shift endophytic populations toward "latent pathogens" that could be detrimental to soybean growth. Continuous applications of high rates of benomyl to both foliage and soils for control of fungal pathogens in ornamental crop production significantly altered the rhizobacterial community, of which a high proportion was DRB (Kremer et al., 1996; Mills et al., 1996). For example, benomyl induced root alterations on leatherleaf fern (*Rumohra adiantiformis*), enhanced growth of DRB with growth suppressive activity, and resulted in various distortions, discolorations and stunting of the fern fronds.

Genetically-modified or transgenic plants that resist non-selective herbicides such as glyphosate or glufosinate or that resist insect attack by expressing *Bacillus thuringiensis* protein toxins (Bt) have been developed in recent years, providing new pest management strategies for conventional agricultural systems. Because of the genetic alterations in these plants and, in the case of herbicide-resistant plants, because applications of herbicides are limited to only one or two compounds, concerns have been raised regarding the potential adverse impacts of any modified root exudation patterns of transgenic crops on soil function due to altered rhizosphere microbial communities. Root exudates of Bt maize differed from non-transgenic maize not only due to presence of the exuded protein toxin but also in chemical composition (Brusetti et al., 2004). The differences in the root exudates determined the selection of different bacterial communities in the rhizosphere. Similarly, differences in rhizosphere bacterial populations and microbial activity between transgenic glufosinate-tolerant and wild-type oilseed rape (*Brassic rapa*) was attributed to altered root exudation patterns associated with the transgenic cultivar (Sessitsch et al., 2004). Because glufonsinate was not applied to either cultivar, it was concluded that the genetic transformation altered the physiology of plants (i.e., root exudation) that led to changes in rhizosphere community structure and bacterial activity. We recently reported that glyphosate-resistant soybean root exudates contained intact glyphosate and significantly higher concentrations of carbohydrates and amino acids compared to a non-transgenic cultivar (Kremer et al., 2005b). The glyphosate combined with the high concentrations of other exudates compounds tended to stimulate specific groups of microorganisms in the rhizosphere. Our data also suggested that plant physiology was altered in soybean during genetic modification for glyphosate resistance.

Much of the foregoing discussion only suggests that DRB might be involved in the shifts of rhizosphere microbial communities due to use of pesticides and transgenic plants in agroecosystems. However, the information to date suggests that management can be a major interacting factor in growth suppression (Figure 2), which often unintentionally creates conditions in the rhizosphere that are ideal for the development and establishment of DRB. Bowen & Rovira (1999) state that application of chemicals can be used as "selective substrates" that cause shifts in microbial community composition. Perhaps this approach will be useful not only in understanding how DRB become dominant under certain management systems but also as a means for manipulating either crop rhizospheres to minimize effects due to DRB or weed rhizospheres to advance the biological control potential of DRB.

7. CONCLUSIONS

From the first report of the impact of deleterious rhizobacteria on plant growth, much information on numerous organisms and their effects on plant species have been collected. No longer can DRBs be relegated to the role of minor pathogens with only subtle effects on plant growth. Research that implicates DRB involvement in numerous plant-microbe interactions continues to be reported in the current literature. The plant-suppressive mechanisms of DRB include the production of phytotoxic substances, plant-growth regulators, or interactions with other microbes, however, it is quite likely that more plant-growth regulating substances produced by DRB have yet to be discovered. Application of new techniques in biotechnology including genome sequencing may provide insight to unknown metabolite production and thereby insight for possible manipulation or control of DRB in agroecosystems (i.e., Paulsen et al., 2005). Root colonization and growth inhibition is often plant species or variety specific, which is intriguing from an ecological and management standpoint. Deleterious rhizobacteria may have potential for use in pest management systems and in biological control. The dynamic and complex nature of DRB and their ecological interactions with other organisms will continue to be investigated to improve our understanding of this group. Strategies might then be developed to reduce crop suppressive DRB populations and enhance weed suppressive strains. Further, rhizobacteria as bioherbicides or natural components of the soil microbial community may receive more consideration as major factors in restoring soil quality and biodiversity to ecosystems degraded through current conventional cropping systems.

8. REFERENCES

Agrios, G.N. (1988). *Plant Pathology*, 3rd ed. Academic Press, San Diego, CA.

Alström, S., & Burns, R.G. (1989). Cyanide production by rhizosphere as a possible mechanism of plant growth inhibition. *Biol. Fert. Soils 7*, 232-238.

Arshad, M., & Frankenberger, W.T., Jr. (1992). Microbial synthesis of ethylene and its influence on plant growth. *Adv. Microb. Ecol. 12*, 69-111.

Barazani, O., & Friedman, J. (1999). Is IAA the major root growth factor secreted from plant-growth-mediating bacteria? *J. Chem. Ecol. 25*, 2397-2406.

Begonia, M.F.T., Kremer, R.J., Stanley, L., & Jamshedi, A. (1990). Association of bacteria with velvetleaf roots. *Trans. Missouri Acad. Sci. 24*, 17-26.

Benizri, E., Piutti, S., Verger, S., Pagès, L., Vercambre, G., Poessel, J.L., & Michelot, P. (2005). Replant diseases: bacterial community structure and diversity in peach rhizosphere as determined by metabolic and genetic fingerprinting. *Soil Biol. Biochem. 37*, 1738-1746.

Berggren, I., Alström, S., & Mårtensson, A.M. (2001a). Deleterious properties of certain rhizosphere bacteria on field pea (*Pisum sativum*) under gnotobiotic and non-sterile conditions. *Appl. Soil Ecol. 16*, 169-177.

Berggren, I., Alström, S., van Vuurde, J.W.L., & Mårtensson, A.M. (2001b). Factors influencing the effect of deleterious *Pseudomonas putida* rhizobacteria on initial infection of pea roots by *Rhizobium leguminosaurum* bv. *viceae*. *Appl. Soil Ecol. 17*, 97-105.

Berggren, I., Alström, S., van Vuurde, J.W.L., & Mårtensson, A.M. (2005). Rhizoplane colonisation of peas by *Rhizobium leguminosarum* bv. *viceae* and a deleterious *Pseudomonas putida*. *FEMS Microbiol. Ecol. 52*, 71-78.

Berner, D.K., Schaad, N.W., & Volksh, (1999). Use of ethylene producing bacteria for stimulation of *Striga* spp. seed germination. *Biol. Contr. 15*, 274-282.

Bolton, H., Jr., & Elliott, L.F. (1989). Toxin production by a rhizobacterial *Pseudomonas* sp. that inhibits wheat root growth. *Plant Soil 114*, 269-278.

Bowen, G.D., & Rovira, A.D. (1999). The rhizosphere and its management to improve plant growth. *Adv. Agron. 66*, 1-102.

Brencic, A., & Winans, S.C. (2005). Detection of and response to signals involved in host-microbe interactions by plant-associated bacteria. *Microbiol. Molec. Biol. Rev. 69*, 155-194.

Brusetti, L., Francia, P., Bertolini, C., Pagliuca, A., et al. (2004). Bacterial communities associated with the rhizosphere of transgenic Bt 176 maize (*Zea mays*) and its non transgenic counterpart. *Plant Soil 266*, 11-21.

Buyer, J.S., Kratzke, M.G., & Sikora, L.J. (1994). Microbial siderophores and rhizosphere ecology. In Manthey, J.A., Crowley, D.E., & Luster, D.G. (Eds), *Biochemistry of Metal Micronutrients in the Rhizosphere* (pp. 67-80), Lewis Publishers: Boca Raton, FL.

Campbell, J.N., Conn, K., Sorlie, L., & Cook, F.D. (1986). Inhibition of growth of canola seedlings caused by an opportunistic *Pseudomonas* sp. under laboratory and field conditions. *Can. J. Microbiol. 32*, 201-207.

Chao, W.L., Nelson, E.B., Harman, G.E., & Hock, H.C. (1986). Colonization of the rhizosphere by biological control agents applied to seeds. *Phytopathology 76*, 60-65.

Cherrington, C.A., & Elliott, L.F. (1987). Incidence of inhibitory pseudomonads in the Pacific Northwest. *Plant Soil 101*, 159-165.

Cleyet-Marel, J.C., Larcher, M., Bertrand, H., Rapior, S., & Pinochet, X. (2001). Plant growth enhancement by rhizobacteria. In Morot-Gaudry, J-F. (Ed), *Nitrogen Assimilation by Plants: Physiological, Biochemical and Molecular Aspects* (pp. 185-197), Science Publishers, Enfield, NH.

Compant, S., Duffy, B, Nowak, J., Clèment, C., & Barka, E.A. (2005). Use of plant growth-promoting bacteria for biocontrol of plant diseases: principles, mechanisms of action, and future prospects. *Appl. Environ. Microbiol. 71*, 4951-4959.

Crowley, D.E., & Gries, D. (1994). Modeling of iron availability in the plant rhizosphere. In Manthey, J.A., Crowley, D.E., & Luster, D.G. (Eds), *Biochemistry of Metal Micronutrients in the Rhizosphere* (pp. 199-223), Lewis Publishers: Boca Raton, FL.

Defago, G., & Haas, D. (1990). Pseudomonads as antagonists of soilborne plant pathogens: modes of action and genetic analysis. In Bollag, J-M., Stotzky, G. (Eds.), *Soil Biochemistry*, vol. 6 (pp. 249-291), Marcel Dekker: New York.

Desai, S., Reddy, M.S., & Kloepper, J.W. (2002). Comprehensive testing of biocontrol agents. In Gnanamanickam, S.S. (Ed.), *Biological Control of Crop Diseases* (pp. 387-420), Marcel Dekker: New York.

Devêre, O., Garbaye, J., & Perrin, R. (1993). Experimental evidence of a deleterious soil microflora associated with Norway spruce decline in France and Germany. *Plant Soil 148*, 145-153.

De Weger, L.A., van der Bij, A.J., Dekkers, L.C., Simons, M., Wijffelman, C.A., & Lugtenberg, B.J. (1995). Colonization of the rhizosphere of crop plants by plant-beneficial pseudomonads. *FEMS Microbiol. Ecol. 17*, 221-228.

Dubeikovsky, A.N., Mordukhova, E.A., Kochetkov, V.V., Polikarpova, F.Y., & Boronin, A.M. (1993). Growth promotion of blackcurrant softwood cuttings by recombinant strain *Pseudomonas fluorescens* BSP53a synthesizing an increased amount of indole-3-acetic acid. *Soil Biol. Biochem. 25*, 1277-1281.

Elliott, L.F., & Lynch, J.M. (1984). Pseudomonads as a factor in the growth of winter wheat (*Triticum aestivum* L.). *Soil Boil. Biochem. 16*, 69-71.

Fett, W.F., Osman, S.F., & Dunn, M.F. (1989). Characterization of exopolysaccharides produced by plant-associated fluorescent pseudomonads. *Appl. Envrion. Microbiol. 55*, 579-583.

Fredrickson, J.K., & Elliott, L.F. (1985a). Effects on winter wheat seedling growth by toxin-producing rhizobacteria. *Plant Soil 83*, 399-409.

Fredrickson, J.K., & Elliott, L.F. (1985b). Colonization of winter wheat roots by inhibitory rhizobacteria. *Soil Sci. Soc. Am. J. 49*, 1172-1177.

Garbeva, P., van Overbeek, L.S., van Vuurde, J.W.L., & van Elsas, J.D. (2001). Analysis of endophytic bacterial communities of potato by plating and denaturing gradient gel electrophoresis (DGGE) of 16S rDNA based PCR fragments. *Microb. Ecol. 41*, 369-383.

Gardner, J.M., Chandler, J.L., & Feldman, A.W. (1985). Growth responses and vascular plugging of citrus inoculated with rhizobacteria and xylem-resident bacteria. *Plant Soil 86*, 333-345.

Gealy, D.R., Gurusiddaiah, S., & Ogg, A.J., Jr. (1996). Isolation and characterization of metabolites from *Pseudomonas syringae* strain 3366 and their phytotoxicity against certain weed and crop species. *Weed Sci. 44*, 383-392.

Grayston, S.J., Wang, S., Campbell, C.D., & Edwards, A.C. (1998). Selective influence of plant species on microbial diversity in the rhizosphere. *Soil Biol. Biochem. 30*, 369-378.

Greaves, M.P., & Sargent, J.A. (1986). Herbicide-induced microbial invasion of plant roots. *Weed Sci. 34 (Suppl. 1)*, 50-53.

Gunner, H.B., Zuckerman, B.M., Walker, R.W., Miller, C.W., Deubert, K.H., & Longley, R.E. (1966). The distribution of diazinon applied to plant and soil and its influence on rhizosphere and soil microflora. *Plant Soil 25*, 249-264.

Gurusiddaiah, S., Gealy, D.R., Kennedy, A.C., & Ogg, A.G., Jr. (1994). Isolation and characterization of metabolites from Pseudomonas fluorescens D7 for control of downy brome (*Bromus tectorum*). *Weed Sci. 42*, 492-501.

Haas, D., & Defago, G. (2005). Biological control of soil-borne pathogens by fluorescent pseudomonads. *Nature Rev. Microbiol. 3*, 307-319.

Hallmann, J. (2001). Plant interactions with endophytic bacteria. In Jeger, M.J. & Spence, N.J. (Eds.), *Biotic Interactions in Plant-Pathogen Associations* (pp. 87-119), Oxon, UK: CAB International.

Handelsman, J., & Stabb, E.V. (1996). Biocontrol of soilborne plant pathogens. *Plant Cell 8*, 1855-1869.

Hasegawa, H., Chatterjee, A., Cui, Y., & Chatterjee, A.K. (2005). Elevated temperature enhances virulence of *Erwinia carotovora* subsp. *carotovora* strain EC153 to plants and stimulates production of the quorum sensing signal, N-acyl homoserine lactone, and extracellular proteins. *Appl. Environ. Microbiol. 71*, 4655-4663.

Hodge, A. (2000). Microbial ecology of the arbuscular mycorrhizae. *FEMS Microbiol. Ecol. 32*, 91-96.

Houston, D.R. (1992). A host-stress-saprogen model for forest dieback-decline diseases. In Manion, P.D., & Lachance, D. (Eds.), *Forest Decline Concepts* (pp. 3-25), APS Press: St. Paul, MN.

Howie, W.J., & Echandi, E. (1983). Rhizobacteria: influence of cultivar and soil type on plant growth and yield of potato. *Soil Biol. Biochem. 15*, 127-132.

Huber, D.M. (2005). The role of nitrogen and sulfur on plant disease incidence and resistance. In Yamada, T. (Ed.), *Proceedings of the Symposium on Relationships Between Plant Nutrition and Disease Incidence.* [CD-ROM], POTAFOS, Piracicaba, S-P, Brazil.

Islam, M.T., Hashidoko, Y., Deora, A., Ito, T., & Tahara, S. (2005). Suppression of damping-off disease in host plants by the rhizoplane bacterium *Lysobacter* sp. Strain SB-K88 is linked to plant colonization and antibiosis against soilborne peronosporomycetes. *App. Environ. Microbiol. 71*, 3786-3796.

Iswandi, A., Bossier, P., Vandenabeele, J., & Verstraete, W. (1987). Relation between soil microbial activity and the effect of seed inoculation with the rhizopseudomonad strain 7NSK2 on plant growth. *Biol. Fertil. Soils 3*, 147-151.

Jones, D.L., Farrar, J., & Giller, K.E. (2003). Associative nitrogen fixation and root exudation – what is theoretical possible in the rhizosphere? *Symbiosis 35*, 19-38.

Kennedy, A.C. (2005). Rhizosphere. In Sylvia, D.M., Fuhrmann, J.J., Hartel, P.G., & Zuberer, D.A. (Eds.), *Principles and Applications of Soil Microbiology* (pp. 242-262), Pearson Prentice Hall: Upper Saddle River, NJ.

Kennedy, A.C., Johnson, B.N., & Stubbs, T.L. (2001). Host range of a deleterious rhizobacterium for biological control of downy brome. *Weed Sci.* 49, 792-797.

Kim, S-J., & Kremer, R.J. (2005). Scanning and transmission electron microscopy of root colonization of morningglory (*Ipomoea* spp.) seedlings by rhizobacteria. *Symbiosis 39*, (in press).

Kirkegaard, J.A., Munns, R., James, R.A., Gardner, P.A., & Angus, J.F. (1995). Reduced growth and yield of wheat with conservation cropping. II. Soil biological factors limit growth under direct drilling. *Aust. J. Agric. Res. 46*, 75-88.

Kloepper, J.W., Zablotowicz, R.M., Tipping, E.M., & Lifshitz, R. (1991). Plant growth promotion mediated by bacterial rhizosphere colonizers. In Keister, D.L. & Cregan, P.B. (Eds), *The Rhizosphere and Plant Growth* (pp. 315-326), Kluwer Academic Publishers: Dordrecht, The Netherlands.

Knowles, C.J., & Bunch, A.W. (1986). Microbial cyanide metabolism. *Adv. Microbiol. Physiol. 27*, 73-111.

Kremer, R.J. (2005). The role of allelopathic bacteria in weed management. In Inderjit (Ed.), *Allelochemicals: Biological Control of Plant Pathogens and Diseases*, Kluwer Academic Publishers: Dordrecht, The Netherlands. (in press).

Kremer, R.J. (2002). Bioherbicides: potential successful strategies for weed control. In Koul, O., & Dhaliwal, G. (Eds.), *Microbial Biopesticides* (pp. 307-323), Taylor & Francis: London.

Kremer, R.J., Begonia, M.F.T., Stanley, L., & Lanham, E.T. (1990). Characterization of rhizobacteria associated with weed seedlings. *Appl. Environ. Microbiol.* 56, 1649-1655.

Kremer, R.J., Caesar, A.J., & Souissi, T. (2005a). Soilborne microorganisms of *Euphorbia* are potential biological control agents of the invasive weed leafy spurge. *Appl. Soil Ecol.*, (in press).

Kremer, R.J., & Kennedy, A.C. (1996). Rhizobacteria as biocontrol agents of weeds. *Weed Technol. 10*, 601-609.

Kremer, R.J., & Li, J. (2003). Developing weed-suppressive soils through improved soil quality management. *Soil Till. Res. 72*, 193-202.

Kremer, R.J., Means, N.E., & Kim, S-J. (2005b). Glyphosate affects soybean root exudation and rhizosphere microorganisms. *Int. J. Environ. Anal. Chem. 85*, (in press).

Kremer, R.J., Sasseville, D.N., & Mills, H.A., Jr. 1996. Promotion of phytotoxic bacteria in rhizospheres of leatherleaf fern by Benlate DF. *J. Plant Nutr. 19*, 939-953.

Kremer, R.J., & Souissi, T. (2001). Cyanide production by rhizobacteria and potential for suppression of weed seedling growth. *Curr. Microbiol. 43*, 182-186.

Kuklinsky-Sobral, J., Welington, L.A., Mendes, R., Pizzirani-Kleiner, A.A., & Azevedo, J.L. (2005). Isolation and characterization of endophytic bacteria from soybean (*Glycine max*) grown in soil treated with glyphosate herbicide. *Plant Soil 273*, 91-99.

Li, J., & Kremer, R.J. (2000). Rhizobacteria associated with weed seedlings in different cropping systems. *Weed Sci. 48*, 734-741.

Li, J., Kremer, R.J., & Ross, L.M., Jr. (2002). Electron microscopy of root colonization of *Setaria viridis* by deleterious rhizobacteria as affected by soil properties. *Symbiosis 32*, 1-14.

Lindow, S.E., & Brandl, M.T. (2003). Microbiology of the phyllosphere. *Appl. Environ. Microbiol. 69*, 1875-1883.

Loper, J.E., & Schroth, M.N. (1986). Influence of bacteria sources of indole-3-acetic acid on root elongation of sugarbeet. *Phytopathology 76*, 386-389.

Lugtenberg, B.J.J. & Dekkers, L.C. (1999). What makes *Pseudomonas* bacteria rhizosphere competent? *Environ. Microbiol. 1*, 9-13.

Manion, P.D., & Lachance, D. (1992). Forest decline concepts: an overview. In Manion, P.D., & Lachance, D. (Eds.), *Forest Decline Concepts* (pp. 181-190), APS Press: St. Paul, MN.

Matschke, J., & Machácková, I. (2002). Changes in the content of indole-3-acetic acid and cytokinins in spruce, fir and oak trees after herbicide treatment. *Boil. Plant. 45*, 375-382.

Maurhofer, M., Keel, C., Schnider, U., Voisard, C., Haas, D., & Defago, G. (1992). Influence of enhanced antibiotic production in *Pseudomonas fluorescens* strain CHA0 on its disease suppressive capacity. *Phytopathology 82*, 190-195.

Mazzola, M. (1998). Elucidation of the microbial complex having a causal role in the development of apple replant disease in Washington state. *Phytopathology 88*, 930-938.

Mills, H.A., Jr., Sasseville, D.N. & Kremer, R.J. (1996). Effects of Benlate on leatherleaf fern growth, root morphology, and rhizosphere bacteria. *J. Plant Nutr. 19*, 917-937.

Nehl, D.B., Allen, S.J., & Brown, J.F. (1997). Deleterious rhizosphere bacteria: an integrating perspective. *Appl. Soil Ecol. 5*, 1-20.

Nelson, E.B. (2004). Microbial dynamics and interactions in the spermosphere. *Annu. Rev. Phytopathol. 42*, 271-309.

Neumann, G., & Römheld, V. (2002). Root-induced changes in the availability of nutrients in the rhizosphere. In Waisel, Y., Eshel, A., & Kafkafi, U. (Eds), *Plant Roots: The Hidden Half* (pp. 617-649), Marcel Dekker: New York.

Owen, A., & Zdor, R.E. (2001). Effect of cyanogenic bacteria on the growth of velvetleaf (*Abutilon theophrasti*) and corn (*Zea mays*) in autoclaved soil and the influence of supplemental glycine. *Soil Biol. Biochem. 33*, 801-809.

Pankhurst, C.E., Magarey, R.C., Stirling, G.R., Blair, B.L., Bell, M.J., & Garside, A.L. (2003). Management practices to improve soil health and reduce the effects of detrimental soil biota associated with yield decline of sugarcane in Queensland, Australia. *Soil Till. Res. 72*, 125-137.

Paulsen, I.T., Press, C.M., Ravel, J., Kobayashi, D.Y., et al. (2005). Complete genome sequence of the plant commensal *Pseudomonas fluorescens* Pf-5. *Nature Biotechnol. 83*, 873-878.

Persello-Cartieaux, F., Nussaume, L., & Robaglia, C. (2003). Tales from the underground: molecular plant-rhizobacteria interactions. *Plant Cell Environ. 26*, 186-199.

Posta, K., Marschner, H., & Römheld, V. (1994). Manganese reduction in the rhizosphere of mycorrhizal and nonmycorrhizal maize. *Mycorrhiza 5*, 119-124.

Preston, G.M. (2004). Plant perceptions of plant-growth promoting *Pseudomonas*. *Phil. Trans. Royal Soc. Lond. B 359*, 907-918.

Probanza, A., Lucas, J.A., Acero, N., & Gutierrez Mañero, F.J. (1996). The influence of native rhizobacteria on European alder (*Alnus glutinosa* (L.) Gaertn.) growth. *Plant Soil 182*, 59-66.

Rengel, Z., Pedler, J.F., & Graham, R.D. (1994). Control of Mn status in plants and rhizosphere: genetic aspects of host and pathogen effects in the wheat take-all interaction. In Manthey, J.A., Crowley, D.E., & Luster, D.G. (Eds), *Biochemistry of Metal Micronutrients in the Rhizosphere* (pp. 125-145), Lewis Publishers: Boca Raton, FL.

Ryu, C-M., Hu, C-H., Locy, R.D., & Kloepper, J.W. (2005). Study of mechanisms for plant growth promotion elicited by rhizobacteria in *Arabidopsis thaliana. Plant Soil 268*, 285-292.

Sarathchandra, S.U., Brown, J.A., & Cox, N.R. (1996). Rhizobacteria harmful to seedling growth in white clover (*Trifolium repens* L.) and perennial ryegrass (*Lolium perenne* L.). *N. Z. J. Agric. Res. 39*, 129-136.

Sarwar, M., & Frankenberger, W.T., Jr. (1994). Influence of L-tryptophan and auxins applied to the rhizosphere on the vegetative growth of *Zea mays* L. *Plant Soil 160*, 97-104.

Sarwar, M., & Kremer, R.J. (1995). Enhanced suppression of plant growth through the production of L-tryptophan-derived compounds by deleterious rhizobacteria. *Plant Soil 172*, 261-269.

Schippers, B. (1988). Biological control of pathogens with rhizobacteria. *Phil. Trans. Royal Soc. Lond. B318,* 283-293.

Schippers, B., Bakker, A.W., & Bakker, P.A.H.M., & van Peer, R. (1990). Beneficial and deleterious effects of HCN-producing pseudomonads on rhizosphere interactions. *Plant Soil 129*, 75-83.

Schippers, B., Bakker, A.W., & Bakker, P.A.H.M. (1987). Interactions of deleterious and beneficial rhizosphere microorganisms and the effect of cropping practices. *Ann. Rev. Phytopathol. 25*, 339-358.

Schroth, M.N., & Hancock, J.G. (1982). Disease-suppressive soil and root-colonizing bacteria. *Science 216*, 1376-1381.

Sessitsch, A., Gyamfi, S., Tscherko, D., Gerzabek, M.H., & Kandeler, E. (2004). Activity of microorganisms in the rhizosphere of herbicide treated and untreated transgenic glufosinate-tolerant and wildtype oilseed rape grown in containment. *Plant Soil 266*, 105-116.

Somers, E., Vanderleyden, J., & Srinivasan, M. (2004). Rhizosphere bacterial signaling: a love parade beneath our feet. *Crit. Rev. Microbiol. 30*, 205-240.

Souissi, T., Kremer, R.J., & White, J.A. (1997). Scanning and transmission electron microscopy of root colonization of leafy spurge (*Euphorbia esula* L.) seedlings by rhizobacteria. *Phytomorphology 47*, 177-193.

Stroo, H.F., Elliott, L.F., & Papendick, R.I. (1988). Growth, survival and toxin production of root-inhibitory pseudomonads on crop residues. *Soil Biol. Biochem. 20*, 201-207.

Sturz, A.V. (1995). The role of endophytic bacteria during seed piece decay and potato tuberization. *Plant Soil 175*, 257-263.

Sturz, A.V., & Christie, B.R. (2003). Beneficial microbial allelopathies in the root zone: the management of soil quality and plant disease with rhizobacteria. *Soil Till. Res. 72*, 107-123.

Sturz, A.V., & Christie, B.R. (1996). Endophytic bacteria of red clover as agents of allelopathic clover-maize syndromes. *Soil Biol. Biochem. 28*, 583-588.

Sturz, A.V., Christie, B.R., & Nowak, J. (2000). Bacterial endophytes: potential role in developing sustainable systems of crop production. *Crit. Rev. Plant Sci. 19*, 1-30.

Surette, M.A., Sturz, A.V., Rajasekaran, R.L., & Nowak, J. (2003). Bacterial endophytes in processing carrots (*Daucus carota* L. var. sativus): their localization, population density, biodiversity and their effects on plant growth. *Plant Soil 253*, 381-390.

Tranel, P.J., Gealy, D.R., & Kennedy, A.C. (1993). Inhibition of downy brome (*Bromus tectorum*) root growth by a phytotoxin from *Pseudomonas fluorescens* strain D7. *Weed Technol. 7*, 134-139.

Turco, R. F., Bischoff, M., Breakwell, D.P., & Griffith, G.R. (1990). Contribution of soil-borne bacteria to the rotation effect in corn. *Plant Soil 122*, 115-120.

Vande Broeke, A. & Vanderleyden, J. (1995). The role of bacterial motility, chemotaxis, and attachment in bacteria-plant interactions. *Molec. Plant-Microbe Interact. 8*, 800-810.

Vessey, J.K. (2003). Plant growth promoting rhizobacteria as biofertilizers. *Plant Soil 255*, 571-586.

Vilich, V. & Sikora, R.A. (1998). Diversity in soilborne microbial communities. In Boland, G.J. & Kuykendall, L.D. (Eds), *Plant-Microbe Interactions and Biological Control* (pp. 1-14), Marcel Dekker: New York.

Waschkies, C., Schropp, A., & Marschner, H. (1994). Relations between grapevine replant disease and root colonization of grapevine (*Vitis* sp.) by fluorescent pseudomonads and endomycorrhizal fungi. *Plant Soil 162*, 219-227.

Widmer, T.L., & Guermache, F. (2004). Filtrates of rhizosphere bacteria suppressive to *Centaurea solstitialis* seed germination. *J. Plant Dis. Protect. 19*, 497-502.

Wilson, G.W.T., Hetrick, B.A.D., & Kitt, D.G. (1988). Suppression of mycorrhizal growth response of big bluestem by non-sterile soil. *Mycologia 80*, 338-343.

Woltz, S.S. (1978). Nonparasitic plant pathogens. *Ann. Rev. Phytopathol. 16*, 403-430.

Yang, C-C., & Leong, J. (1984). Structure of pseudobactin 7SR1, a siderophore from a plant-deleterious *Pseudomonas. Biochemistry 23*, 3534-3540.

Yao, S., Merwin, I.A., Abawi, G.S., & Thies, J.E. (2005). Soil fumigation and compost amendment alter soil microbial community composition but do not improve tree growth or yield in an apple replant site. *Soil Biol. Biochem. 37,* (in press).

Zahir, Z.A., Arshad, M., & Frankenberger, W.T., Jr. (2004). Plant growth promoting rhizobacteria: applications and perspectives in agriculture. *Adv. Agron. 81*, 97-168.

9. AFFILIATIONS

Robert J. Kremer, *United States Department of Agriculture, Agricultural Research Service, Cropping Systems & Water Quality Research Unit, 302 Natural Resources Bldg., University of Missouri, Columbia, MO 65211, U.S.A.*
Tel: 573-882-6408, Fax: 573-884-5070; E-mail: kremerr@missouri.edu

Plant Pathogenic Bacteria

AJITH ANAND AND KIRANKUMAR S. MYSORE

AGROBACTERIUM BIOLOGY AND CROWN GALL DISEASE

Abstract. The crown gall tumors produced in dicotyledonous plants by *Agrobacterium tumefaciens* results from the introduction of a segment of DNA (T-DNA), derived from its tumor-inducing (Ti) plasmid, into plant cells at the infected site. This disease can cause significant economic losses in perennial crops worldwide. Besides their natural hosts, *Agrobacterium* can deliver the T-DNA also to monocotyledonous plants, mammalian cells, yeasts, fungi and other prokaryotes, which makes it one of the most well researched organisms and as well as a promising tool for modern plant breeding. Promising future controls of this disease will include employment of biological control agents; understanding the disease signaling pathways in plants and adopting transgenic approaches for generating crown-gall resistant plants.

1. INTRODUCTION

Agrobacterium tumefaciens is a soil-borne, non-sporulating, motile, rod-shaped phytopathogenic bacterium that elicits neoplastic growth at the site of infection in many dicotyledonous plants causing the "crown-gall" disease. This disease can be traced back to 1850s wherein it was first reported in grapevines, but remained poorly understood until the early 1900s (Smith & Townsend, 1907). This disease by far represents a unique event, involving the transfer of DNA from a prokaryote into the chromosomes of plants, that facilitates its application in modern biology for not only transferring desired genes into plants, but also across other eukaryotic kingdoms including fungi, and mammalian cells (de Groot, Bundock, Hooykaas, & Beijersbergen, 1998; Kunik et al., 2001; Piers, Heath, Liang, Stephens, & Nester, 1996; Rho, Kang, & Lee, 2001). Thus this unique opportunity presented by the bacterium permitting the transfer of foreign genes into many plants and non-plant organisms provides an exceptional system in modern biology research that involves divergent disciplines.

The *Agrobacterium* genus belongs to the family Rhizobiaceae with a number of species, but on the basis of their genetic makeup, chromosome structure, host range, metabolism and relationship with other genera, they are broadly classified into three biovars (Farrand, van Berkum, & Oger, 2003; Goodner et al., 2001; Keane, Kerr, & New, 1970; Tighe et al., 2000). *A. tumefaciens* and *A. rubi* (Tighe et al., 2000) the causative agents for crown gall disease and cane gall belong to biovar I group; *A. rhizogenes* that causes hairy root cultures belongs to biovar II; and *A. vitis* (Kerr & Panagopoulos, 1977) that causes the galls on grapevines belong to the third group, biovar III. There are several excellent reviews detailing various aspects of *Agrobacterium* biology (Anand & Mysore, 2005; Gelvin, 2000, 2003; Tzfira & Citovsky, 2002; Tzfira, Li, Lacroix, & Citovsky, 2004). However, we here in this chapter will focus mainly on the significant developments in *Agrobacterium*

S.S. Gnanamanickam (ed.), Plant-Associated Bacteria, 359–384.

biology, methods towards combating the disease, role of plant genes in *Agrobacterium*-plant interaction and pathogenesis, and the application of *Agrobacterium* as a tool for modern plant breeding.

2. THE CROWN GALL DISEASE

There are several stages in the process of infection by *A. tumefaciens* on dicot plants that requires coordinated responses between an individual viable bacterium and the host cell. The molecular basis for genetic transformation of plant cells by *Agrobacterium* that produces the neoplastic growth at wounded sites, the crown gall, is imprinted on a large tumor-inducing (Ti) plasmid residing in the bacterium. Ti plasmids are in the order of 200 to 800 kb (Gelvin, 2000). The disease is triggered by the compounds exuded at wounded site of the plants including phenolics, simple sugars and amino acids, that are perceived as signals by the bacterium for the induction of *VirA* gene residing in the Ti plasmid. This protein in turn phosphorylates a second constitutive protein *VirG*, followed by a cascade of reactions resulting in the activation of many other *vir* genes. This cascade of events within the bacterium culminates in the nicking and transference of a single stranded DNA molecule from the bacterium into the plant cell (See reviews by Gelvin, 2000, 2003; Tzfira & Citovsky, 2002). A schematic presentation of the various events occurring within the bacterium following wounding is represented in Figure1. The single-stranded DNA molecule, the T-strand resides within the tumor inducing (Ti) plasmid called T-DNA (Chilton et al., 1977; Stachel, Messens, van Montagu, & Zambryski, 1985; Stachel, Timmerman, & Zambryski, 1986). Most wild-type T-DNAs encode enzymes for the synthesis of low molecular weight compounds called opines as well as enzymes involved in the synthesis of plant growth regulators auxin and cytokinins. The uncontrolled proliferation of crown gall cells results from the production of auxins and cytokinins by the bacterium at the wounded sites. The wild-type strains of *Agrobacterium* contain T-DNA strands that carry genes involved in the synthesis of plant growth hormones and opines, while the Ti plasmid carry genes that encode proteins that can catabolize opines.

Opines play a major role in the epidemiology of crown gall and ecology of *Agrobacterium* spp. Opines are low molecular weight carbon compounds (amino acid and sugars phosphate derivatives) that serve as carbon and/or nitrogen sources for the tumor-inducing bacterium. More than 20 different opines have been identified in crown galls and hairy roots, but only a few subset of them are encoded by the T-DNA of any one *Agrobacterium* strain (Dessaux, Petit, Farrand, & Murphy, 1998; Escobar & Dandekar, 2003). The basis of different types of opine(s) metabolized in the infected cells, has been also used for classification of the infecting *Agrobacterium* strains (e.g. octopine, nopaline and agropine-type strains; (Hooykaas & Shilperoort, 1992)). Octopine-type T-DNAs posses four opine synthesis genes that catalyze the production of octopine (*ocs*), agropine (*ags*), and mannopine (*mas1*, *mas2*) and octopine-type Ti plasmid encodes nearly 40 genes related to octopine, agropine and mannopine uptake and use (Zhu et al.,

2000). Opines not only provide a growth substrate but probably are also involved in conjugal Ti plasmid exchange and chemotaxsis (Kim, Baek, Lee, Yang, & Farrand, 2001; Zhu et al., 2000).

The disease that produces the crown gall was first described in grapes in 1882 and was subsequently studied in variety of natural host plants (Braun, 1978; A. Powell & Gordon, 1989). There are several useful reviews describing the major discoveries on crown gall disease in the past (see reviews by (Braun, 1978, 1982; Gelvin, 2000, 2003; Hooykaas & Shilperoort, 1992; Lippincott & Lippincott, 1981; Stafford, 2000; Zupan & Zambryski, 1997). The epidemiology of crown gall disease, identification of the tumor inducing principle (Braun, 1982), bacterial attachment (Braun, 1954; Lippincott & Lippincott, 1981), chemotactic response (Lippincott & Lippincott, 1981), T-DNA transfer (Nester, Gordon, Amasino, & Yanofsky, 1984; Powell & Gordon, 1989; Powell, Rogowsky, & Kado, 1989), development of T-DNA restriction maps (Davey, Curtis, Gartland, & Power, 1994), sequencing of the genome of *A. tumefaciens* C58 strain (Goodner et al., 2001; Wood et al., 2001) and the molecular characterization of events within the plant cell are some of the historical achievements in chronological order leading to better understanding of the pathogen and the disease as well. Apparently, it was not until the early 1990s, the function of chromosomal gene, *chv*, and those of plasmid DNA was described (Sheng & Citovsky, 1996b), subsequently major advances were laid in the area of understanding the roles of *vir* genes and their proteins in the process of infection (see reviews by Gelvin, 2000).

3. GENOME ORGANIZATION OF *AGROBACTERIUM TUMEFACIENS*

The completion of the genome sequencing and annotation of the different genomes of *A. tumefaciens* C58 strain, isolated from cherry tree reaffirmed the presence of four replicons: a circular chromosome, a linear chromosome and plasmids pAtC58 and pTiC58 (Allardet-Servant, Michaux-Charachon, Jumas-Bilak, Karayan, & Ramuz, 1993; Goodner et al., 2001; Wood et al., 2001). The current annotation of the genome of *A. tumefaciens* contains 5,482 predicted protein coding genes. Proteins encoded by 5,415 genes have been assigned a putative function or classified as hypothetical protein, whose functions are unknown in other related genomes. Visit http://depts.washington.edu/agro, for the most recent information on annotations of the *Agrobacterium* genome. The total GC content of the bacterium DNA is roughly about 58%, with two regions of distinctive GC contents in the plasmid pTiC58: the T-DNA with a low GC content of 46% and the *vir* region with 54% (Wood et al., 2001). Similarly the 24-kb segment of pAtC58 has a reduced GC content of 54% that encompasses 17 conserved hypothetical genes, a DNA helicase, and an insertional sequence element (IS). The genes in the plasmid genome (pAtC58 and pTiC58) have different codon usages, when compared to the rest of the genome (Goodner et al., 2001; Wood et al., 2001). The findings from the sequencing studies are in accordance with different evolutionary pressures acting on these elements (IS), reflecting lineage-specific gene loss or acquisitions from other closely-related species.

4. BIOLOGY BEHIND CROWN GALL AND *AGROBACTERIUM*-MEDIATED PLANT TRANSFORMATION

Some of the critical steps involved in the crown gall disease and subsequent plant transformation by *A. tumefaciens* are enlisted as follows; 1) Chemotaxis and activation of bacterial virulence; 2) Bacterial attachment; 3) T-DNA processing and transport into plant cell; and 4) Nuclear import, integration of the T-DNA into the plant genome and expression of oncogenes. A schematic sketch detailing the different steps in the infection process and the putative plant factors that play a critical role in the infection processes are detailed in Figure 1.

4.1. Chemotaxis and activation of bacterial virulence

The phenolics and sugar compounds, extruded at the plant wounded site, is perceived as the signal for infection by *Agrobacterium* and serve as inducers (or co-inducers) for the bacterial virulence (*vir*) genes located on the Ti plasmid. *Agrobacterium* uses a two component sensory systems (Stock, Lukat, & Stock, 1991; Stock, Stock, & Mottonen, 1990) for both sensing and responding to the presence of susceptible host cells. The VirA sensory system perceives the presence of the phenolics at the wounded site either directly or indirectly that is mediated by two chromosomally encoded proteins P10 and P21 (Doty, Yu, Lundin, Heath, & Nester, 1996; Lee, Jin, Sim, & Nester, 1995; Winans, 1991; Winans, 1992). This results in autophosphorylation of the VirA protein and the subsequent trans-phosphorylation of the VirG protein (Jin, Prusti, Roitsch, Ankenbauer, & Nester, 1990; Jin, Roitsch, Christie, & Nester, 1990) cytoplasmic response regulator protein. These interactions within the bacterial cell result in the transcriptional activation of other *vir* genes. Among the various phenolic compounds (e.g., flavonoid precursors) known to induce *vir* gene expression the best categorized inducer is a monocyclic phenolic compound, acetosyringone (Stachel, Nester, & Zambryski, 1986).

Monosaccharides, such as glucose and galactose, can also induce *vir* gene expression in either the absence of acetosyringone or when available at limited quantities (Cangelosi, Ankenbauer, & Nester, 1990). This activation via sugar molecules of the *vir* genes is always indirect; and is under the regulation of a chromosomally encoded protein ChvE. The ChvE, glucose/ galactose binding-protein, interacts with VirA and transmits the signal to VirG, which in turn activates the *vir* gene promoters (Shimoda, Toyoda-Yamamoto, Aoke, & Machida, 1993). Combinations of certain monosaccharides with acetosyringone for maximal *vir* gene activation are therefore routinely employed in protocols involving *Agrobacterium*-mediated plant transformation.

4.2. The vir operon

The autophosphorylation of VirA occurs at a specific histidine residue within the C-terminal half of the protein that also contains the protein kinase domain. Since the phophohistidine VirA complex is very unstable, there is an energy equilibria transfer that occurs in the bacterial cell to an aspartate residue(s) of VirG, a cytoplasmic transcriptional regulator (reviewed by Sheng & Citovsky, 1996a; Winans, Mantis, Chen, Chang, & Han, 1994). The VirG protein shares amino acid similarity with many other transinducer molecules and is relatively stable in the phosphorylated form (Jin, Song, Pan, & Nester, 1993) probably leading to the maximal activation of the '*vir*' gene promoters. Once phosphorylated the VirG activates the expression of the other *vir* operons (*virB*, *virC*, *virD*, *virF*, *virG*, *virH* and *virJ*) that are required for the T-DNA processing and export to the host cell. The known functions of the different Vir proteins is detailed in Table 1.The *vir* region has at least twelve operons, of which eight (*virA, virB, virC, virD, virE, virF, virG and virH*) encode essential proteins required for T-DNA transformation and the rest four operons (*virJ, virK, virL and virM*) encode proteins whose functions are unknown. The number of genes per operon differs, *virA, virG, vir F, virJ, virK, virL, and virM* have only one gene; *virE, virC, virH* have two genes, while *virD* and *virB* have four and eleven genes respectively (Table 1).

4.3. Agrobacterium attachment

The bacterial recognition and attachment is an important step in the infection process. In nature, *Agrobacterium* normally infects wounded plant tissues that are rich in exudates containing phenolic compounds and sugars that trigger chemotaxis of the bacterium for attachment (Ashby, Watson, Ladke, & Shaw, 1988; Lippincott & Lippincot, 1969). Following the loose binding to the host, the bacterium synthesizes cellulose microfilaments, which helps in anchoring the bacterium to the plant cell surface (Matthysse, 1986). The binding of the bacterium to the host cell-surface involves a cell-to-cell recognition step involving a cascade of bacterial attachment proteins that is encoded by chromosomal genes; *chvA*, *chvB*, *pscA* and *att*. The ChvA and ChvB proteins are involved in β-1,2 glucan synthesis (de Iannino & Ugalde, 1989; Zorreguieta et al., 1988). Mutants of these genes are temperature sensitive (Bash & Matthysse, 2002) and lack in their ability to attach and infect host cells (Douglas, Halperin, Gordon, & Nester, 1985a; C. J. Douglas, Staneloni, Rubin, & Nester, 1985b). The chromosomally encoded *pscA* gene (Cangelosi et al., 1987; Uttaro, Cangelosi, Geremia, Nester, & Ugalde, 1990) and *cel* genes are required for the synthesis of exopolysaccharides and cellulose fibrils respectively. Another group of genes involved in the bacterial attachment are the *att* (attachment) genes, which are mapped on the pAtC58 plasmid. Mutations at the *att* locus results in altered attachment phenotype compared to the wild-type. The altered mutant phenotypes for the locus were broadly classified into two groups; (i) those that could complement with wild-type C58 conditioned medium and, (ii) those that could not (Matthysse, 1994; Matthysse, Yarnall, & Young, 1996). A mutation in the *att* (*att*R and *att*D) loci of the pATC58 does not affect tumorigenicity. It was therefore concluded that tumorigenicity is a factor dependent on maximization of *vir* gene expression and not conferred by *att* region alone (Nair, Liu, & Binns, 2003). Even though this complexity of the bacterial genes remains largely resolved, the possible host-cell surface receptors are unexplored.

There are reports suggesting the inefficient infectivity of *Agrobacterium* on plant cells under the absence of wounding by entry through openings in the guard cells (Escudero & Hohn, 1999).

Agrobacterium infection most likely occurs through the formation of an active biofilm and the subsequent attachment of the bacterium to the susceptible host cell. The process of infection is mediated by certain cell wall receptors, which could be protease-sensitive molecules. Two such cell wall proteins are proposed to mediate bacterial attachment, one belonging to a vitronectin-like protein (Wagner & Matthysse, 1992), while the other is a putative plant receptor, rhicadhesin, an adhesion protein encoded by *Agrobacterium* and related phytobacterium (Swart, Logman, Smit, Lugtenberg, & Kijne, 1994). This speculation was based on findings that the attachment of *Agrobacterium* to plant cells exhibited reduced binding when human vitronectin or antibodies raised against vitronectin were applied in cell cultures (Wagner & Matthysse, 1992). Genetic studies and molecular evidences however are currently lacking to support the possible role of these proteins in bacterial attachment.

Few known ecotypes of *Arabidopsis*; B1-1, Petergof and UE-1 are recalcitrant to *Agrobacterium*-mediated transformation and biochemical assays suggested that UE-1 was integration deficient (Nam et al., 1997). The ecotypes Petergof and Bl-1, were found to be deficient in attachment (Nam, Matthysse, & Gelvin, 1997). However, these ecotypes were not recalcitrant to the germ-line transformation method (Mysore, Ranjith Kumar, & Gelvin, 2000). The differences in the ability of the different ecotypes of *Arabidopsis* to be transformed by different methods, lead to the speculation that the transformation capacities of the plant tissue might be dependent upon the differential plant gene expression (Mysore et al., 2000). Solid evidence for involvement of host-cell factors during *Agrobacterium* infection process came from identification of many T-DNA tagged insertional mutants from *Arabidopsis thaliana* that are deficient in *Agrobacterium*-mediated transformation (Nam et al., 1999). These mutants are designated as '*rat*' (resistant to *Agrobacterium*-transformation) and are altered in their efficiency of being transformed by *Agrobacterium*-mediated root transformation (Nam et al., 1999; Zhu et al., 2003). Cut root surfaces of three *rat* mutants (*rat1*, *rat3* and *rat4*), are deficient in binding *Agrobacterium* (Nam et al., 1999; Zhu et al., 2003). Subsequent studies have indicated that the *Rat1* encodes an arabinogalactan-like protein (AGLP), *Rat3* encodes a secreted cell wall protein and, *Rat4* belongs to the family of cellulose-synthase-like genes (*cslA*-09), which is a multi-gene family in *Arabidopsis* (Zhu, Nam, Carpita, Matthysse, & Gelvin, 2003; Zhu et al., 2003).

4.4. T-DNA processing and transmembrane transfer of the T-DNA into the plant cytoplasm

The activation of the *vir* operon leads to the production of a single-stranded (ss) molecules of T-DNA, termed T-strand (Stachel, Timmerman et al., 1986). All T-DNA elements are delimited by a conserved 25-bp homologous sequence that defines their left and right borders (Zambryski, Depicker, Kruger, & Goodman, 1982). The VirD1/VirD2 endonucleases play a critical role in the processing of the T-DNA border sequences. The endonuclease activity of the VirD1 and VirD2 protein complex on the supercoiled Ti plasmid is highly site and strand-specific resulting in the relaxation and nicking of the T-DNA borders between the third and fourth bases of the T-strand borders (Ward & Barnes, 1988; Young & Nester, 1988). After the endonucleotidic cleavage, VirD2 remains attached to the 5'-end of the single stranded T-DNA and the Ti plasmid.

Following the T-strand synthesis, the bacterial repair machinery fills the nicked gap between the right and left borders (Christie, 1997; Zupan & Zambryski, 1995). The VirD2 molecule that is still attached to the left border of the nicked strand may facilitate the ligation of the nicked DNA and restoring intact Ti- plasmid. Whilst the role of the VirD1 protein is limited to T-DNA border nicking, the VirD2 protein remain attached (presumably by covalent linkage) to 5' end of the nicked strand. This attachment is proposed to prevent exonucleolytic cleavage of the 5' end of the T-strand and also to pilot the T-DNA into the plant nucleus. VirE2 is a 69-kDa single stranded DNA binding protein and is believed to bind the T-strand *in vivo* within the *Agrobacterium* cell to protect from nucleases. This cooperative association prevents the attack of nucleases and, in addition, reducing the complex diameter to approximately 2 nm, making the translocation through membrane channels easier (Christie, Ward, Winans, & Nester, 1988; Citovsky, De Vos, & Zambryski, 1988; Citovsky, Wong, & Zambryski, 1989; Das et al., 1986). However, VirE2 is not required for export of T-DNA from the *Agrobacterium*. VirE2 can also be independently exported into the plant cell and is eventually required for protecting the T-strand within the plant cytoplasm (Binns, Beaupre, & Dale, 1995; Lessl & Lanka, 1994). According to the most accepted model, the T-DNA-VirD2 complex is coated by the VirE2 protein once inside the plant cell (Binns et al., 1995; Zhou & Christie, 1999). Complementation studies involving strain mutated in VirE1, with bacterial strains that can export VirE2 indicated that VirE2 can be exported independently and that binding of VirE2 as part of the T-DNA strand is not prerequisite for the transmission (Sundberg, Meek, Carroll, Das, & Ream, 1996; Vergunst, van Lier, den Dulk-Ras, & Hooykaas, 2003). Thus VirE2 could be exported into the plant cytosol by VirE1, wherein these two proteins dissociate, allowing VirE2 to bind to ss-T-strand and assembling the mature T-strand complex including VirE2/ssDNA-VirD2 (Binns et al., 1995; Sundberg et al., 1996; Zhou & Christie, 1999). More recently using the cre/lox recombinase reporter assay system in yeast cells, another product of *virE* locus; VirE3 (Schrammeijer, den Dulk-Ras, Vergunst, Jurado Jacome, & Hooykaas, 2003) was identified to be transported from *Agrobacterium* to the plant cell (Vergunst et al., 2003).

The transport of T-DNA coated with VirD2 requires up to 12 transported proteins involving the *virB* operon (11 proteins) and VirD4 that together constitute the type IV secretion system (see reviews Christie, 2001; Christie & Vogel, 2000;

Zupan, Ward, & Zambryski, 1998). The type IV secretion system is an ancient conjugation system documented for secretion of DNA-protein complexes in many prokaryotes (for more details see Christie, 2001, 2004). VirD4 family of proteins utilizes the energy of the nucleotide triphosphate hydrolysis to link protein complexes required for DNA processing and translocation. A parallel for this has been established with other type IV bacterial transporters (Christie, 1997; Firth, Ippen-Ihler, & Skurray, 1996). It is also postulated that VirD4 is a transmembrane protein but predominantly located at the cytoplasmic side. The role of the other components of the secretion system, proteins encoded from the 9.5 kb *virB* operon, is to generate suitable structures for cell-to-cell adhesion between the bacterium and the host cell wall and also form a channel for the export of DNA-protein complex. These structures also called as pilli and are believed to facilitate the export of the T-strand to the host cell. Of the 11 proteins encoded by the *virB* operon, VirB1* (a processed form of VirB1, Baron, Llosa, Zhou, & Zambryski, 1997), VirB2 and VirB5 are likely to interact with putative host cell receptors, since they reside in the exterior part of the secretion channel (Christie, 1997). VirB2, VirB4 and VirB7 together constitute the structural components of the *Agrobacterium* pilus, which directly interacts with the plant cell surface receptors (Kado, 2000; Zupan et al., 1998). The role of other proteins in the VirB operon is not yet clearly understood. Even though the VirD4-VirB translocation protein interaction is widely accepted as the model for transporting the T-strand complex in to the host cell, a more recent finding suggested the export of VirE2, VirD2 and VirF is independent of the VirB operon (Chen, Li, & Nester, 2000).

4.5. Nuclear import of the T-strand, T-DNA integration and expression of the oncogenes

The final step in the process of successful transformation of the host cell depends on the T-strand targeting to the nucleus by effectively crossing the nuclear membrane barrier. The T-strand coated with the protein complex when synthesized has a very large size (≈12.6 nm; Citovsky, Guralnick, Simon, & Wall, 1997) that by far exceeds the size exclusion limits of the nuclear pore (9 nm, Rout & Wente, 1994) under normal conditions. This exclusion size limit of the nuclear pore can exceed (≈23 nm) during active nuclear uptake (Dworetzky & Feldher, 1988), thus suggesting that the T-DNA complex is most likely actively imported into plant cells (Tzfira & Citovsky, 2000).

As described in the earlier section, the VirD2 and VirE2 proteins are very vital for the nuclear import of the T-DNA into the nucleus of the plant cell. VirD2 has two nuclear localization sequences (NLS), one at each end of the molecule (Howard, Zupan, Citovsky, & Zambryski, 1992). Only one of the NLS at the C-terminal functions in the transport during *Agrobacterium* infection (Rossi, Hohn, & Tinland, 1993; Shurvinton, Hodges, & Ream, 1992). Further mutant analyses using *Agrobacterium* strains carrying NLS-deletion mutants of VirD2 have reaffirmed the role of VirD2 in the nuclear import of the T-DNA (Narasimhulu, Deng, Sarria, & Gelvin, 1996; Shurvinton et al., 1992).

The T-DNA is believed to be coated with a large number of VirE2 molecules (~ 600 molecules per 20-kb T-DNA) in the plant cytoplasm. VirE2 contains two separate bipartite NLS regions that can target linked reported proteins to the plant cell nuclei (Citovsky, D. Warnick, & P. C. Zambryski, 1994; Citovsky, Zupan, Warnick, & Zambryski, 1992). The coating of the T-DNA with VirE2 is suggested to protect it from degradation in the host cell (Rossi, Hohn, & Tinland, 1996). Such a coating is important for efficient transfer of the T-strand into the plant nucleus (Ziemienowicz, Merkle, Schoumacher, Hohn, & Rossi, 2001; Zupan & Zambryski, 1997). The involvement of VirE2 in T-DNA transfer and its nuclear import in the host cell was demonstrated by microinjection of fluorescently labeled ssDNA coated with VirE2, that resulted in efficient nuclear localization of the labeled ssDNA, while the naked DNA (minus VirE2) remained in the cytoplasm (Zupan, Citovsky, & Zambryski, 1996). Furthermore, using the above assay it was demonstrated that VirE2 did not import fluorescently labeled ds-DNA indicating the need for the formation of nucleoprotein complexes for effective transport and nuclear localization. Interestingly the nuclear-import of the VirE2 was shown to be a plant-specific mechanism as it fails to localize to the cell nucleus of yeast and animal cells (Rhee, Gurel, Gafni, Dingwall, & Citovsky, 2000; Tzfira & Citovsky, 2001). Virulence studies on plants with *Agrobacterium* strains mutated in the VirE2 gene were found avirulent on most plant species, but produced tumors on transgenic plants over expressing VirE2 proteins (Citovsky et al., 1992; Gelvin, 1998). VirE2 nuclear localizing ability is sufficient for transport of ssDNA into the plant cell in the absence of VirD2 (Zupan et al., 1996) or for transformation of plant cells by *Agrobacterium* strains lacking the VirD2 nuclear localization signal (Gelvin, 1998). VirE2 is also required for efficient T-DNA integration in plants (Rossi et al., 1996). *In vitro* studies further support the above findings that both VirD2 and VirE2 are required for optimal nuclear uptake and they are actively involved in the nuclear import of the T-strand complex (Ziemienowicz, Gorlich, Lanka, Hohn, & Rossi, 1999; Ziemienowicz et al., 2001).

Using yeast translocation experiments, the role of another novel effector protein VirE3 (Schrammeijer et al., 2003) was identified, which has similar C-terminal signature to VirE2 and VirF. VirE3 is transferred from *Agrobacterium* to the plant cell and then imported into its nucleus via the karyopherin α-dependent pathway. In addition to binding plant importin-α, VirE3 interacts with VirE2, a major bacterial protein that directly associates with the T-DNA and facilitates its nuclear import (Lacroix, Vaidya, Tzfira, & Citovsky, 2005). VirF, an F-box protein, is another important protein that plays a role in the process of plant transformation (Schrammeijer et al., 2000). Studies involving *Agrobacterium* strains mutated in the *virF* gene were found to be avirulent in certain hosts (Hooykaas, Hofker, den Dulk-Ras, & Schilperoort, 1984; Iyer, Klee, & Nester, 1982). The attenuated virulence of VirF mutants is further aggravated by mutations in the *virE3* gene (Schrammeijer et al., 2003). The functional role of VirF protein in *Agrobacterium*-mediated plant transformation is not clearly established. VirF contains an F-box motif and interacts with yeast Skp1 *in vitro*, and is possibly associated in destabilizing the T-DNA complex and its targeted proteolysis inside the plant cell (Schrammeijer et al., 2003; Tzfira, Vaidya, & Citovsky, 2004). The actual role of VirF protein in plant transformation is largely unknown.

Finally once inserted into the plant genome the crown gall tumors result from overproduction of auxin and cytokinin in plant cells transformed by *A. tumefaciens* (Winans, 1992). These abnormally high phytohormone levels result from expression of three genes transferred stably into the plant genome from the *A. tumefaciens* tumor-inducing (Ti) plasmid: *iaaM* (Trp mono-oxygenase), *iaaH* (indole-3-acetamide hydrolase), and *ipt* (AMP isopentenyl transferase; Garfinkel et al., 1981; Ream, Gordon, & Nester, 1983). IaaM converts Trp into indole-3-acetamide, which IaaH converts into indole-3-acetic acid (auxin; Thomashow, Hugly, Buchholz, & Thomashow, 1986; Van Onckelen et al., 1985). Loss of either enzyme prevents auxin production. Ipt converts AMP into isopentenyl-AMP, a cytokinin (Akiyoshi, Klee, Amasino, Nester, & Gordon, 1984; Barry, Rogers, Fraley, & Brand, 1984; Buchmann, Marner, Schroder, Waffenschmidt, & Schroder, 1985). Inactivation of *ipt* and either one of the two auxin biosynthesis genes abolishes crown gall formation (Escobar & Dandekar, 2003; H. Lee et al., 2003; Ream et al., 1983).

5. ROLE OF PLANT GENES IN *AGROBACTERIUM*-MEDIATED PLANT TRANSFORMATION

Once inside the plant cell, the T-DNA is targeted to the plant nucleus wherein it randomly integrates into the host genomic DNA through an unknown process, which most likely is by non-homologous recombination (van Attikum, Bundock, & Hooykaas, 2001). Although the processing of T-strand and their export into the plant cell is reasonably well established, the role of host cell factors in the nuclear targeting of the T-DNA and its integration is at the tip of the iceberg. Several of the putative plant factors were recently identified and their role in the infection process was described and is summarized in Figure 1. There are several other unknown plant genes which play critical roles in *Agrobacterium* pathogenesis, and probably could be excellent candidates for finding durable control of the disease and for increasing the efficiency of the transformation system in many other crops (Gelvin, 2000; Hwang & Gelvin, 2004; Tzfira & Citovsky, 2002; Tzfira & Citovsky, 2003; Tzfira et al., 2004; Tzfira, Rhee, Chen, Kunik, & Citovsky, 2000; van Attikum et al., 2001; Veena, Doerge, & Gelvin, 2003; Zupan, Muth, Draper, & Zambryski, 2000).

One of most commonly used approach for identification of plant genes involved in *Agrobacterium*-mediated plant transformation is by using Vir proteins as bait in the yeast-two-hybrid system. This approach was recently applied in identifying plant interactor proteins of VirB2. As described earlier, VirB2 is a major component of the T-pilus. Using a processed C-terminal domain of VirB2 protein as bait in yeast two-hybrid system to screen against a cDNA library of *Arabidopsis* resulted in identification of three VirB2- interacting proteins (BTI)

BTI1, 2 and 3, and an membrane-associated GTPases, AtRAB8 (Hwang & Gelvin, 2004). The functional role of identified genes BTI1 and AtRAB8 were complemented by overexpression and antisense expression studies in *Arabidopsis* that affected the transformation process by *Agrobacterium*. This further extrapolates their role in *Agrobacterium* infectivity and plant transformation. Using VirD2 as the bait protein, a plant karyopherin (importin-α; *AtKAP*) was characterized as an interacting partner that is involved in the nuclear translocation of many proteins with NLS sequences (Ballas & Citovsky, 1997). Further, the identification of *rat* mutants of importins (importin α-7 and importin β-3) coupled with the antisense expression of importin α-1 confirms the role of plant karyopherin proteins in nuclear targeting of the cognate T-strand in plant cytoplasm (Gelvin, 2003; Zhu et al., 2003). VirD2 was also found to interact with three protein members of *Arabidopsis* cyclophilin chaperone, RocA, Roc4 and CypA, family (Deng et al., 1998). Inhibitors of plant cyclophilins, like Cyclosporin A resulted in reduced *Agrobacterium*-mediated transformation of both *Arabidopsis* and tobacco cell cultures (Deng et al., 1998). Although the exact function of these proteins in T-DNA transformation is unknown, they are hypothesized to aid the cytoplasmic trafficking of the T-complex. VirD2 was also found as an interactor-protein for type 2C protein phosphatases (PP2C; Tao, Rao, Bhattacharjee, & Gelvin, 2004). Co-expression of PP2C and VirD2-NLS-GUS fusion proteins in tobacco BY2 cells decreased nuclear GUS expression, suggesting the phosphorylation of VirD2 is critical in the nuclear targeting of the cognate T-DNA complex (Tao et al., 2004). VirD2 tightly associates with a TATA box-binding protein (TBP) in transformed *Arabidopsis* cells and supports the active role of VirD2 in T-DNA integration into the plant chromosome (Bako, Umeda, Tiburcio, Schell, & Koncz, 2003). In plant cells, VirD2 NLS is phosphorylated by CAK2Ms, a member of cyclin-dependent kinase family (Bako et al., 2003). CAK2Ms play the dual function of phosphorylating cyclin dependent kinase (CDK2) and the C-terminal domain of RNA polymerase II subunit which serves as the TBP-binding domain. These results led to the speculation that both phosphorylation (by unknown kinase) and dephosphorylation steps of VirD2-NLS act as positive and negative regulators in the plant transformation event mediated by *Agrobacterium* (Gelvin, 2003).

The use of VirE2 protein as a bait in the yeast-two-hybrid system led to the identification of additional plant factors involved in this unique plant-microbe interaction. Several independent cDNAs from *Arabidopsis* were identified as VirE2 interactors. Two of these VirE2-interacting proteins (VIPs) were designated as VIP1 and VIP2 (Tzfira, Rhee, Chen, & Citovsky, 2000). VIP1 and VIP2 were found to specifically interact only with VirE2 and not with other non-specific protein interactors in the yeast-two hybrid system including VirD2 supporting their target specificity (Tzfira et al., 2000). The VIP1 protein has structural features similar to the basic-zipper proteins (bZIP) and is hypothesized to be involved in the nuclear import of the T-DNA. It is speculated that VIP1 and VirE2 function as a part of multi-protein complex playing the dual function of nuclear targeting of

VirE2 followed by intranuclear export of T-DNA/VirE2 complex to the site of integration (Tzfira & Citovsky, 2002; Tzfira, Vaidya, & Citovsky, 2001). The *Arabidopsis* Kap114 protein is the only other known protein with similar dual functions involving nuclear import and intranuclear transport of TBP (Pemberton, Rosenblum, & Blobel, 1999). Overexpression of VIP1 in transgenic *Arabidopsis* plants indicated that the protein is imported into the plant cell nucleus by a karyopherin α-dependent pathway and also enhanced significantly the susceptibility to both transient and stable transformation by *Agrobacterium* (Tzfira, Vaidya, & Citovsky, 2002). VIP1 also forms homomultimers and interacts with plant histone H2A and therefore is likely involved in T-DNA integration (Li, Krichevsky, Vaidya, Tzfira, & Citovsky, 2005).

The other VirE2 interacting protein VIP2 has a conserved C-terminal domain that shares homology with the *Drosophila* Rga protein and mammalian *NOT2/NOT3/NOT5* (negative on TATA-less) genes (Tzfira & Citovsky, 2002); Anand et al., unpublished). Transient silencing of *NbVIP2* gene in *N. benthamiana* by virus-induced gene silencing severely affects the tumorogenic capacity of wild-type *A. tumefaciens* strains and blocks the stable transformation event in the silenced plants (Anand et al., unpublished). VIP2 is homologous to yeast Not2p and *Drosophila* Rga proteins, which are thought to mediate intranuclear interactions between chromatin proteins and the transcriptional complex (Collart, 2003; Collart & Struhl, 1993; Frolov, Benevolenskaya, & Birchler, 1998). Furthermore, we speculate that VIP1, VIP2 and VirE2 might function in a multi-protein complex that is best suited for T-DNA import and integration into the host genome (Tzfira & Citovsky, 2002). VirE2 was also found to interact with several of the *Arabidopsis* α-importin proteins in the yeast-two hybrid system proposing similar functions of VirD2 and VirE2 in nuclear import (unpublished data from Gelvin's group).

The role for VirF protein as a host factor was proposed on the molecular basis of host-range differences between the two *Agrobacterium* strains, octopine and nopaline, that differ in their virulence based on the presence/absence of VirF locus. Transgenic tobacco plants over-expressing VirF showed increased susceptibility to infection by a VirF deficient *Agrobacterium* strain, further supporting their role as a host-range factor (Regensburg-Tuink & Hooykaas, 1993). Schrammeijer and colleagues identified an *Arabidopsis* Skp1 (ASK) protein as an F-domain interacting protein of the VirF (Schrammeijer et al., 2001). Plant Skp1 protein is a part of the SCF (Skp1/cdc53/F-box) complexes that target specific protein for proteolysis by an ubiquitin-dependent degradation pathway (Bai et al., 1996; Koepp, Harper, & Elledge, 1999). The specific interaction of VirF with VIP1 but not with VirE2, possibly suggests an unknown mechanism for Vir protein turnover by proteolysis (Tzfira & Citovsky, 2002).

Gelvin and associates have screened T-DNA tagged *Arabidopsis* library to identify mutants that are resistant to root transformation by *Agrobacterium* (*rat* mutants; Nam et al., 1999; Zhu et al., 2003). Plant DNA junction sequences at the

T-DNA insertion site were recovered for many mutants and subsequently sequenced for identifying the mutated plant gene (Zhu et al., 2003). As of date, over 126 *rat* mutants have been identified and many of these mutants can be transiently transformed which suggests that they are disrupted specifically in the step of T-DNA integration (Zhu et al., 2003). Plant factors that were identified using this approach were classified into three broad categories as detailed in Table 1. Although the sequence details are available for the different *rat* mutants, only 13 of these mutants remain successfully complemented.

In addition to identifying *rat* mutants and Vir-interacting proteins, several host DNA polymerases and ligases have shown to play a critical role in creating conducive environment in the plant cell for DNA nicking and filling at the nicked ends to facilitate random T-DNA insertion. (Chilton & Que, 2003) demonstrated the opportunistic exploitation of double stranded breaks in the host genome by the cognate T-DNA strand for random integration. There are several evidences to suggest that plants and animals preferably use the non-homologous end joining (NHEJ) method as an efficient means for repairing double strand DNA breaks (DSDB), while yeast rely on homologous recombination (HR) for DNA repair. Proteins playing crucial role in NHEJ include DNA-dependent protein kinase (DNA-PK), Ku70, Ku80, Lig4 and Xrcc4 (van Attikum et al., 2003; van Attikum & Hooyakaas, 2003). *Arabidopsis* mutants for many of these DNA repair proteins including Rad50, Mre11, Ku70 and Ku80 are now well characterized (Bundock & Hooykaas, 2002; Bundock, van Attikum, & Hooykaas, 2002; Riha, Watson, Parkey, & Shippen, 2002). Most of these mutants were found to be hypersensitive to chemicals inducing DSDB, implicating their role in DNA repair mechanism (Bundock & Hooykaas, 2002; Gallego, Jalut, & White, 2003). However, the role of the DNA repair proteins in T-DNA integration is largely unexplored.

6. GENOMIC APPROACHES FOR IDENTIFICATION OF PLANT FACTORS

To elucidate the early plant responses to *Agrobacterium* infection, both genomic and molecular approaches were applied. Using the cDNA-amplified fragment length polymorphism (AFLP), Ditt, Nester, & Comai, 2001 demonstrated the differential gene expression in the mock inoculated versus virulent *Agrobacterium* strain in suspension BY3 cell cultures. A similar study by Veena, Jiang, Doerge, & Gelvin, 2003 demonstrated the successful utilization of host cellular machinery for genetic transformation purposes. On the basis of the above studies it was inferred that *Agrobacterium* infection induces plant genes necessary for the transformation process while repressing host-defenses for better utilization of the host cellular-machinery during the transformation process.

Our research on identifying plant factors involved in this unique horizontal genetic transfer deploys the modern genomics tool of virus-induced gene silencing (VIGS). VIGS is a RNA-mediated post-transcriptional gene silencing mechanism that can protect plants against foreign gene invasion (Baulcombe, 1999). VIGS in the recent past has emerged as an efficient tool for deciphering the function of many genes (Dinesh-Kumar, Anandalakshmi, Marathe, Schiff, & Liu, 2003; Liu, Schiff, & Dinesh-Kumar, 2002; Liu, Schiff, Marathe, & Dinesh-Kumar, 2002; Peart, Cook, Feys, Parker, & Baulcombe, 2002; Yoshioka et al., 2003). We are currently using the *Tobacco rattle virus* (TRV; Liu, Schiff, & Dinesh-Kumar, 2002) derived vectors for screening several putative plant factors by using both the fast-forward genetics and reverse genetics approaches. As of date we have identified at least 20 cDNA clones that when silenced in *Nicotiana benthamiana* does not produce crown gall tumors incited by a wild-type *Agrobacterium* strain. By the reverse genetics approaches we have revalidated the functional role of many plant genes including *VIP1*, *importin*-α, plant histone *H2A*, and have characterized the functional role of plant genes like expansins (*Exp*), nodulin-like proteins (*nlp*) and plant hisotone *H3*, further demonstrating the versatility of this technique (Anand et al., unpublished). We speculate that this approach will help in the identification of several additional plant factors that could improve the transformation efficiencies of recalcitrant crops and facilitate in finding a durable control of the disease.

7. DISEASE CONTROL

It is likely that a complete understanding of the disease may be achieved in the near future; however the major limitation in combating the disease is in finding durable methods for disease control. Although extensive research efforts were put on better understanding of the epidemiology of crown gall throughout this century, prevention strategies have, for the most part, remained focused on phytosanitation, such as the careful cultural methods, disease free nursery stocks and abandonment of infected soils. Traditional breeding strategies for crown gall disease resistance are possible in some perennial crop species where resistant germplasm is present, but breeding requires decades to achieve resistance in commercially available perennial crop rootstocks Attempts to control this disease have largely failed, with the exception of biological control approach using the *A. radiobacter* strain K84, a non-pathogenic *Agrobacterium* strain that produces agrocin 84 which is toxic to most of the agropine strains of *Agrobacterium* (Burr, Bazzi, Sule, & Otten, 1998; Burr & Otten, 1999). This cross protection is limited to certain *Agrobacterium* strains and is likely to fail in the long run either through conjugal transfer of the agrocin plasmid to more resistant *Agrobacterium* strains or due to spontaneous mutations in the resistant strains. Another strategy currently being developed involves expressing deletion mutants of the Ti plasmid *virE2* gene in plants by genetic engineering. It was previously demonstrated that expression of some of *virE2* mutant genes in tobacco results in plants that are resistant to infection by *A. tumefaciens*, possibly by knocking the T-strand nuclear import (Citovsky, Warnick, & Zambryski, 1994), while producing tumors on transgenic plants over expressing wild-type VirE2 proteins (Gelvin, 1998). Additionally post translation gene silencing (PTGS)/RNA interference (RNAi) technique for silencing the auxin biosynthetic genes including *iaaM* and *ipt* oncogenes gave promising results in developing crown gall resistant stocks in apple (Viss, Humann, Cook, Driver, &

Ream, 2003) and creating gall-resistant plants of *Arabidopsis* and tomato (Escobar, Civerolo, Kristin, Summerfelt, & Dandekar, 2001). Transgenic Arabidopsis plants expressing *VIP1* anti-sense RNA, *importin α*-1anti-sense RNA or *histone H2A* anti-sense RNA also yielded plants resistant to crown gall disease (Tzfira, Vaidya, & Citovsky, 2001; Zhu et al., 2003). It is likely that a better understanding on the role of plant genes in bacterial attachment, chemotaxsis, T-DNA import, integration and subsequent steps leading to production of plant hormones and opines in the plant cytosol will provide new insights in finding durable control for the disease.

8. SIGNIFICANCE OF *AGROBACTERIUM* BIOLOGY AND FUTURE PROSPECTS

Trans-kingdom transfer of the nucleo-protein complex from the bacterial cell to the host cell is achieved through the type IV secretion which is not uncommon for many bacteria infecting plants and animals including humans (Cascales & Christie, 2004; Christie, 2001, 2004). However, what makes it a most valuable tool for genetic engineering of plants, fungi and animals is that *Agrobacterium* represents one of the best characterized Type IV secretion systems (Christie, 2004; Li, Wolf, Elbaum, & Tzfira, 2005; Tzfira, Vaidya, & Citovsky, 2004). Several related genera outside the *Agrobacterium* genus can be modified to mediate gene transfer to a number of diverse plants by acquiring the disarmed Ti plasmid and a suitable binary vector (Broothaerts et al., 2005). This provides a new platform for plant biotechnology and modern plant breeding. Although crown gall continues to a serious disease in horticulture plants and in vineyards, significant progress has been made towards understanding the pathogen and the disease. The breakthrough in the *Agrobacterium* biology and elucidation of the role of certain plant metabolites and plant genes in the infection process has lead to a boom in the application of the *Agrobacterium* system for transforming many crop species. The genetic engineering of plants is one of the most rapidly emerging fields of modern day plant biology and biotechnology. The advances made in the transformation protocols over the last few decades should be termed significant and there is now growing interests in the application of this technology in developing biodegradable plastics in plants, plant-based vaccines that could be administered to humans without significant side-effects and in the use of plants as bioindicators for environmental toxins (Kovalchuk, Telyuk, Kovalchuk, Kovalchuk, & Titov, 2003; Mittendorf et al., 1998; Staub et al., 2000). The current understanding on the roles played by key plant genes in *Agrobacterium* infection could probably help in developing alternate strategies that will interfere with the infection process or disease development in plants. Some of these strategies will include genetic engineering of plants to confer resistance to this bacterial pathogen like the antisense expression of known plant factors that are involved in *Agrobacterium*-infectivity such as *VIP1*, *Histone H2A-1*, *α-importin* and post transcriptional gene silencing of tumor inducing genes *iaaM* and *ipt* (Escobar et al., 2001; Tzfira et al.,

2001; Zhu et al., 2003). This research in future will significantly help in developing appropriate strategies for effective long term control of the crown gall disease.

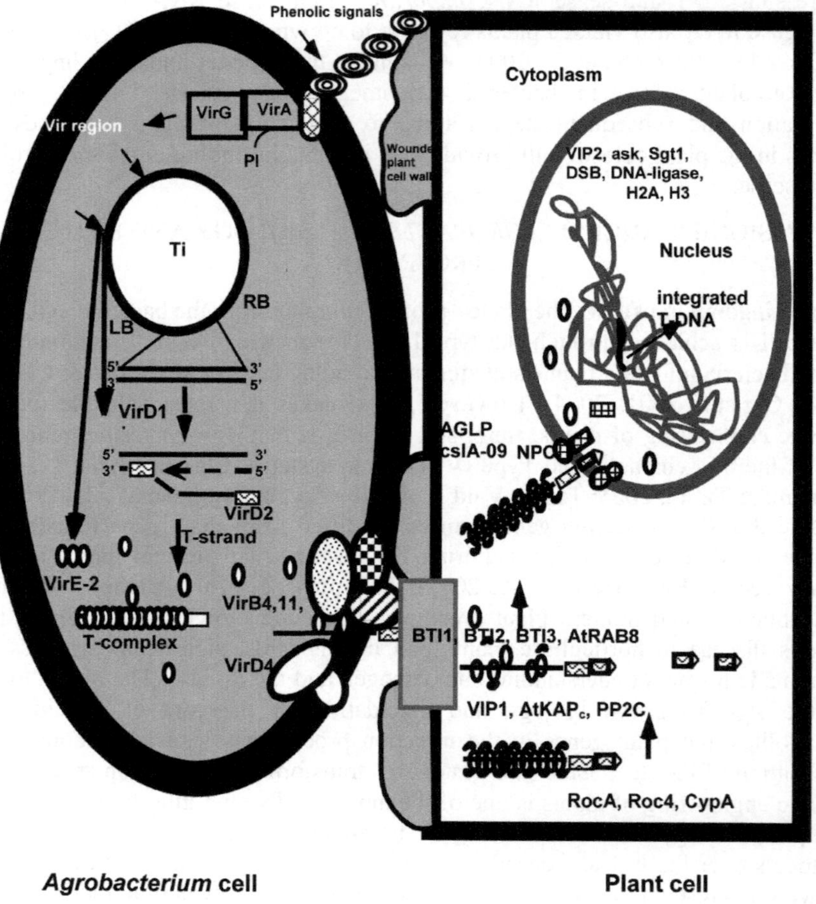

Fig. 1. A schematic sketch of the molecular events that occur between the *Agrobacterium* and the host plant leading to a successful genetic transformation of the host cell. This unique trans-kingdom genetic transfer is facilitated by many host cell factors; while few of the plant factors are well characterized others have been assigned putative functions. The plant factors that are involved in this interaction include genes: facilitating bacterial attachment (Arabinogalactan-like protein; AGLP, cellulose synthase; cslA-09); the cytoplasmic trafficking and nuclear import (plant cyclophilin chaperones; RocA, Roc4, CypA, karyopherin α family; AtKAPα, type 2 protein phosphatase; PP2C, virE2 interacting protein; VIP1; VirB interacting proteins; BTI) and the intranuclear import of the cognate T-DNA and integration (DNA ligases, double strand binding protein; DSB, components of the SCF complex; ask and Sgt1, histone proteins; H2A, H3, virE2 interacting protein; VIP2 etc.)

Table 1. The Vir operon

***vir* loci**	**Known functions**
virB1-11	Type IV transport system to transfer T-DNA and Vir proteins from bacteria to host cytoplasm
virC and *-D*	T-DNA processing. VirD1 and VirD2 nick at T-DNA borders; VirC1 binds overdrive
virE	Nuclear transport of T-DNA. VirE2 binds single-stranded DNA and has nuclear localization sites; VirE1 is a chaperone for VirE2 transport
virF	Host range factor
virH1-2	P450-type oxidases; VirH2 demethylates phenolic inducers
virM, *-L*, *-K*, *-J*, *-F*, *-P*, *-R*, *-D3*, *-D5*, and *-E3*	Other members of the *vir* regulon; VirP resembles phosphatases
virA and *virG*	Two-component regulators of *vir* regulon; VirA is a transmembrane histidine kinase; VirG is an OmpR-type response regulator

Table 2. Putative plant factors and their role in *Agrobacterium* infectivity

Gene family	**Identified Genes**	**Putative role**
Replacement histones	*H2A*, *H2B*, *H3*, *H4* , histone deacetylase and histone acetyl transferases	Chromatin structure and remodeling to facilitate T-DNA integration
Nuclear targeting genes and *vir*-interacting proteins	α-importin (*AtKAP*), *VIP1*, *VIP2*, *PP2C*, Cyclophilin chaperone (*RocA*, *Roc4* and *CypA*), *ask1*	Nuclear targeting of the cognate T-DNA complex and T-DNA integration
Cell wall structural and metabolism genes	*AGLP*, *Actin*, *cslA-09*, gene encoding cell wall protein	Bacterial attachment and cytoplasmic import

9. REFERENCES

Akiyoshi, D. E., Klee, H., Amasino, R. M., Nester, E. W., & Gordon, M. P. (1984). T-DNA of *Agrobacterium* tumefaciens encodes an enzyme of cytokinin biosynthesis Proc. Natl. Acad. Sci., USA, 81, 5994-5998.

Allardet-Servant, A., Michaux-Charachon, S., Jumas-Bilak, E., Karayan, L., & Ramuz, M. (1993). Presence of one linear and one circular chromosome in the *Agrobacterium* tumefaciens C58 genome. J. Bacteriol., 175, 7869-7874.

Anand, A., & Mysore, K. S. (2005). Current advances in *Agrobacterium*-plant interactions and their implications on agricultural biotechnology. In A. Hemantarajan (Ed.), Advances in Plant Physiol: An International Treatise Series, Jodhpur: Scientific publishers (in press).

Ashby, A. M., Watson, M. D., Ladke, G. J., & Shaw, C. H. (1988). Ti plasmid-specified chemotaxis of *Agrobacterium* tumefaciens C58C1 toward vir-inducing phenolic compounds and soluble factores from monocotyledonous and dicotyledonous plants. J. Bacteriol., 170, 4181-4187.

Bai, C., Sen, P., Hofmann, K., Ma, L., Goebl, M., Harper, J. W., et al. (1996). SKP1 connects cell cycle regutlators to the ubiquitin proteolysis machinery through a novel motiff, F- box. Cell, 86, 263-274.

Bako, L., Umeda, M., Tiburcio, A. F., Schell, J., & Koncz, C. (2003). The VirD2 pilot protein of *Agrobacterium*-transferred DNA interacts with the TATA box-binding protein and a nuclear protein kinase in plants. Pro Natl. Acad. Sci., USA, 100, 10108-10113.

Ballas, N., & Citovsky, V. (1997). Nuclear localization signal binding protein from Arabidopsis mediates nuclear import of *Agrobacterium* VirD2 protein. Proc. Natl. Acad. Sci., USA, 94, 10723-10728.

Baron, C., Llosa, M., Zhou, s., & Zambryski, P. C. (1997). VirB1, a component of the T-complex transfer machinery of *Agrobacterium* tumefaciens is processed to a C-terminal secreted product VirB1. J Bacteriol., 179, 1203-1210.

Barry, G. F., Rogers, S. G., Fraley, R. T., & Brand, L. (1984). Identification of a cloned cytokinin biosynthetic gene. Proc. Natl. Acad. Sci., USA, 81, 4776-4780.

Bash, R., & Matthysse, A. G. (2002). Attachment to roots and virulence of a chvB mutant of *Agrobacterium* tumefaciens are temperature sensitive. Mol. Plant Microbe Interact., 15, 160-163.

Baulcombe, D. C. (1999). Fast forward genetics based on virus-induced gene silencing. Curr. Opin. Plant Biol., 2, 109-113.

Binns, A. N., Beaupre, C. E., & Dale, M. (1995). Inhibition of VirB-mediated transfer of diversessubstrate from *Agrobacterium* tumefaciens by the IncQ plasmid RSF100. J. Bacteriol., 177, 4890-4899.

Braun, A. (1954). The physiology of plant tumors. Ann. Rev. Plant Physiol., 5, 133-162.

Braun, A. C. (1978). Plant tumors. Biochim. Biophys. Acata., 516, 167-191.

Braun, A. C. (1982). A history of the crown gall problem. In G. Kahl & J. Schell (Eds.), Molecular biology of plant tumors (pp. 155-210). New York: Academic Press.

Broothaerts, W., Mitchell, H. J., Weir, B., Kaines, S., Smith, L. M. A., Yang, W., et al. (2005). Gene transfer to plants by diverse species of bacteria. Nature, 433, 629-633.

Buchmann, I., Marner, F.-J., Schroder, G., Waffenschmidt, S., & Schroder, J. (1985). Tumour genes in plants: T-DNA encoded cytokinin biosynthesis EMBO J., 4, 853.

Bundock, P., & Hooykaas, P. J. (2002). Severe developmental defects, hypersensitivity to DNA-damaging agents, and lengthened telomeres in Arabidopsis MRE11 mutants. Plant Cell, 14, 2451-2462.

Bundock, P., van Attikum, H., & Hooykaas, P. J. (2002). Increased telomere length and hypersensitivity to DNA damaging agents in an Arabidopsis 70 mutant. Nucl. Acids Res., 30, 3395-3400.

Burr, T. J., Bazzi, C., Sule, S., & Otten, L. (1998). Biology of Agrobactereium vitis and the development of disease control strategies. Plant Dis., 82, 1288-1297.

Burr, T. J., & Otten, L. (1999). Crown gall of grape: biology disease management. Annu. Rev. Phytopathol., 37, 53-80.

Cangelosi, G. A., Ankenbauer, R. G., & Nester, E. W. (1990). Sugars induce the *Agrobacterium* virulence genes through a periplasmic binding protein and a transmembrande signal protein. Proc. Natl. Acad. Sci., USA, 87, 6708-6712.

Cangelosi, G. A., Hung, L., Puvanesarajah, V., Stacey, G., Ozga, D. A., Leigh, J. A., et al. (1987). Common loci for *Agrobacterium* tumefaciens and Rhizobium meliloti exopolysaccharide synthesis and their roles in plant interactions J. Bacteriol., 169, 2086-2091.

Cascales, E., & Christie, P. J. (2004). Definition of a Bacterial Type IV Secretion Pathway for a DNA Substrate. Science, 304, 1170-1173.

Chen, L., Li, C. M., & Nester, E. W. (2000). Transferred DNA (T-DNA)-associated proteins of *Agrobacterium* tumefaciens are exported independently of virB. Pro. Natl. Acad. Sci., USA, 97, 7545-7550.

Chilton, M.-D., Drummond, M. H., Merlo, D. J., Sciaky, D., Montoya, A. L., Gordon, M. P., et al. (1977). Stable incorporation of plasmid DNA into higher plant cells: the molecular basis of crown gall tumorigenesis. Cell, 11, 263-271.

Chilton, M.-D., & Que, Q. (2003). Targeted integration of T-DNA into the tobacco genome at double-stranded breaks: new insights on the mechanism of T-DNA integration. Plant Physiol., 133, 956-965.

Christie, P. J. (1997). *Agrobacterium* tumefaciens T-complex transport apparatus: a paradigm for a new family of multifunctional transporters in Eubacteria. J. Bacteriol., 179, 3085-3094.

Christie, P. J. (2001). Type IV secretion: intercellular transfer of macromolecules by systems ancestrally related to conjugation machines. Mol. Microbiol., 40, 294-305.

Christie, P. J. (2004). Type IV secretion: the *Agrobacterium* VirB/D4 and related conjugation systems. Biochem. Biophys. Acta., 1694, 219-234.

Christie, P. J., & Vogel, J. P. (2000). Bacterial type IV secretion: conjugation systems adapted to deliver effector molecules to host cells. Trends Microbiol., 8, 354-360.

Christie, P. J., Ward, J. E., Winans, S. C., & Nester, E. W. (1988). The *Agrobacterium* tumefaciens VirE2 gene product is a single -stranded-DNA-binding protein that associates with T-DNa. J. Bacteriol., 170, 2659-2667.

Citovsky, V., De Vos, G., & Zambryski, P. C. (1988). Single-stranded DNA binding protein encoded by the virE locus of *Agrobacterium* tumefaciens. Science, 240, 501-504.

Citovsky, V., Guralnick, B., Simon, M. N., & Wall, J. S. (1997). The molecular structure of *Agrobacterium* VirE2-single stranded DNA complexes involved in nuclear import. J. Mol. Biol., 271, 718-727.

Citovsky, V., Warnick, D., & Zambryski, P. C. (1994). Nuclear import of *Agrobacterium* VirD2 and VirE2 proteins in maize and tobacco. Proc. Natl. Acad. Sci., USA, 91, 3210-3214.

Citovsky, V., Wong, M. L., & Zambryski, P. C. (1989). Coooperative interaction of *Agrobacterium* VirE2 protein with single-stranded DNA: Implications for the T-DNA transfer process. Pro. Natl. Acad. Sci., USA, 86, 1193-1197.

Citovsky, V., Zupan, J., Warnick, D., & Zambryski, P. C. (1992). Nuclear localization of *Agrobacterium* VirE2 protein in plant cells. Science, 256, 1802-1805.

Collart, M. A. (2003). Global control of gene expression in yeast by the Ccr4-Not complex. Gene, 313, 1-16.

Collart, M. A., & Struhl, K. (1993). CDC39 an essential nuclear protein that negatively regulates transcription and differentially affects the constitutive and inducible his3 promoters. EMBO J., 12, 177-186.

Das, A., Stachel, S., Ebert, P., Allenza, P., Montoya, A. L., & Nester, E. W. (1986). Promoters of *Agrobacterium* tumefaciens Ti-plasmid virulence genes. Nucl. Acids Res., 14, 1355-1364.

Davey, M. R., Curtis, I. S., Gartland, K. M. A., & Power, J. B. (1994). *Agrobacterium*-induced crown gall and hairy root diseases: Their biology and application to plant genetic engineering. In M. A. J. Williams (Ed.), Plant Galls (Vol. 49, pp 9-55). Systematics Association.

de Groot, M. J., Bundock, P., Hooykaas, P. J., & Beijersbergen, A. G. (1998). *Agrobacterium* tumefaciens-mediated transformation of filamentous fungi. Nat. Biotechnol., 16, 839-842.

de Iannino, N. I., & Ugalde, R. A. (1989). Biochemical characterization of avirulent *Agrobacterium* tumefaciens chvA mutants: synthesis and excretion of beta-(1-2) gulcan. J. Bacteriol., 171, 2842-2849.

Deng, W., Chen, L., Wood, D. W., Metcalfe, T., Liang, X., Gordon, M. P., et al. (1998). *Agrobacterium* VirD2 protein interacts with plant host cyclophilins. Proc. Natl. Acad. Sci., USA, 95, 7040-7045.

Dessaux, Y., Petit, A., Farrand, S. K., & Murphy, P. J. (1998). Opines and opine-like molecules involved in plant-Rhizobiaceae interactions. In H. P. Spaink, A. Kondorosi & P. J. J. Hooykass (Eds.), The Rhizobiaceae: molecular biology of model plant-asscociated bacteria. Dordrecht: Kluwer Academic Publishers.

Dinesh-Kumar, P., Anandalakshmi, R., Marathe, R., Schiff, M., & Liu, Y. (2003). Virus-induced gene silencing in Methods. In E. Grotewold (Ed.), Molecular Biology, Plant Functional Genomics: Methods and Protocols (Vol. 236, pp. 287-293). Totowa: Humana Press.

Ditt, R. F., Nester, E. W., & Comai, L. (2001). Plant gene expression response to *Agrobacterium* tumefaciens. Proc. Natl. Acad. Sci., USA, 98, 10954-10959.

Doty, S. L., Yu, M. C., Lundin, J. I., Heath, J. D., & Nester, E. W. (1996). Mutational analysis of the input domain of the VirA protein of *Agrobacterium* tumefaciens. J. Bacteriol., 178, 961-970.

Douglas, C., Halperin, W., Gordon, M. P., & Nester, E. W. (1985a). Specific attachment of *Agrobacterium* tumefaciens to bamboo cells in suspension cultures. J. Bacteriol., 161, 764-766.

Douglas, C. J., Staneloni, R. J., Rubin, R. A., & Nester, E. W. (1985b). Identification and genetic analysis of an *Agrobacterium* tumefaciens chromosomal virulence region. J. Bacteriol., 161, 850-860.

Dworetzky, S. I., & Feldher, C. M. (1988). Translocation of RNA-coated gold particles through the nuclear pores of oocytes. J. Cell. Biol., 106, 575-584.

Escobar, M. A., Civerolo, E. L., Kristin, R., Summerfelt, K. R., & Dandekar, A. M. (2001). RNAi-mediated oncogene silencing confers resistance to crown gall tumorigenesis. Proc. Natl. Acad. Sci., USA, 98, 13437-13442.

Escobar, M. A., & Dandekar, A. M. (2003). *Agrobacterium* tumefaciens as an agent of disease. Trends Plant Sci,, 8, 380-386.

Escudero, J., & Hohn, B. (1999). Transfer and integration of T-DNA without cell injury in the host plant. Plant Cell, 9, 2135-2142.

Farrand, S. K., van Berkum, P. B., & Oger, P. (2003). *Agrobacterium* is a definable genus of the family Rhizobiaceae. Int. J. Syst. Evol. Microbiol., 53, 1681-1687.

Firth, N., Ippen-Ihler, K., & Skurray, R. A. (1996). Structure and function of the F factor and mechanism of conjugation. In F. C. Neidhardt, R. C. III, J. L. Ingraham, E. C. C. Lin, K. B. Low, B. Magasanik, W. S. Reznifoff, M. Riley, M. Schaechter & H. C. Umbarger (Eds.), Escherichia coli and Salmonella: cellular and molecular biology 2nd Edition. Washington D.C.: American Society for Microbiology.

Frolov, M. V., Benevolenskaya, E. V., & Birchler, J. A. (1998). Regena (Rga), a Drosophila Homolog of the Global Negative Transcriptional Regulator CDC36 (NOT2) from Yeast, Modifies Gene Expression and Suppresses Position Effect Variegation. Genetics, 148, 317-330.

Gallego, M. E., Jalut, N., & White, C. I. (2003). Telomerase dependence of telomere lengthening ku80 mutant Arabidopsis. Plant Cell, 15, 782-789.

Garfinkel, D. J., Simpson, R. B., Ream, L. W., White, F. F., Gordon, M. P., & Nester, E. W. (1981). Genetic analysis of crown gall: Fine structure map of the T-DNA by sit-directed mutagenesis. Cell, 27, 143-153.

Gelvin, S. B. (1998). *Agrobacterium* VirE2 proteins can form a complex with T strands in the plant cytoplasm. J. Bacteriol., 180, 4300-4302.

Gelvin, S. B. (2000). *Agrobacterium* and plant genes involved in T-DNA transfer and integration. Annu. Rev. Plant Physiol. Plant Mol. Biol., 51, 223-256.

Gelvin, S. B. (2003). *Agrobacterium*-mediated plant transformation: the biology behind the "gene-jockeying" tool. Microbiol. Mol. Biol. Rev., 67, 16-37.

Goodner, B., Hinkle, G., Gattung, S., Miller, N., Blanchard, M., Qurollo, B., et al. (2001). Genome sequence of the plant pathogen and biotechnology agent *Agrobacterium* tumefaciens C58. Science, 294, 2323-2328.

Hooykaas, P. J., Hofker, M., den Dulk-Ras, A., & Schilperoort, R. A. (1984). A comparison of virulence determinants in an octopine Ti plasmid, a nopaline Ti plasmid and an Ri plasmid by complementation analysis of *Agrobacterium* tumefaciens mutants. Plasmid, 11, 195-205.

Hooykaas, P. J., & Shilperoort, R. A. (1992). *Agrobacterium* and plant genetic engineering. Plant Mol. Biol., 19, 15-38.

Howard, E., Zupan, J., Citovsky, V., & Zambryski, P. C. (1992). The VirD2 protein of A. tumefaciens contains a C-terminal bipartite nuclear localization signal: implications for nuclear uptake of DNA in plant cells. Cell, 68, 109-118.

Hwang, H.-H., & Gelvin, S. B. (2004). Plant Proteins That Interact with VirB2, the *Agrobacterium* tumefaciens Pilin Protein, Mediate Plant Transformation. Plant Cell, 16, 3148-3167.

Iyer, V. N., Klee, H. J., & Nester, E. W. (1982). Units of genetic expression in the virulence region of a plant tumor-inducing plasmid of *Agrobacterium* tumefaciens. Mol. Gen. Genet., 188, 418-424.

Jin, S., Prusti, R. K., Roitsch, T., Ankenbauer, R. G., & Nester, E. W. (1990). Phosphorylation of the VirG protein of *Agrobacterium* tumefaciens by the autophosphorylated VirA protein: Essential role in biological activity of VirG J. Bacteriol., 172, 4945.

Jin, S., Roitsch, T., Christie, P. J., & Nester, E. W. (1990). The regulatoy VirG protein specifically binds to a cis-acting regulatory sequence involved in transcriptional activation of *Agrobacterium* tumefaciens virulence genes J. Bacteriol., 172, 531.

Jin, S., Song, Y., Pan, S., & Nester, E. W. (1993). Characterization of a virG mutation that confers constitutive virulence gene expression in *Agrobacterium* tumefaciens. Mol. Microbiol., 7, 555-562.

Kado, C. I. (2000). The role of T-pilus in horizontal gene transfer and tumorigenesis. Curr. Opin. Microbiol., 3, 643-648.

Keane, P. J., Kerr, A., & New, P. B. (1970). Crown gall of stone fruit. II. indentification and nomenclature of *Agrobacterium* isolates. Aust. J. Biol. Sci., 23, 585-595.

Kerr, A., & Panagopoulos, C. G. (1977). Biotypes of l*Agrobacterium* radiobacter var. tubefaciens and their biiological control. Phytopathol Z., 90, 172-179.

Kim, C.-H., Baek, J. K., Lee, J., Yang, M., & Farrand, S. K. (2001). Intracellular accumulation of mannopine, an opine produced by crown gall tumors, transiently inhibits growth of *Agrobacterium* tumefaciens. Mol. Plant Microbe Interact., 14, 793-803.

Koepp, D. M., Harper, J. W., & Elledge, S. J. (1999). How the cyclin became a cyclin: regulated proteolysis in the cell cycle. Cell, 97, 431-434.

Kovalchuk, O., Telyuk, P., Kovalchuk, L., Kovalchuk, I., & Titov, V. (2003). Novel plant bioassays for monitoring the genotoxicity of drinking water from the inhabited areas of the Ukraine affected by the Chernobyl accident. Bull. Environ. Contam. Toxicol., 70, 847-853.

Kunik, T., Tzfira, T., Kapulnik, Y., Gafni, Y., Dingwall, C., & Citovsky, V. (2001). Genetic transformation of HeLa cells by *Agrobacterium*. Proc. Natl. Acad. Sci., USA, 98, 1871-1876.

Lacroix, B., Vaidya, M., Tzfira, T., & Citovsky, V. (2005). The VirE3 protein of *Agrobacterium* mimics a host cell function required for plant genetic transformation. EMBO J., 24, 428-437.

Lee, H., Humann, J. L., Pitrak, J. S., Cuperus, J. T., Parks, T. D., Whistler, C. A., et al. (2003). Translation start sequences affect the efficiency of silencing of *Agrobacterium* tumefaciens T-DNA oncogenes. Plant Physiol., 133, 966-977.

Lee, Y.-W., Jin, S., Sim, W.-S., & Nester, E. W. (1995). Genetic evidence for direct sensing of phenolic compounds by the VirA protein of *Agrobacterium* tumefaciens. Proc. Natl. Acad. Sci., USA, 92, 12245-12249.

Lessl, M., & Lanka, E. (1994). Common mechanisms in bacterial conjugation and Ti-mediated T-DNA transfer to plant cells. Cell, 77, 321-324.

Li, J., Krichevsky, A., Vaidya, M., Tzfira, T., & Citovsky, V. (2005). Uncoupling of the functions of the Arabidopsis VIP1 protein in transient and stable plant genetic transformation by *Agrobacterium*. Proc. Natl. Acad. Sci., USA, 102, 5733-5738.

Li, J., Wolf, S. G., Elbaum, M., & Tzfira, T. (2005). Exploring cargo transport mechanics in the type IV secretion systems. Trends in Microbiol., 13, 295-298.

Lippincott, B. B., & Lippincot, J. A. (1969). Bacterial attachment to a specific wound site as an essential stage in tumor initiation by *Agrobacterium* tumefaciens. J. Bacteriol., 97, 620-628.

Lippincott, J. A., & Lippincott, B. B. (1981). Crown gall, a "alignant plant tumor". In H. E. Kaiser (Ed.), Neoplasms: Comparative pathology of growth in animals, plants, and man. Baltimore: Williams & Wilkins.

Liu, Y., Schiff, M., & Dinesh-Kumar, P. (2002). Virus-induced gene silencing in tomato. Plant J., 31, 777-786.

Liu, Y., Schiff, M., Marathe, R., & Dinesh-Kumar, P. (2002). Tobacco Rar1, EDS1 and NPR1/NIMI1 like genes are required for N-mediated resistance to tobacco mosaic virus. Plant J., 30, 415-429.

Matthysse, A. G. (1986). Initial interactions of *Agrobacterium* tumefaciens with plant host cells. Cri. Rev. Microbiol., 13, 281-307.

Matthysse, A. G. (1994). Conditoned medium promotes the attachment of *Agrobacterium* tumefaciens strain NT1 to carrot cells. Protoplasma, 183, 131-136.

Matthysse, A. G., Yarnall, H. A., & Young, N. (1996). Requirement for genes with homology to ABC transport systems for attachment and virulence of *Agrobacterium* tumefaciens. J. Bacteriol., 178, 5302-5308.

Mittendorf, V. V., Robertson, E. J., Leech, R. M., Kruger, N., Steinbuchel, A., & Poirier, Y. (1998). Synthesis of medium-chain-length polyhydroxyalkanoates in Arabidosis thaliana using intermediates of peroxisomal fatty acid betaoxidation. Proc. Natl. Acad. Sci., USA, 95, 13397-13402.

Mysore, K. S., Ranjith Kumar, C. T., & Gelvin, S. B. (2000). Arabidipsis ecotypes and mutants that are recalcitrant to *Agrobacterium* rooot transformation are susceptible to germ-line transformation. Plant J., 21, 9-16.

Nair, G. R., Liu, Z., & Binns, A. N. (2003). Re-examining the role of the cryptic plasmid pAtC58 in the virulence of *Agrobacterium* tumefaciens strain C58. Plant Physiol., 133, 989-999.

Nam, J., Matthysse, A., & Gelvin, S. B. (1997). Differences in susceptibility of Arabidopsis ecotypes to crown gall disease may result from a deficiency in T-DNA integration. Plant Cell, 19, 317-333.

Nam, J., Mysore, K. S., Zheng, C., Knue, M. K., Matthysse, A. G., & Gelvin, G. B. (1999). Identification of T-DNA tagged Arabidopsis mutants that are resistant to transformation by *Agrobacterium*. Mol. Gen. Genet., 261, 429-438.

Narasimhulu, S. B., Deng, X.-B., Sarria, R., & Gelvin, S. B. (1996). Early transcription of *Agrobacterium* T-DNA genes in tobacco and maize. Plant Cell, 8, 873-886.

Nester, E. W., Gordon, M. P., Amasino, R. M., & Yanofsky, M. F. (1984). Crown gall: a molecular and physiological analysis Ann. Rev. Plant Physiol., 35, 387-413.

Peart, J. R., Cook, G., Feys, B. J., Parker, J. E., & Baulcombe, D. C. (2002). An EDS1 orthologue is required for N-mediated resistance against tobacco mosaic virus. Plant J., 29, 569-579.

Pemberton, L. F., Rosenblum, J. S., & Blobel, G. (1999). Nuclear import of the TATA-binding protein: mediation by the karyopherin Kap 114p and a possible mechanism for intranuclear targeting. J. Cell Biol., 145, 1407-1417.

Piers, K. L., Heath, J. D., Liang, X., Stephens, K. M., & Nester, E. W. (1996). *Agrobacterium* tumifaciens-mediated transformation of yeast. Proc. Natl. Acad. Sci., USA, 93, 1613-1618.

Powell, A., & Gordon, M. P. (1989). Tumor formation in plants. In A. Marcus (Ed.), The bio-chemistry of plants (Vol. 15, pp. 617-651). New York: Academic Press.

Powell, B. S., Rogowsky, P. M., & Kado, C. I. (1989). virG of *Agrobacterium* tumefaciens plasmid pTiC58 encodes a DNA-binding protein Mol. Microbiol., 3, 411-419.

Ream, L. W., Gordon, M. P., & Nester, E. W. (1983). Multiple mutations in the T region of the *Agrobacterium* tumefaciens tumor-inducing plasmid Proc. Natl. Acad. Sci., USA, 80, 1660-1664.

Regensburg-Tuink, A. J. G., & Hooykaas, P. J. J. (1993). Transgenic N. glauca plants expressing bacterial virulence gene virF are converted into hosts for nopaline strains of A. tumefaciens Nature, 363, 69-71.

Rhee, Y., Gurel, F., Gafni, Y., Dingwall, C., & Citovsky, V. (2000). A genetic system for detection of protein nuclear import and export. Nat. Biotechnol., 18, 443-437.

Rho, H. S., Kang, S., & Lee, Y. H. (2001). *Agrobacterium* tumefaciens-mediated transformation of plant pathogenic fungus, Magnaporthe grisea. Mol. Cells, 12, 407-411.

Riha, K., Watson, J. M., Parkey, J., & Shippen, D. E. (2002). Telomere length deregulation and enhanced sensitivity to genotoxic stress in Arabidopsis mutants deficient in Ku70. EMBO J., 21, 2819-2826.

Rossi, L., Hohn, B., & Tinland, B. (1993). The VirD2 protein of *Agrobacterium* tumefaciens carries nuclear localization signals important for transfer of T-DNA to plant. Mol. Gen. Genet., 239, 345-353.

Rossi, L., Hohn, B., & Tinland, B. (1996). Integration of completed transferred DNA units is dependent on the activity of virulence E2 protein of *Agrobacterium* tumefaciens. Proc. Natl. Acad. Sci., USA, 93, 126-130.

Rout, M. P., & Wente, S. R. (1994). Pores for thought: nuclear pore complex proteins. Trends Cell. Biol., 4, 357-365.

Schrammeijer, B., Beijersbergen, A. G., Idler, K. B., Melchers, L. S., Thompson, D. V., & Hooykaas, P. J. (2000). Sequence analysis of the vir-region from *Agrobacterium* tumefaciens octopine Ti plasmid pTi15955. J. Exp. Bot., 51, 1167-1169.

Schrammeijer, B., den Dulk-Ras, A., Vergunst, A. C., Jurado Jacome, E., & Hooykaas, P. J. J. (2003). Analysis of Vir protein translocation from *Agrobacterium* tumefaciens using Saccharomyces cerevisiae as a model: evidence for transport of a novel effector proteinVirE3. Nucl. Acids Res., 31, 860-868.

Schrammeijer, B., Risseeuw, E., Pansegrau, W., Regensburg-Tuink, T. J., W.L., C., & Hooykaas, P. J. J. (2001). Interaction of the virulence protein VirF of *Agrobacterium* tumefaciens with plant homologs of the yeast Skp1 protein. Curr. Biol., 11, 258-262.

Sheng, J., & Citovsky, V. (1996). *Agrobacterium*-plant cell DNA transport: have virulence proteins, will travel. Plant Cell, 8, 1699-1710.

Shimoda, N., Toyoda-Yamamoto, A., Aoke, S., & Machida, Y. (1993). Genetic evidence for an interaction between and VirA sensor protein and the ChvE sugar-binding protein of *Agrobacterium* J. Biol. Chem., 268, 26552-26558.

Shurvinton, C. E., Hodges, L., & Ream, W. (1992). A nuclear localization signal and the C-terminal omega sequence in the *Agrobacterium* tumefaciens VirD2 endonuclease are important for tumor formation. Proc. Natl. Acad. Sci., USA, 89, 11837-11841.

Smith, E. F., & Townsend, C. O. (1907). A plant tumor of bacterial orgins. Science, 25, 671-673.

Stachel, S. E., Messens, E., van Montagu, M., & Zambryski, P. C. (1985). Identification of the signal molecules produced by wounded plant cell that activate T-DNA transfer in *Agrobacterium* tumefaciens. Nature, 318, 624-629.

Stachel, S. E., Nester, E. W., & Zambryski, P. C. (1986). A plant cell factor induces *Agrobacterium* tumefaciens vir gene expression Proc. Natl. Acad. Sci., USA, 83, 379-383.

Stachel, S. E., Timmerman, B., & Zambryski, P. C. (1986). Generation of single-stranded T-DNA molecules during the initial stages of T-DNA transfer from *Agrobacterium* tumefaciens to plant cells. Nature, 322, 706-712.

Stafford, H. A. (2000). Crown gall disease and *Agrobacterium* Tumefaciens: A study of the history, present knowlege, missing information, and impact on molecular genetics. In The Botanical Review (Vol. 66, pp. 99-118). Bronx, NY: NYBG Press.

Staub, J. M., Garcia, B., Graves, J., Hajdukiewicz, P. T., Hunter, P., Nehra, N., et al. (2000). High-yield production of a human therapeutic protein in tobacco chloroplasts. Nat. Biotechnol., 18, 333-338.

Stock, J. B., Lukat, G. S., & Stock, A. M. (1991). Bacterial chemotaxis and the molecular logic of intracellular signal transduction networks. Annu. Rev. Biophys. Chem., 20, 109-136.

Stock, J. B., Stock, A. M., & Mottonen, J. M. (1990). Signal transduction in bacteria. Nature, 344, 395-400.

Sundberg, C., Meek, L., Carroll, K., Das, A., & Ream, W. (1996). VirE1 protein mediates export of the single-stranded DNA binding protein VirE2 from *Agrobacterium* tumefaciens into plant cells. J. Bacteriol., 178, 1207-1212.

Swart, S., Logman, T. J. J., Smit, G., Lugtenberg, B. J. J., & Kijne, J. W. (1994). Purfication and partial characterization of a glycoprotein from pea (Pisum sativum) with receptor activity for rhicadhesin, an attachment protein of Rhizobiaceae. Plant Mol. Biol., 24, 171-183.

Tao, Y., Rao, P. K., Bhattacharjee, S., & Gelvin, S. B. (2004). Expression of plant protein phosphatase 2C interferes with nuclear import of the *Agrobacterium* T-complex protein VirD2. Proc. Natl. Acad. Sci., USA, 101, 5164-5169.

Thomashow, M. F., Hugly, S., Buchholz, W. G., & Thomashow, L. S. (1986). Molecular basis for the auxin-independent phenotype of crown gall tumor tissues Science, 231, 616-618.

Tighe, S. W., de Lajudie, P., Dipietro, K., Lindstrom, K., Nick, G., & Jarvis, B. D. W. (2000). Analysis of cellular fatty acids and phenotypic relationships of *Agrobacterium*, Bradyhizobium, Mesorhizobium, Rhizobium and Sinorhizobium species using the Sherlock Microbial Identification System. Int. J. Syst. Evol. Microbiol., 50, 787-801.

Tzfira, T., & Citovsky, V. (2000). From host recognition to T-DNA integration: the function of bacterial and plant genes in the *Agrobacterium*-plant cell interaction. Mol. Plant. Pathol., 1, 201-121.

Tzfira, T., & Citovsky, V. (2001). Comaprision between nuclear import of nopaline-and octopine specific VirE2 protein of *Agrobacterium* on plant and animal cells. Mol. Plant Pathol., 2, 171-176.

Tzfira, T., & Citovsky, V. (2002). Partners-in-infection: host proteins involved in the transformation of plant cells by *Agrobacterium*. Trends Cell Biol., 12, 121-129.

Tzfira, T., & Citovsky, V. (2003). The *Agrobacterium*-plant cell interaction. Taking biology lessons from a bug. Plant Physiol., 133, 943-947.

Tzfira, T., Li, J., Lacroix, B., & Citovsky, V. (2004). *Agrobacterium* T-DNA integration: molecules and models. Trends Genet., 20, 375-383.

Tzfira, T., Rhee, Y., Chen, M.-H., & Citovsky, V. (2000). Nucleic acid transport in plant-microbe interactions: the molecules that walk through the walls. Annu.Rev. Microbiol., 54, 187-219.

Tzfira, T., Rhee, Y., Chen, M.-H., Kunik, T., & Citovsky, V. (2000). Nucleic acid transport in plant-microbe interactions: the molecules that walk through the walls. Annu. Rev. Microbiol., 54, 187-219.

Tzfira, T., Vaidya, M., & Citovsky, V. (2001). VIP1, an Arabidopsis protein that interacts with *Agrobacterium* VirE2, is involved in VirE2 nuclear import and *Agrobacterium* infectivity. EMBO J., 20, 3596-3607.

Tzfira, T., Vaidya, M., & Citovsky, V. (2002). Increasing plant susceptibility to *Agrobacterium* by overexpression of the Arabidopsis nuclear protein VIP1. Proc. Natl. Acad. Sci., USA, 99, 10435-10440.

Tzfira, T., Vaidya, M., & Citovsky, V. (2004). Involvement of targeted proteolysis in plant genetic transformation by *Agrobacterium*. Nature, 431, 87-92.

Uttaro, A. D., Cangelosi, G. A., Geremia, R. A., Nester, E. W., & Ugalde, R. A. (1990). Biochemical characterization of avirulent exoC mutants of *Agrobacterium* tumefaciens J. Bacteriol., 172, 1640-1646.

van Attikum, H., Bundock, P., & Hooykaas, P. J. (2001). Non-homologous end-joining proteins are required for *Agrobacterium* T-DNA integration. EMBO J., 20, 6550-6558.

van Attikum, H., Bundock, P., Overmeer, R. M., Lee, L.-Y., Gelvin, S. B., & Hooyakaas, P. J. J. (2003). The Arabidopsis AtLIG4 gene is required for the repair of DNA damage, but not for the integration of *Agrobacterium* T-DNA. Nucl. Acids Res., 31, 4247-4255.

van Attikum, H., & Hooyakaas, P. J. J. (2003). Genetic requirements for the targeted integration of *Agrobacterium* T-DNA integration. EMBO J., 20, 6550-6558.

Van Onckelen, H., Rudelsheim, P., Inze, D., Follin, A., Messens, E., Horemans, S., et al. (1985). Tobacco plants transformed with the *Agrobacterium* T-DNA gene 1 contain high amounts of indole-3-acetamide FEBS Letters, 181, 373-376.

Veena, J. H., Doerge, R. W., & Gelvin, S. B. (2003). Transfer of T-DNA and Vir proteins to plant cells by *Agrobacterium* tumefaciens induces expression of host genes involved in mediating transformation and suppresses host defense gene expression. Plant J., 35, 219-226.

Vergunst, A. C., van Lier, M. C. M., den Dulk-Ras, A., & Hooykaas, P. J. (2003). Recognition of the *Agrobacterium* VirE2 translocation signal by the VirB/D4 transport system does not require VirE1. Plant Physiol, 133, 978-988.

Viss, W., Humann, J. L., Cook, M., Driver, J., & Ream, W. (2003). Crown-gall-resistant transgenic apple trees that silence *Agrobacterium* tumefaciens T-DNA oncogenes. Mol. Breeding, 12, 283-295.

Wagner, V. T., & Matthysse, A. (1992). Involvement of a vitronectin-like protein in attachment of *Agrobacterium* tumefaciens to carrot suspension culture cells. J. Bacteriol., 174, 5999-6003.

Ward, E. R., & Barnes, W. M. (1988). VirD2 protein of *Agrobacterium* tumefaciens very tightly linked to the 5' end of T-strand DNA. Science, 242, 927-930.

Winans, S. C. (1991). An *Agrobacterium* two-component regulatory system for the detection of chemicals released from plant wounds Mol. Microbiol., 5, 2345.

Winans, S. C. (1992). Two-way chemical signaling in *Agrobacterium*-plant interactions Microbiol. Rev., 56, 12.

Winans, S. C., Mantis, N. J., Chen, C. Y., Chang, C. H., & Han, D. C. (1994). Host recognition by the VirA, VirG two-component regulatory proteins of *Agrobacterium* tumefaciens. Res. Microbiol., 145, 461-473.

Wood, D. W., Setubal, J. C., Kaul, R., Monks, D. E., Kitajima, J. P., Okura, V. K., et al. (2001). Science, 294, 2317-2323.

Yoshioka, H., Numata, N., Nakajima, K., Katou, S., Kawakita, K., Rowland, O., et al. (2003). Nicotiana benthamiana gp91phox homolgs NbrbohA and NbrobohB participate in H_2O_2 accumulation and resistance to Phytophthora infestans. Plant Cell, 15, 706-718.

Young, C., & Nester, E. W. (1988). Association of the virD2 protein with the 5' end of T strands in *Agrobacterium* tumefaciens. J. Bacteriol., 170, 3367-3374.

Zambryski, P. C., Depicker, A., Kruger, K., & Goodman, H. M. (1982). Tumor induction by *Agrobacterium* tumefaciens: analysis of the boundaries of T-DNA. Mol. Appl. Genet., 1, 361-370.

Zhou, X.-R., & Christie, P. J. (1999). Mutagenesis of the *Agrobacterium* VirE2 single-stranded DNA-binding protein identifies regions required for self association and interaction with VirE1 and a permissive site for hybrid protein construction. J. Bacteriol., 181, 4342-4352.

Zhu, J., Oger, P., Schrammeijer, B., Hooyakaas, P. J. J., Farrand, S. K., & Winans, S. C. (2000). The bases of crown gall tumorigenesis. J. Bacteriol., 182, 3885-3895.

Zhu, Y., Nam, J., Carpita, N. C., Matthysse, A. G., & Gelvin, S. B. (2003a). *Agrobacterium*-mediated root transformation is inhibited by mutation of an Arabidopsis cellulose synthase-like gene. Plant Physiol, 133, 1000-1010.

Zhu, Y., Nam, J., Mysore, K. S., Lee, L.-Y., Cao, H., Valentine, L., et al. (2003b). Identification of Arabidopsis rat mutants. Plant Physiol, 132, 494-505.

Ziemienowicz, A., Gorlich, D., Lanka, E., Hohn, B., & Rossi, L. (1999). Import of DNA into mammalian nuclei by proteins originating from a plant pathogenic bacterium. Proc. Natl. Acad. Sci., USA, 96, 3729-3733.

Ziemienowicz, A., Merkle, F., Schoumacher, B., Hohn, B., & Rossi, L. (2001). Import of *Agrobacterium* T-DNA into plant nuclei: two distinct functions of VirD2 and VirE2 proteins. Plant Cell, 13, 369-384.

Zorreguieta, A., Geremia, R. A., Cavaignac, S., Cangelosi, G. A., Nester, E. W., & Ugalde, R. A. (1988). Identification of the product of an*Agrobacterium* tumefaciens chromosomal virulence gene. Mol. Plant-Microbe Interact., 1, 121-127.

Zupan, J., Citovsky, V., & Zambryski, P. C. (1996). *Agrobacterium* VirE2 protein mediates nuclear uptake of ssDNA in plant cells. Proc. Natl. Acad. Sci., USA, 93, 2392-2397.

Zupan, J., Muth, T. R., Draper, O., & Zambryski, P. C. (2000). The transfer of DNA from *Agrobacterium* tumefaciens into plants: a feast of fundamental insights. Plant J., 23, 11-28.

Zupan, J., Ward, D., & Zambryski, P. C. (1998). Assembley of the VirB transport complex for DNA transfer from *Agrobacterium* tumefaciens to plant cells. Curr. Opin. Microbiol., 1, 649-655.

Zupan, J., & Zambryski, P. C. (1995). Transfer of T-DNA from *Agrobacterium* to the plant cell. Plant Physiol., 107, 1041-1047.

Zupan, J., & Zambryski, P. C. (1997). The *Agrobacterium* DNA transfer complex. Crit. Rev. Plant Sci., 16, 279-2995.

10. AFFILIATIONS

Ajith Anand and **Kirankumar S. Mysore**, *Plant Biology Division, The Samuel Roberts Noble Foundation, 2510 Sam Noble Parkway. Ardmore, OK 73401, USA. Author for correspondence: ksmysore@noble.org*

R. EICHENLAUB, K.-H. GARTEMANN, AND A. BURGER

CLAVIBACTER MICHIGANENSIS, A GROUP OF GRAM-POSITIVE PHYTOPATHOGENIC BACTERIA

Abstract. The actinomycete genus *Clavibacter* contains several important pathogens infecting plants of agricultural importance. These Gram-positive bacteria display differences in their mechanism of pathogenicity as compared to the proteobacteria. They are vascular pathogens causing systemic infections after entering the hosts through wounds. The occurence of symptomless latent infections and the invasion of the seeds by the bacteria are widespread. On the molecular level distinct pathogenic mechanisms seem to exist, as no resistant host plants are known and a type III secretion system or avirulence genes have not been identified so far. In this chapter an outline of the biology and the current knowledge on the molecular basis of disease induction by *Clavibacter* is presented

1. INTRODUCTION

Although fungi and viruses may be considered to be the most important plant pathogens, infections by bacterial phytopathogens also are causing significant losses in agriculture. Phytopathogenicity is not a widespread trait amongst bacteria and we find plant pathogenic bacteria only in a few genera. During the last 20 years research has generated extensive information on the molecular mechanisms by which bacteria interact with their host plant, but this progress in the understanding of bacterial plant diseases has been mainly achieved for the Gram-negative phytopathogens. This is due to the fact that most phytopathogenic bacteria are indeed Gram-negatives and often belong to genera for which the methods and tools for a molecular analysis were readily available. Thus the few Gram-positive phytopathogens which are generally more difficult to handle have been neglected although some are quite important by causing great losses in the cultivation of crop plants. One such example are the members of the genus *Clavibacter* which are the subject of this chapter (for recent reviews see also: Metzler et al., 1997; Jahr et al., 1999; Burger and Eichenlaub, 2003; Gartemann et al., 2003).

The bacterial genus *Clavibacter* belongs to the Gram-positive actinomycetes (family *Microbacteriaceae*) (Park et al., 1993). The genus *Clavibacter* consists of only one species *Clavibacter michiganensis* (*Cm*) which is phytopathogenic and is subdivided into five subspecies according to their host specificity. *Clavibacter michiganensis* subsp. *michiganensis* (*Cmm*) causes bacterial wilt and canker of tomato (*Solanum lycopersicum* [*Lycopersicon esculentum*]) (Strider, 1969; Davis et al., 1984) which can be considered to be the most important bacterial disease of tomato. *Clavibacter michiganensis* subsp. *sepedonicus* (*Cms*) is responsible for ring rot of potato (*Solanum tuberosum*; Manzer and Genereux, 1981). *Clavibacter michiganensis* subsp. *insidiosus* (*Cmi*) causes wilting and stunting in alfalfa

S.S. Gnanamanickam (ed.), Plant-Associated Bacteria, 385–421.

(*Medicago sativa*; McCulloch, 1925). The importance of *Cmm*, *Cms*, and *Cmi* as plant pathogens is underlined by the fact that these pathogens are quarantine organisms under the European Union Plant Health Legislation and this is also the case in many other countries. *Clavibacter michiganensis* subsp. *nebraskensis* (*Cmn*) induces wilt and blight of maize (*Zea mays*; Schuster, 1975), and *Clavibacter michiganensis* subsp. *tessellarius* (*Cmt*) causes leaf freckles and leaf spots in wheat (*Triticum aestivum*; Carlson et al., 1982).

Data on the molecular mechanisms of disease induction by *Cm* are mainly available for *Cmm*. For *Cmm* cloning vectors (Laine et al., 1996; Meletzus and Eichenlaub, 1991), an efficient transformation system (Meletzus and Eichenlaub, 1991; Meletzus et al., 2000; Kirchner et al., 2001), and transposons (Gartemann and Eichenlaub, 2001; Kirchner et al., 2001) have been developed and can also be used in the other subspecies (Laine et al., 1996). This led in *Cmm* to the identification of plasmid-encoded virulence factors, *pat*-1 and *celA* (Dreier et al., 1997; Jahr et al., 2000). Furthermore, it could be shown that CelA was also a virulence factor in *Cms* (Laine et al., 1996; Laine et al., 2000).

In the following chapter we try to present an overview on various aspects of the plant pathogenic members of the genus *Clavibacter*, with the hope, that in the future more work will be initiated on molecular aspects of virulence to eventually understand the mechanisms by which these agriculturally important pathogens damage their host plants. The main focus will be on the subspecies *Cmm* and *Cms*, where a lot of research has been done, while for the other subspecies information is relatively scarce.

2. THE GENUS *CLAVIBACTER* - TAXONOMY AND PHYLOGENY

A number of important plant pathogens belongs to the Gram-positive bacteria and display a coryneform morphology. Originally all these Gram-positive bacteria were grouped in the genus *Corynebacterium* mainly on the basis of morphological features (non-sporeforming, more or less irregular Gram-positive rods). With the development of chemotaxonomic and later molecular methods for the differentiation of bacteria it became clear that no close relationship of the plant-pathogenic species to *Corynebacterium* exists and these bacteria were reclassified in several new or different genera (Tab. 1). The current taxonomy is based mainly on menaquinone and cell wall composition and the use of 16SrDNA sequence information.

With the exception of *Corynebacterium fascians* that was reclassified as *Rhodococcus fascians* (Goodfellow, 1984) the plant-pathogenic coryneforms do not contain mycolic acids. *Corynebacterium ilicis*, a pathogen of *Ilex*, has a cell wall type A3*a* containing lysine and was placed in the genus *Arthrobacter* (Collins et al., 1981). However, it remains unclear whether the type strain of *Arthrobacter ilicis* is causing the disease or a *Curtobacterium* strain which was found to be present as contaminant in the culture is the real pathogen (Young et al., 2004).

All the other plant pathogenic coryneforms possess the rare peptidoglycan type B2 (Fig. 1). A new family, the *Microbacteriaceae*, was established to accommodate actinomycetes with this cell wall type (Park et al., 1993). The crosslinkage in this kind of peptidoglycan is between the α-carboxylgroup of a L-glutamate residue in position 2 of one peptide side chain via an interpeptide diamino acid residue to the C-terminal D-alanine in position 4 of the adjacent peptide chain (Schleifer and Kandler, 1972). The *Microbacteriaceae* currently include among others the genera *Agromyces*, *Agrococcus*, *Microbacterium* and the four genera *Clavibacter*, *Curtobacterium*, *Leifsonia*, and *Rathayibacter* containing plant pathogens. Several plant pathogens were classified as pathovars in the species *Curtobacterium flaccumfaciens*. *Curtobacterium* can be differentiated from the three remaining genera by the presence of the diamino acid ornithine in the cell wall (cell wall type B2β), while members of the other genera contain diaminobutyrate (Collins and Jones, 1983).

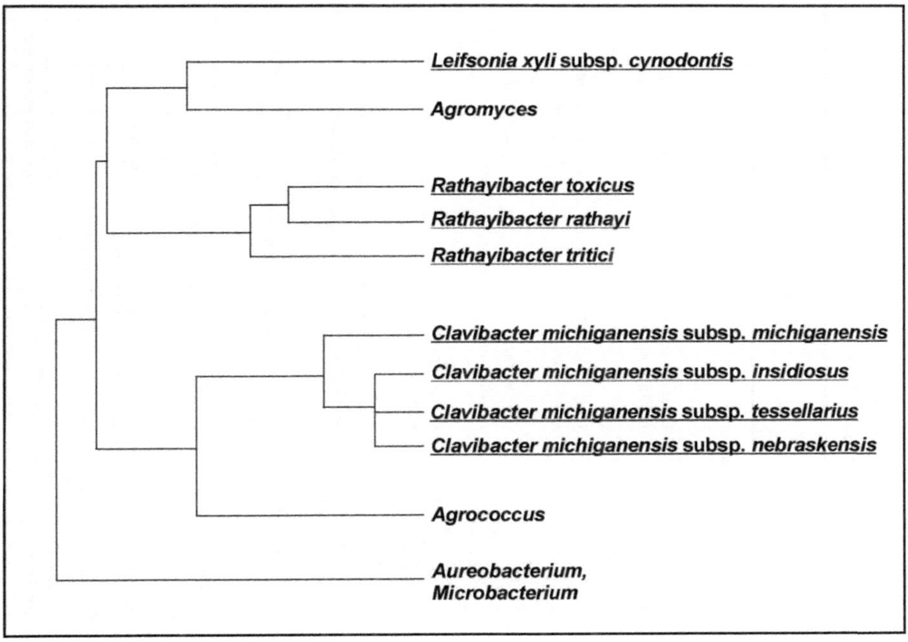

Fig. 1. Phylogenetic tree derived from 16S rDNA sequences of some members of the Microbacteriaceae (modified after Sasaki et al., 1986). The plant pathogens are underlined.

Table 1. Plant-pathogenic coryneform actinomycetes. Not all former names are given. For a list of recent changes in names and synonyms see DSMZ – Bacterial Nomenclature up-to-date, http://www.dsmz.de/bactnom/bactname.htm. DAB – diaminobutyrate, MK – menaquinone

Species/Genus	Former names	Host	Cell wall type	Major quinone	References
Microbacteriaceae					
Clavibacter michiganensis subsp. *michiganensis*	*Corynebacterium michiganense*	tomato (*Solanum lycopersicum*)	B2γ, D,L-DAB	MK-9	Smith, 1910 Davis et al., 1984
Clavibacter michiganensis subsp. *sepedonicus*	*Corynebacterium sepedonicum*	potato (*Solanum tuberosum*)	B2γ, D,L-DAB	MK-9	Spieckermann & Kotthoff, 1914; Davis et al., 1984
Clavibacter michiganensis subsp. *insidiosus*	*Corynebacterium insidiosum*	alfalfa (*Medicago sativa*)	B2γ, D,L-DAB	MK-9	McCulloch, 1925; Davis et al., 1984
Clavibacter michiganensis subsp. *nebraskensis*	*Corynebacterium nebraskense*	corn (*Zea mays*)	B2γ, D,L-DAB	MK-9	Vidaver & Mandel, 1974; Davis et al., 1984
Clavibacter michiganensis subsp. *tessellarius*	*Corynebacterium michiganense tessellarius*	wheat (*Triticum aestivum*)	B2γ, D,L-DAB	MK-9	Carlson & Vidaver, 1982a; Davis et al., 1984
Rathayibacter iranicus	*Corynebacterium iranicum*	wheat (*Triticum aestivum*)	B2γ, L-DAB	MK-10	Carlson & Vidaver, 1982; Davis et al., 1984; Zgurskaya et al., 1993
Rathayibacter rathayi	*Corynebacterium rathayi*	cocksfoot grass (*Dactylis glomerata*)	B2γ, L-DAB	MK-10	Smith, 1913; Davis et al., 1984; Zgurskaya et al., 1993
Rathayibacter tritici	*Corynebacterium tritici*	wheat (*Triticum aestivum*)	B2γ, L-DAB	MK-10	Carlson & Vidaver, 1982; Davis et al., 1984; Zgurskaya et al., 1993
Rathayibacter toxicus	*Clavibacter toxicus*	annual ryegrass (*Lolium rigidum*)	B2γ, L-DAB	MK-10	Riley & Ophel, 1992; Sasaki et al., 1998

Species/Genus	Former names	Host	Cell wall type	Major quinine	References
Microbacteriaceae					
Leifsonia xyli subsp. *xyli*	*Clavibacter xyli* subsp. *xyli*	sugar cane (*Saccharum officinarum*)	B2γ, D,L-DAB	MK-11	Davis et al., 1984; Suzuki et al., 1999; Evtushenko et al., 2000
Leifsonia xyli subsp. *cynodontis*	*Clavibacter xyli* subsp. *cynodontis*	bermudagrass (*Cynodon dactylis*)	B2γ, D,L-DAB	MK-11	Davis et al., 1984; Evtushenko et al., 2000
Curtobacterium flaccumfaciens	*Corynebacterium flaccumfaciens*	several pathovars with different hosts	B2β, ornithine	MK-9	Collins and Jones, 1983
Micrococcaceae					
Arthrobacter ilicis	*Corynebacterium ilicis*	*Ilex opaca*	A3α lysine	MK-9 (H_2)	Collins et al., 1981
Nocardiaceae					
Rhodococcus fascians	*Corynebacterium fascians*	several mono- and dicotyledonous plants	A1γ, diamino-pimelate	MK-9 (H_2) MK-8 (H_2) mycolic acids	Goodfellow and Anderson, 1977

The genus *Clavibacter* was originally proposed by Davis et al. (1984) for all plant pathogenic coryneform bacteria with a cell wall containing diaminobutyrate. *Clavibacter* peptidoglycan is of the B2γ type and contains a diaminobutyrate residue in the interpeptide bridge. Unsaturated menaquinones of 9 to 12 isoprene units are found.

The grass-pathogenic species *C. iranicum, C. rathayi*, *C. toxicus*, and *C. tritici*, were reclassified in the new genus *Rathayibacter* as they possess mainly unsaturated menaquinones with 10 isoprene units, contain L-diaminobutyrate, and no close rela tionship was found in DNA-DNA hybridizations to the remaining *Clavibacter* species (Zgurskaya et al., 1993).

Later, the former *Clavibacter* species *C. xyli* subsp. *cynodontis* and *C. xyli* subsp. *xyli* were placed in the newly established genus *Leifsonia* (Evtushenko et al., 2000; Suzuki et al., 1999). *Leifsonia* species are motile, contain MK-11 and only low amounts of MK-10, D,L-diaminobutyrate in the cell wall and possess putrescine as the predominant polyamine.

So at the moment, only the species *michiganensis* remains in the genus *Clavibacter*. *C. michiganensis* contains five subspecies which infect different host plants. *Clavibacter* is now characterized by non-motility, the presence of D- and L-diaminobutyrate in the cell wall, the major menaquinone MK-9, and the polyamines spermine and spermidine (Altenburger et al., 1997). Glycerol teichoic acids were isolated from both *Cmm* and *Cmi* (Varbanets et al., 1990).

Clavibacter normally forms pleomorphic rods, but depending on the medium coccoid or branched cells may occur. While most of the cells occur singly or in pairs, formation of cell aggregates can also be seen. The characteristic 'snapping division' of coryneform bacteria leading to V-formed arrangements is common. Spores are not formed. The G+C content of the DNA is about 73%.

Two important taxonomic problems remain. Current taxonomy gives the different *Clavibacter* strains subspecies status (Carlson and Vidaver, 1982a), but there was always discussion whether they warrant species status as indicated by DNA/DNA hybridization experiments (Collins and Bradbury, 1986; Döpfer et al., 1982).

Another problem arises from the fact that the subspecies classification is relying heavily on plant-pathological criteria (infection of the host plant is the easiest way to assign *Clavibacter* strains to the subspecies) (Collins and Bradbury, 1992). Identification based on phenotypic characteristics as pigmentation, growth on different media, or use of different carbon sources are often variable depending on the strain or the medium used. This prevents the classification of strains which are either non-pathogenic or for which the host is unknown. Such *Clavibacter* strains which can not be included in the subspecies are known for example from the phyllosphere of peanut (Jacobs and Sundin, 2001) or grasses (Behrendt et al., 2002), from rice seed (Cottyn et al., 2001), and as endophytes not causing any symptoms from several mono- and dicotyledonous prairie plants (Zinniel et al., 2002).

3. LIFESTYLE OF *CLAVIBACTER*

3.1. Physiology

Clavibacter is a mesophile with a temperature optimum of 25-28°C. Generation time is quite high, and it takes three to four days for visible colonies to form on agar plates. Maximum growth temperature is about 35°C. The thermal death point is around 50°C. The pH optimum lies between pH 7 and 8, but growth occurs at pH values as low as pH 5 (Strider, 1969) which is important when the bacteria grow in the xylem vessels.

Clavibacter strains are oxidase-negative obligate aerobes. Neither nitrate nor nitrite reduction occur (Davis et al., 1984). Only menaquinones are found, and unsaturated menaquinones with nine isoprene units (MK-9) predominate. Fatty acids are mainly straight chain saturated, *anteiso-* and *iso*-methyl branched acids. Polar lipids consist of diphosphatidylglycerol, phosphatidylglycerol and some glycosyldiacylglycerols. Cell walls contain rhamnose and fucose. They are non-motile (Collins and Bradbury, 1986).

A wide range of carbohydrates and carbonic acids are used (Collins and Jenkins, 1986; Strider, 1969). Cellulose and some other polysaccharides (e.g. starch) support growth of *Cmm* (Strider, 1969). Organic nitrogen sources, especially urea and peptones are favored by *Cmm* (Strider, 1969). Moderate proteolytic capabilities were reported for *Cmm*. For additional characteristics of *Clavibacter* see Collins and Bradbury (1986, 1992) and Strider (1969). Some important differential characteristics of the subspecies are shown in Table 2.

Table 2. Selected differentiating characteristics of the *C. michiganensis* subspecies

Subspecies	Pigmentation	Gelatin liquefaction	Utilization of acetate	Reducing sugar production	Levan production
Cmm	yellow	+	+	variable	-
Cms	white	-	+	+	-
Cmi	blue-gray (yellow)	-	-	-	-
Cmn	orange	-	+	+	+
Cmt	orange		-		

Plant cell wall degrading exoenzymes are present in the subspecies as far as investigated. *Cmi* produces endocellulase, polygalacturonase, and pectinlyase, but not pectinmethylesterase (Kratka, 1987). For *Cms* has a cellulase and an amylase (Baer and Gudmestad, 1995; Metzler et al., 1997), and for *Cmm* cellulase, polygalacturonase, xylanase, and pectinmethylesterase activities were shown (Beimen et al., 1992; Strider, 1969).

Many strains carry either auxotrophies or are stimulated in growth by the addition of specific compounds to the media (Strider, 1969). Strains of *Cmm*, *Cms*, and *Cmi* require some B vitamins (biotin, nicotinic acid, and thiamine) (Vidaver, 1982). Methionine auxotrophy is also widespread (Collins and Bradbury, 1986; Vidaver, 1982).

3.2. Habitat and ecology

All the *Clavibacter* subspecies are plant-pathogens infecting the vascular system of the host plants and causing characteristic diseases (for a detailed description of the diseases see chapter 4).

3.2.1. Host plants

The subspecies are highly specific for their host plants and a few closely related plant species. A number of other species/genera from the same genus or family can be infected artificially. In some species this leads to the development of disease symptoms, but infections without visible disease symptoms are also common and occur even in the main host plants.

The natural host plant of *Cmm*, *Solanum lycopersicum* (*Lycopersicon esculentum*) is a member of the large *Solanaceae* family of dicotyledonous plants which also includes the agricultural important species, potato (*Solanum tuberosum*), egg plant (*Solanum melongena*), and pepper (*Capsicum frutescens*). The exact taxonomic position of tomato is still debated. The genus *Lycopersicon* is so closely related to *Solanum*, that it is now placed as a section inside of the genus *Solanum* (Spooner et al., 2005). The tomato *S. lycopersicum* is known only as cultivated species. Its origin is unknown, but was probably located in South America. Closely related wild tomato species (e.g. *S. peruvianum*, *S. pimpinellifolium*, *S. habrochaites* [*hirsutum*]) all occur in the mountainous regions of South America (Spooner et al., 2005). These wild tomatoes may constitute natural reservoirs for *Cmm*. Other hosts (at least 13 genera and 35 species of the *Solanaceae*) known to be infected by *Cmm* are described by Strider (1969) and Thyr et al. (1975). However, for most of these species only artificial infection was shown. Disease symptoms are generally less severe than in *S. lycopersicum*. In some hosts (e.g. the potato *S. tuberosum* and *Datura stramonium*) the bacteria live as endophytes without causing symptoms (Thyr et al., 1975).

Cms has potato as natural host (*S. tuberosum*), but many other *Solanum* species including tomato and eggplant (*S. melongena*) can be infected artificially and show disease symptoms (Collins and Bradbury, 1986).

Besides of *Medicago sativa Cmi* was reported to cause disease naturally in other leguminous plants as *Trifolium* spp., other *Medicago* spp., *Onobrychis sativa*, and *Lotus corniculatus* (Collins and Bradbury, 1986). For *Cmn* and *Cmt* no other hosts were described.

3.2.2. Infection mode

The *Clavibacter* subspecies are probably mainly wound pathogens (Strider, 1969). Infection of *Cmm* through stomata (reviewed in Strider, 1969) and through hydathodes was described early, but always discussed. Carlton et al. (1998) finally proved that *Cmm* is able to infect tomato leaves from guttation droplets through hydathodes. Another important mode of transmission is through contaminated seeds (Strider, 1969). Though there are a lot of conflicting results in the literature, *Clavibacter* seems to be a very effective pathogen. For *Cmm*, artificial infection methods like cutting the first true petiole which bring the bacteria directly into the xylem vessels lead to successful infections of about 60% of the plants using as few as about 10 bacteria. The symptoms are only slight with these low inoculum sizes (Thyr, 1968; Chang et al., 1992a). Using titers of 8 x 10^4 bacteria all plants were infected and show severe symptoms. The same correlation between the severity of symptoms and inoculum size was found for *Cms* (Westra and Slack, 1994). Infection through leaves required at least 10^6 bacteria (Strider, 1969). For contaminated seeds low infection rates were reported but secondary infections apparently can take place (Strider, 1969). These data suggest that plant defenses are either not induced or effectively circumvented by the pathogen.

Knowledge about the structure and diversity of natural *Clavibacter* populations is lacking. Thus, the relative importance of the different modes of infection remains unclear. The relative importance of disease outbreaks caused by contaminated seeds distributed by seed companies compared to outbreaks caused by strains originating from the site e.g., in the soil or from latent infections of other host plants is unknown. It is also unknown if mayor outbreaks are caused by only one strain spreading in the site by secondary infections or if several different strains participate.

Infection through animal vectors was never described for the *Clavibacter* subspecies (contrasting with the infection mode of *Rathayibacter* which is associated with different nematode species like *Anguina*) although the severity of the disease can be influenced by the presence of nematodes (Hunt et al., 1971).

In conclusion, all *Clavibacter* subspecies are wound pathogens, but other infection routes are known, and transmission through contaminated seeds is very important in industrialized agriculture.

3.2.3. Life inside the plant

All *Clavibacter* subspecies are vascular plant pathogens causing systemic infections. The life cycle of *Cmm* is shown as an example in Figure 2. The main habitat is the nutrient-poor, slightly acidic xylem fluid. After infection *Clavibacter* spreads in the xylem vessels. *Clavibacter* is biotrophic and mainly living and multiplying inside of the host plant. Only at later stages of the infection, both *Cmm* and *Cms* leave the xylem vessels and enter the surrounding tissues after degradation of plant cell walls leading to tissue macerations (Marte, 1980; Benhamou, 1991; Baer and Gudmestad, 1995). This seems to be caused by the production of cell wall degrading enzymes (e.g., cellulase, polygalacturonase, xylanase) (Beimen et al., 1992). *Clavibacter* is able to produce some plant hormones. For *Cmi* the production of both IAA and cytokinins was shown (Kratka, 1987). IAA synthesis was independent of the presence of tryptophan in the medium. For *Cmm* the production of IAA was also shown (Bermpohl and Eichenlaub, unpubl. data). The pathway used for IAA production is unknown. However, the related plant pathogen *Rhodococcus fascians* used the indolepyruvate pathway (Vandeputte et al., 2005). Whether the phytohormones play a role in symptom development as it is the case for *Rhodococcus fascians* was not investigated, but the occurrence of stunting symptoms in *Medicago* and as atypical symptom in potato (Laurila et al., 2003) and corn suggests a participation of them.

A general feature of *Clavibacter* caused diseases is the occurrence of latently infected symptomless plants. These plants can be a source of reinfections or the spreading of the diseases especially if transplants are used (Nelson, 1982; van Steekelenburg, 1985; Gitaitis et al., 1991; Frank, 1999).

Also the infection of seeds seems to be a general feature of *Clavibacter*. It has been reported for *Cmi* (Samac et al., 1998), *Cmm* (Strider, 1969; Chang et al., 1992b), *Cmn* (Biddle et al., 1990), and *Cmt* (McBeath and Adelman, 1986). *Clavibacter* persists inside of the seed between chalaza and endosperm and close to the embryo (Biddle et al., 1990). Transmission of the disease through contaminated seeds was demonstrated though with a low efficiency if not high bacterial titers were present (Biddle et al., 1990; Chang et al., 1992b; Strider, 1969). Nevertheless, a high a rate of infection can be obtained due to secondary infections (Chang et al., 1992b).

3.2.4. Extracellular polysaccharides (EPS)

Many *Clavibacter* strains form slimy to fluid colonies and produce exopolysaccharides (EPS). EPS are known to have number of diverse functions some of which are related to pathogenicity (Denny, 1995). All subspecies investigated produce several different types of EPS. The amounts vary depending on the strain, the medium and the growth conditions. EPS is produced also *in planta* (van Alfen et al., 1987b; van den Bulk et al., 1991; Bermpohl et al., 1996). EPS production is an unstable trait which is frequently lost giving rise to rough nonfluidal colonies (Fulkerson, 1960; Paschke and van Alfen, 1993).

A direct role of EPS in virulence has been proposed which the EPS acting as a toxin as the work of Strobel and colleagues suggested a direct participation of EPS in virulence. EPS was isolated as high molecular weight glycopeptides from *Cmi*, *Cmm*, and *Cms* which were proposed to act as a toxin (Ries and Strobel, 1972; Rai and Strobel, 1969; Strobel, 1970). Phytotoxic effects like wilt induction in epicotyl cuttings caused by the glycopeptides were reported (Rai and Strobel, 1969; Straley et al., 1974). However, isolation and characterization of the EPS by other groups led to a revised structure as they found no glycopeptides, only some non-covalently bound protein components of the EPS (Bermpohl et al., 1996; Gorin and Spencer, 1980; Henningson and Gudmestad, 1992; van Alfen et al., 1987a; van den Bulk et al., 1991;). For *Cms* there is a discrepancy in sugar composition between the two studies where EPS was isolated (Henningson and Gudmestad, 1992; Westra and Slack, 1992) as Henningson & Gudmestad found no mannose in the EPS. However, both found that no glycoproteins are formed. The composition of characterized EPS fractions is shown in Table 3.

As mutants containing lower amounts of EPS and/or changed sugar compositions were not affected in symptom development or colonization (Bermpohl et al., 1996; Pasche and van Alfen, 1993; Westra and Slack, 1992), a direct participation of EPS in virulence was ruled out. Also, a natural avirulent isolate and a plasmid-free curing derivative of *Cmm* were still able to colonize tomato as the *Cmm* wild-type but displayed no symptoms although the EPS amount and composition was unchanged (Bermpohl et al., 1996).

So the model of van Alfen et al. (1987b) that EPS interferes with water transport due to physical plugging of the xylem vessels leading to a biomass reduction in the otherwise healthy plants, but is not directly involved in virulence is generally accepted now.

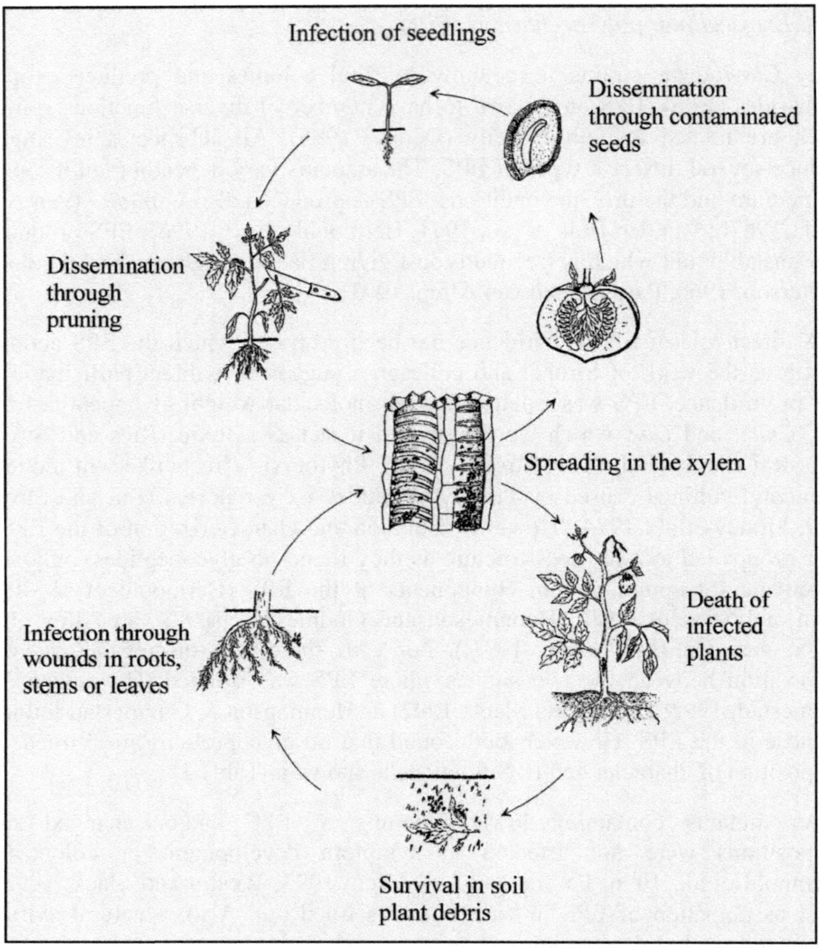

Fig. 2. Infection cycle of *Cmm* in tomato

Analysis of the *Cmm* genome sequence showed the presence of at least four gene clusters which may be involved in the production of different EPS (Gartemann and Eichenlaub, unpubl. data). One of these clusters includes all the genes required to produce the already characterized EPS of *Cmm*. The generation of mutants in this cluster converted the mucoid colony phenotype into a rough one (Schauer and Eichenlaub, unpubl. data). The role of the three other clusters remains unknown, but it is tempting to speculate that they are required for special conditions (e.g. host recognition, evasion of plant defenses).

Table 3. Characterized EPS compounds of the *C. michiganensis* subspecies. Abbreviations: gal - galactose, glu – glucose, fuc –fucose. *Cms*[a] and *Cms*[b] are from Henningson & Gudmestad (1992) and Westra and Slack (1992) respectively

	EPSI	EPSII	EPSIII	EPSIV
Cmi	>10 Mda; acidic; aggregate of EPSII + nine proteins	6-7 MDa; acidic; 2 fuc: 1 gal: 1 glu, pyruvate	21.8 kDa; 3 gal: 1 fuc: 1 rhamnose, glucose, mannose	-
Cmm	-	~5 MDA; acidic; 2 fuc: 1 gal: 1 glu, pyruvate, acetate	-	-
Cms[a]	-	~1 MDa; 2 gal: 1 glu: 1 fuc	~5 kDA; 5 gal: 1 glu: 2 mannose: 2 rhamnose	-
Cms[b]	>20 MDa; acidic; probable aggregate of EPSIII	~4.5 MDa; acidic; probable aggregate of EPSIII	~2.1 kDa; acidic; 7 mannose: 2 gal: 2 gluc: 1 fuc	~2 kDa; neutral; mannan

Other functions like recognition of host plants or evasion/suppression of defense reactions are possible but were not investigated as no mutants are available which do not produce any EPS. Also the function of the single EPS fractions is unclear up to now. For *Cms* the production of biofilms *in planta* in the xylem vessels was shown (Marques et al., 2003). EPS may participate in the establishment of those biofilms.

3.2.5. Strain variability

At least for some traits such as pigmentation and EPS production a high variability and/or the frequent occurrence of variants/mutants is known (Fulkerson, 1960;. Also for *Cmm*, *Cmn*, and *Cmi* a gradual loss in virulence was reported after prolonged cultivation *in vitro* (Carroll and Lukezic, 1971; Schuster et al., 1975; Smidt and Vidaver, 1987; Strider, 1969; Vidaver, 1977). Whether this is a genetic effect or a phenotypic one depending on different artificial media and growth conditions is unclear. Passages of the pathogen through the host plant have been reported to restore the original phenotypes in some cases indicating a phenotypic effect. A genetic basis for the high variability is unknown, but may be due to the presence of active insertion elements (IS*1121*, IS*1122*) occurring in high copy number in *Cmi* and *Cms*.

3.2.6. Clavibacter in soil and as epiphyte

Clavibacter seems to be able to survive in soil for only relatively short periods of time (Carroll and Lukezic, 1971). Longer survival times in the range of 1-3 years were described when the bacteria were associated with host plant material (Chang et al., 1992b; Fatmi and Schaad, 2002; Gleason et al., 1991; Smidt and Vidaver,

1986). The plant debris can form a source of reinfection in the following year. Survival rates in water seem to be even less than in soil. For *Cms* it is possible only for a maximum of 21 days depending on the environmental conditions (van der Wolf and van Beckhoven, 2004). Additionally, the storage in water gave rise to a relatively high proportion of variants and a loss of virulence (Vidaver, 1977).

Other natural reservoirs of *Clavibacter* exist but are poorly investigated. Leaf surface populations of *Clavibacter* were found on several plant species (Behrendt et al., 2002; Carlton et al., 1994; Jacobs and Sundin, 2001; Smidt and Vidaver, 1986). Thus, an epiphytic lifestyle of *Clavibacter* is suggested. The presence of pigments in *Clavibacter* is probably related to this epiphytic lifestyle on leaves where it confers an advantage as UV resistance of the pigmented bacteria is higher (Jacobs et al., 2005). Apart of *Cms*, which form white or pale yellow colonies, the *Clavibacter* subspecies are characterized by colored colonies due to the presence of carotenoids (Collins and Bradbury, 1986). Most *Cmm* strains are yellow containing lycopene and cryptoxanthine. Mutants are known which are red (containing only lycopene), pink (spirilloxanthin and lycopene), or orange (β-carotene, cryptoxanthine, and canthaxanthin) (Saperstein and Starr, 1954; Saperstein et al., 1954). No correlation between pigmentation and virulence was found (Saperstein et al., 1954). The carotenoid composition of the other subspecies is unknown. Most *Cmn* and *Cmt* strains are orange. *Cmi* also forms yellow colonies, but additionally a typical blue-gray pigment (the bipyridyl indigoine) is produced on some media (Starr, 1958).

Additional reservoirs of *Clavibacter* are due to the persistence of *Clavibacter* in infected but symptom free plants (Gitaitis et al., 1991; Franc, 1999), alternative hosts (Chang et al., 1992b) or even completely unrelated plant species growing near the natural host (van der Wolf et al., 2005) which can lead to infections of neighboring main host plants.

In conclusion, *Clavibacter* can be considered as mainly biotrophic and only weakly necrotrophic, but not as soil inhabiting bacterium.

3.3. Bacteriocins

Inhibition of closely related strains by other *Clavibacter* subspecies was reported early (Echandi, 1976; Nelson and Semeniuk, 1964). Gross and Vidaver (1979a) showed, that 85% of the tested strains of the subspecies *Cmi*, *Cmm*, and *Cmn* produced one or more bacteriocins. They were not inducible by mitomycin C. The size of inhibition zones varied with strains, incubation temperature, and media used. In most cases all the other *Clavibacter* subspecies and additionally some *Rathayibacter* and *Curtobacterium* strains were affected by the bacteriocin (Gross and Vidaver, 1979a). A few were also active against *Micrococcus luteus* (Gross and Vidaver, 1979a). Only one bacteriocin from a *Cmm* strain was purified (Huang and Echandi, 1981). It was probably a protein, as it was shown to be trypsin- and pronase-sensitive. In the investigation of Gross and Vidaver (1979a) most of the bacteriocins were found to be heat resistant and protease-sensitive, but some compounds were resistant to protease treatment. The chemical structure of the compounds is still unknown.

The data from the genome projects indicate that *Cmm* might be able to produce a lantibiotic related to type B lantibiotics like mersacidin (Gartemann et al., 2003) which may be one of the bacteriocins. At the moment, experiments are underway to inactivate these genes to prove or reject this hypothesis.

3.4. Phages of Clavibacter

A few phages have been isolated for *Cmi*, *Cmm*, and *Cmn* (Cook and Katznelson, 1960; Echandi and Sun, 1973; Shirako and Vidaver, 1981; Shirako et al., 1986; Wakimoto et al., 1969). All are virulent and subspecies-specific. They have isometric heads and long non-contractile tails. CMP1 from *Cmm* was shown to be specific for virulent *Cmm* strains (Echandi and Sun, 1973). It is a DNA phage with a size of about 60 kb (Gartemann and Eichenlaub, unpubl. results). No lysogenic phage was reported up to now.

3.5. Plasmids of Clavibacter

The presence of ccc-plasmids in most strains of *Cmi*, *Cmm*, *Cmn*, and *Cms* was shown by Gross et al. (1979b). However, no phenotypes could be associated with any of these plasmids. At the moment, only pCM1 and pCM2 from *Cmm* NCPPB382 and pCS1 from *Cms* are better characterized. These plasmids are directly involved in the virulence of *Clavibacter*.

The *Cms* plasmid pCS1 is a 50.5-kb ccc-plasmid present in many *Cms* strains in its free form (Mogen and Oleson, 1987). pCS1 is able to integrate into the chromosome and in nearly all apparently plasmid-free strains the plasmid was found to be integrated into the chromosome (Mogen and Oleson, 1987, 1988). The integration is probably mediated by a 1.3-kb repeated sequence (Mogen and Oleson, 1987, 1990). This repeated sequence, the insertion element IS*1121* (Lee et al., 1997a), is present in two copies on pCS1 and with about 50 copies in the chromosome of *Cms*. The closely related IS*1122* occurs with a high copy number in *Cmi*, while in *Cmm* homologues to these IS elements are absent (Lee et al., 1997a). The copy number of pCS1 in strains containing the free form was estimated to be between one and two (Mogen and Oleson, 1988). A restriction map of pCS1 was established and *celA* encoding an endo-β-1,4-glucanase which is involved in virulence could be localized on the plasmid (Laine et al., 2000).

Each of the two plasmids from *Cmm* NCPPB382 carries a gene encoding proteins necessary for induction of bacterial wilt and canker of tomato (Meletzus et al., 1993). In case of the 27-kb plasmid pCM1 this is an endo-β-1,4-glucanase (Jahr et al., 2000) homologous to the one encoded on pCS1. On the 72-kb plasmid pCM2 the putative serine protease Pat-1 is encoded (Dreier et al., 1997). Curing of both plasmids led to an avirulent derivative of *Cmm* NCPPB382, which is still able to colonize tomato effectively and behaves as an endophyte (Meletzus et al., 1993). Both pCM1 and pCM2 were used for the construction of cloning vector families for *Clavibacter* (see 5.1.1.).

A number of *Cmm* field isolates originating from Israel was recently investigated for their plasmid status. Most of these strains carried one or two plasmids. In contrast to the nearly universally occurring pCS1 of *Cms*, most of these plasmids clearly differ from pCM1 and pCM2. However, the virulence determinants CelA and Pat-1 were shown to be plasmid-encoded in the field isolates, too (Eichenlaub, unpubl. data).

The plasmid backbones of pCM1 and pCS1 from *Cmm* and *Cms* are closely related. Analysis of the data from the genome projects on *Cmm* and *Cms* showed that the 27-kb pCM1 displays homology and a similar gene order over large parts of the plasmid backbone with the 50-kb pCS1, while the remaining part of the plasmid was found to be different (Gartemann and Eichenlaub, unpubl. results).

In addition to pCS1 *Cms* strains carry a linear plasmid, pCSL1, of about 90 kb (Brown et al., 2002a). pCSL1 has a copy number of about five but no phenotype conferred by the plasmid is known. In *Cmm* no linear plasmid could be shown in several strains investigated by PFGE analysis (Gartemann and Eichenlaub, unpubl. results).

4. *CLAVIBACTER* INDUCED PLANT DISEASES

All five subspecies of *Clavibacter michiganensis* induce specific diseases on their respective host plants as summarized in Table 4.

C. m. michiganensis, C. m. sepedonicus, and *C. m. insidiosus* are the target of international quarantine regulations as they cause severe crop failure all over the world. In this chapter the single diseases will be described in detail.

Table 4. *Clavibacter michiganensis* induced plant diseases

Subspecies	Main host plant	Disease
C. m. michiganensis	tomato (*Lycopersicon esculentum)*	wilt and canker
C. m. sepedonicus	potato (*Solanum tuberosum*)	wilt and ring rot of tuber
C. m. nebraskensis	maize (*Zea mays*)	wilt and blight
C. m. insidiosus	alfalfa (*Medicago sativa)*	wilt and stunting
C. m. tessellarius	wheat (*Triticum aestivum*)	leaf freckles and spots

4.1. C. m. michiganensis

The main host of *C. m. michiganensis* (Strider, 1969) is tomato (*Solanum lycopersicum* [*Lycopersicon esculentum*]), but there are also some reports of other *Solanum* subspecies and of the wild plants *Solanum douglasii*, *S. nigrum* and *S. triflorum* as natural hosts. Several different solanaceous plants can be infected artificially with *Cmm* (Thyr et al., 1975).

Infected tomato seeds grow up and build contaminated seedlings. A spread of the disease between single plants occurs via water (rainsplash, irrigation) or by cultural practices as trimming. The bacteria infect the host plants via roots or wounds through stomata and other natural openings and invade the xylem vessels, followed by a systemic infection of the host. Under natural conditions, it takes a long latent period until the first symptoms appear (Strider, 1969). In the field, the first symptom of *Cmm* infection is desiccation on the tips of the leaflets at lower parts of the young plant. Usually without showing the typical wilting symptoms, the plant slowly desiccates. At later stages, stunting, wilting, vascular discoloration, development of open stem cankers and fruit lesions follow. 'Bird' s-eye spots', which are relatively small and surrounded by a white halo, develop on the fruits. These fruits often fail to develop and fall or ripen unevenly.

4.2. C. m. sepedonicus

The only natural host on which *C. m. sepedonicus* (Manzer and Genereux, 1981) causes disease is potato. However, as a natural symptomless host also sugarbeet (Bugbee and Gudmestad, 1988) and in artificial inoculation tests many members of the *Solanaceae* have been described.

Natural infection of the tuber occurs via the stolon and leads to the typical ring rot symptoms. Bacteria divide and multiply very rapidly and spread over vascular strands into the stem and the petioles of the growing plant. About 8 weeks after planting a diseased potato, bacteria can reach the roots and maturing daughter tubes, which are sometimes used as new seed and lead to a distribution of the disease.

In the field, symptoms occur usually late in the growing season and can easily be misunderstood as potatoblight caused by *Phytophthora infestans*, as wilt caused by *Verticillium*, or as stem canker caused by *Thanatephorus cucumeris*. As is also the case for *Cmm*, the disease begins with first wilting symptoms in lower leaves. The margins roll inwards and upwards, and the color changes from green over yellow and brown to necrotic. The symptoms develop stronger under hot and dry weather conditions.

4.3. C. m. insidiosus

The main natural host plant of *C. m. insidiosus* (McCulloch, 1925) is lucerne (*Medicago sativa*) and some other leguminosae, but also different perennial *Medicago* species are susceptible when inoculated through wounded roots.

The bacteria invade the plant through small wounds and then multiply in the vascular tissues of the stem and pods. The bacteria can also be detected in intercellular spaces of seed parenchyma. The wilting disease is induced by an extracellular polysaccharide which is produced by *C. m. insidiosus.* The nutrition of the host strongly influences the development of the disease: the better the nutrition, the stronger develop the wilt symptoms. Mild symptoms appear as leaf mottling, upward curling of the leaf margins, and reduction of total plant height. Moderate infections additionally result in a proliferation of the stem, which is called witches broom effect. In severe infections, the plants grow up only to a few centimeters, the stems are thin, the leaflets small and thicker and often distorted and with marginal or entire bleaching. These symptoms usually lead to the death of the infected plant.

Severely infected plants are the basis for the spread of infected seed. The bacterium can remain infective for more than 3 years. As laboratory experiments showed, *C. m. insidiosus* can remain in dried tissue or seed for 10 years (Erwin, 1990).

4.4. C. m. nebraskensis

C. m. nebraskensis (Vidaver and Mandel, 1974) is a causative agent of leaf spot ('freckles'), leaf blight and wilt of maize or corn (*Zea mays*). This disease is also called Goss`s disease. Infection of the plant with *C. m. nebraskensis* needs wounds and may be direct in leaves or via the roots and stems. The wounding may result from sand, blasting hail, rainstorms, or strong wind.

The symptoms of *C. m. nebraskensis* induced maize disease can be easily confused with Stewart`s bacterial wilt. Characteristic are discrete lesions with water-soaked streaks which are parallel to the leaf veins. Drops of bacterial exsudate appear on the surface of the leaves when the streaks enlarge. After drying these droplets leave a crystalline sheen.

Wilt, whither, and death of the seedling are the result of an early infection. In contrast, infections at a later stage become visible as stunting, wilting, or various degrees of leaf blight. Leaf blight appears as gray to greenish-yellow stripes, which may be reddish on certain lines with wavy margins that follow the leaf veins and sometimes with spots along the leaf veins. Sometimes the entire leaf dries up.

If plants are infected systemically bundles are discolored and discharge an orange colored exsudate after cutting of the stalk.

Infection, wilt or death of the plant can occur at any stage of development but seedlings are more susceptible than older plants. The bacteria can survive in infected corn, but also in infected kernels or in irrigation water.

4.5. C. m. tessellarius

C. m. tessellarius was first described in Nebraska by Carlson and Vidaver (1982b) and is the causative agent of bacterial mosaic of numerous wheat cultivars. The disease is characterized by small yellow lesions with undefined margins which are distributed over the whole leaf. Single lesions can resemble the hypersensitive reaction to rust. The economic significance of *Cmt* is unknown (Carlson and Vidaver, 1982b), but as it can destroy flag leaves in severely infected plants, significant losses may be due to bacterial mosaic.

5. MOLECULAR ANALYSIS OF *CM*-HOST INTERACTION

Most of the work on the molecular level of *Cm*-host interactions has been done with *C. m. michiganensis* and with *C. m. sepedonicus*. Accordingly, these two systems will be described here in some detail. As the availability of genetic tools is a prerequisite for molecular and genetic analyses, the development of an effective host-vector system will be described in the first part of this chapter.

5.1. Genetic tools for Cmm

As known from preliminary experiments (Meletzus and Eichenlaub, 1991) *Cmm* wildtype strain NCPPB382 harbors two endogenous plasmids pCM1 and pCM2. These plasmids were cured out of the cells by incubation of the bacteria at sublethal temperature of 33°C resulting in a plasmid free strain (*CMM*100) and two strains with only one plasmid left, *CMM*101 (pCM1) and *CMM*102 (pCM2) (Meletzus and Eichenlaub, 1991). All strains are viable and can colonize the tomato plant efficiently. Interestingly, the strains *CMM*101 and *CMM*102 containing only one of the plasmids are reduced in virulence, whereas the plasmidfree strain *CMM*100 is non-virulent on the host plant tomato. Thus, it was concluded that important pathogenicity genes were located on the plasmids pCM1 and pCM2 (Meletzus et al., 1993). These findings were the basis for the establishment of an efficient genetic system to study the pathogenicity of *Cmm*.

5.1.1. Establishment of a host-vector-system

The plasmids of *Cmm* NCPPB382 were the origin for the construction of two vector families, the pDM-familiy (Meletzus and Eichenlaub, 1991) based on plasmid pCM1, and the pHN-family (Laine et al., 1996) based on plasmid pCM2. Plasmid borne fragments carrying the origin of replication of the native plasmids were fused with the *E. coli* vector pBR325, resulting in a set of *Cmm-E. coli* shuttle vectors which can replicate in both strains and facilitate the work, as all cloning experiments can be done in *E. coli*. For selection the different vectors were provided with antibiotic resistance cassettes with resistance genes against neomycin (Beck et al., 1992), gentamicin (Wohlleben et al., 1989), or chloramphenicol (Tauch et al., 1998) which as functional in *E. coli* as well as in Gram-positive bacteria.

As the plasmid free *Cmm* strain *CMM*100 was able to colonize the host-plant tomato without causing disease symptoms, this strain was suitable as host strain for cloning and analyzing wilt-inducing genes mapping on the two plasmids pCM1 and pCM2.

The *Cmm-E. coli* shuttle vectors were also shown to stably replicate in *C. michiganensis* subsp. *sepedonicus* (Laine et al., 1996). Different strains of *Cms* carry one single plasmid pCS1 (Clark and Lawrence, 1986; Mogen and Oleson, 1987) which is also involved in virulence. This plasmid can exist in an autonomously replicating form or integrated into the chromosome via IS*1121* copies (Mogen et al., 1987; Mogen et al., 1988; Mogen et al., 1990). For Cms, problems with the use of the pDM vector family may be expected, as the backbones of pCM1 and pCS1 are closely related and recombination or incompatibility may result.

5.1.2. Establishment of a transformation system

First experiments to establish foreign DNA in *Cmm* were performed with polyethylene glycol-mediated transfection of spheroplasts (Meletzus et al., 1991) using DNA of the *Cmm* specific bacteriophage CMP1 (Echandi and Sun, 1973). But the amount of transfectants was poor ($3x10^3$ transfectants per µg of DNA) and plasmid DNA could hardly be transformed using this method. As a first approach to transform plasmid DNA, an electroporation protocol was established giving yields of $2x10^3$ transformants per µg of host specific modified plasmid DNA for *Cmm* and $4,6x10^4$ transformants per µg of host specific modified plasmid DNA for *Cms* (Laine et al., 1996; Meletzus and Eichenlaub, 1991; Meletzus et al., 2000). Kirchner et al. (2001) optimized this electroporation protocol for *Cmm* by varying several parameters in a stepwise fashion: growth phase, amount of cells per transformation assay, amount of DNA, methylation status of DNA, electrical parameters, treatment of competent cells with cell wall-damaging reagents (glycine, lysozyme, penicillin G) during growth or directly prior to electroporation, heat shock after electroporation, different regeneration times, different composition of the regeneration medium. A dramatic increase of the electroporation efficiency was obtained by using non-nmethylated DNA isolated from a *E. coli* *dam$^-$/dcm$^-$*strain and a treatment of the cells with 2.5% glycine prior to electroporation. With this protocol $5x10^6$ transformants per µg plasmid DNA can be generally achieved for *Cmm* (Kirchner et al., 2001). Furthermore, it could be shown that the protocol is also suitable for the transformation of the other *Clavibacter michiganensis* subspecies *insidiosus*, *nebraskensis*, and *sepedonicus*, but further optimization is necessary to improve the transformation rate for these strains.

5.1.3. Mutagenesis of Clavibacter

The high transformation rate for *Cmm* allowed transposon mutagenesis experiments using transposon Tn*1409* which was constructed on the basis of IS*1409* from *Arthrobacter* sp. with resistance genes for chloramphenicol and gentamicin (Gartemann and Eichenlaub, 2001). Electroporation of *Cmm* with a suicide vector carrying transposon Tn*1409* resulted in approximately $1x10^3$ transposon mutants per µg DNA and thus was shown to be suitable for saturation mutagenesis. The analysis of the insertion sites of Tn*1409* suggested a random mode of transposition and under the used conditions single transposition events. A library of randomly created transposon mutants in chromosomal and plasmidal genes is an important tool to identify genes which are involved in the pathogenic bacteria-plant-interaction (Kirchner et al., 2001).

Tn*1409* can also be used for mutagenesis in the other subspecies of *C. michiganensis* as *nebraskensis*, *sepedonicus*, and *insidiosus*. However, the efficiency to obtain transposon mutants depends on an efficient transformation protocol. Thus an optimized electroporation protocol must be worked out for each single subspecies to get sufficient transposition rates (Kirchner et al., 2001).

Besides the random generation of mutants by transposon mutagenesis, the efficient electroporation protocol also allows the defined mutagenesis of single genes. A defined wild-type gene can be replaced by the respective insertionally inactivated gene which requires a double cross-over (Kieser et al., 2000). This method of gene-replacement is one of the most important tools for a functional analysis of genes which are of putative interest and were identified e.g. in a genome project.

5.2. Molecular analysis of the pathogenic Cmm-tomato interaction

C. michiganensis subsp. *michiganensis* causes a vascular disease of the tomato and reaches titers up to 10^9-10^{10} bacteria per g plant tissue in infected plants. As a first approach to elucidate the pathogenic *Cmm*-tomato interaction, the role of the exopolysaccharides (EPS) was analyzed (Bermpohl et al., 1996). As many bacteria living in the soil or in association with plants *Clavibacter* produces great amounts of exopolysaccharides. These exopolysaccharides may be important in the natural environment as they provide a hydrated milieu around the cell, thus facilitating the uptake of nutrients, facilitating adhesion to soil particles and preventing dehydration (Leigh and Coplin, 1992). For many phytopathogenic bacteria a correlation between the amount of EPS production and the degree of virulence has been reported. It was proposed that EPS may prevent bacterial immobilization by host lectins and so allows spreading of the bacteria in the plant xylem (Bradshew-Rouse et al., 1981) or that EPS effects the water transport in the xylem and thereby induces water stress and wilting (van Alfen et al., 1987). However, analysis of the amount and composition of EPS of several strains of *Cmm* which differ in virulence and the pathogenic phenotype indicated that EPS is not a crucial pathogenicity factor in *Cmm* (Bermpohl et al., 1996).

Extracellular enzymes, which might be involved in an enzymatic attack on xylem vessels as well as adjacent parenchymatic cells may also play a role in pathogenicity (Benhamou, 1991; Wallis, 1977). Several extracellular enzymes were found *in vivo* and *in vitro* which have cell wall degrading activity: an endocellulase (Jahr et al., 2000; Meletzus et al., 1993), a polygalacturonase (Beimen et al., 1992), a pectinmethylesterase (Strider, 1969), and a xylanase (Beimen et al., 1992). After enzymatic degradation of plant tissue a lateral spread of *Cmm* within the host plant can become possible, along with the release of nutrients necessary for bacterial proliferation and further colonization of the plant.

An extracellular tomatinase which belongs to the glycosyl hydrolase family 10 (Henrissat, 1991) has been identified recently after cloning the chromosomal insertion site of a transposon mutant of *Cmm* (Kaup et al., 2005). The TomA protein is composed of a predicted signal peptide and an N-terminal catalytic domain as well as a C-terminal fibronectin III-like domain. Tomatinases from fungi are involved in the detoxification of the tomato saponin α-tomatine which acts as a growth inhibitor (Roldan-Arjona et al., 1999; Sandrock and Van Etten, 1998). A *Cmm tomA* gene replacement mutant was constructed and showes a stronger inhibition in growth by α-tomatine than the wildtype. Furthermore, tomatinase activity could be shown in concentrated culture supernatants of *Cmm*, but not in the *tomA* mutant. However, neither the transposon mutant nor the *tomA* gene disruption mutant were affected in virulence on tomato (Kaup et al., 2005).

The wildtype strain which was chosen for the analysis of virulence carries the two endogenous plasmids pCM1 and pCM2 which can be cured out of the cell resulting in strains with a reduced virulence (if one of the plasmids is left) or in a strain which is avirulent (if both plasmids are lost). As all curing derivatives are able to colonize the host plant tomato as well as the plasmid bearing wild type, it was concluded that the ability for effective host plant colonization must be encoded on the bacterial chromosome. Furthermore, the wild-type as well as strain *CMM*100, which does not contain any plasmid, produce EPS in nearly identical quantities and composition (Bermpohl et al., 1996). This indicates that virulence is not correlated to EPS production and is encoded by the chromosome.

5.2.1. The virulence gene celA, encoding an endocellulase

Analysis of plasmid pCM1 led to the identification of a 3.2.kb *Bgl*II fragment which carries the gene for an endoglucanase *celA* (Meletzus et al., 1993). The *celA* gene encodes a protein of 78 kDa which is composed of three domains (Jahr et al., 2000). The first 56 aa residues of the N-terminus resemble a signal sequence which is cleaved off during secretion, resulting in a mature enzyme of 690 aa with a molecular weight of 72 kDa. This protein has a modular structure which is typical for cellulases. The 350 aa N-terminal catalytic domain belongs to the family A_1 of cellulases and is connected via a 27 aa linker region to a 84 aa nearly perfect type IIa bacterial cellulose binding domain. The linker region is rich in serine, proline and glycine residues. Surprisingly, and as a unique feature, the cellulose binding domain of CelA is followed by a third domain, a 200 aa C-terminus that is connected to the cellulose binding domain through a 28 aa linker. While the first two domains show extensive similarity to other bacterial cellulases, the third domain only shows weak homology to α-expansins (Jahr et al., 2000), plant proteins mediating cell wall extension (Cosgrove, 1998). Interestingly, this atypical structure has also been identified for the endocellulase CelA of the potato pathogen *Cms* (Laine et al., 2000). It can be postulated that the atypical composition of the cellulase with the expansin domain comprises the basis for the finding that this enzyme is a primary pathogenicity factor directly involved in the development of disease. After sequence analysis of *celA*, the functional expression of the protein was demonstrated by SDS-page and activity staining (Jahr et al., 2000). The supernatant of *CMM*101(pCM1) gave rise to three bands with hydrolytic activity on carboxymethylcellulose, with molecular masses of about 75, 54, and 43 kDa. The band of 75 kDa represents the native protein after cleavage of the signal peptide whereas the other two bands represent truncated enzymes which result from cleavage within the hinge regions. Activity against crystalline cellulose as well as the pathogenic phenotype was only demonstrated with the intact enzyme contsisting of all three domains. A *celA* gene replacement mutant constructed in strain *CMM*101 had lost the ability to induce wilting in infected tomato plants. Complementation of this mutant with an intact *celA* gene restored the pathogenic phenotype. This indicates that *celA* is the only pathogenicity determinant carried by plasmid pCM1 (Jahr et al., 2000).

5.2.2. The virulence factor Pat-1, a serine protease

Analysis of plasmid pCM2 led to the identification of a 3.75 kb *Bgl*II fragment carrying a pathogenicity locus called *pat*-1 which could be mapped by deletion analysis to a 1.5 kb *Bgl*II/*Sma*I subfragment. Introduction of this *pat*-1 region into the endophytic plasmid-free avirulent *CMM*100 strain converted these bacteria into virulent pathogens (Dreier et al., 1997). The nucleotide sequence of this region revealed one open reading frame encoding a protein of 280 aa with a molecular mass of 29.7 kDa. Northern hybridization of total RNA isolated from the wildtype strain *Cmm* NCPPB382 against a *pat*-1 derived DNA probe identified a 1.5 kb transcript of the *pat*-1 gene. As the introduction of a frame shift mutation into the *pat*-1 coding region led to the complete loss of the pathogenic phenotype, it was concluded that the Pat-1 protein represents the pathogenicity determinant (Dreier et al., 1997). The Pat-1 protein sequence showed that the deduced Pat-1 protein is preceded by a signal sequence which spans the first 33 aa and should be cleaved off during secretion. Database showed similarity of the Pat-1 protein to serine proteases from *Lysobacter enzymogenes* (Epstein and Wensink, 1988), *Streptomyces griseus* (Henderson et al., 1987) and *Staphylococcus* sp. (Genbank accession number S25140). Two motifs, one around a conserved serine at aa position 231 (GDSGG) and the other around a conserved histidine at aa position 96, comprise the consensus sequences of serine proteases of the trypsin type with serine and histidine as part of the catalytic triad (Brenner, 1988; Burger et al., 2005). As a further important feature, Pat-1 possesses a putative sortase motif LPGSG starting at position 251. This motif might be involved in anchoring the secreted Pat-1 protein to the bacterial cell wall (Paterson and Mitchell, 2004).

Up to now it was not possible to demonstrate proteolytic activity of Pat-1 although different substrates were used, and also overexpression in different systems failed. But site specific mutagenesis the exchange of the serine in the conserved motif GDSGG led to a complete loss of the virulent phenotype of the mutated Pat-1 protein. This underlines the function of Pat-1 as a putative serine protease (Burger et al., 2005). Further experiments must elucidate which role this putative protease plays in the pathogenic interaction. It might be involved in the activation or inactivation of other proteins that exist as inactive pre-proteins. Processing might release a peptide which is toxic or has signal function. It is also possible that this protease might serve as a virulence factor and interfere with host defense or signaling systems, thus causing damage to host cells or tissues as it is common in several pathogen-animal-systems (Miyoshi and Shinoda, 2000).

5.2.3. Genes which influence colonization efficiency of Cmm

Hybridization of chromosomal DNA of the wildtype against *pat*-1 as a probe indicated two further bands of *pat*-1 homologous genes in *Cmm*, one located on the chromosome (*chpA*, chromosomal homology of *pat*-1) and the other located on plasmid pCM2 (*php*, plasmid homology of *pat*-1) (Burger et al., 2005). Sequence information obtained by the *Cmm*-genome project extended the *pat*-1 homologous genes to a family with 10 members sharing some common characteristics. All have a signal peptide indicating processing and export, they show the typical motifs of serine proteases, and they possess six cysteine residues at conserved positions indicating a tight folding structure as necessary for a strong substrate specificity. Three of the genes are located on plasmid pCM2, but except for *pat*-1 an involvement of the two *php*-genes in the pathogenic mechanism could not be shown (Burger et al., 2005). The seven chromosomal *chp* genes are clustered in a region of about 50 kb. Three of them, *chpA*, *chpB,* and *chpD,* contain frame shifts and/or in-frame stop codons and thus comprise non-functional pseudogenes. Gene replacement mutants were constructed for *chpC* and *chpG.* While the inactivation of *chpG* did not show an effect on virulence of the resulting mutant strain, the *chpC* mutant was unable to colonize tomato plants effectively and accordingly the development of disease symptoms was drastically reduced (Gräfen and Eichenlaub, unpubl. data).

An different experimental approach to find chromosomal genes involved in pathogenicity led to the identification of the same region carrying the *pat*-1 homologous genes (Abt and Eichenlaub, unpubl. data). Using transposon Tn*1409*Cβ from *Arthrobacter* (Gartemann and Eichenlaub, 2001), transposon mutants from the virulent wildtype strain *Cmm* NCPPB382 were constructed and screened on young tomato plants to find strains with reduced virulence. Two mutants were isolated which colonize tomato plants only up to about $4x10^8$ bacteria per g fresh weight tomato. After complementation with the respective cloned intact wildtype gene the colonization titers increased up to $7x10^8$ or $3x10^9$ bacteria per g fresh weight tomato and virulence was reconstituted, indicating that there is a critical minimal bacterial titer which is necessary for the development of disease symptoms. The loci where the transposon had inserted were in a distance of about 5 kb. Gene bank comparison showed that in both cases genes with similarity to serine proteases, but not to members of the Pat-1 family, were affected (Abt and Eichenlaub, unpubl. data).

5.3. Curing and transfer of plasmids in Cmm

The endogenous plasmids pCM1 and pCM2 of *Cmm* are of great relevance as they carry the pathogenicity determinants *celA* and *pat*-1. However, as shown by experimental curing, each of the plasmids can be lost if *Cmm* is cultivated under sublethal growth temperatures of 30-32°C (Meletzus et al., 1991). Analysis of field isolates by hybridization against the *celA* and the *pat-1* gene as probe, showed that about 80% of the isolates still possessed both pathogenicity genes whereas 20% had the phenotype $celA^+$ pat^-, indicating a relative instability of plasmid pCM2 under natural conditions. These results suggested that transfer functions may be present on the plasmids and that mechanisms exiest to regain a plasmid from an appropriate donor if it is lost. For transfer experiments in the laboratory there existed suitable donor strains with labeled plasmids, a pCM1 with a gentamicin resistance gene cassette (Jahr et al., 2000) and a pCM2 containing a neomycin resistance gene cassette, as well as a transposon mutant (Kirchner et al., 2001) from the plasmid free *Cmm* strain *CMM*100 with a chromosomally located chloramphenicol resistance gene cassette which could be used as a recipient. Tomato plants were infected with one of the donors and the recipient, and two weeks later bacteria were reisolated from the plants and plated on selective media to identify donor, recipient and transconjugant strains. Although there was a strong variation between the frequency of plasmid transfer in the single plants, transfer was observed for pCM1 as well as for pCM2. In case of pCM1 up to 75% of the recipients had obtained the labeled plasmid, and in case of pCM2 up to 97%. These experiments demonstrate that a segregational loss of one of the plasmids can be compensated by reacquisition of the respective plasmid from an appropriate donor strain. This mechanism provides a possibility for pathogens to modulate virulence by varying the plasmid status within the population.As a result of such a modulation, survival of the host can be ensured and an equilibrium between host and pathogen can be established (Burger and Eichenlaub, unpubl. results).

6. DIAGNOSIS AND BIOCONTROL

For Gram-negative phytopathogenic bacteria the gene-for-gene concept has been established, in which a pathogen is virulent on one line of host plants (compatible interaction leading to disease) but avirulent on another line of host plants (incompatible interaction by resistant plant) (Flor, 1971). In the incompatible interaction the gene product of the avirulence gene (Avr-Protein) is transported into the plant cell by the bacterial type-III-secretion system and here it is recognized by the corresponding resistance protein (product of the plant R-gene). As a consequence a hypersensitive reaction is induced in the host plant which abolishes spreading and growth of the pathogen and thus confers resistance against this avirulent race of the pathogenic bacterium.

Infection of the specific host plants by *Clavibacter michiganensis* subspecies inevitably leads to disease. Although there seem to exist differences in sensitivity to *Clavibacter* infections no really resistant cultivars of the respective host plants are in agricultural use. This suggests that the system of avirulence genes and corresponding resistance genes in the host plant as found in Gram-negative phytopathogenic bacteria may not exist in the *Clavibacter michiganensis*-host plant interaction. Recently, two quantitative trait loci, Rcm 2.0 and Rcm 5.1, controlling resistance of *Lycopersicon hirsutum* to *Cmm* have been described (Coaker et al., 2004). We have to wait for the eventual identification of the genes responsible for this resistant phenotype and their functional introduction into *Solanum lycopersicum* which may bring us hopefully closer to a *Cmm*-resistant tomato plant. However, while breeding efforts using related species may give rise to varieties with higher yields displaying less severe or no symptoms, the species used (e.g. the wild potato species *S. acaule* and wild tomato species) can themselves be successfully infected by *Clavibacter* without expressing disease symptoms (Strider, 1969; van Steekelenburg, 1985; Laurila et al., 2003) and thus may not be used for seed production. The construction of really immune varieties may be impossible using such classical breeding approaches.

Therefore, in biocontrol of *Clavibacter michiganensis* we have to employ an efficient diagnosis to prevent spreading of the pathogen by infected seed and other contaminated plant material. Consequently, *Cmm* and *Cms* are quarantine organisms under the European Union Plant Health legislation (Anonymous, 1995). *Cmm* contaminated tomato seed and *Cms* infected seed potatoes are considered to be the primary source of outbreaks of disease in these important crop plants (Thyr, 1969; Tsiantos, 1987). The best assay for detection and identification of a pathogen is of course its isolation on semiselective media (Fatmi and Schaad, 1988) and consecutive virulence testing on the host plant or test for induction of a hypersensitive reaction on a non-host plant (Gitaitis, 1990; Van Steekelenburg, 1985). However, such a procedure is very time consuming and not suitable for high through-put diagnostics. For a while enzyme-linked immuno-sorbent assays (ELISA) and immunofluorescence assays employing mono- and polyclonal antibodies were widely used but were not very reliable due to cross-reactions either with other *Cm* subspecies or other plant associated bacteria (Franken et al., 1993; Nemeth et al., 1991; Riley, 1987; De Boer et al., 1994; De Boer et al., 1984). Also, the Biolog method widely used for species determination which is based on use of different growth substrates was found to be unreliable for the plant-pathogenic actinomycetes (Harris-Baldwin and Gudmestad, 1996; Kokoskova et al., 2005). Meanwhile, PCR (polymerase chain reaction) using primers which are generally designed based on sequence information of the respective pathogen has been established as the standard method (Firrao et al., 1994; Henson et al., 1993; Lee et al., 1997a+b; Schneider et al., 1993; Slack et al., 1996; Smith et al., 2001). PCR is a fast and sensitive procedure to identify the presence of the DNA of a pathogen in a sample. PCR can also very specifically identify a pathogen depending on the specificity of the primer pair used (Dreier et al., 1995). A large number of primers have been described in the literature and some were recently evaluated by Arahal et al. (2004) but one can never completely exclude the generation of false signals. In such cases a classical biotest has to follow to exclude a false positive. Microarrays are currently tested, which promise a more specific detection of a pathogen since presence of mRNA's transcribed from various genes are simultaneous assayed (Fessehaie et al., 2002; Van Beckhoven et al., 2002). It has to be seen whether this more costly techniques will be generally applied.

A still unresolved issue is the epidemiology and population structure of *Cm* in areas where the host plants are regularly or exclusively grown. Here may exist an endemic population of various *Cm* strains either in the soil or as epiphytes on non-host plants. These can be characterized by genomic DNA fingerprinting methods (Brown et al., 2002b; Guimaraes et al., 2003) but as far as we are aware no studies of natural populations of *Cm* have been conducted. In case of an outbreak of a *Cm* induced disease it can be determined whether the bacteria found on diseased plants were introduced from outside for example by contaminated seed or caused by *Cm* strains which contaminated the fields already at the time of sowing.

In addition to an efficient diagnosis of *Cm* biocontrol also includes phytosanitary measures including treatment of seed with hydrochloric acid and solarization of fields and green houses to inactivate *Cm* in the soil (Antoniou et al., 1995; Shlevin et al., 2004). For the prevention of outbreaks of *Cmm* infections in green houses the use of lysozyme gave promising results (Utkhede and Koch, 2004), while streptomycin, copper hydroxide and mancozeb had no effect in protecting tomatoes against *Cmm* (Hausbeck et al., 2000). In a different approach plant resistance to *Clavibacter* was induced by pretreatment with azibenzolar-S-methyl (Bion) leading to a reduction of both symptom severity and titers of *Cmm* (Werner et al., 2002; Soylu et al., 2003). Another strategy employs antagonistic bacteria for biocontrol of *Cms* (Gamard and De Boer, 1995). But a general successful use of this method remains doubtful since the interaction between antagonist and pathogen depends on many parameters which include climate, type of soil, humidity and others.

Since *Cm* is not a typical soil borne pathogen due to various auxotrophies it can survive only for a limited time in soil and is also depended on plant material in the soil which provides the required nutrients. Thus after an outbreak of *Cm* induced diseases the plant material should be completely removed from the fields and burnt. Furthermore, for a period of 3-5 years different crops should be grown on fields potentially contaminated by *Cm* strains (Carrol and Lukeciz, 1971; Fatmi and Schaad, 2002; Gleason et al., 1993).

7. FINAL REMARKS

The development of tools and methods for molecular and genetic analysis of the interaction of *Cm* with its host plant are available which is an essential prerequisite for future more work on this interesting and agriculturally relevant group of phytopathogenic bacteria. Another important development are the recently finished genome projects on *Cmm* and *Cms* (*Cmm*: https://www.genetik.uni-bielefeld.de/GenoMik/partner/bi_eichen.html; Cms: http://www.sanger.ac.uk/Projects/C_michiganensis/). The information available now from the genome data are expected to provide new insights into the genetic relationship of these two *Cm* subspecies and also help to identify genes required for interaction with the host plant. It is our hope that this development will attract more attention to this group of Gram-positive phytopathogens. Only when we understand all aspects of the interaction of the bacterium with the host it will eventually be possible to find a way to generate plants that are less sensitive or even resistant to *Cm* infection and disease development.

8. REFERENCES

Altenburger, P., Kampfer, P., Akimov, V. N., Lubitz, W., & Busse, H. J. (1997). Polyamine distribution in actinomycetes with group B peptidoglycan and species of the genera *Brevibacterium*, *Corynebacterium*, and *Tsukamurella*. *Int. J. Syst. Bacteriol.*, *47*, 270-277.

Anonymous, 1995. Commission Directive 95/4/EC amendment of 21 Feb 1995 to the European Community Plant Health Directive (77/93/EEC). *Official Journal of the European Communities*, *L44*, 56-60.

Arahal, D. R., Llop, P., Alonso, M. P., & Lopez, M. M. (2004). *In silico* evaluation of molecular probes for detection of *Ralstonia solanacearum* and *Clavibacter michiganensis* subsp. *sepedonicus*. *System. Appl. Microbiol.*, *27*, 581-591.

Antoniou, P. P., Tjamos, E. C., & Pnangopoulos, C. G. (1995). Use of solarization for controlling bacterial canker of tomato in plastic houses in Greece. *Plant Pathol.*, *44*, 438-447.

Baer, D., & Gudmestad, N. C. (1995). In vitro cellulolytic activity of the plant pathogen *Clavibacter michiganensis* subsp. *sepedonicus*. *Can. J. Microbiol.*, *41*, 877-888.

Beck, E., Ludwig, G., Auerswald, E. A., Reiss, B., & Schaller, H. (1982). Nucleotide sequence and exact localization of the neomycin phosphotransferase gene from Tn*5*. *Gene*, *19*, 327-336.

Behrendt, U., Ulrich, A., Schumann, P., Naumann, D., & Suzuki, K.-I. (2002). Diversity of grass-associated *Microbacteriaceae* isolated from the phyllosphere and litter layer after mulching the sward; polyphasic characterization of *Subtercola pratensis* sp. nov., *Curtobacterium herbarum* sp. nov. and *Plantibacter flavus* gen. nov., sp. nov. *Int. J. Syst. Evol. Microbiol.*, *52*, 1441-1454.

Beimen, A., Bermpohl, A., Meletzus, D., Eichenlaub, R., & Barz, W. (1992). Accumulation of phenolic compounds in leaves of tomato plants after infection with *Clavibacter michiganense* subsp. *michiganense* strains differing in virulence. *Z. Naturforsch.*, *47C*, 898-909.

Benhamou, N. (1991). Cell surface interactions between tomato and *Clavibacter michiganense* subsp. *michiganense*: localization of some glycoproteins in infected host leaf tissues. *Phys. Mol. Plant Pathol.*, *38*, 15-38.

Bermpohl, A., Dreier, J., Bahro, R., & Eichenlaub, R. (1996). Exopolysaccharides in the pathogenic interaction of *Clavibacter michiganensis* subsp. *michiganensis* with tomato plants. *Microbiol. Res.*, *151*, 391-399.

Biddle, J. A., McGee, D. C., & Braun, E. J. (1990). Seed transmission of *Clavibacter michiganense* subsp. *nebraskense* in corn. *Plant Disease*, *74*, 908-911.

Bradshew-Rouse, J. J., Whatley, M. H., Coplin, D. L., Woods, A., Sequeira, L., & Kelman, A. (1981). Agglutination of *Erwinia stewartii* strain, with a corn agglutinin: correlation with extracellular polysaccharide production and pathogenicity. *Appl. Environ. Microbiol.*, *42*, 344-350.

Brenner, S. (1988). The molecular evolution of genes and proteins: a tale of two serines. *Nature*, *334*, 528-530.

Brown, S. E., Knudson, D. L., & Ishimaru, C. A. (2002a) Linear plasmid in the genome of *Clavibacter michiganensis* subsp. *sepedonicus*. *J. Bacteriol.*, *184*, 2841-2844.

Brown, S. E., Reilley, A. A., Knudson, D. L., & Ishimaru, C. A. (2002b) Genomic fingerprinting of virulent and avirulent strains of *Clavibacter michiganensis* subsp. *sepedonicus*. *Curr. Microbiol.*, *44*, 112-119.

Bugbee, W. M., & Gudmestad, N. C. (1988). The recovery of *Corynebacterium sepedonicum* from sugarbeet seed. *Phytopathology*, *78*, 205-208.

Burger, A., & Eichenlaub, R. (2003). Genetics of phytopathogenic bacteria. *Prog. Bot.*, *64*, 98-114.

Burger, A., Gräfen, I., Engemann, J., Niermann, E., Pieper, M., Kirchner, O., Gartemann, K.-H., & Eichenlaub, R. (2005). Identification of homologues to the pathogenicity factor Pat-1, a putative serine protease of *Clavibacter michiganensis* subsp. *michiganensis*. *Microbiol. Res.* (in press)

Carlson, R. R., & Vidaver, A. K. (1982a). Taxonomy of *Corynebacterium* plant pathogens, including a new pathogen of wheat, based on polyacrylamide gel electrophoresis of cellular proteins. *Int. J. Syst. Bacteriol.*, *32*, 315-326.

Carlson, R. R., & Vidaver, A. K. (1982b). Bacterial mosaic, a new corynebacterial disease of wheat. *Plant Disease*, *66*, 76-79.

Carlton, W. M., Braun, E. J., & Gleason, M. L. (1998). Ingress of *Clavibacter michiganensis* subsp. *michiganensis* into tomato leaves through hydathodes. *Phytopathology*, *88*, 525-529.

Carlton, W. M., Gleason, M. L., & Braun, E. J. (1994). Effects of pruning on tomato plants supporting epiphytic populations of *Clavibacter michiganensis* subsp. *michiganensis*. *Plant Diesease*, *78*, 742-745.

Carroll, R. B., & Lukezic, F. L. (1971). Preservation of *Corynebacterium insidiosum* in a sterile soil mix without loss of virulence. *Phytopathology*, *61*, 688-690.

Chang, R. J., Ries, S. M., & Pataky, J. K. (1992a). Effects of temperature, plant age, inoculum concentration, and cultivar on the incubation period and severity of bacterial canker of tomato. *Plant Disease*, *76*, 1150-1155.

Chang, R. J., Ries, S. M., & Pataky, J. K. (1992b). Local sources of of *Clavibacter michiganensis* ssp. *michiganensis* in the development of bacterial canker on tomatoes. *Phytopathology*, *82*, 553-560.

Clark, N. C., & Lawrence, C. H. (1986). Characterization of a plasmid in isolates of *Corynebacterium sepedonicum*. *Can. J. Microbiol.*, *32*, 617-622.

Coaker, G. L., Willard, B., Kinter, M., Stockinger, E. J., & Francis, D. M. (2004). Proteomic analysis of resistance mediated by Rcm 2.0 and Rcm 5.1, two loci controlling resistance to bacterial canker of tomato. *Mol. Plant-Microbe Interact.*, *17*, 1019-1028.

Collins, M. D., & Bradbury, J. F. (1986). Plant pathogenic species of *Corynebacterium*. In J. P. Butler et al. (Eds.), *Bergey's manual of systematic bacteriology*. (pp. 1276-1283). 2nd edition. Baltimore, U.S.A.: Wilkins & Wilkins.

Collins, M. D., & Bradbury, J. F. (1992). The genera *Agromyces, Aureobacterium, Clavibacter, Curtobacterium,* and *Microbacterium*. In A. Balows et al., (Eds.), *The prokaryotes*. (pp. 1355-1368). 2nd edition. New York: Springer.

Collins, M. D., Jones, D., & Kroppenstedt, R. (1981). Reclassification of *Corynebacterium ilicis* (Mandel, Guba and Litsky) in the genus *Arthrobacter*, as *Arthrobacter ilicis* comb. nov. *Zentralbl. Bakteriol. Parasitenkd. Infektionskr. Hyg. Abt. 1 Orig. Reihe C*, *2*, 318-323.

Collins, M. D., & Jones, D. (1983). Reclassification of *Corynebacterium flaccumfaciens*, *Corynebacterium betae*, *Corynebacterium oortii*, *Corynebacterium poinsettiae* in the genus *Curtobacterium*, as *Corynebacterium flaccumfaciens* comb. nov. *J. Gen. Microbiol.*, *129*, 3545-3548.

Cottyn, B., Regalado, E., Lanoot, B., De Cleene, M., Mew, T. W., & Swings, J. (2001). Bacterial populations associated with rice seed in the tropical environment. *Phytopathology*, *91*, 282-292.

Cook, F. D., & Katznelson, H. (1960). Isolation of bacteriophages for the detection of *Corynebacterium insidiosum*, agent of bacterial wilt of alfalfa. *Can. J. Microbiol.*, *6*, 121-125.

Cosgrove, D. J. (1998). Cell wall loosening by expansins. *Plant Physiol.*, *118*, 333-339.

Cosgrove, D .J. (1999). Enzymes and other agents that enhance cell wall extensibility. *Annu. Rev. Plant Physiol. Mol. Biol.*, *50*, 391-417.

Davis, M. J., Gillespie Jr., A. G., Vidaver, A. K., & Harris, R. W. (1984). *Clavibacter*: a new genus containing some phytopathogenic coryneform bacteria, including *Clavibacter xyli* subsp. *xyli* sp. nov., subsp. nov. and *Clavibacter xyli* subsp. *cynodontis* subsp. nov., pathogens that cause ratoon stunting disease of sugarcane and bermudagrass stunting disease. *Int. J. Syst. Bacteriol.*, *34*, 107-117.

De Boer, S. H., & Wieczorek, A. (1984). Production of monoclonal antibodies to *Corynebacterium sepedonicum*. *Phytopathology*, *74*, 1431-1434.

De Boer, S. H., Stead, D. E., Alivizatos, A. S., Janse, J. D., Van Vaerenbergh, J., De Haan, T. L., & Mawhinney, J. (1994). Evaluation of serological tests for detection of *Clavibacter michiganensis* subsp. *sepedonicus* in composite potato stem and tuber samples. *Plant Disease*, *78*, 725-729.

Denny, T. P. (1995). Involvement of bacterial polysaccharides in plant pathogenesis. *Ann. Rev. Phytopathol.*, *33*, 173-197.

Döpfer, H., Stackebrandt, E., & Fiedler, F. (1982). Nucleic acid hybridization studies on *Microbacterium*, *Curtobacterium*, *Agromyces*, and related genera. *J. Gen. Microbiol.*, *128*, 1697-1708.

Dreier, J., Bermpohl, A., & Eichenlaub, R. (1995). Southern hybridization and PCR for specific detection of phytopathogenic *Clavibacter michiganensis* subsp. *michiganensis*. *Phytopathology, 85*, 462-468.

Dreier, J., Meletzus, D., & Eichenlaub, R. (1997). Characterization of the plasmid encoded virulence region *pat*-1 of phytopathogenic *Clavibacter michiganensis* subsp. *michiganensis*. *Mol. Plant-Microbe Interact., 10*, 195-206.

Echandi, E., & Sun, M. (1973). Isolation and characterization of a bacteriophage for the identification of *Corynebacterium michiganense*. *Phytopathology*, *63*, 1398-1401.

Echandi, E. (1976). Bacteriocin production by *Corynebacterium michiganense*. *Phytopathology*, *66*, 430-432.

Epstein, D. M., & Wensink, P. C. (1988). The alpha-lytic protease gene of *Lysobacter enzymogenes*. *J. Biol. Chem.*, *263*, 16586-16590.

Erwin, D. C. (1990). Bacterial wilt. In, D. L. Stuteville & D. C. Erwin. (Eds.). *Compendium of alfalfa diseases*. (pp. 5-6). 2nd edition, St. Paul, Minnesota, USA: American Phytopathological Society Press.

Evtushenko, L. I., Dorofeeva, L. V., Subbotin, S. A., Cole, J. R., & Tiedje, J. M. (2000). *Leifsonia poae* gen.nov., sp. nov., isolated from nematode galls on *Poa annua*, and reclassification of '*Corynebacterium aquaticum*' Leifson 1962 as *Leifsonia aquatica* (*ex* Leifson 1962) gen. nov., nom. rev., comb. nov. and *Clavibacter xyli* Davis et al., 1984 with two subspecies as *Leifsonia xyli* (Davis et al. 1984) gen. nov., comb. nov. *Int. J. Syst. Evol. Microbiol.*, *50*, 371-380.

Fatmi, M., & Schaad, N. W. (2002). Survival of *Clavibacter michiganensis* subsp. *michiganensis* in infected tomato stems under natural field conditions in California, Ohio and Morocco. *Plant Pathology*, *51*, 149-154.

Fessehaie, A., De Boer, S. H., & Levesque, C. A. (2002). New diagnostic assay based on DNA array for identification and detection of five bacterial pathogens of potato. *Phytopathology*, *92*, S25.

Firrao, G., & Locci, R. (1994). Identification of *Clavibacter michiganensis* subsp. *sepedonicus* using polymerase chain reaction. *Can. J. Microbiol.*, *40*,148-151.

Flor, H. H. (1971) Current status of the gene-for-gene concept. *Ann. Rev. Phytopathol.*, *9*, 275-296.

Frank, G. D. (1999). Persistence and latency of *Clavibacter michiganensis* subsp. *sepedonicus* in field-grown seed potatoes. *Plant Disease*, *83*, 247-250.

Franken, A. A. J. M., Kamminga, G. C., Snyders, W., van der Zouwen, P. S., & Birnbaum, Y. E. (1993). Detection of *Clavibacter michiganensis* ssp. *michiganensis* in tomato seeds by immunofluorescence microscopy and dilution plating. *Neth. J. Plant Pathol.*, *99*, 125-137.

Fulkerson, J. F. (1960). Pathogenicity and stability of strains of *Corynebacterium insidiosum*. *Phytopathology*, *50*, 377-380.

Gamard, P., & De Boer, S. H. (1995). Evaluation of antagonistic bacteria for suppression of bacterial ring rot of potato. *Eur. J. Plant Pathol.*, *101*, 519-525.

Gartemann, K.-H., & Eichenlaub, R. (2001). Isolation and characterization of IS*1409*, an insertion element of 4-chlorobenzoate-degrading *Arthrobacter* sp. strain TM1 and development of a system for transposon mutagenesis. *J. Bacteriol., 183*, 3729-3736.

Gartemann, K.-H., Kirchner, O., Engemann, J., Gräfen, I., Eichenlaub, R., & Burger, A. (2003). *Clavibacter michiganensis* subsp. *michiganensis*: first steps in the understanding of virulence of a Gram-positive phytopathogenic bacterium. *J. Biotechnol.*, *106*, 179-191.

Gitaitis, R. D., Beaver, R. W., & Voloudakis, A. E. (1991). Detection of *Clavibacter michiganensis* subsp. *michiganensis* in symptomless tomato transplants. *Plant Disease*, *75*, 834-838.

Gitaitis, R. D. (1990). Induction of a hypersensitive-like reaction in four-o'clock by *Clavibacter michiganensis* subsp. *michiganensis*. *Plant Disease*, *74*, 834-838.

Gleason, M. L., Braun, E. J., Carlton, W. M., & Peterson, R. H. (1991). Survival and dissemination of *Clavibacter michiganensis* subsp. *michiganensis* in tomatoes. *Phytopathology*, *81*, 1519-1523.

Gleason, M. L., Gitaitis, R. D., & Ricker, M. (1993). Recent progress in understanding and controlling bacterial cancer of tomato in Eastern North America. *Plant Disease*, *75*, 834-838.

Goodfellow, M., & Anderson, G. (1977). The actinomycetes genus *Rhodococcus*: a home for the '*rhodochrous*' complex. *J. Gen. Microbiol.*, *100*, 99-120.

Goodfellow, M. (1984). Reclassification of *Corynebacterium fascians* (Tilford) Dowson in the genus *Rhodococcus*, as *Rhodococcus fascians* comb. nov. *Syst. Appl. Microbiol.*, *5*, 225-229.

Gorin, P. A., & Spencer, J. F. T. (1980). Structure of the extracellular polysaccharide from *Corynebacterium insidiosum*. *Carbohydrate Research*, *79*, 313-315.

Gross, D. C., Vidaver, A. K., & Keralis, M. B. (1979a). Bacteriocins of phytopathogenic *Corynebacterium* species. *Can. J. Microbiol.*, *25*, 367-374.

Gross, D. C., & Vidaver, A. K. (1979b). Indigenous plasmids from phytopathogenic *Corynebacterium* species. *J. Gen. Microbiol.*, *115*, 479-489.

Guimaraes, P. M., Smith, J. J., Palmano, S., & Saddler, G. S. (2003). Characterisation of *Curtobacterium flaccumfaciens* pathovars by AFLP, rep-PCR and pulsed-field gel electrophoresis. *Eur. J. Plant Pathol.*, *109*, 817-825.

Hausbeck, M. K., Bell, J., Medina-Mora, C., Podolsky, R., & Fulbright, D. W. (2000). Effect of bactericides on population sizes and spread of *Clavibacter michiganensis* subsp. *michiganensis* on tomatoes in the greenhouse and on disease development and crop yield in the field. *Phytopathology*, *90*, 38-44.

Harris-Baldwin, A., & Gudmestad, N. C. (1996). Identification of phytopathogenic coryneform bacteria using the Biolog automated microbial identification system. *Plant Disease*, *80*, 874-878.

Henderson, G., Krygsman, P., Liu, C. J., Davey, C. C., & Malek, L. T. (1987). Characterization and structure of genes for protease A and B from *Streptomyces griseus*. *J. Bacteriol.*, *169*, 3778-3784.

Henningson, P. J., & Gudmestad, N. C. (1992). Comparison of exopolysaccharides from mucoid and nonmucoid strains of *Clavibacter michiganensis* subsp. *sepedonicus*. *Can. J. Microbiol.*, *39*, 291-296.

Henrissat, B. (1991). A classification of glycosyl hydrolases based on amino acid sequence similarities. *Biochem. J.*, *280*, 309-316.

Henson, J. M., & French, R. (1993). The polymerase chain reaction and plant disease diagnosis. *Annu. Rev. Phytopathol.*, *31*, 81-109.

Huang, J. S., & Echandi, E. (1981). Purification and characterization of a bacteriocin from *Corynebacterium michiganense* strain 15-2. *Phytopathology*, *71*, 227.

Hunt, O. J., Griffin, G. D., Murray, J. J., Pedersen, M. W., & Peaden, R. N. (1971). The effects of root knot nematodes on bacterial wilt in alfalfa. *Phytopathology*, *61*, 256-259.

Jahr, H., Bahro, R., Burger, A., Ahlemeyer, J., & Eichenlaub, R. (1999). Interactions between *Clavibacter michiganensis* subsp. *michiganensis* and its host plants. *Environ. Microbiol.*, *1*, 113-118.

Jahr, H., Dreier, J., Meletzus, D., Bahro, R., & Eichenlaub, R. (2000). The Endo-β-1,4-glucanase CelA of *Clavibacter michiganensis* subsp. *michiganensis* is a pathogenicity determinant required for induction of bacterial wilt of tomato. *Mol. Plant-Microbe Interact.*, *13*, 703-714.

Jacobs, J. L., Carroll, T. L., & Sundin, G. W. (2005). The role of pigmentation, ultraviolet radiation tolerance, and leaf colonization strategies in the epiphytic survival of phyllosphere bacteria. *Microbial Ecology*, *49*, 104-113.

Jacobs, J. L., & Sundin, G. W. (2001). Effect of solar UV-B radiation on a phyllosphere bacterial community. *Appl. Environ. Microbiol.*, *67*, 5488-5496.

Kaup, O., Gräfen, I., Zellermann, E.-M., Eichenlaub, R., & Gartemann, K.-H. (2005). Identification of a tomatinase in the tomato-pathogenic actinomycete *Clavibacter michiganensis* subsp. *michiganensis* NCPPB382. *Mol. Plant-Microbe Interact.*, in press.

Kieser, T., Bibb, M. J., Buttner, M. M. J., Chater, K. F., & Hopwood, D. A. (2000). Practical *Streptomyces* Genetics. John Innes Foundation, Norwich, UK.

Kirchner, O., Gartemann, K.-H., Zellermann, E.-M., Eichenlaub, R., & Burger, A. (2001). A highly efficient transposon mutagenesis system for the tomato pathogen *Clavibacter michiganensis* subsp. *michiganensis*. *Mol. Plant-Microbe-Interact.*, *14*, 1312-1318.

Kokoskova, B., Mraz, I., Janse, J. D., Fousek, J., & Jerabkova, R. (2005). Reliability of diagnostic techniques for determination of *Clavibacter michiganensis* subsp. *sepedonicus*. *J. Plant Diseases Prot.*, *112*, 1-16.

Kratka, J. (1987). Activity of cell wall destroying enzymes and growth regulators in *Corynebacterium michiganense* subsp. *insidiosum*. *Zentralbl. Mikrobiol.*, *142*, 527-534.

Laine, M. J., Nakhei, H., Dreier, J., Lehtilä, K., Meletzus, D., Eichenlaub, R., & Metzler, M. C. (1996). Stable transformation of the Gram-positive phytopathogenic bacterium *Clavibacter michiganensis* subsp. *sepedonicus* with several cloning vectors. *Appl. Environ. Microbiol.*, *62*, 1500-1506.

Laine, M. J., Haapalainen, M., Wahlroos, T., Kankare, K., Nissinen, R., Kassuwi, S., & Metzler, M. (2000). The cellulase encoded by the native plasmid of *Clavibacter michiganensis* subsp. *michiganensis* plays a role in virulence and contains an expansin-like domain. *Physiol. Mol. Plant Pathology*, *57*, 221-233.

Laurila, J., Metzler, M. C., Ishimaru, C. A., & Rokka, V.-M. (2003). Infection of plant material derived from *Solanum acaule* with *Clavibacter michiganensis* ssp. *sepedonicus*: temperature as a determining factor in immunity of *S. acaule* to bacterial ring rot. *Plant Pathol.*, *52*, 496-504.

Lee, I.-M., Bartoszyk, I. M., Gundersen, D. E., Mogen, B., & Davis, R. E. (1997a). Nested PCR for ultrasensitive detection of the potato ring rot bacterium, *Clavibacter michiganensis* subsp. *sepedonicus*. *Appl. Environ. Microbiol.*, *63*, 625-2630.

Lee, I.-M., Bartoszyk, I. M., Gundersen-Rindal, D. E., & Davis, R. E. (1997b). Phylogeny and classification of bacteria in the genera *Clavibacter* and *Rathayibacter* on the basis of 16S rRNA gene sequence analyses. *Appl. Environ. Microbiol.*, *63*, 631-2636.

Leigh, J. A., & Coplin, D. L. (1992). Exopolysaccharides in plant-bacterial interactions. *Ann. Rev. Microbiol.*, *46*, 307-346.

Manzer, F. & Genereux, H. (1981). Ring rot. In W. J. Hobson, (Ed.) *Compendium of potato disease*. (pp. 31-32). St Paul, MN, USA: American Phytopathological Society Press.

Marques, L. L. R., De Boer, S. H., Ceri, H., & Olson, M. E. (2003). Evaluations of biofilms formed by *Clavibacter michiganensis* subsp. *sepedonicus*. *Phytopathology*, *93*, S57.

Marte, M. (1980). Alcuni aspetti istologici e istochimici dell'infezione da *Corynebacterium michiganense* nel fusto del pomodoro. *Phytopath. Z.*, *97*, 52-271.

McBeath, J. H., & Adelman, M. (1986). Detection of *Corynebacterium michiganense* subsp. *tessellarius* in seeds and wheat plants. *Phytopathology*, *76*, 1099.

McCulloch, L. (1925). *Aplanobacter insidiosum* n. sp., the cause of an alfalfa disease. *Phytopathology*, *15*, 496-497.

Metzler, M. C., Laine, M. J., & De Boer, S. H. (1997). The status of molecular biological research on the plant pathogenic genus *Clavibacter*. *FEMS Microbiol. Lett.*, *150*, 1-8.

Meletzus, D., & Eichenlaub, R. (1991). Transformation of the phytopathogenic bacterium *Clavibacter michiganense* subsp. *michiganense* by electroporation and development of a cloning vector. *J. Bacteriol.*, *173*, 184-190.

Meletzus, D., Bermpohl, A., Dreier, J., & Eichenlaub, R. (1993). Evidence for plasmid-encoded virulence factors in the phytopathogenic bacterium *Clavibacter michiganensis* subsp. *michiganensis* NCPPB382. *J. Bacteriol.*, *175*, 2131-2136.

Meletzus, D., Jahr, H., & Eichenlaub, R. (2000). *Clavibacter michiganensis* - Transformation of a phytopathogenic Gram-positive bacterium. In N. Eynard & J. Tessie, (Eds.), *Electrotransformation of Bacteria*. (pp. 219-226). Berlin Heidelberg, Germany: Springer.

Mogen, B. D., & Oleson, A. E. (1987). Homology of pCS1 plasmid sequences with chromosomal DNA in *Clavibacter michiganensis* subsp. *sepedonicum*: evidence for the presence of a repeated sequence and plasmid integration. *Appl. Environ. Microbiol.*, *53*, 476-2481.

Mogen, B.D., Oleson, A. E., Sparks, R. B., Gudmestad, N. C., & Secor, G. A. (1988). Distribution and partial characterization of pCS1, a highly conserved plasmid present in *Clavibacter michiganense* subsp. *sepedonicum*. *Phytopathology*, *78*, 1381-1386.

Mogen, B. D., Olson, H. R., Sparks, R. B., Gudmestad, N. C., & Oleson, A. (1990). Genetic variation in strains of *Clavibacter michiganensis* subsp. *sepedonicum*: polymorphisms in restriction fragments containing a highly repeated sequence. *Phytopathology*, *80*, 90-96.

Miyoshi, S., & Shinoda, S. (2000). Microbial metalloproteases and pathogenesis. *Microbes Infect.*, *2*, 91-98.

Nelson, G. A., & Semeniuk, G. (1964). An antagonistic variant of *Corynebacterium insidiosum* and some properties of the inhibitor. *Phytopathology*, *54*, 30-335.

Nelson, G. A. (1982). *Corynebacterium sepedonicum* in potato: effect of inoculum concentration on ring rot symptoms and latent infection. *Can. J. Plant Pathol.*, *4*, 129-133.

Nemeth, J., Laszlo, E., & Emody, L. (1991). *Clavibacter michiganensis* ssp. *insidiosus* in lucerne seeds. *Bull. OEPP*, *21*, 713-718.

Paschke, M. C., & van Alfen, N. K. (1993). Extracellular polysaccharide impaired mutants of *Clavibacter michiganense* subsp. *insidiosum*. *Physiol. Mol. Plant Pathol.*, *42*, 309-319.

Park, Y.-H., Suzuki, K.-I., Yim, D.-G., Lee, K.-C., Kim, E., Yoon, J.-S., Kim, S.-J., Kho, Y.-H., Goodfellow, M., & Komogata, K. (1993). Suprageneric classification of peptidoglycan group B actinomycetes by nucleotide sequencing of 5S ribosomal RNA. *Antonie van Leeuwenhoek*, *64*, 307-313.

Paterson, G. K., & Mitchell, T. J. (2004). The biology of Gram-positive sortase enzymes. *Trends Microbiol.*, *12*, 89-95.

Rai, P. V., & Strobel, G. A. (1969). Phytotoxic glycopeptides produced by *Corynebacterium michiganense* II. Biological properties. *Phytopathology*, *59*, 53-57.

Ries, S. M., & Strobel, G. A. (1972). Biological properties and pathological role of a phytotoxic glycopeptide from *Corynebcterium insidiosum*. *Physiol. Plant Pathol.*, *2*, 133-142.

Riley, I. T. (1987). Serological relationship between strains of coryneform bacteria responsible for annual ryegrass toxicity and other plant pathogenic corynebacteria. *Int. J. Syst. Bacteriol.*, *37*, 153-159.

Riley, I. T., & Ophel, K. M. (1992). *Clavibacter toxicus* sp. nov., the bacterium responsible for annual ryegrass toxicity in australia. *Int. J. Syst. Bacteriol.*, *42*, 64-68.

Roldan-Arjona, T., Perez-Espinosa, A., & Ruiz-Rubio, M. (1999). Tomatinase from *Fusarium oxysporum* f. sp. *lycopersici* defines a new class of saponinases. *Mol. Plant-Microbe Interact.*, *12*, 852-861.

Samac, D. A., Nix, R. J., & Oleson, A. E. (1998). Transmission frequency of *Clavibacter michiganensis* subsp. *insidiosus* to alfalfa seed and identification of the bacterium by PCR. *Plant Disease*, *82*, 1362-1367.

Sandrock, R. W., & Van Etten, H. D. (1998). Fungal sensitivity to and enzymatic degradation of the phytoanticipin alpha-tomatine. *Phytopathology*, *88*, 137-143.

Saperstein, S., & Starr, M. P. (1954). The ketonic carotenoid canthaxanthin isolated from a colour mutant of *Corynebacterium michiganense*. *Biochem. J.*, *57*, 273-275.

Saperstein, S., Starr, M. P., & Filfus. J. A. (1954). Alterations in carotenoid synthesis accompanying mutation in *Corynebacterium michiganense*. *J. Gen. Microbiol.*, *10*, 85-92.

Sasaki, J., Chijimatsu, M., & Suzuki, K.-I. (1998). Taxonomic significance of 2,4-diaminobutyric acid isomers in the cell wall peptidoglycan of actinomycetes and reclassification of *Clavibacter toxicus* as *Rathayibacter toxicus* comb. nov. *Int. J. Syst. Evol. Microbiol.*, *48*, 403-410.

Schleifer, K.-H., & Kandler, O. (1972). Peptidoglycan types of bacterial cell walls and their taxonomic implications. *Bacteriol. Rev.*, *36*, 407-477.

Schneider, B. J., Zhao, J., & Orser, C. S. (1993). Detection of *Clavibacter michiganensis* subsp. *sepedonicus* by DNA amplification. *FEMS Microbiol. Lett.*, *109*, 207-212.

Schuster, M. L., Hoff, B., & Compton, W. A. (1975). Variation, maintenance, loss, and failure to recover virulence in *Corynebacterium nebraskense*. *Plant Disease Reporter*, *59*, 101-105.

Shirako, Y., & Vidaver, A. K. (1981). Characteristics of bacteriophages CON11, CONX and ConXC infecting *Corynebacterium nebraskense*. *Phytopathology*, *71*, 903.

Shirako, Y., Vidaver, A. K., & Ackermann, H.-W. (1986). Partial characterization of bacteriophages for *Clavibacter michiganense* subsp. *nebraskense*. *Ann. Phytopath. Soc. Jpn.*, *52*, 793-800.

Shlevin, E., Mahrer, Y., Kritzman, G., & Katan, J. (2004). Survival of plant pathogens under structural solarization. *Phytoparasitica*, *32*, 470-478.

Slack, S. A., Drenna, J. L., Westra, A. A. G., Gudmestad, N. G., & Oleson, A. E. (1996). Comparison of PCR, ELISA, and DNA hybridization for the detection of *Clavibacter michiganensis* subsp. *sepedonicus* in field-grown potatoes. *Plant Disease*, *80*, 519-524.

Smidt, M., & Vidaver, A. K. (1986). Population dynamics of *Clavibacter michiganense* subsp. *nebraskense* in field-grown dent corn and popcorn. *Plant Disease*, *70*, 1031-1036.

Smidt, M., & Vidaver, A. K. (1987). Variation among strains of *Clavibacter michiganense* subsp. *nebraskense* isolated from a single popcorn field. *Phytopathology*, *77*, 388-392.

Smith, N. C., Hennessy, J., & Stead, D. (2001). Repetitive sequence-derived PCR profiling using the BOX-A1R primer for rapid identification of the plant pathogen *Clavibacter michiganensis* subsp. *sepedonicus*. *Eur. J. Plant Pathol.*, *107*, 739-748.

Soylu, S., Baysal, O., & Soylu, E. M. (2003). Induction of disease resistance by the plant activator, acibencolar-S-methyl (ASM), against bacterial canker (*Clavibacter michiganensis* subsp. *michiganensis*) in tomato seedlings. *Plant Science*, *165*, 1069-1075.

Spooner, D. A., Peralta, I. E., & Knapp, S. (2005). Comparison of AFLPs with other markers for phylogenetic inference in wild tomatoes [*Solanum* L. section *Lycopersicon* (Mill.) Wettst.]. *Taxon*, *54*, 43-61.

Starr, M. P. (1958). The blue pigment of *Corynebacterium insidiosum*. *Arch. Microbiol.*, *30*, 325-334.

Straley, C. S., Straley, M. L., & Strobel, G. A. (1974). Rapid screening for bacterial wilt resistance in alfalfa with a phytotoxic glycopeptide from *Corynebacterium insidiosum*. *Phytopathology*, *64*, 194-196.

Strider, D. L. (1969). Bacterial canker of tomato caused by *Corynebacterium michiganense*. A literature review and bibliography. *North Carolina Agric. Exp. Stn., Tech. Bull.*, *193*.

Strobel, G. A. (1970). A phytotoxic glycopeptide from potato plants infected with *Corynebacterium sepedonicum*. *J. Biol. Chem.*, *245*, 32-38.

Suzuki, K. I., Suzuki, M., Sasaki, J., Park, Y. H., & Komogata, K. K. (1999). *Leifsonia* gen. nov., a genus for 2,4-diaminobutyric acid-containing actinomycetes to accommodate '*Corynebacterium aquaticum*' Leifson 1962 and *Clavibacter xyli* subsp. *cynodontis* Davis et al 1984. *J. Gen. Appl. Microbiol.*, *45*, 253-262.

Tauch, A., Zheng, Z., Pühler, A., & Kalinowski, J. (1998). *Corynebacterium striatum* chloramphenicol resistance transposon Tn*5564*: Genetic organization and transposition in *Corynebacterium glutamicum*. *Plasmid, 40*, 126-139.

Thyr, B. D. (1969). Assaying tomato seeds for *Corynebacterium michiganensis*. *Plant Disease Reporter*, *5*, 858-860.

Thyr, B. D., Samuel, M. J., & Brown, P. G. (1975). New solanaceous host records for *Corynebacterium michiganense*. *Plant Disease Reporter*, *59*, 595-598.

Tsiantos, J. 1987. Transmission of the bacterium *Corynebacterium michiganense* pv. *michiganense* by seeds. *J. Phytopathol.*, *119*, 142-146.

Utkhede, R., & Koch, C. (2004). Biological treatments to control bacterial canker of green house tomatoes. *Biocontrol*, *49*, 305-313.

Van Alfen, N. K., McMillan, B. D., & Dryden, P. (1987a). The multi-component extracellular polysaccharide of *Clavibacter michiganense* subsp. *insidiosum*. *Phytopathology*, *77*, 496-501.

Van Alfen, N. K., McMillan, B. D., & Wang, Y. (1987b). Properties of the extracellular polysaccharide of *Clavibacter michiganense* subsp. *insidiosum* that may affect pathogenesis. *Phytopathology*, *77*, 501-505.

Van Beckhoven, J. R., Stead, D. E., & van der Wolf, J. M. (2002). Detection of *Clavibacter michiganensis* subsp. *sepedonicus* by AmpliDet RNA, a new technology based on real time monitoring of NASBA amplicons with a molecular beacon. *J. Appl. Microbiol.*, *93*, 840-849.

Van den Bulk, R. W., Zevenhuizen, L. P. T. M., Cordewener, J. H. G., & Dons, J. J. M. (1991). Characterization of the extracellular polysaccharide produced by *Clavibacter michiganensis* subsp. *michiganensis*. *Phytopathology*, *81*, 619-623.

Vandeputte, O., Öden, S., Mol, A., Vereecke, D., Goethals, K., El Jaziri, M., & Prinsen, E. (2005). Biosynthesis of auxin by the Gram-positive phytopathogen *Rhodococcus fascians* is controlled by compounds specific to infected plant tissues. *Appl. Environ. Microbiol.*, *71*, 1169-1177.

Van der Wolf, J. M., & van Beckhoven, J. R. C. M. (2004). Factors affecting survival of *Clavibacter michiganensis* subsp. *sepedonicus* in water. *J. Phytopathol.*, *152*, 161-168.

Van der Wolf, J. M., van Beckhoven, J. R. C. M., Hukkanen, A., Karjalainen, R., & Muller, P. (2004). Fate of *Clavibacter michiganensis* ssp. *sepedonicus*, the causal organism of bacterial ring rot of potato, in weeds and field crops. *J. Phytopathol.*, *153*, 358-365.

Van Steekelenburg, N. A. W. (1985). Resistance to *Corynebacterium michiganense* in tomato genotypes. *Euphytica*, *34*, 245-250.

Varbanets, L. D., Shashkov, A. S., & Kochorova, N. A. (1990). Phosphorous-containing glycopolymers of *Clavibacter michiganense* cell walls. *Carbohydrate Research*, *204*, 157-160.

Vidaver, A. K., & Mandel, M. (1974). *Corynebacterium nebraskense*, a new orange-pigmented phytopathogenic species. *Int. J. Syst, Bacteriol.*, *24*, 482-485.

Vidaver, A. K. (1977). Maintenance of viability and virulence of *Corynebacterium nebraskense*. *Phytopathology*, *67*, 825-827.

Vidaver, A. K. (1982). The plant pathogenic corynebacteria. *Ann. Rev. Microbiol.*, *36*, 495-517.

Wakomoto, S., Uematsu, T., & Mizukami, T. (1969). Bacterial canker disease of tomato in Japan. 2. Properties of bacteriophages specific for *Corynebacterium michiganense*. *Ann. Phytopath. Soc. Japan*, *35*, 168-173.

Wallis, F. M. (1977). Ultrastructural histopathology of tomato plants infected with *Corynebacterium michiganense*. *Physiol. Plant Pathol.*, *11*, 333-342.

Werener, N. A., Fulbright, D. W., Podolsky, R., Bell, J., & Hausbeek, M. K. (2002). Limiting populations and spread of *Clavibacter michiganensis* subsp. *michiganensis* on seedling tomatoes in the greenhouse. *Plant Disease*, *86*, 535-542.

Westra, A. A. G., & Slack, S. A. (1992). Isolation and characterization of extracellular polysaccharide of *Clavibacter michiganensis* subsp. *sepedonicus*. *Phytopathology*, *82*, 1193-1199.

Westra, A. A. G., & Slack, S. A. (1994). Effect of inoculum dose, cultivar, and geographic location on the magnitude of bacterial ring rot symptom expression in potato. *Phytopathology*, *84*, 228-235.

Wohlleben, W., Arnold, W., Bissonette, L., Pelletier, A., Tanguay, A., Roy, P. H., Gamboa, G. C., Barry, G. F., Aubert, E., Davies, J., & Kagan, S. A. (1989). On the evolution of Tn*21*-like multiresistance transposons: sequence analysis of the gene (*aacC1*) for gentamicin acetyltransferase (AAC(3)-I), another member of the Tn*21*-based expression cassette. *Mol. Gen. Genet.*, *217*, 202-208.

Young, J. M., Watson, D. R. W., & Dye, D. W. (2004). Reconsideration of *Arthrobacter ilicis* (Mandel et al., 1961) Collins et al., 1982 as a plant-pathogenic species. Proposal to emend the authority and description of the species. Request for an opinion. *Int. J. Syst. Evol. Microbiol.*, *54*, 303-305.

Zinniel, D. K., Lambrecht, P., Harris, N. B., Feng, Z., Kuczmarski, D., Higley, P., Ishimaru, C. A., Arunakumari, A., Barletta, R. G., & Vidaver, A. K. (2002). Isolation and characterization of endophytic colonizing bacteria from agronomic crops and prairie plants. *Appl. Environ. Microbiol.*, *68*, 2198-2208.

Zgurskaya, H. I., Evtushenko, L. I., Akimov, V. N., & Kalakoutskii, L. V. (1993). *Rathayibacter* gen. nov., including the species *Rathayibacter rathayi* comb. nov., *Rathayibacter tritici* comb. nov., *Rathayibacter iranicus* comb. nov., and six strains from annual grasses. *Int. J. Syst. Bacteriol.*, *43*, 143-149.

9. AFFILIATION

R. Eichenlaub, K-H. Gartemann and Annette Burger, *Fakultaet fuer Biologie, Gentechnologie/Mikrobiologie, University of Bielefeld, Universitaetsstr., 33615 Bielefeld, Germany.*

AMY O. CHARKOWSKI

THE SOFT ROT *ERWINIA*

Abstract. The soft rot *Erwinia* are members of the most studied bacterial family, the *Enterobacteriaceae*, which also includes other important plant and animal pathogens such as *Pantoea*, *Escherichia*, *Salmonella*, *Klebsiella*, and *Yersinia* species. Plant pathologists have been studying the soft rot *Erwinia* for nearly 120 years and published thousands of pages on this pathogen. Because the soft rot *Erwinia* are amenable to genetic manipulation and because they are wide spread in the environment, they serve as an important model for studying the ecology and evolution of enterobacterial pathogenesis.

1. TAXONOMY OF THE SOFT ROT *ERWINIA*

"The correct name of the blackleg pathogen has been the subject of much discussion and the cause of considerable confusion" Leach, 1931.

The genus *Erwinia* is named after one of the founders of phytobacteriology, Erwin Frink Smith, and was established by Winslow et al. (1917) to include the plant pathogenic enterobacteria. Like other enterobacteria, the *Erwinia* are gram-negative rod-shaped non-sporeforming facultative anaerobes with peritrichous flagella. However, the plant pathogenic enterobacteria are a diverse group of organisms and since establishment of this genus, many new genera have been split from the *Erwinia*. The soft rot *Erwinia* cause wilt, stem rot, and soft rot diseases in many crops and are distinguished from other plant pathogenic enterobacteria by the numerous plant cell wall degrading enzymes that they produce.

For the past several decades, enterobacteria that caused soft rot diseases were placed into one of a few *Erwinia* species, usually *Erwinia carotovora* or *Erwinia chrysanthemi*, even though there was strong evidence that the soft rot *Erwinia* and the necrogenic *Erwinia* should not be in the same genus. In 1945, Waldee proposed moving the pectolytic phytopathogenic *Erwinia* to a new genus, *Pectobacterium*. This genus name was used off and on in the literature from 1945 through the 1990s. In the late 1990s, 16S rRNA gene sequences were used to examine *Erwinia* phylogeny and one group used this data to resurrect the genus *Pectobacterium* for the soft rot *Erwinia* (Hauben et al., 1998; Kwon et al., 1997) (Table 1). In both cases, 16S rRNA gene sequences did not provide enough information for phylogenetic analysis and the statistical support for the observed clusters was low. Afterward, some subspecies of *E. carotovora* were elevated to species level, based upon DNA-DNA hybridization, phenotypic, serological, and DNA sequence data (Gardan et al., 2003). A new genus, *Dickeya*, and several new species have also been proposed for *E. chrysanthemi* (Samson et al., 2005). A recently initiated multi-locus sequencing project, as well as DNA hybridization data from the 1970s supports the transfer of *E. carotovora* and *E. chrysanthemi* to two separate genera

S.S. Gnanamanickam (ed.), Plant-Associated Bacteria, 423–505.

as well as the elevation of some soft rot *Erwinia* subgroups to the species level (Hibbing et al., unpublished; Brenner et al., 1977). All phylogenetic analyses completed to date have suffered from the small number of strains available for some enterobacterial species, which makes it difficult to determine the relatedness of these taxa.

Unfortunately, the naming and re-naming of species has caused considerable confusion in the literature, resulting in manuscripts being published with names that were in use for only a few years. Since a comprehensive phylogenetic study of the entire group of soft rot enterobacteria remains to be completed and since *Erwinia* has remained the preferred genus name in the literature, *E. carotovora* and *E. chrysanthemi* will be referred to by their traditional names throughout this chapter.

E. chrysanthemi is more diverse than *E. carotovora* and more biochemical and molecular characterization has been completed for this species than for *E. carotovora*. The *E. chrysanthemi* have been divided into pathovars, biovars, and various RFLP groups (Nassar et al., 1996a). In these analyses, *E. chrysanthemi* strains fall into groups that roughly follow the host and geographic origins, thus the pathovar designations follow the genetic relationships of the strains.

E. carotovora subsp. *carotovora* is by every measure the most diverse of the *E. carotovora* subspecies (De Boer et al., 1979, 1987; Ried and Collmer, 1986; Hibbing et al., in preparation) and it has the broadest host range, having been reported on hosts from at least 16 plant families. Even *E. carotovora* subsp. *carotovora* strains isolated from a single host, potato, have diverse genome structures, serological attributes, and enzyme profiles (De Boer et al., 1979, 1987; Ried and Collmer, 1986; Yap et al., 2004). Unfortunately, little molecular analysis has been completed on strains isolated from hosts other than potato. It is likely that a more comprehensive analysis of *E. carotovora* subsp. *carotovora* would result in this group being divided into multiple subspecies or possibly even multiple species.

Most of the initial taxonomic work on *E. carotovora* and some with *E. chrysanthemi* was comprised of serological characterization of *Erwinia* antigens (De Boer et al., 1979). Early attempts at serology were complicated by the toxicity of *Erwinia* when inoculated into animals such as mice, rabbits, rats, or guinea pigs. Treatments of *Erwinia* preparations using methods designed to eliminate endo- and exo-toxins eliminated this toxicity and allow production of antibodies (Savulesçu et al., 1964).

Up to 45 *E. carotovora* serogroups, which could classify more than 90% of strains, have been identified. (De Boer et al., 1979, 1987). However, neither the serotyping reagents nor a set of serotyped control strains are widely available, and although serotyping has aided in epidemiological studies, as *Erwinia* serogroups have not correlated well with other types of strain grouping, such as biovars or pathovars, their usefulness to taxonomy is questionable. For these reasons, this method is not currently in wide use in *Erwinia* research.

Table 1. Soft rot *Erwinia* species, hosts, and recently proposed names

Former Name	Proposed Name	Host	Reference
Erwinia carotovora			
subsp. *atroseptica*	*Pectobacterium atrosepticum*	potato, tomato	Gardan et al. (2003)
subsp. *betavasculorum*	*Pectobacterium betavasculorum*	sugarbeet, sunflower, potato, artichoke	Gardan et al. (2003)
subsp. *brasiliensis*	no new name proposed	potato	Duarte et al. (2004)
subsp. *carotovora*	*Pectobacterium carotovorum*	various hosts	Hauben et al. (1998)
subsp. *odorifera*	*Pectobacterium carotovorum* subsp. *odoriferum*	chicory, leeks, celery	Hauben et al. (1998)
subsp. *wasabiae*	*Pectobacterium wasabiae*	horseradish	Gardan et al. (2003)
E. chrysanthemi	*Pectobacterium chrysanthemum*	various hosts	Hauben et al. (1998)
pv dieffenbachia (2, 3)[a]	*Dickeya dieffenbachiae*	*Dieffenbachia* banana, tomato	Samson et al. (2005)
	Dickeya dadantii	geranium, pineapple, carnation, poinsettia, sweetpotato, banana, maize, *Philodendron*, African violet	Samson et al. (2005)
pv. dianthicola (1, 7, 9)	*Dickeya dianthicola*	carnation, chicory, artichoke, dahlia, kalanchoë, tomato, potato	Samson et al. (2005)
pv. paradisiaca (3)	*Dickeya paradisiaca*	banana, maize	Samson et al. (2005)
pv. zeae (3, 8)	*Dickeya zeae*	maize, pineapple, potato *Brachiaria*, banana rice chrysanthemum, tobacco	Samson et al. (2005)
pv. chrysanthemi (5, 6)	*Dickeya chrysanthemi*	chrysanthemum, guayule, artichoke, chicory,	Samson et al. (2005)
pv. parthenii (5, 6)	*Dickeya chrysanthemi*	sunflower, tomato, *Parthenium*, *Philodendron*	
Erwinia cacticida	*Pectobacterium cacticidum*	cacti	Hauben et al. (1998)
Erwinia cypripedii	*Pectobacterium cypripedii*	orchid	Hauben et al. (1998)

[a]*E. chrysanthemi* biovars. For example, *E. chrysanthemi* pv. *dieffenbachia* includes biovars 2 and 3. Biovar 3 strains have also been identified in pathovars paradisiaca and zeae.

Much of the early cladistics work on plant pathogenic bacteria was completed with the *Erwinia*. This work required a revision in the bias of plant pathologists who had focused on the phytopathogenic aspects of bacterial pathogens, perhaps to the exclusion of other important traits. This is clearly stated by Lockhart and Koenig (1964) in describing an *Erwinia* taxonomy based on biochemical tests that "It is significant that this result was obtained despite the omission of such 'key' features as pathogenicity and pectinase production. In measures of overall similarity, these properties would be considered no more important than any other." However, trees built upon phenotypic traits have not correlated well with trees built from genetic sequence data. More work remains to be done to understand how the soft rot *Erwinia* are related to each other and to the rest of the enterobacteria.

2. ISOLATION AND IDENTIFICATION OF SOFT ROT *ERWINIA*

Methods for pathogen identification and strain typing are developing very quickly and many of the methods described here are not likely to be in long-term use. Indeed, many of these methods have only been used in a single publication. This following section is an attempt to summarize and compare the many methods used to detect soft rot *Erwinia* and to type strains for epidemiological studies as well as to highlight those methods most likely to be useful in the near future. In general, DNA-based methods are most likely to become the most widely because the specific reagents for these assays, oligonucleotides, are commercially available and inexpensive, unlike antibodies for immunoassays or phage for phage typing. In addition, those DNA-based assays that efficiently provide phylogenetic information are most likely to be used for strain typing since it is preferable to be able to place new isolates within a framework of previously characterize strains.

2.1. Methods and Media used to Isolate Soft Rot Erwinia

Soft rot enterobacterial plant pathogens are relatively easy to culture from infected tissue. For the past several decades, most researchers have been using calcium pectate based medium to isolate soft rot *Erwinia*. These semi-selective media, such as crystal violet pectate (CVP) medium (Cuppels and Kelman, 1974) (Table 2), inhibit the growth of many bacterial species. In addition, the calcium pectate used to gel the growth medium is degraded by soft rot *Erwinia*, resulting in pitting colonies. Unfortunately, appropriate sources of calcium pectate for this medium can be difficult to find and the recipe has been modified repeatedly to accommodate new pectate sources (Hyman et al., 2001; Pérombelon and Burnett, 1991).

Erwinia are isolated similarly to most bacterial plant pathogens. For example, diseased plant stems or tubers are surface sanitized with 70% ethanol or 10% bleach, then a small piece of plant tissue is aseptically cut and soaked in sterile water for up to 5 minutes to allow the bacteria to stream out of the plant tissue. Bacteria in the water can then be streaked onto CVP medium. If the diseased tissue is very soft, such as a potato tuber with soft rot symptoms, then *Erwinia* can generally be isolated by swabbing a sterile inoculating loop across the diseased tissue and streaking the bacteria onto CVP medium. Incubation at lower temperatures favors isolation of *E. carotovora* subsp. *atroseptica*, which can be difficult to isolate if high levels of *E. carotovora* subsp. *carotovora* are also present.

In asymptomatic tissue, soft rot *Erwinia* may be found in stems, roots, and tubers. In potato tubers, they are more likely to be found at the stem end than elsewhere on the tuber (De Boer, 2002). If the pathogen populations are low, they may be enriched from homogenized plant tissue or soil samples by incubation in pectate enrichment medium (PEM) for several hours prior to selection on CVP. Bacterial colonies that form pits on CVP are presumptive *Erwinia* isolates, however, pectolytic *Pseudomonas* and *Flavobacterium* species can also grow and form pits on CVP. These species are easily distinguished from *Erwinia* when the colonies are streaked to King's B medium, on which *Erwinia* species form cream colored colonies with no fluorescent pigment, *Pseudomonas* species produce fluorescent pigment, and *Flavobacterium* species form bright yellow colonies. Some modifications may be used to increase the likelihood that *Erwinia* species are isolated. For example, plates may be incubated under anaerobic conditions, which allow the growth of facultative anaerobes, such as *Erwinia* species, but not pectolytic pseudomonads.

Some CVP recipes result in a harsh medium that can inhibit the growth of injured cells. Liao and Shollenberger (2004) found that three log units fewer acid-injured *E. carotovora* cells grew on CVP than on a non-selective rich medium, brain heart infusion agar, suggesting that the majority of stressed cells, such as those from environmental sources, may not form colonies on CVP. The CVP-S2 medium recipe reported by Hyman et al. (2001) made with Genu pectin type X-914-02 has a plating efficiency similar to rich medium (Charkowski, unpublished). However, culturing cells in a rich, non-selective medium or in pectate enrichment broth (PEM) prior to growth on CVP may increase the chance of isolating *Erwinia* from environmental samples where it is present at low levels.

Non-pectin based media have also been used to isolate soft rot *Erwinia*. One of the simplest to use is a modified version of a raffinose-based medium developed by Segall (1971). Both *E. carotovora* and *E. chrysanthemi* form small dark red colonies on this medium. Unlike with CVP, pectolytic pseudomonads do not grow on raffinose medium, eliminating the chance of isolating these common bacteria instead of *Erwinia*. However, many other species do grow on raffinose medium and its plating efficiency for soft rot *Erwinia* is about 1% of rich medium, thus CVP is an easier medium to use to isolate *Erwinia*. Numerous other media for the isolation of soft rot *Erwinia* have been described but they are not widely used (Burr and Schroth, 1977; Logan, 1963; Noble and Graham, 1956; Stewart, 1962; Thorne, 1972; Tsuyamu and Sakamoto, 1952).

Table 2. Media used for isolation of soft rot *Erwinia*

CVP-S2 (Hyman et al., 2001)	**500 ml**
Tryptone	0.5 g
Tri-sodium citrate	2.5 g
$NaNO_3$	1.0 g
10% aqueous solution of $CaCl_2.H_2O$	5.1 ml
0.075% aqueous solution of crystal violet	1.0 ml
5 M NaOH	1.3 ml
Agar	2.0 g
Sodium polypectate	9.0 g
(Slendid type 440 or Genu pectin type X-914-02)	
Cold distilled water	500 ml

Dissolve all ingredients, excluding the polypectate and NaOH, in 250 ml of water. Dissolve the polypectate and NaOH in a separate 250 ml of water. Autoclave the two mixes for 15 min at 120ºC, then slowly pour the two mixes together and pour plates. This medium cannot be easily remelted, so plates must be poured immediately.

PEM	**1000 ml**
Sodium polypectate	1.5 g
10% $(NH_4)_2SO_4$	10.0 ml
10% K_2HPO_4	10.0 ml
5% $MgSO_4{\cdot}7H_2O$	5.0 ml

Modified Raffinose Medium (Segall, 1971)	**1000 ml**
Raffinose	10 g
K_2HPO_4	2 g
$(NH_4)_2SO_4$	5 g
Eosin yellow	0.4 g
Methylene blue	0.065 g
Agar	15 g

King's B Medium (King et al., 1954)	**1000 ml**
Proteose Peptone no. 3	20 g
Glycerol	10 g
K_2HPO_4	1.5 g
$MgSO_4 \cdot 7H_2O$	1.5 g
Agar	15 g

2.2. Phenetic Tests

Biochemical tests were widely used in the past to identify soft rot *Erwinia* species to species and subspecies level. However, there are numerous strains that do not have typical biochemical reactions and atypical isolates are commonly reported (Yap et al., 2004). Since molecular assays provide more definitive identifications and generally require fewer specialized reagents, they are becoming more widely used for species identification. We have found that the simplest and most reliable phenetic tests are to determine if the isolate in question grows at 37ºC and if the isolate can grow in Luria-Bertani medium with 5% NaCl. Soft rot *Erwinia* that do not grow at 37ºC are likely to be *E. carotovora* subsp. *atroseptica* and those that do not grow in high salt medium are likely to be *E. chrysanthemi*. However, some *E. chrysanthemi* strains also do not grow at 37°C, so these tests are not definitive.

2.3. Hybridization Assays

Specific hybridization probes for *Erwinia* have been obtained through subtractive hybridization of genomic DNA (Darrasse et al., 1994a). These probes have not been widely used since PCR-based methods have proved to be simpler for the detection and identification of *Erwinia* strains.

Recently, Fessehaie et al. (2003) described a macroarray method for detection of potato-infecting bacteria, including soft rot *Erwinia*. This array contains 40 different 16 to 24-mer oligonucleotides homologous to the 3' end of the 16S rRNA gene and the 16S-23S spacer region. Macroarrays such as these are relatively inexpensive to produce. The oligos may be manually spotted onto nylon membranes with a pin array, then PCR-amplified genomic DNA fragments are labelled with a chemiluminescent non-radioactive DNA labelling kit and hybridized to the array. The signal can be detected by exposure of the membrane to x-ray film or a documentation system capable of recording light emission. This economical method is currently one of the most attractive detection and typing method since numerous oligos may be tested on a single macroarray, allowing detection of or epidemiological studies of multiple strains or even multiple species.

2.4. PCR Assays

PCR-based assays can be divided into two categories, those used to determine if a diseased plant sample is infected with soft rot *Erwinia* and those used to characterize isolates. These assays were all developed prior to the description of the newest *E. carotovora* subspecies, *E. carotovora* subsp. *brasiliensis*. Thus, the efficacy of these primer sets in detection and diagnosis of this *E. carotovora* subspecies is unknown. Most of these assays were also developed prior to the widespread use of real-time PCR and arrays and it is likely that these technologies will eventually supersede many of the assays described here. However, the real-time PCR and array assays developed to date are not able to type strains as well as more conventional PCR-based assays described below, so these primer sets are likely to remain useful for strain typing in the near future.

Bacterial DNA suitable for PCR is not difficult to isolate from *Erwinia* and common DNA isolation protocols may be used. We have found that a simple boil preparation works well if a high number of cells are available. Essentially, a bacterial colony is scraped from an agar plate into sterile water or bacteria are allowed to stream from a diseased plant sample into sterile water. A 500 μl portion of this suspension is boiled for 5 minutes, centrifuged at high speed for 5 minutes, and 1 ml of the supernatant is added to the PCR mix. If a higher DNA concentration or cleaner DNA is needed for the PCR, the DNA in the supernatant may be precipitated with two volumes of ethanol plus 0.02 vol of 10 M sodium acetate, pelleted by centrifugation, then washed with 70% ethanol. The DNA can then be suspended in 50 μl of water or TE and 1 μl can be used in standard PCR.

Erwinia DNA is more difficult to isolate from water, soil or asymptomatic plant samples where the pathogen levels are likely to be much lower than in diseased plant samples. To increase sensitivity in these cases, some researchers have included an enrichment step in a rich medium, such as nutrient agar, or in a selective medium, such as CVP, prior to detection by a PCR assay (Hyman et al., 1997; Toth et al., 1999b).

2.4.1. Primers Designed to Detect All Soft Rot Erwinia

Toth et al. (1999b) designed two primers, SR3f (5'-GGTGCAAGCGTT AATCGGAATG-3') and SR1cR (5'-AGACTCTAGCCTGTCAGTTTT-3') that amplify a 119-bp fragment of the 16S rRNA gene from all soft rot *Erwinia* species tested, including *E. cacticida* and *E. carnegieana*. In addition, this primer set amplified a 119-bp fragment from two necrogenic *Erwinia*, *E. rubrifaciens* and *E. salicis*. This assay was specifically designed to detect soft rot *Erwinia* cells in plant tissue culture. Therefore, an enrichment step was included in the protocol to increase the sensitivity of this assay.

2.4.2. Primers Designed to Detect E. carotovora Subspecies

Kang et al., (2003) developed a primer set, EXPCCR (5'-GCCGTAATTGCCT ACCTGCTTAAG-3') and EXPCCF (5'-GAACTTCGCACCGCCGACCTTCTA-3') based on a fragment of *Erwinia* sequence amplified by universal rice primers (URPs). The primers EXPCCR and EXPCCF amplify a 550-bp fragment from *E. carotovora* subsp. *carotovora* and *wasabiae* strains. Digestion of this fragment with RsaI resulted in 5 different RFLPs for the *E. carotovora* subsp. *carotovora* strains and 1 RFLP for the *E. carotovora* subsp. *wasabiae* strains. Kang et al. (2003) also designed nested primers, INPCCR (5'-GGCCAAGCAGTGCCTGT ATATCC-3') and INPCCF (5'-TTCGATC ACGCAACCTGCATTACT-3'), that amplify a 400-bp fragment and which can be used to increase the sensitivity of this detection assay.

Darrasse et al. (1994b) developed a primer set, Y_1 (5'-TTACCGGACG CCGAGCTGTGGCGT-3') and Y_2 (5'-CAGGAAGATGTCGTTATCGCGAGT-3') that amplifies a fragment of the *pelY* family pectate lyases from *E. carotovora* subspecies, with the exception of *E. carotovora* subsp. *betavasculorum*. Digestion of the amplified fragment with AluI, HaeII, Sau3AI, and HpaII resulted in 21 RFLP groups, with one group containing just *E. carotovora* subsp. *wasabiae*, two with just *E. carotovora* subsp. *atroseptica*, and two with just *E. carotovora* subsp. *odorifera*. Three of the RFLP groups contained both *E. carotovora* subsp. *odorifera* and *E. carotovora* subsp. *carotovora* and the remaining 13 RFLP groups contained *E. carotovora* subsp. *carotovora* strains.

De Boer and Ward (1995) developed a primer set, Eca1f (5'-CGGCATC ATAAAAACACG-3) and Eca2R (5'-GCACACTTCATCCAGCGA-3') that amplifies a fragment only from *E. carotovora* subsp. *atroseptica*. This primer set was later used by Hyman et al. (2000) to develop a quantitative PCR-based assay that includes a DNA extraction control suitable for measuring *E. carotovora* subsp. *atroseptica* levels on potato tubers. Hyman et al. also provided a key to compare PCR results using this assay to the likelihood of blackleg development by extrapolating from epidemiological data collected by Bain et al. (1990). However, the predictive value of this assay has not been validated.

The primers developed by De Boer and Ward (1995) were also used by van der Wolf et al. (1996) to test immunomagnetic sepearation (IMS) to improve the sensitivity of PCR. van der Wolf et al. (1996) compared multiple antibodies and paramagnetic beads to optimize IMS-PCR for detection of *E. cartovora* subsp. *atroseptica*. They found that polyclonal antibodies conjugated to Advanced Magnetics AM-pA particles were optimal. However, they chose to use AM-AR beads for most of their experiments because they were much less expensive. Their optimized IMS-PCR method reduced the detection threshold in potato peel extracts from 10^5 CFU/ml to 2 x 10^3 CFU/ml. They also found that if the IMS particles were plated onto CVP, they could detect as few as 10^2 CFU/ml from potato peel extract. Since *E. carotovora* subsp. *atroseptica* is difficult to isolate if *E. carotovora* subsp. *carotovora* is present, van der Wolf et al. (1996) used this method to attempt to isolate *E. carotovora* subsp. *atroseptica* from samples also containing *E. carotovora* subsp. *carotovora.* They found that IMS followed by plating on CVP allowed isolation of *E. carotovora* subsp. *atroseptica* in suspensions with a 100-fold excess of *E. carotovora* subsp. *carotovora.*

2.4.3. Primers Designed to Detect E. chrysanthemi

Nassar et al. (1996a) designed a primer set, ADE1 (5'-ATCAGAAAGCCCGCAG CCAGAT-3') and ADE2 (5'-CTGTGGCCGAT CAGGATGGTTTTGTCGTGC-3'), which amplifies a 420-bp pectate lyase gene fragment from all *E. chrysanthemi* strains tested, but not from any other soft rot *Erwinia.* Digestion of this fragment with AluI, HpaII, and Sau3AI revealed 16 RFLP patterns among the 78 strains tested. Dendrograms of these fragments correlated well with groupings derived using genetic and phenotypic markers reporter by other groups demonstrating that a typing method based on this virulence gene correlated well with other markers (Boccara, 1991; Dickey, 1979; 1981; Janse and Ruissen, 1988; Nassar et al., 1994).

2.4.4. Primers Designed to classify Erwinia Isolates

Random amplified fragment polymorphism (RAPD) analysis has been used to type soft rot *Erwinia* strains (Maki-Valkama and Karhalainen 1994; Parent et al., 1996; Waleron et al., 2002). However, as more genome sequences have become available, more specific primers are developed, and arrays become more commonly used, and it is unlikely that RAPDs will be widely used in the future. Therefore, all of these methods will not be described in detail.

An ITS-PCR assay useful for differentiation of *Erwinia* species was developed by Toth et al. (2001). In this assay, primers G1 (5'-GAAGTCGTAACAAGG-3') and L1 (5'-CAAGGCATCCACCGT-3') amplify a portion of the intergenic transcribed spacer regions. Two major fragments of 535-550 bp and 575-580 bp are amplified from *E. carotovora* and three major fragments of 440-450 bp, 575-590 bp, and 690-720 bp are amplified from *E. chrysanthemi.* The amplified DNA may be analyzed directly by gel electrophoresis or treated with restriction endonucleases prior to electrophoresis. We have found this assay particularly useful and reliable for differentiating *E. carotovora* from *E. chrysanthemi.* Minor fragments also amplified with primers G1 and L1 or restriction digests of the amplified bands may be used to further differentiate isolates. Importantly, this assay can only be used with pure cultures since the G1 and L1 primers will amplify fragments from numerous other species and it would be impossible to distinguish the *Erwinia* fragments among those amplified from other bacterial species present in plant or other environmental samples.

This same research group also developed an amplified fragment length polymorphism (AFLP) method for differentiating *Erwinia* isolates (Avrova et al., 2002). This method is useful for the rapid identification of unknown isolates and for epidemiology. Unlike several of the methods described above, it appears to be able to differentiate *E. carotovora* subsp. *atroseptica*, *betavasculorum*, *carotovora*, *odorifera*, and *wasabiae*, as well as multiply *E. chrysanthemi* subgroups.

Although the AFLP method described above is likely to be able to classify all soft rot *Erwinia* strains into groups, it can not be used to place these groups phylogenetically among more distantly related species. For this reason, we developed a multi-locus sequence approach for classification of enterobacterial plant pathogens (Hibbing et al., in preparation). Primer sets were designed to amplify fragments from *acnA*, *gapA*, *icdA*, *mdh*, *mltD*, *pgi*, and *proA*, which are central metabolism genes present in all enterobacterial species. These amplified gene fragments were sequenced and phylogenetic analyses were used to describe the relationships of these enterobacterial strains. The data for these gene fragments are available in the Genbank (www.ncbi.nlm.nih.gov) and ASAP (https://asap.ahabs.wisc.edu/asap/ ASAP1.htm) databases (Glasner et al., 2003) so that these sequence alignments may be used to determine the phylogeny of new isolates.

2.5. Immunoassays

Although they are commercially available, enzyme linked immunosorbant assays (ELISAs) are not commonly used for detection of soft rot *Erwinia* because in most cases they have a high limit of detection, which does not make them useful for epidemiological studies. However, there have been reports of DAS-ELISA detecting as little as 10^3 CFU/ml of *Erwinia* from symptomless carnation stem cuttings (Nassar et al., 1996b), thus it may be possible for ELISAs to detect low levels of soft rot *Erwinia*. In addition to ELISAs, several other types of immunoassays have been used for epidemiological studies of the soft rot *Erwinia*. Ouchterlony double diffussion tests were widely used in the 1970s and 80s to examine *Erwinia* serogroups, but this method is rarely used today.

In 1977, Allan and Kelman first used immunofluorescence staining to detect *E. carotovora*. This method was applied in the early 1980s to detection of *E. carotovora* in insects (Phillips and Kelman, 1982), and immunofluorescence methods have continued to be useful. An immunofluorescence colony-staining assay was described by Schober and van Vuurde (1997) that has high sensitivity for soft rot *Erwinia*. They used a spiral plater, which is capable of plating a 100-fold dilution of cells on one plate, to plate extracts from inoculated chicory.

Top-agar was added to the plates, then the cultures were incubated for 24 to 48 hours to allow the bacteria to grow. After colonies had formed, anti-*Erwinia* antibody conjugated to fluorescein isothiocyanate was plated on the top agar and allowed to diffuse into the plate and bind to the *Erwinia* colonies. The plates were then washed and examined with a fluorescence microscope. This assay has a 100 CFU/ml detection limit and also the added benefit that stained colonies can be isolated from the plates for further characterization. This method also allows subgroups of soft rot *Erwinia* to be examined if appropriate antibodies are available.

2.6. Volatile Metabolic Profiling

Soft rot is sometimes known as a disease that must be diagnosed with binoculars since potato storage facilities that have severe soft rot are very noxious. Researchers have found that volatiles released by diseased plants differ depending on which pathogen is present and have suggested using these volatiles as the basis for sensors for disease. To date, no commercial sensors capable of detecting soft rot by profiling volatiles are available.

Of those researchers examining volatiles released by *E. carotovora* infected potato or onion, not all have found the same volatile compound profiles, but the profiles have overlapped for the most part. The variation in profiles could be due to different potato varieties, bacterial strains, or experimental conditions. In addition, it is difficult to eliminate other bacteria from the test samples and secondary invaders also produce volatile compounds, which could account for some of the variability among reports.

Volatiles reported by more than one group as produced by *E. carotovora*-infected potatoes include ethanol, acetaldehyde, 1-propanol, 2-methyl-1-propanol, acetone, 3-hydroxy-2-butanone, and 2-butanone (Lui et al., 2005; Varns and Glynn, 1979; Waterer and Pritchard, 1984). Many of these volatiles have also been found in uninfected controls, which make it more difficult to use volatile profiling as a detection method. However, the concentration of the volatiles increases, in some cases by as much as 2000-fold, as the tuber decays. This increase in volatiles or perhaps the presence of more distinctive volatiles from secondary invaders may eventually be used for early detection of decaying vegetables in warehouses.

Lui et al. (2005) identified the highest number of pathogen specific volatiles. They found that acetic acid ethenyl ester was unique to *E. carotovora* subsp. *atroseptica*, cyclohexene, diazene, and methoxy-(1,1-dimethyl-2-dihydroxy-ethyl)-amine were unique to *E. carotovora* subps. *carotovora*. Over 20 volatiles were found with both *E. carotovora* subsp. that were not present in control tubers or in tubers infected with *Fusarium sambucinum*. Lui et al. (2005) examined 20 replicates for each pathogen and of the *Erwinia*-specific compounds, only five were detected in more than 10 of the 20 replicates, and none were detected in all of the replicates. Most of the volatiles were only detected in one to five of the replicates. Even though most of the pathogen-specific volatiles were only detected in a few of the replicates, this method still holds promise since in a commercial setting, the sample would consist of volatiles from numerous potatoes, some of which are likely to be emitting pathogen-specific volatiles.

2.7. Conductimetry

Measuring changes in conductance (impedance) has been explored mainly as a method to quickly detect food-borne human pathogens. This method has also been adapted to detect soft rot *Erwinia*. Essentially, the conductivity of a bacterial culture is measured over time and a change in conductance due to bacterial metabolic activity indicates that the pathogen is present. Fraaije et al. (1996a) tested several complex and minimal media to identify one useful for detection of *Erwinia* in potato tuber peel extracts by conductivity. They found that a miminal medium with sodium polypectate as a carbon source allowed detection of soft rot *Erwinia*, but not common pectate-degrading bacteria that might also be present on tuber peels, such as pectolytic *Pseudomonas* or *Klebsiella* species. This is because polygalacturonase is required for the conductance change observed and other, non-target bacteria that may be present on potato periderm, do not produce this enzyme. Using this method, they were able to detect 10^2-10^3 CFU/ml *E. carotovora* subsp. *atroseptica* in potato peel extracts. This method for detection of soft rot bacteria in produce is attractive because it is simple, relatively inexpensive, and can be used to test numerous samples easily. However, there are no reports of this being used in the vegetable industry, therefore, we do not yet know how effective it is under real world conditions.

2.8. Comparison of Detection and Strain Typing Methods

Other methods including phage typing and fatty acid profiling have been used by some groups, but with the advent of high throughput DNA analysis methods, these techniques are not likelty to be in widespread use in the future (Gross et al., 1991; De Boer and Sasser, 1986). Some authors have compared the efficacy of different detection methods and differentiation methods to determine which work best for the soft rot *Erwinia* (Fraaije et al., 1996b; 1997; Toth et al., 1999a). In general, traditional PCR has not been found to be very sensitive in comparison to immunofluorescence staining for detection of soft rot *Erwinia* in potato, mainly because there are PCR-inhibiting compounds present in tuber peel extracts (Fraaije et al., 1996b). If immunomagnetic separation of *Erwinia* cells is performed prior to PCR, this greatly increases the sensitivity. (van der Wolf et al., 1996).

Toth et al. (1999a) compared physiological profiling (Biolog), RFLPs, entero-bacterial repetitive intergenic consensus repeats (ERICs), phage typing, RAPDs, and Ouchterlony double diffusion tests to determine which assay was capable of detecting the highest diversity in *E. carotovora* subsp. *atroseptica*. Using Simpson's index of diversity (Hunter and Gaston, 1988), they found that phage typing was best able to discriminate among strains. They also found that they did not obtain similar groupings with all of different methods. For example, phage typing and serotyping grouped strains in one way, while RFLP and ERIC grouped strains in another. Again, DNA sequence based methods are likely to be used for strain typing in the future due to their ease, availability of reagents, and usefulness of the data for phylogenetic placement.

3. *ERWINIA* ECOLOGY AND EPIDEMIOLOGY

3.1. Spread of Erwinia in the Environment

The majority of the soft rot *Erwinia* epidemiology work has been done with *E. carotovora* subsp. *carotovora* and subsp. *atroseptica* on potato; very little epidemiology has been done with any of the other soft rot *Erwinia*, including *E. chrysanthemi*. This bias is because of the large losses faced by the potato industry each year due to *E. carotovora* and because most of the epidemiology work has been done in temperate climates where *E. chrysanthemi* and other soft rot *Erwinia* are uncommon.

The soft rot *Erwinia* are easily found in the environment. They have been cultured from numerous plant hosts, soil, insects, and both surface and ground water (McCarter-Zorner, 1984, 1985). They can spread long distances on vegetatively propagated plants, ranging from carnations to potato tubers. They are also easily spread in aerosols and may travel long distances in the atmosphere. Quinn et al. (1980) found that *E. carotovora* could be caught in the open air during rainfall from mid to late summer, in autumn, and in early winter. Although cells were not found during every rainfall, they were never found in dry weather, suggesting that rainfall generates *Erwinia* aerosols. The *E. carotovora* cells were found downwind of potato fields, as well as in locations not near potato fields and after the potato harvest, suggesting that there are sources other than potato for these bacteria.

As *Erwinia* decay plant material, they open the door for numerous other species, ranging from bacteria to insect larvae, to colonize the dying plant host. The progression of this community and how different members affect the development of disease is unknown. Studies by Leach in the 1920s and 30s suggested that there were intimate interactions between *Erwinia* and the seed corn maggot (Leach, 1926, 1931, 1933). Leach found that maggots, which are often found in decaying plant material, were colonized by *E. carotovora* prior to invasion into plants. He suggested that the maggots inoculate plant tissue, such as potato tubers, with *E. carotovora* and took advantage of the *Erwinia*-caused decay to be able to more easily enter the plant tissue.

Soft rot *Erwinia* have been found in association with other invertebrates, ranging from fruit flies to snails (Harrison et al., 1977; Molina et al., 1974; Phillips and Kelman, 1982). These insects may play an important role in the spread of soft rot bacteria between plant hosts and may also respond to *Erwinia*. For example, Basset et al. (2000) showed that fruit flies have an immune response when inoculated with *Erwinia*. The *Erwinia* genomes also hint at this relationship; both the *E. carotovora* and *E. chrysanthemi* genomes encode several genes homologous to nematode-colonizing enterobacteria and *E. chrysanthemi* 3937 encodes four genes that are similar to Bt toxin (Glasner et al., submitted).

Probably the most important mechanism for spread is on plant material itself. Soft rot *Erwinia* can survive for several generations in association with vegetatively propagated crops, such as potato (De Boer, 2002). These crops are often transported across long distances, effectively spreading and mixing *Erwinia* populations.

3.2. The Role of Environment in Disease Development

In the laboratory, soft rot *Erwinia* can infect nearly any plant if the plant is stressed and enough bacteria are inoculated. With their pervasive nature and broad host range, it appears at first glance that this species is the cockroach of plant pathogens - a pathogen that is found everywhere and that will eat anything. However, an examination of the epidemiology of the soft rot *Erwinia* shows that under natural conditions, the host range of these pathogens appears to be limited. For example, the various subspecies of *E. carotovora* may be distinguished by DNA and biochemical analysis, and are also found on specific hosts. For example, although *E. carotovora* subsp. *carotovora* has been reported on plant hosts in at least 16 plant orders, *E. carotovora* subsp. *atroseptica* has only been reported on potatoes and tomatoes even thought it is able to rot other species under laboratory conditions. Even *E. carotovora* subsp. *carotovora*, which was divided into numerous serogroups in the 1970s and 80s, may have some host specialization. For example, although over 30 *E. carotovora* subsp. *carotovora* serogroups have been described, only a few are routinely found on potatoes (Maher et al., 1986).

Soft rot *Erwinia* colonize mainly the roots, stems, fruits, and other storage organs of host plants. However, they do not appear to colonize plant leaves until the plant has become quite diseased. This suggests that the soft rot *Erwinia* are not fit on the harsh dry environment of a leaf surface, unlike other plant associated bacteria, most notably *Pseudomonas* species. Like *Pseudomonas*, soft rot *Erwinia* can colonize plant tissue and reach high population levels without causing symptoms. In general, *Erwinia* is able to asymptomatically colonize the vascular system of ornamental plants and the lenticels of potato tubers. The bacterial cells can grow and remain in association with their host for long periods of time and not cause disease. This can be devastating for vegetatively propagated crops since the pathogen is easily spread during propagation and once inside the vascular system of a plant, there are no practical large-scale methods for eliminating the pathogen and curing the plant.

Although stem rot and tuber decay are common in potato fields, multiple workers have found that inoculation of potato seed tubers does not guarantee that disease will develop because there is a large environmental component to this disease, with soil type, temperature, and moisture being critical for disease development in the field and oxygen levels affecting disease development in storage (Molina and Harrison, 1980; Webb and Wood, 1974; Weber, 1984). The experiments of Webb and Wood (1974) and Weber (1984) demonstrate the difficulty in getting disease development in inoculated plants in experimental field plots if the environmental conditions are not favorable. Webb and Wood (1974) found that although *E. carotovora* subsp. *atroseptica* was cultured from 40% of the tubers examined prior to planting, only 0.3% of 8400 plants developed blackleg, demonstrating that even though the incidence of infestation is high, the disease incidence may be very low. Weber (1984) planted tubers that had been inoculated with *E. carotovora* during the previous autumn, winter, or spring and found similarly low levels of blackleg in field plots.

The epidemiological effects of temperature have been explored more than the other environmental components. As might be expected, *E. carotovora* subsp. *atroseptica*, which cannot grow above 33°C, is found mainly in the spring, and in cooler climates, while *E. carotovora* subsp. *carotovora* is found more often during the warmer summer months or in warmer regions (De Mendoça and Stanghellini, 1979; Jorge and Harrison, 1986; Powelson, 1980). Multiple researchers have found that temperature differentially affects the production of *Erwinia* virulence proteins and virulence related phenotypes including production of pectate lyase, pectin lyase, polygalacturonase, biofilms, and AHLs, with the specific effects being strain dependent (Hasegawa et al., 2005; Hugouvieux-Cotte-Pattat et al. 1996; Lanham et al., 1991; Nguyen et al., 2002; Smadja et al., 2004b; Yap et al., 2005). Both H-NS and AHL contribute to temperature-controlled gene regulation in the soft rot *Erwinia*, and there are likely to be additional temperature regulators in addition to these (Hasegawa et al., 2005; Nasser et al., 2001).

The role of temperature on seed potato piece decay was examined by Molina et al. (1980) who found that when tubers were inoculated with *E. carotovora* subsp. *carotovora* or *E. carotovora* subsp. *atroseptica* and planted in cool soil (7.0-18.5°C average min and 16-26°C average max for the first 30 days after planting) in the field or greenhouse, *E. carotovora* subsp. *atroseptica* caused more seed piece decay than *E. carotovora* subsp. *carotovora*. Furthermore, the tubers inoculated with *E. carotovora* subsp. *carotovora* were no different than uninoculated controls. The opposite was seen when soil temperatures were high at planting time, (21.4-24°C average min and 29.6-35°C average max for the first 30 days after planting), with *E. carotovora* subsp. *carotovora* causing more disease. At an intermediate soil temperature, both pathogens were ineffective.

Molina et al. (1980) also inoculated *Erwinia*-free stem cuttings with both pathogens and planted the stem cuttings in a greenhouse. They found that both subspecies could cause blackleg, but, that as with tuber decay, temperature played a large role. Only *E. carotovora* subsp. *atroseptica* caused stem infections at the lowest temperature (7-18°C) while only *E. carotovora* subsp. *carotovora* caused disease at the higher temperatures (24-35°C). Both strains caused blackleg at intermediate temperatures. In the field, there was no correlation found between soil temperature and post emergence blackleg or total disease, but in every diseased stem, both subspecies were found, making it difficult to interpret the data.

3.3. Interactions of soft rot Erwinia with other bacterial species

E. carotovora is often found in plants with other pathogens, including soft rot *Clostridium*, *Clavibacter michiganenesis*, *Verticillium*, and the potato rot nematode, *Ditylenchus destructor*. In some cases, synergism may occur. The prevalence of *E. carotovora* on plants that have also been attacked by another pathogen brings the question of how often soft rot *Erwinia* are the primary pathogen versus being a secondary invader that causes symptoms that make the primary pathogen unrecognizable. This scenario is not unlikely with several of the pathogens that *Erwinia* has been found in association with. For example, phytobacteriologists do not routinely examine samples for anaerobic bacteria such as *Clostridium* and therefore there is little data on how often these two pathogens are present together in diseased samples. However, tubers infected with *Clostridium* alone or *Clostridium* and *Erwinia* are more common than those infected with *Erwinia* alone, suggesting that in many cases, *Clostridium* is the primary pathogen (Campos et al., 1982; Pérombelon et al., 1979). Other pathogens, such as *Clavibacter michiganensis* grow very slowly and are easily overgrown by soft rot *Erwinia* even in plant samples. Thus, the presence of *C. michiganensis* may be overlooked in plants with symptoms of soft rot. The advent of array technologies, such as the array developed by Fessehaie et al. (2003), which allow samples to be more easily examined for multiple pathogens, will allow us to better define the roles that these synergisms play in pathogen ecology and epidemiology and identify common pathogen partners of the soft rot *Erwinia*.

In addition to acting in concert or competition with other pathogens, soft rot *Erwinia* also affect the rest of the microbial community that lives in association with its host plants when it causes disease. Reiter et al. (2002) used both 16S rRNA T-RFLP analysis and culturing to examine how the endophytic microbial community was altered by *E. carotovora* subsp. *atroseptica* infection. They found that the community became more complex and contained bacteria from a broad phylogenetic spectrum. They also found that several of the isolates from potato stems were antagonists of *E. carotovora* and were able to protect potato plants from stem rot, suggesting that in a disease-interaction, E. *carotovora* is in competition with potato endophytes for resources.

A few genes that are likely to affect competition with other microbes on plants, in soil, or in water have been identified in the soft rot *Erwinia.* Llama-Palacios et al. (2002) found that *ybiT,* which is likely to encode for part of an ABC transporter, is not required for virulence, but that it is required to limit the growth of secondary invaders in plants infected with *E. chrysanthemi*, perhaps because it is involved in the secretion of an antibiotic.

Some strains of *E. carotovora* produce the carbapenem antibiotic 1-carbapen-2-em-3-carboxylic acid (Car), another indication that *E. carotovora* is in competition with potato endophytes and secondary invaders for resources (Parker et al., 1982). The gene cluster *carRABCDEFGH* is required for production of Car and these genes are regulated by an AHL quorum sensing system (McGowan et al., 1995, 1996). Many *E. carotovora* strains do not produce Car but they do encode cryptic *car* genes. Holden et al. (1998) found that in many cases, the lack of expression of the *car* operon was due to a non-functional CarR and that expression *in trans* of a funcational CarR restored antibiotic production to these strains.

E. carotovora may also produce multiple forms of carotovoricin, a phage-tail-like bacteriocin which kills closely related strains and species (Nguyen et al., 2001). The expression of carotovoricin is induced by DNA damage, similarly to pectin lyase. It is also regulated by temperature, with the highest level of production between 23 and 26ºC (Nguyen et al., 2002). This protein may also affect competition with other enterobacteria in the environment. Multiple strains of E. *carotovora* are usually found in a field with diseased plants, and sometimes even in a single diseased plant (Yap et al., 2004). The role that these anti-bacterial molecules play in the structure of the *Erwinia* population in a plant or field is unknown.

3.4. Soft Rot as a Storage Disease

Once fruit and vegetable crops, ranging from potatoes to pineapples, are in storage, they become very susceptible to soft rot *Erwinia.* As with other aspects of *Erwinia* epidemiology, most of the work examining *Erwinia* as a storage disease has been completed with *E. carotovora* on potato.

There are several aspects of potato storage that contribute to soft rot decay of tubers, with oxygen levels and storage temperature being foremost in the factors that affect disease development. When potatoes are first harvested, they respire at a high rate, in part because the potato tissues are repairing wounds incurred during harvest. Recently harvested potatoes can quickly use all available oxygen if they are stored in unventilated warehouse. This circumstance occurs when temperatures directly after harvest are high, leaving growers, who may not have refrigeration in their warehouses, forced to choose between allowing their stored crop to warm, which is conducive to many storage pathogens, or keeping the warehouse closed and cooler. Unfortunately, once oxygen levels have fallen, the potato tubers are no longer able to respire. These tubers are more vulnerable to soft rot because they can no longer repair wounded periderm and they may be affected in ability to mount other aspects of an effective pathogen defense. That oxygen is critical for tuber resistance to soft rot was clearly demonstrated by Maher and Kelman (1983) who showed that tubers injected with filter-sterilized filtrates from *Erwinia* cultures did not rot when incubated in air, but that these tubers did rot when they were incubated under anaerobic conditions. In addition to the host being compromised under low oxygen levels, some *Erwinia* virulence genes, including those encoding cell wall degrading enzymes are up-regulated under anaerobic conditions (James and Hugouvieux-Cotte-Pattat, 1996).

4. CONTROL OF *ERWINIA*

Soft rot *Erwinia* are present in numerous environments, making it difficult to control soft rot diseases. Numerous chemicals, including copper, formaldehyde, bleach, mercuric chloride, 8-hydroxyquinoline, anthium dioxide, chlorine dioxide, gluytaraldehyde, benzalkonium chloride, cetalkonium chloride, polymeric biguanide salt, EDTA, paraformaldehyde, sulfur dioxide, methyl 4-hydroxybenzoate, ethyl 4-hydroxybenzoate, peracetic acid, sodium 2-phenylphenoxide and acetaldehyd, as well as plant extracts have been tested for control of soft rot *Erwinia* and some provide limited control of this pathogen (Letal, 1977; Lund and Lyon, 1975; Wyatt and Lund, 1981). However, as with other bacterial plant pathogens, there are no good commercial chemical options for control of soft rot *Erwinia*. Antibiotics, such as streptomycin, tetracylcline, and vancomycin are able to control of soft rot *Erwinia*. However, the use of antibiotics in horticultural crops is banned in many countries mainly because of the risk of increased spread of antibiotic resistance genes in microbial populations. Also, in most cases, antibiotics are too expensive to use for control of bacterial plant diseases. There have been some attempts at biocontrol, including groups that recently showed that both *Bacillus thurigiensis* and *Rhodococcus erthropolis*, which interferes with *Erwinia* cell-cell signalling and induction of virulence genes through degradation of the quorum sensing molecule N-acyl homoserine lactone, can inhibit symptom development on potato (Dong et al., 2004; Uroz et al., 2003). However, there are currently no commercial biocontrol agents available specifically for controlling soft rot.

Rotations, cultural practices, sanitation, removal of cull piles, clean irrigation water, appropriate plant nutrition, soil moisture, insect control, and use of resistant cultivars is more effective in controlling soft rot *Erwinia* than is trying to cause a reduction in inoculum load in the seed since the sources of inoculum are so numerous and ubiquitous (Maher et al., 1986). One noted exception to this is *E. carotovora* subsp. *atroseptica*, which has a narrow host range and appears to mainly survive on potato and potato debris. The use of *Erwinia*-free tissue culture plants as the starting material for potato production has reduced the incidence of *E. carotovora* subsp. *atroseptica* on potato in developed countries. However *E. carotovora* subsp. *carotovora* is still widespread and causes significant losses.

Multiple aspects of plant nutrition affect plant resistance to soft rot diseases. In some cases, it appears that at least some of the nutrition effect is due to induction or inhibition of *Erwinia* enzymes. For example, fertilization of calla lilies with superphosphate increased the susceptibility to soft rot (Gracia-Garza et al., 2004). Since the addition of phosphate to bacterial growth medium also increases the production of plant cell wall degrading enzymes, the increased plant susceptibility in this case may be due to increased expression of bacterial virulence factors. Similarly, plants fertilized with high levels of nitrogen are also more susceptible to soft rot (Schober and Vermeulen, 1999).

For many years, growers have known that addition of calcium to the soil or use in a liquid fertilizer reduces soft rot diseases (Bain et al.,1996; Schober and Vermeulen 1999). Since divalent ions, such as calcium, act as cross-linkers for pectate in plant cell walls, the addition of calcium may help the plant for stronger cell walls and thus be able to better resist pathogen attach. Recently, it has also been shown that calcium affects regulation of *E. carotovora* virulence genes, which may also explain why addition of calcium protects plants from soft rot (Flego, 1997).

The best example of control through plant resistance is with *E. carotovora* subsp. *betavasculorum* on sugarbeet. Prior to the late 1970s, sugarbeet cultivars released in the United States were not routinely tested for resistance to *E. carotovora* subsp. *betavasculorum*. In the 1970s there were significant sugarbeet losses to soft rot because a susceptible cultivar was in widespread use. A single domimant resistance gene was identified and sugarbeet cultivars are now routinely tested for resistance to soft rot *Erwinia* prior to release (Lewellen et al., 1978; Whitney and Lewellen, 1978). As a result, soft rot is no longer aa significant of a problem in sugarbeet production. This sugarbeet resistance gene has not been further characterized.

5. HOST PLANT RESPONSES TO *ERWINIA* INFECTION

For the most part, plant responses to *Erwinia* have been examined in plants inoculated with a high level of bacteria. In most cases, *Erwinia* are surviving asymptomatically on plants and almost nothing is known about if or how the plant is responding to these latent infections.

The cell wall degrading enzymes secreted by soft rot *Erwinia* cause degradation of pectate into unsaturated oligogalacturonates, which themselves trigger numerous plant defense responses including phytoalexin synthesis, proteinase inhibitor synthesis, glucanase synthesis, membrane protein phosphorylation, and a response similar to a hypersensitive response known as an XR (Darvill and Albersheim, 1984; Davis and Ausubel, 1989; Davis and Hahlbrock, 1987; Farmer et al., 1991; Mathieu et al., 1991; Messiaen, et al., 1993; Nothnagal et al., 1983; Palva et al., 1993; Ryan, 1988; Thain et al., 1990). The length of the oligogalacturonate is important, with 7- to 20-mers being the best inducers of most defense responses. However, unsaturated digalacturonate and 5-keto-4-deoxyuronate, both products of *Erwinia* digestion of pectin, can also elicit an induced cell death in potato tuber cells (Weber et al., 1996). It is notable that this induced cell death only occurs in tubers for the first few months after harvest and that it varied by cultivar. It is not clear which plants species and tissues have this response or what percentage of cells respond to these elicitors during an infection. For example, when *E. carotovora* is infiltrated into solanaceous plant leaves under low humidity, it elicits a plant cell death in the leaves that is dependent entirely on the type III secretion system and not the type II secretion system, thus oligogalacturonates that may be released by *E. carotovora* do not induce plant cell death (Yap et al., 2004; Charkowski, unpublished). Under the same conditions, many strains of *E. chrysanthemi* do not induce plant cell death at all, again suggesting that cell death induced by digalacturonate and 5-keto-4-deoxyuronate does not occur in all plant tissues (Charkowski, unpublished).

Because leaf infiltration of *E. carotovora* subsp. *carotovora* culture filtrates induces plant defenses as effectively as infiltration of *E. carotovora* cells, Vidal et al. (1998) examined T2SS-secreted plant cell wall degrading enzymes to determine which ones induce plant defenses. To more easily monitor plant gene induction, they used tobacco transformed with a *Nicotinana plumbaginifolia* β-1,3-glucanase promoter, which is induced to high levels by *E. carotovora* culture filtrates, fused to a *gus* reporter gene (Castresana et al., 1990). They found that treatment with polygalacturonase (PehA), pectate lyase (PelA and PelD) or cellulase (CelS or Cel IV) induced expression of the β-1,3-glucanase promoter to varying levels, none of which were as high as the level induced by culture filtrates. Infiltration of a combination of all of the enzymes resulted in the same level of induction as the culture filtrate, demonstrating that the induction activity is additive and that both pectinases and cellulases can induce cell defense genes.

Several groups have taken advantage of the model plant *Arabidopsis thaliana* to determine how plants respond to *Erwinia* infection. Three endogenous signalling molecules, salicylic acid (SA), jasmonic acid (JA), and ethylene (ET), trigger local and systemic plant defenses against pathogen attack in *Arabidopsis*. There is a considerable level of cross talk between the SA and JA pathways and these two pathways are required for resistance to different types of pathogens, with the SA pathway more important for HR elicitation, systemic acquired resistance, and, hence, resistance to hemi-biotrophic pathogens, while the JA pathway is more important for resistance to necrotrophic pathogens.

In *Arabidopsis*, both the JA and ET pathways are induced when leaves are infiltrated with either bacterial cells or culture filtrates (Norman et al., 1999; Norman-Setterblad et al., 2000). In contrast, induction of the SA pathway is delayed. However, these experiments used high levels of cells, which may not be representative of the level of pathogen generally encountered in nature. Kariola et al. (2003) found slightly different results in experiments with purified HrpN, which they found could elicit cell death in *Arabidopsis* and induce genes in both the SA and JA pathways. As discussed below, experiments with purified cell death elicitors can sometimes provide misleading results, so the apparent contradiction in these results may be due to the use of purified protein by Kariola et al. (2003).

An increase in the production indole glucosinolates via induction of the tryptophan biosynthesis pathway and Trp oxidation genes appears to be one of the consequences of induction of the JA pathway in *Arabidopsis* by secreted *Erwinia* elicitors obtained from culture filtrates (Brader et al., 2001). Curiously, the phytoalexin camalexin, which is also synthesized via the tryptophan biosynthesis pathway, was not induced by *E. carotovora* culture filtrates. Glucosinolates, which are non-toxic, are normally sequestered in the plant vacuole. If the plant cell is lysed, endogenous myrosinases break down the glucosinolates into toxic products that can inhibit the growth of pathogens, including *E. carotovora* or affect insect herbivores (Brader, et al., 2001; Rask, 2000). Because appropriate *Arabidopsis* mutants were unavailable, Brader et al. (2001) were unable to definitively show that indole glucosinolates contribute to resistance to soft rot *Erwinia.*

Although induction of the SA pathway is delayed, this pathway also appears important for resistance to soft rot *Erwinia.* Addition of exogenous SA to plants increases their resistance to soft rot *Erwinia* and analysis of *Arabidopsis* mutants also suggests that this pathway plays a role in resistance (Aguilar et al., 2002; Palva et al., 1994; Lopez et al., 2001; Lopez-Lopez et al., 1995). This could be, however, due to effects of SA on *Erwinia* gene expression since this molecule is known to affect gene regulation in other pathogens (Kunin et al., 1995; Prithiviraj et al., 2005). Li et al. (2004) recently showed that over-expression of a SA-dependent WRKY transcription factor provides resistance to *E. carotovora* subsp. *carotovora*. However, upon infection of *Arabidopsis* with *E. carotovora*, this gene is suppressed via the JA pathway. This suggests that soft rot *Erwinia* may be maniulating plant defense gene regulation to inhibit SA pathway genes, perhaps both to manipulate plant defenses and to keep plant produced SA from affecting *Erwinia* gene regulation. *E. carotovora* subsp. *atroseptica* has also been shown to induce a WRKY protein in potato (Dellagi et al., 2000). This potato gene is not induced by SA, JA, ET or wounding, thus the defense pathway it may participate in is still unknown.

There appears to be ties between primary plant metabolism and defense gene induction. For example, it is possible to induce defense genes by addition of sugar to leaves (Herbers et al., 1996). To investigate the ties between metabolism and pathogen defense, Linke et al. (2002) constructed antisense AATP1 antisense plants, which had decreased activity of the plastidic ATP/ADP transporter AATP1. In addition to other physiological traits, such as lower starch levels and altered tuber morphology, these plants were enhanced in responses to pathogens, including an increased oxidative burst, an increased expression of defense-related genes, and increased resistance to *E. carotovora*. The reason for the increase in response to pathogens is not known, but could be due to an increased ATP to ADP ratio, which would provide the plant cells with more energy to mount a defense against a pathogen.

Another aspect of plant basal defense against *Erwinia* infection appears to be sequestering of iron (Dellagi et al., 2005). Arabidopsis plants unable to produce the iron storage protein ferritin are more susceptible to *E. chrysanthemi* than wild type plants. The ferritin gene *AtFer1* is up-regulated in a biphasic manner reponse to both *E. chrysanthemi* infection and infiltration with *E. chrysanthemi* siderophores, but not when an *E. chrysanthemi* mutant unable to produce siderophores is inoculated onto plants. The first induction of *AtFer1*, at 0.5 hours after leaf infiltration with *E. chrysanthemi*, appears to be regulated by both plant NO production and bacterial siderophore production, while the second induction of *AtFer1*, at 24 hours post-inoculation, is independent of NO production. The mechanism of upregulation of *AtFer1* by bacterial siderophores is unknown and could be either due to iron competition or because the siderophores are recognized by the plant cell.

6. SOFT ROT RESISTANCE

Several methods have been tried for screening potato germplasm for resistance to soft rot *Erwinia*, including tuber slice inoculation, tuber stab inoculations, and stem inoculations (Allefs et al., 1995b; Reeves et al., 1999; Zimnoch-Guzowska et al., 1999). Both anaerobic and aerobic conditions and multiple *Erwinia* species and subspecies have also been tested and they give different results in some cases. However, in general, the results for resistance to different strains of *E. chrysanthemi* and *E. carotovora* subspecies are correlated, at least in potato, thus only one strain needs to be used to test for resistance (Wolters and Collins, 1994). In all cases, there is enough variability in the assays that numerous samples must be tested to determine if the variety is resistant. In addition, for potato tubers, tuber resistance changes as the tuber ages and also is affected by the growing season. Therefore tubers from multiple growing seasons need to be tested when examining germplasm for *Erwinia* resistance.

There are currently no examples of gene-for-gene resistance against soft rot *Erwinia* pathogens. However, there are examples of single dominant resistance genes, but since none of these genes have been cloned, the nature of this resistance is unknown. There are also examples of quantitative trait loci contributing to soft rot resistance. For example, using AFLP, RFLP, and resistance-gene-like markers, Zimnoch-Guzowska et al. (2000) found genetic factors affecting resistance to *E. carotovora* on all 12 potato chromosomes. As with single dominant resistance genes, these plant genes have not been cloned or characterized, so their functions are unknown. Also, just as in many cases the contribution of *Erwinia* virulence genes to disease is tissue dependent, resistance can also be tissue dependent. For example, a potato plant may have tubers resistant to soft rot, but may still be susceptible to stem rot (Zimnoch-Guzowska et al., 1999, 2000). However, there is usually a positive correlation between tuber and stem resistance. As described above, one of the most successful examples of soft rot resistance is the use of a single dominant resistance gene in sugarbeet. Similarly high levels of resistance are found in wild species of other crops, including potato, but they have not yet been widely exploited commercially.

There are numerous wild potato species and accessions of many of these have been screened for stem rot and soft rot resistance. Bains et al. (1999) screened 363 accessions of six wild species and found a high percentage of *Solanum boliviense, Solanum, chacoense,* and *Solanum sancta-rosae* were resistant. In contrast, almost all of the *Solanum canasense*, *Solanum tarijense*, and *Solanum spegazzinii* accession were susceptible. Results from this work also suggest that the resistance in *Solanum brevidens*, another wild potato species with a high level of resistance to *E. carotovora*, is simply inherited. Dorel et al. (1996) examined the expression of *E. chrysanthemi* pectate lyase in susceptible *Solanum tuberosum* and a resistant somatic hybrid of *S. tuberosum* and *S. brevidens*. They found that expression of *pelA* and *pelE* in the resistant tubers was very low compared to in the susceptible tubers. They also found that pectin methyl esterase (*pem*) plays a larger role in disease in tubers resistant to soft rot, perhaps because the resistant plants have more highly methylated pectin, which is not effectively degraded by the *E. chrysanthemi* pectinases. This suggests that a more thorough examination of soft rot resistance could provide new insights into plant cell wall synthesis.

Phytoalexins may provide some protection against soft rot, but the literature is somewhat unclear on this topic. For example, Lacy et al. (1979) showed that 80%

of the *E. chrysanthemi* strains isolated from diseased maize were able to detoxify 2,4-dihydroxy-7-methoxy-2H-1,4-benzoxazin-3(4H)-one (DIMBOA), a maize phytoalexin. However maize lines that do not make DIMBOA were not more susceptible to *E. chrysanthemi* than those that do.

The work of Lyon (1984) may hold some clues to the role of phytoalexins. Lyon (1984) showed that the effects of rishitin, a phytoalexin produced by potato tubers, is strongly temperature dependent. It has little effect at 10°C, but is able to kill a progressively higher percentage of bacteria as the temperature is increased until 30°C, where no viable bacteria were detected. Lacy et al. (1979) did their in vitro assays showing detoxification of DIMBOA at 28-30°C, but their greenhouse maize experiments at 21-26°C, which was possibly too low of a temperature to see the protective effects of DIMBOA.

7. *ERWINIA*-RESISTANT TRANSGENIC PLANTS

There have been numerous reports of transgenic potatoes constructed to be resistant to soft rot. Some strategies have taken advantage of plant defenses, some have targeted *Erwinia* cell-cell communication necessary for induction of virulence genes, and some have relied on toxins or enzymes antagonistic to *Erwinia.* Researchers have also used somatic fusion to introgress soft rot resistance into potato (McGrath et al., 2002).

Because host plants induce disease defense pathways in response to unsaturated oligogalacturonates, Wegener et al. (1996) tried expressing *E. carotovora* subsp. *atroseptica pel3* in potato plants to determine if the release of unsaturated oligogalacturonates would increase the resistance of these transgenic plants to soft rot *Erwinia.* Pel3 was chosen because it has the lowest cell wall degrading activity on potato tissue and because its activity results in oligos of a correct size to induce plant defenses. In these plants, the Pel3 enzyme is produced and remains inside the plant cell. When the plant cells are wounded, the enzyme is released and is able to degrade the plant pectin and produce oligos that induce plant defenses. Tubers from Pel3-producing lines grown in field trials were more resistant to *E. carotovora* when wounded and inoculated than control plants; the macerated area was smaller, the level of cells required to cause symptoms was higher, and the plant necrotic reaction that inhibits pathogen growth on the wound surface was stronger (Wegener, 2001; Wegener, 2002).

To determine if transforming potatoes with other *E. carotovora* pectate lyases with subtly different activities would also result in resistant plants, Wegener and Olsen (2004) tested potato lines expressing *E. carotovora* subsp. *atroseptica* Pel1 and found that tubers from these plants were as resistant as the Pel3-expressing lines. Similarly to the Pel3-lines, the Pel1 transgenic potatoes had increased levels of enzymes involved in plant defense, including increased polyphenol oxidase, phenylalanine ammonia-lyase, and POD activities. Because this pectate lyase-mediated defense response is induced by wounds that release the enzyme, it would be interesting to see if these plants have increased resistance to other pests, such as chewing insects or nematodes, that also wound plant cells.

Since the bacterial autoinducer N-acyl homoserine lactone (AHL) is required for expression of numerous *Erwinia* virulence genes, multiple groups have explored the effect of transgenic plants producing either enzymes capable of degrading AHL or producing AHLs themselves. Dong et al. (2001) constructed transgenic tobacco and potato expressing *aiiA*, an AHL lactonase from Bacillus (Dong et al., 2000). They found that plants expressing *aiiA* were resistant to soft rot, presumably because they inhibit AHL-mediated induction of *E. carotovora* virulence genes.

Mäe et al. (2001) found that plants that express *Erwinia expI*, and that were thus able to produce AHL, were also more resistant to *Erwinia* infection when the bacteria were inoculated onto host plants at low levels. They suggest that this is because the *Erwinia* virulence genes are expressed before a quorum is reached and this premature expression induces plant defenses. Inoculation of *expI*-expressing plants with high levels of *Erwinia* overcomes this resistance. In contrast, Toth et al. (1999c) found the opposite; plants producing AHL were more susceptible to *Erwinia*. This conflict has yet to be resolved. It would be interesting to test these transgenic plants with *E. chrysanthemi*, which does not appear to require AHL for virulence gene expression, and with multiple strains of *E. carotovora*, which is a very diverse pathogen, to see how additional *Erwinia* strains and species interact with plants capable of degrading or producing AHLs.

In related experiments, Smadja et al. (2004a) transformed *E. carotovora* subsp. *atroseptica* with the *aiiA* lactonase and, surprisingly found that the cells were able to colonize potatoes to high levels even though symptoms did not develop. They suggest that transgenic plants capable of degrading AHLs may not provide useful control of soft rot *Erwinia* because these plants allow *Erwinia* populations to reach high levels on the plants in the absence of symptoms and if expression cell wall degrading enzymes can be triggered by another signal, this could result in total loss of a crop.

The more direct approach of engineering plants that produce antimicrobial peptides such as attacin or cecropin SB-37 has had mixed results. Arce et al. (1999) reported that some of the transgenic plants that they constructed were resistant, while other groups using similar strategies and antimicrobial peptides did not observe resistance to attack by bacterial pathogens (Allefs et al., 1995a; Florack et al., 1995). Similarly, tubers from transgenic plants expressing tachyplesin I, an antimicrobial peptide normally produced by horseshoe crabs, were only slightly more resistant to soft rot than the untransformed controls (Allefs et al., 1996).

Resistance to soft rot has also been observed in transgenic plants expressing lysozyme (Düring et al., 1993; Serrano et al., 2000). Because some of these strategies are expected to impact the entire rhizosphere community rather than just the *Erwinia*, these plants have been used in experiments to determine how transgenic plants affect soil microbiology (for a review, see Liu et al., 2005). No large effects on the soil community have yet been described and no clear themes have yet emerged on the effects of these transgenic plants on soil microbes.

8. THE *ERWINIA* GENOME

The enterobacteria are the most intensively studied bacteria and this family currently has the largest number of completed genome sequences. One *E. carotovora* subsp *atroseptica* genome, SCRI1043, and one *E. chrysanthemi* genome, 3937, have been sequenced to date. SCRI1043 was isolated from potato and the genome project was headed by scientists at the Scottish Crop Research Institute (Bell et al., 2004). Strain 3937 was isolated from African violet and the genome project was initiated by Noel Keen and Nicole Perna along with a consortium of *Erwinia* researchers (Glasner et al., submitted). The 3937 genome project was unique in that the consortium annotated the genome via web-based software ASAP (Glasner et al., 2003). These annotations are a living document and are continuously updated by those working in soft rot *Erwinia*. These genome sequences have revealed substantial new information about plant pathogenesis and have energized the *Erwinia* research community.

Both species have cirucular chromosomes of approximately the same size (Table 3). No plasmids and relatively few insertion elements and transposons were found in either genome. Based on comparison with other sequenced enterobacteria, approximately 2300 of the *E. chrysanthemi* genes are likely to have been derived from the ancestor of this group. Only 500 additional genes beyond these 2300 are also found in *Erwinia carotovora* subsp. *atroseptica* and just over 400 of these genes are not found in animal associated enterobacterial. However, many of these genes are also present in other plant-associated microbes. Thus the pool of soft rot *Erwinia* specific genes is not large.

Table 3. Features of the soft rot *Erwinia* genomes

Feature	***E. chrysanthemi***	***E. carotovora***
Genome size	4.9 Mb	5.06 Mb
Protein-coding genes	4,638	4,491
rRbosomal RNA-coding genes	22	25
tRNA-coding genes	75	76

The genome sequences revealed several gene clusters likely to be important for survival in plants that had not been yet been identified by experimental approaches. Probably the most surprising find was a second T2SS and two T4SS in *E. chrysanthemi* (Glasner et al., submitted) and a T4SS in *E. carotovora* (Bell et al., 2004). Both *Erwinia* genomes encode nitrogen assimilation genes similar to those found in *Klebsiella*. Homologs of the *nasFEDBA* operon and *nasR* regulator, which are also involved in *Klebsiella* nitrate and nitrite assimilation were found in both genomes. Both genomes encode a large non-ribosomal peptide synthetase system similar to *syrE*, which is required for synthesis of the toxin syringomycin in *P. syringae*. The TAT secretion system was also found in both genomes, but no *Erwinia* proteins have yet been shown to be secreted via this system.

Both genomes had a high number of genes homologous to secretion systems, transporters, regulators, and methyl-accepting chemotaxis proteins (*mcp*) in comparison to previously sequenced enterobacteria. For example, *E. carotovora* subsp. *atroseptica* encodes 36 putative *mcp* genes, while the animal-infecting enterobacteria encode from five to 10 *mcp* genes. Both genomes are the only sequenced enterobacterial genomes to encode all six known types of protein secretion systems known to occur in enterobacteria. Of these, the type I, type II, type III and a putative two-partner system, which secrete proteases, cell-wall degrading enzymes, effectors, and HecA, respectively, were already known to be important for virulence in soft rot *Erwinia*. Both genomes also encode homologs of *srfABC*, which encode proteins likely to be secreted via the T3SS.

Because the T2SS is required for pathogenicity of the soft rot *Erwinia*, characterization of the regulation and activities of the proteins secreted through system has been a major focus of *Erwinia* research for the past several decades. Most of the *E. chrysanthemi* plant cell wall degrading enzymes were identified prior to completion of the genome sequence. In total, nine pectate lyases, two pectin methylesterases, one pectin acetylesterases, one polygalacturonase and one rhamnogalacturonate lyase were experimentally shown to be secreted via the *E. chrysanthemi* T2SS. Xylanase and xylosidase activity has been found in *E. chrysanthemi* (Keen et al., 1996; Vroemen et al., 1995), and xylosidase homologs were found in the *E. chrysanthemi* 3937 genome sequence, suggesting that strain 3937 can also degrade this polymer.

The *E. carotovora* subsp. *atroseptica* genome sequence revealed several new putative virulence genes not found in *E. chrysanthemi* including a *cfa* gene cluster, which is involved in phytotoxin synthesis in *P. syringae*; *aggA*, a root adhesin found in *Pseudomonas putida*; *avrXca* (*svx*), a defense elicitor first identified in *Xanthomonas campestris*; an *ehp* gene cluster, which is used by *Pantoea agglomerans* to produce the antibiotic phenazine; and pilus homologs. In addition, strain 1043 has 20 putative pectinase genes, only 9 of which were known before the genome sequence was completed.

Like *E. chrysanthemi*, *E. carotovora* appears to encode multiple iron acquisition systems, including genes for achromobactin uptake and transport, enterobactin synthesis, ferric citrate uptake, and a HasA heme-binding protein and secretion system.

9. *ERWINIA* FUNCTIONAL GENOMIC AND PROTEOMICS

Erwinia functional genomics and proteomics are still in the early stages of development. Bell et al. (2004) used a transposon mutation grid strategy to quickly identify strains with transposon insertions in newly identified putative pathogenicity genes. They tested these strains on potato stems to determine if any of the mutants were impaired in virulence and found that both the T4SS and the *cfa* gene cluster contribute to virulence. This suggests that the T4SS, which is used by bacteria both for conjugation and to deliver virulence protein (Seubert et al., 2003), may function to deliver virulence proteins into potato. The targets of this system are currently unknown.

Preliminary gene expression studies done while the *E. chrysanthemi* 3937 genome sequencing project was still in progress revealed several *E. chrysanthemi* genes differentially regulated in leaves compared to rich medium (Okinaka et al., 2002). Almost all of the plant down-regulated genes were homologs of housekeeping genes. Many of the plant up-regulated genes were virulence genes that had been previously identified or were homologous to virulence genes or stress-response genes in other species, including genes involved in anaerobiosis, iron uptake, transporters, permeases, chemotaxis, and proteins that protect against heat or oxidative damage. Although this study provided many new insights into *Erwinia*-plant interactions, it had some weaknesses that were unavoidable at the time this work was done. For example, the array used for this study consisted of 3-kb cloned DNA fragments rather than oligos, thus it measured upregulation of operons rather than specific genes. Also, the isolation of the bacteria from plant leaves prior to mRNA extraction may have resulted in the degradation of mRNA transcripts or induction of additional, non-plant induced genes.

The same group used an IVET screen with a green fluorescent protein reporter to identify genes upregulated in plants (Yang et al., 2004). The up-regulated genes function in processes including metabolism, reguation, infrmation transfer, and transport. Unlike the array assay, T3SS genes were identified in the IVET screen. Notably, only one *Erwinia*-specific gene, which has no known function (ASAP ID 16049), was found in this screen.

Several of the induced genes identified in these screens were mutated and examined for ability to infect African violet. Of the genes tested, the T3SS machinery genes *hrpB*, *hrcJ*, and *hrpD*, the rhamnogalacturonide transporter *rhiT*, a peptide synthase homolog, a phenylalanine synthesis homolog *pheC*, a formyltetrahydrofolate deformylase homolog *purU*, and the ethanolamine operon regulator *eutR* were required for full virulence (Okinaka et al., 2002; Yang et al., 2004).

10. GENETICS OF *ERWINIA* VIRULENCE

Most the work on virulence gene regulation and function has been completed in a small number of strains. *E. carotovora* subsp. *Atroseptica* SCRI 1043, *E. carotovora* subsp. *carotovora* Ecc71 and SCC193, and *E. chrysanthemi* 3937 and EC16 have been used as the model strains. This section will focus mainly on findings from these few strains.

10.1. Type II Secretion and Plant Cell Wall Degredation

The type II secretion system (T2SS), or general secretion pathway (GSP), which secretes a variety of plant cell wall degrading enzymes, is crucial for the pathogenicity of soft rot *Erwinia*. Proteins secreted via the *Erwinia* T2SS, which is also known as the Out system in *Erwinia* and the General Secretion Pathway in other species, are secreted first across the bacterial inner membrane via the TAT system or the Sec system. The proteins are then secreted across the outer membrane via the T2SS. Mutations in the T2SS result in protein accumulation in the periplasm. It is noteworthy that there have not been examples of negative feedback regulation in T2SS mutants that down regulate genes encoding secreted proteins since negative regulation of secreted proteins in the absence of secretion is found in many other secretion systems. The recent *Erwinia* sequencing projects revealed that *E. carotovora* subsp. *atroseptica* 1043 encodes one T2SS while *E. chrysanthemi* encodes two T2SS (Bell, 2004; Glasner et al., submitted). At this time it is unknown if the second *E. chrysanthemi* system is expressed or which, if any, proteins it secretes. Its presence requires the careful re-examination of all work done on the *E. chrysanthemi* 3937 T2SS to determine if this second T2SS affects any of the results obtained when studying the functions of the T2SS proteins.

Although the T2SS secretes proteins across the outer membrane, 14 of the 16 T2SS machinery proteins are localized to the inner membrane or the perplasm (Thomas et al., 1997; Filloux, 2004) This protein secretion system has many similarities with the type IV pilus secretion system (Filloux, 2004). In other bacterial species, overproduction of the the T2SS or the pseudopilin GspG (OutG) results in the formation of large helical extracellular filaments (Durand et al., 2003; Sauvonnet et al., 2000). Protein secretion of the normal targets of the T2SS is blocked when the T2SS forms a pilus, suggesting that occupancy of the outer membrane secretion pore by the pilus blocks secretion and that the pilus does not normally reach the outer membrane secretin (Durand et al., 2003). There are examples of the T2SS proteins contributing to secretion of other molecules via other secretion systems (Davis et al., 2000). This participation in other secretion systems may be an example of the evolution of complex new secretion systems from proteins already present in bacterial cells, similarly to how the T3SS may have evolved from a combination of a flagella secretion system and a GspD-like secretin, and should be examined further.

Both *E. carotovora* and *E. chrysanthemi* posses homologous T2SS with conserved gene orders, yet these species are unable to reciprocally secrete pectate lyases via the T2SS. This is specificity is curious since it limits the ability of soft rot *Erwinia* to easily secrete new enzymes acquired via lateral gene transfer and it is different from the T3SS, for which there are numerous examples of heterologous secretion of proteins (for examples, see Anderson et al., 1999; Ham et al., 1998). This specificity also hints that the T2SSs in *E. carotovora* and *E. chrysanthemi* may have different origins, but comparison of gene order and a preliminary phylogenetic analysis (Perna, personnel communication) provides no evidence for this hypothesis. Also, currently, other than for a few *E. carotovora* and *E. chrysanthemi* proteins from a few strains, it is unknown which *Erwinia* species are able to secrete which heterologous proteins. For example, it is unknown if *Erwinia* pathovars or subspecies can secrete heterologous proteins from within the same species but from a different subgroup.

In a search for the T2SS gatekeeper, Lindeberg et al. (1996) mutated each individual *E. chrysanthemi* T2SS gene and attempted to complement the gene with its *E. carotovora* homolog. They found that all of the *E. carotovora* T2SS proteins could substitute for *E. chrysanthemi* proteins except for the OutD secretin and in inner membrane protein OutC, suggesting that these two proteins contribute to the recognition of the T2SS secretion signal on secreted proteins.

Because Lindeberg et al. (1996) used *E. coli* as a host for cloned *Erwinia* T2SSs in their experiments, there was some question as to whether *E. coli* proteins might have affected the results. Bouley et al. (2001) further explored the functions of the PDZ domains OutD and OutC to determine if these domains, which are involved in protein-protein interactions, affect protein secretion. They found that deletion of the *E. chrysanthemi* OutC PDZ domain eliminated secretion of all of the pectate lyases except PelI, which was present at 50-80% of wild type levels in culture supernatants, and that deletion of the OutC PDZ domain had no effect on the secretion of PemA and Cel5. They also found that replacement of *E. chrysanthemi* OutC with the *E. carotovora* OutC resulted in the same secretion profile as the *E. chrysanthemi* OutC PDZ mutant, with no pectate lyase secretion except for PelI and secretion of PemA and Cel5. Furthermore, an *E. carotovora* hybrid OutC containing the *E. chrysanthemi* PDZ motif functioned like the *E. chrysanthemi* OutC protein, demonstrating that the OutC PDZ motif is required to recognize secretion signals from pectate lyases, but not PemA or Cel5. PelI appears to be partially dependent upon the OutC PDZ motif for secretion.

However, the *E. carotovora* OutC is not sufficient to allow secretion of *E. carotovora* proteins by the *E.chrysanthemi* T2SS.

Similarly, the *E. carotovora* OutD protein could substitute for *E. chrysanthemi* OutD for secretion of PemA and partial secretion of PelD, PelE and PelI, but not PelB, PelC, PelL, PelZ or Cel5. Experiments with hybrid proteins identified the N-terminal portion of OutD as important for secretion sepecificity. The secretion specificity for some, but not all T2SS proteins suggests that multiple different secretion signals target proteins to the T2SS.

Curiously, the importance of at least one of the Out proteins, OutB, in secretion depends on the expression level of OutD. If OutD is expressed at high levels from the chromosome or on a plasmid, OutB is not required for protein secretion. However, if OutD is expressed at lower levels, it is required (Condemine and Shevchik, 2000). It has been hypothesized that the role of OutB is to pilot OutD to the outer membrane. Thus, if enough OutD is produced, a functional porin can still form even with inefficient localization of OutD. It is also possible that genes encoded by the second E. chrysanthemi T2SS aid in OutD localization.

There is considerable interest in identifying the signals that target proteins to the T2SS, both so that T2SS-secretion proteins can be identified in genome sequences and because this secretion system would be a useful tool for purification of recombinant proteins from culture supernatants if the recombinant proteins could be targeted for secretion. To address this question, Lindeberg et al. (1998) made 14 *E. carotovora pel1-E. chrysanthemi pelC* chimeras to attempt to identify the T2SS secretion signal. The mature Pel1 and PelC proteins are 71% identical, yet *E. coli* expressing a cloned *E. chrysanthemi* T2SS does not secrete Pel1. Of these chimeras, all 14 retained enzymatic activity, although the levels of activity differed, but only 3 of these hybrid proteins were secreted. These secretion assays revealed that M118 to D175 and V215 to C329 were required for species-specific secretion. Because the structure *of E. chrysanthemi* PelC had already been solved (Yoder and Jurnak, 1995), the authors suggest that one targeting signal is in the external loops formed by G274 to C329 and that M118 to D170 and V215 to G274 are required for proper positioning of these loops.

Other T2SS targeting signals may not have been identified in their experiments since the two pectate lyases studied by Lindeberg et al., 1998) are so closely related. Palomäki et al. (2002) suggested that if the T2SS signal is a short collinear stretch of residues or a conformational feature, some variability might be allowed and therefore these signals might escape detection by analyses where single mutations in secreted proteins are examined. Palomäki and Saarilahti (1995) had found that the C-terminus of PehA, a polygalacturonase secreted by the *E. carotovora* subsp. *carotovora* T2SS, was sufficient for secretion of β-lactamase (Bla) via the T2SS. To further characterize the secretion signal, they constructed a library of 9000 randomly mutagenized *pehA-bla* hybrids and tested these hybrids for T2SS-dependent secretion. All of the mutations that eliminated secretion were then tested to determine if they affected secretion of PehA and 11 mutants were identified. Similarly to Lindeberg et al. (1998) and other workers, Palomäki et al. (2002) found that three regions were required, an N-terminal and C-terminal targeting region and a central region that is probably required to hold the targets in the proper confirmation.

Some of the proteins secreted through the T2SS, including pectate lyases and polygalacturonases, but not cellulase CelV, require disulfide bonds for activity (Cooper and Salmond, 1993). Vincent-Sealy et al. (1999) mutated the *E. carotovora dsbA*, which is a periplasmic disulfide bond forming protein. The mutant strain was affected in activity and secretion of pectate lyases and polygalacturonase, but not cellulase. Importantly, secretion of cellulase by the *dsbA* mutant demonstrates that disulfide bond formation is not required for formation of a function T2SS.

10.2. Virulence Proteins Secreted by the T2SS

Plant cell walls are composed of a pectin portion, which is composed of linear and ramified regions, and a cellulose portion. The linear pectin portion is composed mainly of polygalacturonate, which may be methylated or acetylated, and which is cross-linked by calcium ions. There are two types of ramified pectin. Rhamnogalacturonan I is composed of alternating L-rhamnose and D-galacturonate residues with long side chains of arabinans and galactans. Rhamnoglacturonan II is a complex mix of D-galacaturonate, L-rhamnose, and several uncommon sugars and is crosslinked by a borate ester (O'eill et al., 1996)

The soft rot *Erwinia* encode multiple enzymes which act synergistically to degrade plant cell walls (Tables 4, 5, 6). These enzymes have been most thoroughly studied in *E. chrysanthemi.* Since pectin is generally acetylated or methylated, and since pectate lyases are inefficient at degrading pectate with these modifications, the pectin acetylesterases, PaeX and PaeY, are important for modifying the plant pectin prior to degradation by the pectate lyases (Shevchik and Hugouvieux-Cotte-Pattat 2003). Mutation of both *paeX* and *paeY* in *E. chrysanthemi* results in a strong reduction in virulence (Shevchik and Hugouvieux-Cotte-Pattat 2003). Genes encoding pectin acetylesterases have not been studied in *E. carotovora.*

Table 4. *E. chrysanthemi* T2SS-secreted proteins

Enzyme	Protein	Strain	Reference
Cellulase	CelY	3665	Guiseppi et al. (1991)
	CelZ		Boyer et al. (1984)
Endo-pectate lyases	PelA	3937	Favey et al. (1992)
	PelB	3937	Hugouvieux-Cotte-Pattat and Robert-Baudouy (1992)
	PelC	3937	
	PelD	3937	
	PelE	3937	Reverchon et al. (1989)
	PelI	3937	Shevchik et al. (1997)
	PelL	3937	Lojkowska et al. (1995)
	PelZ	3937	Pissavin et al. (2003)
Exo-pectate lyases	PelX	3937	Shevchik et al. (1999b)
Pectin acetylesterases	PaeY	3937	Shevchik and Hugouvieux-Cotte-Pattat (1997)
Pectin methylesterase	PemA	3937	Laurent et al. (1993)
	PemB	3937	Shevchik et al. (1996)
Pectin lyase		3937	Glasner et al. (submitted)
Polygalacturonase	PehN	3937	Hugouvieux-Cotte-Pattat et al. (2002)
Rhamnogalacturonate lyase	RhiE	3937	Laatu and Condemine (2003)

Table 5. *E. carotovora* subsp. atroseptica T2SS-secreted proteins

Enzyme	Protein	Strain	Reference
Cellulase	Cel		Lanham et al.(1991)
Pectate lyase	PelA (Pel1)	SCRI1043	Bell et al. (2004)
	PelB (Pel2)		
	PelC (Pel3)		
	Pel1	C18	Bartling et al. (1995)
	Pel2		
	Pel3		

	PelA	EC	Lei et al. (1988)
	PelB		Lei et al. (1987)
Polygalacturonase	PG		McMillan et al. (1992)
			Lanham et al. (1991)
Function unknown	SVX	SCRI1043	Corbett et al. (2005)

Table 6. *E. carotovora* subsp. *carotovora* T2SS-secreted proteins

Enzyme	**Protein**	**Strain**	**Reference**
Cellulase	CelV1	SCC3193	Mäe et al. (1995)
	CelS		Saarilahti et al. (1990b)
Endo-pectate lyases	Pel1	Ecc71	Chatterjee et al. (1995)
	Pel3		Liu et al. (1994)
	PelB	SCRI193	Hinton et al. (1989)
	PelC		Hinton et al. (1989)
	Pel1	ER	Ito et al. (1988)
	PelX		Ito et al. (1988)
	Pel2		Yoshida et al. (1992)
	Pel3		Yoshida et al. (1991)
	PelB	SCC3193	Heikinheimo et al. (1995)
Pectin lyase	PnlA	Ecc71	McEvoy et al. (1990)
	Pnl		Pirhonen et al. (1991)
Endo-polygalacturonase	Peh-1	Ecc71	Liu et al. (1994)
	Peh	SCRI193	Hinton et al. (1990)
	PehA	SCC3193	Saarilahti et al. (1990a)

The numerous pectate lyases encoded by *E. chrysanthemi* play overlapping and synergistic roles in degrading these large polymers. Both plant and *in vitro* assays have demonstrated differences in their enzymatic activity. One possible reason for the maintenance of multiple pectate lyases is that their activity differs depending on the pH and the pH across lesions in diseased plants and across plant species varies. Nachin and Barras (2000) showed that in *E. chrysanthemi*, *pelA* and *pelD* were induced in acidic conditions, while *pelE* was induced in basic conditions. Similarly, the numerous pectate lyases differ in their preference for the amount of pectin esterification and contribution to virulence on different hosts (Shevchik et al., 1997; Beaulieu et al., 1993; Pissavin et al., 1996). For example, *pelA* is important for systemic spread while *pelE* is important for lesion formation (Boccara, 1988) and the clustered genes *pelZ*, *pelB*, and *pelC* make no apparent contribution to virulence of *E. chrysanthemi* 3937 on *Saintpaulia ionantha*, but contribute to virulence on witloof chicory (Beaulieu et al., 1993; Pissavin et al., 1996).

The characterized secreted *E. chrysanthemi* pectate lyases all have a basic pH optimum and an absolute requirement for calcium ions, but they vary in relative activity *in vitro*. However, the enzymatic activity *in vitro* is not necessarily a good indicator of how well a pectate lyase will macerate plant tissue. For example, *E. chrysanthemi* EC16 PelA has high pectate lyase activity, but a low plant cell maceration activity compared to EC16 PelB, PelC, or PelE (Tamaki et al., 1988).

E. chrysanthemi also produces two cellulases, CelZ which is secreted, and CelY, which is less abundant and periplasmic (Boyer et al., 1984; Guiseppi et al., 1991). Deletion of *celZ* had no effect on virulence on African violet, but, suprisingly, deletion of *celY*, the periplasmic enzyme, caused a delay in symptom development (Boccara et al., 1994). The *celY* gene is part of a cellulose synthesis gene cluster in *Erwinia* and may actually play a role in bacterial aggregation and not plant cellulose degradation (Jahn et al., in preparation), which could explain why this periplasmic enzyme plays a greater role in virulence than the secreted cellulase.

In *E. chrysanthemi*, many of the genes encoding the cell wall-degrading enzymes are expressed as monocistronic operons under non-inducing conditions and as polycistronic operons under inducing conditions (Shevchik and Hugouvieux-Cotte-Pattat 1997, 2003). The cell-wall degrading enzymes co-synthesized in this fashion have complementary activities. For example, *paeY*, *pemA*, and *pelD*, which have synergistic activity in pectin digestion, are regulated in this manner (Shevchik and Hugouvieux-Cotte-Pattat 1997). In *Erwinia*, a shift from expression as a monocistronic operon to expression as a polycistronic operon has only been examined for the cell wall degrading enzymes and this phenomenon has been little explored in other plant pathogens. If this switch between mono- and polycistronic operons is common, it will require a re-examination of many gene regulation models and it also makes it that much more important to develop methods for quantifying and modelling complex regulatory cascades.

Recently Corbett et al. (2005) demonstrated that Svx, an *E. carotovora* subsp. *atroseptica* 1043 protein homologous to *Xanthomonas campestris* AvrXca, is secreted by the T2SS and that Svx contributes to virulence. Although Corbett et al. (2005) used proteomic analysis of the *Erwinia* secretome to identify Svx, this gene had been discovered earlier in a cDNA-AFLP screen for *E. carotovora* genes differentially expressed in rich and minimal medium (Dellagi et al., 2000). Unexpectedly, the *svx* gene was expressed at higher levels in rich medium than in pectin minimal medium, which was expected to induce virulence genes. This demonstrates that experiments that define virulence genes as those up-regulated in host-mimicking medium compared to rich medium may miss important virulence genes.

Because Corbett et al. (2005) found that Svx migrates as two distinct spots on two-dimensional polyacrylamide gel electrophoresis, the secreted Svx protein is likely to be modified in a fashion that alters its charge while not substantially altering its mass. The nature of this modification has not yet been identified. To determine if Svx was secreted via the T2SS, Corbett et al. (2005) used difference gel electrophoresis, a method where two protein samples are labeled with Cy3 and Cy5 and separated on the same gel (Unlu et al., 1997). They found that Svx was not in the secretome of an *outD* mutant, nor were PelC, CelV, a newly identified cellulase homology (ECA2220) and a newly identified proteoglycan hydrolase homolog (ECA0852). This method was also used to show that Svx was not secreted in an *expI* mutant, presumably because AHLs produced by ExpI are required for *svx* expression. Experiments with the *svx* promoter fused to a β-glucuronidase reporter gene confirmed that *svx* required *expI* for transcription. Svx makes a significant contribution to virulence on stems and tubers; mutation of this gene results in smaller stem lesions and less tuber maceration.

10.3. Metabolism of Plant Cell Wall Components

The pectic component of plant cell walls is composed of oligomers of acidic and neutral sugars, including galactose and rhamnose. These large macromolecules comprising the plant cell wall are broken into oligomers, and then taken up by specific porters into the *Erwinia* cells where they are further degraded. This process is best understood in *E. chrysanthemi*, but *E. carotovora* encodes homologous genes for nearly every step in this process. Fig. 1 outlines the degradation of plant cell walls by *Erwinia*.

In *E. chrysanthemi*, after degradation of the galactose oligos by the extracellular pectate lyases, pectin acetylases, pectin methylesterases and polygalacturonases, smaller oligomers cross the bacterial outer membrane via the

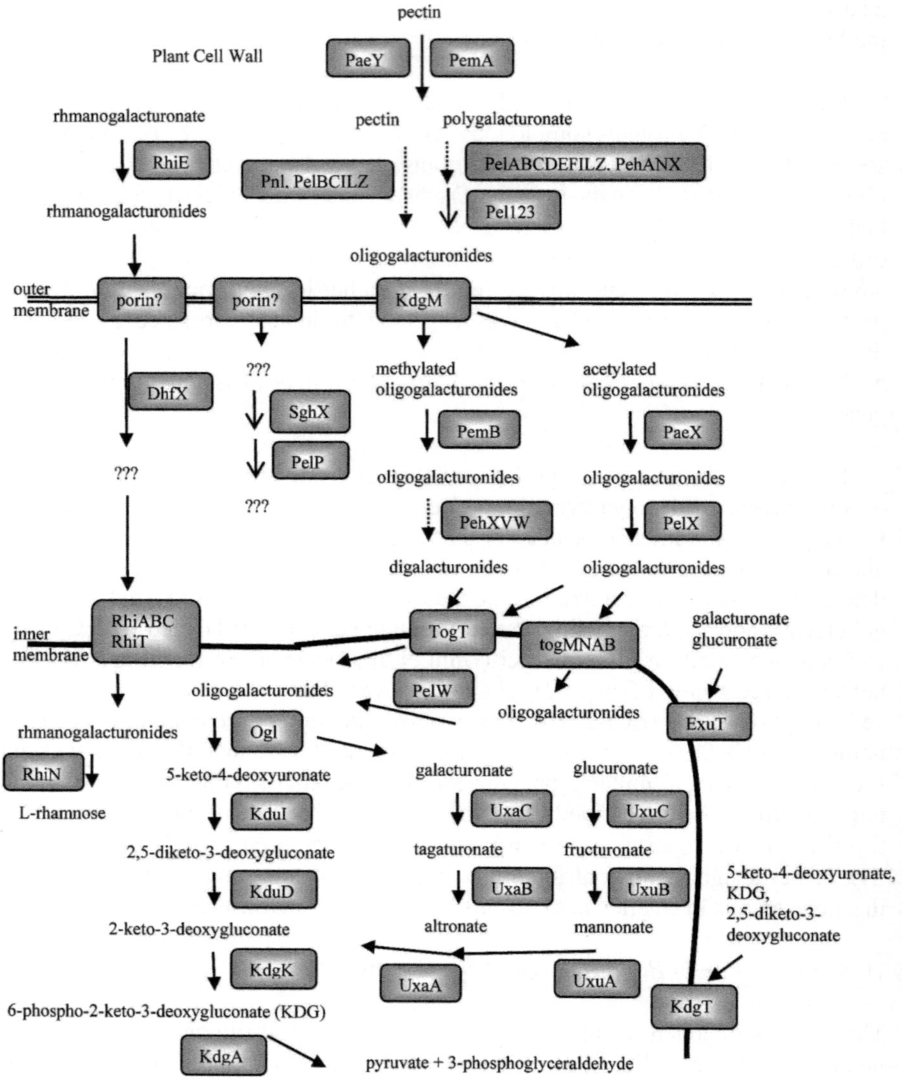

Fig. 1. *Erwinia* degradation of pectin (adapted from Condemine and Robert-Baudouy 1987; Hugouvieux-Cotte-Pattat and Reverchon 2001; Rodionov et al., 2004). Enzymatic steps present only in *E. chrysanthemi* are denoted with dashed arrows and steps present only in *E. carotovora* are indicated with open arrows.

KdgM porin (Blot et al., 2002). Monomers (galacturonic acid) and dimers can enter into the cell via the ExuT porter (Haseloff et al., 1998), the expression of which is regulated by ExuR and probably by KdgR (Valmeekam et al., 2001). The oligos are further degraded in the periplasm by additional enzymes included the exo-pectate lyase PelX, the pectin methylesterase PemB, the polygalacturonase PehX, and the pectin acetylase PaeX. The regulation of these eynzmes is similar to that of the secreted extracellular enzymes. For example, like *pelZ*, *paeX* is transcribed as a monocistronic operon in non-inducing conditions and a polycistronic operon, along with *kdgM*, in inducing conditions. These genes are also under the control of the repressors KdgR, PecS, PecT, and CRP (Shevchik and Hugouvieux-Cotte-Pattat 2003).

Once degraded further, the oligos are transported across the inner membrane via the TogMNAB ABC transporter (Hugouvieux-Cotte-Pattat et al., 2001) or the TogT GPH transporter (Hugouvieux-Cotte-Pattat and Reverchon 2001). Like the cell wall degrading enzymes, the *pelWtogMNAb* operon and *togT* are down-regulated by KdgR and upregulated by CRP via direct interactions of these proteins with the *tog* operon promoter (Hugouvieux-Cotte-Pattat et al., 2001; Hugouvieux-Cotte-Pattat and Reverchon 2001). Interestingly, Hugouvieux-Cotte-Pattat et al. (2001) found that a plasmid containing the genes encoding the TogMNAB transport system and the genes required for oligogalacturonide catabolism *pelW, ogl, kduI* and *kduD*, conferred on *E. coli* the ability to utilize galacturonate or saturated digalactose as a sole carbon source.

Since both the TogMNAB transporter and the TogT transporter can uptake galacturonate, mutation of either *togM* or *togT* does not abolish ability to grow on polygalacturonate or oligos derived from polygalacturonate, including saturated and unsaturated oligogalacturonides (Hugouvieux-Cotte-Pattat and Reverchon, 2001). However, a *togM* mutant is delayed in growth on polygalacturonate apparantly due to delayed induction of the PGA catabolism system (Hugouvieux-Cotte-Pattat et al., 2001). This suggests that TogM is required for induction of these genes, either because inducer molecules can only go through this transporter or because TogM functions as a signal transducer. Hugouvieux-Cotte-Pattat et al. (2001) also found that, like other ABC transporter periplasmic binding proteins, TogM appears to be acting as in chemoreception since a *togM* mutant was less chemotactic toward saturated digalacturonate or unsaturated oligogalacturonides. Chemotaxis toward galacturonate was not affected in the *togM* mutant. TogM probably acts as a receptor for the chemoattractant, then trasfers the signal to a methyl accepting chemotaxis protein, which in turn transfers the signal to the motility apparatus.

Once the oligomers reach the cytoplasm, PelW and Ogl act on these cell wall fragments to break them down further (Shevchik et al., 1999). PelW is homologous to PelY from *Yersinia pseudotuberculosis* and PelB from *E. carotovora*, except that it has no signal sequence, unlike PelY and PelB. Curiously, when a signal peptide was added to a cloned *pelW*, it was secreted to the *E. coli* periplasm, but it was unstable in the periplasm. PelW cleaves tri- and tetragalacturonate, but not digalacturonate, and has the highest affinity for trigalacturonate. PelW generates unsaturated digalacturonate, thus it is an exo-polygalacturonate lyase, like PelX. However, there is no sequence similarity between PelX and PelW. Like the other Pels, PelW requires a cation for activity, but unlike the others Pels, Ca^{2+} works poorly. Instead, PelW is strongly activated by Co^{2+}, Mn^{2+}, and Ni^{2+}. This may be because Ca^{2+} levels are low inside *E. chrysanthemi* cells, necessitating the use of other cations.

In contrast to PelW, Ogl cleaves saturated and unsaturated digalacturonate. Tri- and tetragalacturonate are poor substrates for this enzyme and Ogl is not able to cleave PGA. Ogl also requires divalent cations for activity and Mn^{2+}, Co^{2+}, Fe^{2+}, and Ni^{2+} all can function in this role. Similarly to PelW, Ca^{2+} has no effect on Ogl enzyme activity.

Degradation and metabolism of the rhamnose portion of the plant cell walls is less understood than that of the polygalacturonase portion. The rhamnose is cleaved extracellularly by RhiE (Laatu and Condemine 2003), then the oligos are ported through the outer membrane by an unknown mechanism. The oligos are probably porter across the inner membrane by a dedicated porter encoded in part by *rhiT*, which is homologous to *togT*. Once in the cytoplasm, the rhamnose is further metabolized by RhiN, which acts on unsaturated oligomers produced by RhiE, and by RhaA, RhaB, and RhaD (Hugouvieux-Cotte-Pattat 2004). The rhamnose degradation genes are induced by pectate and rhamnose and are highly induced when *E. chrysanthemi* cells are in plants. The induction of the *rhiTN* operon in the presence of polygalacturonate is mediated by the repressor KdgR, while the induction of the *rhiTN*, *rhaBAD*, and *rhiE* genes by rhamnose is mediated by the *rhaSR* operon. Like most genes involved in pectin degredation, *rhiTN* is induced by CRP, however, it is not regulated by PecT or PecS. This regulatory is importanat for pathogenicity, as a strain with a mutation in *rhaS* is reduced in virulence.

10.4. The Type I Secretion System and Secreted Proteases

The T1SS consists of only three proteins, an MFP protein and ABC-transporter protein in the inner membrane and the TolC protein in the outer membrane. Secretion through this system is usually not dependent on the Sec system. Proteins are targeted to the type I secretion system (T1SS) by an uncleaved C-terminal signal (Holland et al., 2005). Because a C-terminal signal is used to target proteins to this system, they must be interacting with this secretion system after translation has been completed or nearly completed. The best characterized targets of the *Erwinia* T1SS are the metalloproteases that contribute to pathogenicity. In addition to *prtE*, the MFP protein involved in secretion of proteases in E. *chrysanthemi*, eight other MFP homologs are encoded by the *E. chrysanthemi* 3937 genome, including *macA*, and *emrE*. Both *macA* and *emrE* are likely to be required for transport of toxic compounds out of the bacterial cell (Kobayashi et al., 2001; Paulsen et al., 1996).

E. chrysanthemi secretes at least four metalloproteases, PrtA, B, C, and G via a type I secretion system encoded by PrtD, E, and F. (Boyd and Keen, 1993; Dahler et al., 1990; Delepelaire and Wandersman 1989; Ghigo and Wandersman, 1992a, b; Letoffe et al., 1990). Like other type I system-secreted proteins, the secretion signal is located in the C-terminus of the Prt proteins. Also, like many other bacteria that encode proteases, *E. chrysanthemi* also encodes a protease inhibitor, *inh*, linked to the protease genes and their secretion system. Interestingly, the *E. chrysanthemi* proteases modify other secreted proteins and may affect how these proteins interact with host cells. For example, the 97 N-terminal amino acids of PelI are cleaved by secreted *E. chrysanthemi* proteases after secretion. The cleaved PelI is not greatly modified in enzymatic properties, but it is smaller, more basic, and the modified form acts as a defense elicitor in plants (Shevchik et al., 1998).

The *E. carotovora* subsp. *carotovora* SCC3193 secreted protease PrtW, which is homologous to the *E. chrysanthemi* PrtA, PrtB, PrtC, and PrtG proteins, was first identified in a screen for genes upregulated by plant extract (Marits et al., 1999). Like the *E. chrysanthemi* proteases, PrtW is a metalloprotease with two domains that probably bind Ca^{2+} and Zn $^{2+}$. As with *E. carotovora pehA*, *pelB*, *pelC*, and *celVI*, expression of *prtW* is dependent upon *expI*, which is required for production of N-acyl homoserine lactone and the two-component regulatory system GacA-GacS. However, unlike the pectinases, *prtW* is not direcetly regulated by KdgR, although KdgR does appear to have an indirect regulatory effect on this gene (Marits et al., 2002). Unlike with *E. chrysanthemi*, where deletion of the metalloproteases does not appear to affect virulence, deletion of *prtW* in *E. carotovora* caused a significant reduction in virulence (Marits et al., 1999).

Vincent-Sealy et al. (1999) examined *E. carotovora dsbA* and *dsbC* mutants to determine if they were required for protease secretion and found that both *dsbA* and *dsbC* mutants were able to secrete protease activity. However the double mutant was unable to do so, suggesting that the functions of these two genes partially overlap and that cysteine bonds may be important for the function of the T1SS.

10.5. The Type III Secretion System and Effectors

The type III secretion system (T3SS) has been more thoroughly studied in other plant pathogenic bacteria, such as *Pseudomonas syringae* and *Xanthomonas campestris*, where it is more important for pathogenicity than in the soft rot *Erwinia*. There are two classes of T3SS in plant pathogens based on gene organization and regulation (Alfano and Collmer, 1997), with class I being present in both *Pseudomonas* and *Erwnia*. In both cases, an alternative sigma factor, HrpL, is required for expression of the T3SS and its substrates. In *E. chrysanthemi*, expression of *hrpL* requires a sigma 54 enhancer binding protein, HrpS (Chatterjee et al., 2002a; Ham et al., 2004; Yap et al., 2005). The expression of *hrpS* is dependent on the two-component system HrpX-HrpY (Yap et al., 2005). The response regulator HrpY appears to be involved in cross talk with other regulators since *hrpY* is required for T3SS gene expression, while mutation of *hrpX* does not appear to eliminate expression of the T3SS (Yap et al., 2005). The other regulatory proteins that may interact with HrpY, as well as the signal that HrpX responds to is unknown. *E. chrysanthemi* strains differ somewhat in regulation of the T3SS, with the expression of the T3SS being induced in minimal media and by acidic pH in some strains and apparently constitutive in others (Ham et al., 2004)

In *E. carotovora* subsp. *carotovora* Ecc71, the expression of *hrpL* is also under the control of the *rsm* system, and is therefore affected by the many global regulators that control the *rsm* genes, including KdgR, GacA-GacS, HexA, and RsmC (Chatterjee et al., 2002a, b). The contribution of these global regulators to expression of the T3SS has not been explored in *E. chrysanthemi*. In addition to the Hrp regulator cascade and the *rsm* system, the T3SS it is likely to be under the control of negative regulators, such as HrpV, but this too has not been examined in *Erwinia*.

Numerous T3SS substrates have been identified in *Pseudomonas syringae*. In contrast, it appears that very few are present in the soft rot *Erwinia* (Bell et al., 2004; Glasner et al., submitted). However, the T3SS does contribute to growth in association with plants in both *E. chrysanthemi* (Yang et al., 2002) and *E. carotovora* (Holeva et al., 2004). In both *E. chrysanthemi* and *E. carotovora*, *hrpN* contributes to growth (Holeva et al., 2004; Yang et al., 2002) and *dspE* has been shown to be important in *E. carotovora* (Holeva et al., 2004). Yang et al. (2002) saw a greater effect of *hrpN* on virulence in *E. chrysanthemi* 3937 when it was examined in a mutant that also had all major pectate lyase genes deleted, demonstrating that this pectate lyase deletion mutant is useful for examining subtle virulence phenotypes.

The *E. chrysanthemi* T3SS locus appears to be a hotspot for pathogenicity island integration since in addition to the T3SS, a hemoagglutinin-encoding gene and a virulence gene of unknown function (orf11) are also in this locus (Kim et al.,

1998; Rojas et al., 2002; 2004). This region also appears to have undergone recombination in comparison to *E. carotovora* since *dspEF* and *hrpW*, which are located adjacent to the genes encoding the T3SS in other *Erwinia* and in *P. syringae*, are located over 100 kb from the T3SS gene cluster in *E. chrysanthemi* (Glasner et al., submitted).

10.6. The Type IV Secretion System – Conjugation, Effectors or Biofilms?

The bacterial type IV secretion system (T4SS) is very versitile and is used for transfer of DNA between bacteria and into host cells, as well as translocation of effector proteins into host cells and biofilm formation (Cascales and Christie, 2003). *E. carotovora* subsp. *atroseptica* 1043 encodes one type IV secretion system which is required for full virulence on potato (Bell et al., 2004). Similarly, *E. chrysanthemi* 3937 encodes two type IV secretion systems, but their contribution to virulence remains unknown. Whether or not these type IV secretion systems contribute to genetic exchange, deliver effector proteins into host plant cells, or contribute to bacterial adhesion is unknown.

10.7. Erwinia adhesins and biofilm formation

Adhesion to plant surfaces, insects, and surfaces found in agricultural settings, such as greenhouse pots and farm machinery, is likely to be important for the infection cycle of the soft rot *Erwinia*. Both genomes encode T3SS, which may play a role in *Erwinia* biofilm formation. The *E. carotovora* subsp. *atroseptica* 1043 genome, but not *E. chrysanthemi* 3937 genome, encodes pili homologs that may be involved in adhesion to surfaces (Bell et al., 2004). Neither genome encodes aggregative fimbriae, a types of adhesin found in other enterobacteria.

A hemoagglutinin-encoding gene, *hecA*, is located adjacent to the *E. chrysanthemi* T3SS gene cluster (Kim et al., 1998). Both of these gene clusters were likely laterally transferred into the *E. chrysanthemi* genome, thus this location appears to have two separate gene islands important for *Erwinia*-plant interactions. HecA is predicted to be a 3,850-residue protein and it is a member of the *Bordetella pertussis* filamentous hemagglutinin family. Rojas et al. (2002) found that mutation of *hecA* reduced the ability of *E. chrysanthemi* EC16 to attach and form aggregates on leaves and to cause an aggregate-associated killing of epidermal cells. Since production of the cell wall degrading enzymes is upregulated by quorum sensing, this HecA-dependent aggregation may be important for accumulation of acyl-HSL signals to induce macerating enzymes. It appears that homologous hemoagglutinins are present in numerous other plant pathogens, including *Ralstonia*, *Pseudomonas*, and *Xanthomonas* species, suggesting that this class of protein is generally important for virulence on plants. The secretion of HecA has not been examined in soft rot *Erwinia*. However, the homologous *B. pertussis* FHA is secreted by the two-partner secretion system (Locht et al.,

1993). This system requires an outer membrane-associated accessory protein, which is likely to be HecB in *Erwinia*, which interacts with the N-terminal portion of the secreted protein. FHA is proteolytically cleaved after secretion and has multiple substrate-binding domains. Post-secretion processing of HecA has not yet been examined nor is it known if HecA binds to particular substrates.

Many strains of *E. chrysanthemi* swim to the top of stationary cultures and form two distinct types of biofilms when grown in rich medium containing glycerol (Yap et al., 2005). A biofilm ring forms at the air-liquid-solid interface around the top edge of the culture. The genes required for this type of biofilm are unknown. A pellicle may also form across the top of the culture at the air-liquid interface. The T3SS and flagella are required for formation of the pellicle (Yap et al., 2005; Jahn et al., in preparation). Recently Yap et al. (in preparation) also found that HrpN, a T3SS-secreted protein is also required for pellicle formation. This is the first example of bacterial aggregation mediated by the T3SS and of a *in vitro* function for HrpN. Other, as yet unknown, factors also appear to be important for pellicle formation since expression of the T3SS is not sufficient for pellicle formation.

E. chrysanthemi pellicles also contain cellulose synthesized by the bacteria, although cellulose is not absolutely required for pellicle formation (Yap et al., 2005). This was somewhat surprising since *E. chrysanthemi* is better known for degradation of cellulose. Cellulose synthesis in related enterobacteria requires the *bcsABZC* and *bcsEFG* operons, which are divergently transcribed and encode cellulose synthesis genes, the regulatory operon *cgsD*, and *adrA* operon, which encodes a GGDEF protein. *E. chrysanthemi* 3937 only encodes the *bcsABZC* and *adrA* operons. Of these operons, only *bcsABZC* is required for cellulose synthesis (Yap et al., 2005; Jahn et al., in preparation). In addition, the *bcsABZC* operon appears to have been horizontally transferred into *E. chrysanthemi* and it is more similar to *bcs* genes from *Pseudomonas* than from other enterobacteria. Therefore, cellulose synthesis in *E. chrysanthemi* appears to differ in mechanism and regulation compared to other enterobacteria.

E. carotovora does not form pellicles in media. In fact, rather than moving to the top of stationary cultures, they swim to the bottom of the culture tube and form a sediment layer. In both *E. carotovora* and *E. chrysanthemi*, this differing response is likely to be regulated by aerotaxis genes. *E. carotovora* subsp. *atroseptica* 1043 also encodes cellulose synthesis genes *bcsABZC* and *adrA* and it also encodes *bcsEFG*. These genes also appear to have been horizontally transfered into *E. carotovora*.

10.8. Erwinia Necrosis-Inducing Protein

In 2004, two groups reported that a subset of *E. carotovora* subsp. *carotovora* strains lack *hrpN*, and in at least one case, the genes encoding the entire T3SS (Mattinen et al., 2004; Yap et al., 2004). Mattinen et al. (2004) used an *rpoS* mutant constructed in a *hrpN*-minus strain, SCC3193, to identify other secreted virulence factors. The *rpoS* mutant was chosen because it is up regulated in the production of several virulence proteins, probably because RpoS is a positive regulator of *rmsA*, which is itself a negative regulator of several secreted virulence proteins (Mukherjee et al., 1996a, b; Andersson et al., 1999a, b). They found that when the SCC3193 *rpoS* mutant was grown on a solid minimal medium of pH 5.7 at 15°C, then infiltrated into tobacco leaves, it elicited plant cell death similar to an HR. No plant cell death was observed when the cells were grown in liquid medium, at 28°C, in rich medium, or with wild type cells. Mattinen et al. (2004) fractionated bacterial cells, found that the elicitor activity could be purified from the periplasm, and identified the gene encoding the elicitor protien.

This gene encodes a product homologous to the Nep-1 family, which is a group of necrosis- and ethylene-inducing proteins found in many oomycete and fungal pathogens. There are homologs of this gene, which was named *nip* for necrosis-inducing protein, in both *E. carotovora* subsp. *atroseptica* and *E. chrysanthemi* and in all cases appear to encode a Sec secretion signal. Although both purified Nip and HrpN can elicit cell death in tobacco, there is no apparent homology between these two proteins. Nip, however, has two conserved cysteines and a sequence, GHRHDWE, found in all members of the Nep1 family of proteins. An *E. carotovora nip- rpoS-* mutant no longer caused plant cell death and an *E. carotovora nip* deletion mutant was also reduced in virulence on potato tubers, demonstrating that this protein plays a role in soft rot pathogenicity. The *nip* mutant was not reduced in virulence on tobacco seedlings, lettuce, eggplant, cauliflower, broccoli, celery, sweetpotato, or on potato stems, thus the contribution of this virulence factor is both host species and host tissue specific.

The authors did not deposit the sequence of this gene in commonly used DNA databases, therefore this gene is not easily found through homology searches. Oddly enough, Pemberton et al. (2005) reported a homologous protein from *E. carotovora* subsp. *carotovora* found in a search for genes encoding secreted proteins differentially regulated by AHL. They found that AHL reduced expression of a gene encoding a necrosis-inducing protein, which they also named Nip. The effect of AHL on *nip* expression was via the *rsm* system and was not due to a direct interaction. They also found that a mutation in *nip* reduced *E. carotovora* subsp. *carotovora* maceration of potatoes and *E. carotovora* subsp. *atroseptica* lesion formation on potato stems. Oddly enough, Pemberton et al. (2005) did not deposit the Nip sequence in Genbank either.

Several questions remain to be answered about this interesting virulence protein. Is it secreted across the outer membrane, and if so, how? Both groups showed that the Nip protein could cause cell death in tobacco, but they did not determine if this cell death was due to toxic effects of the protein or an elicited active cell death similar to apoptosis. Also, other than its apparent down-regulation by RpoS and AHL / *rsm*, nothing is known about the regulators of this gene nor if and when it is expressed when the bacteria are in association with plants. Finally, leaf infiltration assays with purified proteins can sometimes give very misleading results, thus this data needs to be interpreted with caution. For example, high amounts of purified HrpZ, a T3SS-secreted harpin protein encoded by *P. syringae*, elicits an active plant cell death in tobacco. However, HrpZ does not elicit tobacco plant cell death when produced by bacterial cells in leaves (Alfano et al., 1997). Rather, it assists in the delivery of an effector, HopPsyA (HrmA), into plant cells. Importantly, this shows that when the bacteria are in tobacco leaves, it is the effector, HopPsyA, and not the harpin, HrpZ, that elicits plant cell death. This is probably because HrpZ is not made at high enough levels by the bacteria to elicit the HR. Thus, it is possible that other elicitors, such as HrpN and Nip also do not elicit plant cell death when produced by bacterial cells in leaves.

10.9. Phytotoxins

E. carotovora subsp. *atroseptica* 1043, but not *E. chrysanthemi* 3937, encodes a *cfa* operon, one of two operons required for synthesis of the phytotoxin coronatine in *Pseudomonas syringae*. Coronatine appears to mimic JA and repress expression of the SA-induced plant defense genes (Zhao et al., 2003). The *E. carotovora* subsp. *atroseptica cfa* operon does contribute to virulence, but since it lacks the other operons required for coronatine production, the exact role of this operon is currently unknown (Bell et al., 2004).

10.10. Motility and Chemotaxis

Motility in *Erwinia* is required for full virulence and is co-regulated with other virulence factors via regulators such as PecS, PecT and RsmC (Condemine et al., 1999; Harris et al., 1998; Mulholland et al., 1993; Shih et al., 1999). Although *Erwinia* are motile in soft rot lesions, the co-regulation of protien secretion and motility has been little explored in *Erwinia*. For example, is the T3SS, which is considered a contact-dependent system, repressed in motile cells? Are motile cells secreting cell-wall degrading enzymes? Since the individual bacterial cells would not beneficit from secreted enzymes if they immediately swim away from the location of secretion, is this an example of a cooperative attack on the plant? Finally, does the force of the bacteria swimming in a lesion affect disease development or bacterial growth, perhaps by causing a mixing the substrate and degradative enzymes?

10.11. Iron Acquisition

Iron is essential for the function of multiple bacterial proteins, but iron has low bioavailability in the environment. Like many pathogens, *Erwinia* has multiple systems dedicated to iron acquisition. *E. chrysanthemi* 3937 produces at least two siderophores that contribute to virulence, achromobactin, a citrate/carboxylate siderophore, and chrysobactin, a catechol (Enard et al., 1988; Franza et al., 2005; Münzinger et al., 2000; Neema et al., 1993; Persmark et al., 1989). TonB is required for the uptake of siderophores and for full virulence on plants (Enard and Expert, 2000). The contribution of these siderophores is tissue specific, for example, mutants unable to make either siderophore still macerate potato tubers as well as wild type (Expert 1999).

E. carotovora produces chrysobactin and aerobactin, but no role in disease has been described for either of these siderophores (Bull et al., 1996; Ishimaru and Loper 1992). However, as pointed out in a recent comprehesive review of iron metabolism in *Erwinia,* in these experiments, the bacteria used for plant inoculations were suspended in 0.1 M phosphate buffer, which may represent an important iron supply, in part because *Erwinia* can also acquire iron more easily if phosphate is present (Expert 1999).

Normally siderophore synthesis is repressed if iron is abundant. The corregulation of siderophore regulation in *E. chrysanthemi* was first discovered during characterization of mutants that no longer repressed the chrysobactin operon *fct cbsCEBA* in the presence of iron. This class of mutants was affected in the *cbr* genes, which encode an ABC transporter that is required for uptake of achromobactin (Mahé et al., 1995). Achromobactin and chrysobactin are temporally regulated, with achromobactin produced before chrysobactin. Once chrysobactin production starts, achromobactin production decreases (Franza et al., 2005), suggesting that achromobactin, which has a lower affinity for iron, is produced first and if iron becomes even more limiting, chrysobactin is then produced.

10.12. Resistance to Active Oxygen

Plants produce high levels of several different active oxygen molecules when under attack by pathogens and soft rot *Erwinia* encode enzymes to resist these toxic molecules. These active oxygen molecules are also important for induction of plant defenses, and it is possible that soft rot *Erwinia* activity on active oxygen also has the effect of interfering with plant defense gene induction. *E. chrysanthemi* has long been known to produce a water-insoluble blue pigment, indigoidine, under some growth conditions. Indigoidine production requires the genes *indABC*, which are repressed by PecS, and induced by oxidative stress (Reverchon et al., 1994; 2002). The *ind* genes are required for systemic infection of host plants and may be important for protecting bacterial cells from reactive oxygen species (Reverchon, et al., 2002). *E. carotovora* does not encode *indABC*, nor does it produce indigoidine. Thus, this important function must be performed by another as yet unidentified molecule in *E. carotovora*.

Soft rot *Erwinia* produce several enzymes that detoxify active oxygen and that differ in their contribution to bacterial virulence. One of the first in *Erwinia* was described in 1995, when Favey et al., found that *E. chrysanthemi* 3937 *hmpX*, a flavohaemoglobin homolog, was required for maceration of inoculated African violet leaves and systemic spread in inoculated plants. The role of *hmpX* was not clear until recently when Boccara et al. (2005) demonstrated that HmpX is a NO denitrosylase required for detoxification of NO to nitrate in an oxygen-dependent manner. They also showed that *hmpX* impairs the elicitation of the hypersensitive response, thus apparently interfering with plant defense gene induction, which may better enable *E. chrysanthemi* to cause disease.

Some iron-sulfur clusters in *Erwinia* proteins are sensitive to the defensive oxygen burst that occurs as part of the host plant response to pathogens. The *E. chrysanthemi suf* operon encodes proteins that protect these sensitive proteins and, thus aid in the ability of *Erwinia* to survive the effects of the active oxygen released by its host plants. (Nachin et al., 2001, 2003). This operon is also required for the bacteria to be able to use iron obtained by chrysobactin.

Both *E. carotovora* and *E. chrysanthemi* also posses *msrA*, which encodes a peptide methionine sulfoxide reductase. El Hassouni et al. (1999) showed that this gene, which is in the minimal gene set present in all genomes examined to date and which is important for repairing oxidative damage to proteins, also contributes to virulence. A strain with a mutation in *msrA* was unable to systemically infect plants, suggesting that that repair of oxidative damage is important for surviving the active oxygen species produced by plant leaves under pathogen attack. Curiously, *msrA* is also required for motility on agar plates. It is possible that the role of *msrA* in virulence is due to its affects on motility, which itself is required for full virulence of soft rot *Erwinia* (Charkowski, unpublished) and not because it is required for repair of damaged proteins.

Santos et al. (2001) examined the contribution of an *E. chrysanthemi* 3937 Mn superoxide dismutase, SodA, to pathogenicity. The *sodA* mutant was unable to grow in minimal medium lacking branched chain amino acids, probably because it was unable to inactivate O_2^- molecules, which inactive the enzyme dihydroxyl-acid dehydratase, an enzyme required for branched chain amino acid synthesis (Flint et al., 1993). The *sodA* mutant was able to macerate potato tubers, but was unable to macerate inoculated leaves or cause a systemic infection in African violets. The addition of exogenouse superoxide dismutase and catalase to the inoculum partially restored virulence and resulted in maceration of the inoculated leaves, demonstrating that detoxification of host-produced active oxygen is required for pathogenicity.

However, the specific active oxygen molecule hydrogen peroxide does not appear to be important for plant defense against soft rot *Erwinia*. Miguel et al. (2000) deleted the LysR-family regulatory gene *oxyR* from *E. chrysanthemi* EC16 and found that the mutant strain was able to infect plants as well as wild type even though *in vitro* it was more sensitive to hydrogen peroxide, produced lower levels of catalase and glutathione reductase, and was unable to form colonies on agar plates unless exogenous catalase was added. Because the oxyR mutant was deficient in growth in vitro, but could cause symptoms in plants, the authors hypothesized that a plant-inducible catalase was produced by *E. chrysanthemi*. The authors examined the catalase activity of leaves inoculated with bacterial cells or culture filtrates to explore this idea, but they were unable to detect high catalase activity in the *oxyR* mutant. Therefore, they concluded that hydrogen peroxide is not important for plant defence against *E. chrysanthemi*.

10.13. Resistance to Host Antimicrobial Peptides

Eukaryotic cells produce numerous antimicrobial peptides and some pathogens have evolved methods to detoxify these peptides. The genome sequences of *E. carotovora* subsp. *atroseptica* and *E. chrysanthemi* revealed that both genomes encode *sapABCDF*, which is homologous to a locus required for degradation of host antimicrobial peptides in animal pathogens. The contribution of these genes to virulence was recently examined with *E. chrysanthemi* EC16 (López-Solanilla et al., 1998; 2001). This peptide degradation gene cluster is at least as important as pectate lyases and more important than the T3SS for causing disease on potato tubers and chicory leaves. Importantly, this result demonstrated that production of pectinases is not sufficient for maceration of plant tissue and that soft rot *Erwinia* must also be able to defend against host antimicrobial compounds. The detoxification appears to occur at least partially inside the bacterial cells or tightly associated with the bacterial cells since co-inoculation of wild type and *sap* mutant strains does not recover the growth of the mutant strains.

10.14. Survival of Osmotic Stress

Osmoprotectants are compounds that restore the ability of bacteria to grow under high osmotic stress and *E. chrysanthemi* has at least two osmoprotectant uptake systems (Gouesbet et al., 1996). In addition, osmotic stress also affects production of virulence proteins, specifically pectate lyases, which are upregulated under high osmolarity (Gouesbet et al., 1995; Hugouvieux-Cotte-Pattat et al., 1992; Mildenhall and Prior 1983; Mildenhall et al., 1988; Prior et al., 1994). Gloux et al. (2005) examined the role of osmotolerance in virulence. They found that *E. chrysanthemi* strains varied in their osmotolerance and that virulence, as measured by the ability of *E. chrysanthemi* to decay potato tubers, was inversely correlated with osmotolerance in culture. When an *ousA* mutant strain was examined, it was found to be more virulent on intact potato tubers or on tuber slices incubated under anaerobic conditions than the wild type *E. chrysanthemi* 3937 and also to have increased expression of pectate lyases, but only under anaerobic conditions. Neither virulence nor pectate lyase activity was affected under ambient conditions. This is one of the few examples of a mutation that increases *Erwinia* virulence.

11. *ERWINIA* VIRULENCE GENE REGULATION

Although soft rot *Erwinia* virulence factors are secreted through multiple systems, in many cases the same global regulators control the expression of these diverse secreted proteins. Most of the virulence genes are controlled by multiple regulators, which have varying effects on gene regulation depending upon environmental conditions. In addition, many of these regulators control their own expression and the expression of other regulators. This makes drawing a simple line diagram of gene regulation difficult. A model quantifying the relative importance of the various regulators under diverse conditions is badly needed.

The binding sites for only a few *Erwinia* regulatory proteins have been determined (Table 7), again making predictions about virulence gene regulation difficult. However, these few binding sites have been very useful in bioinformatics approaches for identifying new virulence genes. Determining the binding sites of all of the transcriptional regulators in a model *Erwinia* strain is technically possible and would greatly aid in developing models of virulence gene regulation and function.

Table 7. *Erwinia* regulatory protein binding sites

Protein	Target	Reference
CRP	WWWTGTGATNNNNATCACAWWW	Rodionov et al. (2004)
FruR	GCTGAAWCGWTTCAGC	Rodionov et al. (2004)
KdgR	WAWTRAAAYRnYRTTTYAWTW	Liu et al. (1999) Nasser et al., (1994) Rodionov et al. (2004)
PecS	CGANWTCGTATATTACGANNNCG	Rouanet et al. (2004)
PehR	TTTTTTCGCRATRAAAC (not yet confirmed)	(Nasser et al. 1999)
RdgB[A]	TŦATTAAACTCGATTAATAAGC	Liu et al. (1997)
RhaS	atCTTcgCnntntTGCCAncccCncccnGcngaTGgCaatCncnCAAAGGt	Hugouvieux-Cotte-Pattat (2004)
RhaR	ATCTTGAAAAAnnGACnTn(7)GnGnTnTnnCGnCGnnAnTTTAAGGT	Hugouvieux-Cotte-Pattat (2004)

11.1. The Rsm system – Post-Transcriptional Gene Regulation

The regulator with the most significant effect on protein production in *E. carotovora* is the post-transcriptional regulator RsmA-*rsmB* RNA. RsmA binds to the mRNAs of numerous genes, including those of the cell wall degrading enzymes, *hrpN* and *ohlI* and targets these mRNAs for degradation. The *rsmB* RNA (formerly *aepH*; Murata et al., 1994; Liu et al., 1998) binds to RsmA and inhibits it from binding to other mRNAs, essentially having a negative regulatory effect on RsmA. In effect, the ratio of RsmA to *rsmB* RNA regulates protein production from the target mRNAs by controlling the stability of the target mRNAs. The *rsmB* gene does not appear to encode a protein product and it is post-transcriptionally processed so that two forms, the full-length 479-bp *rsmB* and a 259-bp *rsmB'*, which has its 5' terminus cleaved off, are present in cells (Liu et al., 1998). The smaller *rsmB'* is sufficient to inhibit RNA-degradation of other target mRNAs by RsmA. RsmA and *rsmB* are themselves regulated by multiple proteins, including KdgR, GacA-GacS, RpoS, ExpR, and RsmC (Andersson et al., 1999a, b; Cui et al., 1999, 2001, 2005; Hyytiäinen et al., 2001; Liu et al., 1999).

11.2. The IclR Family Regulators – KdgR, Pir, and RexZ

The KdgR regulator, an IclR-like regulator, is conserved among the enterobacteria, being present in the plant pathogens *E. chrysanthemi*, *E. carotovora*, and *E. amylovora*, as well as animal-associated bacteria, such as *E. coli*. In *E. coli*, this regulator only affects transcription of genes involved in the breakdown of 2-keto-3-deoxygluconate, whereas in the soft rot *Erwinia*, it regulates a wide array of genes and nearly all of the proteins involved in pectin catabolism are down-regulated by KdgR. Even genes involved in the breakdown of non-pectin plant cell wall polymers, such as the *rhi* genes, which are required for degredation of rhamnogalacturonan, are down regulated by KdgR (Hugouvieux-Cotte-Pattat 2004). The binding site of KdgR is also conserved across enterobacterial species, with the *E. coli kdgR* able to complement an *E. chrysanthemi kdgR* mutant (James and Hugouvieux-Cotte-Pattat, 1996).

The KdgR binding site has been identified and is always found near the transcriptional start site of the regulated genes, suggesting that KdgR and RNA polymerase compete for the same location on the promoters of these genes (Nasser et al., 1994). The pectin catabolites 2-keto-3-deoxygluconate (KDG), 5-keto-4-deoxyuronate (DK-I) and 2,5-diketo-3-deoxygluconate (DK-II), are able to bind to KdgR in vivo and in vitro, where they cause KdgR to be unable to bind to DNA, resulting in up-regulation of KdgR-repressed genes. Of these molecules, KDG is the main inducer of KdgR-regulated genes.

Rodionov et al. (2004) used a bioinformatics approach combined with in vivo expression analysis to identify all KdgR-regulated proteins in *E. chrysanthemi*

3937 and to determine the relative strength of each binding site (Table 8). They identified nine additional genes down-regulated by KdgR, *chmX*, *dhfX*, *gntB*, *pykF*, *spiX*, *sotA*, *tpfX*, *yeeO* and *yjgK*, and, surprisingly, two genes, *ppsA* and *ydiA*, up-regulated by KdgR. In general, genes regulated by KdgR are also under catabolite repression and those genes up-regulated by KdgR also appear to be regulated by FruR, another regulator, that like KdgR and CRP, controls genes in response to carbon sources.

To define the functions of these genes, allelic-exchange was used to mutant several of the KdgR-regulated genes. Rodioniv et al. (2004) were unable to obtain *yjgV*, *gntD*, and *gntD2* mutants, suggesting that these genes are required for bacterial growth. They found that *chmX*, *dhfX*, *gntB*, *ppsA*, *pykF*, *sotA*, *spiX*, *tpfX*, *ydiA*, *yeeO* and *yjgK* were not required for growth on polygalacturonic acid or galacturonic acid. Also, *chmX* was not required for chemotaxis towards either of these carbon sources. However, as multiple methyl-accepting chemotaxis receptor proteins are present in *E. chrysanthemi*, there may be redundancy in receptors for plant cell wall oligomers.

In *E. carotovora*, KdgR also regulates *ceV1*, *prtW*, *pehA* and *hrpN* by inhibiting transcription of *rsmB* and by activating RsmA. (Hyytiäinen et al., 2001; Liu et al., 1999; Rodionov et al., 2004). Unlike with the *pel* genes, where KdgR binds to the promoter region, preventing transcription initiation, with *rsmB*, can bind in three places in the coding region for this gene, thus effecting transcriptional elongation (Liu et al., 1999). This novel roadblock mechanism was first described in *E. carotovora*. Since KdgR and the Gac system have opposite effects on the *rsm* genes, mutation of *kdgR* in a *gacA* mutant background nearly restores *E. carotovora* to wild type with respect to production of extracellular proteins.

In *E. chrysanthemi*, the major pectate lyases, PelABCDE, are induced to a greater extent by plant extract than by polypectate and a second set of pectate lyases are only induced in the presence of plant extracts or in plants (Kelemu et al., 1993). Nomura et al. (1998) identified a novel regulator responsible for induction in the presense of plant extract, Pir, by using a DNA-binding assay with the *pelE* promoter. Like KdgR, Pir is a member of the IclR family of transcriptional regulators and it binds to a region of the *pelE* promoter that overlaps with the KdgR binding site. The predicted binding sites of these two proteins are similar and a putative KdgR-binding site was predicted to be upstream of *pir* (Rodionov et al., 2004). However, this site is likely to be a Pir-binding site instead since no regulation of *pir* by KdgR could be demonstrated (Rodionov et al., 2004). A strain with a *pir* mutation does not induce pectate lyase genes in response to plant extracts and it also appears to play a role in induction in reponse to low iron concentrations. However, other enzymes, including cellulases and proteases were not affected by mutation of *pir*.

Nomura et al. (1998) also showed that these Pir and KdgR proteins compete for the same region on the *pelE* promoter, suggesting that the bacteria can only respond to pectate or plant extract and that the response is not additive. This may allow fine tuned modulation of *pel* expression during virulence or may be because KdgR regulation of *pels* is more important during saprophytic growth than during pathogenesis. KdgR-induction of *pels* is not sufficient for full viurlence since *pir* mutants are reduced in ability to cause disease on potato, celery, and cabbage, thus the additional induction of *pels* in response to plants is important for pathogenicity.

Additional IclR-like regulators have been characterized in *E. carotovora*. Thomson et al. (1999) characterized a KdgR homolog in *E. carotovora* SCRI193, RexZ, and found that unlike KdgR, it acts as an activator of extracellular enzymes including pectate lyase, cellulase, and protease. Also unlike KdgR, it does not regulate these enzymes in response to polygalacturonate. The *rexZ* gene has functional binding sites for both KdgR and CRP upstream of its transcriptional start site, suggesting that it is regulated by both of these proteins. However, Thompson et al. (1999) only were able to demonstrate regulation by catabolite repression and not polygalacturonate, suggesting that KdgR does not regulate *rexZ*.

Table 8. Genes regulated by KdgR in soft rot *E. chrysanthemi* 3937 and *E. carotovora* subsp. *atroseptica* 1043 (Rodionov et al., 2004)

	Function	**3937**	**1043**
Regulators			
pecT	Pectate lyase regulator	UC	
pir	Pectate lyase regulator	N	
expI	Regulator of extracellular enzyme production	XC	
rexZ	Regulator of extracellular enzyme production		DC
rsmB	Regulatory RNA		DC
Extracellular pectinases			
pelA, pelB, pelE, pelCZ, pelI	Pectate lyase	DC	
pel1, pel3-pelZ	Pectate lyase		DC
pel2, pelI	Pectate lyase		XC
pehN, pehX, pehW, pehV	Polygalacturonase	DC	
pehA	Polygalacturonase		XC
pehN	Polygalacturonase		DC
pelD-paeY-pemA	Pectate lyase, acetylesterase, methylesterase	DC	
paeY-pemA	Pectin acetylesterase, methylesterase		XC
pemB	Pectin methylesterase	DC	
Secretion systems			
out T2SS	Secretion of cell-wall degrading enzymes	DC	
pel-pehX-out T2SS	Pectate lyase, polygalacturonase, secretion system		XC
Pectin catabolism			
kdgM	OGA porin	C	
kdgT	Transporter of KDG, DK-I and DK-II	DC	
kduI-kduD, kdgF	DK-I catabolism	DC	XC
pelX	Exopectate lyase	DC	XC
togT	Oligogalacturonide transporter	DC	XC
kdgK	KDG kinase	DC	XC

ogl	OGA lyase	DC	XC
kdgN	OGA porin homolog	DC	C
kdgM-paeX	OGA porin, pectate acetylesterase		XC
kdgM3-kdgM4-pelP	OGA porin paralogues, pectate lyase		XC
pelW-togMNAB	Oligogalacturonide catabolism and transport	DC	XC
kdgX	Predicted KDG permease		XC
Other transporters			
rhiT-rhiN	Rhamnogalacturonide transport and catabolism	DC	XC
rhiABC	Predicted transporter for rhamnogalacturonides		XC
yeeO	Multidrug efflux transporter	D	X
sotA	Sugar efflux transporter	DC	XC
Chemotaxis			
chmX13, chmX21	Methyl-accepting chemotaxis protein	XC	
chmX	Methyl-accepting chemotaxis protein	D	
cheX (chmX)[A]	Methyl-accepting chemotaxis protein		XC
Other functions			
ppsA, ydiA	Phosphoenolpyruvate synthase and unknown function	UC	UC
pykF	Pyruvate kinase	DC	X
sghX	Glycosyl hydrolase		XC
spiX	Sugar isomerase	DC	XC
Unknown function			
dhfX, gntDBMNAC, tpfX, yjgK		DC	XC
ygiV		XC	X
gntD2		XC	

D = down-regulated by KdgR; U = up-regulated by KdgR; X = KdgR binding site is present, regulation has not been demonstrated experimentally; N = KdgR binding site is present, but gene is not regulated by KdgR; C = CRP binding site is present

When the operon structure differs between *E. chrysanthemi* and *E. carotovora*, homologous genes are placed in adjacent multiple rows.

[A]*cheX* and *chmX* may be homologous genes

11.3. CRP and Catabolite Repression

The *Erwinia* CRP sequence is very similar to that of *E. coli* and it probably functions in a similar manner (Reverchon et al., 1997). In *E. chrysanthemi*, CRP is a critical inducer since nearly all of the genes down-regulated by KdgR are up-regulated by CRP (Nasser et al., 1997; Reverchon et al., 1997; Rodionov et al., 2004). Because of this, CRP is required by *Erwinia* for growth on pectin and polygalacturonate and for virulence (Reverchon et al., 1997). Just as the regulation of *pelA* differs in many other respects from that of the other *pels*, unlike the other *pels*, *pelA* is not induced by CRP, rather it appears to be repressed. The role that CRP plays in *E. carotovora* has not been examined as closely.

11.4. Signma Factors

Both *E. carotovora* and *E. chrysanthemi* encode multiple sigma factors, but the roles of only a few of these sigma factors in plant-microbe interactions have been examined. The stationary phase sigma factor, RpoS, is unstable in growing cells, but is stabilized during entry into stationary phase or during osmotic stress. RpoS is not required for virulence in *E. carotovora* subsp. *carotovora* Ecc71 and RpoS mutants actually cause more severe symptoms than wild type cells, although they do not grow to higher cell densities (Mukherjee et al., 1998; Andersson et al., 1999a, b). This is probably because RpoS upregulates *rsmA*, which is a negative regulator of secreted virulence proteins. However, RpoS mutants are more sensitive to environmental stresses, including osmotic and oxidative stress, probably in part because they are deficient in accumulation of glycogen, and they are unable to compete with the wild-type either *in planta* or *in vivo* (Andersson et al., 1999a, b). This is one of the few examples of competition experiments with *Erwinia in planta*. This approach should be used more often with *Erwinia* mutants since it can reveal phenotypes for mutations in genes with small contributions to virulence and it can reveal fitness effects in mutants that appear at first to be more virulent.

ExpM, a member of the RssB family, acts negatively on RpoS, decreasing its stability. Andersson et al. (1999b) mutated *expM* and found that it was required for virulence, in part because the mutant had higher levels of RpoS, which resulted in higher levels of RsmA, and hence, lower levels of secreted proteins. ExpM probably also acts on other targets since an *expM-rpoS* mutant was not restored to wild type levels of virulence.

The HrpL sigma factor is required for expression of genes encoding the T3SS and its substrates. It binds to a short sequence known as the *hrp* box upstream of these operons. The Hrp box sequence is conserved among *Erwinia* and *Pseudomonas* species, making it possible to identify putative HrpL-regulated genes through bioinformatics. The *hrpL* gene itself is dependent upon sigma 54 and the enhancer binding protein HrpS (Ham et al., 2004; Yap et al., 2005) and the *hrpS* gene is up-regulated by the two-component system HrpX-HrpY in response to unidentified environmental signals (Yap et al., 2005). In *E. carotovora* subsp. *carotovora*, *hrpL* is also controlled by the *rsm* system, thus its expression is affected by the global regulators that control the *rsm* genes. The *rsm* regulon has not been defined in *E. chrysanthemi*.

11.5. The FUR Family and Regulation in Response to Iron

Genes encoding the proteins that synthesize achromobactin and chrysobactin in *E. chrysanthemi* are regulated by Fur (Franza et al., 1999, 2005). The Fur protein, which acts as a dimer, is an iron-binding negative regulator. When the cellular iron concentration becomes too low, Fur loses its Fe^{2+} and no longer binds to gene operators (Escolar et al., 1999). Although both sets of siderophore genes are regulated by Fur, the achromobactin genes are derepressed at higher iron concentrations than the chrysobactin genes, perhaps because the Fur repressor has a greater affinity for the chrysobactin operator sites (Franza et al., 2005).

In addition to the siderophore genes, several other *E. chrysanthemi* genes are regulated by iron via Fur, including genes encoding PelB, PelC, PelD, and PelE (Masclaux and Expert, 1995; Masclaux, 1996; Sauvage and Expert, 1994). Unlike the other major *pel* genes, *pelA* regulation is not affected by iron concentration and Fur does not bind to the *pelA* promoter (Franza et al., 1999, 2002). In the case of the pectate lyase genes, the Fur binding site overlaps with the CRP-binding site, which would inhibit induction of these genes when iron is available. However, the *pels* are de-repressed in the presence of pectate, suggesting that other iron reponsive regulators are present in *E. chrysanthemi*. Franza et al. (2002) were able to demonstrate that KDG was involved in this derepression, but could not demonstrate a role for KdgR, nor could they construct a *kdgR fur* mutant.

11.6. LuxR Family Regulators and Quorum Sensing

Soft rot *Erwinia* produce a small diffusible molecule, N-acyl homoserine lactone (AHL) that induces several virulence genes including cell wall degrading enzymes, competition-reducing antibiotics, Nip, and Svx, once it reaches a threshold level (Andersson et al., 2000; Corbett et al., 2005; Pemberton et al., 2005). In general, the production of this molecule is higher at lower temperatures, thus AHL contributes in part to the temperature regulation of virulence genes in *Erwinia* (Hasegawa et al., 2005). The contribution of AHL to temperature regulation is at least partially due to the effects of AHL on the *rsm* system.

In many species, binding of AHL to its cognate LuxR-family regulatory protein stabilizes the protein and allows the LuxR homolog to bind to promoters, resulting in the induction of gene expression. Andersson et al. (2000) mutated *expR*, which is located adjacent to *expI* and which is expected to bind to *E. carotovora* subsp. *carotovora* AHL. They found that in contrast to an *expI* mutant where plant cell wall degrading enzymes are not produced, there was little effect on the production of these virulence proteins in an *expR* mutant. Further experimentation with strains over-expressing *expR* suggested that ExpR is actually a negative regulator, but they were not able to rule out a role for cross talk with ExpI-produced AHL and other LuxR family regulators in *E. carotovora*.

An *E. carotovora* subsp. *atroseptica* strain encoding a homoserine lactonase, which degrades AHL, produces only low levels of plant cell wall degrading enzymes and is unable to macerate plant tissue (Smadja et al., 2004a). Surprisingly, this strain was able to grow to high levels in potato tissue, thus neither AHLs nor plant cell maceration are required for *Erwinia* growth in association with plants. It is remarkable that although much is known about the biochemical properties and regulation of the plant cell wall degrading enzymes, their contribution to bacterial growth has not been more thoroughly examined. The homoserine lactonase expressing strain was also unable to elicit the HR on tobacco plants, which is to be expected since the expression of *hrpN* is indirectly controlled by AHLs via their effect on the *rsm* system in *E. carotovora* (Mukherjee et al., 2000).

Nasser et al. (1998) found that regulation of genes by AHL is somewhat different in *E. chrysanthemi* compared to *E. carotovora. E. chrysanthemi* 3937 appears to produce three AHLs, *N*-(3-oxohexanoyl)-homoserine lactone (OHHL) and two others that are probably *N*-hexanoyl-homoserine lactone (HHL) and *N*-decanoyl-homoserine lactone (DHL). Only OHHL and HHL are produced by ExpI, the gene responsible for the production of DHL remains unknown. Nasser et al. (1998) constructed an *E. chrysanthemi expI* mutant and found that, unlike in *E. carotovora*, it was not reduced in total production of pectate lyase activity. They examined the major pectate lyases individually and found that the expression of *pelA* was decreased 10-fold and that of *pelB* was decreased 3-fold under inducing conditions. Since these two enzymes accound for only a small part of the total pectate lyase activity, these changes were not detectable when examining total enzyme activity. Similarly to *E. carotovora*, they did not find any variation in *pel* gene expression in the *expR* mutant, even though ExpR bound to *pe*l gene promoters in an AHL-dependent fashion. Ham et al. (2004) had similar results with *E. chrysanthemi* EC16, where *expI* had no apparent effect on the growth phase regulation of *pelE*, *hrpL*, or *hrpN* transcription, nor contribution to virulence on witloof chicory leaves. This suggests that other LuxR homologs in *E. chrysanthemi* may be able to substitute for ExpR, that synthesis of all three AHLs needs to be eliminated before a large effect on production of extracellular enzymes can be detected, and that possibly, there may be another molecule used for quorum sensing in *Erwnia*.

11.7. LysR Family Regulators

Both *E. carotovora* and *E. chrysanthemi* encode numerous LysR family regulators and the roles of the majority of these genes in pathogenicity are unknown. Of these regulators, the PecT (HexA) regulator has been most thoroughly studied in *Erwinia*. In *E. carotovora* subsp. *carotovora*, HexA negatively regulates extracellular enzymes and *hrpN* mainly by affecting RsmA and *rsmB* levels (Harris et al., 1998; Mukherjee et al., 2000).

In *E. chrysanthemi*, PecT controls the expression of *pelA*, *pelB*, *pelC*, *pelD*, *pelE*, *pelI*, *pelL*, *pelZ*, and *kdgC* (Surgey et al., 1996) by directly interacting with their promoters. Purified PecT varies in affinity for the various *pel* promoters, with the strongest affinity for *pelZ* and *kdgC* (Castillo et al., 1998). The expression of *pelA* is unchanged in a *pecT* mutant and of the promoters tested, PecT had the weakest affinity for this promoter, thus the variation in promoter affinity allow for differential *pel* gene regulation (Surgey et al., 1996; Castillo et al., 1998). Some pectin degrading enzymes, such as PelZ, are encoded by operons that are expressed as monocistronic mRNA under non-inducing conditions and as polycistronic mRNA under inducing conditions (Pissavin, et al., 1996). PecT is the only regulator known to bind to the monocistronic *pelZ* regulatory region and may play a role in controlling whether mono- or polycistronic mRNA is expressed from this locus.

PecT also affects motility and exopolysaccharide synthesis and negatively autoregulates itself (Castillo et al., 1997). There are two PecT binding sites upstream of the *pecT* gene, known as R1 and R2, with R1 being farther upstream, having higher affinity for PecT, and having a lesser effect on *pecT* transcription (Castillo et al., 1998). These two binding sites could help PecT function to quickly turn off and on pectate lyases in response to an as yet unidentified signal. Essentially, low levels of PecT would binds with high affinity to the R1 site, allowing some expression of PecT, which would repress the pectate lyases to varying degrees. When the PecT signal compound is present, PecT would become inactivated and high levels of Pels and PecT would be produced. Once the signal compound is no longer present, the high levels of PecT in the pathogen cells would allow PecT to bind to both sites on its own promoter and to the *pel* regulatory regions, quickly shutting down expression of these virulence genes.

11.8. AraC Family Regulators

There are numerous *Erwinia* AraC family regulators, and in many cases, they are not homologous between *E. carotovora* and *E. chrysanthemi*, suggesting that these genes may provide clues to some of the biological differences between these species. The majority of the AraC family regulators in the soft rot *Erwinia* have not been studied. RhaS, which is in the *rhaSR* operon, which is required for induction of genes required for rhamnose degradation, is an AraC family regulator. In *E. coli*, RhaR activates the *rhaRS* operon and RhaS activates genes required for rhamnose metabolism. To date, examination of this operon has only been done with a polar mutation that should eliminate expression of both *rhaS* and *rhaR*, thus the contributions of these individual genes is unknown (Hugouvieux-Cotte-Pattat 2004). Gel-shift assays with crude soluble extracts from *E. coli* over-expressing the *E. chrysanthemi* 3937 *rhaS* demonstrated that RhaS binds to the *rhiT* promoter only in the presence of rhamnose, confirming a direct interaction between RhaS and one of the rhamnose-inducible operons in *E. chrysanthemi* and demonstrating that rhamnose is an inducer for this operon (Hugouvieux-Cotte-Pattat 2004).

11.9. PecS, a MarR Family Regulators

PecS is a member of the MarR family of regulators, which generally regulate genes in response to toxic compounds, suggesting that *Erwinia* may be using PecS to regulate virulence genes in response to host toxins. In *E. chrysanthemi*, PecS differentially regulates cell wall degrading enzymes, repressing the pectate lyases and cellulases and inducing polygalacturonases (Hugouvieux-Cotte-Pattat et al.,

2002; Nasser et al. 1999; Reverchon et al., 1994, 2002). PecS also suppresses the indigoidine synthesis genes. Because of this, *pecS* mutants are a striking dark blue color. There are numerous other MarR family homologs in *Erwinia*, suggesting that *Erwinia* encounters many types of toxic compounds and is able ot adjust its gene regulation accordingly. None of the other MarR regulators have been examined. PecS also down-regulates itself and a divergently transcribed gene, *pecM*, which is required for full PecS activity (Rouanet and Nasser, 2001). PecM is a membrane protein that may act as a sensor that relays signals to PecS and possible other regulators (Praillet et al., 1997). Since there is no indication that PecS is phosphorylated, this signalling is likely to be via another mechanism.

PecS binds upstream of the genes it down-regulates as a dimer to the PecS consensus binding sequence (Rouanet et al., 2004). Rouanet et al., (2004) used this consensus to search for additional PecS regulated genes and found that *fliE* and *fliFGHIJKLMNOPQR*, which are required for flagella synthesis, are also regulated by PecS, tying together regulation of the cell wall degrading enzymes and motility. This suggests that in *E. chrysanthemi* the *pels* and the flagella are co-ordinately induced, while polygalacturonase is produced by non-motile cells.

11.10. H-NS, a Nucleoid-Associated Protein

H-NS is a global regulatory protein that binds to curved DNA, which is normally found in promoter regions. It generally represses gene expression in the absence of environmental stress. An *E. chrysanthemi hns* mutant is mucoid, non-motile, grows slowly, and is sensitive to high osmolarity, suggesting that H-NS controls the expression of numerous genes in this species (Nasser et al., 2001). This mutant is also reduced in pectate lyase activity and increased in cellulase activity. Nasser et al. (2001) specifically examined the effects of an *hsn* mutation on *pelE*, *celZ*, and *kdgR* expression using a transcriptional reporter gene and found that H-NS appears to be interacting directly with the *pelE* promoter and not through KdgR since *kdgR* expression was not affected in the *hns* mutant. This same group also found that some effects of H-NS were mediated by PecT (Nassar et al., 2002). They examined PelA, B, C, D, and E synthesis in an *hns* mutant under various conditions and found that Pel activity was not affected by osmolarity or temperature, thus the transcriptional regulation by these environmental conditions appears to be regulated by H-NS. They also found that induction of Pel activity occurred in early stationary phase rather than late exponential phase in the *hsn* mutant, suggesting that H-NS also affects growth phase regulation of these virulence genes. As might be expected with a mutant affected in the transcription of multiple virulence genes, the *hns* mutant was reduced in ability to initiate symptoms and spread in African violets.

11.11. RecA, RdgA, and RdgB and Gene Induction in Response to DNA Damage

Pectin lyase (Pnl), carotovoricin synthesis, and cell lysis are co-induced in *E. carotovora* subsp. *carotovora* by DNA-damaging agents, including UV light (Itoh et al., 1980; Kamimiya et al., 1972; Tomizawa and Takahasi 1971; Tsuyumu and Chatterjee, 1984; Tsuyumu et al., 1985; Zink et al., 1985). Unlike the pectate lyases, *pnl* is not under the control of catabolite repression (Chatterjee et al., 1991).

In *E. carotovora* subsp. *carotovora, pnlA* is regulated by RdgA, which is homologous to the LexA repressor, and RdgB, which is homologous to the bacteriophage Mu transcriptional activators (McEvoy et al., 1992; Liu et al., 1994). RdgA represses both *rdgA* and *rdgB* expression. Derepression occurs when, in the presence of single-stranded DNA, RecA interacts with RdgA and probably causes it to undergo autocatalytic cleavage, which disrupts the DNA-binding domain of RdgA. This inactivates RdgA and allows expression of *rdgB*, which binds to the *pnl* promoter and activates expression of pectin lyase (Liu et al., 1996, 1997). Since neither carotovoricin (*ctv*) nor cell lysis genes (*lss*) genes are expressed in an RdgB mutant, RdgB probably controls expression of these operons as well. It is notable that *pnlA* is regulated similarly to a phage gene, both in being induced by DNA damage and in that several phage regulatory genes are homologous to RdgA and RdgB. Because this cell wall degrading enzymes has so few regulatory overlaps with the other pectinases, it suggests that multiple independent mechanisms for pectinase gene regulation are important for *Erwinia* fitness.

11.12. CytR, a LacI Family Regulator

A screen for additional virulence gene regulators in *E. carotovora* subsp. *carotovora* EC1 revealed that a homolog of the LacI-family regulatory protein CytR, which represses nucleoside uptake and catabolism in *E. coli* grown in nucleoside poor environments, was required for expression of polygalacturonase and flagella secretion but not expression of cellulase or pectate lyase (Matsumoto et al., 2003a). The mechanism by which CytR regulates polygalacturonase and flagella genes is unknown, but it does not appear to be through interactions with the FhlD-FhlC flagella regulator. Both *cytR* and *fhlD* mutants were reduced in virulence on cabbage leaves, but a polygalacturonase mutant had no apparent reduction in virulence in this system. Therefore, the effect of the *cytR* mutation on virulence is more likely to be due to its affects on motility or another, unknown system, rather than on polygalacturonase production. This regulator ties pectinases and motility in *E. carotovora*, but this coordinate regulation differs from that in *E. chrysanthemi*, where PecS co-ordinately represses pectate lyases and flagella while inducing polygalacturonases (Rouanet et al., 2004).

11.13. Two-Component Systems

Over 30 two-component systems are present in both *E. carotovora* and *E. chrysanthemi* and several of them regulate virulence genes. The binding sites for none of the transcriptional regulators are known, thus it is not possible at this time to predict from genome sequences which genes are regulated by these two-component systems. In addition, there appears to be a considerable level of cross talk among *Erwinia* two component systems since in most cases, the phenotype of the sensor kinase partner does not match the phenotype of the response regulator. Identifying the binding sites and mapping which genes are likely to be regulated by each of the two component systems is technically feasible and would lead to a more complete picture of how these important pathogens respond to their environment.

The global regulatory proteins, GacA and GacS (also called ExpS and ExpA) are present in *E. carotovora* where they positively regulate Prt, Peh, Pel, Cel, and HrpN levels (Eriksson et al., 1998). This regulation is not via a direct interaction between the Gac system and the virulence proteins. Rather, GacA and GacS are required for wild type expression of the RNA regulator *rsmB* and repression of RsmA (Cui et al., 2001; Hyytiäinen et al., 2001). The phenotypes of *gacS* and *gacA* mutants are distinct, with a gacA mutant being essentially avirulent and a *gacS* mutant being reduced in virulence (Eriksson et al., 1998). The *gacS* and *gacA* genes differ from most two-component regulators in that they are not closely linked in the genome. In addition, as in other species, an *uvrC* homolog is directly downstream of *gacA* and an *E. carotovora* strain with polar insertion into *gacA* is more UV-sensitive than the wild type (Eriksson et al., 1998). However, it has not yet been unequivically shown that this is due to *uvrC* in *E. carotovora.* A gene 92% identical to *gacS* from *E. carotovora* subsp. *carotovora* SCC3193 was found by Frederick et al. (1997) during a transposon mutagenesis screen for *E. carotovora* subsp. *carotovora* genes required for pathogenicity. Unlike GacS, RpfA is required for secretion of proteases and cellulases, but not pectate lyases.

The PehR-PehS system of *E. carotovora* subsp. *carotovora*, which is homologous to PhoP-PhoQ in related animal pathogens, controls production of the PehA polygalacturonase in response to calcium (Flego et al., 1997, 2000). Mutation of either *pehR* or *pehS* essentially eliminates production of polygalacturonase and causes a reduction in virulence, but does not affect secretion of pectate lyases or cellulases. *E. coli* requires *phoP* for expression from the *E. carotovora pehA* promoter, demonstrating that there is functional similarity between the *E. carotovora* PehR-PehS and *E. coli* PhoP-PhoQ.

The PhoPQ system in *E. chrysanthemi* EC16 was first identified in a screen for mutants unable to grow at acidic pH (Llama-Palacios et al., 2003). One of the mutants unable to grow at a pH equal to or lower than 5.5, has a transposon insertion in a *phoQ* homolog. The phenotype of this mutant in association with plants depends on which plant is used as a host. For example, the *phoQ* mutant survived better in chicory, which are at pH 5.5, than in rhubarb stems or tangerines, which have a pH of 3.5 and pH 4.5, respectively. The *phoQ* mutant also is more susceptible to the antimicrobial peptides protamine and thionin, and has decreased polygalacturonase production *in vitro*, unlike with *E. carotovora*. The pectate lyase activity of the *phoQ* mutant was not reduced *in vitro*, but was reduced *in planta*. The *phoQ* gene itself is induced at low Mg^{2+} and in plant tisuse. A *phoP* mutant has similar, but less extreme phenotypes, which is puzzling since PhoQ is expected to be a sensor kinase that activates PhoP by phosphorylation. These phenotypes suggest that PhoQ may be phosphorylating genes in addition to PhoP.

This same group mutated *phoP* and *phoQ* in *E. chrysanthemi* 3937 once the genome sequence became available (Llama-Palacios et al., 2005). For the most part, they found similar phenotypes as in EC16 except that the mutants were not altered in their growth at acid pH and was not altered in sensitivity to protamine. They also found that the mutants respond differently to magnesium; the *phoQ* mutant has diminished virulence at low and high magnesium concentrations, while the *phoP* mutant only has diminished virulence at low magnesium concentrations.

The PmrA-PmrB system responds to extracytoplasmic iron concentrations in related pathogens and is essential for survival when cells are grown in high iron concentration. In *E. carotovora* subsp. *carotovora* SCC3193, mutation of the response regulator *pmrA* and transcriptional activator *pmrB* have opposite effects (Hyytiäinen et al., 2003). For example, *pmrB* is required for transcriptional activation of *celV1*, *pehA*, *pelB*, and *prtW*, but *pmrA* appears to repress these same genes. Similarly, compared to wild type cells, a *pmrB* mutant is more resistant to polymyxin B while a *pmrA* mutant is more susceptible, suggesting that the Pmr system regulates LPS synthesis or modification. Both genes are required for full virulence, although the *pmrB* mutant has a greater reduction in symptom development than the *pmrA* mutant. Both mutant strains were equally impaired in survival in *Arabidopsis* leaves, with the greatest survival defect in the latter half of disease development. The effect on symptom development and survival in plants are likely due to Pmr regulation of extracellular enzymes and cell surface properties, respectively. As in other species, PmrB has a conserved iron-binding site and *pmrA* is required for survival of *E. carotovora* in the presence of excess iron. Mutation of *pmrB* does not appear to affect survival in high iron concentrations, suggesting that PmrA may be phosphorylated by other proteins. This two-component system operon is under auto-repression, with PmrB negatively regulating the *pmr* operon.

The opposite effects of mutations in the *pmr* genes can be explained if unphosphorylated PmrA acts as a negative regulator and if phosphorylation of PmrA relieves this repression. Although the same extracellular enzymes regulated by the Pmr system are also regulated by ExpA and KdgR via the *rmsB*-RsmA system in SCC3193 (Hyytiäinen et al., 2001), mutational analysis demonstrates that the Pmr system does not regulate extracellular enzymes through the Rsm system. PmrA and PmrB homologs are also present in *E. chrysanthemi* and are in a similar operon structure of *pmrCAB*. The function of *pmrC* is unknown.

The T3SS is regulated by HrpX-HrpY, which activate *hrpS*, and HrpS in turn activates *hrpL*, which is an alternate sigma factor that activates the genes encoding the T3SS and its substrates. The signal that HrpX responds to and the HrpY binding site are both unknown. Similarly to other two-component regulators in *Erwinia*, such as PhoP-PhoQ, mutation of HrpX and HrpY does not always result in the same phenotype. For example, as described above, the T3SS is required for pellicle formation in culture. A *hrpX* mutant is still able to form a weak pellicle, while a *hrpY* mutant is unable to form one, suggesting that HrpY is phosphorylated by other proteins in addition to HrpX (Yap et al., 2005). Experiments with overexpression of *hrpX*, which resulted in an *E. carotovora* strain with reduced virulence, suggest that appropriate modulation of the T3SS is important for pathogenicity (Lehtimaki et al., 2003).

12. COMMERCIAL APPLICATIONS OF *ERWINIA* ENZYMES

There are numerous commercial applications for enzymes with similar activities to those produced by the soft rot *Erwinia*, although the sources of these enzymes are ususally fungal species. For example, polygalacturonase is used in the processing of cotton fiber (Hoondal et al., 2002) and for pretreatment of vegetable processing wastewater (Tanabe et al., 1998). Pectate lyase is used in the fruit juice industry, for tea and coffee fermentation, and in processing fiber (Bruhlmann et al., 1994; Carr, 1985; Hoondal et al., 2002; Kapoor et al., 2001). Cellulases are used in plant biotechnology and for cleaning biofilms. Improvements in the catalytic activities and large-scale production of these enzymes would aid all of these industries.

An *E. carotovora* enzyme, L-asparaginase, is used to treat cancer in humans. Some types of tumor cells require high levels of asparagine and absorb it from plasma. L-asparaginase degrades asparagine in plasma, depleting the available supply for tumor cells. Uren and Ragin (1979) used this enzyme for treatment of lymphocytic leukemia, but found that it caused immune reactions in 25% of the patients. The authors added poly-DL-alanyl moieties to the asparaginase enzymes and found that the Km was unchanged and the specific activity was reduced by 50%. The modified *E. carotovora* enzyme was more stable than the native enzyme. The modification resulted in a longer enzyme half-life in mice and a longer effect on reducing asparagine levels in plasma. The modified enzymes also caused less of an immune response in mice. In a later paper, Uren et al. (1982) demonstrated that the modified enzyme elicited a much lower immune response in mice, monkeys, and, in a preliminary trial, humans. Since then, both *E. coli* and *E. carotovora* L-asparaginase has been used to treat cancers (Matsumoto et al., 2003b).

13. REFERENCES

Aguilar, I., Almillo, J. M., Garcia-Olmeda, F., & Rodriguez- Palenzuela, P. (2002) Natural variability in the Arabidopsis response to infection with *Erwinia carotovora* subsp. *carotovora*. Planta 215, 205-209.

Alfano, J. R., & Collmer A. (1997) The type III (Hrp) secretion pathway of plant pathogenic bacteria: trafficking harpins, Avr proteins, and death. J. Bacteriol. 179, 5655-5662.

Alfano, J. R., Kim, H.-S., Delaney, T. P., & Collmer, A. (1997) Evidence that the *Pseudomonas syringae* pv. *syringae hrp*-linked *hrmA* gene encodes an Avr-like protein that acts in an *hrp*-dependent manner within tobacco cells. Mol. Plant-Microbe Interact. 10, 580-588.

Allefs, S. J. H. M., De Jong, E. R., Florack, D. E. A., Hoogendoorn, C., & Stiekema, W. J. (1995a) *Erwinia* soft rot resistant of potato cultivars transformed with a gene construct coding for antimicrobial epeptide cercropin B is not altered. Am. Potato J. 72, 437-445.

Allefs, J. J. H. M., van Dooijeweert, W., De Jong, E. R., Primmel, W., & Hoogendoorn, J. (1995b) Factors affecting potato soft-rot resistance to pectolytic *Erwinia* species in a tuber-slice assay. J. Phytopathol. 143, 705-711.

Allefs, S. J. H. M., De Jong, E. R., Florack, D. E. A., Hoogendoorn, C., & Stiekema, W. J. (1996) *Erwinia* soft rot resistance of potato cultivars expressing antimicrobial peptide tachyplesin I. Mol. Breeding. 2, 97-105.

Allan, E., & Kelman A. (1977) Immunofluorescent stain procedures for detection and identification of *Erwinia carotovora* var. *atroseptica*. Phytopathol. 67, 1305-1312.

Anderson, D. M., Fouts, D. E., Collmer, A., & Schneewind, O. (1999) Reciprocal secretion of proteins by the bacterial type III machines of plant and animal pathogens suggests universal recognition of mRNA targeting signals. Proc. Natl. Acad. Sci. USA 96, 12839-12843.

Andersson, R. A., Kõiv, V., Norman-Setterblad, C., & Pirhonen, M. (1999a) Role of RpoS in virulence and stress tolerance of the plant pathogen *Erwinia carotovora* subsp. *carotovora. Microbiol.* 145, 3547-3556.

Andersson, R. A., Palva, E. T., & Pirhonen, M. (1999b) The response regulator ExpM is essential for the virulence of *Erwinia carotovora* subsp. *carotovora* and acts negatively on the sigma factor RpoS (σ^s). Mol. Plant-Microbe Interact. 12, 575-584.

Andersson, R. A., Eriksson, A. R. B., Keikinheimo, R., Maë, A. Pirhonen, M., Kõiv, V., Hyytiäinen, H., Tuikkala, A., & Palva, E. T. (2000) Quorum sensing in the plant pathogen *Erwinia carotovora* subsp. *carotovora*: the role of $expR_{Ecc}$. Mol. Plant-Microbe Interact. 13, 384-393.

Arce, P., Moreno, M., Gutierrez, M., Gebauer, M., Dell'Orto, P., Torres, H., Acuna, I., et al. (1999) Enhanced resistance to bacterial infection by *Erwinia* carotovora subsp atroseptica in transgenic potato plants expressing the attacin or the cecropin SB-37 genes. Am. J. Potato Res. 76, 169-177.

Avrova, A. O., Hyman, L. J., Toth, R. L., & Toth, I. K. (2002) Application of amplified fragment length polymorphism fingerprinting for taxonomy and identification of the soft rot bacteria *Erwinia carotovora* and *Erwinia chrysanthemi*. Appl. Environ. Microbiol. 68, 1499-1508.

Bain, R. A., Pérombelon, M. C. M., Tsror, L., & Nachmias, A. (1990) Blackleg development and tuber yield in relation to numbers of *Erwinia carotovora* subsp. *atroseptica* on seed potatoes. Plant Pathol. 39, 125-133.

Bain, R. A., Millard, P., & Pérombelon, M. C. M. (1996) The resistance of potato plants to *Erwinia carotovora* subsp. *atroseptica* in relation to their calcium and magnesium content. Potato Res. 39, 185-193.

Bains, P. S., Bisht, V. S., Lynch, D. R., Kawchuk, L. M., & Helgeson, J. P. Identification of stem soft rot (*Erwinia carotovora* subspecies *atroseptica*) resistance in potato. Am. J. Potato Res. 76, 137-141.

Bartling, S., Wegener, C., & Olsen, O. (1995) Synergism between *Erwinia* pectate lyase isoenzymes that depolymerize both pectate and pectin. Microbiol. 141, 873-881.

Bartling, S., Derkx, P., Wegener, C., & Olsen, O. (1996) *Erwinia* pectate lyase differences revealed by action pattern analyses. In: J. Visser & A. G. J Voragen, Editors, *Pectins and pectinases*, *Progress in biotechnology* vol. 14, Elsevier, Amsterdam pp. 282–293.

Basset, A., Khush, R. S., Braun, A., Gardan, L., Boccard, F., Hoffmann, J. A., & Lemaitre, B. (2000) The phytopathogenic bacteria *Erwinia carotovora* infects *Drosophila* and activates an immune response. Proc. Natl. Acad. Sci. USA 97, 3376-3381.

Beaulieu, C., Boccara, M., & Van Gijsegem, F. (1993) Pathogenic behavior of pectinase defective *Erwinia chrysanthemi* mutants on different plants. Mol. Plant-Microbe Interact. 6, 197-202.

Bell, K. S., Sebaihia, M., Pritchard, L., Holden, M. T. G., Hyman, L. J., Holeva, M. C., & Thomson, N. R., et al. (2004) Genome sequence of the enterobacterial phytopathogen *Erwinia carotovora* subsp. *atroseptica* and characterization of virulence factors. Proc. Natl. Acad. Sci. USA 101, 11105-11110.

Blot, N., Berrier, C., Hugouvieux-Cotte-Pattat, N., Ghazi, A., & Condemine, G. (2002) The oligogalacturonate-specific porin KdgM of *Erwinia chrysanthemi* 3937 belongs to a new porin family. J. Biol. Chem. 277, 7936-7944.

Boccara, M., Diolez, A., Rouve, M., & Kotoujansky, A. 1988. The role of individual pectate lyases of *Erwinia chrysanthemi* strain 3937 in pathogenicity on *Saintpaulia* plants. Physiol. Mol. Plant Pathol. 33, 95-104.

Boccara, M., Vedel, R., Lalo, D., Lebrun, M.-H., & Lafay., J. F. (1991) Genetic diversity and host range in strains of *Erwinia chrysanthemi* Mol. Plant-Microbe Interact. 4, 293-299.

Boccara, M., Aymeric, J. L., & Camus, C. (1994) Role of endoglucanases in *Erwinia chrysanthemi* 3937 virulence on *Saintpaulia ionantha*. J. Bacteriol. 176, 1524-1526.

Boccara, M., Mills, C. E., Zeier, J., Anzi, C., Lamb, C., Poole, R. K., & Delledonne, M. (2005) Flavohaemoglobin HmpX from *Erwinia chrysanthemi* confers nitrosative stress tolerance and affects the plant hypersensitive reaction by intercepting nitric oxide produced by the host. Plant J. 43, 226-237.

Bouley, J., Condemine, G., & V. E. Shevchik. (2001) The PDZ domain of OutC and the N-terminal region of OutD determine the secretion specificity of the type II out pathway of *Erwinia chrysanthemi*. J. Mol. Biol. 308, 205-219.

Boyd, C., & Keen, N. T. (1993) Characterization of the *prtA* gene and *prtB* gene of *Erwinia chrysanthemi* EC16. Gene 133,115-118.

Boyer, M.-H., Chambost, J.-P., Magnan, M., & Cattanéo, J. (1984) Carboxymethly-cellulase from *Erwinia chrysanthemi*. II. Purification and partial characterisation of an endo-1,4-glucanase. J. Biotechnol. 1, 241-252.

Boyer, M.-H., Cami, B., Chambost, J.-P., Magnan, M., & Cattanéo, J. (1987) Characterisation of a new endoglucanase from *Erwinia chrysanthemi*. *Eur. J. Biochem.* 162, 311-316.

Brader, G., Tas, E., & Palva, E. T. (2001) Jasmonate-dependent induction of indole glucosinolates in Arabidopsis by culture filtrates of the nonspecific pathogen *Erwinia carotovora*. Plant Physiol. 126, 849-860.

Bruhlmann, F., Kim, K. S., Zimmerman, W., & Fiechter, A. (1994) Pectolytic enzymes from Actinomycetes from the degumming of ramie bast fibres. Appl. Environ. Microbiol. 60, 2107-2112.

Brenner, D. J., Fanning, G. R., & Steigerwalt, A. G. (1977) Deoxyribonucleic acid relatedness among *erwiniae* and other enterobacteria II. corn stalk rot bacterium and *Pectobacterium chrysanthemi*. Int. J. Syst. Bacteriol. 27, 211-221.

Bull, C. T., Ishimaru, C. A., & Loper, J. E. (1994) Two genomic regions involved in catchol siderophore production by *Erwinia carotovora*. Appl. Environ. Microbiol. 60, 662-669.

Burr, T. J., & Schroth, M. N. (1977) Occurrence of soft-rot *Erwinia* spp. in soil and plant material. Phytopathol. 67, 1382-1387.

Campos, E., Maher, E. A., & A. Kelman. 1982. Relationship of pectolytic clostridia and *Erwinia carotovora* strains to decay of potato tubers in storage. Plant Dis. 66, 543-546.

Cascales E., & Christie, P. J. (2003) The versatile bacterial type IV secretion systems. Nature Rev. 1, 137-149.

Castillo, A., & Reverchon, S. J. (1997) Characterization of the *pec T* control region from *Erwinia chrysanthemi* 3937. Bacteriol. 179, 4909-4918.

Castillo, A., Nasser, W., Condemine, G., & Reverchon, S. (1998) The PecT repressor interacts with regulatory regions of pectate lyase genes in *Erwinia chrysanthemi*. Biochemica et Biophysica Acta 1442, 148-160.

Castresana, C., de Carvalho, F., Gheysen, G., Habets, M., Inzé, D., & van Montagu, M. (1990) Tissue-specific and pathogen-induced regulation of a *Nicotiana plumbaginifolia* β-1,3-glucanase gene. Plant Cell 2, 1131-1143.

Chatterjee, A., McEvoy, J. L., Chambost, J. P., Blasco, F., & Chatterjee, A. K. (1991) Nucleotide sequence and molecular characterization of *pnlA*, the structural gene for damage-inducible pectin lyase of *Erwinia carotovora* subsp. *carotovora* 71. J. Bacteriol. 173, 1765-1769.

Chatterjee, A., Liu, Y., & Chatterjee, A. K. (1995). Nucleotide sequence of a pectate lyase structural gene, pel1 of *Erwinia carotovora* subsp. *carotovora* strain 71 and structural relationship of *pel1* with other *pel* genes of *Erwinia* species. Mol. Plant-Microbe Interact. 8, 92-95.

Chatterjee, A., Cui, Y., & Chatterjee, A. K. (2002a) Regulation of *Erwinia carotovora* $hrpL_{Ecc}$ ($sigma\text{-}L_{Ecc}$), which encodes an extracytoplasmic function subfamily of sigma factor required for expression of the HRP regulon. Mol. Plant-Microbe Interact. 15, 971-980.

Chatterjee, A., Cui, Y., & Chatterjee, A. K. (2002b) RsmA and the quorum-sensing signal, N-[3-oxohexanoyl]-L-homoserine lactone, control the levels of rsmB RNA in *Erwinia* carotovora subsp carotovora by affecting its stability. J. Bacteriol. 184, 4089-4095.

Condemine, G., & Robert-Baudouy, J. (1987) 2-keto-3-deoxygluconate transport system in *Erwinia chrysanthemi*. J. Bacteriol. 169, 1972-1978.

Condemine, G., Castillo, A., Passeri, F., & Enard, C. (1999) The PecT repressor coregulates synthesis of exopolysaccharides and virulence factors in *Erwinia chrysanthemi*. Mol. Plant-Microbe Interact. 12, 45-52.

Condemine, G., & Shevchik, V. E. (2000) Overproduction of the secretin OutD suppresses the secretion defect of an *Erwinia chrysanthemi outB* mutant. Microbiol. 146, 639-647.

Cooper, V. J. C. & Salmond, G. P. C. (1993). Molecular analysis of the major cellulase (*celV*) of *Erwinia carotovora*. Evidence for an evolutionary mix and match of enzyme domains. Mol. Gen. Genet. 241, 341-350.

Corbett, M., Virtue, S., Bell, K., Birch, P., Burr, T., Hyman, L., Lilley, K., Poock, S., Toth, I., & Salmond, G. (2005) Identification of a new quorum-sensing-controlled virulence factor in *Erwinia carotovora* subsp *atroseptica* secreted via the type II targeting pathway. Mol. Plant-Microbe Interact. 18, 334-342.

Cui, Y., Mukherjee, A., Dumenyo, C. K., Liu, Y., & Chatterjee, A. K. (1999) *rsmC* of the soft-rotting bacterium *Erwinia carotovora* subsp. *carotovora* negatively controls extracellular enzyme and $harpin_{(Ecc)}$ production and virulence by modulating levels of regulatory RNA (*rsmB*) and RNA-binding protein (RsmA). J. Bacteirol. 181, 6042-6052.

Cui, Y., Chatterjee, A., & Chatterjee, A. K. (2001) Effects of the two-component system comprising GacA and GacS of *Erwinia carotovora* subsp. *carotovora* on the production of global regulatory *rsmB* RNA, extracellular enzymes, and $Harpin_{Ecc}$. Mol. Plant-Microbe Interact. 14, 516-526.

Cui, Y. Y., Chatterjee, A., Hasegawa, H., Dixit, V., Leigh, N., & Chatterjee, A. K. (2005) ExpR, a LuxR homolog of *Erwinia carotovora* subsp. *carotovora*, activates transcription of *rsmA*, which specifies a global regulatory RNA-binding protein. J. Bacteriol. 187, 4792-4803.

Cuppels, D., & Kelman, A. (1974) Evaluation of selective media for isolation of soft rot bacteria from soil and plant tissue. Phytopathol. 64, 468-475.

Dahler, G. S., Barras, F., & Keen, N. T. (1990) Cloning of genes encoding extracellular metalloproteases from *Erwinia chrysanthemi* EC16. J. Bacteriol. 172, 5803-5815.

Darrasse, A., Kotoujanski, A., & Bertheau, Y. (1994a) Isolation by genomic subtraction of DNA probes specific for *Erwinia carotovora* subsp. *atroseptica*. Appli. Environ. Microbiol. 60, 298-306.

Darrasse, A., Priou, S., Kotoujansky, A., & Bertheau, Y. (1994b) PCR and restriction fragment length polymorphism of a *pel* gene as a tool to identify *Erwinia carotovora* in relation to potato diseases. Appl. Environ. Microbiol. 60, 1437-1443.

Darvill, A. G., & Albersheim, P. (1984) Phytoalexins and their elicitors - a defence against microbial infection in plants. Annu. Rev. Plant Physiol. 35, 243-275.

Davis, K. R., & Hahlbrock, K. (1987) Induction of defense responses in cultured parsley cells by plant cell wall fragments. Plant Physiol. 85, 1286-1290.

Davis, K. R., & Ausubel, F. M. (1989) Characterization of elicitor-induced defense responses in suspension cultured cells of *Arabidopsis*. Mol. Plant-Microbe Interact. 2, 363-368.

Davis, B. M., Lawson, E. H., Sandkvist, M., Ali, A., Sozhamanna, S., & Waldor, M. K. (2000) Convergence of the secretory pathways for cholera toxin and the filmentous phage, CTXphi. Science 288, 333-335.

De Boer, S. H., Copeman, R. J., & Vruggink, H. (1979) Serogroups of *Erwinia carotovora* potato strains determined with diffusible somatic antigens. Phytopathol. 69, 316-319.

De Boer, S. H., & Sasser, M. (1986) Differentiation of *Erwinia carotovora* ssp. *carotovora* and *E. carotovora* ssp. *atroseptica* on the basis of cellular fatty acid composition. Can. J. Microbiol. 32, 796-800.

De Boer, S. H., Verdonck, L., Vruggink, H., Harju, P., Bang, H. O., & De Ley, J. (1987) Serological and biochemical variation among potato strains of *Erwinia carotovora* subsp. *atroseptica* and their taxonomic relationship to other *Erwinia carotovora* strains. J. Appl. Bacteriol. 63, 487-495.

De Boer, S. H., & Ward, L. J. (1995) PCR detection of *Erwinia carotovora* subsp. *atroseptica* associated with potato tissue. Phytopathol. 85, 854-858.

De Boer, S. H. (2002) Relative incidence of *Erwinia carotovora* subsp. *atroseptica* in stolen end and peridermal tissue of potato tubers in Canada. Plant Dis. 86, 960-964.

De Mendoça, M., & Stanghellini, M. E. (1979) Endemic and soilborne nature of *Erwinia carotovora* var. *atroseptica*, a pathogen of mature sugar beets. Phytopathol. 69, 1096-1099.

Delepelaire, P., & Wandersman C. (1989) Protease secretion by *Erwinia chrysanthemi*. Protease B and C are secreted as zymogens without a signal peptide. J. Biol. Chem. 264, 9083-9089.

Dellagi, A., Heilbronn, J., Avrova, A. O., Montesano, M., Palva, E. T., Stewart, H. E., Toth, I. K., et al. (2000) A potato gene encoding a WRKY-like transcription factor is induced in interactions with *Erwinia carotovora* subsp. *atroseptica* and *Phytophthora infestans* and is corregulated with class I endochitinase expression. Mol. Plant-Microbe Interact. 13, 1092-1101.

Dellagi, A., Rigauly, M., Segond, D., Roux, C., Kraepil, Y., Cellier, F., Briat, J.-F., Gaymard, F., & Expert, D. (2005) Siderophore-mediated upregulation of Arabidopsis ferritin expression in response to *Erwinia chrysanthemi* infection. Plant J. 43, 262-272.

Dickey, R. S. (1979) *Erwinia chrysanthemi*: A comparative study of phenotypic propoerties of strains from several hosts and other *Erwinia* species. Phytopathol. 69, 324-329.

Dickey, R. S. (1981) *Erwinia chrysanthemi*: Reaction of eight plant species to strains from several hosts and to strains of other *Erwinia* species. Phytopathol. 71, 23-29.

Dong, Y.-H., Xu, J.-L., Li, X. C., & L.-H. Zhang. (2000) AiiA, a novel enzyme inactivates acyl homosering-lactone quorum-sensing signal and attenuates the virulence of *Erwinia carotovora*. Proc. Natl. Acad Sci. USA 97, 3526-3531.

Dong, Y.-H., Wang, L. H., Xu, J.-L., Zhang, H. B., Zhang, X.-F., & Zhang, L.-H. (2001) Qunneching quorum-sensing-dependend bacterial infection by an N-acyl homoserine lactonase. Nature 411, 813-817.

Dong, Y.-H., Zhang, X.-F., Xu, J.-L., & Zhang, L.-H. (2004) Insecticidal *Bacillus thuringiensis* silences *Erwinia carotovora* virulence by a new form of microbial antagonism, signal interference. Appl. Environ. Microbiol. 70,954-960.

Dorel, C., Hugouvieux-Cotte-Pattat, N., Robert-Baudouy, J., & Lojkowska, E. (1996) Production of *Erwinia chrysanthemi* pectinases in potato tubers showing high or low level of resistance to soft rot. Eur. J. Plant Pathol. 102, 511-517.

Durand, E., Bernadac, A., Ball, G., Lazdunski, A., Sturgis, J. N., & Filloux, A. (2003) Type II proteins secretion in *Pseudomonas aeruginosa*: the pseudopilus is a multifibrillar and adhesive structure. J. Bacteriol. 185, 2749-2758.

Düring, K., Porsche, P., Fladung, M., & Lorz, H. (1993) Transgenic potato plants resistant to the phytopathogenic bacterium *Erwinia carotovora*. Plant J. 3, 587-598.

El Hassouni, M., Chambost, J. P., Expert, D., Van Gijsegem, F., & Barras, F. (1999) The minimal gene set member *msrA,* encoding peptide methionine sulfoxide reductase, is a virulence determinant of the plant pathogen *Erwinia chrysanthemi*. Proc. Natl. Acad. Sci. USA 96, 887-892.

Eriksson, A. R. B., Andersson, R. A., Pirhonen, M., & Palva, E. T. (1998) Two-component regulators involved in the global control of viurlence in *Erwinia carotovora* subsp. *carotovora*. Mol. Plant-Microbe Interact. 11, 743-752.

Enard, C., Diolez, A., & Expert, D. (1988) Systemic virulence of *Erwinia chrysanthemi* 3937 requires a functional iron assimilation system. J. Bacteriol. 170, 2419-2426.

Enard, C. & Expert, D. (2000) Characterization of a tonB mutation in *Erwinia chrysanthemi* 3937: $TonB_{Ech}$ is a member of the enterobacterial TonB family. Microbiol. 146, 2051-2058.

Escolar, L., Perez-Martin, J., & de Lorenzo, V. (1999) Opening the iron box: trancriptional metalloregulation by the Fur protein. J. Bacteriol. 181, 6223-6229.

Expert, D. (1999) Withholding and exchanging iron: interactions between *Erwinia* spp. and their plant hosts. Ann. Rev. Phytopathol. 37, 307-334.

Farmer, F. E., Moloshok, T. D., Saxton, M. J., & Ryan, C. A. (1991) Oligosaccharide signaling in plants. Specificity of oligouronide-enhanced plasma membrane protein phosphorylation. J. Biol. Chem. 266, 3140-3145.

Favey, S., Bourson, C., Bertheau, Y., Kotoujansky, A., & Boccara, M. (1992). Purification of the acidic pectate lyase and nucleotide sequence of the corresponding gene (*pelA*) of *Erwinia chrysanthemi* strain 3937. J. Gen. Microbiol. 138, 499-508.

Favey, S., Labesse, G., Vouille, V., & Boccara, M. (1995) Flavohaemoglobin HmpX: a new pathogenicity determinant in *Erwinia chrysanthemi* strain 3937. Microbiol 141, 863-871.

Fessehaie, A., De Boer, S. H., & Levesque, C. A. (2003) An oligonucleotide array for the identification and differentiation of bacteria pathogenic on potato. Phytopathol. 93, 262-269.

Filloux, A. (2004) The underlying mechanisms of type II protein secretion. Biochim. Biophys. Acta 1694, 163-179.

Flego, D., Pirhonen, M., Saarilahti, H., Palva, T. K., & Palva, E. T. (1997) Control of virulence gene expression by plant calcium in the phytopathogen *Erwinia carotovora*. Mol. Microbiol. 25, 831-838.

Flego, D., Marits, R., Eriksson, A. R. B., Kõiv, V., Karlsson, M.-B., Heikinheimo, R., & Palva, E. T. (2000) A two-component regulatory system, *pehR-pehS*, controls endopolygalacturonase production and virulence in the plant pathogen *Erwinia carotovora* subsp. *carotovora*. Mol. Plant-Microbe Interact. 13, 447-455.

Flint, D. H., Turninello, J. F., & Emptage, M. H. (1993). The inactivation of Fe-S containing hydro-lyase by superoxide. J. Biol. Chem. 268, 22369-22376.

Florack, D., Allefs, S., Bollen, R., Bosch, D., Visser, B., & Stiekema, W. (1995) Expression of giant silkmoth cecropin B genes in tobacco. Transgenic Res. 4, 132-141.

Fraaije, B. A., Birnbaum, Y., Franken, A. A. J. M., & VandenBulk R. W. (1996a) The development of a conductimetric assay for automated detection of metabolically active soft rot *Erwinia* spp in potato tuber peel extracts J. Appl. Microbiol. 81, 375-382.

Fraaije, B. A., Birnbaum, Y., & Vandenbulk, R. W. (1996b) Comparison of methods for detection of *Erwinia carotovora* ssp *atroseptica* in progeny tubers derived from inoculated tubers of Solanum tuberosum L. J. Phytopathol. 144, 551-557.

Fraaije, B. A., Appels, M., De Boer, S. H., van Vuurde, J. W. L., & Vandenbulk, R. W. (1997) Detection of soft rot *Erwinia* spp. on seed potatoes: Conductimetry in comparison with dilution plating, PCR and serological assays. Eur. J. Plant Pathol. 103, 183-193.

Franza, T., Sauvage, C., & Expert, D. (1999) Iron regulation and pathogenicity in *Erwinia chrysanthemi* strain 3937: role of the Fur repressor protein. Mol. Plant-Microbe Interact. 12, 119-129.

Franza, T., Michaud-Soret, I., Piquerel, P., & Expert, D. (2002) Coupling of iron assimilation and pectinolysis in *Erwinia chrysanthemi* 3937. Mol. Plant-Microbe Interact. 15, 1181-1191.

Franza, T., Mahé, B., & Expert, D. (2005) *Erwinia chrysanthemi* requires a second iron transport route dependent of the siderophore achromobactin for extracellular growth and plant infection. Mol. Microbiol. 55, 261-275.

Frederick, R. D., Chiu, J., Bennetzen, J. L., & Handa, A. K. (1997) Identification of a pathogenicity locus, *rpfA*, in *Erwinia carotovora* subsp. *carotovora* that encodes a two-component sensor-regulator protein. Mol. Plant-Microbe Interact. 10, 407-415.

Gardan, L., Gouy, C., Christen, R., & Samson, R. (2003) Elevation of three subspecies of *Pectobacterium carotovorum* to species level: *Pectobacterium atrosepticum* sp. nov., *Pectobacterium betavasculorum* sp. nov. and *Pectobacterium wasabiae* sp. nov. Int. J. Syst. Evol. Microbiol. 53, 381-391.

Ghigo, J.-M. & Wandersman, C. (1992a) Cloning, nucleotide sequence and characterization of the gene encoding the *Erwinia chrysanthemi* B374 PrtA metalloprotease: a third metalloprotease secreted via a C-terminal secretion signal. Mol. Gen. Genet. 236, 135-144.

Ghigo, J.-M. & Wandersman, C. (1992b) A fourth metalloprotease gene in *Erwinia chrysanthemi*. Res. Microbiol. 143, 857-867.

Glasner, J. D., Liss, P., Plunkett, G. 3rd, Darling, A., Prasad, T., Rusch, M., Byrnes, A., Gilson, M., Biehl, B., Blattner, F. R., & Perna, N. T. (2003) ASAP, a systematic annotation package for community analysis of genomes. Nucleic Acids Res. 31, 147-51.

Gloux, K., Touze, T., Pagot, Y., Jouan, B., & Blanco, C. (2005) Mutations of *ousA* alter the virulence of *Erwinia chrysanthemi*. Mol. Plant-Microbe Interact. 18, 105-157.

Gouesbet, G., Jebbar, M., Bonnassie, S., Hugouvieux-Cotte-Pattat, N., Hirndikabbab, S., & Blanco, C. (1995) *Erwinia chrysanthemi* at high osmolarity - influence of osmoprotectants on growth and pectate lyase production. Microbiol. 141, 1407-1412.

Gouesbet, G., Trautwetter, A., Bonnassie, S., Wu, L. F., & Blanco, C. (1996) Characterization of the *Erwinia chrysanthemi* osmoprotectant gene *ousA*. J. Bacteriol. 178, 447-455.

Gracia-Garza, J. A., Blom, T. J., Brown, W., Roberts, D. P., Schneider, K., Freisen, M. & Gomber, D. (2004) Increased incidence of *Erwinia* soft-rot on calla lilies in the presence of phosphorous. Eur. J. Plant Pathol. 110, 293-298.

Gross, D. C., Powelson, M. L., Regner, K. M., & Radamaker, G. K.. (1991) A bacteriophage typing system for surveying the diversity and distribution of strains of *Erwinia carotovora* in potato fields. Phytopathol. 81, 220-226.

Guiseppi, A., Cami, B., Aymeric, J.-L., Ball, G., & Creuzet, N. (1991) Sequence analysis of the cellulase-encoding *celY* gene of *Erwinia chrysanthemi*: a possible case of interspecies gene transfer. Gene 106, 109-114.

Ham, J. H., Bauer, D. W., Fouts, D. E., & Collmer, A. (1998) A cloned *Erwinia chrysanthemi* Hrp (type III protein secretion) system functions in *Escherichia coli* to deliver *Pseudomonas syringae* Avr signals to plant cells and to secrete Avr proteins in culture. Proc. Natl. Acad. Sci. USA 17, 10206-10211.

Ham, J. H., Cui, Y., Alfano, J. R., Rodríguez-Palenzuela, P., Rojas, C. M., Chatterjee, A. K., & Collmer, A. (2004) Analysis of *Erwinia chrysanthemi* EC16 *pelE::uidA*, *pelL::uidA*, and *hrpN::uidA* mutants reveals strain-specific atypical regulation of the Hrp type III secretion system. Mol. Plant-Microbe Interact. 17, 184-194.

Harris, S. J., Shih, Y.-L., Bentley, S. D., & Salmond, G. P. C. (1998) The *hexA* gene of *Erwinia carotovora* encodes a LysR homologue and regulates motility and the expression of multiple virulence determinants. Mol. Microbiol. 28, 705-717.

Harrison, M. D., Quinn, C. E., Sells, A., & Graham, D. C. (1977) Waste potato dumps as sources of insects contaminated with soft rot coliform bacteria in relation to re-contamination of pathogen-free potato stocks. Potato Res. 20, 37-52.

Hasegawa, H., Chatterjee, A., Cui, Y., & Chatterjee, A. K. (2005) Elevated temperature enhances virulence of *Erwinia carotovora* subsp. *carotovora* strain EC153 to plants and stimulates production of the quorum sensing signal, N-acyl homoserine lactone, and extracellular proteins. Appl. Environ. Microbiol. 71, 4655-4663.

Haseloff, B. J., Freeman, T. L., Valmeekam, V., Melkus, M. W., Oner, F., Valachovic, M. S., & San Francisco, M. J. D. (1998) The *exuT* gene of *Erwinia chrysanthemi* EC16: nucleotide sequence, expression, localization, and relevance of the gene product. Mol. Plant-Microbe Interact. 11, 270-276.

Hauben, L., Moore, E. R. B., Vauterin, L., Steenackers, M., Mergaert, J., Verdonck, L., & Swings, J. (1998) Phylogenetic position of phytopathogens within the *Enterobacteriaceae*. Syst. Appl. Microbiol. 21, 384-397.

Heikinheimo, R., Flego, D., Pirhonen, M., Karlsson, M. B., Eriksson, A., Mäe, A., Kõiv, V., & Palva, E. T. (1995) Characterization of a novel pectate lyase from *Erwinia carotovora* subsp. *carotovora*. Mol. Plant-Microbe Interact. 8, 207-217.

Herbers, K., Meuwly, P., Métraux, J.-P., Sonnewald, U. (1996) Salicylic acid-indepedent induction of pathogenesis-related protein transcripts by sugars is dependent on leaf developmental stage. FEBS Lett. 397, 239-244.

Hinton, J. C., Sidebotham, J. M., Gill, D. R., & Salmond, G. P. (1989) Extracellular and periplasmic isoenzymes of pectate lyase from *Erwinia carotovora* subspecies *carotovora* belong to different gene families. Mol. Microbiol. 3, 1785-1795.

Hinton, J. C. D., Gill, D. R., Lalo, D., Plastow, G. S., & Salmond, G. P. (1989) Sequence of the peh gene of *Erwinia carotovora*: homology between *Erwinia* and plant enzymes. Mol. Microbiol. 4, 1029-1036.

Holden, M. T. G., McGowan, S. J., Bycroft, B. W., Stewart, G. S. A. B., Williams, P., & Salmond, G. P. C. (1998) Cryptic carbapenem antibiotic production genes are widespread in *Erwinia carotovora*: facile *trans* activation by the *carR* transcriptional regulator. Microbiol. 144, 1495-1508.

Holeva, M. C., Bell, K. S., Hyman, L. J., Avrova, A. O., Whisson, S. C., Birch, P. R. J., & Toth, I. K. (2004) Use of a pooled transposon mutation grid to demonstrate roles in disease development for *Erwinia carotovora* subsp. *atroseptica* putative type III secreted effector (DspE/A) and helper (HrpN) proteins. Mol. Plant-Microbe Interact. 17, 943-950.

Hoondal, G. S., Tewari, R. P., Tewari, R., Dahiya, N., & Beg, Q. K. (2002) Microbial alkaline pectinases and their industrial applications a review. Appl. Microbiol. Biotechnol. 59, 409-418.

Hugouvieux-Cotte-Pattat, N., Dominguez, H., & Robert-Baudouy, J. (1992) Environmental conditions affect transcription of the pectinase genes of *Erwinia chrysanthemi* 3937. J. Bacteriol. 174, 7807-7818.

Hugouvieux-Cotte-Pattat, N. & Robert-Baudouy, J. (1992) Analysis of the regulation of the *pelBC* genes in *Erwinia chrysanthemi* 3937. Mol. Microbiol. 6, 2363-2376.

Hugouvieux-Cotte-Pattat, N., Condemine, G., Nasser, W., & Reverchon, S. (1996) Regulation of pectinolysis in *Erwinia chrysanthemi*. Annu. Rev. Microbiol. 29, 213-257.

Hugouvieux-Cotte-Pattat, N., Blot, N., & Reverchon, S. (2001) Identification of TogMNAB, an ABC transporter which mediates the uptake of pectic oligomers in *Erwinia chrysanthemi* 3937. Mol. Microbiol. 41, 1113-1123.

Hugouvieux-Cotte-Pattat, N. & Reverchon, S. (2001) Two transporters, TogT and TogMNAB, are responsible for oligogalacturonide uptake in *Erwinia chrysanthemi* 3937. Mol. Microbiol. 41, 1125-1132.

Hugouvieux-Cotte-Pattat, N., Shevchik, V. E., & Nasser, W. (2002) PehN, a polygalacturonase homologue with a low hydrolase activity, is coregulated with other *Erwinia chrysanthemi* polygalacturonases. J. Bacteriol. 184, 2664-2673.

Hugouvieux-Cotte-Pattat, N. (2004) The RhaS activator controls the *Erwinia chrysanthemi* 3937 genes *rhiN*, *rhiT* and *rhiE* involved in rhamnogalacturonan catabolism. Mol. Microbiol. 51, 1361-1374.

Hunter, P. P. & Gaston, M. A. (1988) Numerical index of the discriminatory ability of typing systems: an application of Simpson's index of diversity. J. Clin. Microbiol. 26, 2465-2466.

Hyman, L. J., Dewasmes, V., Toth, I. K., & Pérombelon, M. C. M. (1997) Improved PCR detection sensitivity of *Erwinia carotovora* subsp. *atroseptica* in potato tuber peel extract by prior enrichment on a selective medium. Letts. Appl. Microbiol. 25, 143-147.

Hyman, L. J., Birch, P. R. J., Dellagi, A., Avrova, A. O., & Toth, I. K. (2000) A competitive PCR-based method for the detection and quantification of *Erwinia carotovora* subsp. *atroseptica* on potato tubers. Letts. Appl. Microbiol. 30, 330-335.

Hyman, L. J., Sullivan, L., Toth, I. K., & Pérombelon, M. C. M. (2001) Modified crystal violet pectate medium (CVP) based on a new polypectate source (Slendid) for the detection and isolation of soft rot *erwinias*. Potato Res. 44, 265-270.

Hyytiäinen, H., Montesano, M., & Palva, E. T. (2001) Global regulators ExpA (GacA) and KdgR modulate extracellular enzyme gene expression through RsmA-*rsmB* system in *Erwinia carotovora* subsp. *carotovora*. Mol. Plant-Microbe Interact. 14, 931-938.

Hyytiäinen, H., Sjöblom, S., Palomäki, T., Tuikkala, A., & Palva, E. T. (2003) The PmrA-PmrB two-component system responding to acidic pH and iron controls virulence in the plant pathogen *Erwinia carotovora* ssp. *carotovora*. Mol. Microbiol. 50, 795-807.

Ishimaru, C. & Loper, J. (1992) High-affinity iron uptake systems present in *Erwinia carotovora*. subsp. *carotovora* include the hydroxamate siderophore aerobactin. J. Bacteriol.174, 2993-3003.

Ito, K., Kobayashi, R., Nikaido, N., & Izaki, K. (1988) DNA structure of pectate lyase I gene cloned from *Erwinia carotovora*. Agric. Biol. Chem. 52, 479-487.

Itoh, Y., Izaki, K., & Takahashi, H. (1980) Similtaneous synthesis of pectin lyase and carotovoricin induced by mitomycin C, nalidixic acid or ultraviolet light irradiation in *Erwinia carotovora*. Agric.Biol. Chem 44, 1135-1140.

James, V. & Hugouvieux-Cotte-Pattat, N. (1996). Regulatory systems modulating the transcription of the pectinase genes of *Erwinia chrysanthemi* are conserved in *Escherichia coli*. Microbiology 142, 2613-2619.

Janse, J. D. & Ruissen, M. A. (1988) Characterization and classification of *Erwinia chrysanthemi* strains from several hosts in the Netherlands. Phytopathol. 78, 800-808.

Jorge, P. E. & Harrison, M. D. (1986) The association of *Erwinia carotovora* with surface water in Northeastern Colorado. I. The presence and population of the bacterium in relation to location, season and water temperature. Am. Potato J. 63, 517-531.

Kamimiya, S., Izaki, K., & Takahashi, H. (1972) A new pectolytic enzyme in *Erwinia aroideae* formed in the presence of nalidixic acid. Agric. Biol. Chem. 36, 2367-2372.

Kang, H. W., Kwon, S. W., & Go, S. J. (2003) PCR-based specific and sensitive detection of *Pectobacterium carotovora* subsp. *carotovora* by primers generated from a URP-PCR fingerprinting-derived polymorphic band. Plant Pathol. 52, 127-133.

Kapoor, M., Beg, Q. K., Bhusan, B., Singh, K., Dadhich, K. S., & Hoondal, G. S. (2001) Application of an alkaline and thermostable polygalacturonase from *Bacillus* sp. MG-cp-2 in degumming of ramie (*Boehmeria nivea*) and sun hemp (*Crotalaria juncea*) bast fibers. Proc. Biochem. 36, 803-807.

Kariola, T., Palomaki, T. A., Brader, G., & Palva, E. T. (2003) *Erwinia carotovora* subsp. *carotovora* and *Erwinia*-derived elicitors HrpN and PehA trigger distinct but interacting defense responses and cell death in *Arabidopsis*. Mol. Plant-Microbe Interact. 16, 179-187.

Keen, N. T., Boyd, C., & Henrissat, B. (1996). Cloning and characterization of a xylanase gene from corn strains of *Erwinia* chrysanthemi. Mol. Plant-Microbe Interact. 9, 651-657.

Kelemu, S. & Collmer, A. (1993) *Erwinia chrysanthemi* Ec16 produces a second set of plant inducible pectate lyase isozymes. Appl. Environ. Microbiol. 59, 1756-1761.

Kim, J. F., Ham, J. H., Bauer, D. W., Collmer, A., & Beer, S. V. (1998) The *hrpC* and *hrpN* operons of *Erwinia chrysanthemi* EC16 are flanked by *plcA* and homologs of hemolysin/adhesin genes and accompanying activator/transporter genes. Mol Plant-Microbe Interact. 11, 563-567.

King, E. O., Ward, M. K., & Raney, D.E. (1954) Two simple media for the demonstratoin of pyocyanin and fluorescein. J. Lab Clin. Med. 44, 301-307.

Kobayashi, N., Nishino, K., & Yamaguchi, A. (2001) Novel macrolide-specific ABC-type efflux transporter in *Escherichia coli*. J. Bacteriol. 183, 5639-5644.

Kunin, C. M., Hua, T. H., & Bakaletz, L. O. (1995) Effect of salicylate on expression of flagella by *Escherichia coli* and *Proteus*, *Providencia*, and *Pseudomonas* spp. Infect. Immun. 63, 1796-1799.

Kwon, S.-W., Go, S.-J., Kang, H.-W., Ryu, J.-C., & Jo, J.-K. 1997. Phylogenetic analysis of *Erwinia* species based on 16S rRNA gene sequences. Internat. J. Syst. Bacteriol. 47, 1061-1067.

Laatu, M. & Condemine, G. (2003) Rhamnogalacturonate lyase RhiE is secreted by the Out system in *Erwinia chrysanthemi*. J Bacteriol 185, 1642-1649.

Lacy, G. H., Hirano, S. S., Victoria, J. I., Kelman, A., Upper, C. D. (1979) Inhibition of soft-rotting *Erwinia* spp. strains by 2,4-dihydroxy-7-methoxy-2H-1,4-benzoxazin-3(4H)-one in relation to their pathogenicity on *Zea mays*. Phytopathol. 69, 757-763.

Lanham, P. G., McIlravey, K. I., & Pèrombelon, M. C. M. (1991) Production of cell-wall dissolving enzymes by *Erwinia carotovora* subsp. *atroseptica* in vitro at 27°C and 30.5°C. J. Appl. Bacteriol. 70, 20-24.

Laurent, F., Kotoujansky, A., Labesse, G., & Berthau, Y. (1993) Characterization and overexpression of the *pem* gene encoding pectin methylesterase of *Erwinia chrysanthemi* strain 3937. Gene 131, 17-25.

Leach, J. G. (1926) The relation of the seed-corn maggot (*Phorbia fuciceps* Zett.) to the spread and development of potato blackleg in Minnesota. Phytopathol. 16, 149-176.

Leach, J. G. (1931) Further studies on the seed-corn maggot and bacteria with special reference to potato blackleg. Phytopathol. 21, 387-406.

Leach, J. G. (1933) The method of survival of bacteria in the puparia of the seed-corn maggot (Hylemyia cilicrura Rond.). Zeitschr. fur angewondte Entomologie. 20, 150-161.

Lehtimaki, S., Rantakari A., Routtu J., Tuikkala, A., Li, J., Virtaharju, O., Palva, E. T., Romantschuk, M., & Saarilahti, H. T. (2003) Characterization of the *hrp* pathogenicity cluster of *Erwinia carotovora* subsp. *carotovora*, high basal level expression in a mutant is associated with reduced virulence. Mol. Gen. Genom. 270, 263-272.

Lei, S. P., Lin, H. C., Wang, S. S., Callaway, J., & Wilcox, G. (1987) Characterization of the *Erwinia carotovora pelB* gene and its product pectate lyase. J. Bacteriol. 169, 4379-4383.

Lei, S. P., Lin, H. C., Wang, S. S., & Wilcox, G. (1988) Characterization of the *Erwinia carotovora pelA* gene and its product pectate lyase A. Gene 62, 159-164.

Letal, J. R. (1977) Efficacy of disinfestants against potato ring rot and blackleg bacteria. Am. Potato J. 54, 405-409.

Letoffe, S., Delepelaire, P. & Wandersman, C. (1990) Protease secretion by *Erwinia chrysanthemi*: the specific secretion functions are analogous to those of *Escherichia coli* alpha-haemolysin. EMBO J. 9, 1375-1382.

Lewellen, R. T., Whitney, E. D., & Goulas, C. K. (1978) Inheritance of resistance to *Erwinia* root rot in sugarbeet. Phytopathol. 68, 947-950.

Li, J., Brader, G., & Palva, E. T. (2004) The WRKY70 transcription factor: a node of convergence for jasmonate-mediated and salicylate-mediated signals in plant defense. Plant Cell 16, 319-331.

Liao, C.-H. & Shollenberger, L. M. (2004) Enumeration, resuscitation, and infectivity of the sublethally injured *Erwinia* cells induced by mild acid treatment. Phytopathol. 94, 76-81.

Lindeberg, M., Salmond, G. P. C., & Collmer, A. (1996) Complementation of deletion mutations in a cloned functional cluster of *Erwinia chrysanthemi out* genes with *Erwinia carotovora out* homologues reveals OutC and OutD as candidate gatekeepers of species-specific secretion of proteins *via* the type II pathway. Mol. Microbiol. 20, 175-190.

Lindeberg, M., Boyd, C. M., Keen, N. T., & Collmer, A. (1998) External loops at the C terminus of *Erwinia chrysanthemi* pectate lyase C are required for species-specific secretion through the Out type II pathway. J. Bacteriol. 180, 1431-1437.

Linke, C., Conrath, U., Jeblick, W., Betsche, T., Mahn, A., Düring, K., & Neuhaus, H. E. (2002) Inhibition of plastidic ATP/ADP transporter protein primes potato tubers for augmented elicitation of defense responses and enhances their resistance against *Erwinia carotovora*. Plant Physiol. 129, 1607-1615.

Liu, B., Zeng, Q., Yan, F. M., Xu, H. G., & Xu, C. R. (2005) Effects of transgenic plants on soil microorganisms. Plant and Soil 271, 1-13.

Liu, Y., Chatterjee, A., & Chatterjee, A. K. (1994) Nucleotide sequence, orgnaization and expression of *rdgA* and *rdgB* genes that regulate pectin lyase production in the plant pathogenic bacterium *Erwinia carotovora* subsp. *carotovora* in response to DNA-damaging agents. Mol. Microbiol. 14, 999-1010.

Liu, Y., Wang, X., Mukherjee, A., & Chatterjee, A. K. (1996) RecA relieves negative autoregulation of *rdgA* which specifies a component of the RecA-Rgd regulatory circuit controlling pectin lyase production in *Erwinia carotovora* subsp. *carotovora*. Mol. Microbiol. 22, 909-918.

Liu, Y., Cui, Y., Mukherjee, A., & Chatterjee, A. K. (1997) Activation of the *Erwinia carotovora* subsp. *carotovora* pectin lyase structural gene *pnlA*: a role for RdgB. Microbiol. 143, 705-715.

Liu, Y., Cui, Y., Mukherjee, A., & Chatterjee, A. K. (1998) Characterization of a novel RNA regulator of *Erwinia carotovora* subsp. *carotovora* that controls production of extracellular enzymes and secondary metabolites. Mol. Microbiol. 29, 219-234.

Liu, Y., Jiang, G., Cui, Y., Mukherjee, A., Ma, W. L., & Chatterjee, A. K. (1999) $kdgR_{Ecc}$ negatively regulates genes for pectinases, cellulas, protease, $harpin_{Ecc}$, and a global RNA regulator in *Erwinia carotovora* subsp. *carotovora*. J. Bacteriol. 181, 2411-2422.

Llama-Palacios, A., López-Solanilla, E., & Rodríguez-Palenzuela, P. (2002) The *ybiT* Gene of *Erwinia chrysanthemi* codes for a putative ABC transporter and is involved in competitiveness against endophytic bacteria during infection. Appl. Environ. Microbiol. 68, 1624-1630.

Llama-Palacios, A., López-Solanilla, E., Poza-Carrión, C., García-Olmedo, F., & Rodríguez-Palenzuela, & P. (2003) The *Erwinia chrysanthemi phoP-phoQ* operon plays an important role in growth at low pH, virulence and bacterial survival in plant tissue. Mol. Microbiol. 49, 347-357.

Llama-Palacios, A., López-Solanilla, E., & Rodríguez-Palenzuela P. (2005) Role of the PhoP-PhoQ system in the virulence of *Erwinia chrysanthemi* strain 3937: Involvement in sensitivity to plant antimicrobial peptides, survival at acid pH, and regulation of pectolytic enzymes. J. Bacteriol. 187, 2157-2162.

Locht, C., Bertin, P., Menozzi, F. D., & Renauld, G. (1993) The filamentous haemagglutinin, a multifaceted adhesion produced by virulent *Bordetella* spp. Mol. Microbiol. 9, 653-660.

Lockhart, W. R. & Koenig. K. (1965) Use of secondary data in numerical taxonomy of the genus *Erwinia*. J. Bacteriol. 90, 1638-1644.

Logan, C. (1963) A selective medium for the isolation of soft-rot coliforms from soil. Nature 199, 263.

Lojkowska, E., Masclaux, C., Boccara, M., Robert-Baudouy, J., & Hugouvieux-Cotte-Pattat, N. (1995) Characterization of the *pelL* gene encoding a novel pectate lyase of *Erwinia chrysanthemi* 3937. Mol. Microbiol. 16, 1183-1195.

Lopez, M. M., Lopez-Lopez, M. J., Marti, R. Zamora, J., Lopez-Sanchez, J., & Beltra, R. (2001) Effect of acetylsalicylic acid on soft rot produced by *Erwinia carotovora* subsp. *carotovora* in potato tubers under greenhouse conditions. Potato Res. 44, 197-206.

Lopez-Lopez, M. J., Liebana, E., Marcilla, P., & Beltra, R. (1995) Resistance induced in potato tubers by treatment with acetylsalicylic acid to soft rot produced by *Erwinia carotovora* subsp. *carotovora*. J. Phytopathol. 143, 719-724.

López-Solanilla, E., García-Olmedo, F., & Rodríguez-Palenzuela. P. (1998) Inactivation of the *sapA* to *sapF* locus of *Erwinia chrysanthemi* reveals common features in plant and animal bacterial pathogenesis. Plant Cell 10, 917-924.

López-Solanilla, E. Llama-Palacios, A., Collmer, A. García-Olmedo, F., & Rodríguez-Palenzuela. P. (2001) Relative effects on virulence of mutations in the *sap*, *pel*, and *hrp* loci of *Erwinia chrysanthemi*. Mol. Plant-Microbe Interact. 14, 386-393.

Lui, L. H., Vikram, A., Abu-Nada, Y., Kushalappa, A. C., Raghavan, G. S. V., & Al-Mughrabi K. (2005) Volatile metabolic profiling for discrimination of potato tubers inoculated with dry and soft rot pathogens. Am. J. Potato Res. 81, 1-8.

Lund, B. M., & G. D. Lyon. (1975) Detection of inhibitors of *Erwinia carotovora* and *E. herbicola* on thin-layer chromatograms. J. Chrom. 110, 193-196.

Lyon, G. D. (1984) A note on the effect of temperature on the sensitivity of *Erwinia carotovora* subsp. *atroseptica* to the phytoalexin rishitin. J. App. Bacteriol. 57, 191-192.

Mäe, A., Heikinkeimo, R., & Palva, E. T. (1995) Structure and regulation of the *Erwinia carotovora* subspecies *carotovora* SCC3193 cellulase gene *celVI* and the role of cellulase in phytopathogenicity. Mol. Gen. Genet. 247, 17-26.

Mäe, A., Montesano, M., Kõiv, V., & Palva, E. T. (2001). Transgenic plants producing the bacterial pheromone N-acyl-homoserine lactone exhibit enhanced resistance to the bacterial phytopathogen *Erwinia carotovora*. Mol. Plant-Microbe Interact. 14, 1035-1042.

Mahé, B., Masclaux, C., Rauscher, I., Enard, C., & Expert, D. (1995) Differential expression of two siderophore-dependent iron-acquisition pathways in *Erwinia chrysanthemi* 3937: characterization of a novel ferrisiderophore permease of the ABC transporter family. Mol. Microbiol. 18, 33-43.

Maher, E. A. and Kelman, A. (1983) Oxygen status of potato tuber tissue in relation to maceration by pectic enzymes of *Erwinia carotovora*. Phytopathol 73, 536-539.

Maher, E. A., De Boer, S. H., & Kelman, A. (1986) Serogroups of *Erwinia carotovora* involved in systemic infection of potato plants and infestation of progeny tubers. Am. Potato J. 63, 1-11.

Maki-Valkama, T., & Karjalainen., R. (1994) Differentiation of *Erwinia carotovoraa* subsp. *atroseptica* and *carotovora* by RAPD-PCR. Ann. Appl. Biol. 125, 301-309.

Marits, R., Kõiv, V., Laasik, E., & Mäe, A. (1999) Isolation of an extracellular protease gene of *Erwinia carotovora* subsp. *carotovora* strain SCC3193 by transposon mutagenesis and the role of protease in phytopathogenicity. Microbiol. 145, 1959-1966.

Marits, R., Tshuikina, M., Pirhonen, M., Laasik, E., & Mäe, A. (2002) Regulation of the expression of *prtW::gusA* fusions in *Erwinia carotovora* subsp. *carotovora*. Microbiology 148, 835-842.

Masclaux C. & Expert, D. (1995) Signaling potential of iron in plant-microbe interactions: the pathogenic switch of iron transport in *Erwinia chrysanthemi*. Plant J. 7, 121-128.

Masclaux, C., Hugouvieux-Cotte-Pattat, N., & Expert, D. (1996) Iron is a triggering factor for differential expression of *Erwinia chrysanthemi* 3937 pectate lyases during pathogenesis of African violets. Mol. Plant-Microbe Interact. 9, 198-205.

Mathieu, Y., Kurkdjian, A., Xia, H., Guern, J., Koller, A., Spiro, M. D., O'Neill, M., Albersheim, P., & Darvill, A. (1991) Membrane responses induced by oligogalacturonides in suspension-cultured tobacco cells. Plant J. 1, 333-343.

Matsumoto, H., Muroi, H., Umehara, M., Yoshitake, Y., & Tsuyumu, S. (2003a) Peh production, flagellum synthesis, and virulence reduced in *Erwinia carotovora* subsp. *carotovora* by mutation in a homologue of *cytR*. Mol. Plant-Microbe Interact. 16, 389-397.

Matsumoto, Y., Nomura, K., Kanda-Akano, Y., Fujita, Y., Nakao, M., Ueda, K., Horiike, S., Yokota, S., Kusuzaki, K., Kitoh, T., Watanabe, A., & Taniwaki, M. (2003b). Succesful treatment with *Erwinia* L-asparaginase for recurrent natural killer T cell lymphoma. Leuk. Lymphoma 44, 879-882.

Mattinen, L., Tshuikina, M., Mëe, A., & Pirhonen, M. (2004) Identification and characterization of Nip, necrosis-inducing virulence protein of *Erwinia carotovora* subsp. *carotovora*. Mol. Plant-Microbe Interact. 17, 1366-1375.

McCarter-Zorner, N. J., Franc, G. D., Harrison, M. D., Michaud, J. E., Quinn, C. E., Sells, I. A., & Graham, D. C. (1984). Soft rot *Erwinia* in surface and underground waters in Southern Scotland and in Colorado, United States. J. Appl. Bacteriol. 57, 95-105.

McCarter-Zorner, N. J., Harrison, M. D., Franc, G. D., Quinn, C. E., Sells, I. A., & Graham, D. C. (1985) Soft rot *Erwinia* bacteria in the rhizosphere of weeds and crop plants in Colorado, United States and Scotland. J. Appl. Bacteriol. 59, 357-368.

McGowan, S., Sebaihia, M., Jones, S., Yu, B., Bainton, N., Chan, P. F., Bycroft, B., Stewart, G. S., Williams, P., & Salmond, G. P. (1995) Carbapenem antibiotic production in *Erwinia carotovora* is regulated by CarR, a homologue of the LuxR transcriptional activator. Microbiol. 141, 541-550.

McGowan, S., Sebaihia, M., Porter, L. E., Stewart, G. S. A. B., Williams, P., Bycroft, B. W., & Salmond, G. P. C. (1996) Analysis of bacterial carbapenem antibiotic genes reveals a novel β-lactam biosynthesis pathway. Mol. Microbiol. 22, 415-426.

McGrath, J. M., Williams, C. E., Haberlach, G. T., Wielgus, S. M., Uchytil, T. F., & Helgeson, J. P. (2002) Introgression and stabilization of *Erwinia* tuber soft rot resistance into potato after somatic hybridization of *Solanum tuberosum* and *S. brevidens*. Amer. J. Potato Res. 79, 19-24.

McEvoy, J. L., Murata, H., & Chatterjee, A. K. (1990) Molecular cloning and characterization of an *Erwinia carotovora* subsp. *carotovora* pectin lyase gene that responds to DNA-damaging agents. J. Bacteriol. 172, 3284-3289.

McEvoy, J. L., Murata, H., & Chatterjee, A. K. (1992) Genetic evidence for an activator required for induction of pectin lyase in *Erwinia carotovora* subsp. *carotovora* by DNA-damaging agents. J. Bacteriol. 174, 5471-5474.

McMillan, G. P., Johnston, D. J., & Pèrombelon, M. C. M. (1992) Purification to homogeneity of extracellular polygalacturonase and isozymes of pectate lyase of *Erwinia carotoovra* subsp. *atroseptica* by column chromatography. J. Appl. Bacteriol. 73, 83-86.

Messiaen, J., Read, N. D., Van Cutsem, P., & Trewavas, A. J. (1993) Cell wall oligogalacturonides increase cytosolic free calcium in carrot protoplasts. J. Cell Sci. 104, 365-371.

Mildenhall, J. P. & Prior, B. A. (1983) Water relations of *Erwinia chrysanthemi*: growth and extracellular pectic acid lyase production. J. Gen. Microbiol. 127, 27-34.

Mildenhall, J. P., Lindner, W. A., Prior, B. A., & Tutt, K. (1988) Elevation and release of cell-associated pectate lyase in *Erwinia chrysanthemi* by lithium and sodium chloride. Phytopathol. 78, 213-217.

Miguel, E., Poza-Carrion, C., Lopez-Solanilla, E., Aguilar, I., Llama-Palacios, A., Garcia-Olmeda, F., & Rodriguez-Palenzuela, P. (2000) Evidence against a direct antimicrobial role of H_2O_2 in the infection of plants by *Erwinia chrysanthemi*. Mol. Plant-Microbe Interact. 13, 421-429.

Molina, J. J., Harrison, M. D., & Brewer, J. W. (1974) Transmission of *Erwinia carotovora* var. *atroseptica* by *Drosophila melanogaster* Meig. I. Acquisition and transmission of the bacterium. Am. Potato J. 51, 245-250.

Molina, J. J. & Harrison, M. D. (1980) The role of *Erwinia carotovora* in the epidemiology of potato blackleg. II. The effect of soil temperature on disease severity. Am. Potato J. 57, 351-363.

Mukherjee, A., Cui, Y., Liu, Y., Dumenyo, C. K., & Chatterjee, A. K. (1996a) Global regulation in *Erwinia* species by *Erwinia carotovora rsmA*, a homologue of *Escherichia coli csrA*: repression of secondary metabolites, pathogenicity and hypersensitive reaction. Microbiology (U. K.) 142, 427-434.

Mukerjee, A., Cui, Y., Ma, W., Liu, Y., Ishihama, A., Eisenstark, A., & Chatterjee, A. K. (1996b) RpoS (Sigma-S) controls expression of *rsmA*, a global regulator of secondary metabolites, harpin, and extracellular proteins in *Erwinia carotovora*. J. Bacteriol 180, 3629-3624.

Mukherjee, A., Cui, Y., Ma, W.-L., Liu, Y., Ishiama, A., Eisenstark, A., & Chatterjee, A. K. (1998) RpoS (β^S) controls expression of *rsmA*, a global regulator of secondary metabolites, harpin, and extracellular proteins in *Erwinia carotovora*. J. Bacteriol. 180, 3629-3634.

Mukherjee, A., Cui, Y., Ma, W.-L., Liu, Y., & Chatterjee, A. K. (2000) *hexA* of *Erwinia carotovora* subsp. *carotovora* strain Ecc71 negatively regulates production of RpoS and *rsmB* RNA, a global regulator of extracellular proteins, plant virulence and quorum sensing signal, *N*-(3-oxohexanoyl)-L-homoserine lactone. Environ. Microbiol. 2, 203-215.

Mulholland, V., Hinton, J. C., Sidebotham, J., Toth, I. K., Hyman, L. J., Pérombelon, M. C., Reeves, P. J., & Salmond, G. P. C. (1993) A pleiotropic reduced virulence (Rvi^{-}) mutant of *Erwinia carotovora* subspecies *atroseptica* is defective in flagella assembly proteins that are conserved in plant and animal bacterial pathogens. Mol. Microbiol. 9, 343-356.

Münzinger, M., Budzikiewicz, H., Expert, D., Enard, C., & Meyer, J. M. (2000) Achromobactin, a new citrate siderophore of *Erwinia chrysanthemi*. Z. Naturforsch. C. 55, 328-332.

Murata, H., Chatterjee, A., Liu, Y., & Chatterjee, A. K. (1994) Regulation of the production of extracellular pectinase, cellulase, and protease in the soft rot bacterium *Erwinia carotovora* subsp. *carotovora*: evidence that *aepH* of *Erwinia carotovora* subsp. *carotovora* 71 activates gene expression in *Erwinia carotovora* subsp. *carotovora*, *E. carotovora* subsp. *atroseptica*, and *Escherichia coli*. Appl. Environ. Microbiol. 60, 3150-3159.

Nachin, L. & Barras, F. (2000) External pH: an environmental signal that helps to rationalize *pel* gene duplication in *Erwinia chrysanthemi*. Mol. Plant-Microbe Interact. 13, 882-886.

Nachin, L., El Hassouni, M., Loiseau, L., Expert, D., & Barras, F. (2001) SoxR-dependent response to oxidative stress and virulence of *Erwinia chrysanthemi*: the key role of SufC, an orphan ABC ATPase. Mol. Microbiol. 39, 960-972.

Nachin, L., Loiseau, L., Expert, D., & Barras, F. (2003) SufC: an unorthodox cytoplasmic ABC/ATPase required for [Fe-S] biogenesis under oxidative stress. EMBO J. 22, 427-437.

Nassar, A., Bertheau, Y., Dervin, C., Narcy, J. P., & Lemattre, M. (1994) Ribotyping of *Erwinia chrysanthemi* strains in relation to their pathogenic and geographic distribution. Appl. Environ. Microbiol. 60, 3781-3789.

Nassar, A., Darrasse, A., Lemattre, M., Kotoujansky, A., Dervin, C., Vedel, R., & Bertheau, Y. (1996a) Characterization of *Erwinia chrysanthemi* by pectinolytic isozyme polymorphism and restriction fragment length polymorphism analysis of PCR-amplified fragments of *pel* genes. Appl. Environ. Microbiol. 62, 2228-2235.

Nassar, A., Narcy, J. P., & Lemattre, M. (1996b) Detection of *Erwinia chrysanthemi* pv. *dianthicola* (Ech) by the DAS-ELISA method in symptomless carnation cuttings. Agronomie 16, 143-151.

Nasser, W., Reverchon, S., Condemine, G., & Robert-Baudouy, J. (1994) Specific interactions of *Erwinia chrysanthemi* KdgR repressor with different operators of genes involved in pectinolysis. J. Mol. Biol. 236, 427-440.

Nasser, W., Robert-Baudouy, J., Reverchon, S. (1997) Antagonistic effect of CRP and KdgR in the transcription control of the *Erwinia chrysanthemi* pectinolysis gene. *Mol Microbiol* 29, 1071-1082.

Nasser, W., Shevchik, V. E., & Hugouvieux-Cotte-Pattat, N. (1999) Analysis of three clustered polygalacturonase genes in *Erwinia chrysanthemi* 3937 revealed an anti-repressor function for the PecS regulator. Mol. Microbiol. 34, 641-650.

Nasser, W., Faelen, M., Hugouvieux-Cotte-Pattet, N., & Reverchon, S. (2001) Role of the nucleoid-associated protein H-NS in the synthesis of virulence factors in the phytopathogenic bacterium *Erwinia chrysanthemi*. Mol. Plant-Microbe Interact. 14, 10-20.

Nasser, W. & Reverchon, S. (2002) H-NS-dependent activation of pectate lyases synthesis in the phytopathogenicy bacterium *Erwinia chrysanthemi* is mediated by the PecT repressor. Mol. Microbiol. 43, 733-748.

Neema, C., Laulhere, J.-P., & Expert, D. (1993) Iron deficiency induced by chrysobactin in *Saintpaulia* leaves inoculated with *Erwinia chrysanthemi*. Plant Physiol. 102, 967-973.

Nguyen, H. A., Tomita, T., Hirota, M., Kaneko, J., Hayashi, T., & Kamio, Y. (2001) DNA inversion in the tail fiber gene alters the host range specificity of carotovoricin Er, a phage-tail-like bacteriocin of phytopathogenic *Erwinia carotovora* subsp. *carotovora* Er. J. Bacteriol. 183, 6274-6281.

Nguyen, H. A., Kaneka, J., & Kamio, Y. (2002) Temperature-dependent production of carotovoricin Er and pectin lyase in phytopathogenic *Erwinia carotovora* subsp. *carotovora* Er. Biosci. Biotech. Biochem. 66, 444-447.

Noble, M. & Graham, D. C. (1956) A selective medium for the isolation of coliform soft-rot bacteria from plant tissue. Nature 178, 1479-1480.

Nomura, K., Nasser, W., Kawagishi, H., & Tsuyumu, S. (1998) The *pir* gene of *Erwinia chrysanthemi* EC16 regulates hyperinduction of pectate lyase virulence genes in response to plant signals. Proc. Natl. Acad. Sci. 95,14034-14039.

Norman, C., Vidal, S., & Palva, E. T. (1999) Oligogalacturonide-mediated induction of a gene involved in jasmonic acid synthesis in response to the cell-wall-degrading enzymes of the plant pathogen *Erwinia carotovora*. Mol. Plant-Microbe Interact. 12, 640-644.

Norman-Setterblad, C., Vidal, S., & Palva, E. T. (2000) Interacting signal pathways control defense gene expression in *Arabidopsis* in response to cell wall-degrading enzymes from *Erwinia carotovora*. Mol. Plant-Microbe Interact. 13, 430-438.

Nothnagel, E. A., Holstrom, K.-O., Heino, P., & Palva, E. T. (1993) Induction of plant defense response by exoenzymes of *Erwinia carotovora* subsp. *carotovora*. Mol. Plant-Microbe Interact. 6, 190-196.

Okinaka, Y., Yang, C.-H., Perna, N. T., & Keen, N. T. (2002) Microarray profiling of *Erwinia chrysanthemi* 3937 genes that are regulated during plant infection. Mol. Plant-Microbe Interact. 15, 619-629.

O'Neill, M., Warrenfeltz, D., Kates, K., Pellerin, P., Doco, T., Darvill, A., & Albersheim, P. (1996) Rhamnogalacturonan-II, a pectic polysaccharide in the walls of growing plant cell, forms a dimer that is covalently cross-linked by a borate ester. *In vitro* conditions for the formation and hydrolysis of the dimer. *J. Bio.l Chem.* 271, 22923-22930.

Palomäki, T. & Saarilahti, H. T. (1995) The extreme C-terminus is required for secretion of both the native polygalacturonase (PehA) and PehA-BlaA hybrid proteins in *Erwinia carotovora* subsp. *carotovora*. Mol. Microbiol. 43, 449-459.

Palomäki, T., Pickersgill, R., Riekki, R., Romantschuk, M., & Saarilahti, H. T. (2002) A putative three-dimensional targeting motif of polygalacturonase (PehA), a protein secreted through the type II (GSP) pathway in *Erwinia carotovora*. Mol. Microbiol. 43, 585-596.

Palva, T. K., Holmstrõm, K.-O. Heino, P., & Palva, E. T. (1993) Induction of plant defense respons by exoenzymes of *Erwinia carotoovra* subsp. *carotovora*. Mol. Plant-Microbe Interact. 6, 190-196.

Palva, T. K., Hurtig, M., Saindrenan, P., & Palva, E. T. (1994) Salicylic acid induced resistance to *Erwinia carotovora* subsp. *carotovora* in tobacco. Mol. Plant-Microbe Interact. 7, 356-363.

Parent, J.-G., Lacroix, M., Page, D., & Vezina, L. (1996) Identification of *Erwinia carotovora* from soft rot diseased plants by random amplified polymorphic DNA (RAPD) analysis. Plant Dis. 80, 494-499.

Parker, W. L., Rathnum, M. L., Wells, J. S., Trejo, W. H., Principe, P. A., & Sykes, R. B. (1982) SQ27850, a simple carbapenem produced by species of *Serratia* and *Erwinia*. J. Antibiot. 35 ,653-660.

Paulsen, I. T., Skurray, R. A., Tam, R., Saier, M. H. Jr., Turner, R. J., Weiner, J. H., Goldberg, E. B., & Grinius, L. L. (1996) The SMR family, a novel family of multidrug efflux proteins involved with the efflux of lipophilic drugs. Mol Microbiol. 19, 1167-75.

Pemberton, C. L., Whitehead, N. A., Sebalhia, M., Bell, K. S., Hyman, L. J., Harris, S. J., Matlin, A. J., Robson, N. D., Birch, P. R. J., Carr, J. P., Toth, I. K., & Salmond, G. P. C. (2005) Novel quorum-sensing-control led genes in *Erwinia carotovora* subsp. *carotovora*: Identification of a fungal elicitor homologue in a soft-rotting bacterium. Mol. Plant-Microbe Interact. 18, 343-353.

Pérombelon, M. C. M., Gullings-Handley, J., & Kelman, A. 1979. Population dynamics of *Erwinia carotovora* and pectolytic *Clostridium* spp. in relation to decay of potatoes. Phytopathol. 69, 167-173.

Pérombelon, M. C. M. & Burnett, E. M. (1991) Two modified crystal violet pectate (CVP) media for the detection, isolation, and enumeration of soft rot *erwinia*s. Potato Res. 34, 79-85.

Persmark, M., Expert, D., & Neilands, J. B. (1989) Isolation, characterization and synthesis of chrysobactin, a compound with a siderophore activity from *Erwinia chrysanthemi*. J. Biol. Chem. 264, 3187-3193.

Phillips J. A. & Kelman, A. (1982) Direct fluorescent antibody stain procedure applied to insect transmission of *Erwinia carotovora*. Phytopathol. 72, 898-901.

Pirhonen, M, Saarilahti, H., Karlsson, M.-B., & Palva, E. T. (1991) Identification of pathogenicity determinants of *Erwinia carotovora* subsp. *carotovora* by transposon mutagenesis. Mol. Plant-Microbe Interact. 4, 276-283.

Pirhonen, M., Flego, D., Heikinheimo, R., & Palva, E. T. (1993) A small diffusible signal molecule is responsible for the global control of virulence and exoenzyme production in the plant pathogen *Erwinia carotovora*. EMBO J. 12, 2467-2476.

Pissavin, C., Robert-Baudouy, J., & Hugouvieux-Cotte-Pattat, N. (1996) Regulation of *pelZ*, a gene of the *pelB-pelC* cluster encoding a new pectate lyase in *Erwinia chrysanthemi* 3937. J. Bacteriol. 178, 7187-7196.

Powelson, M. L. (1980) Seasonal incidence and cause of blackleg and a stem soft rot of potatoes in Oregon. Am. Potato J. 57, 301-306.

Praillet, T., Reverchon, S., & Nasser, W. (1997) Mutual control of the PecS/PecM couple, two proteins regulating virulence-factor synthesis in *Erwinia chrysanthemi* Mol. Microbiol. 24, 803-814.

Prithiviraj, B., Bais, H. P., Weir, T., Suresh, B., Najarro, E. H., Dayakar, B. V., Schweizer, H. P., & Vivanco, J. M. (2005) Down regulation of virulence factors of *Pseudomonas aeruginosa* by salicylic acid attenuates its virulence on *Arabidopsis thaliana* and *Caenorhabditis elegans*. Infect. Immun. 73, 5319-5328.

Prior, B. A., Hewitt, E. V., Brandt, E. V., Clarke, A., & Mildenhall, J. P. (1994) Growth, pectate lyase production and solute accumulation by *Erwinia chrysanthemi* under osmotic stress: effect of osmoprotectants. J. Bacteriol. 77, 433-439.

Quinn, C. E., Sells, I. A., & Graham, D. C. (1980) Soft rot *Erwinia* bacteria in the atmospheric bacterial aerosol. J. Appl. Bacteriol. 49, 175-181.

Rask, L., Andréasson, E., Ekbom, B., Ericksson, S., Pontoppidan, B., & Meijer, J. (2000) Myrosinase: gene familiy evolution and herbivore defense in *Brassicaceae*. Plant Mol. Biol. 42, 93-113.

Reeves, A. F., Olanya, O. M., Hunter, J. H., & Wells, J. M. (1999) Evaluation of potato varieties and selections for resistance to bacterial soft rot. Amer. J. of Potato Res. 76, 183-189.

Reiter, B., Pfeifer, U., Schwab, H., & Sessitsch, A. (2002) Response of endophytic bacterial communities in potato plants to infection with *Erwinia carotovora* subsp. *atroseptica*. Appl. Environ. Microbiol. 68, 2261-2268.

Reverchon, S., Huang, Y., Bourson, C., Robert-Baudouy, J. (1989) Nucleotide sequences of the *Erwinia chrysanthemi ogl* and *pelE* genes negatively regulated by the *kdgR* gene product. Gene 85, 125-134.

Reverchon, S., Nasser, W., & Robert-Baudouy, J. (1994) *pecS*: a locus controlling pectinase, cellulase and blue pigment production in *Erwinia chrysanthemi*. Mol. Microbiol. 11, 1127-1139.

Reverchon, S., Expert, D., Robert-Baudouy, J., Nasser, W. (1997) The cyclic AMP receptor protein is the main activator of the pectinolysis genes in *Erwinia chrysanthemi*. J Bacteriol 29, 3500-3508.

Reverchon, S., Rouanet, C., Expert, D., & Nasser, W. (2002) Characterization of indigoidine biosynthetic genes in *Erwinia chrysanthemi* and role of this blue pigment in pathogenicity. J. Bacteriol. 184, 654-665.

Ried, J. L. & Collmer A. (1986) Comparison of pectic enzymes produced by *Erwinia chrysanthemi*, *Erwinia carotovora* subsp. *carotovora*, and *Erwinia carotovora* subsp. *atroseptica*. Appl. Environ. Microbiol. 52, 305-310.

Rodionov, D. A., Gelfand, M. S., & Hugouvieux-Cotte-Pattat, N. (2004) Comparative genomics of the KdgR regulon in *Erwinia chrysanthemi* 3937 and other gamma-proteobacteria. Microbiol. 150, 3571-3590.

Rojas, C. M., Ham, J. H., Deng, W.-L., Doyle, J. J., & Collmer, A. (2002) HecA, a member of a class of adhesins produced by diverse pathogenic bacteria, contributes to the attachment, aggregation, epidermal cell killing, and virulence phenotypes of *Erwinia chrysanthemi* EC16 on *Nicotiana clevelandii* seedlings. Proc. Natl. Acad. Sci. USA 99, 13142-13147.

Rojas, C. M., Ham, J. H., Schechter, L. M., Kim, J. F., Beer, S. V., & Collmer.,A. (2004) The *Erwinia chrysanthemi* EC16 *hrp*/*hrc* gene cluster encodes an active Hrp type III secretion system that is flanked by virulence genes functionally unrelated to the Hrp system. Mol. Plant-Microbe Interact. 17, 644-653.

Rouanet, C. & Nasser, W. (2001). The PecM protein of the phytopathogenic bacterium *Erwinia chrysanthemi*, membrane topology and possible involvement in the efflux of the blue pigment indigoidine. J. Mol. Microbiol. Biotechnol. 3, 309-318.

Rouanet, C., Reverchon, S., Rodionov, D. A., & Nasser, W. (2004) Definition of a consensus DNA-binding site for PecS, a global regulator of virulence gene expression in *Erwinia chrysanthemi* and identification of new members of the PecS regulon. J. Biol. Chem. 279, 30158-30167.

Ryan, C. A. (1988) Oligosaccharides as recognition signals for the expression of defensive genes in plants. Biochem. 27, 8879–8898.

Saarilahti, H. T., Heino. P., Pakkanen, R., Kalkkinen, N., Palva, I., & Palva, E. T. (1990a) Structural analysis of the *pehA* gene and characterization of its protein product, endopolygalacturonase. of *Erwinia carotovora* subspecies *carotovora*. Mol. Microbiol. 4,1037-1044.

Saarilahti, H. T., Henirssat, B., & Palva, E. T. (1990b) CelS: a novel endoglucanase identified from *Erwinia carotovora* subsp. *carotovora*. Gene 90, 9-14.

Samson, R., Legendre, J. B., Christen, R., Fischer-Le Saux, M., Achouak, W., & Gardan, L. (2005) Transfer of *Pectobacterium chrysanthemi* (Burkholder et al. 1953) Brenner et al. 1973 and *Brenneria paradisiaca* to the genus *Dickeya* gen. nov as *Dickeya chrysanthemi* comb. nov and *Dickeya paradisiaca* comb. nov and delineation of four novel species, *Dickeya dadantii* sp nov., *Dickeya dianthicola* sp nov., *Dickeya dieffenbachiae* sp nov and *Dickeya zeae* sp nov. J. Syst. Evol. Microbiol. 55, 1415-1427.

Santos, R., Franza, T., Laporte, M.L., Sauvage, C., Touati, D., & Expert, D. (2001) Essential role of superoxide dismutase on the pathogenicity of *Erwinia chrysanthemi* strain 3937. Mol. Plant-Microbe Interact. 14, 758-767.

Sauvage, C. & Expert, D. 1994. Differential regulation by iron of *Erwinia chrysanthemi* pectate lyases: pathogenicity of iron transport regulatory (*cbr*) mutants. Mol. Plant-Microbe Interact. 7, 71-77.

Sauvonnet, N., Vignon, G., Pugsley, A. P., & Gounnon, P. (2000) Pilus formation and protein secretion by the same machinery in *Escherichia coli* EMBO J. 19, 2221-2228.

Savulesçu, A. & Lazar. I. (1964) Studies on the relation between the pathogenicity of certain species on *Erwinia* parasitic on plants and their toxic action on animals. Proc. Symp. Host-Parasite Relations in Plant Path. Budepest. 151-155.

Schober, B. M. & van Vuurde, J. W. L. (1997). Detection and enumeration of *Erwinia carotovora* subsp. *atroseptica* using spiral plating and immunofluorescence colony staining. Can. J. Microbiol. 43, 847-853.

Schober, B. M. & Vermeulen, T. (1999) Enzymatic maceration of witloof chicory by the soft rot bacteria *Erwinia carotovora* subsp. *carotovora*: the effect of nitrogen and calcium treatments of the plant on pectic enzyme production and disease development. Eur. J. Plant Pathol. 105, 341-349.

Segall, R. H. (1971) Selective medium for enumerating *Erwinia* species commonly found in vegetable packinghouse waters. Phytopathol. 61, 425-426.

Serrano, C., Arce-Johnson, P., Torres, H., Gebauer, M., Gutierrez, M., Moreno, M., Jordana, X., Venegas, A., Kalazich, J., & Holuigue, L. (2000) Expression of the chicken lysozyme gene in potato enhances resistance to infection by *Erwinia carotovora* subsp. *atroseptica*. Amer. J. Potato Res. 77, 191-199.

Seubert, A., Hiestand, R., de la Cruz, F., & Dehio, C. (2003) A bacterial conjugation machinery recruited for pathogenesis Mol. Microbiol. 49, 1253-1266.

Shevchik, V. E., Condemine, G., Hugouvieux-Cotte-Pattat, N., & Robert-Baudouy, J. (1996) Characterization of pectin methylesterase B, an outer membrane lipoprotein of *Erwinia chrysanthemi*. Mol. Microbiol. 19, 455-466.

Shevchik, V. E. & Hugouvieux-Cotte-Pattat, N. (1997) Identification of a bacterial pectin acetyl esterase in *Erwinia chrysanthemi* 3937. Mol. Microbiol. 24, 1285-1301.

Shevchik, V. E., Robert-Baudouy, J., & Hugouvieux-Cotte-Pattat, N. (1997) Pectate lyases PelI of *Erwinia chrysanthemi* 3937 belongs to a new family. J. Bacteriol. 179, 7321-7330.

Shevchik, V. E., Boccara, M., Vedel, R., & Hugouvieux-Cotte-Pattat, N. (1998) Processing of the pectate lyase PelI by extracellular proteases of *Erwinia chrysanthemi* 3937. Mol. Microbiol. 29, 1459-1469.

Shevchik, V. E., Condemine, G., Robert-Baudouy, J., & Hugouvieux-Cotte-Pattat, N. (1999a) The exopolygalacturonate lyase PelW and the oligogalacturonate lyase Ogl, two cytoplasmic enzymes of pectin catabolism in *Erwinia chrysanthemi* 3937. J. Bacteriol. 181, 3912-3919.

Shevchik, V. E., Kester, H. C., Benen, J. A., Visser, J., Robert-Baudouy, J., & Hugouvieux-Cotte-Pattat, N. (1999b) Characterization of the exopolygalacturonate lyase PelX of *Erwinia chrysanthemi* 3937. J. Bacteriol. 181, 1652-1663.

Shevchik, V. E. & Hugouvieux-Cotte-Pattat, N. (2003) PaeX, a second pectin acetylesterase of *Erwinia chrysanthemi* 3937. J. Bacteriol. 185, 3091-3100.

Shih, Y. L., Harris, S. J., Borner, G., Rivet, M. M., & Salmond, G. P. C. (1999) The *hexY* genes of *Erwinia carotovora* ssp. *carotovora* and ssp. *atroseptica* encode novel proteins that regulate virulence and motility coordinately. Env. Micrbiol. 1, 535-547.

Smadja, B., Latour, X., Fauer, D., Chevalier, S., Dessaux, Y., & Orange, N. (2004a) Involvement of N-acylhomoserine lactones throughout plant infection by *Erwinia carotovora* subsp. *atroseptica* (*Pectobacterium atrosepticum*). Mol. Plant-Microbe Interact. 17, 1269-1278.

Smadja, B., Latour, X., Trigui, S., Burini, J. F., Chevalier, S., & Orange, N. (2004b) Thermodependence of growth and enzymatic activities implicated in pathogenicity of two *Erwinia carotovora* subspecies (*Pectobacterium* spp.). Can. J. Microbiol. 50, 19-27.

Stewart, D. J. (1962) A selective-diagnostic medium for the isolation of pectinolytic organisms in the *Enterobacteriaceae*. Nature 195, 1023.

Surgey, N., Robert-Baudouy, J., & Condemine, G. (1996) The *Erwinia chrysanthemi pecT* gene regulates pectinase gene expression. J Bacteriol 178, 1593-1599.

Tamaki, S. J., Gold, S., Robeson, M., Manulis, S., & Keen, N. T. (1988) Structure and organization of the *pel* genes from *Erwinia chrysanthemi* EC16. J. Bacteriol. 170, 3468-3478.

Tanabe, H., Kobayashi, Y., & Akamatsu, T. (1998) Pretreatment of pectic wastewater with pectate lyase from an alkalophilic *Bacillus* sp. Agric. Biol. Chem. 52, 1853-1856.

Thain, J. F., Doherty, H. M., Bowles, D. J., & Wildon, D. C. (1990) Oligosaccharides that induce proteinase inhibitor activity in tomato plants cause depolarisation of tomato leaf cells. Plant Cell Environ. 13, 569-574.

Thomas, J. D., Reeves, P. J., & Salmond, G. P. C. (1997) The general secretion pathway of *Erwinia carotovora* subsp. *carotovora*: analysis of the membrane topology of OutC and OutF. Microbiol. 143, 713-720.

Thomson, N. R., Nasser, W., McGowan, S., Sebaihia, M., & Salmond, G. P. C. (1999) *Erwinia carotovora* has two KdgR-like proteins belonging to the IclR family of transcriptional regulators: identification and characterization of the RexZ activator and the KdgR repressor of pathogenesis. Microbiol. 145, 1531-1545.

Thorne, S. N. (1972) A pectate gel medium for the isolation and diagnosis of pectinolytic bacteria. J. Appl. Bact. 35, 357-358.

Toth, I. K., Bertheau, Y., Hyman, L. J., Laplaze, L., López, M. M., McNicol, J. Niepold, F. et al. (1999a) Evaluation of phenotypic and molecular typing techniques for determining diversity in *Erwinia carotovora* subspecies *atroseptica*. J. Appl. Microbiol. 87, 770-781.

Toth, I. K., Hyman, L. J., & Wood, J. R. (1999b) A one step PCR-based method for the detection of economically important soft rot *Erwinia* species on micropropagated potato plants. J. Appl. Microbiol. 87, 158-166.

Toth, I. K., Newton, J. A., Hyman, L. J., Lees, A. K., Daykin, M., Ortori, C., Williams, P., & Fray, R. G. (1999c) Potato plants genetically modified to produce N-acylhomoserine lactones increase susceptibility to soft rot *erwiniae*. Mol. Plant-Microbe Interact. 12, 499-507.

Toth, I. K., Avrova, A. O., & Hyman, L. J. (2001) Rapid identification and differentiation of the soft rot *erwinias* by 16S-23S intergenic transcribed spacer-PCR and restriction fragment length polymorphism analyses. Appl. Environ. Microbiol. 67, 4070-4076.

Tsuyama, H. & Sakamoto, M. (1952) Isolation methods of the soft-rot causing bacteria from the soil. Sci. Rep. Res. Inst. Tohoku, Univ. 3, 29-34.

Tsuyumu, S. & Chatterjee, A. K. (1984) Pectin lyase production in *Erwinia chrysanthemi* and other soft-rot *Erwinia* species. Physiol. Plant Pathol. 24, 291-302.

Tsuyumu, S., Funakubo, T., & Chatterjee, A. K. (1985) Purification and partial characterization of a mitomycin C-induced pectin lyase of *Erwinia chrysanthemi* strain EC183. Physiol. Plant Pathol. 27, 119-130.

Tomizawa, H. & Takahashi, H. (1971) Stimulation of pectolytic enzyme formation of *Erwinia aroideae* by nalidixic acid, mitomycin C and bleomycin. Agric. Biol. Chem. 35, 191-200.

Uren, J. R. & Ragin., R. C. 1979. Improvement in the therapeutic, immunological, and clearance properties of *Escherichia coli* and *Erwinia carotovora* L-asparaginases by attachment of poly-DL-alanyl peptides. Cancer Res. 39, 1927-1933.

Uren, J. R. Hargis, B. J., & Beardsley. P., 1982. Immunological and Pharmacological characterization of poly-DL-alanyl modified *Erwinia carotovora* L-asparaginase. Cancer Res. 42,4068-4071.

Uroz, S., D'Angelo-Picard, C., Carlier, A., Elasri, M., Sicot, C., Petit, A., Oger, P., et al. (2003) Novel bacteria degrading N-homoserine lactones and their use as quenchers of quorum-sensing-regulated functions of plant-pathogenic bacteria. Microbiol. 149, 1981-1989.

Valmeekam, V., Loh, Y.-L., & San Francisco, M. J. D. (2001) Control of *exuT* activity for galacturonate transport by the negative regulator ExuR in *Erwinia chrysanthemi* EC16. Mol. Plant-Microbe Interact. 14, 816-820.

van der Wolf, J. M., Hyman, L. J., Jones, D. A. C., Grevess, C., van Beckhoven, J. R. C. M., van Vuurde, J. W. L., & Pérombelon, M. C. M. (1996) Immunomagnetic separation of *Erwinia carotovora* subsp. *atroseptica* from potato peel extracts to imporve detection sensitivity on a crystal violet pectate medium or by PCR. J. Appl. Bacteriol. 80, 487-495.

Varns, J. L. & Glynn, M. T. (1979) Detection of disease in stored potatoes by volatile monitoring. Am. Potato J. 56, 185-197.

Vidal, S., Eriksson, A. R. B., Montesano, M., Denecke, J., & Palva, E. T. (1998) Cell wall-degrading enzymes from *Erwinia carotovora* cooperate in the salicylic acid-independent induction of a plant defense response. Mol. Plant-Microbe Interact. 11, 23-32.

Vincent-Sealy, L. V., Thomas, J. D., Commander, P., & Salmond, G. P. C. (1999) *Erwinia carotovora* DsbA mutants: evidence for a periplasmic-stress signal transduction system affecting transcription of genes encoding secreted proteins. Micrbiol. 145, 1945-1958.

Vroemen, S., Heldens, J., Boyd, C., Henrissat, B., & Keen, N. T. (1995) Cloning and characterization of the *bgxA* gene from *Erwinia chrysanthemi* D1 which encodes a b-glucosidase/xylosidase enzyme. Mol. Gen. Genet. 246, 465-477.

Waldee, E. L. (1945) Comparative studies of some peritrichous phytopathogenic bacteria. Iowa State J. Sci. 19, 435-484.

Waleron, M., Waleron, K., Podhajska, A. J., & Lojkowska, E. (2000) Genotyping of bacteria belonging to the former *Erwinia* genus by PCR-RFLP analysis of a *recA* gene fragment. Microbiol. 148, 583-595.

Waterer, D. R. & Pritchard, M. K. (1984) Monitoring of volatiles: a technique for detection of soft rot *Erwinia carotovora* in potato tubres. Can J. Plant Pathol. 6, 165-171.

Webb L. E. & Wood R. K. S. (1974) Infection of potato tubers with soft rot bacteria. Ann. Appl. Biol. 76, 91-98.

Weber, J., Olsen, O., Wegener, C., & von Wettstein, D. (1996) Digalacturonates from pectin degradation induce tissue responses against potato soft rot. Physiol. Mol. Plant Pathol. 48, 389-401.

Weber, Z. (1984) Position and date of seed tuber inoculation by *Erwinia carotovora* ssp. *atroseptica* and appearance of potato blackleg. Potato Res. 27, 441-443.

Wegener, C., Bartling, S., Olsen, O., Weber, J., & von Wettstein, D. (1996) Pectate lyase in transgenic potatoes confers preactivation of defence against *Erwinia carotovora*. Physiol. Mol. Plant Pathol. 49, 359-376.

Wegener, C. B. (2001) Transgenic potatoes expressing an *Erwinia* pectate lyase gene - results of a 4-year field experiment. Potato Res. 44, 401-410.

Wegener, C. B. (2002) Induction of defence responses against *Erwinia* soft rot by an endogenous pectate lyase in potatoes. Physiol. Mol. Plant Pathol. 60, 91-100.

Wegener, C. & Olsen, O. (2004) Heterologous pectate lyase isoenzymes are not different in their effects on soft rot resistance in transgenic potatoes. Physiol. Mol. Plant Pathol. 65, 59-66.

Whitney, E. D. & Lewellen, R. T. (1978) Bacterial vascular necrosis and rot of sugarbeet: genetic vulnerability and selecting for resistance. Phytopathol. 68, 657-661.

Winslow, C.-E. A., Broadhurst, J., Buchanan, R. E., Krumwiede, C. Jr., Rogers, L. A., & Smith, G. H. (1917) The families and genera of the bacteria. Preliminary report of the committee of the Society of American Bacteriologists on characterization and classification of bacterial types. J. Bacteriol. 2, 505-566.

Wolters, P. J. & Collins, W. W. (1994). Evaluation of diploid potato clones for resistance to tuber soft rot induced by strains of *Erwinia carotovora* subsp. *atroseptica*, *Erwinia carotovora* subsp. *carotovora* and *Erwinia chrysanthemi*. Potato Res. 37, 143-149.

Wyatt, G. M. & B. M. Lund, B. 1981. The effect of antibacterial products on bacterial soft rot of potatoes. Potato Res. 24, 315-329.

Yang, C.-H., Gavilanes-Ruiz, M., Okinaka, Y., Vedel, R., Berthuy, I., Boccara, M. Chen, J. W. T., Perna, N. T., & Keen, N. T. (2002) *hrp* genes of *Erwinia chrysanthemi* 3937 are important virulence factors. Mol Plant-Microbe Interact. 15, 472-480.

Yang, S., Perna, N., Cooksey, D. A., Okinaka, Y., Lindow, S. E., Ibekwe, A. M., Keen, N. T., & Yang, C.-H. (2004) Genome-wide identification of plant-upregulated genes of *Erwinia chrysanthemi* 3937 using a GFP-based IVET leaf array. Mol. Plant-Microbe Interact. 17, 999-1008.

Yap, M.-N., Barak, J. D., & Charkowski, A. O. (2004) Genomic diversity of *Erwinia carotovora* subsp. *carotovora* and its correlation with virulence. Appl. Environ. Microbiol. 70, 3013-3023.

Yap, M.-N., Yang, C.-H., Barak, J. D., Jahn, C. E., & Charkowski, A. O. (2005) The *Erwinia chrysanthemi* type III secretion system is required for multicellular behavior. J. Bacteriol. 187, 639-648.

Yoder, M. D. & Jurnak, F. (1995) The refined 3-dimensional structure of pectate lyase C from *Erwinia chrysanthemi* at 2.2-angstrom resolution. Implications for an enzymatic mechanism. Plant Physiol. 107, 349-364.

Yoshida, A., Izuta, M., Ito, K., Kamio, Y., & Izaki, K. (1991) Cloning and characterization of the pectate lyase III gene of *Erwinia carotovora* Er. Agric. Biol. Chem. 55, 933-940.

Yoshida, A., Matsuo, Y., Kamio, Y., & Izaki, K. (1992) Molecular cloning and sequencing of the extracellular pectate lyase II gene from *Erwinia carotovora* Er. Biosci. Biotechnol. Biochem. 56, 1596-1600.

Zhao, Y. F., Thilmony, R., Bender, C. L., Schaller, A., He, S. Y., & Howe, G. A. (2003) Virulence systems of *Pseudomonas syringae* pv. tomato promote bacterial speck disease in tomato by targeting the jasmonate signalling pathway. Plant J. 36, 485-499.

Zimnoch-Guzowska, E., Lebecka, R., & Pietrak, J. (1999) Soft rot and blackleg reactions in diploid potato hybrids incoulated with *Erwinia* spp. Amer. J. Potato Res. 76, 199-207.

Zimnoch-Guzowska, E., Marczewski, W., Lebecka, R., Flis, B., Schafer-Pregl, R., Salamini, F., & Gebhardt, C. (2000) QTL analysis of new sources of resistance to *Erwinia carotovora* ssp. *atroseptica* in potato done by AFLP, RFLP, and resistance-gene-like markers. Crop Sci. 40, 1156-1167.

Zink, R. T., Engwall, J. K., McEvoy, J. L. & Chatterjee, A. K. (1985) *recA* is required in the induction of pectin lyase and carotovoricin in *Erwinia carotovora* subsp. *carotovora*. J. Bacteriol. 164, 390-396.

14. AFFILIATION

Amy O. Charkowski, *Department of Plant Pathology, University of Wisconsin-Madison, Russell Laboratories, 1630 Linden Dr., Madison, WI 53706, United States. Tel: (608) 262-7911; Fax: (608) 263-2626; amyc@plantpath.wisc.edu http://plantpath.wisc.edu~amyc*

MONICA HÖFTE AND PAUL DE VOS

PLANT PATHOGENIC *PSEUDOMONAS* SPECIES

Abstract. In the current taxonomy, plant pathogenic *Pseudomonas* species are restricted to rRNA group I organisms belonging to the Gamma subclass of *Proteobacteria*. Currently, about 21 validly described plant pathogenic *Pseudomonas* species are known. The most important species is *P. syringae* with more than 50 described pathovars. The pathovar concept is confusing and the taxonomy of *P. syringae* needs revision. *P. syringae* pv. *tomato* has become an important model organism to study host reactions to pathogen infection, mainly because this pathovar can infect the model plant Arabidopsis. Plant pathogenic Pseudomonads cause important diseases on a variety of crops and symptoms include cankers, leaf and stem spots, blight, soft rot and galls. Important pathogenicity and virulence factors are the type III secretion system, ice nucleation activity, the production of secondary metabolites such as phytotoxins, pectolytic enzymes, exopolysaccharides, and hormone production. Complete genome sequences are available for three important *P. syringae* pathovars. Molecular methods are becoming increasingly important in the diagnosis of plant pathogenic *Pseudomonas* species and specific detection techniques aimed at genes involved in pathogenicity are being developed. No single control strategy is effective against the plant pathogenic *Pseudomonas*. Control should be based on a combination of chemical, biological and cultural strategies.

1. INTRODUCTION

In this overview we will deal with plant pathogenic Pseudomonas species *sensu stricto*, this means organisms that belong to the rRNA group I as described by Palleroni et al. (1973). The often confusing taxonomy of this group of organisms will be discussed together with some information about molecular phylogeny. A brief overview will be given of important pathogenicity and virulence factors, together with some more information about the complete genome sequences that are now available for three important *P. syringae* pathovars. The overview will be concluded with recent information about diagnosis and control.

2. TAXONOMY AND MOLECULAR PHYLOGENY

The nomenclature of bacteria in the genus *Pseudomonas* has changed considerably during the last decennia. The genus *Pseudomonas* is currently restricted to those species related to the type species *Pseudomonas aeruginosa*, i.e. the genuine pseudomonads of the rRNA group I (Palleroni et al., 1973), that belong to the Gamma subclass of the *Proteobacteria* (Kersters et al., 1996). Up to now (2005), the genus *Pseudomonas* comprises 18 validly described plant pathogenic species and 3 species that are pathogenic to mushrooms (Tables 1 and 2). An overview of oxidase positive and oxidase negative plant (and mushroom) pathogenic *Pseudomonas* species is given in Table 1 and Table 2.

S.S. Gnanamanickam (ed.), Plant-Associated Bacteria, 507–533.

Table 1. Validly published names of oxidase positive plant pathogenic pseudomonads[a]

Name	*Host*	*Disease or symptoms* [b]
Pseudomonas agarici	*Agaricus bisporus*	drippy gill
Pseudomonas asplenii	*Asplenum nidus*	leaf spot and blight
Pseudomonas cichorii	wide host range	leaf and stem spots
Pseudomonas constantinii	*Agaricus bisporus*	brown blotch
Pseudomonas corrugata	tomato	pith necrosis
Pseudomonas fuscovaginae	*Oryzae sativa*	leaf sheath brown rot
Pseudomonas marginalis pv. *alfalfae*	*Medicago sativa*	root browning, stunting
Pseudomonas marginalis pv. *marginalis*	wide host range	marginal leaf necrosis, soft rot brown rot of roots, sof rot
Pseudomonas marginalis	*Pastinaca sativa*	
pv. *pastinacea*		pith necrosis
Pseudomonas mediterranea	*Lycopersicon*	weakly pathogenic to rice
Pseudomonas palleroniana	*esculentum*	café au lait disease
Pseudomonas salomonii	*Oryzae sativa*	brown blotch
Pseudomonas tolaasii	*Allium sativum*	
	Agaricus spp.	

[a] Information based on Young et al. (1996) and Young et al. (2004)
[b] Information mainly based on Bradbury (1986) and Smith et al. (1988)

Among the oxidase positive species, *P. chicorii* has a wide host range and causes necrotic leaf and stem lesions. Pectinolytic *P. fluorescens* strains that can cause soft rot on a wide range of hosts, are commonly named *P. marginalis*. *P. marginalis* strains, however, are phenotypically indistinguishable from saprophytic strains of *P. fluorescens* biovar II. The name *P. marginalis* has commonly been used for soft rot bacteria that resemble *P. fluorescens* biovar2 (= *P. marginalis sensu stricto*) or for all fluorescent oxidase positive soft rot bacteria (= *P. marginalis sensu lato*). Janse et al. (1992), however, have shown that within the group of fluorescent oxidase positive pseudomonads various other bacteria show soft rot activity including most biovars of *P. fluorescens*, and isolates identified as *P. putida*, *P. aureofaciens*, and *P. tolaassii*. Since soft rot activity can be demonstrated in so many diverse fluorescent pseudomonads, Janse et al. (1992) concluded that it no longer makes sense to classify all fluorescent oxidase soft rot bacteria in the artificial species *P. marginalis*. All these bacteria appear to belong to the *P. fluorescens* supercluster (Janse et al., 1992).

P. tolaasii, *P. agaraci* and the recently described species *P. constantinii* (Munsch et al., 2002) are pathogenic on the cultivated mushroom *Agaricus*. Also *P. tolaasii* is taxonomically closely related to *P. fluorescens*. *P. corrugata* and the newly established species *P. mediterranea* (Catara et al., 2002) cause pith necrosis on tomato. *P. asplenii* is the causal agent of bacterial leaf blight of bird' nest firn (*Asplenum nidus*) and shows a high similarity to *P. fuscovaginae*, causing leaf sheath brown rot on rice (*Oryzae sativa*) and other grasses. Based on SDS-PAGE of whole cell proteins, a possible synonymy between *P. asplenii* and *P. fuscovaginae* was suggested (Vancanneyt et al., 1996b).

Table 2. Validly published names of oxidase negative plant pathogenic pseudomonads[a]

Name	*Host*	*Disease or symptoms* [b]
P. amygdali	*Prunus amygdalus*	bacterial canker
P. avellanae	*Corylus avellana*	bacterial canker
P. cannabina	*Cannabis sativa*	
P. caricapapayae	*Carica papaya*	leaf spot
P. ficuserectae	*Ficus erectae*	leaf spot, shoot blight
P. meliae	*Melia azadarach*	galls
P. savastanoi (various pathovars)	see Table 3	
P. syringae (various pathovars)	see Table 4	
P. tremae	*Trema orientalis*	
P. viridiflava	wide host range	leaf necrosis, necrotic spots, stem and root rots

[a] Information based on Young et al. (1996) and Young et al. (2004)
[b] Information mainly based on Bradbury (1986) and Smith et al. (1988)

Among the oxidase negative species, *Pseudomonas syringae* is economically the most important with more than 50 pathovars (see Table 3). The pathovar concept was introduced in order to distinguish among bacteria within the species that exhibit different pathogenic abilities (Dye et al., 1980). The term "pathovar" is not part of the taxonomic hierarchy and should therefore be eliminated as the primary name of an organism as soon as sufficient data are obtained to justify species and subspecies ranking (Braun-Kiewnick and Sands, 2001). Within *P. syringae*, the most important and best-studied pathovars are *coronafaciens*, *glycinea*, *lachrymans*, *morsprunorum*, *persicae*, *phaseolicola*, *pisi*, *syringae*, *tabaci* and *tomato*. In recent years, *P. syringae* pv. *tomato* and the closely related pv. *maculicola* have become important model organisms to study molecular mechanisms of host responses to infection, mainly because many strains of these pathovars are pathogenic on the model plant *Arabidopsis thaliana*. Certain strains exhibit race-cultivar specificity on Arabidopsis, thus providing a model pathosystem for studying both compatible and incompatible host-pathogen ineractions (Preston, 2000). *Pseudomonas savastanoi* is an important tumor or gall inducing species on olive, ash and oleander. Currently, various pathovars are distinguished within this species (see Table 4). Schaad et al. (2000), however, have rejected the inclusion of pv. *phaseolicola* and pv. *glycinea* in the species *P. savastanoi* and propose that pv. *phaseolicola* and pv. *glycinea* should remain as pathovars of the species *P. syringae*. The pectinolytic species *P. viridiflava* has a wide host range and causes necrotic leaf and stem lesions and basal stem and root rots. *P. avellana* caused bacterial canker on hazelnut and has been reported in Greece and Italy (Scortichini et al., 2002). *P. cannabina* and *P. tremae* (Gardan et al., 1999) are newly established species, but they are of little economic importance. Other species of minor importance are *P. amygdali* on almond trees, *P. caricapapaya* causing leaf spot on *Carica papaya; P. ficuserectae* causing leaf spot on *Ficus erectae* and *P. meliae* causing bacterial galls on chinaberry.

Table 3. Validly described pathovars of *Pseudomonas syringae*[a]

Name	*Host*	*Disease or symptoms*[b]
P.s. pv. *aceris*	*Acer macrophyllum*	leaf spot
P.s. pv. *actinidiae*	kiwi fruits	bacterial canker
P.s. pv. *alisalensis*	*Brassica* spp.	bacterial blight
P.s. pv. *aesculi*	*Aesculus indica*	leaf spot
P.s. pv. *antirrhini*	*Antirrhinum majus*	leaf spot, stem lesions
P.s. pv. *apii*	*Apium graveolens*	leaf spot
P.s. pv. *aptata*	*Beta vulgaris*	leaf spot, foliar blight
P.s. pv. *atrofaciens*	cereals	leaf spot, basal glume rot
P.s. pv. *berberidis*	*Berberis* spp.	leaf spot
P.s. pv. *broussonetiae*	*Broussonetia*	bacterial blight
P.s. pv. *castaneae*	*Castanea crenata*	bacterial canker
P.s. pv. *cerasicola*	*Prunus yedoensis*	galls
P.s. pv. *ciccaronei*	*Ceratonia siliqua*	leaf spot
P.s. pv. *coriandricola*	*Coriandrum sativum*	umbel blight, seed decay
P.s. pv. *coronafaciens*	cereals	halo blight
P.s. pv. *cunninghamiae*	*Cunninghamia lanceolate*	
P.s. pv. *daphniphylli*	*Daphniphyllum*	galls
P.s. pv. *delphinii*	*Delphinium* spp.	leaf spot
P.s. pv. *dendropanacis*	*Dendropanax trifidus*	
P.s. pv. *dysoxyli*	*Dysoxylum spectabile*	leaf spot, shot hole
P.s. pv. *eriobotryae*	*Eriobotrya japonica*	bud blight, twig canker
P.s. pv. *garcae*	*Coffea arabica*	halo blight
P.s. pv. *helianthi*	*Helianthus* spp.	leaf spot
P.s. pv. *hibisci*	*Hibiscus japonica*	leaf spot
P.s. pv. *lachrymans*	cucurbits	angular leaf spot
P.s. pv. *lapsa*	maize, sorghum	stalk rot
P.s. pv. *maculicola*	*Brassica* spp.	bacterial spotting
P.s. pv. *mellea*	*Nicotiana tabacum*	Wisconsin tobacco disease
P.s. pv. *mori*	*Morus* spp.	leaf spots, shoot blight
P.s. pv. *morsprunorum*	*Prunus* spp.	leaf spot and stem canker
P.s. pv. *myricae*	*Myricae rubra*	galls
P.s. pv. *oryzae*	*Oryzae sativa*	halo blight
P.s. pv. *papulans*	*Malus pumila, Pyrus*	blister spot, blister canker
P.s. pv. *passiflorae*	*Passiflora edulis*	necrotic spots

P.s. pv. *persicae*	*Prunus persicae*	leaf spots, cankers, dieback
P.s. pv. *philadelphi*	*Philadelphus* spp.	leaf spot
P.s. pv. *photiniae*	*Photinia glabra*	leaf spot and blight
P.s. pv. *pisi*	*Pisum, Vicia*	bacterial blight
P.s. pv. *porri*	*Allium porrum*	bacterial blight
P.s. pv. *primulae*	*Primula* spp.	leaf spot
P.s. pv. *raphiolepidis*	*Rhaphiolepis umbellata*	galls
P.s. pv. *ribicola*	*Ribes aureum*	leaf spot, defoliation
P.s. pv. *sesami*	*Sesamum indicum*	leaf spot
P.s. pv. *solidagae*	*Solidago altissima*	leaf spot
P.s. pv. *spinaceae*	*Spinacea oleracea*	leaf spot
P.s. pv. *striafaciens*	*Avena sativa*, triticale	stripe blight
P.s. pv. *syringae*	very wide	leaf spots, cankers, dieback
P.s. pv. *tabaci*	*Glycine max, Nicotiana*	wildfire, angular leaf spot
P.s. pv. *tagetis*	*Ambrosia, Helianthus, Tagetes*	leaf spot
P.s. pv. *theae*	*Camellia sinensis*	shoot blight, stem blight
P.s. pv. *tomato*	tomato, *Arabidopsis*	bacterial speck, leaf spot
P.s. pv. *ulmi*	*Ulmus* spp.	leaf and shoot blight
P.s. pv. *viburni*	*Viburnum* spp.	leaf and stem spot
P.s. pv. *zizaniae*	*Zizania aquatica*	

[a] Information based on Young et al. (1996) and Young et al. (2004)
[b] Information mainly based on Bradbury (1986) and on original references in Young et al. (1996, 2004)

Table 4. Validly described pathovars of *Pseudomonas savastanoi*[a]

Name	*Host*	*Disease or symptoms* [b]
P.sav. pv. *fraxini*	*Fraxinus excelsior*	galls
P.sav. pv. *glycinea*[c]	*Glycine max*	bacterial blight
P.sav. pv. *nerii*	*Nerium oleander*	galls
P.sav. pv. *phaseolicola*[c]	*Phaseolus, Pisum, Vigna*	halo blight
P.sav. pv. *retacarpa*	*Retama sphaerocarpa*	galls
P.sav. pv. *savastanoi*	*Olea europaea*	galls

[a] Information based on Young et al. (1996) and Young et al. (2004)
[b] Information mainly based on Bradbury (1986) and on original references in Young et al. (1996, 2004)
[c] Inclusion of these pathovars in the species *P. savastanoi* was rejected by Schaad et al. (2000)

Gardan et al. (1999) studied a total of 48 pathovars of *P. syringae* and eight related species by DNA-DNA hybridisation and ribotyping and proposed the delineation of nine discrete genomospecies (see Table 5). Each genomospecies, which may eventually be regarded as a new species, contains strains exhibiting at least 70% DNA/DNA homology with the relevant type strain(s). As a consequence, several pathovars of *P. syringae* and related species have already been, or will be renamed. Genomospecies 1 corresponds to *P. syringae sensu stricto* and also corresponds to the DNA-DNA group "*syringae*" of Pecknold and Grogan (1973). All but one strain of genomospecies 1 belonged to ribogroup A. It has been suggested that the nine pathovars that are grouped in genomospecies 1 may be synonyms of pv. *syringae*. Genomospecies 2 includes 16 different pathovars of *P. syringae* and type strains of the related species *P. savastanoi*, *P. ficuserectae*, *P. meliae* and *P. amygdali*. According to Gardan et al. (1999) *P. amygdali* should be the correct name for this species. This genomospecies corresponds to DNA group "*morsprunorum*" of Pecknold and Grogan (1973). The 20 strains of genomospecies 2 were distributed within four ribogroups, B to E. Genomospecies 3 includes 14 strains of different pathovars of *P. syringae* that demonstrated relatedness to the pathotype strain of *P. syringae* pv. *tomato*. This genomospecies corresponds to the DNA-DNA group "tomato" of Pecknold and

Table 5. Genomospecies in *P. syringae* pathovars and related species (Gardan et al., 1999)

Genomospecies	*Taxon*
Genomospecies 1	*P. syringae, P.s.* pv. *aptata, P.s.* pv. *lapsa, P.s.* pv. *papulans, P.s.* pv. *pisi, P.s.* pv. *atrofaciens, P.s.* pv. *aceris, P.s.* pv. *panici, P.s.* pv. *dysoxyli, P.s.* pv. *japonica*
Genomospecies 2	*P. savastanoi, P. ficuserectae, P. meliae, P. amygdali, P.s.* pv. *phaseolicola, P. s.* pv. *ulmi, P.s.* pv. *mori, P. s.* pv. *lachrymans, P.s.* pv. *sesami, P.s.* pv. *tabaci, P.s.* pv. *morsprunorum, P.s.* pv. *glycinea, P.s.* pv. *ciccaronei, P. s.* pv. *eriobotryae, P.s.* pv. *mellea, P.s.* pv. *aesculi, P.s.* pv. *hibisci, P.s.* pv. *myricae, P.s.* pv. *photiniae, P.s.* pv. *dendropanacis*
Genomospecies 3	*P. s.* pv. *tomato, P.s.* pv. *persicae, P.s.* pv. *antirrhini, P.s.* pv. *maculicola, P.s.* pv. *viburni, P.s.* pv. *berberidi, P.s.* pv. *apii,P.s.* pv. *delphinii, P.s.* pv. *passiflorae, P.s.* pv. *philadelphi, P.s.* pv. *ribicola, P.s.* pv. *primulae*
Genomospecies 4	*P. coronafaciens, P.s.* pv. *porri, P.s.* pv. *garcae, P.s.* pv. *striafaciens, P.s.* pv. *atropurpurea, P. s.* pv. *oryzae, P. s.* pv. *zizaniae*
Genomospecies 5	*P. tremae*
Genomospecies 6	*P. viridiflava*
Genomospecies 7	*P. s.* pv. *tagetis, P.s.* pv. *helianthi*
Genomospecies 8	*P.s.* pv. *theae, P. avellanae, P. s.* pv. *actinidiae*
Genomospecies 9	*P. cannabina*

Grogan (1973). Gardan et al. (1999) recommended *P. syringae* pv. *tomato* CFBP 2212 as the type strain for this genomospecies. Genomospecies 4 includes the type strain of *P. coronafaciens* and seven strains of different pathovars of *P. syringae*. This genomospecies represents "*coronafaciens*". The strains of genomospecies 4 constituted ribogroup F. Genomospecies 5 includes only the pathotype strain of *P. syringae* pv. *tremae* and comprises the species *P. tremae* sp. nov. as described by Gardan et al. (1999). The single strain in this genomotype represents ribogroup K. Genomospecies 6 includes the type strain of *P. viridiflava* and represents *P. viridiflava*. The 3 strains of the genomospecies tested all belonged to ribogroup J. Genomospecies 7 includes *P. syringae* pv. *tagetis* and *P. syringae* pv. *helianthi*; *P. syringae* pv. *tagetis* strain CFBP 1694 is recommended to serve as the reference strain. Both strains belong to ribogroup G. Genomospecies 8 included the type strain of *P. avellanae* (Janse et al., 1996) and thus represents *P. avellanae*. *P. syringae* pv. *actinidiae* strains were not included in the study of Gardan et al. (1999). *P. syringae* pv. *actinidiae* is the causal agent of bacterial canker of kiwifruit and has been reported in Japan, Italy and South Korea. Scortichini et al. (2002) showed that *P.s.* pv. *actinidiae* is genetically related to *P.s.* pv. *theae* and *P.s.* pv. *avellanae* and proposed to include this pathovar in genomospecies 8. Genomospecies 9 includes only the pathotype strain of *P. syringae* pv. *cannabina* and thus comprises *P. cannabina* sp. nov. as described by Gardan et al. (1999). This strain represented ribogroup L. Genomospecies 3 and 8 could not clearly be distinguished by ribotyping.

Sawada et al. (1999) conducted a phylogenetic analysis of *P. syringae* using 56 strains belonging to 19 pathovars. *gyrB* and *rpoD* were adopted as the index genes to determine the course of bacterial genome evolution, while *hrpL* and *hrpS* were selected as the representatives of the pathogenicity-related genes located on the chromosome. The data on all four genes were used to create a genomic tree that showed three distinct monophyletic groups: Group 1, 2 and 3. Pathovar *tomato*, pv. *morsprunorum*, pv. *syringae*, pv. *actinidiae* and pv. *theae* were located in Group 1; pv. *aceris*, pv. *aptata*, pv. *japonica*, pv. *syringae*, pv. *pisi* were in Group 2; and pv. *myricae*, pv. *eriobotryae*, pv. *morsprunorum*, pv. *tabaci*, pv. *lachrymans*, pv. *castaneae*, pv. *phaseolicola*, pv. *glycinea*, pv. *mori* and pv. *broussonetiae* were in Group 3. Three pathovars, pv. *lachrymans*, pv. *morsprunorum* and pv. *syringae*, were distributed over two groups. The pathotype strains of pv. *lachrymans* and pv *morsprunorum* were included in Group 1, but all other strains of these pathovars belonged to Group 3. For *P. syringae*, only a Japanese citrus strain belonged to Group 1, while all other strains were included in Group 2. Group 1 largely corresponds to genomospecies 3 and genomospecies 8 as defined by Gardan et al. (1999). As mentioned before, Gardan et al. (1999) were unable to differentiate these two groups by ribotyping. Group 2 largely corresponds to genomospecies 1 as defined by Gardan et al. (1999), while Group 3 corresponds to genomospecies 2. Sarkar and Guttman (2004) studied the population structure and dynamics of the core genome of *P. syringae* via multilocus sequencing typing (MLST) of 60 strains, representing 21 pathovars and 2 nonpathogens, isolated from a variety of plant hosts. MLST is a recently developed strain-typing system that focuses strictly on the core genome. In this approach, the DNA sequences from seven housekeeping genes are used to differentiate strains and clonal lineages. The phylogenetic analysis of *P. syringae* revealed four major groups of strains, three of which largely correspond to those identified by Sawada et al. (1999). The fourth group contained only pathogens of monocots (rice, oats and onions) and includes pathovars that correspond with genomospecies 4 as defined by Gardan et al. (1999). An analysis of molecular variance found that host association explained only a small proportion of the total genetic variation in the sample. With respect to the core genome, *P. syringae* is a highly clonal and stable species that is endemic within plant populations. Sarkar and Guttman (2004) concluded that factors outside of the core genome must be maintaining the cohesion of the species and must play very significant roles in determining host suitability.

3. SYMPTOMS

Plant pathogenic Pseudomonas species can cause a variety of symptoms such as cankers, dieback, blossom, twig, leaf or kernel blight and leaf spots caused by *P. syringae* pathovars; soft or brown rot caused by *P. viridiflava* and pectinolytic *P. fluorescens* strains (*P. marginalis*); tumors or galls caused by *P. savastanoi* and mushroom blights caused by *P. tolaasii* and *P. agarici*. See Tables 1 to 4 for a detailed overview.

4. PATHOGENICITY AND VIRULENCE FACTORS

4.1. TTSS and effectors

The type III protein secretion system (TTSS) is key to the plant parasitism of *P. syringae* pathovars and has been found in all of the *P. syringae* strains examined. The TTSS was first discovered in the mammalian pathogenic bacterium *Yersinia* and has since been found in taxonomically diverse Gram-negative bacterial pathogens of plants and animals and in a few non-pathogenic plant-associated bacteria. Most of the *hrp* (hypersensitive response and pathogenicity) and *hrc* (hrp conserved) genes encoding the TTSS system are essential for pathogenicity, which indicates the collective importance of the effector proteins that are injected into plant cells by the system. The reader is referred to several excellent and detailed reviews about TTSS in bacterial plant pathogens (Alfano and Collmer, 2004; Mudgett, 2005). The TTSS in *Pseudomonas syringae* has recently been reviewed by Jin et al. (2003). Functional analysis of the genome of *P. syringae* pv. *tomato* DC3000 has yielded genes for 31 effectors and 7 additional proteins secreted by the Hrp system that likely have a role in translocating effectors across the plant cell wall and plasma membrane (Buell et al., 2003). Effector genes can be dispersed throughout the genome, clustered in a pathogenicity island located on a plasmid, or associated with the *hrp* gene cluster (Oguiza and Asensio, 2005). Type III effectors are believed to contribute to pathogenesis in two ways: by eliciting the release of water and/or nutrients from the host cell in the apoplastic space; and by suppressing and/or evading plant host defense responses. The type III effectors produced and secreted by the pathogen interact with plant molecules known as virulence targets. In resistant plants, effectors function as avirulence determinants that activate the hypersensitive response (HR), a primary defense response triggered by recognition of the effector-virulence target complex by plant resistance genes. In susceptible plants, effectors avoid specific recognition by the plant host surveillance mechanisms and function as virulence determinants that facilitate pathogenesis and modulate host defense responses and physiology to the benefit of the pathogen. (Oguiza and Asensio, 2005). Based on their localization in plants, type III effector proteins of *P. syringae* can be grouped into two classes: extracellular type III effectors such as HrpZ and HrpW and intracellular type III effectors that are directly transported from the bacterial cell into the plant cytosol. The extracellular type III effectors are glycine-rich, cysteine-lacking, heat-stable proteins, also called harpins, that elicit a hypersensitive-like response when infected into the intercellular space of plant leaves. It is suggested that HrpZ and HrpW may function either in the release of nutrients from the host cell or possibly as mediators of the translocation process, contributing to the effective delivery of effectors in the host cell cytoplasm. Recent studies have firmly established the concept that the suppression of various plant defenses, including basal defense, gene-for-gene resistance, and nonhost resistance, is a major virulence function of intracellular TTSS effectors. This topic has recently been reviewed by Nomura et al. (2005).

4.2. Phytotoxins

Pseudomonas spp. produce a wide spectrum of phytotoxic compounds. Among the most well-characterized bacterial phytotoxins are those produced by *Pseudomonas syringae* (see Bender et al., 1999 for an extensive review). The toxins produced by *P. syringae* include monocyclic lactam (tabtoxin), sulfodiaminophosphinyl peptide (phaseolotoxin), lipodepsipeptide (syringomcins, syringopeptins) and polyketide (coronatine) structures. Although phytotoxins are not required for pathogenicity in *P. syringae*, they generally function as virulence factors and their production results in increased disease severity. *P. syringae* phytotoxins can contribute to systemic movement of bacteria in planta, lesion size, and multiplication of the pathogen in the host. Tagetitoxin is a cyclic hemithioketal molecule that is only produced by strains of *P. syringae* pv. *tagetis*. The toxin interferes with RNA polymerase in protein biosynthesis of chloroplasts. The toxin can rapidly be detected by its ability to elicit apical chlorosis in plant tissues.

Toxins produced by other *Pseudomonas* species include the lipodepsipeptides corpeptin, fuscopeptin, tolaasin and viscosin produced by *P. corrugata*, *P. fuscovaginae*, *P. tolaasii* and *P. fluorescens* (*marginalis*), respectively.

The best studied phytotoxins are coronatine, syringomycin, tabtoxin and phaseolotoxin.

4.2.1. Coronatine

Coronatine is produced by *P. syringae* pv. *atropurpurea, glycinea, maculicola, morsprunorum*, and *tomato* (Bender et al., 1999). The structure of coronatine has two distinct components: the polyketide coronafacic acid and coronamic acid, an ethylcyclopropyl amino acid derived from isoleucine. Coronatine shows a remarkable structural and functional homology to methyl jasmonate, a plant hormone which activates the jasmonate (JA) signalling pathway that is involved in plant defense against herbivores and certain pathogens. It has been shown that coronatine activates JA-signalling responses and suppresses salicylic acid (SA) dependent plant defenses. Inoculation of a local leaf of Arabidopsis with the coronatine producing *P. syringae* pv. *maculicola* induced increased susceptibility to secondary infections in systemic leaves (Cui et al., 2005). The action of coronatine in triggering systemic induced susceptibility is probably mediated by cancelling out SA-dependent defenses in systemic leaves through antagonistic cross-talk between the SA and JA signalling pathways (Cui et al., 2005). Recent evidence suggests, however, that coronatine has the ability to stimulate both SA and jasmonates (Block et al., 2005).

4.2.2. Syringomycin and other lipodepsipeptides

Production of syringomycins has been shown in *P. syringae* pv. *syringae*, pv. *aptata*, pv. *atrofaciens* and *P. fuscovaginae* (Bender et al., 1999). Syringomycin is a member of the cyclic lipodepsinonapeptide class of phytotoxins, which are composed of a polar peptide head and a hydrophobic 3-hydroxy fatty acid tail. The amphipathic syringomycin molecule exhibits potent biosurfactant activity. The surface active properties of syringomycin are similar to those of other biosurfactants produced by fluorescent pseudomonads such as viscosin and tolaasin. All strains of *P. syringae* pv. *syringae* analysed produce both syringomycin and syringopeptin. Both compounds are pore-forming cytotoxins that cause necrosis in plants by similar mechanisms.

4.2.3. Tabtoxin

Tabtoxin is a monocyclic β-lactam produced by *P. syringae* pv. *tabaci*, *coronafaciens*, and *garcae*, which cause wildfire on tobacco, and halo blight of oats and coffee, respectively. *P. syringae* pv. *striafaciens*, the causal agent of bacterial stripe of oats, is tabtoxin-deficient, but further indistinguishable from *P. syringae* pv. *coronafaciens* and pv. *garcae*. Recent evidence suggests that *P. syringae* pv. *coronafaciens, garcae* and *striafaciens* are likely the same pathovar. Introduction of the tabtoxin biosynthetic region in *P. syringae* pv. *striafaciens* resulted in the production of lesions on oat leaves there were indistinguishable from those caused by *P. syringae* pv. *coronafaciens* (Barta and Willis, 2005). Tabtoxin contains tabtoxin-β-lactam linked by a peptide bond to threonine. The chlorosis-inducing activity occurs only after hydrolysis of the peptide bond by aminopeptidases of plant or bacterial origin. Cleavage of the peptide bond releases tabtoxin-β-lactam, the toxic moiety. Tabtoxin-β-lactam irreversibly inhibits glutamine synthetase. This enzyme is the only way to efficiently detoxify ammonia.

4.2.4. Phaseolotoxin

Phaseolotoxin is produced by *P. syringae* pv. *phaseolicola* and *P. syringae* pv. *actinidiae*, which cause halo blight on beans and bacterial canker on kiwifruit, respectively. Phaseolotoxin consists of a sulfodiaminophosphinyl moiety linked to a tripeptide. Phaseolotoxin competitively inhibits ornithine carbamoyl transferase (OCTase), a critical enzyme in the urea cycle, which converts ornithine and carbamoyl phosphate to citrulline. Phaseolotoxin is hydrolysed in planta by peptidases to produce octicidine. Octicidine is an irreversible inhibitor of OCTase and the predominant form of the toxin in infected tissues. Inhibition of OCTase causes an accumulation of ornithine and a deficiency in intracellular pools of arginine, leading to chlorosis.

4.3. Auxin production

Glickmann et al. (1998) studied auxin production by pathovars of *P. syringae* and related species. Most of the analysed strains produced IAA, especially in the presence of tryptophan. The strains *P. syringae* pv. *syringae* 1392 and *P. syringae* pv. *aceris* 2339 (genomospecies 1); P. *savastanoi* pv. *savastanoi* 1670, *P. syringae* pv. *myricae* 2897 and *P. syringae* pv. *photiniae* 2899 (genomospecies 2); *P. syringae* pv. *maculicola* 1657 and *P. syringae* pv. *ribicola* 10971t (genomospecies 3); and *P. syringae* pv. *cannabina* 2341 (genomospecies 9) synthesized IAA at concentrations over 2 μg/ml when grown in modified King B medium without tryptophan and produced high amounts of IAA in the presence of tryptophan. These strains harbor genes homologous to the *iaaM/iaaH* genes of *P. savastanoi.*

The involvement of IAA in pathogenicity has been unambiguously demonstrated for *P. savastanoi* pv. *savastanoi*. For *P. savastanoi* strains, pathogenicity implies biosynthesis of plant growth regulators. The synthesis of hormones such as cytokinins and indole-3-acetic acid (IAA) leads to the formation of the characteristic knots on olive and oleander. *P. syringae* pv. *amygdali* and

P. syringae pv. *myricae* also induce proliferation of plant tissues and also harbor the *iaaM/iaaH* genes. IAA production has also been associated with epiphytic survival or with toxin production as demonstrated for *P. syringae* pv. *syringae* strains on *Phaseolus vulgaris*. There are also indications that IAA may inhibit plant defense mechanisms (Robinette and Matthysse, 1990).

4.4. Ethylene production

Ethylene production has been demonstrated in various pathovars of *P. syringae*, including pvs. *glycinea, pisi* (Weingart and Volksch, 1997a), *cannabina* and *sesami* (Sato et al., 1997). In addition strains of *P. syringae* pv. *phaseolicola* isolated from kudzu (*Pueraria lobata*) also produce ethylene unlike *P.s.* pv. *phaseolicola* strains isolated from bean (Volksch and Weingart, 1997). Volksch and Weingart (1997) have shown that *P.s.* pv. *phaseolicola* strains from kudzu can be clearly differentiated from strains isolated from bean. They utilize mannitol, produce ethylene, and are strongly pathogenic to kudzu, bean, and soybean. It was suggested by Volksch and Weingart (1997) that the strains from kudzu should be separated from the pathovar *phaseolicola* and should represent their own pathovar. The *efe* gene encoding the ethylene-forming enzyme appears to be plasmid-encoded (Watanabe et al., 1998). The role of ethylene production in virulence of *P. syringae* pvs. *glycinea* and *phaseolicola* was studied. Virulence of *P. syringae* pv. *phaseolicola* was not affected by disruption of the *efe* gene, while *efe* mutants of *P. syringae* pv. *glycinea* were significantly reduced in their ability to grow in planta (Weingart et al., 2001).

4.5. Exopolysaccharides

The production of exopolysaccharide polymers by phytopathogenic bacteria has been implicated in several symptoms, including wilting induced by vascular pathogens and the water soaking associated with foliar pathogens (see Denny, 1995 for a review). *P. syringae* pathovars generally produce two EPS molecules: levan, a fructofuranan polymer, and alginate, a co-polymer of O-acetylated β-1,4-linked D-mannuric acid and L-guluronic acid (Gross and Rudolph, 1987). When grown on media with excess sucrose, many *P. syringae* pathovars produce levan (Hettwer et al., 1998). However, alginate appears to be the major EPS produced in water-soaked lesions (Fett and Dunn, 1989; Rudolph et al., 1989). The alginate biosynthetic gene cluster of *P. syringae* pv. *syringae* FF5 was cloned and characterized (Peñaloza-Vàzquez et al., 1997). The arrangement of the alginate gene cluster in *P. syringae* was virtually identical to that described for the human pathogen *P. aeruginosa*. However, the regulation and signals for transcriptional activation of alginate biosynthesis differed in the two species, presumably because of their adaptation to plant and animal hosts, respectively (Peñaloza-Vàzquez et al., 1997). An alginate deficient mutant of *P. syringae* pv. *syringae* was significantly impaired in its ability to colonize tomato leaves (a non-host) compared with the wild type strain, indicating that alginate plays a role in epiphytic fitness. The mutant retained the ability to form lesions on bean leaves, but symptoms were less severe and the mutant population was significantly reduced in comparison with the wild type. Apparently, alginate contributes to the virulence of *P. syringae* pv. *syringae*, perhaps by facilitating colonization or dissemination of the bacterium in planta (Yu et al., 1999).

P. syringae pv. *ciccaronei*, which causes leaf spots on carob plants produces a mannan exopolysaccharide. The pure polysaccharide showed phytotoxic effects, i.e., chlorosis and necrosis on tobacco leaves (Corsaro et al., 2001).

4.6. Pectinolytic enzymes

Soft-rotting *Pseudomonas fluorescens* (*marginalis*) strains are capable of degrading pectic components of plant cell walls by producing a wide variety of pectolytic enzymes, including pectin methyl esterase, pectin lyase, polygalacturonase and two pectate lyase isozymes. *P. viridiflava* produces a single pectate lyase (PelV), which has a very alkaline PI, like the major Pel enzyme of *P. fluorescens* (*marginalis*) (Liao et al., 1994). *pel* genes have been cloned from *P. fluorescens* and *P. viridiflava* and DNA sequence analysis has revealed that *P. fluorescens* and *P. viridiflava* Pels are members of the *Erwinia chrysanthemi* PelADE family (Liao et al., 1996). The *P. viridiflava pelV* gene has been mutated revealing it to be essential for soft-rot pathogenesis (Liao et al., 1988). At least some of the *P. syringae* pathovars also produce pectic enzymes and *pel* gene sequences are available in the database for *P. syringae* pv. *lachrymans*, *P. syringae* pv. *phaseolicola*, *P. syringae* pv. *tabaci*, and *P. syringae* pv. *glycinea*. *P. syringae* pv. *glycinea* produces two alkaline pectate lyase isozymes with pIs of 9.0 and 9.5 and an alkaline polygalacturonase (Magro et al., 1994). *P. syringae* pv. *lachrymans* produces a single pectate lyase enzyme with a pH optimum between 8.0 and 8.5 which is encoded by the *pelS* gene (Bauer and Collmer, 1997).

4.7. Ice nucleation

The ability of bacteria to nucleate supercooled water to form ice is uniquely limited to *P. syringae* and a few other bacterial species including strains of *P. fluorescens* and *P. viridiflava*. Ice nucleation-active (INA) bacteria are usually phyllosphere inhabitants. Ice nucleation has been used as a trait to distinguish strains among some of the *P. syringae* pathovars. For example strains within pv. *syringae* frequently exhibit the ice phenotype, while none of the strains tested thus far within pv. *tomato* or *morsprunorum* are ice nucleation active. The presence of INA bacteria on leaf surfaces can destroy leaf habitats at subzero temperatures. Frost-sensitive plants are injured when ice forms within plant tissues. In the absence of heterogeneous ice nuclei, water associated with leaves will supercool. Supercooling in the temperature range of 0 to about –5°C is primarily limited by the presence of INA bacteria. Below –5°C, other heterogeneous ice nuclei, probably also limit supercooling. Thus, INA bacteria are responsible for ice formation, and hence injury to plants, mainly in the range from 0 to –5°C (Hirano and Upper, 2000). Within the plant pathogenic pseudomonads, *ice* (or *ina*) genes have been cloned and sequenced from strains of *P. syringae* (Green and Warren, 1985) and *P. fluorescens* (Warren et al., 1986). Recently it was shown that *P. syringae* pv. *syringae* B728a also has an unlinked gene encoding an antifreeze protein (Feit et al., 2005). Antifreeze proteins are secreted into the medium, where they inhibit the growth of external ice by adsorbing onto the ice surface and lowering the temperature at which it can grow. It is possible that the ice nucleation activity of *P. syringae* strains, which are quantitatively quite variable, is modulated by the acivity of such antifreeze proteins.

5. COMPLETE GENOME SEQUENCES

Within the group of plant-associated fluorescent pseudomonads, complete genome sequences are available of the plant commensal *P. fluorescens* Pf-5 (Paulsen et al., 2005), the Arabidopsis and tomato pathogen *P. syringae* pv. *tomato* DC3000 (Buell et al., 2003), the epiphytic pathogen *P. syringae* pv. *syringae* B728a (Feil et al., 2005) and the bean pathogen *P. syringae* pv. *phaseolicola* 1448A (Joardar et al., 2005). As expected, *P. fluorescens* Pf-5 lacks a number of virulence factors found in plant pathogens. There is no evidence in the *P. fluorescens* Pf-5 genome for the biosynthesis of known *P. syringae* phytotoxins or enzymes associated with degradation of plant cell walls and cell wall components. In addition, no evidence was found for a type III protein secretion system, frequently found in bacterial pathogens of animals and plants (Paulsen et al., 2005).

The *P. syringae* pv. *tomato* DC3000 genome is composed of one circular chromosome of 6,397,126 bp and two plasmids of 73,661 bp and 67,473 bp. DC3000 encodes a wide range of proteins that are implicated in virulence. Buell et al. (2003) identified 298 genes (5% of the total) in the virulence category. Functional analysis of the DC3000 genome has yielded genes for 31 effectors and 7 additional proteins secreted by the Hrp system that likely have a role in translocating effectors across the plant cell wall and plasma membrane. Genes for the phytotoxin coronatine are chromosomally encoded in DC3000. There is no evidence in de DC3000 genome for the biosynthesis of any of the known *P. syringae* lipodepsinonapeptide phytotoxins. DC3000 contains two genes *iaaH* and *iaaM* required for IAA production. All of the genes required for alginate biosynthesis in P. aeruginosa are present in DC3000. Three genes encoding levansucrases, required for the biosynthesis of the polysaccharide levan, were also identified in DC3000. Genes encoding cell-wall-degrading enzymes are present in DC3000 and include a pectin lyase, a polygalacturonase, and three enzymes predicted to have cellulolytic activity. DC3000 lacks a gene encoding the outer-membrane ice nucleation protein.

P. *syringae* pv. *syringae* B728a is distinct from DC3000 because it exhibits a very pronounced epiphytic phase on plants, while DC3000 is a poor colonizer of the exterior of plants and may be considered as an "endophyte" (Feil et al., 2005). P. *syringae* pv. *syringae* is composed of one circular chromosome of 6,093,698 bp harboring 5,127 genes. *P. syringae* pv. *syringae* B728a has 27 type III secretion effectors, five of which are not found in DC3000. B728a is known to synthesize two syringopeptins and syringostatin and gene clusters for both phytotoxins as well as a gene encoding an ABC transporter for export of boh metabolites are present in the genome of B728a. P. *syringae* pv. *syringae* strains are also capable of producing a family of peptide derivatives called syringolins. Syringolins have no known impact on the interaction of bacteria with their host plants, but they are recognized by nonhost plants, where these peptides activate defense-related genes and induce resistance to fungal pathogens. Orthologs of the genes participating in biosynthesis and export of syringolin A are present in the B728a genome. B728a has an operon for the biosynthesis of IAA, which includes *iaaM* (Psyr1536) and *iaaH* (Psyr1537).

P. *syringae* pv. *phaseolicola* 1448A encodes 5,353 open reading frames on one circular chromosome (5,928,787 bp) and two plasmids (131,950 and 51,711 bp). Searches of the 1448A genome using the DC3000 virulence ORFs revealed that 81% of the DC3000 virulence ORFs are present in 1448A, including genes for many Hop effectors, secretion pathways I, II and III, and cell wall-degrading enzymes (Joardar et al., 2005).

6. DIAGNOSIS

A combination of using (semi)selective media, biochemical/nutritional, pathogenicity, and genetic tests is recommended for the precise identification of phytopathogenic pseudomonads. The reader is referred to Lopez et al. (2003) and Alvarez et al. (2004) for a general overview about detection and diagnosis of plant pathogenic bacteria and to Braun-Kiewnick and Sands (2001) for a detailed overview of diagnostic techniques useful for plant pathogenic pseudomonads.

Usually, a preliminary identification can be made by the use of semiselective media in combination with disease symptoms and host of origin. A semiselective media useful for the isolation of Pseudomonas is the iron-limiting modified King B medium on which fluorescent pseudomonads produce their characteristic yellow-green pigment pyoverdine. Pathovars of *P. syringae* usually produce less pigment then saprophytic *Pseudomonas* strains. Soft-rotting pseudomonads such as *P. marginalis* and *P. viridiflava* can be identified by pit production on Crystal Violet Pectate medium. In addition, MP medium is a general purpose agar for detecting pectate lyases. Some selective media for specific *P. syringae* pathovars can be found in Braun-Kiewnick & Sands (2001). LOPAT characters (Leliott et al., 1966) are still very useful for species identification within the fluorescent pseudomonads and include **Levan** production on sucrose medium, **Oxidase** reaction, **Pectolytic** activity on potato slices or pectate gel, **Arginine** dihydrolase activity, and hypersensitive reaction on **Tobacco** leaves. Additional tests can be found in Braun-Kiewnick and Sands (2001). Some characteristics useful for the differentiation of the most important plant pathogenic *Pseudomonas* species can be found in Table 6.

Biolog and Biotype-100 systems can also be used to differentiate *Pseudomonas* species (Grimont et al., 1996). These methods, however, do not allow correct identification at the pathovar level. Other identification methods use a chemotaxonomic approach such as whole-cell fatty acid composition, which is useful for differentiation of major phylogenetic groups (Vancanneyt et al., 1996b) and SDS-PAGE of whole-cell proteins, which yield species-specific protein profiles (Vancanneyt et al., 1996a). These methods, however, do not give differentiation at the pathovar level.

Pathovar identification is more complicated than species identification, since it relies on more tests and host specificity. Braun-Kiewnick and Sands (2001) have listed a series of tests that can be used to distinguish the most important pathovars of *P. syringae* with good accuracy. Identification, however, should always be confirmed by a pathogenicity test. In addition, the use of toxin bioassays as described by Braun-Kiewnick and Sands (2001) can be helpful to differentiate toxin producing pathovars of *P. syringae*. Presumptive pathovar identification can also be based on serological tests by using specific antibodies raised against the lipopolysaccharides of bacterial cell walls. Antibody-based diagnostic kits and reagents are commercially available for various *P. syringae* pathovars including pv. *glycinea*, pv. *lachrymans*, pv. *phaseolicola*, pv. *tomato*, pv. *pisi* and pv. *syringae*.

Molecular techniques are rapidly overtaking serology, enzymology and metabolic analyses for the identification of plant pathogenic bacteria (Louws et al., 1999; Lopez et al., 2003). Most used in taxonomy and detection are the conserved ribosomal genes (Widmer et al., 1998; Miller et al., 2002). Other target sequences are situated in non-coding regions such as the ribosomal spacers, short repetitive regions, insertion sequences or sequences with unknown function such as determined by SCAR. Using rep-PCR primers, several species and/or pathovars can be identified (Louws et al., 1994; Manceau and Horyais, 1997; Weingart and Volksch, 1997b). In some cases plasmid-based sequences are targeted (Takahashi et al., 1996). When genes involved in pathogenicity are known, species or pathovar species sequences can be determined. Several PCR-based detection techniques are based on *hrp* gene sequences and have been developed for pathogens such as *P. syringae* pv *tomato* (Zaccardelli et al., 2005), *P. syringae* pv. *papulans* (Kerkoud et al., 2002), and *P. avellanae* (Loreti & Gallelli, 2002). DNA-analysis of toxin genes can be helpful to differentiate pathovars of *P. syringae*. Some PCR primers that can be used to detect toxin genes are listed in Braun-Kiewnick and Sands

(2001). PCR detection techniques have been developed for pathogens that produce coronatine (Ullrich et al., 1993; Bereswill et al., 1994; Zhao et al., 2002), lipodepsinonapeptides (Sorensen et al., 1998; Bultreys and Gheysen, 1999), phaseolotoxin (Prosen et al., 1993; Schaad et al., 1995), tabtoxin (Lydon and Patterson, 2001) and tagetitoxin (Kong et al., 2004). In some cases, however, non-toxigenic strains that escape detection based on toxin genes, can cause disease as has been reported for *P. syringae* pv. *phaseolicola* (Rico et al., 2003). Production of indole acetic acid (IAA) is useful for identification of the gall producing pseudomonad *P. savastonoi* and can be identified by serological or molecular techniques (Braun-Kiewnick and Sands, 2001; Penyalver et al., 2000).

Table 6. Differentiation of some important plant pathogenic *Pseudomonas* species[d]

	PHB[a]	Levan	oxidase	pectolytic activity	arginine dihydrolase	fluorescent pigment	Growth at 37°C
P. syringae	-	+[b]	-	-	-	+	-
P. savastanoi	-	-	-	V[c]	-	+	+
P. viridiflava	-	-	-	+	-	+	-
P. cichorii	-	-	+	-	-	+	-
P. marginalis	-	+	+	+	+	+	-
P. corrugata	+	+	+	-	-	-	+
P. agarici	-	-	+	-	-	+	-

+, 80% or more strains positive; -, 80% or more strains negative

[a] Poly β hydroxybutyrate

[b] pathovars *delphinii, papulans* and *passiflorae* are negative

[c] between 21 – 79% of strains positive

d information based on Braun-Kiewnick and Sands (2001) and Smith et al. (1988)

6.1. Pyoverdins

A common characteristic of almost all phytopathogenic pseudomonads is the production of Fe(III)-chelating siderophores, called pyoverdins that are fluorescent under UV light and are typically produced on iron-limiting media such as King's Medium B (King et al., 1954). The only exceptions are the species *P. corrugata,* some strains of *P. amygdali*, *P. cannabina*, *P. meliae*, and *P. fuscovaginae*, and strains of the *P. syringae* pathovars *persicae, morsprunorum, sesami,* and *garcae* (Bultreys et al., 2003). Pyoverdins have a quinoline chromophore, responsible for the colour of the molecule that is bound to a peptide chain and to a dicarboxylic acid or to a dicarboxylic amide. Pyoverdins can be useful in systematics and identification because of the variation found in the peptide part of the molecule. Pyoverdins typically contain three iron-binding ligands; one ligand is located in a catechol moiety in the chromophore, while the other two are located in the peptide chain and are hydroxamic acids derived from ornithine or β-hydroxyaspartic acid. Interestingly, more than 40 pyoverdin peptide chain compositions have been identified in the group containing the arginine dihydrolase-positive, saprophytic or opportunistic animal-pathogenic fluorescent *Pseudomonas* species, but only one composition has been found in the group containing the arginine dihydrolyase-negative phytopathogenic fluorescent *Pseudomonas* species. Arginine dihydrolyase-negative, phytopathogenic, fluorescent Pseudomonas species produce atypical pyoverdins in which two β-hydroxyaspartic acid residues and no derivates of ornithine are involved in iron chelation. Bultreys et al. (2003) have shown that within the group of arginine dihydrolase negative pseudomonads, the oxidase negative species *P. viriflava*, *P. ficuserectae* and the 38 pathovars of *P. syringae* tested, produce the same atypical pyoverdin. The oxidase positive species *P. cichorii* produces a similar atypical pyoverdin that contains a glycine instead of a serine. The more distantly related species *P. asplenii* and *P. fuscovaginae* both produced a less similar atypical pyoverdin. In contrast arginine dihydrolase positive species such as *P. fluorescens*, *P. putida*, *P. chlororaphis*, *P. marginalis* and *P. agarici* produce different typical pyoverdins.

7. CONTROL

7.1. Antibiotics

Antibiotic use in plant agriculture has recently been reviewed by McManus et al. (2002). The use of antibiotics for the treatment of bacterial diseases on plants is modest relative to applications in human and veterinary medicine. Because they are relatively expensive, antibiotics are used primarily on high-value fruit and vegetable crops and ornamental plants. Streptomycin, an aminoglycoside antibiotic, has been the major antibiotic used on plants in the USA. In Europe, streptomycin is either not permitted, only used on an emergency basis, or used regularly, depending on the country. Streptomycin is used to control various pathovars of *Pseudomonas syringae*, which cause fruit-spotting or blossom-blast symptoms on apple, pear and related landscape trees. On tobacco streptomycin is used to control wildfire, caused by *Pseudomonas syringae* pv. *tabaci*. Another *Pseudomonas* pathogen that is targeted is *P. cichorii* on celery, where it causes bacterial blight (McManus et al., 2002). Oxytetracycline, a tetracycline antibiotic and gentamcyin, an aminoglycoside antibiotic, are used to control *Pseudomonas* spp. on several vegetable crops in Latin American countries.

The emergence of streptomycin-resistent plant pathogens has complicated the control of bacterial diseases of plants. Resistance to streptomycine has been reported in *P. cichorii*, *P. syringae* pv. *lachrymans*, *P. syringae* pv. *papulans* and *P. syringae* pv. *syringae* (McManus et al., 2002). Resistance to streptomycin in *Pseudomonas* bacteria is plasmid/transposon determined. The linked *strA-strB* genes that encode streptomycin-inactivating phosphotransferases are located on variants of transposon Tn5393 which are present in *P. syringae* pv. *syringae*. The streptomycin resistance transposon Tn5393a, which carries a *strA-strB* determinant

(Psyr2669-2670) is found in the *P. syringae* pv. *syringae* B728a genome (Feil et al., 2005).

7.2. Copper-based fungicides

Since the use of antibiotics is restricted in most European countries, copper-based fungicides are the only effective compounds available to the farmer to control bacterial plant diseases. Copper-based fungicides such as Bordeaux mixtures are used extensively to control bacterial pathogens on fruit trees such as *P. syringae* pv. *syringae* and *P. syringae* pv. *morsprunorum* on stone fruit trees. The use of copper, however, has several disadvantages. Phytotoxicity can occur and resistance to copper develops rapidly in bacteria. Hwang et al. (2005) have recently shown that most *P. syringae* strains are copper resistant. Copper resistance genes, including the *copABCD* operon and a *copRS* two-component regulatory system are present in the genome of *P. syringae* pv. *syringae* B728a (Feil et al., 2005). These proteins appear to be 92-96% identical to plasmid-encoded CopABCDRS proteins found in other strains of *P. syringae.*

7.3. Plant activators

Plant activators such as 1,2,3-benzothiadiazole (or acibenzolar-S-methyl; also known as Actigard or Bion) and probenazole (Yoshioka et al., 2001; Nakashita et al., 2002; which induce systemic resistance in plants (Sticher et al., 1997; Vallad and Goodman, 2004) can be used to control bacterial leaf pathogens. Louws et al. (2001) have shown that Acibenzolar-S-methyl can be integrated as a viable alternative to copper-based bactericides for field management of bacterial speck, caused by *P. syringae* pv. *tomato*, particularly where copper-resistant populations predominate. Actigard was also used to control P. *syringae* pv. *tabaci* on tobacco in field trials (Cole, 1999).

7.4. Seed treatment

Various pathovars of *P. syringae* are seedborne including *P.s.* pv. *coronafaciens* on cereals, *P.s.* pv. *glycinea* on soybean, *P.s.* pv. *lachrymans* on cucurbits, *P.s.* pv. *maculicola* on brassicas, *P.s.* pv. *phaseolicola* on bean, *P.s.* pv. *pisi* on pea, *P.s.* pv. *porri* on leek, *P.s.* pv. *tabaci* on tobacco, and *P.s.* pv. *tomato* on tomato (Smith et al., 1988). The first consideration in controlling these pathogens is to obtain pathogen-free seed. This can be achieved by seed production in arid regions, seed certification by serological or molecular techniques, chemical treatment of seeds with antibiotics or copper-based compounds or heat treatment of seeds (Kritzman, 1993; Bashan and de Bashan, 2002).

7.5. Biological control

Perhaps the best known example of biological control against plant pathogenic bacteria, including plant pathogenic Pseudomonads, is the use of ice nucleation-deficient deletion mutants of *P. syringae* and *P. fluorescens* to prevent or reduce the growth of frost-forming bacteria on leaves and blossoms (Lindemann and Suslow, 1987; Wilson and Lindow, 1993, Skirvin et al., 2000, Lindow & Brandl, 2003). This research has led to commercial products such as Frostban that can be used on fruit crops, almond, potato, and tomato crops.

There are various examples of plant growth promoting rhizobacteria (PGPR) such as *Pseudomonas* and *Bacillus* spp. (Kloepper et al., 2004) that can control leaf pathogens including pathovars of *P. syringae* via induced systemic resistance (van Loon et al., 1998, Vallad and Goodman, 2004, Meziane et al., 2005). Mixtures of PGPRs, mainly *Bacillus* strains, have been used in field trials to control angular leaf spot caused by *P. syringae* pv. *lachrymans* on cucumber (Raupach and Kloepper, 2000). Induced systemic resistance used in combination with other strategies was effective in controlling bacterial speck on tomato (Bashan and de Bashan, 2002).

Mainly bacterial antagonists, have been tested to control pathovars of *P. syringae* under field conditions. Volksch and May (2001) describe the use of near isogenic or ecologically similar antagonistical strains to target *P. syringae* pv. *glycinea* under field conditions. Strains of *Pantoea agglomerans* suppressed the development of basal kernel blight of barley, caused by *Pseudomonas syringae* pv. *syringae*, Under field conditions, 45 to 74% kernel blight disease reduction was observed (Braun-Kiewnick et al., 2000). A non-pathogenic *P. syringae* strain gave some control in field trials at various locations in the USA and Canada against bacterial speck (Wilson et al., 2002).

7.6. Genetic resistance

Host-specific *P. syringae* pathovars show a typical gene-for-gene interaction with their host and resistance against them is generally mediated by major resistance genes. Breeding programmes and tolerant or resistant host cultivars have been developed for various economic important pathovars of *P. syringae* including pv. *morsprunorum* (Garrett, 1979; Santi et al., 2004), pv. *phaseolicola* (Taylor et al., 1978; Zaiter and Coyne, 1984), pv. *pisi* (Taylor et al., 1989), pv. *tabaci*, pv. *tomato* and pv. *glycinea* (Smith et al.,1988). Resistance genes against *P. syringae* pathovars have been mapped or cloned in tomato (Pedley and Martin, 2003), bean (Ariyarathne et al., 1999), pea (Hunter et al., 2001) and soybean (Ashfield et al., 2003). In addition, various resistance genes against *P. syringae* pathovars have been cloned in Arabidopsis (see Hammond-Kosack and Parker, 2003 for an overview).

7.7. Cultural practices

Free water on susceptible leaves and optimal temperature for bacterial growth are the best combination for promoting disease by plant bacteria. It is possible to reduce bacterial diseases in greenhouses by controlling the environment, by maintaining low relative humidity values using periodic aeration of the greenhouse and drip irrigation, and by holding suboptimal temperatures for pathogen proliferation. In Israel, farmers have been able to reduce damage caused by *P. syringae* pv. *lachrymans* on cucumber and other vegetable crops by using rounded greenhouse structures made of plastic treated with anticondense chemicals. These plastics avoid that water drops fall on leaves (Okon, 1990).

In general disease incidence can be lowered by avoiding excess use of nitrogen fertilizers. Other sensible practices are crop rotation, disinfection of pruning tools and destruction of possibly affected host debris (Okon, 1990; Smith et al., 1988).

7.8. Integrated control

Control of plant pathogenic bacteria is difficult and there is not one strategy that is 100% effective. However, sanitary measures, combined with cultural, chemical and/or biological strategies may lead to satisfactory disease control (Bashan and de Bashan, 2002).

8. REFERENCES

Alvarez, A.M. (2004). Integrated approaches for detection of plant pathogenic bacteria and diagnosis of bacterial diseases. *Annual Review of Phytopathology*, 42, 339-366.

Ariyarathne, H.M. Coyne, D.P., Jung, G., Skroch, P.W., Vidaver, A.K., Steadman, J.R., Miklas, P.N., & Bassett, M.J. (1999). Molecular mapping of disease resistance genes for halo blight, common bacterial blight, and bean common mosaic virus in a segregating population of common bean. *Journal of the American Society for Horticultural Science*, 124, 654-662.

Ashfield, T., Bocian, A., Held, D., Henk, A.D., Marek, L.F., Danesh, D., Penuela, S., Meksem, K., Lightfoot, D.A., Young, N.D., Shoemaker, R.C., & Innes, R.W. (2003). Genetic and physical localization of the soybean Rpg1-b disease resistance gene reveals a complex locus containing several tightly linked families of NBS-LRR genes. *Molecular Plant-Microbe Interactions*, 16, 817-826.

Barta, T.M., & Willis, D.K. (2005). Biological and molecular evidence that *Pseudomonas syringae* pathovars *coronafaciens, striafaciens* and *garcae* are likely the same pathovar. *Journal of Phytopathology*, 153, 492-499.

Bashan, Y., & de Bashan, L.E. (2002). Reduction of bacterial speck (*Pseudomonas syringae* pv. *tomato*) of tomato by combined treatments of plant growth-promoting bacterium, Azospirillum brasilense, streptomycin sulfate, and chemo-thermal seed treatment. *European Journal of Plant Pathology*, 108, 821-829.

Bauer, D.W. & Collmer, A. (1997). Molecular cloning, characterization, and mutagenesis of a *pel* gene from *Pseudomonas syringae* pv. *lachrymans* encoding a member of the *Erwinia chrysanthemi* PelADE family of pectate lyases. *Molecular Plant Microbe Interactions,* 10, 369-379.

Bender, C.L., Alarcon-Chaidez, F., & Gross, D.C. (1999). *Pseudomonas syringae* phytotoxins: mode of action, regulation, and biosynthesis by peptide and polyketide synthetases. *Microbiology and Molecular Biology Reviews*, 63, 266-292.

Bereswill, S., Bugert, P., Volksch, B., Ullrich, M., Bender, C.L., & Geider, K. (1994). Identification and relatedness of coronatine-producing *Pseudomonas syringae* pathovars by PCR analysis and sequence determination of the amplification products. *Applied and Environmental Microbiology*, 60, 2924-2930.

Block, A., Schmelz, E., Jones, J.B., & Klee, H. (2005). Coronatine and salicylic acid: the battle between Arabidopsis and *Pseudomonas* for phytohormone control. *Molecular Plant Pathology*, 6, 79-83.

Bradbury, J.F. (1986). *Guide to plant pathogenic bacteria.* Kew, United Kingdom: CAB International Mycological Institute.

Buell, C.R., Joardar, V., Lindeberg, M., Selengut, J., Paulsen, I.T., Gwinn, M.L. et al. (2003). The complete genome sequence of the Arabidopsis and tomato pathogen *Pseudomonas syringae* pv. *tomato* DC3000. *Proceedings of the National Academy of Sciences of the USA*, 100, 10181-10186.

Braun-Kiewnick, A., & Sands, D.C. (2001). Gram-negative bacteria. Pseudomonas. In Schaad, N.W., Jones, J.B. and Chun, W. (Eds.) (pp. 84-120) *Laboratory guide for identification of plant pathogenic bacteria. Third Edition*. APS Press.

Bultreys, A., & Gheysen, I. (1999). Biological and molecular detection of toxic lipodepsipeptide-producing *Pseudomonas syringae* strains and PCR identification in plants. *Applied and Environmental Microbiology*, 1904-1909.

Bultreys, A., Gheysen I., Wathelet, B., Maraite, H., & de Hoffman, E. (2003). High-performance liquid chromatography analyses of pyoverdin siderophores differentiate among phytopathogenic fluorescent *Pseudomonas* species. *Applied and Environmental Microbiology*, 69, 1143-1153.

Catara, V., Sutra, L., Morineau, A., Achouak, W., Christen, R., & Gardan, L. (2002). Phenotypic and genomic evidence for the revision of *Pseudomonas corrugata* and proposal of *Pseudomonas mediterranea* sp. nov. *International Journal of Systematic and Evolutionary Microbiology*, 52, 1749-1758.

Cole, D.L. (1999). The efficacy of acibenzolar-S-methyl, an inducer of systemic acquired resistance, against bacterial and fungal diseases of tobacco. *Crop Protection*, 18, 267-273.

Corsaro, M.M., Evidente, A., Lanzetta, R., Lavermicocca, P., & Molinaro, A. (2001). Structural determination of the phytotoxic mannan exopolysaccharide from *Pseudomonas syringae* pv. *ciccaronei*. *Carbohydrate Research*, 330, 271-277.

Cui, J., Bahrami, A.K. Pringle, E.G., Hernandez-Guzman, G., Bender, C., Pierce, N.E., & Ausubel, F.M. (2005). *Pseudomonas syringae* manipulates systemic plant defenses against pathogens and herbivores. *Proceedings of the National Academy of Sciences of the USA*, 102, 1791-1796.

Denny, T.P. (1995). Involvement of bacterial polysaccharide in plant pathogenesis. *Annual Review of Phytopathology*, 33, 173-197.

Dye, D.W., Bradbury, J.F., Goto, M., Hayward, A.C., Lelliot, R.A., & Schroth, M.N. (1980). International standards for naming pathovars of phytopathogenic bacteria and a list of pathovar names and pathotype strains. *Review of Plant Pathology*, 59, 153-168.

Feil, H., Feil, W.S., Chain, P., Larimer, F., Dibartolo, G., Copeland, A. et al. (2005). Comparison of the complete genome sequences of *Pseudomonas syringae* pv. *syringae* B728a and pv. *tomato* DC3000. *Proceedings of the National Academy of Sciences of the USA*, 102, 11064-11069.

Fett, W.F., & Dunn, M.F. (1989). Exopolysaccharides produced by phytopathogenic *Pseudomonas syringae* pathovars in infected leaves of susceptible hosts. *Plant Physiology*, 89, 5-9.

Gardan L., Bella, P., Meyer, J.M., Christen, R., Rott, P., Achouak, W., & Samson, R. (2002). *Pseudomonas salomonii* sp. nov., pathogenic on garlic, and *Pseudomonas palleroniana* sp. nov., isolated from rice. *International Journal of Systematic and Evolutionary Microbiology*, 52, 2065-2074.

Gardan, L., Shafik, H., Belouin, S., Broch, R., Grimont, F., & Grimont, P.A.D. (1999). DNA relatedness among the pathovars of *Pseudomonas syringae* and description of *Pseudomonas tremae* sp. nov. and *Pseudomonas cannabina* sp. nov. (ex Sutic and Dowson 1959). *International Journal of Systemic Bacteriology*, 49, 469-478.

Garrett, C.M.E. (1979). Screening *Prunus* rootstocks for resistance to bacterial canker, caused by *Pseudomonas morsprunorum*. *Journal of Horticultural Science*, 54, 189-193.

Glickmann, E., Gardan, L., Jacquet, S., Hussain, S., Elasri, M., Petit, A., & Dessaux, Y. (1998). Auxin production is a common feature of most pathovars of *Pseudomonas syringae*. *Molecular Plant Microbe Interactions*, 11, 156-162.

Green, R.L., & Warren, G.J. (1985). Physical and functional repetition in a bacterial ice nucleation gene. *Nature*, 317, 645-648.

Grimont, P.A.D., Vancanneyt, M., Lefevre, M., Vandemeulebroecke, K., Vauterin, L., Brosch, R., Kersters, K., & Grimont, F. (1996). Ability of the biolog and biotype-100 systems to reveal the taxonomic diversity of the Pseudomonads. *Systematic and Applied Microbiology*, 19, 510-527.

Gross, M., & Rudolph, K. (1987). Demonstration of levan and alginate in bean plants (*Phaseolus vulgaris*) infected by *Pseudomonas syringae* pv. *phaseolicola*. *Journal of Phytopathology*, 120, 9-19.

Hammond-Kosack, K.E., & Parker, J.E. (2003). Deciphering plant-pathogen communication: fresh perspectives for molecular resistance breeding. *Current Opinion in Biotechnology*, 14, 177-193.

Hettwer, U., Jaeckel, F.R., Boch, J., Meyer, M., Rudolph, K., & Ullrich, M.S. (1998). Cloning, nucleotide sequence, and expression in *Escherichia coli* of levansucrase genes from the plant pathogens *Pseudomonas syringae* pv. *glycinea* and *P. syringae* pv. *phaseolicola*. *Applied and Environmental Microbiology*, 64, 3180-3187.

Hirano, S., & Upper, C.D. (2000). Bacteria in the leaf ecosystem with emphasis on Pseudomonas *syringae* – a pathogen, ice nucleus, and epiphyte. *Microbiology and Molecular Biology Reviews*, 64, 624-653.

Hunter, P.J., Ellis, N., & Taylor, J.D. (2001). Association of dominant loci for resistance to *Pseudomonas syringae* pv. *pisi* with linkage groups II, VI and VII of *Pisum sativum*. *Theoretical and Applied Genetics*, 103, 129-135.

Hwang, M.S.H., Morgan, R.L., Sarkar, S.F., Wang, P.W., & Guttman, D.S. (2005). Phylogenetic characterization of virulence and resistance phenotypes of *Pseudomonas syringae*. *Applied and Environmental Microbiology*, 71, 5182-5191.

Janse, J.D., Derks, J.H.J., Spit, B.E., 1 and Van Der Tuin, W.R. (1992). Classification of fluorescent soft rot Pseudomonas bacteria, including *P. marginalis* strains, using whole cell fatty acid analysis. *Systematic and Applied Microbiology*, 15, 538-553.

Jin, Q., Thilmony, R., Zwieser-Vollick, J., & He, S.H. (2003). Type III protein secretion in *Pseudomonas syringae*. *Microbes and Infection*, 5, 301-310.

Joardan, V., Lindeberg M., Jackson, R.W., Selengut, J., Dodson, R., Brinkac L.M. et al. (2005). Whole-genome sequence analysis of *Pseudomonas syringae* pv. *phaseolicola* 1448A reveals divergence among pathovars in genes involved in virulence and transposition. *Journal of Bacteriology*, 187, 6488-6498.

Kerkoud, M., Manceau, C., & Paulin, J.P. (2002). Rapid diagnosis of *Pseudomonas syringae* pv. *papulans*, the causal agent of blister spot of apple, by polymerase chain reaction using specifically designed *hrpL* gene primers. *Phytopathology*, 92, 1077-1083.

Kersters, K., Ludwig, W. Vancanneyt, M., De Vos, P., Gillis, M., & Schleifer, K.H. (1996). Recent changes in the classification of the pseudomonads: an overview. *Systematic and Applied Microbiology*, 19(4), 465-477.

King, R.E., Ward, M.K., & Rainey, D.E. (1954). Two simple media for the demonstration of pyocyanin and fluorescein. *Journal of Laboratory Clinical Medicine*, 44, 301-307.

Kloepper, J.W., Ryu, C.M., & Zhang, S.A. (2004). Induced systemic resistance and promotion of plant growth by *Bacillus* spp. *Phytopathology*, 94, 1259-1266.

Lindemann, J., & Suslow, T.W. (1987). Competition between ice-nucleation-active wild type and ice nucleation-deficient deletion mutant strains of *Pseudomonas syringae* and *Pseudomonas fluorescens* biovar I and biological control of frost injury on strawberry blossoms. *Phytopathology*, 77, 882-886.

Kong, H., Patterson, C.D., Zhang, W., et al. (2004). A PCR protocol for the identification of *Pseudomonas syringae* pv. *tagetis* based on genes required for tagetitoxin production. *Biological control,* 30, 83-89.

Kritzman, G. (1993). A chemi-thermal treatment for control of seed-borne bacterial pathogens of tomato. *Phytoparasitica*, 21, 101-109.

Lelliot, R.A., Billing, E., & Hayward, A.C. (1966). A determinative scheme for the fluorescent plant pathogenic pseudomonads. *Journal of Applied Bacteriology,* 29, 470-489.

Liao, C.H., Hung, H.Y., & Chatterjee, A.K. (1988). An extracellular pectate lyase is the pathogenicity factor of the soft-rotting bacterium *Pseudomonas viridiflava. Molecular Plant Microbe Interactions*, 1, 199-206.

Liao, C.H., McCallus, D.E., & Fett, W.F. (1994). Molecular characterization of two gene loci required for production of the key pathogenicity factor pectate lyase in *Pseudomonas viridiflava. Molecular Plant Microbe Interactions,* 7, 391-400.

Liao, C.H., Gaffney, T.D., Bradley, S.P., & Wong, L.J.C. (1996). Cloning of a pectate lyase gene from *Xanthomonas campestris* pv. *malvacearum* and comparison of its sequence relationship with pel genes of soft rot *Erwinia* and Pseudomonas. *Molecular Plant Microbe Interactions,* 9, 14-21.

Lindow, S.E., & Brandl, M.T. (2003). Microbiology of the phyllosphere. *Applied and Environmental Microbiology*, 69, 1875-1883.

Loreti, S., & Gallelli, A. (2002). Rapid and specific detection of virulent *Pseudomonas avellanae* strains by PCR amplification. *European Journal of Plant Pathology*, 108, 237-244.

Louws, F.J., Fulbright, D.W., Stephens, C.T., & de Bruijn, F.J. (1994). Specific genomic fingerprints of phytopathogenic *Xanthomonas* and *Pseudomonas* pathovars and strains generated with repetitive sequences and PCR. *Applied and Environmental Microbiology*, 60, 2286-2295.

Louws, F.J., Rademaker, J..L.W., & de Bruijn, F.J. (1999). The three Ds of PCR-based genomic analysis of phytobacteria: Diversity, detection, and disease diagnosis. *Annual Review of Phytopathology*, 37, 81-125.

Louws, F.J., Wilson, M., Campbell, H.L., Cuppels, D.A., Jones, J.B., Shoemaker, P.B., Sahin, F., & Miller, S.A. (2001). Field control of bacterial spot and bacterial speck of tomato using a plant activator. *Plant Disease*, 85, 481-488.

Lopez, M.M., Bertolini, E., Olmos, A., Caruso, P., Gorris, M.T., Llop, P., Penyalver, R., & Cambra, M. (2003). Innovative tools for detection of plant pathogenic viruses and bacteria. *International Microbiology*, 6, 233-234.

Lydon, J., & Patterson, C.D. (2001). Detection of tabtoxin-producing strains of Pseudomonas *syringae* by PCR. *Letters in Applied Microbiology*, 32, 166-170.

Magro, P., Varvaro, L., Chilosi, G., Avanzo, C., & Balestra, G.M. (1994). Pectolytic enzymes produced by *Pseudomonas syringae* pv. *glycinea. FEMS Microbiology Letters*, 117, 1-5.

Manceau, C., & Horvais, A. (1997). Assessment of genetic diversity among strains of *Pseudomonas syringae* by PCR-restriction fragment length polymorphism analysis of rRNA operons with special emphasis on P. *syringae* pv. tomato. *Applied and Environmental Microbiology*, 63, 498-505.

McManus, P.S., Stockwell, V.O., Sundin, G.W., & Jones, A.L. (2002). Antibiotic use in plant agriculture. *Annual Review of Phytopathology*, 40, 443-465.

Meziane, H., Van der Sluis, I., van Loon, L.C., Hofte, M., Bakker, P.A.H.M. (2005). Determinants of Pseudomonas putida WCS358 involved in inducing systemic resistance in plants. *Molecular Plant Pathology,* 6, 177-185.

Munsch, P., Alatossava, T., Marttinen, N., Meyer, J.M., Christen, R., & Gardan, L. (2002). *Pseudomonas constantinii* sp. nov., another causal agent of brown blotch disease, isolated from cultivated mushroom sporophores in Finland. *International Journal of Systematic and Evolutionary Microbiology*, 52, 1973-1983.

Nakashita, H., Yoshioka, K., Yasuda, M., Nitta, T., Arai, Y., Yoshida, S., & Yamaguchi, I. (2002). Probenazole induces systemic acquired resistance in tobacco through salicylic acid accumulation. *Physiological and Molecular Plant Pathology*, 61, 197-203.

Nomura, K., Melotto, M., & He, S.Y. (2005). Suppression of host defense in compatible plant-Pseudomonas *syringae* interactions. *Current Opinion in Plant Biology*, 8, 361-368.

Oguiza, J.A., & Asensio, A.C. (2005). The VirPphA/AvrPtoB family of type III effectors in *Pseudomonas syringae*. *Research in Microbiology*, 156, 298-303.

Okon, J. (1990). Methods in agronomy to reduce bacterial diseases. In Klement, Z. Rudolph, K., & Sands, D.C. (Eds.) (pp. 301-306). *Methods in Phytobacteriology*. Budapest: Akadémiai Kiadó.

Palleroni, N.J., Kunisawa, R., Contopoulou, R., & Doudoroff, M. (1973). Nucleic acid homologies in the genus *Pseudomonas*. *International Journal of Systematic Bacteriology*, 23, 333-339.

Paulsen, I.T., Press, C.M., Ravel, J., Kobayashi, D.Y., Myers, G.S.A., Mavrodi, D.V. et al. (2005). Complete genome sequence of the plant commensal *Pseudomonas fluorescens* Pf-5. *Nature Biotechnology*, 23, 873-878.

Pecknold, P.C., & Grogan, R.G. (1973). Deoxyribonucleic acid homology groups among phytopathogenic Pseudomonas species. *International Journal of Systematic Bacteriology*, 23, 111-121.

Pedley, K.F., & Martin, G.B. (2003). Molecular basis of Pto-mediated resistance to bacterial speck disease in tomato. *Annual Review of Phytopathology*, 41, 215-243.

Peñaloza-Vazquez, A., Kidambi, S.P., Chakrabarty, A.M., & Bender, C. (1997). Characterization of the alginate biosynthetic gene cluster in *Pseudomonas syringae* pv. *syringae*. *Journal of Bacteriology*, 179, 4464-4472.

Penyalver, R., Garcia, A., Ferrer, A., Bertolini, E., & Lopez, M.M. (2000). Detection of *Pseudomonas savastanoi* pv. *savastanoi* in olive plants by enrichment and PCR. *Applied and Environmental Microbiology*, 66, 2673-2677.

Preston, G. (2000). *Pseudomonas syringae* pv. *tomato*: the right pathogen, of the right plant, at the right time. *Molecular Plant Pathology*, 1, 263-275.

Prosen, D., Hatziloukas, E., Schaad, N.W., & Panopoulos, N.J. (1993). Specific detection of *Pseudomonas syringae* pv. *phaseolicola* DNA in bean seed by polymerase chain reaction-based amplification of a phaseolotoxin gene region. *Phytopathology,* 83, 965-970.

Raupach, G.S., & Kloeper, J.W. (2000). Biocontrol of cucumber diseases in the field by plant growth-promoting rhizobacteria with and without methyl bromide fumigation. *Plant Disease*, 84, 1073-1075.

Rico, A., Lopez, R., Asensio, C., Aizpun, M.T., Asensio-S-Manzanera, M.C., & Murillo, J. (2003). Non-toxigenic strains of *Pseudomonas syringae* pv. *phaseolicola* are a main cause of halo blight of beans in Spain and escape current detection methods. *Phytopathology,* 93, 1553-1559.

Robinette, D., & Matthysse, A.G. (1990). Inhibition by *Agrobacterium tumefaciens* and *Pseudomonas savastanoi* of development of the hypersensitive response elicited by *Pseudomonas syringae* pv. *phaseolicola*. *Journal of Bacteriology*, 172, 5742-5801.

Rudolph, K.W.E., Gross, M., Neugebauer, M., Hokawat, S., Zachowski, A., Wydra, K. et al. (1989). Extracellar polysaccharides as determinants of leaf spot diseases caused by pseudomonads and xanthomonads. In Graniti, A. Durbin, R.D. & Ballio, A. A. (Eds). *Phytotoxins and Plant Pathogenesis* (pp. 177-218). Berlin: Springer.

Santi, F., Russell, K., Menard, M., & Dufour, J. (2004). Screening wild cherry (*Prunus avium*) for resistance to bacterial canker by laboratory and field tests. *Forest Pathology*, 34, 349-362.

Sarkar, S.F., & Guttman, D.S. (2004). Evolution of the core genome of *Pseudomonas syringae*, a highly clonal, endemic plant pathogen. *Applied and Environmental Microbiology*, 70, 1999-2012.

Sato, M., Watanabe, K., Yazawa, M., Takikawa, Y., & Nishiyama, K. (1997). Detection of new ethylene-producing bacteria, *Pseudomonas syringae* pv. *cannabina* and *sesami*, by PCR amplification of genes for the ethylene-forming enzyme. *Phytopathology*, 87, 1192-1196.

Sawada, H., Suzuki, F., Matsuda, I., & Saitou, N. (1999). Phylogenetic analysis of *Pseudomonas syringae* pathovars suggests the horizontal gene transfer of *argK* and the evolutionary stability of *hrp* gene cluster. *Journal of Molecular Evolution*, 49, 627-644.

Schaad, N.W., Cheong, S.S., Tamaki, S., Hatziloukas, E., & Panopoulos, N.J. (1995). A combined biological and enzymatic amplification (Bio-PCR) technique to detect Pseudomonas *syringae* pv. phaseolicola in been seed extracts. *Phytopathology*, 85, 243- 248.

Schaad, N.W., Vidaver, A.K., Lacy, G.H., Rudolph, K., & Jones, J.B. (2000). Evaluation of proposed amended names of several pseudomonads and xanthomonads and recommendations. *Phytopathology*, 90, 208-213.

Scortichini, M., Marchesi, U., & Di Prospero, P. (2002). Genetic relatedness among *Pseudomonas avellanae*, *P. syringae* pv. *theae* and *P.s.* pv. *actinidiae*, and their identification. *European Journal of Plant Pathology*, 108, 269-278.

.Skirvin, R.M., Kohler, E., Steiner, H., Ayers, D., Laughnan, A., Norton, M.A., & Warmund, M. (2000). The use of genetically engineered bacteria to control frost on strawberries and potatoes. Whatever happened to all of that research? *Scientia Horticulturae*, 84, 179-189.

Smith, I.M., Dunez, J., Lelliott, R.A., Phillips, D.H., & Arch, S.A. (Eds.) (1988). *European Handbook of Plant Diseases*. Oxford, UK: Blackwell Scientific Publications.

Sorensen, K.N., Kim, K.H., & Takemoto, J.Y. (1998). PCR detection of cyclic lipodepsinonapeptide-producing *Pseudomonas syringae* pv. *syringae* and similarity of strains. *Applied and Environmental Microbiology*, 64, 226-230.

Sticher, L., Mauch-Mani, B., & Metraux, J.P. (1997). Systemic acquired resistance. *Annual Review of Phytopathology*, 35, 235-270.

Takahashi, Y., Omura, T., Hibino, H., & Sato, M. (1996). Detection and identification of *Pseudomonas syringae* pv. *atropurpurea* by PCR amplification of specific fragments from an indigenous plasmid. *Plant Disease*, 80, 783- 788.

Taylor, J.D., Bevan, J.R., Crute, I.R., & Reader, S.L. (1989). Genetic relationship between races of *Pseudomonas syringae* pv. *pisi* and cultivars of *Pisum sativum*. *Plant Pathology*, 38, 364-375.

Taylor, J.D., Innes, N.L., Dudley, C.L., & Griffiths, W.A. (1978). Sources and inheritance of resistance to halo blight of *Phaseolus* beans. *Annals of Applied Biology*, 90, 101-110.

Ullrich, M., Bereswill, S., Volksch, B., Fritsche, W., & Geider, K. (1993). Molecular characterization of field isolates of *Pseudomonas syringa* pv. *glycinea* differing in coronatine production. *Journal of General Microbiology*, 139, 1927-1937.

Vallad, G.E., & Goodman, R.M. (2004). Systemic acquired resistance and induced systemic resistance in conventional agriculture. *Crop Science*, 44, 1920-1934.

Vancanneyt, M., Torck, U., Dewettinck, D., Vaerewijck, M., & Kersters, K. (1996a). Grouping of Pseudomonads by SDS-PAGE of whole-cell proteins. *Systematic and Applied Microbiology*, 19, 556-568.

Vancanneyt, M., Witt, S., Abraham, W.R., Kersters, K., & Fredrickson, H.L. (1996b). Fatty acid content in whole-cell hydrolysates and phospholipid fractions of Pseudomonads: a taxonomic evaluation. *Systematic and Applied Microbiology*, 19, 528-540.

Van Loon, L.C., Bakker, P.A.H.M., and Pieterse, C.M.J. (1998). Systemic resistence induced by rhizosphere bacteria. *Annual Review of Phytopathology*, 36, 453-483.

Volksch, B., & May, R. (2001). Biological control of *Pseudomonas syringae* pv. *glycinea* by epiphytic bacteria under field conditions. *Microbial Ecology*, 41, 132-139.

Volksch, B., & Weingart, H. (1997). Comparison of ethylene-producing *Pseudomonas syringae* strains isolated from kudzu (*Pueraria lobata*) with *Pseudomonas syringae* pv. *phaseolicola* and *Pseudomonas syringae* pv. *glycinea*. *European Journal of Plant Pathology*, 103, 795-802.

Warren, G., Corotto, L., & Wolber, P. (1986). Conserved repeats in diverged ice nucleation structural genes from two species of *Pseudomonas*. *Nucleic Acids Research*, 14, 8047-8060.

Watanabe, K., Nagahama, K., & Sato, M. (1998). A conjugative plasmid carrying the *efe* gene for the ethylene-forming enzyme isolated from *Pseudomonas syringae* pv. *glycinea*. *Phytopathology*, 88, 1205-1209.

Weingart, H., Ullrich, H., Geider, K., et al. (2001). The role of ethylene production in virulence of *Pseudomonas syringae* pvs. *glycinea* and *phaseolicola*. *Phytopathology*, 91, 511-518.

Weingart, H., & Volksch, B. (1997). Ethylene production by *Pseudomonas syringae* pathovars in vitro and in planta. *Applied and Environmental Microbiology*, 63, 156-161.

Weingart, H., & Volksch, B. (1997). Genetic fingerprinting of Pseudomonas *syringae* pathovars using ERIC-, REP-, and IS50-PCR. *Journal of Phytopathology*, 145, 339-345.

Widmer, F., Seidler, R.J., Gillevet, P.M., Watrud, L.S., & DiGiovanni, G.D. (1998). A highly selective PCR protocol for detecting 16S rRNA genes of the genus *Pseudomonas* (*sensu stricto*) in environmental samples. *Applied and Environmental Microbiology*, 64, 2545-2553.

Wilson, M., & Lindow, S.E. (1993). Release of recombinant microorganisms. *Annual Review of Microbiology*, 47, 913-944.

Wilson, M., Campbell, H.L., Ji, P., Jones, J.B., & Cuppels, D.A. (2002). Biological control of bacterial speck of tomato under field conditions at several locations in North America. *Phytopathology*, 92, 1284-1292.

Yoshioka, K., Nakashita, H., Klessig, D.F., & Yamaguchi, I. (2001). Probenazole induces systemic acquired resistance in Arabidopsis with a novel type of action. *Plant Journal*, 25, 149-157.

Young, J.M., Saddler, G.S., Takikawa, Y., De Boer, S.H., Vauterin, L., Gardan, L., Gvozdyak, R.I., & Stead, D.E. (1996). Names of plant pathogenic bacteria 1864-1995. *Review of Plant Pathology*, 75, 721-763.

Young, J.M., Bull, C.T., De Boer, S.H., Firrao, G., Gardan, L., Saddler, G.E., Stead, D.E., & Takikawa, Y. (2004). Names of Plant Pathogenic Bacteria Published Since 1995. International Society of Plant Pathology.

Yu, J., Peñaloza-Vazquez, A., Chakrabarty, A.M., & Bender C.L. (1999). Involvement of the exopolysaccharide alginate in the virulence and epiphytic fitness of *Pseudomonas syringae* pv. *syringae*. *Molecular Microbiology*, 33, 712-720.

Zaccardelli, M., Spasiano, A., Bazzi, C., & Merighi, M. (2005). Identification and in planta detection of *Pseudomonas syringae* pv. *tomato* using PCR amplification of *hrpZ*(Pst). *European Journal of Plant Pathology,* 111, 85-90.

Zaiter, H.Z., & Coyne, D.P. (1984). Testing inoculation methods and sources of resistance to the halo blight bacterium (*Pseudomonas syringae* pv. *phaseolicola*) in *Phaseolus vulgaris*. *Euphytica*, 33, 133-141.

Zhao, Y.F., Damicone, J.P., & Bender, C.L. (2002). Detection, survival, and sources of inoculum for bacterial diseases of leafy crucifers in Oklahoma. *Plant Disease*, 86, 883-888.

9. AFFILIATIONS

Monica Höfte, *Lab. Phytopathology, Department of Crop Protection, Faculty of Bioscience Engineering, Ghent University, Coupure Links 653, B-9000 Gent, Belgium. Tel. 0032 9 2646017, e-mail: monica.hofte@ugent.be*

Paul De Vos, *Lab. Microbiology, Department of Biochemistry, Physiology and Microbiology, Faculty of Sciences, Ghent University, K.L. Ledeganckstraat 35, B-9000 Gent, Belgium. Tel. 0032 9 2645110, e-mail: paul.devos@ugent.be*

SUJATHA SUBRAMONI, GOPALJEE JHA &
RAMESH V. SONTI

VIRULENCE FUNCTIONS OF XANTHOMONADS

Abstract. Bacteria belonging to the genus Xanthomonas cause diseases on a wide variety of plant species. They elaborate a number of virulence factors to successfully infect and cause disease on their hosts. These include surface polysaccharides, putative adhesins, various secretion systems and their effectors, toxins, functions involved in detoxification of host derived anti-microbial compounds, nutrient acquisition, etc. Over and above these are the regulatory functions that direct the expression of the various virulence factors. In addition, mutations in genes involved in the biosynthesis of certain amino acids and purines cause virulence deficiency, suggesting an inability to source these metabolites from the host. Some genes involved in sugar metabolism have been shown to be required for virulence, pointing towards the importance of these sugars as carbon sources during in planta growth. The genome sequences of several xanthomonads have been determined during the last few years. These studies have uncovered a number of candidate virulence functions whose role in virulence needs to be systematically analysed. A number of species and strain specific xanthomonad genes have also been identified, some of which may have a role in determining host range of the pathogens.

1. INTRODUCTION

The genus Xanthomonas consists of ϒ-proteobacteria that cause disease on at least 124 monocotyledonous and 268 dicotyledonous plant species (Chan & Goodwin, 1999). Although the genus itself has a very broad host range, individual members are, in general, specialized for causing disease on a limited number of taxonomically related hosts. Many xanthomonads are also found associated with plants, as epiphytes or saprophytes, without causing obvious symptoms of disease. A number of specialised virulence factors are employed by xanthomonads to successfully enter into their hosts, multiply inside them and cause disease. In recent years, there has been a lot of progress in our understanding of xanthomonad virulence functions. This chapter is an overview of this field with emphasis on recent studies that have involved the application of molecular genetic techniques.

The xanthomonads are rod shaped, non-spore forming bacteria that are motile by a single polar flagellum (Dowson, 1939; Vauterin et al., 1995). Two characteristic features of most members of this genus are the production of yellow, outer membrane-bound brominated pigments called xanthomonadins and copious amounts of extracellular polysaccharide. These bacteria enter their hosts through natural openings like stomata and hydathodes or through wounds. In general, the xanthomonads that are specialized for entry through the hydathodes colonize the xylem vessels while those that enter through the stomata colonize the intercellular spaces in plant tissues. They employ a combination of virulence factors to gain entry into and attach to host tissues, obtain essential nutrients, and overcome and suppress preformed/induced host defenses. The expression of the virulence factors

S.S. Gnanamanickam (ed.), Plant-Associated Bacteria, 535–571.

is controlled by regulatory functions that monitor the external environment as the bacterium enters the host and multiplies within it's different compartments. These regulatory functions act to ensure that the virulence factors are produced in the right amounts, at the right time and the right place. Eventually, the activity of these various functions leads to loss of host tissue integrity and cell death. Symptoms may range from leaf blight (*X. oryzae* pv. *oryzae*; Xoo, a pathogen of rice), leaf spot (*X. campestris* pv. *vesicatoria*, Xcv, a pathogen of pepper and tomato plants), leaf rot (*X. campestris* pv. *campestris*, Xcc, a pathogen of crucifer plants) to fruit canker (*X. axonopodis* pv. *citri*, Xac, a pathogen of citrus plants).

The last 15 years have seen a remarkable progress in identifying the virulence determinants of plant pathogenic bacteria in general and more specifically of xanthomonads. The completion of the sequencing of the genomes of Xac, Xcc and Xoo as well as the initiation of many genome-wide studies have greatly contributed to this progress (da Silva et al., 2002; Lee et al., 2005; Qian et al., 2005). Studies on virulence factors unique to each bacterium and their role in infection might help to understand distinct features of their pathogenicity and host specificity. This knowledge of virulence factors may ultimately lead to novel methods of disease control in the form of effective and ecofriendly antibacterials as well as disease resistant crops.

2. VIRULENCE DETERMINANTS OF XANTHOMONADS

In the following sections we overview current knowledge about the virulence functions that are elaborated by xanthomonads.

Adhesins

A critical step in host colonization is the ability of the bacterium to adhere to the host cell surface. Bacterial surface components that mediate adhesion are called adhesins. Bacterial adhesins maybe polysaccharide in nature, like extracellular polysaccharide (EPS), but they are often proteinaceous pilus-like fimbriae (Fimbrial adhesins) or other surface proteins (Non-fimbrial adhesins) (Soto & Hultgren, 1999). Although their role in virulence has been well characterized in animal pathogens, very little information is available regarding the role of proteinaceous adhesins in pathogenicity of plant pathogens. In *Erwinia chrysanthemi* (Ech), *hecA*, a protein similar to filamentous haemaglutinin family of adhesins has been reported to be required for optimum virulence and attachment on tobacco leaves, although the target receptor in the host plant is not yet known (Rojas et al., 2002). In the same study it has been demonstrated that *hecA* mutants are also defective in auto aggregation i.e. the capacity of the bacteria to attach to each other.

In Xoo, mutations in the *xadA* gene which encodes Xanthomonas Adhesin like Protein A, result in reduced lesion lengths following infection at low concentrations of inoculum (Ray et al., 2002). The *xadA* mutants also exhibit a reduced efficiency in epiphytic infection. XadA is homologous to non fimbrial adhesins of animal pathogenic bacteria like YadA (Yersinia Adhesin A) of *Yersinia enterocolitica*, a mammalian pathogen. The YadA protein is required for virulence in a mouse model by mediating binding of the bacterium to collagen (El Tahir et al., 2000). An interesting feature of XadA is the presence of particular sequence motifs that are distributed throughout the length of the protein. The presence of such repetitive sequence motifs has been described in other adhesins of animal pathogenic bacteria, including YadA (Hoiczyk et al., 2000). The genome sequence of a Korean strain of Xoo reveals that this bacterium encodes, besides XadA, a paralog called XadB (Xanthomonas Adhesin like Protein B) (Lee et al., 2005). Orthologs of XadA are encoded in the genomes of Xac, Xcc and the closely related *Xylella fastidiosa* (Xf), which causes the citrus variegated chlorosis disease (Simpson et al., 2000; da Silva et al., 2002). The genome sequence of Xcc has a single XadA ortholog while the Xac genome encodes two orthologs of XadA. A readily identifiable difference between the various orthologs and paralogs of XadA lies in the number of sequence repeats, a feature which ultimately results in proteins of different lengths and, possibly, specificities. Nöel et al. (2001) reported that HrpG* (a constitutively active form of HrpG, a regulator of type III secretion components and effectors) down-regulates a Xcv *xadA* ortholog, suggesting that *xadA* may be required during the initial stages of infection, just prior to induction of the type III system.

Besides XadA like proteins, the genomes of Xoo, Xac and Xcc encode homologs of pilus/fimbrial proteins, filamentous haemaglutinin and certain glycine-rich outer membrane proteins that are reported to function as adhesins in animal pathogenic bacteria. Although a role in virulence has not been demonstrated for many of these proteins in xanthomonads, it is conceivable that they are involved in promoting adhesion to plant tissues as well as in autoaggregation. Mutations in *pilB* and *pilC* genes, which are predicted to function in pilus assembly, have been shown to result in reduced virulence of Xcc (Qian et al., 2005).

Extracellular polysaccharide (EPS)

Many gram-negative bacteria have a coating of EPS that is external to the outer membrane but is attached to it. As indicated above, many xanthomonads produce copious amounts of EPS leading to the characteristic mucoid appearance of colonies. The structure of Xcc EPS (commercially known as xanthan gum) consists of a chain of a repeating pentamer, that is made up of 2 molecules each of glucose and mannose with one molecule of glucuronic acid which are in turn modified by acetylation and pyruvylation (Jansson et al., 1975). Hence, the synthesis and transport of EPS would involve many functions and the disruption of any one of them can be expected to result in an EPS deficient phenotype. Tseng et al. (1999) have reported that 8 chromosomal loci may be involved in biosynthesis of xanthan gum. One of these loci is the *gum* operon of Xcc which contains 12 genes, *gumB* to *gumM,* which are implicated in the first stages of biosynthesis and assembly of the pentamer repeats of xanthan gum (Katzen et al., 1996). The *gum* operon of Xoo is similar to Xcc and Xac except for the presence of *gumN*, which encodes a hypothetical protein as the last gene of the operon (Lee et al., 2005).

A mutation in the *gumD* gene of Xcc caused a severe virulence deficiency when inoculated on broccoli seedlings (Chou et al., 1997). Also, non-polar transposon insertion mutations in *gumL*, *gumF*, *gumG*, *gumI* and *gumD* resulted in varying degrees of virulence deficiency on cabbage leaves (Katzen et al., 1998). In planta expression of the Xcc *gum* operon was studied using β-glucuronidase as reporter (Vojnov et al., 2001). They found that bacteria from inoculated turnip leaves exhibited increased expression of *gum* genes at later stages of growth suggesting the requirement of EPS in later phases of infection.

In Xoo, a mTn*5* insertion in the *gumGXo* gene, which encodes an acetyl transferase, leads to EPS and virulence deficiency. Complementation with a genomic clone containing the *gum* operon of Xoo restored the wildtype phenotype for both characteristics (Dharmapuri & Sonti, 1999). Spontaneous EPS mutants arising due to the insertion of an endogenous IS element in the *gumMXo* gene were also found to be virulence deficient (Rajeshwari & Sonti, 2000). A novel genomic locus of Xoo, different from the *gum* cluster was found to be important for production of EPS and lipopolysaccharide (LPS) as well as virulence on rice plants (Dharmapuri et al., 2001). As discussed below, this locus encodes six genes which appear to function in LPS biosynthesis. Mutations in two of these genes, *wxoA* and *wxoB*, affect LPS as well as EPS production. It is unclear how these genes might function in EPS biosynthesis.

The *xanA* and *xanB* genes of Xcc are involved in the biosynthesis of UDP-D-glucose and GDP-D-mannose, which are precursors of xanthan gum (Köplin et al., 1992). Mutations in these genes result in loss of EPS production and also appear to have an affect on LPS biosynthesis. An insertion mutation in the *xanA* gene causes virulence deficiency in Xcc (Qian et al., 2005).

Lipopolysaccharide (LPS)

Gram-negative bacteria have an inner membrane and an outer membrane separated by a periplasmic space. LPS is an important constituent of the outer membrane and is present in the outer leaflet of this membrane. LPS is made of a lipid portion called LipidA, which is embedded in the outer membrane, and a polysaccharide part that is exposed on the bacterial surface. The polysaccharide part is made up of an inner core, outer core and an O-polysaccharide (also called O-antigen). Vörholter et al. (2001) have identified a cluster of 15 Xcc genes called *wxc* genes that are required for LPS biosynthesis. Analysis of the LPS profiles of various *wxc* mutants led to the conclusion that these genes are involved in the following functions; O-antigen biosynthesis, formation of LPS core, modification of nucleotide sugars (the building blocks in LPS synthesis) and translocation of LPS to the outer membrane. Mutations in the *wxcA*, *wxcB*, *wxcC*, *wxcD* and *wxcM* genes were reported to cause loss of virulence on cabbage (Qian et al., 2005). The *rmlABCD* genes of Xcc are also involved in LPS biosynthesis (Köplin et al., 1993) and a mutation in the *rmlA* gene has been shown to cause reduced virulence (Qian et al., 2005).

Interestingly, different gene clusters that are involved/likely to be involved in LPS biosynthesis are present in Xoo and Xac at the locus that is occupied by the *wxc* genes in Xcc. The gene clusters present at this locus in Xcc, Xoo and Xac are found to differ from each other with respect to the number and nature of the encoded genes. For eg. six genes are encoded at this locus in Xoo as opposed to 15 in Xcc (Patil and Sonti, 2004). The genes at this locus in Xoo include *wxoA* (predicted epimerase), *wxoB* and *wxoC* (predicted glycosyl transferases), *wxoD* (predicted acetyl transferase), *wzt* (predicted ATPase) and *wzm* (predicted integral membrane protein), the first four of which are predicted to be involved in LPS biosynthesis and the last two are predicted to be components of an ABC type of bacterial lipopolysaccharide transport system. All six genes have characteristics like atypical G+C content and altered codon usage pattern that are suggestive of inheritance by horizontal gene transfer. A point to note is that although these six genes are present in the vast majority of Xoo strains from India and other countries, they are absent from two Xoo strains, one of which is from India (BXO8, a variant Indian pathotype) and another from Nepal (Nepal 624) (Patil and Sonti, 2004). In place of this gene cluster, these two Xoo strains appear to have a gene cluster that is homologous to the gene cluster that is present at this locus in Xac. This kind of variation that occurs due to gene cluster replacement has been detected earlier in LPS biosynthetic gene clusters of animal pathogenic bacteria, where it has been attributed to help in evasion of the host immune response. The variation in LPS biosynthetic gene clusters in plant pathogenic bacteria may also serve a similar function in evasion of the host defense response.

LPS is considered to be one of the major pathogen derived molecules recognized by the host plants in order to mount defense responses (Dow et al., 2000a). Prior treatment of pepper leaves with LPS (isolated from either Xcc or enteric bacteria like *Escherichia coli*) before inoculation with Xcc or Xcv cultures, suppresses HR, alters expression of defense genes and delays bacterial growth (Newman et al., 1997; Newman et al., 2002). These results suggest that prior treatment of plants with LPS may induce defense responses and limit the spread of the pathogen without visible HR. The LPS from Xcc has also been shown to induce an oxidative burst reaction in cell-suspension cultures of the tobacco plant, which is a non-host for this pathogen (Meyer et al., 2001). The oxidative burst is considered to be an important part of the plant defense response against microbial pathogens (Wojtasek, 1997).

Interest in LPS also stems from the fact that LPS modulates permeability of the bacterial membrane. Alterations in LPS can therefore influence susceptibility to antimicrobial compounds. An Xcc mutant in one of the LPS biosynthetic genes (*rfaX*) was reported to be non-pathogenic, suggesting a role for LPS in pathogenicity (Dow et al., 1995). As mentioned above, mutations in the *wxoA* and *wxoB* genes of Xoo, lead to a drastic reduction in virulence on rice. In

Xanthomonas campestris pv. *citromelo* (causes leaf spot disease of citrus plants), mutation in the *opsX* gene (that is homologous to the *rfaQ* gene involved in the assembly of LPS core in *E. coli*) resulted in altered EPS and LPS levels and a reduction in pathogenicity on citrus but not on bean plants (Kingsley et al., 1993). It is not clear why mutation in the *opsX* gene would result in loss of virulence on citrus but not on bean plants. It is possible that the change in LPS composition would lead to increased susceptibility against antimicrobial compounds produced by citrus plants but not to those produced by bean plants.

These various studies indicate that LPS has a dual role in the plant-microbe interaction. It is an inducer of plant defense responses and at the same time appears to be involved in promoting virulence. The latter role may stem from its function as a barrier against antimicrobial compounds that are produced by the host. Infiltration of purified Xcv LPS into leaves of pepper plants has been shown to result in increased callose deposition in plant cell walls, a defense response that might result in strengthening of cell walls against microbial attack (Keshavarzi et al., 2004). Co-infiltration of pepper leaves with Xcv and purified LPS has been shown to result in suppression of callose deposition. These results suggest that Xcv, and possibly other xanthomonads, are capable of suppressing the plant defense responses that are induced by LPS. The manner in which these bacteria bring about the suppression of the LPS induced defense response is unclear. As indicated below, one possibility is that the suppression may be brought about by effectors secreted through the bacterial Type III secretion system.

Flagellar apparatus

Flagella are filamentous structures attached to bacterial cell surface that help in motility. Many bacteria sense the presence of certain chemicals in the environment and exhibit chemotactic movement with the aid of their flagella. Motility is considered to be an important factor for pathogenicity in several bacteria (Josenhans & Suerbaum, 2002).

Analysis of the genome sequences of Xac, Xcc and Xoo reveals the presence of a complete set of genes required for flagellar apparatus synthesis, regulation and chemotaxis. In Xac and Xcc these genes are present in four gene clusters that total to a length of 150 Kb in their genomes (da Silva et al., 2002). On the other hand, in Xoo these genes are organized into two clusters spanning 62 Kb (Lee et al., 2005). All the three bacteria have multiple nearly identical copies (ten in Xac, eight in Xcc and two in Xoo) of a methyl-accepting chemotaxis protein gene (*mcp*). In addition to this, both Xac and Xcc have several paralogs of *mcp* distributed randomly throughout their genomes. This kind of gene content suggests a very important role for motility genes in the biology, and possibly the virulence, of xanthomonads. Compounds in rice hydathodal exudates serve as chemoattractants for Xoo and it has been suggested that motility is important for bacterial entry into rice leaves through the hydathodes (Feng & Guo, 1975; Shen et al., 2001).

Role of flagella in virulence has been investigated in Xoo (Shen et al., 2001). Studies on the genes coding for flagellar apparatus were initiated when FlhF (a flagellar component) was identified as an interacting protein with the kinase domain of rice Xa21 resistance protein as bait (Shen et al., 2001). Further characterization of the genomic region encoding the FlhF protein revealed the presence of six orfs namely, *flhB*, *flhA*, *flhF*, *fliA* as well as orfs encoding two proteins of unknown function in an operon. Insertional inactivation of Xoo *flhF* significantly reduced motility but had no effect on virulence or avirulence phenotype. In a yeast two hybrid screen, FlhF was found to interact with itself, Xa21 and also with the Xoo homolog of the *Pseudomonas aeruginosa* PilL protein. The significance of these observations is not yet known.

Pigment

Members of the Xanthomonas group of bacteria produce yellow, brominated, outer membrane located aryl polyene pigments called xanthomonadins. A cluster of genes (*igA* to *pigG*) that are required for pigment biosynthesis was first cloned from Xcc (Poplawsky et al., 1993). The pigment deficient phenotype caused by mutations in one of the genes, *pigB*, appears to be due to loss of production of a pheromone called DF (diffusible factor). This factor is required for normal levels of pigmentation as well as EPS (Poplawsky & Chun, 1998). Mutations in the other *pig* genes affect pigmentation without affecting EPS. The *pigB* mutant was deficient for survival on leaf surfaces and was also severely reduced for infection of cauliflower leaves, when infection was through the hydathodes. The *pigB* mutant was as virulent as the wild-type strain following wound inoculation indicating that DF is required for epiphytic infection but not required for growth within the host plant. Mutations in all the *pig* genes, *pigA* to *pigG,* were found to be virulence proficient following wound inoculation of cauliflower, indicating that xanthomonadin is not required for growth within the host plant. The *pigC* and *pigG* mutants were hypersensitive to UV irradiation in the presence of the photosensitizer, toluidine blue (Poplawsky et al., 2000). Pigment deficient mutants of Xoo and *Xanthomonas juglandis*, a walnut pathogen have also been shown to be hypersensitive to photooxidative damage in the presence of light and photosensitizers (Rajagopal et al., 1997; Jenkins & Starr, 1982). In addition, methanolic extracts of wild type Xoo but not pigment deficient mutants were shown to protect lipids against oxidative damage (Rajagopal et al., 1997). These observations indicate that xanthomonadins can provide protection against oxidative damage, under *in vivo* and *in vitro* conditions. The *pigC* mutant of Xcc was further shown to be defective for survival on cauliflower leaves only under conditions of high light intensities and also showed a small but significant reduction in the efficiency of infection through hydathodes (Poplawsky et al., 2000). This suggests that xanthomonadins have a role in promoting survival of bacteria on leaf surfaces, prior to leaf entry, by providing protection against photooxidative damage.

The Xcc *pig* gene cluster was used as a probe to identify a homologous gene cluster from Xoo. Sequence characterization of the Xoo gene cluster indicated that these genes were homologous to functions involved in polyketide biosynthesis (Goel et al., 2002). Mutations in an unlinked Xoo gene, *aroE*, which encodes shikimate dehydrogenase, were found to be deficient for aromatic amino acid biosynthesis as well as for pigment production (Goel et al., 2001). These results suggest that the aryl ring in xanthomonadin may be derived from the aromatic amino acid biosynthetic pathway and that the polyene tail is derived from a polyketide pathway. A putative membrane transporter, with homology to multidrug efflux proteins, is encoded in the Xoo *pig* cluster and has been shown to be required for outermembrane localization of xanthomonadin (Goel et al., 2002). The *aroE* mutant of Xoo was virulence deficient, but virulence deficiency was attributed to the aromatic amino acid auxotrophy, as the vast majority of Pig$^-$ mutants of Xoo are virulence proficient and prototrophs (Goel et al., 2001).

Type I secretion and AvrXa21 activity

The rice resistance gene *Xa21*, which encodes a receptor kinase, provides resistance against a number of Xoo strains (Song et al., 1995; Wang et al., 1996). However, the nature of the *avrXa21* activity is not known. The Xoo strain PXO99 is unable to cause disease on rice lines containing the *Xa21* resistance gene. In order to identify the *avrXa21* activity, Tn*5* induced mutants of PXO99 were isolated that are able to cause disease on *Xa21* containing rice lines (Shen et al., 2002). The genes that are mutated in these strains were called *rax* genes (required for *avr Xa21* activity). Two genes that were identified in this manner, called *raxP* and *raxQ,* are located among a cluster of sulphur assimilation genes. In another approach, clones from a genomic library of PXO99 were mobilized into another Xoo strain called DY89031 which lacks *avrXa21* activity and causes disease on rice plants carrying *Xa21*. A genomic clone of PXO99 was identified that prevented DY89031 from causing disease on *Xa21* containing rice lines. The Xoo strain carrying this genomic clone retains the ability to cause disease on the rice line IR24, which lacks *Xa21*. Four genes namely *raxST, raxA, raxB* and *raxC* were identified as being required for *avrXa21* activity in this manner (da Silva et al., 2004). These genes encode proteins that are involved in sulfation and type I secretion in bacteria.

In gram –ve bacteria, type I secretion system (T1S) is involved in the export of proteins across both the bacterial membranes to the extracellular environment (Holland et al., 2005). The T1S is composed of an ABC transporter (which provides energy for transport by ATP hydrolysis and forms a channel across the inner membrane), a Membrane Fusion Protein (MFP) (which spans the periplasm and extends upto the outer membrane) and the outer membrane protein, usually a protein called TolC (which completes the channel and forms the point of exit of the type 1 substrate). Three of the Xoo *rax* genes identified namely *raxA*, *raxB* and *raxC* encode components of a T1S. RaxA and RaxB exhibit similarity to MFP and ABC transporter components of the T1S, respectively, while RaxC is similar to the *E. coli* outer membrane protein TolC (da Silva et al., 2004). The *raxA* and *raxB* genes are absent from the genomes of Xac and Xcc although a number of genes encoding components of several type 1 secretion systems are present (da Silva et al., 2002). Marker exchange mutants in PXO99 *raxA*, *raxB* and *raxC* genes result in loss of *avrXa21* activity and lead to a 100-fold increase in bacterial survival within *Xa21* containing plants. Mutations in these genes do not affect virulence on the IR24 rice line (da Silva et al., 2004).

Of the other *rax* genes, *raxP* encodes an ATP sulfurylase, *raxQ* encodes an adenosine phosphosulfate kinase and *raxST* shares similarity to sulfotransferases (Shen et al., 2002; da Silva et al. 2004). RaxP and RaxQ are similar to NodP and NodQ of *Sinorhizobium meliloti*, which together with the sulfotransferase NodH modify the Nod factor that is specifically recognized by its host plant, alfafa (Ehrhardt et al., 1995; Roche et al., 1991). In Xoo, these three genes may be functioning in the modification of the *avrXa21* molecule by sulfation, which may be critical for recognition by the host plant.

raxSTAB genes are suggested to be part of an operon. Downstream of the *raxSTAB* operon, two genes namely, *raxR* (which shows similarity to response regulators) and *raxH* (which shows similarity to histidine protein kinases) were identified (Burdman et al., 2004). Mutations in either of the two genes led to a reduction in *avrXa21* activity. The *raxR* and *raxH mutations* also resulted in reduced expression of the *raxSTAB* genes (Burdman et al., 2004). Orthologs of *raxR* and *raxH* are found in Xac and Xcc, although they lack the *raxSTAB* genes (da Silva et al., 2002).

These studies show that a bacterial T1S and modification of the predicted AvrXa21 molecule by sulfation are necessary for *avrXa21* activity and recognition by the *Xa21* carrying rice plants. Also, the expression of AvrXa21 activity is regulated by a two component regulatory system, which possibly also regulates the expression of other genes in Xoo and other xanthomonads.

Type II protein secretion system and its effectors

The bacterial type II protein secretion system (T2S) mediates the secretion of proteins, from the periplasmic space, across the outer membrane to the extracellular surroundings (Pugsley, 1993). These proteins first cross the inner membrane and reach the periplasm, either through the sec pathway (GSP) or the Twin Arginine pathway, depending on the nature of the signal peptide in the preprotein (Pugsley, 1993; Voulhoux et al., 2001). In xanthomonads, the components of the T2S are encoded by a cluster of about 11-12 genes, depending on the nature of the system and the organism (Jha et al., 2005).

Genome sequence analysis has revealed that two complete T2S gene clusters called *xps* and *xcs* are encoded in the genomes of Xcc and Xac (da Silva et al., 2002). These two gene clusters do not exhibit nucleotide sequence similarity with each other, indicating that they are not paralogs. The *xps* and *xcs* gene clusters of Xcc exhibit a high degree of nucleotide similarity to their counterparts in Xac, indicating that they are orthologs. Interestingly, the genome sequence of Xoo reveals the presence of an *xps* ortholog and the absence of an ortholog of *xcs* (Lee et al., 2005). It is not clear why a single T2S is sufficient for Xoo while Xcc and Xac apparently need two different T2Ss. Mutations in the *xps* gene clusters of Xcc and Xoo result in virulence deficiency (Dow et al., 1987; Hu et al., 1992; Ray et al., 2000). Mutations have not yet been isolated in the *xps* and *xcs* gene clusters of Xac and in the *xcs* gene cluster of Xcc.

A number of T2S secreted proteins have been identified in Xcc and Xoo (Table 1). The interesting feature is that the vast majority, if not all, of these proteins are involved in degrading different components of plant cell walls. These proteins include cellulases, pectinases, xylanases, proteases, lipases, etc. What is the role of these individual proteins in virulence of xanthomonads? A protease deficient mutant of Xcc exhibits reduced virulence when infection is through cut vein endings of turnip but not in other tests of pathogenicity like infiltration into mature leaves or infection of seedlings (Dow et al., 1990). Polygalacturonase isoenzyme I mutants of Xcc are virulence proficient while a slight delay in development of symptoms is associated with mutations in the gene for the major secreted endoglucanase of this bacterium (Dow et al., 1989; Gough et al., 1988). Mutations in genes for either a secreted lipase/esterase (LipA) or xylanase (XynB) of Xoo have slight effects on virulence, while *lipA xynB* double mutants have much

Table 1. T2S secreted proteins in *Xanthomonas*

Name of the Bacterium	Known effectors	References
Xanthomonas campestris pv. *campestris*	Polygalacturonate lyases	1,2,3
	Cellulase/Endoglucanase	1,2,3
	Proteases	1,2,3
	α- Amylase	1,2,3
Xanthomonas oryzae pv. *oryzae*	Xylanase	4
	Cellulase/Endoglucanase	5,6,7
	Putative cysteine protease	5
	Putative cellobiosidase	7
	Lipase/Esterase	8

1 = Dow et al., 1987; 2 = Dow et al., 1989; 3 = Hu et al., 1992; 4 = Ray et al., 2000; 5 = Furutani et al., 2004; 6 = Sun et al., 2005; 7 = Gopaljee Jha, R. Rajeshwari and R.V. Sonti (Unpublished data); 8 = Rajeshwari et al., 2005.

more severe effects on virulence (Rajeshwari et al., 2005). A cosmid clone containing the *xynB* gene restores the virulence deficiency of the *lipA xynB* double mutant. Complementation activity is abolished when a mutation is introduced into the *xynB* gene on the cosmid. This indicates that the XynB protein is important for virulence of Xoo. Anti-XynB antibodies were used to show that XynB is indeed expressed during growth of Xoo within rice xylem vessels.

These results indicate that mutations in individual T2S secreted proteins have slight effects on virulence and that mutating more than one T2S secreted protein leads to more severe effects on virulence. This suggests that there is functional redundancy amongst T2S secreted proteins with regard to their role in virulence. The redundancy might arise from the fact that individual T2S secreted proteins are targeting different components of plant cell walls and an inability to degrade one of these components might not seriously affect virulence. Alternatively, the redundancy might also arise from the fact that xanthomonads can produce multiple isoforms of various cell wall degrading enzymes. Genome sequence analysis predicts the presence of eight cellulases, five xylanases and four pectate lyases in Xcc, while seven cellulases, four xylanases and two pectate lyases are predicted to be encoded in the genomes of Xac and Xoo (da Silva et al., 2002; Lee et al., 2005).

Infiltration of rice leaves with purified T2S secreted proteins of Xoo like cellulase, cellobiosidase or LipA results in a browning reaction, accompanied by lignin accumulation, which is akin to the hypersensitive reaction (HR) (Jha et al., 2005). These responses are similar to those that occur when Xoo is infiltrated into leaves of resistant rice cultivars (Reimers & Leach, 1991; Yang et al., 2000). Pre treatment of the mid-veinal regions of rice leaves with any of these three purified T2S secreted proteins results in resistance against subsequent (twenty hours later) infection with Xoo. Co-infiltration of rice leaves with T2S secreted protein and wild-type Xoo results in suppression of the HR like reactions that are induced by the T2S secreted protein. A Xoo mutant that is defective in the type III secretion system (T3S) is unable to suppress the HR like reactions that are induced by T2S effectors. Taken together, these results indicate that Xoo T2S secreted proteins induce very potent host defense responses that are suppressed by proteins that are secreted through the T3S.

Type III protein secretion system and its effectors

Pathogenic bacteria employ the T3S, the so-called needle apparatus, to deliver virulence factors directly into host cells. Initially, the T3S was identified and studied in animal pathogenic bacteria like *Yersinia*, *Shigella* and *Salmonella* where they were found to secrete into host cells a number of virulence determinants called Yops (Yersinia outer proteins), Ipas (Invasion plasmid-associated proteins) and Sips (Salmonella invasion proteins) respectively. Gough et al. (1992) reported the presence of a homologue of T3S in *Ralstonia solanacearum*, a pathogen of plants of Solanaceae as well as among several other plant pathogens including Xcc. As in other plant pathogenic bacteria, the genes that encode the T3S are organized in a gene cluster in Xcc, Xcv, Xoo, Xac and *X. axonopodis* pv. *glycines* (Arlat et al., 1991; Bonas et al., 1991; da Silva et al., 2002; Zhu et al., 2000a; Kim et al., 2003). Mutations in these genes result in loss of pathogenicity on host plants and an inability to induce the hypersensitive response (HR) on non-host plants. About nine genes in this cluster are referred to as *hrc* (hypersensitive response conserved) genes as they are conserved among plant and animal pathogenic bacteria. Mutations in several other genes, which are located in this cluster as well as at other genomic locations, also result in loss of pathogenicity in hosts and HR on non-hosts. Clearly identifiable homologs of these genes are not present in animal pathogenic bacteria and they are referred to as *hrp* (hypersensitive response and pathogenicity) genes. The genes coding for the T3S have been found to be, by and large, functionally conserved and related to the flagellar apparatus (Bogdanove et al., 1996; Van Gijsegem et al., 1995). Structurally conserved chaperones of the T3S secreted proteins also play an important role in regulated secretion (Karlinsey et al., 2000).

Among xanthomonads, extensive work has been carried out on the *hrp* and *hrc* genes of Xcv. In this bacterium, these genes are organized in 6 different transcription units designated *hrpA* to *hrpF*. These include nine *hrc* genes (*hrcC*, *hrcT*, *hrcN*, *hrcJ*, *hrcU*, *hrcV*, *hrcQ*, *hrcR*, and *hrcS*), which possibly code for the core components of the T3S, and a number of *hrp* genes (Fenselau et al., 1992; Fenselau & Bonas, 1995). As indicated above, mutations in the *hrc* genes lead to loss of virulence in a compatible host and loss of HR in a non-host. Studies to understand the role of the *hrp* genes have revealed that *hrpF* is necessary for translocation of effectors into plant cells whereas *hrpB1*, *hrpB2*, *hrpB4*, *hrpB5*, *hrpD5* and *hrpD6* are required for secretion (Rossier et al., 2000). Non polar mutations in each of the above genes result in a defect in virulence and HR. An exception is the *hpaA* (*hrp* associated) gene in the *hrpD* operon, mutations in which result in reduced virulence in host plants but have no effect on the induction of HR in non-host plants. The ability of the *hpaA* mutant to induce the HR indicates that it is proficient for translocation of proteins into plant cells and that the HpaA protein is therefore not essential for T3S secretory activity (Huguet et al., 1998).

The *hrp* genes of Xcv were found to be induced during growth *in planta* as well as in the plant mimic medium (Schulte & Bonas, 1992; Wengelnik & Bonas, 1996). The *hrp* genes are positively regulated by an AraC type regulator called HrpXv, which is orthologous to the earlier described HrpXo and HrpXc proteins (Wengelnik & Bonas, 1996; Kamdar et al., 1993; Kamoun & Kado, 1990). An ortholog of HrpXv is also present in Xac (Iwamoto & Oku, 2000). The HrpXv protein, and its orthologs, are not encoded in the main *hrp* cluster (Wengelnik & Bonas, 1996). The *hrpXv* mutants were found to produce no symptoms on susceptible pepper and tomato plants and no HR on a resistant host. The promoter regions of HrpXv regulated genes contain a conserved sequence called the plant inducible promoter (PIP) box with TTCGC-N15-TTCGC, as consensus sequence. Genes containing at least two substitutions in the conserved PIP box sequence have also been shown to be regulated by the HrpX ortholog of Xoo (Tsuge et al., 2005). HrpG, a transcriptional activator of the OmpR family of two-component response regulators, controls a wide variety of genes including *hrp* genes and *hrpXv* (Wengelnik et al., 1996). Recent genome-wide studies in Xcv have identified new genes belonging to the HrpG regulon, several of which are secreted through the T3S and have been designated as Xops (Xanthomonas outer proteins) (Wengelnik et al., 1999; Nöel et al., 2001; Nöel et al. 2002).

Zhu et al. (2000a) first reported cloning of the *hrp* gene cluster from Xoo and by genetic analysis showed that strains with mutations in *hrpB*, *hrpC* or *hrpD* operons are virulence deficient and fail to elicit HR in resistant rice lines. Xoo has two genes *hpa1* (similar to harpin-like *popA* of *R. solanacearum*) and *hpa2* (encodes a lysozyme like protein) that are located immediately upstream of *hrpA*. Strains that are deleted for the *hpa1* and *hpa2* genes were virulence deficient and exhibit reduced HR. The *hrpF* deficient mutants of Xoo are virulence deficient but are able to induce HR in rice cultivars containing either *Xa7* or *Xa10* resistance genes (Sugio et al., 2005). The authors suggest that secretion through T3S is reduced but not eliminated in the *hrpF* mutants and that HR is induced even with reduced translocation of T3S effectors.

A number of T3S effector genes identified so far in xanthomonads belong to either the *avrBs3* family or the *avrRxv/YopJ* family. The *avrBs3* family members share 90-97% amino acid sequence identity and have certain distinguishing features like presence of a central domain composed of a variable number (13.5-25.5) of almost identical 34 amino acid repeats and a C-terminal region with Nuclear Localization Signals (NLSs) and an Acidic transcriptional Activation Domain (AAD) (Lahaye & Bonas, 2001). Various mutational studies on *avrBs3*, and two other family members, *avrXa7* and *avrXa10* have revealed that the presence of AAD and at least one NLS is required for eliciting HR (Van den Ackerveken et al., 1996; Zhu et al., 1998). Herbers et al. (1992) demonstrated that the avirulence specificity (cultivar specificity in pepper host plants) of Xcv strains can be changed by altering the number of repeats in the central domain of AvrBs3. Besides *avrBs3*, several other avirulence genes identified in Xcv are *avrBs1*, *avrBs2*, and *avrBs4* (Ronald and Staskawicz, 1988; Kearney & Staskawicz, 1990; Bonas et al., 1989; Bonas et al., 1993; Ballvora et al., 2001). The *avrBs4* gene belongs to the *avrBs3* family. Mutations in the *avrBs2* gene cause reduced growth of Xcv on susceptible pepper plants (Kearney & Staskawicz, 1990). Bacteria with mutations in *avrBs1* alone or both *avrBs1* and *avrBs3* were able to cause disease as well as the wild type strain (Wichmann and Bergelson, 2004). However, *avrBs1avrBs2* double mutants or *avrBs1avrBs2avrBs3* triple mutants were more virulence deficient than the *avrBs2* mutants. These results are indicative of complex additive interactions between these avirulence genes.

A number of *avrBs3* family members have been cloned from *X. campestris* pv. *malvacearum* (Xcm; which infects cotton). Some of them are *avrB4*, *avrb6*, *avrb7*, *avrBIn*, *avrB101* and *avrB102* (De Feyter & Gabriel, 1991; De Feyter et al., 1993). Besides a role in avirulence, members of the gene family are reported to be required for eliciting water soaking symptoms in succeptible cultivars (Yang et al.,

1996). The *pthA* gene is a member of the *avrBs3* family from Xac, the citrus pathogen. This gene was identified as being critical for pathogenesis on citrus but has also been shown to function as an avirulence gene on the non-host bean plant (Swarup et al., 1992). The repeats in the central domain of the *avrb6* and *pthA* have been shown to be required for eliciting symptoms and increased release of Xcm and Xac (Yang et al., 1994).

Members of the *avrBs3* family of avirulence genes cloned from Xoo include *avrXa5*, *avrXa7*, *avrXa10 and avrXa27* (Hopkins et al., 1992; Gu et al., 2005). Lee et al. (2005) report the presence of 8 homologues of *avrBs3* in the genome of Xoo strain KACC 10331. Bai et al. (2000) reported that Xoo strains with mutations in *avrXa5*, *avrXa7* and 4 other proteins of the *avrBs3* family exhibited shorter lesions when inoculated on otherwise susceptible rice plants. The *avrXa7* gene was found to contribute maximum to pathogen aggressiveness on rice plants. Yang & White (2004) showed by combining the middle repeat region of *avrBs3* with the C-terminal domain of *avrXa7* that virulence on susceptible hosts is lost, while avirulence specificity for *Xa7* rice plants is retained. Also, *avrXa7* has been shown to be a T3S dependent, nuclear localized, double stranded DNA binding protein (Yang et al., 2000).

Members of the *avrRxv / yopJ* family are present not only in plant pathogens but also conserved in mammalian pathogens. These proteins have sequence similarity to cysteine proteases. The *avrRxv* gene product of Xcv is similar to *yopJ* of *Yersinia pseusotuberculosis* and was one of the first genes identified in this family. Four YopJ homologues have been identified in Xcv; AvrXv4, AvrBsT, AvrRxv and XopJ (Astua-Monge et al., 2000; Ciesiolka et al., 1999; Nöel et al., 2001; Whalen et al., 1988). Xcv secretes AvrXv4 in a T3S dependent manner and mutational studies indicate that elicitation of HR in resistant tobacco leaves requires the presence of the protease catalytic domain (Roden et al., 2004a). This family of proteins is thought to function like Small ubiquitin-like modifier (SUMO) proteases of eukaryotic cells and modulate host protein functions during infection (Innes, 2003). Xoo does not have an avirulence gene belonging to the *avrRxv/yopJ* family (Lee et al., 2005).

In a genetic screen for T3S effectors of Xcv that are translocated into pepper cells, 7 genes were identified encoding the following Xops; XopC, XopF1, XopF2, XopN, XopO, XopP and XopQ (Roden et al., 2004b). A XopN mutant exhibited reduced growth in susceptible pepper and tomato leaves, indicating its importance in virulence. XopN exhibits similarity to hypothetical proteins from Xanthomonas sp. (da Silva et al., 2002; Lee et al., 2005) and is conserved in Xac, Xcc and Xoo. Mutations in genes for the other Xops identified in the study did not affect growth in pepper plants. Some other T3S secreted proteins of Xcv are XopA, XopB, XopD and XopX. The XopA protein is homologous to the Hpa1 protein of Xoo and HpaG protein of *X. axonopodis*. pv. *glycines* (Nöel et al., 2002). The *xopA* mutants exhibit reduced virulence (Nöel et al., 2002). The XopB protein is homologous to AvrPphD, an avirulence protein of *P. syringae* pv. *phaseolicola* while XopD is homologous to the virulence factor PsrA from *P. syringae* pv. *eriobotryae* (Nöel et al., 2001; Nöel et al., 2002). XopD protein has been shown to be a SUMO protease (Hotson et al., 2003). Mutations in the gene for XopB and XopD proteins do not affect Xcv virulence. XopX is a 74 kDa protein that has been postulated to act as a Xcv virulence factor by suppressing host plant immune response (Metz et al., 2005). Homologs of XopX are present in Xcv, Xac and Xoo. In conjunction with an as yet unidentified T3S effector, XopX has been found to also induce HR in the non-host plant, *Nicotiana benthamiana*.

What is the explanation for the dual role of T3S secreted proteins in virulence and in inducing host defense responses? It is likely that the innate plant defense responses that are induced by the action of cell wall degrading enzymes, LPS as well as other bacterial elicitors like flagellin are potent enough to stop bacterial infection. For infection to proceed, these defense responses must be inhibited. The effectors secreted through the T3S are postulated to suppress the defense responses. Non-host plants, and resistant cultivars of host plants, recognize the presence of these T3S effectors and mount defense responses including a HR. Although recognition might occur through direct interaction between a resistance gene product and the T3S effector, recent results indicate that this is more likely to be due to detection of certain biochemical changes that are brought by the activities of T3S effectors within plant cells (Innes, 2003; Alfano & Collmer, 2004).

The type IV secretion system

A translocator apparatus designated as the type IV secretion system (T4S) is an important virulence factor in certain animal and plant pathogenic bacteria where it is involved in the secretion of either proteins or DNA into host cells (Cascales & Christie, 2003). The T4S stretches from the inner membrane across the periplasm and outer membrane terminating in a pilus like structure that protrudes from the bacterial cell surface. Among phytopathogens, extensive studies on the *virB* locus of *Agrobacterium tumefaciens* have revealed that 11 proteins comprise the T4S (Zhu et al., 2000b).

Xac has two clusters of genes that can each potentially encode components of the T4S; one in the chromosome and the other in the plasmid, pXac64 (da Silva et al., 2002). The Xcc T4S gene cluster is similar to the chromosomally encoded Xac T4S gene cluster except for the presence of *virB6* outside the cluster. Both Xcc and Xac encode multiple paralogs of *virB6* (da Silva et al., 2002; Qian et al., 2005). Interestingly, a T4S gene cluster is not encoded in the Xoo genome (Lee et al., 2005). It is not clear why Xac and Xcc need a T4S while Xoo can apparently do without this system.

A severely virulence deficient mutant of Xcc 8004, isolated from a collection of transposon induced mutants, carried an insertion in the *virB8* gene, which codes for a component of the T4S apparatus (Qian et al., 2005). This suggests that the T4S is likely to play an important role in the virulence of Xcc and other xanthomonads that encode such a secretion system. Further studies are required to elucidate the substrates, either proteins or DNA, that are secreted through the T4S of xanthomonads and to elucidate their role in the virulence of these bacteria. Genes encoding components of the T4S comprise one of the several regions with an altered GC content in the genomes of Xac and Xcc (Van Sluys et al., 2002). The altered GC content of the T4S gene clusters of Xac and Xcc suggests that these regions may have been acquired by horizontal transfer and maintained due to their importance in virulence of these xanthomonads.

Cell-Cell Signaling system

Tang et al. (1991) identified and characterized a cluster of Xcc genes called the *rpf* gene cluster (for regulation of pathogenicity factors) which includes *rpfA-I.* This gene cluster is required for full virulence of Xcc. Some members of this gene cluster code for a cell signaling system involving a diffusible signal factor (DSF). Interestingly, the *rpf* cluster of genes are specific to the xanthomonad group of pathogens, and the closely related Xylella sp. and have not yet been identified in the genome of any other bacteria whose sequence is available (Moreira et al., 2004). From the studies done so far it is clear that RpfB (sequence similarity to long chain fatty acyl CoA ligase) and RpfF (homologous to enoyl CoA hydratase) are involved in the production of DSF (Barber et al., 1997) while RpfC (a complex sensor histidine kinase with fused regulator and phospho transfer domain) and RpfG (a CheY like regulator) constitute a two-component regulatory system, that might be involved in DSF perception (Slater et al., 2000). The RpfG/C system appears to positively regulate EPS and extracellular enzyme production. Mutations in *rpf* genes result in the downregulation of extracellular enzymes like endoglucanase, protease and polygalacturonate lyase as well as EPS (Barber et al., 1997; Tang et al., 1996). Supplementation of DSF to *rpfF* mutants however restores normal levels of extracellular enzymes and EPS. Wang et al. (2004) identified the DSF in Xcc as cis-11-methyl-2-dodecenoic acid which is an α, β unsaturated fatty acid. They reported that several other pathogenic bacteria such as *P. aeruginosa* and *Mycobacterium sp*. also display DSF like activity. Structurally similar extracellular signaling molecules are also produced by *Candida albicans*, making these the first such group of signaling molecules to be conserved in prokaryotes and eukaryotes.

In a recent paper, the *rpf* signaling cluster has been implicated in the regulation of biofilm dispersal in Xcc (Dow et al., 2003). The cells in 24 hour old cultures of *rpfF*, *rpfC*, *rpfG* and *rpf GHC* mutants occur as aggregates. Supplementation of DSF was seen to either prevent or disperse the already formed aggregates only for *rpfF* mutants but not for the other mutants indicating the role of DSF acting through the *rpf G/C* system in biofilm dispersal. DSF supplementation is expected to complement the phenotype of the *rpfF* mutant because mutations in this gene result in loss of DSF production. DSF supplementation does not complement the phenotype of the *rpfG*, *rpfC* or *rpfGC* mutants because these mutants are proficient for DSF production but defective in DSF mediated signaling. Further it has been demonstrated that the enzyme endo β-1,4-mannosidase encoded by *manA* gene of

Xcc is capable of dispersing the Xcc *rpfF* mutant aggregates. However, future experiments are required to understand the role of biofilm formation and dispersal in the virulence of Xcc (Crossman & Dow, 2004).

The other genes namely, *rpfADEI* appear to have no role in this cell signaling system (Crossman & Dow, 2004). The *rpfA* gene encodes an aconitase (Wilson et al., 1998). Mutations in this gene affect EPS and extracellular enzyme production, cause virulence deficiency and also result in reduced levels of intracellular iron. The *rpfD* gene encodes a novel type of transcriptional regulator and insertions in this gene have minor effects on EPS production and extracellular enzymes (Dow et al., 2000b). The *rpfE* and *rpfI* genes are annotated as encoding regulatory functions (da Silva et al., 2002; Lee et al., 2005). Mutations in these genes affect production of EPS and certain extracellular enzymes but not to the same extent as the defects observed in *rpfA, B, C, F* and *G* mutants (Dow et al., 2000b). Mutations in the *rpfE* and *I* genes do not effect either DSF production or virulence.

In Xoo, the *rpfF* mutant was first identified in a screen for virulence defective mutants (Chatterjee & Sonti, 2002). The organization of the *rpf* gene cluster in Xoo is similar to Xcc. However, unlike in Xcc, the Xoo *rpfF* mutant is proficient for EPS and extracellular enzyme production. The virulence defect of the Xoo *rpfF* mutant could be complemented by either the wild-type or *hrp*$^-$ Xoo in co-inoculation experiments. This suggests that the restoration of DSF in trans corrects the virulence deficiency of *rpfF* mutants. Interestingly, the *rpfF* mutants of Xoo overproduce siderophores and exhibit a defect in expressing a tetracycline resistance phenotype, both indicative of iron starvation. Consistent with this observation, the *rpfF* mutant exhibits a growth defect in iron depleted medium. The siderophore overproduction phenotype of *rpfF* mutants could be reversed by cell-free supernatant derived from wild-type Xoo but not from *rpfF* mutant bacteria, indicating the role of DSF. Also, exogenous iron supplementation in a detached leaf assay promotes migration of *rpfF* mutants in rice leaves. This suggests that the virulence deficiency of the Xoo *rpfF* mutants may be due to an inability to source sufficient iron during growth within rice leaves. Also in Xoo, the *rpfC* gene has been shown to be required for virulence (Tang et al., 1996). Interestingly, *rpfC* mutants exhibit reduced EPS but no change in extracellular enzymes. The full set of genes under the *rpf* regulon is not known in either Xcc or Xoo.

Regulatory proteins – Catabolite activator-like protein (Clp)

In Xcc the production of many virulence factors is controlled by a global transcription factor, Catabolite activator-like protein (Clp) which is a homologue of the cyclic AMP receptor protein (CRP) of *E. coli*. Xcc Clp is believed to regulate target genes in a cyclic AMP independent manner as no *cya* homologue (a gene responsible for the synthesis of cyclic AMP) has been found in the Xcc genome sequence. In one of the early studies, mutations in the *clp* gene were found to affect the production of EPS, xanthomonadin pigment and extracellular enzymes (de Crecy-lagard et al., 1990). The Xcc *clp* mutant exhibited reduced levels of cellulase and pectinase activity. Acetone extraction of pigment revealed a 50% increase in xanthomonadin production in the mutant when compared to wild-type strain. Also, the amount of xanthan gum produced by the mutant was less than wild type. Inoculation of the Xcc *clp* mutant into either seedlings or mature leaves of turnip indicated that it is reduced for virulence (de Crecy-lagard et al., 1990). The Xcc *clp* mutant also exhibited reduced disease symptoms on radish (de Crecy-lagard et al., 1990).

Among the extracellular enzymes, the level of endoglucanase activity is decreased by 85% in the Xcc *clp* mutant (de Crecy-lagard et al., 1990). It has been established that Clp activates transcription of the *engA* gene, which encodes an endoglucanase, by direct binding to consensus Clp-binding sites in the promoter region (Hsiao et al., 2005). In an earlier study the expression of the major protease gene *prt1* was shown to be upregulated by Clp in an indirect manner, which does not involve the binding of Clp to the promoter (Hsiao & Tseng, 2002).

Another gene regulated by Clp is the flagellin gene *fliC*, where the exact mode of regulation is still unclear (Lee et al., 2003). Proteomic analysis revealed that certain extracellular proteins are reduced in the *clp* mutant as compared to the wild type strain (Lee et al., 2003). Among the proteins that exhibit a reduction is the flagellin protein encoded by *fliC*. Xcc *clp* mutants are slightly reduced for motility although there is no alteration in the flagellation. On the other hand, Xcc *fliC* mutants lack motility and flagellation but are virulence proficient (Lee et al., 2003). These studies suggest that the Clp regulator modulates the expression of target traits in a complex manner and that some of these traits may be more important for virulence than others.

Virulence deficiency is also caused by a mutation in the XC2827 gene of Xcc 8004 which encodes a *marR* type of transcriptional factor that possesses a helix-turn-helix motif (Qian et al., 2005). In the same study, mutation of a cytoplasmic car-like transducer protein that might be involved in arginine chemotaxis was also found to affect the virulence of Xcc 8004.

Iron uptake functions

Iron is essential for growth as it plays a crucial role in the redox reactions of intermediary metabolism, being a co-factor of several important enzymes. However, iron availability is limited by its low solubility (10^{-18} M) and dissolution kinetics in aerobic conditions at neutral and alkaline pH. Bacteria need 10^{-6} to 10^{-7} M iron for proper growth. For pathogenic bacteria, the concentration of available iron is even lesser within the host due to iron sequestration by iron binding proteins. Under conditions of iron starvation, many bacteria produce low molecular weight (500-1000 daltons) iron chelators called siderophores. The siderophores bind to Fe^{3+} with high affinity, the formation constant being 10^{30} or higher (Neilands, 1995). Siderophores are non-ribosomal peptides synthesized by Non Ribosomal Peptide Synthetases (NRPSs), which are multi-modular enzymes that synthesize peptides of a particular sequence without an RNA template (Crosa &

Walsh, 2002). Among plant pathogens, studies done on *Erwinia chrysanthemi* strain 3937 (causes soft-rot of African violets) and *Erwinia amylovora* strain CFBP 1430 (causes Fireblight of apple) have shown that siderophore-mediated iron uptake is critical for causing disease (Expert, 1999).

Wiggerich et al. (1997) reported that Xcc secretes siderophores upon iron starvation. The nature and structure of these siderophores and the genes involved in their biosynthesis are not characterized in any xanthomonad. Moreira et al. (2004) reported in a comparative study of the genomes of *Xylella fastidiosa* and Xac that although genes involved in siderophore biosynthesis are not readily identifiable, they are probably annotated as hypothetical genes. In another in silico study, Etchegaray et al. (2004) report the identification of genes of the NRPS family in two regions of the Xac genome; one of these regions is also found in the Xcc genome. These proteins are related to polyketide synthetases and may be involved in siderophore biosynthesis. Homologues of the vibrioferrin (a polycarboxylate type siderophore of *Vibrio parahaemolyticus*) biosynthetic gene cluster (Tanabe et al., 2003) have been identified, in the genomes of Xcc, Xac and Xoo (S.Sujatha, Alok Pandey & Ramesh Sonti, unpublished data). However, the role of these genes in siderophore biosynthesis has not yet been established.

The uptake of ferric-siderophore complex into the periplasm is an energy dependent process that requires a complex of three inner membrane proteins, TonB-ExbB-ExbD (Larsen et al., 1996; Stojiljkovic & Srinivasan, 1997) as well as TonB-dependent outer membrane located receptors specific for the ferric-siderophore complex. Amongst plant pathogenic bacteria, virulence deficiency has been associated with the disruption of a siderophore uptake pathway in *E. amylovora* wherein a *foxR* (TonB dependent receptor for desferrioxamine) mutant exhibited reduced colonization on apple flowers (Dellagi et al., 1999). In *E. chrysanthemi*, a *tonB* mutant was able to macerate infected tissue like the wild type strain but was impaired in the spread of infection (Enard & Expert, 2000). In Xcc, the *tonB-exbB-exbD1* cluster of genes have been found to be essential for eliciting a HR on the non-host plant (pepper) and for induction of black-rot symptoms on the host plant (cauliflower) (Wiggerich & Pühler, 2000). The Xcc *tonB* mutants caused only small dry spots on cauliflower leaves instead of the typical black rot symptoms. However, the *tonB* mutant was able to spread within infected leaves. How does the *tonB* mutant manage to obtain iron within the host? One possibility is that the *tonB* mutant may be able to source its iron requirements from the host through uptake of ferrous iron (Wiggerich & Pühler, 2000). Mutations in the gene for FeoB, a major ferrous transporter of gram-ve bacteria, have been shown to cause defects in the virulence phenotype of pathogenic *E. coli*, *Salmonella typhimurium, Legionella pneumophila* and *Helicobacter pylori* (Stojiljkovic et al., 1993; Tsolis et al., 1996; Robey and Cianciotto 2002; Velayudhan et al., 2000). Homologs of FeoB are encoded in the genomes of Xac, Xcc and Xoo (da Silva et al., 2002; Lee et al., 2005) although mutations in these genes have not been reported.

Excess iron inside cells can be toxic due to the ability of Fe^{2+} to catalyze the Fenton reaction, which generates cell damaging hydroxyl radicals (Touati, 2000). Hence, the expression of iron uptake functions is tightly controlled and this function is performed in many bacteria by the Ferric Uptake Regulator (Fur) protein. Homologs of the *fur* gene have been identified in a number of xanthomonads (Loprasert et al., 1999). Mutations in the *fur* gene of Xoo result in constitutive production of siderophores (Subramoni & Sonti, 2005). The Xoo *fur* mutant is hypersensitive to hydrogen peroxide and exhibits reduced catalase activity. The *fur* mutant grows very poorly within rice leaves and it was demonstrated that supplementation with exogenous ascorbic acid (an anti-oxidant) rescues this growth deficiency. It appears that the growth deficiency of the Xoo *fur* mutant within the host is, at least in part, due to an inability to survive oxidative stress during infection. Interestingly, the Xoo *fur* mutant also grows poorly in peptone sucrose (PS) medium but ascorbic acid supplementation only partially alleviates this slow growth phenotype. It is possible that ascorbic acid is not taken up very well by *X. oryzae* pv. *oryzae*. In such a situation, ascorbic acid supplementation might provide better protection to the *fur* mutant against oxidative stress emanating from an external source (as would be experienced during growth in the leaf) as compared to oxidative stress from within the cell (as would be experienced during growth in PS medium). Alternatively, enhanced oxidative stress may only be partially responsible for the slow growth phenotype of the *fur* mutant in PS medium. More studies on the *fur* regulon of xanthomonads are needed to understand how Fur controls iron uptake, response to oxidative stress, growth in laboratory conditions and within the host.

Zinc uptake regulator

A mutation in the Zinc Uptake Regulator (Zur) protein gene of Xcc leads to impaired growth at high zinc concentrations (110 – 400 μM) (Tang et al., 2005). However, the *zur* mutant does not exhibit any growth deficiencies at concentrations of zinc (11 μM) that are similar to those estimated to be present in the intercellular spaces of radish leaves. The *zur* mutant exhibits reduced virulence and EPS production. The virulence deficiency of the *zur* mutant maybe due to the defect in the production of either EPS or some other virulence factor whose expression is regulated by Zur.

Phosphate metabolism

Mutations in the Xoo *phyA* (putative phytase A) gene cause virulence deficiency and an inability to use phytic acid as a sole phosphate source (Chatterjee & Sonti, 2003). Phytic acid is inositol hexaphosphate, a major storage form of phosphate in plants like cereals and legumes. Phytases are enzymes involved in phytic acid degradation. The *phyA* gene encodes a secreted protein which is homologous to Bacillus phytases. In a detached leaf assay, supplementation with exogenous phosphate promotes growth of the *phyA* mutant in rice leaves, indicating that the virulence deficiency of this mutant may be due to an inability to use phytic acid as a phosphate source. Orthologs of *phyA* are present in Xcc and Xac, indicating that this gene might be important for virulence of other xanthomonads, besides Xoo.

Xanthomonas toxins

Albicidins produced by *Xanthomonas albilineans* (Xab), the causal agent of sugarcane leaf scald disease, are some of the best studied toxins of xanthomonads. Albicidins block prokaryotic DNA replication and hence are both bactericidal and phytotoxic due to their effect on plastid replication (Birch & Patil, 1987a). These toxins are thought to function as pathogenicity factors and to provide a growth advantage to *X. albilineans* at the site of infection by inhibiting the growth of competing microbes (Birch & Patil, 1985).

At least two gene clusters involved in albicidin biosynthesis have been identified in the genomes of Florida and Queensland strains of *X. albilineans* (Rott et al., 1996; Wall & Birch, 1997). Amongst the proteins that are encoded in the larger biosynthetic cluster, XALB1 (55,839 bp), are three large proteins with modular features that are characteristic of polyketide synthase (PS) and non-ribosomal peptide synthetase (NRPS) activities (Royer et al., 2004). All three proteins are required for albicidin biosynthesis. One of these proteins, AlbI, is 6879 amino acids long and is similar to the previously described PS and NRPS containing Xabb protein that is required for albicidin biosynthesis in a Queensland strain of *X. albilineans* (Huang et al., 2001). However, the Xabb protein is 4,801 amino acids long and is missing some of the NRPS modules that are present in AlbI. Downstream of *xabB* is *xabC (albII)*, which encodes a protein necessary for methylation of an intermediate in albicidin biosynthesis (Huang et al., 2000a). Another gene, *xabA* located elsewhere in the Xab genome, encodes a phosphopantetheinyl transferase required for activation of XabB synthetase (Huang et al., 2000b).

Several lines of evidence indicate that albicidins induce chlorosis (a typical symptom of leaf scald disease) and promote pathogenesis. The albicidin deficient mutants of *X. albilineans* fail to induce chlorosis in sugarcane plants whereas the revertants for albicidin production are able to do so (Birch & Patil, 1987b; Wall & Birch, 1997). Similarly, expression of a gene for albicidin detoxification, *albD*, in *X. albilineans* abolishes toxin release and affects pathogen growth in inoculated sugarcane leaves (Zhang & Birch, 1997). Inoculation of wild type *X. albilineans* in *albD* expressing transgenic sugarcane resulted in 3-4 log reduction in bacterial cell numbers, when compared to that in inoculated control scald susceptible sugarcane plants (Zhang et al., 1999). These and other studies helped to conclude that albicidins are important pathogenicity factors. They appear to cause chlorosis by diffusing into cells surrounding infected xylem vessels and inhibiting plastid DNA replication. Blockage of plastid DNA replication also leads to inhibition of the biosynthesis of several compounds whose site of synthesis is in the chloroplast. These include certain aromatic compounds which are precursors of molecules involved in plant defense reactions (Buchanan et al., 2000). In this manner, albicidin production might also interfere with the plant defense response and facilitate spread of the pathogen within the host.

Two genes have been identified in Xoo, *rtxA* and *rtxC,* whose products are homologous to Bradyrhizobial proteins that are involved in biosynthesis of rhizobitoxine, which is postulated to be a plant toxin and an inhibitor of ethylene biosynthesis. Orthologs of these genes are not encoded in the Xac and Xcc genome (Lee et al., 2005). The role of *rtxA* and *rtxC* in virulence of Xoo remains to be established.

Detoxification of host derived anti-microbial compounds

Reactive oxygen species as well as phytoalexins and other anti-microbial compounds are produced by plants to counter microbial infections (Wojtaszek, 1997). Mutations in two Xcc 8004 catalase genes (*katE* and *catB*) have been shown to result in reduced virulence (Qian et al., 2005). It is possible that this phenotype is due to a reduction in the ability of the mutants to detoxify reactive oxygen species that are produced by the host. As discussed elsewhere in this review, the in planta growth defect of the Xoo Fur (Ferric Uptake Regulator) mutant can be rescued by supplementation with ascorbic acid, an anti-oxidant. These results provide evidence that the ability to detoxify reactive oxygen species produced by the host is an important aspect of the pathogenesis of xanthomonads on susceptible hosts. The Xoo Fur mutant exhibits reduced catalase activity but the exact mechanism by which the Xoo Fur protein controls catalase expression is not clear; it may be direct or indirect (Subramoni & Sonti, 2005; Wilderman et al., 2004).

Qian et al. (2005) also report that transposon insertion in the *uptB* gene affects virulence of Xcc 8004 on cabbage. This gene is predicted to encode a glutathione S-transferase involved in degradation of xenobiotic molecules. The authors suggest that this phenotype may be due to an inability of the mutant to cope with anti-bacterial compounds that are produced by the host during infection. In the same study, mutations in two genes that are involved in fatty acid degradation were found to be affected for virulence. One of these genes is *fadB* which is predicted to encode a *p*-hydroxycinnamoyl CoA hydratase / lyase. The other gene, *cypC,* is predicted to encode a fatty acid α hydroxylase. It is possible that these functions provide protection to Xcc against toxic fatty acids that are produced by the host. Alternatively, the virulence deficiency may be due to a growth defect caused by an alteration in fatty acid metabolism/degradation.

Recombination functions

Martinez et al. (1997) constructed a *recA* mutant of Xcc using an interrupted *Rhizobium etli recA* gene that had been introduced into the Xcc genome using a non-replicating vector and subsequent selection for double recombinants. Presumably, the high degree of conservation in the sequence of RecA proteins permits recombination between the sequences of the *recA* genes from the two organisms. The Xcc *recA* mutants are very sensitive to the DNA damaging agent methyl-methane-sulfonate (MMS), exhibiting a reduction in viability of about three orders of magnitude as compared to the wild type strain. The *recA* mutants are also hypersensitive to ultra violet (UV) radiation. These phenotypes are complemented by the *recA* gene from *R. etli* indicating that the hypersensitivity to MMS and UV are due to mutations in the *recA* gene. The authors have also shown that the *recA* mutants exhibit a reduction in the frequency of homologous recombination of at least three orders of magnitude as compared to the wild type strain. The Xcc *recA* mutants are partially virulence deficient. Reactive oxygen species are known to cause DNA damage. As the *recA* mutants are hypersensitive to DNA damaging agents, it is possible that the reduced virulence of the Xcc *recA* mutants is due to a hypersensitivity to the reactive oxygen species that are produced by the host. However, it is also possible that the virulence deficiency of the *recA* mutants is due to a reduced growth rate or some non-specific metabolic defect.

Metabolism

Mutations in genes for certain metabolic pathways can also lead to loss of virulence in xanthomonads. Many of these mutations lead to auxotrophy and it is likely that the virulence defect is due to an inability to grow within the host. However, it is also possible that in some of these cases, the mutation leads to loss in production of a pathogenicity factor like a toxin. A few examples of mutations in metabolic genes that lead to loss of virulence in xanthomonad pathogens are given below;

a. Purine biosynthesis: In Xoo, a transposon insertion in the *purH* gene (which encodes a bifunctional purine biosynthesis protein having formyltransferase and hydrolase activities) leads to virulence deficiency. As expected, the Xoo *purH* mutant also exhibited a growth deficiency on minimal medium, which could be rescued by purine supplementation (Chatterjee & Sonti, in press). However, the *purH* mutant was proficient for HR in non-host plant tomato. In Xcc8004, transposon insertions in two genes of the purine biosynthetic pathway, *purC* and *purM,* resulted in severe virulence deficiency (Qian et al., 2005).

b. Amino acid biosynthesis: Ten different amino acid biosynthetic genes were found to be mutated in a collection of seventy five non-redundant virulence deficient mutants of Xcc 8004 (Qian et al., 2005). These include the following: *aroA* and *pheA* genes that are involved in the shikimate-chorismate pathway; *leuA* and *leuC* genes involved in leucine biosynthesis; *hisF* gene involved in histidine biosynthesis; *argB* and *argC* genes involved in arginine biosynthesis; *metA* gene involved in methionine biosynthesis; *trpD* and *trpE* genes involved in tryptophan biosynthesis. It is possible that, during growth on cabbage, Xcc cannot source adequate amounts of amino acids like tryptophan, leucine, histidine, arginine, etc. from the host and is dependent on it's own biosynthetic capabilities for this purpose. In a similar manner, mutations that cause glutamate auxotrophy have been shown to result in loss of virulence in Xac

(Tung & Kuo, 2000). Mutations in the Xoo *aroE* gene, which encodes an enzyme in the shikimate-chorismate pathway, were found to result in auxotrophy for all three aromatic amino acids (Tyrosine, Phenylalanine and Tryptophan) as well as deficiency for virulence and xanthomonadin pigment production (Goel et al., 2001). The virulence deficiency appears to be due to the aromatic amino acid auxotrophy.

c. Gluconeogenesis: A transposon induced Xoo mutant which is reduced for virulence but proficient for inducing the HR in resistant rice varieties and tomato had a mutation in the phosphoglucose isomerase (*pgi*) gene (Tsuge et al., 2004). Pgi is an enzyme involved in gluconeogenesis that catalyses the reversible isomerisation of glucose 6 phosphate and fructose 6 phosphate. The Xoo *pgi* mutant exhibited a reduced growth in synthetic medium containing xylose or fructose as the sole carbon source. It is estimated that rice cell walls contain 60% xylan (Takeuchi et al., 1994), a polysaccharide whose backbone is made of xylose residues. The virulence deficiency of the *pgi* mutant may be due to an inability to use xylose as a carbon source. A *pgi* mutant of Xcc has also been shown to be defective for virulence on citrus plants (Tung & Kuo, 1999). Tsuge et al. (2001) also report that an Xoo mutant that is deficient in using glucose as a carbon source retains full virulence on rice. This result suggests that Xoo is not dependent on glucose as a sole carbon source when it is growing within rice leaves.

d. Sucrose metabolism: Sucrose is considered to be a major carbon source for bacteria that live in the intercellular spaces of leaves. Mutational inactivation of the sucrose hydrolase (*suh*) gene of Xag resulted in the loss of the ability to use sucrose as a carbon source and in reduced growth within the host plant (Kim et al., 2004). The residual in planta growth of the *suh* mutant maybe due to the presence of other carbon sources, besides sucrose, in the intercellular spaces of soybean leaves.

e. Sugar-nucleotide biosynthesis: One of the first steps in the biosynthesis of Xanthan gum in Xcc is the synthesis of sugar-nucleotide precursors. UDP-glucose dehydrogenase (encoded by *udgH*) is an enzyme that catalyses the conversion of UDP-glucose to UDP-glucuronic acid, an essential precursor of Xanthan gum. A *udgH* mutant of Xcc is deficient for EPS biosynthesis (Chang et al., 2001). *udgH* mutants of Xcc and Xcv were virulence deficient when inoculated on cabbage and pepper leaves respectively.

Function unknown genes

A significant proportion of the genes in the genomes of Xac, Xcc and Xoo have been annotated as encoding conserved hypothetical (or function unknown) proteins. Therefore, it is not surprising that insertions in nine such genes were found to lead to virulence deficiency in a very large and systematic screen for Xcc virulence deficient mutants (Qian et al., 2005). Some of these genes have recognizable domains like a helix-turn-helix motif (XC1021 gene) or haemolysin type calcium-binding domain (XC3932). Insertions that affect virulence were also isolated in three genes of this category that are encoded in strain specific regions of the genome of Xcc8004 strain that are absent from the genome of the Xcc ATCC 33913 strain. One gene (XC2068) does not have any recognizable domains but homologs are present in a number of animal and plant pathogenic bacteria indicating that it may be an important virulence determinant. Two other genes (XC2055 and XC2416) have domains which suggest that they may be involved in cell signaling. The XC2055 gene has significant similarity with two genes from the human pathogen, *Bordetella bronchiseptica.* Interestingly, insertion mutations in these three strain specific genes led to different virulence phenotypes on different host plants. For eg. An insertion mutation in the XC2416 gene affected virulence on radish cultivar Huaye without affecting virulence on radish cultivar Xiaojingzhong. This suggests the possibility that the functions of such strain specific xanthomonad genes may be to facilitate infection in particular sub-groups of host plants.

Genome analysis

The genome sequences of one strain each of Xoo and Xac as well as two strains of Xcc have been completed and are available in the public domain (Lee et al., 2005; da Silva et al., 2002; Qian et al., 2005). The genomes of several other xanthomonads including Xcv (Büttner et al., 2003) are being sequenced. Besides the above, the sequences of the *Xylella fastidiosa* 9a5c (causal agent of citrus variegated chlorosis; Simpson et al., 2000) and *Xylella fastidiosa* Temecula 1 (causal agent of Pierce's disease of grapevine; Van Sluys et al., 2003) strains are also available for comparison with the genomes of xanthomonads. These genome sequences have been analysed to identify genes that are specific to particular species or strains (da Silva et al., 2002; Lee et al., 2005; Qian et al., 2005) of xanthomonads. These studies have revealed that 800 Xac ORFs are absent from Xcc ATCC 33913 and that 646 ORFs that are present in Xcc ATCC 33913 are absent from Xac (da Silva et al., 2002). A number of these species specific ORFs are present in clusters (Moreira et al., 2005). Some of these species specific clusters are associated with transposases, phage/plasmid related sequences and tRNAs that are indicative of inheritance by horizontal gene transfer (Lima et al., 2005).

In a comparison of the two Xcc strains whose genomes have been completely sequenced, Qian et al. (2005) report that 108 ORFs are specific to the Xcc 8004 strain and 62 ORFs are unique to the Xcc ATCC 33913. Many of these ORFs were found to localize to certain strain specific chromosomal segments. Genes that are encoded within one of the strain specific chromosome segments of Xcc 8004 exhibit characteristics like altered codon usage pattern, as compared to the rest of the genome, that are indicative of introduction by horizontal gene transfer. Yet another strain specific segment of Xcc8004 contains many genes that are orthologs of sequences present in Xac. This suggests that this region of the genome might have been in the ancestor of Xac and Xcc and that it might have been deleted during the evolution of the Xcc ATCC 33913 strain. As indicated in an earlier section, transposon insertions in some strain specific genes of Xcc8004 result in virulence deficiency indicating that these genes are involved in pathogenicity. Detailed studies are required for analyzing the role of species and strain specific genes in promoting virulence of xanthomonads. It is likely that these studies will shed light on the functions and mechanisms that promote host plant specificity amongst xanthomonads.

In a related study, the partial genome sequence of another Xcc strain, Xcc B100, was obtained by transposon tagging and compared with the complete genome sequence of Xcc ATCC 33913 (Vörholter et al., 2003). At least 27 genes present in Xcc B100 were absent in Xcc ATCC 33913, although similar genes were identified in Xac, *X. fastidiosa* and other bacteria indicating variation at the strain level. Moreover, four genomic regions present in Xcc ATCC 33913 appear to be absent in Xcc B100. One such region consists of a 11kb gene cluster that appears to be involved in biosynthesis of cell surface polysaccharides.

The contribution of mobile genetic elements such as insertion sequences and phages as well as horizontally acquired genomic islands to the genomes of Xac and Xcc have been analysed (Monteiro-Vittorello et al., 2005; Lima et al., 2005). The work of Lima et al. (2005) has revealed that almost 40% of genes in Xac and Xcc exhibit highest similarity to genes from non-ϒ-proteobacteria, which are phylogenetically distant organisms. Many of these genes are present in clusters, forming genomic islands (da Silva et al., 2002; Lima et al., 2005). Of these, five islands are unique to each genome while another thirty islands are represented, at least partially, in both the genomes. An example of such an island that is present in both Xac and Xcc is the *xcs* T2S gene cluster that exhibits similarity to genes in *Caulobacter crescentus*, an α-proteobacterium.

The availability of whole genome sequence information paves the way for research on functional genomics including gene expression profiling, proteomics and large scale screening for mutants. The work of Qian et al. (2005) is an excellent example of a large scale mutant screen that has led to the identification of a number of new genes that are involved in promoting Xcc pathogenesis. Studies of this nature need to be conducted on other xanthomonad pathogens. It is also anticipated that microarray/gene chip analysis will be used to map the transcriptomes of xanthomonads under various growth conditions; for eg. within and outside the host; within different tissue compartments of the host (for eg; fruits vs. leaves for Xac); limitation for nutrients like iron, phosphate, etc. Whole genome expression profiling will also shed light on the complete set of genes that are controlled by master regulatory genes like *rpfF*/*rpfB*, *rpfG*/*C*; *hrpG*; *hrpX, clp*, *fur*, *zur*, etc. A number of candidate virulence factors like putative adhesins, toxins, protein secretion systems, etc. have been uncovered through genome sequence analysis and it is anticipated that their roles in virulence will be systematically analysed. All in all, the coming years promise to be a very exciting period as new insights will be obtained about the virulence functions of xanthomonads and the mechanisms that govern their host plant specificity.

3. SUMMARY AND FUTURE PROSPECTS

It is clear that a number of factors contribute to determine the virulence and host specificity of the xanthomonad group of pathogens. These include the following: adhesins; surface polysaccharides like EPS and LPS; secretion systems like T1S, T2S, T3S and T4S and their effectors; regulatory factors like HrpG, HrpX, DSF, Clp, Zur and Fur; functions like PhyA which are required for nutrient acquisition; functions such as catalase which are required for detoxification of host derived anti-microbial agents; toxins such as albicidin; etc. Mutations in genes involved in the biosynthesis of certain amino acids and purines also result in loss of virulence, most likely because these nutrients cannot be obtained from the host. Mutations involved in the catabolism of certain sugars also result in loss of virulence, most likely because the particular sugars are available as carbon sources for the bacterium during growth within the host. Although the latter two categories of functions cannot strictly be considered as virulence factors, they can serve, along with the virulence factors, as targets for the development of newer generations of antibacterial compounds.

Although the list of virulence factors that have been identified to date is impressive, this is only the beginning and the complete picture of the pathogenesis pathway is yet to emerge. In particular, we know very little about how these pathogens attach to host tissues and the mechanisms by which they suppress plant defense responses and acquire nutrients. Also, there is minimal information about the functions that xanthomonads use to survive as epiphytes. The complete genome sequences of three xanthomonads and two of their host plants (Rice and Arabidopsis) are already available and the genome sequences of several other xanthomonads are currently being determined. This information would greatly accelerate the characterization of different genes required for host infection. Future studies using genomic and proteomic approaches would provide comprehensive information regarding bacterial functions that contribute to host and tissue specificity, as well as host signal transduction pathways that respond to the pathogen. This information should help in the development of new strategies for reducing yield losses due to these pathogens, either by way of effective and ecofriendly anti-bacterial compounds or resistant host plants that are obtained using the techniques of transgenesis and marker assisted breeding.

4. REFERENCES

Alfano, J. R., & Collmer, A. (2004). Type III secretion system effector proteins: double agents in bacterial disease and plant defense. *Annu. Rev. Phytopathol. 42*, 385-414.

Arlat, M., Gough, C. L., Barber, C. E., Boucher, C., & Daniels, M. J. (1991). *Xanthomonas campestris* contains a cluster of *hrp* genes related to the larger *hrp* cluster of *Pseudomonas solanacearum*. *Mol. Plant Microbe Interact. 4*, 593-601.

Astua-Monge, G., Minsavage, G. V., Stall, R. E., Vallejos, C. E., Davis, M. J., & Jones, J. B. (2000). Xv4-vrxv4: a new gene-for-gene interaction identified between *Xanthomonas campestris* pv. *vesicatoria* race T3 and the wild tomato relative *Lycopersicon pennellii. Mol. Plant Microbe Interact. 13,* 1346-1355.

Bai, J., Choi, S. H., Ponciano, G., Leung, H., & Leach, J. E. (2000). *Xanthomonas oryzae* pv. *oryzae* avirulence genes contribute differently and specifically to pathogen aggressiveness. *Mol. Plant Microbe Interact. 13,* 1322-1329.

Ballvora, A., Pierre, M., van den Ackerveken, G., Schornack, S., Rossier, O., Ganal, M., et al. (2001). Genetic mapping and functional analysis of the tomato *Bs4* locus governing recognition of the *Xanthomonas campetris* pv. *vesicatoria* AvrBs4 protein. *Mol. Plant Microbe Interact. 14,* 629-638.

Barber, C. E., Tang, J. L., Feng, J. X., Pan, M. Q., Wilson, T. J., Slater, H., et al. (1997). A novel regulatory system required for pathogenicity of *Xanthomonas campestris* is mediated by a small diffusible signal molecule. *Mol. Microbiol. 24,* 555-566.

Birch, R. G., & Patil, S. S. (1985). Preliminary characterization of an antibiotic produced by *Xanthomonas albilineans* which inhibits DNA synthesis in *Escherichia coli. J. Gen. Microbiol. 131,* 1069-1075.

Birch, R. G., & Patil, S. S. (1987a). Evidence that an albicidin-like phytotoxin induces chlorosis in sugarcane leaf scald disease by blocking plastid DNA replication. *Physiol. Mol. Plant Pathol. 30,* 207-214.

Birch, R. G., & Patil, S. S. (1987b). Correlation between albicidin production and chlorosis induction by *Xanthomonas albilineans*, the sugarcane leaf scald pathogen. *Physiol. Mol. Plant Pathol. 30,* 199-206.

Bogdanove, A. J., Beer, S. V., Bonas, U., Boucher, C. A., Collmer, A., Coplin D. L., et al. (1996). Unified nomenclature for broadly conserved *hrp* genes of phytopathogenic bacteria. *Mol. Microbiol. 20,* 681-683.

Bonas, U., Conrads-Strauch, J., & Balbo, I. (1993). Resistance in tomato to *Xanthomonas campestris* pv. *vesicatoria* is determined by alleles of the pepper-specific avirulence gene *avrBs3*. *Mol. Gen. Genet. 238,* 261-269.

Bonas, U., Schulte, R., Fenselau, S., Minsavage, G. V., Staskawicz, B. J., & Stall, R. E. (1991). Isolation of a gene cluster from *Xanthomonas campestris* pv. *vesicatoria* that determines pathogenicity and the hypersensitive response on pepper and tomato. *Mol. Plant Microbe Interact. 4,* 81-88.

Bonas, U., Stall, R. E., & Staskawicz, B. (1989). Genetic and structural characterization of the avirulence gene *avrBs3* from *Xanthomonas campestris* pv. *vesicatoria. Mol. Gen. Genet. 218,* 127-136.

Buchanan, B. B., Gruissen, W., & Jones, R. L. (2000). *Biochemistry and Molecular Biology of Plants.* Rockville, Maryland: Am. Soc. Plant Physiol.

Burdman, S., Shen, Y., Lee, S. W., Xue, Q., & Ronald, P. (2004). RaxH/RaxR: a two-component regulatory system in *Xanthomonas oryzae pv. oryzae* required for AvrXa21 activity. *Mol. Plant Microbe Interact. 17,* 602-612.

Büttner, D., Nöel, L., Thieme, F., & Bonas, U. (2003). Genomic approaches in *Xanthomonas campestris* pv. *vesicatoria* allow fishing for virulence genes. *J. Biotechnol. 106,* 203-214.

Cascales, E., & Christie, P. J. (2003). The versatile bacterial type IV secretion systems. *Nat. Rev. Microbiol. 1,* 137-149.

Chan, J. W. Y. F., & Goodwin, P. H. (1999). The molecular genetics of virulence of *Xanthomonas campestris*. *Biotechnology Advances, 17,* 489-508.

Chang, K. W., Weng, S. F., & Tseng, Y. H. (2001). UDP-glucose dehydrogenase gene of *Xanthomonas campestris* is required for virulence. *Biochem. Biophys. Res. Commun. 287,* 550-555.

Chatterjee, S., Sankaranarayanan, R., & Sonti, R. V. (2003). PhyA, a secreted protein of *Xanthomonas oryzae* pv. *oryzae*, is required for optimum virulence and growth on phytic acid as a sole phosphate source. *Mol. Plant Microbe Interact. 16,* 973-982.

Chatterjee, S., & Sonti, R. V. (2002). *rpfF* mutants of *Xanthomonas oryzae* pv. *oryzae* are deficient for virulence and growth under low iron conditions. *Mol. Plant Microbe Interact. 15,* 463-471.

Chatterjee, S., & Sonti, R. V. (2005). Virulence deficiency caused by a transposon insertion in the *purH* gene of *Xanthomonas oryzae* pv. *oryzae*. *Canadian Journal of Microbiology*. CJM File NO. V04-032, 04-008002 (in press).

Chou, F. -L., Chou, H. -C., Lin, Y. -S., Yang, B. -Y., Lin, N. -T., Weng, S. -F., & Tseng, Y. -H. (1997). The *Xanthomonas campestris gumD* gene required for synthesis of xanthan gum is involved in normal pigmentation and virulence in causing black rot. *Biochem. Biophys. Res. Commun. 233,* 265–279.

Ciesiolka, L. D., Hwin, T., Gearlds, J. D., Minsavage, G. V., Saenz, R., Bravo, M. et al. (1999). Regulation of expression of avirulence gene avrRxv and identification of a family of host interaction factors by sequence analysis of *avrBsT*. *Mol. Plant Microbe Interact. 12,* 35-44.

Crosa, J. H., & Walsh, C. T. (2002). Genetics and assembly line enzymology of siderophore biosynthesis in bacteria. *Microbiol. Mol. Biol. Rev. 66,* 223-249.

Crossman, L., & Dow, J. M. (2004). Biofilm formation and dispersal in *Xanthomonas campestris*. *Microbes Infect. 6,* 623-629.

da Silva, A. C., Ferro, J. A., Reinach, F. C., Farah, C. S., Furlan, L. R., Quaggio, R. B., et al. (2002). Comparison of the genomes of two *Xanthomonas* pathogens with differing host specificities. *Nature, 417,* 459-463.

da Silva, F. G., Shen, Y., Dardick, C., Burdman, S., Yadav, R. C., de Leon, A. L., et al. (2004). Bacterial genes involved in type I secretion and sulfation are required to elicit the rice Xa21-mediated innate immune response. *Mol. Plant Microbe Interact. 17,* 593-601.

de Crecy-Lagard, V., Glaser, P., Lejeune, P., Sismeiro, O., Barber, C. E., Daniels, M. J., et al. (1990). A *Xanthomonas campestris* pv. *campestris* protein similar to catabolite activation factor is involved in regulation of phytopathogenicity. *J. Bacteriol. 172*, 5877-5883.

De Feyter, R., & Gabriel, D. W. (1991). At least six avirulence genes are clustered on a 90-kilobase plasmid in *Xanthomonas campestris* pv. *malvacearum*. *Mol. Plant Microbe Interact. 4,* 423-432.

De Feyter, R., Yang, Y., & Gabriel, D. W. (1993). Gene-for-genes interactions between cotton R genes and *Xanthomonas campestris* pv. *malvacearum avr* genes. *Mol. Plant Microbe Interact. 6,* 225-237.

Dellagi, A., Reis, D., Vian, B., & Expert, D. (1999). Expression of the ferrioxamine receptor gene of *Erwinia amylovora* CFBP 1430 during pathogenesis. *Mol. Plant Microbe Interact. 12,* 463-466.

Dharmapuri, S., & Sonti, R. V. (1999). A transposon insertion in the *gumG* homologue of *Xanthomonas oryzae* pv. *oryzae* causes loss of extracellular polysaccharide production and virulence. *FEMS Microbiology Letters, 179,* 53-59.

Dharmapuri, S., Yashitola, J., Vishnupriya, M. R., Sonti, R. V. (2001). Novel genomic locus with atypical G+C content that is required for extracellular polysaccharide production and virulence in *Xanthomonas oryzae* pv. *oryzae*. *Mol. Plant Microbe Interact. 14,* 1335-1339.

Dow, J. M., Clarke, B. R., Milligan, D. E., Tang, J. -L., & Daniels, M. J. (1990). Extracellular proteases from *Xanthomonas campestris* pv. *campestris*, the black rot pathogen. *Appl.. Environ. Microbiol. 56,* 2994-2998.

Dow, J. M., Crossman, L., Findlay, K., He, Y. Q., Feng, J. X., & Tang, J. L. (2003). Biofilm dispersal in *Xanthomonas campestris* is controlled by cell-cell signaling and is required for full virulence to plants. *Proc Natl Acad Sci USA. 100,* 10995-11000.

Dow, J. M., Feng, J. –X., Barber, C. E., Tang, J. –L., & Daniels, M. J. (2000b). Novel genes involved in the regulation of pathogenicity factor production within the *rpf* gene cluster of *Xanthomonas campetris*. *Microbiology, 146,* 885-891.

Dow, J. M., Milligan, D. E., Jamieson, L., Barber, C. E., & Daniels, M. J. (1989). Molecular cloning of a polygalacturonate lyase gene from *Xanthomonas campestris* pv. *campestris* and role of the gene product in pathogenicity. *Physiol. Mol. Plant Pathol. 35,* 113-120.

Dow, M., Newman, M. A., & von Roepenack, E. (2000a). The induction and modulation of plant defense responses by bacterial lipopolysaccharides. *Annu. Rev. Phytopathol. 38,* 241-261.

Dow, J. M., Osbourn, A. E., Wilson, T. J., Daniels, M. J. (1995). A locus determining pathogenicity of *Xanthomonas campestris* is involved in lipopolysaccharide biosynthesis. *Mol. Plant Microbe Interact. 8,* 768-777.

Dow, J. M., Scofield, G., Trafford, K., Turner, P. C., & Daniels, M. J. (1987). A gene cluster in *Xanthomonas campestris* pv. *campestris* required for pathogenicity controls the excretion of polygalacturonate lyase and other enzymes. *Physiol. Mol. Plant Pathol. 31*, 261-271.

Dowson, W. J. (1939). On the systematic position and generic names of the gram negative bacterial plant pathogens. *Zentr. Bakteriol. Parasitenk. Abt. II, 100*, 177-193.

Ehrhardt, D. W., Atkinson, E. M., Faull, K. F., Freedberg, D. I., Sutherlin, D. P., Armstrong, R., et al. (1995). In vitro sulfotransferase activity of NodH, a nodulation protein of *Rhizobium meliloti* required for host-specific nodulation. *J. Bacteriol. 177*, 6237-6245.

El Tahir, Y., Kuusela, P., & Skurnik, M. (2000). Functional mapping of the *Yersinia enterocolitica* adhesin YadA. Identification of eight NSVAIG – S motifs on the amino-terminal half of the protein involved in collagen binding. *Mol. Microbiol. 37,* 192-206.

Enard, C., & Expert, D. (2000). Characterization of a *tonB* mutation in *Erwinia chrysanthemi* 3937: $TonB_{Ech}$ is a member of the enterobacterial TonB family. *Microbiology, 146,* 2051-2058.

Etchegaray, A., Silva-Stenico, M. E., Moon, D. H., & Tsai, S. M. (2004). In silico analysis of nonribosomal peptide synthetases of *Xanthomonas axonopodis* pv. *citri*: identification of putative siderophore and lipopeptide biosynthetic genes. *Microbiol. Res. 159,* 425-437.

Expert, D. (1999). Withholding and exchanging iron: Interactions Between *Erwinia spp*. and their plant hosts. *Annu. Rev. Phytopathol. 37,* 307-334.

Feng, T. Y., & Guo, T. T. (1975). Bacterial leaf blight of rice plant. VI. Chemotactic response of *Xanthomonas oryzae* pv. *oryzae* to water droplets exudated form water pores on the leaf of rice plants. *Bot. Bull. Acad. Sin. (Taipei).* 16, 126-136.

Fenselau, S., Balbo, I., & Bonas, U. (1992). Determinants of pathogenicity in *Xanthomonas campestris* pv. *vesicatoria* are related to proteins involved in secretion in bacterial pathogens of animals. *Mol. Plant Microbe Interact. 5*, 390-396.

Fenselau, S., & Bonas, U. (1995). Sequence and expression analysis of the *hrpB* pathogenicity operon of *Xanthomonas campestris* pv. *vesicatoria* which encodes eight proteins with similarity to components of the Hrp, Ysc, Spa, and Fli secretion systems. *Mol. Plant Microbe Interact. 8*, 845-854.

Furutani, A., Tsuge, S., Ohnishi, K., Hikichi, Y., Oku, T., Tsuno, K., et al. (2004). Evidence for HrpXo-dependent expression of type II secretory proteins in *Xanthomonas oryzae* pv. *oryzae*. *J. Bacteriol. 186,* 1374-1380.

Goel, A. K., Rajagopal, L., Nagesh, N., Sonti, R. V. (2002). Genetic locus encoding functions involved in biosynthesis and outer membrane localization of xanthomonadin in *Xanthomonas oryzae* pv. *oryzae*. *J. Bacteriol. 184*, 3539-3548.

Goel, A. K., Rajagopal, L., & Sonti R.V. (2001). Pigment and virulence deficiencies associated with mutations in the *aroE* gene of *Xanthomonas oryzae* pv. *oryzae*. *Appl. Environ. Microbiol. 67*, 245-250.

Gough, C. L., Dow, J. M., Barber, C. E., & Daniels, M. J. (1988). Cloning of two endoglucanase genes of *Xanthomonas campestris* pv. *campestris*: analysis of the role of the major endoglucanase in pathogenesis. *Mol. Plant Microbe Interact. 1,* 275-281.

Gough, C. L., Genin, S., Zischek, C., & Boucher, C. A. (1992). *hrp* genes of *Pseudomonas solanacearum* are homologous to pathogenicity determinants of animal pathogenic bacteria and are conserved among plant pathogenic bacteria. *Mol. Plant Microbe Interact. 5*, 384-389.

Gu, K., Yang, B., Tian, D., Wu, L., Wang, D., Sreekala, C., et al. (2005). R gene expression induced by a type-III effector triggers disease resistance in rice. *Nature, 435*, 1122-1125.

Herbers, K., Conrads-Strauch, J., & Bonas, U. (1992). Race-specificity of plant resistance to bacterial spot disease determined by repetitive motifs in a bacterial avirulence protein. *Nature*, *356*, 172-174.

Hoiczyk, E., Roggenkamp, A., Reichenbecher, M., Lupas A., & Heesemann, J. (2000). Structure and sequence analysis of Yersinia YadA and Moraxella UspAs reveal a novel class of adhesins. *EMBO J. 19*, 5989-5999.

Holland, I. B., Schmitt, L., & Young, J. (2005). Type 1 protein secretion in bacteria, the ABC-transporter dependent pathway. *Mol. Membr. Biol. 22*, 29-39.

Hopkins, C. M., White, F. F., Choi, S. H., Guo, A., & Leach, J. E. (1992). Identification of a family of avirulence genes from *Xanthomonas oryzae* pv. *oryzae*. *Mol. Plant Microbe Interact. 5*, 451-459.

Hotson, A., Chosed, R., Shu, H., Orth, K., & Mudgett, M. B. (2003). *Xanthomonas* type III effector XopD targets SUMO-conjugated proteins *in planta*. *Mol. Microbiol. 50,* 377-389.

Hsiao, Y. M., Liao, H. Y., Lee, M. C., Yang, T. C., & Tseng, Y. H. (2005). Clp upregulates transcription of *engA* gene encoding a virulence factor in *Xanthomonas campestris* by direct binding to the upstream tandem Clp sites. *FEBS Lett. 579*, 3525-3533.

Hsiao, Y. M., & Tseng, Y. H. (2002). Transcription of *Xanthomonas campetris prtI* gene encoding protease I increases during stationary phase and requires global transcription factor Clp. *Biochem. Biophy. Res. Commun. 295,* 43-49.

Hu, N. -T., Hung, M. -N, Chiou, S. -J., Tang, F., Chiang, D. -C., Huang, H. -Y. et al. (1992). Cloning and characterization of a gene required for the secretion of extracellular enzymes across the outer membrane by *Xanthomonas campestris* pv. *campestris*. *J. Bacteriol. 174,* 2679-2687.

Huang, G., Zhang, L., & Birch, R. G. (2000a). Analysis of the genes flanking *xabB*: a methyltransferase gene is involved in albicidin biosynthesis in *Xanthomonas albilineans*. *Gene, 255*, 327-333.

Huang, G., Zhang, L., & Birch, R. G. (2000b). Albicidin antibiotic and phytotoxin biosynthesis in *Xanthomonas albilineans* requires a phosphopantetheinyl transferase gene. *Gene, 258*, 193-199.

Huang, G., Zhang, L., & Birch, R. G. (2001). A multifunctional polyketide-peptide synthetase essential for albicidin biosynthesis in *Xanthomonas albilineans*. *Microbiology, 147,* 631-642.

Huguet, E., Hahn, K., Wengelnik, K., & Bonas, U. (1998). *hpaA* mutants of *Xanthomonas campestris* pv. *vesicatoria* are affected in pathogenicity but retain the ability to induce host-specific hypersensitive reaction. *Mol. Microbiol. 29,* 1379-1390.

Innes R. (2003). New effects of type III effectors. *Mol. Microbiol. 50*, 363-365.

Iwamoto, M., & Oku, T. (2000). Cloning and molecular characterization of *hrpX* from *Xanthomonas axonopodis* pv. *citri*. *DNA Seq. 11*, 167-173.

Jansson, P. E., Kenne, L., & Lindberg, B. (1975). Structure of extracellular polysaccharide from *Xanthomonas campestris*. *Carbohydr. Res. 45,* 275-282.

Jenkins, C. L., & Starr, M. P. (1982). The brominated aryl polyene (Xanthomonadin) pigments of *X. juglandis* protect against photobiological damage. *Curr. Microbiol. 7,* 323-326.

Jha, G., Rajeshwari, R., & Sonti, R. V. (2005). Bacterial type two secretion system secreted proteins: double-edged swords for plant pathogens. *Mol. Plant Microbe Interact. 18,* 891-898.

Josenhans, C., & Suerbaum, S. (2002). The role of motility as a virulence factor in bacteria. *Int. J. Med. Microbiol. 291*, 605-614.

Kamdar, H. V., Kamoun, S., & Kado, C. I. (1993). Restoration of pathogenicity of avirulent *Xanthomonas oryzae* pv. *oryzae* and *X. campestris* pathovars by reciprocal complementation with the *hrpXo* and *hrpXc* genes and identification of HrpX function by sequence analyses. *J. Bacteriol. 175*, 2017-2025.

Kamoun, S., & Kado, C. I. (1990). A plant-inducible gene of *Xanthomonas campestris* pv. *campestris* encodes an exocellular component required for growth in the host and hypersensitivity on nonhosts. *J. Bacteriol. 172*, 5165-5172.

Karlinsey, J. E., Lonner, J., Brown, K. L., & Hughes, K. T. (2000). Translation/secretion coupling by type III secretion systems. *Cell, 102*, 487-497.

Katzen, F., Becker, A., Zorreguieta, A., Pühler, A., Ielpi, L. (1996). Promoter analysis of the *Xanthomonas campestris* pv. *campestris gum* operon directing biosynthesis of the xanthan polysaccharide. *J. Bacteriol. 178*, 4313-4318.

Katzen, F., Ferreiro, D. U., Oddo, C. G., Ielmini, M. V., Becker, A., Pühler, A. et al. (1998). *Xanthomonas campestris* pv. *campestris gum* mutants: effects on xanthan biosynthesis and plant virulence. *J. Bacteriol. 180*, 1607-1617.

Kearney, B., & Staskawicz, B. J. (1990). Widespread distribution and fitness contribution of *Xanthomonas campestris* avirulence gene *avrBs2*. *Nature*, *346*, 385-386.

Keshavarzi, M., Soylu, S., Brown, I., Bonas, U., Nicole, M., Rossiter, J., et al. (2004). Basal defenses induced in pepper by lipopolysaccharides are suppressed by *Xanthomonas campestris* pv. *vesicatoria*. *Mol. Plant Microbe Interact. 17*, 805-815.

Kim, H. -S., Park, H. -J ., Heu, S., & Jung, J. (2004). Molecular and functional characterization of a unique sucrose hydrolase from *Xanthomonas axonopodis* pv. *glycines*. *J. Bacteriol. 186,* 411-418.

Kim, J. -G., Park, B. K., Yoo, C. -H., Jeon, E., Oh, J., & Hwang, I. (2003). Characterization of the *Xanthomonas axonopodis* pv. *glycines* Hrp pathogenicity island. *J. Bacteriol. 185*, 3155-3166.

Kingsley, M. T., Gabriel, D. W., Marlow, G. C., & Roberts, P. D. (1993). The *opsX* locus of *Xanthomonas campestris* affects host range and biosynthesis of lipopolysaccharide and extracellular polysaccharide. *J. Bacteriol. 175*, 5839–5850.

Köplin, R., Arnold, W., Hötte, B., Simon, R., Wang, G., & Pühler, A. (1992). Genetics of xanthan production in *Xanthomonas campestris*: the *xanA* and *xanB* genes are involved in UDP-glucose and GDP-mannose biosynthesis. *J. Bacteriol. 174*, 191-199.

Köplin, R., Wang, G., Hötte, B., Priefer, U. B., & Pühler, A. (1993). A 3.9-kb DNA region of *Xanthomonas campestris* pv. *campestris* that is necessary for lipopolysaccharide production encodes a set of enzymes involved in the synthesis of dTDP-rhamnose. *J. Bacteriol. 175*, 7786-7792.

Lahaye, T., & Bonas, U. (2001). Molecular secrets of bacterial type III effector proteins. *Trends Plant Sci. 6*, 479-485.

Larsen, R. A., Myers, P. S., Skare, J. T., Seachord, C. L., Darveau, R. P., & Postle, K. (1996). Identification of TonB homologs in the family Enterobacteriaceae and evidence for conservation of TonB-dependent energy transduction complexes. *J. Bacteriol. 178*, 1363-1373.

Lee, B. -M., Park, Y. -J., Park, D. -S., Kang, H. -W., Kim, J. -G., Song, E. -S., et al. (2005). The genome sequence of *Xanthomonas oryzae* pathovar *oryzae* KACC10331, the bacterial blight pathogen of rice. *Nucl. Acids Res. 33,* 577-586.

Lee, M. C., Weng, S. F., & Tseng, Y. H. (2003). Flagellin gene *fliC* of *Xanthomonas campestris* is upregulated by transcription factor Clp. *Biochem. Biophys. Res. Commun. 307*, 647-652.

Lima, W. C., Van Sluys, M. A., & Menck, C. F. (2005). Non-gamma-proteobacteria gene islands contribute to the Xanthomonas genome. *OMICS. 9*, 160-172.

Loprasert, S., Sallabhan, R., Atichartpongkul, S., & Mongkolsuk, S. (1999). Characterization of a ferric uptake regulator (*fur*) gene from *Xanthomonas campestris* pv. *phaseoli* with unusual primary structure, genome organization, and expression patterns. *Gene, 239,* 251-258.

Martinez, S., Martinez-Salazar, J., Camas, A., Sanchez, R., & Soberon-Chavez, G. (1997). Evaluation of the role of *recA* protein in plant virulence with *recA* mutants of *Xanthomonas campestris* pv. *campestris*. *Mol. Plant Microbe Interact. 10,* 911-916.

Metz, M., Dahlbeck, D., Morales, C. Q., Al Sady, B., Clark, E. T., & Staskawicz, B. J. (2005). The conserved *Xanthomonas campestris* pv. *vesicatoria* effector protein XopX is a virulence factor and suppresses host defense in *Nicotiana benthamiana*. *Plant J. 41*, 801-814.

Meyer, A., Pühler, A., & Niehaus, K. (2001). The lipopolysaccharides of the phytopathogen *Xanthomonas campestris* pv. *campestris* induce an oxidative burst reaction in cell cultures of *Nicotiana tabacum*. *Planta, 213*, 214-222.

Monteiro-Vitorello, C. B., de Oliveira, M. C., Zerillo, M. M., Varani, A. M., Civerolo, E., & Van Sluys, M. A. (2005). Xylella and Xanthomonas Mobil'omics. *OMICS, 9*, 146-159.

Moreira, L. M., de Souza, R. F., Almeida, N. F. Jr., Setubal, J. C., Oliveira, J. C., Furlan, L. R., Ferro, J. A., & da Silva, A. C. (2004). Comparative genomics analyses of citrus-associated bacteria. *Annu. Rev. Phytopathol. 42,* 163-184.

Moreira, L. M., De Souza, R. F., Digiampietri, L. A., Da Silva, A. C. R., & Setubal, J. C. (2005). Comparative analyses of *Xanthomonas* and *Xylella* complete genomes. *OMICS, 9,* 43-76.

Neilands, J. B. (1995). Siderophores: structure and function of microbial iron transport compounds. *J. Biol. Chem. 270,* 26723-26726.

Newman, M. -A., Daniels, M. J., & Dow, J. M. (1997). The activity of lipid A and core components of bacterial lipopolysaccharides in the prevention of the hypersensitive response in pepper. *Mol. Plant Microbe Interact. 10,* 926-928.

Newman, M. -A., von Roepenack-Lahaye, E., Daniels, M. & Dow, M. (2000). Lipopolysaccharides and plant responses to phytopathogenic bacteria. *Molecular Plant Pathology*, *1*, 25-31.

Newman, M. -A., von Roepenack-Lahaye, E., Parr, A., Daniels, M. J., & Dow, J. M. (2002). Prior exposure to lipopolysaccharide potentiates expression of plant defenses in response to bacteria. *Plant J. 29,* 487-495.

Nöel, L., Thieme, F., Nennstiel, D., & Bonas, U. (2001). cDNA-AFLP analysis unravels a genome-wide *hrpG*-regulon in the plant pathogen *Xanthomonas campestris* pv. *vesicatoria*. *Mol. Microbiol. 41*, 1271-1281.

Nöel, L., Thieme, F., Nennstiel, D. & Bonas, U. (2002). Two novel type III system-secreted proteins of *Xanthomonas campestris* pv. *vescicatoria* are encoded within the *hrp* pathogenicity island. *J. Bacteriol. 184*, 1340-1348.

Patil, P. B., & Sonti, R. V. (2004). Variation suggestive of horizontal gene transfer at a lipopolysaccharide (*lps*) biosynthetic locus in *Xanthomonas oryzae* pv. *oryzae*, the bacterial leaf blight pathogen of rice. *BMC Microbiol. 4*, 40. doi: 10.1186/1471-2180-4-40.

Poplawsky, A. R., & Chun, W. (1998). *Xanthomonas campestris* pv. *campestris* requires a functional *pigB* for epiphytic survival and host infection. *Mol. Plant Microbe Interact. 11,* 466-475.

Poplawsky, A. R., Kawalek, M. D., & Schaad, N. W. (1993). A xanthomonadin-encoding gene cluster for the identification of pathovars of *Xanthomonas campestris*. *Mol. Plant Microbe Interact. 6*, 545–552.

Poplawsky, A. R., Urban, S. C., & Chun, W. (2000). Biological role of xanthomonadin pigments in *Xanthomonas campestris* pv. *campestris*. *Appl. Environ. Microbiol. 66*, 5123-5127.

Pugsley, A. P. (1993). The complete general secretory pathway in gram-negative bacteria. *Microbiol. Rev. 57*, 50-108.

Qian, W., Jia, Y., Ren, S. X., He, Y. Q., Feng, J. X., Lu, L. F., et al. (2005). Comparative and functional genomic analyses of the pathogenicity of phytopathogen *Xanthomonas campestris* pv. *campestris*. *Genome Res. 15*, 757-67.

Rajagopal, L., Sundari, C. S., Balasubramanian, D., Sonti, R. V. (1997). The bacterial pigment xanthomonadin offers protection against photodamage. *FEBS Lett. 415*, 125-128.

Rajeshwari, R., Jha, G., & Sonti, R. V. (2005). Role of an *in planta* expressed xylanase of *Xanthomonas oryzae* pv. *oryzae* in promoting virulence on rice. *Mol. Plant Microbe Interact. 18,* 830-837.

Rajeshwari, R., & Sonti, R. V. (2000). Stationary-phase variation due to transposition of novel insertion elements in *Xanthomonas oryzae* pv. *oryzae*. *J. Bacteriol. 182,* 4797-4802.

Ray, S. K., Rajeshwari, R., & Sonti, R. V. (2000). Mutants of *Xanthomonas oryzae* pv. *oryzae* deficient in general secretory pathway are virulence deficient and unable to secrete xylanase. *Mol. Plant Microbe Interact. 13,* 394-401.

Ray, S. K., Rajeshwari, R., Sharma, Y., & Sonti, R. V. (2002). A high-molecular-weight outer membrane protein of *Xanthomonas oryzae* pv. *oryzae* exhibits similarity to non-fimbrial adhesins of animal pathogenic bacteria and is required for optimum virulence. *Mol. Microbiol. 46,* 637-647.

Reimers, P. J., & Leach, J. E. (1991). Race-specific resistance to *Xanthomonas oryzae* pv. *oryzae* conferred by bacterial blight resistance gene *Xa-10* in rice (*Oryza sativa*) involves accumulation of a lignin-like substance in host tissues. *Physiol. Mol. Plant Pathol. 38,* 39-55.

Robey, M., & Cianciotto, N. P. (2002). *Legionella pneumophila feoAB* promotes ferrous iron uptake and intracellular infection. *Infect. Immun. 70,* 5659-5669.

Roche, P., Debelle, F., Maillet, F., Lerouge, P., Faucher, C., Truchet, G., et al. (1991). Molecular basis of symbiotic host specificity in *Rhizobium meliloti*: *nodH* and *nodPQ* genes encode the sulfation of lipo-oligosaccharide signals. *Cell, 67,* 1131-1143.

Roden, J. A., Belt, B., Ross, J. B., Tachibana, T., Vargas, J., & Mudgett, M. B. (2004b). A genetic screen to isolate type III effectors translocated into pepper cells during *Xanthomonas* infection. *Proc. Natl. Acad. Sci. USA. 101,* 16624-16629.

Roden, J., Eardley, L., Hotson, A., Cao, Y., & Mudgett, M. B. (2004a). Characterization of the *Xanthomonas* AvrXv4 effector, a SUMO protease translocated into plant cells. *Mol. Plant Microbe Interact. 17,* 633-643.

Rojas, C. M., Deng, W. -L., Doyle, J. J., & Collmer, A. (2002). HecA, a member of a class of adhesins produced by diverse pathogenic bacteria, contributes to the attachment, aggregation, epidermal cell killing, and virulence phenotypes of *Erwinia chrysanthemi* EC16 on *Nicotiana clevelandii* seedlings. *Proc. Natl. Acad. Sci. USA 99,* 13142-13147.

Ronald, P. C., & Staskawicz, B. J. (1988). The avirulence gene *avrBs1* from *Xanthomonas campestris* pv. *vesicatoria* encodes a 50-kD protein. *Mol. Plant Microbe Interact. 1,* 191-198.

Rossier, O., Van den Ackerveken, G., & Bonas, U. (2000). HrpB2 and HrpF from Xanthomonas are type III-secreted proteins and essential for pathogenicity and recognition by the host plant. *Mol. Microbiol. 38,* 828-838.

Rott, P. C., Costet, L., Davis, M. J., Frutos, R., & Gabriel, D. W. (1996). At least two separate gene clusters are involved in albicidin production by *Xanthomonas albilineans*. *J. Bacteriol. 178,* 4590-4596.

Royer, M., Costet, L., Vivien, E., Bes, M., Cousin, A., Damais, A., et al. (2004). Albicidin pathotoxin produced by *Xanthomonas albilineans* is encoded by three large PKS and NRPS genes present in a gene cluster also containing several putative modifying, regulatory, and resistance genes. *Mol. Plant Microbe Interact. 17,* 414-427.

Schulte, R., & Bonas, U. (1992). Expression of the *Xanthomonas campestris* pv. *vesicatoria hrp* gene cluster, which determines pathogenicity and hypersensitivity on pepper and tomato, is plant inducible. *J. Bacteriol. 174,* 815-823.

Shen, Y., Chern, M., Silva, F. G., & Ronald, P. (2001). Isolation of a *Xanthomonas oryzae* pv. *oryzae* flagellar operon region and molecular characterization of flhF. *Mol. Plant Microbe Interact. 14,* 204-213.

Shen, Y., Sharma, P., da Silva, F. G., & Ronald, P. (2002). The *Xanthomonas oryzae* pv. *oryzae raxP* and *raxQ* genes encode an ATP sulphurylase and adenosine-5'-phosphosulphate kinase that are required for AvrXa21 avirulence activity. *Mol. Microbiol. 44,* 37-48.

Simpson, A. J., Reinach, F. C., Arruda, P., Abreu, F. A., Acencio, M., Alvarenga, R., et al. (2000). The genome sequence of the plant pathogen *Xylella fastidiosa*. *Nature, 406,* 151-157.

Slater, H., Alvarez-Morales, A., Barber, C. E., Daniels, M. J., & Dow, J. M. (2000). A two-component system involving an HD-GYP domain protein links cell-cell signalling to pathogenicity gene expression in *Xanthomonas campestris*. *Mol. Microbiol. 38,* 986-1003.

Song, W. -Y., Wang, G. -L., Chen, L. -L., Kim, H. -S., Pi, L. -Y., Holsten T, et al. (1995). A receptor kinase-like protein encoded by the rice disease resistance gene, *Xa21*. *Science*, *270*, 1804-1806.

Soto, G. E., & Hultgren, S. J. (1999). Bacterial adhesins: common themes and variations in architecture and assembly. *J. Bacteriol. 181*, 1059-1071.

Stojiljkovic, I., Cobeljic, M., & Hantke, K. (1993). *Escherichia coli* K-12 ferrous iron uptake mutants are impaired in their ability to colonize the mouse intestine. *FEMS Microbiol. Lett. 108,* 111-115.

Stojiljkovic, I., & Srinivasan, N. (1997). *Neisseria meningitidis tonB, exbB*, and *exbD* genes: Ton-dependent utilization of protein-bound iron in *Neisseriae*. *J. Bacteriol. 179*, 805-812.

Subramoni, S., & Sonti, R. V. (2005). Growth deficiency of a *Xanthomonas oryzae* pv. *oryzae fur* mutant in rice leaves is rescued by ascorbic acid supplementation. *Mol. Plant Microbe Interact. 18*, 644-651.

Sugio, A., Yang, B., & White, F. F. (2005). Characterization of the *hrpF* pathogenicity peninsula of *Xanthomonas oryzae* pv. *oryzae*. *Mol. Plant Microbe Interact. 18*, 546-554.

Sun, Q. H., Hu, J., Huang, G. X., Ge, C., Fang, R. X., & He, C. Z. (2005). Type-II secretion pathway structural gene *xpsE*, xylanase- and cellulase-secretion and virulence in *Xanthomonas oryzae* pv. *oryzae*. *Plant Pathology*, *54*, 15-21.

Swarup, S., Yang, Y., Kingsley, M. T., & Gabriel, D. W. (1992). A *Xanthomonas citri* pathogenicity gene, *pthA*, pleiotropically encodes gratuitous avirulence on nonhosts. *Mol. Plant Microbe Interact. 5*, 204-213.

Takeuchi, Y., Tohbaru, M., & Sato, A. (1994). Polysaccharide in primary cell walls of rice cells in suspension culture. *Phytochemistry, 35*, 361-363.

Tanabe, T., Funahashi, T., Nakoa, H., Miyoshi, S. -I., Shinoda, S., & Yamamoto, S. (2003). Identification and characterization of genes required for biosynthesis and transport of the siderophore vibrioferrin in *Vibrio parahaemolyticus*. *J. Bacteriol. 185,* 6938-6949.

Tang, D. -J., Li, X. -J., He, Y. -Q., Feng, J. -X., Chen, B., Tang, J. -L. (2005). The zinc uptake regulator Zur is essential for the full virulence of *Xanthomonas campestris* pv. *campestris*. *Mol. Plant Microbe Interact. 18*, 652-658.

Tang, J. -L., Feng, J. -X., Li, Q. -Q., Wen, H. -X., Zhou, D. -L., Wilson, T. J., et al. (1996). Cloning and characterization of the *rpfC* gene of *Xanthomonas oryzae* pv. *oryzae*: involvement in exopolysaccharide production and virulence to rice. *Mol. Plant Microbe Interact. 9,* 664-666.

Tang, J. -L., Liu, Y. -N., Barber, C. E., Dow, J. M., Wootton, J. C., & Daniels, M. J. (1991). Genetic and molecular analysis of a cluster of *rpf* genes involved in positive regulation of synthesis of extracellular enzymes and polysaccharide in *Xanthomonas campestris* pathovar *campestris*. *Mol. Gen. Genet. 226,* 409-417.

Touati, D. (2000). Iron and oxidative stress in bacteria. *Arch. Biochem. Biophys. 373*, 1-6.

Tseng, Y. -H., Choy, K. -T., Hung, C. -H., Lin, N. -T., Liu, J.- Y., Lou, C. -H., et al. (1999). Chromosome map of *Xanthomonas campestris* pv. *campestris* 17 with locations of genes involved in xanthan gum synthesis and yellow pigmentation. *J. Bacteriol. 181,* 117-125.

Tsolis, R. M., Baumler, A. J., Heffron, F., & Stojiljkovic, I. (1996). Contribution of TonB- and Feo-mediated iron uptake to growth of *Salmonella typhimurium* in the mouse. *Infect. Immun. 64,* 4549-4556.

Tsuge, S., Furutani, A., Kubo, Y., & Horino, O. (2001). Identification of a H+/glucose and galactose symporter gene *glt* from *Xanthomonas oryzae* pv. *oryzae*. *Microbiol. Immunol. 45*, 543-547.

Tsuge, S., Ochiai, H., Inoue, Y., Oku, T., Tsuno, K., & Kaku, H., et al. (2004). Involvement of Phosphoglucose Isomerase in Pathogenicity of *Xanthomonas oryzae pv. oryzae*. *Phytopathology, 94*, 478-483.

Tsuge, S., Terashima, S., Furutani, A., Ochiai, H., Oku, T., Tsuno, K., et al. (2005). Effects on promoter activity of base substitutions in the cis-acting regulatory element of HrpXo regulons in *Xanthomonas oryzae* pv. *oryzae*. *J. Bacteriol.* *187*, 2308-2314.

Tung, S. Y., & Kuo, T. T. (1999). Requirement for phosphoglucose isomerase of *Xanthomonas campestris* in pathogenesis of citrus canker. *Appl. Environ. Microbiol.* *65*, 5564-5570.

Van den Ackerveken, G., Marois, E., & Bonas, U. (1996). Recognition of the bacterial avirulence protein AvrBs3 occurs inside the host plant cell. *Cell*, *87*, 1307-1316.

Van Gijsegem, F., Gouch, C., Zischek, C., Niqueux, E., Arlat, M., Genin, S., et al. (1995). The *hrp* gene locus of *Pseudomonas solanacearum*, which controls the production of a type III secretion system, encodes eight proteins related to components of the bacterial flagellar biogenesis complex. *Mol. Microbiol.* *15,* 1095-1114.

Van Sluys, M. A., de Oliveira, M. C., Monteiro-Vitorello, C. B., Miyaki, C. Y., Furlan, L. R., Camargo, L. E., et al. (2003). Comparative analyses of the complete genome sequences of Pierce's disease and citrus variegated chlorosis strains of *Xylella fastidiosa*. *J. Bacteriol.* *185,* 1018-1026.

Van Sluys, M. A., Monteiro-Vitorello, C. B., Camargo, L. E., Menck, C. F., Da Silva, A. C., Ferro, J. A., et al. (2002). Comparative genomic analysis of plant-associated bacteria. *Annu. Rev. Phytopathol.* *40*, 169-189.

Vauterin, L., Hoste, B., Kersters, K., & Swings, J. (1995). Reclassification of Xanthomonas. *Int. J. Syst. Bacteriol.* *45*, 472-489.

Velayudhan, J., Hughes, N. J., McColm, A. A., Bagshaw, J., Clayton, C. L., Andrews, S. C., & Kelly, D. J. (2000). Iron acquisition and virulence in *Helicobacter pylori*: a major role for FeoB, a high-affinity ferrous iron transporter. *Mol. Microbiol.* *37,* 274-286.

Vojnov, A. A., Slater, H., Daniels, M. J., & Dow, J. M. (2001). Expression of the *gum* operon directing xanthan biosynthesis in *Xanthomonas campestris* and its regulation *in planta*. *Mol. Plant Microbe Interact.* *14*, 768-774.

Vörholter, F. J., Niehaus, K., & Pühler, A. (2001). Lipopolysaccharide biosynthesis in *Xanthomonas campestris* pv. *campestris*: a cluster of 15 genes is involved in the biosynthesis of the LPS O-antigen and the LPS core. *Mol. Genet. Genomics*. *266*, 79-95.

Vörholter, F. J., Thias, T., Meyer, F., Bekel, T., Kaiser, O., Pühler, A., et al. (2003). Comparison of two *Xanthomonas campestris* pathovar *campestris* genomes revealed differences in their gene composition. *J. Biotechnol.* *106*, 193-202.

Voulhoux, R., Ball, G., Ize, B., Vasil, M. L., Lazdunski, A., Wu, L. F. & Filloux, A. (2001). Involvement of the twin-arginine translocation system in protein secretion via the type II pathway. *EMBO J.* *20,* 6735-6741.

Wall, M. K., & Birch, R. G. (1997). Genes for albicidin biosynthesis and resistance span at least 69 kb in the genome of *Xanthomonas albilineans*. *Lett. Appl. Microbiol.* *24*, 256-260.

Wang, G. -L., Song, W. -Y., Ruan, D. L., Sideris, S., & Ronald, P. C. (1996). The cloned gene, *Xa21*, confers resistance to multiple *Xanthomonas oryzae* pv. *oryzae* isolates in transgenic plants. *Mol. Plant Microbe Interact.* *9*, 850-855.

Wang, L. H., He, Y., Gao, Y., Wu, J. E., Dong, Y. H., He, C. et al. (2004). A bacterial cell-cell communication signal with cross-kingdom structural analogues. *Mol. Microbiol.* *51,* 903-912.

Wengelnik, K., & Bonas, U. (1996). HrpXv, an AraC-type regulator, activates expression of five out of six loci in the *hrp* cluster of *Xanthomonas campestris* pv. *vesicatoria*. *J. Bacteriol.* *178*, 3462-3469.

Wengelnik, K., Rossier, O. & Bonas, U. (1999). Mutations in the regulatory gene *hrpG* of *Xanthomonas campestris* pv. *vesicatoria* result in constitutive expression of all *hrp* genes. *J. Bacteriol.* *181*, 6828-6831.

Wengelnik, K., Van den Ackerveken, G., & Bonas, U. (1996). HrpG, a key *hrp* regulatory protein of *Xanthomonas campestris* pv. *vesicatoria* is homologous to two-component response regulators. *Mol. Plant Microbe Interact.* *9,* 704-712.

Whalen, M. C., Stall, R. E., & Staskawicz, B. J. (1988). Characterization of a gene from a tomato pathogen determining hypersensitive resistance in non-host species and genetic analysis of this resistance in bean. *Proc. Natl. Acad. Sci. USA. 85*, 6743-6747.

Wichmann, G., & Bergelson, J. (2004). Effectors genes of *Xanthomonas axonopodis* pv. *vesicatoria* promote transmission and enhance other fitness traits in the field. *Genetics, 166,* 693-706.

Wiggerich, H. G., & Pühler, A. (2000). The *exbD2* gene as well as the iron-uptake genes *tonB*, *exbB* and *exbD1* of *Xanthomonas campestris* pv. *campestris* are essential for the induction of a hypersensitive response on pepper (*Capsicum annuum*). *Microbiology, 146,* 1053-1060.

Wiggerich, H. G., Klauke, B., Köplin, R., Priefer, U. B., & Pühler, A. (1997). Unusual structure of the *tonB-exb* DNA region of *Xanthomonas campestris* pv. *campestris*: *tonB*, *exbB*, and *exbD1* are essential for ferric iron uptake, but *exbD2* is not. *J. Bacteriol. 179,* 7103-7110.

Wilderman, P. J., Sowa, N. A., FitzGerald, D. J., FitzGerald, P. C., Gottesman, S., Ochsner, U. A., & Vasil, M. L. (2004). Identification of tandem duplicate regulatory small RNAs in *Pseudomonas aeruginosa* involved in iron homeostasis. *Proc. Natl. Acad. Sci. USA. 101,* 9792-9797.

Wilson, T. J., Bertrand, N., Tang, J. -L., Feng, J. -X., Pan, M. -Q., Barber, C. E., et al. (1998). The *rpfA* gene of *Xanthomonas campestris* pathovar *campestris*, which is involved in the regulation of pathogenicity factor production, encodes an aconitase. *Mol. Microbiol. 28*, 961-970.

Wojtaszek, P. (1997). Oxidative burst: an early plant response to pathogen infection. *Biochem. J. 322*, 681-692.

Yang, B., & White, F. F. (2004). Diverse members of the AvrBs3/PthA family of type III effectors are major virulence determinants in bacterial blight disease of rice. *Mol. Plant Microbe Interact. 17*, 1192-1200.

Yang, B., Zhu, W., Johnson, L. B., & White, F. F. (2000). The virulence factor AvrXa7 of *Xanthomonas oryzae* pv. *oryzae* is a type III secretion pathway-dependent nuclear-localized double-stranded DNA-binding protein. *Proc. Natl. Acad.. Sci. USA. 97,* 9807-9812.

Yang, Y., De Feyter, R., & Gabriel, D. W. (1994). Host-specific symptoms and increased release of *Xanthomonas citri* and *X. campestris* pv. *malvacearum* from leaves are determined by the 102-bp tandem repeats of *pthA* and *avrb6*, respectively. *Mol. Plant Microbe Interact. 7*, 345-355.

Yang, Y., Yuan, Q., & Gabriel, D. W. (1996). Watersoaking function(s) of XcmH1005 are redundantly encoded by members of the *Xanthomonas avr/pth* gene family. *Mol. Plant Microbe Interact. 9*, 105-113.

Zhang, L., & Birch, R. G. (1997). The gene for albicidin detoxification from *Pantoea dispersa* encodes an esterase and attenuates pathogenicity of *Xanthomonas albilineans* to sugarcane. *Proc. Natl. Acad. Sci. USA. 94*, 9984-9989.

Zhang, L., Xu, J., & Birch, R. G. (1999). Engineered detoxification confers resistance against a pathogenic bacterium. *Nat. Biotechnol. 17*, 1021-1024.

Zhu, J., Oger, P. M., Schrammeijer, B., Hooykaas, P. J. J., Farrand, S. K., & Winans, S. C. (2000b). The bases of crown gall tumorigenesis. *J. Bacteriol. 182,* 3885-3895.

Zhu, W., MaGbanua, M. M., & White, F. F. (2000a). Identification of two novel hrp-associated genes in the *hrp* gene cluster of *Xanthomonas oryzae* pv. *oryzae*. *J. Bacteriol. 182*, 1844-1853.

Zhu, W., Yang, B., Chittoor, J. M., Johnson, L. B., & White, F. F. (1998). AvrXa10 contains an acidic transcriptional activation domain in the functionally conserved C terminus. *Mol. Plant Microbe Interact. 11*, 824-832.

5. AFFILIATION

Sujatha Subramoni, Gopaljee Jha & Ramesh V. Sonti, *Centre for Cellular and Molecular Biology, Uppal Road, Hyderabad-500007, India.*

TIMOTHY P. DENNY

PLANT PATHOGENIC *RALSTONIA* SPECIES

Abstract. *Ralstonia solanacearum* is an important phytopathogen that attacks many crops and native plants in warm and moist climates. In contrast, the closely related *Ralstonia syzygii* and blood disease bacterium attack primarily clove and banana, respectively, in and around Indonesia. This review should be useful both for scientist unfamiliar with these organisms and specialists in the field. It briefly examines the taxonomy of these organisms and summarizes the phenotypic and genetic diversity known to exist within the *R. solanacearum* species complex. The strategies used to detect and identify these pathogens are discussed. General aspects of the pathogens' life cycle, the disease symptoms on representative hosts, and approaches to disease control are described. The biochemical and genetic mechanisms underlying pathogenesis, including results from recent genomic analyses, and host responses to infection are summarized.

1. INTRODUCTION

Bacterial wilts of tomato, pepper, eggplant and Irish potato caused by *Ralstonia solanacearum* were among the first diseases that E. F. Smith proved to be caused by a bacterial pathogen. *R. solanacearum* is now known as one of the world's most important phytopathogenic bacteria due to its lethality, persistence, wide host range, and broad geographic distribution (Elphinstone, 2005). Although most troublesome in the tropics and subtropics, *R. solanacearum* is a continuing threat in temperate climates. Consequently, *R. solanacearum* is one of the more intensively studied phytopathogenic bacteria, and bacterial wilt of tomato is a model system for investigating mechanisms of pathogenesis.

Research on *R. solanacearum* up to the early 1950's, much of it in journals that are now difficult or impossible to access, was summarized by Kelman in a seminal monograph (Kelman, 1953). Since then, short reviews have usually focused on selected aspects of *R. solanacearum* pathobiology (Boucher, Gough, & Arlat, 1992; Buddenhagen & Kelman, 1964; Genin & Boucher, 2004; Hayward, 1991; Hayward, 1995; Hayward, 2000; Schell, 2000). In contrast, the international conferences held since 1985 generated six books that summarized most of the contemporaneous *R. solanacearum* research (Persley, 1986a; Hartman & Hayward, 1993; Hayward & Hartman, 1994; Prior, Allen, & Elphinstone, 1998; Allen, Prior, & Hayward, 2005). Although valuable for scientists specializing in this pathogen, their detail and depth make these reviews and books less useful for non-specialists. Consequently, I have endeavored to provide a wide-ranging review of *R. solanacearum* biology and pathology suitable for a broad audience.

S.S. Gnanamanickam (ed.). Plant-Associated Bacteria, 573–644.

2. TAXONOMY

Taxonomic groups should follow the natural divisions that are apparent after characterizing the phenotype and genotype of related organisms and determining their relationship to known taxa. Until recently, however, the knowledge necessary to discern natural groups was often unavailable and many groups were 'incorrectly' classified in phylogenetic terms. This was certainly true for the organisms now considered members of the *Burkholderia* group, which were long classed as non-fluorescent *Pseudomonas* species (Anzai, Kim, Park, Wakabayashi, & Oyaizu, 2000; Palleroni, 1984). However, recent genetic analyses have revealed many new relationships and prompted renaming many bacteria in this and other groups (one Web site with official nomenclature is http://www.bacterio.cict.fr/).

Rank	Name	Comments
Class:	β proteobacteria	one of four classes of 'purple bacteria'
'Family'	*Burkholderia* group	has nine genera (three shown)
Genus:	*Ralstonia*	has five species (three shown)
Species:	*pickettii*	human pathogen; common outgroup
	solanacearum complex	plant pathogen, wide host range
	syzygii	plant pathogen of cloves in Indonesia
	Burkholderia	has human and plant pathogens and symbionts
	Cupriavidus (*Wautersia*)	
	metallidurans	many orthologous genes in *R. solanacearum*
	necator	originally *Alcaligenes eutrophus*
	taiwanensis	nodulating, nitrogen-fixing *Mimosa* symbiont

Fig. 1. Abbreviated taxonomic list of the *Burkholderia* group.

2.1. Introduction to the Burkholderia Group

Figure 1 presents an abbreviated taxonomic list for the *Burkholderia* group, showing only the plant-associated or other relevant species. A complete list is available on several Web sites (e.g., TaxBrowser on the NCBI site (http://www.ncbi.nlm.nih.gov/)).

The relevant taxonomy of this group began in the early 1970's when nucleic acid hybridization revealed that fluorescent *Pseudomonas* species comprising RNA homology group I are unrelated to the non-fluorescent pseudomonads in group II (Hayward, 2000; Palleroni, 1984). However, group II organisms were not officially separated until Yabuuchi et al. (1992) established the genus *Burkholderia*, which included *B. cepacia*, *B. pickettii*, and *B. solanacearum* among its seven species. A few years later, the latter two species were transferred into the new genus *Ralstonia* along with *Alcaligenes eutrophus*, which was renamed *R. eutropha* (Yabuuchi, Kosako, Yano, Hotta, & Nishiuchi, 1995). Very recently, 16S rDNA sequence analyses and phenotypic differences (Table 1) stimulated Vaneechoutte et al. (2004) to divide the genus *Ralstonia* into two groups: the *R. pickettii* lineage and the *R. eutropha* lineage. In addition to three species found in human clinical samples, the genus *Ralstonia* now contains the plant pathogens *R. solanacearum* and the newly renamed *R. syzygii*. Species in the *R. eutropha* lineage were reassigned to the new genus *Wautersia*, with *W. eutropha* as the type species. Unexpectedly, only a few months later, Vandamme and Coeyne (2004) showed that *W. eutropha* is a later synonym of *Cupriavidus necator*, which prompted renaming the genus and eliminating *eutropha* as a specific epithet. The phylogenetic relationships of *Ralstonia* and *Cupriavidus* species based on a 16S rDNA gene sequence comparison are shown in Figure 2.

Table 1. Phenotypic differences between genus *Ralstonia* and genus *Cupriavidus* (adapted from Vaneechoutte et al., 2004).

Character	*Ralstonia*	*Cupriavidus*
flagellation	polar, 1-4[a]	peritrichous
colistin resistance	yes	no
viability on TSA at 25° C[b]	<6 days	>9 days
acid from glucose	yes	No

[a] In motile species. *R. syzygii* is nonmotile (Roberts, Eden-Green, Jones, & Ambler, 1990). Wild-type *R. solanacearum* is most motile in a rich broth medium at about 10^8 cfu/ml (Clough et al., 1997).
[b] Not applicable to *R. syzygii*, which does not grow on TSA (tryptic soy agar).

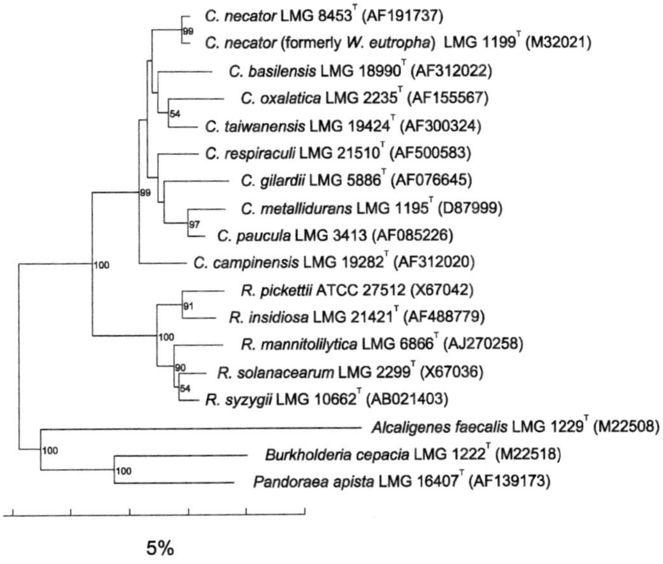

Fig. 2. Phylogenetic tree based on 16S rDNA sequence similarity. Reproduced from Vandamme & Coenye, 2004 (with permission).

2.2. *Ralstonia* Species

Ralstonia contains only five recognized species, many fewer than other genera in the *Burkholderia* group. *R. insidiosa*, *R. mannitolilytica*, and *R. pickettii* have been isolated from human clinical samples, and the latter two can be pathogenic (Coenye, Goris, De Vos, Vandamme, & LiPuma, 2003; De Baere et al., 2001; Ralston, Palleroni, & Doudoroff, 1973). *R. pickettii* has often been used as an outgroup in genetic studies of *R. solanacearum*, but *R. mannitolilytica* is more closely related to *R. solanacearum* (Fig. 2) (De Baere et al., 2001; Vandamme & Coenye, 2004; Vaneechoutte, Kampfer, De Baere, Falsen, & Verschraegen, 2004). They will not be discussed further.

This chapter focuses on *R. solanacearum*, which causes lethal wilting diseases on many crops and wild plants. Many of its most important hosts are in the nightshade family (*Solanaceae*), hence the specific epithet. This pathogen was briefly known as *Bacillus solanacearum* and then for most of the last century as *Pseudomonas solanacearum* (Saddler, 2000). *R. solanacearum* is an aerobic, Gram-negative rod with a polar flagella tuft. It is oxidase positive, arginine dihydrolase negative, and accumulates poly-β-hydroxybutyrate intracellularly. Most strains denitrify and also produce a diffusible brown pigment on rich medium. It does not grow at 4 or 40°C, and there is little or no growth in 2% NaCl. Its G + C content is 66.5-68 mol%. Numerical taxonomy, serology, and phylogenetic analyses, revealed that the blood disease (BD) bacterium, an Indonesian banana pathogen, is closely related to *R. solanacearum*, but its taxonomic standing remains undefined (Baharuddin, Rudolph, & Niepold, 1994; Coenye et al., 2003; Eden-Green, 1994; Taghavi, Hayward, Sly, & Fegan, 1996; Villa, Tsuchiya, Horita, Opina, & Hyakumachi, 2005).

Ralstonia syzygii, previously known as *Pseudomonas syzygii* (Roberts et al., 1990), is a pathogen of clove trees in Indonesia. Sumatra disease of cloves was first reported from western Sumatra over 80 years ago, and the pathogen may have evolved there. Numerical taxonomy (Eden-Green, 1994; Roberts et al., 1990) and sequence analysis of rDNA and other genes (De Baere et al., 2001; Poussier et al., 2000b; Poussier, Prior, Luisetti, Hayward, & Fegan, 2000a; Taghavi et al., 1996; Villa et al., 2005) showed that it is closely related to *R. solanacearum*, but is more distant than the BD bacterium. Unlike *R. solanacearum*, strains of *R. syzygii* grow poorly or not at all on standard media, are less nutritionally versatile, do not grow at 37°C, are more sensitive to NaCl, and are nonmotile and nonflagellate (Roberts et al., 1990). Its G + C content is 66-67 mol%, and DNA-DNA homology studies show that *R. syzygii* should be retained as a species rather than being incorporated into *R. solanacearum* or established as a subspecies (Roberts et al., 1990; Vaneechoutte et al., 2004).

3. DIVERSITY IN THE *R. SOLANACEARUM* SPECIES COMPLEX

R. solanacearum is present on all continents and many islands between the Tropics of Cancer and Capricorn. As a group, it causes disease on over 200 plant species in over 50 families (Hayward, 1994b; Kelman, 1953). In light of its geographic and pathogenic diversity, Ivan Buddenhagen, a veteran of the campaign to control bacterial wilt (BW) on banana in Central America, opined that there are many bacterial wilts and there are many *'Pseudomonas solanacearums'* They have originated and evolved in widely different places and they have different capabilities with both native flora and introduced hosts, and presumably with different soils and environmental conditions (Buddenhagen, 1986, p. 126).

The most recent genetic studies have proven that Dr. Buddenhagen was correct, and the BW pathogens are currently considered to be members of a species complex.

3.1. Races

Early attempts to codify the diversity present in the *R. solanacearum* species complex resulted in separate race and biovar systems, which greatly influenced thinking about *R. solanacearum* pathobiology during the last four decades. Buddenhagen et al. (1962) recognized three races based largely on host range at the plant species level using strains collected in the Americas (North, Central and South) and Caribbean up to the early 1960's. Unfortunately, this research was only published as an abstract, so a full description of this work is now unavailable. Buddenhagen later admitted that a formal paper was never prepared because their subsequent work revealed more variation among strains than originally observed (Buddenhagen & Kelman, 1964; Buddenhagen, 1986). Nevertheless, most *R. solanacearum* biologists and regulatory agencies have used (and sometimes misused) the system up to the present day.

Race 1 strains were originally described as affecting tobacco, tomato, many solanaceous weeds, some other weeds, and diploid bananas (e.g., *Musa* with AA and BB genotypes) (Buddenhagen, Sequeira, & Kelman, 1962). Over time, the acceptable host range was gradually expanded to include many other plants, including but not limited to groundnut, potato, pepper, eggplant, olive, ginger, strawberry, geranium, and *Eucalyptus*. One long recognized problem with the definition of race 1 is that some strains are highly virulent on tomato and eggplant but low in virulence on tobacco (Granada & Sequeira, 1975; Kelman & Person, 1961) and most of these induce a hypersensitive response (HR; a rapid defensive reaction (Klement, 1982)) when infiltrated into tobacco leaves (Granada & Sequeira, 1975; Robertson, Wechter, Denny, Fortnum, & Kluepfel, 2004). In contrast, strains virulent on tobacco almost always cause necrosis that appears 48-72 h after infiltration. Therefore, tobacco is not a good host for differentiating races. Another problem is that race 1 strains are phenotypically diverse and not part of a natural taxonomic group (Table 3) (Hayward, 1964; Hayward, 1994a).

Race 2 strains were described as pathogenic on either triploid bananas (AAA, AAB, ABB genotypes), *Heliconia* species or both (Buddenhagen et al., 1962; Thwaites, Eden-Green, & Black, 2000). Buddenhagen and Sequeira established that *R. solanacearum* is endemic on *Heliconia* species native to Central America and, although very rarely pathogenic on triploid *Musa* cultivars, the establishment of large commercial plantations apparently provided the biological filter necessary to select the rare strains that cause BW of banana (Moko disease) (Buddenhagen, 1986; Sequeira, 1993; Sequeira, 1998). Subgroups of strains from Central and South America were recognized based on pathogenicity, cultural characteristics, and whether they were insect transmitted (Fegan, 2005; French, 1986). When artificially inoculated, race 2 strains isolated from diseased triploid banana are highly virulent to *Musa acuminata* and *M. balbisiana* and are often at least moderately virulent to either eggplant, pepper, potato or tomato (French & Sequeira, 1970; Janse, 1991; Raymundo, Orlina, Lavina, & Opina, 2005). When infiltrated into tobacco leaves, race 2 strains typically induce a HR (Janse, 1991; Lozano & Sequeira, 1970), but some strains can be moderately virulent on tobacco (French & Sequeira, 1970).

Race 3 strains originally were described as pathogenic on potato and tomato but weakly virulent on other solanaceous crops (Buddenhagen et al., 1962). Race 3 was considered to have a narrow host range compared to race 1 (Persley, 1986b), and it has sometimes been referred to as the 'potato race' (Buddenhagen & Kelman, 1964; French, 1994; Hayward, 1991). Race 3 is thought to have evolved in the Andes mountains, and it appears to be well adapted to cool-temperate climates (Elphinstone, 1996; French, 1986; Hayward, 1991). Most strains are not pathogenic on tobacco and when infiltrated into tobacco leaves only induce yellowing within the inoculated zone (Lozano & Sequeira, 1970). Recently, race 3 (biovar 2) strains have naturally infected cultivated geraniums (*Pelargonium* spp.) in Central America and Africa, and asymptomatic cuttings were subsequently shipped to the United States and Europe (Janse et al., 2004; Swanson, Yao, Tans-Kersten, & Allen, 2005; Williamson, Nakaho, Hudelson, & Allen, 2002). In Europe, bittersweet nightshade (*Solanum dulcamara*), a semi-aquatic weed, is also an asymptomatic host (Elphinstone, 1996). A variety of solanaceous and nonsolanaceous herbaceous weeds can harbor race 3strains, but grasses seem to remain free of the pathogen (Janse et al., 2004; Pradhanang, Elphinstone, & Fox, 2000a; Tusiime, Adipala, Opio, & Bhagsari, 1998). In addition, artificial inoculation of race 3 (biovar 2) strains from Columbia and Peru showed them to be moderately or highly virulent on tomato, eggplant, *Datura stramonium*, *Cyphomamdra betacea*, and *Solanum nigrum* (Marín & El-Nashaar, 1993; Swanson et al., 2005; Thurston, 1963). Therefore, race 3 has a much larger host range than was originally envisioned (Buddenhagen et al., 1962).

Race 4 strains are particularly aggressive on ginger (Anon., 2004; Persley, Batugal, Gapasin, & Vander Zaag, 1986). I found no paper in the primary literature designating such strains as race 4, but this terminology came into general use in the mid 1980's (Buddenhagen, 1986; Persley et al., 1986) and is now the defacto standard (Anon., 2004; Denny & Hayward, 2001). *R. solanacearum* strains that cause a rapid wilting of edible ginger (*Zingiber officinale*) and ornamental gingers have been isolated in Australia, China, Hawaii, India, Indonesia, Japan, Malaysia, Mauritius, the Philippines, and Thailand (Alvarez et al., 2005; Hayward, 1994b; Kumar, Sarma, & Anandaraj, 2004). A host range test in a naturally infested Australian field determined that race 4 (biovar 4) strains also wilted tomato, pepper, eggplant and some native weeds (e.g., *Solanum nigrum* and *Crassocephalum crepidioides*) (Pegg & Moffett, 1971). The few race 4 strains tested are low in virulence on tobacco (Pegg & Moffett, 1971; Quinon, Aragaki, & Ishii, 1964; Zehr, 1970a) and when infiltrated into tobacco leaves they induce a slow necrosis similar to that caused by race 1 strains virulent on tobacco (Janse, 1991). The host range of race 4 (biovar 3) strains from India has not been reported. Some race 1 strains may cause a slow wilting of ginger, but other races are not pathogenic (Hayward, 1994b; Janse, 1991; Quinon et al., 1964; Zehr, 1970b).

Race 5 is known only from China, where it causes wilt of mulberry trees (*Morus alba*) (He, Sequeira, & Kelman, 1983). Artificial inoculations show that race 5 strains are weakly virulent on eggplant and potato, and not virulent on tomato, pepper, groundnut or tobacco. Strains from mulberry were originally designated as race 4 due to the confusion surrounding the unofficial prior designation of ginger strains as race 4.

Several caveats regarding race designation are worth mentioning. First, although pathogenicity assays with *R. solanacearum* are relatively easy, artificial inoculations are influenced by a many variables (e.g., host cultivar, growth conditions, inoculum concentration, and inoculation method) and may overestimate the natural host range. In addition, because host range assays require substantial time and resources, only a few strains have usually been included in any one experiment. There are no *in vitro* tests that reliably predict the race or host range of *R. solanacearum*. Although Janse (1991) reported that whole cell fatty acid analysis supports differentiation of races 1, 2, and 3, confirmation of this correlation using a larger, more diverse set of strains has not been reported. Thus, it understandable why Cook et al. (1989, page 113) stated that Ahost range is often an ambiguous and unreliable taxonomic character.@

A second caveat is that the *R. solanacearum* race system is confusing to most scientists not familiar with this pathogen. For almost all other host-pathogen systems, races are determined by the matrix of resistance responses generated when two or more strains are tested on a set of differential cultivars of a single host species (i.e., cultivar-level specificity or gene-for-gene specificity) (Keen, 1990). In contrast, the unavailability of highly resistant cultivars of crop plants precludes cultivar-level specificity testing of *R. solanacearum*. Only recently did several authors (Alvarez, 2005; Prior & Fegan, in press) finally state the obvious: that the races of *R. solanacearum* resemble pathovars in other species of phytopathogenic bacteria.

After considering the ambiguities and problems with the existing *R. solanacearum* race system, I propose that it be abandoned. It generally has not lived up to its original purpose of 'aiding in the evaluation of the pathogenic potential' of *R. solanacearum* (Buddenhagen et al., 1962). On the contrary, the race system has become increasingly unwieldy and unreliable as a predictor of pathogenic potential as the known diversity of this species complex has increased. Consequently, mistakes could be made when the race system is used by regulatory agencies to define quarantine agents. Even worse, the race system's imprecision and inaccuracies impede our developing an accurate picture of *R. solanacearum* pathobiology.

3.2. Biovars

In contrast to the pathocentric race system, the biovar system originally proposed by Hayward (1964) groups strains by their ability to acidify media containing one each of three disaccharides or three sugar alcohols (Table 2). The system was later expanded to include additional substrates, production of nitrite from nitrate, and production of gas from nitrate (Hayward, 1994a). This is a special-purpose classification system that is only useful once a strain has already been identified as *R. solanacearum* using other methods. Nevertheless, because biovar determination is easy, inexpensive, and reproducible, it has been widely adopted as an essential trait in strain characterization.

Table 2. Differentiation of *Ralstonia solanacearum* biovars[a].

Test	*Biovars*					
	1	*2*[b]	*2-T*	*3*	*4*	*5*
Acidification of medium						
Mannitol	-	-	-	+	+	+
Sorbitol (Glucitol)	-	-	-	+	+	-
Dulcitol (Galactitol)	-	-	-	+	+	-
myo-Inositol[c]	+	+	+	+	+	+
D-Ribose[c]	d	-	+	+	+	+
Trehalose	+	-	+	+	+	+
Lactose	-	+	+	+	-	+
Maltose	-	+	+	+	-	+
D-(+)-Cellobiose	-	+	+	+	-	+
Nitrite from nitrate	+	+	+	+	+	+
Gas from nitrate	-	-	-	+	+	+

[a] +, 80% or more strains positive; -, 80% or more strains negative; d, 21-79% of strains positive.
[b] Some biovar 2 strains from Chile and Columbia (RFLP 27) are negative for inositol, positive for trehalose, and most are negative for nitrite production (Hayward, 1994a).
[c] For biovars 1, 3, 4, and 5 the results are from nine or fewer strains and should be considered preliminary

Four biovars (numbers 1 to 4) were recognized initially and they have accommodated almost all *R. solanacearum* strains isolated. Some strains isolated from potato in the Amazon basin can be differentiated from the archetypal biovar 2 using the extended panel of carbohydrates (French, Aley, Torres, & Nydegger, 1993; Hayward, 1994a). This new group is known as 2-T (T for tropical) in recognition of its lowland origin or N2 (new biovar 2). Consequently, the original biovar 2 is sometimes now referred to as biovar 2-A (A for Andean) in recognition of its highland origin (see below). Biovar 5 was established to accommodate strains isolated from mulberry in China.

There is no general correlation between biovars and races, but biovar 2 strains are almost always race 3 (and the reverse is always true) and biovar 5 strains are usually race 5 (and vice versa). Until recently, race 4 strains were biovar 4, but strains highly virulent on ginger in India are biovar 3 (Kumar et al., 2004). Race assignment to biovar 2-T strains is problematic, because they exhibit a wider host range than do the archetypal biovar 2 (race 3) strains. Biovar 1 and 3 strains are isolated from many different plants in many locations, so little can be predicted about the biology of such strains based only on biovar typing.

Biovars are based on a few traits, each of which probably requires only one or a few genes (some of which may have been acquired by horizontal gene transfer (Gabriel et al., in press)), and preliminary work indicates that one gene is essential for utilization of both sorbitol and dulcitol (Denny, unpublished results). Therefore, spontaneous mutation, in nature or in storage, could change a strain's reaction profile and confuse biovar assignment. For example, a single frame-shift mutation could change a biovar 3 strain into one classified as biovar 5. However, experience has demonstrated that biovars are stable in culture. In addition, recent phylogenetic studies indicate that biovars 3, 4, and 5 are part of a separate genetic lineage distinct from the other biovars (see below) and these strains also can utilize more varied carbon sources than can those in other biovars (Hayward, 1994a; Palleroni & Doudoroff, 1971). Consequently, biovar type appears to be relatively stable in nature.

3.3. Phylogenetic Groups

Although pathogenicity and biovar typing have genetic underpinnings, these traits are one or more steps removed from the DNA encoding them. Strictly genetic characterization has dramatically increased our knowledge and understanding of the diversity, relationships, and evolution of the *R. solanacearum* species complex. The trend over the last decade has been to more finely divide this group of related organisms as more strains from around the globe are studied in greater depth, and it may not be long before there is a proposal to split the complex into two or more subspecies or species.

The seminal phylogenetic work by Cook et al. (1989) used classical restriction fragment length polymorphism (RFLP) analysis to examine 62 *R. solanacearum* strains representing 4 races and 5 biovars isolated in the Americas, Asia and Oceania. Southern blots of restriction endonuclease-digested genomic DNAs were probed with each of nine DNA fragments cloned from *R. solanacearum* that hybridized to regions encoding biosynthesis of tryptophan, the high molecular mass extracellular polysaccharide (EPS1), the core lipopolysaccharide, three regions of the HR and pathogenicity (*hrp*) gene cluster, and three regions of unknown function. Twenty eight unique RFLP groups (later called multilocus genotypes (MLGs)) were recognized and a similarity coefficient matrix showed that they formed two distinct groups. These divisions were apparent in the RFLP patterns with each probe individually, suggesting that the loci sampled had coevolved as a part of the same genome.

Follow-up studies examined 102 additional *R. solanacearum* strains that better represented the same races and localities and also included biovar 2-T strains (Cook & Sequeira, 1994). RFLP data generated using eight of the original nine DNA probes produced a total of 44 MLGs and hierarchical cluster analysis confirmed the presence of the same Divisions I and II (Table 3). Division I contains all strains characterized as race 1 biovars 3, 4, 5, and they are mostly from Asia and Oceania. Division II contains strains characterized as race 1 biovar 1, race 2, and race 3 biovar 2, and they are mostly from the Americas. Division II also has many biovar 2-T strains (which produced many new MLGs) and comprises two sub-clusters: the biovar 2 and 2-T strains are in one subcluster and almost all the biovar 1 strains are in the other subcluster. Strains from banana (race 2) are in both subclusters. Most strains from potato are in MLG 26, despite being isolated from many locations around the world, which strongly suggests that a clonal population has been distributed by humans in latently infected potatoes.

A variety of modern techniques have been used since the early 1990's to investigate the phylogeny of *R. solanacearum*. These include total genomic RFLP fingerprinting (Gillings & Fahy, 1993; van der Wolf et al., 1998), polymerase chain reaction (PCR)-RFLP (Gillings, Fahy, & Davies, 1993; Poussier et al., 2000b; Poussier, Vandewalle, & Luisetti, 1999), amplified fragment length polymorphism (AFLP) (Poussier et al., 2000b; van der Wolf et al., 1998), tRNA-anchored PCR (Seal, Jackson, & Daniels, 1992) and repetitive element (rep)-PCR (Horita & Tsuchiya, 2001; Thwaites, Mansfield, Eden-Green, & Seal, 1999; van der Wolf et al., 1998).

DNA sequencing of desired loci that have been PCR-amplified from many strains has become increasingly popular as the price of this method has fallen (Fegan & Prior, 2005; Fegan, Taghavi, Sly, & Hayward, 1998; Li et al., 1993; Pastrik, Elphinstone, & Pukall, 2002; Poussier et al., 2000b; Poussier et al., 2000a; Prior & Fegan, in press; Taghavi et al., 1996; Villa et al., 2003; Villa et al., 2005). Computer software is used to identify polymorphic nucleotides, calculate genetic distances and perform cluster analyses. Relationships between strains that are distantly related can be determined by analyzing loci that evolve slowly, such as the 16S ribosomal RNA gene (rDNA). Conversely, relationships between more closely related strains require targeting loci that are evolving more rapidly. Genes that encode important proteins may evolve at a moderate rate, but non-coding intergenic regions should evolve most rapidly. The *R. solanacearum* loci examined to date include genes encoding an endopoly-galacturonase (*pehA*), an endoglucanase (*egl*), the AraC-type regulatory protein in the *hrp* gene cluster (*hrpB*), the DNA mismatch repair system (*mutS*), the intergenic region adjacent to the *lpxC* gene, and the 16S-23S intergenic spacer region (also called the ITS region).

Regardless of the method, all of the genetic analyses support the existence of two major groups that are essentially the same as Divisions I and II described by Cook et al. (1994). Studies from the mid 1990's found only the same two divisions, because they only examined strains representing populations similar to those sampled previously. Later studies included strains isolated in Africa and/or Indonesia, which had not been examined previously, and the greater diversity resulted in at first three and then four major divisions being recognized (Table 3). The most recent study compared the 16S rDNA, *egl*, and *hrpB* sequences from the most representative set of strains studied to date and consistently found four major divisions that correlated with geographic origin (Villa et al., 2005). The data also indicate that the group of strains originating in the Americas has two subclusters (phylotypes IIa and IIb) (Prior & Fegan, in press; Villa et al., 2005). The same relationships are observed regardless of the locus examined, suggesting that these regions have co-evolved as part of a conserved *R. solanacearum* genome. This is not the case for all loci, because the genome of strain GMI1000 is a mosaic of genes apparently acquired from related and unrelated bacteria (see below) (Salanoubat et al., 2002).

The new classification scheme proposed by Fegan and Prior (2005) denotes the major groups as phylotypes, defined as a monophyletic cluster of strains revealed by phylogenetic analysis of sequence data (Table 3). A set of multiplex PCR primers is available to determine a strain's phylotype (Fegan & Prior, 2005). Each phylotype is comprised of a number of sequevars, or sequence variants, and single strains that have a highly conserved sequence within the target locus. Sequevars are comparable to RFLP MLGs and the two schemes often produce the same or similar groups of strains (Table 3). However, both schemes suffer from the same potential problem, which is that the recognized clusters may be dependent on the particular gene(s) sequenced or the set of polymorphic fragments examined (e.g., sequevar groups might be different when based on *egl* and *hrpB* sequence data (Villa et al., 2005)). Additional sequevars probably will be recognized as more strains are examined (especially strains not isolated from diseased plants), but few if any new phylotypes are likely to be described (Fegan & Prior, 2005). This classification scheme has a distinct advantage over other measures of genetic diversity, such as RFLPs, because sequences entered into an on-line database can easily be 'reused' in future, increasingly comprehensive phylogenetic studies.

Table 3. Diversity within the *Ralstonia solanacearum* species complex assessed using different methods[a].

Phylotype & Origin	*RFLP div.*	*16S group*	*Sequevar (egl)*[b]	*RFLP MLG*[c]	*Biovar or species*	*Host of origin*	*Comments*
I Asia and Oceania	I	1	12-18	8-14	3	many	Race 1
				11, 15-18, 42	4	many	
				22, 23	4	Ginger	Race 4
				19, 20	5	Mulberry	Race 5
					2-T	Pepper, Tomato	Japan (one strain from each)
IIa Americas	II	2A	3, 4	24, 25	1	Musaceous	Race 2
			1[e], 2	26[e], 27, 34	2	Potato mostly	Race 3
			none yet	29-31, 33, 36, 37, 39	2-T	Potato, Eggplant	only single strains sequenced
IIb Americas			5, 7	1-7, 38	1	many	Race 1, type strain
			6	28, 46	1	Musaceous	Race 2
III Africa	nd	nd	19-23	nd	1	many	unrelated to Bv.1 in Div. II
					2-T	Potato	
IV Indonesia and Asia	nd	2B		nd	1	Clove	unrelated to Bv. 1 in Div. II
			11		2	Clove, Potato	Bv. 2 also in Australia
			8		2-T	Potato	also in Japan & the Philippines
			10		2-T, BDB[d]	Musaceous	
			9		1, *R. syzygii*	Clove	

[a] Sources of data are as follows: Phylotype (Fegan & Prior, 2005; Prior & Fegan, 2005; Villa et al., 2005); RFLP (restriction fragment length polymorphism) Division (Cook et al., 1994); 16s rDNA group (Poussier et al., 2000b; Taghavi et al., 1996); RFLP MLG (multilocus genotype) (Cook et al., 1994). Biovars and host ranges are compiled from many references. nd, no data.

[b] Sequevar designations are from analysis of the *egl* gene. Analysis of other genes might not cluster strains into the same groups.

[c] Cook et al. (1994) did not define MLGs 32 or 35 and merged MLGs 16 and 21 with MLGs 11 and 8, respectively.

[d] BDB, Blood Disease Bacterium.

[e] Strains isolated from potato worldwide are most often in sequevar 1 (MLG 26).

The robustness of the phylotypes strongly suggests that they reflect true evolutionary lineages within the *R. solanacearum* species complex. These lineages presumably arose when progenitors became geographically isolated and subsequently adapted to different environments and potential host plants. Therefore, unlike the races and biovars, phylogenetic analyses and phylotyping classification should make possible a stable and meaningful taxonomy that defines subspecific groups of *R. solanacearum* that are at least related to geographic origin. Available DNA-DNA hybridization data (Palleroni & Doudoroff, 1971; Roberts et al., 1990) indicate low homology (<70%, the threshold for speciation) between some strains in different biovars, but the data are too preliminary to conclude that the species complex should be divided into subspecies or new species. There is also insufficient information on the biological, epidemiological, and ecological properties of strains that would make the phylotype system more useful to plant breeders, plant pathologists and quarantine officials (Fegan & Prior, 2005).

4. DETECTION AND IDENTIFICATION

Detection and identification are conceptually and methodologically intertwined processes. When studying the pathobiology and epidemiology of *R. solanacearum* it is usually necessary to first detect its presence (based on a tentative identification) before isolating and rigorously identifying a strain. Substantial international effort has been focused on developing better detection methods, because soil and water samples typically have low populations of *R. solanacearum* and not all cells may grow *in vitro*. General aspects of detection methods will be reviewed and their relative advantages and disadvantages discussed. Technical details and access to the earlier literature are available elsewhere (Anon., 2004; Alvarez, 2005; Denny et al., 2001; Elphinstone, Hennessy, Wilson, & Stead, 1996; Saddler, 2000; Seal & Elphinstone, 1994).

Pure cultures of *R. solanacearum* are not difficult to identify. Cultural and physiological tests can quickly rule out related organisms (Anon., 2004). There are also commercially available fatty acid methyl ester (FAME) analyses and BIOLOG™ kits, but they will not be described here (see (Black & Sweetmore, 1993; Janse, 1991; Li & Hayward, 1993; Stead, Sellwood, Wilson, & Viney, 1992)). There are also many suitable serological and nucleic acid based methods (see (Alvarez, 2005; Seal et al., 1994)), but only those useful for detecting the pathogen are discussed below.

4.1. General Considerations

The first consideration is that the purpose of a diagnosis should dictate its thoroughness (Black & Elphinstone, 1998). If only advice concerning an ongoing pest management problem is needed, then presumptive evidence of *R. solanacearum* will probably suffice. In contrast, quarantine issues dictate an unequivocal, well documented identification of the pathogen. Second, no single detection method is suitable for all samples or meets all the requirements for sensitivity, specificity, and speed. Plant tissue, seeds, soil, and water each present different challenges that vary with the detection method and increase as the pathogen population decreases. Third, the amount of material sampled affects relative sensitivity. For example, method A may detect 10x more cells per unit volume than method B, but if the latter uses 100x more sample volume then it may be the more sensitive overall. Fourth, the sampling protocol prior to detection can influence the results. For example, it is a statistical certainty that a random sample of 200 tubers from a ton lot will contain at least one infected tuber when 10% of the lot is infected, but the probability of a sample having an infected tuber drops to only 63.3% when 0.5% of the lot is infected (Elphinstone et al., 1996). Fifth, cost effectiveness will be important for diagnostic labs that process many samples, but the least expensive method may be different in developed and less developed countries (Seal, 1998). Availability, reliability and price of reagents and equipment are more important in developing countries, whereas labor costs are generally higher in developed countries.

4.2. Traditional Methods

A variety of pathogens can cause wilting, so symptoms alone are not definitive for BW. However, plants wilted by *R. solanacearum* have $>10^8$ colony forming units (cfu)/g of tissue, so a milky white ooze often forms on the cut surface of a stem left after decapitating a wilted plant as root pressure forces out xylem fluid. Ooze also may accumulate on the cut surface of infected tubers or rhizomes. Even if ooze does not form spontaneously, a 'streaming test' may be positive (Fig. 3) (Allen, Kelman, & French, 2001). Other wilt-inducing pathogens do not produce comparable ooze. The ooze is usually an almost pure culture of *R. solanacearum*, which can be cultured on standard, low ionic strength bacteriological media. Detecting *R. solanacearum* in soil and water samples is more difficult, because the pathogen population is usually small ($<10^4$ cfu/g soil or ml water) and saprophytic bacteria are present in equal or greater numbers. Such samples are best cultured on a semi-selective medium, such as modified SMSA (Denny et al., 2001; Elphinstone et al., 1996), which usually suppresses contaminants well enough to permit detection of *R. solanacearum* down to 100 to 500 cfu/g soil or ml water, and about 10-fold lower in tissue extracts (Elphinstone et al., 1996; Poussier, Cheron, Couteau, & Luisetti, 2002; Pradhanang, Elphinstone, & Fox, 2000b). Other semi-selective media may work better in some locations or with particular soils (see (Denny et al., 2001; Pradhanang et al., 2000b)). Surprisingly, none of the common serological or nucleic acid based techniques described below has a substantially lower threshold of detection. Even when other methods are used for detecting *R. solanacearum*, culturing is required to confirm pathogen viability and provide a pure strain for confirmation of identity and pathogenicity bioassays.

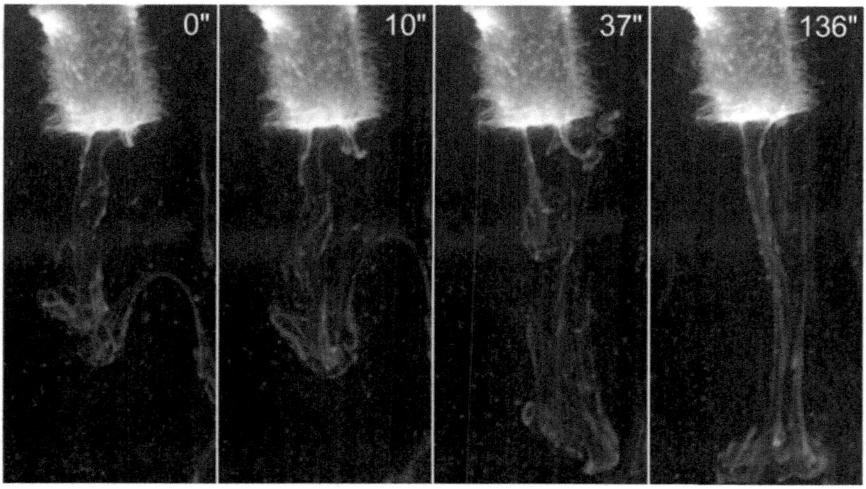

Fig. 3. Streaming test to observe bacteria oozing from an infected tomato stem in water.

It is also possible to use a susceptible plant as the enrichment 'medium'. Typically, a crude extract from tissue or soil is introduced into the stem of young tomato plants (3 to 4 leaf stage) that are incubated at 25-30°C for up to 4 weeks. Plants are monitored for symptom development and any unwilted plants are sampled for latent infection by plating tissue extracts on SMSA. Tomato seedlings also can be transplanted into infested soil. Stem inoculation reliably detects *R. solanacearum* when it is present at $\geq 10^4$ cfu/ml of sample extract (Elphinstone et al., 1996; van der Wolf et al., 2000), whereas for the transplantation test there must be $>10^5$ cfu/g soil (Pradhanang et al., 2000b).

4.3. Serological Methods

Serological methods are popular for primary screening of samples because they are generally quick and reliable. These methods have thresholds of detection that range from 10^3-10^4 bacteria/ml of sample. Each serological method has different strengths, but they all suffer from problems with either specificity, sensitivity or both. Polyclonal antisera (PAb) usually react more strongly with *R. solanacearum* than do monoclonal antibodies (MAb), but this greater sensitivity is counterbalanced by reduced specificity. False positives become more common as the proportion of contaminants in samples increases, whereas false negatives occur when the pathogen lacks the targeted epitope(s). Serological methods also have the disadvantage that they do not discriminate between live and dead cells. Knowledge that a sample has living *R. solanacearum* is essential before making a decision to quarantine or destroy a shipment of produce.

Rapid, virtually fool-proof serological kits are commercially available (e.g., Agdia (Elkhart, IN, USA), Neogen Europe Ltd. (Auchincruive, Scotland), and the Central Science Laboratory (Sand Hutton, York, UK)). Most kits use a MAb to *R. solanacearum* coated onto colored latex beads that are preloaded onto a dipstick or horizontal flow unit. Results can be read in several minutes: positive samples produce two colored bands, negative samples produce a single control band, and no band indicates a method failure. The Pocket Diagnostic™ kit from the CSL is approved by the United States Department of Agriculture for testing of tissue from symptomatic potatoes and geraniums. A rapid agglutination test kit also is available form Neogen Europe. However, speed and convenience come at a price, because each test costs \$3 to \$6 USD.

Enzyme-linked immunosorbent assays (ELISA) are still used in many labs, because they are relatively inexpensive, easy, fairly fast, and tolerate foreign material in the sample (e.g., plant tissue, soil). Direct ELISA, indirect ELISA, and double antibody sandwich (DAS) assays have been developed. ELISA kits are commercially available from Agdia and Neogen Europe. A variation on this theme is to bind the samples to a nitrocellulose membrane, which can be stored for long periods before and after the ELISA (Priou, Aley, & Gutarra, in press). This method has been used in many countries for pathogen surveys and testing propagative tissues (Elphinstone, 2005; Priou et al., in press). ELISA has a relatively high detection threshold; under ideal conditions, PAb cannot detect fewer than 10^4 bacteria/ml sample and MAb are about 100-fold less sensitive (Pradhanang et al., 2000b; Seal, 1998). Values are 10- to 100-fold higher when using samples from plants or soil. Sensitivity can be improved substantially by adding the sample to broth medium and incubating 24 to 72 h prior to running the ELISA on the enriched sample (Elphinstone et al., 1996; Pradhanang et al., 2000b). Caruso et al. (2002) used a non-selective medium and reported detecting *R. solanacearum* at 1 to 10 cfu/ml of potato extract and Priou et al. (in press) used a new semi-selective medium and detected about 100 cfu/g of naturally infested soil.

Alternatives to ELISA are immunofluorescence-antibody staining (IFAS) and immunofluorescence-colony staining (IFCS) (Elphinstone et al., 1996; Janse, 1988; van der Wolf et al., 2000). These methods have most of the same advantages and disadvantages as ELISA, but require special microscopes that may not be readily available. Both have a detection threshold of 10^2 to 10^4 bacteria/g soil or ml tissue extract (Elphinstone et al., 1996; Seal, 1998; van der Wolf et al., 2000). The specificity of IFAS can be improved by simultaneously running fluorescent *in situ* hybridization (FISH) using a probe to the 23S rDNA gene sequence (Wullings, van Beuningen, Janse, & Akkermans, 1998). The specificity of IFCS can be verified by recovering viable bacteria from the stained colonies for either direct PCR testing or subculturing on SMSA and subsequent IFAS testing (van der Wolf et al., 2000).

4.4. Nucleic Acid-Based Methods

Nucleic acid-based methods to detect *R. solanacearum* rely on hybridization of probes or PCR primers to targeted pathogen sequences. FISH is the only current method that still uses labeled probes, because greater sensitivity can be achieved with amplification of target sequences by PCR. Direct PCR, nested PCR, co-operational PCR, real-time PCR and multiplex PCR reactions have all been developed to detect *R. solanacearum*. When the target sequence is present in multiple copies (e.g., tRNA genes) and reaction conditions are ideal, PCR can detect a single pathogen cell or its equivalent in purified DNA. With this level of sensitivity, though, extra care is needed to avoid false positives due to sample contamination by *R. solanacearum* cells or DNA.

Proper selection of target sequences is essential when developing a PCR assay. The target should be conserved within the species or the desired subspecies group (e.g., phylotype or race) and absent in all other bacteria. Targeting more than one pathogen sequence is also recommended to guard against unexpected failures in specificity. The primers should be tested on as many *R. solanacearum* strains and related species as possible to determine the likelihood of false positive or negative reactions. The primers also should be tested in multiple laboratories to assess the effect of small differences in sample preparation and reaction conditions.

PCR methods are theoretically more specific and sensitive than serological approaches. However, similar to serological methods, PCR detects both living and dead cells, because DNA can remain intact for long periods after death. More importantly, unlike serological methods, PCR is often partially or completely inhibited by compounds introduced into the reaction along with the sample, resulting in false negatives. For example, extracts from potato tubers interfere strongly with PCR (Elphinstone et al., 1996; Seal, 1998; Weller, Elphinstone, Smith, Boonham, & Stead, 2000a). Extracts from other plants, seeds, soil and water usually have different, unidentified PCR inhibitors. Multiplex PCR addresses this issue by including primers to co-amplify the sequence of an internal standard (e.g., a conserved plant or eubacterial rDNA gene) that is consistently present in (or is added to) all samples (Glick, Coffey, & Sulzinski, 2002; Pastrik et al., 2002; Schönfeld, Heuer, van Elsas, & Smalla, 2003; Weller et al., 2000a). The internal standard tests the 'amplification competence' of each sample to reveal those that may give a false negative.

A variety of approaches can reduce or eliminate PCR inhibitors. Diluting a sample will sometimes reduce the inhibitor concentration enough to allow amplification (Weller et al., 2000a), as can washing the bacteria by centrifugation, filtration or during immunocapture (Dittapongpitch & Surat, 2003; Poussier et al., 2002).

However, the more reliable solution to this problem is to extract DNA from the sample. Extraction buffer additives, in particular polyvinyl pyrrolidone (PVP) and polyvinyl polypyrrolidone (PVPP), can help to remove PCR inhibitors (Llop, Caruso, Cubero, Morente, & Lopez, 1999; Poussier et al., 2002). In the few reports where different extraction methods were compared, the commercial QIAamp DNA (QIAGEN, Inc) and the Easy-DNA (Invitrogen) kits were more reliable and produced DNA that worked better in PCR than other commercial kits and conventional recipes (Pastrik & Maiss, 2000; Poussier et al., 2002). PCR buffer additives, such as bovine serum albumin and trehalose (Ozakman & Schaad, 2003; Poussier et al., 2002), may also be helpful. Another common strategy is to enrich the population of *R. solanacearum* by culturing the environmental sample on SMSA broth or SMSA plates prior to PCR (Elphinstone et al., 1996; Ozakman & Schaad, 2003; Pradhanang et al., 2000b; Weller, Elphinstone, Smith, & Stead, 2000b). This approach has the added advantages of reducing the detection threshold by providing a biological amplification before the detection step and eliminating the need for DNA purification, because amplifiable DNA is prepared simply by heating the cells at 100° C for several minutes.

Other PCR variations can enhance sensitivity of detecting *R. solanacearum*. Two-stage nested PCR is the simplest method (Elphinstone et al., 1996; Poussier & Luisetti, 2000; Pradhanang et al., 2000b), but the added sample manipulation step increases the risk that contamination will generate a false positive. Co-operational PCR is related to nested PCR, but the reaction is performed in a single tube, so contamination is expected to be less of a problem (Caruso, Bertolini, Cambra, & Lopez, 2003). Real-time PCR using TaqMan fluorescent probes can detect 10 to 30 cfu/ml of potato tuber extract after the *R. solanacearum* population has been enriched by growth on SMSA (Ozakman & Schaad, 2003; Weller et al., 2000a; Weller et al., 2000b). This method also eliminates post-PCR analyses, provides a quantitative estimate of the target sequence in a sample, and demonstrates viability of *R. solanacearum* cells when the amount of target sequence increases in samples removed at intervals while incubating the enrichment culture.

Another approach to detect viable cells is to target RNA, which degrades quickly after death. Nucleic acid sequence based amplification (NASBA), based on isothermal amplification of *R. solanacearum* 16S rRNA, detects only live cells (Bentsink et al., 2002). NASBA was converted to a real-time detection method (called AmpliDet RNA) by incorporating a fluorescent molecular beacon into the reaction (van der Wolf, 2004). Similar to a TaqMan probe, the fluorescence of a molecular beacon increases linearly with the amount of homologous sequence in the reaction mixture. In a 90 minute reaction, AmpliDet RNA detected 10^4 cfu/ml in potato tuber extract and 10 cfu/ml in a water sample previously concentrated 200-fold.

5. THE GENERAL PATHOGEN CYCLE

Before describing diseases caused by *R. solanacearum* on representative hosts, general aspects of the pathogen's life cycle (disease cycle) will be reviewed. Susceptible hosts suffering from BW contain massive populations of *R. solanacearum* (10^8 to 10^{10} cfu/g tissue) and usually die from the disease. Therefore, the first hurdle for the pathogen is to survive until it can infect another host plant (Coutinho, 2005; Hayward, 1991; Persley, 1986b). The large number of pathogen cells shed from roots of symptomatic and nonsymptomatic plants (Elphinstone, 1996; Swanson et al., 2005) and bacterial ooze on plant surfaces (Buddenhagen & Kelman, 1964; Kelman, 1953) enter the surrounding soil or water, contaminate farming equipment, or may be acquired by insect vectors.

Survival of *R. solanacearum* in water can range from weeks to years depending on the interacting abiotic and biotic factors. The pathogen can survive for >40 years when stored in pure water at 20-25°C, and this is still an acceptable method for long term storage. Temperature extremes (especially storage at 4°C), pH extremes, and the presence of salts or other contaminants directly reduce survival (Caruso et al., 2005; van Elsas, Kastelein, de Vries, & van Overbeek, 2001). These factors can also affect the number, type, and activity of other microorganisms present in non-sterile water that compete with or prey upon *R. solanacearum*. Contaminated irrigation water also is an important route for pathogen dispersal and inoculation (Elphinstone, 1996; Janse, 1996; Swanson et al., 2005).

Survival in soil is less well understood, due to the greater complexity of this substrate and the difficulty in detecting pathogen populations <100 cfu/g soil. In the short term (up to 2 years), survival in soil is largely controlled by moisture (matric potential) and temperature. *R. solanacearum* can survive in moderately dry soil (e.g., -0.3 MPa ≈ 20% moisture), but lower moisture levels drastically reduce survival. Other factors may include the soil type, soil depth, host plant debris, organic matter content, nutrient level, and the microflora and microfauna (Coutinho, 2005; Hayward, 1991; Persley, 1986b). Increased amounts of organic material often correlates with decreased pathogen survival (Gorissen, van Overbeek, & van Elsas, 2004; Hayward, 1991; Schönfeld et al., 2003), but why some soils are 'conducive to disease' while others are 'suppressive' remains unclear. For long term survival (>2 years) in the absence of a true host, the pathogen must either colonize the more nutrient rich soil near roots or latently infect roots of plants that remain asymptomatic (Coutinho, 2005; Elphinstone, 1996; Janse et al., 2004; Pradhanang et al., 2000a). After multiplying in these sheltered sites, the pathogen returns to the bulk soil to repeat the cycle.

Recently, several research groups have claimed that *R. solanacearum* can enter a state where it is viable but not culturable (VBNC) as a result of incubation at 4°C or exposure to copper ions (Caruso et al., 2005; Grey & Steck, 2001; van Elsas, van Overbeek, & Trigalet, 2005b). Cells in the VBNC state do not form colonies on most laboratory media but remain metabolically active, as determined by their enlargement in response to nutrients or their retention of an intact cytoplasmic plasma membrane (van Elsas et al., 2005b). The existence of the VBNC state is controversial, because (for example) nonculturability may be a laboratory artifact and 'resuscitation' of VBNC cells implies that they were never nonculturable. In practical terms, however, the potential for some or most of a pathogen population to escape detection methods that require *in vitro* multiplication (i.e., plating on SMSA, detection by IFCS, enrichment prior to PCR) means that our understanding of the ecology and epidemiology of *R. solanacearum* may be incomplete and could compromise quarantine practices. It is possible that cold-stressed viable cells are not a threat to agriculture, because in a controlled test strain 1609 (biovar 2 race 3) held at 4°C for 100 to 125 days became avirulent on tomato (van Overbeek, Bergervoet, Jacobs, & van Elsas, 2004). However, this may not always be the case, because *R. solanacearum* strains recovered during the winter from a Spanish river retained virulence on tomato (Caruso et al., 2005).

R. solanacearum also survives in hosts that, due to resistance or cool temperatures, are latently infected and do not develop symptoms. Pathogen populations are often low enough to make detection difficult, but can be as high as 10^8 cfu/g tissue in parts of some latently infected plants (Elphinstone et al., 1996; Swanson et al., 2005). Latently infected propagating tissues also are a major route for long-distance dispersal of *R. solanacearum*. The most important plants in this regard are potato, banana (and other musaceous spp.), ginger, and geranium. It is widely believed that cool-temperature adapted biovar 2 strains pathogenic on potato were repeatedly introduced into Europe via infected seed and ware potatoes (Janse, 1996) and that strains pathogenic to banana were moved from Central America to the Philippines (Buddenhagen, 1986; Fegan & Prior, in press). There is no doubt that biovar 2 strains recently were introduced onto the USA and Europe on latently infected geranium cuttings (Janse et al., 2004; Williamson et al., 2002). There are only a few studies of the survival and dispersal on true seed, and these are rare events (Coutinho, 2005; Martins, Nabizadeh-Ardekani, & Rudolf, 2005).

After survival and/or dispersal, the next step in the *R. solanacearum* life cycle is to invade a potential host. Like most other phytopathogenic bacteria, *R. solanacearum* usually enters via a wound that exposes internal tissues. Unlike many phytopathogenic bacteria, *R. solanacearum* potentially requires only one entry site to establish a systemic infection that results in BW. Normal agricultural practices frequently wound plants, either accidentally or intentionally, and infested tools may simultaneously inoculate the pathogen. Growth in natural soil also exposes roots to wounding by a variety of other agents, foremost of which are nematodes. In the case of Sumatra disease of clove trees caused by *R. syzygii*, xylem-feeding spittlebugs (*Hindola* spp.) both disperse the pathogen and introduce it directly into a suitable infection court (Bennet, Hunt, & Asman, 1985; Roberts et al., 1990).

However, 'natural wounds' also appear during normal plant development. For example, in the case of bananas and plantains, abscission of male flowers creates a moist site with open xylem vessels that can be inoculated by bees and other insects that inadvertently vector the pathogen from diseased plants that are oozing bacteria (Buddenhagen & Kelman, 1964). A more general example is the genesis of lateral roots, because a meristem growing outward from the central cylinder breaks through the endodermis, root cortex and epidermis. Before it heals, the 'wound' at the lateral root axil provides a route for small molecules to reach the xylem (Peterson, Emanuel, & Humphreys, 1981) and, at least for tomato plants, this site also can be invaded by *R. solanacearum* (Araud-Razou, Vasse, Montrozier, Etchebar, & Trigalet, 1998; Kelman & Sequeira, 1965; Schmit, 1978; Vasse, Frey, & Trigalet, 1995).To access 'wounded' tomato roots, the pathogen benefits from both flagellar-mediated swimming motility (Tans-Kersten, Brown, & Allen, 2004; Tans-Kersten, Huang, & Allen, 2001) and chemotaxic attraction to root exudates (C. Allen, personal communication), but these attributes were not essential at the high soil populations used in controlled tests.

After invading a susceptible host, *R. solanacearum* must multiply and move systemically within the plant before BW symptoms occur. The pathogen's goal is to maximize its population size by using its rare ability to exploit the ecological niche within a plant. Consequently, wilting should be considered as the most visible side affect that usually, but not always, occurs after extensive pathogen colonization. There are only a few histological studies of the systemic colonization of host plants by *R. solanacearum* and most of these examined tomato plants. When plants are grown in liquid nutrient medium, bacteria that enter naturally via a lateral root axil multiply in the root cortex for 2 to 6 days before breaching the endodermis (Schmit, 1978; Vasse, Danoun, & Trigalet, 2005; Vasse et al., 1995). The pathogen then colonizes the intercellular spaces within the central cylinder, invades the xylem vessels, and begins rapidly migrating through the vascular tissue (Vasse et al., 1995). In contrast, Wallis and Truter (1978) reported that, after inoculating the severed end of the tomato taproot, the pathogen first multiplies within cells (maybe tracheids or xylem parenchyma) adjacent to vessels. How bacteria invade these cells was not determined. The pathogen migrates into some of the many tyloses that form within the xylem vessels, and bacteria were observed in vessel lumena 3-4 cm above the root tip beginning 3 days after inoculation.

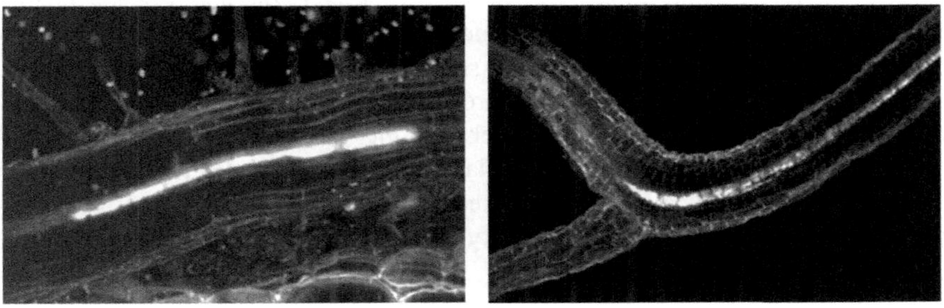

Fig. 4. Confocal microscopy of fluorescent strain AW1-gfp38 colonizing the central cylinder of tomato lateral roots 2 days after soil-drench inoculation.

A somewhat different scenario was observed when unwounded roots of young tomato plants growing in a soil-less potting mixture were inoculated by drenching with strain AW that constitutively expresses green fluorescent protein (GFP) (Denny, unpublished results). The site of pathogen entry was not observed, but within the first day or two the central cylinder of a few lateral roots became brightly fluorescent due to a high density of pathogen cells (Fig. 4). Bacteria also were observed in xylem vessels in the taproot and lower stem 2 or 3 days after inoculation (Fig. 5A) and plants began to wilt after 4 days. Similarly rapid colonization of plants with unwounded roots was observed based on viable cell recovery (Araud-Razou et al., 1998; McGarvey, Denny, & Schell, 1999; Saile, Schell, & Denny, 1997). These results indicate that in soil the pathogen invades a very low percentage of lateral roots almost immediately after inoculation rather than first colonizing the root cortex as observed for roots in liquid culture. Thus, roots of tomato plants in hydroponic culture probably are not well suited for studying natural invasion processes.

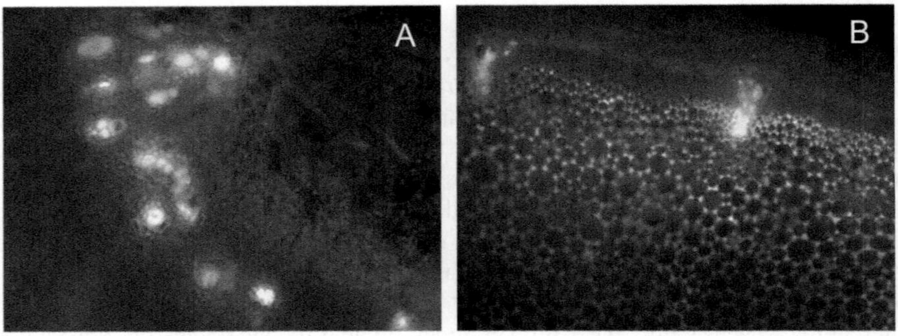

Fig. 5. Epifluorescent microscopy of fluorescent AW1-gfp38 colonizing tomato plants after soil-drench inoculation. A. Bacteria in tap root xylem vessels 3 days post inoculation. B. Bacteria in stem pith intercellular spaces 6 days post inoculation.

Once the pathogen has invaded tomato root xylem vessels there are relatively few physical barriers preventing it from systemically colonizing the remaining plant body. As in most plants, the end walls of functional vessels have completely degenerated, so there is nothing blocking the axial movement of bacteria unless tyloses develop from adjacent parenchyma cells. However, in susceptible tomato plants tylose formation occurs too infrequently or too slowly to prevent pathogen migration, and may instead contribute to vascular dysfunction by obstructing uncolonized vessels (Grimault, Gélie, Lemattre, Prior, & Schmit, 1994; Nakaho, Hibino, & Miyagawa, 2000). Radial movement of bacteria is initially blocked by the pit membranes that separate a colonized vessel from adjacent vessels, xylem parenchyma, or ray parenchyma. The pathogen tends to congregate near the pits and adjacent parenchyma cells often die. The pit membranes eventually break down and allow bacteria to invade adjacent vessels or the space occupied by dead parenchyma cells (Grimault et al., 1994; Nakaho et al., 2000). Immunofluorescence microscopy and tracking of GFP+ cells revealed that after leaving xylem vessels the pathogen migrates radially through intercellular spaces to colonize primarily the pith, but also the cortex, of roots and stems (Fig. 5B) (McGarvey et al., 1999; Denny, unpublished results).

Wilting begins to appear as pathogen density increases throughout a plant. In a susceptible tomato variety that was root-inoculated, the onset of wilt was correlated with a bacterial density exceeding 4 x 10^7 cfu/g tissue at the midstem (McGarvey et al., 1999) and with one of the major vascular bundles becoming heavily colonized from the taproot to near the shoot tip (Denny, unpublished results). At this time, extracellular polysaccharide (EPS1) content is about 10 μg/g tissue in the taproot, hypocotyl and midstem; EPS1 concentrations later are ≥100 μg/g tissue in fully wilted plants (Denny & Baek, 1991; McGarvey et al., 1999). Wilting is due to vascular dysfunction that prevents sufficient water from reaching the leaves (Buddenhagen & Kelman, 1964; Denny, Carney, & Schell, 1990). There is no evidence for excessive transpiration due to loss of stomatal control as might result from a systemic toxin (Buddenhagen & Kelman, 1964; Van Alfen, 1989). How *R. solanacearum* colonization reduces water flow is not completely clear; the primary factor is probably plugging of pit membranes in the petioles and leaves by the high molecular mass EPS1 (Van Alfen, 1989), but high bacterial cell densities, byproducts of plant cell wall degradation, and plant-produced tyloses and gums may be contributing factors.

6. DISEASES

In 1953 the list of 'true' hosts included plants in 35 families that are naturally infected or highly susceptible when inoculated (Kelman, 1953). Hayward (1994b) updated this host list, and susceptible plants are now known in over 50 families. Some widely cultivated plants, such as sweet potato, strawberry, and groundnut, are only attacked by *R. solanacearum* when they have been moved outside of their normal range (Hayward, 1994b). Many more dicots (Class Magnoliopsida) suffer from BW than do monocots (Class Liliopsida), and five of the nine monocot families that include hosts (*Cannaceae*, *Heliconniaceae*, *Musaceae*, *Strelitziaceae* and *Zingerberfloraceae*) are in the order *Zingiberales*. Why some families have more species susceptible to BW is not known. This section describes disease symptoms on representative hosts and includes some relevant aspects of pathogen diversity and epidemiology.

6.1. BW of Tomato and Tobacco

Tomato is probably the crop most often affected by BW, because it is grown world wide and is susceptible to almost all pathogen strains. The youngest leaves are usually the first to become flaccid, and in the field this often does not happen until after flowering has begun (McCarter, 1991). Onset of wilt is rapid at warm temperatures (>28°C); it may at first be unilateral, but wilting of the whole plant usually ensues (Fig. 6). The rate of wilt is slower at cooler temperatures, and there may be enough time before the plant dies for adventitious root primordia to appear on the lower and middle stem. Wilting plants are stunted and, if cut open, browning of the vascular system is evident. The stem pith rots during the later stages of the disease either before or after the plant dies. A majority of plants may be affected by BW in fields heavily infested by *R. solanacearum*.

BW of tobacco (also known as Granville Wilt in the USA) occurs in many tobacco-growing countries with moist tropical to warm-temperate climates. Symptoms in tobacco are similar to those on tomato, but unilateral wilt is often more obvious due to the large leaves (Echandi, 1991). During disease onset it is not uncommon for only half of one or two leafs to become flaccid. When the disease progresses slowly leaves on infected plants become light green and then yellow and necrotic areas may appear between veins and at leaf margins (scorch symptoms). Similar to tomato, the vascular system becomes discolored brown and then black, and the pith eventually rots leaving a hollow stem. Lesions may appear on the stem surface.

Fig. 6. Bacterial wilt of tomato. An uninoculated control plant is on the left.

6.2. BW of Potato

BW of potato, often called brown rot, can be caused by strains in biovars 1, 3, 4 and 2-T in warm climates and by biovar 2 (race 3) strains in both warm and cool climates. BW is a limiting factor in production of potato in both lowland and highland tropics (Allen et al., 2001; French, 1994; Lemaga, Kakuhenzire, Kassa, Ewell, & Priou, 2005). In warm climates there may initially be transient wilting only when transpiration rates are high, but leaves and stems soon wilt permanently (although remaining green) and then die and desiccate. Petiole epinasty may occur. The stem above the soil line may become streaked brown due to discoloration and necrosis of vascular tissues, and extensive secondary rotting may develop. Tubers may not show external symptoms before they begin to rot, but in well established infections vascular discoloration may be visible through the periderm. There may be external signs of the pathogen when bacteria ooze from eyes and the stolon-end attachment site, which may then retain a soil coating (Fig. 7). Tuber internal symptoms usually include distinct grayish brown discoloration of the vascular tissue and necrosis of the vascular ring and adjacent tissues that can be confused with ring rot caused by *Clavibacter michiganesis* subsp. *sepidonicus*.

Unlike most *R. solanacearum* strains, some biovar 2 strains are virulent on potato even when temperatures are ≤24°C (French, 1986; Swanson et al., 2005). These conditions are common at higher elevations in the tropics and at higher latitudes. Thurston (1963) reported epidemics of potato brown rot in Columbian highlands when temperatures averaged ~13°C (23°C maximum). He also demonstrated that potato plants inoculated by stem wounding slowly develop wilt symptoms when incubated constantly at 18°C and more rapidly at 24 and 30°C. Other researchers reported that following inoculation of wounded stems or roots some biovar 2 strains latently infect and cause at least limited wilt symptoms at 16°C and are highly virulent at ≥20°C (Ciampi & Sequeira, 1980; Swanepoel, 1990). In contrast, biovar 1 and 3 strains cause no wilt at <20°C and are highly virulent only at ≥24°C. Recently, Swanson et al. (2005) inoculated unwounded roots of potato and tomato and found that a biovar 2 strain from geranium is more virulent than a biovar 1 strain from tomato at 24°C, but the reverse is true at 28°C. Nothing is known about the genetics or physiology responsible for the ability of biovar 2 strains to cause disease at cool temperatures, except that it is not strongly correlated with the ability to multiply *in vitro* at 16°C (Ciampi & Sequeira, 1980; French, 1986). It is also unclear if all biovar 2 strains exhibit this trait, because relatively few strains have been tested and in one case only 2 of 13 strains from potato were highly virulent at cool temperatures (Ciampi & Sequeira, 1980).

Fig. 7. Bacterial wilt of potato. Tubers with dirt adhering to bacterial ooze emerging from the eyes. Courtesy of Dr. Joe Thurston, Cornell University.

In cooler climates yellowing of the foliage and stunting may appear before wilt or there may be no foliar symptoms. Apparently healthy plants can produce tubers that are latently infected by biovar 2 strains. Processing of infected ware potatoes can release the pathogen into streams and rivers where alternate hosts like *S. dulcamara* can become infected and further infest the watercourse. Irrigating with contaminated water will often result in fields becoming infested by *R. solanacearum* biovar 2 strains (Elphinstone et al., 1996; Janse, 1996).

6.3. BW of Geranium

During the last decade it became apparent that in addition to biovar 1, biovar 2 strains from potato also affect cultivated zonal geranium (*Pelargonium* x *hortorum*) (Janse, 1996; Williamson et al., 2002). Several companies produce geranium cuttings in regions of Central America and Africa were biovar 2 strains are endemic. Irrigation with infested water or other phytosanitary failure results in plants latently infected by *R. solanacearum* and infected cuttings have been shipped to the US and Europe were they are rooted and grown for resale. If the plants are maintained at cool temperatures they may show no disease symptoms, but unknown stresses or incubation at warmer temperatures (>25°C) are conducive to disease. Early symptoms are upward curling and then wilting of lower leaves and subsequent sectorial chlorosis and necrosis. Stems may show external brown to black discoloration and internal discolored vascular tissue. The disease sometimes progresses and the whole plant desiccates and dies, but other times plants 'out grow' the disease (but not the infection). BW of geranium would not be a serious problem for horticulturalists except that detection of biovar 2 infection triggers quarantine restrictions that usually result in destruction of many thousands of plants.

6.4. BW of Banana and Related Plants

Cultivated *Musa* species are hybrids between the diploid species *Musa acuminata* (genome A) and *Musa balbisiana* (genome B). Sterile triploid plants with different proportions of the A and B genomes determine the sweetness or starchiness of the fruit: dessert bananas (commonly Cavendish type) are AAA, whereas cooking bananas may be either AAB (e.g. Plantain type) or ABB (e.g. Bluggoe type) (Jones, 2000). These and other triploid cultivars are propagated vegetatively throughout the humid tropics and provide the main carbohydrate consumed by many of the world's poor. A few varieties are grown on large plantations for export, and they are a major cash crop in some countries (Jones, 2000; Sequeira, 1998). *R. solanacearum* and the BD bacterium cause wilt diseases of *Musa* and *Heliconia* species in Latin America and the Caribbean, the Philippines, and Indonesia (Thwaites et al., 2000). Relevant *R.*

solanacearum strains are in phylotype II, whereas BD bacteria are in phylotype IV (Table 3). Colonies of the BD bacterium are smaller than those of *R. solanacearum* and are non-fluidal. When first isolated they do not use glucose, but readily metabolize galactose and glycerol, and do not reduce nitrate to nitrite (Eden-Green, 1994).

BW of bananas in Latin America and the Caribbean is usually referred to as Moko disease, the name having originated in the early 1900's during an epidemic in Trinidad on the local variety called Moko (a Bluggoe type) that is very susceptible (Thwaites et al., 2000). Serious epidemics of Moko disease in Central and South America during the 1950's and 1960's in commercial dessert banana plantations and in Bluggoe bananas planted in gardens and smallholdings attracted substantial scientific attention. The original literature described two subgroups of *R. solanacearum* stains causing Moko disease based on bacterial cultural characteristics and whether the pathogen is vectored by insects (French, 1986; Sequeira, 1998). Colonies of strains in the 'B' subgroup were described as elliptical with lace-like EPS1 slime and a light pink center on medium containing triphenyl tetrazolium chloride (TZC). This subgroup is synonymous with sequevar 3 (RFLP MLG 24) (Fegan, 2005; Prior & Fegan, 2005). These strains are mainly soil-borne and are transmitted by root-to-root contact, cultivation or pruning. Infected dessert bananas exhibit rapid yellowing and wilting of leaves and suckers, vascular discoloration in the pseudostem, premature fruit ripening and blackening, and dry rot of fruit pulp (Fig. 8). 'B' strains may be transmitted by insects, but this is infrequent because infected plants exude relatively little bacterial ooze.

Colonies of strains in the 'SFR' (small, fluidal, round) and 'A' subgroups were described as being nearly round with little to plentiful faintly lace-like EPS1 slime and a light pink center (in a faint spiral pattern for A types). More importantly, strains in these subgroups are readily insect-transmitted and enter plants via natural wounds on male flowers (Buddenhagen & Kelman, 1964). Bluggoe type cultivars are particularly susceptible, because male flowers abscise every day for several months, whereas flower infection is uncommon in cultivars with persistent bracts. Infected flower buds and peduncles (flower stalks) become blackened and shriveled and there is internal blackening and rotting of fruit of the affected bunch. Copious bacterial ooze exudes from the bases of bracts or abscission sites on diseased peduncles, and some of the many insects that visit these sites become infested. During the 1961 epidemic in Costa Rica, the pathogen was frequently transmitted from diseased inflorescences to uninfected plants nearby and infrequently to plants one mile or more away. Typical Moko disease wilt symptoms may appear months later after the pathogen has systemically colonized the pseudostem and rhizome. Nearby plants not susceptible to flower infection by insects may then exhibit typical Moko disease symptoms after being infected via soil, flood water, root contact, or pruning tools. Genetic analyses show that SFR strains are found in sequevars 4 and 6 (MLGs 25 and 28) (Fegan, 2005; Prior & Fegan, 2005; Raymundo, Aves-Ilagan, & Denny, 1998; Thwaites et al., 1999).

Surprisingly, sequevar 4 is closely related to B strains that cause typical Moko disease (sequevar 3) in the phylotype IIa subcluster, whereas sequevar 6 is in the phylotype IIb subcluster. Therefore, although SFR strains share unusual pathogenic traits, they are not a clonal.

Fig. 8. Moko disease of banana. A. Diseased raceme with severe symptoms on fruit (some have been cut open). B. Cross sections of pseudostem (top middle) and fruit (lower three sections) from a diseased plant. Fruit in upper corners are from a healthy plant. Courtesy of Dr. Joe Thurston, Cornell University.

In the Philippines the local cooking banana varieties Saba (ABB or BBB), Cardaba (ABB or BBB) and Latundan (AAB) suffer from a disease known as Bugtok (or Tapurok), which was already widespread in 1965 when the first description was published (Soguilon, Magnaye, & Natural, 1994; Thwaites et al., 2000). Symptoms are largely confined to the inflorescence, with the peduncle becoming blackened, dry and distorted, and fruit pulp becoming discolored grayish black to yellowish red and later becoming hard. There may be an associated reddish brown discoloration of the vascular tissue of the peduncle and pseudostem, but rarely does discoloration extend into the rhizome. Because the pathogen is never fully systemic there are no wilt symptoms and the plant appears relatively normal to the untrained eye. Bugtok remains a common disease in bananas cultivated by smallholders in the Philippines. Unexpectedly, numerical taxonomy and genetic analyses determined that the *R. solanacearum* strains responsible for Bugtok are indistinguishable from those causing Moko disease on Cavendish type dessert bananas in the Philippines and are very similar to the 'B' subgroup of Moko strains from Honduras (Eden-Green, 1994; Fegan & Prior, in press; Ilagan, Lavina, Natural, & Raymundo, 2003; Raymundo et al., 2005; Thwaites et al., 1999). In addition, the very low genetic diversity among *R. solanacearum* strains isolated from *Musa* spp. in the Philippines (Fegan & Prior, in press; Raymundo et al., 2005) suggests that a single genotype of the pathogen was introduced from Central America. There probably are several reasons why the 'same' strains can cause two apparently different diseases in the Philippines. First, Bugtok starts with infection of male flowers by insects (similar to infection by SFR strains) rather than via stems or roots like in Moko disease. Only Moko disease is observed in the commercial banana plantations, because the standard control practices of removing male flowers and protecting fruit in insect-proof bags prevent infection of the inflorescence. Second, the cooking banana varieties cultivated in the Philippines may resist systemic infection better than the Bluggoe type (ABB) varieties popular in the Americas. Third, unique environmental factors in the Philippines may make cooking bananas more resistant to systemic infection than they are in the Americas.

In Indonesia cooking and dessert bananas and native *Heliconia* plants suffer from another wilting disease known as blood disease (BD). The pathogen may have originated on Salayar Island near Sulawesi, and BD was first reported in the early 1900's after the introduction of dessert bananas (Eden-Green, 1994; Thwaites et al., 2000). Genetic analyses indicate that there is little diversity among strains of the BD bacterium (Fegan & Prior, in press; Thwaites et al., 1999). BD was 'rediscovered' in Java in the late 1980's (Thwaites et al., 2000) and was common on local cooking banana cultivars in Sulawesi (Stover & Espinoza, 1992). Unfortunately, the pathogen has spread to most of the larger Indonesian islands and average yield losses exceed 35% (Supriadi, 2005). It has also been reported on the island of New Guinea (see (Fegan & Prior, in press)). Symptoms of BD are quite similar to Moko disease caused by insect-transmitted strains in the SFR/A subgroup, namely discoloration and shriveling of the male flower bud and peduncle, reddish dry rot of the fruit pulp, and reddish discoloration of vascular tissue throughout the plant, which emits reddish-brown bacterial ooze when cut. Older leaves become yellow, followed by wilting, necrosis and collapse; younger leaves turn bright yellow before becoming necrotic and dry. The pathogen rapidly colonizes the entire plant, and suckers will also wilt and die. Insect transmission of the BD bacterium is strongly implicated by the sequence of symptoms and the rapid dispersal of the pathogen. However, poor hygienic practices of banana traders may also be a contributing factor. Unlike *R. solanacearum* strains causing Moko and Bugtok, the BD bacterium is not pathogenic on tomato and eggplant seedlings (Eden-Green, 1994; Supriadi, 2005).

6.5. BW of Ginger

Edible ginger is a perennial herb cultivated in the warm, humid tropics. BW of ginger was first reported in Australia and Hawaii during the 1960's (Hayward, 1994b), and it is now a serious problem in Hawaii, Japan, China, India, and several southeast Asian countries (Alvarez et al., 2005; Elphinstone, 2005). In some countries BW is the most important disease of ginger. Typical BW symptoms start with the yellowing and wilting of lower leaves, but the whole plant soon becomes golden brown and wilted. The base of the pseudostem may become watersoaked and will easily break away from the rhizome (the edible root). Diseased rhizomes are grayish brown with transparent patches covering milky white tissue below. Copious white, milky exudate oozes out after pseudostems or rhizomes are cut.

In Oceania, some biovar 4 strains (race 4) are responsible for BW of ginger (Hayward, 1994b), and genetic analysis showed that they cluster as a relatively homogenous group separate from selected strains from tomato, banana, and potato (Alvarez et al., 2005; Cook et al., 1994). Until recently, the few biovar 3 strains found that attack ginger were weakly aggressive (Alvarez et al., 2005; Hayward, 1994b), but in India a clonal group of biovar 3 strains are more aggressive than biovar 4 strains from the some localities (Kumar et al., 2004). These contrasting findings indicate the necessity for a comprehensive genetic survey of ginger strains from many countries.

6.6. Sumatra Disease of Clove

On the Indonesian islands of Sumatra and Java *R. syzygii* causes a lethal wilting disease of clove trees (*Syzygium aromaticum*), a high value crop grown for manufacture of kretek cigarettes (Bennet et al., 1985; Eden-Green, 1994; Roberts et al., 1990). *R. solanacearum* can colonize the roots and lower trunk, but only *R. syzygii* systemically colonizes and kills clove trees. Observed since the 1920's, Sumatra disease is a recurring problem in the western coastal region of Sumatra, especially on trees planted near forests 400 to 1000 meters above sea level. Seedlings and saplings <2 years old are unaffected, but susceptibility increases with age and most of the trees >10 years old are killed. The first symptom is unseasonal yellowing and leaf-drop from tips of branches high in the crown. Leaves may also wilt suddenly and turn brown, but stay attached (fire blight symptoms). Affected twigs turn reddish brown and progressively die back. Internally, there is a pale grayish-brown discoloration of the newly-formed wood adjacent to the cambium (in an arc or complete ring) that is diagnostic for Sumatra disease. Infected branches often release a milky white to pale brown bacterial ooze from cut surfaces. The discolored xylem can be traced down the trunk into one or more major roots. Symptoms typically progress to lower branches until the whole crown is affected, and the tree dies within 6 to 18 months. Insect transmission of the pathogen by xylem-feeding spittlebugs (*Hindola* spp.) combined with the greater susceptibility of older trees often results in death of all the mature trees in a region. The disease then disappears for years until young trees mature and the cycle repeats. The host range of *R. syzygii* has not been determined. It has been experimentally transmitted to some other myrtaceous species, but if it follows the pattern of some other xylem-limited bacteria, then its host range (including symptomless hosts) might be substantially larger (Purcell & Hopkins, 1996).

7. DISEASE CONTROL

No single strategy can reduce the incidence and/or severity of BW in regions where the pathogen is endemic (Hartman & Elphinstone, 1994; Saddler, 2005). Nevertheless, losses due to BW can be greatly reduced by following a holistic approach, often referred to as Integrated Disease Management (IDM), which employs multiple disease control strategies. For BW, all successful IDM packages include use of pathogen-free planting material, planting less susceptible host varieties, and rotation of susceptible crops with those resistant or immune to BW (Akiew & Trevorrow, 1994; French, 1994; Lemaga et al., 2005; Saddler, 2005).

7.1. Host Resistance

Planting disease resistant cultivars is almost always the most successful, most economical and most environmentally benign disease control strategy (Boshou, 2005). Unfortunately, for most crops susceptible to BW, there are almost no sources of high level, gene-for-gene type resistance encoded by single dominant genes. Instead, available sources of resistance are usually polygenic and it has been difficult or impossible to transfer all the identified quantitative trait loci (QTL) into desirable cultivars due to their number or linkage to undesirable traits. The only exception is groundnut, where dominant resistance genes were identified long ago and introduced into all four botanical types of *Arachis hypogaea* to produce cultivars with high-level resistance. Remarkably, Schwarz 21, a groundnut cultivar developed over 80 years ago, is still resistant in different regions of the world (Boshou, 2005).

The best that normal breeding has achieved for most solanaceous crops is tolerance of BW (i.e., satisfactory yield despite infection) on a regional level when conditions are not excessively hot or wet. Some potato cultivars are less susceptible to BW at least on a regional scale, and they are useful as part on an IDM package (French, 1994; Lemaga et al., 2005). Due to potato's importance as a subsistence crop, there are still active resistance breeding programs, some of which are focusing on resistance to latent infection (Priou et al., in press). Three distinctly different cultivars of eggplant very tolerant of BW have been released for cultivation in India, and might perform well elsewhere (Gopalakrishnan et al., 2005). A small number of tolerant tobacco cultivars with multiple recessive genes have been developed and some have been widely planted (Akiew et al., 1994). However, even the most tolerant tobacco cultivars bred in the USA must be used as part of an IDM package, because they still suffer serious losses when disease pressure is high (Fortnum & Kluepfel, 2005). In contrast, despite extensive international research that has produced some highly resistant tomato breeding lines, such as Hawaii 7996, there are still no acceptable large-fruited cultivars generally available (Prior, Grimault, & Schmit, 1994; Wang, Hanson, & Barnes, 1998).

There are several reports where genetic engineering has increased BW tolerance of tomato and tobacco. One approach is to constitutively activate defense responses (e.g., pathogenesis-related (PR genes (Van Loon & Van Strien, 1999)) and other defense-response genes) that are normally induced only after pathogen infection. For example, transgenic tomato lines constitutively expressing the *Arabidopsis NPR1* gene (nonexpressor of *PR* genes) are outwardly normal looking but overexpress a subset of PR genes (Lin et al., 2004). NPR1 protein is a conserved and essential intermediate regulator in salicylic acid and jasmonic acid/ethylene-dependent activation of PR genes and acquired resistance. The *NPR1* overexpressing lines are almost as resistant to BW after soil drench inoculation as the control resistant line Hawaii 7997, and also suppress *R. solanacearum* multiplication *in planta*. Constitutive expression of the tomato stress response factor gene *TSRF1* also enhances BW tolerance in transgenic tobacco and tomato (Zhang et al., 2004). TSRF1 protein is a member of the family of ethylene-responsive factors (ERFs), which are transcription factors that help regulate plant pathogen resistance, abiotic stress tolerance and plant development. Like other ERFs, over production of TSRF1 activates expression PR genes that have promoters containing a GCC box, probably by binding to the GCC motif. It is not known in either example whether resistance is due to over production of PR proteins or to activation of uncharacterized responses.

The second approach for engineering tolerance has been to constitutively express foreign proteins or peptides with antimicrobial activity. *R. solanacearum* is sensitive to some peptides *in vitro* (e.g., potato pseudothionin-St1, lactoferrin, and bovine lactoferricin), but is unusually tolerant of others (e.g. Cecropin B, a cationic lytic peptide, and Shiva-1, a synthetic analog of Cecropin B) (Alan & Earle, 2002; Jaynes et al., 1993; Segura, Moreno, Madueno, Molina, & García-Olmedo, 1999; Zhang, Coyne, Vidaver, & Mitra, 1998). Tobacco expressing Shiva-1 from a wound-inducible promoter are moderately more tolerant to BW than normal, especially when inoculated via a stem wound (Jaynes et al., 1993). Some transgenic potato plants expressing a different Cecropin B analog are similarly more BW tolerant (Montanelli, Stefanini, Chiari, Chiari, & Nascari, 1995). Tobacco and tomato plants that constitutively produce iron-binding human lactoferrin protein are more tolerant to BW than controls and inhibited multiplication of the pathogen *in planta* (Lee, Coyne, Clemente, & Mitra, 2002; Zhang et al., 1998). Tolerance could be due to bacteriostatic sequestration of iron by lactoferrin, the bactericidal lactoferricin peptide released from lactoferrin by proteolysis, or to unknown side effects of the transgenes. However, regardless of the approach used to genetically engineer enhanced BW resistant crops, there are major scientific and societal hurdles that must be cleared before they are ready for field production (Denny, 2005).

7.2. Cultural Control

A variety of cultural practices help to reduce losses due to bacterial wilt. In regions or fields where *R. solanacearum* is not present, the fist line of defense is to avoid introducing the pathogen by using pathogen-free propagative tissue (e.g., tubers and rhizomes) and good sanitation. In some developed countries, regulatory agencies have promulgated quarantine regulations for biovar 2 (race 3) strains. These may include testing all lots of seed potatoes for latent infection, surveying ware and starch potatoes and destroying loads containing any infected tubers, monitoring surface water and prohibiting use of contaminated waterways for irrigation, and requiring that affected farms disinfest machinery, storage facilities, etc. and plant grasses in infested fields for 4 or 5 years (Janse, Araluppan, Schans, Wenneker, & Westerhuis, 1998). Planting clean seed potatoes is helpful even in regions where the pathogen is endemic, but in developing countries there may not be enough certified seed tubers or they are too expensive for subsistence farmers. An innovative on-farm seed-plot technique pioneered in eastern Africa has the potential to help satisfy farmer's needs for high quality, pathogen-free seed potatoes (Kinyua et al., 2005).

Where *R. solanacearum* is already endemic, the best cultural control is crop rotation. Several grasses are especially effective in reducing BW incidence (Akiew et al., 1994; French, 1994; Hartman et al., 1994; Saddler, 2005). Other crops that are locally desirable can also be effective, with sweet potato in Africa being a good example (Lemaga et al., 2005). Elimination of volunteer plants and weeds that promote survival of *R. solanacearum* is an important component of successful rotation. Rotation may provide an additional advantage if it reduces nematode populations that attack the susceptible crop. The number of years that a rotation crop must be grown depends on the level of infestation, the survival capacity of the pathogen in local soils and climate, and other factors. In some cases a single year of rotation can significantly reduce BW (at least in the short term), but the usual recommendation is for a two or three year rotation schedule. Bare fallowing and flooding can serve the same purpose (Hartman et al., 1994), but generally are not feasible. Unfortunately, farmers often do not adopt effective rotations because of limited land availability and pressures to produce a subsistence crop or one with high cash value.

A variety of other cultural controls are available, some of which are crop or region specific. One of the simplest is to shift cropping dates to avoid the peak periods of warm and/or moist conditions conducive to disease. Soil amendments, such as organic matter (compost, bagasse, rice husk powder), inorganic fertilizers, or other material like oyster shell powder, may modify native soil microbial communities to suppress the size or activity of the *R. solanacearum* population, but have not been widely studied or generally applied (Lemaga et al., 2005; Saddler, 2005; Schönfeld et al., 2003; van Elsas, van Overbeek, Bailey, Schönfeld, & Smalla, 2005a). Solarization appears to be marginally effective at best, and in one case the *R. solanacearum* population increased in treated beds (Saddler, 2005; Sharma, Rajesh, Dohroo, & Rajesh, 2004; van Elsas et al., 2005a). Grafting susceptible, horticulturally desirable scions of eggplant and tomato onto BW-resistant rootstocks reduces disease and increases yield (Grimault & Prior, 1994a; Nakaho et al., 2000), and has been used successfully in Japan, India and southeast Asia. For banana and plantain, a combination of sanitation, roguing of diseased plants and those nearby (and soil disinfestation), removing the male flower and bagging fruit clusters all help to control BW.

7.3. Chemical Control

Commercial chemicals and antibiotics generally are ineffective in controlling BW (Hartman et al., 1994; Saddler, 2005). One exception may be fumigation with chloropicrin either alone or combined with other fumigants (Enfinger, McCarter, & Jaworski, 1979; Fortnum & Martin, 1998) or with solarization. However, for tobacco in the southeastern USA, fumigation was not as effective as rotation or use of tolerant cultivars. Moreover, fumigation is environmentally destructive and the cost effectiveness of fumigation has not been examined, so this strategy is unlikely to be part of an IDM package in developing countries.

7.4. Biological Control

There has been a great deal of interest in finding bacteria that can be coated on propagating tissues or added to soil to reduce infection of susceptible crops or reduce *R. solanacearum* populations, respectively (Prior, Allen, & Elphinstone, 1998; Akiew et al., 1994; Hartman et al., 1994; Saddler, 2005; Trigalet, Frey, & Trigalet-Demery, 1994). Theoretically, biological control agents (BCAs) may work directly by competing with the pathogen for limited resources in the soil, the rhizosphere, or within the plant, or by producing antibiotics, bacteriocins, or bacteriophage. BCAs may also work indirectly by stimulating plant defense capabilities. BCAs often are nonpathogenic bacteria, but the greatest effort on developing a BCA for *R. solanacearum* has focused on non-pathogenic *hrpO* mutants of the pathogen. Unfortunately, although showing promise when tested in controlled conditions, none of the potential BCAs have effectively or consistently reduced BW in field conditions. Development of a useful, affordable BCA is unlikely in near term, but progress is being made and there is still hope for success in the future.

8. MECHANISMS OF PATHOGENESIS

The biochemical and genetic mechanisms underlying *R. solanacearum* pathogenesis have been studied for over 50 years (Denny, 2005). In this literature, virulence factors are defined as those which contribute to the incidence, rate or severity of wilt symptoms, whereas pathogenicity factors are essential for disease. Most virulence factors enhance the pathogen's ability to colonize host tissues systemically and to reach populations $>10^8$ cfu/g tissue; disease symptoms appear only after, and largely because of, this extensive colonization. The expectation is that basic research into *R. solanacearum*'s pathogenic mechanisms will result in new or improved approaches for disease control. This topic has been recently reviewed (Denny, 2005; Hayward, 1995; Schell, 2000), so work published in the last five years will be emphasized.

Compared to most other aspects of *R. solanacearum* pathobiology, the mechanisms of pathogenesis have been studied by relatively few research groups using only a few pathogen strains. Strain K60 (the type strain; phylotype II, biovar 1), isolated from a wilted tomato in North Carolina, was first studied by A. Kelman, then by L. Sequeira, and most recently by C. Allen. Strain AW, which was isolated from tomato in Alabama and is genetically and biologically similar to K60, has been studied primarily by T. Denny and M. Schell. Strain GMI1000 (phylotype I, biovar 3) was isolated from tomato in French Guyana and used primarily to investigate pathogenicity factors by a French consortium lead by C. Boucher. The complete genomic sequence of GMI1000 is available (Salanoubat et al., 2002). All three strains are highly virulent on tomato, the primary host for pathogenesis research, but only K60 causes BW of tobacco.

8.1. Virulence Factors

8.1.1. Extracellular Polysaccharides

A major virulence factor of *R. solanacearum* is its acidic extracellular polysaccharide EPS1, a long ($>10^6$ Da) polymer with a repeating unit of three unusual *N*-acetylated monosaccharides (Orgambide et al., 1991). EPS1 is produced in massive amounts by *R. solanacearum* on various laboratory media and in plants (Araud-Razou et al., 1998; McGarvey, Bell, Denny, & Schell, 1998). Tests with EPS1-specific antibodies show that diverse *R. solanacearum* strains produce EPS1 (or a very similar polymer) and that 85% of the EPS1 is released as a cell-free slime (McGarvey et al., 1998). Many proteins comprising the EPS1 biosynthetic pathway are encoded by the 16-kb *eps* operon. Mutants of *R. solanacearum* unable to produce EPS1 rarely wilt or kill plants, even when bacteria are injected directly into the stem (Araud-Razou et al., 1998; Saile et al., 1997). Potted plants inoculated by soil-drenching, the most natural laboratory method available, showed that EPS1 promotes rapid systemic colonization of tomato plants, because EPS1$^-$ mutants generally colonize only the roots and lower stems (Saile et al., 1997). In contrast, EPS1$^-$ mutants do not move into the xylem vessels of hydroponically-grown plants, but instead multiply in the intercellular spaces of the root cortex (Araud-Razou et al., 1998).

Lipopolysaccharide (LPS), consisting of lipid A, core polysaccharide, and O-antigen polysaccharide, is a major component of the outer leaflet of the outer membrane of Gram-negative bacteria, making it the primary environmental interface. Most *R. solan-acearum* strains characterized have an O-antigen repeating unit that contains three rhamnose and one acetylglucosamine (Kocharova et al., 1993; Varbanets, Kocharova, Knirel, & Moskalenko, 1996). Studying the role of LPS has been difficult, because LPS mutants are usually pleiotropic. For example, mutations that interfere with synthesis of the core polysaccharide in K60 reduce virulence (Kao

& Sequeira, 1991; Titarenko, López-Solanilla, García-Olmedo, & Rodríguez-Palenzuela, 1997), but they also reduce production of EPS1 and increase sensitivity to plant antimicrobial peptides. Many years of research in Sequeira's lab suggested that the LPS could play several roles in host-pathogen interactions (Sequeira, 1985), but without well-characterized LPS-minus mutants the overall results were inconclusive. Recently, Kang et al. (2004) specifically blocked synthesis of the O-antigen side-chain without affecting production of EPS1 by inactivating the gene (*waaL*; RSc2204) in GMI1000 that appears to encode the lipid A core:O-antigen ligase, an enzyme that couples the completed O-antigen oligosaccharide subunits to the core polysaccharide (Whitfield, 1995). The mutant is resistant to an LPS-specific phage and analysis of purified LPS indicated that it lacks the O-antigen moiety. Although the mutant is normal for multiple other traits, it is dramatically less virulent than the wild type. These results suggest that the O-antigen contributes to *R. solanacearum* virulence, but the mechanism remains undetermined.

8.1.2. Protein Appendages: Flagella and Pili (Fimbriae)

R. solanacearum can produce two or three lophotrichous flagella, but in a rich broth culture cells are motile only at cell densities between 10^7 and 10^9 cells ml^{-1} (Clough, Flavier, Schell, & Denny, 1997). Similarly, bacteria recovered from within tomato plants are overwhelmingly nonmotile (Tans-Kersten et al., 2001) and microscopic observations of GFP+ *R. solanacearum* in tomato plants confirms this observation (Denny, unpublished results). K60 Mutants that are nonflagellated due to inactivation of the flagellin structural gene (*fliC*) are reduced in virulence when applied to potted tomato plants in a soil drench, but exhibit normal virulence when inoculated via a severed petiole (Tans-Kersten et al., 2001). This suggests that flagella are not important for pathogenesis once bacteria are inside a tomato plant. Non-chemotactic K60 mutants are similarly reduced in virulence when inoculated by soil drench (Allen, personal communication). Unlike the flagellin protein from some bacteria (Asai et al., 2002), *R. solanacearum* FliC does not trigger an innate defense response by *Arabidopsis thaliana* or tobacco (Pfund et al., 2004).

Polar, retractable, type 4 pili (Tfp) are produced by diverse bacteria, and give them the ability to migrate over solid surfaces, a process called twitching motility (Strom & Lory, 1993). *R. solanacearum* strains make Tfp composed of PilA protein, and Tfp are essential for twitching motility and virulence on tomato when plants were inoculated either by a soil drench or via severed petioles (Kang, Liu, Genin, Schell, & Denny, 2002). A Tfp-minus strain also is reduced in autoaggregation and biofilm formation in broth culture, and does not exhibit polar attachment to cultured tobacco cells or to tomato roots. The Hrp pili made by *R. solanacearum* will be discussed below.

8.1.3. Plant Cell Wall Degrading Enzymes (CWDEs)

The involvement of *R. solanacearum* CWDEs in virulence has been examined in all three strains, usually with tomato as the host (González & Allen, 2003; Liu, Zhang, Schell, & Denny, in press; Schell, 2000). Six extracellular enzymes have been identified: a β-1,4-endoglucanase (Egl), an exoglucanase (ChbA), an endopolygalacturonase (PehA or PglA), two exopolygalacturonases (PehB and PehC), and a pectin methylesterase (Pme). Inactivation of single genes has shown that none of these exoenzymes is essential and that their relative contribution to disease may vary with the strain. In K60, PehA and PehB, but not PehC or Pme, contribute to pathogen colonization and wilt severity (González & Allen, 2003; Huang & Allen, 2000; Tans-Kersten, Guan, & Allen, 1998), but the cellulolytic enzymes have not been studied. The PehA-PehB double mutant was less virulent than either single mutant (Huang & Allen, 2000), but eliminating PehC in addition partially restored virulence (González & Allen, 2003). This counterintuitive result was attributed to the absence of pectic breakdown products that stimulate plant defenses (see also Jha, Rajeshwari, & Sonti, 2005). In GMI1000, however, where mutants lacking one to all six enzymes were recently studied, no combination of the four pectolytic enzymes contribute significantly to disease when using a soil drench inoculation (Liu et al., in press). Both Egl and CbhA contribute to virulence of GMI1000 regardless of the inoculation method, and the Egl-CbhA double mutant is the least virulent CWDE mutant strain. Similar to the results for K60, simultaneous elimination of Peh enzymes consistently enhances virulence of GMI1000. Why the CWDEs do not contribute equally to virulence of K60 and GMI1000 is not known.

8.1.4. Other Potential Virulence Factors

In *R. solanacearum* most of the CWDEs and many other extracellular proteins transit the inner membrane via the Sec-dependent general export pathway and the outer membrane via the type II secretion system (T2SS) (Preston, Studholme, & Caldelari, 2005; Schell, 2000). The one known exception is PehC, which is exported by the twin-arginine translocation (Tat) system (González and Allen, personal communication) and secreted by the T2SS. Inactivation of the T2SS results in proteins that normally are secreted accumulating in the periplasm or cytoplasm. A T2SS mutant of strain AW does not secrete CWDEs, multiplies less than the wild type in tomato stems, and does not cause wilt symptoms (Kang, Huang, Mao, He, & Schell, 1994). A more thorough examination of a GMI1000 T2SS mutant lacking its outer membrane secretin (SdpD), found that it does not secrete any of the six CWDEs and is much less virulent than either the Egl-CbhA double mutant or a mutant lacking all six CWDEs (Liu et al., in press). The T2SS mutant also colonized the lower stem of about 50% fewer plants than the wild type. These results indicate that extracellular proteins in addition to Egl and CbhA contribute to the ability of GMI1000 to systemically colonize tomato plants.

It has long been suspected that phytohormones produced by *R. solanacearum* might contribute to virulence (Buddenhagen & Kelman, 1964; Hayward, 1995).

Infected plants may exhibit leaf epinasty and over production of adventitious root primordia, suggesting an imbalance in indoleacetic acid (IAA) or auxin. K60 produces relatively little IAA in culture and increased IAA concentrations in plants may be due the pathogen reducing its degradation. *R. solanacearum* can make cytokinin and ethylene. Inactivation of the *tzs* gene responsible for production of cytokinin (Akiyoshi, Regier, & Gordon, 1989) reduces the virulence of strain AW, but inactivation of the gene encoding the putative ethylene-forming enzyme (RSp1529), which makes ethylene and succinate from oxoglutarate, does not affect virulence (Ratnayake, 2002).

Many bacteria produce siderophores (low-molecular mass extracellular iron-scavenging compounds) to help them acquire essential iron, which is often in forms that are biologically unavailable to aerobes (Andrews, Robinson, & Rodriguez-Quinones, 2003). Some phytopathogenic bacteria need their iron acquisition system for full virulence whereas others do not. *R. solanacearum* strain K60 had been reported to produce the dihydroxamate siderophore called schizokinen, but more recent work showed that this strain (and probably all seven others tested, including AW and GMI1000) produce the polycarboxylate siderophore staphyloferrin B (Bhatt & Denny, 2004). However, tomato xylem fluid may have sufficient iron to repress expression of the pathogen's iron-acquisition system, and a mutant that does not make staphyloferrin B is fully virulent on tomato.

8.1.5. Regulated Production of Virulence Factors

Spontaneous loss of virulence, EPS1 and other traits by *R. solanacearum* in culture was until recently a vexing and perplexing problem for scientists studying BW (Denny, 2005; Kelman, 1953; Sequeira, 1985). Although there are ways to cope with this propensity (Buddenhagen & Kelman, 1964), the genetics of phenotype conversion (PC) remained unclear until the discovery of PhcA, a LysR-type transcriptional regulator that controls expression of many virulence genes (Fig. 9) (Brumbley, Carney, & Denny, 1993; Schell, 2000). There is no evidence that *phcA* spontaneously mutates at a high frequency; instead, *phcA* mutants (PC-types) accumulate because they are selected for during some stressful conditions (e.g., prolonged stationary phase in culture or *in planta*, high salt or low oxygen concentrations) (Denny, Brumbley, Carney, Clough, & Schell, 1994). DNA replication errors and transposition of IS elements can inactivate *phcA* (Brumbley et al., 1993; Jeong & Timmis, 2000; Poussier et al., 2003). Traits in strain AW currently known to be positively regulated directly or indirectly by PhcA are: (i) production of EPS1, (ii) production of Egl, (iii) production of Pme, (iv) competence for natural transformation by DNA, and (v) an acyl-homoserine lactone quorum sensing system (Kang et al., 2002; Schell, 2000). Negatively regulated traits (i.e., those expressed better in PC-types) are: (i) production of PehA, (ii) production of staphyloferrin B siderophore, (iii) production of type 4 pili and thereby twitching motility, autoaggregation and biofilm formation, (iv) flagellar motility at high cell density, (v) salt tolerance, and (vi) activity of the HrpG transcriptional regulator (Bhatt & Denny, 2004; Genin, Brito, Denny, & Boucher,

2005; Schell, 2000). There are likely to be additional genes controlled by PhcA. The acyl-homoserine lactone quorum sensing system does not regulate virulence, but it likely controls more than just *aidA* (whose function is unknown). Although some PC-types are stimulated to revert to wild type by plant compounds (Poussier et al., 2003), spontaneous mutation of *phcA* is usually just a genetic mistake and a pathological dead end for *R. solanacearum*. However, wild type PhcA plays a critical role as part of a complex regulatory network that enables *R. solanacearum* to cycle between two very different phenotypic states (i.e., reversible PC) in response to nutrient availability and cell density (Denny, 2005; Schell, 2000). Levels of functional PhcA are controlled by a unique confinement-sensing system encoded by the *phcBSR* operon. PhcB appears to be a small-molecule methyltransferase that synthesizes 3-OH palmitic acid methyl ester (3-OH PAME), which accumulates in the extracellular environment when bacteria are growing rapidly within a confined space (Fig. 9). PhcS and PhcR comprise a two-component regulatory system that senses and responds to threshold concentrations of 3-OH PAME by elevating the level of functional PhcA. It is likely that 3-OH PAME stimulates PhcS to phosphorylate PhcR, and that PhcR ~ P lacks the ability to post-transcriptionally inhibit production of functional PhcA. In other words, cells at low-density have little functional PhcA and, like *phcA* mutants, exhibit a low virulence phenotype. These cells are hypothesized to be optimized for survival and invasion of plant tissues, because low densities of *R. solanacearum* routinely occur in soil and in plants at the leading edge of infection. In contrast, cells at high densities, like those in well colonized xylem vessels, have abundant functional PhcA and so produce multiple virulence factors (while suppressing production of survival/invasion factors) that promote multiplication and further colonization of tissues. Functional PhcA activates expression of some genes, like *egl* and *xpsR*, directly by binding to their promoter, while traits like EPS1 are controlled indirectly (Fig. 9) (Schell, 2000). Most research on confinement-sensing was done in culture, but the *eps* operon is similarly regulated in tomato plants during pathogenesis (Kang, Saile, Schell, & Denny, 1999; McGarvey et al., 1998).

PhcA has a central role in regulating virulence, but there are additional regulatory proteins in the supporting cast (Fig. 9) (Schell, 2000). Transcription of the *eps* operon is the best example, because downstream of PhcA is a complex cascade that may respond to two additional signals. Functional PhcA first activates transcription of *xpsR*, an intermediate regulator whose expression is also enhanced by the VsrA/VsrD two-component system. XpsR, a unique and very basic protein, then works in conjunction with VsrC, a response regulator paired with the VsrB sensor, to activate *eps* transcription (Garg, Huang, Yindeeyoungyeon, Denny, & Schell, 2000). The signals, if any, sensed by VsrA and VsrB are not known. VsrC binds to a 20 nucleotide region of the *eps* promoter that is also essential for activation by XpsR. The biochemical role of XpsR remains unknown due to poor solubility of the purified protein. EpsR, an atypical response regulator-type protein (Chapman & Kao, 1998), inhibits EPS1 production when borne on a multicopy plasmid, and binds to the same 20 nucleotide region. However, since inactivation of *epsR* in the genome has no major effect on *eps* expression or EPS1 production, the physiological role of EpsR remains obscure.

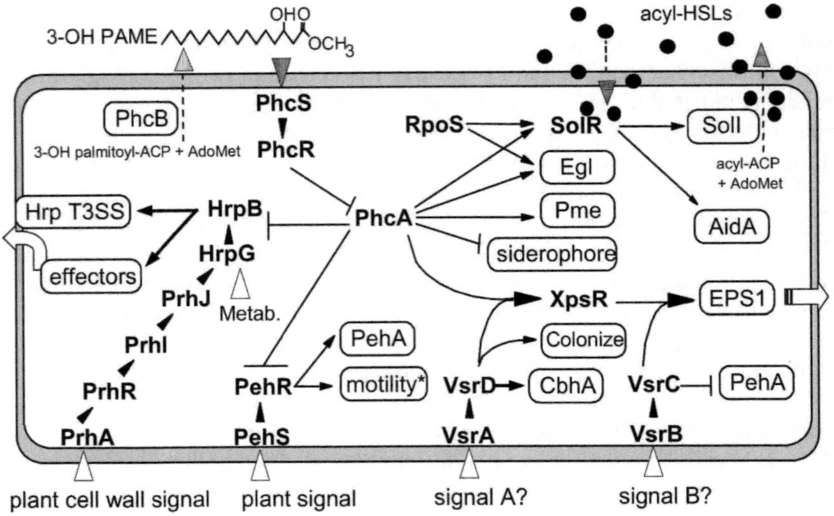

Fig. 9. Model of the regulatory networks in *R. solanacearum*. The major circuits are shown, but there is cross talk between many of these pathways (Brown and Allen, 2004). Within this network the four known two-component regulatory systems are PhcS-PhcR, PehS-PehR, VsrA-VsrD, and VsrB-VsrC. Except for 3-OH PAME, the novel autoinducer sensed by PhcS-PhcR, the signals for the two-component systems are not known. Other transcriptional regulators are PhcA, SolR (that responds to acyl-homoserine lactones (AHLs)), and PrhI, PrhJ, HrpG, and HrpB in the Prh/Hrp signal cascade. Also involved with regulating transcription are RpoS (an alternative sigma factor) and XpsR (a signal integrator). Proteins essential for biosynthesis of extracellular signal molecules are PhcB (predicted to be an S-adenosylmethionine-dependent methyltransferase) and SolI (an AHL synthase). Predicted substrates for these enzymes are 3-OH palmitoyl-acyl carrier protein (3-OH palmitoyl-ACP), C6- and C8-acylated-acyl carrier proteins (acyl-ACPs), and S-adenosyl-methionine (AdoMet). AidA is a protein of unknown function; its structural gene is regulated by SolR in response to sufficient concentrations of acyl-HSLs. Dashed lines with hatched arrowheads represent diffusion of signal compounds into and out of the cell. The metabolic/nutritional signal affecting hrp gene expression is not known; also unknown is whether this signal affects HrpG activity, the step between HrpG and HrpB, or HrpB directly. Motility□refers to both flagellar swimming and type 4 pili-mediated twitching. Open arrowheads represent perception of extracellular signals by two-component sensors; filled arrowheads represent presumed transfer of phosphate from sensor proteins to response regulators. Lines with filled arrowheads or bars represent positive or negative control, respectively.

Both the VsrBC and VsrAD two-component systems regulate other traits. Inactivation of either *vsrB* or *vsrC* increases PehA production about seven fold by an undetermined mechanism. Inactivation of *vsrA* or *vsrD* strongly reduces production of CbhA and largely eliminates the ability of *R. solanacearum* to cause disease symptoms. The poor ability of a *vsrAD* mutant to colonize stems and multiply *in planta* is not due to the reduced EPS1 production that accompanies loss of this regulator (McGarvey, 1999), so it is likely that VsrAD activates expression of some genes that promote rapid *in planta* growth and colonization.

Unexpectedly, pectinolytic CWDEs are not coordinately controlled, since functional PhcA activates production of Pme, represses production of PehA, and has little effect on PehB and PehC (Schell, 2000). PhcA similarly controls both flagellar and twitching motility, because unlike in the wild type, *phcA* mutants express both swimming and twitching motility at high cell density. Functional PhcA indirectly controls PehA production and motility by reducing the function of the PehSR two-component system that positively controls their expression (Fig. 9). Expression of *pehR* is activated by an uncharacterized plant signal (Tans-Kersten et al., 2004), and its inactivation results in loss of PehA and flagellar and twitching motility. In culture, PehR controls flagellar motility by activating expression of *flhDC*, which encode a tetrameric regulatory protein, but *in planta* there appear to be other activators of *flhDC* expression and repressors of motility (Tans-Kersten et al., 2004). PehR controls twitching motility by activating expression of *pilA*, which encodes the major pilin protein (Kang et al., 2002).

8.2. Hypersensitive Response and Pathogenicity (Hrp) System

Most bacteria that kill plant cells during pathogenesis have a type III secretion system (T3SS) similar to that in some animal pathogens (He, Nomura, & Whittam, 2004; Preston et al., 2005). *R. solanacearum* was among the first phytopathogenic bacteria found to harbor a T3SS, and mutants lacking this system have the typical *hrp* phenotype: HR-negative on nonhosts (usually tobacco) and nonpathogenic on hosts (usually tomato) (Schell, 2000; van Gijsegem, Vasse, De Rycke, Castello, & Boucher, 2002). Similar to other phytopathogenic bacteria, *R. solanacearum* T3SS mutants multiply very little after being infiltrated into a host's leaves. They retain the ability to invade unwounded roots of tomato and to colonize the tap root and lower stem as rapidly as the wild type, but they poorly colonize the upper half of infected tomatoes. The population size of T3SS mutants in infected tissues is typically 100 to 1000-fold lower than for the wild type (Etchebar, Trigalet-Demery, van Gijsegem, Vasse, & Trigalet, 1998; Frey et al., 1994; Vasse, Genin, Frey, Boucher, & Brito, 2000), but in young plants grown in conducive conditions the mutant population can equal that of the wild type and cause transient wilt (Denny, unpublished results).

The primary purpose of the T3SS probably is to secrete 'effector' proteins into the cytoplasm of plant cells, where they facilitate nutrient release and/or suppress basal defense responses (Alfano & Collmer, 2004; Büttner & Bonas, 2002a; He et al.,

2004). Individual effectors usually have subtle and nonessential biochemical functions, so determining their role in pathogenicity has been difficult. However, some plants have evolved (or been bred for) proteins that recognize an effector and trigger a rapid, vigorous defense response, often manifested as a HR (Keen, 1990; Nimchuk, Eulgem, Holt, & Dangl, 2003; White, Yang, & Johnson, 2000). In such cases, the effector is called an avirulence protein because it restricts the pathogen's host range by making it nonpathogenic (incompatible) on a resistant host. Thus, a single effector protein can be a 'double agent' (Alfano & Collmer, 2004) by promoting pathogenesis in a susceptible cultivar, but triggering HR in a resistant cultivar.

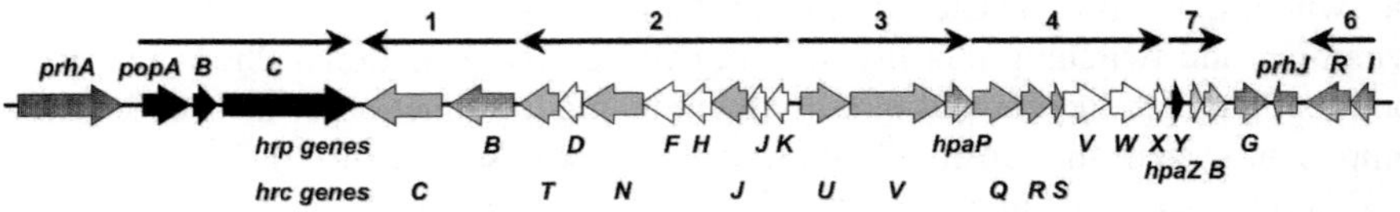

Fig. 10. Genetic organization of the *R. solanacearum* hrp/hrc gene cluster. Thin arrows indicate operons; thick arrows indicate genes. Conserved hrc genes are represented by filled grey arrows, genes encoding T3SS secreted proteins by filled black arrows, hpa genes by stippled arrows, and regulatory genes by hatched arrows. Modified from Van Gijsegem et al., 2002 (with permission).

8.2.1. The hrp/hrc *Gene Cluster and the T3SS*

In *R. solanacearum* the T3SS is encoded by a gene cluster containing five transcriptional units located in a 23-kb region of the megaplasmid (Fig. 10) (van Gijsegem et al., 1995; van Gijsegem et al., 2002). Most genes in the GMI1000 T3SS are very similar to and nearly syntenic with those in several *Xanthomonas* species (the hrp2 group (He et al., 2004)), but the gene cluster is arranged quite differently in *P. syringae* and *Erwinia* spp. (the hrp1 group). Available evidence indicates that the GMI1000 T3SS co-evolved with the rest of the genome rather than being recently acquired by horizontal gene transfer (Salanoubat et al., 2002). The nine *hrc* (hrp conserved) genes encode proteins essential for the core secretion apparatus (the secreton) of all T3SS. Most Hrc proteins localize to the inner membrane of Gram-negative bacteria, with the exceptions being the HrcN ATPase, which is cytoplasmic, and the HrcC secretin, which creates a pore in the outer membrane. In contrast, the *hrp* genes encode proteins with more species-specific roles in translocating effector proteins into the plant cell cytoplasm. In GMI1000, *hrpB*, *hrpF*, *hrpK*, *hrpW*, *hrpX* and *hrpY* are essential for the HR in tobacco, virulence on tomato, production of a Hrp pilus, and secretion of the PopA protein (see below), but *hrpD*, *hrpH*, *hpaP* (hrp associated), *hrpV*, and *hrpJ* are variably impaired in these traits. (Schell, 2000; van Gijsegem, Vasse, Camus, Marenda, & Boucher, 2000; van Gijsegem et al., 2002). Mutants of *hpaZ*, which encodes a HrpY-like protein, and *hpaB* have not been studied in GMI1000. However, in *R. solanacearum* strain RS1000 (biovar 4, from Japan), *hpaZ* mutants exhibit wild-type virulence and HR induction, but *hpaB* mutants are negative for both traits despite production of Hrp pili (Mukaihara, Tamura, Murata, & Iwabuchi, 2004). *hrpY*, *hpaZ* and *hpaB* are all regulated by HrpB (see below) and constitute operon 7 in RS1000 (Fig. 10) (Cunnac, Occhialini, Barberis, Boucher, & Genin, 2004b; Mukaihara et al., 2004).

Only a few Hrp proteins have well characterized functions. The best example is HrpB, which is an AraC-type transcriptional regulator that activates expression of the other transcriptional units in the *hrp/hrc* gene cluster as well as a majority of the known and putative effector proteins (see below) (Cunnac, Boucher, & Genin, 2004a; Cunnac et al., 2004b; Mukaihara et al., 2004). It is also clear that HrpY is the major subunit of the *R. solanacearum* Hrp pilus (Kang et al., 2002; van Gijsegem et al., 2000), and probably forms a hollow conduit through which these proteins move (He et al., 2004). However, HrpY pili are not required for the polar adherence of *R. solanacearum* bacteria to cultured plant cells (van Gijsegem et al., 2000). HrpX is essential for assembly of Hrp pili (but not for secretion of HrpY pilin), but how is unknown (van Gijsegem et al., 2002). HrpJ is suggested to function in translocation of a subset of effectors, because a *hrpJ* mutant has a very low HR-inducing potential but is almost normally virulent on tomato. HrpV and HpaB may have a similar functions in effector 'exit control', because they are orthologous to the better characterized HpaA and HpaB proteins in *X. campestris* pv. vesicatoria (Büttner, Gurlebeck, Noel, & Bonas, 2004; Huguet, Hahn, Wengelnik, & Bonas, 1998) and *hpaB* mutants of RS1000 are nonpathogenic and HR-negative despite making Hrp pili (Mukaihara et al., 2004). HrpV is also needed for efficient display of Hrp pili (van Gijsegem et al., 2002).

The *hrp/hrc* gene cluster in *X. campestris* pv. vesicatoria also has *hrpF*, which is not present in the 23-kb *hrp/hrc* region of GMI1000. *hrpF* is not required for secretion of effectors into the culture supernatant, but it is required for translocation of a the AvrBs3 avirulence protein into the plant cytosol (Rossier, Van den, & Bonas, 2000). HrpF also exhibits *in vitro* lipid-binding activity and pore-formation potential (Büttner, Nennstiel, Klusener, & Bonas, 2002), so it is proposed to be part of an oligomeric protein channel (the translocon) in the plant cell membrane (Büttner & Bonas, 2002b). GMI1000 has two genes, *popF1* and *popF2* (*Pseudomonas* outer protein), that encode proteins about 50% identical to HrpF. These genes are located on the megaplasmid >30 kb away from the *hrp/hrc* cluster. Similar to HrpF, PopF1 and PopF2 are both secreted by the T3SS and, although essential for translocation of the AvrA avirulence protein into plant cells, they are not required for assembly of Hrp pili or secretion of proteins to the supernatant (M. Arlat, personal communication). If additional work proves that PopF1 and PopF2 are part of the translocon, then they are members of the *hrp* subset of T3SS genes and probably should be renamed to reflect this status.

8.2.2. Secreted Proteins and Effectors

The definition of an effector is in flux now, but I use it to refer only to proteins that are translocated via the T3SS directly into the host cell (Alfano & Collmer, 2004). Although ten GMI1000 proteins are known to transit the T3SS, only six have been proven to be effectors. Nevertheless, bioinformatics and transcriptomic analyses suggest that GMI1000 may produce 60-80 effectors (Genin & Boucher, 2004).

Avirulence Proteins. The best known effectors are those which trigger a HR in plants that have a effector-specific recognition system, commonly called a gene-for-gene interaction (Keen, 1990). There are only a few gene-for-gene type interactions between *R. solanacearum* and a host, so unlike phytopathogenic pseudomonads and xanthomonads, there are few known avirulence effectors. The first avirulence gene (*avrA*) was found in AW and encodes a protein necessary for this strain to elicit a HR on 27 cultivars of tobacco (*Nicotiana tabacum*) (Carney & Denny, 1990). Thus, unlike most avirulence proteins that determine cultivar-level specificity, AvrA determines host specificity of AW on *N. tabacum* at the species level. GMI1000 has an AvrA ortholog that is 60% identical to AvrA in AW, and despite lacking 33 amino acids (from four locations within the protein) it is responsible for this strain eliciting a HR on *N. tabacum* cultivars (Carney & Denny, 1990; S. Genin, personal communication). Furthermore, GMI1000 translocates AvrA fused in frame to the *Bordetella pertussis* calmodulin-dependent adenylate cyclase reporter (CyaA=) into plant cells (S. Genin, personal communication), and transient expression of *avrA* from GMI1000 in *N. tabacum* cells elicits a HR and. In K60, which is virulent on tobacco, *avrA* has been inactivated by insertion of a miniature inverted transposable element (Robertson et al., 2004). Inactivation of *avrA* is at least partly responsible for the virulence on tobacco of K60 and most of the *R. solanacearum* strains isolated from tobacco in North and South Carolina, USA. AvrA is not homologous to any protein in the GenBank nonredundant database, so bioinformatics has provided no clues as to its potential biochemical function.

PopP1 and PopP2 in GMI1000 are encoded by genes in chromosomal regions with atypical codon usage and both are secreted by the T3SS into the culture medium (Deslandes et al., 2003; Lavie, Shillington, Equiluz, Grimsley, & Boucher, 2002). GMI1000 also carries an inactivated *popP3* gene (Lavie, Seunes, Prior, & Boucher, 2004). Most of the strains in phylotypes I and III have one or more of the *popP1*, *popP2* or *popP3* genes, but they are all absent in K60 and most other phylotype II strains with the exception of strains in MLG 25 (sequevar 4; mostly from banana) (Lavie et al., 2004). Although PopP1 and PopP2 are only slightly related, amino acid sequence analysis shows that they both are members of the YopJ/AvrBsT family of ubiquitin-like cysteine proteases (Alfano & Collmer, 2004; Deslandes et al., 2003; Lavie et al., 2004; Lavie et al., 2002). Neither protein has been demonstrated to have the predicted enzymatic activity *in vitro* or *in planta*. *popP1* and *popP2* behave like avirulence genes, because mutants are pathogenic on *Petunia* cultivars and *Arabidopsis* ecotypes, respectively, that normally resist BW caused by GMI1000. In addition, a PopP2-CyA' fusion protein is translocated from bacteria into plant cells (Cunnac et al., 2004b). PopP2 appears to interact with the RRS1-R resistance protein in *Arabidopsis* (Deslandes et al., 2003) and, due to its functional nuclear localization signal, it directs the movement of RRS1-R into the plant nucleus. Like most effector genes in *R. solanacearum* and other phytopathogenic bacteria, *popP2* is positively regulated along with the *hrp/hrc* operons (by HrpB in this case); however, *popP1* is expressed constitutively.

Other Proteins. Seven other genes activated by HrpB in GMI1000 encode proteins that transit the T3SS in culture. Four *rip* genes (for *Ralstonia* effector injected into plant cells) encode effectors, because their adenylate-cyclase fusion proteins are translocated from bacteria into plant cells (Cunnac et al., 2004b). RipA is a member of the *R. solanacearum* AWR family (five genes in GMI1000 that have a conserved Alg-Trp-Arg motif). RipB is homologous to the *P. syringae* effector HopPtoQ and has a putative nucleoside N-ribohydrolase domain. RipG has 18 leucine-rich repeats (LRRs) and is a member of the LRR-GALA subfamily in GMI1000. RipT is related to the YopT family of cysteine proteases. The functions of these proteins have not been demonstrated.

Relatively large amounts of PopA1 protein are secreted into culture medium, where it was identified by its ability to elicit a HR when an unnaturally high protein concentration is infiltrated into tobacco and some BW-resistant Petunia cultivars (Arlat, van Gijsegem, Huet, Pernollet, & Boucher, 1994). Tomato cultivars are unaffected by PopA proteins. This HR-inducing activity is similar to the 'harpins' that *P. syringae* and *Erwinia* spp. secrete when cultured in *hrp*-inducing minimal media (MM) (Alfano & Collmer, 2004; He et al., 2004). Like other harpins, PopA is glycine rich and heat stable. In GMI1000, *popA* is the first gene in the HrpB-activated *popABC* operon that is adjacent to the *hrp/hrc* gene cluster (Fig. 10). Removal of 93 amino acids from the N-terminus of PopA1, which occurs in culture and *in planta*, creates PopA3 that retains HR-inducing activity. PopA is not required by GMI1000 either for pathogenicity on susceptible tomato and *Petunia* cultivars or for eliciting a HR on tobacco (Arlat et al., 1994). Interestingly, a mutant of strain OE1-1 constitutively expressing *popA* is not virulent on tobacco, tomato, eggplant and pepper (Kanda et al., 2003); however, since PopA1 is inactive in tomato, the reason for this mutant's reduced virulence is unclear. No biochemical data or sequence motif indicates that PopA is translocated into plant cells, and it may normally be targeted to the plant apoplast.

In contrast, PopB and PopC have amino acid sequence motifs that suggest they have evolved to function inside eukaryotic cells (Gueneron, Timmers, Boucher, & Arlat, 2000). PopB is a basic protein with a functional bipartite nuclear localization signal and PopC carries 22 tandem LRR repeats that match the predicted eukaryotic cytoplasmic LRR consensus present in some resistance gene products. These proteins are detected in culture supernatants only when Congo Red is added to the *hrp*-inducing growth medium, because this dye appears to stabilize some extracellular proteins. PopB or PopC are not required for pathogenicity or HR induction, and whether they are translocated into plant cells has not been reported.

8.2.3. Regulated Production of the T3SS

Expression of genes for the T3SS and most effectors is environmentally regulated. Like in all other phytopathogenic bacteria, transcription of these genes in *R. solanacearum* is coordinately increased *in planta* and in minimal media (MM), which mimics some conditions in the apoplast (Genin & Boucher, 2004; He et al., 2004; Schell, 2000). Addition of a complex nitrogen source (e.g., peptone) to MM strongly represses gene expression. In parallel with the two *hrp/hrc* evolutionary lineages in phytopathogenic bacteria, *R. solanacearum* and *X. campestris* pv. vesicatoria (hrp2 group) have a regulatory cascade completely different from *P. syringae* and *Erwinia* spp. (hrp1 group).

In GMI1000, *hrpB* and *hrpG* are essential for activating *hrp/hrc* operons in minimal medium (Schell, 2000), but how they sense nitrogen status is unknown. HrpB mostly controls genes with promoters containing a hrp_{II} box (TTCGn16TTCG) (Cunnac et al., 2004a). This motif is very similar to the putative PIP box motif identified by sequence analysis of *X. campestris* pv. vesicatoria promoters regulated by HrpX (the ortholog of HrpB) (Büttner & Bonas, 2002a). Deletion and mutation analyses of promoters showed that the direct repeat sequence, the distance (but not the sequence) between the repeats, and the distance of the hrp_{II} box from the transcriptional start site are all important for activation by HrpB (Cunnac et al., 2004a). However, HrpB/HrpX have not been shown to interact physically with the hrp_{II} box (e.g., by footprinting), and other aspects of HrpB/HrpX function are not understood. For example, in RS1000 a few HrpB-regulated *hpx (hrpB-dependent expression)* genes lack a hrp_{II} box, have an imperfect hrp_{II} box, or have a hrp_{II} box abnormally far upstream (Mukaihara et al., 2004). Similarly, in *X. campestris* pv. vesicatoria, HrpX activates some promoters that lack a PIP box, and also does not activate some promoters with a PIP box (Büttner & Bonas, 2002a). Furthermore, in *R. solanacearum* strain OE1-1, HrpB activation of *hrpY* occurs more quickly than activation of *popABC* after bacteria are infiltrated into tobacco leaves, which suggests that factors in addition to HrpB regulate the *popABC* operon (Kanda et al., 2003).

The HrpG transcriptional regulator is a member of the OmpR subclass of two-component response regulators (and is orthologous to HrpG in *X. campestris* pv. vesicatoria) (Brito, Marenda, Barberis, Boucher, & Genin, 1999). A cognate two-component sensor kinase has not been identified. Transcription of *hrpG* does not increase in MM, but is stimulated >15-fold by cocultivation with *Arabidopsis* or tomato cells. Nevertheless, HrpG is required for activating expression of *hrpB* and downstream *hrp/hrc* operons both in MM and in cocultivation with *Arabidopsis* and tomato cells, so it acts as a convergence point for separate metabolic/nutritional and plant signals that affect expression of genes in the T3SS. Contrary to the original report, *hrpG* mutants are nonpathogenic on tomato and HR-negative on tobacco. HrpG also appears to regulate additional genes important for pathogenesis, because a *hrpG* mutant colonizes tomato plants less aggressively than a *hrpB* mutant (Vasse et al., 2000).

Contact with plant cell walls is another environmental signal that induces expression of *hrp/hrc* and coregulated genes in GMI1000. Four additional genes (*prhA*, *prhR*, *prhI*, and *prhJ* (plant regulator of hrp genes)) in a linear pathway upstream of *hrpG* sense and transduce the plant signal (Fig. 9) (Aldon, Brito, Boucher, & Genin, 2000; Brito, Aldon, Barberis, Boucher, & Genin, 2002; Brito et al., 1999; Marenda et al., 1998), but they are not required for induction of *hrpB* in MM. PrhA is similar to TonB-dependent outer membrane siderophore receptor proteins, PrhR is similar to signal transducer proteins that span the inner membrane, and PrhI is similar to members of the ECF subset of σ^{70} factors that induce genes in response to extracytoplasmic stimuli. These three proteins may comprise a signal transduction module analogous to the FecA-FecR-FecI system that controls expression of genes for ferric citrate transport in *E. coli*, except that the *R. solanacearum* module does not respond to low iron concentrations. Thus, PrhA (a putative outer membrane protein) may respond to wall contact by stimulating PrhR, which spans the inner membrane. PrhR in turn stimulates the PrhI σ factor to activate expression of *prhJ*, which encodes a member of the LuxR/UhpA family of transcriptional activators that then activates expression of *hrpG*. Mutation of *prh* genes results in varying degrees of reduction in HR elicitation on tobacco and virulence on tomato or *Arabidopsis*, with mutations toward the beginning of the signal cascade having progressively less impact on T3SS function.

A third environmental signal controlling *hrp/hrc* gene expression is the concentration of 3-OH PAME autoinducer (see above). As culture density increases in a confined space, the PhcSR sensing system reacts to increasing autoinducer concentration by enhancing PhcA activity. Genin et al. (2005) found that inactivation of *phcA* in GMI1000 increased *hrpB* expression up to 60-fold in normally repressive rich culture medium and up to 6-fold in normally inductive MM. That a *phcA* mutant of AW overproduces Hrp pili in rich medium (Kang and Denny, unpublished results), and a GMI1000 strain constitutively expressing *phcA* elicits partial and delayed HR on tobacco both support the conclusion that active PhcA depresses functionality of the T3SS. Genetic evidence indicates that PhcA represses *hrpB* transcription by modifying the activity of the HrpG response regulator post-transcriptionally, possibly by affecting its phosphorylation state (Genin et al., 2005). Thus, in the wild type, HrpG integrates signals from two environmental pathways and uses quorum sensing to further modulate the transduced signal (Fig. 9). During pathogenesis, *R. solanacearum* is envisioned as maximally producing its T3SS and effectors when at low cell density in the apoplast but fully repressing them in highly colonized tissue of a wilted plant, presumably to conserve resources and to prepare for surviving after the host dies.

9. PLANT RESPONSES TO INFECTION

In comparison to the research on pathogenic mechanisms, there have been relatively few reports describing the genetics, physiology, or histology of host responses to infection by *R. solanacearum*. Most of the genetic research has emphasized aspects related to breeding for resistance (Hartman & Hayward, 1993; Prior, Allen, & Elphinstone, 1998; Allen, Prior, & Hayward, 2005), and will not be reviewed. I will instead focus on basic research using tomato and *Arabidopsis* as models for understanding the nature of resistance loci and subsequent plant responses. Additionally, only studies in which roots were inoculated (with or without wounding) will be considered, because responses to other inoculation methods, especially leaf infiltration, may be artifactual.

9.1. Genetics of Susceptibility

Susceptibility has been studied much less than resistance, but it is logical to consider this aspect first, because it is the most common outcome of *R. solanacearum* infection of host plants. Feng et al. (2004) isolated a single fast-neutron-induced *nws1* (no wilt symptoms) mutant of *Arabidopsis* that, unlike the wild type glabrous Col-0 parent, does not develop wilt symptoms after inoculation with GMI1000 or other strains. Pathogen multiplication in *nws1* plants is reduced, similar to that in a well-characterized resistant ecotype. Reduced susceptibility is only expressed under high light intensity, which also makes the plants smaller and have shorter roots than normal. Unlike many resistant mutants, *nws1* plants do not spontaneously develop leaf lesions (due to abnormal constitutive defense responses), abnormally express pathogenesis-related (PR) genes associated with defense responses (Van Loon & Van Strien, 1999) prior to or after inoculation, or become more resistant to other pathogens. Since *nws1* is recessive, this suggests that *NWS1* encodes a 'susceptibility factor' required for normal disease development. *Arabidopsis* genes also have been found that are required for susceptibility to powdery mildew (Vogel, Raab, Somerville, & Somerville, 2004) and downy mildew (Van Damme et al., 2005). Therefore, BW disease does not occur just due to an absence of resistance, but requires active plant metabolism before or after inoculation. *NWS1* has not been characterized and its function is unknown, but theoretical functions include production of a cell wall binding site necessary for stimulating expression of *hrpB* or a metabolite that stimulates pathogen genes required for multiplication or virulence (see IVET research below).

The ethylene signal transduction pathway may also contribute to symptom development in *Arabidopsis*, because *ein2-1* mutant plants (ethylene insensitivity;

EIN2 encodes a signal transducer downstream of the ethylene receptors (Chang & Shockey, 1999)) wilt more slowly than susceptible Col-0 plants inoculated with GMI1000 (Hirsch, Deslandes, Feng, Balague, & Marco, 2002). Bacterial colonization is less than normal in the *ein2-1* plants, but greater than in a resistant ecotype. The ethylene responsive genes *PR-3* and *PR-4* that are strongly induced by infection of Col-0 are not expressed in the *ein2-1* mutant. In contrast, wilt symptoms develop normally in Col-0 plants lacking either of two ethylene receptors or a transcriptional factor downstream of *EIN2*. Therefore, it is not clear if ethylene itself is required for expression of wilt symptoms. It is worth noting that Col-0 plants expressing *NahG* (which do not accumulate salicylic acid required for resistance), and Col-0 mutants constitutively expressing PR genes or insensitive to jasmonic acid are normally susceptible to *R. solanacearum* (Hirsch et al., 2002).

9.2. Genetics of Resistance in Tomato

Despite decades of work, researchers in public and private institutions have had only limited success in developing a tomato cultivar with large fruit and high level resistance to BW in diverse geographical locations. Useful levels of resistance exist in *L. esculentum* var. cerasiforme (e.g. L285) and in *L. pimpinellifolium* (especially PI 127805 A), and a number of breeding lines and a few named cultivars have been created by introgressing one or more loci into *L. esculentum*. When 35 tomato lines were assessed for BW resistance at 11 locations world wide, L285 exhibited intermediate resistance while Hawaii 7997 (and related lines with PI 127805 A as a parent) performed well at all sites and had the highest average survival rate (97%) (Wang et al., 1998). Crosses between L285 and H7997 (or three additional resistant lines) did not produce any F1 progeny that were significantly more resistant to BW than the parents (Hanson, Licardo, Hanudin, Wang, & Chen, 1998). Unfortunately, H7997 is susceptible to some *R. solanacearum* strains (e.g. two isolated in Taiwan (Jaunet & Wang, 1999)), so even its resistance is not 'universal' .

Modern marker-assisted genetics methods revealed that BW resistance in both L285 and H7997 is polygenic and carried on two or more chromosomes. Danesh et al. (1994) crossed L285 and susceptible line C286 and developed an RFLP linkage map with 79 markers that cover about 75% of the genome. Root inoculation of F_2 progeny with strain UW364 (biovar 4 from China) revealed a quantitative trait locus (QTL) on chromosome 6 centered on markers CT184 and TG240 that accounts for 77% of the variation in disease response. This locus may be strain specific, because it is not associated with resistance to strain Pss4 (biovar 3 from Taiwan) (see Wang et al. (2000)). A less effective locus on chromosome 10 between RFLP markers CT225b and TG230 accounts for 24% of the variation. Both QTLs are also found when F_3 plants are inoculated with UW364, and they are partially dominant. The presence of a major QTL on chromosome 6 is interesting, because some genes for resistance to other plant pests are also on this chromosome.

A series of studies examining the progeny of a cross between the highly resistant line H7997 and the very susceptible *L. pimpinellifolium* line WVa700 identified multiple QTLs using a linkage map covering about 75% of the genome. Growth chamber tests of plants inoculated with strain GMI8217 (biovar 1 from Guadeloupe) revealed a major locus on chromosome 6 spanning a large region 10-20 cM distal to the QTL observed in L285 (Thoquet et al., 1996a). This QTL and weaker ones on chromosomes 10 and 11 account for up to 56% of the variation in resistance. Two QTLs originally identified on chromosome 4 were later determined to be statistical artifacts (Wang et al., 2000). Field tests of the F_2 clones and additional F_3 families inoculated with GMI8217 in Guadeloupe identified four QTLs (Thoquet et al., 1996b). Besides the previously identified loci on chromosomes 6 and 11, two new QTLs were found on chromosomes 3 and 8. Temporal analysis of BW development in a field test of an F_3 population inoculated with GMI8217 revealed that the large QTL on chromosome 6 contains two linked loci about 30 cM apart on the upper arm of chromosome 6 (Mangin, Thoquet, Olivier, & Grimsley, 1999). One locus is near the end of the chromosome close to the *Cf-2* gene for *Cladosporium fulvum* resistance and the *Mi* gene for nematode resistance. The close proximity of the BW QTL and *Mi* makes recombination between these loci very difficult and explains why introgression of the *Mi* gene from *L. peruvianum* into *L. esculentum* results in progeny that are more susceptible to BW (Deberdt, Olivier, Thoquet, Queneherve, & Prior, 1999). The second locus on chromosome 6 is in the same region as the QTL in L285 (near marker TG240), but likely is a separate locus or different allele. In contrast, a greenhouse test of the F_3 population root-inoculated with strain Pss4 (biovar 3) identified a new major QTL on chromosome 12 that appears to be active specifically against this strain (Wang et al., 2000). In this case, the QTL on chromosome 6 near marker TG240 contributes a little to resistance to Pss4 and the QTL near *Mi* is inactive. Wang et al. (2000) speculated that BW QTLs may act like single resistance genes that determine race-cultivar specificity, but that they provide partial resistance only to selected strains. Tomato lines like H7997 may have a collection of such loci on chromosome 6 that make it resistant to almost all *R. solanacearum* strains (Jaunet & Wang, 1999; Wang et al., 1998).

9.3. Genetics of Resistance in Arabidopsis thaliana

Arabidopsis has multiple genes for BW resistance, and the facile genetics of this model organism has resulted in some of them being well characterized. A single dominant locus for resistance is present in ecotype S96, which develops a HR-like response when leaves are infiltrated with *R. solanacearum* strain Ps95, but the gene's identity has not been reported (Ho & Yang, 1999). Similarly, resistance is dominant in F_1 progeny from a cross between resistant accession Col-0 and highly susceptible L*er* when inoculated with *R. solanacearum* strain 14.25 (Godiard et al., 2003). Analysis of 100 F_9 recombinant inbred lines revealed that resistance involves the major QTL loci *QRS1* (quantitative resistance to *R. solanacearum*) and *QRS2* on chromosome 2 and the weaker *QRS3* on chromosome 5, which together explain about 90% of the variation in resistance. Unexpectedly, analysis of both an L*ER* line and two L*er* lines transformed with the wild-type *ERECTA* gene, which is located near to QRS1 and normally affects development of aerial organs, showed that *ERECTA* also contributes quantitatively to BW resistance.

In contrast, BW resistance to GMI1000 in accession Nd-1 is inherited as a single recessive gene in crosses with the susceptible Col-5 accession (Deslandes et al., 1998). The DNA sequences of the dominant *RRS1-S* and recessive *RRS1-R* alleles are 98% identical up to where a stop codon in the former results in a protein lacking the C-terminal 90 amino acids present in the latter (Deslandes et al., 2002). Normally susceptible Col-5 plants carrying the cloned *RRS1-R* allele are resistant and prevent normal pathogen multiplication *in planta*, so this allele encodes a resistance *R* gene. The biochemically nonfunctional but genetically dominant *RRS1-S* is proposed to encode a protein that interferes with the function of RRS1-R (a dominant negative effect). Analysis of the putative *RRS1* alleles show that they are members of the large TIR-NBS-LRR subclass of *R* genes (Lahaye, 2002), but are unique in having a group III conserved C-terminal WRKY transcriptional activation domain. WRKY transcriptional regulators are known to interact with the W box in many pathogen-responsive plant promoters (Genin & Boucher, 2004). Another unusual feature of RRS1-R is that it physically interacts with the PopP2 avirulence effector secreted by GMI1000, which is responsible for triggering resistance in accession Nd-1 (Deslandes et al., 2003). PopP2 has a functional nuclear localization signal and co-expression of fluorescent PopP2 and RRS1-R fusion proteins in *Arabidopsis* cells results in their colocalizing in the nucleus. It therefore seems likely that BW resistance in Nd-1 plants is manifested when PopP2 is injected into the plant cytoplasm by the T3SS and carries RRS1-R 'piggyback' into the nucleus where it activates defense-response genes. A separate study found that a homozygous *ein2-1* mutant of Nd-1 is normally resistant, so ethylene sensing is not required for resistance (Hirsch et al., 2002).

9.4. Manifestation of Resistance

The mechanisms of BW resistance are even less well understood than its genetics. Most of our knowledge has come from spatial-temporal analyses of pathogen colonization and microscopic comparisons of susceptible and resistant tomato lines (especially Hawaii 7996, 7997 or 7998) infected with *R. solanacearum*. The pathogen multiplies equally well in the xylem sap collected from H7996 and a susceptible cultivar, so there appear to be no pre-existing inhibitory compounds in the xylem (McGarvey et al., 1999). Instead, resistance is probably due to rapid development of physical or chemical 'barriers' in response to pathogen infection.

Several papers by Grimault and associates showed that, after wounded roots are inoculated, resistance in H7996 is most clearly manifested by a low percentage of plants being colonized at the lower and mid-stem levels (Grimault, Anais, & Prior, 1994; Grimault & Prior, 1993; Grimault & Prior, 1994b). Despite most H7996 plants having 10^5 to 10^6 cfu/gram fresh tissue in their taproot and collar tissue, all the plants remained healthy looking. Grafting experiments showed that resistance is associated with limited *R. solanacearum* invasion of the lower stem, because resistant scions (CRA66 or Caraïbo) grafted 10 cm above the collar of a susceptible root stock (Floradel) wilted, but the susceptible scion on a resistant root stock did not (Grimault & Prior, 1994a). However, in plants inoculated by soil drench without intentionally wounding the roots, the pathogen also multiplies much less well in the taproot of H7996 than in a susceptible cultivar (McGarvey et al., 1999). *R. solanacearum* also inefficiently invades the unwounded roots of H7996 within 5 hours after inoculation. Microscopic examination of H7997 growing in hydroponic culture similarly found that invasion of unwounded roots of this resistant line is less efficient than in a susceptible cultivar, and that tissue browning near infection sites and deposition of polyphenolic-like material on vessel walls and in the lumen of some vessels was only observed in H7997 (Vasse et al., 2005).

The resistance response was examined using light and electron microscopy of vascular tissue from the lower stem of two cultivars. Unlike in the susceptible cultivar Floradel, many tyloses occluded colonized xylem vessels and adjacent vessels in resistant Caraïbo plants 15 days after inoculation (Grimault et al., 1994). Deposits of electron-dense material (gums) were also common only in Caraïbo. In contrast, no tyloses were seen in vessels from the lower stem of either susceptible cultivar Ponderosa or resistant line LS-89 (a selection from H7998) 14 days after inoculation (Nakaho et al., 2000). Nevertheless, pathogen movement between vessels is restricted in LS-89, which correlates with its pit membranes being thicker and more electron dense than in Ponderosa and the development of a more conspicuous electron-dense layer near the pits and along vessel walls. In both experiments, however, plants were examined only once late in pathogenesis, so it is unknown whether the responses observed occurred before or after resistance was manifested. The timing of defense responses is crucial, because it is often the rapidity and not that nature of the response that differentiates resistant and susceptible cultivars. Because the impaired colonization of resistant tomato lines by wild-type *R. solanacearum* is very similar to that of susceptible cultivars by *hrp* mutants (Etchebar et al., 1998; Frey et al., 1994; Vasse et al., 2000), it may be that effector proteins secreted by the T3SS delay induction of defense responses in susceptible cultivars but not in resistant plants (Alfano & Collmer, 2004).

10. GENOME ANALYSIS

GMI1000 was one of the first phytopathogenic bacteria to have its complete genome sequenced (Salanoubat et al., 2002), and this accomplishment ushered in a new phase of research on *R. solanacearum* (Denny, 2005). In addition, by the time this chapter is published, draft sequences of three additional strains should also be available (Gabriel et al., in press) (S. Genin and C. Boucher, personal communication). With these resources, anyone with an Internet connection can access in minutes knowledge about a specific locus that previously would have required months of laboratory work. It then is easy to create mutants necessary to test the gene's function experimentally. However, rather than just providing shortcuts in standard experimental strategies, the genomic sequence also makes possible novel approaches to study gene content, function, and expression on a large scale.

10.1. Genome Structure and Predicted Gene Functions

At 5.81 megabases (Mb), GMI1000 has a genome about the same size as other phytopathogenic proteobacteria. There are two circular replicons of 3.716 and 2.094 Mb that have almost the same G + C content. The larger replicon, which has an origin of replication typical of bacterial chromosomes and carries genes for all essential life functions, is referred to as the chromosome. The smaller replicon has an origin of replication characteristic of plasmids and it is well known in the literature as a megaplasmid. However, three observations indicate that the megaplasmid is an essential part of the genome. First, no derivative of GMI1000 has been found that completely lacks the megaplasmid, and most other strains have a megaplasmid (Genin & Boucher, 2004). Second, it carries the *hrp* gene cluster, many virulence-associated genes (e.g., CWDEs, flagellar motility, EPS1 biosynthesis), and genes that allow the pathogen to exploit diverse environments. Third, the megaplasmid probably co-evolved with the chromosome rather than being acquired recently by horizontal gene transfer (Coenye & Vandamme, 2003). *R. solanacearum* is, therefore, considered to have a bipartite genome. Unlike many phytopathogenic bacteria, only a few *R. solanacearum* strains are known to carry small plasmids (<100 kb).

The most significant information on GMI1000 gene content and genome structure that can be gleaned from DNA sequence analysis was recently reviewed (Genin & Boucher, 2004). The chromosome and megaplasmid together are estimated to have approximately 5,120 protein-coding open reading frames (ORFs), about half of which have a functional assignment based on homology with genes in other organisms. Among these are multiple genes and protein secretion systems that may contribute to virulence, but other than those already described above, they have not been studied experimentally. Some of the conserved genes explain the metabolic versatility of GMI1000 (typical of biovar 3), because they should encode for utilization of a wide range of amino acids, carbohydrates, and fatty acids and some phenolic compounds. Less versatile biovar 1 and 2 strains presumably lack these genes. Unexpectedly, GMI1000 also has multiple genes possibly involved with detoxification of noxious compounds. For example, eight gene clusters are related to those for heavy metal tolerance exhibited by *Cupriavidus metallidurans* strain CH34, which was isolated from a waste tank at a zinc factory (Mergeay et al., 2003). If functional, such detoxification loci may improve the fitness of GMI1000 in soil. These and many other genes probably were acquired by horizontal gene transfer, which may be facilitated by the natural ability of *R. solanacearum* to take up DNA from its environment (Bertolla, van Gijsegem, Nesme, & Simonet, 1997). Many of these genes are present in 93 ACURs (alternative codon usage regions) that also have significantly different G + C content from the rest of the genome. In addition, there are an unusually large number of mobile genetic elements (e.g., insertion sequences, defective prophage, and a potential conjugative transposon).

I want to emphasize that only a predicted function is available for the vast majority of genes in GMI1000 and other sequenced genomes. Prediction of a new gene's function is most reliable when deduced from homology with a gene whose function has been demonstrated experimentally. Unfortunately, non-curated databases (e.g., GenBank) contain many genes whose functions are, for one or more reasons, incorrectly predicted (Valencia, 2005). Furthermore, even when orthologous proteins in two organisms appear to have a similar biochemical function (e.g., enzyme activity), their biological function may be different. Therefore, it is important to know when an inference is based only on sequence analysis so as to avoid over-interpreting or misinterpreting the data.

10.2. Utilization of Genomic Sequence Data

Promoters of genes encoding proteins in the T3SS and many of its co-regulated secreted effectors in phytopathogenic bacteria have one of several conserved motifs (Preston et al., 2005). After analysis of the *hrpY* promoter in GMI1000 demonstrated that the hrp_{II} box is essential for induction by the HrpB transcriptional regulator (see above), computer-assisted analysis of the GMI1000 genomic sequence identified 95 transcriptional units (estimated to comprise 110 genes and 4 pseudogenes) with a properly positioned hrp_{II} box (Cunnac et al., 2004a). Nineteen of these genes encode proteins that may transit the T3SS; among these are the putative translocon proteins PopF1 and PopF2, and 14 proteins previously identified as having domains suggestive of their functioning within eukaryotic cells. Cunnac et al. (2004b) subsequently created transcriptional *lacZ*-reporter fusions to 71 of these candidate promoters and found 48 genes that are positively regulated by HrpB. Among these *brg* (HrpB-regulated) genes are those encoding the avirulence proteins AvrA and PopP2. Six *brg* genes appear to encode accessory components of the T3SS, eight genes encode proteins homologous to known harpins or effectors, and the remaining genes encode hypothetical proteins. A complementary study by Mukaihara et al. (2004) examined random transposon-induced *lacZ*-reporter fusions in strain RS1000 and found 25 of the 48 *brg* genes (which they called *hpx* genes) and 3 genes predicted to be HrpB-regulated (Cunnac et al., 2004a). Four other *hpx* genes are unusual, because two don't have a hrp_{II} box, one has an imperfect hrp_{II} box, and one has a hrp_{II} box far upstream from its normal position. Therefore, strategies in addition to sequence analysis are necessary to identify all of the genes regulated by HrpB.

Fortunately, the genomic sequence also facilitates production of microarrays and other methods that make it possible to evaluate transcription of all genes simultaneously (the transcriptome) in different strains or in one strain subjected to different conditions (Rhodius, Van Dyk, Gross, & LaRossa, 2002). A microarray with one oligonucleotide for each of 5047 predicted ORFs in GMI1000 was generated and used to determine the HrpB transcriptome (Occhialini, Cunnac, Reymond, Genin, & Boucher, 2005). Microarray slides were hybridized to cDNAs prepared from pairs of strains cultured under *hrp*-inducing conditions: GMI1000 and the *hrpB* mutant GMI1525, or GMI1000 overexpressing a plasmid-borne *hrpB* and GMI1525 with an empty vector. A stringent threshold of differential gene expression detected 143 *brg* genes positively controlled by HrpB and 50 genes that are negatively controlled by HrpB; regulated expression was confirmed for 91.4% of the representative genes tested. However, 81 genes may be within operons, so only 112 HrpB-regulated promoters were identified. Among the *brg* genes, 64 (or 92%) had been previously identified as being HrpB-regulated (i.e., known *hrp*, *brg*, and *pop* genes), and a disproportionate number are on the megaplasmid. Four known and 36 new *brg* genes do not have a hrp_{II} box within 500 bases of the start codon of the cognate gene or the first gene of the operon and none of the negatively regulated promoters have a hrp_{II} box. Promoters lacking a hrp_{II} box may be controlled indirectly by other transcriptional regulators, some of which may be directly controlled by HrpB. Many of the new HrpB-regulated genes with annotated function appear to be involved in chemotaxy or intermediate metabolism. Thus, HrpB appears to be the master regulator in a developmental program to up and down regulate many functions associated with the shift from saprophytic to parasitic life.

Analysis of the 50 N-terminal amino acids in five secreted Pop proteins, four Rip proteins, and many Brg proteins revealed that they are rich in serine and proline, low in leucine, and most lack acidic residues in the first 12 positions. Similar sequence features have been identified previously in *P. syringae* T3SS effectors (Preston et al., 2005), although what role they play during protein secretion is not known. The microarray results identified 26 new *brg* genes that also have these characteristics, so GMI1000 may have a total of 70-80 effectors or T3SS-accessory proteins (Occhialini et al., 2005). It is unlikely that all of these *R. solanacearum* proteins will be true effectors, because recent results showed that a minority of the *P. syringae* proteins predicted to be T3SS substrates are translocated into plant cells (Chang et al., 2005).

In vivo expression technology (IVET) provides an experimental approach to identify promoters that are induced by the environment within an animal or plant host, which acts as a selective 'medium'. Brown and Allen (2004) used IVET to screen a library of potential promoter-containing DNA fragments from *R. solanacearum* K60 for induction in tomato plants. They cloned the DNA fragments into a custom-made vector upstream of promoterless, tandem copies of the K60 *trpEG* locus and a β-glucuronidase gene, introduced the library into the genome of strain K909 (Δ*trpEG*), and inoculated pools of exconjugants into tomato plants. Cells with promoters active *in planta* produced tryptophan, multiplied, and were recovered from wilting plants three days later. Assaying for β-glucuronidase activity *in vitro* during growth on a rich medium differentiated uninteresting strains containing constitutively expressed promoters from strains containing inducible promoters, and the cloned DNA in 307 of the latter was sequenced. Homology searches in the GMI1000 genomic DNA database revealed 153 unique *in planta*-expressed (*ipx*) genes. Based on the GMI1000 genome annotation, the *ipx* genes can be organized in eight categories: phage and transposases (5 genes), transport/ secretion systems (8), signal peptide motif (12), stress response (15), regulators (19), novel with no motif (21), transmembrane motif (26), and metabolism (31). Fifteen have the identifiable promoter aligned backwards with respect to the *trpEG* locus. *In planta* expression was confirmed for 44 *ipx* promoters and ranged from 2-fold to over 35-fold. Nineteen of 32 fusions are not expressed in minimal medium, suggesting that their promoters are specifically triggered by *in planta* conditions or signals. Seven known virulence-associated genes are among the *ipx* genes (*pehR*, *vsrB*, *vsrD*, *rpoS*, *hrcC*, *pme*, and *gspK*). The diverse functions of *ipx* genes indicate that *R. solanacearum* significantly modifies its physiology to adapt to a stressful, low-nutrient environment *in planta*. Surprisingly, only three *ipx* genes are also positively regulated by HrpB (Occhialini et al., 2005), suggesting that, because bacteria were recovered from plants after extensive pathogen colonization, this IVET screen may have identified mostly genes expressed late in pathogenesis.

11. REFERENCES

Akiew, E. & Trevorrow, P. R. (1994). Management of bacterial wilt of tobacco. In A. C. Hayward & G. L. Hartman (Eds.), *Bacterial Wilt: The Disease and Its Causative Agent, Pseudomonas solanacearum* (pp. 179-198). Wallingford, UK: CAB International.

Akiyoshi, D. E., Regier, D., & Gordon, M. P. (1989). Nucleotide sequence of the *tzs* gene from *Pseudomonas solanacearum* strain K60. *Nucleic Acids Research, 17,* 8886.

Alan, A. R. & Earle, E. D. (2002). Sensitivity of bacterial and fungal plant pathogens to the lytic peptides, MSI-99, magainin II, and cecropin B. *Molecular Plant-Microbe Interactions, 15,* 701-708.

Aldon, D., Brito, B., Boucher, C., & Genin, S. (2000). A bacterial sensor of plant cell contact controls the transcriptional induction of *Ralstonia solanacearum* pathogenicity genes. *EMBO Journal, 19,* 2304-2314.

Alfano, J. R. & Collmer, A. (2004). Type III secretion system effector proteins: double agents in bacterial disease and plant defense. *Annual Review of Phytopathology, 42,* 385-414.

Allen, C., Kelman, A., & French, E. R. (2001). Brown rot. In W. R. Stevenson, R. Loria, G. D. Franc, & D. P. Weingartner (Eds.), *Compendium of Potato Diseases* (2 ed., pp. 11-13). St. Paul: APS Press.

Allen, C., Prior, P., and Hayward, A. C. (2005). *Bacterial Wilt Disease and the Ralstonia solanacearum Species Complex*. St. Paul, MN: APS Press.

Alvarez, A. (2005). Diversity and diagnosis of *Ralstonia solanacearum*. In C. Allen, P. Prior, & A. C. Hayward (Eds.), *Bacterial Wilt Disease and the Ralstonia solanacearum Species Complex* (pp. 437-447). St. Paul, MN: APS Press.

Alvarez, A. M., Trotter, K. J., Swafford, M. B., Berestecky, J. M., Yu, Q., Ming, R. et al. (2005). Characterization and detection of *Ralstonia solanacearum* strains causing bacterial wilt of ginger in Hawaii. In C. Allen, P. Prior, & A. C. Hayward (Eds.), *Bacterial Wilt Disease and the Ralstonia solanacearum Species Complex* (pp. 471-477). St. Paul, MN: APS Press.

Andrews, S. C., Robinson, A. K., & Rodriguez-Quinones, F. (2003). Bacterial iron homeostasis. *FEMS Microbiology Reviews, 27,* 215-237.

Anon. (2004). *Ralstonia solanacearum. EPPO Bulletin, 34,* 173-178.

Anzai, Y., Kim, H., Park, J. Y., Wakabayashi, H., & Oyaizu, H. (2000). Phylogenetic affiliation of the pseudomonads based on 16S rRNA sequence. *International Journal of Systematic and Evolutionary Microbiology, 50,* 1563-1589.

Araud-Razou, I., Vasse, J., Montrozier, H., Etchebar, C., & Trigalet, A. (1998). Detection and visualization of the major acidic exopolysaccharide of *Ralstonia solanacearum* and its role in tomato root infection and vascular colonization. *European Journal of Plant Pathology, 104,* 795-809.

Arlat, M., van Gijsegem, F., Huet, J. C., Pernollet, J. C., & Boucher, C. A. (1994). PopA1, a protein which induces a hypersensitivity-like response on specific *Petunia* genotypes, is secreted via the Hrp pathway of *Pseudomonas solanacearum. EMBO Journal, 13,* 543-553.

Asai, T., Tena, G., Plotnikova, J., Willmann, M. R., Chiu, W. L., Gómez-Gómez, L. et al. (2002). MAP kinase signalling cascade in *Arabidopsis* innate immunity. *Nature, 415,* 977-983.

Baharuddin, B., Rudolph, K., & Niepold, F. (1994). Production of monospecific antiserum against the blood disease bacterium affecting banana and plantain. *Phytopathology, 84,* 570-575.

Bennet, C. P. A., Hunt, P., & Asman, A. (1985). Association of a xylem-limited bacterium with Sumatra disease of cloves in Indonesia. *Plant Pathology, 34,* 487-494.

Bentsink, L., Leone, G. O. M., van Beckhoven, J. R. C. M., van Schijndel, H. B., van Gemen, B., & van der Wolf, J. M. (2002). Amplification of RNA by NASBA allows direct detection of viable cells of *Ralstonia solanacearum* in potato. *Journal of Applied Microbiology, 93,* 647-655.

Bertolla, F., van Gijsegem, F., Nesme, X., & Simonet, P. (1997). Conditions for natural transformation of *Ralstonia solanacearum. Applied and Environmental Microbiology, 63,* 4965-4968.

Bhatt, G. & Denny, T. P. (2004). *Ralstonia solanacearum* iron scavenging by the siderophore staphyloferrin B is controlled by PhcA, the global virulence regulator. *Journal of Bacteriology, 186,* 7896-7904.

Black, R. & Elphinstone, J. (1998). Developing appropriate detection methods for developing countries. In Ph. Prior, C. Allen, & J. Elphinstone (Eds.), *Bacterial Wilt Disease: Molecular and Ecological Aspects* (pp. 123-127). Berlin: Springer-Verlag.

Black, R. & Sweetmore, A. (1993). Identification and characterization of *Pseudomonas solanacearum* using metabolic profiles. In G. L. Hartman & A. C. Hayward (Eds.), *Bacterial Wilt. ACIAR Proceedings No. 45* (pp. 32-44). Canberra, Australia: Australian Centre for International Agricultural Research.

Boshou, L. (2005). A broad review and perspective on breeding for resistance to bacterial wilt. In C. Allen, P. Prior, & A. C. Hayward (Eds.), *Bacterial Wilt Disease and the Ralstonia solanacearum Species Complex* (pp. 225-238). St. Paul, MN: APS Press.

Boucher, C. A., Gough, C. L., & Arlat, M. (1992). Molecular genetics of pathogenicity determinants of *Pseudomonas solanacearum* with special emphasis on *hrp* genes. *Annual Review of Phytopathology, 30,* 443-461.

Brito, B., Aldon, D., Barberis, P., Boucher, C., & Genin, S. (2002). A signal transfer system through three compartments transduces the plant cell contact-dependent signal controlling *Ralstonia solanacearum hrp* genes. *Molecular Plant-Microbe Interactions, 15,* 109-119.

Brito, B., Marenda, M., Barberis, P., Boucher, C., & Genin, S. (1999). *prhJ* and *hrpG*, two new components of the plant signal-dependent regulatory cascade controlled by PrhA in *Ralstonia solanacearum. Molecular Microbiology, 31,* 237-251.

Brown, D. G. & Allen, C. (2004). *Ralstonia solanacearum* genes induced during growth in tomato: an inside view of bacterial wilt. *Molecular Microbiology, 53,* 1641-1660.

Brumbley, S. M., Carney, B. F., & Denny, T. P. (1993). Phenotype conversion in *Pseudomonas solanacearum* due to spontaneous inactivation of PhcA, a putative LysR transcriptional activator. *Journal of Bacteriology, 175,* 5477-5487.

Buddenhagen, I. & Kelman, A. (1964). Biological and physiological aspects of bacterial wilt caused by *Pseudomonas solanacearum. Annual Review of Phytopathology, 2,* 203-230.

Buddenhagen, I., Sequeira, L., & Kelman, A. (1962). Designation of races in *Pseudomonas solanacearum. Phytopathology, 52,* 726.

Buddenhagen, I. W. (1986). Bacterial wilt revisited. In G. J. Persley (Ed.), *Bacterial Wilt Disease in Asia and the South Pacific. ACIAR Proceedings No. 13* (pp. 126-143). Canberra: Australian Centre for International Agricultural Research.

Büttner, D. & Bonas, U. (2002a). Getting across - bacterial type III effector proteins on their way to the plant cell. *EMBO Journal, 21,* 5313-5322.

Büttner, D. & Bonas, U. (2002b). Port of entry - the type III secretion translocon. *Trends in Microbiology, 10,* 186-192.

Büttner, D., Gurlebeck, D., Noel, L. D., & Bonas, U. (2004). HpaB from *Xanthomonas campestris* pv. *vesicatoria* acts as an exit control protein in type III-dependent protein secretion. *Molecular Microbiology, 54,* 755-768.

Büttner, D., Nennstiel, D., Klusener, B., & Bonas, U. (2002). Functional analysis of HrpF, a putative type III translocon protein from *Xanthomonas campestris* pv. vesicatoria. *Journal of Bacteriology, 184,* 2389-2398.

Carney, B. F. & Denny, T. P. (1990). A cloned avirulence gene from *Pseudomonas solanacearum* determines incompatibility on *Nicotiana tabacum* at the host species level. *Journal of Bacteriology, 172,* 4836-4843.

Caruso, P., Bertolini, E., Cambra, M., & Lopez, M. M. (2003). A new and sensitive co-operational polymerase chain reaction for rapid detection of *Ralstonia solanacearum* in water. *Journal of Microbiological Methods, 55,* 257-272.

Caruso, P., Gorris, M. T., Cambra, M., Palomo, J. L., Collar, J., & Lopez, M. M. (2002). Enrichment double-antibody sandwich indirect enzyme-linked immunosorbent assay that uses a specific monoclonal antibody for sensitive detection of *Ralstonia solanacearum* in asymptomatic potato tubers. *Applied and Environmental Microbiology, 68,* 3634-3638.

Caruso, P., Palomo, J. L., Bertolini, E., Alvarez, B., Lopez, M. A., & Biosca, E. G. (2005). Seasonal variation of *Ralstonia solanacearum* biovar 2 populations in a Spanish river: Recovery of stressed cells at low temperatures. *Applied and Environmental Microbiology, 71,* 140-148.

Chang, C. & Shockey, J. A. (1999). The ethylene-response pathways: signal perception to gene regulation. *Current Opinion in Plant Biology, 2,* 352-358.

Chang, J. H., Urbach, J. M., Law, T. F., Arnold, L. W., Hu, A., Gombar, S. et al. (2005). A high-throughput, near-saturating screen for type III effector genes from *Pseudomonas syringae. Proceedings of the National Academy of Sciences, USA, 102,* 2549-2554.

Chapman, M. R. & Kao, C. C. (1998). EpsR modulates production of extracellular polysaccharides in the bacterial wilt pathogen *Ralstonia* (*Pseudomonas*) *solanacearum. Journal of Bacteriology, 180,* 27-34.

Ciampi, L. & Sequeira, L. (1980). Influence of temperature on virulence of race 3 strains of *Pseudomonas solanacearum* potatoes. *American Potato Journal, 57,* 307-317.

Clough, S. J., Flavier, A. B., Schell, M. A., & Denny, T. P. (1997). Differential expression of virulence genes and motility in *Ralstonia* (*Pseudomonas*) *solanacearum* during exponential growth. *Applied and Environmental Microbiology, 63,* 844-850.

Coenye, T., Goris, J., De Vos, P., Vandamme, P., & LiPuma, J. J. (2003). Classification of *Ralstonia pickettii*-like isolates from the environment and clinical samples as *Ralstonia insidiosa* sp. nov. *International Journal of Systematic and Evolutionary Microbiology, 53,* 1075-1080.

Coenye, T. & Vandamme, P. (2003). Simple sequence repeats and compositional bias in the bipartite *Ralstonia solanacearum* GM11000 genome. *Bmc Genomics, 4,* 10.

Cook, D., Barlow, E., & Sequeira, L. (1989). Genetic diversity of *Pseudomonas solanacearum*: Detection of restriction fragment length polymorphisms with DNA probes that specify virulence and the hypersensitive response. *Molecular Plant-Microbe Interactions, 2,* 113-121.

Cook, D. & Sequeira, L. (1994). Strain differentiation of *Pseudomonas solanacearum* by molecular genetic methods. In A. C. Hayward & G. L. Hartman (Eds.), *Bacterial Wilt: The Disease and its Causative Agent, Pseudomonas solanacearum* (pp. 77-93). Wallingford, UK: CAB International.

Coutinho, T. A. (2005). Introduction and prospectus of the survival of *R. solanacearum*. In C. Allen, P. Prior, & A. C. Hayward (Eds.), *Bacterial Wilt Disease and the Ralstonia solanacearum Species Complex* (pp. 29-38). St. Paul, MN: APS Press.

Cunnac, S., Boucher, C., & Genin, S. (2004a). Characterization of the *cis*-acting regulatory element controlling HrpB-mediated activation of the type III secretion system and effector genes in *Ralstonia solanacearum*. *Journal of Bacteriology, 186,* 2309-2318.

Cunnac, S., Occhialini, A., Barberis, P., Boucher, C., & Genin, S. (2004b). Inventory and functional analysis of the large Hrp regulon in *Ralstonia solanacearum*: identification of novel effector proteins trans-located to plant host cells through the type III secretion system. *Molecular Microbiology, 53,* 115-128.

Danesh, D., Aarons, S., McGill, G. E., & Young, N. D. (1994). Genetic dissection of oligogenic resistance to bacterial wilt in tomato. *Molecular Plant-Microbe Interactions, 7,* 464-471.

De Baere, T., Steyaert, S., Wauters, G., De Vos, P., Goris, J., Coenye, T. et al. (2001). Classification of *Ralstonia pickettii* biovar 3/'thomasii' strains (Pickett 1994) and of new isolates related to nosocomial recurrent meningitis as *Ralstonia mannitolytica* sp. nov. *International Journal of Systematic and Evolutionary Microbiology, 51,* 547-558.

Deberdt, P., Olivier, J., Thoquet, P., Queneherve, P., & Prior, P. (1999). Evaluation of bacterial wilt resistance in tomato lines nearly isogenic for the *Mi* gene for resistance to root-knot. *Plant Pathology, 48,* 415-424.

Denny, T. P. (2005). A short history of the biochemical and genetic research on *Ralstonia solanacearum* pathogenesis. In C. Allen, P. Prior, & A. C. Hayward (Eds.), *Bacterial Wilt Disease and the Ralstonia solanacearum Species Complex* (pp. 323-334). St. Paul, MN: APS Press.

Denny, T. P. & Baek, S. R. (1991). Genetic evidence that extracellular polysaccharide is a virulence factor of *Pseudomonas solanacearum*. *Molecular Plant-Microbe Interactions, 4,* 198-206.

Denny, T. P., Brumbley, S. M., Carney, B. F., Clough, S. J., & Schell, M. A. (1994). Phenotype conversion of *Pseudomonas solanacearum*: its molecular basis and potential function. In A. C. Hayward & G. L. Hartman (Eds.), *Bacterial Wilt: The Disease and its Causative Agent, Pseudomonas solanacearum* (pp. 137-143). Wallingford, UK: CAB International.

Denny, T. P., Carney, B. F., & Schell, M. A. (1990). Inactivation of multiple virulence genes reduces the ability of *Pseudomonas solanacearum* to cause wilt symptoms. *Molecular Plant-Microbe Interactions, 3,* 293-300.

Denny, T. P. & Hayward, A. C. (2001). *Ralstonia*. In N. W. Schaad, J. B. Jones, & W. Chun (Eds.), *Laboratory Guide for Identification of Plant Pathogenic Bacteria, 3rd Ed.* (pp. 165-189). St. Paul, MN: APS Press.

Deslandes, L., Olivier, J., Peeters, N., Feng, D. X., Khounlotham, M., Boucher, C. et al. (2003). Physical interaction between RRS1-R, a protein conferring resistance to bacterial wilt, and PopP2, a type III effector targeted to the plant nucleus. *Proceedings of the National Academy of Sciences, USA, 100,* 8024-8029.

Deslandes, L., Olivier, J., Theulieres, F., Hirsch, J., Feng, D. X., Bittner-Eddy, P. et al. (2002). Resistance to *Ralstonia solanacearum* in *Arabidopsis thaliana* is conferred by the recessive RRS1-R gene, a member of a novel family of resistance genes. *Proceedings of the National Academy of Sciences, USA, 99,* 2404-2409.

Deslandes, L., Pileur, F., Liaubet, L., Camut, S., Can, C., Williams, K. et al. (1998). Genetic characterization of *RRS1*, a recessive locus in *Arabidopsis thaliana* that confers resistance to the bacterial soilborne pathogen *Ralstonia solanacearum*. *Molecular Plant-Microbe Interactions, 11,* 659-667.

Dittapongpitch, V. & Surat, S. (2003). Detection of *Ralstonia solanacearum* in soil and weeds from commercial tomato fields using immunocapture and the polymerase chain reaction. *Journal of Phytopathology, 151,* 239-246.

Echandi, E. (1991). Root diseases caused by bacteria: bacterial wilt. In H. D. Shew & G. B. Lucas (Eds.), *Compendium of Tobacco Diseases* (pp. 33-35). St. Paul, MN: APS Press.

Eden-Green, S. J. (1994). Diversity of *Pseudomonas solanacearum* and related bacteria in south east Asia: new directions for Moko disease. In A. C. Hayward & G. L. Hartman (Eds.), *Bacterial Wilt: The Disease and its Causative Agent, Pseudomonas solanacearum* (pp. 25-34). Wallingford, UK: CAB International.

Elphinstone, J. G. (1996). Survival and possibilities for extinction of *Pseudomonas solanacearum* (Smith) Smith in cool climates. *Potato Research, 39,* 403-410.

Elphinstone, J. G. (2005). The current bacterial wilt situation: A global view. In C. Allen, Ph. Prior, & A. C. Hayward (Eds.), *Bacterial Wilt Disease and the Ralstonia solanacearum Species Complex* (pp. 9-28). St. Paul, Minnesota: APS Press.

Elphinstone, J. G., Hennessy, J., Wilson, J. K., & Stead, D. E. (1996). Sensitivity of different methods for the detection of *Pseudomonas solanacearum* in potato tuber extracts. *EPPO/OEPP Bulletin, 26,* 663-678.

Enfinger, J. M., McCarter, S. M., & Jaworski, C. A. (1979). Evaluation of chemicals and application methods for control of bacterial wilt of tomato transplants. *Phytopathology, 69,* 637-640.

Etchebar, C., Trigalet-Demery, D., van Gijsegem, F., Vasse, J., & Trigalet, A. (1998). Xylem colonization by an HrcV$^-$ mutant of *Ralstonia solanacearum* is a key factor for the efficient biological control of tomato bacterial wilt. *Molecular Plant-Microbe Interactions, 11,* 869-877.

Fegan, M. (2005). Bacterial wilt diseases of banana: evolution and ecology. In C. Allen, P. Prior, & A. C. Hayward (Eds.), *Bacterial Wilt Disease and the Ralstonia solanacearum Species Complex* (pp. 379-386). St. Paul, MN: APS Press.

Fegan, M. & Prior, P. Diverse members of the *Ralstonia solanacearum* species complex cause bacterial wilt of banana. *Australasian Plant Pathology,* (in press).

Fegan, M. & Prior, P. (2005). How complex is the *Ralstonia solanacearum* species complex? In C. Allen, P. Prior, & A. C. Hayward (Eds.), *Bacterial Wilt Disease and the Ralstonia solanacearum Species Complex* (pp. 449-461). St. Paul, MN: APS Press.

Fegan, M., Taghavi, M., Sly, L. I., & Hayward, A. C. (1998). Phylogeny, diversity and molecular diagnostics of *Ralstonia solanacearum*. In Ph. Prior, C. Allen, & J. Elphinstone (Eds.), *Bacterial Wilt Disease: Molecular and Ecological Aspects* (pp. 19-33). Berlin: Springer-Verlag.

Feng, D.-X., Deslandes, L., Keller, H., Revers, F., Favery, B., Lecomte, P. et al. (2004). Isolation and characterization of a novel *Arabidopsis thaliana* mutant unable to develop wilt symptoms after inoculation with a virulent strain of *Ralstonia solanacearum*. *Phytopathology, 94,* 289-295.

Fortnum, B. A. & Kluepfel, D. (2005). Mechanization has contributed to the spread of bacterial wilt on flue-cured tobacco in the southeastern USA. In C. Allen, P. Prior, & A. C. Hayward (Eds.), *Bacterial Wilt Disease and the Ralstonia solanacearum Species Complex* (pp. 51-59). St. Paul, MN: APS Press.

Fortnum, B. A. & Martin, S. B. (1998). Disease management strategies for control of bacterial wilt of tobacco in the southeastern USA. In Ph. Prior, C. Allen, & J. Elphinstone (Eds.), *Bacterial Wilt Disease: Molecular and Ecological Aspects* (pp. 394-402). Berlin: Springer-Verlag.

French, E. R. (1986). Interaction between strains of *Pseudomonas solanacearum*, its hosts and the environment. In G. J. Persley (Ed.), *Bacterial Wilt Disease in Asia and the South Pacific. ACIAR Proceedings No. 13* (pp. 99-104). Canberra: Australian Centre for International Agricultural Research.

French, E. R. (1994). Strategies for integrated control of bacterial wilt of potatoes. In A. C. Hayward & G. L. Hartman (Eds.), *Bacterial Wilt: The Disease and Its Causative Agent, Pseudomonas solanacearum* (pp. 199-207). Wallingford, UK: CAB International.

French, E. R., Aley, P., Torres, E., & Nydegger, U. (1993). Diversity of *Pseudomonas solanacearum* in Peru and Brazil. In G. L. Hartman & A. C. Hayward (Eds.), *Bacterial Wilt. ACIAR Proceedings No. 45* (pp. 70-77). Canberra: Australian Centre for International Agricultural Research.

French, E. R. & Sequeira, L. (1970). Strains of *Pseudomonas solanacearum* from Central and South America: a comparative study. *Phytopathology, 60,* 506-512.

Frey, P., Prior, P., Marie, C., Kotoujansky, A., Trigalet-Demery, D., & Trigalet, A. (1994). Hrp^- mutants of *Pseudomonas solanacearum* as potential biocontrol agents of tomato bacterial wilt. *Applied and Environmental Microbiology, 60,* 3175-3181.

Gabriel, D. W., Allen, C., Schell, M. A., Denny, T. P., Greenberg, J. T., Duan, Y. P. et al. Identification of open reading frames unique to a Select Agent: *Ralstonia solanacearum* race 3 biovar 2. *Molecular Plant-Microbe Interactions,* (in press).

Garg, R. P., Huang, J., Yindeeyoungyeon, W., Denny, T. P., & Schell, M. A. (2000). Multicomponent transcriptional regulation at the complex promoter of the exopolysaccharide I biosynthetic operon of *Ralstonia solanacearum*. *Journal of Bacteriology, 182,* 6659-6666.

Genin, S. & Boucher, C. (2004). Lessons learned from the genome analysis of *Ralstonia solanacearum*. *Annual Review of Phytopathology, 42,* 107-134.

Genin, S., Brito, B., Denny, T. P., & Boucher, C. (2005). Control of the *Ralstonia solanacearum* type III secretion system (Hrp) genes by the global virulence regulator PhcA. *FEBS Letters, 579,* 2077-2081.

Gillings, M. & Fahy, P. (1993). Genetic diversity of *Pseudomonas solanacearum* biovar-2 and biovar-N2 assessed using restriction endonuclease analysis of total genomic DNA. *Plant Pathology, 42,* 744-753.

Gillings, M., Fahy, P., & Davies, C. (1993). Restriction analysis of an amplified polygalacturonase gene fragment differentiates strains of the phytopathogenic bacterium *Pseudomonas solanacearum*. *Letters in Applied Microbiology, 17,* 44-48.

Glick, D. L., Coffey, C. M., & Sulzinski, M. A. (2002). Simultaneous PCR detection of the two major bacterial pathogens of geranium. *Journal of Phytopathology, 150,* 54-59.

Godiard, L., Sauviac, L., Torii, K. U., Grenon, O., Mangin, B., Grimsley, N. H. et al. (2003). ERECTA, an LRR receptor-like kinase protein controlling development pleiotropically affects resistance to bacterial wilt. *Plant Journal, 36,* 353-365.

González, E. T. & Allen, C. (2003). Characterization of a *Ralstonia solanacearum* operon required for polygalacturonate degradation and uptake of galacturonic acid. *Molecular Plant-Microbe Interactions, 16,* 536-544.

Gopalakrishnan, T. R., Singh, P. K., Sheela, K. B., Shankar, M. A., Kutty, P. C. J., & Peter, K. V. (2005). Development of bacterial wilt resistant varieties and basis of resistance in eggplant (*Solanum melongena* L.). In C. Allen, P. Prior, & A. C. Hayward (Eds.), *Bacterial Wilt Disease and the Ralstonia solanacearum Species Complex* (pp. 293-300). St. Paul, MN: APS Press.

Gorissen, A., van Overbeek, L. S., & van Elsas, J. D. (2004). Pig slurry reduces the survival of *Ralstonia solanacearum* biovar 2 in soil. *Canadian Journal of Microbiology, 50,* 587-593.

Granada, G. A. & Sequeira, L. (1975). A hypersensitive reaction induced in tobacco leaves by a compatible isolate of *Pseudomonas solanacearum*. *Phytopathology, 65,* 731-733.

Grey, B. E. & Steck, T. R. (2001). The viable but nonculturable state of *Ralstonia solanacearum* may be involved in long-term survival and plant infection. *Applied and Environmental Microbiology, 67,* 3866-3872.

Grimault, V., Anais, G., & Prior, P. (1994). Distribution of *Pseudomonas solanacearum* in the stem tissues of tomato plants with different levels of resistance to bacterial wilt. *Plant Pathology, 43,* 663-668.

Grimault, V., Gélie, B., Lemattre, M., Prior, P., & Schmit, J. (1994). Comparative histology of resistant and susceptible tomato cultivars infected by *Pseudomonas solanacearum*. *Physiological and Molecular Plant Pathology, 44,* 105-123.

Grimault, V. & Prior, P. (1993). Bacterial wilt resistance in tomato associated with tolerance of vascular tissues to *Pseudomonas solanacearum*. *Plant Pathology, 42,* 589-594.

Grimault, V. & Prior, P. (1994a). Grafting tomato cultivars resistant or susceptible to bacterial wilt: analysis of resistance mechanisms. *Journal of Phytopathology, 141,* 330-334.

Grimault, V. & Prior, P. (1994b). Invasiveness of *Pseudomonas solanacearum* in tomato, eggplant and pepper: a comparative study. *European Journal of Plant Pathology, 100,* 259-267.

Gueneron, M., Timmers, A. C., Boucher, C., & Arlat, M. (2000). Two novel proteins, PopB, which has functional nuclear localization signals, and PopC, which has a large leucine-rich repeat domain, are secreted through the Hrp-secretion apparatus of *Ralstonia solanacearum*. *Molecular Microbiology, 36,* 261-277.

Hanson, P. M., Licardo, O., Hanudin, Wang, J.-F., & Chen, J. (1998). Diallel analysis of bacterial wilt resistance in tomato derived from different sources. *Plant Disease, 82,* 74-78.

Hartman, G. L. & Elphinstone, J. G. (1994). Advances in the control of *Pseudomonas solanacearum* race 1 in major food crops. In A. C. Hayward & G. L. Hartman (Eds.), *Bacterial Wilt: The Disease and Its Causative Agent, Pseudomonas solanacearum* (pp. 157-177). Wallingford, UK: CAB International.

Hartman, G. L. and Hayward, A. C. (1993). *Bacterial Wilt. Proceedings of an international conference held at Kaohsiung, Taiwan, 28-31 October 1992. ACIAR Proceedings No. 45.* Canberra: Australian Centre for International Agricultural Research.

Hayward, A. C. (1964). Characteristics of *Pseudomonas solanacearum*. *Journal of Applied Bacteriology, 27,* 265-277.

Hayward, A. C. (1991). Biology and epidemiology of bacterial wilt caused by *Pseudomonas solanacearum*. *Annual Review of Phytopathology, 29,* 65-87.

Hayward, A. C. (1994a). Systematics and phylogeny of *Pseudomonas solanacearum* and related bacteria. In A. C. Hayward & G. L. Hartman (Eds.), *Bacterial Wilt: The Disease and its Causative Agent, Pseudomonas solanacearum* (pp. 123-135). Wallingford, UK: CAB International.

Hayward, A. C. (1994b). The hosts of *Pseudomonas solanacearum*. In A. C. Hayward & G. L. Hartman (Eds.), *Bacterial Wilt: The Disease and its Causative Agent, Pseudomonas solanacearum* (pp. 9-24). Wallingford, UK: CAB International.

Hayward, A. C. (1995). *Pseudomonas solanacearum*. In U. S. Singh, R. P. Singh, & K. Kohmoto (Eds.), *Pathogenesis and host specificity in plant diseases: histopathological, biochemical, genetic and molecular bases. vol. I. Prokaryotes* (pp. 139-151). Tarrytown, N.Y.: Elsevier Science, Inc.

Hayward, A. C. (2000). *Ralstonia solanacearum*. In J. Lederberg (Ed.), *Encyclopedia of Microbiology, Vol. 4* (Second ed., pp. 32-42). San Diego, CA: Academic Press.

Hayward, A. C. and Hartman, G. L. (1994). *Bacterial Wilt: The Disease and its Causative Agent, Pseudomonas solanacearum*. Wallingford, UK: CAB International.

He, L. Y., Sequeira, L., & Kelman, A. (1983). Characteristics of strains of *Pseudomonas solanacearum* from China. *Plant Disease, 67,* 1357-1361.

He, S. Y., Nomura, K., & Whittam, T. S. (2004). Type III protein secretion mechanism in mammalian and plant pathogens. *Biochimica et Biophysica Acta-Molecular Cell Research, 1694,* 181-206.

Hirsch, J., Deslandes, L., Feng, D. X., Balague, C., & Marco, Y. (2002). Delayed symptom development in *ein2-1*, an *Arabidopsis* ethylene-insensitive mutant, in response to bacterial wilt caused by *Ralstonia solanacearum*. *Phytopathology, 92,* 1142-1148.

Ho, G.-D. & Yang, C.-H. (1999). A single locus leads to resistance of *Arabidopsis thaliana* to bacterial wilt caused by *Ralstonia solanacearum* through a hypersensitive-like response. *Phytopathology, 89,* 673-678.

Horita, M. & Tsuchiya, K. (2001). Genetic diversity of Japanese strains of *Ralstonia solanacearum*. *Phytopathology, 91,* 399-407.

Huang, Q. & Allen, C. (2000). Polygalacturonases are required for rapid colonization and full virulence of *Ralstonia solanacearum* on tomato plants. *Physiological and Molecular Plant Pathology, 57,* 77-83.

Huguet, E., Hahn, K., Wengelnik, K., & Bonas, U. (1998). *hpaA* mutants of *Xanthomonas campestris* pv. *vesicatoria* are affected in pathogenicity but retain the ability to induce host-specific hypersensitive reaction. *Molecular Microbiology, 29,* 1379-1390.

Ilagan, Y. A., Lavina, W. A., Natural, M. P., & Raymundo, A. K. (2003). Genetic homogeneity of the banana-infecting strains of *Ralstonia solanacearum* (Smith) Yabuuchi *et al.* in the Philippines. *Philippine Agricultural Scientist, 86,* 394-402.

Janse, J. D. (1988). A detection method for *Pseudomonas solanacearum* in symptomless potato tubers and some data on its sensitivity and specificity. *EPPO/OEPP Bulletin, 18,* 343-351.

Janse, J. D. (1991). Infra- and intraspecific classification of *Pseudomonas solanacearum* strains, using whole cell fatty acid analysis. *Systematic and Applied Microbiology, 14,* 335-345.

Janse, J. D. (1996). Potato brown rot in western Europe - history, present occurrence and some remarks on possible origin, epidemiology and control strategies. *EPPO/OEPP Bulletin, 26,* 679-695.

Janse, J. D., Araluppan, F. A. X., Schans, J., Wenneker, M., & Westerhuis, W. (1998). Experiences with bacterial brown rot, *Ralstonia solanacearum* biovar 2 race 3, in the Netherlands. In Ph. Prior, C. Allen, & J. Elphinstone (Eds.), *Bacterial Wilt Disease: Molecular and Ecological Aspects* (pp. 146-154). Berlin: Springer-Verlag.

Janse, J. D., van den Beld, H. E., Elphinstone, J., Simpkins, S., Tjou-Tam-Sin, N. N. A., & van Vaerenbergh, J. (2004). Introduction to Europe of *Ralstonia solanacearum* biovar 2, race 3 in *Pelargonium zonale* cuttings. *Journal of Plant Pathology, 87,* 147-155.

Jaunet, T. X. & Wang, J.-F. (1999). Variation in genotype and aggressiveness of *Ralstonia solanacearum* race 1 isolated from tomato in Taiwan. *Phytopathology, 89,* 320-327.

Jaynes, J. M., Nagpala, P., Destéfano-Beltrán, L., Huang, J. H., Kim, J. H., Denny, T. et al. (1993). Expression of a Cecropin B lytic peptide analog in transgenic tobacco confers enhanced resistance to bacterial wilt caused by *Pseudomonas solanacearum*. *Plant Science, 89,* 43-53.

Jeong, E.-L. & Timmis, J. N. (2000). Novel insertion sequence elements associated with genetic heterogeneity and phenotype conversion in *Ralstonia solanacearum*. *Journal of Bacteriology, 182,* 4673-4676.

Jha, G., Rajeshwari, R., & Sonti, R. V. (2005). Bacterial type two secretion system secreted proteins: double-edged swords for plant pathogens. *Molecular Plant-Microbe Interactions, 18,* 891-898.

Jones, D. R. (2000). Introduction to banana, abacá and enset. In D. R. Jones (Ed.), *Diseases of Banana, Abacá and Enset* (pp. 1-36). Wallingford, UK: CABI Publishing.

Kanda, A., Yasukohchi, M., Ohnishi, K., Kiba, A., Okuno, T., & Hikichi, Y. (2003). Ectopic expression of *Ralstonia solanacearum* effector protein PopA early in invasion results in loss of virulence. *Molecular Plant-Microbe Interactions, 16,* 447-455.

Kang, Y., Huang, J. Z., Mao, G. Z., He, L. Y., & Schell, M. A. (1994). Dramatically reduced virulence of mutants of *Pseudomonas solanacearum* defective in export of extracellular proteins across the outer membrane. *Molecular Plant-Microbe Interactions, 7,* 370-377.

Kang, Y., Liu, H., Carlson, R. W., & Denny, T. P. (2004). Lipopolysaccharide O-antigen of *Ralstonia solanacearum* contributes to virulence on tomato plants. *Phytopathology, 94,* S49.

Kang, Y., Liu, H., Genin, S., Schell, M. A., & Denny, T. P. (2002). *Ralstonia solanacearum* requires type 4 pili to adhere to multiple surfaces, and for natural transformation and virulence. *Molecular Microbiology, 46,* 427-437.

Kang, Y., Saile, E., Schell, M. A., & Denny, T. P. (1999). Quantitative immunofluorescence of regulated *eps* gene expression in single cells of *Ralstonia solanacearum. Applied and Environmental icrobiology, 65,* 2356-2362.

Kao, C. C. & Sequeira, L. (1991). A gene cluster required for coordinated biosynthesis of lipopolysaccharide and extracellular polysaccharide also affects virulence of *Pseudomonas solanacearum. Journal of Bacteriology, 173,* 7841-7847.

Keen, N. T. (1990). Gene-for-gene complementarity in plant-pathogen interactions. *Annual Review of Genetics, 24,* 447-463.

Kelman, A. (1953). *The bacterial wilt caused by Pseudomonas solanacearum.* North Carolina Agricultural Experiment Station, Technical Bulletin No. 99.

Kelman, A. & Person, L. H. (1961). Strains of *Pseudomonas solanacearum* differing in pathogenicity to tobacco and peanut. *Phytopathology, 51,* 158-161.

Kelman, A. & Sequeira, L. (1965). Root-to-root spread of *Pseudomonas solanacearum. Phytopathology, 55,* 304-309.

Kinyua, Z. M., Olanya, M., Smith, J. J., El-Bedewy, R., Kihara, S. N., Kakuhenzire, R. K. et al. (2005). Seed-plot technique: empowerment of farmers in production of bacterial wilt-free seed potato in Kenya and Uganda. In C. Allen, P. Prior, & A. C. Hayward (Eds.), *Bacterial Wilt Disease and the Ralstonia solanacearum Species Complex* (pp. 167-175). St. Paul, MN: APS Press.

Klement, Z. (1982). Hypersensitivity. In M. S. Mount & G. H. Lacy (Eds.), *Phytopathogenic Prokaryotes* (pp. 149-177). New York: Academic Press.

Kocharova, N. A., Knirel, Y. A., Shashkov, A. S., Nifantev, N. E., Kochetkov, N. K., Varbanets, L. D. et al. (1993). Studies of O-specific polysaccharide chains of *Pseudomonas solanacearum* lipopolysaccharides consisting of structurally different repeating units. *Carbohydrate Research, 250,* 275-287.

Kumar, A., Sarma, Y. R., & Anandaraj, M. (2004). Evaluation of genetic diversity of *Ralstonia solanacearum* causing bacterial wilt of ginger using REP-PCR and PCR-RFLP. *Current Science, 87,* 1555-1561.

Lahaye, T. (2002). The *Arabidopsis RRS1-R* disease resistance gene - uncovering the plant's nucleus as the new battlefield of plant defense? *Trends in Plant Science, 7,* 425-427.

Lavie, M., Seunes, B., Prior, P., & Boucher, C. (2004). Distribution and sequence analysis of a family of type III-dependent effectors correlate with the phylogeny of *Ralstonia solanacearum* strains. *Molecular Plant-Microbe Interactions, 17,* 931-940.

Lavie, M., Shillington, E., Equiluz, C., Grimsley, N., & Boucher, C. (2002). PopP1, a new member of the YopJ/AvrRxv family of type III effector proteins, acts as a host-specificity factor and modulates aggressiveness of *Ralstonia solanacearum. Molecular Plant-Microbe Interactions, 15,* 1058-1068.

Lee, T. J., Coyne, D. P., Clemente, T. E., & Mitra, A. (2002). Partial resistance to bacterial wilt in transgenic tomato plants expressing antibacterial lactoferrin gene. *Journal of American Society of Horticultural Science, 127,* 158-164.

Lemaga, B., Kakuhenzire, R., Kassa, B., Ewell, P. T., & Priou, S. (2005). Integrated control of potato bacterial wilt in eastern Africa: the experience of African Highlands Initiative. In C. Allen, P. Prior, & A. C. Hayward (Eds.), *Bacterial Wilt Disease and the Ralstonia solanacearum Species Complex* (pp. 145-157). St. Paul, MN: APS Press.

Li, X., Dorsch, M., del Dot, T., Sly, L. I., Stackebrandt, E., & Hayward, A. C. (1993). Phylogenetic studies of the rRNA group II pseudomonads based on 16S rRNA gene sequences. *Journal of Applied Bacteriology, 74,* 324-329.

Li, X. & Hayward, A. C. (1993). The use of the Biolog identification system for the rapid identification of plant pathogenic pseudomonads. In G. L. Hartman & A. C. Hayward (Eds.), *Bacterial Wilt. ACIAR Proceedings No. 45* (pp. 45-48). Canberra, Australia: Australian Centre for International Agricultural Research.

Lin, W. C., Lu, C. F., Wu, J. W., Cheng, M. L., Lin, Y. M., Yang, N. S. et al. (2004). Transgenic tomato plants expressing the *Arabidopsis NPR1* gene display enhanced resistance to a spectrum of fungal and bacterial diseases. *Transgenic Research, 13,* 567-581.

Liu, H., Zhang, S., Schell, M. A., & Denny, T. P. Pyramiding unmarked mutations in *Ralstonia solanacearum* shows that secreted proteins in addition to plant cell wall degrading enzymes contribute to virulence. *Molecular Plant-Microbe Interactions,* (in press).

Llop, P., Caruso, P., Cubero, J., Morente, C., & Lopez, M. M. (1999). A simple extraction procedure for efficient routine detection of pathogenic bacteria in plant material by polymerase chain reaction. *Journal of Microbiological Methods, 37,* 23-31.

Lozano, J. C. & Sequeira, L. (1970). Differentiation of races of *Pseudomonas solanacearum* by a leaf infiltration technique. *Phytopathology, 60,* 833-838.

Mangin, B., Thoquet, P., Olivier, J., & Grimsley, N. H. (1999). Temporal and multiple quantitative trait loci analyses of resistance to bacterial wilt in tomato permit the resolution of linked loci. *Genetics, 151,* 1165-1172.

Marenda, M., Brito, B., Callard, D., Genin, S., Barberis, P., Boucher, C. et al. (1998). PrhA controls a novel regulatory pathway required for the specific induction of *Ralstonia solanacearum hrp* genes in the presence of plant cells. *Molecular Microbiology, 27,* 437-453.

Marín, J. E. & El-Nashaar, H. M. (1993). Pathogenicity of the new phenotypes of *Pseudomonas solanacearum* from Peru. In G. L. Hartman & A. C. Hayward (Eds.), *Bacterial Wilt. ACIAR Proceedings No. 45* (pp. 78-84). Canberra, Australia: Australian Centre for International Agricultural Research.

Martins, O. M., Nabizadeh-Ardekani, F., & Rudolf, K. (2005). Seeds from infected tomato plants appear to be free from contamination by *Ralstonia solanacearum* when tested by PCR and microbiological assays. In C. Allen, P. Prior, & A. C. Hayward (Eds.), *Bacterial Wilt Disease and the Ralstonia solanacearum Species Complex* (pp. 95-101). St. Paul, MN: APS Press.

McCarter, S. M. (1991). Root diseases caused by bacteria: bacterial wilt. In J. B. Jones, J. P. Jones, R. E. Stall, & T. A. Zitter (Eds.), *Compendium of Tomato Diseases* (pp. 28-29). St. Paul, MN: APS Press.

McGarvey, J. A. (1999). *In vivo and in vitro growth and virulence factor production by Ralstonia solanacearum.* Ph.D. dissertation, University of Georgia, Athens.

McGarvey, J. A., Bell, C. J., Denny, T. P., & Schell, M. A. (1998). Analysis of extracellular polysaccharide I in culture and *in planta* using immunological methods: new insights and implications. In Ph. Prior, C. Allen, & J. Elphinstone (Eds.), *Bacterial Wilt Disease: Molecular and Ecological Aspects* (pp. 157-163). Berlin: Springer-Verlag.

McGarvey, J. A., Denny, T. P., & Schell, M. A. (1999). Spatial-temporal and quantitative analysis of growth and EPS I production by *Ralstonia solanacearum* in resistant and susceptible tomato cultivars. *Phytopathology, 89,* 1233-1239.

Mergeay, M., Monchy, S., Vallaeys, T., Auquier, V., Benotmane, A., Bertin, P. et al. (2003). *Ralstonia metallidurans*, a bacterium specifically adapted to toxic metals: towards a catalogue of metal-responsive genes. *FEMS Microbiology Reviews, 27,* 385-410.

Montanelli, C., Stefanini, F. M., Chiari, A., Chiari, T., & Nascari, G. (1995). Variability in the response to *Pseudomonas solanacearum* of transgenic lines of potato carrying a cecropin gene analogue. *Potato Research, 38,* 371-378.

Mukaihara, T., Tamura, N., Murata, Y., & Iwabuchi, M. (2004). Genetic screening of Hrp type III-related pathogenicity genes controlled by the HrpB transcriptional activator in *Ralstonia solanacearum*. *Molecular Microbiology, 54,* 863-875.

Nakaho, K., Hibino, H., & Miyagawa, H. (2000). Possible mechanisms limiting movement of *Ralstonia solanacearum* in resistant tomato tissues. *Journal of Phytopathology, 148,* 181-190.

Nimchuk, Z., Eulgem, T., Holt, B. E., & Dangl, J. L. (2003). Recognition and response in the plant immune system. *Annual Review of Genetics, 37,* 579-609.

Occhialini, A., Cunnac, S., Reymond, N., Genin, S., & Boucher, C. (2005). Genome-wide analysis of gene expression in *Ralstonia solanacearum* reveals that the *hrpB* gene acts as a regulatory switch controlling multiple virulence pathways. *Molecular Plant-Microbe Interactions, 18,* 938-949.

Orgambide, G., Montrozier, H., Servin, P., Roussel, J., Trigalet-Demery, D., & Trigalet, A. (1991). High heterogeneity of the exopolysaccharides of *Pseudomonas solanacearum* strain GMI 1000 and the complete structure of the major polysaccharide. *Journal of Biological Chemistry, 266,* 8312-8321.

Ozakman, M. & Schaad, N. W. (2003). A real-time BIO-PCR assay for detection of *Ralstonia solanacearum* race 3, biovar 2, in asymptomatic potato tubers. *Canadian Journal of Plant Pathology, 25,* 232-239.

Palleroni, N. J. (1984). Family I. *Pseudomonadaceae*. Genus I. *Pseudomonas*. In N. R. Krieg & J. G. Holt (Eds.), *Bergey's Manual of Systematic Bacteriology*, vol. 1 (pp. 141-219). Baltimore: Williams and Wilkins.

Palleroni, N. J. & Doudoroff, M. (1971). Phenotypic characterization and deoxyribonucleic acid homologies of *Pseudomonas solanacearum*. *Journal of Bacteriology, 107,* 690-696.

Pastrik, K.-H., Elphinstone, J. G., & Pukall, R. (2002). Sequence analysis and detection of *Ralstonia solanacearum* by multiplex PCR amplification of 16S-23S ribosomal intergenic spacer region with internal positive control. *European Journal of Plant Pathology, 108,* 831-842.

Pastrik, K.-H. & Maiss, E. (2000). Detection of *Ralstonia solanacearum* in potato tubers by polymerase chain reaction. *Journal of Phytopathology, 148,* 619-626.

Pegg, K. G. & Moffett, M. L. (1971). Host range of the ginger strain of *Pseudomonas solanacearum* in Queensland. *Australian Journal of Experimental Agriculture and Animal Husbandry, 11,* 696-698.

Persley, G. J. (1986a). *Bacterial Wilt Disease in Asia and the South Pacific. Proceedings of an international workshop held at PCARRD, Los Baños, Philippines, 8-10 October 1985. ACIAR Proceedings No. 13.* Canberra: Australian Centre for International Agricultural Research.

Persley, G. J. (1986b). Ecology of *Pseudomonas solanacearum*, the causal agent of bacterial wilt. In G. J. Persley (Ed.), *Bacterial Wilt Disease in Asia and the South Pacific. ACIAR Proceedings No. 13* (pp. 71-76). Canberra: Australian Centre for International Agricultural Research.

Persley, G. J., Batugal, P., Gapasin, D., & Vander Zaag, P. (1986). Summary of discussion and recommendations. In G. J. Persley (Ed.), *Bacterial Wilt Disease in Asia and the South Pacific. ACIAR Proceedings No. 13* (pp. 7-13). Canberra: Australian Centre for International Agricultural Research.

Peterson, C. A., Emanuel, M. E., & Humphreys, G. B. (1981). Pathway of movement of apoplastic fluorescent dye tracers through the endodermis at the site of secondary root formation in corn (*Zea mays*) and broad bean (*Vicia faba*). *Canadian Journal of Botany, 59,* 618-625.

Pfund, C., Tans-Kersten, J., Dunning, F. M., Alonso, J. M., Ecker, J. R., Allen, C. et al. (2004). Flagellin is not a major defense elicitor in *Ralstonia solanacearum* cells or extracts applied to *Arabidopsis thaliana*. *Molecular Plant-Microbe Interactions, 17,* 696-706.

Poussier, S., Cheron, J. J., Couteau, A., & Luisetti, J. (2002). Evaluation of procedures for reliable PCR detection of *Ralstonia solanacearum* in common natural substrates. *Journal of Microbiological Methods, 51,* 349-359.

Poussier, S. & Luisetti, J. (2000). Specific detection of biovars of *Ralstonia solanacearum* in plant tissues by nested-PCR-RFLP. *Journal of Plant Pathology, 106,* 255-265.

Poussier, S., Prior, P., Luisetti, J., Hayward, C., & Fegan, M. (2000a). Partial sequencing of the *hrpB* and endoglucanase genes confirms and expands the known diversity within the *Ralstonia solanacearum* species complex. *Systematic and Applied Microbiology, 23,* 479-486.

Poussier, S., Thoquet, P., Trigalet-Demery, D., Barthet, S., Meyer, D., Arlat, M. et al. (2003). Host plant-dependent phenotypic reversion of *Ralstonia solanacearum* from non-pathogenic to pathogenic forms via alterations in the *phcA* gene. *Molecular Microbiology, 49,* 991-1003.

Poussier, S., Trigalet-Demery, D., Vandewalle, P., Goffinet, B., Luisetti, J., & Trigalet, A. (2000b). Genetic diversity of *Ralstonia solanacearum* as assessed by PCR-RFLP of the *hrp* gene region, AFLP and 16S rRNA sequence analysis, and identification of an African subdivision. *Microbiology-Sgm, 146,* 1679-1692.

Poussier, S., Vandewalle, P., & Luisetti, J. (1999). Genetic diversity of African and worldwide strains of *Ralstonia solanacearum* as determined by PCR-restriction fragment length polymorphism analysis of the *hrp* gene region. *Applied and Environmental Microbiology, 65,* 2184-2194.

Pradhanang, P. M., Elphinstone, J. G., & Fox, R. T. V. (2000a). Identification of crop and weed hosts of *Ralstonia solanacearum* biovar 2 in the hills of Nepal. *Plant Pathology, 49,* 403-413.

Pradhanang, P. M., Elphinstone, J. G., & Fox, R. T. V. (2000b). Sensitive detection of *Ralstonia solanacearum* in soil: a comparison of different detection techniques. *Plant Pathology, 49,* 414-422.

Preston, G. M., Studholme, D. J., & Caldelari, I. (2005). Profiling the secretomes of plant pathogenic Proteobacteria. *FEMS Microbiology Reviews, 29,* 331-360.

Prior, P. & Fegan, M. (2005a). Diversity and molecular detection of *Ralstonia solanacearum* race 2 strains by multiplex PCR. In C. Allen, P. Prior, & A. C. Hayward (Eds.), *Bacterial Wilt Disease and the Ralstonia solanacearum Species Complex* (pp. 4405-414). St. Paul, MN: APS Press.

Prior, P. & Fegan, M. Recent developments in the phylogeny and classification of *Ralstonia solanacearum*. *Acta Horticulturae,* (in press).

Prior, P., Grimault, V., & Schmit, J. (1994). Resistance to bacterial wilt (*Pseudomonas solanacearum*) in tomato: present status and prospects. In A. C. Hayward & G. L. Hartman (Eds.), *Bacterial Wilt: The Disease and Its Causative Agent, Pseudomonas solanacearum* (pp. 209-223). Wallingford, UK: CAB International.

Prior, Ph., Allen, C., and Elphinstone, J. (1998). *Bacterial Wilt Disease: Molecular and Ecological Aspects*. Berlin: Springer-Verlag.

Priou, S., Aley, P., & Gutarra, L. (2005). Assessment of resistance to bacterial wilt in CIP advanced potato clones. In C. Allen, P. Prior, & A. C. Hayward (Eds.), *Bacterial Wilt Disease and the Ralstonia solanacearum Species Complex* (pp. 261-267). St. Paul, MN: APS Press.

Priou, S., Gutarra, L., & Aley, P. An improved enrichment broth of the highly sensitive and reliable detection of *Ralstonia solanacearum* (biovar 1 and 2A) in soil using double-antibody sandwich enzyme-linked immunosorbent assay. *Plant Pathology,* (in press).

Purcell, A. H. & Hopkins, D. L. (1996). Fastidious xylem-limited bacterial plant pathogens. *Annual Review of Phytopathology, 34,* 131-151.

Quinon, V. L., Aragaki, M., & Ishii, M. (1964). Pathogenicity and serological relationship of three strains of *Pseudomonas solanacearum* in Hawaii. *Phytopathology, 54,* 1096-1099.

Ralston, E., Palleroni, N. J., & Doudoroff, M. (1973). *Pseudomonas pickettii*, a new species of clinical origin related to *Pseudomonas solanacearum*. *International Journal of Systematic Bacteriology, 23,* 15-19.

Ratnayake, A. S. (2002). *Screening for physiological characteristics that contribute to the virulence of Ralstonia solanacearum GMI1000.* M.S. Thesis, University of Georgia, Athens.

Raymundo, A. K., Aves-Ilagan, Y., & Denny, T. P. (1998). Analysis of genetic variation in a population of banana-infecting strains of *Ralstonia solanacearum*. In Ph. Prior, C. Allen, & J. Elphinstone (Eds.), *Bacterial Wilt Disease: Molecular and Ecological Aspects* (pp. 56-60). Berlin: Springer-Verlag.

Raymundo, A. K., Orlina, M. E., Lavina, W. A., & Opina, N. L. (2005). Comparative genome plasticity of tomato and banana strains of *Ralstonia solanacearum* in the Philippines. In C. Allen, P. Prior, & A. C. Hayward (Eds.), *Bacterial Wilt Disease and the Ralstonia solanacearum Species Complex* (pp. 387-393). St. Paul, MN: APS Press.

Rhodius, V., Van Dyk, T. K., Gross, C., & LaRossa, R. A. (2002). Impact of genomic technologies on studies of bacterial gene expression. *Annual Review of Microbiology, 56,* 599-624.

Roberts, S. J., Eden-Green, S. J., Jones, P., & Ambler, D. J. (1990). *Pseudomonas syzygii* sp. nov., the cause of Sumatra disease of cloves. *Systematic and Applied Microbiology, 13,* 34-43.

Robertson, A. E., Wechter, W. P., Denny, T. P., Fortnum, B. A., & Kluepfel, D. A. (2004). Relationship between avirulence gene (*avrA*) diversity in *Ralstonia solanacearum* and bacterial wilt incidence. *Molecular Plant-Microbe Interactions, 17,* 1376-1384.

Rossier, O., Van den, A. G., & Bonas, U. (2000). HrpB2 and HrpF from *Xanthomonas* are type III-secreted proteins and essential for pathogenicity and recognition by the host plant. *Molecular Microbiology, 38,* 828-838.

Saddler, G. S. (2000). IMI description of fungi and bacteria No. 1220: *Burkholderia solanacearum. Mycopathologia, 128,* 61-63.

Saddler, G. S. (2005). Management of bacterial wilt disease. In C. Allen, P. Prior, & A. C. Hayward (Eds.), *Bacterial Wilt Disease and the Ralstonia solanacearum Species Complex* (pp. 121-132). St. Paul, MN: APS Press.

Saile, E., Schell, M. A., & Denny, T. P. (1997). Role of extracellular polysaccharide and endoglucanase in root invasion and colonization of tomato plants by *Ralstonia solanacearum. Phytopathology, 87,* 1264-1271.

Salanoubat, M., Genin, S., Artiguenave, F., Gouzy, J., Mangenot, S., Arlat, M. et al. (2002). The genome sequence of the wide host-range plant pathogen *Ralstonia solanacearum. Nature, 415,* 497-502.

Schell, M. A. (2000). Control of virulence and pathogenicity genes of *Ralstonia solanacearum* by an elaborate sensory array. *Annual Review of Phytopathology, 38,* 263-292.

Schmit, J. (1978). Microscopic study of early stages of infection by *Pseudomonas solanacearum* E.F.S. on *in "vitro"* grown tomato seedlings. In *Proceedings of the Fourth International Conference on Plant Pathogenic Bacteria* (pp. 841-857). Anger, France.

Schönfeld, J., Gelsomino, A., van Overbeek, L. S., Gorissen, A., Smalla, K., & van Elsas, J. D. (2003). Effects of compost addition and simulated solarisation on the fate of *Ralstonia solanacearum* biovar 2 and indigenous bacteria in soil. *FEMS Microbiology Ecology, 43,* 63-74.

Schönfeld, J., Heuer, H., van Elsas, J. D., & Smalla, K. (2003). Specific and sensitive detection of *Ralstonia solanacearum* in soil on the basis of PCR amplification of *fliC* fragments. *Applied and Environmental Microbiology, 69,* 7248-7256.

Seal, S. E. (1998). Molecular methods for detection and discrimination of *Ralstonia solanacearum*. In Ph. Prior, C. Allen, & J. Elphinstone (Eds.), *Bacterial Wilt Disease: Molecular and Ecological Aspects* (pp. 103-109). Berlin: Springer-Verlag.

Seal, S. E. & Elphinstone, J. G. (1994). Advances in identification and detection of *Pseudomonas solanacearum*. In A. C. Hayward & G. L. Hartman (Eds.), *Bacterial Wilt: The Disease and its Causative Agent, Pseudomonas solanacearum* (pp. 35-57). Wallingford, UK: CAB International.

Seal, S. E., Jackson, L. A., & Daniels, M. J. (1992). Use of tRNA consensus primers to indicate subgroups of *Pseudomonas solanacearum* by polymerase chain reaction amplification. *Applied and Environmental Microbiology, 58,* 3759-3761.

Segura, A., Moreno, M., Madueno, F., Molina, A., & García-Olmedo, F. (1999). Snakin-1, a peptide from potato that is active against plant pathogens. *Molecular Plant-Microbe Interactions, 12,* 16-23.

Sequeira, L. (1985). Surface components involved in bacterial pathogen-plant host recognition. *Journal of Cell Science Supplement, 2,* 301-316.

Sequeira, L. (1993). Bacterial wilt: past, present, and future. In G. L. Hartman & A. C. Hayward (Eds.), *Bacterial Wilt. ACIAR Proceedings No. 45* (pp. 12-21). Canberra, Australia: Australian Centre for International Agricultural Research.

Sequeira, L. (1998). Bacterial wilt: the missing element in international banana improvement programs. In Ph. Prior, C. Allen, & J. Elphinstone (Eds.), *Bacterial Wilt Disease: Molecular and Ecological Aspects* (pp. 6-14). Berlin: Springer-Verlag.

Sharma, S. K., Rajesh, K., Dohroo, N. P., & Rajesh, K. (2004). Effect of soil solarization on *Ralstonia solanacearum* population and bacterial wilt of tomato in Himachal Pradesh. *Indian Phytopathology, 57,* 200-204.

Soguilon, C. E., Magnaye, L. V., & Natural, M. P. (1994). "Bugtok" disease of cooking bananas: etiology and diagnostic symptoms. *Philippine Phytopathology, 30,* 26-34.

Stead, D. E., Sellwood, J. E., Wilson, J., & Viney, I. (1992). Evaluation of a commercial microbial identification system based on fatty acid profiles for rapid, accurate identification of plant pathogenic bacteria. *Journal of Applied Bacteriology, 72,* 315-321.

Stover, R. H. & Espinoza, A. (1992). Blood disease of bananas in Sulawesi. *Fruits, 47,* 611-613.

Strom, M. S. & Lory, S. (1993). Structure-function and biogenesis of the type IV pili. *Annual Review of Microbiology, 47,* 565-596.

Supriadi. (2005). Present status of blood disease in Indonesia. In C. Allen, P. Prior, & A. C. Hayward (Eds.), *Bacterial Wilt Disease and the Ralstonia solanacearum Species Complex* (pp. 395-404). St. Paul, MN: APS Press.

Swanepoel, A. E. (1990). The effect of temperature on the development of wilting and on progeny tuber infection of potatoes inoculated with South African strains of biovar 2 and biovar 3 of *Pseudomonas solanacearum. Potato Research, 33,* 287-290.

Swanson, J. K., Yao, J., Tans-Kersten, J., & Allen, C. (2005). Behavior of *Ralstonia solanacearum* race 3 biovar 2 during latent and active infection of geranium. *Phytopathology, 95,* 136-143.

Taghavi, M., Hayward, C., Sly, L. I., & Fegan, M. (1996). Analysis of the phylogenetic relationships of strains of *Burkholderia solanacearum, Pseudomonas syzygii,* and the blood disease bacterium of banana based on 16S rRNA gene sequences. *International Journal of Systematic Bacteriology, 46,* 10-15.

Tans-Kersten, J., Brown, D., & Allen, C. (2004). Swimming motility, a virulence trait of *Ralstonia solanacearum,* is regulated by FlhDC and the plant host environment. *Molecular Plant-Microbe Interactions, 17,* 686-695.

Tans-Kersten, J., Guan, Y. F., & Allen, C. (1998). *Ralstonia solanacearum* pectin methylesterase is required for growth on methylated pectin but not for bacterial wilt virulence. *Applied and Environmental Microbiology, 64,* 4918-4923.

Tans-Kersten, J., Huang, H. Y., & Allen, C. (2001). *Ralstonia solanacearum* needs motility for invasive virulence on tomato. *Journal of Bacteriology, 183,* 3597-3605.

Thoquet, P., Olivier, J., Sperisen, C., Rogowsky, P., Laterrot, H., & Grimsley, N. (1996a). Quantitative trait loci determining resistance to bacterial wilt in tomato cultivar Hawaii7996. *Molecular Plant-Microbe Interactions, 9,* 826-836.

Thoquet, P., Olivier, J., Sperisen, C., Rogowsky, P., Prior, P., Anais, G. et al. (1996b). Polygenic resistance of tomato plants to bacterial wilt in the French West Indies. *Molecular Plant-Microbe Interactions, 9,* 837-842.

Thurston, H. D. (1963). Bacterial wilt of potatoes in Columbia. *American Potato Journal, 40,* 381-390.

Thwaites, R., Eden-Green, S. J., & Black, R. (2000). Diseases caused by bacteria. In D. R. Jones (Ed.), *Diseases of Banana, Abacá and Enset* (pp. 213-239). Wallingford, UK: CABI Publishing.

Thwaites, R., Mansfield, J., Eden-Green, S., & Seal, S. (1999). RAPD and rep PCR-based fingerprinting of vascular bacterial pathogens of *Musa* spp. *Plant Pathology, 48,* 121-128.

Titarenko, E., López-Solanilla, E., García-Olmedo, F., & Rodríguez-Palenzuela, P. (1997). Mutants of *Ralstonia* (*Pseudomonas*) *solanacearum* sensitive to antimicrobial peptides are altered in their lipopolysaccharide structure and are avirulent in tobacco. *Journal of Bacteriology, 179,* 6699-6704.

Trigalet, A., Frey, P., & Trigalet-Demery, D. (1994). Biological control of bacterial wilt caused by *Pseudomonas solanacearum*: state of the art and understanding. In A. C. Hayward & G. L. Hartman (Eds.), *Bacterial Wilt: The Disease and Its Causative Agent, Pseudomonas solanacearum* (pp. 225-233). Wallingford, UK: CAB International.

Tusiime, G., Adipala, E., Opio, F., & Bhagsari, A. S. (1998). Weeds as latent host of *Ralstonia solanacearum* in highland Uganda: implications to development of an integrated control package for bacterial wilt. In Ph. Prior, C. Allen, & J. Elphinstone (Eds.), *Bacterial Wilt Disease: Molecular and Ecological Aspects* (pp. 413-419). Berlin: Springer-Verlag.

Valencia, A. (2005). Automatic annotation of protein function. *Current Opinion in Structural Biology, 15,* 267-274.

Van Alfen, N. K. (1989). Reassessment of plant wilt toxins. *Annual Review of Phytopathology, 27,* 533-550.

Van Damme, M., Andel, A., Huibers, R. P., Panstruga, R., Weisbeek, P. J., & van den Ackerveken, G. (2005). Identification of *Arabidopsis* loci required for susceptibility to the downy mildew pathogen *Hyaloperonospora parasitica. Molecular Plant-Microbe Interactions, 18,* 583-592.

van der Wolf, J. M. (2004). Specific detection of *Ralstonia solanacearum* 16S rRNA sequences by AmpliDet RNA. *European Journal of Plant Pathology, 110,* 25-33.

van der Wolf, J. M., Bonants, P. J. M., Smith, J. J., Hagennar, M., Nijhuis, E., van Beckhoven, J. R. C. M. et al. (1998). Genetic diversity of *Ralstonia solanacearum* race 3 in western Europe determined by AFLP, PC-PFGE, and rep-PCR. In Ph. Prior, C. Allen, & J. Elphinstone (Eds.), *Bacterial Wilt Disease: Molecular and Ecological Aspects* (pp. 44-49). Berlin: Springer-Verlag.

van der Wolf, J. M., Vriend, S. G. C., Kastelein, P., Nijhuis, E. H., van Bekkum, P. J., & van Vuurde, J. W. L. (2000). Immunofluorescence colony-staining (IFC) for detection and quantification of *Ralstonia* (*Pseudomonas*) *solanacearum* biovar 2 (race 3) in soil and verification of positive results by PCR and dilution plating. *European Journal of Plant Pathology, 106,* 123-133.

van Elsas, J. D., Kastelein, P., de Vries, P. M., & van Overbeek, L. S. (2001). Effects of ecological factors on the survival and physiology of *Ralstonia solanacearum* bv. 2 in irrigation water. *Canadian Journal of Microbiology, 47,* 842-854.

van Elsas, J. D., van Overbeek, L. S., Bailey, M. J., Schönfeld, J., & Smalla, K. (2005a). Fate of *Ralstonia solanacearum* biovar 2 as affected by conditions and soil treatments in temperate climate zones. In C. Allen, P. Prior, & A. C. Hayward (Eds.), *Bacterial Wilt Disease and the Ralstonia solanacearum Species Complex* (pp. 39-49). St. Paul, MN: APS Press.

van Elsas, J. D., van Overbeek, L. S., & Trigalet, A. (2005b). The viable but non-culturable state in *Ralstonia solanacearum*: is there a realistic threat to our strategic concepts? In C. Allen, P. Prior, & A. C. Hayward (Eds.), *Bacterial Wilt Disease and the Ralstonia solanacearum Species Complex* (pp. 103-119). St. Paul, MN: APS Press.

van Gijsegem, F., Gough, C., Zischek, C., Niqueuz, E., Arlat, M., Genin, S. et al. (1995). The *hrp* gene locus of *Pseudomonas solanacearum*, which controls the production of a type III secretion system, encodes eight proteins related to components of the bacterial flagellar biogenesis complex. *Molecular Microbiology, 15,* 1095-1114.

van Gijsegem, F., Vasse, J., Camus, J. C., Marenda, M., & Boucher, C. (2000). *Ralstonia solanacearum* produces Hrp-dependent pili that are required for PopA secretion but not for attachment of bacteria to plant cells. *Molecular Microbiology, 36,* 249-260.

van Gijsegem, F., Vasse, J., De Rycke, R., Castello, P., & Boucher, C. (2002). Genetic dissection of the *Ralstonia solanacearum hrp* gene cluster reveals that the HrpV and HrpX proteins are required for Hrp pilus assembly. *Molecular Microbiology, 44,* 935-946.

Van Loon, L. C. & Van Strien, E. A. (1999). The families of pathogenesis-related proteins, their activities, and comparative analysis of PR-1 type proteins. *Physiological and Molecular Plant Pathology, 55,* 85-97.

van Overbeek, L. S., Bergervoet, J. H. H., Jacobs, F. H. H., & van Elsas, J. D. (2004). The low-temperature-induced viable-but-nonculturable state affects the virulence of *Ralstonia solanacearum* biovar 2. *Phytopathology, 94,* 463-469.

Vandamme, P. & Coenye, T. (2004). Taxonomy of the genus *Cupriavidus*: a tale of lost and found. *International Journal of Systematic and Evolutionary Microbiology, 54,* 2285-2289.

Vaneechoutte, M., Kampfer, P., De Baere, T., Falsen, E., & Verschraegen, G. (2004). *Wautersia* gen. nov., a novel genus accommodating the phylogenetic lineage including *Ralstonia eutropha* and related species, and proposal of *Ralstonia* [*Pseudomonas*] *syzygii* (Roberts et al. 1990) comb. nov. *International Journal of Systematic and Evolutionary Microbiology, 54,* 317-327.

Varbanets, L. D., Kocharova, N. A., Knirel, Y. A., & Moskalenko, N. V. (1996). Structure of O-specific polysaccharide chains of lipopolysaccharides of *Burkholderia* (*Pseudomonas*) *solanacearum* ICMP 750, 8093, 8115, 7864, and 7945. *Biochemistry, English Translation, 61,* 580-585.

Vasse, J., Danoun, S., & Trigalet, A. (2005). Microscopic studies of root infection in resistant tomato cultivar Hawaii 7996. In C. Allen, P. Prior, & A. C. Hayward (Eds.), *Bacterial Wilt Disease and the Ralstonia solanacearum Species Complex* (pp. 285-291). St. Paul, MN: APS Press.

Vasse, J., Frey, P., & Trigalet, A. (1995). Microscopic studies of intercellular infection and protoxylem invasion of tomato roots by *Pseudomonas solanacearum. Molecular Plant-Microbe Interactions, 8,* 241-251.

Vasse, J., Genin, S., Frey, P., Boucher, C., & Brito, B. (2000). The *hrpB* and *hrpG* regulatory genes of *Ralstonia solanacearum* are required for different stages of the tomato root infection process. *Molecular Plant-Microbe Interactions, 13,* 259-267.

Villa, J., Tsuchiya, K., Horita, M., Natural, M., Opina, N., & Hyakumachi, M. (2003). DNA analysis of *Ralstonia solanacearum* and related bacteria based on 282-bp PCR-amplified fragment. *Plant Disease, 87,* 1337-1343.

Villa, J. E., Tsuchiya, K., Horita, M., Opina, N., & Hyakumachi, M. (2005). Phylogenetic relationships of *Ralstonia solanacearum* species complex strains from Asia and other continents based on 16S rDNA, endoglucanase, and *hrpB* gene sequences. *Journal of General Plant Pathology, 71,* 39-46.

Vogel, J. P., Raab, T. K., Somerville, C. R., & Somerville, S. C. (2004). Mutations in *PMR5* result in powdery mildew resistance and altered cell wall composition. *Plant Journal, 40,* 968-978.

Wallis, F. M. & Truter, S. J. (1978). Histopathology of tomato plants infected with *Pseudomonas solanacearum*, with emphasis on ultrastructure. *Physiological Plant Pathology, 13,* 307-317.

Wang, J.-F., Hanson, P., & Barnes, J. A. (1998). Worldwide evaluation of an international set of resistance sources to bacterial wilt in tomato. In Ph. Prior, C. Allen, & J. Elphinstone (Eds.), *Bacterial Wilt Disease: Molecular and Ecological Aspects* (pp. 269-275). Berlin: Springer-Verlag.

Wang, J.-F., Olivier, J., Thoquet, P., Mangin, B., Sauviac, L., & Grimsley, N. H. (2000). Resistance of tomato line Hawaii7996 to *Ralstonia solanacearum* Pss4 in Taiwan is controlled mainly by a major strain-specific locus. *Molecular Plant-Microbe Interactions, 13,* 6-13.

Weller, S. A., Elphinstone, J. G., Smith, N. C., Boonham, N., & Stead, D. E. (2000a). Detection of *Ralstonia solanacearum* strains with a quantitative, multiplex, real-time, fluorogenic PCR (TaqMan) assay. *Applied and Environmental Microbiology, 66,* 2853-2858.

Weller, S. A., Elphinstone, J. G., Smith, N. C., & Stead, D. E. (2000b). Detection of *Ralstonia solanacearum* from potato tissue by post-enrichment TaqMan PCR. *EPPO/OEPP Bulletin, 30,* 381-383.

White, F. F., Yang, B., & Johnson, L. B. (2000). Prospects for understanding avirulence gene function. *Current Opinion in Plant Biology, 3,* 291-298.

Whitfield, C. (1995). Biosynthesis of lipopolysaccharide O-antigens. *Trends in Microbiology, 3,* 178-185.

Williamson, L., Nakaho, K., Hudelson, B., & Allen, C. (2002). *Ralstonia solanacearum* race 3, biovar 2 strains isolated from geranium are pathogenic on potato. *Plant Disease, 86,* 987-991.

Wullings, B. A., van Beuningen, A. R., Janse, J. D., & Akkermans, A. D. L. (1998). Detection of *Ralstonia solanacearum*, which causes brown rot of potato, by fluorescent in situ hybridization with 23S rRNA-targeted probes. *Applied and Environmental Microbiology, 64,* 4546-4554.

Yabuuchi, E., Kosako, Y., Oyaizu, H., Yano, I., Hotta, H., Hashimoto, Y. et al. (1992). Proposal of *Burkholderia* gen. nov. and transfer of seven species of the genus *Pseudomonas* homology group II to the new genus, with the type species *Burkholderia cepacia* (Palleroni and Holmes 1981) comb. nov. *Microbiology and Immunology, 36,* 1251-1275.

Yabuuchi, E., Kosako, Y., Yano, I., Hotta, H., & Nishiuchi, Y. (1995). Transfer of two *Burkholderia* and an *Alcaligenes* species to *Ralstonia* gen. nov.: proposal of *Ralstonia pickettii* (Ralston, Palleroni and Doudoroff 1973) comb. nov., *Ralstonia solanacearum* (Smith 1986) comb. nov. and *Ralstonia eutropha* (Davis 1969) comb. nov. *Microbiology and Immunology, 39,* 897-904.

Zehr, E. I. (1970a). Cultural, physiological and biochemical properties of isolates of Philippine *Pseudomonas solanacearum*. *Philippine Phytopathology, 6,* 29-43.

Zehr, E. I. (1970b). Strains of *Pseudomonas solanacearum* in the Philippines as determined by cross-inoculation of hosts at different temperatures. *Philippine Phytopathology, 6,* 44-54.

Zhang, H., Zhang, D., Chen, J., Yang, Y., Huang, Z., Huang, D. et al. (2004). Tomato stress-responsive factor TSRF1 interacts with ethylene responsive element GCC box and regulates pathogen resistance to *Ralstonia solanacearum*. *Plant Molecular Biology, 55,* 825-834.

Zhang, Z., Coyne, D. P., Vidaver, A. K., & Mitra, A. (1998). Expression of human lactoferrin cDNA confers resistance to *Ralstonia solanacearum* in transgenic tobacco plants. *Phytopathology, 88,* 730-734.

12. AFFILIATION

Tim Denny, *University of Georgia, Department of Plant Pathology, Plant Science Bldg., Athens, GA 30602-7274, Tel: 706-542-1282; Fax: 706-542-1262; E-mail: TDenny@uga.edu*

RONALD GITAITIS AND CLAUDIA NISCHWITZ

BURKHOLDERIA

Abstract. *Burkholderia* are bacteria that were recognized as members of the "nonfluorescent pseudomonads". As classification of bacteria evolved over time and as improvements in molecular tools for bacterial identification occurred, they became members of the rRNA homology group II. They were reclassified under the new genus *Burkolderia* with *B. cepacia* as the type species. As of the date of preparing this chapter, a total of 41 species and four proposed species (*candidatus*) have been placed in the genus Burkholderia. We describe in detail, the six known species of this genus that are pathogenic to plants.

1. INTRODUCTION

The origins of the genus *Burkholderia* can be traced back to the early reports of the onion (*Allium cepa*) pathogens *B. gladioli* and *B. cepacia* and the carnation (*Dianthus caryophyllus*) pathogen, *B. caryopyilli* (17,18). Walter Burkholder of Cornell University first published the description of the slippery skin of onion pathogen, *Phytomonas alliicola*, and the bacterial wilt of carnation pathogen, *P. caryophylii*, in 1942 (17), and later he described the sour skin of onion pathogen, *Pseudomonas cepacia*, in 1950 (18). Eventually *Phytomonas alliicola* and *Phytomonas caryophylli* were reclassified within the genus *Pseudomonas*. In the past, the genus *Pseudomonas* was somewhat of a dumping ground for bacteria that were grouped on the basis of the visual examination of characteristics such as cell morphology, Gram reaction, flagella type, and if it were a strict aerobe.

Bacteriologists recognized that there was great diversity and numerous species within the genus *Pseudomonas* and over the years began sorting the different groups into new genera and eventually came to grips with defining a bacterial species. Some of the earliest and most definitive work clarifying the situation was done by a group of researchers from the University of California at Berkeley (10,98,122). Stanier et al. (122) published a comprehensive and thorough study of many phenotypic traits of members of the genus *Pseudomonas*. Ballard et al. (10) indicated that *P. pseudomallei, P. mallei, P. cepacia, P. marginata* (syn. *P. gladioli*), and *P. caryophylli* comprised a major and isolated cluster within the pseudomonads and tentatively termed these species as the "pseudomallei" group. Later, the genus *Pseudomonas* was subdivided into five rRNA homology groups, which were based on rRNA and DNA homologies (98). They further subdivided the groups based on DNA homology. Group I contained the true fluorescent pseudomonads and possessed three subgroups. Groups II and III contained many of the nonfluorescent pseudomonads. Group II contained the species that eventually would be classified within the genus *Burkholderia,* and contained two subgroups. One subgroup contained the bacteria *P. solanacearum* and *P. pickettii*, destined to become *Ralstonia*, and the other DNA subgroup contained members of the

S.S. Gnanamanickam (ed.), Plant-Associated Bacteria, 645–670.

pseudomallei cluster, namely *P. caryophylli*, *P. cepacia*, *P. gladioli*, *P. mallei*, and *P. pseudomallei*. For a number of years these bacteria continued to be recognized as members of the "nonfluorescent" pseudomonads. To identify a nonfluorescent pseudomonad, it first had to fit the general requirements for the genus *Pseudomonas*, i.e., gram-negative rod, aerobic, catalase positive, "usually" oxidase positive, and a G+C content of DNA between 58-70 moles %. The non-fluorescent subgroup, in addition to not producing a water-soluble, diffusible, fluorescent pigment, was largely determined by the ability to accumulate poly-beta-hydroxybutyrate (PBHB) as a carbon reserve, (96). Most strains within the DNA subgroup containing the *pseudomallei* cluster could grow at 41°C, and use arginine and betaine as a sole carbon source (96). On the other hand, strains within the DNA subgroup containing *P. solanacearum* did not grow at 41°C, and could not use arginine or betaine. Sands et al.(108) studied just the phytopathogenic pseudomonads and divided them into five groups. In their classification, the nonfluorescent pseudomonads capable of accumulating PBHB were placed in groups III and IV. Based on their criteria for separation, *B. caryophyli* was placed in group III and *B. cepacia* in group IV.

As was the case in many instances when phenotypic traits were used to define bacterial species characteristics, there were exceptions to the rule and misconceptions to avoid. For example, there were nonfluorescent strains of *Pseudomonas* species classified as fluorescent pigment producers (110). Also, there were some non-fluorescent strains, e.g. *P. cepacia*, that produced pigments in some growth media that appeared similar in color to the fluorescent pigments but when examined under ultra-violet light failed to fluoresce. The 8th edition of *Bergey's Manual of Determinative Bacteriology* (57) attempted to simplify *Pseudomonas* taxonomy. Many bacteria formerly recognized as *Pseudomonas* nomenspecies, including some destined to be classified as *Burkholderia*, were not included in the principal key. In that version of *Bergey's Manual,* one of the criteria for the genus *Pseudomonas* was that it did not produce sheathed flagella, but in a later publication of *Bergey's Manual of Systematic Bacteriology* in 1984 they classified *P. andropogonis* (syn. *P. stizolobii*) within the genus *Pseudomonas* even though it possessed a sheathed flagellum (36,96). Additional confusion could be generated regarding the accumulation of PBHB. Many of the organisms in RNA homology group II cannot hydrolyze PBHB extracellularly and thus cannot use PBHB as a carbon source. Different enzymes are required for the extracellular use of PBHB as a substrate than those for use of PBHB synthesized as a cellular reserve. Thus, a bacterium could accumulate PBHB and use it as an internal reserve but might not be able to use it as an external carbon source. The demonstration of the accumulation of PBHB generally was by microscopic examination of bacterial smears stained with Sudan black B and counter-stained with Safranin O. The PBHB would appear as a black granule inside of the pink bacterial cell.

The classification of bacteria evolved over time as science and technology progressed. A generally accepted definition of a species is that it is a category of taxonomic classification ranking below genus and consists of related organisms capable of successful interbreeding, i.e., with the production of fertile offspring. Since bacteria reproduce by an asexual process, cell fission, the concept of classifying bacteria into genera and species faced difficulties. Over time, using morphological, physiological and biochemical characteristics, taxonomists decided that out of hundreds of characteristics evaluated, 80% of the characters should be common to a species. As improvements in molecular biology occurred, the identification of a species began to be based on homology of the 16S rDNA gene and DNA-DNA hybridization, where ≥ 70% binding was considered as the value defining a species. Using both the 16S sequences, DNA-DNA hybridization, fatty acid analysis, and several phenotypic characteristics, Yabuuchi et al. (151) proposed that members of the rRNA homology group II should be reclassified under the new genus *Burkholderia* and designated *B. cepacia* as the type species. As of the date of preparing this chapter, a total of 41 species and four proposed species (Table 1) have been placed in the genus *Burkholderia*.

Table 1. Recognized species within the genus *Burkholderia*.

Burkholderia *species*	*Reference*
Burkholderia ambifaria	29
B. andropogonis	27, 117
B. anthina	135
B. brasilensis	9
B. caledonica	28
B. caribensis	1
B. caryophylli	17
B. cenocepacia	136
B. cepacia	18, 151
B. cocovenenans	156
B. dolosa	143
B. fungorum	28
B. gladioli	17
B. glathei	144
B. glumae	75
B. graminis	144
B. hospita	44
B. kururiensis	155
B. mallei	151
B. multivorans	137
B. norimbergensis	147
B. phenazinium	144
B. phenoliruptrix	25
B. phymatum	134
B. phytofirmans	113

B. plantarii	6
B. pseudomallei	10
B. pyrrocinia	144
B. sacchari	14
B. singaporensis	120
B. sordidicola	80
B. stabilis	138
B. terricola	44
B. thailandensis	15
B. tropica	104
B. tuberum	134
B. ubonensis	150
B. unamae	20
B. vandii	26, 134
B. vietnamiensis	42
B. xenovorans	45
Candidatus Burkholderia calva	141
Candidatus Burkholderia kirkii	140
Candidatus Burkholderia nigropunctata	141
Candidatus Burkholderia verschuerenii	139

Candidatus is placed before the genus of a bacterium that has not been formally named. It is commonly used when a species or genus is well characterized, but has not been cultured.

The scope of this chapter encompasses only the plant pathogenic species within the genus *Burkholderia* and does not cover the many other species within the genus. The following are known to be plant pathogens: *B. andropogonis, B. caryophylli, B. cepacia, B. gladioli, B. glumae and B. plantarii.*

2. BURKHOLDERIA ANDROPOGONIS (SMITH 1911) STAPP 1928 GILLIS *ET AL.*, 1995

Historical Names and Synonyms

Aplanobacter stizolobii Wolf 1920

Bacterium andropogonis Smith 1911
Bacterium stizolobii (Wolf) McCulloch 1920
Bacterium woodsii Smith 1911
Phytobacterium andropogonis (Smith) Magrou & Prévot 1948
Phytobacterium stizolobii (Wolf) Magrou & Prévot 1948
Phytobacterium woodsii (Smith) Magrou & Prévot 1948
Phytomonas andropogonis (Smith) Bergey et al., 1930
Phytomonas stizolobii (Wolf) Bergey et al., 1930
Phytomonas woodsii (Smith) Bergey et al., 1930

Pseudomonas andropogonis pv. stizolobii (Wolf) Palleroni 1984
Pseudomonas andropogonis (Smith 1911) Stapp 1928
Pseudomonas stizolobii (Wolf) Stapp 1935
Pseudomonas woodsii (Smith 1911) Stevens 1925

Burkholderia andropogonis probably has the widest host range of all the *Burkholderia* species, and was first described as causing a disease of sorghum (*Sorghum bicolor*), also known as broom corn, in 1905 (117). They indicated that the disease began on lower leaves as reddish-colored stripes that coalesced causing large blighted areas of the leaf. If conditions remained favorable, the disease progressed to the top of the plant and caused defoliation of the lower portion of the plant. The causal organism was a strictly aerobic, non-spore-forming, motile, gram-negative, rod-shaped bacterium. They further stated that the red 'stain' of the sorghum leaves was a host reaction, as the bacterium produced white colonies on all the substrates they tested. Smith (116) further characterized the causal agent and eventually named it *Bacterium andropogoni.* The species name was derived from the name of the host, because at the time, the scientific name of broom corn was *Andropogon sorghum*, and the name *Andropogon* refers to a bearded man. Other synonyms of sorghum included, *Sorghum saccharatum*, *Holcus sorghum* and *Holcus bicolor.*

In 1920, Wolf (148) described a bacterial leaf spot of velvet bean (*Stizolobium deeringianum* Bort) that was first observed in 1916 in North Carolina. Initial symptoms were described as small, point-like, translucent areas, then as lesions developed, the center areas became dark brown and the translucent character disappeared except at the lesion periphery. However, Wolf went on to state that mature lesions lacked water-soaked borders that are so characteristic of bacterial leaf spot diseases(148). Some chlorosis was observed in areas surrounding the lesions and as disease progressed, the chlorosis more or less encompassed the leaf. Mature spots were angular, limited by veins for the most part and ranged from 2 mm to 8 mm in diameter. He characterized the bacterium as non-motile and named it *Aplanobacter stizolobii.* In 1957, Burkholder (19) indicated that *P. stizolobii* in addition to causing a leaf spot on velvet bean also caused a leaf spot disease of white clover (*Trifolium repens*). Added to the host range of diseases caused by *P. stizolobii* were *Bougainvillea* spp. (105) and common vetch (*Vicia sativa*) (3) in Rhodesia in 1964 and in Australia in 1970, respectively. A comparative study of *P. andropogonis, P. alborpreciptans* (syn. *A. avenae*), and *P. stizolobii* by Goto and Starr (46) indicated that *P. andropogonis* and *P. stizolobii* were synonymous and recommended that *P. andropogonis* (syn. *B. andropogonis*) be retained as the proper epithet. Results from Gitaitis et al. (43) supported the data that *P. andropogonis* and *P. stizolobii* were synonymous when they characterized a bacterial pathogen from white clover in Georgia and found that it was pathogenic to both sorghum and velvet bean. They also found that arrowleaf clover (*T. vesiculosum*), crimson clover (*T. incarnatum)*, red clover (*T. pratense*) and subterraneum clover (*T. subterraneum*) were suitable hosts for *B. andropogonis.*

Ullstrup (132) reported a bacterial stripe disease of corn (*Zea mays*) caused by *B. andropogonis* in Indiana and compared it to the bacterial blight and stalk rot caused by *A. avenae* and concluded that the two corn pathogens were distinct. He described the disease as being associated with periods following high temperatures and "rainy weather" which increased levels of water congestion in the leaves, similar to conditions that favor *A. avenae*. Furthermore, the symptoms *B. andropogonis* produces in corn are not reddish as in sorghum and are easy to confuse with symptoms caused by *A. avenae* subsp. *avenae*. Primary symptoms in corn are amber-colored to tan stripes, which initially appear translucent and water-soaked. In a few rare cases, highly susceptible varieties develop chlorotic stripes and bleaching of the upper leaves. In addition to corn and sorghum, other hosts within the grasses include Sudangrass (*Sorghum X drummondii*) and Johnsongrass (*Sorghum halepense*).

Both *B. andropogonis* and *A. avenae* subsp. *avenae* were members of the nonfluorescent, pseudomonad, RNA homology group II and appear quite similar on standard microbiological growth media such as nutrient agar. Prior to using nucleic acid homologies and 16S sequencing, the primary means to separate these two bacteria were by the inability of *B. andropogonis* to reduce nitrates to nitrites or have a positive oxidase reaction. Hale and Wilkie (50) also compared *B. andropogonis* with *A. avenae* subsp. *avenae*, as well as with the sugarcane pathogens *P. rubrisubalbicans* (syn. *Herbaspirillum rubrisubalbicans*) and *P. rubrilineans* and another sorghum pathogen, *P. sorghicola*. They concluded that *B. andropogonis* was a distinct species from the others. Based on their tests they also concluded that *P. rubrilineans* was distinct from *P. alboprecipitans*, but we now know that they are both synonymous with *A. avenae* subsp. *avenae*. The name *P. sorghicola* has been lost and no longer recognized as a viable culture representing this species is not known to exist, thus the name was not included on the *Approved Lists Of Bacterial Names* (115).

Although most frequently a pathogen of members of the Gramineae and the Leguminosae, other hosts have been reported for *B. andropogonis*. A carnation leaf spot pathogen *P. woodsii* was described in 1911 (116) that eventually was placed in the rDNA group II. This pathogen was distinct from *B. caryophylli* and eventually was shown to be synonymous with *B. andropogonis* (27). The Belgian research group used both phenotypic data (42) and a polyphasic approach (27) comparing whole-cell proteins by electrophoresis, cellular fatty acid composition by gas chromatography, DNA–DNA binding values and 16S rDNA sequences to show that *P. woodsii* and *B. andropogonis* were synonymous. Nishiyama et al. (93) demonstrated that *P. woodsii* and *B. andropogonis* had an indistinguishable host range when isolates from tulip (*Tulipa gesneriana*), carnation, and sorghum demonstrated reciprocal cross-infectivity. In carnation, the bacterium causes lesions that are difficult to distinguish from several fungal pathogens. Lesions generally are pale brown with purplish concentric rings. Lesion margins frequently display water-soaking, which are more readily visible with transmitted light. Bacteria may ooze out on to the leaf surface under conditions of high humidity. As in corn, the progress of the disease is from the lower leaves upwards. Under favorable conditions, lesions can occur on carnation stems and flower buds.

Terete *Vanda* orchids were also reported as hosts for *B. andropogonis* in Hawaii in 1964 where it produced a firm, dark-brown rot that progressed from the leaf tip to the sheath and eventually caused defoliation (95). In their host range study, Oshiro et al. (95) demonstrated that in addition to the terete *Vanda* orchid, their strain of *B. andropogonis* was pathogenic to *Phalaenopsis* spp., *Dendrobium* spp., strap leaf *Vanda*, sugarcane (*Saccharum officinarum*) and corn. More recently, Takahashi et al. (125) reported isolating *B. andropogonis* from *Odontoglossum*, *Odontioda*, *Odontocidium*, and *Vuylstekeara* orchids in Japan. They described typical symptoms as being dark brown or black leaf spots with yellow haloes. Interestingly, they found that their strains were also pathogenic to *Phalaenopsis* orchids and tulips but not on white clover or corn. Other hosts reported include coffee (*Coffea arabica*), carob (*Ceratonia siliqua*), statice (*Limonium sp*), white-flowered wandering Jew (*Tradescantia* fluminensis), purple heart (*Tradescantia pallida 'Purpurea'*), hormigo (*Triplaris felipensis*), queen's spiderwort (*Dichorisandra reginae*), blueberry (*Vaccinium* sp), baby's breath (*Gypsophila paniculata*), annual baby's breath (*G. elegans*), inch plant (*Zebrina pendula*), and *Strelitzia sp.*

This bacterium is characterized as being a gram-negative rod with one and sometimes two polar flagella and the GC content of the DNA is 57.8 moles % (13). A sheath surrounding the flagellum has been reported (36). It is a strict aerobe, PBHB is accumulated as a cellular reserve, and the bacterium produces no extracellular fluorescent pigments. The majority of the strains are negative for the following: starch hydrolysis, gelatin hydrolysis, reduction of nitrates to nitrites, lipolysis, lecithinase, arginine dihydrolase, pectolytic activity, and oxidase. Whereas, phosphatase, urease, phenylalanine deaminase and catalase are usually positive. Acid is produced when the bacterium is grown in the presence of adonitol, arabinose, fructose, galactose, glucose, glycerol, inositol, mannitol, mannose, rhamnose, sorbitol, and xylose. Acid is not produced when it is stabbed into cellobiose, dulcitol, erythritol, glycogen, inulin, maltose, melibiose, raffinose salicin, sorbose, and sucrose (13,43).

In addition to characterizing the bacterium by classical methods listed above, *B. andropogonis* has been identified by a dot-immunobinding assay, immuno-fluorescence, enzyme-linked immunosorbent assay, cellular protein profiles, monoclonal antibodies and commercial identification systems such as substrate utilization patterns by Biolog (Hayward, CA) and fatty acid analysis by MIDI, Inc. (Newark, DE) (2,27,28,79,109). Species-specific primers for PCR that produce an amplicon of 410-bp have been reported (7). Ramundo and Claflin (103) reported success in differentiating 30 strains of *B. andropogonis* from closely related bacteria using BOX primers and rep-PCR.

Principal means of controlling diseases caused by *B. andropogonis* include crop rotation, using clean seed and plants, using resistant cultivars, avoiding handling of plants in the presence of free moisture, and in some instances spraying with copper bactericides.

3. BURKHOLDERIA CEPACIA (PALLERONI AND HOLMES 1981) YABUUCHI *ET AL.*, 1993

Historical Names and Synonyms

Pseudomonas cepacia Burkholder 1950

Pseudomonas multivorans *Stanier* et al., *1966*

Pseudomonas kingii Jonsson, 1970

As stated above, *B. cepacia* was first isolated from onions and was shown to cause the disease sour skin. By happenstance the name was not included in the *Approved List of Bacterial Names* and lost official recognition by the rules of bacterial nomenclature (115). The name was revived by Palleroni and Holmes (97) a year later. Other names in the literature include *P. multivorans* (122) and *P. kingii* (65), but both were found to be synonymous with *P. cepacia* (10,114).

There are probably more publications about *B. cepacia* than all of the other *Burkholderia* species combined, and several excellent reviews (30,48,56, 87,100,102) provide more extensive coverage of the species then what will be presented in this chapter, which is primarily concerned with plant pathogenic strains. However, this species is extremely diversified and is considered both a pest and a resource for mankind, thus a brief discussion about the roles of *B. cepacia* other than that as a plant pathogen are warranted. It is one of the few plant pathogenic bacteria considered truly soilborne, but it is also considered a rhizosphere organism. It also has been known to inhabit natural water sources. In addition to being a plant pathogen, other strains have been reported as effective biocontrol agents (61,73,99,100,101,102), as plant growth promoting bacteria (21,32,91), and at least in one instance, a promoter of the rot of orange fruit by *Penicillium digitatum* (58). Then still, there are other strains categorized as bioremediation strains and have the potential to be used in cleansing the environment of hazardous waste (70,74,155). Furthermore, clinical strains were reported to be often associated with nosocomial infections and as a common pathogen of cystic fibrosis (CF) patients (29) and has been described as the 'cepacia syndrome' (60). A primary concern with CF patients contracting *B. cepacia* is that pulmonary colonization reduces survival by 50% due to a necrotizing pneumonia, which is sometimes accompanied by bacteraemia (47). Outbreaks of *B. cepacia* infections in the 1980's and 1990's were responsible for a significant number of deaths in cystic fibrosis patients world-wide. In all of these studies, the bacterium had the phenotypic characteristics highly similar to those of

B. cepacia, originally described as an onion pathogen. The importance of *B. cepacia* as a human pathogen has resulted in extensive research regarding the organism, including its taxonomic standing.

As the strains were studied and compared, patterns began to unfold which indicated there were groups within the species that generally correlated with their environmental niche. At first the different types were referred to as the *B. cepacia* complex and then eventually the different strains were grouped into nine different "genomovars", with the type species and the original onion pathogen categorized as genomovar I and clinical species classified as genomovar III. At present, each genomovar has been assigned species status (Table 2).

Table 2. Nine phylogenetically similar but genomically distinct species (genomovars) comprise the *Burkholderia cepacia* complex.

Species/(genomovar)	*Reference*
B. cepacia I)	151
B. multivorans (II)	137
B. cenocepacia (III)	136
B. stabilis (IV)	138
B. vietnamiensis (V)	42
B. dolosa (VI)	143
B. ambifaria (VII)	29
B. anthina (VIII)	135
B. pyrrocinia (IX)	135

Nonetheless, some studies (12) indicate that all clinical strains tested had the ability to macerate onion tissues to some degree. In that same study, the majority of clinical isolates were assigned to genomovar III (*B. cenocepacia*), but a few were assigned to genomovar I (*B. cepacia*), which contains the type species originally isolated from onion. Another study (137) showed that clinical strains were found to come from at least five different genomonvars, although the majority belonged to genomovar III. Balandreau et al. (8) and LiPuma et al. (82,83) found that despite the developing picture that human pathogenic and environmental strains of *B. cepacia* complex are distinguishable, e.g. clinical strains primarily belong to genomovar III, that members of genomovar III are frequently associated with plants and the environment. Several studies have attempted to group clinical strains and environmental strains into distinct categories based on phenotypic characterization. Optimum temperature for growth, products synthesized, substrates utilized (12), 16S ribosomal rDNA sequences, restriction endonuclease cutting of DNA, random amplified polymorphic DNA fingerprinting, Southern hybridization with gene probes (124), multilocus isoenzyme analysis, BIOLOG patterns (153), pectolytic activity and pathogenicity on onion slices (154) all have been used to differentiate strains within the *B. cepacia* complex. In a 2-year study (124), using *recA*- and 16S rDNA-targeted PCR assays, a number of genomovar III strains were found to survive in organic soils in onion fields. Analyzing the rep-PCR and randomly amplified polymorphic DNA (RAPD) data showed that several of these were genomovar III isolates and were identical to the strain most frequently recovered from sputum culture of the majority of patients with cystic fibrosis in the mid-Atlantic region of the US. They concluded that human pathogenic strains are not always distinct from environmental strains, and reiterated the concern over the commercial use of *B cepacia* complex species in agriculture and bioremediation. That debate stresses the need for further studies to better define the preferred and specific natural habitats of each *B cepacia* complex species.

The extensive diversity found in the *B cepacia* complex might be related to its large genome, as an individual cell can contain up to four large circular pieces of DNA (bacterial "chromosomes") and one or more plasmids (78). The total genome size can range from 4 to 9 Mb, which is more than twice the size of most bacteria, which commonly contain only one large bacterial chromosome. In addition to having a large genome, *B. cepacia* contains many insertion sequences (IS). The IS in *B. cepacia* have been implicated in genetic rearrangement, modification of neighboring genes, and to make use of foreign genes through the fusion of replicons (11,39). The huge genome, presence of plasmids and IS provide a basis for spontaneous evolutionary bursts accounting for rapid adaptation, the vast substrate utilization profile and extensive ecological diversity exhibited by the *B. cepacia* complex. The genome complexity may also cause difficulties in interpreting epidemiologic studies. While assessing the risk of rhizosphere strains to humans, Tabacchioni et al. (124) found that clinical strains used in their study had identical 16S rDNA sequences but were different from agricultural strains isolated from the rhizospheres of rice and corn. These results were interpreted as evidence of evolutionary divergence of the rhizophere isolates from their clinical counterparts. Similar conclusions were made by Yohalem and Lorbeer (154) when they concluded that clinical and plant pathogenic strains could be distinguished by correlating pathogenicity to onions with pectolytic activity. In contrast, they observed that clinical strains recovered from hospital environments were not pathogenic to onion. However, Vandamme et al. (137) reported that a small number of strains recovered from cystic fibrosis patients belonged to genomovar I, which is the home of the plant pathogenic strains. In a similar study comparing environmental and clinical strains from Thailand and Japan, Seo and Tsuchiya (112) concluded that 81% of the environmental strains were clustered in genomovar I. The remaining 19% fell in genomovar III (*B. cenocepacia*). When they grouped the strains from clinical sources they found that they grouped into four different genomovars, namely *B. cepacia, B. cenocepacia, B. stabilis*, and *B. vietnamiensis*. Holmes et al. (56) reported diversity among 16S rDNA sequences among a larger sample size of clinical strains and theorized that multiple genomes could account for varying sequences of the 16S gene. They further speculated that the transfer of genetic material between closely related species, e.g. the *B. cepacia* complex, is not only possible but is highly probable. This adds additional complexities to interpreting epidemiologic data. This was supported by the work of Mack et al. (86) who found *B. cepacia* IS in *B. pseudomallei* that belonged to the CF highly-transmissible lineage. Consequently, they concluded that even if environmental strains incapable of infecting humans could be identified, the distribution of clinical and environmental strains and the potential for gene interaction poses a threat and that *B. cepacia* strains in agriculture, beneficial or pest, pose a risk to human health.

As a plant pathogen, *B. cepacia* has a narrow host range. Mark et al. (87) list Chinese cabbage (*Brassica pekinensis*), leek (*Allium porrum*), onion (*A. cepa*) and shallot (*A. cepa var ascalonicum*) as susceptible hosts. Bradbury (13) adds tomato (*Lycopersicon esculentum*) as a natural host and indicates that corn, bean (*Phaseolus vulgaris*) and pea (*Pisum sativum*) pods were susceptible upon artificial inoculation. Other reported hosts include, banana (76), mushroom (40) and *Cymbidium* orchid (128). As with the reported host range of any plant pathogenic bacterium, host range studies prior to the understanding of a hypersensitive response (71) or the accurate identification of bacterial species by molecular techniques should be viewed within the context of when they were conducted.

The main plant disease associated with *B. cepacia* is sour skin of onion. Primarily observed as a post-harvest disease, sour skin has been reported from onion-growing regions world-wide. Symptoms include a breakdown of one or more inner bulb scales, while adjacent outer scales and inner portions of the bulb may remain intact. Rotting scales appear pale yellow, reddish-brown to a darker brown decay. Outer scales, while remaining dry, may become a darker brown. In advanced stages, outer scales may slide off while handling. As scales are teased apart, a yellow "cheesy" mass of bacteria and onion tissue can be seen on the underside of the scale. As the name sour skin suggests, there is a distinct, pungent, "fermenty" odor associated with this disease.

As with many bacterial diseases, water-congested tissues may be predisposed to infection. In the case with sour skin, warm temperatures favor rapid disease development, but under marginal conditions the disease can develop slowly and require weeks for complete destruction of the bulb. Generally, when temperatures exceed 30°C sour skin progresses quite rapidly and losses can be severe.

For the most part, onion leaves are relatively resistant to *B. cepacia*, but bacterial cells can translocate through water-congested intercellular spaces and move as either free cells or as small aggregates (69). The pattern of inner scales decaying with outer, fleshy scales remaining intact may be related to bacterial ingress through the neck or leaf blades (68). This led Burkholder (18) to speculate that the harvest practice of "topping", the clipping of foliage above the bulb, is a primary means of bacterial infection. However, wounds caused by insects, windblown soil particles or hail occurring in the field prior to harvest may also serve as an infection court (68). Lorbeer et al. (84) described a foliar phase in the field that they described as canker-like. These lesions were bleached, dry and restricted on the leaf-blade axil of the outer most leaves. Under conditions of high humidity and free moisture, the bacteria will continue to colonize the onion tissues and work their way into the bulb. However, when conditions are dry and humidity is low, the "canker" remains restricted. Differences in when and how the onion becomes infected will dictate if outer bulb scales can become infected and rot.

Kawamoto and Lorbeer (67) demonstrated a clear relationship between time for symptom expression and inoculum density. When they infiltrated high concentrations of bacterial inoculum in to the lacunar cavity of the leaf blade, they observed rapid wilting and death of the leaf within 24 hrs. They hypothesized that this may be a hypersensitive response. However, upon further analysis they inoculated leaves with inoculum suspensions adjusted so that initial intracellular populations were $\sim 1 \times 10^4$ CFU/ 0.5 cm leaf disc and found that *B. cepacia* multiplied to a population of $\sim 4.8 \times 10^9$ CFU/0.5 cm leaf disc. These levels would indicate that they were observing a typical compatible pathogenic relationship as the final bacterial populations were 100 to 1000 times higher than what would have been expected from a typical hypersensitive response (121).

Initial inoculum has been associated with soil and contaminated irrigation water. Teviotdale et al. (127) demonstrated that under the dry conditions found in California, furrow irrigation substantially reduces disease levels when compared to overhead sprinkler irrigation. While this may be an effective disease management strategy in arid or semi-arid regions, use of furrow irrigation is unlikely to be a significant option for onion growers in areas with abundant and frequent rain events. Other control measures include proper maturing, field-curing prior to topping, and rapid drying after harvest. There may be some benefit in using copper bactericides to reduce the foliar, "canker" phase of the disease. Preliminary studies using crop rotation (51) with certain specific crops and solarization (111) show promise in reducing bacterial levels in soils used with continuous onion cultivation. Haudenshield et al. (51) also showed that populations of *B. cepacia* increased in soils following a rotation with corn. It is interesting to note that *B. cepacia* has also been described as a plant growth promoting bacterium that improve corn productivity (142). Further studies showing that crop rotations have an effect on soilborne populations of *B. cepacia* are warranted, as several researchers have found (32,34) different genomovar types associated with the rhizosphere of corn roots. Different corn cultivars and soil types also influenced the genetic diversity within *B. cepacia* populations. It needs to be determined if crop rotations are impacting mutualistic and plant pathogenic strains equally.

Identification of a bacterium from a symptomatic plant, is always easier than when it comes from an environmental source, because the pathogen usually is present in high concentrations and the disease symptoms are a key starting point for identification. Classical biochemical and physiological tests that are helpful in identifying *B. cepacia* include: utilization of cellobiose, orthophthalate, penicillin

G, D-serine, L-threonine, and trehalose. Other soft-rot bacteria isolated from rotting onions are either members of the Enterobacteriacae such as *Erwinia* or *Pantoea* and can easily be distinguished from *B. cepacia* by their facultative use of glucose under anaerobic conditions. *Burkholderia cepacia* can be separated from *P. marginalis* or *P. viridiflava* by its lack of production of a water-soluble, fluorescent pigment when grown on King's medium B. Since, many strains of *B. cepacia* will produce a nonfluorescent yellow-green, diffusible pigment in that medium, it is always wise to check for fluorescence using an ultraviolet light source. Other useful tests include checking for pectolytic activity on a pectin gel medium and for rot in onion slices. The most difficult organism to differentiate it from is the slippery skin pathogen, *B. gladioli* pv. *alliicola*. However, most strains of *B. cepacia* will utilize gluterate, putrescine, levulinate and tryptamine but not mesaconate, nicotinate or D(-) tartrate within 7 days, whereas *B. gladioli* pv. *alliicola* usually will test the opposite when exposed to those substrates.

Recently developed methods to identify *B. cepacia* include whole cell fatty acid analysis (109), PCR (23,66) and hybridization with species-specific rRNA gene probes (77). Sasser (109) not only was able to identify the bacterium to species by fatty acid analysis but was also able to track specific strains in the environment and identify a source of industrial contamination. A number of different labs (23,66,82) have published specific oligionucleotide primers, mostly targeting the 16S or 23S rDNA genes, that are useful for identifying different genomovars of *B. cepacia*. Salles et al. (107) was successful in identifying strains of *B. cepacia* using a DNA-based PCR-denaturing gradient gel electrophoresis (DGGE). DGGE analyses of the PCR products obtained using specific primers for the 16S region showed that there were sufficient differences in migration behavior to distinguish the majority of the *Burkholderia* species tested.

Finally, growth and appearance on semi-selective media can also be useful in identifying *B. cepacia* (41,49,52,85,119,146,149). Choice of the most appropriate medium in environmental studies depends on the researcher's goals and may have to be empirically derived. For example, Tabacchioni et al. (123) found that a higher degree of genetic diversity was observed among strains isolated from PCAT medium (16,85) than among those isolated from TB-T medium (49). In their study, all strains isolated from TB-T medium were assigned to the *B. cepacia* species, whereas among PCAT isolates only 74% were assigned to the *B. cepacia* species. PCAT medium inhibits background microflora incapable of using azelaic acid and TB-T selects against bacteria that cannot use L-asparagine and are sensitive to tetracycline. In contrast, PCAT medium has been used successfully and has been the media of choice in several studies (51,111) with plant pathogenic strains.

4. BURKHOLDERIA CARYOPHYLLI (BURKHOLDER 1942) YABUUCHI *ET AL.*, 1993

Historical Names and Synonyms

Phytomonas caryophylli

Pseudomonas caryophylli

Burkholderia caryophylli causes bacterial wilt of carnation (*Dianthus caryophyllus*) and crown rot of statice (*Limonium* sp.). It was first reported in Washington State in 1941 (64). Infected carnation plants develop grayish-green leaves before they turn yellow and die. Vascular discoloration and root rotting can also be observed (94,64). Under cool conditions and low soil temperatures, the carnation stems crack longitudinally between the nodes on the lower part of the stem (33). Symptoms on statice plants infected with *B. caryophylli* include chlorotic and necrotic leaves, root decay, crown rot and eventual death of the plant (63). The pathogen is soilborne and infects through the roots and colonizes vascular tissues (94).

Differentiation of *B. caryophylli* from other *Burkholderia* species and bacterial genera can be done by substrate utilization patterns, using a commercially available kit, Biolog GN MicroPlate System (Biolog, Hayward, CA). Each kit consists of a microplate with 96 wells with each well containing a different carbon source, and computer software can be used to help differentiate *B. caryophylli* from other *Burkholderia* or *Pseudomonas* species.

Management of the disease is limited. The best control is to use non-infected cuttings and soil disinfestation by steam or chemicals. No cultivars have been found to be completely resistant to bacterial wilt but some cultivars seem to be less susceptible than others. In Japan, 277 carnation cultivars were tested for their resistance against bacterial wilt and three cultivars showed resistance with ≤20% of the plants wilting (94).

5. BURKHOLDERIA GLADIOLI (SEVERINI 1913) YABUUCHI *ET AL.,* 1993

Historical Names and Synonyms

Pseudomonas gladioli

Pseudomonas antimicrobica

Pseudomonas marginata

Bacterium marginatum

Burkholderia gladioli was first described in 1921 as the causal agent of a new disease on *Gladiolus* sp. (89). It has a diverse range of interests. Currently three pathovars are recognized. *Burkholderia gladioli* pv. *gladioli* infects among others *Gladiolus* spp., *Crocus* spp., *Freesia* spp., and *Iris* spp., as well as several ferns and orchids. On *Gladiolus* leaves, it causes necrotic, reddish lesions that enlarge and turn dark brown to black. On the corms, sunken lesions appear yellowish in color (106). Under moist and warm conditions, the disease spreads rapidly, resulting in a rot of leaves and corms and their subsequent collapse (89,106). On ferns, the bacterium causes leaf spots and blight (106). The bacterium is found all over the world including the United States, South Africa, Zimbabwe, Thailand, Australia, Germany, Italy, Finland, Japan and China.

Burkholderia gladioli pv. *alliicola* causes the disease slippery skin of onion. The disease was first reported in 1899 in the United States but has since been found in countries around the world including India, Indonesia, Thailand, Hungary, Spain and Australia (92). Except for a softening of the neck, symptoms may be absent at the beginning of the disease. When bulbs are cut longitudinally some of the fleshy scales may appear water-soaked. The rot progresses downward in the bulb and will eventually affect the entire bulb. As the disease progresses, the infected tissues may dry out and the bulb shrivels. The disease derives its name from the fact that the core of an infected bulb will slip out when pressure is applied to the outer scales when holding the bulb. The bacterium enters through wounds on leaves or bulbs in the field or after harvest. Hail or strong winds that damage the plants can lead to severe disease outbreaks (92).

Burkholderia gladioli pv. *agaricicola* is the newest pathovar described. It distinguishes itself from the other *Burkholderia* species in the fact that it does not attack plants but is a pathogen of cultivated mushrooms (*Agaricus bitorquis*) causing a soft rot of the sporocarp. At 28°C symptoms start to appear 13 hours after inoculation. Initial symptoms are small, depressed lesions on the pileus which progress fast. After 48 hours an extensive browning and a soft rot of the sporocarp are visible (4,81). There are other bacteria causing diseases on cultivated mushrooms but none cause a wet or soft rot. However, this *Burkholderia* species differs from other bacteria except for *Steptomyces* in the fact that it produces a family 19 chitinase (ChiB gene) (72). Chitin is an essential component of fungal cell walls.

Its potential as a biological control agent against fungal plant pathogens such as *Botrytis* spp. (22,145) and *Fusarium oxysporum* is interesting but may not be practical. The production of chitinase makes *Burkholderia gladioli* a successful competitor on the leaf surface. However, its application as a biological control agent is probably very limited. It is a plant pathogen that can cause severe losses in commercial crops and in addition has been found as a human pathogen in cystic fibrosis patients (24).

Differentiation between *B. gladioli* and other *Burkholderia* or *Pseudomonas* species and between the three pathovars is not easy. Several different approaches are possible depending on time and resources available. *Burkholderia gladioli* is the only known bacterial species other than *Streptomyces* where many strains produce family 19 chitinase (72). Using specific primers for the chiB gene, PCR could distinguish *B. gladioli* pathovars from other bacterial species and genera. PCR targeting the 16S-23S rDNA spacer region to distinguish *B. gladioli* from other *Burkholderia* sp. and other genera has also been effective (38).

Host range testing in addition to other methods can help distinguish between the three different pathovars. All three *B. gladioli* pathovars caused a soft rot of mushroom blocks whereas *B. gladioli agaricicola* caused no rot on onion slices or whole bulbs. Some strains of caused rotting of gladiolus corm slice but not whole corms (81). This again shows that there is variation between strains and species and a combination of tests is necessary to identify different *Burkholderia* species and pathovars.

Nutritional tests on the ability to utilize organic compounds as their sole nitrogen or carbon source can be used together with other methods such as host range testing to differentiate between *Burkholderia* species and between the three pathovars of *B. gladioli*. When comparing bacteria in several publications, variations between strains of the same species with regard to utilization of carbon and nitrogen sources were observed. Biolog and fatty acid analysis can be used to help differentiate *B. gladioli* from other *Burkholderia* or *Pseudomonas* species.

Siderophore typing is another potential tool for the identification of bacterial species. So far, only a few siderophores have been attributed to a few *Pseudomonas* species. To identify bacteria from other siderophore producing genera, including *Burkholderia*, a database or library of isoelectrofocusing (IEF) profiles of known siderophores needs to be developed. Bacteria produce siderophores to capture iron when they are iron-deprived. The siderophores/iron complexes are then taken up by the bacteria through specific membrane receptors. The receptors are specific to the structure of the siderophore molecule produced by a particular bacterial species. Siderophore molecules can belong to a particular group of chemical compounds whose molecular structure differs by the number and types of amino acids in the peptide chain depending on the bacterial species or they can belong to a group of compounds that are unique to a bacterial species (35,90).

A new technique, rapid extraction-HPLC, shows distinct peaks for *Burkholderia gladioli* and *B. plantarii* that are absent in B. *glumae* allowing the differentiation between these three species (88).

6. *BURKHOLDERIA GLUMAE* (KURITA AND TABEI 1967) URAKAMI *ET AL.*, 1994

Historical Names and Synonyms

Pseudomonas glumae

Burkholderia glumae causes bacterial seedling and grain rot in rice (*Oryza sativa*) and other grasses, resulting in severe losses in Japan, Korea and China (106). The bacteria reduce the germination rate by causing a soft rot of soaked seeds. Seedlings which emerge from infected seeds are smaller and have brown rotting sheaths. Symptoms on the grain range from small brown spots on the hull to the discoloration of the entire grain kernels (31). The bacteria produce the phytotoxin toxoflavin responsible for characteristic symptoms of bacterial grain rot on rice. Symptoms include a brown band on the palea and lemma, and stunted rice seedlings. *B. glumae* isolates that did not produce the toxin did not cause bacterial seedling or grain rot (59). In addition to the toxin, it is assumed that *B. glumae* produces enzymes that result in the rotting of invaded plant tissue.

Transmission occurs through latent infected seed. Rice seeds are soaked in water for several days and incubated for a day at 100% humidity to hasten germination before being sown in nursery boxes. The bacteria population on infected seeds increases during this time, and when sown in the nursery beds bacterial seedling rot occurs (53,54). Primary infection of panicles in the field occurs when the bacteria present on the leaf sheaths colonize the spikelets. Rice plants are susceptible for about 11 days after heading, and the spikelets are highly susceptible during flowering and until three days after flowering. For successful infection to occur, relative humidity has to exceed 95%. The bacteria enter the lemmata and paleae through stomata, and seedlings through wounds that occur when the first leaves and secondary roots emerge and multiply in intercellular spaces. Secondary infection in the field spreads from an infected plant to neighboring plants creating a focus in the field. Studies have shown that the further away plants are from the inoculum source, disease severity is less (5,129,130). The disease was first reported in Japan but is now found in several other rice growing countries in Asia including Korea, the Philippines, China and Sri Lanka as well as in Colombia (106).

Several different approaches for identification of *Burkholderia glumae* exist depending on time and resources available: An ELISA test kit (Agdia, Elkhart, IN, USA) is available for detection. Samples are dispensed into a microplate with 96 wells. A *B. glumae* specific coating antibody is added and the samples dry overnight. After following additional steps in the protocol and adding another antibody and enzyme conjugate, the addition of a substrate for the enzyme will result in a color reaction if the bacteria are *B. glumae*.

PCR, using primers specific for *Burkholderia glumae* that have been developed, is a good tool to distinguish *B. glumae* from other *Burkholderia* species and bacterial genera (38,126). A selective medium (S-PG) developed by Tsushima et al. (131) can be used together with other techniques to identify *B. glumae*. It cannot be used as the only tool because there are closely related bacteria, such as *A. avenae*, that also grow on this medium. Biolog can be used to help differentiate *B. glumae* from other *Burkholderia* or *Pseudomonas* species. Rapid extraction-HPLC, shows distinct peaks for *Burkholderia gladioli* and *B. plantarii* that are absent in B. *glumae* allowing the differentiation between these three species (88).

There are a few control strategies being investigated. Biological control using avirulent *B. glumae* strains as a seed treatment is being studied. The seeds are dipped into a bacterial solution before being sown into nursery boxes. The success of avirulent *B. glumae* to control bacterial seedling and grain rot depends on the avirulent strain used and the virulent strain present on the seed (37). Oxolinic acid, a quinoline derivative, shows good success against bacterial seedling rot when the seed are dipped in the chemical before being soaked in water (55). Spray applications 2 days after heading resulted in good control of the disease. However, Oxolinic acid-resistant *B. glumae* strains have already been recovered from rice seedlings (56) and other treatments such as biological control should be considered.

7. *BURKHOLDERIA PLANTERII* (AZEGAMI *ET AL., 1987*) URAKAMI *ET AL.,* 1994

Historical Names and Synonyms
Pseudomonas planterii

Burkholderia planterii causes bacterial seedling blight of rice and was first found in Japan in 1983 and described in 1987 (6). In the early stages of infection, the symptoms resemble those caused by *B. glumae* but then the seedlings become reddish-brown and dry without becoming rotted. Root growth of *B. planterii*-infected seedlings in contrast to *B. glumae* is strongly reduced causing the seedlings to lodge. *B. planterii* produces the toxin tropolone which is responsible for the symptoms observed on rice seedlings (6). The bacteria enter the seedlings on leaf sheaths and coleoptiles through stomata and wounds that occur when the first leaves and secondary roots emerge and multiply in intercellular spaces similar to *B. glumae* (5). Differentiation of *B. planterii* from other *Burkholderia* species and bacterial genera is possible with different techniques. Biolog, rapid extraction-HPLC and PCR allowing the differentiation of B. planteria from similar species (88.126). Physiological characteristics can be used to identify *B. planterii* but should used with caution as different strains from the same species can differ in some characteristics and lead to the wrong identification. The characters described here are based on *B. planterii* strains from one study. In contrast to other *Burkholderia* and *Pseudomonas* spp. tested, none of the *B. planterii* strains grew at 40°C and produced tropolone in a shake culture as well as reddish brown granules, which are thought to be tropolone derivatives, in Ayers agar supplemented with 1% glucose and 100ppm iron. In contrast to *B. glumae*, *B. planterii* was oxidase positive, and grew in Cohn's solution (6).

8. REFERENCES

Achouak, W., R. Christen, M. Barakat, M. H. Martel and T. Heulin. 1999. *Burkholderia caribensis* sp. nov., an exopolysaccharide-producing bacterium isolated from vertisol microaggregates in Martinique. *Int. J. Syst. Bacteriol. 49*: 787-794.

Akhtar, M.A. and L.E. Claflin. 1986. Detection of *Pseudomonas andropogonis* in plant tissue employing fluorescent antibody stain procedure. Pakistan *J Agric Res 7:* 319-322.

Allen, R.N., A.C. Hayward, W.J. Halliday and J. Fulcher. 1970. Bacterial blight of *Vicia sativa*: Aetiology of the disease and identification of the pathogen. *Aust. J. Biol. Sci. 23*: 597-606.

Atkey, P.T., T.R. Fermor and S.P. Lincoln. 1992. Electron microscopy of the infection process of rapid soft rot disease of the edible mushroom *Agaricus bitorquis*. *Mycol. Res. 96*: 717-722.

Azegami, K., K. Nishiyama and H. Tabei. 1988. Infection courts of rice seedlings with *Pseudomonas planterii* and *Pseudomonas glumae*. *Ann. Phytopathol. Soc. Jpn. 54*: 337-341.

Azegami, K., K. Nishiyama, Y. Watanabe, I. Kadota, A. Ohuchi, and C. Fukazawa. 1987. *Pseudomonas plantarii* sp. nov., the causal agent of rice seedling blight. *Int. J. Syst. Bacteriol. 37*: 144-152.

Bagsic, R.D., M. Fegan, X. Li and A.C. Hayward. 1995. Construction of species-specific primer for *Pseudomonas andropogonis* based on 16s rDNA sequences. *Lett Appl Microbiol 21*: 87-92.

Balandreau, J., V. Viallard, B. Cournoyer, T. Coenye, S. Laevens, and P. Vandamme. 2001. *Burkholderia cepacia* genomovar III is a common plant-associated bacterium. *Appl. Environ. Microbiol. 67:* 982-985.

Baldani, V.D.L., E. Oliveira, E. Balota, J.I. Baldani, G. Kirchhof and J. Döbereiner. 1997 *Burkhoderia brasilensis* sp. nov., uma espécie de bactéria diazotrófica endofítica. *An. Acad. brasi. Ci. 69*: 116.

Ballard, R.W., N.J. Palleroni,, M. Doudoroff, R.Y. Stanier and M. Mandel. 1970. Taxonomy of the aerobic pseudomonads: *Pseudomonas cepacia, P. marginata, P. allicola*, and *P. caryophylli. J. Gen. Microbiol. 60*: 199-214.

Barsomian, G. and T.G. Lessie. 1986. Replicon fusions promoted by insertion sequences on *Pseudomonas cepacia* plasmid pTGL 6. *Molecular and General Genetics 204*: 273-280.

Bevivino, A., C. Dalmastri, S. Tabacchioni, L. Chiarini, M.L. Belli, S. Piana, A. Materazzo, P. Vandamme and G. Manno. 2002. *Burkholderia cepacia* complex bacteria from clinical and environmental sources in Italy: Genomovar status and distribution of traits related to virulence and transmissibility. *J. Clin. Microbiol. 40*: 846-851.

Bradbury, J.F. 1986. P. 114-115 *In:* Guide to Plant Pathogenic Bacteria. C.A.B. International, Farnham House, Slough, UK.

Brämer, C.O., P. Vandamme, L.F. da Silva, J.G.C. Gomez, and A. Steinbüchel. 2001. *Burkholderia sacchari* sp. nov., a polyhydroxyalkanoate-accumulating bacterium isolated from soil of a sugar-cane plantation in Brazil. *Int. J. Syst. Evol. Microbiol. 51*: 1709-1713.

Brett, P.J., D. DeShazer, and D. E. Woods. 1998. *Burkholderia thailandensis* sp. nov., a *Burkholderia pseudomallei*-like species. *Int. J. Syst. Bacteriol. 48*: 317-320.

Burbage, D.A., M. Sasser and R.D. Lumsden. 1982. A medium selective for *Pseudomonas cepacia. Phytopathology 72*: 706.

Burkholder, W.H. 1942. Three bacterial plant pathogens: *Phytomonas caryophilli*, sp. n., *Phytomonas alliicola* sp. n., and *Phytomonas manihotis* (Arthaud-Berthet et Bondar) Viegas. *Phytopathology 32:* 141-149.

Burkholder, W.H. 1950. Sour skin, a bacterial rot of onion bulbs. *Phytopathology 40*: 115-117.

Burkholder, W.H. 1957. A bacterial disease of clover and velvet beans. *Phytopathology 47*: 48-50.

Caballero-Mellado, J., L. Martinez-Aguilar, G. Paredes-Valdez and P. Estrada-de los Santos. 2004. *Burkholderia unamae* sp. nov., an N2-fixing rhizospheric and endophytic species. *Int. J. Syst. Evol. Microbiol. 54*: 1165-1172.

Chiarini, L., A. Bevivino, S. Tabacchioni and C. Dalmastri. 1998. Inoculation of *Burkholderia cepacia, Pseudomonas fluorescens* and *Enterobacter* sp. on *Sorghum bicolor*: root colonization and plant growth promotion of dual strain inocula. *Soil Biol. Biochem. 30*: 81-87.

Chiou, A.L. and W.S. Wu. 2001. Isolation, identification and evaluation of bacterial antagonists against *Botrytis elliptica* on Lily. *J. Phytopath. 49*: 319-324.

Clode, F.E., M.E. Kaufman, H. Malnick and T.L. Pitt. 1999. Evaluation of three oligonucleotide primer sets in PCR for the identification of *Bukholderia cepacia* and their differentiation from *Burkholderia gladioli. J. Clin. Pathol. 52*: 173-176.

Coenye, T., M. Gillis and P. Vandamme. 2000. (*Pseudomonas antimicrobica* Attafuah and Bradbury 1990 is a junior synonym of *Burkholderia gladioli* (Severini 1913) Yabuuchi et al. 1993. *Int. J. Syst. Evol. Microbiol. 50*: 2135-2139.

Coenye, T., D. Henry, D.P. Speert, and P. Vandamme. 2004. *Burkholderia phenoliruptrix* sp. nov., to accommodate the 2,4,5-trichlorophenoxyacetic acid and halophenol-degrading strain AC1100. *Syst Appl Microbiol 27*: 623-627.

Coenye, T., B. Holmes, K. Kersters, J.R.W. Govan, and P. Vandamme. 1999. *Burkholderia cocovenenans* (van Damme et al. 1960) Gillis et al. 1995 and *Burkholderia vandii* Urakami et al., 1994 are junior subjective synonyms of *Burkholderia gladioli* (Severini 1913) Yabuuchi et al., 1993 and *Burkholderia plantarii* (Azegami et al., 1987) Urakami et al. 1994, respectively. *Int. J. Syst. Bacteriol. 49*: 37-42.

Coenye, T., Laevens, S., Gillis, M. & Vandamme, P. 2001. Genotypic and chemotaxonomic evidence for the reclassification of *Pseudomonas woodsii* (Smith 1911) Stevens 1925 as *Burkholderia andropogonis* (Smith 1911) Gillis et al. 1995. *Int J. Syst Evol Microbiol 51*: 183-185.

Coenye, T., S. Laevens, A. Willems, M. Ohlen, W. Hannant, J.R.W. Govan, M. Gillis, E. Falsen, and P. Vandamme. 2001. *Burkholderia fungorum* sp. nov. and *Burkholderia caledonica* sp. nov., two new species isolated from the environment, animals and human clinical samples. *Int. J. Syst. Evol. Microbiol. 51*: 1099-1107.

Coenye, T., E. Mahenthiralingam, D. Henry, J. J. LiPuma, S. Laevens, M. Gillis, D. P. Speert, and P. Vandamme. 2001. *Burkholderia ambifaria* sp. nov., a novel member of the *Burkholderia cepacia* complex including biocontrol and cystic fibrosis-related isolates. *Int. J. Syst. Evol. Microbiol. 51*: 1481-1490.

Coenye, T., E., P. Vandamme, J.R.W. Govan and J.J. LiPuma. 2001. Minireview: Taxonomy and Identification of the Burkholderia cepacia complex. *J. Clin. Microbiol. 39*: 3427-3436.

Cottyn, B., M.T. Cerez, M.F. Van Outryve, J. Barroga, J. Swings and T.W. Mew. 1996. Bacterial diseases of rice. I. Pathogenic bacteria associated with sheath rot complex and grain discoloration of rice in the Phillipines. *Plant Dis. 80*: 429-437.

Dalmastri, C., L. Chiarini, C. Cantale, A. Bevivino, and S. Tabacchioni. 1999. Soil type and maize cultivar affect the genetic diversity of maize root-associated *Burkholderia cepacia* populations. *Microb. Ecol. 38*: 273-284.

Dickey, R.S. and P.E. Nelson. 1963. *Pseudomonas caryophylli* in carnation. I. Relation of soil temperature to symptom expression. *Phytopathology 53*: 1237-1238.

Fiore, A., S. Laevens, A. Bevivino, C. Dalmastri, S. Tabacchioni, P. Vandamme and L. Chiarini. 2001. *Burkholderia cepacia* complex: Distribution of genomovars among isolated from the maize rhizosphere in Italy. *Environ Microbiol. 3*: 137-43.

Fuchs, R., M. Schäfer, V. Geoffroy and J.-M. Meyer. 2001. Siderotyping – A powerful tool for the characterization of pyoverdines. *Curr. Top. Med. Chem. 1*: 31-57.

Fuesrt, J.A. and A.C. Hayward. 1969. The sheathed flagellum of *Pseudomonas stizolobii*. *J. Gen. Microbiol. 58*: 239-245.

Furuya, N., T. Okamato, Y. Kori, N. Matsuyama and S. Wakimoto. 1991. Control of bacterial seedling rot of rice by avirulent strains of *Pseudomonas glumae*. *Ann. Phytopathol. Soc. Jpn. 57*: 371-376.

Furuya, N., H. Ura, K. Iiyama, M. Matsumoto, M. Takeshita and Y. Takanami. 2002. Specific oligonucleotide primers based on sequences of the 16S-23S rDNA spacer region for the detection of *Burkholderia gladioli* by PCR. *J. Gen. Plant Pathol. 68*: 220-224.

Gaffney, T.D. and T.G. Lessie. 1987. Insertion-sequence dependent rearrangement of *Pseudomonas cepacia* plasmid pTGL 1. *J. Bacteriol. 169*: 224-230.

Gill, W.M. and A.L.J. Cole. 1992. Cavity disease of *Agaricus bitorquis* caused by *Pseudomonas cepacia*. *Can. J. Microbiol. 38*: 394-397.

Gilligan, P.H., P.A. Gage, L.M. Bradshaw, D.V. Schidlow and B.T. DeCicco. 1985. Isolation medium for the recovery of *Pseudomonas cepacia* from respiratory secretions of patients with cystic fibrosis. *J. Clin. Microbiol. 22*: 5-8.

Gillis, M., V. Tran Van, R. Bardin, M. Goor, P. Hebbar, A. Willems, P. Segers, K. Kerstens, T. Heulin, and M. P. Fernadez. 1995. Polyphasic taxonomy in the genus *Burkholderia* leading to an amended description of the genus and proposition of *Burkholderia vietnamiensis* sp. nov. for N_2-fixing isolates from rice in Vietnam. *Int. J. Syst. Bacteriol. 45*: 274-289.

Gitaitis, R.D., J. Miller and H.D. Wells. 1983. Bacterial leaf spot of white clover in Georgia. *Plant Dis. 67*: 913-914.

Goris, J., W. Dejonghe, E. Falsen, E. De Clerck, B. Geeraerts, A. Willems, E. M. Top, P. Vandamme, and P. De Vos. 2002. Diversity of transconjugants that acquired plasmid pJP4 or pEMT1 after inoculation of a donor strain in the A- and B-horizon of an agricultural soil and description of *Burkholderia hospita* sp. nov. and *Burkholderia terricola* sp. nov. *Syst. Appl. Microbiol. 25*: 340-352.

Goris, J., P. De Vos, J. Caballero-Mellado, J. Park, E. Falsen, J.F. Quensen, III, J.M Tiedje, P. Vandamme. 2004. Classification of the biphenyl- and polychlorinated biphenyl-degrading strain $LB400_T$ and relatives *as Burkholderia xenovorans* sp. nov. *Int. J. Syst. Evol. Microbiol 54*: 1677-1681.

Goto, M. and M.P. Starr. 1971. A comparative study *of Pseudomonas andropogonis*, *P. stizolobii* and *P. alboprecipitans*. *Ann. Phytopathol. Soc. Jpn. 37*: 233-241.

Govan J.R.W. and V. Deretic. 1996. Microbial pathogenesis in cystic fibrosis: mucoid *Pseudomonas aeruginosa* and *Burkholderia cepacia*. *Microbiol. Rev. 60*: 539-574.

Govan, J., Hughes, J., Vandamme, P. 1996. *Burkholderia cepacia*: medical, taxonomic, and ecological issues. *J. Med. Microbiol. 45*: 395-407.

Hagedorn, C., W.D. Gould, D.R. Bardinelli and D.R. Gustavson. 1987. A selective medium for enumeration and recovery of *Pseudomonas cepacia* biotypes from soil. *Appl. Environ. Microbiol. 53*: 2265-2268.

Hale, C.N. and J.P. Wilkie. 1972. A comparative study of *Pseudomonas* species pathogenic to sorghum. N.Z. *J. Agric. Res. 15*: 448-456.

Haudenshield, J.S., N.A. Gundersheim, G.L. Mark and J.W. Lorbeer. 2003. Multi-seasonal population dynamics of *Burkholderia cepacia* in organic soils cropped to onion, and the influence of non-onion crops. *Phytopathology 93*: S34.

Henry, D.A., M.E. Campbell, J.J. LiPuma and D.P. Speert. 1997. Identification of *Burkholderia cepacia* from patients with cystic fibrosis and the use of a simple new selective medium. *J. Clin. Microbiol. 35*: 614-619.

Hikichi, Y. 1995. The spread of bacterial grain rot of rice and its control in the paddy fields. *J. Pestic. Sci. 20:* 329-331.

Hikichi, Y., H. Egami, Y. Oguri and T. Okuno. 1998. Fitness for survival of *Burkholderia glumae* resistant to oxolinic acid in rice plants. *Ann. Phytopathol. Soc. Jpn. 64*: 147-152.

Hikichi, Y., T. Okuno and I. Furusawa. 1995. Mode of action of oxolinic acid against bacterial seedling rot of rice caused by *Pseudomonas glumae*. III. Infection with *P. glumae* into plumules. *Ann. Phytopathol. Soc. Jpn. 61*: 134-136.

Holmes, A., J. Govan and R. Goldstein. 1998. Agricultural Use of *Burkholderia* (*Pseudomonas*) *cepacia*: A Threat to Human Health? *Emerging Infectious Diseases 4*: 221-227.

Holt, J.G. (ed). 1977. The Shorter Bergey's Manual of Determinative Bacteriology. Eighth Edition. Williams & Wilkins. *Baltimore*. 356 pp.

Huang, Y., B.J. Deverall and S.C. Morris. 1991. Promotion of infection of orange fruit by *Penicillium digitatum* with a strain of *Pseudomonas cepacia*. *Phytopathology 81*: 615-618.

Iiyama, K., N. Furuya, Y. Takanami and N. Matsuyama. 1995. A role of phytotoxin in virulence of *Pseudomonas glumae* Kurita et Tabei. *Ann. Phytopathol. Soc. Jpn. 61*: 470-476.

Isles, A., I. Maclusky, M. Corey. R. Gold, C. Prober, P. Fleming, and H. Levison. 1984. *Pseudomonas cepacia* infection in cystic fibrosis: an emerging problem. *J. Pediatr. 104*: 206-210.

Janisiewicz W. J. and J. Roitman. 1988. Biological control of blue mold and gray mold on apple and pear with *Pseudomonas cepacia*. *Phytopathology 78*: 1697-1700.

Janisiewicz, W.J., L. Yourman, J. Redman and L. Mahone. 1991. Postharvest control of blue mold and gray mold of apples and pears by dip treatment with pyrrolnitrin, a metabolite of *Pseudomonas cepacia. Plant Dis. 75*: 490-494.

Jones, J.B. and A.W. Engelhard 1984. Crown and leaf rot of statice incited by a bacterium resembling *Pseudomonas caryophylli. Plant Dis. 68*: 338-340.

Jones, L.K. 1941 Bacterial wilt of carnation. *Phytopathology 31*: 199.

Jonsson, V. 1970. Proposal of a new species *Pseudomonas kingii. Int. J. Syst. Bact. 20*: 255-257.

Karpati, F. and J. Jonasson. 1996. Polymerase chain reaction for the detection of *Pseudomonas aeruginosa, Stenotrophomonas maltiphilia* and *Burkholderia cepacia* in the sputum of patients with cystic fibrosis. *Molecular and Cellular Probes 10*: 397-403.

Kawamoto, S.O. and J.W. Lorbeer. 1972a. Multiplication of *Pseudomonas cepacia* in onion leaves. *Phytopathology 62*: 1263-1265.

Kawamoto, S.O. and J.W. Lorbeer. 1972b. Histology of onion leaves infected with *Pseudomonas cepacia. Phytopathology 62*: 1266-1271.

Kawamoto, S.O. and J.W. Lorbeer. 1974. Infection of onion leaves by *Pseudomonas cepacia. Phyotopathology 64*: 1440-1445.

Kilbane, J.J., D. K. Chatterjee, J.S. Karns, S.T. Kellogg and A.M. Chakrabarty. 1982. Biodegradation of 2:4,5-trichlorophenoxyacetic acid by a pure culture of *Pseudomonas cepacia. Appl. Environ. Microbiol. 44*: 72-78.

Klement, Z., G.L. Farkas and L. Lovrekovich. 1964. Hypersensitive reaction induced by phytopathogenic bacteria in the tobacco leaf. *Phytopathology 54*: 474-477.

Kong, H., M. Shimosaka, Y. Ando, K. Nishiyama, T. Fujii and K. Miyashita. 2001. Species-specific distribution of a modular family 19 chitinase gene in *Burkholderia gladioli. FEMS Microbiol. Ecol. 37*: 135-141.

Knudsen, G.R. and H.W. Spurr. 1987. Field persistence and efficiency of five bacterial preparations for control of peanut leaf spot. *Plant Dis. 71*: 442-445.

Krumme, M.L., K.N. Timmis, and D.F. Dwyer. 1993. Degradation of trichloroethylene by *Pseudomonas cepacia* G4 and the constitutive mutant strain G4 5223 PR1 in aquifer microcosms. *Appl. Environ. Microbiol. 59*: 2746-274.

Kurita, T., Tabei, H. (1967) [On the pathogenic bacterium of bacterial grain rot of rice.] *Annals of the Phytopathological Society, Japan 33*: 111.

Lee, Y.-A., Y.-Y. Shiao and C.-P. Chao. 2003. First report of *Burkholderia cepacia* as a pathogen of banana finger-tip rot in Taiwan. *Plant Dis. 87*: 601.

Leff, L.G., R.M. Kernan, J.V. McArthur, and L.J. Shimkets. 1995. Identification of aquatic *Burkholderia (Pseudomonas)* cepacia by hybridization with species-specific rRNA gene probes. *Applied and Environmental Microbiology 61*: 1634-1636.

Lessie, T.G., W. Hendrickson, B.D. Manning and R. Devereux. 1996. Genomic complexity and plasticity of *Burkholderia cepacia. Fems Microbiol. Lett. 144*: 117-28.

Li, X., W.C. Wong and A.C. Hayward. 1993. Production and use of monoclonal antibodies to *Pseudomonas andropogonis*. *J Phytopathol 138*: 21-30.

Lim, Y. W., K.S. Baik, S.K. Han, S.B. Kim and K.S. Bae. 2003. *Burkholderia sordidicola* sp. nov., isolated from the white-rot fungus *Phanerochaete sordida*. *Int. J. Syst. Evol. Microbiol. 53*: 1631-1636.

Lincoln, S.P., T.R. Fermor, D.E. Stead and J.E. Sellwood. 1991. Bacterial soft rot of *Agaricus bitorquis*. *Plant Path. 40*: 136-144.

LiPuma, J.J., B.J. Dulaney, J.D. McMenamin, P.W. Whitby, T.L. Stull, T. Coenye and P. Vandamme. 1999. Development of a RNA-based PCR assays for the identification of *Burkholderia cepacia* complex isolates recovered from cystic fibrosis patients. *J. Clin. Microbiol. 327*: 3167-3170.

LiPuma, J.J., T. Spilker, T. Coenye and C.F. Gonzalez. 2002. An epidemic complex strain of *Burkholderia cepacia* identified in soil. *The Lancet 359*: 2002-2003.

Lorbeer, J.W., N.A. Gundersheim and G.L. Mark. 1998. Present status of bacterial diseases of onion and their control in New York. P. 209-213 In: Proceedings of the 1998 National Onion (and other *Allium*) Research Conference. 10-12 December, Sacramento, California, USA. University of California, Davis, California.

Lumsden, R.D., M.D. Bowie and M. Sasser. 1986. Medium for isolation of *Pseudomonas cepacia* biotype from soil and the isolated biotype. U.S. Patent number 4,588,584. Issued 13 May, 1986. U.S. Patent Office, Washington, D.C.

Mack, K. and R.W. Titball 1998. The detection of insertion sequences within the human pathogen *Burkholderia pseudomallei* which have been identified previously in *Burkholderia cepacia. FEMS Microbiol. Letters 162*: 69-74.

Mark, G.L., R.D. Gitaitis and J.W. Lorbeer. 2002. Bacterial Diseases of Onion. p. 267-292 *In:* (Ed.) H.D. Rabinowitch and L. Currah. *Allium Crop Science: Recent Advances*. CABI Publishing, Oxon, U.K.

Matsuyama, N., Y. Ueda, K. Iiyama, N. Furuya, H. Ura, A.A. Khan. and M. Matsumoto. 1998. Rapid extraction-HPLC as a tool for presumptive identification of *Burkholderia gladioli, B. glumae* and *B. planterii*, causal agents of various rice diseases. *J. Fac. Agric., Kyushu Univ. 42*: 265-272.

McCulloch, L. 1921. A bacterial disease of *Gladiolus*. Science 54: 115-116.

Meyer, J.-M., V. Geoffroy, N. Baida, L. Garden, D. Izard, P. Lemanceau, W. Achouak and N.J. Palleroni. 2002. Siderophore typing, a powerful tool for the identification of fluorescent and non-fluorescent pseudomonads. *Appl. Environ. Mirobiol. 68*: 2745-2753.

Minerdi, D., R. Fani, R. Gallo, A. Boarino, and P. Bonfante. 2001. Nitrogen fixation genes in an endosymbiotic *Burkholderia* strain. *Appl. Environ. Microbiol. 67*: 725-732.

Mohan, S.K. 1995. Slippery skin. In: *Compendium of Onion and Garlic Diseases. p. 32*. APS Press.

Nishiyama, K., T. Kusaba, K. Ohta, K. Nahata and K. Ezuka. 1979. Bacterial black rot of tulip caused by *Pseudomonas andropogonis*. *Annals of the Phytopathol. Soc. Japan 45*: 668-674.

Onozaki, T., T. Yamaguchi, M. Mimeno and H. Ikeda. 1999. Evaluation of 277 carnation cultivars for resistance to bacterial wilt (*Pseudomonas caryophylli*). *J. Jpn. Soc. Hortic. Sci. 68*: 546-550.

Oshiro, L.S., R.B. Hine and S. Goto. 1964. The identification of *Pseudomonas andropogonis* as the cause of a firm rot disease of the terete *Vanda* orchid in Hawaii. *Plant Dis. Rep. 48*: 736-740.

Palleroni, N. J. 1984. Genus I. *Pseudomonas* Migula 1894, p. 141-199. *In* N. R. Krieg and J. G. Holt (ed.), Bergey's manual of systematic bacteriology, vol. 1. The Williams & Wilkins, Co., Baltimore, Md.

Palleroni, N.J. and B. Holmes. 1981. *Pseudomonas cepacia*, sp. nov., nom. rev. *Int. J. Syst. Bacteriol. 31*: 479-481.

Palleroni, N.J., R. Kunisawa, R. Contopoulou and M. Doudoroff. 1973. Nucleic acid homologies in the genus *Pseudomonas*. *Int. J. Syst. Bacteriol. 23*: 333-339.

Park, K.S, S.W. Jang, C.H. Kim and E.J. Lee. 1989. Studies on biological control of *Phytophthora* blight of red pepper: III Formulations of *Trichoderma harzianum* and *Pseudomonas cepacia* antagonistic to *Phytophthora capsici* and their preservation. Korean *J Plant Pathol.5*: 131-138.

100.Parke, J.L. 1998. *Burkholderia cepacia*: Friend or Foe? APSnet Feature, October 1 thru October 31, 1998. http://www.apsnet.org/education/feature/*Burkholderia cepacia.*

Parke, J.L., R. E. Rand, A. E. Joy, and E. B. King. 1991. Biological control of *Pythium* damping-off and *Aphanomyces* root rot of peas by application of *Pseudomonas cepacia* or *P. fluorescens* to seed. *Plant. Dis. 75*: 987-992.

Parke, J.L. and D. Gurian-Sherman. 2001. Diversity of the *Burkholderia cepacia* complex and implications for risk assessment of biological control strains. *Annu. Rev. Phytopathol. 39*: 225-258.

Ramundo, B.A. and L.E. Claflin. 2005. Identification of *Burkholderia andropogonis* with a repetitive sequence BOX element and PCR. *Current Microbiology 50*: 52-56.

Reis, V. M., P.E. de los Santos, S. Tenario-Salgado, J. Vogel, S. Guyon, M. Schmid, J.I. Baldani, A. Hartmann, and J. Caballero-Mellado. 2004. *Burkholderia tropica* sp. nov., a novel nitrogen-fixing, plant-associated bacterium. *Int. J. Syst. Evol. Microbiol. 54*: 2155-2162.

Rothwell, A. and A.C. Hayward. 1964. A bacterial disease of bougainvillea. Rhodesian *Journal of Agricultural Research 2*: 97-99.

Saddler, G.S. 1994. *Burkholderia gladioli* pv. *gladioli. Mycopathologia 128*: 57-58.

Salles, J.F., F.A. De Souza and J.D. van Elsas. 2002. Molecular method to assess the diversity of *Burkholderia* species in environmental samples. *Appl. Environ. Microbiol. 68*: 1595-1603.

Sands, D.C., M.N. Schroth and D.C. Hildebrand. 1970. Taxonomy of phytopathogenic pseudomonads. *J. Bacteriol. 101*: 9-23.

Sasser. M. 2001. "Tracking" a strain using the Sherlock Microbial Identification System (MIS). MIDI Technical Note #102. www.midi-inc.com

Schroth, M.N., D.C. Hildebrand and M.P. Starr. 1983. Phytopathogenic members of the genus *Pseudomonas*. p. 701-718 *In*: M.P. Starr (ed.). Selections From *The Prokaryotes*: *A Handbook On Habits, Isolation, And Identification Of Bacteria. Phytopathogenic Bacteria*. Springer-Verlag New York.

Seebold, K.W., H.F. Sanders, J.C. Diaz-Perez and R.D. Gitaitis. 2004. Effect of solarization on soilborne pathogens and weeds in short-day, sweet onion in Georgia. 2004 National Allium Research Conference. December 9-10, 2004. Two Rivers Convention Center, Grand Junction, Colorado.

Seo, S.-T. and K. Tsuchiya. 2004. PCR-based identification and characterization of *Burkholderia cepacia* complex bacteria from clinical and environmental sources *Letters Appl. Microbiol. 39*: 413-419.

Sessitsch, A. T. Coenye, A.V. Sturz, P. Vandamme, E.A. Barka, J.F. Salles, J.D. Van Elsas, D. Faure, B. Reiter, B.R. Glick, G. Wang-Pruski and J. Nowak. 2005. *Burkholderia phytofirmans* sp. nov., a novel plant-associated bacterium with plant-beneficial properties. *Int. J. Syst. Evol. Microbiol. 55*: 1187-92.

Sinsabaugh, H.A. and G. W. Howard. 1975. Emendation of the description of *Pseudomonas cepacia* Burkholder (synonyms *Pseudomonas multivorans* Stanier et al., *Pseduomonas kingii* Jonsson; EO-1 Group). *Int. J. Syst. Bacteriol. 25*: 187-201.

Skerman, V.B.D., V. McGowan and P.H.A. Sneath. 1980. Approved Lists Of Bacterial Names. *Int. J. Syst. Bacteriol. 30*: 225-420.

Smith, E.F. 1905-1911. *Bacteria In Relation To Plant Diseases*. V. 1-2, Washington, D.C. (Carnegie Inst. Wash. Publ. 27).

Smith, E.F. and F. Hedges. 1905. Burrill's bacterial disease of broom corn. (Abstr.) *Science 21*: 502-503.

Snell, J.J.S., L.R. Hill, S.P. Lapage, and M.A. Curtis. 1972. Identification of *Pseudomonas cepacia* Burkholder and its synonymy with *Pseudomonas kingii* Jonsson. *Int. J. Syst. Bacteriol. 22*:127-138.

Speert, D.P., D.A. Henry, M.E. Campbell and J.J. LiPuma. 1997. Identification of *Burkholderia cepacia* isolates from patients with cystic fibrosis and use of a simple new selective medium *J. Clin. Microbiol. 35*: 614-619.

Stabnikova, O., Chen, X., Ivanov, V. and Wang, J.-Y. 2003. AY197330 *Burkholderia singaporensis* strain Gs 16S ribosomal RNA gene, partial sequence. Direct Submission NCBI GenBank.

http://www.ncbi.nlm.nih.gov/entrez/viewer.fcgi?db=nucleotide&val=28629816.

Stall, R.E. and A.A. Cook. 1966. Multiplication of *Xanthomonas vesicatoria* and lesion development in resistant and susceptible pepper. *Phytopathology 56*: 1152-1154.

Stanier, R.Y., N.J. Palleroni and M. Doudoroff. 1966. The aerobic pseudomonads: A taxonomic study. *J. Gen. Microbiol. 43*: 159-271.

Tabacchioni, S., L. Chiarini, A. Bevivino, C. Cantale and C. Dalmastri. 2000. Bias caused by using different isolation media for assessing the genetic diversity of a natural microbial population. *Microb Ecol. 40*: 169-176.

Tabacchioni S, P. Visca, L. Chiarini, A. Bevivino, C. Di Serio, S. Francelli and R. Fani. 1995. Molecular characterization of rhizosphere and clinical isolates of *Burkholderia cepacia*. *Res. Microbiol. 146*: 531-42.

Takahashi, Y.K., Takahashi, K. Watanabe and T. Kawano. 2004. Bacterial black spot caused by *Burkholderia andropogonis* on *Odontoglossum* and intergeneric hybrid orchids. *J. Gen. Plant Pathol. 70*: 285-287.

Takeuchi, T., H. Sawada, F. Suzuki and I. Matsuda. 1997. Specific detection of *Burkholderia plantarii* and *B. glumae* by PCR using primers selected from the 16S-23S rDNA specer region. *Ann. Phytopathol. Soc. Jpn. 63*: 455-462.

Teviotdale, B.L., R.M. Davis, J.P. Guerard and D.H. Harper. 1989. Effect of irrigation management on sour skin of onion. *Plant Dis. 73*: 819-822.

Tsuchiya, K., T. Akhter, T. Takasaki, S. Tanaka. and S. Wakimoto. 1986. Bacterial brown spot disease of *Cymbidium* spp. caused by *Pseudomonas cepacia*. *Ann. Phytopathol. Soc. Jpn. 52*: 825-834.

Tsushima, S. 1996. Epidemiology of bacterial grain rot of rice caused by *Pseudomonas glumae*. *JARQ 30*: 85-89.

Tsushima, S., H. Naito and M. Koitabashi. 1995. Change in panicle susceptibility associated with flowering rate of spikelets in bacterial grain rot of rice caused by *Pseudomonas glumae*. *Ann. Phytopathol. Soc. Jpn. 61*: 109-113.

Tsushima, S., S. Wakimoto and S. Mogi. 1986. Selective medium for detecting *Pseudomonas glumae* Kurita et Tabei, the causal bacterium of grain rot of rice. *Ann. Phytopathol. Soc. Jpn. 52*: 253-259.

Ullstrup, A.J. 1960. Bacterial stripe of corn. *Phytopathology 50*: 906-910.

Urakami, T., C. Ito-Yoshida, H. Araki, T. Kijima, K. Suzuki, and K. Komagata. 1994. Transfer of *Pseudomonas plantarii* and *Pseudomonas glumae* to *Burkholderia* as *Burkholderia* spp. and description of *Burkholderia vandii* sp. nov. *Int. J. Syst. Bacteriol. 44*: 235-245.

Vandamme, P., J. Goris, W.-M. Chen, P. de Vos, and A. Willems. 2002. *Burkholderia tuberum* sp. nov. and *Burkholderia phymatum* sp. nov., nodulate the roots of tropical legumes. *Syst Appl Microbiol 25:* 507-512.

Vandamme, P., Henry, D., Coenye, T., Nzula, S., Vancanneyt, M., LiPuma, J. J., Speert, D. P., Govan, J. R. W. & Mahenthiralingam, E. 2002. *Burkholderia anthina* sp. nov. and *Burkholderia pyrrocinia*, two additional *Burkholderia cepacia* complex bacteria, may confound results of new molecular diagnostic tools. *FEMS Immun Med Microbiol 33*: 143-149.

Vandamme, P., B. Holmes, T. Coenye, J. Goris, E. Mahenthiralingam, J. J. LiPuma, and J. R. Govan. 2003. *Burkholderia cenocepacia* sp. nov.—a new twist to an old story. *Res. Microbiol. 154* : 91-96.

Vandamme, P., B. Holmes, M. Vancanney, T. Coenye, B. Hoste, R. Coopman, H. Revets, S. Lauwers, M. Gillis, K. Kersters, and J.-R. W. Govan. 1997. Occurrence of multiple genomovars of *Burkholderia cepacia* in cystic fibrosis patients and proposal of *Burkholderia multivorans* sp. nov. *Int. J. Syst. Bacteriol. 47*: 1188-1200.

Vandamme, P., E. Mahenthiralingam, B. Holmes, T. Coenye, B. Hoste, P. De Vos, D. Henry, and D. P. Speert. 2000. Identification and population structure of *Burkholderia stabilis* sp. nov (formerly *Burkholderia cepacia* genomovar IV). *J. Clin. Microbiol. 38*: 1042-1047.

Van Oevelen,S., R. De Wachter and E. Prinsen. 2005. AY277699 *Candidatus Burkholderia verschuerenii* 16S ribosomal RNA gene, complete sequence. http://www.ncbi.nlm.gov/entrez/viewer.fcgi?db=nucleotide&val=33622538.

Van Oevelen, S., R. De Wachter, P. Vandamme, E. Robbrecht and E. Prinsen, E. 2002. Identification of the bacterial endosymbionts inleaf galls of *Psychotria* (Rubiaceae, angiosperms) and proposal of '*Candidatus Burkholderia kirkii'* sp. nov. *Int. J. Syst. Evol. Microbiol. 52*: 2023-2027.

Van Oevelen, S., R. De Wachter, P. Vandamme, E. Robbrecht and E. Prinsen, E. 2004. *'Candidatus Burkholderia calva'* and *'Candidatus Burkholderia nigropunctata'* as leaf gall endosymbionts of African *Psychotria*. *Int. J. Syst. Evol. Microbiol. 54*: 2237-2239.

Velazquez, M., E. Ventura, A. Hernandez, S. Aguilar and A.N. Hernandez. 1999. The interaction between corn and *Burkholderia* (*Pseudomonas*) *cepacia*. *Rev. Latinam. Microbiol. 41*: 17-23.

Vermis, K.,T. Coenye, J.J. LiPuma, E. Maehenthiralingam, H.J. Nelis and P. Vandamme. 2004. Proposal to accommodate *Burkholderia cepacia* genomovar VI as *Burkholderia dolosa* sp. nov. *Int.J. Sys. Evol. Microbiol. 54*: 689-691.

Viallard, V., I. Poirier, B. Cournoyer, J. Haurat, S. Wiebkin, K. K. Ophel and J. Balandreau. 1998. *Burkholderia graminis* sp. nov., a rhizospheric *Burkholderia* species, and reassessment of *Pseudomonas phenazinium*, *Pseudomonas pyrrocinia* and *Pseudomonas glathei* as *Burkholderia*. *Int. J. Syst. Bacteriol. 48*: 549-563.

Walker, R., K.A. Emslie and E.J. Allan. 1996. Bioassay methods for the detection of antifungal activity by *Pseudomonas antimicrobica* against the grey mould pathogen *Botrytis cinerea*. *J. Appl. Microbiol. 81*: 531-537.

Welch, D.F., M.J. Muszynski, C.H. Pai, M.J. Marcon, M.M Hribar, P.H. Gilligan, J.M. Matsen, P.A. Ahlin, B.C. Hillman and S.A. Chartrand. 1987. Selective and differential medium for recovery of *Pseudomonas cepacia* from the respiratory tracts of patients with cystic fibrosis. *J. Clin. Microbiol. 25*: 1730-1734.

Wittke, R., W. Ludwig, S. Peiffer and D. Kleiner. 1997. Isolation and characterisation of *Burkholderia norimbergensis* sp. nov., a mildly alkaliphilic sulfur oxidizer. *Syst. Appl. Microbiol. 20*: 549-553.

Wolf, F.A. 1920. A bacteria leaf spot of velvet bean. Phytopathology 10: 73-80.

Wu, B.J. and S.T. Thompson. 1984. Selective medium for *Pseudomonas cepacia* containing 9-chloro-9-(4-diethylaminophenyl)-10-phenylacridan and polymyxin B sulfate. *Appl Environ Microbiol. 48*: 743-746.

Yabuuchi, E., Y. Kawamura, T. Ezaki, M. Ikedo, S. Dejsirilert, N. Fujiwara, T. Naka, and K. Kobayashi. 2000. *Burkholderia uboniae* sp. nov., L-arabinose-assimilating but different from *Burkholderia thailandensis* and *Burkholderia vietn amiensis*. *Microbiol. Immunol. 44*: 307-313.

Yabuuchi E., Y. Kosako, H. Oyaizu, I. Yano, H. Hotta, Y. Hashimoto, T. Ezaki, and M. Arakawa.1992. Proposal of *Burkholderia* gen. nov. and transfer of seven species of the genus *Pseudomonas* homology group II to the new genus, with the type species *Burkholderia cepacia* (Palleroni and Holmes 1981) comb. nov. *Microbiol. Immunol. 36*: 1251-1275.

Yabuuchi, E., Y. Kosako, I. Yano, H. Hotta, and Y. Nishiuchi. 1995. Transfer of two *Burkholderia* and an *Alcaligenes* species to *Ralstonia* gen. nov.: proposal of *Ralstonia pickettii* (Ralston, Palleroni and Doudoroff 1973) comb. nov., *Ralstonia solanacearum* (Smith 1896) comb. nov. and *Ralstonia eutropha* (Davis 1969) comb. nov. *Microbiol. Immunol. 39*: 897-904.

Yohalem, D. S. and J. W. Lorbeer. 1994. Multilocus isoenzyme diversity among strains of *Pseudomonas cepacia* isolated from decayed onions, soils, and clinical sources. *Syst. Appl. Microbiol. 17*: 116-124.

Yohalem, D. S., and J.W. Lorbeer. 1997. Distribution of *Burkholderia cepacia* phenotypes by niche, method of isolation and pathogenicity to onion. *Ann. Appl. Biol. 130*: 467-479.

Zhang, H., S. Hanada, T. Shigematsu, K. Shibuya, Y. Kamagata, T. Kanagawa and R. Kurane. 2000. *Burkholderia kururiensis* sp. nov., a trichloroethylene (TCE)-degrading bacterium isolated from an aquifer polluted with TCE. *Int. J. Syst. Evol. Microbiol. 50*: 743-749.

Zhao, N., C. Qu, E. Wang, and W. Chen. 1995. Phylogenetic evidence for the transfer of *Pseudomonas cocovenenans* (van Damme et al. 1960) to the genus *Burkholderia* as *Burkholderia cocovenenans* (van Damme et al. 1960) comb. nov. *Int. J. Syst. Bacteriol. 45*: 600-603.

9. AFFILIATION

Ronald Gitaitis and Claudia Nischwitz, *Department of Plant Pathology, University of Georgia, Coastal Plain Experiment Station – Tifton Campus, Tifton, GA 31793-0748 USA.*

MARK FEGAN

PLANT PATHOGENIC MEMBERS OF THE GENERA *ACIDOVORAX* AND *HERBASPIRILLUM*

Abstract. The genera *Acidovorax* and *Herbaspirillum* contain organisms that were previously classified as part of the genus *Pseudomonas*. Together with many plant pathogenic pseudomonads, as the genus *Pseudomonas* has been revised over the past twenty years, organisms within these two genera have been taxonomically reclassified. The taxonomy, host range, ecology, control and detection of the plant pathogenic members of the genera *Acidovorax* and *Herbaspirillum* are discussed.

1. INTRODUCTION

The genera *Acidovorax* and *Herbaspirillum* are members of the *β-Proteobacteria* (*Fig. 1*), which primarily cause leaf spot and streak diseases of a variety of host plants (Table 1). The most significant pathogens within the genus *Acidovorax* genus belong to the species *A. avenae*. *A. avenae* subsp. *avenae* is a significant pathogen of graminaceous species and *A. avenae* subsp. *citrulli* is an important pathogen of *cucurbit* species, especially watermelon, and is arguably the most important plant pathogen in the genus. *A. avenae* subsp. *cattleyae* causes disease of orchids and is of lesser significance that the other two subspecies of *Acidovorax avenae*. *A. konjaci* causes a disease of the konjac plant and has only been found in Japan. The other members of the genus *Acidovorax*, which cause diseases of plants, have been described relatively recently. *A. anthurii* was formally described in the year 2000 (Gardan, Dauga, Prior, Gillis, & Saddler, 2000) and *A. valerianella* was formally described in the year 2003 (Gardan, Stead, Dauga, & Gillis, 2003). An unclassified *Acidovorax* sp. causing disease of geranium and petunia has also recently been identified.

The genus *Herbaspirillum* contains only a single plant pathogenic member, *H. rubrisubalbicans*, which has been described as a "mild plant pathogen" (Baldani et al., 1996). Members of this genus are endophytic diazotrophic organisms, including *H. rubrisubalbicans*. Due to the mild nature of the disease caused by *H. rubrisubalbicans* little research has been conducted into the disease

This review covers the taxonomic history of members of the genera *Acidovorax* and *Herbaspirillum* including a discussion of variation below the level of the species or subspecies. The host range and methods employed to detect and identify these organisms is discussed as is the ecology and control of the disease they cause.

2. TAXONOMY OF THE GENUS *ACIDOVORAX*

The genus *Acidovorax* was created in 1990 and initially contained only the non-phytopatogenic species *A. facilis*, *A. temperans* and *A. delafeldii* (Willems et al.,

S.S. Gnanamanickam (ed.), Plant-Associated Bacteria, 671–702.

1990). The plant pathogenic members of the genus, originally classified in the genus *Pseudomonas*, were transferred into the genus *Acidovorax* in 1992 by Willems et al. (1992). The genus *Acidovorax* belongs to the beta subdivision of the *Proteobacteria* (Woese et al., 1984) within the family *Comamonadaceae* (Willems, Deley, Gillis, & Kersters, 1991) and contains four described plant pathogenic species; *A. anthurii*, *A. avenae*, *A. konjaci* and *A. valerianellae*. *A. avenae* includes three subspecies; *A. avenae subsp. avenae*, *A. avenae subsp. citrulli* and *A. avenae subsp. cattleyae*. Not all members of the genus *Acidovorax* are plant pathogens with the species *A. facilis*, *A. defluvii*, *A. temperans* and *A. delafeldii* commonly found in soil, water, activated sludge or clinical environments.

Table 1. Plant pathogenic members of the genera *Acidovorax* and *Herbaspirillum*.

Pathogen	*Hosts*	*Disease common name*
A. anthurii	*Anthurium* spp.	Bacterial leaf-spot of Anthurium
A. avenae subsp. *avenae*	Wide host range including *Saccharum officianarum*, *Oryzae sativa* and *Sorghum bicolor*	Bacterial leaf blight of maize and sorghum; brown stripe of rice; red stripe of sugarcane etc.
A. avenae subsp. *citrulli*	*Cucurbitaceae*	Watermelon fruit blotch
A. avenae subsp. *cattleyae*	*Cattleya* spp., *Phalaenopsis* spp.	Bacterial leaf spot of orchids
A. konjaci	*Amorphophallus konjac*	Bacterial leaf spot of konjac
A. valerianellae	*Valerianella locusta*	Bacterial spot of lamb's lettuce
Acidovorax sp.	Geranium and petunia	Bacterial leaf spot of geranium and petunia
Herbaspirillum rubrisubalbicans	Sugarcane and Sorghum	Mottled stripe of sugar cane, bacterial leaf stripe of sorghum

Phylogenetic analysis of the 16S rRNA sequences of the members of the genus *Acidovorax* has shown that the plant pathogenic members of the genus cluster together on one branch within the genus with the other, non-plant pathogenic, organisms clustering together on a separate branch (Gardan et al., 2003; Wen, Fegan, Hayward, Chakraborty, & Sly, 1999) (*Fig. 1*). This clustering of plant pathogenic and non-plant pathogenic species is in agreement with the early rRNA cistron analysis of Willems et al. (1992) where the three subspecies of *A. avenae* form a tight cluster based upon rRNA:DNA hybridisation data and *A. konjaci* was located on a deeper sub-branch of the avenae rRNA branch (Willems et al., 1992).

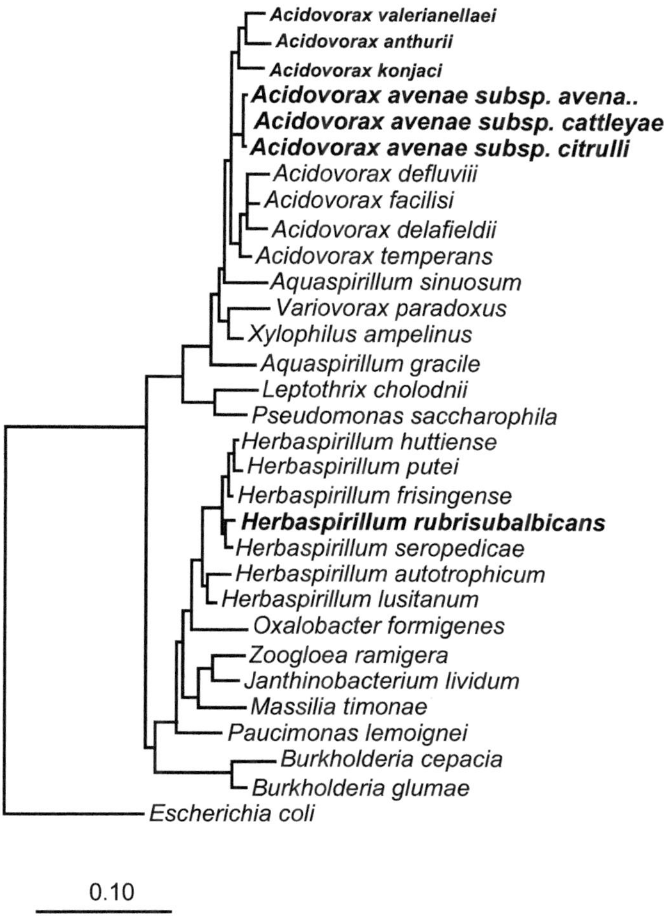

Fig. 1. Phylogenetic tree inferred from 16S rRNA sequence information using the ARB software program (Ludwig et al., 2004) from sequences available in the GenBank database. The tree shows the phylogenetic position of plant pathogenic *Acidovorax* sp. and *Herbaspirillum* sp. (indicated in bold text) within the β-proteobacteria.

ACIDOVORAX AVENAE

The three subspecies of *A. avenae* were transferred from the genus *Pseudomonas* to the genus *Acidovorax* by Willems et al. (1992). The DNA relatedness among the three subspecies of *A. avenae* ranges from 54 to 68 % (Willems et al., 1992). Phenotypically the three subspecies of *A. avenae* have been shown to form separate phena in a numerical analysis of carbon assimilation tests by Willems et al. (1992). Differences occur in the carbon source assimilation patterns of the subspecies;

A. avenae subsp. *avenae* and *A. avenae* subsp. *cattleyae* can utilise mannitol, D-arabitol, sorbitol, L-threonine, and L-histidine, but *A. avenae* subsp. *citrulli* does not. Similar to the relationship found using phenotypic tests gel electrophoresis fingerprinting patterns of total protein extractions of *A. avenae* subsp. *avenae* and *A. avenae* subsp. *cattleyae* have been found to be similar (Willems et al., 1992), but the fingerprinting pattern of *A. avenae* subsp. *citrulli* strains are different.

The three subspecies of *A. avenae* exhibit differences in host range. *A. avenae* subsp. *avenae* is pathogenic on members of the *Gramineae* (Claflin, Ramundo, Leach, & Erinle, 1989; F. P. Hu, Young, & Triggs, 1991; F.-P. Hu, Young, Triggs, & Wilkie, 1997; Ramundo & Claflin, 1990; Schaad, Kado, & Sumner, 1975; Schaad, Sowell, Goth, Colwell, & Webb, 1978), but not on *Cucurbitaceae* (Schaad et al., 1978) or *Cattleya* sp. (F.-P. Hu et al., 1997). *A. avenae* subsp. *citrulli* is pathogenic on members of *Cucurbitaceae* but not *Gramineae* (F.-P. Hu et al., 1997; Schaad et al., 1978), and *A. avenae* subsp. *cattleyae* is pathogenic on *Cattleya* and *Phalaenopsis* only (Ark & Thomas, 1946). However, a recent study by Hu et al. (1997) demonstrated that the existing strains of *A. avenae* subsp. *cattleyae* do not cause symptoms on *Cattleya* sp. under the conditions used to assess pathogenicity

2.1. A. avenae subsp. avenae

2.1.1. Taxonomic history of A. avenae *subsp.* avenae

A. avenae subsp. *avenae* contains strains that were previously classified in the species *Pseudomonas avenae*, *P. setariae* and *P. rubrilininans*. Incomplete descriptions and the lack of original cultures for comparison have resulted in a confused nomenclatural history of *A. avenae* subsp. *avenae* and only *P. avenae* and *P. rubrilineans* were included in the Approved Lists of Bacterial Names (Skerman, McGowan, & Sneath, 1980). Manns (1909) first described *Pseudomonas avenae* as the cause of leaf blight on oats in Ohio. Rosen (1922) observed a similar disease on foxtail (a common weed in corn fields) in Arkansas but, due to phenotypic differences between the foxtail pathogen and the pathogen originally described by Manns (1909), he gave the pathogen a new name, "*P. alboprecipitans*". However, Schaad et al. (1975) concluded that "*P. alboprecipitans*" was a synonym of *P. avenae* when comparing isolates from diseased corn, in Georgia and Florida, with the previous descriptions of *P. avenae* and "*P. alboprecipitans*". The pathogen causing red stripe disease on sugarcane was initially named *Phytomonas rubrilineans* (Lee, Purdy, Barnum, & Martin, 1925). Stapp (1928) re-assigned *Phytomonas rubrilineans* to *Pseudomonas rubrilineans*. In 1942 this species was transferred to the genus *Xanthomonas* (Starr & Burkholder, 1942), and subsequently transferred back into the genus *Pseudomonas* and the species description extended (Hayward, 1962). *P. setariae* was initially described by Okabe (1934) as "*Bacterium setariae*" the causal agent of brown stripe of rice. The pathogen was renamed as *P. setariae* (Savulescu, 1947) and although considered to be a separate species (Young, Dye, Bradbury, Panagopoulos, & Robbs, 1978) it was not included in the Approved Lists of Bacterial Names (Skerman et al., 1980).

2.1.2. A. avenae *subsp.* avenae *infrasubspecific variation*

Genetic fingerprinting employing rep-PCR has shown that *A. avenae* subsp. *avenae* strains are heterogeneous (D. Stead, Stanford, Aspin, & Heeney, 2003). The different genotypes of strains recognised by rep-PCR fingerprinting may be related to the hosts the strains were isolated from (eg rice, pelargonium and canna) (D. Stead et al., 2003). Based upon DNA:DNA homology studies *A. avenae* subsp. *avenae* strains that cause bacterial streak of corn and strains causing bacterial strip of rice have been shown to have a mean DNA:DNA homology of only 45%, much less that the 70% required for these strains to belong to the same bacterial species (Postnikova & Schaad, 2002). These results suggest that *A. avenae* subsp. *avenae* contains subgroups of strains that may represent more that one bacterial species. Sequence variation in the intergenic spacer (ITS) region between the 16S and 23S rRNA genes of *A. avenae* subsp. *avenae* strains from different hosts has also indicated that variation exists between *A. avenae* subsp. *avenae* strains (Song, Kim, Hwang, & Schaad, 2004). A phylogenetic analysis of ITS region sequences of *A. avenae* subsp. *avenae* strains also shows that *A. avenae* subsp. *avenae* strains are polyphyletic (*Fig. 2*). Strains of *A. avenae* subsp. *avenae* from rice and maize form a separate cluster from *A. avenae* subsp. *avenae* strains isolated from rescue and wheat grasses (*Fig. 2*).

It has been reported that DNA/DNA reassociation assays, ITS sequencing and amplified fragment length polymorphism (AFLP) data show that the corn and rice strains may be considered as separate species or subspecies (E. Postnikova, L. E. Claflin, I. Agarkova, A. Sechler, B. A. Ramundo, A. K. Vidaver and N. W. Schaad, unpublished data referenced in (Song et al., 2004)). However, further taxonomic work on a representative set of strains from all known hosts of the *A. avenae* subsp. *avenae* is required to clarify the taxonomic significance of this genetic variation. Phenotypic variation between corn and rice strains of *A. avenae* subsp. *avenae* has also been identified. The use of d-sorbitol as the sole carbon source in SNR medium (Sumner & Schaad, 1977) was shown to be useful for the isolation of strains of *A. avenae* subsp. *avenae* from corn, but not strains from rice.

Variation in pathogenicity and virulence among strains of *A. avenae* subsp. *avenae* has also been observed by Hu et al. (1997). Generally, strains of *A. avenae* subsp. *avenae* have been shown to be more virulent to *Zea mays* var. rugosa Bonaf. cv. Golden Treasure (sweetcorn) than to *Zea mays* L. cv. 'XP74' (maize) or sugarcane, and only weakly virulent to oats (F.-P. Hu et al., 1997). Che et al. (1999) also showed pathogenic specialisation of certain strains. Strains isolated from rice were shown to only infect rice and a strain isolated from finger millet could not infect rice (Che et al., 1999; Kadota, 1996). However, Ramundo and Claflin (1990) have report no differences in pathogenicity of strains.

Using the agar gel double diffusion method serological variation in *A. avenae* subsp. *avenae* strains from rice have been identified by Kadota et al. (1991) who grouped 77 isolates from rice into four serovars based on the main precipitin bands produced from four antisera to *A. avenae* subsp. *avenae*. The serological properties of isolates from different hosts were distinguishable.

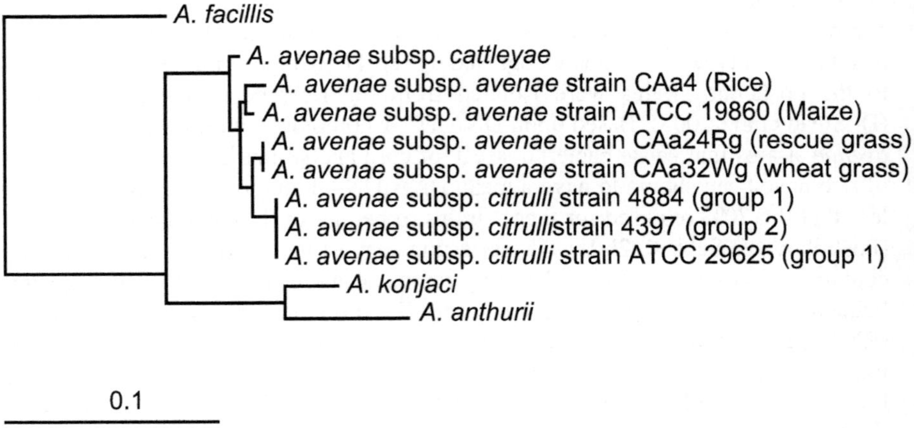

Fig. 2. Phylogenetic tree inferred from the analysis of sequences of the 16S-23S rRNA gene spacer region of *Acidovorax* sp. strains. The tree was generated using the ARB software program (Ludwig et al., 2004) from sequences available in the GenBank database.

2.1.3. A. avenae *subsp.* avenae *Hosts and symptoms*

The primary economically important hosts of *A. avenae* subsp. *avenae* are rice (*Oryza sativia*) (K. Goto & Ohata, 1961; Kadota & Ohuchi, 1983; Kadota et al., 1991; Tominaga, Kimura, & Goh, 1983; Zeigler & Alvarez, 1989), sugarcane (*Saccharum officinarum*) (Christopher & Edgerton, 1932; Hayward, 1962), Common sorghum (*Sorghum bicolour*) (Akhtar & Aslam, 1986; Rott, Frossard, Vuong, & Katile, 1989) and maize (*Zea mays*) (Hsu, Chang, Tzeng, & Leung, 1991; Johnson, Robert, & Cash, 1949; Karganilla & Cabauatan, 1974; Sumner & Schaad, 1977; Ullstrup, 1960; White, Pataky, & Stall, 1994). There are many hosts secondary importance including oats (*Avena sativa*) (Manns, 1909), foxtail millet

Table 2. Natural hosts and symptoms caused by *A. avenae* subsp. *avenae.*

Host plant species	*Symptom*	*Reference(s)*
Primary Hosts		
Oryza sativa [rice]	brown stripe	(K. Goto & Ohata, 1961; Kadota & Ohuchi, 1983; Kadota et al., 1991; Tominaga et al., 1983; Zeigler & Alvarez, 1989)
Saccharum officinarum [sugarcane]	red stripe and top rot	(Christopher & Edgerton, 1932; Hayward, 1962)
Sorghum bicolor [sorghum]	leaf blight	(Akhtar & Aslam, 1986; Rott et al., 1989; White et al., 1994)
Zea mays [maize]	leaf blight and stalk rot	(Hsu et al., 1991; Johnson et al., 1949; Karganilla & Cabauatan, 1974; Sumner & Schaad, 1977; White et al., 1994)
Secondary Hosts		
Avena sativa [oats]	blade blight	(Manns, 1909)
Camellia sinensis [tea]	leaf spot	(Takikawa et al., 1988)
Eleusine coracana [finger millet]	brown stripe on sheath and leaf blade	(Billimoria & Hegde, 1971; Mudingotto, Veena, & Mortensen, 2002; Nishiyama, Nishihara, & Ezuka, 1979)
Panicum miliaceum [millet]	leaf, sheath, and culm brown streaks, leaf blight	(Elliott, 1923; Ullstrup, 1960)
Pennisetum glaucum [pearl millet]	leaf stripe	(Claflin et al., 1989)
Setaria italica [foxtail millet]	leaf stripe	(M. Goto & Okabe, 1952; Okabe, 1934)
Setaria lutescens	leaf blight	(Rosen, 1922)
Setaria viridis [green foxtail millet]	leaf stripe	(Shakya, Vinther, & Mathur, 1985)
Wild Hosts		
Agropyron intermedium [intermediate pubescent wheat-grass]	unknown	(Bradbury, 1986)
Bromus catharticus [rescue brome]	unknown	(Bradbury, 1986)
Bromus marginatus [mountain brome]	unknown	(Bradbury, 1986)

Host plant species	*Symptom*	*Reference(s)*
Caryota mitis [fishtail palm]	brown to black foliage blight	(Knauss, Miller, & Virgona, 1978)
Digitaria sanguinalis [hairy finger grass, lance crab grass or hairy crab grass]	leaf stripe	(Shakya, 1981)
Echinochloa crus-galli [Japanese millet, barnyard grass]	leaf stripe	(Shakya, 1981)
Paspalum dilatatum [dallisgrass]	brown leaf stripe	(Kadota et al., 1991)
Paspalum notatum [bahiagrass]	unknown	(Saddler, 1994)
Paspalum paniculatum [arrocillo]	unknown	(Saddler, 1994)
Paspalum urvillei [Vasey's grass]	leaf blight and stalk rot	(Gitaitis, Stall, & Strandberg, 1978)
Agropyron tricophorum (syn. *Elytrigia intermedia*) [intermediate pubescent wheat-grass]	unknown	(Bradbury, 1986)
Chaetochloa italica [foxtail millet]	unknown	(Rosen, 1922)
Euchlaena mexicana (syn. *Zea mexicana*) [teosinte]	leaf stripe	(Dange & Payak, 1972; M. Goto & Starr, 1971)
Panicum hirsutum [hairy panicgrass]	unknown	(Bradbury, 1986)
Pennisetum americanum [American fountaingrass]	leaf blight	(Rott et al., 1989)
Setaria glauca [yellow foxtail]	leaf stripe	(Claflin et al., 1989)

(*Setaria italica*) (M. Goto & Okabe, 1952; Okabe, 1934) and tea (*Camellia sinensis*) (Takikawa, Ando, Hamaya, Tsuyumu, & Goto, 1988). There are also many wild hosts, a list of hosts and common symptoms is provided in Table 2.

Although the pathogen has caused losses in the production of corn, oats, rice and millet the economic impact of the disease on rice (Cottyn, Cerz, & Mew, 1994) and maize (Dange, Payak, & Renfro, 1978) are considered minor. In a study of bacterial leaf blight on sweet corn hybrids with the *shrunken-2* endosperm mutation Pataky et al. (1997) showed that despite the presence of moderately severe foliar symptoms the disease did not decrease the yield substantially even in the most susceptible plants. However, heavy infections in rice nurseries have been reported in Japan (K. Goto & Ohata, 1961; Kadota & Ohuchi, 1983; Kadota et al., 1991; Tominaga et al., 1983).

The symptoms commonly produced on host plants infected by *A. avenae* subsp. *avenae* consist of leaf streaks and stripes, which may extend into the leaf sheaths with occasional development of a stalk rot (Saddler, 1994). Symptoms are more severe on seedlings and immature plants than mature plants (Shakya, Chung, & Vinther, 1986; Sumner & Schaad, 1977).

On sugarcane, *A. avenae* subsp. *avenae* (formerly *P. rubrilineans*) causes red stripe and top rot (Christopher & Edgerton, 1932; Hayward, 1962). The disease occurs on leaves and leaf sheaths, producing red coloured stripes, and a stem rot which normally begins near the growing point.

In rice *A. avenae* subsp. *avenae* infects the primary leaves of rice seedlings in nurseries producing brown stripes on the sheaths which frequently extend out to the leaf blade (Tominaga et al., 1983). Four types of symptoms have been observed in infected rice seedlings: inhibition of germination, brown stripes on leaves, curving of the leaf sheath, and abnormal elongation of the mesocotyl (Kadota & Ohuchi, 1983). Severely affected seedlings may become stunted and die (Sato et al., 1983; Shakya et al., 1985). Discoloration of rice seeds has also been described (Cottyn, Outryve et al., 1996; Zeigler & Alvarez, 1989). Inoculation of rice seed with large numbers of the bacterium (10^8 cfu/mL) has been shown to affect seed germination and seed vigour (Xie, Sun, & Mew, 1998).

Bacterial leaf spot of tea caused by *A. avenae* subsp. *avenae* has so far only been observed in Japan (Takikawa et al., 1988) and it is possible that *A. avenae* subsp. *avenae* together with *P. syringae* pv. *theae* may form a disease complex which gives rise to symptoms. Typical symptoms are leaf spots with irregular shape and apparent necrosis of the spongy parenchyma tissue (Takikawa et al., 1988).

Severe infection of sorghum has been reported to occur under greenhouse conditions (M. Goto & Starr, 1971) and natural infection of sorghum in the USA (White et al., 1994) and Mali (Rott et al., 1989) has been reported. The leaves of sorghum infected by the pathogen develop a greyish-green colour with red borders or small, irregular and reddish necrotic stripes which may coalesce to form extensive necrotic areas covering a large proportion of leaf. When severe infection occurs, seedlings may be stunted, and eventually die.

On maize typical symptoms of infection are shredding of infected leaves and rotting of the stalk above the ear (Johnson et al., 1949; Karganilla & Cabauatan, 1974), symptoms may occur on both young and old plants. The disease was named as bacterial leaf blight and stalk rot by Johnson (1949), and bacterial leaf stripe by Karganilla and Cabauatan (1974). Natural infections have been noted on various types of maize (*Zea mays* L.) grown in central Illinois, USA (White et al., 1994).

A. avenae subsp. *avenae* causes disease in several species of millet (Table 2) producing a leaf stripe. On pearl millet (Claflin et al., 1989; Rott et al., 1989) affected plants show interveinal lesions varying from several millimetres to more than 25 cm in length. On ragi millet brown stripes are produced on the sheaths and leaf blades (Nishiyama et al., 1979) which may be greater than 20 cm in length. On proso millet, a stripe disease was described for the symptom of water-soaked streaks on leaves, sheaths, and culms (Elliott, 1923). In Australia, leaf blight disease outbreaks on proso and Italian millet caused by *A. avenae* subsp. *avenae* have been recorded in experimental field plots (Fahy, Gillings, Bradley, Diatloff, & Singh, 1989).

2.1.4. Epidemiology and Ecology of A. avenae subsp. avenae

A. avenae subsp. *avenae* has a worldwide distribution (CMI, 1995; Shakya et al., 1985). The bacterium is seed-borne (Claflin et al., 1989; Fahy et al., 1989; Gitaitis et al., 1978; Shakya et al., 1986; Shakya et al., 1985) and has been found to be viable in 8-yr old rice seed samples stored at 5°C (Shakya et al., 1985). In rice, the bacterium has been reported to be located between the glumes and the pericarp, or potentially deeper in the seed (Shakya et al., 1986). Shakya et al. (1985) concluded that infected seed may act as an important means of dissemination of the bacterium from one geographical area to another. In bacterial leaf blight of maize alternative, or wild, hosts such as *Paspalum urvillei* (Vasey grass) in Florida, USA, have also been reported to act as an inoculum reservoir for outbreaks (Gitaitis et al., 1978). The bacterium survives between seasons in association with perennial Vasey grass (*Paspalum urvillei*) seed as well as in leaf tissue but is not thought to survive well in soil or in plant debris (Sumner & Schaad, 1977).

Transmission of *A. avenae* subsp. *avenae* from rice seed to seedling through coleoptile infection during germination and from plant to seed has been demonstrated (Shakya et al., 1986). The bacterium invades the plant through the stomata of coleoptile and the first leaf where it multiplies in substomatal chambers and invades intercellularly reaching the lacunae which run along the full length of the leaves, the bacterium is not present in xylem and phloem vessels (Shakya et al., 1986). A similar observation by Gitaitis et al. (1981) showed that the ingress of *A. avenae* subsp. *avenae* occurred through stomata on corn leaves and due to the nature of the symptoms on Pearl millet Claflin et al. (1989) postulated that hydathodes may be the portals of entry for the bacterium.

Kadota and Ohuchi (1990) observed that in rice plants the bacteria invade the seeds during several days around flowering and heading. There is evidence that mature plants, which survive infection at the seedling stage, harbour latent infections, and that the bacteria can be internally transmitted from plant to seed in latently infected plants (Shakya et al., 1986; Shakya et al., 1985). Dissemination may also occur via farm equipment (Gitaitis et al., 1978). In general, conditions of high temperature and humidity favour symptom development (Akhtar & Aslam, 1986; Johnson et al., 1949; Shakya et al., 1985; Tominaga et al., 1983; White et al., 1994).

Control of diseases caused by *A. avenae* subsp. *avenae* is most effectively achieved by using pathogen-free seed. The pathogen free state of seeds is commonly determined by the use of a germination test before planting and maintaining a temperature and humidity which are favourable for symptom development in growing rice seedlings (Kadota & Ohuchi, 1990).

There is little information on the use of chemicals to control diseases caused by *A. avenae* subsp. *avenae*. However, the bactericide kasugamycin has been used to control brown stripe disease in rice (Kadota & Ohuchi, 1990) and coper sulphate and streptocycline (streptomycin and tetracycline) has been used to control infection in maize (Thind, Randhawa, & Soni, 1984).

To date no varieties of rice resistant to bacterial brown stripe have been developed, but, varietal differences in symptom expression in laboratory tests have been observed (Sato et al., 1983). Dange et al. (1978) have found that there is variation in the susceptibility of maize cultivars to *A. avenae* subsp. *avenae.* Sumner and Schaad (1977) suggest that by using available resistant cultivars of maize the impact of bacterial leaf blight will be negligible. Pataky et al. (1997) have found that, under the environmental conditions studied, all sweetcorn hybrids tested were resistant to bacterial leaf blight although cosmetic symptoms on leaves were present.

2.1.5. Detection/Diagnosis of A. avenae subsp. avenae

Phenotypic and chemotaxonomic techniques. Characterization and identification of *Acidovorax avenae* subsp. *avenae* is commonly based on phenotypic properties such as colony morphology, physiological and biochemical properties, disease symptoms, host range and hypersensitivity reaction in tobacco (Akhtar & Aslam, 1986; Christopher & Edgerton, 1932; Claflin et al., 1989; Elliott, 1923; Hsu et al., 1991; Johnson et al., 1949; Knauss et al., 1978; Nishiyama et al., 1979; Ramundo & Claflin, 1990; Rott et al., 1989; Schaad et al., 1975; Shakya et al., 1985; Takikawa et al., 1988; Ullstrup, 1960).

The Biolog GN Microplate automated identification system has been used to identify *A. avenae* subsp. *avenae* by several authors (Cottyn, Outryve et al., 1996; Xie et al., 1998; Xie, Zhu, & Ren, 2002) and fatty acid analysis has been used to identify pure cultures of *A. avenae* subsp. *avenae* isolated from diseased planting materials (Fahy et al., 1989; White et al., 1994). Fahy et al. (1989) used fatty acid methyl ester (FAME) profiles and restriction fragment length polymorphism (RFLP) to trace a quarantine outbreak of *A. avenae* subsp. *avenae* on French and Italian millet in Australia.

Selective media have been developed to aid the isolation of *A. avenae* subsp. *avenae* to aid in the isolation of the pathogen from infected seed which may be difficult due to overgrowth by *Pseudomonas* sp. (Cottyn, Van Outryve et al., 1996). Sorbitol neutral red agar (Sumner & Schaad, 1977) has been employed for the isolation of the pathogen from soil and plant debris, the semi-selective liquid medium, SP medium, developed by Song et al. (2004) and the *P. avenae*-selective medium (PASM) (Kadota, 1996) have been developed to assay rice seeds for the pathogen. The SP medium (Song et al., 2004) is based upon the differentially utilised substrates d-sorbitol and l-pyroglutamic acid which are used by *A. avenae* subsp. *avenae* strains.

Techniques such as serology (Kadota, 1996; Shakya, 1987) and plant grow-out tests (Rice-blotter test of (Shakya & Chung, 1983)) have been employed to identify the presence of *A. avenae* subsp. *avenae* in rice seeds. Dange et al. (1978) describe a grow out test for Maize (Maize-incubation). Although these tests have been successfully employed they lack specificity and sensitivity and suffer from a lack of speed for use as routine detection methods.

Molecular Techniques. Molecular tests for the diagnosis of A. avenae subsp. *avenae* have been developed. *A. avenae* subsp. *avenae* strain NCPPB 1392 could be differentiated from other phytopathogenic pseudomonads when genomic DNA was digested with the endonuclease *Spe* I in a followed by pulsed field gel electrophoresis analysis (Grothues & Rudolph, 1991) but the applicability of this method to differentiate all *A. avenae* subsp. *citrulli* strains is questionable. Employing size polymorphisms in amplification products from the spacer regions between the 16S and 23S rRNA genes Kim et al. (1996) differentiated *A. avenae* subsp. *avenae* (*Pseudomonas avenae*) strains isolated from rice from other common seedborne rice pathogens (*Pseudomonas glumae*, *P. fuscovaginae*, *P. syringae* pv. *syringae*, *Xanthomonas oryzae* pv. *oryzae*, *X. oryzae* pv. *oryzicola*, and *Erwinia herbicola*). The three *A. avenae* subsp. *avenae* (*Pseudomonas avenae*) strains tested produced one major (950 bp in size) and two secondary (600 and 1,400 bp in size) amplification products which allowed differentiation of *A. avenae* subsp. *avenae*. However, Kim et al. (1996) did not include other subspecies of *A. avenae* (*A. avenae* subsp. *citrulli* and *A. avenae* subsp. *cattleyae*), therefore the specificity of the technique for *A. avenae* subsp. *avenae* is unknown. Song et al. (2003) subsequently cloned and sequenced the 950 bp fragment produced by amplification of the spacer regions between the 16S and 23S rRNA genes, and designed primers (Oafl and Oarl; *Table 3*). This primer pair amplifies both *A. avenae* subsp. *avenae* and *A. avenae* subsp. *citrulli*, producing a 550-bp product. Recently Song et al. (2004) have developed a BIO-PCR protocol for the detection of *A. avenae* subsp. *avenae* from rice seed. The nested-PCR described by Song et al. (2004) (primers presented in *Table 3*) is specific for strains of *A. avenae* subsp. *avenae* from rice and does not amplify template DNA from *A. avenae* subsp. *avenae* strains from other hosts or any other bacterium tested.

2.2. A. avenae subsp. citrulli

2.2.1. Taxonomic history of A. avenae *subsp.* citrulli

A. avenae subsp. *citrulli* is arguably the most important pathogen within the genus *Acidovorax*. The pathogen was first reported by Webb and Goth (1965) as the causal agent of bacterial leaf spot in commercial watermelon fields in Florida, USA. Webb and Goth (1965) isolated a bacterium that produced white colonies on potato-dextrose agar and beef extract agar, but did not identify the pathogen. Later, Crall and Schenck (1969) described a fruit rot disease on watermelon with photographs of the symptoms of bacterial fruit blotch of watermelon (BFB) as large, firm, dark green, water-soaked lesion with irregular margins. Goth and Webb (1975) observed a bacterial wilt of watermelon seedlings and, although several of their observations were not consistent with currently established characteristics of bacterial fruit blotch, they concluded that the pathogen of bacterial wilt of watermelon seedlings was a non-fluorescent pseudomonad that could be seed transmitted.

The name "fruit blotch" was first used as the common name for a disease described in a plant disease handbook published in Queensland, Australia (Vock, 1978). The description was very brief, and the pathogen was described as a *Pseudomonas* species. However, a photograph of an affected watermelon fruit included clearly shows that the symptoms are similar to the disease now attributed to the bacterium *Acidovorax avenae* subsp. *citrulli*.

Schaad et al. (1978) assigned the name *P. pseudoalcaligenes* subsp. *citrulli* to a bacterial pathogen associated with water-soaked lesion on watermelon cotyledons, although the mol% G+C of the DNA (65-67%) was higher than that of *P. pseudoalcaligenes* (62-64). Schaad et al. (1978) also showed that the bacterium was similar to the unidentified non-fluorescent bacterium described by Webb and Goth (1965).

The connection between the seedling disease and the fruit blotch symptoms was not made until Wall and Santos (1988) associated *P. pseudoalcaligenes* subsp. *citrulli*, with the characteristic symptoms of BFB of watermelon fruit.

2.2.2. Infrasubspecific variation of A. avenae *subsp.* citrulli

On the basis of both phenotypic and genotypic tests two groups of *A. avenae* subsp. *citrulli* strains have been identified. Phenotypically, gas chromatography-fatty acid methyl ester (GC-FAME) profile analysis (Walcott, Langston, Sanders, & Gitaitis, 2000), carbon substrate utilisation profiles (O'Brien & Martin, 1999; Walcott, Fessehaie, & Castro, 2004) and pathogenicity on different host plants (O'Brien & Martin, 1999; Walcott et al., 2004) have defined two groups of strains. The same two groups of strains have also been identified by using the genomic fingerprinting techniques, rep-PCR (Walcott et al., 2004; Wen et al., 1997) and pulsed field gel electrophoresis (Walcott et al., 2004; Walcott et al., 2000). These two groups have been defined as *A. avenae* subsp. *citrulli* Group I and Group II by Walcott et al. (2004). The distinguishing phenotypic characteristics of *A. avenae* subsp. *citrulli* group I and Group II strains are presented in *Table 4*.

Table 3. Oligonucleotide primer pairs for the detection of *Acidovorax avenae* strains.

Organism detected (specificity)	*Primer designation*	*Primer sequence (5'-3')*	*Reference*
Acidovorax sp. and *Delftia acidovorans*	WFB1	GACCAGCCACACTGGGAC	(Walcott & Gitaitis, 2000)
	WFB2	CTGCCGTACTCCAGCGAT	
Acidovorax sp.	RST63	TCCGGCGGCGCGCTCACCGTGGTGCTG	(Jones, Gitaitis, & Schaad, 2001)
	RST64	AGCGCGGCGGCGTAGGCGCGCGAG	
A. avenae	Oaf1	GTCGGTGCTAACGACATGG	(Song et al., 2003)
	Oar1	AGACATCTCCGCTTTCTTTCAA	
A. avenae	RST49	GATGGCCGTGCCCTTCTTCATCCTCG	(Jones et al., 2001; Minsavage, Hoover, Kucharek, & Stall, 1995)
	RST51	CATGGCCACGATGAGGATCG	
A. avenae	Oaf1	GTCGGTGCTAACGACATGG	(Song et al., 2003)
	Oar2	TCCTCGCATCTTATGTTCGGAA	
	AaP1(Probe)	6FAM-TCAGCTGGTTAGAGCACCGTCTAGA-TAMARA	
A. avenae subsp. *citrulli*	Aacf2	GGAAGAATTCGGTGCTACCC	(Song et al., 2003)
	Aacr2	TCGTCATTACTGAATTTCAACA	
A. avenae subsp. *citrulli*	Aacf3	CCTCCACCAACCAATACGCT	(Song et al., 2003)
	Aacr2	TCGTCATTACTGAATTTCAACA	
	Aap2 (Probe)	6FAM-CGGTAGGGCGAAGAAACCAACACC-TAMARA	
A. avenae subsp. *avenae* strains from rice	Aaaf3 (SEQ10)	GTCATCCTCCACCAACCAAG	(Song et al., 2004)
	Aaar2 (SEQ 12)	AGAACAATTCGTCATTACTGAAC	
	Aaaf5 (SEQ 11)[a]	TGC CCT GCG GTA GGG CG	

[a] to be used in conjunction with the primer Aaar2 in a hemi-nested PCR reaction using amplicon from application of primer pair Aaaf3/Aaar2 as template

Table 4. Distinguishing phenotypic characteristics of *A. avenae* subsp. citrulli group I and Group II strains.

Phenotypic characteristic	*Group I*	*Group II*
GC-FAME profile		
11:0, 3:OH	-	+
12:0, 3:OH	-	+
17:0 *cyclo*	-	+
Carbon Substrate Utilisation		
L-Leucine	- (96%)	+ (96%)
Disease severity		
Watermelon	Moderate	High
cantaloupe	Moderate	Low

A. avenae subsp. *citrulli* strains belonging to Group I exhibit greater genetic diversity than group II strains (Walcott et al., 2004; Walcott et al., 2000) this is also reflected in the variation of GC-FAME profiles (Walcott et al., 2000).

It has been suggested that the introduction of *A. avenae* subsp. *citrulli* group I isolates into growing areas of the United States corresponds to the increase in outbreaks of the disease on cantaloupes and other cucurbit hosts (Walcott et al., 2004). Similarly, O'Brien and Martin (O'Brien & Martin, 1999) suggested that introduction of *A. avenae* subsp. *citrulli* strains potentially on infected seed, later identified as belonging to Group I, were responsible for an outbreak of the disease on melons in Queensland Australia. In contrast to the situation for *A. avenae* subsp. *avenae* where the different groups of strains have variable sequences within the 16S-23S rRNA gene spacer region there is no variation in sequence between group I and II strains of *A. avenae* subsp. *citrulli* in this region of the genome (*Fig. 2*).

2.2.3. Diseases caused by A. avenae *subsp.* citrulli

A. avenae subsp. *citrulli* is pathogenic to plants of the family *Cucurbitaceae*. Watermelon [*Citrullus lanatus* (Thunb.) Matsum & Nakai] was the first reported natural host (Webb & Goth, 1965). Subsequently, honeydew (Isakeit, Black, Barnes, & Jones, 1997), muskmelon (Sowell, 1981), citronmelon (Isakeit, Black, & Jones, 1998), Hami melon (Zhao, Sun, Wang, & Hui, 2001), pumpkin (Langston, Walcott, Gitaitis, & Sanders, 1999), squash (Schaad et al., 1978), anguria fruit (de Oliveira & Moura, 1994), cucumber (H. L. Martin, O'Brien, & Abbott, 1999), prickly paddy melon (O'Brien & Martin, 1999) and gramma (*Cucurbita moschata*)

(H. L. Martin & Horlock, 2002) have been noted as natural hosts. *A. avenae* subsp. *citrulli* has also been isolated from diseased eggplant seedlings (Assouline, Milshtein, Mizrahi, Levy, & BenZeev, 1997), and tomato seeds (Assouline et al., 1997) Other members of *Cucurbitaceae* (e.g. zucchini) and tomato seedlings are susceptible following inoculation (Assouline et al., 1997; O'Brien, 1996).

On watermelon, seedlings, leaves, and fruit can be affected (Latin & Hopkins, 1995). Two distinct symptoms have been observed: leaf spots, forming water-soaked lesions on the cotyledons of seedlings and watermelon fruit blotch, large, firm, water-soaked lesions with irregular margins form on the fruit. As lesions on fruit age the periderm can crack and bacterial ooze is produced. The pathogen is thought to enter the fruit through stomata; immature fruits in particular are infected (Frankle, Hopkins, & Stall, 1993).

On rockmelon or muskmelon, angular spots appear on the leaf (Sowell, 1981) and affected fruit show water-soaked pits on the fruit surface (Latin & Hopkins, 1995; O'Brien, 1996). O'Brien (O'Brien, 1996) observed that the spots on leaves of rockmelon are lighter in colour than those on watermelon, and fruit spots are also smaller and darker in colour. Isakeit et al. (1997) reported that the circular lesions on diseased honeydew fruits were either water-soaked or had a scabby centre with water-soaked edge, lesions were 3-10 mm diameter that did not extend into the flesh of the fruit. Citron-melon, a common weed in south Texas, USA and in Queensland, Australia, is often infected by *A. avenae* subsp. *citrulli* (Isakeit et al., 1998). The lesions on citron fruit are circular and necrotic, or water-soaked. Symptoms on prickly paddy melon (*Cucumis myriocarpus*) are similar to those on watermelon (O'Brien & Martin, 1999) although *A. avenae* subsp. *citrulli* Group II strains are more pathogenic for prickly paddy melon than Group I strains.

Infected cucumber shows necrotic lesions on leaves, but not on fruit or stems (H. L. Martin et al., 1999). Initially angular, chlorotic, water-soaked lesions form on the leave and these later dry to produce necrotic areas of light brown dead tissue. Lesions are delimited by leaf veins and are distributed uniformly over the leaf surface.

2.2.4. Epidemiology and control of A. avenae *subsp.* citrulli

A. avenae subsp. *citrulli* has been reported in Australia (O'Brien, 1996; O'Brien & Martin, 1999; Vock, 1978), Brazil (Assis, Mariano, Silva Hanlin, & Duarte, 1999; Silveira, Mariano, & Michereff, 2003), Guam and the Northern Mariana Islands (Wall, 1991; Wall & Santos, 1988; Wall, Santos, Cruz, & Nelson, 1990), South Africa (Serfontein, Uys, & Goszczynska, 1997), Turkey (Demir, 1996a, 1996b), and the USA (Black et al., 1994; Evans & Mulrooney, 1991; Hamm, Spink, Clough, & Mohan, 1997; Hopkins, 1994; Walcott et al., 2000). Severe epidemics of the disease have been reported in the USA (Hopkins, 1989, 1993, 1994; Somodi et al., 1991), Guam (Wall, 1991; Wall & Santos, 1988; Wall et al., 1990), and Australia (O'Brien, 1996; O'Brien & Martin, 1999).

The primary inoculum for fruit blotch epidemics comes from contaminated seed, where the pathogen remains on the seed coat and embryos (Rane & Latin, 1992). Infected seeds result in seedlings with symptoms of BFB under conducive conditions (*i.e.* a warm, humid environment commonly found in glasshouses with overhead irrigation) (Hopkins, 1994). Secondary spread in the transplant glasshouse can be responsible for significant proportions of infected seedlings reaching the field (Hopkins, 1994). As plants grow in the field, the pathogen spreads to new leaves and neighbouring plants. Infected plants may not be killed if the infection is not severe, but lesions on foliage provide a source of inoculum for infection of immature fruit. Before fruit ripen, the pathogen penetrates the stomata of fruit producing the characteristic blotch symptoms. The characteristic lesions on fruit expand very rapidly, and may cover the entire upper surface of the fruit within a 2-week period (Frankle et al., 1993). As diseased fruit decay in the field, seeds associated with affected fruit slip into the soil serving as a source of inoculum. Although seed is contaminated both internally and externally, seed transmission does not take place systemically through the vascular system (Rane & Latin, 1992). The openings in the seed coat near the hilum are probably the ports of entry into the seed during the seed extraction process (Rane & Latin, 1992). Infestation of seed probably occurs during the seed harvesting process, with the method of separation of seeds from the fleshy fruit providing an opportunity for the bacteria to contaminate the seed (Lovic & Hopkins, 2003). Recently Walcott et al. (Walcott, Gitaitis, & Castro, 2003) have shown that watermelon blossoms are also a potential site for the ingress of *A. avenae* subsp. *citrulli* into watermelon fruit and seed. These authors found that the inoculation of blossoms with *A. avenae* subsp. *citrulli* led to the production of infested seed within symptomless fruit. It has also been suggested that bees pollinating flowers may be a vector of the bacterium to the watermelon blossoms (Fessehaie, Hopkins, Gitaitis, & Walcott, 2005). However, the significance of these findings to the epidemiology of the disease in the field requires further study.

Wild cucurbit host in the vicinity of crops and volunteer plants that grow from a previously planted crop represent a potentially important source of infection in proceeding crops (Latin & Hopkins, 1995). Any cucurbit crop can potentially act as a potential source of inoculum of *A. avenae* subsp. *citrulli* and it has been suggested that seedlots of all curcurbits should be tested for *A. avenae* subsp. *citrulli* and that transplant houses should only be used to grow a single cucurbit type to prevent cross contamination (Hopkins & Thompson, 2002). *A. avenae* subsp. *citrulli* has been isolated from diseased eggplant seedlings, and tomato seeds imported into Israel from the USA (Assouline et al., 1997). Solanaceous seeds (eggplant and tomato) infested by *A. avenae* subsp. *citrulli* may also assist the movement of the pathogen between countries (Assouline et al., 1997).

Control measures recommended include the use of disease-free seeds, early field spray of copper to protect the young fruit, careful field hygiene, including crop rotation, burning of infected plant residues, and elimination of wild and volunteer cucurbits (de Oliveira & Moura, 1994; O’Brien, 1996).

Treatment of seed prior to planting to remove *A. avenae* subsp. *citrulli* has also been investigated as a method of control. O'Brien and Martin (1999) tested hot water, chlorine and acid treatments, and found that hot water treatment is the most effective treatment to eradicate the bacterium from the seed. Heat treatment of infested seed at 50°C for 20 min effectively controlled disease development in seedlings in a completely randomized experiment by Wall (1989), and O'Brien and Martin (1999) found that 50°C for 30 min gave excellent control. Nomura and Shirakawa (2001) showed that hot water treatment of watermelon seed for 10 to 30 min at 54 to 56°C effectively reduced the incidence of disease but did not totally eradicate the pathogen. Hopkins et al. (1996) describe the use of wet seed treatments, including fermentation of seeds in watermelon juice and debris, followed by chemical treatment. Fermentation of seeds for 24 to 48h followed by treatment with 1% HCl or 1% $CaOCl_2$ for 15 min prior to washing and drying were the most effective treatments either eliminating the pathogen or reducing seed transmission to less than 1% (Hopkins et al., 1996). However, this treatment does reduce the germination of triploid seed.

Chemical control of the disease in the field by the use of copper sprays prior to fruit set has been successfully employed to reduce the incidence of the disease (Hopkins, 1991, 1995). Recently Hopkins (2005) has shown that the spread of bacterial fruit blotch in greenhouses can be controlled by including iodised copper (1-1.5ppm) and peroxyacetic acid (80ppm) in combination in the irrigation water. This application does not cause phytotoxicity, a common problem with copper application, and is particularly useful in controlling spread of the disease when used in combination with bi-weekly applications of acibenzolar-S-methyl (a systemic compound which induces host plant resistance). The potential for biological control of bacterial fruit blotch has been demonstrated by Fessehaie and Walcott (2005). Treatment of seed with a strain of *A. avenae* subsp. *avenae* was shown to effectively reduce transmission of the disease in growth chamber and green house conditions. Treatment of watermelon blossom with the *A. avenae* subsp. *avenae* strain was also successful in reducing seed infestation via the watermelon blossom. However, as noted by the authors, the *A. avenae* subsp. *avenae* strain used in these experiments is pathogenic to maize and therefore could not be commercialised as a biocontrol agent (Fessehaie & Walcott, 2005).

No varieties resistant to bacterial fruit blotch exist and under disease conducive conditions significant losses will occur. Fruit involvement is the reason for major economic losses because of bacterial fruit blotch. It has been reported that fruit with a darker rind, in comparison to fruit with a lighter rind, are more resistant to fruit involvement (Hopkins, 1993). Triploid watermelons have been reported to be more resistant than diploid watermelons (Rhodes, Zhang, Garrett, & Fang, 1996).

2.2.5. Detection of A. avenae *subsp.* citrulli

As seed borne transmission is one of the most common mechanisms leading to infection of crops with *A. avenae* subsp. *citrulli* testing of seed for the pathogen is the best way to control the spread of the disease. At present the "gold-standard" test for the presence of the bacterium is the seedling grow out assay conducted under conditions favourable for symptom development (Lovic & Hopkins, 2003). After inspection of seedlings for the presence of symptoms isolations of the causative organism are made and *A. avenae* subsp. *citrulli* identified using biochemical, serological or DNA based methods (see below). The seedling grow out test suffers from being slow, taking a month to complete, is very expensive (Lovic & Hopkins, 2003) and requires the destructive use of large numbers of seeds (Walcott, 2003).

Characterization and identification of *A. avenae* subsp. *citrulli* is commonly based on morphological, physiological and biochemical properties, disease symptoms, host range and hypersensitivity reaction in tobacco (Assouline, 1996; Assouline
et al., 1997; Rane & Latin, 1992; Schaad et al., 1978; Somodi et al., 1991; Wall et al., 1990). Biochemical techniques employing the Biolog GN Microplate automated identification system as well as the Microbial Identification System (MIS) fatty acid profiles (Assouline, 1996; Assouline et al., 1997; Black et al., 1994; Evans & Mulrooney, 1991; Hamm et al., 1997; Serfontein et al., 1997) are often combined with pathogenicity tests on watermelon for identification of the pathogen (Isakeit et al., 1997; Isakeit et al., 1998; Jones, Chase, & Harris, 1993). Serological based techniques are commercially available from various sources and are commonly employed to confirm that bacteria isolated from symptomatic seedlings in a grow out test are *A. avenae* subsp. *citrulli*.

Various molecular tests to detect the pathogen in infected seed have been developed primarily based upon the polymerase chain reaction the sequences of most of the published primer pairs are presented in *Table 3*. Minsavage et al. (1995) developed a set of oligonucleotide primers for the detection of *A. avenae* subsp. *citrulli* on seeds using PCR. However, Song et al. (1998) have questioned the reliability of this PCR and the primer pair also amplifies other plant pathogens in the genus *Acidovorax*, although *A. avenae* subsp. *citrulli* could be differentiated from other species using a restriction enzyme analysis of the amplified product. An immunomagnetic separation (IMS) method has also been developed for the detection of *A. avenae* subsp. *citrulli* in watermelon seeds (Walcott & Gitaitis, 2000). While this method is very sensitive and removes potential inhibitory substances which may prevent the PCR from working, as with all PCR based methods, the efficiency of target cell recovery is dependent on the levels of *A. avenae* subsp. *citrulli* in the seed wash. Song et al. (1998) have cloned and sequenced the internal transcribed spacer region between the 16S and 23S rRNA genes from a strain of *A. avenae* subsp. *citrulli* and used this sequence to design specific primers. Primers Aacf2/Aacr2 (*Table 3*) specific to *A. avenae* subsp. *citrulli* amplified a product of 450-bp in size using PCR (Song et al., 1998). This primer pair, together with other specific primer pairs of varying specificities, has been patented (Song, Kim, & Schaad, 2001). Song et al. (2003) have recently also described a real-time PCR assay for the specific identification of *A. avenae* subsp. *citrulli* (*Table 3*).

A. avenae *subsp.* cattleyae

2.2.6. Taxonomic history of A. avenae *subsp.* cattleyae

"*Bacterium cattleyae*" was first isolated from diseased orchids with leaf spot by Pavarino (1911) in Italy together with three other bacteria, "*Bacillus farnetianus*", "*Bacillus pollacii*", and "*Bacterium krameriani*". Ark and Thomas (1946) identified *Phytomonas cattleyae* as the causal agent of leaf spot and bud rot disease on *Cattleya* spp. and *Phalaenopsis* spp. They concluded that "*Phytomonas cattleyae*" was probably the bacterium "*Bacterium cattleyae*" described by Pavarino (1911), due to the similarity of phenotypic properties including the symptoms of the disease. One year later, Savulescu (1947) transferred "*Phytomonas cattleyae*" to *Pseudomonas cattleyae.*

In Bergey's Manual of Systematic Bacteriology, the pathogen was not assigned to any of the rRNA homology groups, and was placed within Section V (Palleroni, 1984). However, the close relationship of the type strain, NCPPB 961, to rRNA homology group III pseudomonads and *A. avenae* in particular was demonstrated by De Vos et al. (1985) using DNA:rRNA hybridisation, by Stead (1992) based on cellular fatty acid profiles, and by Hu et al. (1991) on the basis of polyphasic characterisation. The taxon was subsequently placed in the genus *Acidovorax* as *A. avenae* subsp. *cattleyae* (Willems et al., 1992).

2.2.7. Diseases caused by A. avenae *subsp.* cattleyae

The natural hosts of *A. avenae* subsp. *cattleyae* are *Cattleya* spp., *Phalaenopsis* spp. and their hybrids (Ark & Thomas, 1946; A. D. Ding, Liu, & Qi, 1993; Quimio & Tabei, 1979). When artificially inoculated, the following orchids are susceptible: *Epidendrum O'breinianum*, *Dendrobium* sp., *Cypripedium* sp., *Phalaenopsis amabilis,* and Vanilla (Ark & Thomas, 1946).

The typical symptom of the disease is leaf spot. Lesions initially appear water-soaked on the leaves. These lesions increase in size rapidly, and colour changes from light brown to dark brown with age. A considerable amount of exudate may also be observed. Lesions can occur on any part of the leaf, older spots may be surrounded by a light green or yellow halo. Spots may coalesce and form large irregular patches of dark brown to black areas of tissues. The disease can kill seedlings and mature plants, if the infections reach the growing point (Ark & Thomas, 1946; A. D. Ding et al., 1993; Quimio & Tabei, 1979).

2.2.8. Epidemiology and control of A. avenae *subsp.* cattleyae

The disease caused by *A. avenae* subsp. *cattleyae* on *Cattleya* spp., *Phalaenopsis* spp. and hybrids has been reported in Australia (Duff, 1997; Stovold, Bradley, & Fahy, 2001), the Philippines (Quimio & Tabei, 1979), Italy (Pavarino, 1911), China (A. D. Ding et al., 1993), Taiwan (Wey, 1988), USA (Ark & Thomas, 1946; Frank, 1988), and Portugal (Martins, 1981). Heavy losses have been caused to the orchid industry in USA, China, and the Philippines (Ark & Thomas, 1946; A. D.

Ding et al., 1993; Quimio & Tabei, 1979; Tabei & Quimio, 1978). However, the disease has occurred only sporadically since first reported in 1911.

Bacterial exudate from heavily infected plants may act as the source of inoculum (Ark & Thomas, 1946). The bacterium is thought to enter the plant through the stomata (Ark & Thomas, 1946).

Removal of diseased material and avoidance of overhead watering can achieve control of the disease (Ark & Thomas, 1946). Reduction of prolonged wetness on leaves, by increasing air circulation, rinsing with a quaternary ammonium compound such as Physan; or carefully swabbing the diseased area with a sponge soaked in 1 in 1000 corrosive sublimate ($HgCl_2$), can also reduce symptoms (Ark & Thomas, 1946; Frank, 1988).

2.2.9. Detection of A. avenae *subsp.* cattleyae

Characterization and identification of *A. avenae* subsp. *cattleyae* is commonly based on morphological, physiological and biochemical properties, disease symptoms, host range and hypersensitivity reaction in tobacco (Ark & Thomas, 1946; A. D. Ding et al., 1993; Quimio & Tabei, 1979; Tabei & Quimio, 1978).

3. PATHOGENICITY/VIRULENCE FACTORS AND HOST RESPONSE TO INFECTION WITH *A. AVENAE*

Little is known about the pathogenicity or virulence factors produced by *A. avenae* subspecies. However, *A. avenae* subsp. *avenae* some *A. avenae* subsp. *citrulli* and *A. avenae* subsp. *cattleyae* have antifungal activity, which may help these organisms compete in their natural environment (F. P. Hu & Young, 1998). *A. avenae* subsp. *avenae* shows a weak hybridisation signal to probes for some of the genes for alginate biosynthesis a known virulence factor for *Pseudomonas syringae* pv. *phaseolicola* (Fett, Osman, Dunn, & Panopoulos, 1992; Koopmann, Rollwage, Nollenburg, & Rudolph, 2001). However, *A. avenae* subsp. *avenae* does not produce alginate (Fett et al., 1992) and the relevance of the weak hybridisation signal to probes for some of the genes for alginate biosynthesis is unknown. *A. avenae* subsp. *avenae* and *A. avenae* subsp. *citrulli* have homologues of genes involved in type III secretion indicated by sequences submitted to the GenBank database (accession numbers AB053454 and AY898625 respectively). Although an active type III secretion system has not been confirmed in *A. avenae* it would be expected that given this type III secretion system is active it would contribute to the virulence of these organisms. It has recently been found that motility of *A. avenae* subsp. *citrulli* strains may be required for the migration of the pathogen through the ovaries of female watermelon blossoms (Walcott, Lessl, & Fessehaie, 2005) a potentially important mechanism of ingress of the pathogen into fruit and therefore seed infestation.

The response of plants to infection with *A. avenae* has received little attention. However, the interaction of a compatible strain and an incompatible strain of *A. avenae* subsp. *avenae* with rice led to the identification of differentially expressed rice genes including a gene potentially involved in repair of the plant cell wall (Che et al., 2002).

4. OTHER PLANT PATHOGENIC *ACIDOVORAX* SP.

Acidovorax konjaci

The species *A. konjaki* contains strains originally described as *Pseudomonas psedoalcaligenes* subsp. *konjaci* (M. Goto, 1983). *A. konjaci* strains cause bacterial leaf blight of konjac (*Anorphophallus konjac*) a root vegetable. The major symptoms of the disease are leaf spots and leaf blight but under severe conditions the petioles become infected and the plant will wilt and the roots may rot (M. Goto, 1983). The disease is only known in Japan. The bacterium enters the plant through stomatal openings and wounds and spreads by wind, rain and contact with diseased material (Hayashi, 1991b). The organism can survive for extended periods of time in plant debris and soil and may serve as an inoculation source for future crops (Hayashi, 1989). The bacterium may be splashed directly onto leaflets of newly emerging plants from the soil by rain splash (Hayashi, 1991a)

Acidovorax anthurii

A. anthurii causes bacterial leaf-spot of anthurium which is a serious limiting factor in commercial anthurium production in the French West Indies and Trinidad and Tobago (Gardan et al., 2000; Prior & Rott, 1989; Saddler, O'Grady, & Spence, 1995) and is becoming more widespread in these countries (Gardan et al., 2000). The disease begins as necrotic lesions close to the veins and leaf margins which become chlorotic and water soaked, blacken and turn grey. The bacteria become systemic and which results in tissue discolouration and finally plant death (Gardan et al., 2000; Saddler, 2000).

There is evidence from DNA fingerprinting analysis of *A. anthurii* using raqndom amplified polymorphic DNA analysis that there is variation below the species level with two groups of isolates being recognised (Saddler et al., 1995). However, it is unknown if these subgroups relate to any phenotypic characteristic or virulence of these isolates (Saddler et al., 1995).

Problems with the phytotoxicity of copper for *Anthurium* spp. and the lack of resistant varieties mean that the only way to control the disease is strict sanitation (Prior & Rott, 1989).

Acidovorax valerianellae

A. valerianellae causes black spot of corn salad (also known as lambs lettuce) (*Valerianella locusta*) in Europe (Gardan et al., 2003; Grondeau, Cerceau, Bureau, & Samson, 2003b; Moltmann et al., 2000). The pathogen was described as a new species on the basis of phenotypic and genotypic properties (Gardan et al., 2003).

The disease is considered the most important disease of corn salad in France as crops can become unmarketable (Grondeau, Cerceau, Bureau, & Samson, 2003a). The disease is thought to be soil transmitted with the bacterium being detected in contaminated soil and root debris (Grondeau et al., 2003a). It is also possible that seed transmission occurs. Consistent cultivation of only *Valerianella locusta* is advised against due to the soil-borne nature of the disease.

Acidovorax *sp., a pathogen of geranium and petunia*

An *Acidovorax* sp. has been isolated from leaf spots on geranium (Simone, Cullen, & Hodge, 1996) and petunia (Jones et al., 2003). Based upon DNA:DNA hybridisation data the *Acidovorax* sp. strains causing these diseases potentially represent a new species within the genus (Jones et al., 2003).

5. *HERBASPIRILLUM RUBRISUBALBICANS*

Taxonomic history of H. rubisubalbicans

The genus *Herbaspirillum* was established by Baldani et al (1986) and belongs to the β-subclass of the *Proteobacteria* (*Fig. 1*). The genus contains seven described species, *H. autotrophicum*, *H. lusitanum*, *H. frisingense*, *H. putei*, *H. huttiense*, *H. seropedicae* and *H. rubrisubalbicans* (L. X. Ding & Yokota, 2004). The only known plant pathogenic member of this genus is *H. rubisubalbicans* (Baldani et al., 1996). However, *H. seropedicae* has been reported to cause symptoms of red stripe disease when artificially inoculated into sorghum (James, Olivares, Baldani, & Döbereiner, 1997). Many of species in this genus, including *Herbaspirillum rubisubalbicans*, exhibit the ability to fix nitrogen (Baldani et al., 1996; L. X. Ding & Yokota, 2004).

Diseases caused by H. rubisubalbicans

H. rubisubalbicans has been described as a "mild plant pathogen" causing disease of sugar cane and sorghum. *H. rubisubalbicans* (syn. *Pseudomonas rurisubalbicans*) was originally described by Christopher and Edgerton (1932) as the causative agent of mottled stripe disease of sugar cane and is also known to cause bacterial leaf stripe of sorghum (Hale & Wilkie, 1972). Typical disease symptoms on sugar cane are the presence of mottled patches and red stripes on leaves. The stripes on the leaf blade are primarily red in colour but white margins commonly occur giving the appearance of a red stripe on a white background (Christopher & Edgerton, 1932). On Sorghum *H. rubisubalbicans* causes dark red lesions on the leaves and leaf sheaths, young lesions appear as streaks between veins whereas older lesions may cover a large area of the leaf surface. In severe cases the disease may lead to premature senescence of the leaf (Hale & Wilkie, 1972). The xylem tissue of infected plants becomes almost completely occluded by bacterial cells (Olivares, James, Baldani, & Döbereiner, 1997)

Epidemiology and control of H. rubisubalbicans

H. rubisubalbicans has a world wide distribution being recorded in Asia, Africa, North America, Central America, South America, and Australia (Steindl & Edgerton, 1964). The epidemiology of the disease has not been widely investigated, probably due to the mild nature of the disease. However, Steindl and

Edgerton (Steindl & Edgerton, 1964) note that the disease is probably spread in the field by wind and rain. Most agronomically important cultivars of sugar cane appear to be resistant to infection (Olivares et al., 1997; Pimentel et al., 1991) and although symptoms may become plentiful in summer sufficient leaf area is seldom affected to cause crop losses (Steindl & Edgerton, 1964). *H. rubisubalbicans* can survive for some time on, and colonise, leaf surfaces which allows the organism to enter leaves through the stomata (Olivares et al., 1997).

Although, as previously mentioned, *H. seropedicae* causes a pathogenic response in sugarcane on artificial inoculation it is not naturally found in the leaves of plants being confined to the stems and roots (Olivares et al., 1997). *H. rubisubalbicans*, on the other hand, is naturally found associated with the roots, stems and leaves of plants (Olivares et al., 1997). Olivares et al. (1997) have speculated that *H. rubisubalbicans* and *H. seropedicae* may be partly responsible for the nitrogen fixation attributed to diazotrophic microorganisms in sugarcane. If this is the case then *H. rubisubalbicans* has both positive and negative effects on plant growth.

6. REFERENCES

Akhtar, M. A., & Aslam, M. (1986). Resistance of sugarcane to bacterial red stripe, a new bacterial disease from Pakistan. *Tropical Pest Management,* 32, 118-119.

Ark, P. A., & Thomas, H. E. (1946). Bacterial leaf spot and bud rot of orchids caused by Phytomonas cattleyae. *Phytopathology,* 36, 695-698.

Assis, S. M. P., Mariano, R. L. R., Silva Hanlin, D. M. W., & Duarte, V. (1999). Bacterial fruit blotch caused by Acidovorax avenae subsp. citrulli in melon in the state of Rio Grande do Norte, Brazil. *Fitopatologia Brasileira,* 24, 191.

Assouline, I. (1996). He watermelon fruit blotch disease and other diseases caused by Acidovorax avenae. *Phytoparasitica,* 24, 136-137.

Assouline, I., Milshtein, H., Mizrahi, M., Levy, E., & BenZeev, I. S. (1997). *Acidovorax avenae* subsp. *citrulli* transmitted by solanaceous seeds. *Phytoparasitica,* 25, 117-118.

Baldani, J. I., Baldani, V. L. D., Seldin, L., & Dobereiner, J. (1986). Characterization of *Herbaspirillum seropedicae* gen. nov., sp. nov., a root associated nitrogen fixing bacterium. *International Journal of Systematic Bacteriology,* 36, 86-93.

Baldani, J. I., Pot, B., Kirchhof, G., Falsen, E., Baldani, V. L. D., Olivares, F. L., et al. (1996). Emended description of *Herbaspirillum*; Inclusion of *Pseudomonas rubrisubalbicans*, a mild plant pathogen, as *Herbaspirillum rubrisubalbicans* comb. nov.; And classification of a group of clinical isolates (EF group 1) as *Herbaspirillum* species 3. *International Journal of Systematic Bacteriology,* 46, 802-810.

Billimoria, K. N., & Hegde, R. K. (1971). New Bacterial Disease of Ragi, Eleusine-Coracana (Linn) Gaertn in Mysore State. *Current Science,* 40, 611-612.

Black, M. C., Isakeit, T., Barnes, L. W., Kucharek, T. A., Hoover, R. J., & Hodge, N. C. (1994). First report of bacterial fruit blotch of watermelon in Texas. *Plant Disease,* 78, 831.

Bradbury, J. F. (1986). *Guide to Plant Pathogenic Bacteria*. Wallingford, UK: CAB International.

Che, F. S., Entani, T., Marumoto, T., Taniguchi, M., Takayama, S., & Isogai, A. (2002). Identification of novel genes differentially expressed in compatible and incompatible interactions between rice and *Pseudomonas avenae*. *Plant Science,* 162, 449-458.

Che, F. S., Iwano, M., Tanaka, N., Takayama, S., Minami, E., Shibuya, N., et al. (1999). Biochemical and morphological features of rice cell death induced by Pseudomonas avenae. *Plant and Cell Physiology,* 40, 1036-1045.

Christopher, W. N., & Edgerton, C. W. (1932). Bacterial stripe diseases of sugar cane in Louisiana. *Journal of Agricultural Research,* 41, 259.

Claflin, L. E., Ramundo, B. A., Leach, J. E., & Erinle, I. D. (1989). *Pseudomonas avenae*, causal agent of bacterial leaf stripe of pearl millet. *Plant Disease,* 73, 1010-1014.

CMI. (1995). *Acidovorax avenae subsp. avenae. Distribution Maps of Plant Diseases No 511.* Wallingford, UK: CAB International.

Cottyn, B., Cerz, M. T., & Mew, T. W. (1994). Bacterial Pathogens. In T. W. Mew & J. K. Misra (Eds.), *A manual of rice seed health testing* (pp. 91-96). Manila, The Philippines: International Rice Research Institute.

Cottyn, B., Outryve, M. F. V., Cerez, M. T., Cleene, M. D., Swings, J., & Mew, T. W. (1996). Bacterial diseases of rice. II. Characterization of pathogenic bacteria associated with sheath rot complex and grain discoloration of rice in the Philippines. *Plant Disease,* 80, 438-445.

Cottyn, B., Van Outryve, M. F., Cerez, M. T., De Cleene, M., Swings, J., & Mew, T. W. (1996). Bacterial diseases of rice. II. Characterization of pathogenic bacteria associated with sheath rot complex and grain discoloration of rice in the Philippines. *Plant Disease,* 80, 438-445.

Crall, J. M., & Schenck, N. C. (1969). Bacterial fruit rot of watermelon in Florida. *Plant Disease Reporter,* 53, 74-75.

Dange, S. R. S., & Payak, M. M. (1972). Additional hosts for red stripe disease of sugarcane. *Sugarcane Pathologists' Newsletter,* 9, 8.

Dange, S. R. S., Payak, M. M., & Renfro, B. L. (1978). Seed transmission of *Pseudomonas rubrilineans*, the incitant of bacterial leaf stripe of maize. *Indian Phytopathology,* 31, 523-524.

de Oliveira, J. R., & Moura, A. B. (1994). Diseases caused by bacteria on Cucurbitaceae. *Informe Agropecuario Belo Horizonte,* 17, 54-57.

Demir, G. (1996a). Disease and pest outbreaks. Turkey. A new bacterial disease of watermelon in Turkey: bacterial fruit blotch of watermelon. *Arab and Near East Plant Protection Newsletter*, 33.

Demir, G. (1996b). A new bacterial disease of watermelon in Turkiye: bacterial fruit blotch of watermelon (Acidovorax avenae subsp. citrulli (Schaad et al.) Willems et al.). *Journal of Turkish Phytopathology,* 25, 43-49.

DeVos, P., Goor, M., Gillis, M., & De Ley, J. (1985). Ribonsomal ribonucleic acid cistron similarities of phytopathogenic Pseudomonas species. *International Journal of Systematic Bacteriology,* 35, 169-184.

Ding, A. D., Liu, C. Z., & Qi, P. K. (1993). Identification of the pathogen causing bacterial brown spot of Phalaenopsis aphrodite Reichb. *Journal of South China Agricultural University,* 14, 124-126.

Ding, L. X., & Yokota, A. (2004). Proposals of *Curvibacter gracilis* gen. nov., sp. nov. and *Herbaspirillum putei* sp. nov. for bacterial strains isolated from well water and reclassification of *Pseudomonas huttiensis*, *Pseudomonas lanceolata*, *Aquaspirillum delicatum* and *Aquaspirillum autotrophicum* as *Herbaspirillum huttiense* comb. nov., *Curvibacter lanceolatus* comb. nov., *Curvibacter delicatus* comb. nov. and *Herbaspirillum autotrophicum* comb. nov. *International Journal of Systematic and Evolutionary Microbiology,* 54, 2223-2230.

Duff, J. D. (1997). *Orchid diseases of the Northern Territory*: Darwin, Australia: Department of Industries and Fisheries, Northern Territory.

Elliott, C. (1923). A bacterial stripe disease of Proso millet. *Journal of Agricultural Research,* 26, 151-159.

Evans, T. A., & Mulrooney, R. P. (1991). First report of watermelon fruit blotch in Delaware. *Plant Disease,* 75, 1074.

Fahy, P., Gillings, M., Bradley, J., Diatloff, A., & Singh, S. (1989). Use of fatty acid profiles and restriction fragment length polymophism to trace a quarantine outbreak of *Pseudomonas avenae* on French and Italian millet. In Z. Klement (Ed.), *Proceedings of the 7th International Conference on Plant Pathogenic Bacteria* (pp. 491-496). Budapest, Hungary: Akadémiai Kiado.

Fessehaie, A., Hopkins, D. L., Gitaitis, R. D., & Walcott, R. R. (2005). Role of honey bees in watermelon seed infestation by Acidovorax avenae subsp. citrulli. *Phytopathology,* 95, S29.

Fessehaie, A., & Walcott, R. R. (2005). Biological control to protect watermelon blossoms and seed from infection by *Acidovorax avenae* subsp. *citrulli. Phytopathology,* 95, 413-419.

Fett, W., Osman, S. F., Dunn, M. F., & Panopoulos, N. J. (1992). Cell surface properties of Pseudomonas syringae pv. phaseolicola wild-type and hrp mutants. *Journal of Phytopathology,* 135, 135-152.

Frank, D. W. (1988). Control of bacterial soft rot caused by Pseudomonas cattleya in the culture of Phalaenopsis. *Orchid Digest,* 52, 66-67.

Frankle, W. G., Hopkins, D. L., & Stall, R. E. (1993). Ingress of the Watermelon Fruit Blotch Bacterium into Fruit. *Plant Disease,* 77, 1090-1092.

Gardan, L., Dauga, C., Prior, P., Gillis, M., & Saddler, G. S. (2000). *Acidovorax anthurii* sp. nov., a new phytopathogenic bacterium which causes bacterial leaf-spot of anthurium. *International Journal of Systematic and Evolutionary Microbiology,* 50, 235-246.

Gardan, L., Stead, D. E., Dauga, C., & Gillis, M. (2003). *Acidovorax valerianellae* sp. nov., a novel pathogen of lamb's lettuce [Valerianella locusta (L.) Laterr.]. *International Journal of Systematic and Evolutionary Microbiology,* 53, 795-800.

Gitaitis, R. D., Samuelson, D. A., & Strandberg, J. O. (1981). Scanning electron microscopy of the ingress and establishment of *Pseudomonas alboprecipitans* in sweet corn leaves. *Phytopathology,* 71, 171-175.

Gitaitis, R. D., Stall, R. E., & Strandberg, J. O. (1978). Dissemination and survival of *Pseudomonas alboprecipitans* ascertained by disease distribution. *Phytopathology,* 68, 227-231.

Goth, R. W., & Webb, R. E. (1975). Bacterial wilt of watermelon. *Proceedings of American Phytopathology Society,* 2, 122-123.

Goto, K., & Ohata, K. I. (1961). Bacterial stripe of rice. *Special Publication of College of Agriculture, Taiwan University,* 10, 49-57.

Goto, M. (1983). *Pseudomonas pseudoalcaligenes* subsp. *konjaci* subsp. nov., the causal agent of bacterial leaf blight of konjac (*Amorphophalus konjac* Koch.). *International Journal of Systematic Bacteriology,* 33, 539-545.

Goto, M., & Okabe, N. (1952). Studies on the causal organisms of bacterial stripe disease of millet and the brown stripe disease of Italian millet. *Bulletin, Faculty of Agriculture, Shizuoka University,* 2, 15-24.

Goto, M., & Starr, M. P. (1971). A comparative study of *P. andropogonis*, *P. stizolobii* and *P. alboprecipitans*. *Annals of the Phytopathological Society of Japan,* 37, 233-241.

Grondeau, C., Cerceau, V., Bureau, C., & Samson, R. (2003a). Evidence that *Acidovorax valerianellae*, bacterial black spot of corn salad (*Valerianella locusta*) agent, is soil transmitted. In N. S. Iacobellis, A. Collmer, S. W. Hutcheson, J. W. Mansfield, C. E. Morris, J. Murillo, N. W. Schaad, D. E. Stead, G. Surico & M. S. Ullrich (Eds.), *Pseudomonas syringae pathovars and related pathogens; Biology and Genetic* (pp. 89-91). Dordrecht, Netherlands: Kluwer Academic Publishers.

Grondeau, C., Cerceau, V., Bureau, C., & Samson, R. (2003b). Evidence that *Acidovorax valerianellae*, bacterial black spot of corn salad (*Valerianella locusta*) agent, is soil transmitted. In N. S. Iacobellis, A. Collmer, S. W. Hutcheson, J. W. Mansfield, C. E. Morris, J. Murillo, N. W. Schaad, D. E. Stead, G. Surico & M. S. Ullrich (Eds.), *Pseudomonas syringae and related pathogens* (pp. 89-91). Dordrecht, The Netherlands: Kluwer Academic Publishers.

Grothues, D., & Rudolph, K. (1991). Macrorestriction analysis of plant pathogenic *Pseudomonas* species and pathovars. *FEMS Microbiology Letters,* 79, 83-88.

Hale, C. N., & Wilkie, J. P. (1972). Bacterial leaf stripe of sorghum in New Zealand caused by *Pseudomonas rubrisubalbicans*. *New Zealand Journal of Agricultural Research,* 15, 457-460.

Hamm, P. B., Spink, D. S., Clough, G. H., & Mohan, K. S. (1997). First report of bacterial fruit blotch of watermelon in Oregon. *Plant Disease,* 81, 113.

Hayashi, N. (1989). Survival of Pseudomonas pseudoalcaligenes subsp. konjaci, the causal agent of bacterial leaf blight of Konnyaku (Amorphophallus konjaci), in residues of diseased leaflets, on corms and in soil. *Annals of the Phytopathological Society of Japan,* 55, 609-614.

Hayashi, N. (1991a). Dissemination of Pseudomonas pseudoalcaligenes subsp. konjaci, the causal agent of bacterial leaf blight of konnyaku (Amorphophallus konjac), from primary inoculum source to konnyaku leaflet. *Annals of the Phytopathological Society of Japan,* 57, 49-52.

Hayashi, N. (1991b). Secondary dissemination of konnyaku bacterial leaf blight caused by Pseudomonas pseudoalcaligenes subsp. konjaci. *Annals of the Phytopathological Society of Japan,* 57, 345-350.

Hayward, A. C. (1962). Studies on bacterial pathogens of sugar cane. II. Differentiation, taxonomy and nomenclature of the bacteria causing red stripe and mottled stripe diseases. *Mauritius Sugar Industry Research Institute Occasional Paper,* 13, 13-17.

Hopkins, D. L. (1989). Bacterial fruit blotch of watermelon: a new disease in the eastern USA. *Proceedlings of Cucurbitaceae,* 89, 74-75.

Hopkins, D. L. (1991). Chemical control of bacterial fruit blotch of watermelon. *Proceedings of the Florida State Horticultural Society,* 104, 270-272.

Hopkins, D. L. (1993). Field spread of bacterial fruit blotch of watermelon. *Phytopathology,* 83, 466.

Hopkins, D. L. (1994). Spread of bacterial fruit blotch of watermelon in greenhouse. *Phytopathology,* 84, 775.

Hopkins, D. L. (1995). Copper-containing fungicides reduce the spread of bacterial fruit blotch of watermelon in the greenhouse. *Phytopathology,* 85, 510.

Hopkins, D. L. (2005). Control of greenhouse spread of bacterial fruit blotch of cucurbits with peroxyacetic acid and ionized copper in the irrigation water. *Phytopathology,* 95, S44.

Hopkins, D. L., Cucuzza, J. D., & Watterson, J. C. (1996). Wet seed treatments for the control of bacterial fruit blotch of watermelon. *Plant Disease,* 80, 529-532.

Hopkins, D. L., & Thompson, C. M. (2002). Evaluation of *Citrullus* sp. germ plasm for resistance to *Acidovorax avenae* subsp. *citrulli. Plant Disease,* 86, 61-64.

Hsu, S. T., Chang, R. J., Tzeng, K. C., & Leung, L. K. (1991). Occurrence of bacterial stripe of corn in Taiwan. *Plant Protection Bulletin,* 33, 376-383.

Hu, F. P., & Young, J. M. (1998). Biocidal activity in plant pathogenic *Acidovorax, Burkholderia, Herbaspirillum, Ralstonia* and *Xanthomonas* spp. *Journal of Applied Microbiology,* 84, 263-271.

Hu, F. P., Young, J. M., & Triggs, C. M. (1991). Numerical analysis and determinative tests for nonfluorescent plant-pathogenic *Pseudomonas* spp. and genomic analysis and reclassification of species related to *Pseudomonas avenae* Manns 1909. *International Journal of Systematic Bacteriology,* 41, 516-525.

Hu, F.-P., Young, J. M., Triggs, C. M., & Wilkie, J. P. (1997). Pathogenic relationships of the subspecies of *Acidovorax avenae. Australasian Plant Pathology,* 26, 227-238.

Isakeit, T., Black, M. C., Barnes, L. W., & Jones, J. B. (1997). First report of infection of honeydew with *Acidovorax avenae* subsp. *citrulli. Plant Disease,* 81, 694-694.

Isakeit, T., Black, M. C., & Jones, J. B. (1998). Natural infection of citronmelon with *Acidovorax avenae* subsp. *citrulli. Plant Disease,* 82, 351-351.

James, E. K., Olivares, F. L., Baldani, J. I., & Döbereiner, J. (1997). *Herbaspirillum*, an endophytic diazotroph colonizing vascular tissue in leaves of *Sorghum bicolor* L. Moench. *Journal of Experimental Botany,* 48, 785-797.

Johnson, A. G., Robert, A. L., & Cash, L. (1949). Bacterial leaf blight and stalk rot of corn. *Journal of Agricultural Research,* 78, 719-732.

Jones, J. B., Chase, A. R., & Harris, G. K. (1993). Evaluation of the Biolog GN MicroPlate system for identification of some plant-pathogenic bacteria. *Plant Disease,* 77, 553-558.

Jones, J. B., Gitaitis, R. D., & Schaad, N. W. (2001). *Acidovorax* and *Xylophilus.* In N. W. Schaad, J. B. Jones & W. Chun (Eds.), *Laboratory guide for the identification of plant pathogenic bacteria* (pp. 121-138). St. Paul, Minnisota, USA: APS Press.

Jones, J. B., Stall, R. E., Minsavage, G. V., Dickstein, E. R., Bouzar, H., Roberts, P. D., et al. (2003). Characterisation of an *Acidovorax* sp. associated with geranium and petunia. In N. S. Iacobellis, A. Collmer, S. W. Hutcheson, J. W. Mansfield, C. E. Morris, J. Murillo, N. W. Schaad, D. E. Stead, G. Surico & M. S. Ullrich (Eds.), *Pseudomonas syringae and related pathogens* (pp. 667-673). Dordrecht, The Netherlands: Kluwer Academic Publishers.

Kadota, I. (1996). Studies on the pathogen of bacterial brown stripe of rice and its ecology. *Bulletin of the Hokuriku National Agricultural Experiment Station,* 38, 113-171.

Kadota, I., & Ohuchi, A. (1983). Symptoms of bacterial brown stripe of rice seedlings in nursery boxes. *Annals of the Phytopathological Society of Japan,* 49, 561-564.

Kadota, I., & Ohuchi, A. (1990). Symptoms and ecology of bacterial brown stripe of rice. *Japan Agricultural Research Quarterly,* 24, 15-21.

Kadota, I., Ohuchi, A., & Nishiyama, K. (1991). Serological properties and specificity of *Pseudomonas avenae* Manns 1909, the causal agent of bacterial brown stripe of rice. *Annals of the Phytopathological Society of Japan,* 49, 561-564.

Karganilla, A. D., & Cabauatan, P. Q. (1974). Bacterial leaf stripe of corn in the Philippines. *The Philippine Agriculturist,* 58, 105-114.

Kim, H. M., & Song, W. Y. (1996). Characterization of ribosomal RNA intergenic spacer region of several seedborne bacterial pathogens of rice. *Seed Science and Technology,* 24, 571-580.

Knauss, J. F., Miller, J. W., & Virgona, R. J. (1978). Bacterial blight of fishtail palm, a new disease. *Proceedings of the Florida State Horticultural Society,* 91, 245-247.

Koopmann, B., Rollwage, H., Nollenburg, M., & Rudolph, K. (2001). Isolation and characterization of the *algD* gene of *Pseudomonas syringae* pv. *phaseolicola* and its distribution among other pseudomonads and related organisms. *Journal of Phytopathology-Phytopathologische Zeitschrift,* 149, 511-519.

Langston, D. B., Jr., Walcott, R. D., Gitaitis, R. D., & Sanders, F. H., Jr. (1999). First report of a fruit rot of pumpkin caused by *Acidovorax avenae* subsp. *citrulli* in Georgia. *Plant Disease,* 83, 199.

Latin, R. X., & Hopkins, D. L. (1995). Bacterial fruit blotch of watermelon - the hypothetical exam question becomes reality. *Plant Disease,* 79, 761-765.

Lee, H. A., Purdy, H. A., Barnum, C. C., & Martin, J. P. (1925). A comparison of red-stripe of sugar cane and other grasses. In *Red-stripe disease studies* (pp. 64-74): Bulletin, Experiment Station of Hawaiian Sugar Planters' Association.

Lovic, B. R., & Hopkins, D. L. (2003). Production steps to reduce seed contamination by pathogens of cucurbits. *Horttechnology,* 13, 50-54.

Ludwig, W., Strunk, O., Westram, R., Richter, L., Meier, H., Yadhukumar, et al. (2004). ARB: a software environment for sequence data. *Nucleic Acids Research,* 32, 1363-1371.

Manns, T. F. (1909). The blade blight of oats - a bacterial disease. *Bulletin of Ohio Agricultural Experiment Station,* 210, 91-167.

Martin, H. L., & Horlock, C. M. (2002). First report of Acidovorax avenae subsp. citrulli as a pathogen of gramma in Australia. *Plant Disease,* 86, 1406.

Martin, H. L., O'Brien, R. G., & Abbott, D. V. (1999). First report of *Acidovorax avenae* subsp. *citrulli* as a pathogen of cucumber. *Plant Disease,* 83, 965.

Martins, J. M. S. (1981). Bacterial diseases of ornamental plants in Portugal. III. Pseudomonas spp. form orchids. *Agronomia Lusitana,* 41, 67-75.

Minsavage, G. V., Hoover, R. J., Kucharek, T. A., & Stall, R. E. (1995). Detection of the watermelon fruit blotch pathogen on seeds with the polymerase chain reaction. *Phytopathology,* 85, 1162.

Moltmann, E., Blum, E., Detzel, P., Riesterer, K., Krauss, J., & Schrameyer, K. (2000). Leaf spots in lamb's lettuce due to *Acidovorax valerianellae* bacteria. *Gemuse Munchen,* 36, 10-12.

Mudingotto, P. J., Veena, M. S., & Mortensen, C. N. (2002). First report of bacterial blight caused by *Acidovorax avenae* ssp. *avenae* associated with finger millet seeds from Uganda. *Plant Pathology,* 51, 396.

Nishiyama, K., Nishihara, N., & Ezuka, A. (1979). Bacterial brown stripe of ragi caused by *Pseudomonas alboprecipitans*. *Annals of the Phytopathological Society of Japan,* 45, 25-31.

Nomura, T., & Shirakawa, T. (2001). Efficacy of hot water and bactericide treatments of watermelon seeds infested by Acidovorax avenae subsp. citrulli. *Proceedings of the Kansai Plant Protection Society,* 1-6.

O'Brien, R. G. (1996). *Bacterial fruit blotch of watermelon. Farm Note*. Brisbane: Queensland Department of Primary Industries.

O'Brien, R. G., & Martin, H. L. (1999). Bacterial blotch of melons caused by strains of *Acidovorax avenae* subsp *citrulli*. *Australian Journal of Experimental Agriculture,* 39, 479-485.

Okabe, N. (1934). Bacterial diseases of plant occurring in Formosa. IV. *Journal of the Society of Tropical Agricriculture Taiwan,* 6, 54-63.

Olivares, F. L., James, E. K., Baldani, J. I., & Döbereiner, J. (1997). Infection of mottled stripe disease-susceptible and resistant sugar cane varieties by the endophytic diazotroph *Herbaspirillum*. *New Phytologist,* 135, 723-737.

Palleroni, N. J. (1984). Genus I. Pseudomonas Migula 1894. In N. R. Krieg., & J. G. Holt (Eds.), *Bergey's Manual of Systematic Bacteriology* (Vol. 1, pp. 141-199). Baltimore, USA: Williams & Wilkins.

Pataky, J. K., duToit, L. J., & Kerns, M. R. (1997). Bacterial leaf blight on shrunken-2 sweet corn. *Plant Disease,* 81, 1293-1298.

Pavarino, G. L. (1911). Malattie causate de bacterii nelle Orchidee. *Atti della Reale Accademia dei Lincei,* 20, 233-237.

Pimentel, J. P., Olivares, F., Pitard, R. M., Urquiaga, S., Akiba, F., & Dobereiner, J. (1991). Dinitrogen fixation and infection of grass leaves by *Pseudomonas rubrisubalbicans* and *Herbaspirillum seropedicae*. *Plant and Soil,* 137, 61-65.

Postnikova, E., & Schaad, N. W. (2002). Phylogenetic studies of corn and rice strains of *Acidovorax avenae* subsp. *avenae* by DNA/DNA hybridization. *Phytopathology,* 92, S66.

Prior, P., & Rott, P. (1989). Bacterial leaf-spot of Anthurium (*Anthurium* spp.) caused by a *Pseudomonas* sp. in the French West Indies. *Journal of Phytopathology,* 124, 215-224.

Quimio, A. J., & Tabei, H. (1979). Identity of the bacterium associated with bacterial brown spot of Phalaenopsis orchids. *Philippine Phytopathology,* 15, 76-80.

Ramundo, B. A., & Claflin, L. E. (1990). Demonstration of synonymy between the plant pathogens *Pseudomonas avenae* and *Pseudomonas rubrilineans*. *Journal of General Microbiology,* 136, 2029-2033.

Rane, K. K., & Latin, R. X. (1992). Bacterial fruit blotch of watermelon - association of the pathogen with seed. *Plant Disease,* 76, 509-512.

Rhodes, B. B., Zhang, X. P., Garrett, J. T., & Fang, C. (1996). Watermelon fruit blotch infection rates in diploids and triploids. *Report Cucurbit Genetics Cooperative*, 70-72.

Rosen, H. R. (1922). A bacterial disease of foxtail (Chaetochloa lutescens). *Annals of the Missouri Botanical Garden,* 9, 333-402.

Rott, P., Frossard, P., Vuong, H. H., & Katile, O. (1989). Des raies foliaires sur mil (*Pennisetum americanum*) et sorgho (*Sorghum bicolor*) provoquees par *Pseudomonas rubrilineans* au Mali. *L'agronomie Tropicale,* 44, 233-237.

Saddler, G. S. (1994). IMI descriptions of fungi and bacteria No. 1211: *Acidovorax avenae* subsp. *avenae*. *Mycopathologia,* 128, 41-43.

Saddler, G. S. (2000). *Acidovorax anthurii*. In *IMI descriptions of fungi and bacteria* (pp. No. 1451). Wallingford, UK: CAB International.

Saddler, G. S., O'Grady, E. B., & Spence, J. A. (1995). Characterization of a pseudomonad-like bacterium pathogenic for *Anthurium* species. *Bulletin OEPP,* 25, 211-217.

Sato, Z., Kawakubo, Y., Shirata, A., Matsuda, I., Tsuchiya, Y., & Date, H. (1983). Bending symptoms of young rice seedlings grown in nursery flat caused by *Pseudomonas avenae*, the causal agent of bacterial stripe of rice. I. Symptoms, inoculation method, factors affecting symptom appearance and varietal differences. *Bulletin of the National Institute of Agricultural Sciences, Japan, C,* 38, 161-168.

Savulescu, T. (1947). Contribution à la classification des bactériacées phytopathogènes. *Analele Academiei Romane Series III,* 22, 135-160.

Schaad, N. W., Kado, C. I., & Sumner, D. R. (1975). Synonymy of *Pseudomonas avenae* Manns 1905 and *Pseudomonas alboprecipitans* Rosen 1922. *International Journal of Systematic Bacteriology,* 25, 133-137.

Schaad, N. W., Sowell, G., Goth, R. W., Colwell, R. R., & Webb, R. E. (1978). *Pseudomonas pseudoalcaligenes* subsp. *citrulli* subsp. nov. *International Journal of Systematic Bacteriology,* 28, 117-125.

Serfontein, J. J., Uys, M. D. R., & Goszczynska, T. (1997). First report of bacterial fruit blotch of watermelon in South Africa. In *Proceedings of the 35th of Congress of the Southern African Society for Plant* (pp. 32).

Shakya, D. D. (1981). *Etiology and seed transmission of bacterial leaf stripe of rice.* Unpublished Ph.D Thesis, National University of Republic of Korea, Seoul.

Shakya, D. D. (1987). Rapid diagnosis of Pseudomonas avenae by pathogenicity and serology. *Korean Journal of Plant Pathology,* 3, 300.

Shakya, D. D., & Chung, H. S. (1983). Detection of *Pseudomonas avenae* in rice seed. *Seed Science and Technology,* 11, 583-587.

Shakya, D. D., Chung, H. S., & Vinther, F. (1986). Transmission of *Pseudomonas avenae*, the cause of bacterial stripe of rice. *Journal of Phytopathology-Phytopathologische Zeitschrift,* 116, 92-96.

Shakya, D. D., Vinther, F., & Mathur, S. B. (1985). World wide distribution of a bacterial stripe pathogen of rice identified as *Pseudomonas avenae*. *Phytopathologische Zeitschrift-Journal of Phytopathology,* 114, 256-259.

Silveira, E. B., Mariano, R. L. R., & Michereff, S. J. (2003). Variability of strains of Acidovorax avenae subsp. citrulli from melon produced in the Rio Grande do Norte State, Brazil. *Summa Phytopathologica,* 29, 255-261.

Simone, G. W., Cullen, R. E., & Hodge, N. C. (1996). A new leaf spot disease of geranium caused by *Acidovorax* sp. *Phytopathology,* 86, S50.

Skerman, V. B. D., McGowan, V., & Sneath, P. H. A. (1980). Approved lists of bacterial names. *International Journal of Systematic Bacteriology,* 30, 225-420.

Somodi, G. C., Jones, J. B., Hopkins, D. L., Stall, R. E., Kucharek, T. A., Hodge, N. C., et al. (1991). Occurrence of a bacterial watermelon fruit blotch in Florida. *Plant Disease,* 75, 1053-1056.

Song, W. Y., Hatziloukas, E., Randhawa, P., & Schaad, N. W. (1998). Development of PCR primers for identification of *Acidovorax avenae* subsp. *citrulli*, the causal agent of watermelon fruit blotch. *Phytopathology,* 88, S84.

Song, W. Y., Kim, H. M., Hwang, C. Y., & Schaad, N. W. (2004). Detection of *Acidovorax avenae* ssp. *avenae* in rice seeds using BIO-PCR. *Journal of Phytopathology,* 152, 667-676.

Song, W. Y., Kim, H. M., & Schaad, N. W. (2001). PCR primers for detection and identification of plant pathogenic species, subspecies and strains of *Acidovorax*. USA.

Song, W. Y., Sechler, A. J., Hatziloukas, E., Kim, H. M., & Schaad, N. W. (2003). Use of PCR for the rapid identification of *Acidovorax avenae* and *A. avenae* subsp. *citrulli*. In N. S. Iacobellis, A. Collmer, S. W. Hutcheson, J. W. Mansfield, C. E. Morris, J. Murillo, N. W. Schaad, D. E. Stead, G. Surico & M. S. Ullrich (Eds.), *Pseudomonas syringae and related pathogens* (pp. 531-544). Dordrecht, The Netherlands: Kluwer Academic Publishers.

Sowell, G., Jr. (1981). A bacterial disease causing severe damage to susceptible plant introductions of muskmelon. *Plant Disease,* 65, 609-610.

Stapp, C. (1928). *Schizomycetes* (Spaltpilze oder Bakterien). In P. Sorauer (Ed.), *Handbuch der Pflanzenkrankheiten* (5th ed., Vol. 2, pp. 1-295). Berlin, Germany: Paul Parey.

Starr, M. P., & Burkholder, W. H. (1942). Lipolytic activity of phytopathogenic bacteria determined by means of spirit blue agar and its taxonomic significance. *Phytopathology,* 32, 598-604.

Stead, D., Stanford, H., Aspin, A., & Heeney, J. (2003). Current status of some new and some old plant pathogenic pseudomonads. In N. S. Iacobellis, A. Collmer, S. W. Hutcheson, J. W. Mansfield, C. E. Morris, J. Murillo, N. W. Schaad, D. E. Stead, G. Surico & M. S. Ullrich (Eds.), *Pseudomonas syringae and related pathogens* (pp. 561-572). Dordrecht, The Netherlands: Kluwer Academic Publishers.

Stead, D. E. (1992). Grouping of plant-pathogenic and some other Pseudomonas spp. by using cellular fatty acid profiles. *International Journal of Systematic Bacteriology,* 42, 281-295.

Steindl, D. R. L., & Edgerton, C. W. (1964). Mottled Stripe. In C. G. Hughes, E. V. Abbott & C. A. Wismer (Eds.), *Sugar-cane diseases of the world* (Vol. 2, pp. 13-16). Amsterdam. Netherlands: Elsevier.

Stovold, G. E., Bradley, J., & Fahy, P. C. (2001). *Acidovorax avenae* subsp *cattleyae* (*Pseudomonas cattleyae*) causing leafspot and death of *Phalaenopsis* orchids in New South Wales. *Australasian Plant Pathology,* 30, 73-74.

Sumner, D. R., & Schaad, N. W. (1977). Epidemiology and Control of Bacterial Leaf-Blight of Corn. *Phytopathology,* 67, 1113-1118.

Tabei, H., & Quimio, A. J. (1978). Bacterial brown spot of Phalaenopsis orchids in the Philippines. *Jarq,* 12, 241-242.

Takikawa, Y., Ando, Y., Hamaya, E., Tsuyumu, S., & Goto, M. (1988). Identification of the pathogens responsible for bacteriosis of tea plant occurred in 1983. *Annals of the Phytopathological Society of Japan,* 54, 224-228.

Thind, B. S., Randhawa, P. S., & Soni, P. S. (1984). Chemical control of bacterial stalk rot (*Erwinia chrysanthemi* pv. *zeae*) and leaf stripe (*Pseudomonas rubrilineans*) of maize. *Zeitschrift fur Pflanzenkrankheiten und Pflanzenschutz.,* 19, 424-430.

Tominaga, T., Kimura, K., & Goh, N. (1983). Bacterial brown stripe of rice in nursery box, caused by Pseudomonas avenae. *Annals of the Phytopathological Society of Japan,* 49, 463-466.

Ullstrup, A. J. (1960). Bacterial Stripe of Corn. *Phytopathology,* 50, 906-910.

Vock, N. T. (1978). *A Handbook of Plant Diseases in Colour* (2 ed. Vol. 1 Fruit and Vegetables). Brisbane: Government Printer, Queensland, Australia.

Walcott, R. R. (2003). Detection of seedborne pathogens. *Horttechnology,* 13, 40-47.

Walcott, R. R., Fessehaie, A., & Castro, A. C. (2004). Differences in pathogenicity between two genetically distinct groups of *Acidovorax avenae* subsp. *citrulli* on cucurbit hosts. *Journal of Phytopathology,* 152, 277-285.

Walcott, R. R., & Gitaitis, R. D. (2000). Detection of *Acidovorax avenae* subsp *citrulli* in watermelon seed using immunomagnetic separation and the polymerase chain reaction. *Plant Disease,* 84, 470-474.

Walcott, R. R., Gitaitis, R. D., & Castro, A. C. (2003). Role of blossoms in watermelon seed infestation by *Acidovorax avenae* subsp *citrulli. Phytopathology,* 93, 528-534.

Walcott, R. R., Langston, D. B., Sanders, F. H., & Gitaitis, R. D. (2000). Investigating intraspecific variation of *Acidovorax avenae* subsp *citrulli* using DNA fingerprinting and whole cell fatty acid analysis. *Phytopathology,* 90, 191-196.

Walcott, R. R., Lessl, J. T., & Fessehaie, A. (2005). Blossom colonization by *Acidovorax avenae* subsp. *citrulli* and its role in watermelon seed infection. *Phytopathology,* 95, S129.

Wall, G. C. (1989). Control of watermelon fruit blotch by seed heat-treatment. *Phytopathology,* 79, 1191.

Wall, G. C. (1991). Plant diseases of recent introduction to Guam. *Micronesica,* 41-45.

Wall, G. C., & Santos, V. M. (1988). A new bacterial disease of watermelon in the Mariana Islands. *Phytopathology,* 78, 1605.

Wall, G. C., Santos, V. M., Cruz, F. J., & Nelson, D. A. (1990). Outbreak of watermelon fruit blotch in the Mariana Islands. *Plant Disease,* 74, 80.

Webb, R. E., & Goth, R. W. (1965). A seedborne bacterium isolated from watermelon. *Plant Disease Reporter,* 49, 818-821.

Wen, A. M., Fegan, M., Hayward, C., Chakraborty, S., & Sly, L. I. (1999). Phylogenetic relationships among members of the *Comamonadaceae*, and description of *Delftia acidovorans* (den Dooren de Jong 1926 and Tamaoka et al. 1987) gen. nov., comb. nov. *International Journal of Systematic Bacteriology,* 49, 567-576.

Wen, A. M., Hayward, A. C., Sly, L. I., Chakraborty, S., Fegan, M., & O'Brien, R. G. (1997). *Rapid identification and detection of Acidovorax avenae in melon.* Paper presented at the 11th Biennial Conference of the Australasian Plant Pathology Society, Perth, Western Australia, Australia.

Wey, G. C. (1988). Occurrence and investigation of important diseases on Phalaenopsis in Taiwan. *Reports of the Taiwan Sugar Research Institute,* 122, 31-41.

White, D. G., Pataky, J. K., & Stall, R. E. (1994). Unusual occurrences of bacterial leaf-blight on maize and sorghum in central Illinois. *Plant Disease,* 78, 640-640.

Willems, A., Deley, J., Gillis, M., & Kersters, K. (1991). *Comamonadaceae*, a new family encompassing the acidovorans ribosomal RNA complex, including *Variovorax paradoxus* gen. nov., comb. nov., for *Alcaligenes paradoxus* (Davis 1969). *International Journal of Systematic Bacteriology,* 41, 445-450.

Willems, A., Falsen, E., Pot, B., Jantzen, E., Hoste, B., Vandamme, P., et al. (1990). *Acidovorax*, a new genus for *Pseudomonas facilis*, *Pseudomonas delafieldii*, E-falsen (Ef) group 13, Ef group 16, and several clinical isolates, with the species *Acidovorax facilis* comb. nov., *Acidovorax delafieldii* comb. nov., and *Acidovorax temperans* sp. nov. *International Journal of Systematic Bacteriology,* 40, 384-398.

Willems, A., Goor, M., Thielemans, S., Gillis, M., Kersters, K., & Deley, J. (1992). Transfer of several phytopathogenic *Pseudomonas* species to *Acidovorax* as *Acidovorax avenae* subsp. *avenae* subsp. nov., comb. nov., *Acidovorax avenae* subsp. *citrulli*, *Acidovorax avenae* subsp. *cattleyae*, and *Acidovorax konjaci. International Journal of Systematic Bacteriology,* 42, 107-119.

Woese, C. R., Weisburg, W. G., Paster, B. J., Hahn, C. M., Tanner, R. S., Krieg, N. R., et al. (1984). The phylogeny of purple bacteria - the Beta-Subdivision. *Systematic and Applied Microbiology,* 5, 327-336.

Xie, G. L., Sun, X. L., & Mew, T. W. (1998). Characterization of *Acidovorax avenae* subsp. *avenae* from rice seeds. *Chinese Journal of Rice Science,* 12, 165-171.

Xie, G. L., Zhu, G. N., & Ren, X. P. (2002). Diversity of pathogenic bacteria from rice seeds. *Acta Phytopathologica Sinica,* 32, 114-120.

Young, J. M., Dye, D. W., Bradbury, J. F., Panagopoulos, C. G., & Robbs, C. F. (1978). A proposed nomenclature and classification for plant pathogenic bacteria. *New Zealand Journal of Agricultural Research,* 21, 153-177.

Zeigler, R. S., & Alvarez, E. (1989). Pseudomonas species causing rice sheath rot (ShR) and grain discoluoration (GID). *International Rice Research Newsletters,* 14, 26-27.

Zhao, T. C., Sun, F. Z., Wang, B. W., & Hui, W. G. (2001). Pathogen identification of Hami melon bacterial fruit blotch. *Acta Phytopathologica Sinica,* 31, 357-364.

7. AFFILIATION

Mark Fegan, *School of Molecular and Microbial Sciences, Molecular Bioscience Building, Room 156, University of Queensland, Australia.*
E-mail:mfegan@uq.edu.au

INDEX

GPSR Compliance

The European Union's (EU) General Product Safety Regulation (GPSR) is a set of rules that requires consumer products to be safe and our obligations to ensure this.

If you have any concerns about our products, you can contact us on ProductSafety@springernature.com

In case Publisher is established outside the EU, the EU authorized representative is:

Springer Nature Customer Service Center GmbH
Europaplatz 3
69115 Heidelberg, Germany

Batch number: 10755320

Printed by Printforce, the Netherlands